Handbook of Biomarkers and Precision Medicine

Handbook of Biomarkers and Precision Medicine

Edited by
Claudio Carini
Mark Fidock
Alain van Gool

CRC Press is an imprint of the
Taylor & Francis Group, an **informa** business
A CHAPMAN & HALL BOOK

CRC Press
Taylor & Francis Group
6000 Broken Sound Parkway NW, Suite 300
Boca Raton, FL 33487-2742

First issued in paperback 2020

ISBN-13: 978-1-4987-6258-8 (hbk)
ISBN-13: 978-0-367-73005-5 (pbk)

Visit the Taylor & Francis Web site at
http://www.taylorandfrancis.com

and the CRC Press Web site at
http://www.crcpress.com

The editors greatly acknowledge Charlotte Bryant-Turner for excellent administrative support and fellow contributors for their excellent contributions without whom this book would never had happened; a BIG thank you!

Contents

Preface

Today, the practice of medicine remains largely empirical. Physicians generally rely on matching patterns in patient history, symptoms, and laboratory data to make the diagnosis. The choice for a given treatment is often based on physicians' past experience with patients affected by that disease. In this paradigm, blockbuster medicines are prescribed for a "typical patient" with a specific disease. As a consequence, patients may be exposed to unnecessary drug side effects by taking a medicine with poor or no efficacy for them, even though it works well in the majority of patients—thus the need for a more patient-focused, precise, therapeutic approach.

There is consensus that biological markers (biomarkers) are valuable drug development tools that help validate potential drug targets, allow us to better understand the mechanisms of action and to identify earlier compounds with the highest potential for efficacy. Biomarkers are also essential for eliminating compounds with an unacceptable therapeutic index enabling the strategy to "fail fast, fail early."

Companion diagnostics are medical devices that provide information for the safe and effective use of a linked targeted drug or biological product. These tools can help identify patients most likely to benefit from a particular medicine as well as those who may present serious side effects if they use the therapeutic product. The development of a companion diagnostic enables the identification of the patient population that will benefit from the new drug and allows those patients with a negative test result to have an alternative treatment decision. Indeed, early diagnosis and the right therapy for diseases such as cancer are crucial for patients to have a favorable clinical outcome.

Greater use of biomarkers and companion diagnostics should better enable a shift from empirical medicine to personalized medicine, or more appropriately, precision medicine, whereby treatment is tailored for the individual. Today, medical practice is moving away from the concept of "one size fits all" and shifting, instead, to a medical approach that uses the right medicine, for the right patient, at the right dose, at the right time. Precision medicine is targeting medical treatment to the individual characteristics, needs, and preferences of a patient during all stages of care, including prevention, diagnosis, treatment, and follow-up.

Precision medicine highlights the importance of coupling established clinical indexes with molecular profiling to create diagnostic, prognostic and therapeutic strategies regarding each patient's needs, and it has the potential to provide a rapid acceleration in the development of next generation therapeutics.

Regulatory and government agencies in the United States and abroad have recognized the value of biomarkers and companion diagnostics to guide decision making about drug safety, efficacy and patients' stratification. The magnitude and seriousness of the U.S. Food and Drug Administration (FDA) and European Medicines Agency (EMA) commitment to precision medicine is reflected in several initiatives encouraging the use of biomarkers and, in collaboration with the Foundation for the National Institutes of Health (FNIH), fostering public–private partnerships.

In recent years, the topic of biomarkers and companion diagnostics has been featured at many specialized scientific conferences and has received extensive media coverage. The co-editors recognized that a book that approached the topic with an emphasis on the practical aspects of biomarkers identification and use, as well as their strategic implementation, was missing and essential to improve the application of biomarkers and companion diagnostics in precision medicine. The co-editors also recognized that the specialized knowledge of a diverse group of experts was necessary to create the kind of comprehensive book needed. Therefore, contributions were invited from renowned experts in their respective fields and included scientists from government, academia, research hospitals, biotechnology and pharmaceutical companies, and contract research organizations. The result is a book that will appeal broadly to pharmaceutical research scientists, clinical and academic investigators, regulatory scientists, managers, students, and professionals engaged in drug development.

As discussed in Chapter 1, biomarkers are not new, they have been used for several years to help physicians diagnose and treat disease. The importance, promises, and pitfalls of biomarkers are described. Chapter 1 discusses the biomarkers regulatory strategies, why biomarkers are important to payers, and the pros and cons of immunotherapy.

Chapter 2 highlights the limitation of animal models to predict cardio-, hepatic, and renal toxicity of novel drugs and safety parameters like immunogenicity and stresses the role of system biology in identifying new biomarkers.

In Chapter 3, the authors describe the importance of translating preclinical to clinical data and back and the use of molecular testing for clinical applications and patient stratification. Importantly, Chapter 3 also addresses the need for global collaboration in translational medicine.

Chapter 4 analyzes the use of statistics, including a practical overview of concepts, study design, common mistakes, and helpful tips to ensure credible biomarkers that can address their intended use, novel adaptive designs and improving the efficacy of drug development. This chapter also provides a comprehensive review of classical clinical designs leveraging biomarker information. Statistical considerations related to biomarker discovery are also discussed.

Chapter 5 highlights the importance of identifying the hurdles and solutions for the implementation of precision medicine, taking into consideration the crucial role played by high-quality biosamples in biomarkers analysis. This chapter also addresses the value of computational biology and the value of public–private partnerships.

Chapters 6 focuses on "lessons from the past and pioneers of the future" reviewing the practical aspects of implementing biomarkers in drug development programs. Examples are drawn from a wide range of target organ toxicities, therapeutic areas, and product types. In this chapter, the authors describe several case studies that stress the importance of integrating drug development with companion diagnostics as well as the relevance of interpreting molecular testing in clinical trials. This chapter addresses the relevant contribution of biomarkers consortia to the discovery and clinical applications of novel biomarkers.

Chapter 7 focuses on important emerging technologies that have direct impact on drug discovery and development and on the conduct of clinical trials and patients' treatment. In it, the authors present concepts and examples of utilizing biomarkers in drug discovery, preclinical safety assessment, clinical trials, and translational and precision medicine.

In the last two chapters (Chapters 8 and 9), the book concludes on the next frontiers of therapeutic targets and future trends analyzing the future challenges in precision medicine.

The field of biomarkers in drug development is rapidly evolving and this book presents a snapshot of some exciting new approaches. The co-editors hope that by presenting a wide range of biomarker applications, discussed by knowledgeable and experienced scientists, readers will develop an appreciation of the scope and breadth of biomarker knowledge and find examples that will help them in their own work.

By utilizing this book as a source of new knowledge, to reinforce or to integrate existing knowledge, the co-editors hope that readers will benefit from a greater understanding and appreciation of the strategic approach of using biomarkers in drug development and companion diagnostics, thus becoming contributors to the global discourse that will allow treatment and prevention to be personalized and precise.

Maria C. Freire
Foundation for the National Institute of Health

About the Editors

Claudio Carini, MD, PhD, FRCPath, has been involved in the application of biomarkers in precision medicine for the past 18 years. He is currently appointed as H. Faculty in the Department of Cancer and Pharmaceutical Sciences, Faculty of Life Sciences & Medicine at King's College, School of Medicine, London. He is a member of the F-NIH Biomarkers Consortium Steering Committees for Cancer, Inflammation & Immunity, and Neuroscience. He is also a senior member of the Scientific Advisory Boards for several biotechnology companies and a member of the New York Academy of Sciences. Claudio was formerly the Global Head of Clinical Immunology and Biomarkers at Pfizer Inc., where he led precision medicine activities across multiple therapeutic areas and facilitated international collaborations with prestigious academic institutions. He also held senior roles in pharmaceutical companies (Wyeth, Novartis, and Roche) and faculty positions at Harvard and Johns Hopkins Medical School. He is a strong believer of open innovation networks, and in his roles he has been collaborating extensively with different stakeholders to speed up the clinical application of scientific research through public–private partnerships. He has served in several national and international scientific boards: F-NIH Inflammation/Immunity; MRC, Inflammation/ Immunology Initiative—Rheumatoid Arthritis Consortium; MRC, "Stratified Medicine, SLE Consortium; IMI, ABIRISK, EU/EFPIA"; The PML Consortium. Claudio has over 200 publications in national and international peer-reviewed journals. He has also co-edited three books in the field of Biomarkers and Precision Medicine: *Biomarkers in Drug Development: A Handbook of Practice Application and Strategy*, (2010), *Clinical and Statistical Considerations in Personalized Medicine*, (2014), and *Handbook of Biomarkers & Precision Medicine* (2019 in press).

Mark Fidock, PhD, is Vice President of Precision Medicine Laboratories within the IMED Biotech Unit, AstraZeneca, Cambridge, UK. He received his PhD from the University of East Anglia, UK, in Molecular Pharmacology. Mark is a trained molecular biologist/pharmacologist and gained extensive experience in drug development early in his career while at Pfizer Ltd, Sandwich, UK. He supported the development and approval of four new therapies and provided leadership for the Biomarker and Translational Medicine activities across multiple therapeutic areas. Mark joined AstraZeneca in 2013, appointed to the role of Head of Precision Medicine Laboratories and a member of the Precision Medicine and Genomics Leadership Team. He has overall accountability for all clinical testing activities that enables diagnostic development and regulatory approval. Mark leads a global multidisciplinary team providing strategy and direction in the development of Companion Diagnostic tests linked to targeted therapeutics across Oncology; Respiratory; and Cardiovascular, Renal, & Metabolism. These laboratory tests are deployed across both early- and late-phase clinical trials with the intent for global regulatory approval. By 2019 Mark had achieved >25 regulatory approved companion diagnostic tests across 3 major markets (USA, EU and Japan) linked to 4 AstraZeneca targeted therapies. Mark has a passion for innovative diagnostic science, driving research such as the use of artificial intelligence across tissue and molecular diagnostics to enable approaches that improve clinical testing outcomes with the ambition to change clinical care pathways. Mark has published more than 30 peer-reviewed articles in scientific journals and an active editor of the Journal of Precision Medicine. As an internationally recognized expert in Precision Medicine he holds an honorary position at the University of Cambridge and a member of the Institute of Translational and Stratified Medicine, Plymouth, scientific advisory board.

Alain van Gool, PhD, is a professor of Personalized Healthcare and heads the Translational Metabolic Laboratory at the Radboud University Medical Center, Nijmegen Netherlands, with a strong passion in the application of biomarkers in translational medicine and personalized healthcare. Being trained as biochemist and molecular biologist, Alain worked at pharmaceutical industry (Organon, Schering-Plough, MSD), academia, applied research institutes, and university medical centers in EU, USA, and Singapore. He has been leading technology-based laboratories, cross-functional expert teams, therapeutic project teams, and public–private consortia, many of which were focused on the discovery, development, and implementation of translational biomarkers in a variety of therapeutic areas. His technical expertise resides most strongly in molecular profiling through x-omics, analytical biomarker development, and applications in translational scientific research. Alain is a strong believer of public–private partnerships and thrives to work with specialists to translate research findings to applications. With that background, he currently also acts as Strategic Advisor to the Executive Board of Radboudumc, coordinates the Radboudumc Technology Centers, is Scientific Lead Technologies of DTL (the Dutch Techcenter for Life Sciences), is Chair Biomarker Platform of EATRIS (the European infrastructure for Translational Medicine), is co-initiator of Health-RI (the Netherlands Personalized Medicine and Health Research Infrastructure), and project leader and PI of the Netherlands X-omics Initiative, thus contributing to the organization and coordination of local, national, and European technology infrastructures. Alain has published 40 peer-reviewed articles in scientific journals and is a frequently invited keynote speaker at scientific conferences and public events. Complementing his daily work, he enjoys contributing as translational biomarker scientist to editorial boards of scientific journals, funding agencies, and scientific advisory boards of start-up enterpreneurs, multinational companies, diagnostic organizations, public–private consortia, and conference organizers.

Contributors

Nurulamin Abu Bakar
Department of Neurology
Radboud University Medical Center
and
Translational Metabolic Laboratory
Radboud University Medical Center
Nijmegen, the Netherlands

Viatcheslav R. Akmaev
BERG LLC.
Framingham, Massachusetts

Alexandre Akoulitchev
Oxford BioDynamics PLC
Oxford, United Kingdom

John L. Allinson
Biomarker Services
Biologics Development Services (BDS)
Tampa, Florida

Antoni L. Andreu
EATRIS ERIC
European Infrastructure for Translational Medicine
Amsterdam, the Netherlands

Joan-Carles Arce
Ferring Pharmaceuticals Inc.
Parsippany, New Jersey

Paolo A. Ascierto
Melanoma, Cancer Immunotherapy and Development Therapeutics Unit
Istituto Nazionale Tumouri Fondazione "G. Pascale"
Napoli, Italy

Rita Azevedo
Lygature
Utrecht, the Netherlands

Robert A. Beckman
Departments of Oncology and of Biostatistics, Bioinformatics, and Biomathematics
Lombardi Comprehensive Cancer Center
and
Innovation Center for Biomedical Informatics
Georgetown University Medical Center
Washington, District of Columbia

Jeroen A.M. Beliën
VU University Medical Center
Amsterdam, the Netherlands

Neil Benson
Unit 43
Certara Quantitative Systems Pharmacology
Canterbury Innovation Centre
Canterbury, United Kingdom

Florence Bietrix
EATRIS ERIC
European Infrastructure for Translational Medicine
Amsterdam, the Netherlands

The Biomarker Development Center Consortium
University of Groningen
Groningen, the Netherlands

Rainer Bischoff
Department of Analytical Biochemistry
Interfaculty Mass Spectrometry Center
Research Institute of Pharmacy
University of Groningen
Groningen, the Netherlands

Jan-Willem Boiten
Lygature
Utrecht, the Netherlands

Corry-Anke Brandsma
Department of Pathology & Medical Biology
University Medical Center Groningen
University of Groningen
Groningen, the Netherlands

Sarah Brockbank
Institute of Cellular Medicine
Newcastle University
and
Freeman Hospital
Newcastle upon Tyne Hospitals NHS Foundation Trust
Newcastle upon Tyne, United Kingdom

Ian N. Bruce
Arthritis Research UK Centre for Epidemiology
Centre for Musculoskeletal Research
School of Biological Sciences
Faculty of Biology, Medicine and Health
The University of Manchester
and
NIHR Manchester Biomedical Research Centre
Central Manchester Foundation Trust
Manchester Academic Health Science Centre
Manchester, United Kingdom

Harrie Buist
TNO Innovation for Life
Unit Healthy Living
Research Group Risk Analysis for Products in Development
Zeist, the Netherlands

Marylou Buyse
Integrated Health Solutions
Quality and Population Health Solutions
Precision for Value
Gladstone, New Jersey

Claudio Carini
Department of Cancer & Pharmaceutical Sciences
School of Medicine
Life Science Medicine
King's College
London, United Kingdom

Marinel Cavelaars
The Hyve
Utrecht, the Netherlands

Jeremy Chaufty
BERG LLC.
Framingham, Massachusetts

Cong Chen
Department of Biostatistics and Research Decision Sciences
Merck & Co., Inc.
Kenilworth, New Jersey

Andrew P. Cope
Centre for Inflammation Biology and Cancer Immunology
School of Immunology and Microbial Sciences
Faculty of Life Sciences and Medicine
King's College London
Guy's Campus
Great Maze Pond
London, United Kingdom

Edwin Cuppen
University Medical Center
Utrecht, the Netherlands

Raymond H. Cypess
ATCC
Manassas, Virginia

Wendy de Leng
University Medical Center
Utrecht, the Netherlands

Simon P. Dearden
Precision Medicine and Genomics
IMED Biotech Unit
AstraZeneca
Cambridge, United Kingdom

André Dekker
Department of Radiation Oncology (MAASTRO)
GROW School for Oncology and Developmental Biology
Maastricht University Medical Centre+
Maastricht, the Netherlands

James Demarest
ViiV Healthcare
Research Triangle Park, North Carolina

Benoit Destenaves
Precision Medicine and Genomics
IMED Biotech Unit
AstraZeneca
Waltham, Massachusetts

E. Dinant Kroese
TNO Innovation for Life
Unit Healthy Living
Research Group Risk Analysis for Products in Development
Zeist, the Netherlands

Nick Edmunds
New Modalities, Drug Safety and Metabolism
IMED Biotech Unit
AstraZeneca
Cambridge, United Kingdom

Mark B. Effron
John Ochsner Heart and Vascular Institute
Ochsner Clinical School
University of Queensland School of Medicine
and
Ochsner Health System
New Orleans, Louisiana

Alessio Fasano
Center for Celiac Research
and
Mucosal Immunology and Biology Research Center and Center
Massachusetts General Hospital
Harvard Medical School
Boston, Massachusetts

Brisa S. Fernandes
IMPACT Strategic Research Centre (Barwon Health)
School of Medicine
Deakin University
Geelong, Victoria, Australia

and

Laboratory of Calcium Binding Proteins in the Central Nervous System
Department of Biochemistry
Federal University of Rio Grande do Sul
Porto Alegre, Brazil

Mark Fidock
Precision Medicine and Genomics
IMED Biotech Unit
Astra Zeneca
Cambridge, United Kingdom

F. Owen Fields
Pfizer Inc.
Inflammation and Immunology
Research and Development Unit
Philadelphia, Pennsylvania

Remond J.A. Fijneman
Netherlands Cancer Institute
Amsterdam, the Netherlands

Leonard P. Freedman
Global Biological Standards Institute
Washington, District of Columbia

Maria C. Freire
President and Executive Director
Foundation for the National Institute of Health
North Bethesda, Maryland

Dominic Galante
Quality and Populations Health Solutions
Precision for Value
Gladstone, New Jersey

María Laura García Bermejo
Ramon y Cajal Health Research Institute
Madrid, Spain

Stephane Gesta
BERG LLC.
Framingham, Massachusetts

Federico Goodsaid
Regulatory Pathfinders, LLC
Pescadero, California

Boris Gorovits
BiomedicineDesign Bioanalytical
Pfizer
Andover, Massachusetts

Francis Hector Grand
Oxford BioDynamics PLC
Oxford, United Kingdom

George A. Green
Translational Medicine
Bristol-Myers Squibb
New York, New York

Bennett P. Greenwood
BERG LLC.
Framingham, Massachusetts

Hanna L. Groen
Top Sector Life Sciences & Health (Health~Holland)
The Hague, the Netherlands

Peter M.A. Groenen
Translational Science
Idorsia Pharmaceuticals Ltd.
Allschwil, Switzerland

Victor Guryev
European Research Institute for the Biology of Ageing
University Medical Center Groningen
Groningen, the Netherlands

Coşkun Güzel
Laboratory of Neuro-Oncology, Clinical & Cancer Proteomics
Department of Neurology
Erasmus MC
Rotterdam, the Netherlands

Marian Hajduch
Institute of Molecular and Translational Medicine
Faculty of Medicine and Dentistry
Palacký University Olomouc and Cancer Research Czech Republic
Olomouc, Czech Republic

Abdel-Baset Halim
Translational Medicine, Biomarkers and Diagnostics
Celldex Therapeutics
Hampton, New Jersey

Trevor T. Hansel
Imperial Clinical Respiratory Research Unit (ICRRU)
National Heart and Lung Institute (NHLI)
St Mary's Hospital
Imperial College
London, United Kingdom

Jayvant Heera
Pfizer Worldwide Research and Development
Groton, Connecticut

Ron M.A. Heeren
Division of Imaging Mass Spectrometry
Maastricht MultiModal Molecular Imaging (M4I) Institute
Maastricht University
Maastricht, the Netherlands

Crispin T. Hiley
Translational Cancer Therapeutics Laboratory
The Francis Crick Institute
London, United Kingdom

John Hinrichs
University Medical Center
Utrecht, the Netherlands

Zsuzsanna Hollander
PROOF Centre of Excellence
and
Centre for Heart Lung Innovation
University of British Columbia
Vancouver, British Columbia, Canada

Stephanie Holst
Center for Proteomics and Metabolomics
Leiden University Medical Center
Leiden, the Netherlands

Robert Holt
Biomarkers and Companion Diagnostics
BerGenBio ASA
Munich, Germany

Jorrit J. Hornberg
AstraZeneca R&D Gothenburg
Mölndal, Sweden

Peter Horvatovich
Department of Analytical Biochemistry
Research Institute of Pharmacy
University of Groningen
Groningen, the Netherlands

Bo Huang
Pfizer Inc.
New York, New York

Glen Hughes
Precision Medicine and Genomics
IMED Biotech Unit
AstraZeneca
Cambridge, United Kingdom

Manon Huibers
University Medical Center
Utrecht, the Netherlands

Martin Hund
Roche Diagnostics International Ltd
Rotkreuz, Switzerland

Ewan Hunter
Oxford BioDynamics PLC
Oxford, United Kingdom

John D. Isaacs
Institute of Cellular Medicine
Newcastle University
and
Freeman Hospital
Newcastle upon Tyne Hospitals NHS Foundation Trust
Newcastle upon Tyne, United Kingdom

Danyel G.J. Jennen
Department of Toxicogenomics
Maastricht University
Maastricht, the Netherlands

Akhilesh Jha
Imperial Clinical Respiratory Research Unit (ICRRU)
National Heart and Lung Institute (NHLI)
St Mary's Hospital
Imperial College
London, United Kingdom

Victor L. Kallen
Department of Microbiology & Systems Biology
The Netherlands Organization for Applied Sciences (TNO)
Zeist, the Netherlands

Hiroshi Kato
Division of Rheumatology
Departments of Medicine, Microbiology and Immunology
and
Biochemistry and Molecular Biology
College of Medicine
State University of New York
Upstate Medical University
Syracuse, New York

Martina Kaufmann
Strategic Consulting
Muellheim, Germany

Zahoor Khan
Ochsner Medical Center
New Orleans, Louisiana

Ji-Young V. Kim
PROOF Centre of Excellence
and
Centre for Heart Lung Innovation
University of British Columbia
and
Department of Medical Genetics
University of British Columbia
Vancouver, British Columbia, Canada

Michael A. Kiebish
BERG LLC.
Framingham, Massachusetts

Wietske Kievit
Department for Health Evidence
Radboud Institute for Health Sciences
Radboud University Medical Center
Nijmegen, the Netherlands

Frank Klont
Department of Analytical Biochemistry
Interfaculty Mass Spectrometry Center
University of Groningen
Groningen, the Netherlands

Charles Knirsch
Pfizer Worldwide Research and Development
Pearl River, New York

Vladimira Koudelakova
Institute of Molecular and Translational Medicine
Faculty of Medicine and Dentistry
Palacký University Olomouc and Cancer Research
Czech Republic
Olomouc, Czech Republic

Lars Kuepfer
Bayer AG
RWTH Aachen University
Aachen, Germany

Penelope Kungl
Diagnostic and Research Center for Molecular BioMedicine
Medical University of Graz
Graz, Austria

Nina Kusch
Joint Research Center for Computational Biomedicine
RWTH Aachen
Aachen, Germany

Karen K.Y. Lam
PROOF Centre of Excellence
and
Centre for Heart Lung Innovation
University of British Columbia
and
Department of Pathology and Laboratory Medicine
University of British Columbia
Vancouver, British Columbia, Canada

Kim-Anh Lê Cao
Melbourne Integrative Genomics
School of Mathematics and Statistics
The University of Melbourne
Melbourne, Victoria, Australia

Dirk J. Lefeber
Department of Neurology
Radboud University Medical Center
and
Translational Metabolic Laboratory
Radboud University Medical Center
Nijmegen, the Netherlands

Richard Lee
School of Immunology & Microbial Sciences
King's College London
London, United Kingdom

Cecilia Lindskog
Department of Immunology, Genetics and Pathology
Science for Life Laboratory
Rudbeck Laboratory
Uppsala University
Uppsala, Sweden

Massimo Loda
Department of Oncologic Pathology
Dana-Farber Cancer Institute
Harvard Medical School
Boston, Massachusetts

Theo M. Luider
Laboratory of Neuro-Oncology, Clinical & Cancer Proteomics
Department of Neurology
Erasmus MC
Rotterdam, the Netherlands

Giuseppe Masucci
Department of Oncology-Pathology
Karolinska Institutet
Stockholm, Sweden

Bruce M. McManus
PROOF Centre of Excellence
and
Centre for Heart Lung Innovation
University of British Columbia
and
Department of Pathology and Laboratory Medicine
University of British Columbia
Vancouver, British Columbia, Canada

Gerrit A. Meijer
Netherlands Cancer Institute
Amsterdam, the Netherlands

and

University Medical Centre Utrecht
Utrecht, the Netherlands

Joseph P. Menetski
Foundation for the National Institutes of Health
Bethesda, Maryland

Sandeep Menon
Pfizer Inc.
New York, New York

Jane Mellor
Department of Biochemistry
University of Oxford
Oxford, United Kingdom

Giovanni Migliaccio
EATRIS ERIC
European Infrastructure for Translational Medicine
Amsterdam, the Netherlands

Michael C. Montalto
Translational Medicine
Bristol-Myers Squibb
New York, New York

Paula P. Narain
BERG LLC.
Framingham, Massachusetts

Niven R. Narain
BERG LLC.
Framingham, Massachusetts

Eric J. Nestler
BERG LLC.
Framingham, Massachusetts

and

Icahn School of Medicine at Mount Sinai
New York, NY

Raymond T. Ng
PROOF Centre of Excellence
and
Department of Computer Science
University of British Columbia
Vancouver, British Columbia, Canada

Steven W.M. Olde Damink
Department of General Surgery
NUTRIM
Maastricht University
Maastricht, the Netherlands

and

Department of General, Visceral and Transplantation Surgery
RWTH University Hospital Aachen
Aachen, Germany

Maria C.M. Orr
Precision Medicine and Genomics
IMED Biotech Unit
AstraZeneca
Cambridge, United Kingdom

Peter J. Parker
King's College London
New Hunt's House
and
Francis Crick Institute
London, United Kingdom

Ayako Pedersen
Institute of Cellular Medicine
Newcastle University
and
Freeman Hospital
Newcastle upon Tyne Hospitals NHS Foundation Trust
Newcastle upon Tyne, United Kingdom

Michela Perani
King's College London
New Hunt's House
London, United Kingdom

Andras Perl
Division of Rheumatology
Departments of Medicine, Microbiology and Immunology
and
Biochemistry and Molecular Biology
College of Medicine
State University of New York
Upstate Medical University
Syracuse, New York

Nelma Pértega-Gomes
Department of Oncologic Pathology
Dana-Farber Cancer Institute
Harvard Medical School
Boston, Massachusetts

Roy H. Perlis
Center for Quantitative Health
Massachusetts General Hospital
Harvard Medical School
Boston, Massachusetts

Suso Platero
Covance Inc.
Princeton, New Jersey

Amy Pointon
Safety and ADME Translational Sciences
IMED Biotech Unit
AstraZeneca
Cambridge, United Kingdom

Tiffany Porta
Division of Imaging Mass Spectrometry
Maastricht MultiModal Molecular Imaging (M4I) Institute
Maastricht University
Maastricht, the Netherlands

Shobha Purushothama
Biogen Inc.
Cambridge, Massachusetts

John A. Reynolds
Arthritis Research UK Centre for Epidemiology
Centre for Musculoskeletal Research
School of Biological Sciences
Faculty of Biology, Medicine and Health
The University of Manchester
and
NIHR Manchester Biomedical Research Centre
Central Manchester Foundation Trust
Manchester Academic Health Science Centre
Manchester, United Kingdom

Tanja Rouhani Rankouhi
TNO
Zeist, the Netherlands

Frans G.M. Russel
Department of Pharmacology and Toxicology
Radboud Institute of Molecular Life Sciences
Radboudumc
Nijmegen, the Netherlands

Matthew Salter
Oxford BioDynamics PLC
Oxford, United Kingdom

George Santis
School of Immunology & Microbial Sciences
King's College London
London, United Kingdom

Anna Sapone
Center for Celiac Research
and
Mucosal Immunology and Biology Research Center and Center
Massachusetts General Hospital
Harvard Medical School
Boston, Massachusetts

Karine Sargsyan
Biobank Graz
Graz, Austria

Rangaprasad Sarangarajan
BERG LLC.
Framingham, Massachusetts

Andreas Scherer
Institute for Molecular Medicine Finland (FIMM)
University of Helsinki
Helsinki, Finland

Andreas Schuppert
Joint Research Center for Computational Biomedicine
RWTH Aachen
Aachen, Germany

Gloria Serena
Center for Celiac Research
and
Mucosal Immunology and Biology Research Center
Massachusetts General Hospital
Harvard Medical School
Boston, Massachusetts

Attila A. Seyhan
Fox Chase Cancer Center
Temple Health
Temple University
Philadelphia, Pennsylvania

and

The Brown Institute for Translational Science
Department of Pathology and Laboratory Medicine
Warren Alpert Medical School
Division of Biology and Medicine
Brown University
Providence, Rhode Island

Casey P. Shannon
Prevention of Organ Failure (PROOF) Centre of Excellence
Vancouver, British Columbia, Canada

Amrit Singh
Department of Pathology and Laboratory Medicine
University of British Columbia
and
Prevention of Organ Failure (PROOF) Centre of Excellence
Vancouver, British Columbia, Canada

Nehmat Singh
Imperial Clinical Respiratory Research Unit (ICRRU)
National Heart and Lung Institute (NHLI)
St Mary's Hospital
Imperial College
London, United Kingdom

Reinier L. Sluiter
Department for Health Evidence
Radboud Institute for Health Sciences
Radboud University Medical Center
Nijmegen, the Netherlands

Rob H. Stierum
TNO Innovation for Life
Unit Healthy Living
Research Group Risk Analysis for Products in Development
Zeist, the Netherlands

Christoph Stingl
Laboratory of Neuro-Oncology, Clinical & Cancer Proteomics
Department of Neurology
Erasmus MC
Rotterdam, the Netherlands

Cornelia Stumptner
Diagnostic and Research Center for Molecular BioMedicine
Medical University of Graz
Graz, Austria

Frank Suits
IBM T.J. Watson Research Center
Yorktown Heights, New York

Roel Tans
Translational Metabolic Laboratory
Department of Laboratory Medicine
Radboud Institute of Molecular Life Sciences
Radboud University Medical Center
Nijmegen, the Netherlands

Scott J. Tebbutt
PROOF Centre of Excellence
and
Centre for Heart Lung Innovation
University of British Columbia
and
Division of Respiratory Medicine
Department of Medicine (Respiratory Division)
University of British Columbia
Vancouver, British Columbia, Canada

Annemette V. Thougaard
H. Lundbeck A/S
Valby, Denmark

Ryan S. Thwaites
Imperial Clinical Respiratory Research Unit (ICRRU)
National Heart and Lung Institute (NHLI)
St Mary's Hospital
Imperial College
London, United Kingdom

Vladimir Tolstikov
BERG LLC.
Framingham, Massachusetts

Lisa Turnhoff
Joint Research Center for Computational Biomedicine
RWTH Aachen
Aachen, Germany

Anton E. Ussi
EATRIS ERIC
European Infrastructure for Translational Medicine
Amsterdam, the Netherlands

Arnoud van der Maas
Radboud University Medical Center
Nijmegen, the Netherlands

Elna van der Ryst
The Research Network
Sandwich, United Kingdom

Gert Jan van der Wilt
Department for Health Evidence
Donders Centre for Neuroscience
Radboud University Medical Center
Nijmegen, the Netherlands

Alain van Gool
Translational Metabolic Laboratory
Department of Laboratory Medicine
Radboud Institute for Molecular Life Sciences
Radboud University Medical Center
Nijmegen, the Netherlands

Nico L.U. van Meeteren
Top Sector Life Sciences & Health (Health~Holland)
The Hague, the Netherlands

and

CAPHRI
Maastricht University
Maastricht, the Netherlands

and

Topcare
Leiden, the Netherlands

Monique van Scherpenzeel
Department of Neurology
Radboud University Medical Center
and
Translational Metabolic Laboratory
Radboud University Medical Center
Nijmegen, the Netherlands

Eugene P. van Someren
TNO Innovation for Life
Unit Healthy Living
Research Group Risk Analysis for Products in Development
Zeist, the Netherlands

Jennifer Venhorst
TNO
Zeist, the Netherlands

Lars Verschuren
TNO
Zeist, the Netherlands

Lisa Villabona
Department of Oncology-Pathology
Karolinska Institutet
Stockholm, Sweden

Vivek Vishnudas
BERG LLC.
Framingham, Massachusetts

Jack W.T.E. Vogels
TNO Innovation for Life
Unit Healthy Living
Research Group Risk Analysis for Products in Development
Zeist, the Netherlands

Jing Wang
Pfizer Inc.
New York, New York

John C. Waterton
Division of Informatics Imaging & Data Sciences
Centre for Imaging Sciences
Faculty of Biology Medicine & Health
School of Health Sciences
University of Manchester
Manchester, United Kingdom

Mike Westby
Centauri Therapeutics
Sandwich, United Kingdom

Rhonda Wideman
PROOF Centre of Excellence
Vancouver, British Columbia, Canada

Jonathan I. Wilde
Kjwilde Consulting
Cambridge, United Kingdom

Martijn J. Wilmer
Department of Pharmacology and Toxicology
Radboud Institute of Molecular Life Sciences
Radboudumc
Nijmegen, the Netherlands

Esther Willems
Translational Metabolic Laboratory
Department of Laboratory Medicine
Radboud Institute of Molecular Life Sciences
Radboud University Medical Center
Nijmegen, the Netherlands

Stefan Willems
University Medical Center
Utrecht, the Netherlands

Manfred Wuhrer
Center for Proteomics and Metabolomics
Leiden University Medical Center
Leiden, the Netherlands

Mary Zacour
PROOF Centre of Excellence
Vancouver, British Columbia, Canada

and

Biozac Consulting
Montreal, Quebec, Canada

Kurt Zatloukal
Diagnostic and Research Center for Molecular BioMedicine
Medical University of Graz
Graz, Austria

Weidong Zhang
Pfizer Inc.
New York, New York

1 What Is a Biomarker and Its Role in Drug Development?

Definitions and Conceptual Framework of Biomarkers in Precision Medicine

1.1

Claudio Carini, Attila A. Seyhan, Mark Fidock, and Alain van Gool

Contents

INTRODUCTION

The word "biomarker" in its medical context is a little over 30 years old, having first been used by Karpetsky, Humphrey and Levy in the April 1977 edition of the *Journal of the National Cancer Institute* where they reported that "serum RNase level… was not a biomarker either for the presence or extent of the plasma cell tumor" [1]. Few new words have proved so popular—a recent PubMed search lists more than 370,000 publications that use it.

Part of this success can no doubt be attributed to the fact that the word gave a long-overdue name to a phenomenon that has been around at least since the seventh century BC. However, while the origins of biomarkers are indeed ancient, the pace of progress over two thousand years was somewhat less than frenetic.

The pace of basic discovery research progress has been profound with the intertwining of innovative technologies. As technology and information advanced, the use of genetic, epigenetics, genomics, metabolomics, proteomics, and imaging in clinical and translational research is now being actively engaged.

Despite the significant growth in investment in biomarker research in private and publicly funded research, the reach of the impact of biomarkers into clinical practice interventions at present is challenging to quantify. The understanding of the clinical implications of disease markers has taken much longer than many had predicted as a consequence of relatively slow evidence development. There have been many challenges with establishing a translational research infrastructure that serves to verify and validate the clinical value of biomarkers as disease end points and their value as independent measures of health conditions. The wide adoption of biomarkers in clinical practice measures has not yet matured.

DEFINITION OF BIOMARKERS

A biomarker is a biological characteristic that can be molecular, physiologic, or biochemical. These characteristics can be measured and evaluated objectively. They act as indicators of a normal or a pathogenic biological process. They allow assessment of the pharmacological response to a therapeutic intervention. A biomarker shows a specific physical trait or a measurable biologically produced change in the body that is linked to a disease.

A biomarker may be used to assess or detect the following:

- A specific disease as early as possible—**diagnostic biomarkers** (Hepatitis C Virus RNA after infection).
- The risk of developing a disease—**susceptibility/risk biomarkers** (BRCA1-breast cancer).
- The evolution of a disease (indolent vs. aggressive)—**prognostic biomarkers** (HER-2-breast cancer)—but it can be predictive too.
- The response and the toxicity to a given treatment—**predictive biomarkers** (EGFRNSCLC/gefitinib, DPD-gastrointestinal cancer/fluoropyrimidines).
- The mechanism(s) of action of a drug—**mechanistic biomarkers**. Sequential gene profiling of basal cell carcinomas treated with imiquimod (a Toll-like receptor-7 agonist capable of inducing complete clearance of basal cell carcinoma.
- The pharmacological response to a drug in order to monitor clinical response—**pharmacodynamic biomarkers**. Biomarkers are the "PD" in pharmacokinetic and pharmacodynamic (PK/PD) and are critical to understanding the relationship between dose and biomarker response providing information during drug development.
- The clinical benefit/survival of a disease—**surrogate end point biomarkers**. If a treatment affects the biomarker, the biomarker will serve as a surrogate end point for evaluating clinical benefit (e.g., hemoglobin A1C).
- To predict toxicity to a specific therapy at the time of patient's enrollment—**safety biomarkers**. Given its significance in ultimate "go" or "no-go" decisions, no other biomarker is more important to drug testing than drug safety biomarkers.

HISTORY OF BIOMARKERS

Biomarkers have been one of the pillars of medicine since the time of Hippocrates and have become even more important in recent times. When clinical laboratories became well established in the middle of the twentieth century, many components of body fluids and tissues were analyzed. The use of blood glucose levels as a marker for diabetes and the later use of Her2 expression levels as the indicator of treatment response for breast cancer were recognized as some of the markers indicative of a pathological state and disease. Hundreds of biomarkers are now used for the prediction, diagnosis, and prognosis of disease and therapeutic interventions. The aim of medicine is now selecting biomarkers that are relevant for monitoring therapeutic intervention. Although we think of genes and proteins as biomarkers, there are many other forms of biomarkers that have expanded our understanding of, for instance, cardiovascular diseases. In 1733, Stephen Hales first measured blood pressure [2]. Over many years, the meaning of blood pressure as a surrogate marker for cardiovascular disease has become ingrained in current clinical practice.

Another early example is the recording of the first electrocardiogram (ECG) by Waller and its refinement by Einthoven in 1887 [3]. These scientists could not have ever imagined the value that the ECG would have in today's cardiology. The ECG is an excellent example of a diagnostic biomarker that is indicative of disease.

A modern concept of biomarkers for vascular diseases started with the correlation of blood cholesterol with coronary disease as reported in the Framingham Heart Study (1961), in which correlations between cholesterol and coronary disease were first reported.

During the last decade of the twentieth century, biomedical research underwent significant changes when the advance of new molecular and imaging technologies started to yield a remarkable body of knowledge for the discovery of new approaches to the management of human health and diseases.

The global economic threat from HIV was one of the first steps toward the development of a targeted therapeutic approach and the use of viral load and immune parameters (e.g., CD4+ count) as indicators of disease. The regulatory authorities have initiated a procedure allowing accelerated approval of medical products using surrogate end points. For example, in cancer therapy the use of clinical laboratory tests to stratify patients into responders and nonresponders allowed for a targeted therapy based on the relation that the Her2/neu tyrosine kinase receptor had with aggressive breast cancer and its response to Trastuzumab (Herceptin®) [4]. Similarly, the association of Imatinib (Gleevac®) responsiveness with the presence of the Philadelphia chromosome translocation involving BCR/Abl genes in chronic myelogenous leukemia [5] represented an example of a biomarker used as a surrogate end point for a patient clinical response.

Over the last decade of the twentieth century, however, a swift movement across the research and development enterprise was underway. It was obvious at the time that biomarker research in the 1990s, and into the early years of the twenty-first century, was driven by the rapid pace of genome mapping and the fall in the cost of large-scale genomic sequencing technology resulting from the Human Genome Project. A decade later, it is now apparent that biomarker research in the realm of clinical application has acquired a momentum of its own and is self-sustaining. Biomarkers and patient stratification will also help to reduce the attrition often observed in phase III clinical studies due to the wrong patient population having been selected or due to the lack of translation from preclinical to clinical studies.

UNDERSTANDING RISK

There is no question that most people who know something about the pharmaceutical industry consider developing a drug product a risky business. Usually, they mean that the investment of time and money is high, and the chance of a successful outcome is low compared with other industries that create new products. Yet the rewards can be great, not only in terms of monetary return on investment (ROI) but also in the social value of contributing an important product to the treatment of human disease.

A risk is defined as "the possibility of loss or injury" (Merriam-Webster's Collegiate Dictionary [6]).

The everyday decisions and actions that people take are guided by their conscious and unconscious assessments of risk, and we are comfortable with compartmentalized schemes where we sense a situation has very high, high, medium, low or very low risk. Some risks can be defined in more absolute terms such as some kind of population measure based on trend analysis of prior incidence statistics (e.g., current risk of postmenopausal white women being diagnosed with breast cancer). These types of population-based risk data affect decision making at the individual level.

As opposed to an individual's skill at assessing risks, decisions and actions taken during drug development require objective and systematic risk assessment by groups of people. There are many stakeholders (e.g., scientists, clinical investigators, ethics committees, regulatory authorities) involved in the development of a drug product. Each stakeholder has his or her own unique perspective of risk. The prime focus is the business risk to progress the drug at each phase of drug development. On the other hand, institutional review boards (IRBs), regulators, investigators, and patient associations are primarily concerned with the safety risks to the patient.

Before risk assessment can be fully appreciated in the context of decision making in drug development, it must be balanced with the perceived benefits of the product. For example, the patient and physician would tolerate a high level of risk to the patient if the potential benefits offered by a novel therapeutic were for a life-threatening disease for which there is no effective treatment.

Ideally, on many occasions during drug development, each stakeholder is asked to assess their view of risk versus benefit based on current data. Their assessment will become part of the decision-making process that drives drug development in a logical and hopefully collaborative way. Effective decision making requires integrating these varying risk–benefit assessments in a balanced way. The "stage gate" approach is a useful way to integrate these needs into the highly competitive and risky drug development process.

BIOMARKERS AND DECISION MAKING

Drug development is a process that proceeds through several high-level stage gates from the identification of a potential therapeutic druggable target to the discovery and development of candidate agents through to marketing a new drug product [7].

In order for a drug candidate to progress further in drug development, it must meet a set of criteria that have been previously agreed upon by the decision makers before they will open the gate. It is a "go/no-go" decision because the future product life of the drug hangs in the balance. Once a new drug candidate has progressed through a decision process, the organization should be committed to expend even greater resources (money and time) to do the studies and address the criteria for the next decision stage along the development path.

Disciplined planning and decision making are required to leverage the value of the decision-stage approach. Initially, a clear set of questions needs to be agreed upon at each decision stage and understood by the decision makers. In later stages of drug development, these questions and criteria are often presented in a target product profile (TPP), which is, in essence, a summary of a drug development plan described in terms of labeling concepts. The U.S. Food and Drug Administration (FDA) has formalized the agency's expectations for a TPP in a guidance document. Many companies start evolving a TPP even at the earliest go/no-go decision stages of drug development. In this way, a common way of thinking is preserved throughout the life of the product. The development program plan is assembled by determining what studies need to be done, as well as their design, in order to provide the information critical to answering the key questions at each go/no-go decision stage.

The value of an effective and successful drug development plan based on a decision-stage approach can only be leveraged if there is discipline in the decision-making process. Stakeholders need to clearly understand what their role is in making the decision at each stage. Are they a decider, a consultant or someone who just needs to know what the decision is in order to effectively do their job? Go/no-go decisions need to be made when all information required to answer the key questions is available. Go/no-go decisions should not be revisited. If this occurs, then either there was a lack of decision discipline or key information was missing when the original decision was made.

Some go/no-go decision stages require agreement to proceed by different groups of stakeholders. This is particularly true when the question "Is the drug candidate safe to give to humans?" is addressed. The sponsor must decide whether to file an Investigational New Drug (IND) application based on the data and information collected so far from animal tests and *in vitro* assays. The same data are also evaluated by the regulator (e.g., FDA) who must have time (30 days) to object if they feel the safety of the subjects in the first few clinical trials will be unduly compromised. Finally, the data are reviewed again by the IRB who look specifically at how safety will be evaluated and managed in the clinical trials.

The proper incorporation of biomarkers in the drug development strategy enables the concept of "fail fast, fail early," allowing early identification of the extremely high proportion of compounds that fail during drug development. The challenge, therefore, is to identify relevant biomarkers early enough to implement them for go/no-go decisions at critical stages of the development process.

Biomarker development forms one of the cornerstones of a new working paradigm in the pharmaceutical industry by increasing the importance of linking diagnostic technologies with the use of drugs. The position of the FDA is rather clear on this. Biomarkers are crucial to generate safe and efficacious drugs and are essential for deciding which patients should receive which treatment. It appears likely that the development of pharmaceuticals also will drive the development of biomarkers as diagnostic probes for clinical decisions and for the stratification of patients.

ROLE OF BIOMARKERS IN DISCOVERY AND PRECLINICAL DEVELOPMENT

Less than 10% of the drug candidates that enter clinical testing result in a marketed product [8]. This means that at some point during clinical development, over 90% of the drug candidates will fail [8–12] and the process of getting a new drug to market is becoming a longer, costlier, and riskier business with high failure rates [13,14].

The scale of the problem could be as simple as differences in methodology, with the improper use of statistical analysis methods and the misuse of p-values resulting in inaccurate conclusions that can result in a complete failure to replicate the results. It may be that the effect seen *in vitro* did not translate into animal studies or that the candidate drug did not have a good safety profile in animals or that the drug had poor pharmacology, a poor safety profile, or efficacy or that it had no effect on the mechanism of the disease. Occasionally, a drug is effective for only a subset of patients but not for everyone. In that case, stratification strategies such as the use of biomarkers can be used to identify responder versus nonresponder patients. Biomarkers that signal correct dosing and whether the specific molecular target has been hit in early proof-of-concept clinical trials can mitigate such attrition risks.

If failure occurs late in clinical development, hundreds of millions of dollars and a great deal of time will have been invested with little or no return. Proper incorporation of biomarkers in drug development strategy enables the fail-fast, fail-early concept. Early failure actually lowers overall risk because it enables one to move resources and to utilize available patients on other promising therapies.

As illustrated in Figure 1.1.1, biomarker data can be critical to a go/no-go decision. By definition, biomarkers may reflect the biology or the progression of the disease and/or the effect of drug treatment.

Therefore, information provided by properly selected biomarkers can greatly influence the decision on whether or not to progress through a go/no-go decision stage. The challenge is to identify relevant biomarkers early enough to implement them for go/no-go decisions at the critical early stages of the discovery–development process.

Biomarkers in discovery are valuable tools for understanding the pathogenesis of a disease and the pharmacology of a target and/or compounds (hits or leads) under investigation. Moreover, even at this very early stage, a biomarker may already be identified as a potential clinical safety or efficacy biomarker. Decision stages in discovery are frequently directed toward selection of a lead or a small number of leads to take forward to preclinical development and prioritization of the leads. For this purpose,

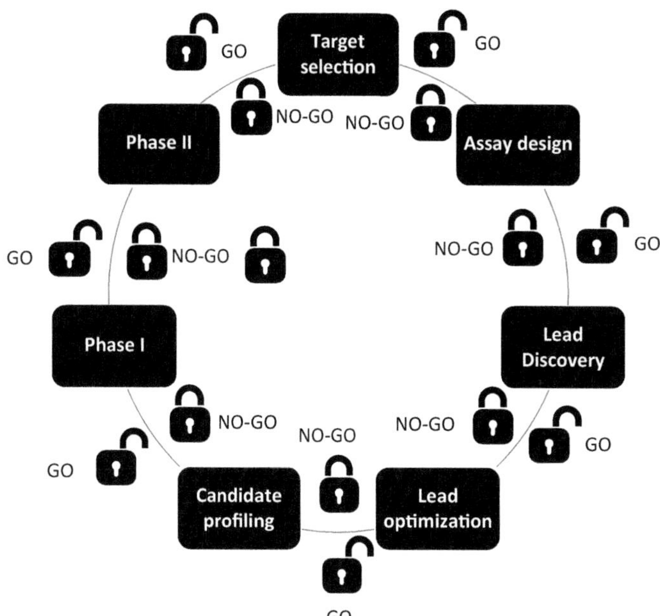

FIGURE 1.1.1 Stage-gate decision-making process in biomedical research. A stage-gate decision process in drug discovery, development, and management of research along milestones such as those indicated above is widely used in the biopharmaceutical industry. As in preclinical and clinical development, go/no-go criteria are defined for each milestone according to the target product profile that must be fulfilled to continue research. Biomedical discovery research is also covered by benchmark initiatives that provide average risk:success rates, timelines, and resources used per milestone. (Adapted from Bode-Greuel, K.M. and Nickisch, K.J., *J. Commer. Biotechnol.*, 14, 307–325, 2008.)

a number of commonly employed screening assays can generate useful biomarker data.

During preclinical development, the focus of studies is on revealing the potential toxicity of the drug candidate. Accordingly, there is great interest in identifying and monitoring safety biomarkers that can be used for go/no-go decisions, particularly for classes of compounds whose development has been plagued by unusual findings in animals that have not translated into human safety. Biomarkers that translate well across species are highly desirable but not always easy to find.

In both discovery and preclinical development, the abundance of new technologies—high content screening, imaging, genomics, proteomics, metabolomics, and systems biology, to mention a few—has opened the door to the identification of theoretically innumerable novel biomarkers along with an expectation that this will result in enhanced efficiency and better decision making with clinical impact. However, the technologies, and the biomarkers that they produce, have to be validated so that they are applicable to a wider patient population and they have to be evaluated in how they provide insight into understanding the biology and pharmacology being investigated and how they can be effectively used by scientists and clinicians in decision making. Typically, the increased volume and complexity of data requires sophisticated data analysis and informatics infrastructure to be in place before the new knowledge generated can be meaningfully applied to decision making.

ROLE OF BIOMARKERS IN EARLY CLINICAL DEVELOPMENT

Biomarkers may be the only way to address the decision stage asking the question, "Is there evidence that the drug is working in humans using the same mechanism of action defined by animal studies?" Can target engagement be demonstrated? Does this engagement deliver good pharmacology effect for a predictive positive outcome? Is the appropriate dose reaching the target organ? Those questions are answered by the clinical proof-of-concept (POC) study, often a phase IIa study in a small number of patients. However, occasionally clinical POC can be addressed during phase I development if the mechanism of action can be demonstrated in healthy volunteers (e.g., effects on blood pressure, body weight, lipids). The changes in the data collected during this study phase do not necessarily have to have achieved statistical significance; rather, there needs to be enough indication of potential efficacy to convince the decision makers to open the gate and spend the needed resources to progress the drug further. Often the minimal result for progression can be defined ahead of time, making the actual decision-making process much easier, more objective and transparent.

Early in clinical development, multiple biomarker assays or technologies may be utilized to more fully understand the safety and actions of the drug in humans. One may use a collection of biomarker methods, most of which are not likely to be fully validated surrogate markers of effect. For early decision making, it is not necessary to have results that meet the same standards of accuracy and quality as those used for drug approval. Indeed, it is during these early trials that promising tests for use in later clinical development or marketing are first identified from among the pool of available experimental biomarkers. Resources can then be put together to further develop the selected biomarker methodology to meet the expected quality standards for continued future use.

ROLE OF BIOMARKERS IN PHASE III AND POST-MARKETING DECISIONS

Traditional clinical trial end points such as morbidity and mortality often require extended time frames and may be difficult to evaluate. Fully validated surrogate markers can be used as primary end points for pivotal phase IIb and III safety and efficacy trials. Lowering blood pressure and lowering low-density lipoprotein cholesterol (LDL-C) are both examples of surrogate biomarker end points upon which several drugs have been approved. It has been reported that effects on these markers directly impact morbidity and mortality several years later. Similar promises have been delivered by the use of new technologies showing efficacy signals earlier in the treatment of certain slowly progressing diseases.

Patient selection is another area where decisions are made using biomarker information. For example, patients who tested negative for the Her2/Neu receptor responded poorly to Herceptin treatment, whereas good responders were positive for Her2/Neu receptor expression. This is consistent with the mechanism of action of the drug and resulted in regulatory approval for patients positive for Her2/Neu. The outcome was that a diagnostic test was required before this drug could be effectively marketed. Thus, this is now a biomarker that drives the treatment decisions of breast cancer oncologists. Her2 is a good example, although it is a rather old one. A current example is represented by programmed death-ligand 1 (PD-L1).

Identifying potential responders or nonresponders prior to phase III clinical trials can have a profound effect on the size of a clinical trial. Eliminating nonresponders can dramatically reduce the intersubject variability in response, thereby reducing the number of subjects in each group of a pivotal clinical trial required to demonstrate the effect. This results in large savings in time and costs. However, if the biomarker is novel, a diagnostic will need to be developed along with the drug before one can effectively market the product.

Nonetheless, the development and use of biomarkers must be dictated by the principle of being associated with how they will be used; that is, are they "fit-for-purpose"? [15].

Because biomarkers provide insightful information during the drug development and decision-making process, the biomarker strategy must be integrated into various phases of the drug development from early discovery to late clinical drug development. Linking biomarkers and mechanistic PK/PD models early in the preclinical phase of drug development can provide information on the underlying mechanism of drug action, establish a relationship between dose and response to treatment, impact dose escalation studies on the time course of response, measure duration of response and lag between plasma drug concentration and response, measure effect of repeated doses (toleration, sensitization, reflex mechanisms), help to identify range of doses and dosing intervals, and form the foundation for clinical trial simulations. Moreover, biomarkers can guide dose selection for phase II/III studies based on biomarker PK/PD and the projected therapeutic index and can help to differentiate the candidate drug from competitors based on the demonstration of a key response (e.g., safety, efficacy).

To have any utility, assays for biomarkers must be validated so that anytime, anywhere, and by anyone, they are reliable and reproducible. After validation, a biomarker can be used to diagnose disease risk, to make a prognosis, or to establish treatments for the patients.

However, the ultimate goal for a biomarker is for it to be used for making informed decisions that contribute to increasing confidence in the biomarker(s) to make go/no-go decisions and help design future studies. It is also worth mentioning that biomarkers used in clinical studies need to be performed under the Clinical Laboratory Improvement Amendments (CLIA) and in accordance with good clinical practice (GCP).

CONCLUSIONS

Biomarkers serve many purposes in drug development and the evaluation of therapeutic intervention and treatment strategies. They can be impactful for the selection of candidate compounds for preclinical and clinical studies. They can play a critical role in characterizing subtypes of diseases, allowing patient stratification for targeting with the most appropriate therapeutic intervention. However, strategies for development have faced many technical challenges, in part due to poor experimental design, small sample size, and patient heterogeneity.

Presumably, a more radical strategy is required to warrant future success of biomarker development and validation. By using the knowledge gained from earlier studies as well as new strategies for study designs and multi-omics technologies to execute them, it is anticipated that we should be able to develop robust biomarkers. Gathering information and careful design and planning of the study are crucial steps toward this goal. As design and planning commence toward the execution phase, other factors and considerations should be evaluated to allow for the optimization of the process until the objectives are achieved. This process should yield candidate biomarkers that can be validated and qualified as surrogate markers substituting for clinical end points and help in the decision-making process. Moreover, the standardization of sample collection and data analysis and an emphasis on computational biology to process large data sets is crucial to facilitate this process.

Finally, the robust linkage of a biomarker with a clinical end point is critical, although this may not be needed for the early clinical development where the aim is the validation of pharmacologic response, dose and regimen optimization. Realization of the potential benefits of biomarkers to facilitate the development of targeted, safe, and precision therapies will require linking biomarkers to clinical end points. To achieve all this will require a network of collaboration among academia, the pharmaceutical industry, governmental regulatory agencies and international networks all working toward the same goal.

REFERENCES

1. Karpetsky TP, Humphrey RL, Levy CC (1977) Influence of renal insufficiency on levels of serum ribonuclease in patients with multiple myeloma. Journal of the National Cancer Institute 58: 875–880.
2. Hamilton WF, Richards DW (1982) The output of the heart. In: Fishman AP, Richards DW (eds.) *American Physiological Society*. Bethesda, MD: American Physiological Society, 83–85.
3. Katz LN, Hellerstein HK (1982) Electrocardiography. In: Fishman AP, Richards DW (eds). *American Physiological Society*. Bethesda, MD: American Physiological Society, 86–89.
4. Ross JS, Fletcher JA, Linette GP et al. (2003) The Her-2/neu gene and protein in breast cancer 2003: Biomarker and target of therapy. *Oncologist* 8: 307–25.
5. Deininger M, Druker BJ. (2003) Specific targeted therapy of chronic myelogenous leukemia with imatinib. *Pharmacological Reviews* 55: 401–423.
6. Merriam-Webster (2008) *Merriam-Webster's Collegiate Dictionary*, 11th ed. Springfield, MA.
7. Pritchard JF, Jurima-Romet M, Reimer MLJ, Mortimer E, Rolfe B, Cayen MN (2003) Making better drugs: Decision gates in non-clinical drug development. *Nature Reviews Drug Discovery* 2: 542–553.
8. Kola I, Landis J (2004) Can the pharmaceutical industry reduce attrition rate? *Nature Reviews Drug Discovery* 3: 711–715.
9. Garner JP, Gaskill BN, Weber EM, Ahloy-Dallaire J, Prickett KP (2017) Introducing therioepistemology: The study of how knowledge is gained from animal research. *Lab Animal* 46: 103–113.
10. Paul SM, Mytelka DS, Dunwiddie CT et al. (2010) How to improve R7D productivity: The pharmaceutical industry's grand challenge. *Nature Reviews Drug Discovery* 9: 203–214.
11. Scannell JW, Blanckley A, Boldon H, Warrington B (2012) Diagnosing the decline in pharma R&D efficiency. *Nature Reviews Drug Discovery* 11: 191–200.
12. Hay M, Thomas DW, Craighead JL et al. (2014) Clinical development success rate for investigational drug. *Nature* 32: 40–51.
13. DiMasi JA, Hansen RW, Grabowski HG (2003) The price of innovation: New estimates of drug development costs. *Health Economy* 22 (2): 151–185.
14. Morgan S, Grootendorst P, Lexchin J, Cunningham C, Greyson D (2011) The cost of drug development: A systematic review. *Health Policy* 100: 4–17.
15. Lee JW, Devanarayan V et al. (2006). Fit-for-purpose method development and validation for successful biomarker measurement. *Pharmaceutical Research* 23 (2): 312–328.

In Search of Predictive Biomarkers in Cancer Immunotherapy

1.2

Establishment of Collaborative Networks for the Validation Process

Giuseppe Masucci, Paolo A. Ascierto, and Lisa Villabona

Contents

INTRODUCTION

Cancer immunotherapy is one of the most promising treatments for patients. Recently, immune checkpoints inhibitors have been introduced successfully in the care of a variety of malignancies. However, there are still limitations in the overall results, and the number of responses are not optimal, neither are the accompanying toxicity and high costs. Therefore, identifying biomarkers to determine which patients are likely to benefit from which immunotherapy and/or be susceptible to adverse side effects is an undeniable clinical and social need. In addition, with several new immunotherapy agents in different phases of development and with approved therapeutics being tested in combination with a variety of different standard-of-care treatments, there is a requirement to stratify patients and select the most appropriate population in which to assess clinical efficacy. The ideal solution could be to design parallel biomarker studies integrated within key randomized clinical trials. Sample collection (e.g., fresh and/or archival tissue, peripheral blood mononuclear cells [PBMCs], serum, plasma, stool) at specific points of treatment is important for evaluating possible biomarkers and studying the mechanisms of responsiveness, resistance, toxicity and relapse. Although numerous candidate biomarkers have been described, currently the only FDA-approved assays, based on PD-L1 expression, which have been clinically validated to select patients who may be more likely to benefit from single-agent anti-PD-1/PD-L1 therapy. Because of the complexity of the immune response and tumor biology, it is unlikely that a single biomarker will be sufficient to predict clinical outcomes in response to immune-targeted therapy. Rather, the integration of multiple tumor and immune response parameters—such as protein expression, genomics, and transcriptomics—may be necessary for accurate clinical prediction. Before a candidate biomarker and/or new technology can be used in a clinical setting, several steps are necessary to demonstrate its clinical validity. Although regulatory guidelines provide general roadmaps for the validation process, their applicability to biomarkers in the cancer immunotherapy field is somewhat limited.

The focus on the validation of clinical efficacy, clinical utility and regulatory considerations for biomarker development provides guidance for the entire biomarker improvement process, with a particular attention on the unique aspects of developing immune-based biomarkers. Specifically, knowledge about the challenges to clinical validation of predictive biomarkers, which has been gained from numerous successes and failures in other contexts, should be considered together with statistical methodological issues related to bias and overfitting. The different trial designs used for the clinical validation of biomarkers, as the selection of clinical metrics and end points, becomes critical to establish the utility of the biomarker during the clinical validation phase of their development. The regulatory aspects of submission of biomarker assays to the FDA as well as regulatory considerations in the European Union have to be considered in this process.

Biomarker Discovery and Validation in Cancer Immunotherapy

Biomarker discovery is a fundamental objective in the design of many clinical trials. Therefore, the incorporation of correlative biomarker studies, using state-of-the-art technologies, is required to maximize data generation. The challenge at this stage is that most completed or ongoing studies have not sufficiently incorporated biomarker assessment into their design.

The complexity of the immune response, tumor heterogeneity and patient diversity makes it unlikely that a single biomarker will be sufficient to predict clinical outcomes in response to the spectrum of immune-targeted therapies. Biomarkers correlated with clinical outcome can be identified at the molecular (e.g., genetics, epigenetics, metagenomics, proteomics, metabolomics), cellular, and tissue levels. Before a candidate biomarker and/or new technology can be used for treatment decisions in a clinical setting, several steps are necessary to demonstrate its clinical validity. The discovery and assessment of biomarkers using cutting-edge technologies across different clinical studies is a fundamental step in maximizing data generation (Figure 1.2.1).

A biomarker with clinical relevance requires rigorous validation that can be separated into several sequential steps: the assessment of basic assay performance (analytical validation), characterization of the assay performance with regard to its intended use (clinical validation), validation in clinical trials that ensures that the assay performs robustly according to predefined specifications (fit-for-purpose), and the establishment of definitive acceptance criteria for clinical use (validation of clinical utility). The fit-for-purpose approach (an umbrella term used to describe distinct stages of the validation process) for biomarker development and validation addresses the proper assay tailored to meet the intended purpose of the biomarker [1,2].

Clinical study design in which biomarker analysis is one of the primary objectives/end points needs to be promoted. As a good example, CA184-004 (NCT00261365) was such a study: a phase II trial to determine predictive markers of response to ipilimumab (MDX-010) [3]. In this study, the primary end point was to identify candidate markers predictive of response and/or serious toxicity to ipilimumab. Tissue and blood samples were collected at different time points from enrolled patients and the subsequent biomarker analyses generated interesting data. The findings of this study could have been considered as the first evidence for biomarker association with outcome and could potentially have been confirmed and prospectively validated in subsequent trials. However, results often cannot sufficiently meet expectations. They may be contradictory, in part due to the constraints of underpowered cohort size, variables in the time and type of sample collection, and differences in procedures, data generation and tools used at different sites. As stated earlier, the solution could be to design parallel biomarker studies that are integrated as key components of important randomized clinical trials. Sample collection at specific stages of treatment is important to evaluate possible biomarkers and to study mechanisms of response, resistance, toxicity and relapse. At present, several research institutions around the world are collecting tumor, blood, serum and stool samples to investigate prognostic and predictive biomarkers and to better understand the complex immunobiology of patients and their cancers.

Biomarker discovery is a fundamental objective in the design of many clinical trials. Therefore, the incorporation of correlative biomarker studies using state-of-the-art technologies within clinical trials in order to maximize data generation is required. The challenge at this stage is that most completed or ongoing clinical trials have not sufficiently incorporated biomarker assessment into their design. There is a need for an international joint effort to maximize data, information and knowledge generation from existing and completed clinical trials and to design clinical trials that will better address these important issues.

International Collaboration for Studying Cancer Immunotherapy Biomarkers

It is fundamental to encourage the creation of a network to facilitate the sharing and coordinating of samples from clinical trials in order to allow more in-depth analyses of correlative biomarkers than is currently possible. The feasibility, logistics, and various stakeholder interests in such a network could also be considered. A high standard of sample collection and storage as well as the exchange of samples and knowledge through collaboration has

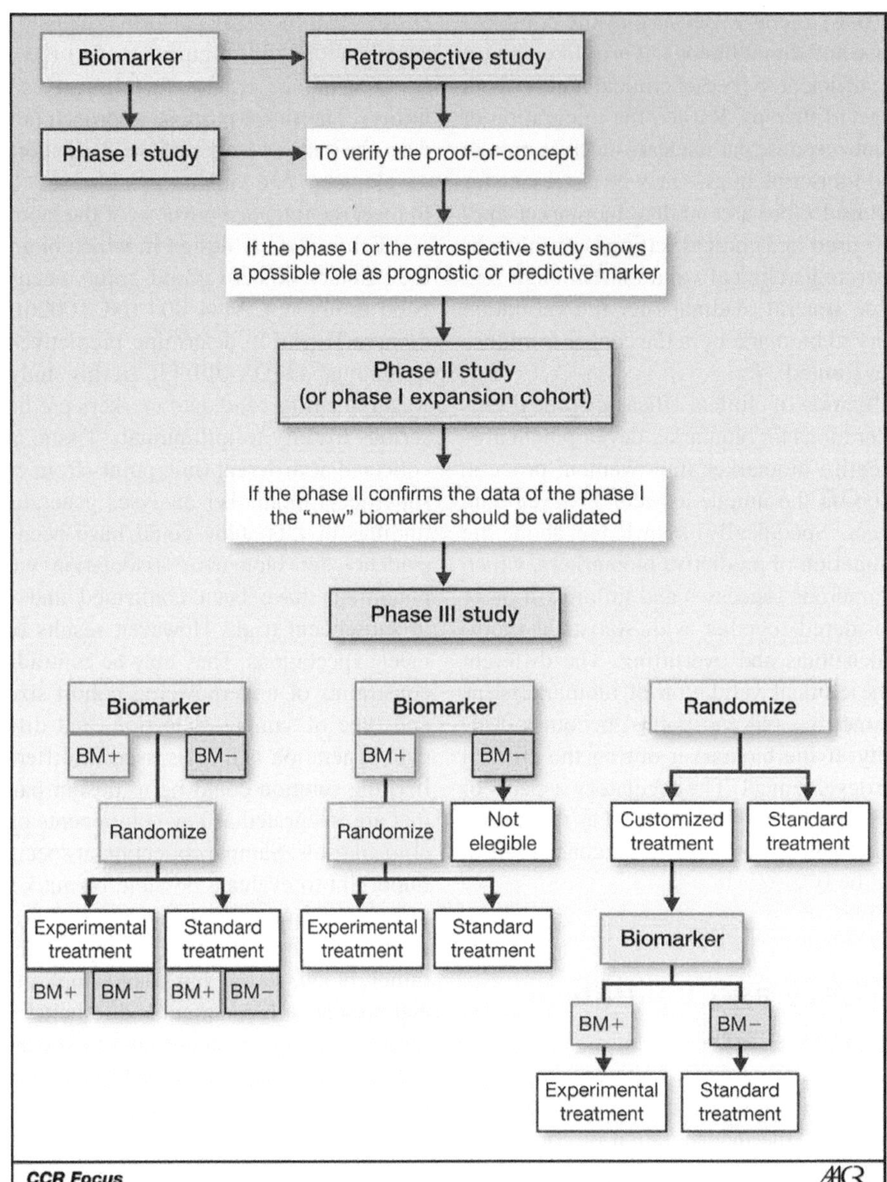

FIGURE 1.2.1 Procedures for introduction and validation of biomarkers (BMs). Alternative solutions for randomized phase III trials in the presence of a potentially predictive marker of efficacy of treatment. (BM+, biomarker positive; BM−, biomarker negative). Bottom left, "randomize-all" solution with determination and prospective stratification of BM+ and BM− patients. Center, "targeted" design. Right, "customized" solution. (From Ascierto, P.A. et al., *Clin. Cancer. Res.*, 19, 1009–1020, 2013; Di Maio, M. et al., *Lung Cancer*, 67, 127–135, 2010.)

to be taken into consideration in addition to how this could move forward using banked samples from completed studies or with prospective planning of ongoing and future clinical trials.

There are a number of compelling arguments for the establishment of international cancer immunotherapy biomarker collaborations to (a) maximize the potential of novel biomarker discovery using samples from multi-institutional clinical practice and clinical studies and will offer a new breadth of experience and expertise; (b) allow streamlined sample access by developing online registration, bio-banking and the inventory and tracking of archived samples in order to best utilize samples that exist after designated trials are completed; (c) set up and share standard operating procedures (SOPs) to harmonize future sample collection processing and banking; (d) improve access to samples by initiating a new patient sample registry or by joining forces with existing international clinical trial patient registries with available clinical data and biological sample collection, including storage conditions and inventory information; (e) provide support/guidelines in correlative study design; (f) engage and leverage with cancer societies, pharma and biotech companies, and government institutes in order to improve the development of biomarkers; and (g) accelerate biomarker development by bringing together groups from around the world that can collaborate from proof-of-concept to validation. Moreover, the experience of the worldwide Immunoscore Validation Project, coordinated by the Society for Immunotherapy of Cancer (SITC), is an important example of an "honest broker" approach for coordinating a specific study and data sharing [4] to reinforce the concept of a stable-standing consortium that would be able to take short, intermediate and long-term views toward biomarker development for more effective care of patients with cancer.

It is relevant to emphasize the importance of proposal (d) above in order to gain data from large cohorts of patients. It is also beneficial to have a stable standing collaborating syndicate among the major contributing institutions, societies and research units. Recently, a consortium organized by the University of Tubingen was able to collect samples from several different institutions and analyze a large number of patients treated with checkpoint inhibitors. [5]. The worldwide Immunoscore project is another good example of an effective collaboration, which has successfully validated a previously described biomarker [4].

Challenges in Biomarker Discovery

A series of challenges are confronted when collaborative projects are proposed:

1. Limited or fragmented resources
2. Insufficient numbers of patients per cohort
3. Inclusion of patients with diverse treatment history (previous antitumor treatment), histological and radiographic conditions
4. Limited and/or suboptimal correlative study design due to funding and/or regulatory constraints
5. Heterogeneity in the types of biological samples with different time points of collection, storage conditions, platforms used for data generation, lab-driven SOPs and data analysis algorithms and tools
6. Heterogeneity in the clinical data collected and the length of follow-up
7. Lack of international joint initiatives
8. Potential intellectual property

Previous programs that utilized treatment stratification biomarkers had higher success rates at each phase of development versus the overall data set [6]. Moreover, the importance of determining biomarkers for both patient selection for treatment and selection of treatment for the patient is of utmost importance. For example, the assessment of tumor PD-L1 status is not critical for selecting patients with metastatic melanoma for treatment with anti-PD-1 inhibitors (pembrolizumab or nivolumab). In fact, patients with PD-L1–negative tumors may receive long-term benefit from anti-PD-1/PD-L1 treatment. However, PD-L1 status might be important for selecting patients for combination treatment (e.g., anti-CTLA-4 and anti-PD-1). Currently available data show that overall survival (OS) in PD-L1–positive patients treated with combined anti-CTLA-4 and anti-PD-1 is not superior to nivolumab monotherapy [7]. PD-L1 expression, therefore, is not the best example of a biomarker for patient selection in the context of checkpoint inhibitors. Although patients with strongly positive PD-L1 tumors (>50% on tumor cells) clearly had an advantage for anti-PD-1 therapy versus chemotherapy in the Keynote 024 study, it is less clear why the same patient population did not demonstrate the same advantage from nivolumab therapy in the Checkmate 026 pivotal study [8]. Anti-PD-1/PD-L1 treatments produced similar efficacy across different clinical trials; the problem may be in the immunohistochemistry (IHC) assay and the immunological characterization/criteria used to evaluate PD-L1 positivity. However, the FDA recently (May 11, 2017) approved the combination of chemotherapy with pembrolizumab for first-line treatment of non-small cell lung cancer (NSCLC) regardless of PD-L1 expression, confirming the need for predictive biomarkers [9].

Because the FDA issued important guidance that all drugs should be accompanied by a companion diagnostic (CDx) [10,11], the industry has applied a thorough understanding of biology and the immune system to develop robust and meaningful biomarkers. While bio-companies may believe that leveraging biomarkers and companion diagnostics are critical to precision medicine, developing them is challenging and expensive and has therefore been less of a priority. On the other hand, the introduction of Pembrolizumab in clinical trials should be an example to others. Industry efforts might be helped by strategic partnerships with academic institutions identifying relevant clinical biomarkers. For example, the processes of detecting reproducible, predictive biomarkers and developing robust companion diagnostics substantially benefit from the correlation between genomics and complex tissue analysis data. This includes the analysis of spatial relationships between the immune cells hosting the tumor environment and the development of highly sensitive, precise, quantifiable and reproducible assays. These issues are challenging, time-consuming and need a large capital investment. Pursuing a thorough, evidence-based and scientifically driven drug discovery and development program is the desired long-term goal. However, funds to support these programs could instead be used for more short-term plans such as recruiting additional patient groups into a clinical trial in order to meet regulatory objectives.

At present, the best example of a truly successful and important immune-oncology predictive biomarker is mismatch repair deficiency: Cancer patients with microsatellite instability high (MSI-H) tumors independent of the tissue of origin benefit from immunotherapy with PD-1 inhibitors [12]. In particular, in colorectal cancer, MSI-H patients experience up to 50% or higher responses, while few if any responses are seen in MSI-low (L) colorectal cancer patients. The clinical demonstration of this concept has led to a recent approval by the FDA [13]. We believe that these two examples of single predictive biomarkers (i.e., PD-L1 and MSI) will be the exception and not the rule and that we will have to search for the integration of several different biomarkers. This amplifies the challenge and will require a highly coordinated effort across multiple institutions.

Patient Inclusion According to the Study Design

The development and implementation of appropriate biomarker assays to study T cells and other cells in the microenvironment is an essential companion objective for clinical trials that seek to evaluate immunotherapeutic agents, particularly when used in combination. In principle, the efficacy of a compound (in this case, an immune-targeted agent) is at least partially dependent on the presence of the target in the tumor. In an ideal scenario, when complete information on predictive factors and proper selection of patients can be obtained in the early phases of drug development (phase I–II studies), the conduct of subsequent phase III studies could be optimized. Unfortunately, this ideal scenario rarely occurs. The clinical immuno-oncology research community is dealing with several key questions, including (a) what metrics are best for biomarker evaluation in phase III studies, (b) what primary and secondary end points should be assessed, and (c) what are the statistical properties of various metrics. When planning a phase III trial comparing an experimental treatment with the standard treatment, we often have evidence supporting the predictive role of a biomarker, whereas patients with the absence of such expression should not respond. In such a scenario, different strategies are theoretically possible (Figure 1.2.2): (a) a "randomize all" strategy, that is, randomization between standard and experimental treatment without selection, possibly with stratification based on biomarker status (in this case, "stratified trial design" or "treatment–marker interaction design"); (b) "targeted" design, that is, randomization between standard and experimental treatment only in patients selected according to the status of the marker (also called "enrichment design"); and (c) "customized"

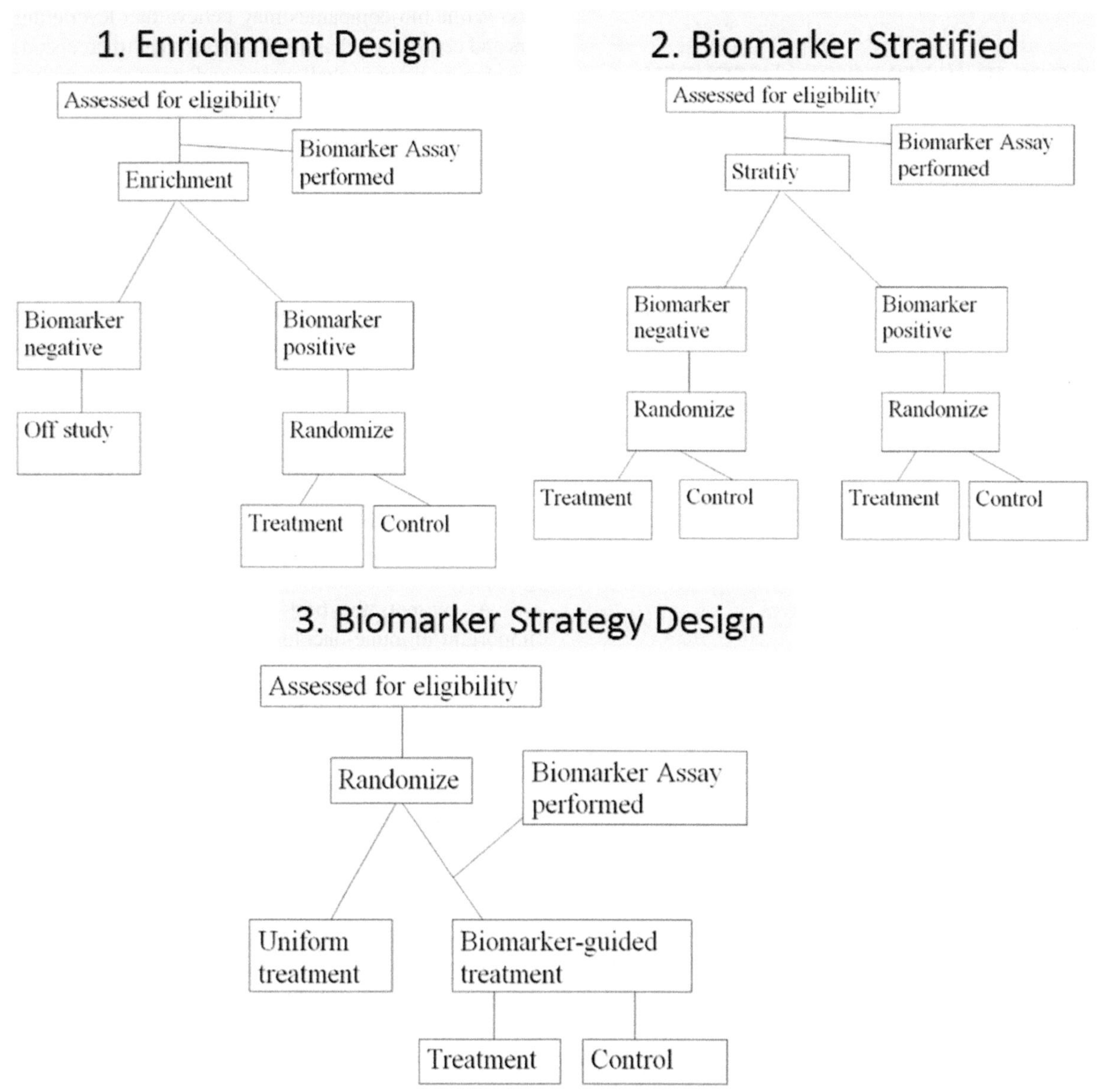

FIGURE 1.2.2 There are three basic phase III design options for assessing the ability of a biomarker. (1) The enrichment design includes only patients who are positive for the biomarker in a study evaluating the effect of a new therapy. (2) In the biomarker stratified design, all patients, independent of biomarker results, are enrolled and randomized to treatment and control groups within each of the biomarker positive and negative groups to ensure balance. Finally, (3) in the biomarker strategy design, patients are randomized between no use of the biomarker (all patients receive standard therapy on that arm) and a biomarker-based strategy where biomarker-negative patients receive standard therapy and biomarker-positive patients receive the new therapy. (From Dobbin, K.K. et al., *J. Immunother. Cancer*, 4, 77, 2016.)

strategy (also called "marker-based strategy"), that is, randomization between a standard arm in which the treatment is the same for all patients, and a personalized arm in which treatment is chosen on the basis of the marker status of each patient.

BIOMARKERS VALIDATION PROCESS

Validation and Regulatory Considerations

With the increasing understanding of the molecular basis of cancer, research and clinical laboratories are developing and implementing a variety of molecular diagnostic tests to guide cancer therapy, including immunotherapy. Before introducing any new test into the market, the analytic and clinical performance characteristics of the assay must be validated. If the assay is developed as an *in vitro* diagnostic (IVD), then it must be approved and cleared by the FDA and by corresponding bodies in other countries; if the assay is developed as a laboratory-developed test (LDT), only analytic validation is needed for commercialization. Understanding the regulatory approval process for IVDs to be used in making health care decisions is important for the development and performance assessment of any clinical diagnostic.

It happens that a subcohort of patients will experience substantial clinical benefit in response to different immunotherapeutic approaches, while the rest of the patients do not benefit but are still exposed to potentially significant drug toxicities. Therefore, a growing critical need for the development and clinical use of predictive biomarkers exists in immune-oncology. Predictive cancer biomarkers can be used to identify *a priori*, that is, before treatment, the patients who are or are not likely to derive benefit from a specific therapeutic approach. In order to be applicable in the clinical setting, predictive biomarkers must be carefully shepherded through a step-wise, highly regulated developmental process. There are documents providing background, examples and "good practice" recommendations on the pre-analytical and analytical phases of the biomarker development process as outlined above. The clinical validation and regulatory considerations provide specific knowledge gained from the numerous successes and failures in other contexts. Because the selection of metrics and end points becomes critical to establishing the clinical utility of the biomarker during the clinical validation phase of the biomarker development, the different prospective trials designed for the clinical validation of biomarkers have to be taken in consideration (Table 1.2.1).

TABLE 1.2.1 Parameters for evaluating clinical validity of a predictive biomarker

CLINICAL PARAMETER	*DEFINITION*
Sensitivity	Ability of a biomarker to predict a meaningful change in a clinical end point. It describes the relationship between the magnitude of change in the biomarker toward the clinical end point. *[For example, a 50-unit increase in OncotypeDX recurrence score (RS-PCT/50) was associated with an estimated increase of 2.87 in hazard ratio (Tang et al., 2011 [17]) of distant recurrence (DRFI end point) in tamoxifen-treated patients.]*
Specificity	Ability of a biomarker to distinguish changes in clinical end points in patients responding from nonresponding to an intervention in terms of changes in clinical endpoints. *[For example, the estimated hazard ratio for chemotherapy (no chemotherapy divided by chemotherapy) in the low OncotypeDX recurrence score (RS) group was 1.31 vs. 0.26 in the high RS group (Tang et al., 2011 [17]), where the outcome is DRFI.]*
Probability of false positives	An expected change in a biomarker does not reflect a positive change in a clinical end point or, even worse, is associated with a negative change in a clinical end point. For instance the detection of elevated levels of the functional or biochemical marker in the absence of clinical response to treatment. *[For example, a tumor that has expressed programmed death ligand-1 (PD-L1) on the tumor cells, but does not respond to targeted anti-PD-L1 immunotherapy, is a false positive.]*
Probability of false negatives	When no change or a small observed change in a biomarker fails to signal a positive, meaningful change in a clinical end point. *[For example, a tumor that does not express PD-L1 but does respond to anti-PD-L1 immunotherapy is a false negative.]*
Receiver operative characteristics (ROC) analysis	A graphical approach for showing accuracy across the entire range of biomarker concentrations. ROC, used to set cut points, is essentially a plot that captures the true positive rate against the false positive rate of an assay.
Area under the curve (AUC)	Area under the ROC curve, AUC is used to compare different tests, If an AUC value is close to 1 it indicates good discrimination, whereas an AUC of 0.5 provides *no* useful information regarding the likelihood of response.
Cut point	The sensitivity and specificity of the assay must be demonstrated through robust ROC curves that provide support for the cut points established to identify responders vs. nonresponders.
Hazard ratio	Chance of an event (e.g., disease recurrence, death) occurring in the treatment arm divided by the chance of the event occurring in the control arm, or vice versa.
Relative risk	Ratio of the probability of an event (e.g., disease recurrence, death) occurring in treated group to the probability of the event occurring in the control group.

Source: Dobbin, K.K. et al., *J. Immunother. Cancer* 4, 77, 2016.

Pre-Analytical and Clinical Validation

Although numerous candidate biomarkers have been described, there are currently only FDA-approved assays based on PD-L1 expression that have been clinically validated to select patients who may be more likely to benefit from single-agent anti-PD-1/PD-L1 therapy. Although regulatory guidelines provide general roadmaps for the validation process, their applicability to biomarkers in the cancer immunotherapy field is somewhat limited. To illustrate the requirements for validation, there are examples of biomarker assays that have shown preliminary evidence of an association with clinical benefit from immunotherapeutic interventions. The scope includes only those assays and technologies that have established a certain level of validation for clinical use (fit-for-purpose). Recommendations are considered to meet challenges and strategies to guide the choice of analytical and clinical validation design for specific assays. Their validation should ultimately qualify them for use in clinical decision making. Examples of assays already or soon to be approved for use in clinical laboratories are presented to exemplify the requirements for analytical validation (Table 1.2.2).

TABLE 1.2.2 Different levels of evidence of clinical validity/utility from biomarker assays predictive for response to immunotherapy

BIOMARKER ASSAY	*BIOMARKER*	*CLINICAL USE*	*STUDY TYPE/LEVEL OF EVIDENCE*	*REFERENCES/ REGULATORY CLEARANCE*
PD-L1 22C3 pharmDx IHC companion diagnostic	PD-L1	Predicting response to anti-PD-1 therapy (pembrolizumab) in NSCLC 50% cutoff	Prospective, phase III clinical trial KEYNOTE-001	[14,15]
PD-L1 28-8 pharmDx IHC complementary test	PD-L1	Risk vs. benefit of anti-PD-1 therapy (nivolumab) in NSCLC and melanoma- continuous correlation of PD-1 expression with magnitude of treatment effect	Prospective, phase III clinical trial CheckMate-057	[14]
PD-L1 SP142 IHC complementary test	PD-L1	Risk vs. benefit of anti-PD-L1 therapy (atezolizumab) for metastatic urothelial bladder cancer	Prospective, phase II clinical trial IMvigor-210	[16]
Immunohistochemistry	Tumor T cell Infiltrate, PD-L1 with spatial resolution	Predictive to anti-PD-1 therapy in melanoma and NSCLC	Retrospective, exploratory analysis	[17–19]
Enzyme-linked immunospot (ELISpot)	IFNγ release	Posttreatment/monitoring, cancer vaccines	Retrospective, exploratory analysis	[20,21]
Multiparametric flow cytometry	MDSC, Tregs, ICOS+ CD4 T cells	Posttreatment/monitoring, cancer vaccines, Predictive of anti-CTLA-4 therapy in RCC and melanoma	Retrospective, exploratory analysis, phase I, II trials	[22–24]
Single cell network profiling	AraC → cPARP AraC→CD34	Predictive of response to induction therapy in elderly patients with *de novo* acute myeloid leukemia	Retrospective, training and validation study establishing clinical utility	[25]
Complete blood count	ALC	Predictive of response to anti-CTLA-4 therapy	Retrospective, small cohort, significant variability among institutions	[26]
TCR sequencing	Limited clonality	Assessments of tumor-infiltrating lymphocytes. Predictive to response with anti-CTLA-4 and anti-PD-1 in melanoma	Retrospective, small cohort	[17] [27]
NanoString	Gene expression profile	Predictive of response to anti-PD-1 therapy in melanoma	Retrospective, training and test sets—prospective validation ongoing on different tumor types	[28]
Next-generation sequencing	Mutational load	Predictive of response to anti-CTLA-4 therapy in melanoma and anti-PD-1 in NSCLC	Retrospective, small cohort, training and test sets	[29,30]
NGS/*in Silico* epitope prediction	MHC class I epitope frequency/specificity	Predictive of response to anti-CTLA-4 and anti-PD-1 in melanoma, NSCLC, and CRC	Retrospective, small cohorts	[29–31]
Microsatellite instability analysis	Mismatch repair status	Predictive of response to anti-PD-1 therapy in CRC	Phase II study, small cohort	[31]

Source: Masucci, G.V. et al., *J. Immunother. Cancer*, 4, 76, 2016.

Prototypes of these assays have been shown in research studies to be potentially valid as tools to enrich selected patient cohorts. Although analytical validation data for each specific platform are available, none of these have yet been clinically validated as a predictive biomarker except for PD-L1.

Although assays for cancer immunotherapy are subject to the same analytical validation requirements as other bioanalytic assays, some basic differences may impact the validation process. Table 1.2.2 highlights the differences between single-analyte bioassays (measuring a single protein or metabolite) versus assays measuring an immune response. Although immune response assays can be singular, most biomarkers will require multiparameter tests that depend on an increased number of controls, complex scoring algorithms, high-throughput performance data analysis, and results output. In addition, when a predictive marker will be used to direct patient enrollment or for patient stratification in clinical trials in the United States, the assay will need to be performed in a CLIA laboratory. CLIA labs follow Clinical and Laboratory Standards Institute (CLSI) guidelines for determination of standard assay parameters such as precision, accuracy, limit of detection, specificity, and reference range. The most well known accreditation bodies are listed below:

1. Clinical Laboratory Improvement Amendments (CLIA)
2. United Kingdom Accreditation Service (UKAS)
3. European Co-operation for Accreditation (EA)
4. International Laboratory Accreditation Cooperation (ILAC)
5. International Accreditation Forum (IAF)
6. The American Association for Laboratory Accreditation (A2LA)
7. International Accreditation New-Zealand (IANZ)

A typical *pre-* and *analytical validation* plan involves several steps in which the assay must be optimized for multiple parameters:

1. Sample-related (pre-analytic parameters)
2. Assay-related (analytical parameters)
3. Data-related (post-analytical parameters)

An important moment in biomarker validation is the evaluation of *pre-analytical factors* that may affect assay performance due to specimen-related variability. The efficacy of immunotherapies needs to be monitored *ex vivo* in phenotypical or functional assays, which require high-quality samples to ensure reliable analytic output. SOPs are essential to ensure that optimal pre-analytic processing regimens are followed to control specific biomarker development moments. To create the best practice metrics, blood collection and storage media optimization protocols are often obtained in conjunction with other pre-analytical parameters. General guidance on pre-analytical quality indicators and their harmonization, including analytical stability and laboratory quality control (QC) have been published. [35] To improve the standardization of specimens, the U.S. National Cancer Institute (NCI) has published best practice guidelines for bio-specimen collections [36]. In addition, specific guidelines for the analytical requirements of biomarkers have been set up [37,38].

Analytical validation confirms that the assay used for the biomarker measurement has established (a) accuracy, (b) precision, (c) analytical sensitivity, (d) analytical specificity, (e) reportable range of test results for the test system, (f) reference intervals (normal values) with controls and calibrators, (g) harmonized analytical performance if the assay is to be performed in multiple laboratories, and (h) appropriate QC measures. These requirements for analytical validation as well as their definitions are summarized in full in Table 1.2.3.

Clinical validity and utility, the final stage in the development of a biomarker predictive of clinical outcome, is the assessment of its clinical validity and utility through the application of the analytically validated assay within a clinical trial, with multiple design options depending on the intended use of the test and availability of specimens from previous clinical trials. Clinical validity relates to the observation that the predictive assay reliably divides the patient population(s) of interest into distinct groups with divergent expected outcomes to a specific treatment [5,6]. The criteria for validation are defined by the nature of the question that the biomarker is intended to address (i.e., fit-for-purpose). A predictive biomarker needs to demonstrate the association with a specific clinical end point

TABLE 1.2.3 Analytical validation

REQUIREMENTS	*DEFINITION*
Sensitivity	Ability of the assay to distinguish the analyte of interest from structurally similar substance
Specificity	Degree of interference by compounds that may resemble but differ from the analyte to be quantified
Linearity	Ability of an assay to give concentrations that are directly proportional to the levels of the analyte following sample dilution
Precision	Agreement between replicate measurements
Limit of detection	Lowest concentration of analyte significantly different from zero; also called the analytical sensitivity
Accuracy	Agreement between the best estimate of a quantity and its true value
Repeatability	Describes measurements made under the same conditions
Reproducibility	Describes measurements done under different conditions
Robustness	Precision of an assay following changes in assay conditions, e.g., ambient temperature, storage condition of reagents

Source: Jennings, L. et al., *Arch. Pathol. Lab. Med.*, 133, 743–755, 2009.

(e.g., survival or tumor response) in pretreatment samples from patients who have been treated or exposed to a uniform treatment intervention.

Validating Mathematic Methods: Biomarker Characteristics—Single-Analyte versus Multivariate Assays

Predictive markers can be defined as a single biomarker or signature of markers that separate different populations with respect to the outcome of interest in response to a particular treatment. A distinguishing characteristic of multivariate assays is that computational methods are applied to the high-dimensional data such as gene expression profiling using NanoString, single cell network profiling (SCNP), or fluorescence-activated cell sorting (FACS), to build mathematical models, often from a subset of the measured variables that have been identified through data-driven selection. This is in contrast to the single-analyte molecular tests based on prespecified, biologically driven variables such as mutations in genes (e.g., BRAF) or protein expression targeted by a specific therapeutic agent (e.g., HER2/neu expression). Single-analyte tests must be based on well-established analytical performance. Similarly, multianalyte assays based on complex computational models must also achieve robust analytical performance but pose additional challenges that are distinct from the single-analyte realm.

One of the most common problems in clinical validation is bias or systematic error that is the source of results unrelated to clinical outcomes and that are not reproducible. Sources of bias can include (a) differences in relevant demographic characteristics between training and testing sets, (b) differences in pre-analytic variables (e.g., sample handling, storage time, and variability arising from different collection protocols), and (c) divergence from assay protocols. These critical issues, which are often overlooked in the biomarker discovery process, are likely to be among the greatest reasons why most biomarker discoveries fail to be clinically validated.

Computational methods are applied to generate functional algorithms for assays that measure multiple variables to predict clinical parameters, such as patient outcome in response to treatment. These algorithms are vulnerable to overfitting, which can occur when large numbers of potential predictors are used to discriminate among a small number of outcome events. It can result in apparent discrimination (e.g., between patients whose tumor responded or did not respond to a certain treatment) that is actually caused by chance and is, therefore, not reproducible. Thus, the importance of rigorously assessing the biological relevance and clinical reproducibility of the predictive accuracy of an assay is higher in the development of the computational model than for a single biomarker–based test.

The high dimensionality of omics data and the complexity of many algorithms used to develop omics-based predictors, including immunomics, present many potential pitfalls if proper statistical modelling and evaluation approaches are not used. Various statistical methods and machine learning algorithms are available to develop models, and each has its strengths and weaknesses. With the development of next-generation sequencing (NGS) and other molecular technologies, the dimensionality and complexity of potential diagnostics has greatly increased; in particular, storing the resulting terabytes of biological data becomes challenging.

Recommendations for Clinical Validation

- For multi-analyte classifiers, internal validation should be performed for the model development, tuning, and validation.
- External validation is critical. In external validation, a fully "nailed down" predictor is applied to a novel data set from a source that is different (typically a different laboratory and clinic) and most critically a non-overlapping set of patients.
- Many modern statistical methods involve extensive resampling of a training set during the model development and complex averaging over a large and varied set of prediction models. These methods include statistical boosting and bagging as well as Bayesian model averaging. The resulting black box nature of these algorithms makes them problematic to evaluate. As they move toward the clinic, these should be simplified into more transparent models, such as linear or generalized linear models.
- Cut points, used for classification and stringency levels and model tuning, need to be specified prior to external validation on independent data sets.

FINAL REMARKS

Cancer immunotherapies are rapidly changing traditional treatment paradigms and resulting in durable clinical responses in patients with a variety of malignancies. However, the overall number of patients who will respond to these therapies is limited. Furthermore, there are significant costs as well as potential toxicities that are associated with these therapies that impede their potential impact. Thus, there is a need to develop predictive biomarkers in order to maximize the clinical benefits of this innovative therapy. Although numerous candidate biomarkers have been described, currently only two assays are FDA-approved (one as a companion and one as a complementary diagnostic) to identify NSCLC patients who would benefit from anti-PD-1 therapies. Because of the complexities of both the immune response and of tumor biology, there are unique aspects to the validation process that must be taken into consideration during the planning and implementation phases of biomarker development.

SUMMARY

Immunotherapy has emerged as one of the most promising approaches to treating patients with cancer. Recently, the entire medical oncology field has been revolutionized by the introduction of inhibitors of immune checkpoint blockers. Despite demonstrated successes in a variety of malignancies, responses typically occur in only a small percentage of patients in any given histology and treatment regimen. There is a concern that immunotherapies are associated with immune-related toxicity and have high costs. Therefore, biomarkers determining which patients would derive clinical benefit from which immunotherapy and/or be susceptible to adverse side effects is a compelling clinical and social question. In addition, with several new immunotherapy agents in different phases of development, and approved therapeutics being tested in combination with a variety of different standard-of-care treatments, there is an urgent need to stratify patients and select the right population for clinical efficacy testing. The opportunity to design parallel biomarker studies inside the most important randomized clinical trials could be the ideal solution. Sample collection (e.g., fresh and/or archival tissue, PBMC, serum, plasma, stool), at specific points of treatment, is important for evaluating possible biomarkers and studying the mechanisms of responsiveness, resistance, toxicity and relapse.

There is a critical requirement of collaborative networks to achieve a fingerprint for each patient to an eligible, ad hoc, highly granted immunotherapy treatment. The care of cancer patients needs the discovery of new biomarkers to improve the possibility of narrowing the best effective immunotherapy, the pre-analytical and analytical aspects and clinical validation and regulatory considerations as they relate to immune biomarker development. Together, these considerations are relevant challenges for the entire biomarker discovery and validation process.

REFERENCES

1. Masucci GV, Cesano A, Hawtin R, Janetzki S, Zhang J, Kirsch I et al. Validation of biomarkers to predict response to immunotherapy in cancer: Volume I—Pre-analytical and analytical validation. *J Immunother Cancer* 2016, 4:76.
2. Dobbin KK, Cesano A, Alvarez J, Hawtin R, Janetzki S, Kirsch I et al. Validation of biomarkers to predict response to immunotherapy in cancer: Volume II—Clinical validation and regulatory considerations. *J Immunother Cancer* 2016, 4:77.
3. Hamid O, Schmidt H, Nissan A, Ridolfi L, Aamdal S, Hansson J et al. A prospective phase II trial exploring the association between tumor microenvironment biomarkers and clinical activity of ipilimumab in advanced melanoma. *J Transl Med* 2011, 9:204.
4. Galon J, Mlecnik B, Marliot F, Ou F-S, Bifulco CB, Lugli A et al. Validation of the Immunoscore (IM) as a prognostic marker in stage I/II/III colon cancer: Results of a worldwide consortium-based analysis of 1,336 patients. *J Clin Oncol* 2016, 34.
5. Weide B, Martens A, Hassel JC, Berking C, Postow MA, Bisschop K et al. Baseline biomarkers for outcome of melanoma patients treated with pembrolizumab. *Clin Cancer Res* 2016, 22:5487–5496.
6. Buyse M, Sargent DJ, Grothey A, Matheson A, de Gramont A. Biomarkers and surrogate end points—the challenge of statistical validation. *Nat Rev Clin Oncol* 2010, 7:309–317.
7. Ugurel S, Rohmel J, Ascierto PA, Flaherty KT, Grob JJ, Hauschild A et al. Survival of patients with advanced metastatic melanoma: The impact of novel therapies. *Eur J Cancer* 2016, 53:125–134.
8. Larkin J, Chiarion-Sileni V, Gonzalez R, Rutkowski P, Grob JJ, Cowey CL et al. Overall Survival Results from a Phase III Trial of Nivolumab Combined with Ipilimumab in Treatment-naïve Patients with Advanced Melanoma (CheckMate 067). *AACR, Annual Meeting* 2017, Abstract Number CT075.
9. Reck M, Rodríguez-Abreu D, Robinson AG, Hui R, Csőszi T, Fülöp A et al. Pembrolizumab versus chemotherapy for PD-L1–positive non–small-cell lung cancer. *N Engl J Med* 2016, 375:1823–1833.
10. Draft Guidance for Industry, Clinical laboratories, and FDA staff. In *Vitro Diagnostic Multivariate Index Assays*. (Services USDoHaH, Administration FaD, Health CFDAR, Safety Ooivddea, Research CFBEA Eds.) 2007.
11. Invitro Diagnostic (IVD) Regulatory Assistance: Overview of IVD Regulation, http://www.fda.gov/MedicalDevices/DeviceRegulationandGuidance/IVDRegulatoryAssistance/ucm123682.htm#1
12. Xiao Y, Freeman GJ. The Microsatellite Instable (MSI) subset of colorectal cancer is a particularly good candidate for checkpoint blockade immunotherapy. *Cancer Discov* 2015, 5:16–18.
13. Approval F. FDA grants accelerated approval to pembrolizumab for first tissue/site agnostic indication. 2017.
14. https://ww.agilent.com/en/product/pharmdx/pd-l1-ihc-22c3-pharmdx-overview
15. https://www.google.com/url?sa=t&rct=j&q=&esrc=s&source=web&cd=1&cad=rja&uact=8&ved=2ahUKEwjhrqSp5_TfAhWJh6YKHXklCWIQFjAAegQIBBAC&url=https%3A%2F%2Fwww.agilent.com%2Fcs%2Flibrary%2Fpackageinsert%2Fpublic%2F124668001.PDF&usg=AOvVaw3N2uFyYRaDra1QxNuzq6A4
16. VENTANA, Ventana PD-L1 (SP142) Assay, http://www.accessdata.fda.gov/cdrh_docs/pdf16/P160002c.pdf, 2016.
17. Tumeh PC, Harview CL, Yearley JH, Shintaku IP, Taylor EJ, Robert L et al. PD-1 blockade induces responses by inhibiting adaptive immune resistance. *Nature* 2014, 515:568–571.
18. Teng MW, Ngiow SF, Ribas A, Smyth MJ. Classifying cancers based on T-cell infiltration and PD-L1. *Cancer Res* 2015, 75:2139–2145.
19. Herbst RS, Soria JC, Kowanetz M, Fine GD, Hamid O, Gordon MS et al. Predictive correlates of response to the anti-PD-L1 antibody MPDL3280A in cancer patients. *Nature* 2014, 515:563–567.
20. Kenter GG, Welters MJ, Valentijn AR, Lowik MJ, Berends-van der Meer DM, Vloon AP et al. Vaccination against HPV-16 oncoproteins for vulvar intraepithelial neoplasia. *N Engl J Med* 2009, 361:1838–1847.
21. Sheikh NA, Petrylak D, Kantoff PW, Dela Rosa C, Stewart FP, Kuan LY et al. Sipuleucel-T immune parameters correlate with survival: An analysis of the randomized phase 3 clinical trials in men with castration-resistant prostate cancer. *Cancer Immunol Immunother* 2013, 62:137–147.
22. Walter S, Weinschenk T, Stenzl A, Zdrojowy R, Pluzanska A, Szczylik C et al. Multipeptide immune response to cancer vaccine IMA901 after single-dose cyclophosphamide associates with longer patient survival. *Nat Med* 2012, 18:1254–1261.
23. Tarhini AA, Edington H, Butterfield LH, Lin Y, Shuai Y, Tawbi H et al. Immune monitoring of the circulation and the tumor microenvironment in patients with regionally advanced melanoma receiving neoadjuvant ipilimumab. *PLoS One* 2014, 9:e87705.

24. Di Giacomo AM, Calabro L, Danielli R, Fonsatti E, Bertocci E, Pesce I et al. Long-term survival and immunological parameters in metastatic melanoma patients who responded to ipilimumab 10 mg/kg within an expanded access programme. *Cancer Immunol Immunother* 2013, 62:1021–1028.
25. Cesano A, Willman CL, Kopecky KJ, Gayko U, Putta S, Louie B et al. Cell signaling-based classifier predicts response to induction therapy in elderly patients with acute myeloid leukemia. *PLoS One* 2015, 10:e0118485.
26. Ku GY, Yuan J, Page DB, Schroeder SE, Panageas KS, Carvajal RD et al. Single-institution experience with ipilimumab in advanced melanoma patients in the compassionate use setting: Lymphocyte count after 2 doses correlates with survival. *Cancer* 2010, 116:1767–1775.
27. Cha E, Klinger M, Hou Y, Cummings C, Ribas A, Faham M, Fong L. Improved survival with T cell clonotype stability after anti-CTLA-4 treatment in cancer patients. *Sci Transl Med* 2014, 6:238ra270.
28. Ribas A, Robert C, Hodi FS, Wolchok JD, Joshua AM, Hwu WJ, et al. Association of response to programmed death receptor 1 (PD-1) blockade with pembrolizumab (MK-3475) with an interferon-inflammatory immune gene signature. *J Clin Oncol* 2015, 33:3001–3001.
29. Snyder A, Makarov V, Merghoub T, Yuan J, Zaretsky JM, Desrichard A, et al. Genetic basis for clinical response to CTLA-4 blockade in melanoma. *N Engl J Med* 2014, 371:2189–2199.
30. Rizvi NA, Hellmann MD, Snyder A, Kvistborg P, Makarov V, Havel JJ et al. Cancer immunology. Mutational landscape determines sensitivity to PD-1 blockade in non-small cell lung cancer. *Science* 2015, 348:124–128.
31. Le DT, Uram JN, Wang H, Bartlett BR, Kemberling H, Eyring AD et al. PD-1 Blockade in tumors with mismatch-repair deficiency. *N Engl J Med* 2015, 372:2509–2520.
32. Jennings L, Deerlin VMV, Gulley ML. Recommended principles and practices for validating clinical molecular pathology tests. *Arch Pathol Lab Med.* 2009, 133:743–755.
33. Ascierto PA, Kalos M, Schaer DA, Callahan MK, Wolchok JD. Biomarkers for immunostimulatory monoclonal antibodies in combination strategies for melanoma and other tumor types. *Clin Cancer Res* 2013, 19:1009–1020.
34. Di Maio M, Gallo C, De Maio E, Morabito A, Piccirillo MC, Gridelli C, Perrone F. Methodological aspects of lung cancer clinical trials in the era of targeted agents. *Lung Cancer* 2010, 67:127–135.
35. Plebani M, Sciacovelli L, Aita A, Chiozza ML. Harmonization of pre-analytical quality indicators. *Bioch Med* 2014, 24(1):105–113. doi:10.11613/BM.2014.012.
36. Burghel GJ, Hurst CD, Watson CM, Chambers PA, Dickinson H, Roberts P et al. Towards a next-generation sequencing diagnostic service for tumour genotyping: A comparison of panels and platforms. *Biomed Res Int* 2015, 2015:478017. doi:10.1155/2015/478017.
37. Chau CH, Rixe O, McLeod H, Figg WD. Validation of analytic methods for biomarkers used in drug development. *Clin Cancer Res* 2008, 14(19):5967–5976. doi:10.1158/1078-0432.ccr-07-4535.
38. Lee JW, Weiner RS, Sailstad JM, Bowsher RR, Knuth DW, O'Brien PJ et al. Method validation and measurement of biomarkers in nonclinical and clinical samples in drug development: A conference report. *Pharm Res* 2005, 22(4):499–511.

Integrating Personalized Medicine in the Health Care 1.3

Payers and Reimbursements

Dominic Galante and Marylou Buyse

Contents

PERSONALIZED MEDICINE AND BIOMARKERS

Over the past two decades, a growing number of pharmaceutical products have been approved in the United States, signaling a new era of drug development. Many of these new therapies employ the use of biomarkers (Figure 1.3.1) [1]. The FDA defines a biomarker as a characteristic that is measured as an indicator of normal biological processes, pathogenic processes, or responses to an exposure or intervention, including therapeutic interventions. Molecular, histologic, radiographic, or physiologic characteristics can function as biomarkers. A biomarker is not an assessment of how an individual feels, functions, or survives.

The use of biomarkers combined with targeted therapies appears to be a growing trend in a number of clinical areas, further developing the field of "personalized medicine." The FDA has approved several cancer drugs for use in patients whose tumors have specific genetic characteristics identified with a companion diagnostic test for a specific biomarker (Figure 1.3.2) [2–7]. An example of this is the antibody trastuzumab, which is targeted to patients whose breast cancer tests positive for HER2, formerly a hard-to-treat form of breast cancer. Since approval for its use in the United States in 1998,

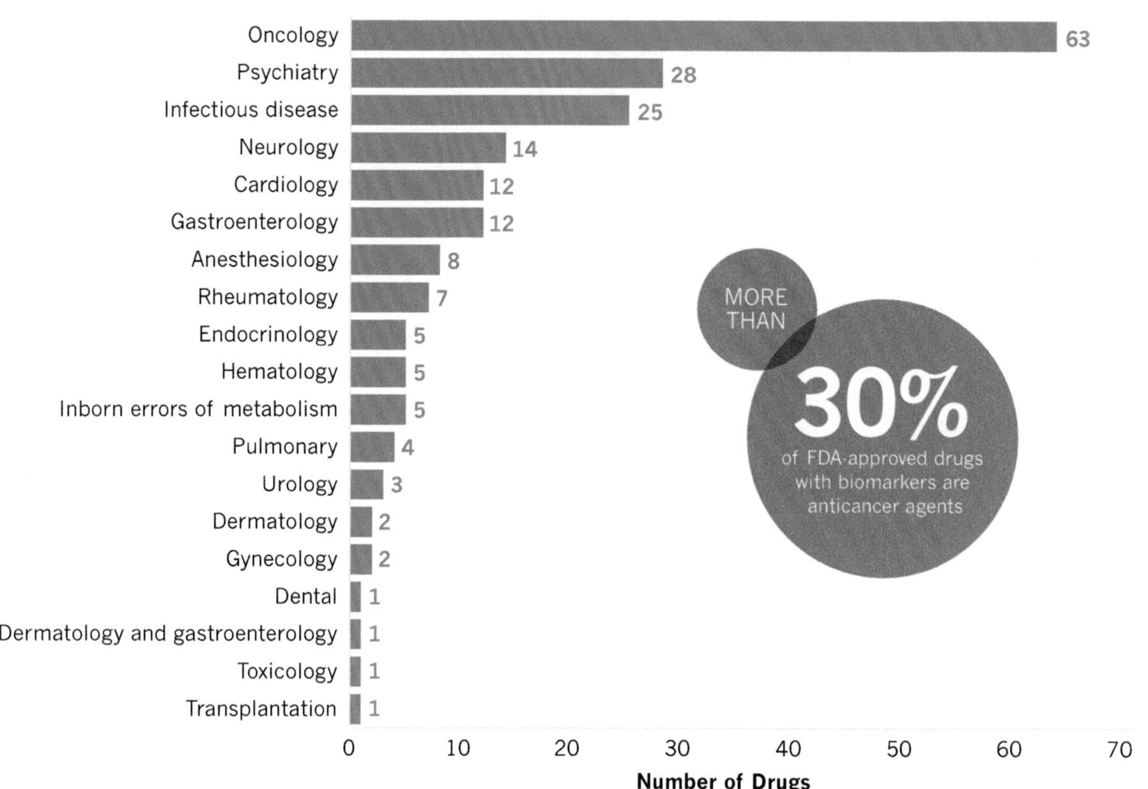

FIGURE 1.3.1 As of October 2017, there were nearly 200 FDA-approved drugs with pharmacogenomic biomarkers in their labels across a wide range of therapeutic areas. More than 30% of these drugs are anticancer agents.

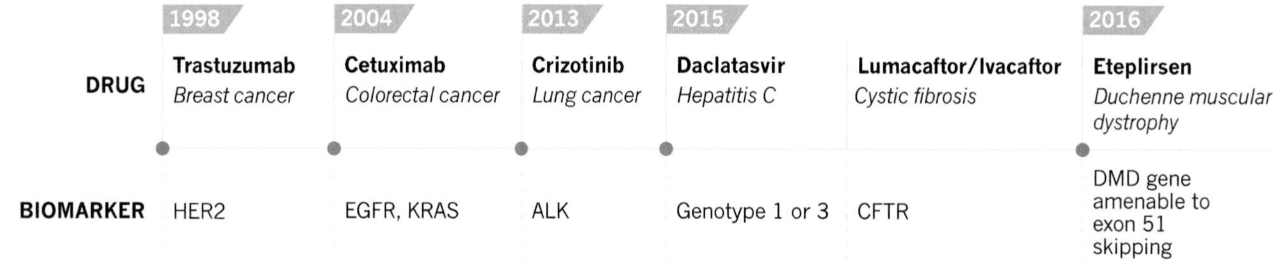

FIGURE 1.3.2 Since the approval of Herceptin in 1998, several other drugs have come to market with companion diagnostic requirements in their label. These products extend beyond oncology.

trastuzumab has transformed the treatment and improved the prognosis of women testing positive for overexpression of HER2 protein in specific tumors.

In 2015, the FDA approved a new therapy for use in certain cystic fibrosis (CF) patients with a specific genetic mutation. Ivacaftor was approved in 2012 by the FDA for treatment of the underlying cause of CF in patients with certain mutations of the CFTR gene; the gene expression in this case served as a biomarker. Also in 2015, CF patients with two copies of the F508del mutation (the biomarker) in their CFTR gene were able to take lumacaftor/Ivacaftor combination therapy to change the course of the disease. Several medications for the treatment of hepatitis C are now on the market for patients who are infected with a specific genotype of hepatitis C.

Personalized medicine as a concept has been around for many years. Clinicians have long known that patients with similar symptoms may have illnesses with different causes and respond differently to appropriate therapeutics.

What is new, however, are advanced diagnostics, such as genomic or proteomic testing for biomarkers. We are now developing therapeutic agents that are targeted to persons who have a specific genotype or whose disease has specific characteristics that allow a more focused and targeted therapy. Biomarkers are critical to the identification of patients who would benefit from these new highly targeted therapeutic agents.

HOW DO PAYERS USE BIOMARKERS?

This section of the chapter reviews how payers use biomarkers and why they are advantageous for payers. Payers want to ensure their members are receiving effective and appropriate care. To accomplish this goal, managed care organizations (MCOs) have set up a number of procedures and controls to avoid overuse and misuse of health care services. In general, this comes under their utilization management (UM) program. UM is a major part of how payers identify high-cost targets, whether a health care service or a pharmaceutical agent or device. Once a payer makes the determination to manage a health care service, the organization may develop a policy that describes the coverage requirements for that service. Medical policies and other UM requirements are regularly communicated to the provider network and are generally reviewed annually. These coverage requirements may be pre-service, concurrent (generally for acute care services), or post-service; when biomarkers are involved, the MCO will specify the test(s) needed to justify insurance coverage.

Prior to determining that it will manage a particular service or therapeutic agent based on the results of a biomarker test, the MCO or health plan will gather sufficient evidence to ascertain that the biomarker test is readily available, accurate, valid, and reliably predicts improved outcomes. In making this determination, the payer is relying on high-quality randomized clinical research studies published in peer-reviewed journals. The MCO will also consider evidence-based clinical guidelines developed by specialty groups or government agencies. Payers may also consider guidance from the FDA, National Institutes of Health (NIH), and Centers for Disease Control and Prevention (CDC).

MCOs and other payers publish coverage criteria detailing circumstances in which a particular service or drug will or will not be covered (generally in a medical policy statement). In many plans, the medical or pharmaceutical policy lays out these coverage criteria. Coverage policies and UM programs can vary from payer to payer. The next section will detail how and why payers use biomarkers to identify their members who are most likely to either benefit or not benefit from a specific health care service. In some cases, companion biomarkers are required by the FDA in order to use a specific therapeutic agent; in others, a payer may inform its network to test and identify patients most likely to benefit from a specific agent. Biomarker results can qualify costly therapeutic agents for insurance coverage. If a health care provider (HCP) does not submit the required biomarker result for that patient, the requested therapeutic agent may be denied for payment.

WHY ARE BIOMARKERS IMPORTANT TO A PAYER?

Patient Selection

When used effectively, biomarkers can reliably determine which patients will and will not benefit from a particular therapy. The biomarker status guides the HCP to direct treatment to those individuals who can benefit from the therapy; hence, the term "personalized medicine" or "precision medicine." Even after accounting for the cost of testing, this approach saves in unnecessary treatment costs and spares patients from adverse events, which may occur even in the absence of a therapeutic benefit. There are numerous examples of this approach to therapy resulting from advances in genetic and other advanced molecular testing. An early example was discussed in the beginning of the chapter, describing the use of HER2 protein biomarker testing of breast cancer tissue to determine whether a patient overexpresses the HER2 protein. If the patient is HER2 positive, the therapeutic antibody trastuzumab can be used to greatly improve outcomes. Prior to trastuzumab therapy, HER2-positive cancers were a difficult-to-treat form of breast cancer. Since its FDA approval, trastuzumab has transformed the treatment and improved the prognosis of women with tumors that overexpress the HER2 protein and thus test positive for the HER2 receptor (Figure 1.3.3) [8].

Assess Disease Risk and Severity

Biomarkers can be indicators for many clinical aspects of a condition or disease process. Payers will indicate which biomarker requirements are needed to justify insurance coverage for a specific health care service or therapeutic agent. Payers/MCOs have an interest in using biomarkers to segment or identify persons in their population who are at risk for a disease. Most payers under specific circumstances will, for example, pay for BRCA testing in certain women who meet defined criteria for BRCA testing to determine if they carry a specific gene that puts them at high risk for both familial breast and ovarian cancer. Currently, payers, including the Centers for Medicare and Medicaid Services (CMS), restrict BRCA testing to women whose clinical histories indicate that they are more likely to test positive for the *BRCA* gene.

There are many examples of biomarker use to assess risk for disease. Payers will generally cover such tests if they are widespread and easily accessible, such as testing for LDL-C. For high-cost tests, the payer may have medical policy or prior authorization requirements in place to ensure only people who meet clearly defined criteria will be tested. As mentioned, payers want to make sure that covered medical services are necessary, reliable, and valid and generally benefit the persons being tested.

Predictive biomarkers have been reported in a wide range of conditions such as Alzheimer's disease, chronic obstructive pulmonary disease, rheumatoid arthritis, cardiovascular disease, and sepsis.

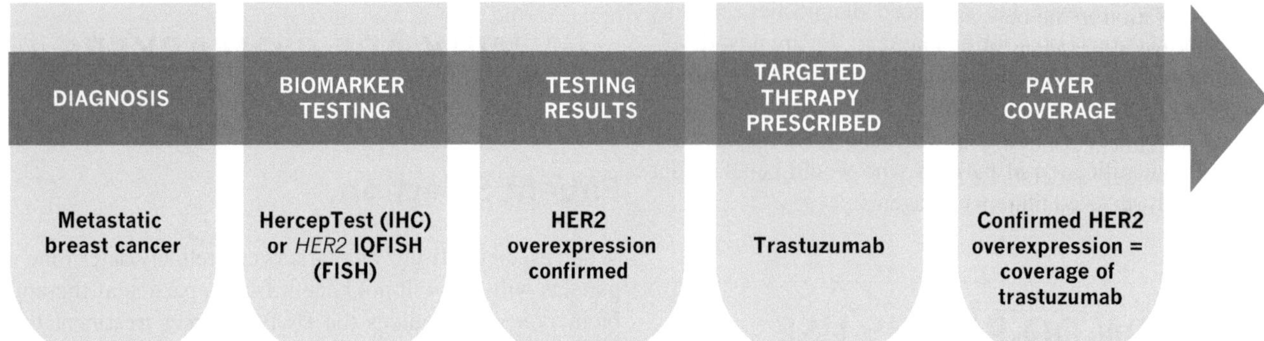

FIGURE 1.3.3 Biomarkers are important to payers because they reliably determine which patients will and will not benefit from a particular therapy, such as Herceptin.

Track Disease Progression

An example of the use of biomarkers from 1991 is the use of tumor necrosis factor-α (TNF-α) or cachectin to determine whether an individual with multiple sclerosis had progressive disease [9].

Predict Events

Biomarkers have been used to predict a wide variety of clinical events such as cardiovascular disease and cancer. BRCA test results can reliably predict who is at very high risk for familial breast and/or ovarian cancer. These tests show only that someone has a high likelihood of developing cancer.

Inform Therapeutics

One of the most promising areas in therapeutics is the use of biomarkers to predict an individual's response to a therapeutic agent. In 2009, the American Society of Clinical Oncology (ASCO) recommended that patients diagnosed with colon cancer be tested for a mutation of the *KRAS* gene before being treated with medicines such as cetuximab or panitumumab. In July 2009, the FDA approved labeling changes based on findings that cetuximab and panitumumab are not effective for patients whose *KRAS* gene has mutated.

Today, numerous therapeutic agents have required companion diagnostic tests to identify patients who may benefit from a specific treatment prior to initiation of therapy. Payers can legitimately deny payment for therapy when such companion diagnostic tests have not been performed prior to the initiation of therapy when required by the FDA.

Prognostic Markers

Biomarker testing can often provide prognostic information. Breast cancer is a good example. Triple-negative breast cancer describes a patient with breast cancer that is negative for all three receptors: estrogen, progesterone, and HER2/NEU. Such patients have a form of breast cancer that is difficult to treat.

Another example in breast cancer is the use of the commercially available biomarker test, Oncotype DX. This test is designed to segment women with newly diagnosed breast cancer of low risk from those who are likely to have a recurrence and thus benefit from chemotherapy.

Diagnostic Markers

Prostate-specific antigen (PSA) testing has long been used as a surrogate marker for prostate cancer risk. While not a definitive test, PSA test results have for decades been used to refer patients for diagnostic biopsy to confirm whether prostate cancer is present.

Dose Selection

Biomarkers are now being used to test for dose selection in the course of drug development to improve treatment of tuberculosis. A study examined the safety, tolerability, and pharmacokinetics of multiple ascending doses of oxazolidinone PNU-100480 in healthy volunteers using biomarkers for safety and efficacy. This randomized, controlled trial showed the agent to be safe at all doses tested, providing a role for biomarkers in accelerating drug development [10].

MEDICAL POLICY APPLICATIONS

Payers develop medical benefit policies based on best available evidence. A policy is intended to inform stakeholders of the details and the circumstances under which a member can receive coverage for a specific benefit. Medical policies are developed when there is nuanced coverage beyond a simple yes or no, or a limitation to the benefit such as covering 30 physical therapy visits per year. Many medicines have a corresponding medical policy or a pharmaceutical policy detailing coverage. Medications distributed through retail pharmacies generally will have a pharmaceutical policy that helps the retail pharmacy and prescribing provider understand coverage requirements for a specific therapeutic agent. Internal heath plan staff responsible for reviewing medications requiring prior authorization (PA) can review the policy before approving or denying coverage for a specific requested drug. Certain medications are paid for under medical benefits. These medicines are generally, but not always, administered in a physician's office or infusion center, or they may be managed and delivered by a specialty pharmacy. They are generally not available at a retail pharmacy where most prescriptions are obtained. If the requested medication falls under the medical benefit, that policy will outline the specific coverage requirements.

Health plans aim to provide the highest value for their policyholders. They are well aware that biomarkers provide valuable information about the effectiveness and appropriateness of a specific drug. Biomarker results can inform physicians and others about choice of therapeutic agents, including risks and benefits, based on the expected response of an individual.

In general, MCOs and health plans do not cover therapies that are considered. The use of biomarkers that predict positive or negative response with a specific therapeutic agent allows health plans to formulate coverage policy. An experimental drug can also be placed into a medical policy detailing the reasoning and the evidence for noncoverage and describing what lines of business fall under the policy.

It helps to understand that commercial, Medicare, and Medicaid lines of business have different coverage requirements, and the medical policies will call these out, or there may be separate policies for each line of business.

TRENDS IN PAYER BIOMARKER ADOPTION AND MANAGEMENT

Specialized Laboratories for Biomarker Testing and Access

There is increasing pressure to provide cost-effective health care based on "best practice." Consequently, new biomarkers are likely to be introduced into routine clinical biochemistry departments only if they are supported by strong evidence showing improved patient management and outcomes. Carefully designed audit and cost–benefit studies in relevant patient groups must demonstrate that introducing the biomarker delivers an improved clinical pathway. Good stability of the biomarker in relevant physiological matrices is essential to avoid the need for special processing. Absence of specific timing requirements for sampling and knowledge of the effect of medications that might be used to treat the patients in whom the biomarker will be measured are also highly desirable. Assays must therefore be robust, fulfilling standard requirements for linearity on dilution, precision, and reproducibility, both within- and between-run. Provision of measurements by a limited number of specialized reference laboratories may be most appropriate, especially when a new biomarker is first introduced into routine practice [11].

Successfully taking a biomarker from the research laboratory into the clinical laboratory ideally requires a four-way collaboration involving the research laboratory (which develops the fundamental concept), the diagnostics industry (which turns the concept into a practical reliable tool), the clinical laboratory (which evaluates the tool in real-life practice), and clinicians (who help identify unanswered clinical questions and needs that the measurement of a new biomarker might usefully address; clinicians also provide the carefully characterized clinical specimens necessary for its assessment).

The decision to introduce a new biomarker will clearly be influenced by different reimbursement policies and other logistical arrangements among health care systems. The introduction of a new biomarker into routine clinical practice requires rigorous assessment from three different perspectives: that of the clinician, the laboratory pathologist, and the health care funding organization. An integrated approach to funding the entire patient-care pathway, including additional tests recommended as a part of other initiatives (e.g., Quality Outcome Framework targets in the United Kingdom), is preferred to piecemeal funding of separate functions (e.g., laboratory, pharmacy, radiology), which is sometimes termed "silo budgeting." However, such an approach is infrequently in place. In the United States, gaining approval and payment rates for new tests can be a limiting factor in determining whether a new test will be performed [12].

Additive Health Care Costs

A survey of approximately 130 oncologists/hematologists and medical oncologists conducted from 2016 to 2017 by H. Jack West, MD, Medical Director, Thoracic Oncology Program, Swedish Cancer Institute, Seattle, Washington [13], found that private health insurance is by far the most common way genomic testing is paid for among cancer patients (85%), followed by research funding (35%) and patient self-pay (29%) (Figure 1.3.4). More than 30% of the oncologists surveyed had clinical concerns with genomic testing, stating that it rarely provides clinically actionable, evidence-based information; more than 60% said that less than a quarter of their patients would benefit from the testing. Eighty-four percent of oncologists also have concerns about insurance coverage of genomic testing.

An average drug on the market today is reported to be effective in only 50% of those who take it [14]. Prescribing medications to those who are unlikely to respond not only unnecessarily inflates our annual $2.5 trillion spending on health care ($1.3 trillion of which is provided by private payers and $378 million is paid for out-of-pocket by patients) [15,16] but also exposes patients to the side effects of medications without the potential therapeutic benefit. Personalized medicine can allow for prediction of nonresponders, thereby avoiding unnecessary exposure to side effects

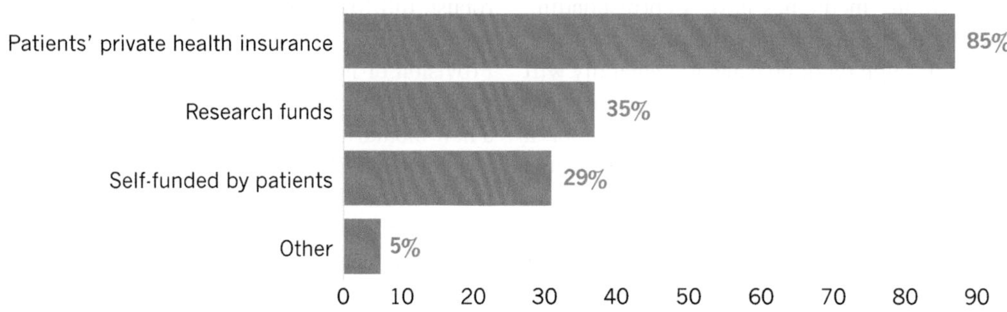

FIGURE 1.3.4 Survey results demonstrate that genomic testing is primarily paid for by private health insurance.

from medications predicted to be ineffective. A classic example of a targeted therapy, trastuzumab, is highly effective in the 15%–25% of breast cancers that overexpress the HER2 protein (a cell growth promoter) and is generally not effective against breast tumors without HER2 overexpression. Drugs like erlotinib are effective therapies in those patients whose non-small cell lung cancer carries specific EGFR mutations but not KRAS mutations [17].

Although payers are generally familiar with the concept of selecting drugs based on genetic targets in several therapeutic areas, a hallmark example of this is the anti-thrombotic clopidogrel and the SNPs of CYP2C19, a CYP gene encoding a key enzyme in the metabolic activation of clopidogrel and associated with pharmacokinetic and pharmacodynamic responses to clopidogrel. Pharmacogenomic research of more commonly prescribed drugs such as clopidogrel, [18,19] warfarin, [20] and statins [21] has been stimulated owing to the enhanced value perceived for targeted medicines.

Health Economic Value of Biomarker Use

Health care payers represent stakeholders who can act as gatekeepers to the translation of personalized medicine into routine clinical practice. To date, the slow realization of the promise of personalized medicine has been partly attributable to the lack of clear evidence supporting the clinical utility of genetic and genomic tests—and the lag in development of clinical guidelines for the use and interpretation of tests. These factors, along with a paucity of clear guidance from health care payers and little clinical experience with genomic tests, serve as impediments to timely and consistent reimbursement decisions. The design of alternative strategies for collaborative evidence generation, clinical decision support, and educational initiatives for health care providers, patients, and the payers themselves are critical to achieve the full benefit of personalized medicine in day-to-day health care settings [22].

Payers recognize that the size of the opportunity and potential return on investment both clinically and economically in personalized medicine is tremendous, yet they are hesitant to embrace the use of molecular diagnostics when many tests' utility is not yet fully proven. Molecular diagnostics are defined as genetic and/or esoteric tests that assay for biomarkers from gene to gene product, including RNA, miRNA, protein, metabolites, antibodies, and all varieties of genomes including human, cancers, and pathogens. Estimates suggest that genetic variation accounts for between 20% and 95% of the variability in individual response to medications [23]. The opportunity to prescribe medications only to those who are most likely to respond, or to avoid life-threatening side effects, will allow physicians to improve care with economic benefits for both patients and payers. Furthermore, although diagnostics represent a relatively modest expense, they inform a huge portion of health care decisions and subsequent expenditures. It has been estimated that diagnostic tests represent less than 5% of health care spending and influence up to 70% of health decision making [24].

Modeling the clinical and economic outcomes of pharmacogenomic interventions suggests that the medical costs potentially avoided by testing 100 patients initiating therapy for which there is an actionable pharmacogenomics test ranges from $6,000 for tests with less common variants such as HLA-B*5701 for abacavir to $50,000 for more common genotypes such as CYP2C19 metabolizer status for clopidogrel (Figure 1.3.5) [25]. At a population level, considering only the medical costs avoided, this represents up to a 200% return on investment for the cost of testing (data presented reflect 17 proprietary unpublished clinical and economic cost-effectiveness models evaluating the potential costs and benefits of pharmacogenomic interventions from a payer's perspective).

Health care payers primarily view the clinical and health economic benefits of personalized medicine and pharmacogenomics as driven by three potential factors: better prediction of responsiveness to treatment, proactive prevention of drug adverse events, and promotion of adherence to therapy through minimization of undesirable side effects.

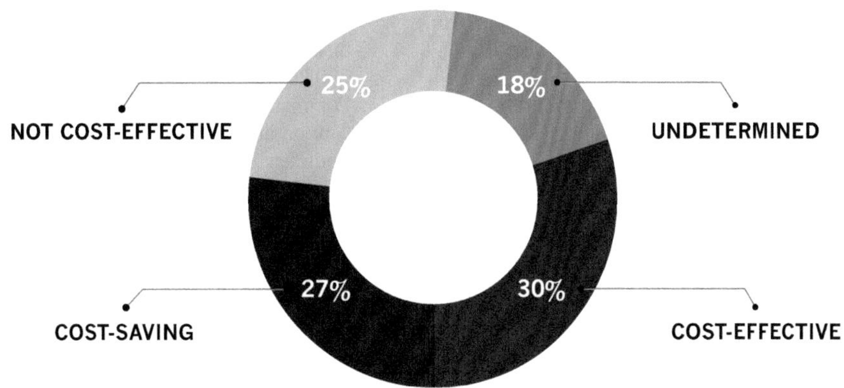

Cost-saving/dominant	Cost-effective	Undetermined	Not cost-effective
PGx was more effective at lower cost	PGx was more effective at acceptable additional cost	Reviewed study did not reach unequivocal conclusion	PGx was not cost-effective

FIGURE 1.3.5 Pharmacogenomic-guided treatment can be a cost-effective and even a cost-saving strategy. These findings are based on economic evaluations for pharmacogenomic associations listed in the FDA Table of Pharmacogenomic Biomarkers in Drug Labeling. Authors determined the proportion of evaluations that found pharmacogenomic-guided treatment to be cost-effective or dominant over the alternative strategies, and estimated the impact on this proportion of removing the cost of genetic testing. Of the 137 pharmacogenomic associations in the FDA table, 44 economic evaluations, relating to 10 drugs, were identified.

REFERENCES

1. U.S. Food and Drug Administration. Table of pharmacogenomics biomarkers in drug labeling. https://www.fda.gov/Drugs/ScienceResearch/ucm572698.htm. Updated October 3, 2017. Accessed October 12, 2017.
2. Herceptin [package insert]. South San Francisco, CA: Genentech.
3. Erbitux [package insert]. Indianapolis, IN: Eli Lilly and Co.
4. Xalkori [package insert]. New York: Pfizer.
5. Daklinza [package insert]. Princeton, NJ: Bristol-Myers Squibb.
6. Orkambi [package insert]. Boston, MA: Vertex Pharmaceuticals.
7. Exondys 51 [package insert]. Cambridge, MA: Sarepta Therapeutics.
8. Agilent. Products; pharmDx. http://www.agilent.com/en-us/products/pharmdx. Accessed October 12, 2017.
9. Sharief MK, Hentges R. Association between tumor necrosis factor-alpha and disease progression in patients with multiple sclerosis. *N Engl J Med*. 1991;325(7):467–472.
10. Wallis RS, Jakubiec W, Kumar V et al. Biomarker-assisted dose selection for safety and efficacy in early development of PNU-100480 for tuberculosis. *Antimicrob Agents Chemother*. 2011;55(2):567–574.
11. Sturgeon C, Hill R, Hortin GL, Thompson D. Taking a new biomarker into routine use—A perspective from the routine clinical biochemistry laboratory. *Proteomics Clin Appl*. 2010;4(12):892–903.
12. Hortin GL, Jortani SA, Ritchie JC Jr, Valdes R Jr, Chan DW. Proteomics: A new diagnostic frontier. *Clin Chem*. 2006;52(7):1218–1222.
13. West HJ, Miller G. Genomic testing and precision medicine in cancer care. *Medscape*. http://www.medscape.com/slideshow/genomics-and-oncology-report-6008655. Accessed October 4, 2017.
14. Spear BB, Heath-Chiozzi M, Huff J. Clinical application of pharmacogenetics. *Trends Mol Med*. 2001;7(5):201–204.
15. Truffer CJ, Keehan S, Smith S et al. Health spending projections through 2019: The recession's impact continues. *Health Aff (Millwood)*. 2010;29(3):522–529.
16. U.S. Department of Labor, Bureau of Labor Statistics (2010). *Consumer Expenditure Survey, 2010: Quarterly Interview Survey and the Diary Survey*. Washington, DC: Department of Labor, Bureau of Labor Statistics [producer and distributor] (2011). http://www.bls.gov/cex.
17. Keedy VL, Temin S, Somerfield MR et al. American society of clinical oncology provisional clinical opinion: Epidermal Growth Factor Receptor (EGFR) mutation testing for patients with advanced non-small-cell lung cancer considering first-line EGFR tyrosine kinase inhibitor therapy. *J Clin Oncol*. 2011:29(15):2121–2127.
18. Mega JL, Simon T, Collet JP et al. Reduced-function CYP2C19 genotype and risk of adverse clinical outcomes among patients treated with clopidogrel predominantly for PCI: A meta-analysis. *JAMA*. 2010;304(16):1821–1830.
19. Roden DM, Shuldiner AR. Responding to the clopidogrel warning by the U.S. Food and Drug Administration: Real life is complicated. *Circulation*. 2010;122(5):445–448.

20. Wadelius M, Chen LY, Lindh JD et al. The largest prospective warfarin-treated cohort supports genetic forecasting. *Blood.* 2009;113(4):784–792.
21. SEARCH Collaborative Group, Link E, Parish S et al. SLCO1B1 variants and statin-induced myopathy–a genomewide study. *N Engl J Med.* 2008;359(8):789–799.
22. Canestaro WJ, Martell LA, Wassman ER, Schatzberg R. Healthcare payers: A gate or translational bridge to personalized medicine? *Pers Med.* 2012;9(1):73–84.
23. Evans WE, McLeod HL. Pharmacogenomics—Drug disposition, drug targets, and side effects. *N Engl J Med.* 2003;348(6):538–549.
24. Forsman RW. Why is the laboratory an afterthought for managed care organizations? *Clin Chem.* 1996;42(5):813–816.
25. Verbelen M, Weale ME, Lewis CM. Cost-effectiveness of pharmacogenetic-guided treatment: Are we there yet? *Pharmacogenomics J.* 2017;17(5):395–402.

1.4 Regulatory Strategies to Accelerate the Implementation of Biomarkers and Companion Diagnostics

F. Owen Fields

Contents

INTRODUCTION

The development of new drugs is one of the most regulated of human activities. In many cases, biomarkers are utilized in various manners and "contexts of use" to support elements of new drug development. Such uses include establishing biochemical/biological proof of principle, informing mechanism of action (MOA), helping to support or even define dose response, selecting patients that are more sensitive to either desirable efficacy-related effects or undesirable safety-related effects, and most challengingly as surrogate end points that are sufficiently predictive of clinical benefit to support approval. Other uses are now well established, such as judging susceptibility/risk of developing a condition, safety monitoring, and prognostic markers that predict the probability of clinically relevant events. As such, regulatory science concepts pertaining to the use of biomarkers must be well understood to optimize the efficiency of using biomarkers in the regulatory development of new drugs.

Before initiating this discussion, it should be noted that biomarkers are best used in a regulatory context when there is no readily available/practical clinically relevant end point that is capable of efficiently serving the same purpose. The discussion below will provide examples of contexts of use in which such clinically relevant end points are either not available or not practical in light of the size of the population or other disease characteristics, such as progressive diseases with a slow rate of progression.

In this section, the concept of context of use will be described and examples provided; understanding the concept of context of use and the elastic nature of this key concept is key to predicting which biomarker-based strategies will be acceptable to global regulatory agencies because it defines the specific use for which the biomarker serves as a useful tool in a regulatory context.

Following this, some basic principles of companion diagnostics co-development (defined as concurrent development of a new drug with the diagnostic method necessary for its appropriate use) will be reviewed. At the close of this section, a look forward to how the next generation of diagnostic/analytical technologies will require regulatory agencies, and new drug developers to modernize their thinking based on the development of such systems. Included among these advanced diagnostic-related technologies are multiplex proteomic and transcriptomic assays, as well as next-generation DNA sequencing.

CONTEXT OF USE OF A BIOMARKER IS A CONTROLLING CONCEPT THAT DEFINES THE APPROPRIATE REGULATORY USE OF BIOMARKERS

In short, biomarkers are tools that are available for use in new drug development; they can comprise various analytes or images or even isolated measures of organ function. The context of use of a biomarker is perhaps best described as a way of defining specifically what one is expecting to do with a biomarker, in terms of the important elements in regulatory development that are to be supported with biomarker-derived results.

Precision/personalized medicine requires the use of biomarkers, but their place in supporting the overall drug development effort in regulatory contexts depends greatly on their context of use. In Figure 1.4.1, the various high-level regulatory contexts of use of various biomarkers are shown schematically.

Beginning on the left are biomarkers used for enriching and diagnosing populations (e.g., HLA genotypes are utilized in supporting the diagnosis and predicting rate of progression of certain autoimmune diseases such as ankylosing spondylitis [1]. Another example is the presence of anti CCP antibodies, which has been utilized in an attempt to define autoimmune disease patients more likely to progress rapidly as well as to respond more strongly to certain therapeutic modalities [2].

Examples of different contexts of use for various biomarkers are provided in the latter sections of this section.

QUALIFICATION OF BIOMARKERS

Another concept that needs to be understood when considering the appropriate regulatory use of a biomarker is the concept of "qualification." The term qualification refers to the process of working with a regulatory agency(s) to formally substantiate the appropriateness of a biomarker for its intended context of use. The requirements and applicability of regulatory qualification of a biomarker is one of the most elastic concepts in regulatory practice simply because there can be an infinite number of individual contexts of use based on various specific fact patterns and resulting delimiters that must be defined. For example, the acceptable use of a biomarker to develop a drug in an ultra-rare disease context (e.g., genetically defined subpopulations of cystic fibrosis—see below) may not be acceptable in a more prevalent and, therefore, less practically challenging population.

There are three broad categories of biomarker qualification. Going from the narrowest/least challenging to broader and more challenging uses, these categories are what we will call "project-specific vetting," formal qualification, and the special case of validation as a surrogate marker.

The Process of Project-Specific Vetting of Use of a Biomarker

When asking a regulatory agency about using any new tool in drug development, or using an extant tool in a novel way, the narrowest, most molecule- and program-specific use would normally be the least challenging to support. This is because the opinion requested from the agency is very narrowly "delimited" by the specific drug

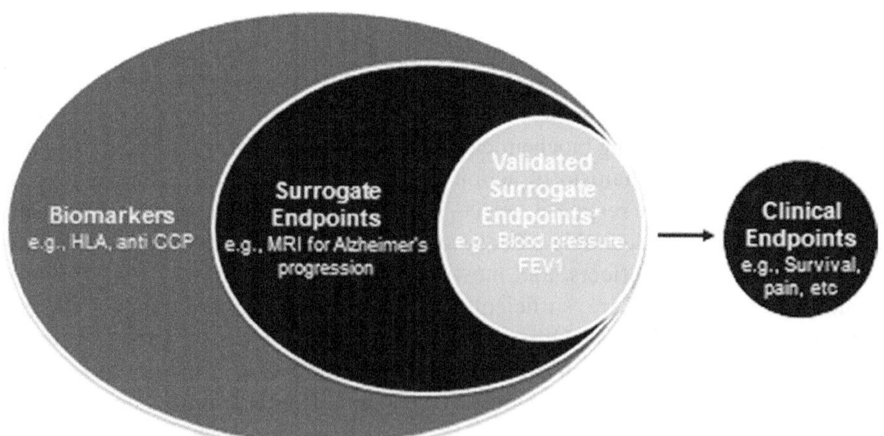

FIGURE 1.4.1 Precision medicine requires biomarkers.

and generally also by the specific and narrow proposed use in the overall individual new drug development program.

It is, in fact, often the case that regulatory agencies permit use of a biomarker in new drug development in such a narrow project-specific fashion well before the same biomarker is formally qualified for less narrowly defined use. An example of this comes from the use of total kidney volume (TKV) to both select patients at a high risk of progression of autosomal dominant polycystic kidney disease and to define dose response. In this disease, there are proliferative somatic lesions (cysts) in the kidneys that arise from a loss of heterozygosity; these innumerable cysts result in a logarithmic increase in TKV. Because the rate of increase is logarithmic, the larger baseline TKV is, the faster TKV is expanding. Due to this, patients with a high baseline TKV are a more sensitive population in which to assess dose response and initial therapeutic activity. In 2015, the FDA issued a formal qualification opinion on this use of TKV as a biomarker [3]. However, examination of public clinical trials databases indicate that at least two different clinical trial sponsors utilized a high initial TKV apparently as a way of "isolating" patients who are progressing rapidly and thus were a more sensitive population to assess initial therapeutic activity and dose–response. Based on the designs of the publicly disclosed programs, dose–response was also solely characterized using TKV [4,5]. This situation also applies to some early attempts at assessing agents to inhibit progression of other slowly progressing diseases [6].

The Process of Formally Qualifying a Biomarker for a Given Context of Use

The term qualification is used to refer for a formal written qualification as described in U.S. and EU agency regulatory guidances [7,8]. This form of qualification is burdensome and can take several years of analysis and compilation of supporting data from various sources. This extra burden is driven by the fact the result is an agency finding of relatively broad utility, rather than of very narrow utility to a single individual program. Under formal qualification, the regulatory agency essentially issues a finding that in any cases that are within the context of the formal qualification opinion, one can rely on the formal qualification finding by the agency.

The FDA has issued opinions on a wide range of biomarkers for specific contexts of use [9]. The EU authorities have issued equivalent opinions, which are partly overlapping [10]. It is recommended that the reader examine these opinions and the evidentiary basis that supported them because this will provide a great deal of insight on the process and evidentiary requirements applying to formal qualification.

Relevant Contexts of Uses of Biomarkers in Regulatory Development

This section addresses the concept of qualification of biomarkers for specific contexts of use as well as the special case of validating a biomarker for use as a surrogate end point. As noted above, in a regulatory context if an efficient, clinically relevant end point is available, one should generally avoid relying on biomarkers for anything other than supportive use.

Biomarkers are now regularly used in new drug development for such purposes as identifying patients at a high risk of a clinical event of interest, diagnosing a disease and/or the extent of disease activity, or monitoring safety (e.g., various enzymes potentially derived from hepatic sources), and such examples are presumably well known to the reader. Other examples of reasonable regulatory contexts of use of biomarkers that fall within the rubric of precision medicine are as follows:

Establishment of dose–response and target engagement. There are several examples of this context of use in settings in which rare populations, very slowly progressing diseases, or settings with insensitive end points limit the ability to use other means to establish these parameters. One will note that in all of the settings provided as examples, the biomarker relates very closely with the underlying pathophysiology of the disease:

- *Use of sweat chloride in defining dose response in cystic fibrosis.* In CF, an ATP-gated ion channel is defective; this physiologic function (which plays a controlling and highly debilitating role in the lung pathophysiology of the disease) is also expressed in sweat because the affected ion channel helps control sweat chloride concentration. Given this very close mechanistic relationship and the insensitivity of lung function end points (considering this rare population), regulatory agencies have accepted sweat chloride for establishing dose response of agents in rare CF populations. In some cases, these are actually genetically defined subpopulations (so-called ultra-orphan populations) within this overall rare population, further increasing the value of the use of a biomarker for this purpose [11].
- As referenced above, TKV has been used in autosomal dominant polycystic kidney disease (ADPKD) for both defining dose–response and enriching for patients that are likely to be rapidly progressing [3]. As noted above, this is another example of a context of biomarker use that is driven by practical considerations, in that development of a new drug for ADPKD without such a biomarker (given the slow progression of the disease) may otherwise be impractical.

Enrichment of a highly responsive population. There are also numerous examples of this context of use. The most prominent context currently would comprise tumors that are expressing, or overexpressing, the targeted biological moiety, such as ectopic expression of HER2 or the ALK kinase translocation [12,13]. This context of use for oncology therapeutics is increasingly common [14].

Elimination/deselection of a sensitive safety population. The classic example of this context of use are those individuals who have "atypical" (sometimes geographically differential) genotypes in drug metabolizing enzymes that lead to atypically robust or atypically low rates of drug metabolism [15].

The special case of surrogate end points. To move to a much higher-level context of use, there has been some success in the use of biomarkers as surrogate markers that support approval based on the likelihood that the biomarker effects confidently predict clinically meaningful benefit. A detailed discussion of so-called validation concepts for surrogate markers is beyond the scope of this section but has been addressed by other authors [16]. Relevant examples of such end points follow:

- FEV_1 in pulmonary indications. This end point has been very important in a series of approvals for drugs that improve pulmonary function. This is based on the very well-established and close relationship of this particular measure of lung function to basic human functions, such as exercise capacity and lack of pulmonary symptoms such as dyspnea. This is an example of a surrogate marker based on isolated organ function.
- Lipid measures have been used by developers of lipid-lowering therapies as validated surrogate end points. This is an example of a biochemical marker.
- Blood-pressure lowering agents have in the past regularly utilized blood pressure as a surrogate end point. This is an example of a biomarker resulting from an instrumented measurement.

All of the above examples involve the concept of what can be called "external validation"—that is, validation of a surrogate end point by data external to a single program.

In areas such as oncology and in some rare diseases, however, there has also been what can be called "internal validation" of surrogate end points; unlike external validation, this validation would generally only apply to that specific drug in the tested population. In the case of oncology settings, the surrogate is often an end point such as radiographic response or progression free survival; internal validation is based on extending the controlled trial to longer time-points in order to address the ability of the drug to establish an effect on clearly clinically meaningful end points such as overall survival. These data are typically submitted after the initial approval application to support internal validation of the initial surrogate end point-based results.

DEVICE/ANALYTICAL SERVICES REGULATION: *IN VITRO* DIAGNOSTICS VERSUS CLIA ASSAYS

Now that we have provided a discussion of biomarker types, context of use, and qualification considerations, it is time to turn to the regulatory requirements for the products that detect and quantitate such biomarkers. These are regulated in the United States generally as either devices (in-vitro diagnostics [IVDs] under the legal authority of the Federal Food, Drug, and Cosmetic Act—FFDCA) or as diagnostic services (under the Clinical Laboratory Improvement Amendments [CLIA]). Those devices regulated under the FFDCA are further divided into those regulated under the 510(k) pathway versus the premarket approval (PMA) pathway. Although the distinction between these two FFDCA pathways is in practice a very complicated matter driven by precedence and nuance, the 510(k) pathway is generally utilized for devices that are well precedented and, therefore, have "predicate devices," whereas those regulated under the PMA procedures are generally based on new analytes or less mature technology.

The division between these major overarching types of regulation (CLIA versus FFDCA) has historically been considered as being based on whether the methods is a so-called laboratory-developed test (in which what is being provided is a service and is, in essence, an analytical result rather than a product) or a so-called IVD in which the product being sold is an actual device that is used in a medical setting.

Laboratory-developed tests regulated under CLIA include many thousands of assays that are run in normally centralized laboratories using custom reagents and custom arrangements of equipment; normally the element of commerce is an analytical result. This is why CLIA assays are sometimes described as analytical services rather than analytical products in contrast to products regulated as IVD products.

Historically, the major division between CLIA and FFDCA regulated contexts, while not subject to a bright-line definition, was generally understood. More recently, the FDA proposed that they could assert broad legal authority over laboratory-developed tests [17]. However, whether they have such legal authority has been subject to substantial debate on the part of legal experts [18]. Late in 2016, the FDA informally announced that they were placing on hold their plans to assert such authority, but a formal announcement is not available at the time of writing [19].

In vitro Diagnostics/Analytical Services Development—High-Level Fundamentals

A full discussion of device development and validation is beyond the scope of this review, but in general *in vitro* device development is scientifically analogous to analytical development in a centralized laboratory. Analytical validation at the development stage begins with feasibility studies as well as assessment of preliminary repeatability, sensitivity, and specificity. Following this, development moves into the preliminary validation stage, with selection of optimal reagent concentrations, determining the optimal temporal and chemical variables in the protocol/device, and assessment of preliminary repeatability, sensitivity, specificity, and cross-reactivity using the near-final device. In the final stage of development (full validation), the commercial or "near commercial" device/method is used and sensitivity, specificity, precision, repeatability (between intra-laboratory replicates for IVDs), and reproducibility (between analysts within a CLIA laboratory) are assessed. Generally, this analysis is conducted versus an established gold standard methodology, if one exists. At this late stage, the cutoffs for a reporting threshold (limit of detection [LOD]) and for the limit of quantitation (LOQ) are generally finally established. In addition, potential interferences between potentially structurally related analytes, robustness to various sample preparation conditions, and clinical sample performance around the reporting cutoffs are typically assessed. In this late phase of development, it is important that data be derived through analysis of samples from the intended human population. Finally, for IVDs, the manufacturing robustness has to be assessed (that is, the ability to consistently manufacturer batches of devices that perform as designed), as does the need for operator training and usability under actual use conditions.

DRUG–*IN VITRO* DIAGNOSTIC CO-DEVELOPMENT

More recently, the application of precision medicine concepts has led to the process of what is now termed drug–companion diagnostic (CDx) co-development. This now typically can involve the custom (or semi-custom) development of an IVD aimed at a biomarker that is targeted by a drug or predicts a drug's likely performance. The biomarker is generally used to mark a biological setting in which the drug will be particularly well-suited; this is often a marker that indicates ectopic expression of protein involved in the underlying pathophysiology (e.g., HER2), high- or low-level expression of a normal biomarker of disease activity, or the presence of a somatic mutation in an oncology setting.

To provide insight on how to manage and coordinate these complexities, the FDA has published a guidance noting that any diagnostic that is critical for the safe and effective use of a new drug would generally be expected to have an FDA-approved IVD product available [20]. This guidance does, however, express some flexibility regarding this general policy based on public health considerations.

The need for such co-development can complicate the development of new precision medicines, as it has led to the need to coordinate the simultaneous development of two products that are regulated by different agency centers under processes that are partly discordant in timelines, scientific standards, and developmental stages. This complexity has been the subject of increased attention by senior FDA leaders. While it is hoped that better coordination will be achieved based on improvements in inter-center coordination, development of two regulated products simultaneously will always remain more complex than development of a product in isolation.

The complexity of drug–companion diagnostic co-development is illustrated in Figure 1.4.2. In essence, such co-development requires that the stage of development of the two products be coordinated closely so that one of the required elements does not delay overall approval of the combination. Development of the drug in an expedited setting (e.g., skipping a traditional phase of development) further compresses timelines and makes this coordination even more challenging.

The above discussion is partially specific to the United States. However, the European Union has recognized the need to modernize the regulation of companion IVDs. Currently IVDs in the EU are regulated under an EU Directive that is implemented nationally, with the intensity of oversight commensurate with increasing risk. Most IVDs are currently self-certified. Under the current EU Directive, the concept of a "companion diagnostic" is not defined. In contrast, under a proposed In Vitro Diagnostic Regulation [21], there will be EU implementation of regulatory requirements based on risk. Companion diagnostics will largely fall into a high-risk category and will require explicit Notified Body involvement. Companion diagnostics will also require submission of relevant clinical data. Further, under the proposed IVD regulation, companion diagnostics will involve

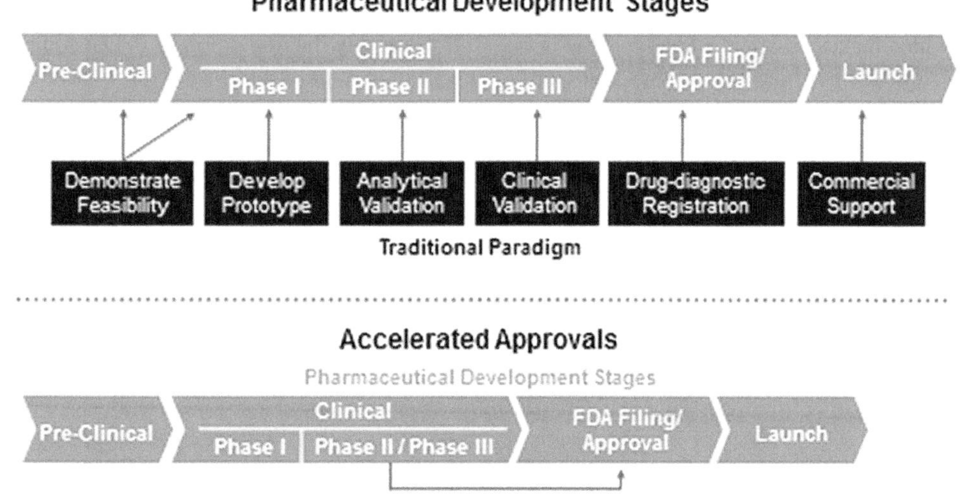

FIGURE 1.4.2 Regulatory process choreography: "Companion diagnostics."

consultation with the European Medicines Agency (EMA) or relevant National Drug Authorities, as needed to review the product's overall use in concert with a new drug. Under the current proposal, existing Conformite europeenne (CE; French for "European conformity")-marked IVD products will eventually require recertification by the end of a transition period.

WHAT DOES THE FUTURE OF PRECISION MEDICINE REGULATORY DEVELOPMENT LOOK LIKE?

More recently, the concept of what has become known as "one IVD–one drug" (a device that is developed for use with, and approved, only for use with a single drug) has increasingly been recognized as unsustainable in the long term. This is based on multiple factors:

- Multi-analyte diagnostics are being developed (capable of reading out multiple analytes in a single assay) that will be able to serve the needs of informing the use of multiple precision medicines.
- The development of IVD assays specific to the use of a single drug within a class with multiple entries has led to confusion on the part of health care professionals. Perhaps the best-known case of this is in the area of assays for immune-oncology–related antigens, which are being used as biomarkers in multiple drug development programs and often have followed the one drug–one IVD concept [22].
- Perhaps the ultimate case of multi-analyte diagnostic technology is raised by next-generation sequencing technology, which is capable of reading out multiple genetic variants within a single somatic tumor sample. Even more fundamentally, next-generation sequencing technology now makes practical whole-genome or whole-exome sequencing of germ-line DNA that can detect thousands of potentially informative genetic variants. Current developments in this general area are reviewed in the next section.

The Coming Era of Multiple Analyte IVDs/Diagnostic Services, Including Next-Generation DNA Sequencing Technology

As noted above, analytical technology has been moving forward, and nowhere is this more evident than in next-generation multiple-analyte and next-generation sequencing technology. Such new technologies allow the detection of multiple protein (proteomics) or transcript analytes (transcriptomics), detection of multiple genetic variants in limited somatic tissue samples, and even detection of thousands or millions of potentially informative variants in whole-genome or whole-exome sequencing. It is clear that such technologies are going to force a reform in current regulatory paradigms.

With regard to the regulation of next-generation sequencing, and specifically undirected whole-genome and whole-exome sequencing, the transformative nature of this new technology has led to two separate groups of authors calling for the FDA to break their general "one approval–one analyte" IVD paradigm. Specifically, in 2016 two publications called for separating the mode of regulation of the technology system leading to the analytical result from the method used to provide biological interpretation of, and assign utility to, the analytical result [23,24].

This concept now appears to have been adopted, in high-level concept, by the FDA based on a two-part FDA workshop held late in 2016. One of these parts focused on a broad systematic validation of the accuracy of next-generation sequencing systems and the other on the development of what could be called a relational database that provides insight on the possible or likely medical and pharmaceutical implications of genetic variants [25]. Late in 2016, the FDA published two relevant draft guidances in this area that provide a clear indication of their long-term policy direction [26,27].

Looking forward, it is now believed that the increasing availability of both multiple-analyte proteomic and transcriptomic technologies, multi-genic analysis of somatic mutations, and whole-genome or whole-exome sequencing will eventually lead to a partial reorientation of the relationship between drug and diagnostic technology. More specifically, what is now called the "companion diagnostic era" (involving the availability of a drug MOA driving the utilization of IVDs) could be supplemented by the "companion drug" paradigm, in which availability of multiple analyte (most often proteomic or transcriptomic methods) and next-generation sequencing services will instead drive utilization of drugs. This overall concept is illustrated in Figure 1.4.3.

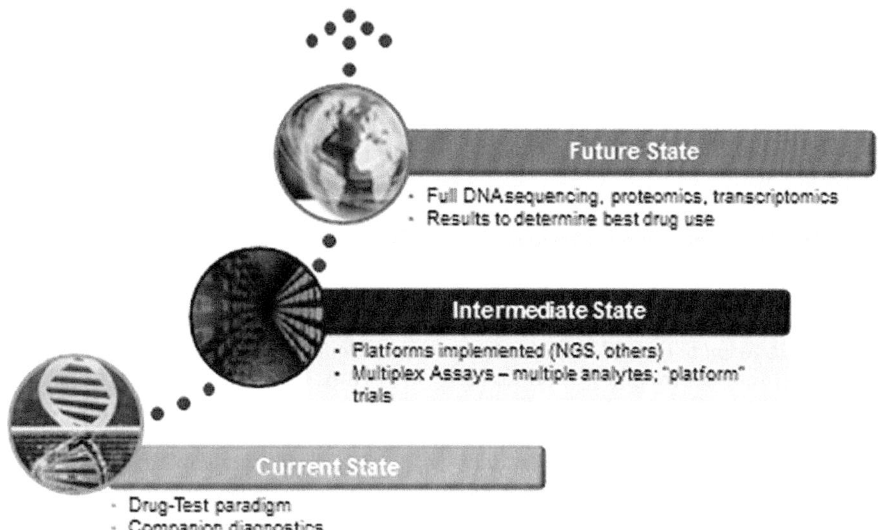

FIGURE 1.4.3 Next-generation sequencing and beyond: "Companion drugs."

REFERENCES

1. Anna Luisa Di Lorenzo, MBBCh. HLA-B27 Syndromes http://emedicine.medscape.com/article/1201027-overview.
2. A. Kastbom, G. Strandberg, A. Lindroos, T. Skogh. Anti-CCP antibody test predicts the disease course during 3 years in early rheumatoid arthritis (the Swedish TIRA project) *Annals of the Rheumatic Diseases* 63, p 1085–1089, 2004.
3. Qualification of Biomarker—Total Kidney Volume in Studies for Treatment of Autosomal Dominant Polycystic Kidney Disease; http://www.fda.gov/downloads/Drugs/GuidanceCompliance RegulatoryInformation/Guidances/UCM458483.pdf.
4. V.E. Torres, A.B. Chapman, O. Devuyst, RT. Gansevoort, J.J. Grantham, E. Higashihara, R.D. Perrone, H.B. Krasa, J. Ouyang, F.S. Czerwiec. Tolvaptan in patients with autosomal dominant polycystic kidney disease. *New England Journal of Medicine* 367(25), p 2407–2418, 2012.
5. Bosutinib For Autosomal Dominant Polycystic Kidney Disease; https://clinicaltrials.gov/ct2/show/NCT01233869?term=autosomal+dominant+bosutinib&rank=1; ClinicalTrials.gov Identifier: NCT01233869.
6. K. Steenland, L. Zhao, F. Goldstein, J. Cellar, J. Lah. Biomarkers for predicting cognitive decline in those with normal cognition. *Journal of Alzheimers Disease* 40(3), p 587–594, 2014.
7. Guidance for Industry and FDA Staff Qualification Process for Drug Development Tools; January 2014 Procedural; http://www.fda.gov/downloads/drugs/guidancecomplianceregulatoryinformation/guidances/ucm230597.pdf.
8. Qualification of novel methodologies for drug development: Guidance to applicants; European Medicines Agency; http://www.ema.europa.eu/docs/en_GB/document_library/Regulatory_and_procedural_guideline/2009/10/WC500004201.pdf.
9. FDA Drug Development Tools (DDT) Qualification Programs; http://www.fda.gov/drugs/developmentapprovalprocess/drugdevelopmenttoolsqualificationprogram/default.htm.
10. Qualification of novel methodologies for medicine development; http://www.ema.europa.eu/ema/index.jsp?curl=pages/regulation/document_listing/document_listing_000319.jsp&mid=WC0b01ac0580022bb0.
11. A. Mishra, R. Greaves, J. Massie. The relevance of sweat testing for the diagnosis of cystic fibrosis in the genomic era. *Clinical Biochemical Review* 26, p 135–153, 2005.
12. J.A. Incorvati, S. Shah, Y. Mu, J. Lu1. Targeted therapy for HER2 positive breast cancer. *Journal of Hematology & Oncology* 6, p 38, 2013.
13. T. Sasaki, S.J. Rodig, L.R. Chirieac, P.A. Jänne. The biology and treatment of EML4-ALK non-small cell lung cancer. *European Journal of Cancer* 46(10), p 1773–1780, 2010.
14. A.J. Vargas and C.C. Harris. Biomarker development in the precision medicine era: Lung cancer as a case study. *Nature Reviews* 16, p 525–537, 2016.

15. S.C. Preissner, M.F. Hoffmann, R. Preissner, M. Dunkel, A. Gewiess, S. Preissner. Polymorphic cytochrome P450 enzymes (CYPs) and their role in personalized therapy. *PLoS One* 8(12), p 1–12, 2013.
16. L.J. Lesko, A.J. Atkinson, Jr. Use of biomarkers and surrogate endpoints in drug development and regulatory decision making: Criteria, validation, strategies. *Annual Review of Pharmacology and Toxicology* 41, p 347–366, 2001.
17. M. Bayefsky, B.E. Berkman. FDA's proposed guidance for laboratory developed tests: How should regulators balance the risks and promise of innovation in clinical genetics? *FDLIs Food Drug Policy Forum* 5(2), p 1–15, 2015.
18. P.D. Clement, L.H. Tribe. *Laboratory Testing Services, As The Practice of Medicine, Cannot Be Regulatory As Medical Devices.* Washington, DC: American Clinical Laboratory Association; http://www.acla.com/wp-content/uploads/2015/01/Tribe-Clement-White-Paper-1-6-15.pdf.
19. Z. Brennan. FDA delays finalization of lab-developed test draft guidance. November 2016; http://www.raps.org/Regulatory-Focus/News/2016/11/18/26218/FDA-Delays-Finalization-of-Lab-Developed-Test-Draft-Regulations/
20. Principles for Co-development of an In Vitro Companion Diagnostic Device with a Therapeutic Product; http://www.fda.gov/downloads/medicaldevices/deviceregulationandguidance/guidancedocuments/ucm510824.pdf.
21. Proposal for a Regulation of the European Parliament and of the Council on In Vitro Diagnostic Medical Devices; http://www.europarl.europa.eu/meetdocs/2009_2014/documents/com/com_com(2012)0541_/com_com(2012)0541_en.pdf.
22. FDA Public Workshop—Complexities in Personalized Medicine: Harmonizing Companion Diagnostics Across a Class of Targeted Therapies; March 24, 2015; http://www.fda.gov/MedicalDevices/NewsEvents/WorkshopsConferences/ucm436716.htm.
23. E.S. Lander. Cutting the Gordian helix—Regulating genomic testing in the era of precision medicine. *The New England Journal of Medicine* 372(13); p 1185–1186, 2015.
24. P. Vicini, O. Fields, E. Lai, E.D. Litwack, A.M. Martin, T.M. Morgan, M.A. Pacanowski et al. Precision medicine in the age of big data: The present and future role of large-scale unbiased sequencing in drug discovery and development. *Journal of Clinical Pharmacology and Therapeutics* 99(2), p 198–207, 2016.
25. FDA Public Workshop—Adapting Regulatory Oversight of Next Generation Sequencing-Based Tests—September 23, 2016; http://www.fda.gov/MedicalDevices/NewsEvents/WorkshopsConferences/ucm514720.htm.
26. Use of Standards in FDA Regulatory Oversight of Next Generation Sequencing (NGS)-Based In Vitro Diagnostics (IVDs) Used for Diagnosing Germline Diseases; July 2016 http://www.fda.gov/downloads/medicaldevices/deviceregulationandguidance/guidancedocuments/ucm509838.pdf.
27. Use of Public Human Genetic Variant Databases to Support Clinical Validity for Next Generation Sequencing (NGS)-Based In Vitro Diagnostics; 2016 http://www.fda.gov/downloads/medicaldevices/deviceregulationandguidance/guidancedocuments/ucm509837.pdf.

2 Biomarkers in Preclinical Sciences

Lost in Translation

The Challenges with the Use of Animal Models in Translational Research

2.1

Attila A. Seyhan

Contents

INTRODUCTION

Ninety percent of the drugs entering human trials fail [1–5], and the process of getting a new drug to market is becoming a longer, costlier, and riskier business [6,7]. As discussed in this book and elsewhere [3], drugs fail the development process for various reasons. One of the problems is the irreproducibility of many preclinical research findings and their poor translatability to human studies. There has been a growing argument that the failure of translation from preclinical findings to human outcomes may in part be due to the problems in the preclinical animal research itself [10–24]. There could be many possible causes for the poor translatability for animal model research [26], in particular, the methodological flaws such as underpowered studies, low group sizes, and lack of blinding [19], or whether appropriate animal models or species have been used (Figure 2.1.1).

Animal models are immensely valuable when used appropriately [30] and a broad range of animal models used for the study of human disease [31]. Each year, more than 100 million animals—including mice, rats, frogs, dogs, cats, rabbits, hamsters, guinea pigs, monkeys, fish, and birds are used in biomedical, agribusiness, veterinary research, cosmetics, chemical and food industry as well as biology and medical training in U.S. laboratories alone [32] and (https://www.peta.org/issues/animals-used-for-experimentation/animals-used-experimentation-factsheets/animal-experiments-overview/).

Scientists have relied on animal models as important research tools to study the pertinent questions in a variety of human diseases including cancer [33], diabetes [34] [35], inflammatory [36] and autoimmune disease and disease of central nervous systems [37].

Various recommendations to improve the translatability of animal model studies have been reported [27,28], including research on relevant, carefully designed, well-characterized and controlled animal models that will remain for a long time an essential step for fundamental discoveries, for testing hypotheses

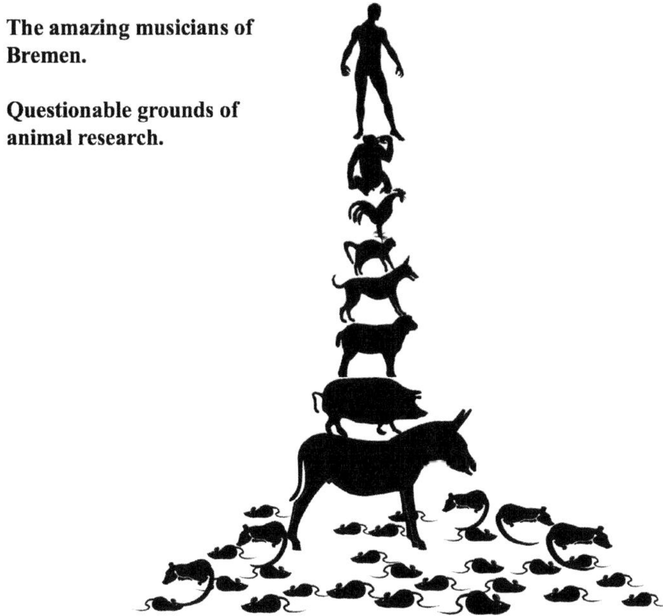

FIGURE 2.1.1 The amazing musicians of Bremen. Questionable grounds of animal research.

at the organism level, and for the validation of human data [25]. However, due to long-standing issues with animal model studies, there is a renewed interest to focus on human studies for early discovery [22]. Recently, biopharmaceutical industry has begun disinvesting in internal animal research, which has been taken up by academic and startup biotech sectors [11,29].

POOR REPRODUCIBILITY OF PRECLINICAL RESEARCH

Despite the enormous benefit of preclinical research findings, many preclinical reports contain irreproducible findings and are dead ends in the search for new drugs for a number of human diseases. This is in part due to the significant differences between humans and animal such as genetic background, developmental span, and organ structure and functions, which make it challenging to translate findings in animals into human disorders and corresponding therapies for human patients.

The research community has been well aware of the issues and merits involving the translatability of various preclinical animal findings to human outcomes [2–5,24,38–44].

It has been estimated that only 6% of animal studies translate to human responses [45] and only about one-third of highly cited animal research findings translate at the level of human trials [19,46] and that only one-tenth of the interventions are subsequently approved for use in humans.

The use of mouse and other animal models of human diseases has been the cornerstone of modern biomedical research [47,48]. The theory is that the findings from current mouse models developed to mimic human diseases translate directly to human conditions. However, this argument has been challenged by many studies including Seok et al. [36]. For example, researchers reported on the existence of 195 published methods that prevented or delayed the development of type 1 and 2 diabetes in mice [34], yet none of these "breakthroughs" ever translated to human medicine.

In addition, mouse models of disease have been extensively used in modern biomedical research to identify and test the efficacy and safety of drug candidates for subsequent human studies [48,49]; however, translating animal studies to human studies has a poor record and only a few of those human trials have been successful [50–52]. In particular, the failure rate is even greater for those trials in the area of inflammation, a condition present in many human diseases. As documented by Seok et al., approximately 150 clinical trials that tested candidate drug compounds that were intended to block the inflammatory response in human patients failed [53–56].

In a recent article, researchers found that medical treatments developed in animals rarely translate to humans, mostly due to poor replication of even high-quality animal studies [46]. Diseases that are artificially induced in animals in a laboratory, whether they are mice or monkeys, are never identical to those that occur naturally in human beings. Because animal species differ from one another biologically in many important ways, it becomes even more unlikely that animal experiments will generate results that will be correctly interpreted and translated to the human studies in a meaningful way.

Systematic studies evaluating how well animal models mimic human diseases are sparse. For example, Seok et al. [36] conducted a systematic analysis on the genomic responses of animal models and humans to similar inflammatory stresses and found that, although acute inflammatory stresses from different etiologies result in highly similar genomic responses in humans, the responses in the corresponding mouse models correlate poorly with the human conditions. For example, among genes that differ significantly between humans and mice, the mouse ortholog genes were almost random in matching their human orthologs (e.g., R^2 between 0.0 and 0.1).

The study by Seok and colleagues [36] supports a higher priority for translational medical research to focus on the more complex human conditions rather than relying on mouse models to study human inflammatory and other complex chronic diseases. They further stated that the complete genetic dissociation in their study challenges the usefulness of mice models in other areas of biomedical research and for drug development. In fact, many other researchers have been voicing their concerns about the shortcomings of the mice models of numerous other human diseases.

For example, according to former National Cancer Institute Director Dr. Richard Klausner, "We have cured mice of cancer for decades, and it simply didn't work in humans".

A recent opinion article [57] argued that even if the research were conducted rigorously, our ability to predict human responses from animal models will be limited by interspecies differences in molecular and metabolic pathways.

Similar concerns were highlighted in a widely cited paper titled "Why most published research findings are false," which highlighted the problems caused by poor experimental and study design and analysis [39].

A seminal paper published by a team at Bayer HealthCare reported that only about 25% of published preclinical studies could be validated [14]. Another paper, published by Glenn Begley [13], has made this problem even more visible. Begley's test involved 53 studies whose results were purported to be groundbreaking. Most of the results could not be reproduced even by the original investigators themselves. Of the 53 original "groundbreaking" studies that Begley put to test, he could reproduce only 6 of them. A recent report published by Begley titled "Six red flags for suspect work" [13] brings forward the six most common, but avoidable, mistakes that he has identified. These included whether (1) experiments were performed blinded, (2) basic experiments were repeated, (3) all results were presented or only cherry-picked results were presented, (4) the appropriate controls were used, (5) the reagents used in experiments were valid and of high purity and quality, and (6) appropriate statistical tests and methods were employed. His conclusion was that the appropriate scientific conduct was often not applied.

Traditional training for biomedical research is often a haphazard process. Few basic scientists receive any formal teaching, most rely on what they learn from their supervisors and whom they often imitate [58,59].

Despite the evidence that questions the usefulness of animals to model human disease [47,48,60], in the absence of significant and successful translation of animal studies to human trials, many scientists and public regulators still assume that results from animal studies mirror human disease.

THE LIMITATIONS OF USING ANIMALS FOR HUMAN DISEASE

What prevents most of the successes in animal experiments from eventually becoming successful findings in human studies? The reason is largely due to the species-specific differences.

Traditionally, many animal models are generated by inducing disease in otherwise healthy animals. Because of this, common comorbidity contexts of human patients in a specific disease may not be replicated in animal models. Animal models are limited in their capacity for self reporting; therefore, the clinical end points must be carefully defined and reliable to distinguish treatment effects.

As discussed recently [14,61], contradictory findings have been found for every biological level of investigation—from genetics to physiology, from life span to metabolism, from biochemistry to immunology. Animals are different from humans.

Furthermore, even deeper differences at the nucleic acid, protein, pathway, cellular, tissue, organ, organism, and population levels, as well as differences in animals' interactions with their environment and their responses to environmental stimuli contribute to the difficulty of developing effective human treatments and to the absence of cures.

Genetic and physiological variation within each species or between closely related species has led to the development of inbred strains, which have highly homogeneous genetic composition, to increase the reproducibility of results and the statistical power of experiments [25].

However, disease clinical presentation and response to therapeutics can vary greatly between mouse strains. A recent report showed that although some mouse strains showed full resistance to Ebola virus, others died without specific symptoms and some other strains developed fatal hemorrhagic fever [62]. Differences in responses in animal strains might also underscore the differences in clinical observations among human patients.

The use of animal models of disease is not the only problem to blame for the high failure rate of biomedical research. It has been revealed recently [24] that many published research findings cannot be reproduced or are not as robust as they were described. Begley has shown that only 11% of 53 seminal publications in preclinical cancer research could be confirmed by Amgen researchers [24,59].

The selection of the specific animal species for an animal model is crucial to the study of human disease. For example, a cholesterol-lowering drug failed in rats and one pharmaceutical company gave up on it entirely; however, a Japanese researcher, Arika Endo, asked a colleague who was using chickens for experiments to test the compound. The results of testing this compound in the birds was a success and marked the emergence of statins, drugs used successfully to manage high cholesterol in millions of people.

However, nobody knows how well an animal model predicts a human response. In fact, tests on mice may not even predict how a drug will work in another rodent model [97].

A previous study that examined the results of rat and mouse LD50 tests for 50 chemicals found that these tests predicted toxicity in humans with only 65% accuracy—while a series of human cell-line tests was found to predict toxicity in humans with 75%–80% accuracy [98].

Moreover, using animal models for research other than the commonly used models such as mice also presents unique challenges for scientists, such as financial feasibility and the usefulness of previous studies that utilized a given species for a specific human disease. Similarly, the unusual biological characteristics of a species and the available molecular techniques available for that particular species must also be considered. The choice of a naturally occurring species model must be identified in which it can be most conveniently studied. One example is the use of the nine-banded armadillo in studies for leprosy due to the armadillo's unique susceptibility to *M. leprae.*

Many studies have confirmed the failure to replicate published results including in areas ranging from physiology to cancer biology [63–65], diabetes [34,35,61,66,67], Alzheimer's [68], stroke [18], and amyotrophic lateral sclerosis [69–71], for example. The reasons for failure are complex and multifaceted. As discussed in a recent popular book published by science journalist Richard Harris [59], reasons range from the complexity of biomedical science to the limitations of the tools and training of scientists as well as incentives pressure to publish and secure funds. In the competitive field of science, several counterintuitive incentives conspire to undermine the scientific method, leading to a literature scattered with unreliable non-replicable data.

Furthermore, animal models are often poor predictors of results in human trials. Sample sizes could be too small to produce reliable results, not to mention that animal models are often too homogeneous or, in fact, genetically identical. Yet, animals still provide valuable information about fundamental biology if not the disease pathobiology itself. Nonetheless, translating that basic biology into medically relevant insights is complicated, and it is even more complicated when cell lines or even primary human cells are used in research. Studies conducted in human cell lines have been shown to be problematic and prone to generate irreproducible data. It is not only the models that fail us, but biology itself is also a factor. Evolution has created many redundant systems to ensure that organisms can survive and pass on their genes to next generation by rapidly adapting to a dynamic environment. Many cancer drugs provide good examples. Humans evolved many different biological signaling pathways and mechanisms to enable cellular growth and proliferation. Targeting a single pathway is not going to block a specific disease phenotype entirely. This is why chemotherapy drugs eventually fail: Tumors often eventually develop resistance to many chemotherapy drugs.

EFFORTS TO AMELIORATE ANIMAL RESEARCH AND REPORTING

To address the significant shortcomings found in the conduct and reporting of preclinical research, Sena et al. [72] proposed minimum quality standards for the range of animal research before findings from these studies are taken to human clinical trials. The ARRIVE (Animal Research: Reporting In Vivo Experiments) guidelines [57,73] were created to further address this problem. However, despite the endorsement of these guidelines by the major journals and funding agencies, the improvement in reporting standards was found to be marginal [74].

The Systematic Review Centre for Laboratory Animal Experimentation (SYRCLE) in the Netherlands has created "a gold standard publication checklist" to promote more rigor in the conduct, not just in the reporting, of animal research [75].

It has also been acknowledged by the National Institutes of Health (NIH) that poor training may in part be responsible for the lack of reproducibility of findings based on animal models [76]. Resources such as workshops (www.3rs-reduction.co.uk) and training (www.syrcle.nl) are available to help preclinical investigators in experimental design and statistical analysis. There is even an online course in experimental design to help investigators learn how to conduct systematic reviews.

It has been reported that basic scientists' motivation comes from scientific discovery rather than the application of their findings to medicine [58,77].

The "co-clinical trial" in which preclinical trials explicitly parallel ongoing human phase I and II trials have also been conducted [78] alongside with the development of a translatability scoring system to identify biomarkers that more accurately predict therapeutic outcomes [79].

USE OF NON-ANIMAL RESEARCH METHODS

Recent developments in non-animal research methods have shown promise. For example, research conducted with human volunteers with the aid of sophisticated computational methods and *in vitro* studies based on human cells and tissues have been shown to be critically more effective and accurate than crude animal model studies.

New technologies such as organ-on-a-chip systems for animal-free toxicity testing are also emerging [45,80–85]. These surrogate non-animal testing systems need not be perfect. If they are just significantly better than animal model systems and *in vitro* human studies, they will be still highly impactful.

As discussed in a recent commentary by Shuler [45], there must be some advantages of the organ-on-a-chip systems for them to have broad acceptance and impact, particularly on drug development. They need to be better than the existing animal model and human *in vitro* systems, which are typically in the static mono-layer single cell–type culture system. While animals present a fully integrated model, significant differences exist at the cellular, genetic and immune levels as well as in size and pharmacokinetics and in true biology and physiological response levels and pathophysiology and disease levels. These differences make the translation of animal findings to human responses difficult. Similarly, *in vitro* human cell models often provide an incomplete assessment of a complex, fully integrated human response. Communication among various cells within an organ or between organs is critical for paracrine and endocrine signaling and to a response to a drug at the organism level.

The development of drugs during preclinical research fails due to both insufficient efficacy and toxicological side effects of the drug or its metabolites. New model systems that can identify in preclinical studies those drugs that will be both efficacious and have a good safety profile could result in and more effective and safer drugs at a relatively lower development cost, not to mention saving animals from unnecessary suffering.

There is no easy solution to this. However, by studying the disease using a combination of human cell cultures and tissues and *in vitro* and stem cell methods with a better understanding of disease mechanisms in patient samples as well as other approaches directly relevant for patients may provide better alternatives and be more ethical and humane than using animal models of disease. The NIH has mandated that applicants specifically report in their grant proposals how they will make sure that their results are robust and reproducible, and many journals now have sections to ensure that a study has followed a list of criteria. There are also new initiatives such as the one championed by Center for Open Science in Charlottesville, Virginia, which encourages open sharing of research processes and resources, including protocols, materials, and data.

Furthermore, the growing field of metascience—the scientific study of science—may facilitate progress in medical research and reduce the failure rate by analyzing large data sets.

EMERGING MODEL SYSTEMS

Novel non-animal–based models are becoming more available and have been shown to be more accurate than the not-so-efficient animal experiments. Organs-on-a-chip that use induced pluripotent stem (iPS) cells taken from humans, or in the future from patients themselves, is one promising model system. Cells obtained from an individual with a particular disease can be induced in a miniaturized lab on a chip microfluidic device [86,87] to become specific cell types or miniature organs with full physiological function, not to mention they are genetically identical to the patient. This is a paradigm shift in early drug discovery and development from what has been done traditionally using cell lines, human primary cell cultures, and animal models of disease. The organ-on-a-chip can also be used for drug safety and efficacy studies. However, the question remains how these organoids truly represent a fully functioning organ such as a liver or heart. Emulate, a spin-off of the Wyss Institute of Harvard, has developed organoids using plastic chips built by three-dimensional (3D) printers with intricate channels. These chips can be used to study organ biology and to screen and test candidate drugs for their efficacy. It is hoped that these model systems will demonstrate their value as a replacement for animal studies and that regulatory agencies will accept them as a model system in place of animal studies in the near future.

It is inevitable that medical research that uses human volunteers, sophisticated computational methods, and advanced *in vitro* studies based on human organoids generated from iPS and tissues will be the future advancement of biomedical research [88,89].

OTHER APPROACHES FOR IMPROVING TRANSLATIONAL EFFICIENCY

Small differences in preclinical studies, including animal model designs, can lead to large differences in the effect size of therapeutics [90]. This is analogous to chaos theory, where small differences in initial conditions or model parameters may lead to large differences in results. Because of this, Ergorul and Levin [90] recommend applying strategies used by meteorologists in dealing with the chaos inherent in forecasting weather [91] to translational research. For example, in translational research, the characteristics of each model can be systematically varied in several dimensions. If the result remains the same despite changing parameters (e.g., the species, the method of inducing disease, the degree of injury, and other factors), it is more likely that the results will translate to humans. These approaches are those of systems biology, and there are examples for their use, for example, in translation initiation in eukaryotes [92] and differentiation of HL-60 cells [93].

As discussed earlier [90], there are other methods for improving translational efficacy. For example, the phase 0 clinical trial is one where receptor binding, mechanism of action, pharmacodynamics, pharmacokinetics, or other biologic principles are studied in humans using microdoses of a drug [94,95] without any therapeutic intent. The goal is to test whether the biological mechanism seen in animals can be affected by treatment at a very low dose as used in a human. Phase 0° studies could be vital in improving the translational research process than at present. This type of study is best done with a biomarker that could be highly sensitive to small concentrations of a drug tested.

Knowledge of a biological mechanism found in a disease can be employed to find new therapies, for example, the aberrant activation of Bcr-Abl tyrosine kinase in chronic myelocytic leukemia led to the development of the highly effective kinase inhibitor, imatinib.

Such bedside-to-bench observations in human disease can help focus the direction of animal research, which in turn will improve the translational process because they are already known to be associated with a clinical end point.

SUMMARY

Data suggest that medical treatments developed in animals rarely translate to humans; thus, researchers should remain cautious about extrapolating the finding of prominent animal research to the care of human disease. Moreover, irreproducibility of many animal studies is an additional concern for the value of this type of research.

Diseases that are artificially induced in animals in a controlled laboratory environment are never identical to those that occur naturally in human beings. Moreover, because animal species differ from one another in biology, in physiology and in many other ways, it becomes even more unlikely that animal experiments will generate results that will be correctly interpreted and applied to the human condition in a meaningful way.

Further complicating the situation is the multifactorial nature of complex diseases such as cancer, cardiovascular diseases, infectious diseases, neurodegenerative disorders, and the pathological consequences of aging. It has been a challenge to develop efficacious therapies rapidly and efficiently.

To overcome the challenges with translational research, some methodological changes are needed in how experiments are designed and conducted, how animal models are used, and how preclinical and clinical teams work together to solve this "lost in translation" conundrum and increase the probability of preclinical findings that can be translated to the clinic.

Because of their complementarity, all experimental approaches —such as genetics, genomics, biochemistry, physiology, *in vitro* and cell culture experiments, *in silico* modeling, animal models of disease, and clinical studies—are indispensable. Therefore, preclinical research on carefully designed, well-characterized and controlled disease-relevant animal models will remain a key stepping-stone for testing early discoveries at the organism level and for the testing of human findings for the foreseeable future.

Other considerations include the applicability of the target, rigor, and validity of the experimental platform and models systems. Possible criteria may include a greater than 50% increase in survival in the case of animal models of cancer (three primary human orthopedic patient-derived xenograft (PDX) models) and a 50% increase in survival of pancreatic ductal adenocarcinoma (PDAC) genetically engineered mouse models (GEMM). Additional considerations are route and schedule of delivery of agents similar to what will be used in humans, demonstration of target inhibition, duration of therapy that approximates anticipated duration in a trial, matched data to the clinical indication and context such as stage of disease and the presence of intact immune response if appropriate (for immune therapeutics), and availability of biomarkers to predict responses.

The better use of the more complex human conditions and data is needed rather than relying on mouse and other animal models to study human diseases. Furthermore, the use of non-animal techniques such as the use of human volunteers, epidemiological and clinical data sets, mathematical and computer (*in silico*) models, *in vitro* human tissue or organ models, and computer-modeling studies will improve overall biomedical research.

As important, "precision medicine" trials to evaluate targeted therapies combined with mechanism-based trials in which eligibility is based on criteria other than traditional disease definitions might help to improve overall biomedical research and the development of more efficacious therapies more efficiently in less time [96].

CONCLUSIONS

It is the author's opinion that the future lies with models that can replace animal models, models that must be constantly improved to go beyond animal models and mimic real human conditions. Likewise, biomedical research can be made more efficient by adopting new technologies and methods of scientific research to better understand disease biology in humans.

REVIEW CRITERIA

Publicly available information such as PubMed and Internet were used for the literature review. The focus was on identifying articles published on translational research, use of animal models for human diseases, preclinical and clinical research, biomedical research, drug development, reproducibility of biomedical research data, failure of drug development, and clinical trials. The search was restricted to the most recent studies in this field and all searches were limited to human studies published in English.

REFERENCES

1. Garner JP, Gaskill BN, Weber EM, Ahloy-Dallaire J, Pritchett-Corning KR. Introducing therioepistemology: The study of how knowledge is gained from animal research. *Lab Anim.* 2017;46: 103–113.
2. Paul SM, Mytelka DS, Dunwiddie CT, Persinger CC, Munos BH, Lindborg SR et al. How to improve R&D productivity: The pharmaceutical industry's grand challenge. *Nat Rev Drug Discov.* 2010;3: 203. doi:10.1038/nrd3078.
3. Kola I, Landis J. Can the pharmaceutical industry reduce attrition rates? *Nat Rev Drug Discov.* 2004;3: 711–715.
4. Scannell JW, Blanckley A, Boldon H, Warrington B. Diagnosing the decline in pharmaceutical R&D efficiency. *Nat Rev Drug Discov.* 2012;11: 191–200.
5. Hay M, Thomas DW, Craighead JL, Economides C, Rosenthal J. Clinical development success rates for investigational drugs. *Nat Biotechnol.* 2014;32: 40–51.
6. DiMasi JA, Hansen RW, Grabowski HG. The price of innovation: New estimates of drug development costs. *J Health Econ.* 2003;22: 151–185.
7. Morgan S, Grootendorst P, Lexchin J, Cunningham C, Greyson D. The cost of drug development: A systematic review. *Health Policy.* 2011;100: 4–17.
8. Moher D, Shamseer L, Cobey KD, Lalu MM, Galipeau J, Avey MT et al. Stop this waste of people, animals and money. *Nature.* 2017;549: 23–25.
9. Shen C, Björk B-C. "Predatory" open access: A longitudinal study of article volumes and market characteristics. *BMC Med.* 2015;13: 230.
10. Sabroe I, Dockrell DH, Vogel SN, Renshaw SA, Whyte MKB, Dower SK. Identifying and hurdling obstacles to translational research. *Nat Rev Immunol.* 2007;7: 77–82.
11. Rosenblatt M. An incentive-based approach for improving data reproducibility. *Sci Transl Med.* 2016;8: 336ed5.
12. Pusztai L, Hatzis C, Andre F. Reproducibility of research and preclinical validation: problems and solutions. *Nat Rev Clin Oncol.* 2013;10: 720–724.
13. Begley CG. Six red flags for suspect work. *Nature.* 2013;497: 433–434.
14. Prinz F, Schlange T, Asadullah K. Believe it or not: How much can we rely on published data on potential drug targets? *Nat Rev Drug Discov.* 2011;10: 712.
15. Mak IW, Evaniew N, Ghert M. Lost in translation: Animal models and clinical trials in cancer treatment. *Am J Transl Res.* 2014;6: 114–118.
16. Cummings JL, Morstorf T, Zhong K. Alzheimer's disease drug-development pipeline: Few candidates, frequent failures. *Alzheimers Res Ther.* 2014;6: 37.
17. Zahs KR, Ashe KH. "Too much good news"—Are Alzheimer mouse models trying to tell us how to prevent, not cure, Alzheimer's disease? *Trends Neurosci.* 2010;33: 381–389.

18. Sena ES, van der Worp HB, Bath PMW, Howells DW, Macleod MR. Publication bias in reports of animal stroke studies leads to major overstatement of efficacy. *PLoS Biol.* 2010;8: e1000344.
19. van der Worp HB, Howells DW, Sena ES, Porritt MJ, Rewell S, O'Collins V et al. Can animal models of disease reliably inform human studies? *PLoS Med.* 2010;7: e1000245.
20. Peers IS, Ceuppens PR, Harbron C. In search of preclinical robustness. *Nat Rev Drug Discov.* 2012;11: 733–734.
21. Macleod MR, van der Worp HB, Sena ES, Howells DW, Dirnagl U, Donnan GA. Evidence for the efficacy of NXY-059 in experimental focal cerebral ischaemia is confounded by study quality. *Stroke.* 2008;39: 2824–2829.
22. Garner JP. The significance of meaning: Why do over 90% of behavioral neuroscience results fail to translate to humans, and what can we do to fix it? *ILAR J.* 2014;55: 438–456.
23. Tricklebank MD, Garner JP. Chapter 20. The possibilities and limitations of animal models for psychiatric disorders. *Drug Discov.* pp. 534–557.
24. Begley CG, Ellis LM. Drug development: Raise standards for preclinical cancer research. *Nature.* 2012;483: 531–533.
25. Barré-Sinoussi F, Montagutelli X. Animal models are essential to biological research: Issues and perspectives. *Future Sci OA.* 2015;1: FSO63.
26. Sabroe I, Dockrell DH, Vogel SN, Renshaw SA, Whyte MKB, Dower SK. Identifying and hurdling obstacles to translational research. *Nat Rev Immunol.* 2007;7: 77–82.
27. Ioannidis JPA. Clinical trials: What a waste. *BMJ.* 2014;349: g7089.
28. Ioannidis JPA, Greenland S, Hlatky MA, Khoury MJ, Macleod MR, Moher D et al. Increasing value and reducing waste in research design, conduct, and analysis. *Lancet.* 2014;383: 166–175.
29. Hunter J. Challenges for pharmaceutical industry: New partnerships for sustainable human health. *Philos Trans A Math Phys Eng Sci.* 2011;369: 1817–1825.
30. Michael Conn P. *Animal Models for the Study of Human Disease.* London, UK: Academic Press; 2017.
31. Tannenbaum J. *Animal Models for the Study of Human Disease: Chapter 1. Ethics in Biomedical Animal Research: The Key Role of the Investigator.* St. Louis, MO: Elsevier; 2013.
32. Newkirk I. *The PETA Practical Guide to Animal Rights: Simple Acts of Kindness to Help Animals in Trouble.* New York: St. Martin's Griffin; 2009.
33. Holen I, Speirs V, Morrissey B, Blyth K. In vivo models in breast cancer research: Progress, challenges and future directions. *Dis Model Mech.* 2017;10: 359–371.
34. Roep BO, Atkinson M, von Herrath M. Opinion: Satisfaction (not) guaranteed: Re-evaluating the use of animal models of type 1 diabetes. *Nat Rev Immunol.* 2004;4: 989–997.
35. Chen Y-G, Mathews CE, Driver JP. The role of NOD Mice in type 1 diabetes research: Lessons from the past and recommendations for the future. *Front Endocrinol.* 2018;9. doi: 10.3389/fendo.2018.00051
36. Seok J, Warren HS, Cuenca AG, Mindrinos MN, Baker HV, Xu W et al. Genomic responses in mouse models poorly mimic human inflammatory diseases. *Proc Natl Acad Sci* . 2013;110: 3507–3512.
37. Cavanaugh SE. Animal models of Alzheimer disease: Historical pitfalls and a path forward. *ALTEX.* 2014;31: 279–302.
38. Geerts H. Of mice and men: Bridging the translational disconnect in CNS drug discovery. *CNS Drugs.* 2009;23: 915–926.
39. Ioannidis JPA. Why most published research findings are false. *Chance.* 2005;18: 40–47.
40. Viktor R. Environmental enrichment and refinement of handling procedures. Wolfe-Coote S, editor, In *The Laboratory Primate,* Academic Press, pp. 209–227; 2005.
41. Richter SH, Garner JP, Würbel H. Environmental standardization: Cure or cause of poor reproducibility in animal experiments? *Nat Methods.* 2009;6: 257–261.
42. Richter SH, Helene Richter S, Garner JP, Auer C, Kunert J, Würbel H. Systematic variation improves reproducibility of animal experiments. *Nat Methods.* 2010;7: 167–168.
43. Insel TR. From animal models to model animals. *Biol Psychiatry.* 2007;62: 1337–1339.
44. Gaskill BN, Garner JP. Stressed out: Providing laboratory animals with behavioral control to reduce the physiological effects of stress. *Lab Anim.* 2017;46: 142–145.
45. Shuler ML. Organ-, body- and disease-on-a-chip systems. *Lab Chip.* 2017;17: 2345–2346.
46. Hackam DG, Redelmeier DA. Translation of research evidence from animals to humans. *JAMA.* 2006;296: 1731–1732.
47. Hayday AC, Peakman M. The habitual, diverse and surmountable obstacles to human immunology research. *Nat Immunol.* 2008;9: 575–580.
48. Woodcock J, Woosley R. The FDA critical path initiative and its influence on new drug development. *Annu Rev Med.* 2008;59: 1–12.
49. Vargas HM, Amouzadeh HR, Engwall MJ. Nonclinical strategy considerations for safety pharmacology: Evaluation of biopharmaceuticals. *Expert Opin Drug Saf.* 2012;12: 91–102.
50. Pound P, Ebrahim S, Sandercock P, Bracken MB, Roberts I. Reviewing Animal Trials Systematically (RATS) Group. Where is the evidence that animal research benefits humans? *BMJ* 2004;328(7438): 514–517. doi:10.1136/bmj.328.7438.514.
51. Pound P. Where is the evidence that animal research benefits humans? *BMJ.* 2004;328: 514–517.
52. Rice J. Animal models: Not close enough. *Nature.* 2012;484: S9–S9.
53. Hotchkiss RS, Opal S. Immunotherapy for sepsis—A new approach against an ancient foe. *N Engl J Med.* 2010;363: 87–89.
54. Hotchkiss RS, Coopersmith CM, McDunn JE, Ferguson TA. The sepsis seesaw: Tilting toward immunosuppression. *Nat Med.* 2009;15: 496–497.
55. Wiersinga WJ. Current insights in sepsis: From pathogenesis to new treatment targets. *Curr Opin Crit Care.* 2011;17: 480–486.
56. Mitka M. Drug for severe sepsis is withdrawn from market, fails to reduce mortality. *JAMA.* 2011;306: 2439–2440.
57. Pound P, Bracken MB. Is animal research sufficiently evidence based to be a cornerstone of biomedical research? *BMJ.* 2014;348: g3387.
58. Festing MFW, Nevalainen T. The design and statistical analysis of animal experiments: Introduction to this issue. *ILAR J.* 2014;55: 379–382.
59. Harris R. *Rigor Mortis: How Sloppy Science Creates Worthless Cures, Crushes Hope, and Wastes Billions.* New York: Hachette; 2017.
60. Davis MM. A prescription for human immunology. *Immunity.* 2008;29: 835–838.
61. Chandrasekera PC, Pippin JJ. Of rodents and men: Species-specific glucose regulation and type 2 diabetes research. *ALTEX.* 2014;31: 157–176.
62. Rasmussen AL, Okumura A, Ferris MT, Green R, Feldmann F, Kelly SM et al. Host genetic diversity enables Ebola hemorrhagic fever pathogenesis and resistance. *Science.* 2014;346: 987–991.
63. Day CP, Merlino G, Van Dyke T. Preclinical mouse cancer models: A maze of opportunities and challenges. *Cell.* 2015;163: 39–53.

64. Gengenbacher N, Singhal M, Augustin HG. Preclinical mouse solid tumour models: Status quo, challenges and perspectives. *Nat Rev Cancer.* 2017;17: 751–765.
65. Zhao X, Li L, Starr TK, Subramanian S. Tumor location impacts immune response in mouse models of colon cancer. *Oncotarget.* 2017;8: 54775–54787.
66. DIAbetes Genetics Replication And Meta-analysis (DIAGRAM) Consortium, Asian Genetic Epidemiology Network Type 2 Diabetes (AGEN-T2D) Consortium, South Asian Type 2 Diabetes (SAT2D) Consortium, Mexican American Type 2 Diabetes (MAT2D) Consortium, Type 2 Diabetes Genetic Exploration by Nex-generation sequencing in multi-Ethnic Samples (T2D-GENES) Consortium, Mahajan A et al. Genome-wide trans-ancestry meta-analysis provides insight into the genetic architecture of type 2 diabetes susceptibility. *Nat Genet.* 2014;46: 234–244.
67. Dupuis J, Langenberg C, Prokopenko I, Saxena R, Soranzo N, Jackson AU et al. New genetic loci implicated in fasting glucose homeostasis and their impact on type 2 diabetes risk. *Nat Genet.* 2010;42: 105–116.
68. Cavanaugh SE. Animal models of Alzheimer disease: Historical pitfalls and a path forward. *ALTEX.* 2014;31: 279–302.
69. Rothstein JD. Preclinical studies: How much can we rely on? *Amyotroph Lateral Scler Other Motor Neuron Disord.* 2004;5 Suppl 1: 22–25.
70. Traynor BJ, Bruijn L, Conwit R, Beal F, O'Neill G, Fagan SC et al. Neuroprotective agents for clinical trials in ALS: A systematic assessment. *Neurology.* 2006;67: 20–27.
71. Benatar M. Lost in translation: Treatment trials in the SOD1 mouse and in human ALS. *Neurobiol Dis.* 2007;26: 1–13.
72. Sena E, van der Worp HB, Howells D, Macleod M. How can we improve the pre-clinical development of drugs for stroke? *Trends Neurosci.* 2007;30: 433–439.
73. Kilkenny C, Browne WJ, Cuthill IC, Emerson M, Altman DG. Improving bioscience research reporting: The ARRIVE guidelines for reporting animal research. *PLoS Biol.* 2010;8: e1000412.
74. Baker D, Lidster K, Sottomayor A, Amor S. Two years later: Journals are not yet enforcing the ARRIVE guidelines on reporting standards for pre-clinical animal studies. *PLoS Biol.* 2014;12: e1001756.
75. Hooijmans CR, Leenaars M, Ritskes-Hoitinga M. A gold standard publication checklist to improve the quality of animal studies, to fully integrate the Three Rs, and to make systematic reviews more feasible. *Altern Lab Anim.* 2010;38: 167–182.
76. Collins FS, Tabak LA. Policy: NIH plans to enhance reproducibility. *Nature.* 2014;505: 612–613.
77. Morgan M, Barry CA, Donovan JL, Sandall J, Wolfe CDA, Boaz A. Implementing "translational" biomedical research: Convergence and divergence among clinical and basic scientists. *Soc Sci Med.* 2011;73: 945–952.
78. Chen Z, Cheng K, Walton Z, Wang Y, Ebi H, Shimamura T et al. A murine lung cancer co-clinical trial identifies genetic modifiers of therapeutic response. *Nature.* 2012;483: 613–617.
79. Wendler A, Wehling M. Translatability scoring in drug development: Eight case studies. *J Transl Med.* 2012;10: 39.
80. Mahler GJ, Esch MB, Stokol T, Hickman JJ, Shuler ML. Body-on-a-chip systems for animal-free toxicity testing. *Altern Lab Anim.* 2016;44: 469–478.
81. Bessems J. Toxicokinetic modelling: A necessary tool for quantitative risk assessment in animal-free toxicity testing. *Toxicol Lett.* 2015;238: S47.
82. Leist M. Consensus report on the future of animal-free systemic toxicity testing. *ALTEX.* 2008; 341–356.
83. Kessler R. Filling a gap in developmental toxicity testing: Neural crest cells offer faster, cheaper, animal-free testing. *Environ Health Perspect.* 2012;120: a320–a320.
84. Materne E-M, Tonevitsky AG, Marx U. Chip-based liver equivalents for toxicity testing—organotypicalness versus cost-efficient high throughput. *Lab Chip.* 2013;13: 3481.
85. Oleaga C, Bernabini C, Smith AST, Srinivasan B, Jackson M, McLamb W et al. Multi-Organ toxicity demonstration in a functional human in vitro system composed of four organs. *Sci Rep.* 2016;6. doi: 10.1038/srep20030
86. Ozcan A. Mobile phones democratize and cultivate next-generation imaging, diagnostics and measurement tools. *Lab Chip.* 2014;14: 3187–3194.
87. Ozcan A. Democratization of next-generation imaging, diagnostics and measurement tools through computational photonics. *Biophys J.* 2015;108: 371a.
88. Bredenoord AL, Clevers H, Knoblich JA. Human tissues in a dish: The research and ethical implications of organoid technology. *Science.* 2017;355: eaaf9414.
89. Xinaris C, Brizi V, Remuzzi G. Organoid models and applications in biomedical research. *Nephron.* 2015;130: 191–199.
90. Ergorul C, Levin LA. Solving the lost in translation problem: Improving the effectiveness of translational research. *Curr Opin Pharmacol.* 2013;13: 108–114.
91. Shukla J. Predictability in the Midst of Chaos: A scientific basis for climate forecasting. *Science.* 1998;282: 728–731.
92. Nayak S, Siddiqui JK, Varner JD. Modelling and analysis of an ensemble of eukaryotic translation initiation models. *IET Syst Biol.* 2011;5: 2.
93. Tasseff R, Nayak S, Song SO, Yen A, Varner JD. Modeling and analysis of retinoic acid induced differentiation of uncommitted precursor cells. *Integr Biol.* 2011;3: 578–591.
94. Kummar S, Rubinstein L, Kinders R, Parchment RE, Gutierrez ME, Murgo AJ et al. Phase 0 clinical trials: Conceptions and misconceptions. *Cancer J Sci Am.* 2008;14: 133–137.
95. Lorusso PM. Phase 0 clinical trials: An answer to drug development stagnation? *J Clin Oncol.* 2009;27: 2586–2588.
96. Woodcock J, LaVange LM. Master protocols to study multiple therapies, multiple diseases, or both. *N Engl J Med.* 2017;377: 62–70.
97. Shanks N, Greek R, Greek J. Are animal models predictive for humans? Philos Ethics Humanit Med. 2009;4:2. doi:10.1186/1747-5341-4-2.
98. Ekwall B. Overview of the final MEIC results: II. The in vitro–in vivo evaluation, including the selection of a practical battery of cell tests for prediction. Toxicol In Vitro. 1999;13(4–5):665–673.

Application of Pharmacokinetic and Pharmacodynamic Modeling of Biomarkers of Efficacy in Translational Research

2.2

Neil Benson

Contents

INTRODUCTION

Translational biomarker pharmacokinetic/pharmacodynamic (PK/PD) modeling utilizes preclinical *in vitro* and *in vivo* data and employs mathematical and statistical models to describe and understand them [1]. The PK/PD models developed can then be used to refine preclinical experiments and extrapolate findings to the clinical setting. One key benefit to this approach is that it facilitates understanding of the often-complex interaction between time- and dose-dependent PK and PD.

Biomarkers can be useful "way markers" for diseases. They can often be measured quantitatively and respond within a time frame suitable for preclinical experiments and early clinical trials. They are a valuable tool to show expression of pharmacology, which can be crucial to performing a clinical trial that can be interpreted in the event of either a positive or a negative effect [2]. Furthermore, by selecting the right biomarkers, a pharmacological audit trail can be obtained, linking pharmacology and disease, which can expedite and optimize drug discovery [3].

The purpose of this review is to evaluate the current corpus of literature on this topic, to identify any useful general conclusions, and to comment on possible future developments.

DISCUSSION

In order to survey and interpret the literature, disease-related topics were categorized into the following subtypes: cancer, endocrine disease, neuropsychiatric, cardiovascular disease, musculoskeletal, neurological, congenital, anti-infectives, sensory organs, respiratory, digestive, genitourinary, skin and other (e.g., aging, autoimmunity).

The articles found describing the use of translational biomarker efficacy PK/PD on these topics are summarised in Tables 2.2.1 to 2.2.7.

In general, some 80 or so articles were identified describing the use of PK/PD to analyze biomarker efficacy data and translate to humans. Plotting the frequency of reports by a 4-year range (Figure 2.2.1), it can be seen that the research reported

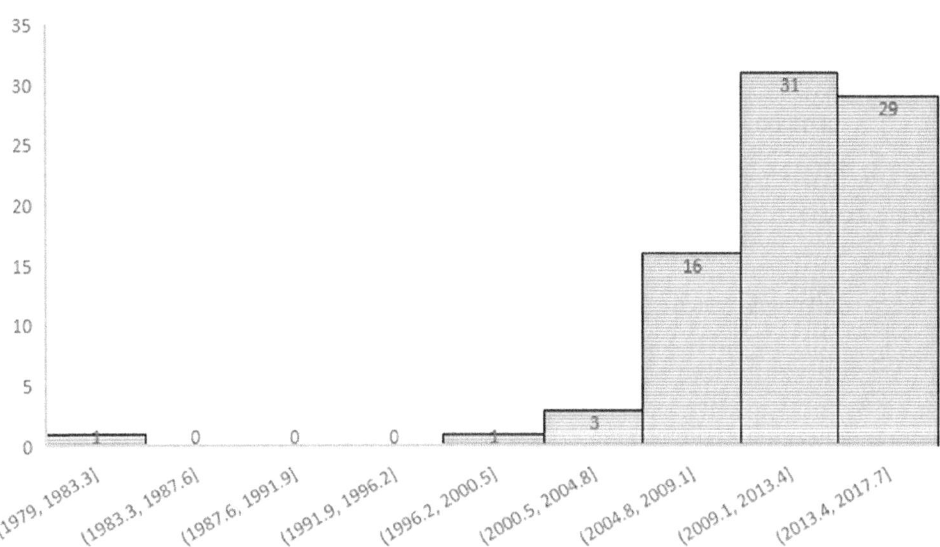

FIGURE 2.2.1 Reference frequency histogram. The plot shows that the translational PK/PD work reported were made in the last 10 years and increased during 2004–2008, 2008–2012 and 2012–20016. To date from 2016, seven publications were found and if this continues, then the trend would seem to be reaching a plateau.

were predominantly carried out in the last 10 years, with very few papers published before 2000. Thus, this approach would seem to be a relatively new development in drug discovery. Interestingly, the numbers of papers grew rapidly between 2000 and 2016, but the trend from 2016 on appears to suggest a plateauing of the number of publications.

The majority of applications were in the area of cancer drug development. Whether this reflects the relative current dominance of cancer in drug discovery or a bias in terms of expected utility cannot be definitively concluded. The frequent use of human tumor cells lines in xenograft models perhaps encourages the use of PK/PD and biomarkers approaches. In addition, there is a tendency toward targets involved in signal transduction, in particular kinases, and since the technology to measure phosphorylated biomarkers has become increasingly robust, the ease of making useful biomarker measures has increased and this may explain some of the increase in signal transduction-related articles.

A significant corpus of literature was also available for endocrine, neurological and anti-infectives disease areas, each with 10 or more publications. The richness of the endocrine literature is mainly due to drug discovery research into diabetes. This disease is both prevalent and good translatable biomarkers such as glucose, insulin and glycosylated hemoglobin are available. Moreover, the relationship between the biomarkers and the disease are relatively well understood, perhaps explaining some of the confidence in using these biomarkers for translational PK/PD research.

Similarly, tissue measures (e.g., in plasma) of viral particle and bacterial counts can be robust and are, in principle, conceptually easy to relate to disease. Hence, perhaps the apparently more prevalent use of translational PK/PD for these diseases.

The relatively large number of publications for neuroscience is interesting. Due to the complexity of the diseases and limited understanding of how animal models can be translated to clinical effects, the confidence in translating model results to humans is arguably less clear than, for example, diabetes and anti-infectives. Nevertheless, evidence was found of the use of preclinical PK/PD biomarker evaluation to aid in the extrapolation of dose to the clinic. This work was predominantly carried out in Alzheimer's disease (AD), probably reflecting the relative proportion of drug discovery work on AD versus other central nervous systems (CNS) diseases. However, other CNS-related topics did receive significant attention, for example, pain, despite the challenge of translating results from preclinical models to humans.

The numbers of papers found in other disease areas was limited.

Most of the papers described the modeling of the preclinical PK biomarker interactions and how the biomarker changes could be related to tumor stasis or regression, typically in a xenograft or syngeneic mouse model (Table 2.2.1). Both small molecules and biologicals have been evaluated, typically using mouse xenograft or syngeneic models. In turn, the authors often aimed to scale the predictions to clinical outcome. Examples of where such scaling has been tested against clinical data include dasatinib, crizotinib, E7820, sunitinib, everolimus, pembrolizumab, A1mcMMAF and motolimod (Table 2.2.1). Overall, it was concluded that, within certain boundaries, preclinical biomarker PK/PD was useful for projecting clinical efficacy. Another interesting observation was that, broadly speaking, tumor stasis or regression was associated with a high percentage of inhibition or occupancy of biomarker responses both in preclinical animal models and in patients (e.g., dasatinib, crizotinib, G-573, pembrolizumab; Table 2.2.1).

TABLE 2.2.1 Summary of references describing the use of biomarker PK/PD models in translational pharmacology for oncology drug discovery

TARGET	*BIOMARKER*	*DRUGS*	*COMMENT*	*REFERENCES*
BCr-Abl	Phospho-BCR-ABL/ phospho-CrkL	Dasatinib	Preclinical and clinical PK/PD showed that 90% of phospho-BCR-ABL *in vivo* was correlated in CML patients who responded to dasatinib treatment in the clinic.	[4]
Mitogen-activated protein kinase/extracellular signal-regulated kinase	p-ERK	G-573	IC_{50} p-ERK inhibition was approximately 1/8th of IC_{50} for TGI.	[5]
α2-Integrin mRNA expression	α2-Integrin mRNA	E7820	Moderate inhibition of α2-integrin expression corresponded to tumor stasis in mice, and similar levels could be reached in patients with the dose level of 100 mg q.d. This analysis showed that only at the highest toxic dose of 200 mg q.d., both targets were met in more than 95% of patients.	[6]
ALK and MET	Phospho-ALK and MET	Crizotinib	The results suggest that 50% ALK inhibition is required for >50% TGI whereas >90% MET inhibition is required for > 50% TGI. Furthermore, >75% ALK inhibition and >95% MET inhibition in patient tumors were projected by PK/PD modeling during the clinically recommended dosing regimen, twice daily doses of crizotinib 250 mg (500 mg/day).	(7)
Multiple receptor tyrosine kinases	Circulating angiogenic factors	Sunitinib	Altered VEGF is more likely to serve as a useful biomarker reflecting tumor responses in cancer patients whose pretreatment VEGF is higher than baseline VEGF in healthy subjects. The findings provide a mechanistic insight into tumor modulation of angiogenic molecules, and may explain the inconsistent results found in previous biomarker studies.	[8]
Ataxia–telangiectasia and Rad3-related kinase	Ser-139 phosphorylated histone 2AX (γH2AX)	AZD6738	The QSP model is being used for P1 clinical trial designs for AZD6738, with the aim of improving patient care through quantitative dose and scheduling prediction.	[9]
mTOR	Inhibition of S6K1	Everolimus	A direct-link PK/PD model predicting the time course of S6K1 inhibition during weekly and daily everolimus administration allowed extrapolation from preclinical studies and first clinical results to select optimal doses and regimens of everolimus to explore in future clinical trials.	[10]
PD-1	PD-1 receptor occupancy	Pembrolizumab	PD-1 receptor occupancy is modeled in the tissue compartment associated with tumor, indicating that the pathway leading to antitumor activity is effectively saturated at clinically relevant pembrolizumab concentrations.	[11]
Cyclin-dependent kinases	Genomic markers of 8 main genes: CD86, CNNB1, EGR1, FLJ4432, HEXIM1, JUN, MK167, PFAAP5 in blood plasma	R547	The paper describes the identification of PD classifiers for the drug to help define clear clinical end points in patients. PK/PD modeling was used to clarify conclusions.	[12]
Microtubule stabilization	TAX resistance factors, e.g., beta-3 tubulin, Pgp; acetylated alpha-tubulin	IXA	PK/PD modeling for a combination therapy for effective translation to a P1 trial.	[13]

(Continued)

TABLE 2.2.1 (*Continued*) Summary of references describing the use of biomarker PK/PD models in translational pharmacology for oncology drug discovery

TARGET	*BIOMARKER*	*DRUGS*	*COMMENT*	*REFERENCES*
TORC1/2	Levels of c-PARP, pSer473-AKT, p-S6	Paclitaxel and AZD2014	PK/PD modeling for a combination therapy to treat lung and ovarian cancer.	[14]
HDAC/Topoisomerase	Chromatin decondensation, cellular differentiation, DNA-strand breaks, presence of topoII-beta in cells	HDACi+ VPA+ epirubicin	The paper gives an overview of PK/PD modeling to test the clinical significance of a combination therapy.	[15]
HER2/cytotoxin	Plasma ADC concentration, tumor volume/stasis	T-M1, A1mcMMAF	The paper outlines the preclinical models (mouse tumor models) evaluated and PK/PD modeling to define clinical end points in humans.	[16]
pEGFR inhibition	Plasma AZD9291 and des-methylated metabolite concentrations	AZD9291	The paper outlines PK/PD analysis in preclinical models to track the change in the specified biomarker to evaluate drug efficacy. A comparison in the change of this biomarker between murine models and humans was also made.	[17]
STAT3 SH2 phosphotyrosine domain-binding site	STAT3 phosphorylation, IFN-α and IFN-ϒ signaling, IL-6 signaling	LY5	The paper highlights that PK/PD modeling of preclinical research shows that the drug lacks efficacy.	[18]
MEK	p-ERK	GDC-0973	Following single doses of GDC-0973, estimated *in vivo* IC_{50} of percentage p-ERK decrease based on tumor concentrations in xenograft mice was 0.78 (WM-266-4) and 0.52 μM (A375). Following multiple doses of GDC-0973, the estimated *in vivo* IC_{50} in WM-266-4 increased (3.89 μM). Human simulations predicted a minimum target plasma concentration of 83 nM and an active dose range of 28–112 mg. The steep relationship between tumor PD (percentage p-ERK decrease) and antitumor efficacy suggests a pathway modulation threshold beyond which antitumor efficacy switches on. Conclusions: Clinical observations of percentage p-ERK decrease and antitumor activity were consistent with model predictions.	[19]
cMET kinase	Phosphorylated cMET	PF02341066	The EC_{90} for the inhibition of cMet phosphorylation corresponded to the EC_{50} for the tumor growth inhibition, suggesting that near-complete inhibition of cMet phosphorylation (>90%) is required to significantly inhibit tumor growth (>50%).	[20]
TLR8 receptor	Plasma levels of IL-6, G-CSF, MCP-1 and MIP1-β	Motolimod	The abstract outlines translation from PK/PD analysis in preclinical models into an FIH trial.	[21]
PI3K/mTOR signaling	Fasting serum insulin and glucose levels, hyperglycemia, 2[18F] fluoro-2-deoxy-D-glucose (PDG)-PET, phosphorylation of AKT, EPK and S6	GSK2126458	The paper describes the translation from PK/PD analysis of the drug and biomarker in preclinical models, into a P1 trial. The need for correct patient pre-selection is highlighted. The conclusions highlight that activating somatic PI3KCA mutations are not a marker of disease or efficacy, hence the need for novel markers to determine how heavily the tumor depends on PI3K/mTOR signaling for survival, which will help predict the success of this therapy.	[22]

This observation is concordant with previous general reviews of efficacy:target occupancy relationships [23] where it was concluded that, in the case of antagonists and enzyme inhibitors, approximately 60%–90% target occupancy is required for clinical efficacy across a range of targets and indications. One hypothesis that might explain the apparent need for approaching or saturating inhibition of the biomarker is that the pathways the biomarker are controlled by are important to the proliferation of a given neoplasm; simultaneously there is some robustness to its attenuation whereby a tumor is insensitive within the linear portion of the sigmoid inhibition concentration curve. According to this theory, redundancies exist in the networks of signal transduction that maintain the proliferative drive for the tumor even if a given pathway is attenuated (see [24] for further discussion) and this may explain why any given individual pathway needs significant attenuation before any effect is evident.

Another notable trend was the use of more mechanistic PK/PD and systems pharmacology models to evaluate biomarker data and make recommendations, for example, for dose and dose regimen (e.g., pembrolizumab; Table 2.2.1). These may utilize data from multiple linked biomarkers, which can help to build confidence in the structure and parameters for a model [25]. It seems likely that this trend toward more mechanistic models using multiple biomarkers will continue.

T-cell therapy represents an emerging and promising modality for the treatment of cancer. Data from recent clinical trials of genetically modified T cells, most notably chimeric antigen receptor T cells, have been interesting. However, continued progress will require the identification of the relevant PK/PD and biomarker strategies to support and guide clinical development of the candidate products (for review see [26]).

One area of difficulty for cancer biomarker research may be that tumor biopsy quality and quantity is currently a technical hurdle that limits the utility of biomarkers in clinical cancer research [27]. In addition, an assumption in many, if not all, biomarker studies was that biomarkers exhibit a flat baseline with respect to time at steady state. Some recent experimental work has called this assumption into question, for example, C-reactive protein (CRP) [28] and phospho signal transducer and activator of transcription (pSTAT3) [29] have been observed to oscillate. On a per case basis, this implies that biomarker data interpretation would be greatly facilitated by the capture of time course data.

Despite the insight the PK/PD understanding bring, the latest metrics show that attrition remains an issue for cancer as for other indications [30]. It may be that the relatively recent use of biomarker PK/PD in cancer drug discovery means that its true impact has yet to be fully seen. It will be interesting to see whether these techniques lead to improved attrition statistics in the future.

Most of the reports of translational PK/PD in the endocrine disease area describe work on diabetes. Diabetes is perhaps the exemplar disease area for mathematical modeling in clinical practice (Table 2.2.2). Furthermore, a large number of mathematical models have been developed to investigate PK/PD and other questions with respect to this disease [42]. Although currently not as rich as the cancer area, there are nevertheless also some detailed

TABLE 2.2.2 Summary of references describing the use of biomarker PK/PD models in translational pharmacology for endocrine disease drug discovery

DISEASE	*TARGET*	*BIOMARKER*	*DRUGS*	*COMMENT*	*REFERENCES*
Diabetes	Reducing appetite, suppressing glucagon release	Glucose	Pramlintide	This model can be used as a platform to optimize dosing of both pramlintide and insulin as a combined therapy for glycemic regulation, and in the development of an artificial pancreas as the kernel for a model-based controller.	[31]
Diabetes	NA	Glucose	Glucose	Dog model initially developed and used in clinical practice.	[32]
Diabetes	NA	Phospho (IRS1, IR, protein kinase B) and glucose		Hierarchical modeling can potentially create bridges between other experimental model systems and the *in vivo* human situation and offers a framework for systematic evaluation of the physiological relevance of *in vitro* obtained molecular/cellular experimental data.	[33]
Diabetes	NA	Insulin, glucose	Insulin	Based on these results, it may be concluded that the indirect pharmacodynamic response model is a more appropriate approach for modeling the PK/PD of insulin than the effect-compartment link model.	[34]

(Continued)

TABLE 2.2.2 (*Continued*) Summary of references describing the use of biomarker PK/PD models in translational pharmacology for endocrine disease drug discovery

DISEASE	*TARGET*	*BIOMARKER*	*DRUGS*	*COMMENT*	*REFERENCES*
Diabetes	Peroxisome proliferator-activated receptor gamma	Mean blood glucose level, percentage of glucose reduction	Effect of curcumin on pioglitazone	The paper outlines a food-drug interaction, tested by PK/PD modeling in preclinical models that could impact upon drug metabolism and clinical efficacy.	[35]
Diabetes	Glucokinase	Glucose		The paper outlines PK/PD modeling of preclinical models to predict human dose and regimen.	[36]
Diabetes	SGLT2	Urinary glucose	Dapagliflozin	Preclinical and clinical translation using a physiologically based PK/PD modeling approach.	[37]
Diabetes/ cancer	Both the IGR-1R and IR-A receptors involved in IGF signaling	IR-A RNA, MEK phosphorylation, AKT phosphorylation, IR:IGF-1R ratio, oral glucose tolerance	BMS-754807	BMS-754807 is currently being tested in a P1 trial for safety (from toxic hyperglycemia as a result of reduced insulin signaling) and efficacy, after preclinical and *in vitro* research highlights that targeting IGF signaling by blocking the IGF-1R alone will not achieve inhibition of tumor growth, due to secondary signaling via the insulin receptor (IR), and the formation of an IGF-1R/IR hybrid receptor as an adoptive resistance mechanism to treatment, causing tumor relapse. Research showing that the IR-A isoform, and a high IR:IGF-1R ratio must be looked for when selecting patients for clinical trials, in addition to hyperinsulinemic patients having underlying NET-associated disease. A combination therapy of BMS-754807 (which only induces tumor-cell apoptosis) and a drug targeting EGFR signaling to shrink the tumor is suggested to maximize therapeutic benefit.	[38]
Hyperal-dosteronism/ Cushing's disease	Selective aldosterone synthase production over cortisol production	Plasma and urinary-aldosterone, cortisol, sodium, potassium; plasma renin activity; and biomarkers of cardiorenal damage including: albuminuria, mean arterial pressure, LV fractional shortening, LV isometric relaxation time (as percentage of R-R interval), blood urea nitrogen, creatinine clearance, LV-myocardial cell size, urine volume.	LCI699	LCI699 is more potent at inhibiting human aldosterone synthase enzymes than those of preclinical species. EC50 and IC_{50} are similar for both human and animal enzymes. However, LCI699 is less selective for human enzyme inhibition (i.e., aldosterone production over cortisol production) as compared with selectivity in preclinical models. Based on dose-dependent specificity, dosage of >1 mg (where specificity is lost) is suggested to treat Cushing's syndrome.	[39]

(*Continued*)

TABLE 2.2.2 (*Continued*) Summary of references describing the use of biomarker PK/PD models in translational pharmacology for endocrine disease drug discovery

DISEASE	*TARGET*	*BIOMARKER*	*DRUGS*	*COMMENT*	*REFERENCES*
Weight loss disorders	TrkB	Weight gain, low ADAs	TAM-163	The pharmacologically active dose is observed to be 1–15 mg/kg upon IV administration in preclinical models (lean cynomolgous monkeys and obese rhesus monkeys). A TMDD model predicts the dose to be administered to the patient depending on their own TrkB levels. The paper suggests 0.05 mg/kg as the starting dose based on the analysis for the MABEL dose. Doses ≥1 mg/kg should yield pharmacological activity. Given the very high target coverage at 15 mg/kg, the need for toxicity assessment for the highest clinically safe dose is highlighted. The model is also useful for diseases with a highly variable or complex pathobiology.	[40]
Benign prostatic hyperplasia	Vitamin D receptors	Markers for prostatic growth, inflammatory activity and smooth muscle activity	Elocalcitol	The paper shows PK/PD modeling for the drug after translation from preclinical models to evaluate the drug's therapeutic potential based on mechanism of action, where clinical end points were also defined to conduct a P1 trial.	[41]

biomarker PK/PD reports. It would appear that some success has been achieved scaling biomarkers such as glucose, insulin and acetylated haemoglobin (HbAC) from preclinical to humans using models. The relative ease of measuring translatable blood biomarkers has no doubt aided the application of biomarker PK/PD methods, as has the clear link between biomarkers such as glucose and HbAc and the disease (e.g., see dapagliflozin; Table 2.2.2).

Most of the papers in the cardiovascular disease section describe the use and scaling of cardiovascular parameters, for example, mean arterial pressure, left ventricular end diastolic pressure, heart rate and electrocardiogram (ECG) data (Table 2.2.3). Typically, these can be measured accurately, easily and with a very rich sampling density with respect to time. Furthermore, the same or very similar parameters can be measured across species. Indeed the basis for scaling these types of cardiac parameters has been the subject of significant investigation [51]. This combination of easily collectable, rich and translatable biomarker data is ideal for building translational PK/PD understanding and encourages research in this area. Consistent with this, there are many more published translational PK/PD biomarker examples aimed at quantifying safety risks for candidate drugs, for example [52–54].

Translational biomarker PK/PD in the musculoskeletal disease area has primarily been used in the development of certain drugs for arthritis (Table 2.2.4). In particular, the use of inflammatory biomarkers in plasma and synovial fluid such as interleukin (IL)-6 was common across some of the studies. A unique key issue for this disease area is the question of the joint as a site of drug action; the relevant inflammatory processes are hypothesized to be local to the joint and, accordingly, inhibition of targets in the joint may be a requirement for full efficacy. In this regard, questions concerning equilibration dynamics and amplitude for both anti-arthritic drugs and inflammatory cytokines, for example, are complex and dependent upon a PK/PD understanding.

It was notable that although monoclonal antibodies (mAbs) and other biologicals are widely used in, for example, Rheumatoid arthritis (RA) [60] relatively few examples of translational PK/PD biomarker study were identified. This may reflect the species specificity of mAbs. In the case where a mAb exhibits human specificity, a preclinical experiment would require a bespoke reagent, and this may be a barrier to preclinical PK/PD study.

The majority of the preclinical translational biomarker PK/PD in neurology has been carried out for AD, with, for example, extensive enquiries made into amyloid-beta (Aβ) species using direct binding or production inhibitors (e.g., beta-secretase [BACE]). Translational modeling has shown that BACE inhibitors (BACEi) can dramatically decrease Aβ in preclinical models and that this PK/PD can be quantitatively scaled to humans (e.g., see [72]). Unfortunately, the biomarker impact did not manifest as an improvement in disease, showing that sometimes caution is required when extrapolating from biomarker effects to disease effects. Similarly, the PK/PD for mAbs that bind Aβ has shown that Aβ can be decreased by a plasma drug. Consistent with the observation for

TABLE 2.2.3 Summary of references describing the use of biomarker PK/PD models in translational pharmacology for cardiovascular disease drug discovery

DISEASE AREA	*TARGET*	*BIOMARKER*	*DRUGS*	*COMMENT*	*REFERENCES*
Myocardial infarction	Mitochondrial permeability transition pore	Levels of free radicals, lipid peroxidation, ascorbate, LVEDP	TRO40303	The paper shows PK/PD modeling was carried out. TRO40303 can be safely administered by the intravenous route in humans at doses expected to be pharmacologically active. These results allowed evaluation of the expected active dose in human at 6 mg/kg, used in a p2 proof-of-concept study currently ongoing.	[43]
Top of form Arrhythmia Bottom of form	IKr	Q-T prolongation, ECG markers, activity levels of hERG, Ica,i, LNav1.5, hKv4.3/ hkCHIP2,2 qnd hKv1.5	Dofetilide	Retrospective PK/PD analysis was carried out to assess the drug's effects on heart rate in dogs, which was translated to humans and the drug's safety was assessed.	[44]
Hypercholesterolemia	apoB-100	Levels of apoB, LDL-C		The paper describes PK/PD modeling carried out for the drug in *in vivo* models to compare its metabolism between preclinical models and humans.	[45]
Atrial fibrillation	IKACh	Left atrial effective refractory period	AZD2927, A7071 separately	The paper describes PK/PD modeling was employed for dog data, and compared with humans. Difference in electrophysiology between dogs and humans is highlighted, which will impact on the translation of the drug's effects seen in dogs to humans.	[46]
Coagulation	Various	Markers for thrombin generation and coagulation	Various	The paper highlights the need to have antidotes to overcome warfarin toxicity. PK/PD profiles of such drugs in both preclinical and clinical trials are outlined.	[47]
Thrombosis	Factor Xa	Anti Factor Xa assay	Apixaban	The PK/PD analyses were performed using an inhibitory E max model for anti-fXa assay and a linear model for PT and HCT assays.	[48]
Various	Various	MAP, cardiac output, peripheral resistance	Various	Mechanism based model could predict mode of action and be scaled to human.	[49]
Hypertension	angiotensin II type 1 receptor	NA	Irbesartan	The paper outlines PK/PD modeling done in dog models to assess drug safety and efficacy.	[50]

BACEi, this decrease in Aβ had limited or no effect on AD cognition end points [79]. Nevertheless, the preclinical and clinical PK/PD work in this area showed that by using PK/PD and biomarkers, clinical trials could be efficiently performed. Clinical evaluations were carried out that quantitatively demonstrated that the expected and required pharmacology was expressed. In the absence of disease impact, the conclusion is arguably, therefore, clear that the mechanistic hypothesis was erroneous and new approaches are required. Although disappointing, the rich translational and clinical PK/PD and biomarker knowledge generated to date will be useful in the search for the next horizon for AD treatments. A focus on optimal start time point of intervention (e.g., to prevent the formation of plaques), the impact of soluble Aβ on plaques in the brain, and the extent of reversibility of any plaque brain damage, together with a wider investigation of the Aβ and tau hypotheses are all warranted. Importantly, due to initiatives at the U.S. Food and Drug Administration (FDA), the increased use of biomarkers is likely to be a feature of future AD drug discovery [80].

Pain has also received some attention (Table 2.2.5) despite the challenges of understanding and extrapolating the pain biology [81].

TABLE 2.2.4 Summary of references describing the use of biomarker PK/PD models in translational pharmacology for musculoskeletal disease drug discovery

DISEASE AREA	*TARGET*	*BIOMARKER*	*DRUGS*	*COMMENT*	*REFERENCES*
Osteoporosis	Dkk-1-binding at the LRP5/6	Dkk-1 levels in both blood and bone marrow plasma	PF-04840082	The paper highlights the need for using a TMDD mechanistic PK/PD model for accurate dose calculation of a mAb MABEL. It also shows that at low doses-high clearance rate and half-life too short to achieve efficacy (nonlinear pharmacokinetics), requiring a TMDD model that takes into account target and target–mAb kinetics. Importantly, levels of Dkk-1 in osteoporotic, osteopenic and postmenopausal patients must be looked at and models designed accordingly, to account for target expression levels and turnover rate.	[55]
Duchenne muscular dystrophy (DMD)	HDACs	Myosin heavy chain-positive myotubes, fatty infiltration, MPO activity	Givinostat	The paper outlines the results of PK/PD modeling in preclinical research, and translation strategies have been suggested. Parallel pharmacokinetic/ pharmacodynamic analysis confirmed the relationship between the effective doses of givinostat and the drug distribution in muscles and blood of treated mice. These findings provide the preclinical basis for an immediate translation of givinostat into clinical studies with DMD patients.	[56]
Osteoarthritis (OA)	ADAMTS-5	ARGS levels, macrophage infiltration	GSK2394002	The paper describes the PK/PD modeling using preclinical models, followed by translation into a FIH trial. ADAMTS-5 is the major aggrecanase involved in cartilage degradation and provides a link between a biological pathway and pharmacology that translates to human tissues, non-human primate models and points to a target OA patient population.	[57]
Rheumatoid arthritis	JAK-STAT signaling	Plasma levels of cytokines: IL-6, KC, MIG, IP-10 and MCP-5/CCL2	Tofacitinib	The paper outlines PK/PD modeling for the drug, to define clinical end points by comparing with human PK/PD profiles. The collective clinical and preclinical data indicated the importance of Cave as a driver of efficacy, rather than C_{max} or C_{min}, where $Cave_{50}$ values were within approximately twofold of each other.	[58]
Rheumatoid arthritis	Bruton's tyrosine kinase	Blood pBTK-Tyr223 levels, inflammatory markers present in the synovial fluid, markers for bone resorption and cartilage damage, ankle diameter	GDC-0834	The paper outlines PK/PD modeling to assess drug efficacy. Simultaneous fitting of data from vehicle- and GDC-0834-treated groups showed that overall 73% inhibition of pBTK was needed to decrease the rate constant describing the ankle swelling increase (kin) by half. These findings suggest a high degree of pBTK inhibition is required for maximal activity of the pathway on inflammatory arthritis in rats.	[59]

TABLE 2.2.5 Summary of references describing the use of biomarker PK/PD models in translational pharmacology for neurological disease drug discovery

DISEASE AREA	*TARGET*	*BIOMARKER*	*DRUGS*	*COMMENT*	*REFERENCES*
Depression	Various	Receptor occupancy	NA	This review outlines the use of receptor occupancy as a tool for accurate PK/PD modeling of preclinical models and how this can be translated to humans.	[61]
Depression	κ-Opioid receptor	Prolactin	PF-04455242	The results illustrate the utility of the proposed PK/PD model in supporting the quantitative translation of preclinical studies into an accurate clinical expectation. As such, the proposed PK/PD model is useful for supporting the design, selection, and early development of novel KOR antagonists.	[62]
Depression	SERT	SERT occupancy	Escitalopram, paroxetine, sertraline	PK/PD modeling of SERT occupancy to predict clinical dose. PK/PD modeling using SERT occupancy and 5-HTP-potentiated behavioral syndrome as response markers in mice may be a useful tool to predict clinically relevant plasma Css values.	[63]
Pain	Various	Target-site binding and downstream signaling	Various	The current gap between animal research and clinical development of analgesic drugs presents a challenge for the application of translational PK/PD modeling and simulation. First, animal pain models lack predictive and construct validity to accurately reflect human pain etiologies and, secondly, clinical pain is a multidimensional sensory experience that cannot always be captured by objective and robust measures. These challenges complicate the use of translational PK/PD modeling to project PK/PD data generated in preclinical species to a plausible range of clinical doses. To date, only a few drug targets identified in animal studies have been shown to be successful in the clinic. PK/PD modeling of biomarkers collected during the early phase of clinical development can bridge animal and clinical pain research.	[64]
Pain	Various	Physiological markers of pain	Various	Analysis of drug development failures indicates that they occur primarily in clinical phases and are mostly due to a lack of translation of efficacy in animal models to patients. Although a comprehensive analysis of problems in analgesic development is beyond the scope of this review, we hypothesize that the preclinical to clinical transition can be facilitated by a rational use, in early drug development stages, of biological markers of activity in human nociceptive pathways.	[65]
Pain	Various	Nociceptive axon-reflex flare, secondary hyperalgesia, Capsaicin-evoked flare.	Various	A seamless boundary between basic preclinical and clinical arms of the discovery process, embodying the concept of translational research is viewed by many as the way forward. The rational application of human experimental pain models in early clinical development is reviewed. Capsaicin, UV-irradiation and electrical stimulation methods have each been used to establish experimental hyperalgesia in p1 human volunteers and the application of these approaches is discussed in the context of several pharmacological examples.	[66]

(*Continued*)

TABLE 2.2.5 (*Continued*) Summary of references describing the use of biomarker PK/PD models in translational pharmacology for neurological disease drug discovery

DISEASE AREA	*TARGET*	*BIOMARKER*	*DRUGS*	*COMMENT*	*REFERENCES*
Pain	NGF	dppERKnuc	Tanezumab	A QSP model used for dose prediction for anti-NGF mAb.	[67]
Pain	Fatty acid amide hydrolase	Anandamide	PF-04457845	Use of PK/PD and SP to extrapolate biomarker effect and dose to clinic.	[68]
AD	BACE1	CSF Aβ levels		The paper describes PK/PD modeling and translation strategies suggested. i.c.v. administration of anti-BACE1 resulted in enhanced BACE1 target engagement and inhibition, with a corresponding dramatic reduction in CNS Aβ concentrations due to enhanced brain exposure to antibody.	[69]
AD	BACE1	Aβ40 and target-binding levels		A mathematical model incorporating PK/PD and safety profiles is developed for bispecific TfR/BACE1 antibodies with a range of affinities to TfR in order to guide candidate selection.	[70]
AD	BACE1	Plasma levels of Aβ1-40, Aβ1-42, Aβx-40 and C99	LY2886721	The paper reports the nonclinical and early clinical development of LY2886721, a BACE1 active site inhibitor that reached P2 clinical trials in AD. LY2886721 has high selectivity against key off-target proteases, which efficiently translates *in vitro* activity into robust *in vivo* amyloid β lowering in nonclinical animal models. Similar potent and persistent amyloid β lowering was observed in plasma and lumbar CSF when single and multiple doses of LY2886721 were administered to healthy human subjects. Collectively, these data add support for BACE1 inhibition as an effective means of amyloid lowering and as an attractive target for potential disease modification therapy in AD.	[71]
AD	BACE1	Aβ in the plasma and CSF, and sAPPβ and sAPPα in the CSF	GNE-629, GNE-892	A PK/PD model was developed to mechanistically describe the effects of BACE1 inhibition on Aβ, sAPPβ, and sAPPα in the CSF, and Aβ in the plasma. This model can be used to prospectively predict *in vivo* effects of new BACE1 inhibitors using their *in vitro* activity and PK data.	[72]
AD	Various	CSF Ab40 and Ab42 levels	BACEi, GSI or GSM	The paper describes a semi-mechanistic PK/PD model that can describe the PK/Aβ data by accounting for Aβ generation and clearance. The modeling characterizes the *in vivo* PD (i.e., Aβ lowering) properties of compounds and generates insights about the salient biological systems. The learning from the modeling enables us to establish a framework for predicting *in vivo* Aβ lowering from *in vitro* parameters.	[73]
AD	NA	CSF Aβ40 levels	NA	The turnover of CSF Aβ40 was systematically examined, for the first time, in multiple species through quantitative modeling of multiple data sets. Our result suggests that the clearance mechanisms for CSF Aβ in rodents may be different from those in the higher species. The understanding of Aβ turnover has considerable implications for the discovery and development of Aβ-lowering therapeutics, as illustrated from the perspectives of preclinical PK/PD characterization and preclinical-to-clinical translation.	[74]

(*Continued*)

TABLE 2.2.5 (*Continued*) Summary of references describing the use of biomarker PK/PD models in translational pharmacology for neurological disease drug discovery.

DISEASE AREA	*TARGET*	*BIOMARKER*	*DRUGS*	*COMMENT*	*REFERENCES*
AD	Various	BOLD-fMRI, pharmaco-EEG, markers of neuro-degeneration and taupathy	NA	The paper outlines the importance of PBPK modeling and the use of QSP as a more accurate translation tool than PK/PD modeling. Proposes QSP is a new "humanized" tool for supporting drug discovery and development in general and CNS disorders in particular.	[75]
Brain metastases	EGFRm	NA	Osimertinib	Osimertinib demonstrated greater penetration of the mouse blood–brain barrier than gefitinib, rociletinib (CO-1686), or afatinib, and at clinically relevant doses induced sustained tumor regression in an EGFRm PC9 mouse brain metastases model; rociletinib did not achieve tumor regression. Under positron emission tomography micro-dosing conditions, [11C] osimertinib showed markedly greater exposure in the cynomolgus monkey brain than [11C] rociletinib and [11C] gefitinib. Early clinical evidence of osimertinib activity in previously treated patients with EGFRm-advanced NSCLC and brain metastases is also reported.	[76]
Neuroblastoma	Topoisomerase	NA	Irinotecan	The paper describes PK/PD modeling using murine models to translate to a P1 trial in children. The protracted schedule is well tolerated in children. The absence of significant myelosuppression and encouraging clinical responses suggest compellingly that irinotecan be further evaluated in children using the (q.d. × 5) × 2 schedule, beginning at a dose of 20 mg/m^2. These results imply that data obtained from xenograft models can be effectively integrated into the design of clinical trials.	[77]
DIPG	Histone deacetylase	H3 acetylation	Panobinostat	The paper outlines PK/PD analysis for the drug to evaluate drug safety and efficacy.	[78]

One view is that preclinical pain models lack sufficient predictive and construct validity to accurately reflect human pain etiologies and that clinical pain is a multidimensional sensory experience that is not easily captured by objective and robust measures. In this context, candidate drugs that are positive preclinically, but fail clinically, for example, [82] often leave open the question of whether the apparent lack of efficacy was due to the target not being sufficiently modulated or whether the mechanistic hypothesis was not correct. Biomarker and PK/PD strategies may be one way forward in this case; by enabling demonstration of significant modulation of pharmacology and biomarkers, an argument can be made that we can quantify the extent of impact of a candidate drug on a given pharmacology. Preclinical technical success in this regard will increase the confidence of achieving a positive clinically. In this way, pharmacology can be shown clinically and, in the event of a negative result, it can be more confidently concluded that the hypothesis was flawed.

Given the extent of the unmet medical need for pain and dementias such as AD, it seems likely that translational biomarker PK/PD will increasingly become of interest. To achieve success, it will be important to focus on the understanding of the validity of translating complex biology from animal models to human disease.

Translating preclinical biomarker PK/PD for anti-infectives to humans has been carried out with arguably some success, at least in specific cases. The biomarker of choice is often the pathogen in blood and, therefore, relatively easy to measure. In addition, although the pathogen can be regarded as a biomarker, the link to disease is often clear and the confidence in biomarker-to-disease effect is generally good.

Translational PK/PD of bacterial infections has been often carried out using mouse models. Traditionally, antimicrobial drug discovery has focused on the use of the minimum inhibitory concentration (MIC), or some derived parameter (maximum concentration of drug in plasma (C_{max})/MIC, area under the curve (AUC/MIC), or time over MIC [95]. In the future, a more mechanistic approach, for example, incorporating microbial behavior parameters, resistance data and concentrations in target tissue may refine clinical dose projections.

Antiviral translational PK/PD can be more difficult to execute due to the species specificity of many viruses. Nevertheless, there are often useful PK/PD insights that can be derived from preclinical species, for example, TLR7 where the interferon alpha (IFNα) immune response was quantified using PK/PD methods in mice and the model used to extrapolate TLR7 agonist effects to Hepatitis C Virus (HCV) patients (Table 2.2.6, BHMA). Preclinical data has also been used in the HIV arena. In this case *in vitro Kd* was used to identify efficacious doses for a CCR5 antagonist. Receptor occupancy was measured as a biomarker as well as viral load, and it was concluded that a maintained, very high receptor occupancy (>99%) was a prerequisite for success (Table 2.2.6, maraviroc). This high level was required due to the committed, irreversible nature of the HIV cell infection process.

One paper (Table 2.2.7) described the application of PK/PD biomarker methods to the development of ribonucleic acid interference (RNAi). A key finding was the ability to measure vbiomarker effects, and drug PK was comparable in many ways to more conventional drug discovery platforms (e.g., it has been shown that measuring RNAi drug PK and inflammatory biomarkers in plasma is possible in principle). In the future, it is reasonable to expect increasing utilization of PK/PD data for dose and regimen prediction from the preclinical data for Ribonucleic acid interference RNAi and comparable exciting new technologies such as gene therapy [103] and clustered regularly interspaced short palindromic repeats clustered regularly interspaced short palindromic repeats (CRISPR) where PK/PD understanding might help optimize dose and regimen selection and the management of immunogenic responses [104].

Arguably, the areas of digestive, skin and respiratory could be regarded in some way as "topical" applications. This additional potential complexity may explain why there are relatively few published examples of biomarker PK/PD analysis (Table 2.2.7). For example, biomarker measures in the location of the disease

TABLE 2.2.6 Summary of references describing the use of biomarker PK/PD models in translational pharmacology for anti-infectives drug discovery

DISEASE AREA	*TARGET*	*BIOMARKER*	*DRUGS*	*COMMENT*	*REFERENCES*
Malaria	*Plasmodium falciparum*	*Plasmodium falciparum*/ *P. berghei* in plasma	Artemisinin	The paper evaluates current PK/PD preclinical models and suggests translation strategies for the drug.	[83]
HIV	CCR5	Target-binding levels, protein concentrations and HIV in plasma	Maraviroc	The paper outlines a PK/PD model to predict drug efficacy and safety. The model parameters were derived from the literature, as well as from a model-based analysis of available p2a clinical data from another investigational antiretroviral drug. The PD component that links the plasma concentrations of maraviroc to the inhibition of virus replication was based on *in vitro* measurements of drug potency and took into account the difference in the *in vitro* and *in vivo* protein binding and the uncertainties regarding the interpretation of the *in vitro* to *in vivo* extrapolation of the 50% inhibitory concentration.	[84]
HIV		NA		The paper outlines translation using a PK/PD model developed by the same team in this paper: J Pharmacokinet Pharmacodyn (2012) 39:357–368 (doi:10.1007/s10928-012-9255-3), after testing in preclinical models.	[85]
HCV	TLR7	IFNα	BHMA	PK/PD model for mouse IFNα data. Discussion on use for extrapolation to man.	[86]
H. influenza antibiotic resistance	Cell wall synthesis	β-Lactamase-positive and -negative isolates	Cefixime	The paper describes the effects of the drug in preclinical models using PK/PD modeling to translate into clinical trials.	[87]
Antifungals	Various	NA	Various	The paper outlines the PK/PD analysis of a range of drug classes and the preclinical models supporting them. The need for improving these models to model newly emerging fungal pathogens causing infection is highlighted, and how PK/PD analysis can be improved accordingly.	[88]
Antibiotics	Bacterial DNA gyrase and topoiso-merase IV	Bacterial count	Pradofloxacin	The paper describes the use of an *in vitro* PK/PD model to determine dosing of the antibiotic required to prevent the emergence of mutant strains in response to drug use.	[89]

(Continued)

TABLE 2.2.6 (*Continued*) Summary of references describing the use of biomarker PK/PD models in translational pharmacology for anti-infectives drug discovery

DISEASE AREA	*TARGET*	*BIOMARKER*	*DRUGS*	*COMMENT*	*REFERENCES*
Antibiotics	The large ribosomal subunit	Bacterial count	Evernimicin	PK/PD was used to integrate preclinical and clinical data for P2/3 doses.	[90]
Antibiotics	Various	Bacterial count	Various	The paper outlines current PK/PD modeling and analysis strategies to allow more successful translation to humans.	[91]
Antibiotics	Cell wall biosynthesis	Bacterial count	Doripenem, meropenem, cilastin, imipenem and cilastatin	PK/PD model to simulate optimal dosing regimens.	[92]
Multidrug-resistant *P. aeruginosa*	Cell wall biosynthesis	CFU	Imipenem	The paper outlines PK/PD modeling for a special murine model and the results used to guide translation to humans.	[93]
Antibiotics	30S ribosomal subunit	Bacterial count	Tigecycline	A preliminary PK/PD analysis in experimental animal models of infection indicate that the efficacy of tigecycline is probably best predicted by the ratio of the area under the concentration–time curve to the minimum inhibitory concentration.	[94]

(skin, lung, gut) may be less technically feasible than plasma measures and tissue-specific PK/PD may not be as well understood. It will be interesting to see whether the database of examples of application to topical questions expands.

NEUROPSYCHIATRIC AND SENSORY ORGANS

No reports of translational biomarker PK/PD were observed for these disease areas. This may reflect the limited confidence in translating neuropsychiatric observations in preclinical models to human patients. It may also be that neuropsychiatric and sensory organ disease areas are, currently at least, relatively limited funding therapeutic areas for drug discovery as a whole and the limited PK/PD reports reflect the broader research investment in these areas.

CONCLUSIONS

The examples discussed of the application of biomarker PK/PD methodology to pain and AD research highlight that, although preclinical disease models can appear to have predictive validity, they are sometimes not an accurate representation of the human disease. Biomarkers and PK/PD can help enable a rational investigation into the source of any discrepancy, for example, by facilitating a better understanding of whether a negative is due to an inadequate pharmacological effect or is due to inaccurate extrapolation of a disease model from animal models.

A potential additional value of translating biomarker PK/PD is that it allows evaluation of the impact of including variability parameters into the extrapolation exercise. Techniques such as nonlinear mixed effects modeling can be used to estimate variability parameters and these, together with any insight into clinical variability, can be used to simulate scenarios and facilitate understanding of the implications of these for clinical development rather than basing decisions only on average representations see, for example, [105].

The demonstration of a better understanding of pharmacology and PK/PD is likely to be of increasing importance for drug discoverers going forward following recent events in first-in-humans (FIH) clinical trials where frank toxicity and a death was observed [106]. As a direct result of this, European Medicines Evaluation Agency (EMEA) updated the guidance for FIH trials [107], recommending, among other things, that a PK/PD modeling approach is useful to inform clinical dose levels and schedules, taking into consideration repeated-dose applications as anticipated in the clinical setting.

Although the study of scaling of PK from preclinical to humans is well established [108], the scaling of biomarker PD is still in its infancy. The examples discussed in this chapter show that PD biomarker end points can, in some cases, be scaled effectively to the clinical response. Although the sample size per biomarker is small, as the database grows it will be useful to survey the examples with the aim of identifying useful general PD scaling principles.

A clear trend in the data was toward more mechanistic models and, in particular, systems pharmacology models [109,110].

TABLE 2.2.7 Summary of references describing the use of biomarker PK/PD models in translational pharmacology for other disease area drug discovery

PATHO-PHYSIOLOGICAL AREA	*DISEASE AREA*	*TARGET*	*BIOMARKER*	*DRUGS*	*COMMENT*	*REFERENCES*	General
RNAi therapeutics	Various	Inflammatory markers (TNFα, IFNα)	Various	The paper outlines PK/PD analysis of preclinical models and how this has been used to translate into ongoing clinical trials.	[96]	Congenital	Gaucher's disease
NA	Blood counts, plasma levels of chitotriosidase, GC	Velaglucerase-α and imiglucerase	Similar *in vitro* enzymatic properties, and *in vivo* PK/PD and therapeutic efficacy of GCase were found with two human GCases, recombinant GCase (CHO cell, imiglucerase, Imig) and gene-activated GCase (human fibrosarcoma cells, velaglucerase alfa, Vela), in a Gaucher mouse, D409V/null	[97]	Respiratory	Chronic respiratory diseases	CXCR2
Sputum levels of inflammatory markers (e.g., neutrophils) and myeloperoxidase	SB-656933	The paper describes the use of PK/PD modeling in preclinical models to translate to a clinical trial, and reports the result of the P1 trial.	[98]	Digestive	Crohn's disease	IL-6	Gut and serum levels of pSTAT3 and IL-6; serum levels of CRP, sIL-6R and sgp130

(*Continued*)

TABLE 2.2.7 (*Continued*) Summary of references describing the use of biomarker PK/PD models in translational pharmacology for other disease area drug discovery

PATHO-PHYSIOLOGICAL AREA	*DISEASE AREA*	*TARGET*	*BIOMARKER*	*DRUGS*	*COMMENT*	*REFERENCES*	General
Monoclonal antibodies	The paper describes a QSP multiscale systems model to predict the outcome of this immunotherapy and its implications for translation to clinical trials.	[99]	Colorectal cancer		Angiogenesis inhibition	DCE-MRI to measure vascular perfusion and permeability, MRI-IAUC, plasma VEGF, bFGF, sTIE-2, sE-SEL	PTK/ZK
PK/PD analyses were carried out with mouse models and human trials, and a comparison made between them.	[100]	Gram-positive bacterial infections	NA		CFU	MCB3581	The paper describes PK/PD analysis to test drug efficacy.
[101]	Skin	Atopic dermatitis	Chymase inhibition	Secretory leucocyte protease inhibitor (cSLPI) levels in the serum	SUN13834	A chymase inhibitor SUN13834 has been shown to improve skin condition in animal models for atopic dermatitis. In the present study, effective dosages of SUN13834 for atopic dermatitis patients were predicted by PK/PD analyses of SUN13834 in NC/Nga mice, which spontaneously develop atopic dermatitis-like skin lesions. The clinical effective dosage predicted in this paper is also discussed in relation to a recently conducted P2a study.	[102]

LIST OF ABBREVIATIONS USED IN TABLES 2.2.1–2.2.7; NA, not applicable; CML, Chronic myeloid leukaemia; p-ERK, phosphorylated extracellular regulated kinase; TGI, tumour growth inhibition; ALK, anaplastic lymphoma kinase; MET, tyrosine-protein kinase Met; VEGF, vascular endothelial growth factor; PD-1, programmed cell death protein 1; QSP; quantitative systems pharmacology; mTOR, mammalian target of rapamycin; PI3K, phosphatidylinositol-4,5-bisphosphate 3-kinase; P1, phase 1 clinical trial; IGF, insulin-like growth factor; IGF-1R, insulin-like growth factor receptor 1; EGFR, epidermal growth factor receptor; TMDD, target-mediated drug disposition; TrkB, Tropomyosin receptor kinase B; PT, prothrombin time; HCT, high haematocrit; LDL, low-density lipoprotein.

As the technology to accurately sample and measure biomarkers improves, together with increasingly better tools for the development of complex models, it seems likely that such models will play an increasingly important part in drug discovery.

It was notable that the application of biomarker and PK/PD techniques to novel technology such as RNAi was recently reported. The lessons learned from small and large molecule drug discovery can, in part, be deployed to support the clinical development of the new wave of medicines such as gene therapy, CRISPR and others. The questions of optimal dose route, dose, dose regimen, dose frequency and immunogenicity are perhaps as important for these as they are for conventional small molecule and biological drugs and biomarkers, and PK/PD will have a role to play in optimizing these.

Overall, it was clear from this enquiry that the publication of examples of utilization of translational biomarker PK/PD is an increasing trend over the last decade or so. By inference, it can be concluded that this is probably more generally true in drug discovery. Some evidence for this and, indeed, of a positive impact has been published recently [111]. However, it is necessary to be conscious that a positive bias may exist in publication, where, for example, examples of good translation between preclinical and clinical biomarker PK/PD are more likely to be reported than those where this was not observed. It will be interesting to see how these methods impact drug discovery and, in particular, attrition in the future.

ACKNOWLEDGMENTS

Acknowledgment is given to Divyanshi Karmani for assistance with the literature search and summary.

REFERENCES

1. Dayneka NL, Garg V, Jusko WJ. Comparison of four basic models of indirect pharmacodynamic responses. *J Pharmacokinet Biopharm*. 1993;21(4):457–478.
2. Morgan P, Van Der Graaf PH, Arrowsmith J, Feltner DE, Drummond KS, Wegner CD et al. Can the flow of medicines be improved? Fundamental pharmacokinetic and pharmacological principles toward improving phase II survival. *Drug Discov Today*. 2012;17(9–10):419–424.
3. Workman P. How much gets there and what does it do?: The need for better pharmacokinetic and pharmacodynamic endpoints in contemporary drug discovery and development. *Curr Pharm Des*. 2003;9(11):891–902.
4. Luo FR, Yang Z, Camuso A, Smykla R, McGlinchey K, Fager K et al. Dasatinib (BMS-354825) pharmacokinetics and pharmacodynamic biomarkers in animal models predict optimal clinical exposure. *Clin Cancer Res*. 2006;12(23):7180–7186.
5. Choo EF, Belvin M, Chan J, Hoeflich K, Orr C, Robarge K et al. Preclinical disposition and pharmacokinetics-pharmacodynamic modeling of biomarker response and tumor growth inhibition in xenograft mouse models of G-573, a MEK inhibitor. *Xenobiotica*. 2010;40(11):751–762.
6. Keizer RJ, Funahashi Y, Semba T, Wanders J, Beijnen JH, Schellens JH et al. Evaluation of alpha2-integrin expression as a biomarker for tumor growth inhibition for the investigational integrin inhibitor E7820 in preclinical and clinical studies. *AAPS J*. 2011;13(2):230–239.
7. Yamazaki S. Translational pharmacokinetic-pharmacodynamic modeling from nonclinical to clinical development: A case study of anticancer drug, crizotinib. *AAPS J*. 2013;15(2):354–366.
8. Sharan S, Woo S. Quantitative insight in utilizing circulating angiogenic factors as biomarkers for antiangiogenic therapy: Systems pharmacology approach. *CPT Pharmacometrics Syst Pharmacol*. 2014;3:e139.
9. Checkley S, MacCallum L, Yates J, Jasper P, Luo H, Tolsma J et al. Bridging the gap between in vitro and in vivo: Dose and schedule predictions for the ATR inhibitor AZD6738. *Sci Rep*. 2015;5:13545.
10. Tanaka C, O'Reilly T, Kovarik JM, Shand N, Hazell K, Judson I et al. Identifying optimal biologic doses of everolimus (RAD001) in patients with cancer based on the modeling of preclinical and clinical pharmacokinetic and pharmacodynamic data. *J Clin Oncol*. 2008;26(10):1596–1602.
11. Lindauer A, Valiathan CR, Mehta K, Sriram V, de Greef R, Elassaiss-Schaap J et al. Translational pharmacokinetic/pharmacodynamic modeling of tumor growth inhibition supports dose-range selection of the anti-PD-1 antibody pembrolizumab. *CPT Pharmacometrics Syst Pharmacol*. 2017;6(1):11–20.
12. DePinto W, Chu XJ, Yin X, Smith M, Packman K, Goelzer P et al. In vitro and in vivo activity of R547: A potent and selective cyclin-dependent kinase inhibitor currently in phase I clinical trials. *Mol Cancer Ther*. 2006;5(11):2644–2658.
13. Lee AL F, Wen M, Ryseck R, Fargnoli J, Poruchynsky M, Fojo T et al., Antiangiogenic (AG) synergy with Ixabepilone (IXA): Translation of preclinical studies to the clinical setting. *Cancer Res*. 2009:206.
14. Thavasu P, Te Fong ACLW, Rodriguez BJ, Basu B, Turner A, Hall E et al., Abstract CT138: Translating preclinical observations to the clinic: Combination of the dual m-TORC1/2 inhibitor AZD2014 and paclitaxel in ovarian and lung cancer. *Cancer Res*. 2015:CT138-CT138.
15. Munster P, Marchion D, Bicaku E, Schmitt M, Lee JH, DeConti R et al. Phase I trial of histone deacetylase inhibition by valproic acid followed by the topoisomerase II inhibitor epirubicin in advanced solid tumors: A clinical and translational study. *J Clin Oncol*. 2007;25(15):1979–1985.
16. Haddish-Berhane N, Shah DK, Ma D, Leal M, Gerber HP, Sapra P et al. On translation of antibody drug conjugates efficacy from mouse experimental tumors to the clinic: A PK/PD approach. *J Pharmacokinet Pharmacodyn*. 2013;40(5):557–571.
17. Ballard P, Ashton S, Cross D, Dimelow R and Yates Y. Abstract B212: Integrating the pre-clinical pharmacokinetic, pharmacodynamics, and efficacy data for AZD9291, an oral, irreversible inhibitor of EGFR activating (EGFRm+) and resistant (EGFRm+/T790M) mutations and an active metabolite to predict the human pharmacokinetics and potential efficacious dose in patients. *Mol Cancer Ther*. 2013:B212–B212.
18. Yu PY, Gardner HL, Roberts R, Cam H, Hariharan S, Ren L et al. Target specificity, in vivo pharmacokinetics, and efficacy of the putative STAT3 inhibitor LY5 in osteosarcoma, Ewing's sarcoma, and rhabdomyosarcoma. *PLoS One*. 2017;12(7):e0181885.
19. Wong H, Vernillet L, Peterson A, Ware JA, Lee L, Martini JF et al. Bridging the gap between preclinical and clinical studies using pharmacokinetic-pharmacodynamic modeling: An analysis of GDC-0973, a MEK inhibitor. *Clin Cancer Res*. 2012;18(11):3090–3099.

20. Yamazaki S, Skaptason J, Romero D, Lee JH, Zou HY, Christensen JG et al. Pharmacokinetic-pharmacodynamic modeling of biomarker response and tumor growth inhibition to an orally available cMet kinase inhibitor in human tumor xenograft mouse models. *Drug Metab Dispos.* 2008;36(7):1267–1274.
21. Dietsch GN, Randall TD, Gottardo R, Northfelt DW, Ramanathan RK, Cohen PA et al. Late-stage cancer patients remain highly responsive to immune activation by the selective TLR8 agonist motolimod (VTX-2337). *Clin Cancer Res.* 2015;21(24):5445–5452.
22. Munster P, Aggarwal R, Hong D, Schellens JH, van der Noll R, Specht J et al. First-in-human phase I study of GSK2126458, an oral pan-class I phosphatidylinositol-3-kinase inhibitor, in patients with advanced solid tumor malignancies. *Clin Cancer Res.* 2016;22(8):1932–1939.
23. Grimwood S, Hartig PR. Target site occupancy: Emerging generalizations from clinical and preclinical studies. Pharmacol Ther. 2009;122(3):281–301.
24. Kitano H. Cancer as a robust system: Implications for anticancer therapy. *Nat Rev Cancer.* 2004;4(3):227–235.
25. Benson N, van der Graaf PH. The rise of systems pharmacology in drug discovery and development. *Future Med Chem.* 2014;6(16):1731–1734.
26. Lacey SF, Kalos M. Biomarkers in T-cell therapy clinical trials. *Cytotherapy.* 2013;15(6):632–640.
27. Pavlou MP, Diamandis EP, Blasutig IM. The long journey of cancer biomarkers from the bench to the clinic. *Clin Chem.* 2013;59(1):147–157.
28. Coventry BJ, Ashdown ML, Quinn MA, Markovic SN, Yatomi-Clarke SL, Robinson AP. CRP identifies homeostatic immune oscillations in cancer patients: A potential treatment targeting tool? *J Transl Med.* 2009;7:102.
29. Yoshiura S, Ohtsuka T, Takenaka Y, Nagahara H, Yoshikawa K, Kageyama R. Ultradian oscillations of Stat, Smad, and Hesl expression in response to serum. *Proc Natl Acad Sci.* 2007;104(27):11292–11297.
30. Moreno L, Pearson AD. How can attrition rates be reduced in cancer drug discovery? *Expert Opin Drug Discov.* 2013;8(4):363–368.
31. Ramkissoon CM, Aufderheide B, Bequette BW, Palerm CC. A model of glucose-insulin-pramlintide pharmacokinetics and pharmacodynamics in type I diabetes. *J Diabetes Sci Technol.* 2014;8(3):529–542.
32. Bergman RN, Ider YZ, Bowden CR, Cobelli C. Quantitative estimation of insulin sensitivity. *Am J Physiol.* 1979;236(6):E667–E677.
33. Nyman E, Brannmark C, Palmer R, Brugard J, Nystrom FH, Stralfors P et al. A hierarchical whole-body modeling approach elucidates the link between in vitro insulin signaling and in vivo glucose homeostasis. *J Biol Chem.* 2011;286(29):26028–26041.
34. Chien SLaYW. Pharmacokinetic-pharmacodynamic modelling of insulin: Comparison of indirect pharmacodynamic response with effect-compartment link models. *J Pharm Pharmacol.* 2010;54(6):791–800.
35. Neerati P. Influence of curcumin on pioglitazone metabolism and Pk/Pd: Diabetes mellitus . *J Diabetes Metab.* 2012;S:6.
36. Zager MG, Kozminski K, Pascual B, Ogilvie KM, Sun S. Preclinical PK/PD modeling and human efficacious dose projection for a glucokinase activator in the treatment of diabetes. *J Pharmacokinet Pharmacodyn.* 2014;41(2):127–139.
37. Maurer TS, Ghosh A, Haddish-Berhane N, Sawant-Basak A, Boustany-Kari CM, She L et al. Pharmacodynamic model of sodium-glucose transporter 2 (SGLT2) inhibition: Implications for quantitative translational pharmacology. *AAPS J.* 2011;13(4):576–584.
38. Carboni JM, Wittman M, Yang Z, Lee F, Greer A, Hurlburt W et al. BMS-754807, A small molecule inhibitor of insulin-like growth factor-1R/IR. *Mol Cancer Ther.* 2009;8(12):3341–3349.
39. Menard J, Rigel DF, Watson C, Jeng AY, Fu F, Beil M et al. Aldosterone synthase inhibition: Cardiorenal protection in animal disease models and translation of hormonal effects to human subjects. *J Transl Med.* 2014;12:340.
40. Vugmeyster Y, Rohde C, Perreault M, Gimeno RE, Singh P. Agonistic TAM-163 antibody targeting tyrosine kinase receptor-B: Applying mechanistic modeling to enable preclinical to clinical translation and guide clinical trial design. *MABS.* 2013;5(3):373–383.
41. Maggi M, Crescioli C, Morelli A, Colli E, Adorini L. Pre-clinical evidence and clinical translation of benign prostatic hyperplasia treatment by the vitamin D receptor agonist BXL-628 (Elocalcitol). *J Endocrinol Invest.* 2006;29(7):665–674.
42. Ajmera I, Swat M, Laibe C, Le Novere N, Chelliah V. The impact of mathematical modeling on the understanding of diabetes and related complications. *CPT Pharmacometrics Syst Pharmacol.* 2013;2:e54.
43. Le Lamer S, Paradis S, Rahmouni H, Chaimbault C, Michaud M, Culcasi M et al. Translation of TRO40303 from myocardial infarction models to demonstration of safety and tolerance in a randomized phase I trial. *J Transl Med.* 2014;12:38.
44. Parkinson J, Visser SA, Jarvis P, Pollard C, Valentin JP, Yates JW et al. Translational pharmacokinetic-pharmacodynamic modeling of QTc effects in dog and human. *J Pharmacol Toxicol Methods.* 2013;68(3):357–366.
45. Yu RZ, Lemonidis KM, Graham MJ, Matson JE, Crooke RM, Tribble DL et al. Cross-species comparison of in vivo PK/PD relationships for second-generation antisense oligonucleotides targeting apolipoprotein B-100. *Biochem Pharmacol.* 2009;77(5):910–919.
46. Walfridsson H, Anfinsen OG, Berggren A, Frison L, Jensen S, Linhardt G et al. Is the acetylcholine-regulated inwardly rectifying potassium current a viable antiarrhythmic target? Translational discrepancies of AZD2927 and A7071 in dogs and humans. *Europace.* 2015;17(3):473–482.
47. Mo Y, Yam FK. Recent advances in the development of specific antidotes for target-specific oral anticoagulants. *Pharmacotherapy.* 2015;35(2):198–207.
48. He K, Luettgen JM, Zhang D, He B, Grace JE, Jr., Xin B et al. Preclinical pharmacokinetics and pharmacodynamics of apixaban, a potent and selective factor Xa inhibitor. *Eur J Drug Metab Pharmacokinet.* 2011;36(3):129–139.
49. Snelder N, Ploeger BA, Luttringer O, Rigel DF, Webb RL, Feldman D et al. PKPD modelling of the interrelationship between mean arterial BP, cardiac output and total peripheral resistance in conscious rats. *Br J Pharmacol.* 2013;169(7):1510–1524.
50. Carlucci L, Song KH, Yun HI, Park HJ, Seo KW, Giorgi M. Pharmacokinetics and pharmacodynamics (PK/PD) of irbesartan in beagle dogs after oral administration at two dose rates. *Pol J Vet Sci.* 2013;16(3):555–561.
51. Lindstedt SL, Schaeffer PJ. Use of allometry in predicting anatomical and physiological parameters of mammals. *Lab Anim.* 2002;36(1):1–19.
52. Dubois VF, de Witte WE, Visser SA, Danhof M, Della Pasqua O, Cardiovascular Safety Project T et al. Assessment of interspecies differences in drug-induced QTc interval prolongation in cynomolgus monkeys, dogs and humans. *Pharm Res.* 2016;33(1):40–51.
53. Jonker DM, Kenna LA, Leishman D, Wallis R, Milligan PA, Jonsson EN. A pharmacokinetic-pharmacodynamic model for the quantitative prediction of dofetilide clinical QT prolongation from human ether-a-go-go-related gene current inhibition data. *Clin Pharmacol Ther.* 2005;77(6):572–582.

54. Gotta V, Cools F, van Ammel K, Gallacher DJ, Visser SA, Sannajust F et al. Inter-study variability of preclinical in vivo safety studies and translational exposure-QTc relationships—A PKPD meta-analysis. *Br J Pharmacol.* 2015;172(17):4364–4379.
55. Betts AM, Clark TH, Yang J, Treadway JL, Li M, Giovanelli MA et al. The application of target information and preclinical pharmacokinetic/pharmacodynamic modeling in predicting clinical doses of a Dickkopf-1 antibody for osteoporosis. *J Pharmacol Exp Ther.* 2010;333(1):2–13.
56. Consalvi S, Mozzetta C, Bettica P, Germani M, Fiorentini F, Del Bene F et al. Preclinical studies in the mdx mouse model of Duchenne muscular dystrophy with the histone deacetylase inhibitor givinostat. *Mol Med.* 2013;19:79–87.
57. Larkin J, Lohr TA, Elefante L, Shearin J, Matico R, Su JL et al. Translational development of an ADAMTS-5 antibody for osteoarthritis disease modification. *Osteoarthr Cartil.* 2015;23(8):1254–1266.
58. Dowty ME, Jesson MI, Ghosh S, Lee J, Meyer DM, Krishnaswami S et al. Preclinical to clinical translation of tofacitinib, a Janus kinase inhibitor, in rheumatoid arthritis. *J Pharmacol Exp Ther.* 2014;348(1):165–173.
59. Liu L, Di Paolo J, Barbosa J, Rong H, Reif K, Wong H. Antiarthritis effect of a novel Bruton's tyrosine kinase (BTK) inhibitor in rat collagen-induced arthritis and mechanism-based pharmacokinetic/pharmacodynamic modeling: Relationships between inhibition of BTK phosphorylation and efficacy. *J Pharmacol Exp Ther.* 2011;338(1):154–163.
60. Campbell J, Lowe D, Sleeman MA. Developing the next generation of monoclonal antibodies for the treatment of rheumatoid arthritis. *Br J Pharmacol.* 2011;162(7):1470–1484.
61. Melhem M. Translation of central nervous system occupancy from animal models: Application of pharmacokinetic/pharmacodynamic modeling. *J Pharmacol Exp Ther.* 2013;347(1):2–6.
62. Chang C, Byon W, Lu Y, Jacobsen LK, Badura LL, Sawant-Basak A et al. Quantitative PK-PD model-based translational pharmacology of a novel kappa opioid receptor antagonist between rats and humans. *AAPS J.* 2011;13(4):565–575.
63. Kreilgaard M, Smith DG, Brennum LT, Sanchez C. Prediction of clinical response based on pharmacokinetic/pharmacodynamic models of 5-hydroxytryptamine reuptake inhibitors in mice. *Br J Pharmacol.* 2008;155(2):276–284.
64. Yassen A, Passier P, Furuichi Y, Dahan A. Translational PK-PD modeling in pain. *J Pharmacokinet Phar.* 2013;40(3):401–418.
65. Chizh BA, Greenspan JD, Casey KL, Nemenov MI, Treede RD. Identifying biological markers of activity in human nociceptive pathways to facilitate analgesic drug development. *Pain.* 2008;140(2):249–253.
66. Chizh BA, Priestley T, Rowbotham M, Schaffler K. Predicting therapeutic efficacy—Experimental pain in human subjects. *Brain Res Rev.* 2009;60(1):243–254.
67. Benson N, Matsuura T, Smirnov S, Demin O, Jones HM, Dua P et al. Systems pharmacology of the nerve growth factor pathway: Use of a systems biology model for the identification of key drug targets using sensitivity analysis and the integration of physiology and pharmacology. *Interface Focus.* 2013;3(2):20120071.
68. Benson N, Metelkin E, Demin O, Li GL, Nichols D, van der Graaf PH. A systems pharmacology perspective on the clinical development of fatty acid amide hydrolase inhibitors for pain. *CPT Pharmacometrics Syst Pharmacol.* 2014;3:e91.
69. Yadav DB, Maloney JA, Wildsmith KR, Fuji RN, Meilandt WJ, Solanoy H et al. Widespread brain distribution and activity following i.c.v. infusion of anti-beta-secretase (BACE1) in nonhuman primates. *Br J Pharmacol.* 2017;174(22):4173–4185.
70. Gadkar K, Yadav DB, Zuchero JY, Couch JA, Kanodia J, Kenrick MK et al. Mathematical PKPD and safety model of bispecific TfR/BACE1 antibodies for the optimization of antibody uptake in brain. *Eur J Pharm Biopharm.* 2016;101:53–61.
71. May PC, Willis BA, Lowe SL, Dean RA, Monk SA, Cocke PJ et al. The potent BACE1 inhibitor LY2886721 elicits robust central Abeta pharmacodynamic responses in mice, dogs, and humans. *J Neurosci.* 2015;35(3):1199–1210.
72. Liu X, Wong H, Scearce-Levie K, Watts RJ, Coraggio M, Shin YG et al. Mechanistic pharmacokinetic-pharmacodynamic modeling of BACE1 inhibition in monkeys: Development of a predictive model for amyloid precursor protein processing. *Drug Metab Dispos.* 2013;41(7):1319–1328.
73. Lu Y. Integrating experimentation and quantitative modeling to enhance discovery of beta amyloid lowering therapeutics for Alzheimer's disease. *Front Pharmacol.* 2012;3:177.
74. Lu Y, Barton HA, Leung L, Zhang L, Hajos-Korcsok E, Nolan CE et al. Cerebrospinal fluid beta-amyloid turnover in the mouse, dog, monkey and human evaluated by systematic quantitative analyses. *Neurodegener Dis.* 2013;12(1):36–50.
75. Geerts H, Spiros A, Roberts P, Carr R. Towards the virtual human patient. Quantitative systems pharmacology in Alzheimer's disease. *Eur J Pharmacol.* 2017;817:38–45.
76. Ballard P, Yates JW, Yang Z, Kim DW, Yang JC, Cantarini M et al. Preclinical comparison of osimertinib with other EGFR-TKIs in EGFR-Mutant NSCLC brain metastases models, and early evidence of clinical brain metastases activity. *Clin Cancer Res.* 2016;22(20):5130–5140.
77. Furman WL, Stewart CF, Poquette CA, Pratt CB, Santana VM, Zamboni WC et al. Direct translation of a protracted irinotecan schedule from a xenograft model to a phase I trial in children. *J Clin Oncol.* 1999;17(6):1815–1824.
78. Hennika T, Hu G, Olaciregui NG, Barton KL, Ehteda A, Chitranjan A et al. Pre-clinical study of panobinostat in xenograft and genetically engineered murine diffuse intrinsic pontine glioma models. *PLoS One.* 2017;12(1):e0169485.
79. Doody RS, Thomas RG, Farlow M, Iwatsubo T, Vellas B, Joffe S et al. Phase 3 trials of solanezumab for mild-to-moderate Alzheimer's disease. *N Engl J Med.* 2014;370(4):311–321.
80. Gottlieb S. How FDA plans to help consumers capitalize on advances in science. https://blogs.fda.gov/fdavoice/index.php/2017/07/how-fda-plans-to-help-consumers-capitalize-on-advances-in-science/. Accessed July 2017.
81. Taneja A, Di Iorio VL, Danhof M, Della Pasqua O. Translation of drug effects from experimental models of neuropathic pain and analgesia to humans. *Drug Discov Today.* 2012;17(15–16):837–849.
82. Huggins JP, Smart TS, Langman S, Taylor L, Young T. An efficient randomised, placebo-controlled clinical trial with the irreversible fatty acid amide hydrolase-1 inhibitor PF-04457845, which modulates endocannabinoids but fails to induce effective analgesia in patients with pain due to osteoarthritis of the knee. *Pain.* 2012;153(9):1837–1846.
83. Patel K, Simpson JA, Batty KT, Zaloumis S, Kirkpatrick CM. Modelling the time course of antimalarial parasite killing: A tour of animal and human models, translation and challenges. *Br J Clin Pharmacol.* 2015;79(1):97–107.
84. Rosario MC, Jacqmin P, Dorr P, van der Ryst E, Hitchcock C. A pharmacokinetic-pharmacodynamic disease model to predict in vivo antiviral activity of maraviroc. *Clin Pharmacol Ther.* 2005;78(5):508–519.
85. Fang J, Jadhav PR. From in vitro EC(5)(0) to in vivo dose-response for antiretrovirals using an HIV disease model. Part II: Application to drug development. *J Pharmacokinet Pharmacodyn.* 2012;39(4):369–381.

86. Benson N, de Jongh J, Duckworth JD, Jones HM, Pertinez HE, Rawal JK et al. Pharmacokinetic-pharmacodynamic modeling of alpha interferon response induced by a Toll-like 7 receptor agonist in mice. *Antimicrob Agents Chemother.* 2010;54(3):1179–1185.
87. Verhoef J, Gillissen A. Resistant Haemophilus influenzae in community-acquired respiratory tract infections: A role for cefixime. *Int J Antimicrob Agents.* 2003;21(6):501–509.
88. Lepak AJ, Andes DR. Antifungal pharmacokinetics and pharmacodynamics. *Cold Spring Harb Perspect Med.* 2014;5(5):a019653.
89. Marcusson LL, Komp Lindgren P, Olofsson SK, Hughes D, Cars O. Mutant prevention concentrations of pradofloxacin for susceptible and mutant strains of Escherichia coli with reduced fluoroquinolone susceptibility. *Int J Antimicrob Agents.* 2014;44(4):354–357.
90. Drusano GL, Preston SL, Hardalo C, Hare R, Banfield C, Andes D et al. Use of preclinical data for selection of a phase II/III dose for evernimicin and identification of a preclinical MIC breakpoint. *Antimicrob Agents Chemother.* 2001;45(1):13–22.
91. Andes DR, Lepak AJ. In vivo infection models in the pre-clinical pharmacokinetic/pharmacodynamic evaluation of antimicrobial agents. *Curr Opin Pharmacol.* 2017;36:94–99.
92. Katsube T, Yamano Y, Yano Y. Pharmacokinetic-pharmacodynamic modeling and simulation for in vivo bactericidal effect in murine infection model. *J Pharm Sci.* 2008;97(4):1606–1614.
93. Yadav R, Bulitta JB, Wang J, Nation RL, Landersdorfer CB. Evaluation of pharmacokinetic/pharmacodynamic model-based optimized combination regimens against multidrug-resistant pseudomonas aeruginosa in a murine thigh infection model by using humanized dosing schemes. *Antimicrob Agents Chemother.* 2017;61(12) e01268–17.
94. Meagher AK, Ambrose PG, Grasela TH, Ellis-Grosse EJ. Pharmacokinetic/pharmacodynamic profile for tigecycline-a new glycylcycline antimicrobial agent. *Diagn Microbiol Infect Dis.* 2005;52(3):165–171.
95. Ambrose PG, Bhavnani SM, Rubino CM, Louie A, Gumbo T, Forrest A et al. Pharmacokinetics-pharmacodynamics of antimicrobial therapy: It's not just for mice anymore. *Clin Infect Dis.* 2007;44(1):79–86.
96. Vaishnaw AK, Gollob J, Gamba-Vitalo C, Hutabarat R, Sah D, Meyers R et al. A status report on RNAi therapeutics. *Silence.* 2010;1(1):14.
97. Xu YH, Sun Y, Barnes S, Grabowski GA. Comparative therapeutic effects of velaglucerase alfa and imiglucerase in a Gaucher disease mouse model. *PLoS One.* 2010;5(5):e10750.
98. Lazaar AL, Sweeney LE, MacDonald AJ, Alexis NE, Chen C, Tal-Singer R. SB-656933, A novel CXCR2 selective antagonist, inhibits ex vivo neutrophil activation and ozone-induced airway inflammation in humans. *Br J Clin Pharmacol.* 2011;72(2):282–293.
99. Dwivedi G, Fitz L, Hegen M, Martin SW, Harrold J, Heatherington A et al. A multiscale model of interleukin-6-mediated immune regulation in Crohn's disease and its application in drug discovery and development. *CPT Pharmacometrics Syst Pharmacol.* 2014;3:e89.
100. Lee L, Sharma S, Morgan B, Allegrini P, Schnell C, Brueggen J et al. Biomarkers for assessment of pharmacologic activity for a vascular endothelial growth factor (VEGF) receptor inhibitor, PTK787/ZK 222584 (PTK/ZK): Translation of biological activity in a mouse melanoma metastasis model to phase I studies in patients with advanced colorectal cancer with liver metastases. *Cancer Chemoth Pharm.* 2006;57(6):761–771.
101. Dalhoff A, Rashid MU, Kapsner T, Panagiotidis G, Weintraub A, Nord CE. Analysis of effects of MCB3681, the antibacterially active substance of prodrug MCB3837, on human resident microflora as proof of principle. *Clin Microbiol Infect.* 2015;21(8):767 e1–e4.
102. Ogata A, Fujieda Y, Terakawa M, Muto T, Tanaka T, Maruoka H et al. Pharmacokinetic/pharmacodynamic analyses of chymase inhibitor SUN13834 in NC/Nga mice and prediction of effective dosage for atopic dermatitis patients. *Int Immunopharmacol.* 2011;11(10):1628–1632.
103. Collins M, Thrasher A. Gene therapy: Progress and predictions. *Proc Biol Sci.* 2015;282(1821):20143003.
104. Chew WL. Immunity to CRISPR Cas9 and Cas12a therapeutics. *Wiley Interdiscip Rev Syst Biol Med.* 2018;10(1):e1408.
105. Langdon G, Davis JD, McFadyen LM, Dewhurst M, Brunton NS, Rawal JK et al. Translational pharmacokinetic-pharmacodynamic modelling; application to cardiovascular safety data for PF-00821385, a novel HIV agent. *Br J Clin Pharmacol* 2010;69(4):336–345.
106. Chaikin P. The bial 10-2474 phase 1 study—A drug development perspective and recommendations for future first-in-human trials. *J Clin Pharmacol.* 2017;57(6):690–703.
107. European Medicines Agency. Guideline on strategies to identify and mitigate risks for first-in-human and early clinical trials with investigational medicinal product. 2017. https://www.ema.europa.eu/documents/scientific-guideline/guideline-strategies-identify-mitigate-risks-first-human-early-clinical-trials-investigational_en.pdf.
108. Rowland M, Benet LZ. Lead PK commentary: Predicting human pharmacokinetics. *J Pharm Sci.* 2011;100(10):4047–4049.
109. van der Graaf PH, Benson N. Systems pharmacology: Bridging systems biology and pharmacokinetics-pharmacodynamics (PKPD) in drug discovery and development. *Pharm Res.* 2011;28(7):1460–1464.
110. Agoram BM, Demin O. Integration not isolation: Arguing the case for quantitative and systems pharmacology in drug discovery and development. *Drug Discov Today.* 2011;16(23–24):1031–1036.
111. Morgan P, Brown DG, Lennard S, Anderton MJ, Barrett JC, Eriksson U et al. Impact of a five-dimensional framework on R&D productivity at AstraZeneca. *Nat Rev Drug Discov.* 2018;17(3):167–181.

2.3 The DIAMONDS Platform

Providing Decision Support for Toxicological Assessment Based on Integration of Structural, Toxicological, and Biomarker Information

Rob H. Stierum, E. Dinant Kroese, Jack W. T. E. Vogels, Harrie Buist, Danyel G. J. Jennen, and Eugene P. van Someren

Contents

INTRODUCTION

The assessment of drug safety is highly important in order to ensure human safety. Before administrating drugs to humans in early clinical trials, toxicity testing is performed in animal models, where acute toxicity end points, including behavior and clinical signs, are monitored. In addition to this, end points evolving from more chronic exposures—such as effects on reproduction, long-term survival and carcinogenicity—are also studied. Animal studies are costly, time consuming and not concordant with the 3Rs principle (replacement, reduction and refinement) that aims at the replacement, reduction and refinement of animal use (Russell and Burch, 1959). In addition, clinical observations and biomarkers of toxicity derived from these animal studies are not necessarily predictive for the human situation. This contributes to the high drug attrition rate. In fact, only ~10% of drugs tested in clinical trials reach ultimate approval (Plenge, 2016), with toxicity in preclinical *and* phase I clinical trials as one of the leading causes (~30%) to cancel the further development of originally conceived drug candidates (Kola and Landis, 2004).

To foster, on one hand, the development of alternatives to animal testing and, on the other hand, the development of methods that yield better predictive outcomes and associated biomarkers for man, innovative approaches are needed. In recent years, fundamental toxicology has been evolving from an animal-based discipline, primarily concerned with histopathological and clinical observations, toward an interdisciplinary science combining *in vivo* animal data with high-content data from molecular biology, computational chemistry, high-throughput cell-based *in vitro* assays, pharmacokinetic and systems biology modeling, toxicogenomics, bioinformatics and *in silico* approaches, to early de-risk drug targets (see also the Chapter "Predicting the safety of drug targets: an *in silico* perspective" elsewhere in this book by Jennifer Venhorst et al.). In addition, public data resources containing toxicogenomics or molecular data are becoming increasingly available [e.g., Open TG-GATEs[1] (Igarashi et al., 2015), DrugMatrix[2] (Ganter et al., 2006), ToxCast[3] (Dix et al., 2007), diXa[4] (Hendrickx et al., 2015), CEBS[5], BioStudies[6] (Sarkans et al., 2018)].

To assist in toxicological assessments, there is a need for data management and data-driven decision support systems that are able to collect and integrate these heterogeneous types of information originating from these very diverse domains. In this chapter, we describe the current development of such a generic data management and data integration platform: Data-warehouse Infrastructure for Algorithms, Models and Ontologies toward Novel Design and Safety (DIAMONDS). Besides collecting and storing various chemical, structural and kinetic information for thousands of compounds, DIAMONDS also contains various transcriptomics data sets. This is augmented with diverse sources of (*in vivo*) toxicological data that are integrated into a single label of toxicity. With traditional read-across approaches, the potential toxicity of a novel compound is based on compounds that are structurally similar to that compound and for which toxicological end points and biomarkers are known. DIAMONDS strengthens traditional read-across by combining structural similarities with similarities based on biological responses, in other words, based on the available processed transcriptomics data. In addition, the potential for biomarker discovery based upon comparison of data from structurally similar molecules is promoted.

To achieve the DIAMONDS approach, an infrastructure platform has been developed consisting of (1) pipelines that store all data into a data-warehouse, knowledge base and file system, which are fully annotated using ontologies; (2) a website that allows users to browse and query the data and visualize the results; and (3) a tooling infrastructure to run algorithms and models (accessible for other systems via an application programming interface). Figure 2.3.1 illustrates how the DIAMONDS approach of integrated toxicity prediction is based on the technical elements of the DIAMONDS infrastructure platform.

While still in development, we provide illustrations on how DIAMONDS may be used to obtain integrated data views to support toxicological assessment. Specifically, we take the drug clofibrate as an exemplar model compound.

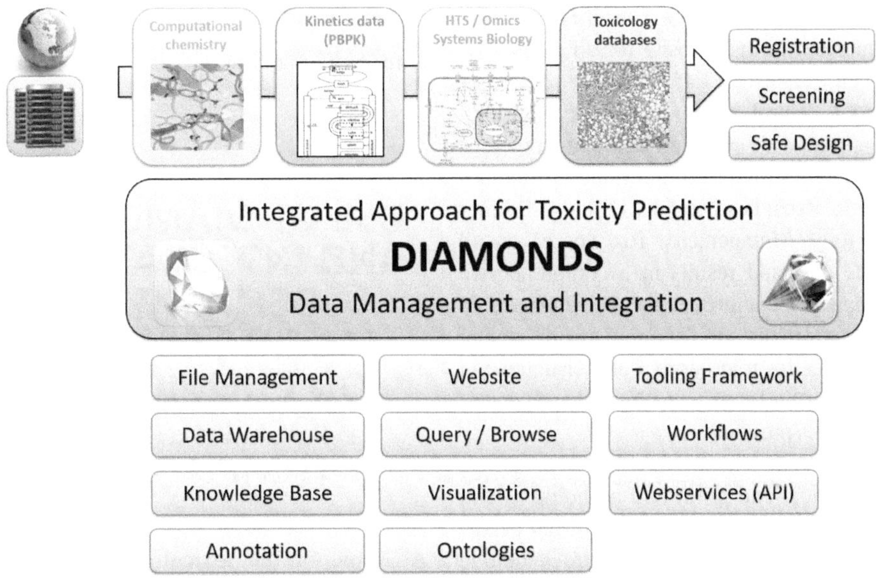

FIGURE 2.3.1 Diagram illustrating schematically how the development of the DIAMONDS platform supports the DIAMONDS approach toward an integrated toxicity prediction. The lower part of the diagram shows the technical elements within the DIAMONDS platform that are essential for an integrated approach. The upper part of the diagram shows how data (some of which originating from public domain websites or databases) are collected from the diverse domains to help with registration, screening or safe design of compounds. To this end, data concerning the chemical structure, kinetics, diverse omics approaches as well as known *in vivo* pathology need to be integrated.

[1] http://toxico.nibiohn.go.jp/english/
[2] https://ntp.niehs.nih.gov/drugmatrix/index.html
[3] https://actor.epa.gov/dashboard/
[4] www.dixa-fp7.eu
[5] tools.niehs.nih.gov/cebs3/ui/
[6] https://www.ebi.ac.uk/biostudies/

DESCRIPTION OF THE DIAMONDS INFRASTRUCTURE AND ASSOCIATED DATA CONTENT

The DIAMONDS platform (https://diamonds.tno.nl/) is a sustainable and secure web-based system that harbors diverse life science projects. The design of the platform is modular, such that a specific group of users connected to a certain project can work together on their specific goal. Different projects may share data and functional modules, but project members may also work with their own specific data, knowledge and tools independent of other projects. Over time, the DIAMONDS platform has been used by different toxicogenomics projects, starting with the Netherlands Toxicogenomics Centre (NTC). DIAMONDS was further developed by The Netherlands Organization for Applied Scientific Research (TNO) and used in many other projects amongst others EU FP7 HEALS, H2020-EU-ToxRisk, H2020 HMB4EU, ZonMw ASAT, CEFIC-LRI funded projects as well as pharma industry and nutritional projects within TNO.

In technical terms, the DIAMONDS platform consists of a Linux filesystem, MySQL data warehouse and a Hypertext Preprocessor (PHP)-based website connecting to tools such as R, Python and KNIME. The infrastructure management and monitoring, together with the provision of daily backups, is performed within the TNO's data center. Secure Sockets Layer (SSL) certificates ensure a secure (encrypted) connection to the website for all users. The website is based on the well-established Laravel framework. User accounts are built around a central group-access approach. All users are members of a group with common access to projects, modules and data. DIAMONDS provides a multitude of modules (e.g., Compounds Module, Casestudies Module, Distance Profiler Module) and viewpoints (e.g., Mutagenicity Tox Space) on all stored data (presenting the data and results in a comprehensive manner) and allows data management through the web interface. In addition, a diverse set of bioinformatics tools ranging from, among others, pre-processing pipelines for microarray data, visualization techniques, pattern recognition tools up to pathway-based meta-analysis and distance profiling is present.

Considering the data content within DIAMONDS, omics, clinical chemistry, toxicological end point and pathology data, high-throughput screening data and extensive meta-data have been collected, curated and pre-processed. Data sources include selected toxicogenomics and molecular data sets from a variety of projects and repositories, including the NTC, ZonMw ASAT, Open TG-GATEs, DrugMatrix, diXa (www.dixa-fp7.eu) and the U.S. Environmental Protection Agency (EPA) ToxCast. DrugMatrix[7] (Ganter et al., 2006) is a molecular toxicology reference database and informatics system. DrugMatrix contains data from standardized toxicological experiments in which rats or primary rat hepatocytes were treated with chemicals at toxic and subtoxic/nontoxic dose levels. The U.S. EPA ToxCast data set[8] consists of data on ~700 high-throughput assays spanning a variety of cellular responses and signaling pathways (Dix et al., 2007). Open TG-GATEs[9] is a repository of *in vivo* and *in vitro* transcriptomics data, together with clinical chemistry, biomarker data and pathology (Igarashi et al., 2015). The Open TG-GATEs database spans 170 well-known reference compounds/drugs and includes omics studies in human as well as rat: Human data consists of *in vitro* primary hepatocyte results, whereas rat data consists of *in vivo* liver and kidney (repeated dose and single dose) as well as hepatocyte results. In addition, similarity in ligand-based biological target predictions based upon fingerprints of chemical structures (Koutsoukas et al., 2011, 2013) is available for interpretation within DIAMONDS. Further, chemoinformatic data has been collected and curated from PubChem.[10] Specifically, two-dimensional (2D) fingerprints were collected using PipelinePilot,[11] Extended Connectivity FingerPrints (ECFP4) (Rogers and Hahn, 2010) and Feature-Connectivity FingerPrints (FCFP4, thereby depicting for each compound which of the 880 structural features are present (presented as 880 bits). Aside from these molecular data, diverse toxicological data sources have been utilized within the data context of DIAMONDS (e.g., ToxRefDb, Registration, Evaluation, Authorisation and Restriction of Chemicals [REACH], U.S. National Toxicology Program; ZEBET; RepDose; CPDB; ISSSTY; ISSCAN; ToxTree; and Open TG-GATEs). For some sources, only the *availability* of compound-specific data is indicated, in relation to that source.

USE OF DIAMONDS TO PROCESS AND INTEGRATE DATA TOWARD DEFINING COMPOUND-SPECIFIC SIMILARITIES BASED ON CHEMICAL STRUCTURES, BIOLOGICAL TARGETS AND TOXICOGENOMICS DATA

As some of the toxicological sources in DIAMONDS may have overlapping toxicity information concerning the same compounds, we first defined a general workflow that ranks toxicological sources according to their inherent strength of evidence (Figure 2.3.2a). For example, as a general principle, if (in light of EU regulation involving REACH) the REACH classifications are available for a compound, there is no need for assessing available *in vivo* or *in vitro* or for *in silico* predictions for that

[7] https://ntp.niehs.nih.gov/drugmatrix/index.html

[8] https://actor.epa.gov/dashboard/

[9] http://toxico.nibiohn.go.jp/english/

[10] https://pubchem.ncbi.nlm.nih.gov/

[11] http://accelrys.com/products/collaborative-science/biovia-pipeline-pilot/

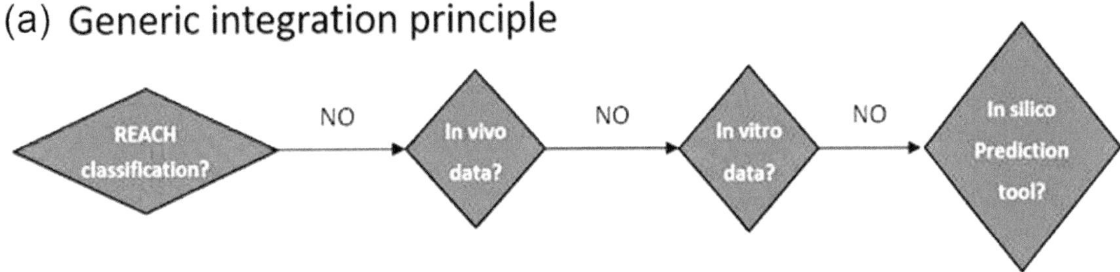

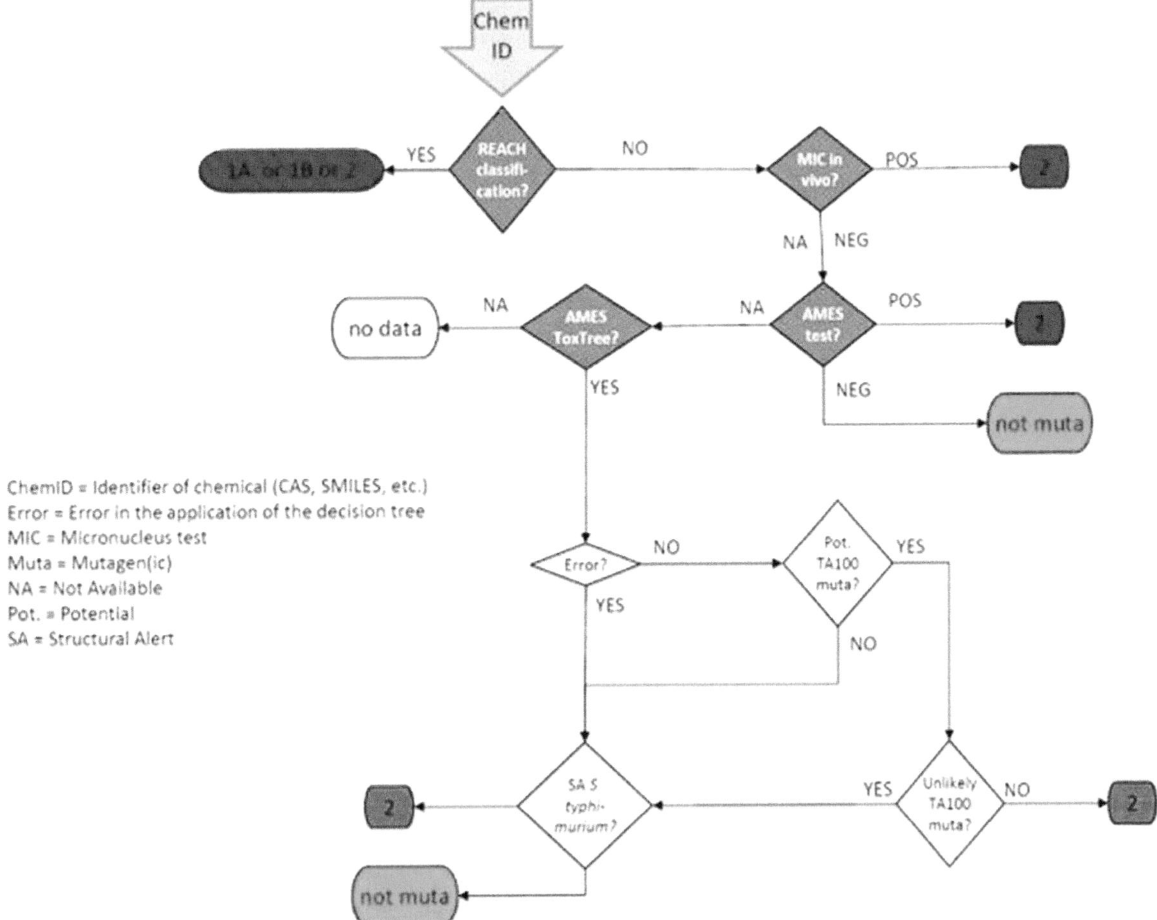

FIGURE 2.3.2 Workflow diagrams illustrating how diverse toxicity data is integrated into a single label of toxicity for implementation into DIAMONDS. (a) Generic workflow to design a decision tree, that follows the principle that a higher ranked toxicological source is used first and if no information is available a subsequently ranked source is used. So, preferably REACH classification is used, then *in vivo* data, then *in vitro* data and finally *in silico* predictions. In DIAMONDS this principle has been used to create specific decision trees for acute toxicity, sensitization, mutagenicity and carcinogenicity. (b) An example of the specific decision tree for mutagenicity. If REACH classification is available, a 1A, 1B or 2 label is assigned according to its REACH classification. If no REACH classification is available, the decision tree checks for an available *in vivo* micronucleus test, and if that is not positive, it continues to the AMES test and so on.

compound. However, if a classification is not available, and if no *in vivo* or *in vitro* study information is available either, then *in silico* predicted values become valuable. Therefore, our general principle was worked out in more detail for four toxicological end points—acute toxicity, sensitisignal, mutagenicity and carcinogenicity—for that specific end point and corresponding particular data sources. This resulted in four decision trees that aim to integrate diverse toxicological sources into a single integrated toxicity classification label. Figure 2.3.2b shows the specific decision tree for mutagenicity.

Importing, storing, annotating and curating of transcriptomics data is a laborious task. Therefore, we have implemented an interactive transcriptomics-processing pipeline within DIAMONDS (partly based on Eijssen et al., 2013). This pipeline can be fed with original raw data files (e.g., CEL files from Affymetrix) and original annotation, allowing the investigator to transform annotation toward a standardized format, normalize data, define ratios, execute limma statistics (Smyth et al., 2004) and compute signature similarities. More specifically, $\log_2$ ratios were computed for each sample, as compared with the average of the set of corresponding controls. Because this was also done for the separate control samples themselves, this ensured the availability of control-corrected replicate values, both for compound (experimental) treatment groups as well as for their corresponding control experiments, enabling subsequent limma statistical analysis to retrieve compound-specific differentially expressed genes.

After normalization (and limma statistics), two types of similarity measures were computed for each of the omics studies between each pair of samples. First, Euclidean distance-based similarity, which is the scaled Euclidean distance between normalized $\log_2$ ratios. This distance is based on comparing the full omics responses (all gene expression values) of both samples. (Note: for this distance measure, $\log_2$ ratios were directly used and limma statistics were ignored.) The distance measure was converted into a similarity measure by first dividing by the highest observed distances (to map between 0 and 1) and then subtracting it from 1 (to map high distances to low similarities). Second, geneset overlap-based similarity, which is the Jaccard index between the sets of significantly regulated genes (obtained by the Benjamini-Hochberg–corrected p-value of $\leq 5\%$ in the limma statistic). This similarity measure is based on computing the overlap of gene sets of both samples. (Note: for this distance, measure $\log_2$ ratios were ignored and the resulting genesets from the limma statistics were directly used.) The Jaccard index is here referred to as the number of common regulated genes present in both sets (shared present; intersection) divided by the total number of genes present in (any of the) both sets (combined present; union). These two approaches were both employed because both have advantages and disadvantages. The similarity measure based on Euclidean distance has the advantage of not requiring statistical pre-processing, but it can be influenced by noisy genes that may not be relevant (although effects are assumed to cancel out over many genes). The similarity measure based on the Jaccard index calculated from geneset overlap has the advantage of focusing on relevant genes, but it sensitive to the limma statistic and the number of replicate samples. In addition, similarities were calculated for chemical structures. For PubChem 2D Fingerprints, ECFP4 as well as FCFP4 structural similarity was defined by calculating the Tanimoto score between each pair of compounds for each of the three fingerprints. The Tanimoto score is defined as the number of structural features that are present in both compounds, divided by the total number of structural features present in at least one of the compounds. Finally, scaled Euclidian distances between compounds were calculated from ligand-based target predictions based on circular fingerprints of chemical structures (Koutsoukas et al., 2011, 2013), using the PIDGIN tool (Mervin et al., 2015).

USE OF DIAMONDS TO INTERROGATE THE DATA FOR CLOFIBRATE—USING THE COMPOUNDS MODULE

After collection within DIAMONDS and processing of the various data sets, the Open TG-GATEs data set was explored in greater detail, together with compound-specific structural information, ligand-based target predictions based on circular fingerprints of chemical structures and toxicological information derived from the decision tree approach (Figure 2.3.2), as described above. In addition, various other toxicological data sets were employed, including drug-induced liver injury scores obtained from Chen et al. (2011, 2013) and pathology descriptions and clinical chemistry data from Open TG-GATEs. These were not processed in terms of establishing similarity measures; however, together with data sources for which similarity measures were calculated, they were considered in the final interpretations as well.

We here consider peroxisome proliferation as the toxicological end point of the drug clofibrate. Within eukaryotic cells, peroxisomes are intracellular organelles that contain enzymes for oxidative reactions, including the beta-oxidation of very-long-chain fatty acids. During beta-oxidation, fatty acids are sequentially broken down to two carbon units, which are subsequently converted to acetyl-CoA, which is then transported back from the peroxisomes to the cytosol for further use. Aside from peroxisomal, mitochondrial beta-oxidation injury can also occur, at least in animal cells. Peroxisome proliferation is evidenced by hepatocellular hypertrophy caused by "cloudy swelling" or granular degeneration of the cells. Specifically, affected cells in the liver appear to be more eosinophilic, with abundant granular cytoplasm. An increase in the number of peroxisomes leads to an increased formation of reactive oxygen species (ROS). As a resultant of this increase in oxidative stress, increased peroxisomal activity may lead to damaged proteins, lipids or nucleic acids and may cause pathogenesis (e.g., carcinogenesis).

Drugs, such as fenofibrate and clofibrate, have been shown to induce peroxisomal proliferation via peroxisome proliferator-activated receptors (PPARs). As a consequence of increased oxidative stress, these compounds increase the risk of somatic mutations resulting in, for example, cancer or impaired metabolism (e.g., thyroid disorders, reproductive dysfunction, skeletal, cardiac myopathies). In addition, it has been known for some time that peroxisome proliferators have a liver tumor-promoting effect in rodents (Reddy and Lalwai, 1983). It is noted though that peroxisome proliferation, as a mode of action leading to carcinogenesis, is considered not to be relevant in humans by most experts or, if relevant, only so under exceptional exposure conditions (Corton et al., 2014; Felter et al., 2018).

In DIAMONDS, the user has the option of obtaining the available information on chemical, physical and toxicological properties of a compound of interest by visiting the Compound module. The user can select one of three tabs with information:

Compound Details, Data Availability and Similar Compounds. The Compound module allows one to start typing any part of the compound name (or one of its identification numbers [IDs]) during which suggestions of existing compounds are given through the autocomplete functionality. Once a compound of interest is chosen, the Compound Details tab shows a list of synonyms, an image of the chemical structure, basic identifying properties (e.g., CAS, Formula, Mass, Einecs, InChiKey, InChi, Smiles, & PubChem identification number), experimental physical and chemical properties as well as computed physical and chemical properties. The Data Availability tab shows for which sources toxicity information is available concerning acute toxicity (AT), sensitization (S), mutagenicity and carcinogenicity (M&C), repeated dose toxicity (RDT), developmental and reproductive toxicity (DART) and whether a registration dossier is available on the ECHA website.

The Similar Compounds tab allows the user to find compounds similar to the compound of interest based on one of the three structural similarity measures (PubChem 2D Fingerprints, ECFP4, FCFP4) or one of the two omics similarity measures (Euclidean distance-based similarity; geneset overlap-based similarity). Subsequently, the user can select what kind of information they want to see for the compound of interest as well as its set of similar compounds. The choices are (1) data availability (see above); (2) decision tree–based toxicity labels for acute toxicity (Acute), sensitization (Sens), mutagenicity (Muta) and carcinogenicity (Carc); (3) Tox Spaces (2D scatter plots) for the same end points as for B and D), an extensive overview of study results concerning repeated dose toxicity, developmental toxicity and reproductive toxicity.

Figure 2.3.3 illustrates what is shown in the Compound module when (1) clofibrate is selected as the compound of interest, (2) the Pubchem 2D fingerprint Tanimoto scores are selected as a similarity measure, and (3) decision tree–based toxicity labels (according to Figure 2.3.2) are used as the information of interest to display. If we would assume that no information was known concerning mutagenicity and/or carcinogenicity of clofibrate, this integrated view may help the toxicologist with assessing these toxicities through a traditional read-across approach. In this example,

Safety table for compounds similar to Ethyl 2-(4-chlorophenoxy)-2-methylpropanoate

TNO ID	Name	Cas	Similarity	Acute	Sens	Muta	Carc
1720	Ethyl 2-(4-chlorophenoxy)-2-methylpropanoate	637-07-0	1.0000		not sens	not muta	2
714263	sodium;2-(4-chlorophenoxy)-2-methylpropanoate	7314-47-8	0.9714			not muta	2
714292	potassium;2-(4-chlorophenoxy)-2-methylpropanoate	26723-02-4	0.9714			not muta	2
78164	butyl 2-(4-chlorophenoxy)acetate	52716-17-3	0.9712			not muta	2
78018	2-Chloroethyl 2-(4-chlorophenoxy)-2-methylpropanoate	52161-12-3	0.9630			2	2
642514	[2-(4-chlorophenoxy)-2-methylpropanoyl]oxyaluminum;dihydrate	14613-28-6	0.9444			not muta	2
1721	2-(4-Chlorophenoxy)-2-methylpropanoic acid	882-09-7	0.9444	4		not muta	2
712204	sodium;2-(4-chlorophenoxy)acetate	13730-98-8	0.9429			not muta	2
14444	2-(4-CHLOROPHENOXY)PROPANOIC ACID	3307-39-9	0.9358			not muta	2
3179	3-[2-(4-chlorophenoxy)-2-methylpropanoyl]oxypropyl 2-(4-chlorophenoxy)-2-methylpropanoate	14929-11-4	0.9286			not muta	2
20118	2-(4-chlorophenoxy)acetic acid	122-88-3	0.9252	4		not muta	2
49001	calcium;2-(4-chlorophenoxy)-2-methylpropanoate	39087-48-4	0.9107			not muta	2
67324	Methyl 2-(2,4-dichlorophenoxy)propanoate	23844-57-7	0.9027			not muta	2
305512	propan-2-yl 2-(2,4-dichlorophenoxy)propanoate	61961-10-2	0.9027			not muta	2
69657	ethyl 2-(2,4-dichlorophenoxy)propanoate	58048-39-8	0.9027			not muta	2
12032	2-methylpropyl 2-(2,4-dichlorophenoxy)acetate	1713-15-1	0.9027			not muta	2
12358	propyl 2-(2,4-dichlorophenoxy)acetate	1928-61-6	0.9018			not muta	2
994	methyl 2-(2,4-dichlorophenoxy)acetate	1928-38-7	0.9018			not muta	2
4814	butyl 2-(2,4-dichlorophenoxy)acetate	94-80-4	0.9018			not muta	not carc
7931	Ethyl 2-(2,4-dichlorophenoxy)acetate	533-23-3	0.9018			not muta	2

FIGURE 2.3.3 Screenshot from DIAMONDS showing an integrated view of structural similarity with integrated toxicity labels for clofibrate. In DIAMONDS, (1) clofibrate was chosen by the user as the compound of interest (top row, ethyl 2-(4-chlorophenoxy)-2-methylpropanoate), (2) structural similarity using the Tanimoto score on PubChem 2D fingerprints (880 bits) was chosen as the similarity of interest, with rows defining similar compounds in decreasing similarity to clofibrate; and (3) integrated toxicity labels resulting from the four specific decision trees (according to Figure 2.3.2) was chosen as the information of interest to depict. Columns 1–3 indicate identification numbers, chemical name, and CAS number, respectively. Column 4 displays the Tanimoto similarity score. Columns 5–8 depict integrated toxicity labels of Acute Toxicity [Acute], Sensitization [Sens], Mutagenicity [Muta] and Carcinogenicity [Carc], respectively.

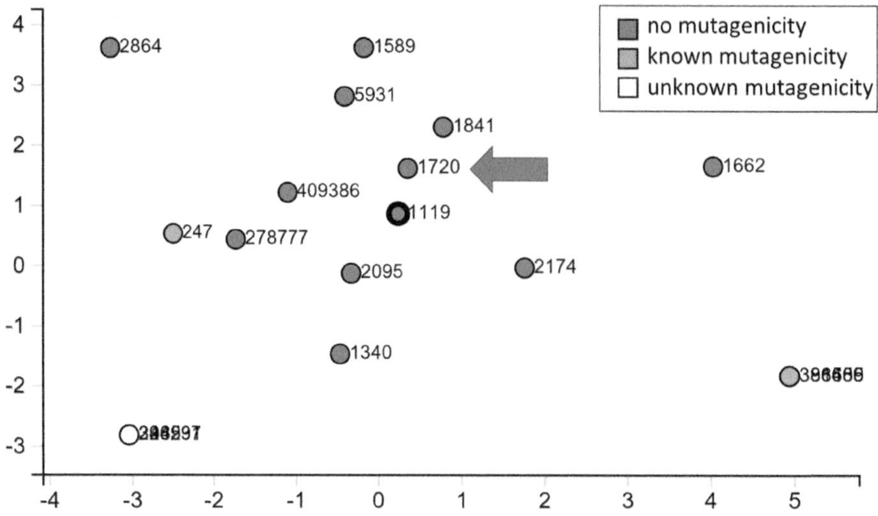

FIGURE 2.3.4 Mutagenicity Tox Space for clofibrate. 2D scatter plot of clofibrate and it is compounds with similar biological response. The position of the compounds (circles) are based on multidimensional scaling, which optimizes the positions to reflect the actual similarities. The similarities were determined based on Euclidean distance from TG-GATEs transcriptomics data. The arrow indicates clofibrate (TNO ID = 1720). The colors indicate known mutagenicity, where green indicates "no mutagenicity," orange indicates "known mutagenicity," and white indicates "unknown." This shows how clofibrate is surrounded by compounds that are not mutagenic and if its mutagenicity would not have been known, biological read-across would suggest "no mutagenicity."

we see that most of the similar compounds are not mutagenic ("not muta"[12] label in "Muta" column), except for one compound, and that most of the similar compounds are causing concern for being carcinogenic ("2"[1] label in "Carc" column), except for one compound. As can be seen in the first row, clofibrate itself is not mutagenic ("not muta") and is also identified as of concern for being carcinogenic ("2").

Because structural similarity does not always reflect toxicological outcome, information from read-across based on structure alone may be strengthened by taking the similarity in biological responses of the compounds, for example, inferred from ligand binding or transcriptome changes, into account as well. This is shown in Figure 2.3.4. This figure shows a different integrative view in DIAMONDS, where (1) again clofibrate was chosen as the compound of interest, but (2) Euclidean-distance–based similarity from transcriptomics responses was chosen to determine similar compounds, and (3) the so-called Mutagenicity Tox Space was chosen as the view of interest. In DIAMONDS, a Tox Space shows the compound of interest and its most similar compounds as small circles in a 2D scatterplot, where the distances in the 2D plot are optimized such that they best reflect the actual distances (e.g., by using multidimensional scaling). By projecting one of the decision tree–based toxicity labels as colors into this plot, the user gains more insight into the associated toxicity of similar compounds as well as an indication of which sets of similar compounds in the same areas share the same toxicity label. In this example, we see that most of the compounds that have a biological response (inferred from transcriptomics) similar to clofibrate are not mutagenic (green circle), except for two single compounds in remote areas (orange). Although Figure 2.3.3 already indicated that structurally similar compounds to clofibrate are non-genotoxic, Figure 2.3.4 strengthens the evidence for clofibrate to be non-genotoxic by indicating that compounds with similar biological response are non-genotoxic as well.

[12] Using principles of Regulation (EC) No 1272/2008 ("CLP Regulation") for mutagenicity and carcinogenicity; note, data used for concluding may be limited.

USE OF DIAMONDS TO INTERROGATE THE DATA FOR CLOFIBRATE—USING THE CASESTUDIES MODULE

To compare the compound information for a set of (presumably related) compounds, the Casestudies module (not shown) allows defining a specific case study and attaching a number of user-selected compounds to this case study. The Casestudies module allows a user to search for or browse through the available stored case studies and select a case study of interest. For this selected case study, similar information as described in the Compounds module can be shown. The only difference is that this information is shown for the set of compounds defined in that case study instead of for a single compound (because a group of compounds is already selected the Similar Compounds tab is not available).

USE OF DIAMONDS TO INTERROGATE THE DATA FOR CLOFIBRATE—USING THE DISTANCE PROFILER MODULE

The Distance Profiler module allows the showing of the different types of available distances measures in a single view in relation to a compound of interest. Using the Distance Profiler module (Figure 2.3.5), diverse types of information can be visualized in different columns for a set of compounds (in rows). Some of the columns can depict the actual distance values, such as distances based on structural- and ligand-based target prediction information (Figure 2.3.5a: "Struct"; "Target"), or distances based on transcriptomics data on human and rat hepatocytes (Figure 2.3.5: "HumHepa," "RatHepa"), or transcriptomics data on rat liver (Figure 2.3.5: "RatVivo"). Other columns present their information directly, for example, toxicity-related information such as mutagenicity (Figure 2.3.5: "MutaGen"), clinical chemistry (Figure 2.3.5. not shown) and pathology here shown as the drug induced liver injuries based upon Chen et al. (2011, 2013) and annotated as "DILI-Label," "DILI severity," "DILI-Concern") within each data source (e.g., Open TG-GATEs, DrugMatrix).

To use the Distance Profiler, for clofibrate, for example, assuming the user would know little about the potential toxicity of clofibrate, the user may select this compound from the examples and choose to visualize a combination of distances based on structural information, ligand-binding target prediction, transcriptomics responses in human hepatocytes, rat hepatocytes and rat *in vivo* liver. In addition, toxicity data are shown for DILI-label, Mutagenicity and Carcinogenicity. The compounds and associated data are ordered from top to bottom based on structural similarity (based upon PubChem 2D Fingerprints). Figure 2.3.5a provides a screenshot of the result. In this view, the user learns that fenofibrate is structurally most similar to clofibrate; however, this similarity is not very strong, as indicated

Distance Profiler for clofibrate

Compound	Struct ↓	Target	HumHepa	RatHepa	RatVivo	MutaGen	Carcino	DILI-Label	DILI-Severity	DILI-Concern
clofibrate	0	0	0	0	0	not muta	2	Less-DILI-Concern	3	Warnings and precautions
fenofibrate	0.26	[illegible]	0.168038342	[illegible]	0.410905514	not muta	2	Less-DILI-Concern	3	Warnings and precautions
coumarin	0.32	0.171524656	0.223473122	0.230211138	0.36830119	2	2	#N/A	#N/A	#N/A
gemfibrozil	0.33	0.106056324	0.197927742	0.128629087	0.294921482	not muta	2	Less-DILI-Concern	3	Warnings and precautions
aspirin	0.37	0.071665542	0.241298145	0.207180689	0.30621428	not muta	not carc	No-DILI-Concern	0	No match
moxisylyte	0.39	0.071575064	0.240955462	0.255311111	0.349961178	not muta	not carc	Most-DILI-Concern	-2	Withdrawn
mexiletine	0.4	0.071714281	0.241819176	0.161442985	0.354505808	not muta	not carc	Most-DILI-Concern	3	Box warning
butylated hydroxyanisole	0.41	0.136330496	0.503558741	0.215492345	0.360135723	not muta	not carc	#N/A	#N/A	#N/A
tamoxifen	0.41	0.182207822	0.186529528	0.160671155	0.371571843	not muta	not carc	Most-DILI-Concern	8	Warnings and precautions
griseofulvin	0.41	0.165127158	0.218273295	0.175906777	0.354180057	not muta	2	Most-DILI-Concern	8	Warnings and precautions
adapin	0.41	0.180713155	0.331398123	0.219521398	0.339948353	?	not carc	Less-DILI-Concern	4	Adverse reactions
phenacetin	0.42	0.071814208	0.181979114	0.173731606	0.371585069	not muta	2	#N/A	#N/A	#N/A
bucetin	0.45	0.07163382	0.176210778	0.170086821	0.35288611	not muta	not carc	#N/A	#N/A	#N/A
pemoline	0.49	0.072148363	0.178653573	0.152385737	0.353781773	not muta	not carc	Most-DILI-Concern	-2	Withdrawn
methyldopa	0.5	0.194674585	0.164089543	0.15654482	0.351858925	not muta		Most-DILI-Concern	8	Warnings and precautions

FIGURE 2.3.5 (a) Illustration of the Distance Profiler module for clofibrate, ordered on structural similarity in combination with diverse toxicity labels. Column 1 depicts compound names starting with clofibrate. Columns 2–6 depict distance values of diverse sources, that is, two-dimensional (2D) PubChem structural distance (Struct), ligand-based target prediction-based distance (Target) and 3 TG-GATEs omics response distances in human hepatocytes (HumHepa), rat hepatocytes (RatHepa) and rat *in vivo* liver data (RatVivo). In addition, known mutagenic (MutaGen), carcinogenic (Carcino) and three columns with Drug Induced Liver Injuries (DILI) information (Label, Severity and Concern) are shown. All columns containing green shadings represent distances values, whereas the columns containing yellow shadings depict toxicity information. The intensity of the green colouring is related to the distance, whereas the yellow coloring is related to the severity of the toxicity label. Rows depict clofibrate (first row) and clofibrate's most similar compounds (based on ordering of column "Struct"). Structural similarity of fenofibrate to clofibrate is supported by target prediction as well as by hepatic transcriptomics response in both human and rats. Nonmutagenicity (not muta) and carcinogenicity (2) appears to be common between clofibrate and its structural similars.

(Continued)

Distance Profiler for clofibrate

Compound	Struct	Target	HumHepa	RatHepa↓	RatVivo	Pathology
clofibrate	0	0	0	0	0	Degeneration, granular, eosinophilic;Swelling
pirinixic acid	0.69	0.074083831	0.205650653	0.110709818	0.444607388	Degeneration, granular, eosinophilic;Hypertrophy
benzbromarone	0.53	0.073890208	0.513471681	0.11534189	0.271256666	Ground glass appearance;Hypertrophy
fenofibrate	0.26	0.064418567	0.168038342	0.119224329	0.410905514	Degeneration, granular, eosinophilic
gemfibrozil	0.33	0.106056324	0.197927742	0.128629087	0.294921482	Degeneration, granular, eosinophilic
sulfasalazine	0.73	0.169314893	0.192003265	0.131978015	0.379231681	Swelling
ethinylestradiol	0.55	0.393558655	0.176593794	0.139336768	0.488205574	Change, eosinophilic;Dilatation;Hypertrophy;Proliferation, Kupffer cell;Vacuolization,
amiodarone	0.51	0.115988718	0.163909014	0.146005302	0.361644387	Change, eosinophilic;Vacuolization, cytoplasmic
pemoline	0.49	0.072148363	0.178653573	0.152385737	0.353781773	NO Pathology found
phenylanthranilic acid	0.61	0.130451639	0.291182195	0.152681721	0.340566235	DEAD
glibenclamide	0.51	0.163151624	0.160563776	0.152846674	0.377987318	NO Pathology found
cimetidine	0.92	0.226730101	0.161002761	0.153379701	0.354378765	Swelling
phenytoin	0.66	0.095023841	0.192738535	0.154568864	0.373974942	Hypertrophy
chlormezanone	0.59	0.128236333	0.186177035	0.154599769	0.359001974	Hypertrophy
tetracycline	0.6	0.585427377	0.215400081	0.154677068	0.348626175	Vacuolization, cytoplasmic

FIGURE 2.3.5 (Continued) (b) Illustration of the Distance Profiler module for clofibrate ordered on rat hepatocyte transcriptome response (RatHepa) distance in combination with pathology. Column 1 depicts compound names starting with clofibrate. Columns 2–6 depict distance values of diverse sources exactly similar to (a), that is, Struct, Target and (3) HumHepa, RatHepa and RatVivo. Column 7 depicts a summary of the pathology descriptions from TG Gates. Rows depict clofibrate (first row) and clofibrate's most similar compounds (based on ordering on rat hepatocyte transcriptome data [column "RatHepa"]). Pathology description "Degeneration, granular, eaosinophilic" appears to be common between clofibrate and those other compounds for which similarity with clofibrate based upon rat hepatocyte transcriptome profile was observed as well.

by the light green background rather than a dark green background, with the intensity of the color being a measure for the similarity (the more intense green, the more similar). If the user would perform read-across of clofibrate based on its structural similar fenofibrate, which is non-genotoxic but carcinogenic and of less DILI concern, this would lead to indications that clofibrate may also be not genotoxic but carcinogenic and have less DILI concern (severity 3). Because clofibrate is a reference compound with known toxicity, we see that the read-across suggestion is consistent with what is known for clofibrate, but without this information, the weak structural similarity would have led to doubts on the validity of grouping these compounds for read-across. The Distance Profiler, however, shows that the weak similarity of fenofibrate with clofibrate is supported by a similarity in ligand-based target prediction, as well as by similarities in transcriptomics responses in human and rat hepatocytes (although not in the liver in rodents *in vivo*). Such information can strengthen the support for grouping these two compounds. Moreover, as a result, it can lead to the postulation of biomarkers for the compound under study, using the information derived from the known reference drugs. Last, this result shows that the two additional (structurally) best-matching compounds (coumarin and gemfibrozil) are also indicative of carcinogenicity and less DILI concern (at least for gemfibrozil) (severity 3) but inconclusive concerning mutagenicity. It also shows that coumarin, which is less similar in terms of structure, is also not similar in any of the other sources, whereas gemfibrozil at least shares a similar response in rat hepatocytes and in rat liver.

Alternatively, if the user would have performed an *in vitro* experiment of clofibrate with rat primary hepatocytes and used these results as the starting point for exploring the DIAMONDS data content, then Figure 2.3.5b illustrates the potential outcome. By ordering on the *in vitro* rat hepatocyte transcriptome response, the user would see, besides fenofibrate and gemfibrozil, that pirinixic acid (WY-14643) and benzbromarone also show similar biological responses. In addition, by requesting to see the known rodent pathological information, a picture emerges of degeneration, eosinophilic changes and hypertrophy/vacuolization, which is the very *in vivo* rodent pathological phenotype for peroxisome proliferation. Note that only 7 of the 170 compounds within TG Gates have "degeneration" as a pathological outcome (one of which is clofibrate). Compared with clofibrate, three of these—fenofibrate, gemfibrozil and pirinixic acid—show a similar transcriptome response in rat hepatocytes. Although not visible in the current tool, the described compounds with similar responses have a commonality in that most are known to activate the PPARα receptor.

Interestingly, the ranking was not supported by the toxicogenomics data from human hepatocytes (column "HumHepa"). This illustrates that the total transcriptome changes were less similar for these structurally similar compounds in the *in vitro* human hepatocyte model, despite similar *in vivo* rodent toxicity (degeneration). Although a specific bioinformatics analysis targeted toward understanding mechanisms of peroxisome proliferation for these compounds is not discussed here, this might indicate that the human hepatocytes are less capable of reflecting a common mechanism of peroxisome proliferation compared to the relevance to rodent data.

CONCLUDING REMARKS

In summary, we have described the DIAMONDS infrastructure to provide a basis for integrative analyses of different toxicological and molecular data types, including omics and *in silico* based target prediction, from different public sources. Although DIAMONDS is still in development, we have shown the initial application to understanding the toxicological effects of a well-known drug clofibrate, in relation to structurally similar drugs and in light of possible biomarker development. From the clofibrate case study, it appears that the combination of the structural and omics similarity could possibly strengthen predictivity over the use of a single chemical similarity parameter. Furthermore, comparison of the omics data on rat and human hepatocytes seems to indicate a difference in mechanism of toxicity between rats and humans. Comparative pathway analysis of both omics profiles may clarify whether this is indeed the case and whether this difference is related to the fact that peroxisome proliferation is a mechanism more relevant in rats compared to humans.

REFERENCES

Chen, M. et al. (2011) FDA-approved drug labeling for the study of drug-induced liver injury. *Drug Discov. Today*, **16**, 697–703.

Chen, M., Hong, H., Fang, H., Kelly, R., Zhou, G., Borlak, J., and Tong, W. (2013) Quantitative structure-activity relationship models for predicting drug-induced liver injury based on FDA-approved drug labeling annotation and using a large collection of drugs. *Toxicol Sci.*, **136**(1), 242–249. doi:10.1093/toxsci/kft189. Epub 2013 Aug 31. PubMed PMID: 23997115.

Corton, J.C. et al. (2014) Mode of action framework analysis for receptor-mediated toxicity: The peroxisome proliferator-activated receptor alpha (PPARα) as a case study. *Crit. Rev. Toxicol.*, **44**, 1–49.

Dix, D.J. et al. (2007) The ToxCast program for prioritizing toxicity testing of environmental chemicals. *Toxicol. Sci.*, **95**, 5–12.

Eijssen, L.M. et al. (2013) User-friendly solutions for microarray quality control and pre-processing on ArrayAnalysis.org. *Nucleic Acids Res.*, **41**, W71–W76.

Felter, S.P. et al. (2018) Human relevance of rodent liver tumors: Key insights from a toxicology Forum workshop on nongenotoxic modes of action. *Regul. Toxicol. Pharmacol.*, **92**, 1–7.

Ganter, B. et al. (2006) Toxicogenomics in drug discovery and development: Mechanistic analysis of compound/class-dependent effects using the DrugMatrix® database. *Pharmacogenomics*, **7**, 1025–1044.

Hendrickx, D.M. et al. (2015) diXa: A data infrastructure for chemical safety assessment. *Bioinformatics*, **31**, 1505–1507.

Igarashi, Y. et al. (2015) Open TG-GATEs: A large-scale toxicogenomics database. *Nucleic Acids Res.*, **43**, D921–D927.

Kola, I. and Landis, J. (2004) Can the pharmaceutical industry reduce attrition rates? *Nat. Rev. Discov.*, **3**, 711–715.

Koutsoukas, A. et al. (2011) From in silico target prediction to multi-target drug design: Current databases, methods and applications. *J. Proteomics*, **74**, 2554–2574.

Koutsoukas, A. et al. (2013) In silico target predictions: Defining a benchmarking data set and comparison of performance of the multiclass naive Bayes and Parzen-Rosenblatt window. *J. Chem. Inf. Model.*, **53**, 1957–1966.

Mervin, L.H. et al. (2015) Target prediction utilising negative bioactivity data covering large chemical space. *J. Cheminform.*, **7**, 51.

Plenge, R.M. (2016) Disciplined approach to drug discovery and early development. *Sci. Transl. Med.*, **8**, 349ps15.

Reddy, J.K. and Lalwai, N.D. (1983) Carcinogenesis by hepatic peroxisome proliferators: Evaluation of the risk of hypolipidemic drugs and industrial plasticizers to humans. *Crit. Rev. Toxicol.*, **12**, 1–58.

Rogers, D. and Hahn, M. (2010) Extended-connectivity fingerprints. *J. Chem. Inf. Model.*, **50**, 742–754.

Russell, W.M.S. and Burch, R.L. (1959) *The Principles of Humane Experimental Technique*. London, UK: Methuen.

Sarkans, U. et al. (2018) The BioStudies database—one stop shop for all data supporting a life sciences study. *Nucleic Acids Res.*, **46**, D1266–D1270.

Smyth, G. K. (2004). Linear models and empirical Bayes methods for assessing differential expression in microarrayexperiments. *Stat. Appl. Genet. Mol.*, **3**(1), Article 3.

Predicting the Safety of Drug Targets

2.4

An In Silico *Perspective*

Jennifer Venhorst, Lars Verschuren, Annemette V. Thougaard, Jorrit J. Hornberg, and Tanja Rouhani Rankouhi

Contents

THE NEED FOR PREDICTIVE (PRE) CLINICAL SAFETY ASSESSMENTS

The high attrition rates in drug development have presented a challenge to the pharmaceutical industry. Only ~10% of drug candidates entering clinical trials reach ultimate approval, and the cost of developing a new drug is ever increasing (Hay et al., 2014; diMasi, 2016; Plenge, 2016). Innovation in the pharmaceutical R&D process has been initiated in order to deal with this challenge and increase the success of developing novel therapies for patients.

The underlying causes of drug failure have been thoroughly analyzed and reviewed in recent years. In the year 2000, termination of post "first-in-human" drug development programs, in 10 major pharmaceutical companies, could be attributed to safety issues in around 30% of cases (Kola and Landis, 2004). A similar rate of failure due to toxicity was reported in the years 2003–2011 for investigational drugs that reached the stage of either new drug application (NDA) or biologic license application (Hay et al., 2014). In both of these studies, the lack of efficacy was the second largest contributor to the discontinuation of drug candidates. Analyses of the success rate of drugs in phase II (2008–2010) and phase III (2007–2010) clinical trials identified safety as the second major cause of failure, after lack of efficacy (Arrowsmith, 2011a, 2011b). Finally, several internal reviews performed by large pharmaceutical companies confirmed that safety and efficacy represent the most significant factors in the failure of candidate drugs (Cook et al., 2014; Hornberg et al., 2014a; Waring et al., 2015).

With safety liabilities being a major cause of drug failure, especially in preclinical and phase I studies, there is a strong case for implementing de-risking strategies as early as possible in the discovery pipeline (Hornberg and Mow, 2014b). Consequently, it is now common practice to perform front loading of toxicity screening assays (Blomme and Will, 2016), but more can be done. It is essential to develop a thorough understanding of potential safety issues associated with that target/concept at the earliest phase of drug discovery, when a novel drug target is being nominated.

We have designed a comprehensive *in silico* target safety assessment (TSA) approach to predict the safety issues that could arise from the modulation of a drug target. The toxicological risks identified can be used to make strategic choices about the progression of target–drug combinations and for the design of toxicity studies in the drug discovery pipeline (Roberts, 2018; Hornberg, 2014a).

CAUSES OF ADVERSE EFFECTS IN DRUG DISCOVERY

The TSA covers three of the four different types of toxicity that may arise from drug target modulation (Table 2.4.1). First, the modulation may be the cause of a direct downstream, on-target effect. These effects are closely related to the biological function of the protein and are often referred to as *exaggerated pharmacological effects* (Rudmann, 2013).

Second, effects of target modulation can combine in unexpected ways to disturb biological systems. The function of a protein is generally not restricted to a singular or linear signaling pathway. A holistic view of the biological complexity, as opposed to a focus on constituent parts, is important when assessing target-related toxicities. By taking into account the biological network of a target—and thus the pathways in which it operates—we can analyze the toxicological outcome arising from the composite effects, including compensatory feedback mechanisms. Although *pathway effects* represent on-target effects, their delineation is more complex and requires additional methodologies.

The third potential cause of toxicity is particularly relevant for targets that are modulated by small molecules and concerns off-target effects. *Off-target* effects are due to nonselective binding of the drug that modulates the intended target. Off-target proteins are able to accommodate the ligand in its binding site even though binding-site similarity with the primary target may be limited (Barelier, 2015). Off-target binding often occurs within the same family or class of proteins, but may involve completely unrelated proteins.

On another level, compound-based toxicity, which is related to the physicochemical characteristics of a drug candidate, may also give rise to adverse effects. This type of toxicity is largely out of the scope of this chapter because it is not related to the target itself and can often be successfully addressed by medicinal chemistry strategies. It is also less relevant for safety assessments of exploratory targets performed at the beginning of the discovery pipeline as such evaluations are performed prior to the start of a hit finding screening campaign. Nevertheless, (pre)clinical data of compounds that cover multiple structural classes can be very informative for assessing whether observed adverse effects are compound-related or mechanism-based (i.e., target-related).

TARGET SAFETY ASSESSMENT

In Figure 2.4.1 the TSA process is shown. It is structured as three pillars: Target characterization, Biological network analysis, and Toxicological evaluation of (pre)clinical studies (Figure 2.4.1, orange boxes). Target characterization takes all the features that are inherent to the target into account, and may contribute to on-target and off-target toxicities. Data mining and *in silico* analyses are particularly useful for this part of the TSA. Network biology places the target in a broader perspective—covering on-target and pathway effects—by applying systems biology approaches. During the toxicological evaluation of (pre)clinical studies, effects of drug interventions are analyzed. All types of toxicity, including compound-based effects, may be at play here. Text mining provides a powerful approach to collect all relevant data of the target and corresponding interventions from scientific literature, conference proceedings, and other sources.

In the following paragraphs, various aspects of the TSA will be highlighted and illustrated using heat shock protein 90 (HSP90) as a case study. The inhibition of HSP90 has been thoroughly investigated as a therapeutic approach for treating various cancers, including breast cancer, renal cancer and multiple myeloma (Neckers and Workman, 2012). More recently, interest has extended to the treatment of neurodegenerative diseases that involve aberrantly folded proteins, such as Alzheimer's disease (Ou et al., 2014) and Huntington's disease (Reis et al., 2016).

TABLE 2.4.1 Causes of toxicities occurring with drug interventions. The target safety assessment analyses toxicities that are target-related: on-target, pathway-related and off-target effects

	BINDING SPECIFICITY	*TOXICITY TYPE*	*EFFECT*
Drug intervention	Selective binding	On-target toxicity Pathway-related toxicity	Exaggerated pharmacology Composite effects disturbed biological system
	Nonselective binding	Off-target toxicity	Related to biological function of off-target(s)
	Nonbinding	Ligand-based toxicity	Related to physicochemical properties

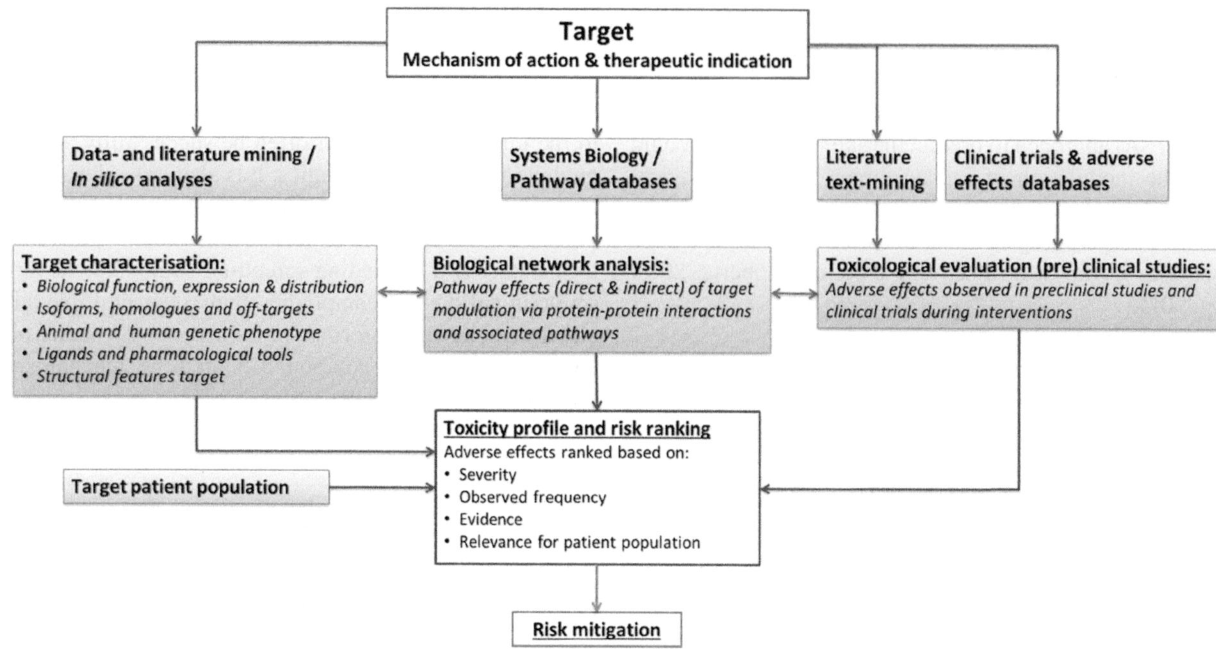

FIGURE 2.4.1 Schematic representation of the target safety assessment (TSA).

Target Characterization

Biological function, expression and distribution

A crucial component of the TSA is the delineation of the biological function of the target under scrutiny, as modulation of the target is likely to perturb its function. Information is mainly derived from text mining of *in vitro* and *in vivo* mechanistic studies, although general information can be obtained from protein databases as well (e.g., UniProt). Another highly informative source constitutes observations made with alternate genetic phenotypes in animals and humans (*vide infra*).

The biological function of a target is closely related to its expression and distribution in cells and organ tissues and to its interaction with other proteins (Villa-Vialaneix et al., 2013). Together, this information provides a first glimpse of the potential organs of toxicity and the associated adverse effects. In terms of (re)sources, both public and commercial informatics tools and databases are commonly used to understand aspects of target distribution and expression in normal and diseased tissues. Examples of public (re)sources include the National Center of Biotechnology Information (NCBI; NCBI Resource Coordinators, 2016), UniProt (UniProt Consortium, 2015), GenAtlas (Frezal, 1998), Protein Atlas (Uhlén et al., 2015) and AceView (Thierry-Mieg and Thierry-Mieg, 2006). An example of a commercial database is NextBio (www.nextbio.com).

A summary of the general features of HSP90 is depicted in Table 2.4.2. As listed, HSP90 acts as a molecular chaperone in the folding and trafficking of client proteins during

TABLE 2.4.2 Summary of some of the target characteristics of HSP90α

HSP90α	
Biological function	HSP90α is an inducible molecular chaperone that functions as a homodimer. The encoded protein aids in the proper folding of specific target proteins by use of an ATPase activity that is modulated by co-chaperones. It performs key roles in the protein-signaling pathway, protein folding, protein degradation, and morphologic evolution and antigen presentation.
Indications	Cancer, CNS diseases
Homologs	HSP90β GRP94 TRAP1
Expression profile	Ubiquitous
Ligand binding site	ATP-binding site, N-terminal domain (e.g., geldanamycin) ATP-binding site, C-terminal domain (e.g., novobiocin)
Mechanism of Action	Inhibition
Downstream effect	Up-regulation of HSP70

normal metabolism. Under stress, it amplifies the levels of repair and refolding of damaged polypeptides. In line with its crucial role in protein homeostasis, HSP90 is expressed by all eukaryotic cells. This ubiquitous expression makes HSP90 modulation potentially prone to more widespread toxicological effects than a target with restricted expression. To exert its function, HSP90 forms a multi-component complex with co-chaperones. These co-chaperones (e.g., Hop [Li et al., 2012]) serve to recognize client proteins and assist their binding to the HSP90 hetero-protein complex (Ernst et al., 2014; Wang et al., 2014).

The interest in HSP90 inhibition as an anticancer therapy originated from its overexpression in various cancer cells. In addition, many of the HSP90 client proteins play crucial roles in establishing cancer cell hallmarks, such as self-sufficiency in growth signals, insensitivity to growth-inhibitory mechanisms and evasion of programmed cell death (Neckers and Workman, 2012; Miyata et al., 2013).

In the context of neurodegenerative disorders involving protein aggregates, two modes of action are of therapeutic interest. First, inhibition of HSP90-mediated stabilization of client proteins inhibits the cycling of the not-yet-unfolded proteins and promotes their degradation. Second, inhibition of HSP90 induces the expression of HSP70, which in turn results in promotion of HSP70-dependent degradation of already-unfolded or misfolded client proteins (Reis, 2016). As HSP70 upregulation is a direct downstream effect of HSP90 inhibition, potential safety liability of enhanced HSP70 activity was also assessed in this TSA. HSP70 plays a role in protein folding and cytoprotection (Table 2.4.3) and may exert protective activity in, for example, gastrointestinal diseases such as gastric ulcers (Ishihara et al., 2011).

Isoforms, homologs and off-targets

Over 90% of human genes undergo alternative splicing resulting in different protein products, that is, isoforms. Although it is not clear to what extent this is the case in general, alternatively processed isoforms have been demonstrated to play distinct or even opposing functions in many studies (Li et al., 2014). The long isoform of caspase-8 (caspase-8L), for example, was found to be an endogenous competitive inhibitor of caspase-8 itself (Himeji et al., 2002).

The isoforms of a target and their potential divergent function are thus mined during a TSA. Homologs/off-targets are similarly identified, often based on sequence similarities determined by a Basic Local Alignment Search Tool (BLAST) search. Off-targets can alternatively be inferred from the nonselective binding profile of the target's known ligands or by three-dimensional (3D)-binding site comparison, ligand-based similarities or docking studies (Patel et al., 2015). However, these approaches require known ligands or pharmacological tools and may, therefore, be prohibitive for more exploratory targets.

Human HSP90 occurs as two cytosolic protein variants, HSP90α and HSP90β, which share about 85% sequence identity as indicated by a BLAST analysis (Altschul, 1990). Although originating from different genes, the term HSP90 used in literature generally refers to both proteins. HSP90α constitutes the stress-inducible form, whereas HSP90β is constitutively expressed. HSP90α furthermore exists as two isoforms, which do not appear to have divergent functions. A comparative study demonstrated that HSP90α and β interact similarly with the major co-chaperone components of the HSP90 chaperone machinery and that heat shock does not alter these interactions. Some differences were described, however, regarding inhibitor or client binding-specificity under stress conditions (Taherian et al., 2008; Prince et al., 2015).

In addition to HSP90α and β, two other ubiquitously expressed homologs exist: 94 kDa glucose-regulated protein (GRP94) and Tumor necrosis factor type 1 receptor-associated protein (TRAP1). GRP94 resides in the endoplasmic reticulum (ER), whereas TRAP1 is predominantly expressed in mitochondria. TRAP1 has no known co-chaperones to date (Johnson, 2012). GRP94 functions as a pivotal regulator of ER proteostasis and calcium storage, being one of the major calcium-binding proteins (Table 2.4.3). Many of the GRP94 client proteins are cell-surface or secreted proteins involved in intercellular communication

TABLE 2.4.3 Summary of the most important biological functions of GRP94, TRAP1 and HSP70

PROTEIN	BIOLOGICAL FUNCTION
GRP94	Cell defense mechanisms Angiogenesis Intestinal homeostasis Platelet activation and aggregation Muscle physiology and myogenic cell differentiation Embryogenesis Calcium storage Intercellular communication and cell adhesion
TRAP1	Mitochondrial integrity and calcium homeostasis Cytoprotective pathways Anti-apoptotic activity
HSP70	Protein folding, degradation and transport Prevention of protein aggregation Cytoprotection

and adhesion (e.g., Toll-like receptors, immunoglobulins). In the absence of GRP94, these proteins are targeted for ER-associated degradation (Arin, 2014; Muth et al., 2014). The function of TRAP1 is not fully understood, but it includes cytoprotection by preventing the opening of the mitochondrial permeability transition pore. It has also been associated with the ER stress response, maintenance of the electron transport chain, selective degradation of mitochondria by autophagy, and mitochondrial morphology (Altieri, 2012). More recently, TRAP1 has also been implicated in metabolism and in the regulation of cancer stem cells (Im, 2016).

Animal and human genetic phenotypes

The biological function of proteins can in part be deduced from genetic phenotypes of animals and humans alike. Natural variants encompass small-scale sequence variations (e.g., point mutations, insertions and deletions), large-scale structural variations (e.g., copy number variation) or numerical variation in whole chromosomes or genomes. These modifications may result in either increased or reduced activities, and can be associated with disease states. Duplication of the amyloid precursor protein (APP) gene, for example, has been linked to Alzheimer's disease (Lee and Lupski, 2006), whereas gain-of-function mutation in the sulfonylurea receptor causes neonatal diabetes (Proks, 2013). In terms of resolving biological function from genetic phenotypes, monogenic targets pose the most straightforward approach. They are also a rich source for the identification of novel drug targets (Brinkman et al., 2006). Known human genetic variants associated with disease are comprehensively listed in the Online Mendelian Inheritance in Man (OMIM®; http://omim.org) database. None of the HSP90 homologs have described allelic variants that are linked to disease. However, numerous studies have implicated overexpression of each of the homologs in cancer (Barrott and Haystead, 2013; Im, 2016; Wu et al., 2016).

Apart from natural variants, reverse genetics involving targeted gene modifications are used *in vivo* to study both function and phenotypes. Knock-out, knock-down, knock-in, conditional, tissue-specific, and overexpression of a protein or silencing of the mRNA, all contribute to the understanding of its role in various organs and in its interplay with other proteins. It may also be a useful tool to assess likely adverse effects associated with target modulation. However, it should be noted that whereas, for example, knock-out models result in a complete lack of protein, small molecule–induced inhibition may only partially reduce the function of a protein, rather than abolishing it completely. Genetically modified animal models may thus display more extreme phenotypes with more severe effects. In addition, early lethality in mice homozygous for a gene deletion does not exclude the corresponding protein as a therapeutic target. Another limitation of genetically engineered models is the fact that environmental and genetic factors may influence the phenotypic outcome (Lin, 2008). Finally, it should be noted that not necessarily all aspects (clinical or morphological) of a genetically modified phenotype are studied, and some effects may remain unnoticed. In some cases, the use of inducible knockout models, in which gene deletion is triggered in the adult mouse, may be useful to study whether the phenotype from a germline knockout is relevant (Liljevald et al., 2016).

A comprehensive resource describing genetically modified mouse models and associated phenotypes is the Mouse Genome Informatics (MGI) database (Eppig et al., 2015). Table 2.4.4 shows the affected phenotypes observed in the MGI database for the various HSP90 homologs. The corresponding observations, illustrated for HSP90α/β, highlight their functions in the investigated organs. The overview shows that—despite high conservation—the HSP90 homologs are functionally nonredundant. Whereas the absence of HSP90α results in incomplete preweaning lethality, HSP90β and GRP94 are essential for survival. HSP90β deficient mice are not viable because they fail to develop a placental labyrinth (Voss et al., 2000). GRP94 null mice do not survive beyond day 7 of gestation (Wanderling et al., 2007). Several other functions appear unique to HSP90α compared with its beta counterpart. For example, HSP90α deficiency in mice reportedly affects the male reproductive system and fertility (e.g., testicular atrophy and azoospermia) and the immune response (e.g., abnormal dendritic cell physiology and antigen presentation).

Ligands and pharmacological tools

Information on known compounds exerting a biological action (inhibition, activation) identical or similar to the intended mechanism of action is collected within a TSA. Subsequently, (pre)clinical studies describing adverse effects associated with those compounds are mined. However, the availability of known ligands and pharmacological tools depends on the novelty of the target under scrutiny. For first-in-class therapies (i.e., very exploratory targets), compound information, and especially toxicological data, is often very limited or lacking altogether. These studies generally focus on compound efficacy, or concern mechanistic studies. For more established targets, such as HSP90, compound information may be abundant. Both commercial and public databases are available to extract relevant compounds. For example, Drugbank (Wishart, 2006) and ChEMBL (Bento, 2014) are public resources and Thomson Reuters Integrity® (https://integrity.thomson-pharma.com/integrity/xmlxsl/), Springers Adis Insight (adisinsight.springer.com) and Informa's Medtrack (oneview.medtrack.com) are commercial databases that provide compound data, mechanistic information, competitive information and more.

Although compound-based toxicity is largely out of the scope of a TSA, clustering of chemical structures to generate compound cluster–oriented toxicity profiles may provide insight into whether toxicities are target or compound related. Compound-based toxicity may be of less concern at early stages of drug discovery because integrated safety assays in lead optimization cycles will ensure the design of compounds with a more favorable efficacy/toxicity profile. Clustering chemical structures can be performed, for example, based on fragment-based similarity (fingerprints) as implemented in the Pipeline Pilot tool (Hassan et al., 2006). Representative structural classes of HSP90 inhibitors are shown in Figure 2.4.2.

TABLE 2.4.4 Overview of affected phenotypes (gray boxes) in transgenic mouse models lacking the mouse orthologs of human HSP90α, HSP90β, GRP94 or TRAP1, and the main phenotypic observations for HSP90α and HSP90β

AFFECTED PHENOTYPE	*HSP90α*	*HSP90β*	*GRP94*	*TRAP1*
Cardiovascular system		■	■	
Behavior/neurological	■			
Cellular	■		■	■
Embryo		■	■	
Endo-/exocrine glands	■			■
Growth/size/body			■	■
Hematopoietic system			■	■
Homeostasis/metabolism				■
Immune system	■		■	■
Limbs/digits/tail	■			
Liver/biliary system				■
Mortality/ageing	■	■	■	
Reproductive system	■			
Skeleton	■			
Neoplasm				■
Main effects due to HSP90 deficiency				
HSP90α				
Effects on male reproductive system				
Preweaning lethality (incomplete penetrance)				
Abnormal tail morphology				
Increased grip strength				
Abnormal immune system physiology				
HSP90β				
Embryonic lethality during organogenesis (complete penetrance)				
Embryo tissue necrosis				
Absent heart beat				
Abnormal extra-embryonic tissue morphology				
Abnormal placenta vasculature				

Structural characteristics and binding selectivity toward off-targets

Both desired and adverse effects can be governed by interactions of a drug with many, often unrelated, targets. Next to broad *in vitro* profiling (Bowes, 2012), structural biology may be used to assess the likelihood of nonselective binding of compounds to protein targets. If available, structures of proteins can be downloaded—in apo- or complexed form—from the Brookhaven Protein Databank (PDB) and complemented by docking and/or molecular dynamics studies. In the absence of a resolved 3D protein structure, homology modeling may offer an alternative to inspect binding sites. When predicting binding modes, protein flexibility—an induced fit—is a complicating factor that deserves special attention (Spyrakis et al., 2011; Antunes, 2015). This is exemplified by the most studied and dramatic conformational changes of tyrosine kinases, the so-called "DFG-flip," notably induced by the anticancer drug Gleevec (Imatinib; Morando et al., 2016). However, as described below, conformational flexibility can also be utilized to force compound selectivity.

All HSP90 homologs consist of three domains: the N-terminal domain, which binds ATP and regulates the turnover of folded client proteins; the highly charged middle domain, which displays high affinity for co-chaperones and client proteins; the C-terminal domain, which contains a second ATP binding site and is believed to regulate homo-dimerization (Donnelly and Blagg 2008). The N-terminal domain of HSP90α/β and its ATP-binding site are well characterized due to the abundant availability of crystal structures. The majority of known inhibitors bind to this domain, competing with ATP. The C-terminal domain is less well described because mammalian crystal structures are lacking. Nevertheless, the C-terminal ATP binding site is known to accommodate small molecule inhibitors such as novobiocin and cisplatin (Marcu et al., 2000). Targeting the C-terminal binding site is of growing interest because this may regulate signaling pathways that are mechanistically different from those targeted by N-terminal HSP90 inhibitors (Shelton et al., 2009; Eskew et al., 2011).

Overall sequence identities of the homologs range from 30% to 86% when compared with HSP90α. Sequence identities for

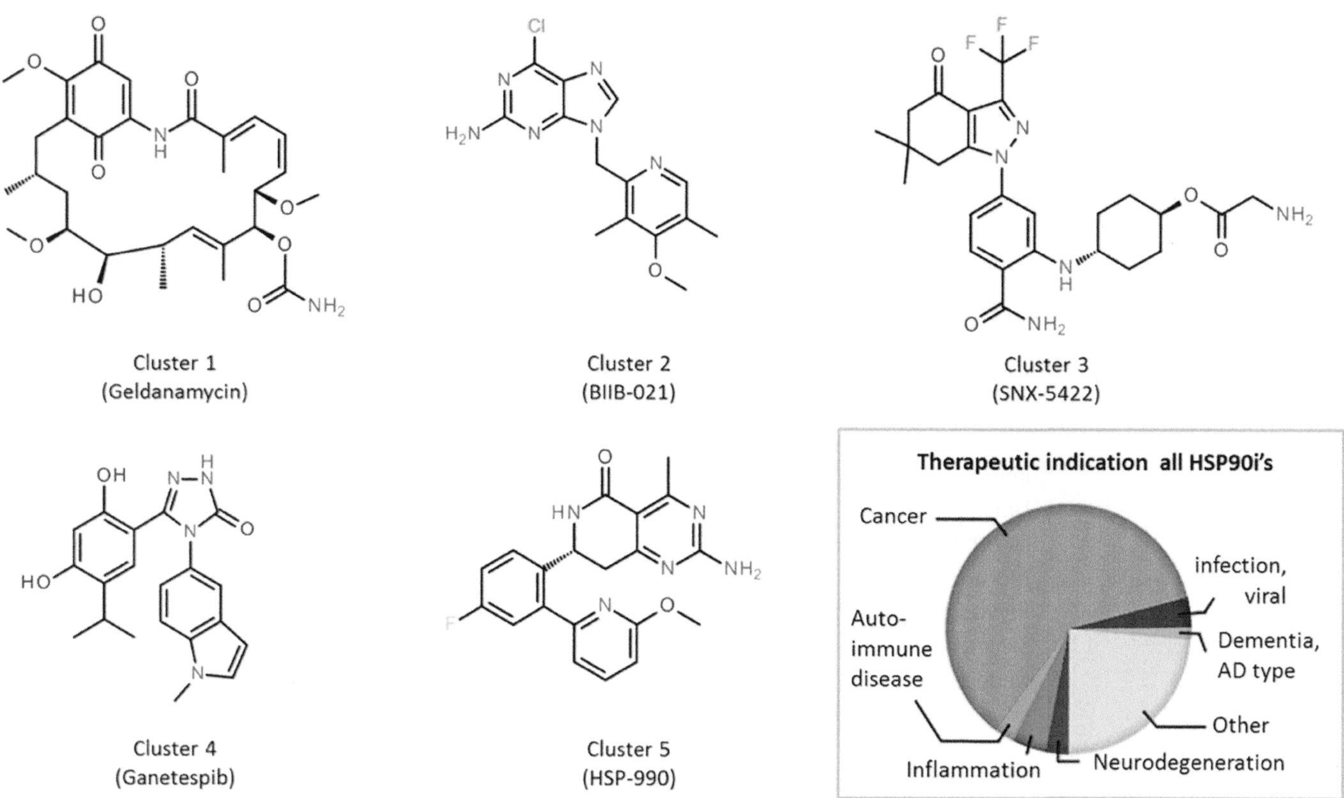

FIGURE 2.4.2 Representative clusters of heat shock protein 90 (HSP90) inhibitors and therapeutic indications of known HSP90 inhibitors in all phases of development. (From Thomson Reuters Integrity®.)

the N-terminal domain are at least 85%. Combined with the high conservation in both the overall fold and in the ATP-binding site (Figure 2.4.3), it can be concluded that homologs represent the most relevant off-targets in the case of HSP90 (Immormino et al., 2010). However, the high conservation in sequence and structure does not prohibit the design of selective inhibitors. All forms of HSP90 exist as dimers, bind and hydrolyze ATP, and cycle between distinct conformational states. These include (1) an open, nucleotide-free form dimerized at the C termini; (2) a closed form that is stabilized by ATP and is dimerized at the N termini; and (3) a distinct ADP-bound conformation (Johnson, 2012; Chiosis, 2013). The conformational flexibility of the N-terminal ATP-binding site can be exploited to enhance isoform-selective inhibition of HSP90α/β versus TRAP1 and/or GRP94. Because not all homologs can adopt similar conformations, protein conformation is a primary driver of selectivity (Figure 2.4.3; Ernst et al., 2014). Interestingly, the middle domain has very recently been implicated in HSP90β-specific inhibition and may offer a first chance to distinguish between the cytosolic HSP90s (Yim et al., 2016). In terms of safety, conformational sampling implies that toxicities due to off-target (homolog) binding may be avoided.

Biological Network Analysis

Proteins exert their function in a complex network of interactions, affecting a multitude of other biological components and regulating corresponding pathways (Cusick, 2005). Consequently, a reductionist approach, which merely includes a single component of a biological system, does not provide sufficient context to comprehend or predict the full spectrum of biological effects occurring downstream due to target modulation. Network biology provides a means of obtaining this essential global view (García-Campos, 2015), by, for example, using a data driven, top-down approach such as enrichment analysis (Figure 2.4.4) (Huang et al., 2009). Enrichment analysis is a statistical method that links regulated or differentially expressed genes/proteins to individual pathways or processes. The holistic nature of network analysis allows a better understanding of the relationships between multiple genes and other biological or chemical entities. In terms of safety, it can be used to predict toxicological outcomes and may contribute to biomarker development (Caberlotto, 2015). Network biology is therefore an important part of a TSA study, especially when few preclinical observations are described in the public domain. The success of network-based methods, however, does rely heavily on the availability, further development, and curation of high-quality databases, containing, for example, protein–protein interactions and corresponding functional annotations. These include, for example, Ingenuity Pathway Analysis (IPA®; QIAGEN Redwood City; www.qiagen.com/ingenuity), StringDB (Jensen et al., 2009), Pathway Commons (Cerami, 2011), Comparative Toxicogenomics Database (CTD; Davis, 2016) and Kyoto Encyclopedia of Genes and Genomes (KEGG; Kanehisam and Goto, 2000). A comprehensive list of resources can be found at http://www.pathguide.org.

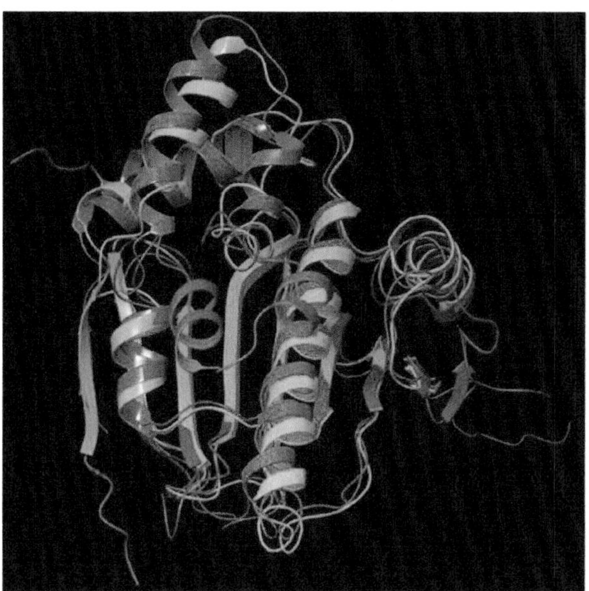

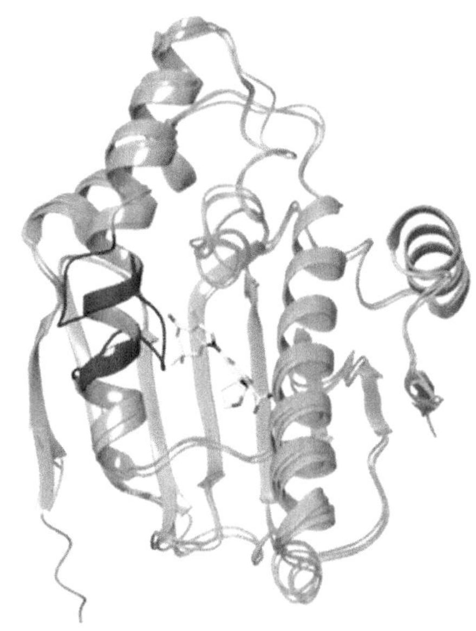

FIGURE 2.4.3 Overlay of the N-terminal domain of HSP90α (PDB 1YET, blue), GRP94 (PDB 1YT2, orange), TRAP1 (PDB 4IVG, green), displaying high conservation of the N-terminal domain and inhibitor binding site (left). However, the ligand-induced fit of HSP90 amino acid residues 104–111 (shown in red) allows for isoform-selective inhibition (right; PDBs: 1YET, 1OSF, 4NH8). As shown, these residues can adopt an open- or closed-loop conformation or a helical structure, depending on distinct chemotypes. (From Ernst, J.T., *Bioorganic Med. Chem. Lett.*, 24, 204–208, 2014.)

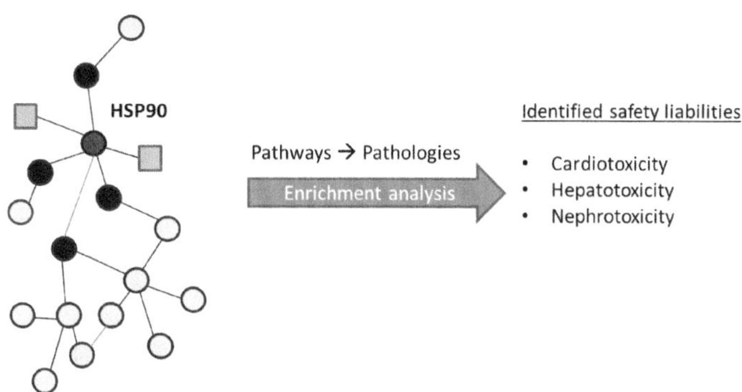

FIGURE 2.4.4 Schematic representation of biological network analysis. Biological (dark blue nodes) and chemical (light blue nodes) entities directly interacting with the target (red node) are extracted from databases, integrated, and used as input for the enrichment analysis of toxicological pathways/processes in which these regulated entities play a role. Mathematical methods such as shortest path algorithms can be used to rank and present results of relevant entities and biological pathways. Top pathways are often ordered by confidence values such as *p*-value and can be translated to pathological outcomes associated with target modulation.

Enrichment analysis of HSP90-regulated proteins with IPA indicated that the heart (e.g., heart failure), liver (e.g., liver necrosis) and kidney (e.g., renal damage) are major organs of toxicity. Apart from ranking toxicological predictions, zooming-in on individual protein–protein interactions may reveal the mechanistic foundation of toxicological effects. For example, when inspecting the interaction partners of HSP90α, the human ether-a-go-go-related gene (hERG) surfaced as a key player in heart toxicity (Figure 2.4.5). hERG constitutes the pore-forming alpha subunit of a voltage-gated potassium channel expressed in the heart and in nervous tissue responsible for repolarization of cardiac action potential (Peterson et al., 2012). Disturbed function of hERG is associated with a prolonged QT interval and arrhythmia that may lead to death. hERG was found to preferentially immunoprecipitate with HSP90α and genetic knockdown of HSP90α, but not HSP90β, resulted in defective maturation and trafficking of hERG (Ficker, 2003; Peterson et al., 2012; Muth et al., 2014).

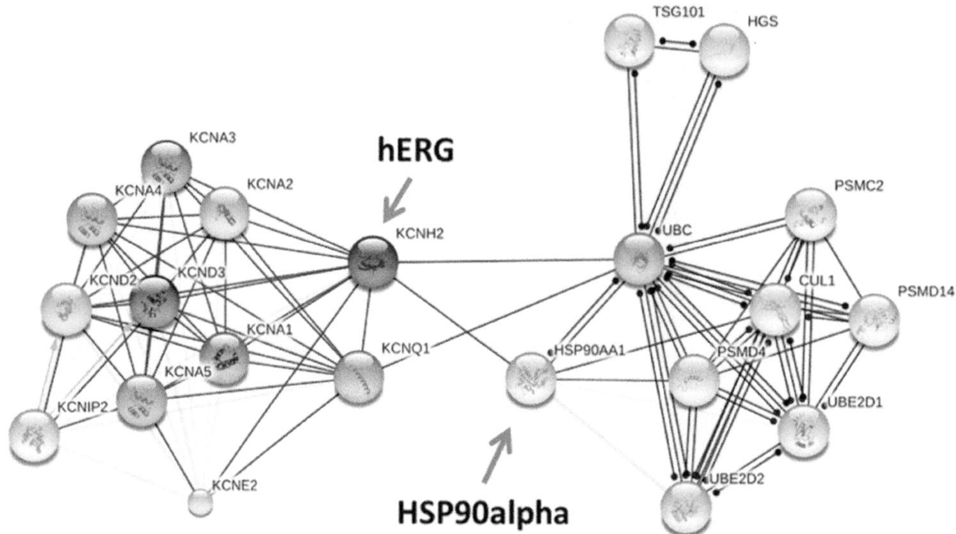

FIGURE 2.4.5 Protein–protein binding plot from StringDB illustrating the direct binding interaction between HSP90α (gene: HSP90AA1) and the cardiac potassium channel hERG (gene: KCNH2). hERG is a HSP90α client protein that is dependent on the chaperone for its maturation and functional trafficking.

Toxicological Evaluation (Pre)Clinical Studies

The retrieval of toxicological information from (pre)clinical studies described in literature constitutes an important part of a TSA. This information can be obtained with searches on competitor drugs and pharmacological tools extracted from databases, as described above. Alternatively, text-mining tools are increasingly applied to efficiently extract data based on specific ontologies, key entities and their relationships (Harpaz et al., 2014; Gonzalez et al., 2016). Apart from the generally used PubMed/Medline databases, other sources can be queried, such as conference proceedings, adverse event reports and clinical trials.

Within the TSA, toxicological information derived from (pre) clinical studies is structured based on the organ system involved and the model system used (*in vitro*/*in vivo*/clinical). Together with corresponding data on expression, biological function, and genetic phenotypes, a comprehensive view on organ-specific toxicity is obtained that allows risk ranking. Table 2.4.5 shows a consolidated view on the major toxicological observations with HSP90 inhibitors. A compound-oriented view may help in building hypotheses regarding the mechanisms of toxicity. The fact that the liver toxicity consistently displayed by the geldanamycin cluster (cluster 1) is absent in many other clusters alludes to this being a compound-based effect. Hepatotoxicity of geldanamycin and analogs has indeed been attributed to their metabolism, forming reactive oxygen species (Samuni et al., 2010). PK may, at least in part, explain adverse visual effects. The geldanamycin analog 17-DMAG accumulated in the retina and caused photoreceptor cell death. In contrast, the geldanamycin analog 17-AAG, showing rapid retinal elimination, did not display photoreceptor injury (Zhou et al., 2013). Thus, retinal toxicity—despite being deemed an on-target effect—was compound-specific as a result of PK properties and biodistribution, varying even within a compound cluster.

Toxicity Profile and Risk Ranking

The ultimate aim of a TSA is to identify those safety liabilities that need immediate monitoring and de-risking during drug discovery. All data collected in the TSA is therefore brought together at the end of the study, analyzed on the organ-system level and weighted for risk ranking. Weighting is performed by expert opinion based on severity of the effects, evidence level (e.g., *in vitro* vs. clinical observation), and observed frequency. Toxicities are then categorized as high, medium or low priority for further investigation.

While ranking, there are a few general considerations. Toxicological effects can be exacerbated in patients already suffering from preexisting disorders that affect the same organ system. Patient population and comorbidities may therefore influence the weighting process. For example, in the case of cancer, comorbidities include cardiovascular illness, obesity and metabolic illness, mental health problems, and musculoskeletal conditions (Sarfati et al., 2016). For the elderly in general, polypharmacy is of concern because it may lead to drug–drug or drug–disease interactions and concomitant adverse effects (Hassan, 2014). Declining renal function (Mühlberg and Platt, 1999) and an increase in the incidence of liver pathologies are also of general concern for the elderly patient. Finally, the extent to which significant adverse effects are perceived acceptable is dependent on the therapeutic indication and also affects risk weighting. In general, for life-threatening diseases, more

TABLE 2.4.5 Major toxicities observed with heat shock protein 90 (HSP90) inhibitor clusters (Figure 2.4.2) in (pre)clinical studies

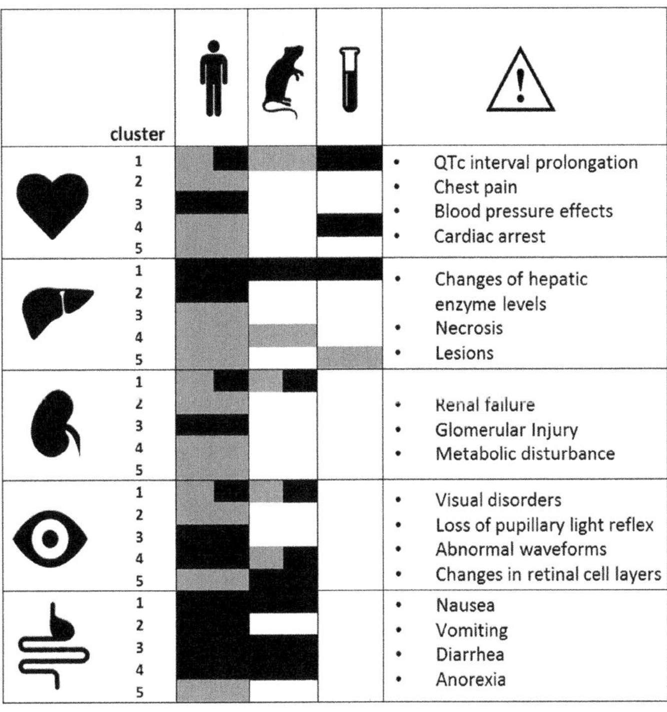

[a] *Note:* Gray: no toxicity found; black: toxicity reported; gray/black: mixed profile in compound cluster.

severe adverse effects may be acceptable. Thus, it is essential to place the TSA into context of the intended patient population (e.g., disease phenotype, patient characteristics, comorbidities, common co-medication) and medical need.

Based on the consolidated data and ranking criteria, cardio- and hepatotoxicity were classified as high priority risks for HSP90 inhibition, followed by retinal, gastrointestinal and renal toxicity. Off-target binding may contribute to cardio- and gastrointestinal toxicity. For example, GRP94 plays an essential role in gastrointestinal homeostasis (Liu et al., 2013), whereas TRAP1 fulfils a protective role in cardiomyocytes (Xiang et al., 2010). Of lower toxicological concern were, for example, central nervous system and immune effects because these were not commonly observed and less severe. It should be noted that some effects may only become evident after long-term treatment. It should also be noted that, with novel compounds being developed and existing ones progressing, the evidence and perception of toxicological risks associated with a drug target is likely to change with time. A TSA should therefore be regularly updated during the corporate lifetime of the drug target.

Risk Mitigation

The final part of the TSA is the design of a risk mitigation strategy, resulting in a toxicological screening funnel for candidate drugs. Nonclinical safety studies may encompass *in silico*, *in vitro*, *ex vivo*, and *in vivo* approaches or concern the development of (translational) *in vivo* safety biomarkers.

Standard toxicological screening cascades generally follow a tiered approach. A standard set of assays is run to test for commonly observed effects that can easily be measured (Dambach et al., 2015). These tests predominantly focus on organs that have historically given rise to the highest drug attrition, for example, the cardiovascular system and the liver (Dambach et al., 2015; Hornberg et al., 2014c). Examples of early safety end points are hERG and other ion channels binding (heart), cytotoxicity (liver) and genotoxicity (Escobar et al., 2013; Hornberg et al., 2014c). Off-target promiscuity is also routinely screened, both *in vitro* and *in silico* (Bowes, 2012; Schmidt et al., 2014). Early—mostly *in vitro*—safety testing is followed by more complex and specialized *in vitro* and *in vivo* models for advancing compounds. For example, compounds positive for hERG can be further inspected in additional cardiotoxicity assays, such as primary heart cells and tissue and immortalized mammalian cardiac cell lines (Clements et al., 2015). In addition to such standard screening strategies that may be considered generic across drug discovery programs, each individual program would encompass a selection of bespoke assays and models that focus on safety concerns that were raised in the TSA. Typical examples are the addition of safety end points to early target validation studies using genetic or pharmacological tools, the application of counter screens to rapidly de-select chemical equity that has intrinsic cross-reactivity to related off-targets, the integration of specific models in which compounds can be benchmarked against tools with a known clinical risk, and the conduction of investigative toxicology work in disease models.

In summary, the *in silico* TSA provides a holistic approach to the prediction of potential safety liabilities associated with the modulation of a drug target. Strategic choices can be made on whether to discontinue a target based on potential safety show-stoppers. For advancing targets and ligands, a strategy can be designed including dedicated screening cascades and studies to monitor and de-risk the toxicities designated high priority. The mechanistic insights gained, for example, by systems biology approaches may aid in identifying candidate safety biomarkers for highly specific toxicological effects.

REFERENCES

Altieri, D. C., G. S. Stein, J. B. Lian, and L. R. Languino. 2012. TRAP-1, the mitochondrial Hsp90. *Biochimica et Biophysica Acta (BBA)—Molecular Cell Research* 1823 (3): 767–773. doi:10.1016/j.bbamcr.2011.08.007.

Altschul, S. F., W. Gish, W. Miller, E. W. Myers, and D. J. Lipman. 1990. Basic local alignment search tool. *Journal of Molecular Biology* 215 (3): 403–410. doi:10.1016/S0022-2836(05)80360-2.

Antunes, D. A., D. Devaurs, and L. E. Kavraki. 2015. Understanding the challenges of protein flexibility in drug design. *Expert Opinion on Drug Discovery* 10 (12): 1301–1313. doi:10.1517/17460441.2015.1094458.

Arin, R. M., Y. Rueda, O. Casis, M. Gallego, A. I. Vallejo, and B. Ochoa. 2014. Basolateral expression of GRP94 in parietal cells of gastric mucosa. *Biochemistry (Moscow)* 79 (1): 8–15. doi:10.1134/S0006297914010027.

Arrowsmith, J. 2011a. Trial watch: Phase II failures: 2008–2010. *Nature Reviews Drug Discovery* 10 (5): 328–329.

Arrowsmith, J. 2011b. Trial watch: Phase III and submission failures: 2007–2010. *Nature Reviews Drug Discovery* 10 (2): 87–87.

Barelier, Sarah, Teague Sterling, Matthew J. O'Meara, and Brian K. Shoichet. 2015. The recognition of identical ligands by unrelated proteins. *ACS Chemical Biology* 10 (12): 2772–2784.

Barrott, J. J., and T. A. J. Haystead. 2013. Hsp90, an unlikely ally in the war on cancer. *FEBS Journal* 280 (6): 1381–1396. doi:10.1111/febs.12147.

Bento, A. P., A. Gaulton, A. Hersey, L. J. Bellis, J. Chambers, M. Davies, F. A. Krüger, et al. 2014. The ChEMBL bioactivity database: An update. *Nucleic Acids Research* 42 (D1): D1083–D1090. doi:10.1093/nar/gkt1031.

Blomme, E. A. G., and Y. Will. 2016. Toxicology strategies for drug discovery: Present and future. *Chemical Research in Toxicology* 29 (4): 473–504. doi:10.1021/acs.chemrestox.5b00407.

Bowes, J., A. J. Brown, J. Hamon, W. Jarolimek, A. Sridhar, G. Waldron, and S. Whitebread. 2012. Reducing safety-related drug attrition: The use of in vitro pharmacological profiling. *Nature Reviews Drug Discovery* 11 (12): 909–922. doi:10.1038/nrd3845.

Brinkman, R. R., M. P. Dube, G. A. Rouleau, A. C. Orr, and M. E. Samuels. 2006. Human monogenic disorders—A source of novel drug targets. *Nature Reviews Genetics* 7 (4): 249–260. doi:10.1038/nrg1828.

Caberlotto, L., and M. Lauria. 2015. Systems biology meets—Omic technologies: Novel approaches to biomarker discovery and companion diagnostic development. *Expert Review of Molecular Diagnostics* 15 (2): 255–265. doi:10.1586/14737159.2015.975214.

Cerami, E. G., B. E. Gross, E. Demir, I. Rodchenkov, O. Babur, N. Anwar, N. Schultz, G. D. Bader, and C. Sander. 2011. Pathway commons, a web resource for biological pathway data. *Nucleic Acids Research 39 (Database):* D685–D690. doi:10.1093/nar/gkq1039.

Chiosis, G., C. A Dickey, and J. L Johnson. 2013. A global view of Hsp90 functions. *Nature Structural and Molecular Biology* 20 (1): 1–4. doi:10.1038/nsmb.2481.

Clements, M., V. Millar, A. S. Williams, and S. Kalinka. 2015. Bridging functional and structural cardiotoxicity assays using human embryonic stem cell-derived cardiomyocytes for a more comprehensive risk assessment. *Toxicological Sciences* 148 (1): 241–260. doi:10.1093/toxsci/kfv180.

Cook, D., D. Brown, R. Alexander, R. March, P. Morgan, G. Satterthwaite, and M. N. Pangalos. 2014. Lessons learned from the fate of AstraZeneca's drug pipeline: A five-dimensional framework. *Nature Reviews Drug Discovery* 13 (6): 419–431.

Cusick, M. E. 2005. Interactome: Gateway into systems biology. *Human Molecular Genetics* 14 (suppl 2): R171–81. doi:10.1093/hmg/ddi335.

Dambach, D. M., D. Misner, M. Brock, A. Fullerton, W. Proctor, J. Maher, D. Lee, K. Ford, and D. Diaz. 2016. Safety lead optimization and candidate identification: Integrating new technologies into decision-making. *Chemical Research in Toxicology* 29 (4): 452–472. doi:10.1021/acs.chemrestox.5b00396.

Davis, A. P., C. J. Grondin, R. J. Johnson, D. Sciaky, B. L. King, R. McMorran, J. Wiegers, T. C. Wiegers, and C. J. Mattingly. 2016. The comparative toxicogenomics database: Update 2017. *Nucleic Acids Research*, 45(D):D972–D978. doi:10.1093/nar/gkw838.

DiMasi, J. A., H. G. Grabowski, and R. W. Hansen. 2016. Innovation in the pharmaceutical industry: New estimates of R&D costs. *Journal of Health Economics* 47:20–33. doi:10.1016/j.jhealeco.2016.01.012.

Donnelly, A., and B. S. J. Blagg. 2008. Novobiocin and additional inhibitors of the Hsp90 C-terminal nucleotide-binding pocket. *Current Medicinal Chemistry* 15 (26): 2702–2717.

Eppig, J. T., J. A. Blake, C. J. Bult, J. A. Kadin, J. E. Richardson, and The Mouse Genome Database Group. 2015. The Mouse Genome Database (MGD): Facilitating mouse as a model for human biology and disease. *Nucleic Acids Research* 43 (D1): D726–D736. doi:10.1093/nar/gku967.

Ernst, J. T., M. Liu, H. Zuccola, T. Neubert, K. Beaumont, A. Turnbull, A. Kallel, B. Vought, and D. Stamos. 2014. Correlation between chemotype-dependent binding conformations of HSP90α/β and isoform selectivity—Implications for the structure-based design of HSP90α/β selective inhibitors for treating neurodegenerative diseases. *Bioorganic and Medicinal Chemistry Letters* 24 (1): 204–208. doi:10.1016/j.bmcl.2013.11.036.

Ernst, J. T., T. Neubert, M. Liu, S. Sperry, H. Zuccola, A. Turnbull, B. Fleck et al. 2014. Identification of novel HSP90α/β isoform selective inhibitors using structure-based drug design. Demonstration of potential utility in treating CNS disorders such as Huntington's disease. *Journal of Medicinal Chemistry* 57 (8): 3382–3400. doi:10.1021/jm500042s.

Escobar, P.A., R.A. Kemper, J. Tarca, J. Nicolette, M. Kenyon, S. Glowienke, S.G. Sawant et al. 2013. Bacterial mutagenicity screening in the pharmaceutical industry. *Mutation Research/Reviews in Mutation Research* 752 (2): 99–118. doi:10.1016/j.mrrev.2012.12.002.

Eskew, J. D., T. Sadikot, P. Morales, A. Duren, I. Dunwiddie, M. Swink, X. Zhang et al. 2011. Development and characterization of a novel C-Terminal inhibitor of Hsp90 in androgen dependent and independent prostate cancer cells. *BMC Cancer* 11: 468.

Ficker, E. 2003. Role of the cytosolic chaperones Hsp70 and Hsp90 in maturation of the cardiac potassium channel hERG. *Circulation Research* 92 (12): e87–e100. doi:10.1161/01.RES.0000079028.31393.15.

Frezal, J. 1998. Genatlas database, genes and development defects. *Comptes Rendus de l'Academie Des Sciences. Serie III, Sciences de La Vie* 321 (10): 805–817.

García-Campos, M. A., J. Espinal-Enríquez, and E. Hernández-Lemus. 2015. Pathway analysis: State of the art. *Frontiers in Physiology* 6: 383. doi:10.3389/fphys.2015.00383.

Gonzalez, G. H., T. Tahsin, B. C. Goodale, A. C. Greene, and C. S. Greene. 2016. Recent advances and emerging applications in text and data mining for biomedical discovery. *Briefings in Bioinformatics* 17 (1): 33–42. doi:10.1093/bib/bbv087.

Harpaz, R., A. Callahan, S. Tamang, Y. Low, D. Odgers, S. Finlayson, K. Jung, P. LePendu, and N. H. Shah. 2014. Text mining for adverse drug events: The promise, challenges, and state of the art. *Drug Safety* 37 (10): 777–790. doi:10.1007/s40264-014-0218-z.

Hassan, M., R. D. Brown, S. Varma-O'Brien, and D. Rogers. 2006. Cheminformatics analysis and learning in a data pipelining environment. *Molecular Diversity* 10 (3): 283–299. doi:10.1007/s11030-006-9041-5.

Hay, M., D. W. Thomas, J. L. Craighead, C. Economides, and J. Rosenthal. 2014. Clinical development success rates for investigational drugs. *Nature Biotechnology* 32 (1): 40–51. doi:10.1038/nbt.2786.

Himeji, D., T. Horiuchi, H. Tsukamoto, K. Hayashi, T. Watanabe, and M. Harada. 2002. Characterization of caspase-8L: A novel isoform of caspase-8 that behaves as an inhibitor of the caspase cascade. *Blood* 99 (11): 4070–4078.

Hornberg, J. J., and T. Mow. 2014b. How can we discover safer drugs? *Future Medicinal Chemistry* 6 (5): 481–483. doi:10.4155/fmc.14.15.

Hornberg, J. J., M. Laursen, N. Brenden, M. Persson, A. V. Thougaard, D. B. Toft, and T. Mow. 2014a. Exploratory toxicology as an integrated part of drug discovery. Part I: Why and how. *Drug Discovery Today* 19 (8): 1131–1136. doi:10.1016/j.drudis.2013.12.008.

Hornberg, J. J., M. Laursen, N. Brenden, M. Persson, A. V. Thougaard, D. B. Toft, and T. Mow. 2014c. Exploratory toxicology as an integrated part of drug discovery. Part II: Screening strategies. *Drug Discovery Today* 19 (8): 1137–1144. doi:10.1016/j.drudis.2013.12.009.

Huang, D. W., B. T. Sherman, and R. A. Lempicki. 2009. Bioinformatics enrichment tools: Paths toward the comprehensive functional analysis of large gene lists. *Nucleic Acids Research* 37 (1): 1–13. doi:10.1093/nar/gkn923.

Im, C. 2016. Past, present, and emerging roles of mitochondrial heat shock protein TRAP1 in the metabolism and regulation of cancer stem cells. *Cell Stress and Chaperones* 21 (4): 553–562. doi:10.1007/s12192-016-0687-3.

Immormino, R. M., L. E. Metzger, P. N. Reardon, D. E. Dollins, B. S.J. Blagg, and D. T. Gewirth. 2009. Different poses for ligand and chaperone in inhibitor-bound Hsp90 and GRP94: Implications for paralog-specific drug design. *Journal of Molecular Biology* 388 (5): 1033–1042. doi:10.1016/j.jmb.2009.03.071.

Ishihara, T., S. Suemasu, T. Asano, K. Tanaka, and T. Mizushima. 2011. Stimulation of gastric ulcer healing by heat shock protein 70. *Biochemical Pharmacology* 82 (7): 728–736. doi:10.1016/j.bcp.2011.06.030.

Jensen, L. J., M. Kuhn, M. Stark, S. Chaffron, C. Creevey, J. Muller, T. Doerks et al. 2009. STRING 8—A global view on proteins and their functional interactions in 630 organisms. *Nucleic Acids Research* 37: D412–D416. doi:10.1093/nar/gkn760.

Johnson, J. L. 2012. Evolution and function of diverse Hsp90 homologs and cochaperone proteins. *Biochimica et Biophysica Acta (BBA)—Molecular Cell Research* 1823 (3): 607–613. doi:10.1016/j.bbamcr.2011.09.020.

Kanehisa, M., and S. Goto. 2000. KEGG: Kyoto Encyclopedia of Genes and Genomes. *Nucleic Acids Research* 28 (1): 27–30.

Kola, I., and J. Landis. 2004. Can the pharmaceutical industry reduce attrition rates? *Nature Reviews Drug Discovery* 3 (8): 711–716.

Lee, J. A., and J. R. Lupski. 2006. Genomic rearrangements and gene copy-number alterations as a cause of nervous system disorders. *Neuron* 52 (1): 103–121. doi:10.1016/j.neuron.2006.09.027.

Li, H.D., R. Menon, G. S. Omenn, and Y. Guan. 2014. The emerging era of genomic data integration for analyzing splice isoform function. *Trends in Genetics* 30 (8): 340–347. doi:10.1016/j.tig.2014.05.005.

Li, J., J. Soroka, and J. Buchner. 2012. The Hsp90 chaperone machinery: Conformational dynamics and regulation by co-chaperones. *Biochimica et Biophysica Acta (BBA)—Molecular Cell Research* 1823 (3): 624–635. doi:10.1016/j.bbamcr.2011.09.003.

Liljevald, M., M. Rehnberg, M. Söderberg, M. Ramnegård, J. Börjesson, D. Luciani, N. Krutrök et al. 2016. Retinoid-Related Orphan Receptor γ (RORγ) adult induced knockout mice develop lymphoblastic lymphoma. *Autoimmunity Reviews* 15 (11): 1062–1070. doi:10.1016/j.autrev.2016.07.036.

Lin, J. H. 2008. Applications and limitations of genetically modified mouse models in drug discovery and development. *Current Drug Metabolism* 9 (5): 419–438.

Liu, B., M. Staron, F. Hong, B. X. Wu, S. Sun, C. Morales, C. E. Crosson et al. 2013. Essential roles of grp94 in gut homeostasis via chaperoning canonical wnt pathway. *Proceedings of the National Academy of Sciences* 110 (17): 6877–6882. doi:10.1073/pnas.1302933110.

Marcu, M. G., A. Chadli, I. Bouhouche, M. Catelli, and L. M. Neckers. 2000. The heat shock protein 90 antagonist novobiocin interacts with a previously unrecognized ATP-Binding domain in the carboxyl terminus of the chaperone. *Journal of Biological Chemistry* 275 (47): 37181–37186. doi:10.1074/jbc. M003701200.

Miyata, Y., H. Nakamoto, and L. Neckers. 2013. The therapeutic target Hsp90 and cancer hallmarks. *Current Pharmaceutical Design* 19 (3): 347–365.

Morando, M. A., G. Saladino, N. D'Amelio, E. Pucheta-Martinez, S. Lovera, M. Lelli, B. López-Méndez, M. Marenchino, R. Campos-Olivas, and F. Luigi Gervasio. 2016. Conformational selection and induced fit mechanisms in the binding of an anticancer drug to the c-Src kinase. *Scientific Reports* 6: 24439. doi:10.1038/srep24439.

Mühlberg, W., and D. Platt. 1999. Age-dependent changes of the kidneys: Pharmacological implications. *Gerontology* 45 (5): 243–253.

Muth, A., V. Crowley, A. Khandelwal, S. Mishra, J. Zhao, J. Hall, and B. S. J. Blagg. 2014. Development of radamide analogs as Grp94 inhibitors. *Bioorganic and Medicinal Chemistry* 22 (15): 4083–4098. doi:10.1016/j.bmc.2014.05.075.

NCBI Resource Coordinators. 2016. Database resources of the national center for biotechnology information. *Nucleic Acids Research* 44 (D1): D7–D19. doi:10.1093/nar/gkv1290.

Neckers, L., and P. Workman. 2012. Hsp90 molecular chaperone inhibitors: Are we there yet? *Clinical Cancer Research* 18 (1): 64–76. doi:10.1158/1078-0432.CCR-11-1000.

Ou, J. R., M. S. Tan, A. M. Xie, J. T. Yu, and L. Tan. 2014. Heat shock protein 90 in Alzheimer's disease. *BioMed Research International* 2014: 1–7. doi:10.1155/2014/796869.

Patel, H., X. Lucas, I. Bendik, S. Günther, and I. Merfort. 2015. Target fishing by cross-docking to explain polypharmacological effects. *ChemMedChem* 10 (7): 1209–1217. doi:10.1002/cmdc.201500123.

Peterson, L. B., J. D. Eskew, G. A. Vielhauer, and B. S. J. Blagg. 2012. The hERG channel is dependent upon the Hsp90α isoform for maturation and trafficking. *Molecular Pharmaceutics* 9 (6): 1841–1846. doi:10.1021/mp300138n.

Plenge, R. M. 2016. Disciplined approach to drug discovery and early development. *Science Translational Medicine* 8 (349): 349ps15–349ps15.

Prince, T. L., T. Kijima, M. Tatokoro, S. Lee, S. Tsutsumi, K. Yim, C. Rivas et al. 2015. Client proteins and small molecule inhibitors display distinct binding preferences for constitutive and stress-induced HSP90 Isoforms and their conformationally restricted mutants. *PLoS One* 10 (10): e0141786. doi:10.1371/journal.pone.0141786.

Proks, P. 2013. Neonatal diabetes caused by activating mutations in the sulphonylurea receptor. *Diabetes and Metabolism Journal* 37 (3): 157. doi:10.4093/dmj.2013.37.3.157.

Reis, S. D., B. R. Pinho, and J. M. A. Oliveira. 2016. Modulation of molecular chaperones in Huntington's disease and other polyglutamine disorders. *Molecular Neurobiology* 54 (8) 5829–5854. doi:10.1007/s12035-016-0120-z.

Roberts, R. A. 2018. Understanding drug targets: No such thing as bad news. *Drug Discovery Today* 23 (12) 1925–1928. doi:10.1016/j.drudis.2018.05.028.

Rudmann, D. G. 2013. On-target and off-target-based toxicologic effects. *Toxicologic Pathology* 41 (2): 310–314. doi:10.1177/0192623312464311.

Samuni, Y., H. Ishii, F. Hyodo, U. Samuni, M. C. Krishna, S. Goldstein, and J. B. Mitchell. 2010. Reactive oxygen species mediate hepatotoxicity induced by the Hsp90 inhibitor geldanamycin and its analogs. *Free Radical Biology and Medicine* 48 (11): 1559–1563. doi:10.1016/j.freeradbiomed.2010.03.001.

Sarfati, D., B. Koczwara, and C. Jackson. 2016. The impact of comorbidity on cancer and its treatment. *CA: A Cancer Journal for Clinicians* 66 (4): 337–350. doi:10.3322/caac.21342.

Schmidt, F., H. Matter, G. Hessler, and A. Czich. 2014. Predictive *in silico* off-target profiling in drug discovery. *Future Medicinal Chemistry* 6 (3): 295–317. doi:10.4155/fmc.13.202.

Shelton, S. N., M. E. Shawgo, S. B. Matthews, Y. Lu, A. C. Donnelly, K. Szabla, M. Tanol, et al. 2009. KU135, a novel novobiocin-derived C-terminal inhibitor of the 90-kDa heat shock protein, exerts potent antiproliferative effects in human leukemic cells. *Molecular Pharmacology* 76 (6): 1314–1322. doi:10.1124/mol.109.058545.

Spyrakis, F., A. BidonChanal, X. Barril, and F. Javier Luque. 2011. Protein flexibility and ligand recognition: Challenges for molecular modeling. *Current Topics in Medicinal Chemistry* 11 (2): 192–210.

Taherian, A., P. H. Krone, and N. Ovsenek. 2008. A comparison of Hsp90alpha and Hsp90beta interactions with cochaperones and substrates. *Biochemistry and Cell Biology = Biochimie et Biologie Cellulaire* 86 (1): 37–45. doi:10.1139/o07-154.

Thierry-Mieg, D., and J. Thierry-Mieg. 2006. AceView: A comprehensive cDNA-Supported gene and transcripts annotation. *Genome Biology* 7 (1): 1.

Uhlén, M., L. Fagerberg, B. M. Hallström, C. Lindskog, P. Oksvold, A. Mardinoglu, Å. Sivertsson et al. 2015. Tissue-Based map of the human proteome. *Science* 347 (6220). 1260419. doi:10.1126/science.1260419.

UniProt Consortium. 2015. UniProt: A hub for protein information. *Nucleic Acids Research* 43 (D1): D204–D212. doi:10.1093/nar/gku989.

Villa-Vialaneix, N., L. Liaubet, T. Laurent, P. Cherel, A. Gamot, and M. San Cristobal. 2013. The structure of a gene co-expression network reveals biological functions underlying eQTLs. Edited by Marinus F.W. te Pas. *PLoS One* 8 (4): e60045. doi:10.1371/journal.pone.0060045.

Voss, A. K., T. Thomas, and P. Gruss. 2000. Mice lacking HSP90beta fail to develop a placental labyrinth. *Development* 127 (1): 1–11.

Wanderling, S., B. B. Simen, O. Ostrovsky, N. T. Ahmed, S. M. Vogen, T. Gidalevitz, and Y. Argon. 2007. GRP94 is essential for mesoderm induction and muscle development because it regulates insulin-like growth factor secretion. *Molecular Biology of the Cell* 18 (10): 3764–3775.

Wang, H., M. S. Tan, R. C. Lu, J. T. Yu, and L. Tan. 2014. Heat shock proteins at the crossroads between cancer and Alzheimer's disease. *BioMed Research International* 2014: 1–9. doi:10.1155/2014/239164.

Waring, M. J., J. Arrowsmith, A. R. Leach, P. D. Leeson, S. Mandrell, R. M. Owen, G. Pairaudeau et al. 2015. An analysis of the attrition of drug candidates from four major pharmaceutical companies. *Nature Reviews Drug Discovery* 14 (7): 475–486. doi:10.1038/nrd4609.

Wishart, D. S. 2006. DrugBank: A comprehensive resource for in silico drug discovery and exploration. *Nucleic Acids Research* 34 (90001): D668–D672. doi:10.1093/nar/gkj067.

Wu, Bill X., F. Hong, Y. Zhang, E. Ansa-Addo, and Z. Li. 2016. GRP94/gp96 in cancer. In *Advances in Cancer Research*, 129:165–190. Amsterdam the Netherlands: Elsevier. http://linkinghub.elsevier.com/retrieve/pii/S0065230x15000937.

Xiang, F., Y. S. Huang, X. H. Shi, and Q. Zhang. 2010. Mitochondrial chaperone tumour necrosis factor receptor-associated protein 1 protects cardiomyocytes from hypoxic injury by regulating mitochondrial permeability transition pore opening: TRAP1 protects cells from hypoxic injury by MPTP. *FEBS Journal* 277 (8): 1929–1938. doi:10.1111/j.1742-4658.2010.07615.x.

Yim, K. H., T. L. Prince, S. Qu, F. Bai, P. A. Jennings, J. N. Onuchic, E. A. Theodorakis, and L. Neckers. 2016. Gambogic acid identifies an isoform-specific druggable pocket in the middle domain of Hsp90beta. *Proceedings of the National Academy of Sciences of the United States of America* 113 (33): E4801–E4809. doi:10.1073/pnas.1606655113.

Zhou, D., Y. Liu, J. Ye, W. Ying, L S. Ogawa, T. Inoue, N. Tatsuta et al. 2013. A rat retinal damage model predicts for potential clinical visual disturbances induced by Hsp90 inhibitors. *Toxicology and Applied Pharmacology* 273 (2): 401–409. doi:10.1016/j.taap.2013.09.018.

Unwanted Immunogenicity Responses to Biotherapeutics

2.5

Shobha Purushothama and Boris Gorovits

Contents

INTRODUCTION

All biotherapeutics (proteins and peptides) have the potential to induce an unwanted immune response resulting in variable clinical impact—from no effect to potentially life threatening. Reporting of unwanted immunogenicity is one of the key elements in the regulatory submissions for product approval. The current regulatory expectation is that immunogenicity is assessed using a risk-based approach with a goal of mitigating safety consequences. Immunogenicity is typically assessed as anti-drug antibody (ADA) responses and detected/characterized in an ADA assay. In this chapter, we discuss the factors that must be considered in conducting the risk assessment as well as some of the considerations for assay design and validation and conclude with the need to understand the impact of ADA on clinical safety and efficacy.

IMMUNOGENICITY RISK ASSESSMENT—A REGULATORY EXPECTATION

Biotherapeutics have the inherent potential to be immunogenic and the nature of the immune response could be complex [1–3]. This potential to raise an unwanted immune response has consequences that range from no effect [4,5] to devastating [6]. There is a regulatory expectation that immunogenicity is assessed using a risk-based approach with an emphasis on understanding of the potential safety consequences of the immune response [1,3,7] when designing the assessment strategy. Hence, understanding the immunogenicity of a biologic from a safety perspective has become a critical aspect of regulatory filings [8]. The immune responses typically result in production of drug-specific antibodies (described as ADAs). Based on the specificity and binding affinity of the ADAs, these may have various impacts on biotherapeutic exposure (i.e., PK), pharmacological activity (efficacy) or may in some cases lead to serious clinical sequelae (safety, such as hypersensitivity reactions, anaphylaxis or immune complex disease). A careful risk-based evaluation of immunogenicity potential is the basis for clinical study relevant anti-drug antibody response monitoring.

What Is Immunogenicity Risk?

Immunogenicity risk assessment is a balanced evaluation of the ability of the biologic to induce an immune response and the consequence of said immune response. Hence, this assessment should be done on a case-by-case basis. There are several factors to be considered in the conduct of such an assessment: target, mechanism of action (MOA), published data, posttranslation modification, expression systems, formulation, disease indication, patient population, co-medication, and clinical trial design [8,9]. These factors are summarized in Table 2.5.1.

TABLE 2.5.1 Factors impacting immunogenicity risk of the molecule

TARGET RELATED	*BIOTHERAPEUTIC RELATED*	*PATIENT/STUDY DESIGN RELATED*
Location—soluble or membrane bound	Mechanism of action—agonist/ antagonist	HLA type, immune status
Expressed on dendritic cells or other professional antigen presenting cells	Protein sequence/ structure—size, sequence homology to endogenous proteins, contains neo epitopes, repeat structures	Disease status (life threatening), preexisting antibodies, prior exposure to a similar class of biologic
Internalized on binding to drug	Modifications for half-life extension (e.g., pegylation, addition of albumin binding domains) and posttranslational modifications (e.g., glycosylation, deamidation, oxidation)	Age (adult vs. pediatric), sex, ethnicity
Expression level/ turnover	Host cell proteins, aggregates and other impurities	Route of administration, frequency of administration, dose, drug holiday and re-exposure
Endogenous counterpart—sequence homology	Formulation, administration devices, leachables Sole therapy	Concomitant medication, pre-medication tapering

The immunogenicity risk potential for a biotherapeutic should be considered throughout the product life cycle. The early phase of evaluation may start at the point of identification of the biotherapeutics' molecular target and may include considerations of the modality type, protein primary sequence and possible homology to an endogenous protein. Other product-specific characteristics often listed as impactful for the immunogenicity potential are possible production cell line protein contaminants (e.g., Chinese Hamster Ovary cell proteins) and protein aggregates, small molecular contaminants originating from the container or its components [10].

Patient-related factors include prior medical history, prior treatment with biologic of a similar class and concomitant medications leading to enhanced or abrogated immune response potential. Treatment details such as frequency of administration, dose level, duration of the treatment, drug formulation and route of administration are also highly critical. Attempts have been made to predict immunogenicity potential based on biotherapeutic sequence and ability to induce T cell and other components of immune system. These remain in development.

Preexisting Anti-drug Antibodies

ADAs have been broadly reported for various types of biotherapeutics. A topic of current interest given the multitude of biotherapeutics in development is the impact of preexisting antibodies. It was proposed that preexisting antibodies may be a cause of concern because they could increase safety- and efficacy-related risks; however, a recent review indicated that preformed antibodies generally have little impact [11]. Typically, preexisting drug-reactive antibodies are defined as immunoglobulins reactive with protein or glycan domains on the biotherapeutic compound that are present in the matrix of patients who have no prior exposure to the biotherapeutic [11,12]. Evidence of preexisting drug-binding antibodies is commonly explained by prior exposure to proteins, glycans or other compounds with similar structural characteristics. In addition, exposure to xenobiotic environmental antigens or homologous biologics has also been implicated. Potential clinical sequelae of preexisting antibodies may vary greatly, are highly dependent on the biotherapeutic modality, the nature of the disease, patient history and demographics [13]. Preexisting antibodies may or may not be boosted after treatment with the biotherapeutic. Overall, the exact nature of the potential impact of preexisting antibody on the biotherapeutic varies and is challenging to predict [12].

RISK CATEGORIZATION

There is no one prescribed method for risk determination. As alluded to above, the risk determination should take into account the varied factors that could potentially cause an immune response and the consequences of such a response. As a result, this assessment is of necessity multidisciplinary in nature. Typically, molecules are classified into "low," "medium" or "high" risk categories [8] based on the severity of the clinical sequelae of the immune response. However, there is no consistency in the published literature on how these risk categories translate to a testing strategy [9]. There is also considerable latitude (perhaps deliberate) in the regulatory guidance [1,3]. This lack of clarity, particularly for biologics where the induction of an immune response rarely has serious safety consequences, has translated into a one-size-fits-all approach where testing is defined based on high-risk molecules. An alternate approach based on the philosophy of anticipated clinical consequences of ADA categorizes the molecules into two classes: category 1 (no severe ADA mediated clinical sequelae) and category 2 (severe ADA mediated clinical sequelae) [9].

Risk Assessment-Driven Testing Strategy

The risk assessment serves to drive the testing strategy that is appropriate to the development stage of the biologic. Hence, assessment and categorization should be done continuously through the biologic life cycle to ensure that any potential impact of ADA on safety and efficacy is reduced, mitigated and monitored. It is well understood and generally accepted that in nonclinical species, the immune response is not predictive of that expected in humans [14]. The purpose of assessing ADA in nonclinical studies is to help with the interpretation of exposure/response (i.e., PK/PD) data. Hence, routine ADA testing in nonclinical studies is not warranted (regardless of risk). Even when testing is not planned prospectively, it is recommended that samples be collected to allow for retrospective testing [15]. It is recommended to test for ADA when there is an expectation that the consequences of ADA seen in the preclinical studies may translate into the clinic, for example, with immune complex–mediated reactions [1,3,9]. In the clinic, a tiered approach to immunogenicity testing is most commonly practiced and expected by the health authorities irrespective of the immunogenicity risk [1]. In terms of assessing and understanding the immune response, in the early clinical development phase (phase I/II), it may be prudent to collect samples for ADA assessment more frequently regardless of risk [8,9]. For the low-to-medium risk (category 1) molecules, less frequent sampling may be considered at the later stage (phase III). For the high-risk (category 2) molecules, frequent sampling should be considered across all development stages [8,9]. Other elements of the testing strategy that could be informed by the risk assessment is the extent and timing of characterization of the immune response (e.g., neutralizing antibody, isotyping, epitope mapping). In the tiered approach, the current expectation [1] is to have a neutralizing antibody assay in place for pivotal studies. However, this is a point of ongoing discussion and debate, particularly for biotherapeutics where the expected impact is limited to a reduced exposure. In addition, given the drawbacks of these assays such as poor sensitivity and drug tolerance [16], a standalone neutralizing antibody assay may not provide additional information. Instead, neutralizing activity may be inferred by the integration of PK/PD data sets [9].

ANTI-DRUG ANTIBODY DETECTION

Monitoring for ADA development continues to be the main method for the evaluation of biotherapeutic's ability to induce immune response in patients. Assessment of a cellular (i.e., T-cell) response is also possible and may be warranted in some cases, for example, for viral gene therapy and cellular modalities [17].

Assessment of immunogenicity potential is typically expected for both nonclinical and clinical studies. However, the goals, value and conclusions made are very different. Nonclinical immune responses are typically expected as the majority of modern-day biotherapeutics have human or highly humanized primary sequences. This makes a typical biotherapeutic foreign by nature in nonhuman species, frequently resulting in a high percent of ADA-positive animals, often resulting in high titer responses. Because nonclinical immune responses are not expected to be predictive of human immunogenicity, nonclinical evaluations focusing on the exposure (PK) impact assessed using a free PK assay are considered adequate. In some cases, toxicity observations linked to the presence of ADA may suggest possible safety outcomes, should a similar degree of immunogenicity be observed in the clinic.

Immune responses to biotherapeutics in clinical studies are important to understand in the assessment of compound safety and efficacy and are reflected in the product label. For example, in the instance of Krystexxa® (pegloticase) see at drugs at FDA, https://www.accessdata.fda.gov/drugsatfda_docs/label/2012/125293s034lbl.pdf a higher incidence of infusion reactions in patients with high anti-pegloticase antibody titer was reported. In instances where there has been an ADA impact on exposure and the ability to induce various types of adverse reactions, it has resulted in prolonged compound development or termination. Detection, characterization and evaluation of the impact of ADAs in the clinic are therefore key and an integral component of a majority of clinical investigations in the event of safety events.

Anti-Drug Antibody Assay Considerations

Biotherapeutics may induce a variety of binding ADAs, including differences in immunoglobulin isotype (IgM, IgG, IgE and others) as well as varying by the specificity of binding. Hence, ADA responses are polyclonal; including an array of antibody isotypes as well as specificity to multiple epitopes on the therapeutic molecule. Current regulatory guidelines require detection of the entire spectrum of responses. Most common ADA assays are semiquantitative, reporting sample titer value or, in rare cases, mass unit–based value based on the positive control (PC) dilution curve interpolation.

It is critical to understand that a PC sample will not truly represent the collection of isotypes and epitope specificities found in an incurred study sample and, therefore, should be viewed as a system suitability control only. Most typically, PC material is generated by hyperimmunizing nonclinical species such as rabbits or mouse to generate polyclonal or monoclonal drug-specific antibody. The resulting material often contains high-affinity immunoglobulins, frequently of the IgG isotype. PC material is frequently purified by using either Protein A or a ligand-affinity step making it even more homogenous in nature. Such PC material can therefore be very different from a polyclonal ADA mix expected in an ADA-positive sample, particularly early after the drug exposure, where the majority of response is of the IgM subclass. The PC material should therefore not be viewed as assay reference material. In cases where mass unit is reported, values should be viewed as semiquantitative at best. Several assessments based on the PC performance are requested and applied to determine assay quality, including assay drug tolerance and sensitivity. These will be discussed later in the chapter. The semiquantitative nature of reported ADA values makes it inappropriate to compare results across various biotherapeutics or even between studies for a given compound when analysis is done at different labs or using different methods [18].

Commonly, a tier-based assessment of ADA is conducted to include initial screen test, followed by confirmatory and characterization tier. These are sometimes referred to as Tier 1, Tier 2 and Tier 3. The most common ADA characterization assessment focuses on the evaluation of ADA ability to neutralize biotherapeutic activity. Such anti-drug immunoglobulins are referred to as neutralizing antibodies (NAb). Other ADA specificity characterizations such as understanding domain specificity may be relevant, particularly in cases of multi-domain biotherapeutics.

As with PK assays used during support of clinical studies, ADA and NAb assays are expected to undergo development, qualification and validation phases. Assay validation is particularly important for regulated nonclinical (good laboratory practice [GLP]) and clinical study support activities. Methodologies applied during ADA and NAb assay development, qualification and validation have been described in great detail in several industry white papers and regulatory guideline documents [1,3,19–21]. The initial screening test (Tier 1) identifies putatively positive samples, which are confirmed for specificity against the biotherapeutic in the confirmatory test (Tier 2). The most common approach to determine specificity of ADA binding to the biotherapeutic is based on the ability of the unlabeled biotherapeutic to inhibit an assay signal by competing with assay reagents. In both screen and confirmatory tests, an assay-specific cut point value is applied to determine whether the signal (or percentage of inhibition in the confirmatory test) is statistically significantly different from the noise level. The cut point values therefore establish background noise level and are defined by analysis of a treatment-naïve study-relevant population of samples. Generated data sets are analyzed to determine an appropriate confidence interval, for example, 95% in the case of clinical screen cut point and 99% for confirmatory cut point. Produced values are compared with the performance of the assay negative control to generate cut point factors, which are applied during the assay production phase. Other important details relevant to the cut point definition can be found in industry white papers [21]. As a result, a sample is defined as putatively positive in the screen test if the sample produces a signal greater than or equal to the predefined screen assay cut point value. Putatively positive samples are then defined as specific to the biotherapeutic if, after pre-incubation with the specific drug, percentage signal inhibition is greater than or equal to the predefined confirmatory assay cut point value.

Neutralizing Antibody Assay Considerations

Similar concepts are applied during NAb assay development. Typically, NAb assessment is performed on ADA-confirmed (Tier 2) samples; hence, most NAb assays incorporate the screen and titer tiers. In addition, the NAb assay screening cut point is set to have a 1% or 0.1% false-positive rate because the goal here is to characterize the specific nature of the immune response.

It is important to point out that NAb assays are expected to reflect the biotherapeutic mode of action (MOA). Currently, cell-based assays (CBAs) are frequently viewed as functional and are often the go-to methods, particularly for molecules with agonistic MOAs and those that target cell-surface targets. In other cases where the compound's MOA is complex and includes multiple steps, more than one NAb assay may be required. However, CBAs suffer from poor sensitivity and drug tolerance, resulting in false-negative results even if NAbs are present. Hence, non–cell-based methodologies may be applicable, particularly when the MOA is based on binding to a humoral target or is antagonistic by nature [16].

ADA and NAb Assay Validation Considerations

The following assay parameters are commonly evaluated during the validation phase:

- Critical reagents identification and generation, including assay positive and negative controls
- Assay matrix identification
- Selection of an appropriate assay format and platform
- Assessment of assay performance at various conditions with the goal to obtain most appropriate assay characteristics, such as
    - Assay sensitivity
    - Assay drug tolerance
- Analysis of PC performance at various storage conditions (e.g., freeze/thaw or room temperature stability evaluations)
- Assay robustness and variability based on performance of PC and negative control materials

Assay sensitivity and drug tolerance parameters are evaluated using assay PC material. The assay sensitivity is defined as the PC concentration at the assay cut point value and often is an interpolated value based on the PC dilution curve profile.

The previous FDA draft guideline had called for a 250-ng/mL sensitivity, while a recently issued draft guideline is suggesting a 100 ng/mL value using the assay PC. It must be emphasized that the PC is not reflective of the sample for the reasons listed above. Assay drug tolerance is defined as the unlabeled drug concentration at which PC material spiked at a given concentration remains positive. For example, a 250-ng/mL PC material spiked with various unlabeled drug concentrations will result in a decrease in assay signal until the signal drops below the cut point. Concentration of the drug at which PC signal remains at or above the cut point is defined as the assay drug tolerance value. Importantly, assay drug tolerance is highly dependent on the binding properties of the PC material, for example, immunoglobulin isotype and affinity of interaction with the biotherapeutic. The assay drug tolerance, therefore, should be viewed as a relative value and should be an important consideration when evaluating different assay formats and platforms during method development.

CONSIDERATIONS IN INTERPRETING ADA DATA

Importantly, many of the methods employed in quantification of biotherapeutic in human matrices can be impacted by the presence of ADAs. For example, neutralizing ADAs will interfere with the ability of an anti-idiotype antibody assay to detect a given drug [22]. The apparent PK of a biotherapeutic, therefore, needs to be evaluated for the evidence of direct impact of ADAs on compound clearance versus ADA impact on the PK assay. Similarly, residual circulating biotherapeutic is typically expected to impact the ability of the assay to detect ADA, potentially leading to false-negative or inconclusive results due to the potential lack of drug tolerance of the assay. Evaluating the impact of ADA on exposure (efficacy) and safety should be a data-driven decision. For example, for low-risk molecules (category 1), in particular, an *in vivo* PD assay can provide a better assessment of neutralizing activity [9]. A need to evaluate PK and ADA as well as the PD data set in an integrated manner has become evident in recent years. This is particularly important because assays to detect ADA have become more sensitive, leading to a higher reported incidence of ADA [23]. In order to understand the clinical relevance (i.e., impact of ADA on efficacy and safety), the integrated approach is particularly helpful [24,25].

CONCLUSION

- A risk-based approach to immunogenicity testing is a regulatory expectation, should be conducted throughout the product life cycle, taking into account the target, MOA, production and formulation considerations, patient and disease considerations as well as posology and is, in effect, a multidisciplinary exercise. The goal of a risk assessment is to balance the probability of the generation of an immune response with the clinical sequelae of said immune response. The extent of testing at the different phases of drug development is driven by the risk category of the molecule.
- Immunogenicity testing is done in a tiered fashion: detection followed by characterization. There are several considerations to keep in mind during the development and validation of the methods to detect and characterize ADA. The key is to realize that these assays have no reference standard and are quasi-quantitative. In addition, one should be aware that the level of drug present in the sample may impact the ability of the method to accurately detect these ADA. Assessment is done in a tiered fashion using statistically set cut points.
    - Typically, ADA responses are characterized by understanding the neutralizing capability of the ADA. The choice of NAb assay format should be one that best reflects the MOA. Often times, CBAs are used to assess neutralizing activity. However, these are often associated with poor sensitivity and drug tolerance, possibly resulting in false-negative responses. In addition, CBAs may not be appropriate for drugs that do not impact signaling events.
    - Competitive ligand binding assays may be considered as an alternative for molecules that have an antagonistic MOA; however, these assays are not necessarily reflective of the *in vivo* MOA.
    - CBAs are often *ex vivo* modifications of PD assays. Hence, PD assays could be used as a surrogate to inform on neutralizing activity. For low risk molecules, exposure assessed with a free PK assay may be used as a surrogate for the presence of NAbs.
- ADA data should never be interpreted in isolation; rather an integrated approach to PK/PD/efficacy along with ADA data will provide a meaningful interpretation of clinical relevance.

REFERENCES

1. European Medicines Agency, Guideline on immunogenicity assessment of biotechnology-derived therapeutic proteins. Draft. 2016.
2. Rosenberg, A.S., Immunogenicity of biological therapeutics: A hierarchy of concerns. *Dev Biol (Basel)*, 2003. **112**: 15–21.
3. Guidance for Industry, Immunogenicity Assessment for Therapeutic Protein Products. U.S. Department of Health and Human Services. Food and Drug Administration. Center for Drug Evaluation and Research (CDER). Center for Biologics Evaluation and Research (CBER). August 2014.

4. Green, D., Spontaneous inhibitors to coagulation factors. *Clin Lab Haematol*, 2000. **22**: 21–25.
5. Schellekens, H. and N. Casadevall, Immunogenicity of recombinant human proteins: Causes and consequences. *J Neurol*, 2004. **251 Suppl 2**: ii4–ii9.
6. Schellekens, H., Immunogenicity of therapeutic proteins: Clinical implications and future prospects. *Clin Ther*, 2002. **24**(11): 1720–1740; discussion 1719.
7. Rosenberg, A.S. and A. Worobec, A risk-based approach to immunogenicity concerns of therapeutic protein products—Part 1—Considering consequences of the immune response to a protein. *Biopharm Int*, 2004. **17**: 22–26.
8. Shankar, G., C. Pendley, and K. Stein, A risk-based bioanalytical strategy for the assessment of antibody immune responses against biological drugs. *Nat Biotechnol*, 2007. **25**(5): 555–561.
9. Kloks, C., et al., A fit-for-purpose strategy for the risk-based immunogenicity testing of biotherapeutics: A European industry perspective. *J Immunol Methods*, 2015. **417**: 1–9.
10. Pollock, C., et al., Pure red cell aplasia induced by erythropoiesis-stimulating agents. *Clin J Am Soc Nephrol*, 2008. **3**(1): 193–199.
11. van Schie, K.A., G.-J. Wolbink, and T. Rispens, Cross-reactive and pre-existing antibodies to therapeutic antibodies—Effects on treatment and immunogenicity. *mAbs*, 2015. **7**(4): 662–671.
12. Gorovits, B., et al., Pre-existing antibody: Biotherapeutic modality-based review. *AAPS J*, 2016. **18**(2): 311–320.
13. Xue, L. and B. Rup, Evaluation of pre-existing antibody presence as a risk factor for posttreatment anti-drug antibody induction: Analysis of human clinical study data for multiple biotherapeutics. *AAPS J*, 2013. **15**(3): 893–896.
14. Bugelski, P.J. and G. Treacy, Predictive power of preclinical studies in animals for the immunogenicity of recombinant therapeutic proteins in humans. *Curr Opin Mol Ther*, 2004. **6**(1): 10–16.
15. ICH. International Conference on Harmonization (ICH) of Technical Requirements for Registrsation of Pharmaceutical for Human Use. Preclinical safety evaluation of Biotechnology Derived Pharmaceuticals S6(R1) 2011 [cited August 24, 2017]; Available from: http://www.ich.org/fileadmin/Public_Web_Site/ICH_Products/Guidelines/Safety/S6_R1/Step4/S6_R1_Guideline.pdf.
16. Wu, B., et al., Strategies to determine assay format for the assessment of neutralizing antibody responses to biotherapeutics. *AAPS J*, 2016. **18**(6): 1335–1350.
17. Mingozzi, F. and K.A. High, Immune responses to AAV vectors: overcoming barriers to successful gene therapy. *Blood*, 2013. **122**(1): 23–36.
18. Gunn, G.R., et al., From the bench to clinical practice: Understanding the challenges and uncertainties in immunogenicity testing for biopharmaceuticals. *Clin Exp Immunol*, 2016. **184**(2): 137–146.
19. Koren, E., et al., Recommendations on risk-based strategies for detection and characterization of antibodies against biotechnology products. *J Immunol Methods*, 2008. **333**(1–2): 1–9.
20. Guidance for the Industry. Assay development for Immunogenicity Testing of Therapeutic Proteins. Draft Guidance. U.S. Department of Health and Human Services. Food and Drug Administration. Center for Drug Evaluation and Research (CDER). Center for Biologics Evaluation and Research (CBER). December 2009.
21. Shankar, G., et al., Recommendations for the validation of immunoassays used for detection of host antibodies against biotechnology products. *J Pharm Biomed Anal*, 2008. **48**(5): 1267–1281.
22. Sailstad, J.M., et al., A white paper—Consensus and recommendations of a global harmonization team on assessing the impact of immunogenicity on pharmacokinetic measurements. *AAPS J*, 2014. **16**(3): 488–498.
23. Song, S., et al., Understanding the supersensitive anti-drug antibody assay: Unexpected high anti-drug antibody incidence and its clinical relevance. *J Immunol Res*, 2016. **2016**: 3072586.
24. Wang, Y.-M.C., et al., Evaluating and reporting the immunogenicity impacts for biological products—A clinical pharmacology perspective. *AAPS J*, 2016. **18**(2): 395–403.
25. Chirmule, N., V. Jawa, and B. Meibohm, Immunogenicity to therapeutic proteins: Impact on PK/PD and efficacy. *AAPS J*, 2012. **14**(2): 296–302.

Biomarker Use for Advanced Screening of Drug-Induced Kidney Injury

2.6

Martijn J. Wilmer and Frans G. M. Russel

Contents

INTRODUCTION

The kidney plays a pivotal role in the elimination of drugs from the human body. Glomerular filtration is the initial step in renal clearance, whereas active secretion in the renal tubular epithelium is especially important for many drugs that are protein bound. Because of its concentrative and excretory function, it is a frequent target of drug-induced kidney injury (DIKI). In 18%–27% of hospitalized patients, acute kidney injury (AKI) is linked to their medication [1–3]. Importantly, repeated AKI can lead to progressive chronic kidney disease or end-stage renal disease, which is associated with high mortality rates [4,5]. DIKI is consequently a serious burden for society and is frequently encountered by clinicians. Many currently used drugs are known nephrotoxicants for which monitoring renal function is common clinical practice. These include aminoglycosides (e.g., gentamycin, tobramycin), antiretrovirals (e.g., tenofovir), antineoplastics (e.g., cisplatin, methotrexate, ifosfamide) and immunosuppressants (e.g., cyclosporine A) [6]. Limiting the dose of a nephrotoxic drug could decrease the occurrence of DIKI, but that inevitably negatively affects its effectivity. Early detection of DIKI via highly sensitive and specific biomarkers is an important line of investigation to limit adverse effects. Although creatinine-based measurements are still common practice in monitoring renal function, early and specific biomarkers are required to protect the kidney from toxic events and irreversible loss of functionality. Recent developments have resulted in the discovery of more specific and sensitive biomarkers for renal toxicity that have the potential to reduce the occurrence of DIKI in combination with therapeutic drug monitoring.

If detection of such highly sensitive and specific biomarkers is compatible with predictive preclinical models, development of safer drugs without renal adverse events is an attractive perspective that could become reality. Safety programs consisting of *in vitro* screening tools with high predictivity for the clinical phases of drug development can benefit from specific biomarkers that allow translation of *in vitro* to *in vivo* findings. Although current models are inadequate, encouraging new developments emerge from functional human renal cells cultured in a physiological-relevant environment enabled by organ-on-a-chip technology [7].

In this chapter, we describe the anatomy of the kidney with an emphasis on the pathogenesis of DIKI. Further, we discuss recent advancements in sensitive and specific biomarker detection for renal toxicity, both in clinical as well as *in vitro* settings. This overview will be extended toward applications of biomarkers in next-level drug screening techniques in order to prevent DIKI for novel marketed compounds and to increase drug safety.

KIDNEY ANATOMY AND CAUSES FOR KIDNEY INJURY

Due to its unique anatomical structure and concentrative function, the kidneys are highly susceptible to the harmful effects of drugs. The mechanisms that contribute to renal drug exposure are closely related to the physiological processes that take place along the nephrons, that is, glomerular filtration, passive back diffusion, and transporter-mediated secretion and reabsorption. The kidneys receive 20%–25% of the cardiac output and are among the best-perfused organs in the body. This leads to a relatively high exposure of drugs present in the general circulation. Drugs can be further concentrated in the proximal tubules, which are equipped with an array of influx and efflux transporters that mediate active drug uptake from the blood across the basolateral membrane, followed by efflux across the apical membrane into urine. In addition, drugs can be reabsorbed from the urine by carrier processes, including receptor-mediated endocytosis (e.g., aminoglycosides). At the basolateral membrane, separate transporters are located for the influx of mainly hydrophilic, small-molecular-weight (MW<400–500 Da) anionic and cationic drugs. Because these transport systems have high uptake capacity and broad substrate specificity, many drugs tend to accumulate in the proximal tubules, sometimes causing harmful effects (e.g., tenofovir, methotrexate, Non-Steroidal Anti-Inflammatory Drugs (NSAIDs), cisplatin). During water reabsorption down the nephron, the tubular concentration of drugs that do not diffuse back to the circulation is gradually increased to potentially harmful levels. In particular, the loop of Henle and collecting duct are nephron segments at greater risk for nephrotoxicity (e.g., amphotericin B) because these cells are highly metabolically active and reside in a microenvironment with relatively low oxygen levels [8]. Although DIKI is most commonly associated with injury of the tubular compartments of the nephron, there is increasing evidence for the potential of glomerular lesions induced by drugs (e.g., bisphosphonates, nonsteroidal anti-inflammatory drugs, antiplatelet agents, antiangiogenesis drugs) [9]. DIKI is a complex condition because the pathophysiological mechanisms include changes in renal blood flow (i.e., increased vasoconstriction), tubular and glomerular dysfunction, metabolic (i.e., mitochondrial) alterations, cell death, and inflammation. The delicate balance between renal injury and tissue repair determines the ultimate outcome of a nephrotoxic insult.

BIOMARKERS FOR KIDNEY INJURY

The current basis for studying kidney function in the clinic is the biochemical analysis of metabolites like serum creatinine and blood urea nitrogen (BUN), which give an indication of a decrease in glomerular filtration rate (GFR) suggestive of renal failure. The GFR describes the flow rate of filtered fluid in the glomerular capillaries and is usually calculated by measuring serum creatinine and applying the modification of diet in renal disease (MDRD) formula [10]. However, creatinine and BUN levels only increase with a substantial delay, when more than 40% of renal function is already lost. Serum levels of the low-molecular-weight protein cystatin C are considered a more precise marker for renal injury as represented by GFR [11]. Because these serum GFR markers are not specific for the mechanism or location of kidney injury, there is a great need for the development of earlier biomarkers that are more sensitive and histologically defined to specific regions of the nephron. Such biomarkers could enable therapeutic interventions before renal function is significantly and, more importantly, irreversibly, decreased. Novel biomarkers representative of the pathophysiological mechanism at the injured tubular site are mostly found in urine, the number of which is rapidly increasing due to improved analytical methods (Table 2.6.1). Currently, the most promising urinary biomarkers for DIKI include neutrophil gelatinase-associated lipocalin (NGAL), kidney injury molecule-1 (KIM-1), interleukin 18 (IL-18), L-type fatty acid binding protein (L-FABP), angiotensinogen, tissue inhibitor of metalloproteinase-2 (TIMP-2), and insulin-like growth factor-binding protein 7 (IGFB-7). There is not yet available a specific marker for glomerular injury, which means that only by excluding damage to other nephron segments, can the origin of the injury be attributed to the glomerulus.

Biomarkers detected in urine can be categorized with respect to a specific nephron segment, whether they represent function or injury, and if expression is constitutive or induced (Figure 2.6.1).

TABLE 2.6.1 Biomarkers for drug-induced kidney injury

BIOMARKER	*AFFECTED NEPHRON SEGMENT*	*DETECTION METHOD*	*COMMENTS*	*REFERENCES*
BUN	Glomerulus	Enzymatic colorimetric assay	Kidney- and liver-related biomarker; chronic injury	[36]
Cystatin C	Glomerulus	Immune detection/ ELISA	Qualified kidney marker in rat; glomerular filtration rate	[17,37]
Total protein	Glomerulus	Enzymatic colorimetric assay	Qualified kidney marker in rat; functional compromise	[37]
Albumin	Glomerulus and/or tubules		Qualified kidney marker in rat; functional compromise	[37]
B2m	Glomerulus and/or tubules	Immune detection/ ELISA	Qualified kidney marker in rat; functional compromise	[17,37]
Serum creatinine	Glomerulus and/or tubules	Enzymatic colorimetric assay	Kidney-related biomarker; functional compromise	[36,38]
α-GST	Proximal tubules	Immune detection	Kidney-related biomarker; tubular injury	[39]

(Continued)

TABLE 2.6.1 (Continued) Biomarkers for drug-induced kidney injury

BIOMARKER	*AFFECTED NEPHRON SEGMENT*	*DETECTION METHOD*	*COMMENTS*	*REFERENCES*
GGT	Proximal tubules	Immune detection/ enzymatic colorimetric assay	Kidney-related biomarker; tubular injury	[39]
KIM-1	Proximal tubules	Immune detection/ ELISA/qPCR	Qualified kidney marker in rat; early/acute tubular toxicity	[17,36,37,40]
L-FABP	Proximal tubules	Immune detection/ ELISA	Kidney-related biomarker; functional compromise	[12]
NGAL	Proximal tubules	Immune detection/ ELISA	Kidney-related biomarker; functional compromise	[12,41]
NAG	Proximal tubules	Enzymatic colorimetric assay	Kidney-related biomarker; tubular injury	[36]
Calbindin	Distal tubules	Immunoassay; qPCR	Kidney-related biomarker; tubular injury	[42,43]
μ-GST	Distal tubules	Immune detection	Kidney related biomarker; tubular injury	[16,37]
RPA1	Collecting duct	Immune detection	Qualified kidney marker in rat; tubular injury	
Clusterin	Tubules	Immune detection/ ELISA	Qualified kidney marker in rat; tubular damage/regeneration	[17,37]
IL-18	Tubules	Immune detection/ ELISA	Kidney-related biomarker; acute tubular injury	[12,44]
Osteopontin	Tubules	Immunoassay; qPCR	Kidney-related biomarker; inflammation; regeneration	[43]
TFF3	Tubules	Immune detection/ ELISA	Qualified kidney marker in rat; tubular damage	[17,37]
TIMP-1	Tubules	Immunoassay; qPCR	Kidney-related biomarker; tubular injury	[42,43]
nrf-2	Nonspecific	qPCR	Genomic marker; stress response related; nonspecific	[40]
HIF-1α	Nonspecific	qPCR	Genomic marker; stress response related; nonspecific	[40]
ATF4	Nonspecific	qPCR	Genomic marker; stress response related; nonspecific	[40]
MTF1	Nonspecific	qPCR	Genomic marker; stress response related; nonspecific	[40]
HO-1	Nonspecific	qPCR	Genomic marker; stress response related; nonspecific	[40]
miR-21	Not yet determined	qPCR	Genomic marker; specificity not yet determined	[45]
miR-127	Not yet determined	qPCR	Genomic marker; ischemia/reperfusion	[46]
miR-155	Not yet determined	qPCR	Genomic marker; inflammation	[45]
miR-18	Not yet determined	qPCR	Genomic marker; specificity not yet determined	[45]
miR-29	Not yet determined	qPCR	Genomic marker; kidney fibrosis	[47]

Note: ATF4, activating transcription factor 4; B2m, β2-microglobin; BUN, blood urea nitrogen; GGT, gamma glutamyl transferase; HIF-1α, hypoxia-inducible factor 1-alpha; HO-1, heme oxygenase; IL, interleukin; KIM-1, kidney injury marker; L-FABP, liver-type fatty acid-binding protein; miR, microRNA; MTF1, metal-regulatory transcription factor; NAG, N-acetyl-β-D-glucosaminidase; NGAL, lipocalin-2; nrf-2, nuclear factor (erythroid-derived 2)-like 2; qPCR, qualitative polymerase chain reaction; RPA1, renal papillary antigen; TFF3, trefoil factor 3; TIMP-1, tissue inhibitor of metalloproteinase-1; α-GST, α-glutathione-*S*-transferase; μ-GST, μ-glutathione-*S*-transferase.

Impaired tubular reabsorption of low-molecular-weight proteins freely filtered in the glomerulus result in increased urine levels of biomarkers such as cystatin C or β2-microglobulin, indicating tubular dysfunction or overload of the uptake machinery. Alternatively, a toxic exposure can cause induction of biomarker synthesis in the kidney (e.g., KIM-1, clusterin, interleukins). Multiple sources are possible for lipocalin-2 (NGAL) and this constitutively expressed low-molecular-weight protein can be elevated in urine as a result of extrarenal tissue injury and incomplete megalin/cubilin-mediated endocytosis in the proximal

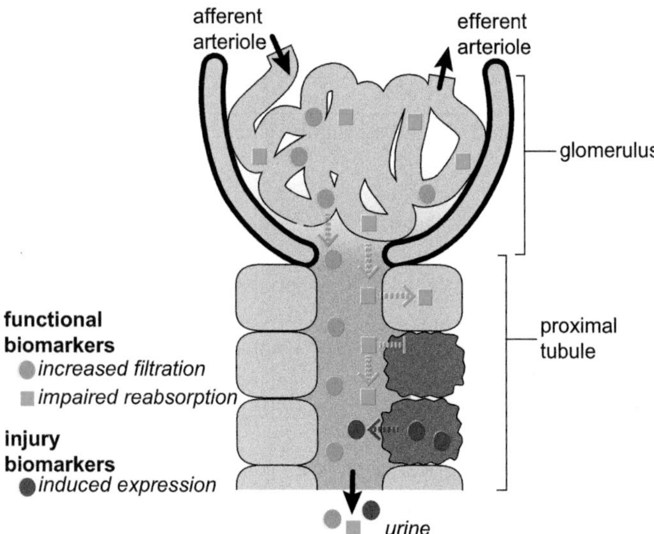

FIGURE 2.6.1 Urinary biomarkers for kidney injury. Kidney injury biomarkers observed in urine can originate from increased glomerular filtration (green circles) or they can have a tubular source. Impaired tubular reabsorption of biomarkers from the glomerular filtrate lead to increased urinary levels as a cause of tubular dysfunction (blue squares). In addition, drug-induced toxicity affecting tubular cells can stimulate cellular injury markers that end up in the urine (red circles).

tubule epithelium, while the distal nephron has the capacity to increase renal gene expression levels of NGAL [12,13].

Recent advances in analytical methodology in genomics, proteomics and metabolomics have led to an era of novel potential biomarkers with improved sensitivity and specificity that can be applied clinically. Analyses of metabolites and proteins in urine, serum or tissue provide insights into cellular and systemic processes that are induced by drugs [14]. The techniques in this field are rapidly evolving, including quantification of an array of metabolites, which will benefit a personalized medicine approach. Alternatively, the detection of microRNAs (miRNAs), a class of noncoding RNAs, is opening a plethora of novel potentially valuable cell type specific biomarkers to diagnose kidney injury at an early stage [6]. Because miRNAs are highly conserved among species, they seem particularly suitable for translational research. Recent studies report that miRNAs have a regulatory role in gene expression and signaling pathways, making them also a potential target for interventional approaches [15].

Although clinical biomarkers are mainly used to diagnose renal dysfunction, drug development will benefit from highly sensitive translational biomarkers that improve predictive capacity during preclinical studies [16]. In this perspective, species differences as well as intraindividual differences need further recognition and extensive research. In addition, the importance of reducing animal experiments has gained broad acceptance in academia and pharmaceutical industry in the last decades, but there is still a long way to go for drug safety research to develop reliable and validated advanced *in vitro* models that fulfill the 3Rs criteria.

PRECLINICAL SAFETY TESTING SYSTEMS

The newly developed biomarkers show promise for early *in vivo* detection of nephrotoxicity in animal models, but need further validation in clinical studies and have yet to be successfully applied for *in vitro* prediction of drug-induced nephrotoxicity. Animal experiments are the current gold standard for organ toxicity testing, despite the widespread awareness of the 3Rs and their limited predictive value due to species differences. For kidney-related toxicity, eight biomarkers in the rat have been qualified by regulatory agencies and are generally screened by pharmaceutical companies before approval of a new drug is obtained [17]. A paradigm shift toward *in vitro* models requires meticulous validation demonstrating improved drug safety in order to convince all stakeholders.

A reliable *in vitro* screening tool for DIKI is based on a physiologically relevant model that recapitulates the microenvironment of a functional nephron. Many widely used renal cell models are derived from human kidney 2 (HK-2), pork kidney (LLC-PK) or dog kidney (MDCK) and lack relevant (human) drug transporter expression [18]. Most likely, the optimal cell model for predictive drug toxicity screening are primary cells derived from the proximal tubule because this segment is the predominant target for nephrotoxicants and freshly isolated cells are probably closest to the *in vivo* situation [19]. To enable high-content screening, reproducibility is an important asset that can be accomplished via cell lines generated by immortalization techniques. Introducing human telomerase (hTERT) to primary proximal tubule cells, either with or without co-transfection with temperature sensitive simian virus (SV40T ts A58), was demonstrated to be an attractive alternative to primary cells [20,21]. In RPTEC/TERT1, a repeated compound dose screening was performed and transcriptomic analysis demonstrated increased expression of biomarkers IL-19 and NGAL, among others [22]. Using overexpression of OAT1 in conditionally immortalized proximal tubule epithelial cells (ciPTEC), decreased cell viability was observed upon antiviral cidofovir, tenofovir and adefovir exposure, which correlated with OAT1 function [7].

To circumvent dedifferentiation of isolated cells in culture and sparse availability of fresh renal cortical tissue, the field of regenerative medicine opens new avenues with the generation of human kidney cells from embryonic stem cells or from human induced pluripotent stem cells (hiPSC) [23,24]. Functional proximal tubule-like cells were developed from neonatal mouse kidney with an intact receptor-mediated endocytosis apparatus [25]. Using IL-6 and IL-8 as end points for renal toxicity, human proximal tubular-like cells derived from embryonic stem cells were used in a toxicity screen [26]. The predictive value of this study was high because the cells demonstrated expression of a range of drug transporters and proximal tubule characteristics. The combination of hiPSC-derived proximal tubule cells with machine-based learning techniques to evaluate a set of end-point measurements, even further increased the predictive capacity [27].

An additional feature of regenerative medicine is the differentiation of human pluripotent stem cell (hPSC) into 3D functional nephron organoids, with nephron structures equivalent to podocytes, proximal tubule cells, loop of Henle and distal tubule markers [28]. These self-organized kidney organoids were exposed to cisplatin and gentamycin, which resulted in enhanced KIM-1 expression. Although the organoids have a 3D structure, they lack a tubular luminal flow that induces fluid sheer stress. Emerging microfluidic technologies enable cultured renal cells that are exposed to fluid sheer stress, which was demonstrated to impact drug transport and cellular sensitivity to a toxic event [29]. Many microfluidic systems are currently available and generally called kidney-on-a-chip, with various characteristics in terms of controlled flow, materials, multi-compartmental organization and throughput [30]. In the context of drug screening, a certain throughput in a system manufactured from nonabsorptive material is pivotal. Complex systems with multiple tube connections in a polydimethylsiloxane (PDMS) chip are less suitable because concentrations to which the cells are exposed can be highly variable.

Together, these technological and biological advances provide a challenging perspective in drug development to improve preclinical prediction, decrease attrition during expensive clinical trials and reduce animal experiments. The missing link that should convince the stakeholders is a thorough validation of next-level systems, in which translation of *in vitro* to *in vivo* data via relevant biomarkers is crucial. Validation of *in vitro* models for preclinical safety studies would require a large set of compounds with multiple read-outs, ideally including a data set derived from *in vivo* or clinical data. Considering the complexity and multiple pathways involved, it seems likely that a panel of complimentary biomarkers will be necessary to reliably detect and eliminate potentially nephrotoxic compounds at an early stage in the drug development process. Recently, nine compounds were tested in human primary PTEC followed by a multiplex gene expression profile, demonstrating HO-1 as a potent biomarker that also has potential in translating rat renal toxicity studies to a clinical setting [19]. Interestingly, the researchers could translate their findings to a kidney-on-a-chip system in a 3D microfluidic structure. Advanced multicellular systems like a kidney-on-a-chip could bridge the gap with human clinical studies and provide a reliable basis for kidney toxicity marker and assay development.

FUTURE ADVANCEMENTS IN DRUG DISCOVERY

Translation from complex *in vitro* systems to the clinic opens the door for investigating the molecular mechanisms underlying DIKI via specific biomarkers that pinpoint to a certain pathologic process. Elucidation of affected intracellular pathways, often called adverse outcome pathways, can hold the key to decrease the incidence of DIKI by linking *in vitro* data with a patients' genetic background. Revolutions in genome sequencing during the last two decades have led to methods that allow genotyping of an individual, where targeted sequencing of a single gene was the initial step in finding functional mutations. The enormous amount of genetic information gathered in databases via whole genome or whole exome sequencing can be linked to population-based pharmacokinetics and could be used to predict individual drug responses prior to monitoring sensitive biomarker profiles. Significant associations between genotype and renal pharmacokinetic data have already been demonstrated, especially for cyclosporine A used as immunosuppressant in renal transplantation [31]. Renal elimination of cyclosporine A is mediated via P-glycoprotein at the apical membrane of proximal tubules. The TT genotype of the 3435C3T polymorphism in the ABCB1 gene of the donor kidney leads to low P-glycoprotein expression levels, which in turn increases the risk for cyclosporine A nephrotoxicity in the recipient. In renal transplantation literature, additional clinically relevant polymorphisms have been demonstrated that affect the clinical response of immunosuppressants. For example, an increased rejection risk was linked to polymorphisms in the metabolizing enzyme UGT1A9, causing a reduced mycophenolic acid exposure, the active metabolite of the pro-drug mycophenolate mofetil [32]. However, further implementation of pharmacogenetics is still modest because improved clinical outcomes have been only sparsely reported and therapeutic drug monitoring is rapid and efficient in many cases, such as monitoring mycophenolic acid levels [33]. To make personalized medicine a success, variations in the human genome that affect a drug response need more research [34]. Highly sensitive and specific biomarkers are the stepping-stone to linking genetic differences in renal transporters or metabolic enzymes with potential adverse effects impeding kidney function. Further developments in genomics and electronic patient records are expected to boost the predicted clinical implementation of personalized medicine. At the same time, the pharmaceutical industry can benefit from associations between genotype and pharmacokinetic data because this opens the opportunity to increase drug safety for subpopulations with a specific genotype, known to affect renal elimination of drugs. As discussed above, the use of proximal tubule-like cells derived from hiPSC isolated from donors with a different genetic background could aid in the implementation of personalized renal *in vitro* models in drug toxicity screening programs, provided that such advanced systems allow a certain throughput [27].

CONCLUSIVE REMARKS

The occurrence of AKI in hospitalized patients is often drug induced. Improved insights into the mechanism of DIKI can predict the variability in responses observed between patients with a different genotype. Detection of specific and sensitive biomarkers provide a stepping-stone to improve our understanding of these mechanisms and, moreover, can be used to stop potentially toxic therapeutic interventions before irreversible damage occurs and leads to chronic renal failure. Innovations in biomarker analytics during drug discovery and preclinical renal safety testing will be pivotal for validation and implementation of advanced early screening models with potentially improved predictive capacity.

Discovery of sensitive biomarkers that pinpoint the specific nephron segment affected by a toxic event will aid the development of safer drugs. Improved biomarker analysis can specifically accelerate the development of drugs designed for a subpopulation with a genetic background linked to increased risks for DIKI [35]. Altogether, recent developments in biomarker detection can ultimately limit drug-induced renal toxicity, improve drug safety and pave the way to a personalized medicine approach.

REFERENCES

1. Choudhury, D. and Z. Ahmed, Drug-induced nephrotoxicity. *Med Clin North Am*, 1997. **81**(3): 705–717.
2. Nash, K., A. Hafeez, and S. Hou, Hospital-acquired renal insufficiency. *Am J Kidney Dis*, 2002. **39**(5): 930–936.
3. Leape, L.L., et al., The nature of adverse events in hospitalized-patients—Results of the Harvard Medical-Practice Study-II. *N Engl J Med*, 1991. **324**(6): 377–384.
4. Chawla, L.S. and P.L. Kimmel, Acute kidney injury and chronic kidney disease: An integrated clinical syndrome. *Kidney Int*, 2012. **82**(5): 516–524.
5. Coca, S.G., S. Singanamala, and C.R. Parikh, Chronic kidney disease after acute kidney injury: A systematic review and meta-analysis. *Kidney Int*, 2012. **81**(5): 442–448.
6. Pavkovic, M. and V.S. Vaidya, MicroRNAs and drug-induced kidney injury. *Pharmacol Ther*, 2016. **163**: 48–57.
7. Nieskens, T.T.G. and M.J. Wilmer, Kidney-on-a-chip technology for renal proximal tubule tissue reconstruction. *Eur J Pharmacol*, 2016. **790**: 46–56.
8. Pazhayattil, G.S. and A.C. Shirali, Drug-induced impairment of renal function. *Int J Nephrol Renovasc*, 2014. **7**: 457–468.
9. Markowitz, G.S., A.S. Bomback, and M.A. Perazella, Drug-induced glomerular disease: Direct cellular injury. *CJASN*, 2015. 10(7): 1291–1299.
10. Levey, A.S., et al., A more accurate method to estimate glomerular filtration rate from serum creatinine: A new prediction equation. Modification of Diet in Renal Disease Study Group. *Ann Intern Med*, 1999. **130**(6): 461–470.
11. Ozer, J.S., et al., A panel of urinary biomarkers to monitor reversibility of renal injury and a serum marker with improved potential to assess renal function. *Nat Biotechnol*, 2010. **28**(5): 486–494.
12. Devarajan, P., Review: Neutrophil gelatinase-associated lipocalin: A troponin-like biomarker for human acute kidney injury. *Nephrology*, 2010. **15**(4): 419–428.
13. Charlton, J.R., D. Portilla, and M.D. Okusa, A basic science view of acute kidney injury biomarkers. *Nephrology, Dialysis, Transplantation: Official Publication of the European Dialysis and Transplant Association—European Renal Association*, 2014. **29**(7): 1301–1311.
14. Hocher, B. and J. Adamski, Metabolomics for clinical use and research in chronic kidney disease. *Nat Rev Nephrol*, 2017. **13**(5): 269–284.
15. Bhatt, K., M. Kato, and R. Natarajan, Mini-review: Emerging roles of microRNAs in the pathophysiology of renal diseases. *Am J Physiol Renal Physiol*, 2016. **310**(2): F109–F118.
16. Ennulat, D. and S. Adler, Recent successes in the identification, development, and qualification of translational biomarkers: The next generation of kidney injury biomarkers. *Toxicol Pathol*, 2015. **43**(1): 62–69.
17. Xie, H.G., et al., Qualified kidney biomarkers and their potential significance in drug safety evaluation and prediction. *Pharmacol Ther.*, 2013. **137**(1): 100–107.
18. Jenkinson, S.E., et al., The limitations of renal epithelial cell line HK-2 as a model of drug transporter expression and function in the proximal tubule. *Pflugers Arch*, 2012. **464**(6): 601–611.
19. Adler, M., et al., A quantitative approach to screen for nephrotoxic compounds in vitro. *J Am Soc Nephrol*, 2015. 27(4): 1015–1028.
20. Wieser, M., et al., hTERT alone immortalizes epithelial cells of renal proximal tubules without changing their functional characteristics. *Am J Physiol Renal Physiol*, 2008. **295**(5): F1365–F1375.
21. Wilmer, M.J., et al., A novel conditionally immortalized human proximal tubule cell line expressing functional influx and efflux transporters. *Cell Tissue Res.*, 2010. **339**(2): 449–457.
22. Aschauer, L., et al., Application of RPTEC/TERT1 cells for investigation of repeat dose nephrotoxicity: A transcriptomic study. *Toxicol In Vitro*, 2015. **30**(1 Pt A): 106–116.
23. Lam, A.Q., B.S. Freedman, and J.V. Bonventre, Directed differentiation of pluripotent stem cells to kidney cells. *Semin Nephrol*, 2014. **34**(4): 445–461.
24. Little, M.H., Generating kidney tissue from pluripotent stem cells. *Cell Death Dis*, 2016. **2**: 16053.
25. Ranghini, E., et al., Stem cells derived from neonatal mouse kidney generate functional proximal tubule-like cells and integrate into developing nephrons in vitro. *PloS One*, 2013. **8**(5): e62953.
26. Li, Y. et al., Identification of nephrotoxic compounds with embryonic stem-cell-derived human renal proximal tubular-like cells. *Mol Pharm*, 2014. **11**(7): 1982–1990.
27. Kandasamy, K. et al., Prediction of drug-induced nephrotoxicity and injury mechanisms with human induced pluripotent stem cell-derived cells and machine learning methods. *Sci Rep.*, 2015. **5**: 12337.
28. Morizane, R. et al., Nephron organoids derived from human pluripotent stem cells model kidney development and injury. *Nat Biotechnol*, 2015. **33**(11): 1193–1200.
29. Jang, K.J., et al., Human kidney proximal tubule-on-a-chip for drug transport and nephrotoxicity assessment. *Integr Biol (Camb)*, 2013. **5**(9): 1119–1129.
30. Wilmer, M.J., et al., Kidney-on-a-chip technology for drug-incuced nephrotoxiciy screening. *Trends Biotechnol*, 2016. **34**(2): 156–170.
31. Hauser, I.A. et al., ABCB1 genotype of the donor but not of the recipient is a major risk factor for cyclosporine-related nephrotoxicity after renal transplantation. *J Am Soc Nephrol*, 2005. **16**(5): 1501–1511.
32. van Schaik, R.H.N., et al., UGT1A9 -275T>A/-2152C>T polymorphisms correlate with low MPA exposure and acute rejection in MMF/tacrolimus-treated kidney transplant patients. *Clin Pharmacol Ther.*, 2009. **86**(3): 319–327.
33. van Gelder, T., R.H. van Schaik, and D.A. Hesselink, Pharmacogenetics and immunosuppressive drugs in solid organ transplantation. *Nat Rev Nephrol.*, 2014. **10**(12): 725–731.
34. Ingelman-Sundberg, M., Personalized medicine into the next generation. *J. Intern Med.*, 2015. **277**(2): 152–154.
35. Sim, S.C. and M. Ingelman-Sundberg, Pharmacogenomic biomarkers: New tools in current and future drug therapy. *Trends Pharmacol Sci*, 2011. **32**(2): 72–81.
36. Vaidya, V.S., et al., Kidney injury molecule-1 outperforms traditional biomarkers of kidney injury in preclinical biomarker qualification studies. *Nat Biotechnol*, 2010. **28**(5): 478–485.
37. Dieterle, F., et al., Renal biomarker qualification submission: A dialog between the FDA-EMEA and predictive safety testing consortium. *Nat Biotechnol*, 2010. **28**(5): 455–462.

38. Kasiske, B.L.K., W.F., Laboratory assessment of renal disease: Clearance, urinalysis, and renal biopsy, in The Kidney, B.M. Brenner, Editor. 2000, WB Saunders Company: Philadelphia, PA. 1129–1170.
39. Westhuyzen, J., et al., Measurement of tubular enzymuria facilitates early detection of acute renal impairment in the intensive care unit. *Nephrol Dial Transplant*, 2003. **18**(3): 543–551.
40. Jennings, P., et al., An overview of transcriptional regulation in response to toxicological insult. *Arch Toxicol*, 2013. **87**(1): 49–72.
41. Hvidberg, V., et al., The endocytic receptor megalin binds the iron transporting neutrophil-gelatinase-associated lipocalin with high affinity and mediates its cellular uptake. *FEBS Lett*, 2005. **579**(3): 773–777.
42. Sohn, S.J., et al., In vitro evaluation of biomarkers for cisplatin-induced nephrotoxicity using HK-2 human kidney epithelial cells. *Toxicol Lett*, 2013. **217**(3): 235–242.
43. Hoffmann, D., et al., Evaluation of a urinary kidney biomarker panel in rat models of acute and subchronic nephrotoxicity. *Toxicology*, 2010. **277**(1–3): 49–58.
44. Ferguson, M.A., et al., Urinary liver-type fatty acid-binding protein predicts adverse outcomes in acute kidney injury. *Kidney Int*, 2010. **77**(8): 708–714.
45. Saikumar, J., et al., Expression, circulation, and excretion profile of microRNA-21, -155, and -18a following acute kidney injury. *Toxicol Sci*, 2012. **129**(2): 256–267.
46. Aguado-Fraile, E., et al., miR-127 protects proximal tubule cells against ischemia/reperfusion: Identification of kinesin family member 3B as miR-127 target. *PLoS One*, 2012. **7**(9): e44305.
47. Wang, B., et al., Suppression of microRNA-29 expression by TGF-beta1 promotes collagen expression and renal fibrosis. *J Am Soc Nephrol*, 2012. **23**(2): 252–265.

Soluble Biomarkers for Drug-Induced Cardiotoxicity 2.7

Amy Pointon and Nick Edmunds

Contents

INTRODUCTION

Cardiotoxicity can be a major challenge for the safe development of novel drugs, and there are significant gaps in our translational understanding of preclinical findings to clinical and disease settings. Findings of this type can halt the development of potentially exciting medications completely or restrict their potential clinical impact by capping maximum dose levels or lead to patient withdrawal. Drug-induced cardiotoxicity can affect all components of the cardiovascular system and can be manifest as reversible or irreversible changes in function (acute alteration of the mechanical and or electrical function of the myocardium), structure (morphological damage to cardiomyocytes and/or loss of viability, changes in cell signaling and intracellular organelles) and/or metabolism (Cross et al., 2015; Laverty et al., 2011). Moreover, due to the intricate balance between these critical elements of the cardiovascular system, identification of initiating molecular interactions responsible for drug-induced cardiotoxicity can be very difficult. In this regard, a given cardiotoxic agent may modulate structure, function, or metabolism of the heart via a direct effect on one specific element that can subsequently lead to an indirect modulation of another element altogether. An example of this is the chronic elevation of heart rate caused by agents such as isoprenaline, which can result in the secondary development of changes in cardiac structural pathology with time (Arnold et al., 2008). Consequently, when assessing cardiotoxicity, a holistic view of the cardiovascular system and recognition of these interdependencies is necessary.

The molecular mechanisms responsible for drug-induced cardiotoxicity of many agents are yet to be fully elucidated, and this is particularly relevant for those drugs that cause structural damage to the myocardium. This is well illustrated by the current understanding of doxorubicin-induced cardiotoxicity. Doxorubicin is an anticancer agent used since the 1960s that results in a dose-dependent irreversible cardiac pathology clinically. Despite numerous years of research, multiple mechanisms are still proposed to be responsible for the observed cardiotoxicity (Minotti et al., 2004). These include the generation of reactive oxygen species (Arola et al., 2000), binding to cardiolipin (Oliveira et al., 2004; Wallace, 2003), changes in iron regulation (Minotti et al., 1999), DNA damage (Arola et al., 2000; L'Ecuyer et al., 2006), inhibition of topoisomerase 2b (Vejpongsa and Yeh, 2014; Zhang et al., 2012), alteration of transcription factors such as GATA-4 (Aries et al., 2004; Kim et al., 2003) and energy depletion (Wallace, 2003). Regardless of the precise initiating mechanism responsible, doxorubicin treatment can ultimately lead to apoptosis and/or necrosis in the myocardium, resulting in the clinical development of heart failure.

The lack of definitive molecular insight upstream of structural damage of many cardiotoxic agents limits our ability to develop strategies to monitor and detect these liabilities before the occurrence of gross morphological abnormalities. Unlike drug-induced functional cardiovascular changes that can be monitored directly (e.g., heart rate), structural damage cannot be easily detected in a noninvasive manner. Moreover, this inadequate mechanistic knowledge has made it very difficult to establish predictive *in vitro* strategies to identify structural cardiotoxicity of agents, and there is a large reliance on histopathology from rodent and non-rodent toxicology studies to identify this liability. Hence, efforts have focused on identification of soluble biomarkers of structural cardiotoxicity for both preclinical and clinical application. An ideal soluble biomarker of cardiotoxicity would have the following characteristics: (1) be specific to the heart, (2) have a high tissue-to-serum ratio, (3) have a low baseline expression relative to the change induced by toxicity, (4) be released prior to or immediately following insult with a sufficiently long half-life in serum to enable detection, (5) permit a robust assessment across species and (6) be predictive or reflective of cardiotoxicity (Wallace et al., 2004). Additionally, although not a requirement, understanding at a molecular level how a given biomarker relates to the drug-induced toxicity would be greatly advantageous.

ESTABLISHED CARDIAC BIOMARKERS, APPLICATION TO CARDIOTOXICITY

Many biomarkers of cardiovascular disease exist and are applied routinely in the clinic for disease diagnosis and have been explored for their potential to serve as biomarkers of drug-induced cardiotoxicity. These include those primarily associated with acute coronary syndrome, myocardial infarction and ischemia, for example, cardiac troponin (cTn), creatine-kinase muscle brain (CK-MB), fatty acid binding protein-3 (FABP3) and myoglobin (Collinson et al., 2014) as well as those associated with excessive stretching and cardiac hypertrophy, for example, natriuretic peptides (atrial natriuretic peptide [ANP] and B-type natriuretic peptide [BNP]) and galectin-3 (Gardner, 2003). Each of these biomarkers of cardiac disease have differing levels of acceptance and both sensitivity and specificity. However, the majority are reflective of the damage that is caused by the condition and largely lack discrimination to give any insight into the etiology at a subcellular level. Many of these biomarkers have been investigated for their potential as biomarkers of cardiotoxicity preclinically and clinically, and these will be described in the following sections.

cTn as a Cardiac Biomarker

The application of cTn to drug-induced cardiotoxicity has evolved from its clinical use as a gold-standard biomarker in the setting of acute coronary syndrome and myocardial infarction (Collinson et al., 2014). Tn is predominantly bound within the actin filaments in striated muscle, cardiac (i.e., cTn) and skeletal muscle (i.e., sTn), where it plays a central role in controlling calcium dependent actin–myosin cross-bridge cycling. The appearance in the circulation is indicative of cardiomyocyte injury via membrane rupture and loss of sarcolemma integrity (Park et al., 2017; Wallace et al., 2004). cTn consists of three subunits: cTnT, cTnC and cTnI. The three subunits together provide an integrated regulation of cardiac proteins controlling calcium-mediated interactions between actin and myosin filaments in cardiomyocytes. Specifically, cTnT binds to tropomyosin, cTnC is the calcium-responsive element facilitating cardiac contractions, and cTnI maintains the necessary troponin–tropomyosin complex (Figure 2.7.1) (Park et al., 2017; Sharma et al., 2004). At the cellular level, two pools of cTn are present; the majority of cTn is bound to structural proteins, but a small free pool is also present within the cytosol representing approximately 3.5% of the tool pool for cTnI and 6%–8% for cTnT in humans (Korff et al., 2006). At the organ level, cTnI is only expressed in the myocardium, and although cTnT has a low level of expression in skeletal muscle, it is considered relatively specific for the myocardium (Sharma et al., 2004). At the organism level, low circulatory levels of cTn in plasma are present, but it is rapidly released following injury irrespective of the etiology (Korff et al., 2006). Increases in both cTnT and cTnI correlate well with the severity of myocardial damage as determined by light microscopy in the rat, dog and nonhuman primate (Bertinchant et al., 2000; Bertsch et al., 1997; Bleuel et al., 1995; Christiansen et al., 2002; Herman et al., 1998, 1999; Nie et al., 2016; O'Brien et al., 2006; Reagan et al., 2017). These characteristics at the level of the cell, organ and organism make cTn a potential ideal biomarker for cardiotoxicity. Indeed, cTn is used in the clinical detection of cardiotoxicity, particularly in relation to anticancer agents such as doxorubicin and trastuzumab (Kitayama et al., 2017; Shafi et al., 2017). However, application within the preclinical setting of cardiotoxicity has in contrast been more protracted.

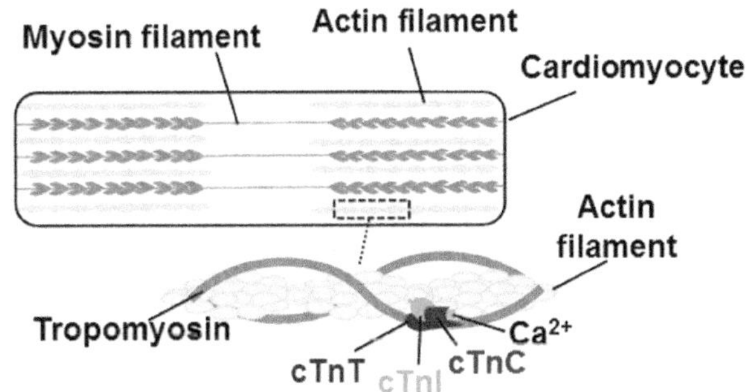

FIGURE 2.7.1 Schematic representation of the cardiac troponin (cTn) complex on the actin filament of cardiomyocytes.

Preclinical Use of cTn as a Biomarker of Drug-Induced Cardiotoxicity

Despite its accepted clinical use in the diagnosis of cardiovascular disease in 1998, it was not until 2012 that cTn was accepted as a biomarker for drug-induced cardiotoxicity. This work was first initiated in 2000 via the commission by the FDA of a working group to support the identification of predictive biomarkers (Pierson et al., 2013; Wallace et al., 2004). Nonetheless, it took a further 12 years until the use of cTn in the rat, mouse and dog was qualified as a biomarker by the FDA for the use in cardiotoxicity. One of the major reasons for this protracted qualification was the analytical difficulties of detecting cTn across multiple species. Specifically, important differences in assay reactivity and precision were identified resulting in the requirement of individual assay characterization by animal species to minimize misinterpretation of data (Apple et al., 2008; Pierson et al., 2013); this learning and the amount of time invested must be consisted when beginning to evaluate other biomarkers.

Evidence supporting application of cTn as a preclinical biomarker for cardiotoxicity is manifold, with the majority of mechanistic and validation studies using sympathomimetic agents (e.g., isoprenaline) and/or cardiotoxic chemotherapeutic agents (e.g., doxorubicin) as positive controls (Figure 2.7.2) (Clements et al., 2010; Herman et al., 1999). Specifically, following 50 mg/kg isoprenaline in the rat, serum cTn values reflected the development of histopathologic lesions, with peak cTn levels temporally preceding the maximal lesion severity (York et al., 2007). Complementary studies using lower doses of isoprenaline have reported similar findings (Brady et al., 2010; Zhang et al., 2008). Following 1 mg/kg per week doxorubicin for 2–12 weeks in a spontaneously hypertensive rat model, Herman et al. (1999) reported a correlation of $r = 0.92$ between cardiomyopathy scores and serum cTnT concentrations. In addition, a dose-response relationship between the cumulative dose of doxorubicin and serum cTnT was observed (Herman et al., 1999). In slight contrast, Reagan et al. (2013) reported that the magnitude of cTn absolute values did not always match directly the grade of histological change following 1, 2 or 3 mg/kg per week doxorubicin for 2 or 4 weeks in the rat; however, cTn release was still able to discriminate doxorubicin-induced injury as indicated by histological changes (vacuolation of cardiomyocytes). These apparent discrepancies could be partly attributed to the histopathological evaluations conducted or differences in cTn assay sensitives and timings at which plasma samples were taken (Clements et al., 2010; Reagan 2010). More recently, a correlation ($r = 0.81$) between maximal cTnT and end-diastolic left ventricular diameters was determined following doxorubicin and daunorubicin in the rat (Adamcova et al., 2007; Bertinchant et al., 2003). In support of this, a detailed evaluation of serological (including cTn), pathological and functional parameters following 1.25 mg/kg per week doxorubicin for 8 weeks in the rat again found a positive correlation with histopathological grading and diastolic dysfunction. However, subcellular cardiomyocyte degeneration was evident via electron microscopy and progressive functional decline

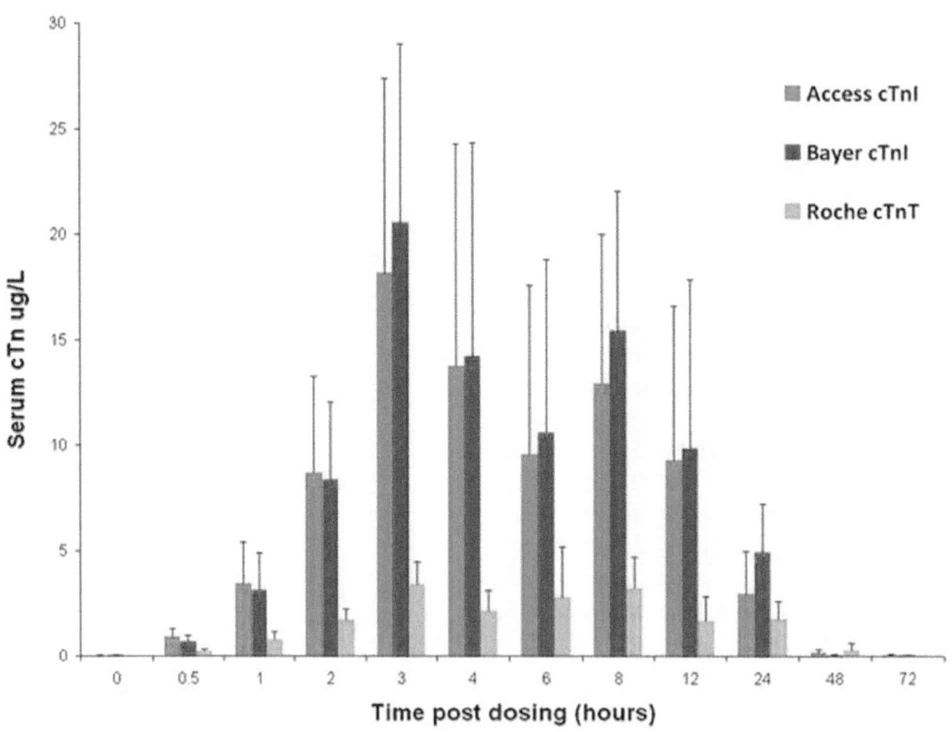

FIGURE 2.7.2 Serum cardiac troponin (cTn) levels in rats treated with single subcutaneous dose of isoprenaline (4,000 µg/kg) measured using three different cTn bioanalytical kits. Myocardial degeneration was apparent from 0.5 h post dosing which progressed to myocardial necrosis (minimal to moderate) after 6 h post dose. (Reproduced from Clements, P. et al., *Toxicol. Pathol.*, 38, 703–714, 2010.)

TABLE 2.7.1 Considerations on the use of cardiac troponin (cTn) in preclinical cardiotoxicity assessments

CONSIDERATIONS	*DETAILS*	*REFERENCE*
Selectivity: cTn is not released into the circulation during all forms of cardiotoxicity	Primarily restricted to the detection of cardiac necrosis, hence limiting the scope when assessed in isolation.	Reagan (2010)
Small changes in cTn without overt histopathology cannot be interpreted	The rational for small increases is currently unknown, but could represent cTn release from the cytosolic pool only. Blood sampling methodology and anesthesia are both known to transiently increase cTn release, further confounding data interpretation.	Berridge et al. (2009); Nagata et al. (2017); Verbiest et al. (2013); Walker (2006)
Relatively small amounts of cTn are released into the circulation following drug-induced damage	Drug-induced cardiac necrosis often effects small areas of the heart in contrast to the damage observed during myocardial infarction, presenting both interpretation and analytical challenges.	O'Brien (2008)
Lack of reference ranges in preclinical species	Presents difficulties when determining if a change in cTn is of biological consequence.	Berridge et al. (2009)
Specificity: occurrence of spontaneous cardiomyopathy in Sprague-Dawley rats	The occurrence of spontaneous cardiomyopathy in Sprague-Dawley rats, gives raise to false positive increases and when taken together with the lack of reference ranges makes interpretation of drug-induced cardiac damage difficult.	Chanut et al. (2013); O'Brien (2008); Friden et al. (2017); Nie et al. (2016)
Specificity: external stimuli impacting cTn levels	cTn increases can be observed with exercise, and therefore could be secondary to an effect on behavior. cTn is to some degree cleared renally, and changes in renal physiology and pathology could interfere with cTn measurements.	
Species differences in kinetics of cTn clearance	Clearance in both the rat and dog is relatively rapid with a half-life of around 1–2 h, this is significantly shorter than human and most notable cTn levels in both the rat and dog can return to baseline 24 h after an insult resulting in a myocardial pathology. This can result in a lack of cTn but a histopathology finding due to inappropriate timing of samples for cTn analysis.	Dunn et al. (2011)

(left ventricular ejection fraction) prior to an increase in cTn, suggesting that sensitivity of cTn as a biomarker may be limited under certain conditions (Cove-Smith et al., 2014).

These studies highlight the correlation of cTn release to not only histopathological changes but also to the ultimate decrease in function of the heart. Despite this, and the acceptance by the FDA of cTn as a marker of drug-induced cardiac damage, there remains some controversy regarding how useful cTn can truly be in a purely preclinical setting. Many agents that show cardiotoxicity preclinically do not have a path to clinical development and cTn is a biomarker of damage that has already happened or is in the process of happening and is, therefore, often used to support histopathological evaluation of the heart where damage is assessed directly. This lack of routine adoption is additionally influenced by a number of other considerations detailed in Table 2.7.1.

Taken together, we believe that these data provide a weight of evidence for the utility of cTn in the preclinical arena to detect myocyte injury and loss of sarcolemma integrity and assessment of cTn can provide useful supplementary evidence to support histopathology findings, however, limitations in application must be carefully considered. The true utility of cTn is that it can provide a synergetic translational biomarker applicable through both drug discovery and development, with its clinical application for therapies such as doxorubicin allowing the physician to titrate/cease administration if increases in cTn levels are observed. However, cTn is unable to provide any mechanistic insight to initiating events of a cardiotoxic agent nor does it give predictive insight into future cardiac damage that a given agent might cause. These evaluations also highlight the importance of fully characterizing the performance of a given biomarker regarding the concentration and time-course of exposure and how this relates to injury before fully defining its application. Additionally, they demonstrate limitations in cTn application and show the value of identifying of alternative biomarkers that could be predictive of future cardiac damage, being measurable prior to both gross histological changes and significant functional decline.

Natriuretic Peptides as Biomarkers of Drug-Induced Cardiotoxicity

The natriuretic peptides, ANP and BNP, and their respective amino-terminal fragments (NT-proANP and NT-proBNP) are cardiac hormones involved in plasma volume homeostasis, cardiovascular remodeling and counter regulation of the renin-angiotensin-aldosterone system (Kim et al., 2016). Both are synthesized and secreted from differing anatomical areas of the heart, ANP is generated within the atria, whereas BNP is generated in the ventricles. They are synthesized and stored as full-length prohormones and on secretion, are cleaved to yield equimolar amounts of N-terminal pro NP and the active NP hormone (Colton et al., 2011). These active fragments are secreted from the heart in response to volume expansion (cardiac hypertrophy), wall stretch, and pressure overload (Levin et al., 1998; Maeda et al., 1998) and bind NP receptors activating cGMP resulting in downstream decreases in blood volume and pressure (Colton et al., 2011; Levin et al., 1998). Consequently, they are indicators of cardiac homeostatic responses and dysfunction and not cardiac damage per se (Levin et al., 1998). They are rapidly gaining acceptance as the biomarker of choice for the diagnosis of congestive heart failure, septic shock and myocardial infarction (Jarai et al., 2009; Meyer et al., 2007; Mukoyama et al., 1991; Wallace et al., 2004). Specifically, BNP is now included in the European guidelines for the diagnosis of chronic heart failure (Remme and Swedberg, 2002). In addition, clinically plasma levels of BNP have correlated with the severity of left ventricular dysfunction and remodeling and served as a predictor of future cardiovascular events in asymptomatic patients (Kim et al., 2016; Morrow and Braunwald, 2003). Clinically recent meta-analysis studies have provided strong evidence for a correlation between serum BNP levels and drug-induced doxorubicin cardiotoxicity and, importantly, changes in BNP observed prior to any measurable changes in traditional echocardiography have been observed (D'Errico et al., 2015; Wang et al., 2016c). Based on this promising clinical application, like for cTn, work is underway to back-translate natriuretic-peptide biomarkers from the clinical to the preclinical situation; however, supporting data is limited. Positive correlations between plasma BNP and left ventricular function in the rat following doxorubicin have been reported (Koh et al., 2004); nevertheless, changes in cTn preceded BNP release. However, a study comparing physiological and drug-induced cardiac hypertrophy in rats identified increases in NT-proANP and NT-proBNP were only associated with drug-induced hypertrophy (Dunn et al., 2017). Further investigations of the utility of BNP and ANP are required before adoption to preclinical cardiotoxicity as a single biomarker, but the potential to detect changes in cardiac homeostatic responses and dysfunction prior to gross pathological damage raises the possibilities that prognostic biomarkers of cardiotoxicity could be identified.

Additional Soluble Cardiac Biomarkers

In comparison with cTn and BNP, other clinical cardiac biomarkers have not received the same level of interest as potential biomarkers of drug-induced cardiotoxicity. This lack of adoption can be to some extent be attributed to four main areas: (1) lack of tissue specificity, in particular the cross release from skeletal muscle as is the case for both CK-MB and myoglobin (Apple, 1999; Kanatous and Mammen, 2010); (2) interspecies variation in the basal tissue content as is the case for myoglobin (Wallace et al., 2004); (3) rapid tissue release kinetics and clearance resulting in limited window for detection as is the case of FABP3 (Clements et al., 2010); and (4) the relative lack of sensitivity to drug-induced changes. These characteristics are not ideal for a biomarker that can be applied across both drug discovery and development; however, the value as part of a panel of biomarkers should not be dismissed. Within *in vitro* models applied early in drug discovery some of these biomarkers that lack tissue specificity could potentially have utility due to the limited cell types present in these models (Kopljar et al., 2017). However, like cTn, all these biomarkers are reflective of cardiomyocyte damage rather than prognostic in value. Overall, improved biomarkers that are applicable for use both preclinically and clinically are required.

NEXT-GENERATION CARDIAC BIOMARKERS

Recent years have seen investment and interest in new and different forms of biomarkers that are distinct from more traditional soluble peptide and protein markers such as cTn. Although many of these still require validation as translational biomarkers of drug-induced cardiotoxicity, clinical and preclinical studies have investigated associations with cardiac function, dysfunction and disease.

MicroRNA

MicroRNAs (miRNAs) are defined as single stranded, noncoding RNA molecules of approximately 22 nucleotides in length. They were first described in *C. elegans* in 1993 and have subsequently been shown to be critical mediators of cellular function through messenger RNA silencing and posttranscriptional regulation of gene expression (Lee et al., 1993). Approximately 60% of messenger RNA is thought to be regulated by miRNA. Indeed, so effective are miRNA in controlling translational pathways that new-generation therapeutics have been developed that either mimic miRNA mechanisms of action, or target specific miRNA for cellular destruction (Chakraborty et al., 2017). miRNA can be released from cells either via active processes or during cellular injury and can be stabilized within extracellular vesicles (EVs) (Valadi et al., 2007) or by associations with RNA-binding proteins (Arroyo et al., 2011) and high-density lipoproteins (Vickers et al., 2011). This makes them attractive candidates as biomarkers of ongoing cardiovascular disease and drug-induced cardiotoxicity, with the potential to give a level of mechanistic understanding that is not available with traditional cardiac biomarkers such as cTn. As such, miRNA have been shown to control a variety of cellular pathological pathways in the heart, initiating cardiac remodeling and fibrosis, apoptosis, necrosis, autophagy and cardiogenesis (Wang et al., 2016a).

Although an in-depth review of the role of miRNA in cardiac function and disease is beyond the scope of this chapter, it is useful to consider how these regulators interact with a disease processes to understand the potential mechanistic insight they may give when employed as biomarkers of cardiotoxicity. Cardiac hypertrophy is an important physiological and pathological process for heart and is, therefore, a useful example. From a physiological perspective, hypertrophy enables the heart to adapt to situations of functional stress by increasing myocyte size and subsequent mechanical force. Such functional stress can occur as the result of normal physiological responses to exercise but can also result from pathological processes like hypertension and myocardial injury. Although initially protective, prolonged and exaggerated hypertrophy can result in heart failure characterized by fibrosis and increased cardiac stiffness (Frey and Olson, 2003). miRNAs play a coordinating role in the development and progression of hypertrophy, integrating both pro- and antihypertrophic pathways, with multiple miRNAs associated with each. miR-155, miR-199am miR-199b, miR-19, miR-208a, miR-21, miR-221 and miR-22 have all been associated with prohypertrophic pathways (Wang et al., 2016a). As an example, miR-21 targets phosphatase and tensin homolog (Pten) modulating the AKT/mTOR pathway to increase hypertrophic related gene expression (Adam et al., 2012). Moreover, miR-21 has been shown to target extracellular regulated kinase inhibitor sprout homolog 1 (Spryl) to induce mitogen-activated protein kinase (MAPK) signaling in cardiac fibroblasts (Thum et al., 2008). miR-208a and miR-208b are encoded within introns of α- and β-cardiac muscle myosin heavy chain genes (Myh6 and Myh7), and can signal the myosin heavy chain isoform switching that occurs during hypertrophy, by activating and repressing myofiber gene programs (van Rooij et al., 2009). Other miRNAs may be considered actively prohypertrophic by targeting and down-regulating antihypertrophic genes (miR-19) (Song et al., 2014) and impairing autophagy (miR-199a) (Li et al., 2017). Conversely, a number of other miRNAs are thought to protect against hypertrophy: miR-1, miR-101, miR-133, miR-145, miR-150, miR-185, miR-223, miR-26 and miR-9 (Wang et al., 2016a). For example, miR-1 has been shown to down-regulate target genes known to be important in hypertrophic progression, such as cyclin-dependent kinase 9, Ras GTPase-activating protein and fibronectin (Sayed et al., 2007). In a similar manner, miR-133 shows an inverse relationship with cardiac hypertrophy and overexpression of miR-133 can inhibit increases in cardiomyocyte size, and inhibition of miR-133 can evoke cardiomyocyte hypertrophy in the absence of any other stimulus (Carè et al., 2007). Mechanistically, miR-133 targets Rhoa and Cdc42, which are involved in cytoskeletal and myofibrillar rearrangements during hypertrophy.

Many of the above-mentioned miRNA are detectable in plasma, and therefore have the potential to serve as biomarkers for both early and late development of hypertrophy, giving both dynamic and mechanistic insight. However, an important consideration when reflecting on the suitability of these miRNAs as biomarkers for cardiac processes is how specific they are for the heart. For example, miR-21 is one of the most highly expressed miRNAs in the cardiovascular system and plasma levels have been associated with cardiovascular diseases such as coronary artery disease (Han et al., 2015) and heart failure (Zhang et al., 2017). However, miR-21 has also been extensively explored as a biomarker for diagnosis and progression of a variety of cancers (Peng et al., 2017), allergic inflammation (Sawant et al., 2015), hypertensive and diabetic kidney disease (Chen et al., 2017; Wang et al., 2016b) and diabetes (Zampetaki et al., 2010), demonstrating the potential ubiquitous stress–response miRNAs are involved in, in a number of diseases and cell types. Although this lack of selectivity may not completely negate the value of miR-21 as a biomarker of cardiovascular disease, it demonstrates that care must be taken in interpreting data. On the other hand, miR-208a is known to be a cardiac muscle-specific miRNA; it is increased in experimental myocardial injury and increased clinically during acute myocardial infarction, with apparent higher diagnostic value than troponin (Wang et al., 2010).

miRNA as Biomarkers of Drug-Induced Cardiotoxicity

As with other soluble biomarkers, much of the work investigating the potential of miRNAs as biomarkers for drug-induced cardiotoxicity has also focused on doxorubicin due to the wealth of existing data on the clinical and preclinical deleterious effects of this anthracycline. One of the first and most comprehensive publications on cardiac miRNAs during doxorubicin treatment was published by Vacchi-Suzzi et al. (2012), who detailed cardiac tissue miRNA levels following 2, 4 and 6 weeks doxorubicin administration in the rat. The resulting cardiotoxicity was manifested as atrial and ventricular vacuolation and was associated with the up-regulation of miR-208b, miR-215, mir-216b, miR-34c and miR-367 in cardiac tissue. Moreover, miR-216b actually increased at the lowest dose in the absence of histopathological findings. In a similar study in mice, miR-34a, miR-21, miR-221, miR-222 and miR-208b expression was increased by doxorubicin treatment (Desai et al., 2014). Taken together these studies demonstrate that cardiac miRNA levels are associated with doxorubicin-induced cardiotoxicity but that different miRNAs may be identified in quite similar studies and that miRNA responses to cardiotoxic agents may be different in different studies. Currently, it is unclear whether this limited concordance across species is due to true biological differences or to variations in the experimental approach, for example, analysis of the different anatomical areas of the heart or targeted miRNA assessment. This then raises the question whether any single miRNA would have sufficient specificity and sensitivity to function as a useful biomarker in this context.

Despite this interest in miRNAs, relatively few studies have investigated their potential as circulating biomarkers of drug-induced cardiotoxicity. The miR-34 family of miRNAs appear to be emerging as the most promising circulating biomarkers associated with doxorubicin-induced cardiac dysfunction preclinically (Piegari et al., 2016). Mechanistically, miR-34 appears to actually contribute to doxorubicin-induced cardiotoxicity and apoptosis by increasing mitochondrial depolarization, Bax expression, caspase-3 activation and attenuating Bcl2 expression (Zhu et al., 2017). Clinically, interest in miR-1 as a biomarker of cardiotoxicity has also received attention. Rigaud et al. (2017) showed a strong correlation between left ventricular ejection fraction and miR-1 levels in patients being treated with doxorubicin and statistically demonstrated

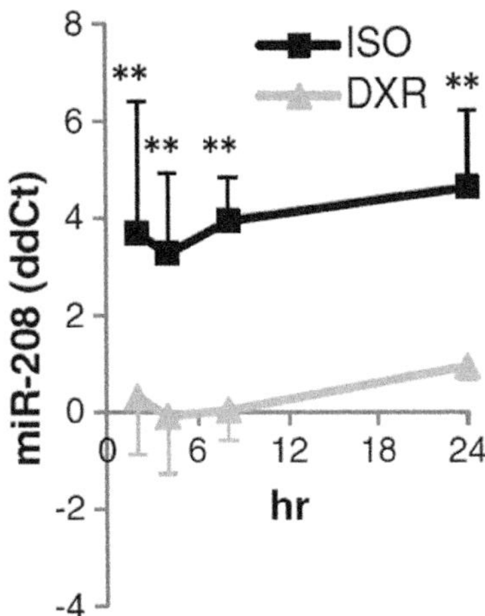

FIGURE 2.7.3 Plasma profile of miR-208 following single intravenous dose of either isoprenaline (1 mg/kg) or doxorubicin (30 mg/kg). ddCT, algorithm approach to determine relative expression with qPCR. (Reproduced from Nishimura, Y. et al., *J. Appl. Toxicol.*, 35, 173–180, 2015.)

that miR-1 was superior to cTnI in its ability to predict development of cardiotoxicity. In fact, a recent review of all miRNA studies investigating doxorubicin-induced cardiotoxicity only identified miR-1 as modulated in both preclinical models and clinical samples (Rigaud et al., 2017; Ruggeri et al., 2018).

As discussed previously, miR-208 plays a role in cardiac remodeling and myosin heavy chain isoform switching, and miR-208 has consistently been shown to be elevated in plasma during isoprenaline-induced cardiac injury with a similar time course to cTnI (Figure 2.7.3) (Gilneur et al., 2016; Ji et al., 2009; Nishimura et al., 2015) and could actually be a better biomarker of injury than cTnI (Liu et al., 2014). However, miR-208a does not appear to be consistently elevated during doxorubicin-induced cardiotoxicity either preclinically (Figure 2.7.3) (Nishimura et al., 2015) or clinically (Oliveira-Carvalho et al., 2015; Rigaud et al., 2017).

The sheer number of potential miRNAs make these an important source of potential novel biomarkers of cardiotoxicity both clinically and preclinically. However, the scale of this opportunity and the complexity of their functional biology in both the cardiac and noncardiac tissues and cells, underwrites the importance of conducting hypothesis-lead, well-controlled validation and mechanistic studies to support their more routine application in drug discovery.

Extracellular Vesicles

It has been recognized for some time that most cells release vesicles from their plasma membranes that are measurable in biological fluids such as blood. EVs are small lipid-bilayer membrane-surrounded structures, and it is now recognized that they are released by cells through active processes and can contain proteins, lipids and miRNA. Furthermore, it is now thought that EVs form the basis of a complex intercellular communication system that is able to signal origin cell status (van Niel et al., 2018). EVs are therefore being considered as potential biomarkers of various cellular processes with measures of EV content and/or origin being used to provide insight into organ function and damage. Much of the work investigating EVs as biomarkers of cardiovascular disease have focused on the characterization of miRNA content similar to the preceding passage. However, recently Yarana et al. (2018) demonstrated that doxorubicin treatment caused a significant increase in overall circulating EVs and that EVs from treated animals contained a lipid peroxidation product that has been implicated in doxorubicin-induced cardiotoxicity. Moreover, they demonstrated brain/heart glycogen phosphorylase in EVs specific to a cardiac source. Dynamic analysis revealed that glycogen phosphorylase-containing EVs were released prior to cTnI release (Yarana et al., 2018). Further work is required to determine the full utility of EVs as biomarkers and whether the phenotype of the EVs, the content of the EVs, or both when considered together have the best potential as markers of drug-induced cardiotoxicity.

CONCLUSIONS

To date, clinical biomarkers of cardiovascular disease have offered a valuable opportunity to explore potential biomarkers for utility in preclinical cardiotoxicity. However, cTn is the only biomarker with FDA qualification in this area. Novel biomarkers of drug-induced cardiotoxicity would have enormous value, especially if they could offer mechanistic insight into the liability. The effort required to identify such biomarkers should not be underestimated; however, much the clinical cardiovascular biomarker landscape is continually evolving and will provide a source of potential new biomarkers of drug-induced cardiotoxicity. The emergence of both miRNAs and EVs offer the alternative avenues to identify mechanistic biomarkers. Nevertheless, the lack of any form of prognostic biomarker of preclinical or clinical cardiotoxicity remains a major gap (Moazeni et al., 2017). Advancements in systems-based approaches, for example, metabolomics, transcriptomics and proteomics, offers one opportunity to facilitate the next generation of cardiac biomarkers that have the potential to fill this void (Figure 2.7.4) (Holmgren et al., 2018; Messinis et al., 2018). Furthermore, the development of human-based *in vitro* models with improved predictive capabilities offers the opportunity to identify translatable biomarkers and to allow detection of cardiotoxicity early during drug discovery. Such advancements would allow temporal changes in cardiac pathology and physiology to be detected prior to the occurrence of gross cardiomyocyte destruction.

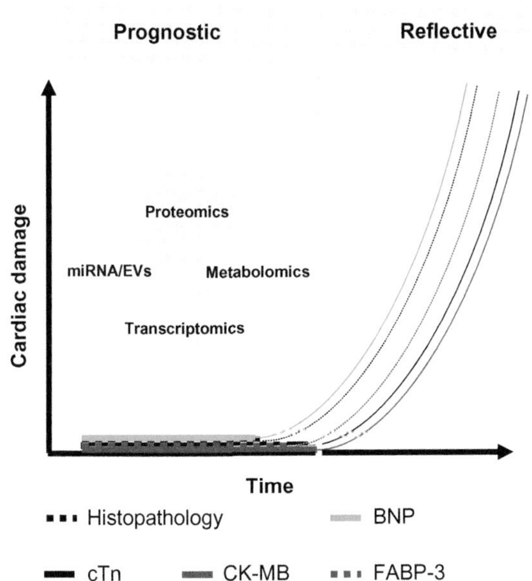

FIGURE 2.7.4 Illustration performance of current biomarkers of cardiotoxicity and the potential of novel emerging biomarker platforms that have the potential to provide prognostic value to predict future risk of cardiac damage.

REFERENCES

Adam O, Lohfelm B, Thum T, Gupta SK, Puhl SL, Schafers HJ et al. (2012). Role of miR-21 in the pathogenesis of atrial fibrosis. *Basic Res Cardiol* 107: 278.

Adamcova M, Simunek T, Kaiserova H, Popelova O, Sterba M, Potacova A et al. (2007). In vitro and in vivo examination of cardiac troponins as biochemical markers of drug-induced cardiotoxicity. *Toxicology* 237: 218–228.

Apple FS (1999). Tissue specificity of cardiac troponin I, cardiac troponin T and creatine kinase-MB. *Clin Chim Acta* 284: 151–159.

Apple FS, Murakami MM, Ler R, Walker D, York M et al. (2008). Analytical characteristics of commercial cardiac troponin I and T immunoassays in serum from rats, dogs, and monkeys with induced acute myocardial injury. *Clin Chem* 54: 1982–1989.

Aries A, Paradis P, Lefebvre C, Schwartz RJ, and Nemer M (2004). Essential role of GATA-4 in cell survival and drug-induced cardiotoxicity. *Proc Natl Acad Sci* 101: 6975–6980.

Arnold JM, Fitchett DH, Howlett JG, Lonn EM, and Tardif JC (2008). Resting heart rate: A modifiable prognostic indicator of cardiovascular risk and outcomes? *Can J Cardiol* 24 Suppl A: 3A–8A.

Arola OJ, Saraste A, Pulkki K, Kallajoki M, Parvinen M, and Voipio-Pulkki LM (2000). Acute doxorubicin cardiotoxicity involves cardiomyocyte apoptosis. *Cancer Res* 60: 1789–1792.

Arroyo JD, Chevillet JR, Kroh EM, Ruf IK, Pritchard CC, Gibson DF et al. (2011). Argonaute2 complexes carry a population of circulating microRNAs independent of vesicles in human plasma. *Proc Natl Acad Sci* 108: 5003–5008.

Berridge BR, Pettit S, Walker DB, Jaffe AS, Schultze AE, Herman E et al. (2009). A translational approach to detecting drug-induced cardiac injury with cardiac troponins: Consensus and recommendations from the cardiac troponins biomarker working group of the health and environmental sciences institute. *Am Heart J* 158: 21–29.

Bertinchant JP, Polge A, Juan JM, Oliva-Lauraire MC, Giuliani I, Marty-Double C et al. (2003). Evaluation of cardiac troponin I and T levels as markers of myocardial damage in doxorubicin-induced cardiomyopathy rats, and their relationship with echocardiographic and histological findings. *Clin Chim Acta* 329: 39–51.

Bertinchant JP, Robert E, Polge A, Marty-Double C, Fabbro-Peray P, Poirey S, Aya, G et al. (2000). Comparison of the diagnostic value of cardiac troponin I and T determinations for detecting early myocardial damage and the relationship with histological findings after isoprenaline-induced cardiac injury in rats. *Clin Chim Acta* 298: 13–28.

Bertsch T, Bleuel H, Aufenanger J, and Rebel W (1997). Comparison of cardiac troponin T and cardiac troponin I concentrations in peripheral blood during orciprenaline induced tachycardia in rats. *Exp Toxicol Pathol* 49: 467–468.

Bleuel H, Deschl U, Bertsch T, Bolz G, and Rebel W (1995). Diagnostic efficiency of troponin T measurements in rats with experimental myocardial cell damage. *Exp Toxicol Pathol* 47: 121–127.

Brady S, York M, Scudamore C, Williams T, Griffiths W, and Turton J (2010). Cardiac troponin I in isoproterenol-induced cardiac injury in the Hanover Wistar rat: Studies on low dose levels and routes of administration. *Toxicol Pathol* 38: 287–291.

Carè A, Catalucci D, Felicetti F, Bonci D, Addario A, Gallo P *et al* (2007). MicroRNA-133 controls cardiac hypertrophy. *Nat Med* 13: 613–618.

Chakraborty C, Sharma AR, Sharma G, Doss CGP, and Lee SS (2017). Therapeutic miRNA and siRNA: Moving from bench to clinic as next generation medicine. *Mol Ther Nucleic Acids* 8: 132–143.

Chanut F, Kimbrough C, Hailey R, Berridge B, Hughes-Earle A, Davies R et al. (2013). Spontaneous cardiomyopathy in young Sprague-Dawley rats: Evaluation of biological and environmental variability. *Toxicol Pathol* 41: 1126–1136.

Chen C, Lu C, Qian Y, Li H, Tan Y, Cai L et al. (2017). Urinary miR-21 as a potential biomarker of hypertensive kidney injury and fibrosis. *Sci Rep* 7: 17737.

Christiansen S, Redmann K, Scheld HH, Jahn UR, Stypmann J, Fobker M et al. (2002). Adriamycin-induced cardiomyopathy in the dog—an appropriate model for research on partial left ventriculectomy? *J Heart Lung Transplant* 21: 783–790.

Clements P, Brady S, York M, Berridge B, Mikaelian I, Nicklaus R et al. (2010). Time course characterization of serum cardiac troponins, heart fatty acid-binding protein, and morphologic findings with isoproterenol-induced myocardial injury in the rat. *Toxicol Pathol* 38: 703–714.

Collinson P, Gaze D, and Goodacre S (2014). Comparison of contemporary troponin assays with the novel biomarkers, heart fatty acid binding protein and copeptin, for the early confirmation or exclusion of myocardial infarction in patients presenting to the emergency department with chest pain. *Heart* 100: 140–145.

Colton HM, Stokes AH, Yoon LW, Quaile MP, Novak PJ, Falls JG et al. (2011). An initial characterization of N-terminal-proatrial natriuretic peptide in serum of Sprague Dawley rats. *Toxicol Sci* 120: 262–268.

Cove-Smith L, Woodhouse N, Hargreaves A, Kirk J, Smith S, Price SA et al. (2014). An integrated characterization of serological, pathological, and functional events in doxorubicin-induced cardiotoxicity. *Toxicol Sci* 140: 3–15.

Cross MJ, Berridge BR, Clements PJ, Cove-Smith L, Force TL, Hoffmann P et al. (2015). Physiological, pharmacological and toxicological considerations of drug-induced structural cardiac injury. *Br J Pharmacol* 172: 957–974.

D'Errico MP, Petruzzelli MF, Gianicolo EA, Grimaldi L, Loliva F, Tramacere F et al. (2015). Kinetics of B-type natriuretic peptide plasma levels in patients with left-sided breast cancer treated with radiation therapy: Results after one-year follow-up. *Int J Radiat Biol* 91: 804–809.

Desai VG, J CK, Vijay V, Moland CL, Herman EH, Lee T et al. (2014). Early biomarkers of doxorubicin-induced heart injury in a mouse model. *Toxicol Appl Pharmacol* 281: 221–229.

Dunn ME, Coluccio D, Hirkaler G, Mikaelian I, Nicklaus R, Lipshultz SE et al. (2011). The complete pharmacokinetic profile of serum cardiac troponin I in the rat and the dog. *Toxicol Sci* 123: 368–373.

Dunn ME, Manfredi TG, Agostinucci K, Engle SK, Powe J, King NM et al. (2017). Serum natriuretic peptides as differential biomarkers allowing for the distinction between physiologic and pathologic left ventricular hypertrophy. *Toxicol Pathol* 45: 344–352.

Frey N, and Olson EN (2003). Cardiac hypertrophy: The good, the bad, and the ugly. *Annu Rev Physiol* 65: 45–79.

Friden V, Starnberg K, Muslimovic A, Ricksten SE, Bjurman C, Forsgard N et al. (2017). Clearance of cardiac troponin T with and without kidney function. *Clin Biochem* 50: 468–474.

Gardner DG (2003). Natriuretic peptides: Markers or modulators of cardiac hypertrophy? *Trends Endocrinol Metab* 14: 411–416.

Glineur SF, De Ron P, Hanon E, Valentin JP, Dremier S, and Nogueira da Costa A (2016). Paving the route to plasma miR-208a-3p as an acute cardiac injury biomarker: Preclinical rat data supports its use in drug safety assessment. *Toxicol Sci* 149: 89–97.

Han H, Qu G, Han C, Wang Y, Sun T, Li F et al. (2015). MiR-34a, miR-21 and miR-23a as potential biomarkers for coronary artery disease: A pilot microarray study and confirmation in a 32 patient cohort. *Exp Mol Med* 47: e138.

Herman EH, Lipshultz SE, Rifai N, Zhang J, Papoian T, Yu ZX et al. (1998). Use of cardiac troponin T levels as an indicator of doxorubicin-induced cardiotoxicity. *Cancer Res* 58: 195–197.

Herman EH, Zhang J, Lipshultz SE, Rifai N, Chadwick D, Takeda K et al. (1999). Correlation between serum levels of cardiac troponin-T and the severity of the chronic cardiomyopathy induced by doxorubicin. *J Clin Oncol* 17: 2237–2243.

Holmgren G, Sartipy P, Andersson CX, Lindahl A, and Synnergren J (2018). Expression profiling of human pluripotent stem cell-derived cardiomyocytes exposed to doxorubicin—integration and visualization of multi omics data. *Toxicol Sci* 163(1):182–195.

Jarai R, Fellner B, Haoula D, Jordanova N, Heinz G, Karth GD et al. (2009). Early assessment of outcome in cardiogenic shock: Relevance of plasma N-terminal pro-B-type natriuretic peptide and interleukin-6 levels. *Crit Care Med* 37: 1837–1844.

Ji X, Takahashi R, Hiura Y, Hirokawa G, Fukushima Y, and Iwai N (2009). Plasma miR-208 as a biomarker of myocardial injury. *Clin Chem* 55: 1944–1949.

Kanatous SB, and Mammen PP (2010). Regulation of myoglobin expression. *J Exp Biol* 213: 2741–2747.

Kim K, Chini N, Fairchild DG, Engle SK, Reagan WJ, Summers SD et al. (2016). Evaluation of cardiac toxicity biomarkers in rats from different laboratories. *Toxicol Pathol* 44: 1072–1083.

Kim Y, Ma AG, Kitta K, Fitch SN, Ikeda T, Ihara Y et al. (2003). Anthracycline-induced suppression of GATA-4 transcription factor: Implication in the regulation of cardiac myocyte apoptosis. *Mol Pharmacol* 63: 368–377.

Kitayama H, Kondo T, Sugiyama J, Kurimoto K, Nishino Y, Kawada M et al. (2017). High-sensitive troponin T assay can predict anthracycline- and trastuzumab-induced cardiotoxicity in breast cancer patients. *Breast Cancer* 24: 774–782.

Koh E, Nakamura T, and Takahashi H (2004). Troponin-T and brain natriuretic peptide as predictors for adriamycin-induced cardiomyopathy in rats. *Circ J* 68: 163–167.

Kopljar I, De Bondt A, Vinken P, Teisman A, Damiano B, Goeminne N et al. (2017). Chronic drug-induced effects on contractile motion properties and cardiac biomarkers in human induced pluripotent stem cell-derived cardiomyocytes. *Br J Pharmacol* 174: 3766–3779.

Korff S, Katus HA, and Giannitsis E (2006). Differential diagnosis of elevated troponins. *Heart* 92: 987–993.

L'Ecuyer T, Sanjeev S, Thomas R, Novak R, Das L, Campbell W et al. (2006). DNA damage is an early event in doxorubicin-induced cardiac myocyte death. *Am J Physiol Heart Circ Physiol* 291: H1273–H1280.

Laverty H, Benson C, Cartwright E, Cross M, Garland C, Hammond T et al. (2011). How can we improve our understanding of cardiovascular safety liabilities to develop safer medicines? *Br J Pharmacol* 163: 675–693.

Lee RC, Feinbaum RL, and Ambros V (1993). The C. elegans heterochronic gene lin-4 encodes small RNAs with antisense complementarity to lin-14. *Cell* 75: 843–854.

Levin ER, Gardner DG, and Samson WK (1998). Natriuretic peptides. *N Engl J Med* 339: 321–328.

Li Z, Song Y, Liu L, Hou N, An X, Zhan D et al. (2017). miR-199a impairs autophagy and induces cardiac hypertrophy through mTOR activation. *Cell Death Differ* 24: 1205–1213.

Maeda K, Tsutamoto T, Wada A, Hisanaga T, and Kinoshita M (1998). Plasma brain natriuretic peptide as a biochemical marker of high left ventricular end-diastolic pressure in patients with symptomatic left ventricular dysfunction. *Am Heart J* 135: 825–832.

Messinis DE, Melas IN, Hur J, Varshney N, Alexopoulos LG, and Bai JPF (2018). Translational systems pharmacology-based predictive assessment of drug-induced cardiomyopathy. *CPT Pharmacometrics Syst Pharmacol* 7: 166–174.

Meyer B, Huelsmann M, Wexberg P, Delle Karth G, Berger R, Moertl D et al. (2007). N-terminal pro-B-type natriuretic peptide is an independent predictor of outcome in an unselected cohort of critically ill patients. *Crit Care Med* 35: 2268–2273.

Minotti G, Cairo G, and Monti E (1999). Role of iron in anthracycline cardiotoxicity: New tunes for an old song? *FASEB J* 13: 199–212.

Minotti G, Menna P, Salvatorelli E, Cairo G, and Gianni L (2004). Anthracyclines: Molecular advances and pharmacologic developments in antitumor activity and cardiotoxicity. *Pharmacol Rev* 56: 185–229.

Moazeni S, Cadeiras M, Yang EH, Deng MC, and Nguyen KL (2017). Anthracycline induced cardiotoxicity: Biomarkers and "omics" technology in the era of patient specific care. *Clin Transl Med* 6: 17.

Morrow DA, and Braunwald E (2003). Future of biomarkers in acute coronary syndromes: Moving toward a multimarker strategy. *Circulation* 108: 250–252.

Mukoyama M, Nakao K, Hosoda K, Suga S, Saito Y, Ogawa Y et al. (1991). Brain natriuretic peptide as a novel cardiac hormone in humans: Evidence for an exquisite dual natriuretic peptide system, atrial natriuretic peptide and brain natriuretic peptide. *J Clin Invest* 87: 1402–1412.

Nagata K, Sawada K, Minomo H, Sasaki D, Kazusa K, and Takamatsu K (2017). Effects of repeated restraint and blood sampling with needle injection on blood cardiac troponins in rats, dogs, and cynomolgus monkeys. *Comp Clin Path* 26: 1347–1354.

Nie J, George K, Duan F, Tong TK, and Tian Y (2016). Histological evidence for reversible cardiomyocyte changes and serum cardiac troponin T elevation after exercise in rats. *Physiol Rep* 4(24).

Nishimura Y, Kondo C, Morikawa Y, Tonomura Y, Torii M, Yamate J et al. (2015). Plasma miR-208 as a useful biomarker for drug-induced cardiotoxicity in rats. *J Appl Toxicol* 35: 173–180.

O'Brien PJ (2008). Cardiac troponin is the most effective translational safety biomarker for myocardial injury in cardiotoxicity. *Toxicology* 245: 206–218.

O'Brien PJ, Smith DE, Knechtel TJ, Marchak MA, Pruimboom-Brees I, Brees DJ et al. (2006). Cardiac troponin I is a sensitive, specific biomarker of cardiac injury in laboratory animals. *Lab Anim* 40: 153–171.

Oliveira PJ, Bjork JA, Santos MS, Leino RL, Froberg MK, Moreno AJ et al. (2004). Carvedilol-mediated antioxidant protection against doxorubicin-induced cardiac mitochondrial toxicity. *Toxicol Appl Pharmacol* 200: 159–168.

Oliveira-Carvalho V, Ferreira LR, and Bocchi EA (2015). Circulating mir-208a fails as a biomarker of doxorubicin-induced cardiotoxicity in breast cancer patients. *J Appl Toxicol* 35: 1071–1072.

Park KC, Gaze DC, Collinson PO, and Marber MS (2017). Cardiac troponins: From myocardial infarction to chronic disease. *Cardiovasc Res* 113: 1708–1718.

Peng Z, Pan L, Niu Z, Li W, Dang X, Wan L et al. (2017). Identification of microRNAs as potential biomarkers for lung adenocarcinoma using integrating genomics analysis. *Oncotarget* 8: 64143–64156.

Piegari E, Russo R, Cappetta D, Esposito G, Urbanek K, Dell'Aversana C et al. (2016). MicroRNA-34a regulates doxorubicin-induced cardiotoxicity in rat. *Oncotarget* 7: 62312–62326.

Pierson JB, Berridge BR, Brooks MB, Dreher K, Koerner J, Schultze AE et al. (2013). A public-private consortium advances cardiac safety evaluation: Achievements of the HESI cardiac safety technical committee. *J Pharmacol Toxicol Methods* 68: 7–12.

Reagan WJ (2010). Troponin as a biomarker of cardiac toxicity: Past, present, and future. *Toxicol Pathol* 38: 1134–1137.

Reagan WJ, Barnes R, Harris P, Summers S, Lopes S, Stubbs M et al. (2017). Assessment of cardiac troponin I responses in nonhuman primates during restraint, blood collection, and dosing in preclinical safety studies. *Toxicol Pathol* 45: 335–343.

Reagan WJ, York M, Berridge B, Schultze E, Walker D, and Pettit S (2013). Comparison of cardiac troponin I and T, including the evaluation of an ultrasensitive assay, as indicators of doxorubicin-induced cardiotoxicity. *Toxicol Pathol* 41: 1146–1158.

Remme WJ, Swedberg K, and European Society of C (2002). Comprehensive guidelines for the diagnosis and treatment of chronic heart failure: Task force for the diagnosis and treatment of chronic heart failure of the European Society of Cardiology. *Eur J Heart Fail* 4: 11–22.

Rigaud VO, Ferreira LR, Ayub-Ferreira SM, Avila MS, Brandao SM, Cruz FD et al. (2017). Circulating miR-1 as a potential biomarker of doxorubicin-induced cardiotoxicity in breast cancer patients. *Oncotarget* 8: 6994–7002.

Ruggeri C, Gioffre S, Achilli F, Colombo GI, and D'Alessandra Y (2018). Role of microRNAs in doxorubicin-induced cardiotoxicity: An overview of preclinical models and cancer patients. *Heart Fail Rev* 23: 109–122.

Sawant DV, Yao W, Wright Z, Sawyers C, Tepper RS, Gupta SK et al. (2015). Serum microRNA-21 as a biomarker for allergic inflammatory disease in children. microRNA 4: 36–40.

Sayed D, Hong C, Chen IY, Lypowy J, and Abdellatif M (2007). MicroRNAs play an essential role in the development of cardiac hypertrophy. *Circ Res* 100: 416–424.

Shafi A, Siddiqui N, Imtiaz S, and Din Sajid MU (2017). Left ventricular systolic dysfunction predicted by early troponin I release after anthracycline based chemotherapy in breast cancer patients. *J Ayub Med Coll Abbottabad* 29: 266–269.

Sharma S, Jackson PG, and Makan J (2004). Cardiac troponins. *J Clin Pathol* 57: 1025–1026.

Song DW, Ryu JY, Kim JO, Kwon EJ, and Kim DH (2014). The miR-19a/b family positively regulates cardiomyocyte hypertrophy by targeting atrogin-1 and MuRF-1. *Biochem J* 457: 151–162.

Thum T, Gross C, Fiedler J, Fischer T, Kissler S, Bussen M et al. (2008). MicroRNA-21 contributes to myocardial disease by stimulating MAP kinase signalling in fibroblasts. *Nature* 456: 980–984.

Vacchi-Suzzi C, Bauer Y, Berridge BR, Bongiovanni S, Gerrish K, Hamadeh HK et al. (2012). Perturbation of microRNAs in rat hear during chronic doxorubicin treatment. *PLoS One* 7; e40395.

Valadi H, Ekstrom K, Bossios A, Sjostrand M, Lee JJ, and Lotvall JO (2007). Exosome-mediated transfer of mRNAs and microRNAs is a novel mechanism of genetic exchange between cells. *Nat Cell Biol* 9: 654–659.

van Niel G, D'Angelo G, and Raposo G (2018). Shedding light on the cell biology of extracellular vesicles. *Nat Rev Mol Cell Biol* 19: 213–228.

van Rooij E, Quiat D, Johnson BA, Sutherland LB, Qi X, Richardson JA, Kelm RJ et al. (2009). A family of microRNAs encoded by myosin genes governs myosin expression and muscle performance. *Dev Cell* 17: 662–673.

Vejpongsa P, and Yeh ET (2014). Prevention of anthracycline-induced cardiotoxicity: Challenges and opportunities. *J Am Coll Cardiol* 64: 938–945.

Verbiest T, Binst D, Waelbers T, Coppieters E, and Polis I (2013). Perioperative changes in cardiac troponin I concentrations in dogs. *Res Vet Sci* 94: 446–448.

Vickers KC, Palmisano BT, Shoucri BM, Shamburek RD, and Remaley AT (2011). MicroRNAs are transported in plasma and delivered to recipient cells by high-density lipoproteins. *Nat Cell Biol* 13: 423–433.

Walker DB (2006). Serum chemical biomarkers of cardiac injury for nonclinical safety testing. *Toxicol Pathol* 34: 94–104.

Wallace KB (2003). Doxorubicin-induced cardiac mitochondrionopathy. *Pharmacol Toxicol* 93: 105–115.

Wallace KB, Hausner E, Herman E, Holt GD, MacGregor JT, Metz AL et al. (2004). Serum troponins as biomarkers of drug-induced cardiac toxicity. *Toxicol Pathol* 32: 106–121.

Wang GK, Zhu JQ, Zhang JT, Li Q, Li Y, He J et al. (2010). Circulating microRNA: A novel potential biomarker for early diagnosis of acute myocardial infarction in humans. *Eur Heart J* 31: 659–666.

Wang J, Duan L, Tian L, Liu J, Wang S, Gao Y et al. (2016b). Serum miR-21 may be a potential diagnostic biomarker for diabetic nephropathy. *Exp Clin Endocrinol Diabetes* 124: 417–423.

Wang J, Liew OW, Richards AM, and Chen YT (2016a). Overview of microRNAs in cardiac hypertrophy, fibrosis, and apoptosis. *Int J Mol Sci* 17(5): 749.

Wang YD, Chen SX, and Ren LQ (2016c). Serum B-type natriuretic peptide levels as a marker for anthracycline-induced cardiotoxicity. *Oncol Lett* 11: 3483–3492.

Yarana C, Carroll D, Chen J, Chaiswing L, Zhao Y, Noel T et al. (2018). Extracellular vesicles released by cardiomyocytes in a doxorubicin-induced cardiac injury mouse model contain protein biomarkers of early cardiac injury. *Clin Cancer Res* 24(7): 1644–1653.

York M, Scudamore C, Brady S, Chen C, Wilson S, Curtis M et al. (2007). Characterization of troponin responses in isoproterenol-induced cardiac injury in the Hanover Wistar rat. *Toxicol Pathol* 35: 606–617.

Zampetaki A, Kiechl S, Drozdov I, Willeit P, Mayr U, Prokopi M et al. (2010). Plasma microRNA profiling reveals loss of endothelial miR-126 and other microRNAs in type 2 diabetes. *Circ Res* 107: 810–817.

Zhang J, Knapton A, Lipshultz SE, Weaver JL, and Herman EH (2008). Isoproterenol-induced cardiotoxicity in Sprague-Dawley rats: Correlation of reversible and irreversible myocardial injury with release of cardiac troponin T and roles of iNOS in myocardial injury. *Toxicol Pathol* 36: 277–278.

Zhang J, Xing Q, Zhou X, Li J, Li Y et al. (2017). Circulating miRNA-21 is a promising biomarker for heart failure. *Mol Med Rep* 16: 7766–7774.

Zhang S, Liu X, Bawa-Khalfe T, Lu LS, Lyu YL, Liu LF et al. (2012). Identification of the molecular basis of doxorubicin-induced cardiotoxicity. *Nat Med* 18: 1639–1642.

Zhu JN, Fu YH, Hu ZQ, Li WY, Tang CM, Fei HW et al. (2017). Activation of miR-34a-5p/Sirtl/p66shc pathway contributes to doxorubicin-induced cardiotoxicity. *Sci Rep* 7: 11879.

3

Biomarkers in Translational Medicine

Lost in Translation
Bridging the Preclinical and Clinical Worlds Concepts, Examples, Successes, and Failures in Translational Medicine

3.1

Attila A. Seyhan and Claudio Carini

Contents

INTRODUCTION

There is growing pressure from the general public, funding agencies, and policy makers for scientists and drug companies to improve biomedical research, accelerate drug development process and reduce the cost.

Because of this, significant efforts have been made in recent years to reduce the efficacy- and safety-related failures of clinical trials by analyzing possible links to the physicochemical properties of drug candidates; however, the results have been inconclusive due to the limited size of data sets or various other reasons including a few discussed here [1].

Evidence suggests that many published research findings in biomedical research are misleading, not as robust as they claim, or cannot be reproduced. Because of this, the problem with the reproducibility of findings from preclinical findings and their translatability to human studies is now widely recognized by both biopharmaceutical industry and academic researchers. Scientists are expected to be objective and diligent critics of not only others' but also of their own findings. In reality, scientists are influenced by natural susceptibility to look for evidence and to see patterns in randomness to support their preconceived ideas, beliefs and overall scientific hypotheses. In the competitive field of scientific research, other various contrary incentives collude to compromise the scientific method, resulting in literature scattered with unreliable data.

The scale of the problem could be as plain as the simple differences in methodological differences (for example, type of coating on tubes, the temperature cells are grown at, how cells are stirred in culture, or subtle differences in medium such as pH or the ingredients on which they are cultured) can result in complete failure to replicate the results.

Additional factors may include the model systems that have been used in scientific research where human studies cannot be conducted. For example, basic scientific and biomedical research using animal models has been considered to be necessary for preclinical research for developing new drug targets and therapies for a number of human diseases as well as for elucidating underlying molecular and pathophysiological mechanisms that lead to or are associated with the disease. Almost 90% of the animal studies the industry conducts are legally required or justified by scientific reasons. However, the poor translation of animal studies to human studies has led to significant amounts of wasted resources. It has been claimed that just only 6% of animal studies are translatable to human responses [2].

Furthermore, the improper use of statistical analysis methods and the misinterpretation or misuse of *p*-values appears to have a significant effect, leading to inaccurate conclusions that add to the reproducibility crisis.

To address several of the many possible reasons leading to the high attrition rate of drug development, this article reviews literature on (1) the possible causes of bias on the reproducibility and translatability of preclinical research findings, (2) poor experimental designs, (3) the limitations of using animal models for human diseases, and (4) the emerging systems and methods as models to improve biomedical research and drug discovery and development.

We then provide recommendations and alternative model systems to use in place of animal models to improve the reproducibility, robustness, and reliability of such studies and thereby to improve their translation to human studies.

CHALLENGES IN NEW DRUG DEVELOPMENT

The process of getting a new drug, from first testing to final FDA approval, and ultimately to market, is a long, costly, and risky process with high failure rates [3,4]. However, the failure rates of human clinical trials are high and continue to grow [5]. Almost 90% of the drugs entering human trials fail [6–9], mostly due to the failure of efficacy or safety tests [6–9]. Recent years have shown no substantial improvement in drug development process, in spite of an increasing commitment of resources from the biopharmaceutical industry and regulatory agencies [8]. The majority of projects fail for problems unrelated to a therapeutic hypothesis, and this may be due to unexpected side effects and tolerability [10,11]. What is often not so apparent to the general public, or even to scientists, is the number of drugs that do not make it through all the hoops of the drug development and approval process. For every drug that gains FDA approval, more than 1,000 were developed but failed. Thus, the odds of moving new compounds from preclinical research into human studies and approved as the marketable drug are only about 0.1%.

The development of a newly approved drug takes more than 10 years, the cost of developing a newly approved drug is about $2.6 billion [12,13], a 145% increase, correcting for inflation, over the estimate made in 2003. The analysis presented in a recent report [13] was based on the data obtained from 10 biopharmaceutical companies on 106 randomly selected drugs tested in human trials between 1995 and 2007 [12,13].

As illustrated in Figure 3.1.1, the efficiency of biomedical research and development of new drugs in the United States halves every 9 years; hence, this is sometimes referred to as Eroom's law—the reverse of Moore's law for microprocessors [14]. In other words, the cost of getting a drug developed and approved will double every 9 years. If business continues to be conducted in the usual way, the biopharmaceutical industry would have to spend $16 billion on drug development in the year 2043. This would force the industry to develop only the most profitable drugs—not the ones most needed (http://ecorner.stanford.edu/videos/4224/Moores-Law-for-Pharma).

Basically, in the late 1980s through the late 1990s, the cost of drug R&D spending tended to level out. However, that is where the first wave of biotech drugs started coming through the market. Biotech drugs such as antibody-based drugs commonly referred as to "biologics" are different in that they involve a completely different mode of action—a completely different platform from traditional small-molecule pharmaceuticals—and they are costly.

The traditional process of new drug development often, but not necessarily, begins with the discovery of a putative dysregulated or disrupted/mutated target(s) [15] or a pathway(s) assumed to be the key factor(s) causing the disease state in humans, although the identification and early validation of disease-modifying targets is an essential first step in the drug discovery pipeline [16]. For example, sometimes, the drug target or mechanism of action (MOA) is discovered much later after its first use in humans (e.g., aspirin). After being used for so many years, aspirin's MOA was only discovered in 1972 by John Robert Vane, who showed that aspirin suppressed the production of prostaglandins and thromboxanes [17,18]. For this discovery Vane along with Bergström and Bengt Ingemar Samuelsson were awarded the 1982 Nobel Prize in Physiology or Medicine [19].

Target identification can be approached by direct biochemical methods, genetic manipulations, or functional genomic or computational inference of patient data compared with healthy

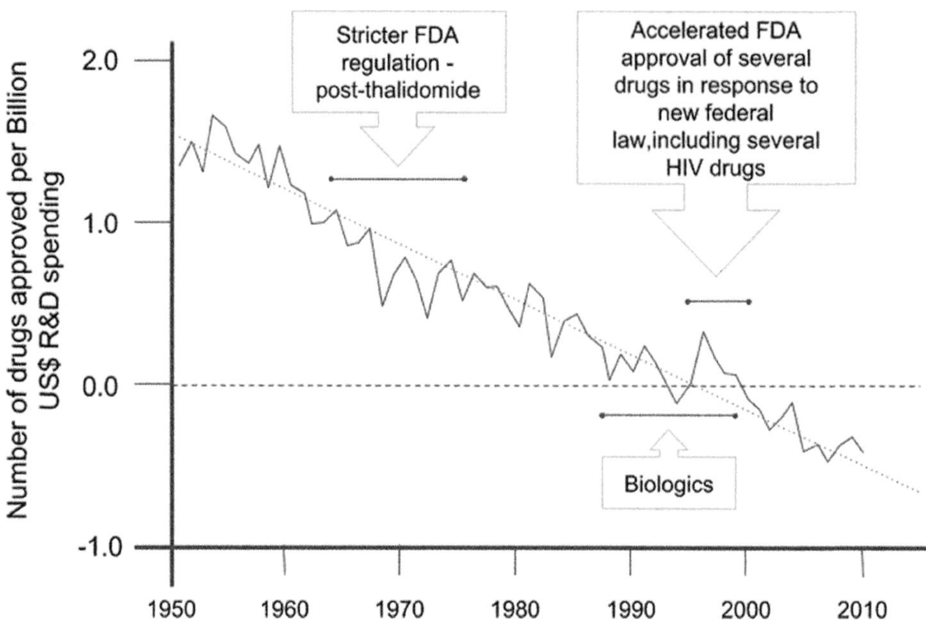

FIGURE 3.1.1 Eroom's Law. The overall trend in R&D efficiency (adjusted for inflation) of new drugs in the United States halves every 9 years. The biopharmaceutical industry refers to this phenomenon as Eroom's law—Moore's law for microprocessors in reverse. (From Seyhan, A.A. *Hum. Genet.*, 130, 583–605, 2011.) In other words, the cost of getting a drug developed and approved will double every 9 years. If the cost of drug development goes like this, the biopharmaceutical industry would have to spend $16 billion on drug development in the year 2043. This would force the industry to develop only the most profitable drugs—not the ones most needed for many newly emerging age-related diseases. However, from approximately the late 1980s through the late 1990s, the only point in this graph where the regression line kind of levels out. That is where the first wave of biotech (antibody-based) drugs started coming through the market. Antibody-based drugs are different in that they have a completely different mode of action and require a completely different platform compared with traditional small-molecule pharmaceuticals. Thus, biologics may be a key solving this problem. (Adapted from Nosengo, N., *Nature*, 534, 314–316, 2016.)

individuals. However, in most cases, combinations of approaches may be required, including searching the literature [20] and available genetic data, to fully characterize disease-associated targets and to understand the mechanisms of candidate drug action [15,16,21–23].

The biopharmaceutical industry continuously screens thousands of compounds to identify ones of therapeutic usefulness. During the preclinical phase (6–7 years), the company evaluates the efficacy of several of these compounds in limited animal testing.

A variety of strategies is used to identify potential drug candidates, including as target-based screening, phenotypic screening, modification of natural substances and biologic-based approaches [24].

As illustrated in Figure 3.1.2, beginning with the selection of a novel target for regulatory approval, the overall probability of success is <8% (fdareview.org). In a traditional drug candidate screening process, thousands of compounds are screened, leading to the identification of ~5,000 candidates. Of 5,000 compounds tested, approximately 5 will appear promising enough to induce the company to file an investigational new drug application (IND). If the IND is approved by the U.S. Food and Drug Administration (FDA) and by an institutional review board, the drug company may begin the first phase of development. The IND phase comprises three phases of human clinical trials. In phase I, the drug's basic properties and safety profile are determined in a small group of healthy humans (20–80 people, this period takes 1–2 years) to evaluate safety, determine safe dosage ranges, and begin to identify side effects, whereas in phase II, the efficacy of the drug is evaluated in 100–300 volunteers of the target patient population. If the drug meets the anticipated outcomes, the FDA gives permission for phase III drug testing, which targets a much larger target patient population (1,000–3,000) to confirm its efficacy, evaluate its effectiveness, monitor side effects, compare it to commonly used treatments, and collect information that will allow it to be used safely. Once phase III is complete, the biopharmaceutical company files a new drug application (NDA). Review of the NDA might take 1–2 years, resulting in the total drug development and approval process taking more than 10 years.

The high attrition rate of potential new drugs in biomedical research costs the industry both time and money. If a drug fails during clinical trials, then the industry ends up losing 7–8 years of R&D time and approximately $900 million. Repositioning drugs could help counter this loss [8,14].

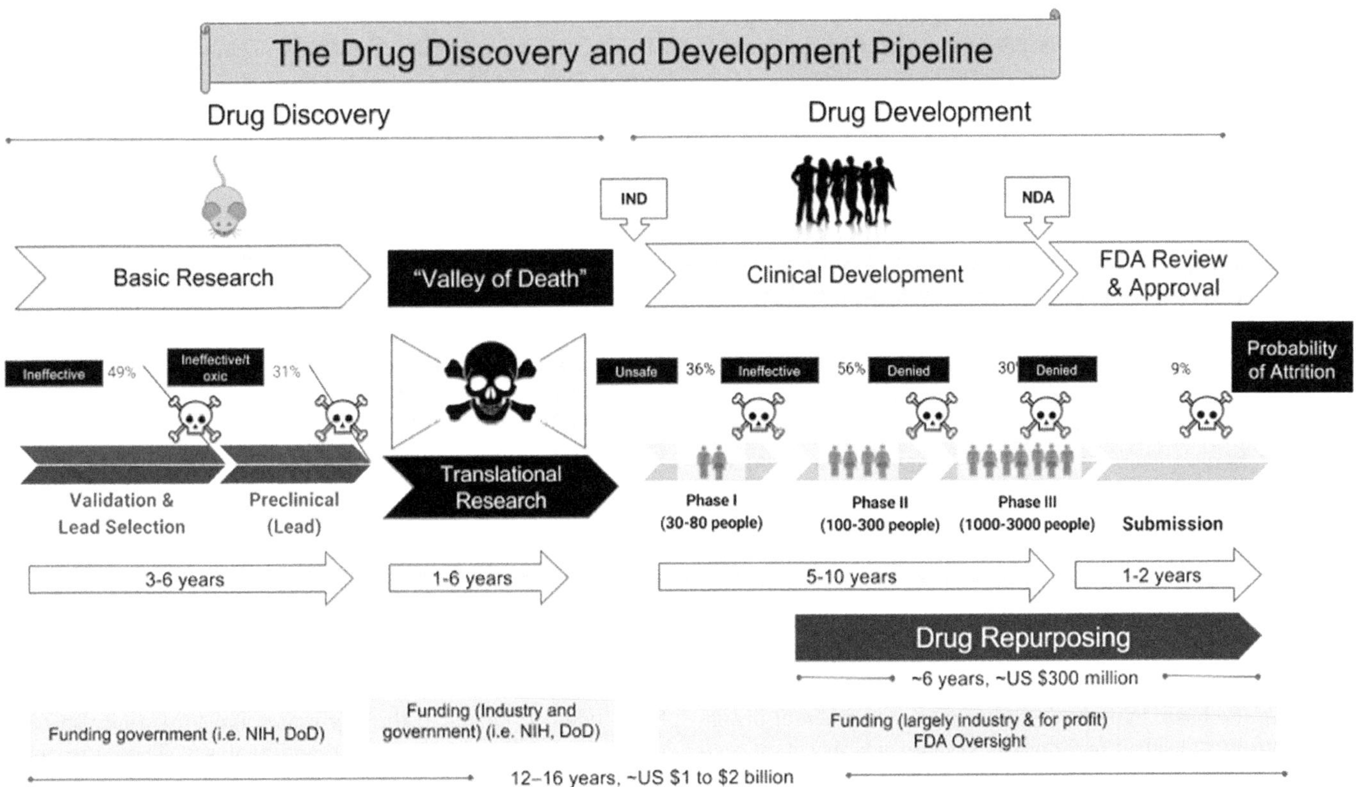

FIGURE 3.1.2 Crossing over the "Valley of Death." The distinct stages of drug development from target discovery to human studies, with timeline and cost for each stage and potential roadblocks that result in high failure rates in this process. Discovering and developing new drugs can take more than 12 years and can cost as much as $2.5 billion. For every drug that gains FDA approval, many thousands enter the pipeline. The process is long and entrenched with major roadblocks. Drug repurposing (repositioning) is another strategy to accelerate drug development and skip early phases of roadblocks. This is because many repurposed drugs have already passed the early phases of development, clinical safety, and bioavailability testing and, thus, can potentially gain FDA approval in less than half the time and at one-quarter of the cost. DoD, Department of Defense; FDA, U.S. Food and Drug Administration; IND, investigational new drug; NDA, new drug application; NIH, National Institutes of Health.

WHY DRUGS FAIL THE DEVELOPMENT PROCESS

As discussed here and elsewhere [7], there are many reasons why drugs fail the development process. According to the literature, this is partly due to the increased complexity of clinical trials, an increased focus on chronic and degenerative diseases, and tests that use comparative drug effectiveness. It may be that the effect seen in cultured cells did not translate into an effect in animal studies, or that the candidate drugs did not have a good safety profiles in animals for unexpected reasons, or that they had a poor pharmacokinetic (PK)/bioavailability and pharmacodynamic (PD) profile, or they may fail due to efficacy and safety issues in human trials [1,7,9,11,25,26] or they were no more effective than the current drugs, or they proved to have intolerable side effects in humans.

A variety of other reasons may include the lack of relevance of the target or mechanism to the disease and, hence, leading to a lack of safety and a high heterogeneity in the patient population in which the target exists in only a subset of patients. In this case, the drug is effective only for a subset of patients but not for everyone. In that case, strategies such as biomarkers can be used to identify the patients that respond and, thus, a new trial may be run. Biomarkers that signal correct dosing and whether the specific molecular target has been hit in early proof-of-concept clinical trials can mitigate such attrition risks.

Drugs that target novel targets with new mechanisms appear to have higher attrition rates, [7,27] but a combination of better-validated preclinical targets that have significant preclinical efficacy and safety profiles and proof of principle data can reduce such attrition risks.

As for why clinical trials fail, this has been discussed eloquently by David Grainger (www.forbes.com/sites/davidgrainger/2015/01/29/why-too-many-clinical-trials-fail-and-a-simple-solution-that-could-increase-returns-on-pharma-rd/#607c0b8edb8b) and others [7,28], who point to a major, but simple, cause for unexpected late-stage clinical trial failure: a fundamental misunderstanding of the *p*-value. Colqhoun [28] proposes simple steps that could materially increase the success rate of late-stage trials.

However, ascribing single reasons for candidate drug failure could be an oversimplification because there may be multiple factors contributing to the failure of the drug candidate.

REPRODUCIBILITY BLUES—POOR REPRODUCIBILITY OF PRECLINICAL RESEARCH

Growing attention is being paid to the poor reproducibility and translation of preclinical findings to human studies. This is evidenced by an emerging literature that has begun to build a consensus around a set of repeating themes that is accompanied by new suggestions, which in themselves represent a paradigm shift.

There are many reports [5,6,7,9,29–32] in the literature attempting to address the same set of questions: (1) why the failure rates of human trials is progressively increasing; (2) why there is a reproducibility crisis; (3) are the animal studies to blame for the failure; or (4) is there a more systematic problem involving the setting up the basic hypothesis and proper experimental design with controls, the selection of model systems whether it is a cell line or animal models, or the statistical analysis.

Recent advances in biomedical research and the rapid introduction of new biological concepts, measurement and analytical methods involving genomes, gene products, biomarkers, and their interactions have resulted in new analysis methods that are complex and poorly understood by many researchers. Further complicating the situation is the inappropriate use of statistical methods, which is magnified by the inadequate training of researchers and scientists in research and laboratory methods.

Further complicating this is the multidisciplinary nature of the biomedical research that uses methods and involves scientists from other disciplines, emphasizing the need for careful study design, conduct, and analysis.

As discussed recently [5], in most of the studies that could not be reproduced because (1) the principles of experimental design and statistics were ignored, for example, because there was no primary hypothesis and primary outcome measure; (2) because the assumption was that every single thing in the experiment was controlled and those controls have worked; (3) because the false discovery rate (FDR) was not considered, especially for unexpected results; (4) because randomization, blinding, and other cornerstones of good practice were disregarded; or (5) because there was a tendency to misinterpret confirmatory experiments and technical replicates as supporting evidence. It is important to recognize that many of these issues have been brought forward by many authors for a number of years with no simple solution to the problem.

As discussed recently [5], the failure rates of human clinical trials continue to grow (90% of the drugs entering human trials fail) [6,7,9,10], leading to the recognition that most drugs fail in human trials due to a failure in efficacy or safety [6,7,9].

Therefore, there is a heightened interest in the so-called "the reproducibility crisis" in biomedical research [30–32]. These failures in biomedical research have led to a growing suspicion that the failure of translation from preclinical findings to human outcomes may be in part due to problems in animal research itself [30–42]. Because of this high attrition rate in translating animal findings to human outcomes, the biopharmaceutical industry has begun disinvesting in internal animal research in the first half of the 2000s, which was taken up by academic and startup biotech sectors [29,43].

There could be many possible causes for the poor translatability for preclinical results [44], in particular, methodological flaws such as underpowered studies, low group sizes, and a lack of blinding [37] or whether appropriate animal models or species were used. Awareness is increasing in the research community that preclinical studies can be flawed in multiple ways, and many recommendations to improve the translatability of these studies have been reported [45,46]. As discussed by Ioannidis et al. [46], even small flaws in the design, conduct, and analysis of biomedical research studies often result in misleading data and a wasteful use of precious resources.

In the chaos theory, "the butterfly effect [42] is the sensitive dependence on initial conditions in which a small change in one state of a deterministic nonlinear system can lead to large differences in a later state" [43]. These small effects can be difficult to detect in the beginning and to distinguish from bias introduced by study design and analyses. Therefore, as in the chaos theory, small causes in highly complex biomedical research can create a significantly different outcome [41].

Adding to this is the poorly described protocols of research and low-quality data presented that might not be useful or important; moreover, the poor statistical precision or inappropriately power further decreases the quality of information.

Further compounding this is the failure to involve biostatisticians and experienced clinical researchers and laboratory scientists in research methods and design and an inadequate emphasis that is placed on the reproducibility of research, which further degrades the quality of findings.

In addition, current rewarding systems, and requirements to secure tenureship, encourage quantity more than quality, and novelty more than reliability and reproducibility. Because there is more at stake in terms of career development and financial gains and securing grants, this approach has been more problematic as biopharmaceutical companies often cannot replicate the results of published work from academia [29,32,35]. Because of this, there is a heightened interest to focus on human studies for early discovery [40].

SIX RED FLAGS FOR SUSPECT SCIENTIFIC WORK

Furthermore, the crisis of reproducibility in biomedical research goes as far as back as the 1960s. Back then, scientists raised the concerns about many problems including the use of human cells used in research. Many of the cell lines used in biomedical research were often not at all what they are supposed to be. Similar concerns were highlighted in a widely cited paper, titled "Why most published research findings are false," that highlighted the problems caused by poor experimental and study design and analysis [44].

Another seminal paper published by a team at Bayer HealthCare reported that only about 25% of published preclinical studies could be validated [28].

Then, a paper published by Glenn Begley [27] made the problem even more visible. Begley's test involved 53 studies whose results were purported to be groundbreaking. Most often, the results could not be reproduced, even by the original investigators themselves. Of the 53 original groundbreaking studies that Begley examined, he could reproduce only 6 of them. A recent report published by Begley titled "Six red flags for suspect work" [27] brings forward the six most common but avoidable mistakes that he identified during his analysis of these studies. These included (1) whether experiments were performed blinded, (2) whether basic experiments repeated, (3) whether all results were presented or whether only the selected hypotheses-supporting results were presented, (4) whether the appropriate controls were used, (5) whether the reagents used in experiments were valid and of high purity and quality, and (6) whether appropriate statistical tests and methods were employed.

His conclusion was that the appropriate scientific conduct was often not applied. There is a term in the U.S. scientific community: "quick and dirty." Unfortunately, results obtained from a "quick and dirty" experiment may lead the researcher in a wrong and often dead-end path, wasting resources and costing time and money and, sometimes, ending up with dead-end careers. Traditional training methods for biomedical research is often a haphazard process. Scientists mostly learn the research process and methodologies from their mentors and often imitate them [46].

THE VALLEY OF DEATH—POOR TRANSLATION OF PRECLINICAL RESEARCH

The high attrition rates of human clinical trials and the magnitude of the reproducibility and translatability problems with the preclinical research findings to human studies have been widely recognized by scientists and biomedical researchers.

The sharp increase in costs comes in spite of significant efforts made in recent years to improve the translatability of preclinical research findings to human studies and ultimately to an approved and marketable drug status.

The difficulty in translating discoveries from basic and preclinical research to human studies and getting them approved by the FDA is well known, and this is often referred to as the "Valley of Death."

In biomedical research, the main objective of translational science is to have an in-depth understanding of the health and disease state of the organism that is studied and the underlying molecular factors contributing to disease and mechanism(s) associated with it, with the goal of "carrying across" this knowledge to treat disease and condition.

To be translational in biomedical research, information is often gathered from the wide range of molecular and cellular biology research—primarily using *in vitro* cell culture, patient primary cell, and *in vivo* animal models—and is then applied to restore a healthy state [44].

Of course, the quality of research findings (i.e., a change in pathophysiological changes leading to disease incidence and/or the development of a therapy) is predominantly contingent on the quality of the input data and the methods for their processing and interpretation. The acquisition of information that may be less relevant than anticipated further corrupts the process, as does the use of models, including animal models, that are irrelevant to human disease.

There are many reasons why translational research might fail. These could be as simple as insufficient understanding of the nature of the translational process, failure to effectively integrate the data coming from different technological approaches to disease, the use of models, access to tissues and appropriate materials, and the need for support in increasingly complex areas such as ethics, bioinformatics and biostatistics and personal privacy in relation to carrying out research in humans [44]. These greatly hinder progress in translational science.

The value of preclinical research mainly conducted in animal model experiments for predicting the effectiveness of therapies and treatment strategies in human trials has remained contentious. This is in part due to a failure of animal research findings to translate to the human studies. Translational failure may be explained in part by methodological flaws and poor experimental designs in preclinical *in vitro* and *in vivo* animal studies, leading to systematic bias and thus leading to irreproducible and unreliable data and inaccurate conclusions.

Of course, improving the quality of hypotheses before testing them will save a lot of time, resources, and grievances and, potentially warrant success. This, of course, requires careful thinking and searching the literature and conducting a proper assessment of the likelihood of success before committing to testing a hypothesis.

Crossing the "Valley of Death" is not only about the science. In addition to the scientific, design, and other reasons, this gap arises because of a systematic funding gap and also because of a knowledge and reward gap. The ability of research teams involved in the drug development process as well as project management and negotiation skills, the track record of the organization, intellectual property, market opportunity and a demonstrable competitive advantage all impact upon the ability to move the drug candidates toward FDA approval and marketed drug status.

STRATEGIES TO IMPROVE TRANSLATIONAL RESEARCH TO REDUCE ATTRITION

There is a bias to view translational research as a linear process in which mice and rats are the bridge between basic science and human clinical studies. In particular, the traditional method of identifying genes *in vitro*, followed by generating experimental animal models of human disease *in vivo*, has been challenging because, in general, the targets and drugs developed in animal models have been poorly translatable to humans [44]. Although this approach has been helpful for a better understanding of disease biology and the role of new candidate drug targets, the predictive utility of animal model experimentation is less than desired, especially in the context of studies of single knockouts in specific disease models and mouse strains [44].

Perhaps, it may be the first step to improve translational research by improving the quality of hypotheses before testing them. This will save a lot of time, resources, and agony, and potentially increase the chances of success. This, of course, necessitates meticulous thinking, planning, searching the literature and conducting a proper assessment of the likelihood of success before committing to testing a hypothesis.

For example, a more comprehensive integration of evidence coming from various *in vitro*, *in vivo* and human studies might allow the refinement of objectives and target relevance and might improve translatability of preclinical findings to humans, hence increasing the chance of successful drug development.

Another emerging strategy is the identification of candidate drugs by screening the compound libraries in less complex biological systems instead of in cell lines or primary human cells that represent disease phenotype that has long been the workhorse of the pharmaceutical industry. Currently, there is interest in the use of model organisms such as *Caenorhabditis elegans* and zebrafish in high-throughput screens for new drugs [44,47,48]. It is becoming increasingly evident that these systems can be used to identify therapeutic targets in the immune system [49] and to enable understanding of the biology of pathways and gene products and to directly assess the ability of a compound to exert its effect in complex biological processes.

In addition, there are several approaches to improve the current models. These include better phenotyping of patients and generating comprehensive genomic descriptions in patient studies to define the human disease as well as allowing access to tissues and overcoming ethical anxieties [44].

Drug repurposing (repositioning) is another strategy used to accelerate drug development and skip early phases of roadblocks. Drug repurposing evaluates drugs that are already approved to treat one indication or condition to see if they are safe and effective for treating other indication [50–61]. Repurposing existing drugs for different indications or targets drugs can be developed in 4–5 years without much risk of failure, by virtue of reverse engineering and licensing Intellectual property rights. This is because, many repurposed drugs have already passed the early phases of development, clinical safety, and bioavailability testing, thus can potentially gain FDA approvals in less than half the time and at one-quarter of the cost. However, side effects that would be acceptable for a life-threatening disease might not be acceptable for a chronic disease. Because the argument that repositioning a drug reduces the cost of drug development because safety tests are not needed because they already exist—works only if the dose and mode of administration remain similar. If the new indication necessitates a significantly higher dose, the drug will have to go through phase I trials again.

FUTURE PERSPECTIVES

Back in the twentieth century, when drug development was progressing rapidly, physicians were not trying to create a new drug entity by understanding the MOA of the drug as well as deep understanding of the biology of a disease state. The studies were conducted simply on patients—not mice or any other animal models of disease—to observe what worked. This is not to defend this historical approach, but rather to emphasize that when a proper model system is present, the drug discovery and development could be fast and efficient. It is, of course, unethical to revisit this approach; however, we can learn from this to set new strategies; that is, identify the appropriate models for a specific disease instead of solely focusing on mice.

For example, a researcher in the Philippines accidentally identified a glucose-lowering effect of an herbal extract that was originally developed to treat flu and malaria. In 1957, a French scientist tried this drug in animals and in the 1960s, British researchers tried the drug in diabetes patients and found that it lowered blood glucose levels. Jean Sterne was the first to try this drug, (i.e., metformin) on humans for the treatment of diabetes; he coined the name "Glucophage" (glucose eater) for the drug and published his results in 1957 [62,63].

Similar accidental discoveries were made for some other drugs such as minoxidil, which was initially developed for blood pressure control and later found to stimulate unusual hair growth, or more recently, a drug initially developed for high blood pressure that resulted in the blockbuster drug Viagra® (sildenafil). The lessons learned here are that the drugs do not work with pinpoint precision; rather they work as shotguns, with some off-target effects, some of which are beneficial and some clearly dangerous. It is worth remembering that many important discoveries begin with human observational studies—the top-down approach—rather than with mice—the bottom-up approach.

The recent development in biologics, nevertheless, shows the success of the bottom-up approach. Biologics are drugs designed to hit their target very specifically, and many of these drugs have translated basic biological concepts into successful treatments. Checkpoint inhibitors are one example, confirming that targeting specific pathways in the immune system to stimulate the patients' own tumor-fighting T lymphocytes is raising hopes to control cancer. These drugs have also been shown to be helpful in treating different types of cancer, including bladder cancer, non-small cell lung cancer, and Merkel cell skin cancer as well as some other types of cancer. Several drugs that target either programmed cell death protein 1 (PD-1) or programmed death-ligand 1 (PD-L1) are currently being evaluated in clinical trials either as mono- or combo-therapy with other drugs. These are drugs that target only a subset of patients with specific genetic alterations, hence the approach is called targeted therapies or precision medicine. However, these drugs can allow the immune system to attack some healthy organs in the body, leading to serious side effects in some people.

CONCLUSIONS

The future of biomedical research requires a different outlook, one that is creative and embraces the underlying philosophy of ethical science. Adopting new technologies and methods of scientific research to better understand disease biology in humans can help to facilitate this transition.

Therefore, we may need to reevaluate the idea of studying human disease in humans: The consensus is building around the need to refocus and adapt new methodologies for use in humans in order to better understand disease biology in humans. These new methodologies may include the better use of the more complex human conditions and data rather than relying on mouse and other animal models to study human diseases. In addition, other methods such as the use of human volunteers, epidemiological and clinical data sets, mathematical and computer (*in silico*) models, *in vitro* human tissue or organ models, and computer-modeling studies will improve overall biomedical research.

Collectively, it may be that better phenotypic screening, more useful model studies, and careful observations of patient populations might be the way to go in prioritizing and developing new, more efficacious, safer, and cost-effective drugs perhaps faster than what it takes now.

Finally, the successful development and implementation of alternative strategies, next-generation models systems, and methods that limit or even avoid the use of animals require cooperation with multiple stakeholders including the biopharmaceutical industry, universities and other research institutions, local and international regulatory agencies and associations, such as the International Life Science Institute.

REVIEW CRITERIA

Publicly available information such as PubMed and Internet were used for the literature review. We focused on identifying articles published on translational research, the use of animal models for human diseases, preclinical and clinical research, biomedical research, drug development, reproducibility of biomedical research data, failure of drug development and clinical trials. The search was restricted to the most recent studies in this field, and all searches were limited to human studies published in English.

REFERENCES

1. Waring MJ, Arrowsmith J, Leach AR, Leeson PD, Mandrell S, Owen RM et al. An analysis of the attrition of drug candidates from four major pharmaceutical companies. *Nat Rev Drug Discov.* 2015;14: 475–486.
2. Shuler ML. Organ, body and disease-on-a-chip systems. *Lab Chip.* 2017;17: 2345–2346.
3. DiMasi JA, Hansen RW, Grabowski HG. The price of innovation: New estimates of drug development costs. *J Health Econ.* 2003;22: 151–185.
4. Morgan S, Grootendorst P, Lexchin J, Cunningham C, Greyson D. The cost of drug development: A systematic review. *Health Policy.* 2011;100: 4–17.
5. Garner JP, Gaskill BN, Weber EM, Ahloy-Dallaire J, Pritchett-Corning KR. Introducing Therioepistemology: The study of how knowledge is gained from animal research. *Lab Anim.* 2017;46: 103–113.
6. Paul SM, Mytelka DS, Dunwiddie CT, Persinger CC, Munos BH, Lindborg SR et al. How to improve R&D productivity: The pharmaceutical industry's grand challenge. *Nat Rev Drug Discov.* 2010; doi:10.1038/nrd3078.
7. Kola I, Landis J. Can the pharmaceutical industry reduce attrition rates? *Nat Rev Drug Discov.* 2004;3: 711–715.
8. Scannell JW, Blanckley A, Boldon H, Warrington B. Diagnosing the decline in Pharmaceutical R&D efficiency. *Nat Rev Drug Discov.* 2012;11: 191–200.
9. Hay M, Thomas DW, Craighead JL, Economides C, Rosenthal J. Clinical development success rates for investigational drugs. *Nat Biotechnol.* 2014;32: 40–51.
10. DiMasi JA, Feldman L, Seckler A, Wilson A. Trends in risks associated with new drug development: Success rates for investigational drugs. *Clin Pharmacol Ther.* 2010;87: 272–277.
11. Arrowsmith J, Miller P. Trial watch: Phase II and phase III attrition rates 2011–2012. *Nat Rev Drug Discov.* 2013;12: 569.
12. Mullin R Cost to develop new pharmaceutical drug now exceeds $2.5B: A benchmark report estimates that the cost of bringing a drug to market has more than doubled in the past 10 years. *Scientific American,* November 24, 2014. [cited September 28, 2017]. Available: https://www.scientificamerican.com/article/cost-to-develop-new-pharmaceutical-drug-now-exceeds-2-5b/
13. DiMasi JA, Grabowski HG, Hansen RW. Innovation in the pharmaceutical industry: New estimates of R&D costs. *J Health Econ.* 2016;47: 20–33.
14. Nosengo N. Can you teach old drugs new tricks? *Nature.* 2016;534: 314–316.
15. Seyhan AA. RNAi: A potential new class of therapeutic for human genetic disease. *Hum Genet.* 2011;130: 583–605.
16. Lindsay MA. Target discovery. *Nat Rev Drug Discov.* 2003;2: 831–838.
17. Vane JR. Inhibition of prostaglandin synthesis as a mechanism of action for aspirin-like drugs. *Nat New Biol.* 1971;231: 232–235.
18. Vane JR, Botting RM. The mechanism of action of aspirin. *Thromb Res.* 2003;110: 255–258.
19. Oates J. The 1982 Nobel Prize in physiology or medicine. *Science.* 1982;218: 765–768.
20. Santos R, Ursu O, Gaulton A, Bento AP, Donadi RS, Bologa CG et al. A comprehensive map of molecular drug targets. *Nat Rev Drug Discov.* 2017;16: 19–34.
21. Schenone M, Dančík V, Wagner BK, Clemons PA. Target identification and mechanism of action in chemical biology and drug discovery. *Nat Chem Biol.* 2013;9: 232–240.
22. Seyhan AA, Rya TE. RNAi screening for the discovery of novel modulators of human disease. *Curr Pharm Biotechnol.* 2010;11: 735–756.
23. Seyhan AA, Varadarajan U, Choe S, Liu Y, McGraw J, Woods M et al. A genome-wide RNAi screen identifies novel targets of neratinib sensitivity leading to neratinib and paclitaxel combination drug treatments. *Mol Biosyst.* 2011;7: 1974–1989.
24. Swinney DC, Anthony J. How were new medicines discovered? *Nat Rev Drug Discov.* 2011;10: 507–519.

25. Bajorath J. Faculty of 1000 evaluation for clinical development success rates for investigational drugs [Internet]. F1000—Post-publication peer review of the biomedical literature. 2015. doi:10.3410/f.718236299.793509111
26. Bunnage ME. Getting pharmaceutical R&D back on target. *Nat Chem Biol.* 2011;7: 335–339.
27. Ma P, Zemmel R. From the analyst's couch: Value of novelty? *Nat Rev Drug Discov.* 2002;1: 571–572.
28. Colquhoun D. An investigation of the false discovery rate and the misinterpretation of p-values. *R Soc Open Sci.* 2014;1: 140216.
29. Rosenblatt M. An incentive-based approach for improving data reproducibility. *Sci Transl Med.* 2016;8: 336ed5.
30. Pusztai L, Hatzis C, Andre F. Reproducibility of research and preclinical validation: Problems and solutions. *Nat Rev Clin Oncol.* 2013;10: 720–724.
31. Begley CG. Six red flags for suspect work. *Nature* 2013;497: 433–434.
32. Prinz F, Schlange T, Asadullah K. Believe it or not: How much can we rely on published data on potential drug targets? *Nat Rev Drug Discov.* 2011;10: 712.
33. Mak IW, Evaniew N, Ghert M. Lost in translation: Animal models and clinical trials in cancer treatment. *Am J Transl Res.* 2014;6: 114–118.
34. Cummings JL, Morstorf T, Zhong K. Alzheimer's disease drug-development pipeline: Few candidates, frequent failures. *Alzheimers Res Ther.* 2014;6: 37.
35. Sena ES, van der Worp HB, Bath PMW, Howells DW, Macleod MR. Publication bias in reports of animal stroke studies leads to major overstatement of efficacy. *PLoS Biol.* 2010;8: e1000344.
36. Zahs KR, Ashe KH. "Too much good news"—Are Alzheimer mouse models trying to tell us how to prevent, not cure, Alzheimer's disease? *Trends Neurosci.* 2010;33: 381–389.
37. Van der Worp HB, Howells DW, Sena ES, Porritt MJ, Rewell S, O'Collins V et al. Can animal models of disease reliably inform human studies? *PLoS Med.* 2010;7: e1000245.
38. Peers IS, Ceuppens PR, Harbron C. In search of preclinical robustness. *Nat Rev Drug Discov.* 2012;11: 733–734.
39. Macleod MR, Van der Worp HB, Sena ES, Howells DW, Dirnagl U, Donnan GA. Evidence for the efficacy of NXY-059 in experimental focal cerebral ischaemia is confounded by study quality. *Stroke.* 2008;39: 2824–2829.
40. Garner JP. The significance of meaning: Why do over 90% of behavioral neuroscience results fail to translate to humans, and what can we do to fix it? *ILAR J.* 2014;55: 438–456.
41. Tricklebank MD, Garner JP. Chapter 20. The possibilities and limitations of animal models for psychiatric disorders. *Drug Discov.* pp. 534–557.
42. Begley CG, Ellis LM. Drug development: Raise standards for preclinical cancer research. *Nature.* 2012;483: 531–533.
43. Hunter J. Challenges for pharmaceutical industry: New partnerships for sustainable human health. *Philos Trans A Math Phys Eng Sci.* 2011;369: 1817–1825.
44. Sabroe I, Dockrell DH, Vogel SN, Renshaw SA, Whyte MKB, Dower SK. Identifying and hurdling obstacles to translational research. *Nat Rev Immunol.* 2007;7: 77–82.
45. Ioannidis JPA. Clinical trials: What a waste. *BMJ* 2014;349: G7089–G7089.
46. Ioannidis JPA, Greenland S, Hlatky MA, Khoury MJ, Macleod MR, Moher D et al. Increasing value and reducing waste in research design, conduct, and analysis. *Lancet* 2014;383: 166–175.
47. Kwok TCY, Ricker N, Fraser R, Chan AW, Burns A, Stanley EF et al. A small-molecule screen in C. Elegans yields a new calcium channel antagonist. *Nature* 2006;441: 91–95.
48. Zon LI, Peterson RT. In vivo drug discovery in the zebrafish. *Nat Rev Drug Discov.* 2005;4: 35–44.
49. Renshaw SA, Loynes CA, Trushell DMI, Elworthy S, Ingham PW, Whyte MKB. A transgenic zebrafish model of neutrophilic inflammation. *Blood.* 2006;108: 3976–3978.
50. Strittmatter SM. Overcoming drug development bottlenecks with repurposing: Old drugs learn new tricks. *Nat Med.* 2014;20: 590–591.
51. Yang YS, Marder SR, Green MF. Repurposing drugs for cognition in schizophrenia. *Clin Pharmacol Ther.* 2016;101: 191–193.
52. Sleigh SH, Barton CL. Repurposing strategies for therapeutics. *Pharmaceut Med.* 2010;24: 151–159.
53. Pollak M. Overcoming drug development bottlenecks with repurposing: Repurposing biguanides to target energy metabolism for cancer treatment. *Nat Med.* 2014;20: 591–593.
54. Zhang L, Kebebew E. Repurposing existing drugs for the treatment of thyroid cancer. *Expert Rev Endocrinol Metab.* 2012;7: 369–371.
55. Ranjan A, Srivastava SK. Abstract 1251: Repurposing antipsychotic drug Penfluridol for cancer treatment. *Cancer Res.* 2014;74: 1251–1251.
56. Fagan SC. Drug repurposing for drug development in stroke. *Pharmacotherapy.* 2010;30: 51S–54S.
57. Mercado G, Hetz C. Drug repurposing to target proteostasis and prevent neurodegeneration: Accelerating translational efforts. *Brain* 2017;140: 1544–1547.
58. M. Telleria C, Telleria CM. Drug repurposing for cancer therapy. *J Cancer Sci Ther.* 2012;4; doi:10.4172/1948-5956.1000e108.
59. Sharlow ER. Revisiting repurposing. *Assay Drug Dev Technol.* 2016;14: 554–556.
60. Zheng W, Sun W, Simeonov A. Drug repurposing screens and synergistic drug-combinations for infectious diseases. *Br J Pharmacol.* 2017; doi:10.1111/bph.13895.
61. Mucke HAM, Mucke E. Sources and targets for drug repurposing: Landscaping transitions in therapeutic space. *DRRR.* 2015;1: 22–27.
62. Campbell IW. Metformin—Life begins at 50: A symposium held on the occasion of the 43rd Annual meeting of the European Association for the Study of Diabetes, Amsterdam, The Netherlands, September 2007. *Br J Diabetes Vasc Dis.* 2007;7: 247–252.
63. Bailey CJ, Day C. Metformin: Its botanical background. *Practical Diabetes International.* 2004;21: 115–117.

Necessary Characterization of Biomarkers to Be "Fit for Purpose"

3.2

Suso Platero

Contents

From preclinical research to the clinic, biomarkers play a significant role in building greater efficiencies into the drug development process. High rates of failure for new therapeutics and an ongoing need for stronger translational models have continued to drive biomarker advancement as a tool to demonstrate the safety and efficacy of new therapeutics while reducing development time and costs. Biomarkers hold a key role in advancing personalized medicine strategies and targeted therapies by stratifying patients, matching the right patient to the right clinical trials, and minimizing the potential for adverse effects.

Whether driving early development go/no-go decisions or identifying an individual's response to a treatment in the clinic, each fit-for-purpose application of a biomarker must present sufficient sensitivity and specificity for its context of use (COU). Applying the appropriate scientific rigor to characterize a biomarker assay can vary based on the type of biomarker and its role in the development process, ranging from driving internal decisions to supporting label claims.

The fit-for-purpose approach has been discussed in numerous publications as tiered or iterative methods that are "tailored to meet the intended purpose of the biomarker study" (Lee et al., 2006). From a regulatory guidance perspective, fit for purpose was mentioned in the 2013 FDA Draft Guidance for Bioanalytical Method Validation, where sponsors are advised to "incorporate the extend of method validation they deem appropriate" to support early drug development and fully validate bioanalytical assays that support safety, effectiveness or dosing instructions (U.S. Food and Drug Administration, 2013). Because it relates to biomarker assays, many drug development stakeholders have cited the need for greater clarity and a generalized framework to build greater consistency into the process for qualifying and validating biomarker assays across a wide range of applications.

This section focuses on the types of novel biomarkers, described in Table 3.2.1, following the terminology recommended by the FDA in the Biomarkers, EndpointS, and other Tools (BEST) Resource (FDA-NIH 2016). Based on these biomarker types, this section also discusses a published framework that can help determine the most appropriate, fit-for-purpose method for biomarker characterization.

In preclinical research, **PD biomarkers** can show a biological response or change after treatment with the compound, providing crucial evidence to assess that the agent modulates its desired target. Other PD biomarkers can provide insights on a molecule's mechanism of action (MOA). These exploratory biomarkers can demonstrate that a drug hits its target and appropriately affects the biochemical pathway. After homing in on lead compounds in preclinical models, first-in-human trials can help validate the proposed pathway of activity of the drug to elucidate the pathway that the drug is inhibiting, uncover possible secondary effects and provide information on the optimum dose. The PD biomarkers that were run in preclinical models need to be run in the human trials. New assays need to be developed that are human specific. Unfortunately, a lot of times what works well in models do not translate to humans.

PD biomarkers are well suited for preclinical and phase I clinical trials, but often they do not translate into the later stages

TABLE 3.2.1 Types of biomarkers and how they are used in drug development

MARKER	*FUNCTION*	*TEST*	*EXAMPLE*
PD response	• Determine if a drug hit the target and has impact on the biological pathway • Evaluate the MOA • PK/PD correlations and determine the dose and schedule • Determine the biologically effective dose	• Research text used during drug development • Not developed as a CDx	HbA1c
Predictive	• Identify patients most likely to respond (or least likely to suffer an adverse event) to the treatment	Complementary/ CDx	HER2/neu, ALK translocation, PD-L1
Safety	• Indicate the presence or extent of toxicity related to an intervention or exposure to treatment	Approved test	Kim-1
Susceptibility/risk	• Indicate the potential for developing a disease or sensitivity to an exposure	Approved test	BRCA
Diagnostic	• Confirm a disease or subset of a disease state	Approved tests	Sweat chloride testing for cystic fibrosis
Prognostic	• Predict course of disease independent of any specific treatment modality	Approved tests	CellSearch, MammaPrint
Monitoring	• Detect a change in a disease or provide evidence of exposure to a treatment	Toxicity or safety testing	HCV-RNA for chronic hepatitis C

of the drug development process or serve as a companion diagnostic. Similarly, the data generated by these biomarkers are not reviewed by regulatory agencies and may only need minimum qualification, not extensive method validation.

This fit-for-purpose approach with PD biomarkers helps conserve early development resources. For example, patients in early phase II proof-of-concept clinical trials for nonalcoholic steatohepatitis (NASH) often undergo liver biopsies in long-term (>48 week) studies (Williams et al., 2017) to assess disease progression and the effect of a treatment. Noninvasive biomarkers, based on the MOA of the treatment, can be preselected for these NASH studies to determine whether the treatment has the desired impact expected on histological outcomes. Not only is this use of noninvasive evaluation easier for patients, it can significantly reduce the early clinical development timeline.

Predictive biomarkers are used to guide therapy choices or determine patient eligibility by providing information about likely outcomes of responding to, or resisting, a specific intervention or treatment. These markers typically target a single genetic or proteomic change, such as gene mutation or alteration in the structure or expression of a protein.

The staining of breast or gastric cancers for human epidermal growth factor receptor 2 (HER2) using immunohistochemistry (IHC) is one example of a sensitive predictive biomarker. Overexpression of HER2 indicates a higher likelihood for a treatment response and helps patients without the biomarker avoid unnecessary treatment. Predictive biomarkers can also help identify mechanisms driving acquired drug resistance. For example, panitumumab (an anti-EGFR antibody approved for treatment of colorectal cancer) is contraindicated if the person's tumor tissue has a KRAS mutation, a feature that allows clinicians to more accurately prescribe the targeted treatment and optimize care (Amado et al., 2008).

In a clinical trial for targeted therapies, predictive markers can help inform the trial design by studying only groups of patients that are expected to benefit or forecast the extent to which the drug can be effective, based on the presence or absence of a biomarker. A randomized clinical trial (RCT) is an ideal setting for establishing the utility of a predictive biomarker for an experimental targeted therapy (Polley et al., 2013). The biomarker status is obtained on the patients but does not inform treatment decisions. Several alternative trial designs can also evaluate the treatment effect in biomarker-defined subpopulations, even trials that only accrue positive biomarker patients. The risk with this later approach is that if the end point difference is not large with regard to the control arm, the drug may not be approved.

Assessing clinical utility of biomarkers can also be performed retrospectively, for example, as in the case of KRAS mutations predicting nonresponse to epidermal growth factor receptor-directed antibodies (Garcia et al., 2011).

Precision medicine in immuno-oncology is one area that has highlighted the utility of predictive biomarkers for targeting the right patients. The expression of PD-L1 has shown which patients would likely respond in some human malignancies, particularly for melanoma, non-small cell lung cancer, renal cell carcinoma, and bladder cancer (Ohaegbulam et al., 2015). Interestingly, we are starting to see a trend in which the drug with no predictive biomarker (e.g., Opdivo [PD1 made by BMS]) is slowly loosing market to the drug with a predictive biomarker (e.g., Keytruda [PD1 made by Merck]).

Safety biomarkers indicate the likelihood, presence, or extent of toxicity as an adverse effect by measuring before or after exposure to the product or agent. Preclinical studies can use urinary kidney biomarkers such as Kim-1 and urinary cystatin C, which have been qualified by the FDA to detect drug-induced nephrotoxicity in nonclinical studies. With period monitoring, safety biomarkers can also signal the need for adjusting a dose, such as hyperkalemia with an aldosterone antagonist, or show the need for an additional treatment, such as a potassium supplementation when detecting hypokalemia with a diuretic. Safety biomarkers can also be used as eligibility criteria in a clinical protocol to identify patients who may be likely to experience adverse events or not respond to a critical treatment.

Diagnostic biomarkers can confirm the presence of a disease or condition or identify individuals with a subtype of the disease. As a diagnostic biomarker, the commonly used sweat chloride test can confirm cystic fibrosis, but certain mutations on the cystic fibrosis transmembrane conductance regulator (CFTR) can serve as a predictive biomarker to determine the likelihood of treatment response. In kidney disease, estimated glomerular filtration rate (eGFR) is a biomarker used to assess disease stage because eGFR typically decreases with disease progression. In patients with heart failure, a diagnostic biomarker can categorize heart failure with reduced ejection fraction (HFrEF) or preserved ejection fraction (HFpEF) because the treatment recommendations may be different between these subgroups.

Prognostic biomarkers can inform about the likely disease outcome, independent of the treatment received; they are typically used in a treatment setting. At the most basic level in oncology, this can include the tumor size or presence of metastasis. In early phases of drug development, identifying and validating a novel prognostic biomarker can serve as a valuable tool for establishing target patient populations for a clinical trial.

Susceptibility/risk biomarkers can guide preventative strategies by indicating an increased or even decreased chance to develop a disease or medical condition in an individual who does not currently have clinically apparent disease or the medical condition. Elevated low-density lipoprotein cholesterol (LDL-C) levels can signal an increased risk of coronary artery disease; apolipoprotein E (APOE) gene variations may identify a predisposition to develop Alzheimer's disease (Genin et al., 2011).

Beyond evaluating the likelihood of a disease or medical condition, susceptibility/risk biomarkers can also function as prognostic biomarkers for clinical trial enrichment by defining subsets of clinical trial participants that may be more likely to develop the disease or medical condition. In turn, this can also inform the study design, including preventative interventions in the case of possible side effects.

Monitoring biomarkers are measured serially to assess the status of a disease or medical condition. They can also provide evidence of exposure to, or the effect of, a treatment or agent. In prostate cancer, the commonly used prostate-specific antigen (PSA) has been used as a monitoring biomarker to assess disease status or burden in patients (Freedland et al., 2007). Another one, the symphysis-fundal height (SFH) serves as a monitoring biomarker to screen for fetal growth disturbances (Papageorghiou et al., 2016).

Surrogate end points are related to these types of biomarkers because they are intended to be used as a substitute for clinically meaningful end points. They are faster and easier to study, especially when observing the effect on a direct end point may not be feasible. For example, trials with events such as stroke or a myocardial infarction could use the surrogate end point of hypertension; intraocular pressure can serve as a surrogate end point for the loss of vision in glaucoma patients.

Surrogates can be used as end points and provide early indicators of drug efficacy if they correlate with a true clinical outcome—and if the treatment effect on the surrogate captures the full effect of the treatment on the clinical end point. The changes induced on a surrogate end point from the treatment are expected to reflect changes in a meaningful end point. Demonstrating the treatment effect with a surrogate can be difficult to validate. These biomarkers require extensive preclinical evaluation, analytical validation and clinical qualification.

CREATING A STRUCTURED PROCESS

The Biomarkers Consortium Evidentiary Standards Writing Group published a "Framework for Defining Evidentiary Criteria for Biomarker Qualification" (2016), outlining a proposed structure for aligning multiple stakeholders on how to best support biomarker qualification for regulatory use. The framework is intended to improve the overall quality of submissions and "enhance the clarity, predictability, and harmonization of the process" to assist the development of biomarkers for qualification.

Starting with a needs statement, researchers should identify the gap that the proposed biomarker will address, along with the current understanding of the biomarker's role, and how the biomarker will be used in the drug development context. This can include the biomarker's ability to significantly improve clinical outcomes, for example, and how this compares with current practices. The needs statement also feeds into the COU statement as part of biomarker qualification. A well-defined COU describes the proposed category of biomarker and the information it is intended to provide. The COU may evolve as greater understanding of the biomarker develops over time.

Next, the framework recommends assessing the risk and benefit and the risk mitigation strategy. This section broadly describes the qualities of improved sensitivity or selectivity, for example, and can also include addressing the consequences of false positives or negatives as it relates to the impact on the patient.

With the overall risk and benefits documented, the evidentiary criteria for biomarker qualification can be developed. This involves documenting the relationship of the biomarker to its desired clinical outcome, the biological rational for use of the biomarker, the study design and type of data to collect, the use of independent data sets, comparisons with the current standards, assay performance and the statistical methods to use.

With all the elements of evidentiary criteria for qualification gathered, stakeholders are ideally aligned and empowered to make informed decisions on level of assay validation required and then outline a fit-for-purpose method.

Fit-for-purpose validation can be thought of as an overarching umbrella term that describes the stages of the validation process. The often-cited position paper by Lee et al. (2006) outlines the biomarker assay validation process and organizes it into iterative, interconnected fit-for-purposes approaches: prevalidation, exploratory validation, in-study validation and advanced validation.

Once a biomarker is evaluated within the framework, validation can determine if the assay meets its intended purposed, provides reliable data and is suited for its proposed application (Ray, 2005). The level of analytical validation required, such as the methods, calibration standards, accuracy, precision, reproducibility, linearity, quality specifications, is influenced by the stage of development.

The evidence required and degree of biological understanding needed can greatly vary depending on the question that the biomarker is intended to address and the level of acceptable risk or reliability provided by the biomarker. Exploratory biomarkers used to inform internal decision making may not need extensive validation. Biomarkers that support PD end points or primary or secondary end points, such as therapeutic efficacy and dose selection, require extensive validation to support regulatory filings and prove their reliability and accuracy.

Based on the level of validation needed, there are several properties that can be evaluated in the process. Analytical validation is the process of evaluating the performance characteristics of an assay. The test must be reproducible, and it must accurately and reliably measure the analyte of interest. Clinical *qualification*, a term used to avoid confusion with the process of the method *validation*, describes how the assay results correlate with the desired clinical outcome, for example, linking a biomarker to both biologic and clinical end points.

The availability of a reference standard and its limits of data produced must also be taken into account in the assay development process. Lee and colleagues (2003) describe these categories of biomarker assay data as *definitive quantitative*, *relative quantitative*, *quasi-quantitative* and *qualitative*. A definitive quantitative assay can use a well-defined reference standard that is "fully representative of the endogenous biomarker," such as steroids. A relative quantitative assay relies on calibration curve for the reference standard, but this reference may not fully represent the endogenous or heterogeneous biomarker; therefore, accuracy can only be estimated. A quasi-quantitative assay does not use calibration standards but can demonstrate precision, for example, through the use of flow cytometry for single-cell analysis such as the activation status. Finally, a qualitative assay may provide a positive or negative response, such as the presence or absence of a genomic mutation.

OPERATIONAL CONSIDERATIONS IN PREVALIDATION

Pre-analytical considerations and method development involves evaluating the potential impact of factors on the resulting data, such as obtaining specimens in sufficient quantities and maintaining stability to successfully assay the specimen and product reliable results. Crafting and following standard operating procedures (SOPs) to document development steps helps form best practices and ensures the analyte remains stable from collection through processing, thereby reducing a potentially major source of bias.

Reagents used in the assay should also undergo qualification because they can affect the stability of an assay. Qualification should involve reproducing situations that mimic actual sample collection, storage and analysis, rather than relying on values and recommendations in literature. The stability windows and expiration dates of reagents should also be documented to ensure reproducibility of an assay.

EXPLORATORY VALIDATION

With the prevalidation steps defined and documented, exploratory validation can help evaluate a drug candidate at a high level and inform internal decision making. Following the fit-for-purpose model, a "quick and dirty" development approach may be appropriate for identifying large-scale changes in new biomarker research. At a more detailed level, a higher level of reliability is required to inform go/no-go decisions, thus requiring a slightly greater level analytical rigor and validation. However, for the most part, exploratory biomarkers are not translated to the clinic; therefore, a less thorough investigation is required for validation and documentation as compared with biomarkers supporting regulatory filings.

To confirm relative accuracy for exploratory biomarkers, running an additional, complementary experiment and/or an orthogonal method comparison may help improve confidence in the results. More sophisticated, in-depth technologies such as mass spectrometry and protein separation can provide a greater characterization of the biomarker, but sponsors need to balance the factors such as time, expertise, and cost against their tolerance for risk and uncertainty along with the level of reliability needed. With these factors defined and weighed against each other, sponsors can create the most appropriate fit-for-purpose method for the exploratory validation of their biomarkers.

CONSIDERATIONS FOR FULL VALIDATION

If the biomarker is intended for use in the clinic, proactive planning is key. The target population should be defined in advance during early development to help guide clinical evaluation and implementation. In addition, it is a helpful exercise to define possible confounding factors, determine how to connect to clinically valid surrogate end points and list the potential limitations in interpreting the assay. Discussing and documenting this level of detail can help illuminate potential pitfalls and reduce the risk of failure in development.

After internal validation, researchers should consider how candidate biomarkers can be validated and adapted to a variety of assay platforms and technologies, each with their own utility, depending on the purpose of the biomarker. Many development processes rely on genomic technologies, which have continued to advance and offer more powerful, rapid identification of genetic aberrations, enhancing the industry's ability to leverage biomarker targets. Multiplex technologies, such as next-generation sequencing (NGS) and gene expression profiling, can examine genomic variations and discover clinically relevant biomarkers with increasing power and speed, including whole exome sequencing (WES) and whole genome sequencing (WGS). Iterative applications of NGS in a study could also serve as a monitoring biomarker to pinpoint changes of total mutational loads over time (McShane et al., 2013).

Considerations in the sample collection method may also factor into choosing the most appropriate technology to employ. Liquid biopsy technologies also have gained recognition as noninvasive methods to provide insights on drug resistance and potentially help determine appropriate, targeted treatments by identifying and analyzing circulating tumor cells (CTCs) and circulating free tumor DNA (cfDNA) from blood.

Given that a biomarker is also described as "a response to an exposure or intervention" per the FDA, the measurement of the rate at which a cell gains mass can be considered one such response. One emerging technology is a single-cell functional assay, which can determine mass accumulation rate of a single cancer cell, using a microfluidic device. Researchers can observe the reduction in mass in a single cell after exposure to targeted small-molecule therapies, as compared with cells with a resistance mutation that maintain a mass accumulation rate matching that of the control conditions (Cermak et al., 2016). As with the case of exploratory biomarker validation, sponsors need to evaluate what will be measured, the level of sensitivity and precision required and determine which technology best meets those needs.

PARAMETERS FOR FULL VALIDATION

Regardless of the technology used to support biomarker development, fully validated biomarkers need to consider the following assay performance characteristics: the calibration/standard curve, validation samples and quality controls, precision and accuracy, limit of quantification, biomarker range assessment, parallelism, dilutional linearity, specificity and interference, and stability. Rather than going into the intricate procedural and statistical guidelines provided by Lee et al. (2006) and Ray (2005), this section provides a high-level overview of the purpose of each biomarker validation attribute.

Calibration/standard curve: The nonlinear calibration curve is used to investigate the concentration–response relationship between the calibrators and test sample analyte. These matrices can be complicated by endogenous analyte, which can be removed or a substitute calibrator matrix can be used.

Validation samples and quality controls: Validation samples serve as quality controls to estimate accuracy within and between runs. They can also serve as quality control samples during in-study validation. Both can "assess the ability of the assay to measure the biomarker of interest for its intended use" (Lee et al., 2006) and, therefore, should closely mimic the study samples. The reference materials vary depending on the technology or assay used.

Precision and accuracy: According to ISO 5725-1, Accuracy (Trueness and Precision) of Measurement Methods and Results (1994), precision is "the closeness of agreement between independent test results obtained under stipulated conditions" and includes repeatability (the variability observed within-run or within-day), intermediate precision (repeatability between-run or between-day), and reproducibility (varied factors and measurements showing reproducible conditions). Precision is usually quantified by the standard deviation and coefficient of variation (CV) as an estimate of imprecision.

Limit of quantification: The range of a method is bracketed by the lower limits of quantification (LLOQ) and upper limits of quantification (ULOQ), each specific for the lowest and highest analyte concentration, respectively, that can be quantified with acceptable precision and accuracy, where an "acceptable" value can vary based on the defined context of use for the biomarker.

Biomarker range assessment: This assessment evaluates the concentration range of the biomarker to understand the distribution of the concentration among a select population. Understanding this value highlights biological variation and can feed into study design. For example, LabCorp (Laboratory Corporation of America Holdings) often queries its database of test results from de-identified patients to analyze real-world data, leveraging biomarkers values to determine the likelihood that a select population will meet eligibility criteria. These data can help shape a clinical protocol and optimize trial design.

Parallelism: When a test sample is serially diluted, is should demonstrate parallelism, meaning that the sample dilution response curve is parallel to the standard concentration response curve and does not result in biased measurements of the analyte concentration (Miller et al., 2001). The goal of investigating the parallelism is to ascertain that the binding characteristic of the endogenous analyte to the antibodies is the same as for the calibrator (Andreasson et al., 2015). This step can be completed in the pre-study validation stage to document any issues that may arise in final evaluation. Acceptance criteria for the CV depend on the context of use for the assay.

Dilution linearity: Related to the concept of parallelism, dilution linearity uses samples spiked with a high concentration of the analyte above the ULOQ. Then it is diluted back down within the working range to determine whether a reliable result has been achieved. This process also evaluates if high concentrations above the ULOQ suppress the signal in what is called the "hook effect."

Specificity and interference: Specificity measures the accuracy of a method in the presence of interfering molecules that may resemble, but differ, from the analyte.

Stability: Also called robustness, this measures the precision of the assay by following changes in conditions such as temperature or the storage condition of reagents. Both short-term and long-term stability must be collected to better understand the potential conditions at the clinic study site.

UNDERSTANDING THE NEED FOR FULLY VALIDATED BIOMARKERS

Whether exploratory or fully validated, ensuring a fit-for-purpose approach for biomarker assay development requires accounting for many considerations about the current and future context of use. The development of a predictive biomarker for an approved treatment illustrates one example of the biomarker journey from the bench to the clinic and post-marketing trials.

The immunotherapy Keytruda® (pembrolizumab) has been associated with many historical advances as it relates to biomarkers. First, it used an *in vitro* diagnostic device with a qualitative immunohistochemical assay to guide therapy decisions for non-small cell lung cancer (NSCLC) patients; this indication was later expanded to gastric or gastroesophageal junction adenocarcinoma tumors that express PD-L1. Next, pembrolizumab became the first drug to be approved based on the results of different biomarkers-guided indication—agnostic of the origin of tumor. Using either a mismatch repair deficient (dMMR) solid tumor biomarker or a microsatellite instability-high (MSI-H) tumor as a predictive biomarkers, the treatment can now be applied to many additional types of cancers that have those biomarkers.

MSI testing is regularly performed in most diagnostic laboratories by evaluating selected microsatellite sequences using a simple polymerase chain reaction (PCR) assay; an IHC method can be used to detect dMMR. As a predictive biomarker, MSI and/or dMMR status can predict which patients would likely derive clinical benefit from PD-1/PD-L1 pathway inhibitors such as pembrolizumab.

Interestingly, the FDA approved the drug but asked the company to conduct further studies to assure the biomarkers were fully validated, and since the trial was run, several assays were used for each of those biomarkers.

Among other studies to characterize the safety and effectiveness in pediatric patients, the FDA asked for the following:

> *Commitment to support the availability through an appropriate analytical and clinical validation study using clinical trial data that will support labeling of an* ***immunohistochemistry based*** **in vitro** ***diagnostic device*** *that is essential to the safe and effective use of pembrolizumab for patients with tumors that are mismatch repair deficient.*
>
> *Commitment to support the availability through an appropriate analytical and clinical validation study using clinical trial data that will support labeling of a* ***nucleic acid-based*** **in vitro** ***diagnostic device*** *that is essential to the safe and effective use of pembrolizumab for patients with tumors that are microsatellite instability high (FDA, 2017).*

Even though MMR testing with IHC is common in the clinic, this example highlights the need for analytical and clinical validation steps. In what is often called a "rear-view mirror" approach, cost and complexity is added to the development process when a companion diagnostic is introduced in the clinic or even after the therapy is on the market.

Today, many sponsors are evaluating novel pathways in personalized medicine to better tailor their compounds to targeted populations (Agarwal et al., 2015), signaling the growing inclusion of a companion diagnostic with a treatment. Integrating the biomarker validation process, starting in early development, can not only help guide scientific conclusions but can also influence the design of more strategic clinical trials and, ultimately, advance personalized medicine programs faster.

REFERENCES

Agarwal A, Ressler D, Snyder G. The current and future state of companion diagnostics. *Pharmacogenomics Person Med.* 2015;8:99–110. doi:10.2147/PGPM.S49493.

Amado RG, Wolf M, Peeters M et al. Wild-type KRAS is required for panitumumab efficacy in patients with metastatic colorectal cancer. *J Clin Oncol.* 2008;26(10):1626–1634.

Andreasson U, Perret-Liaudet A, van Waalwijk van Doorn LJC et al. A practical guide to immunoassay method validation. *Front Neurol.* 2015;6:179. doi:10.3389/fneur.2015.00179.

Cermak N, Olcum S, Delgado FF et al. High-throughput measurement of single-cell growth rates using serial microfluidic mass sensor arrays. *Nat Biotechnol.* 2016;9;34(10):1052–1059. doi: 10.1038/nbt.3666. Epub 2016 Sep 5.

Evidentiary Criteria Writing Group, Framework for Defining Evidentiary Criteria for Biomarker Qualification: Final Version (2016).

FDA correspondence for BLA 125514/S-14. https://www.accessdata.fda.gov/drugsatfda_docs/appletter/2017/125514Orig1s014ltr.pdf. Accessed December 11, 2017.

FDA-NIH Biomarker Working Group. 2016. *BEST (Biomarkers, EndpointS, and other Tools) Resource* [Internet]. Silver Spring, MD: Food and Drug Administration. Available from: https://www.ncbi.nlm.nih.gov/books/NBK326791/.

Freedland SJ, Moul JW. Prostate specific antigen recurrence after definitive therapy. *J Urol.* 2007;177(6):1985–1991.

Garcia, V. M., Cassier, P. A. and de Bono, J. Parallel anticancer drug development and molecular stratification to qualify predictive biomarkers: Dealing with obstacles hindering progress. *Cancer Discov.* 2011;1, 207–212.

Genin E, Hannequin D, Wallon D et al. APOE and Alzheimer disease: A major gene with semi-dominant inheritance. *Mol Psychiatry* 2011;16(9):903–907.

ISO 5725-1. Accuracy (Trueness and Precision) of Measurement Methods and Results—Part 1: General Principles and Definitions (1994).

Lee JW, Smith WC, Nordblom GD, Bowsher RR. Validation of assays for the bioanalysis of novel biomarkers. In C. Bloom, R. A. Dean (eds.), *Biomarkers in Clinical Drug Development*, Marcel Dekker, New York, 2003, pp. 119–149.

Lee JW, Devanarayan, V, Barret YC et al. Fit-for-purpose method development and validation for successful biomarker measurement. *Pharm Res.* 2006;23(2):312–328.

McShane LM, Cavenagh MM, Lively TG et al. Criteria for the use of omics-based predictors in clinical trials. *Nature* 2013;502(7471):317–320.

Miller KJ et al., Workshop on bioanalytical methods validation for macromolecules: Summary report. *Pharm Research* 2001;18(9):1373–1381.

Ohaegbulam KC, Assal A, Lazar-Molnar E, Yao Y, Zang X. Human cancer immunotherapy with antibodies to the PD-1 and PD-L1 pathway. *Trends Mol Med.* 2015;21(1):24–33.

Papageorghiou A, Ohuma E, Gravett M et al. International standards for symphysis-fundal height based on serial measurements from the fetal growth longitudinal study of the INTERGROWTH-21st Project: Prospective cohort study in eight countries. *BMJ* 2016;355:i5662.

Polley MYC, Freidlin B, Korn EL, Conley BA, Abrams JS, and McShane LM. Statistical and practical considerations for clinical evaluation of predictive biomarkers. *J Natl Cancer Inst.* 2013;105:1677–1683

Ray C. Fit-for-purpose validation. Weiner R, Kelley M (eds.), *Translating Molecular Biomarkers into Clinical Assays: Techniques and Applications,* AAPS Advance in the Pharmaceutical Sciences Series 21 (2005).

U.S. Food and Drug Administration, Center for Drug Evaluation and Research. Guidance for industry, bioanalytical method validation. 2013. http://www.fda.gov/downloads/Drugs/GuidanceCompliance RegulatoryInformation/Guidances/UCM368107.pdf. Accessed December 6, 2017.

Williams AS, Slavin DE, Wagner JA, Webster CJ. A cost-effectiveness approach to the qualification and acceptance of biomarkers. *Nat Rev Drug Discov.* 2006;5:897–902.

Williams RN, Filozof C, Goldstein BJ et al. Structure of proof of concept studies that precede a nonalcoholic steatohepatitis development program. *Clin Pharmacol Ther.* 2017;101(4): 444–446.

Crucial Role of High Quality Biosamples in Biomarker Development

3.3

Cornelia Stumptner, Karine Sargsyan, Penelope Kungl, and Kurt Zatloukal

Contents

INTRODUCTION

Developments in personalized medicine have divided classical disease entities into smaller and smaller subentities requiring specific therapies. In order to identify relevant disease subentities and to develop targeted therapies specific for these subentities, molecular insight into the characteristic features of individual diseases is needed. This requires investigating molecular and environmental determinants of individual diseases in a variety of biological samples, such as tissues and body fluids, by applying the latest analytical technologies (e.g., NGS, proteomics, metabolomics and various *in situ* detection and imaging technologies) (National Research Council 2011). Due to the fact that in personalized medicine biomarkers have to be able to identify differences in individual diseases, new quality requirements for biosamples and analytical technologies were generated. For example, in personalized medicine variations in measurements cannot be normalized by pooling large data sets, which would hide relevant features of individual diseases. Furthermore, even large medical centers often do not have access to sufficient numbers of biosamples from certain disease subentities, which makes international collaboration mandatory. This need for joint analysis of samples and data obtained from different centers generates several new challenges in biomedical research. First, biosamples have to be collected, processed and stored according to internationally harmonized standards. Otherwise, differences in pre-analytical processes might introduce major variation in the results generated, making integrated data analysis unrealistic (Moore et al. 2011). The importance of standardization of pre-analytical procedures that may affect biosample quality has been highlighted and quantified in a study performed by Freedman and co-workers who showed that approximately half of the $56 billion spent by the U.S. pharmaceutical industry per year in preclinical research is lost because of irreproducible data (Freedman et al. 2015). Furthermore, they found that approximately one-third of this loss can be attributed to inappropriate quality of biosamples and reference materials data. The impact of pre-analytical factors on biomarker assay performance is also underlined by observations that 46%–68% of diagnostic testing process errors are in the pre-analytical phase (Plebani 2006).

This error rate gained particular relevance in the context of companion diagnostics where the development and prescription of very expensive drugs (50.000–100.000 $ per patient per year) depends on the performance of a biomarker assay. Examples of FDA-listed companion diagnostic devices and their biomarker features are shown in Table 3.3.1. NGS is the predominant assay type, and formalin-fixed, paraffin-embedded (FFPE) tissue the most commonly used biosample (Figure 3.3.1).

Second, international collaboration in biomarker research faces the challenge of dealing with the heterogeneous ethical

TABLE 3.3.1 List of FDA-cleared or -approved companion diagnostic devices (*in vitro* and imaging tools)

DRUG	*DISEASE*	*TARGET*	*BIOSAMPLE*	*ASSAY*
Ado-trastuzumab emtansine	Breast cancer	HER2	DNA/protein from FFPE tissue	IHC/FISH
Ado-trastuzumab emtansine	Gastric cancer	HER2	DNA/protein from FFPE tissue	IHC/FISH
Afatinib	NSCLC	EGFR	DNA from FFPE tissue	NGS/PCR
Alectinib	NSCLC	ALK	DNA from FFPE tissue	NGS
Ceritinib	NSCLC	ALK	DNA/Protein from FFPE tissue	NGS/IHC
Cetuximab (1)	CRC	EGFR	Protein in FFPE tissue	IHC
Cetuximab (2)	mCRC	KRAS	DNA from FFPE tissue	NGS/PCR
Cobimetinib+ vemurafenib	Melanoma	BRAF	DNA from FFPE tissue	NGS
Crizotinib	NSCLC	ALK	DNA from FFPE tissue	NGS/FISH
Crizotinib	NSCLC	ROS1	RNA from FFPE tissue	NGS
Crizotinib	NSCLC	ALK	Protein/DNA in FFPE tissue	IHC
Dabrafenib	Melanoma	BRAF	DNA from FFPE tissue	NGS/PCR
Dabrafenib+trametinib	NSCLC	BRAF	DNA/RNA from FFPE tissue	NGS
Deferasirox	Thalassemia	Iron	Liver imaging	MRI
Enasidenib	AML	IDH2	DNA from blood or bone marrow	PCR
Erlotinib	NSCLC	EGFR	DNA from FFPE tissue or cfDNA from blood	PCR/NGS
Gefitinib	NSCLC	EGFR	DNA from FFPE tissue	PCR/NGS
Imatinib mesylate	GIST	c-Kit	Protein in FFPE tissue	IHC
Imatinib mesylate	MDS, MPD	PDGFRB	Fresh bone marrow	FISH
Imatinib mesylate	ASM	c-Kit	Fresh bone marrow	PCR
Midostaurin	AML	FLT3	DNA from blood or bone marrow	PCR
Nilotinib	CML	BCR-ABL1	RNA from blood	RT-PCR
Olaparib	Breast cancer	BRCA1/2	DNA from blood	PCR, Sanger seq.
Osimertinib	NSCLC	EGFR	DNA from FFPE tissue or cfDNA from blood	PCR/NGS
Panitumumab (1)	CRC	EGFR	Protein in FFPE tissue	IHC
Panitumumab (2)	CRC	KRAS	DNA from FFPE tissue	PCR
Panitumumab (3)	mCRC	KRAS/NRAS	DNA from FFPE tissue	NGS
Pembrolizumab	NSCLC/gastric or GEJ Adenoca	PD-L1	FFPE tissue	IHC
Pertuzumab	Breast cancer	HER2/NEU	DNA/protein from FFPE tissue	NGS/IHC/FISH
Rucaparib	Ovarian cancer	BRCA1/2	DNA from FFPE tissue	NGS
Trametinib	Melanoma	BRAF	DNA from FFPE tissue	NGS/PCR
Trastuzumab	Breast/gastric cancer	HER2/NEU	DNA from FFPE tissue	NGS/FISH/IHC/CISH
Vemurafenib	Melanoma	BRAF	DNA from FFPE tissue	NGS/PCR
Venetoclax	CLL	LSI TP53	Blood	FISH

Note: Data were retrieved on April 27, 2018, from https://www.fda.gov/MedicalDevices/ProductsandMedicalProcedures/InVitroDiagnostics/ucm301431.htm. ALK, alkaline phosphatase; cfDNA, circulating-free tumor DNA; CISH, chromogenic *in situ* hybridization; FFPE, formalin-fixed, paraffin-embedded; FISH, fluorescence *in situ* hybridization; IHC, immunohistochemistry; NGS, next-generation sequencing; PCR, polymerase chain reaction.

and legal requirements for cross-border sample and data transfer (Knoppers et al. 2014; Dove et al. 2016; Warner et al. 2017). Although there are major efforts by large organizations and initiatives, such as the European Biobanking and Biomolecular Resources research infrastructure (BBMRI-ERIC) (Yuille et al. 2008), P3G (Knoppers et al. 2013), GA4GH (Lawler et al. 2017) or the BioSHaRE project (Gaye et al. 2014), to develop solutions facilitating international collaboration, there are differences in the legal requirements that place a major burden on international collaboration. In this context, compliance with data protection legislation became a specific challenge in personalized medicine because the detailed data items describing individual disease phenotypes makes anonymization of sample and data sets impossible or impractical (Eder et al. 2012; Holub et al. 2018). To improve this situation, biobanks have developed tools and provide services that reduce the risks of re-identification of sample donors and data subjects and facilitate sharing of samples as well as sensitive data (Eder et al. 2012; Gaye et al. 2014).

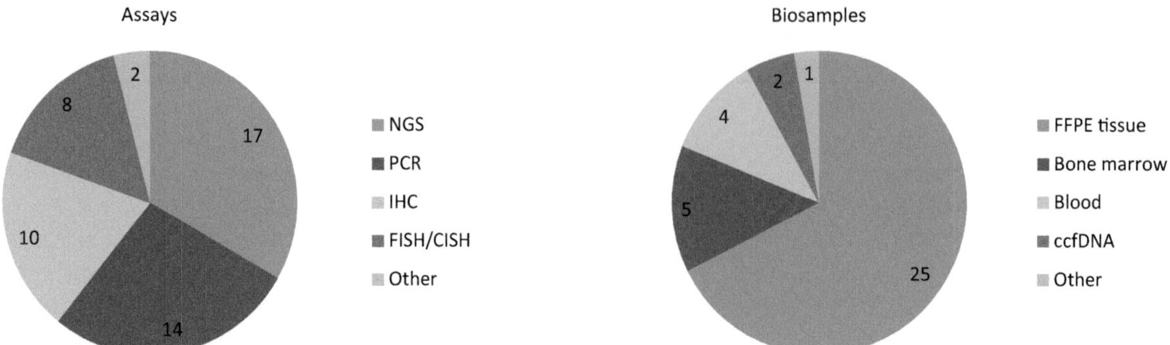

FIGURE 3.3.1 Biosamples and assays for companion diagnostics (FDA-listed). Note: NGS, Next generation sequencing; PCR, Polymerase chain reaction; IHC, Immunohistochemistry; FISH, Fluorescence in situ hybridisation; CISH, Chromogenic in situ hybridisation; FFPE, Formalin-fixed, paraffin-embedded; ccfDNA, circulating cell-free DNA.

STANDARDIZATION OF BIOSAMPLE QUALITY

During the past years, there has been growing evidence of the importance of biosample (biospecimen) quality in academic and industrial research as well as in patient care that has led to an increased awareness of the need for high quality biospecimens that are fit for a specific diagnostic and/or research purpose. This accounts also for biomarker discovery and validation.

A major denominator of sample quality is the pre-analytical phase, which comprises all steps of the sample collection and processing workflow preceding the analysis. It includes, for example, documentation of patient-related information such as patient anamnesis, medical and/or surgical procedure, biospecimen collection, transport, processing, and storage as well as the isolation of an analyte (e.g., RNA, DNA, protein) or the sectioning of tissue, deparaffinization/rehydration and pretreatment of sections for *in situ* detection techniques. As major sources of error or variation occur in the pre-analytical phase, a test result is not just a function of the analytical procedure (e.g., performance of PCR machine or sequencer), but is also influenced by the phase before the actual examination (analysis), which consequently has to be taken into consideration when defining and standardizing biosample quality.

Numerous pre-analytical variables exist that can negatively affect the reliability and reproducibility of analyses results (Table 3.3.2). The nature and extent of the impact may vary according to the biospecimen type and biomarker (i.e., analyte or biomolecule) of interest and the subsequent analytical procedure. Also in biomarker discovery and validation, it is imperative that the measured concentration of the analyte (such as DNA, RNA, proteins, metabolites or other biomolecules) in biospecimens is as close as possible to the actual analyte concentration at the time of sample collection in the body in order to effectively utilize biomarkers to draw meaningful conclusions. Following the principle "garbage in–garbage out" (Compton 2007), the quality of the biospecimen is a major precondition of the quality of a subsequent analysis.

TABLE 3.3.2 Examples of major pre-analytical variables affecting analytical results

FACTORS (CATEGORY)	*FACTORS (EXAMPLES)*
Patient/donor	Disease condition, medication and treatment, nutrition, stress, physical activity, demographics, etc.
Sample collection	Specimen type, surgical procedure, biopsy device, warm ischemia, cold ischemia, blood collection device, collection tube, etc.
Sample stabilization	Stabilization method and solution: freezing, fixation (type of fixative [e.g., formalin], duration, temperature, size of biosample, etc.)
Sample transport	Transport duration, temperature, container, mechanical stress (e.g., pneumatic dispatch)
Sample preparation	Centrifugation, aliquoting, grossing of tissues, tissue processing and embedding in paraffin
Storage	Duration, temperature, humidity
Preprocessing for analysis	Isolation of biomolecules (e.g., RNA, DNA, protein), sectioning of tissues, antigen retrieval)

In response to the growing concern from the scientific community and from industry for safeguarding biospecimen quality and controlling pre-analytical variations, several organizations have developed best practice and standardization guidelines. These include for example Biospecimen Reporting for Improved Study Quality (BRISQ) Guidelines (Moore, Compton, Alper, and Vaught 2011), College of American Pathologists (CAP) Guidelines (Barnes, Shea, and Watson 2017), Clinical & Laboratory Standards Institute (CLSI) Standards, International Agency for Research on Cancer (IARC) Common minimum technical standards (Mendy 2017), International Society for Biological and Environmental Repositories (ISBER) Best Practice Recommendations (Campbell et al. 2018), National Cancer Institute (NCI) Best Practices (https://biospecimens.cancer.gov/

bestpractices/2016-NCIBestPractices.pdf), Organisation for Economic Co-operation and Development (OECD) Best Practice Guidelines (http://www.oecd.org/sti/biotech/38777417.pdf), and a plethora of other guidelines for specific biomarkers or assays.

In addition to the many voluntary guidelines, the European Committee for Standardization (CEN) published in 2015/2016 nine evidence-based CEN Technical Specifications (CEN/TS) for "pre-examination processes" (i.e., pre-analytical workflows). These are currently being developed into international standards by the International Organization for Standardization's Technical Committee 212 for "Clinical Laboratory Testing and *In Vitro* Diagnostic Test Systems" (ISO/TC 212) under the Vienna Agreement, an agreement between ISO and CEN to share information, attend each other's meetings and collaborate on standards at international and European levels (www.iso.org) (Table 3.3.3). Five standards were published in 2018 and the remaining four are expected to be published in 2019 and will then have global impact. Further CEN/TS and ISO Standards on pre-analytical sample processing are planned in the context of the European H2020 project SPIDIA4P (http://www.spidia.eu/) (Oelmüller 2017).

Importantly, all these standards refer to the existing ISO accreditation standard ISO 15189:2012, *Medical laboratories—Requirements for quality and competence* and are based on large-scale studies performed in the context of the European FP7-funded research project SPIDIA in Europe and on studies performed by the National Cancer Institute in the United States to provide sound scientific evidence on which pre-analytical parameters influence analytical results and which need to be standardized.

Both CEN/TS and ISO standards exist for different types of biospecimens (matrices) and analytes (e.g., RNA, DNA, protein, metabolites). They give recommendations on the handling, storage and processing of biospecimens/samples and the documentation of sample-associated data along the whole pre-analytical workflow for a specific purpose (e.g., for the handling of venous whole blood for analyses based on isolated circulating cell free DNA [ccfDNA]) (Table 3.3.3) (ISO/DIS 20186-3).

The CEN/TS and respective ISO standards list in the form of requirements, recommendations, permissions and possibilities, which parameters related to pre-analytical processes "shall", "should," "can" or "may" be standardized and their proper fulfillment documented. This information helps sample users to interpret their analytical results obtained with these samples and to provide regulators with data on assay performance generated with samples complying with international quality standards. The CEN/TS and ISO standards do not per se resemble standard operating procedures (SOPs) because only in some very critical steps, do they give explicit advice about sample handling or reagents to use (e.g., centrifugation speed/temperature/time of blood samples for ccfDNA, or type/concentration/pH of formalin fixative solution). However, they are the basis for standardizing a laboratory's sample management processes and for developing internationally harmonized SOPs. Notably, the series of pre-analytical standards represent a standard family with similar structure and requirements. This is important because in the context of patient care it is often not feasible to establish several different pre-analytical workflows and collect and process samples in parallel in different ways.

TABLE 3.3.3 Overview on CEN/TS and ISO standards for pre-examination processes (published or under drafting/development)

MOLECULAR IN VITRO DIAGNOSTIC EXAMINATIONS—SPECIFICATIONS FOR PRE-EXAMINATION PROCESSES FOR ….	
CEN/TS (published)	• Snap frozen tissue—Part 1: Isolated RNA (CEN/TS 16826-1:2015) • Snap frozen tissue—Part 2: Isolated proteins CEN/TS 16826-2:2015) • FFPE tissue—Part 1: Isolated RNA (CEN/TS 16827-1:2015) • FFPE tissue—Part 2: Isolated proteins (CEN/TS 16827-2:2015) • FFPE tissue—Part 3: Isolated DNA (CEN/TS 16827-3:2015) • Venous whole blood—Part 1: Isolated cellular RNA (CEN/TS 16835-1:2015) • Venous whole blood—Part 2: Isolated genomic DNA (CEN/TS 16835-2:2015) • Venous whole blood—Part 3: Isolated circulating cell free DNA from plasma (CEN/TS 16835-3:2015) • Metabolomics in urine, venous blood serum and plasma (CEN/TS 16945:2016)
CEN/TS (under drafting)	• Saliva—Isolated DNA (WI=00140116) • Circulating tumor cells (CTCs) in venous whole blood—Part 1: Isolated RNA (WI=00140123) • Circulating tumor cells (CTCs) in venous whole blood—Part 2: Isolated DNA (WI=00140125) • Circulating tumor cells (CTCs) in venous whole blood—Part 3: Preparations for analytical CTC staining (WI=00140124)
ISO Standards (published)	• FFPE tissue—Part 1: Isolated RNA (ISO 20166-1:2018) • FFPE tissue—Part 2: Isolated proteins (ISO 20166-2:2018) • FFPE tissue—Part 3: Isolated DNA (ISO 20166-3:2018) • Frozen tissue—Part 1: Isolated RNA (ISO 20184-1:2018) • Frozen tissue—Part 2: Isolated proteins (ISO 20184-2:2018)
ISO Standards (under development)	• Venous whole blood—Part 1: Isolated cellular RNA (ISO 20186-1) • Venous whole blood—Part 2: Isolated genomic DNA (ISO 20186-2) • Venous whole blood—Cellular RNA—Part 3: Isolated circulating cell free DNA from plasma (ISO/DIS 20186-3) • FFPE tissue—*In situ* detection techniques (ISO/AWI 20166-4)

Although these standards primarily address molecular *in vitro* diagnostics, they are also relevant for biobanking and any laboratory or company working in the field of biomarker research and development because the same quality criteria should apply for all biosamples used for research and development as well as for medical diagnostics.

BIOBANKS AS A SOURCE FOR QUALITY-CONTROLLED BIOSAMPLES AND ASSOCIATED DATA

Biobanking, as already discussed, is a rapidly increasing science field. A huge number of biobanks and biorepositories are documented around the world, and several emerging biobanking activities are currently being established. Traditionally, collected specimens have demonstrated a vital role in investigation of the etiology and pathogenesis of a disease as well as its treatment. Due to deep understanding of the research and development process in modern biomedical research, biospecimens gained an essential role as a resource for biomarker research and drug development (Zatloukal and Hainaut 2010). Usage of human biological samples has facilitated the development of diagnostic as well as prognostic biomarkers for clinical practice. One of the current illustrations of the effective usage of biobanks in biomarker research is the Oncotype DX 21-gene test in breast cancer and the MammaPrint assays. These two predictive biomarker tests give information about the recurrence risk of early-stage breast cancer (Newman and Freitag 2011). Based on many effective examples like these, we can assume that standardized high quality biobanking can significantly assist and facilitate the discovery and large-scale validation of biomarkers.

Biobanks as Biomarker Research Partners

The development of biobanking addresses critical points in technical, procedural and scientific forecasting practices in order to facilitate the best possible improvement in biomarker research. Biomarker scientists together with biobanking scientists are involved in collaborative efforts for strategic and successful implementation of the value chain of research-based biomarker-biobank cooperation.

Independent of the type, purpose, automation grade and scope of biobanks, biobanked specimens and data collections are critical resources for biomarker research. However, different quality criteria may apply according to the research requirements.

There are several standards and guidelines relevant for quality in biobanks (see also above). Before the development of standards there were several evidence/experience–based best practice documents available for biobanks (both for technical and operational issues). They could be used by biobanks for improving quality and to some extent harmonizing standardization of specimen and data collections, such as the above-mentioned ISBER Best Practices for Repositories 4.0 (Campbell et al. 2018), OECD Best Practice Guidelines for Biological Resource Centers and/or WHO/IARC Common Minimum Standards and Protocols for Biological Resource Centers Dedicated to Cancer Research (IARC Working Group Reports 2007). These and other guidelines have been used as effective tools for biobanks to support biomarker research.

Biobanking Challenges with Impacts on Biomarker Research

Biospecimen researchers in cooperation with biomarker researchers can develop the competence to improve the quality as well as quantity of biobanking capabilities and enhance the impact of biobanks on significant biomarker research by addressing some common challenges. The main challenges have been described by many authors both in biobanking and biomarker research (Bevilacqua et al. 2010; Zatloukal and Hainaut 2010; Vaught and Lockhart 2012).

The main biobanking-related challenges for biomarker researchers include the following:

- Finding the best biobank containing the required material (currently there is no comprehensive global biobank directory or inventory). This challenge is approached in Europe by cataloging biobanks in BBMRI-ERIC and the National nodes.
- Gaining rapid access to standardized medical data, which is still hampered for multicenter studies by lack of common biospecimen and data standards (Holub et al. 2018).
- Providing quality related meta-data (e.g., warm and cold ischemia times), which could be solved by broad implementation of pre-analytical CEN/TS and ISO standards in biobanks.
- Availability of appropriate control biospecimens for a study.
- Heterogeneity and fragmentation of collections at different levels (e.g., late-stage vs. early-stage diseases; different biospecimens tissue vs. serum or DNA, data stored in different formats and databases). This heterogeneity can be markedly reduced when biobanks are established as a centralized infrastructure in medical centers. Such centralized biobanks usually collect a broad spectrum of sample types and data sets from the same patient. However, still more emphasis has to be placed in the future on longitudinal collections documenting the whole time course of a disease and to complement biospecimens with more detailed clinical data sets (especially the medical follow-up data, treatment response and outcome).
- Finding biobanks with installed quality management systems that are open for external audits. With the increasing demand on quality and implementation of CEN/TS and ISO standards, this challenge will be well addressed in the future. BBMRI-ERIC already

highlights biobanks that meet the CEN/TS requirements in its directory (https://directory.bbmri-eric.eu).

- Accessing the biobank samples in an acceptable time period including efficient management of administrative requirements (e.g., approval procedures and signing of material transfer agreements). This challenge depends on multiple factors such as on established and standardized procedures in biobanks, regulations of the carrier organization and national law.

An important aspect is that biobanking and biomarker scientists are working together to define and find "fit-for-purpose" specimens and data as well as to benchmark the required level of biobanking and sample management. The term "fit for purpose" is adopted in biobanking research to match the proposed research goal with the specific required features of specimens and data to achieve this goal. SOPs for data collection, sample collection, processing, and storage, for example, should be mandatory for biomarker researchers and are required to evaluate whether data and samples are "fit for purpose." The selection of "fit-for-purpose" biospecimens for a biomarker study should be an interactive cooperation between biobankers and biomarker scientists in which the biobank specialists provide the data on specimens (especially quality relevant information and SOPs) for the choice of best fit specimens and data for the given research. This procedure should also involve medical expertise in project relevant fields, such as pathologists, clinicians, data scientists and/or also laboratory scientists. A specific challenge in this interactive process relates to the definition and selection of the appropriate control cohort, which may be matched healthy subjects or other patient groups. In the last two decades more data have been published about biomarkers than on any other topic in biomedicine.

CONCLUSION

Plenty of new publications are issued every day with a potential new biomarker (protein, gene, or other markers) for diagnostics and monitoring. At the same time, there is inefficient translation of these findings to the bedside, often because of a lack of reproducibility and clinical proof-of-concept (van Gool et al. 2017). Some of these problems are related to issues of quality of sample and data collection (e.g., poor quality in collection process and documentation, incomplete clinical data, combination of different sample cohorts with different quality). In fact, strong correlation was found between the success percentages in biomarker research with the quality of biobanking (Ioannidis and Panagiotou 2011). This highlights the key role of biobanks in biomarker research because biobanks can provide an environment that is required to provide high quality biospecimens and data as well as professional governance, ensuring efficient procedures and ethical and legal compliance (Riegman et al. 2008; Hewitt 2011).

FUNDING

This work was supported by the Austrian BMWFW (GZ 10.470/0016-II/3/2013, BBMRI.at) and the H2020-funded project SPIDIA4P (grant agreement no. 733112).

REFERENCES

Barnes, R. O., K. E. Shea, and P. H. Watson. 2017. The Canadian Tissue Repository Network biobank certification and the College of American Pathologists biorepository accreditation programs: Two strategies for knowledge dissemination in biobanking. *Biopreserv Biobank* 15 (1):9–16. doi:10.1089/bio.2016.0021.

Bevilacqua, G., F. Bosman, T. Dassesse et al. 2010. The role of the pathologist in tissue banking: European Consensus Expert Group report. *Virchows Arch* 456 (4):449–454. doi:10.1007/s00428-010-0887-7.

Campbell, L. D., J. J. Astrin, Y. DeSouza et al. 2018. The 2018 revision of the ISBER best practices: Summary of changes and the editorial team's development process. *Biopreserv Biobank* 16 (1):3–6. doi:10.1089/bio.2018.0001.

Compton, C. 2007. Getting to personalized cancer medicine: Taking out the garbage. *Cancer* 110 (8):1641–1643. doi:10.1002/cncr.22966.

Dove, E. S., B. Thompson, and B. M. Knoppers. 2016. A step forward for data protection and biomedical research. *Lancet* 387 (10026):1374–1375. doi:10.1016/S0140-6736(16)30078-2.

Eder, J., H. Gottweis, and K. Zatloukal. 2012. IT solutions for privacy protection in biobanking. *Public Health Genomics* 15 (5):254–262. doi:10.1159/000336663.

Freedman, L. P., I. M. Cockburn, and T. S. Simcoe. 2015. The economics of reproducibility in preclinical research. *PLoS Biol* 13 (6):e1002165. doi:10.1371/journal.pbio.1002165.

Gaye, A., Y. Marcon, J. Isaeva et al. 2014. DataSHIELD: Taking the analysis to the data, not the data to the analysis. *Int J Epidemiol* 43 (6):1929–1944. doi:10.1093/ije/dyu188.

Hewitt, R. E. 2011. Biobanking: The foundation of personalized medicine. *Curr Opin Oncol* 23 (1):112–119. doi:10.1097/CCO.0b013e32834161b8.

Holub, P., F. Kohlmayer, F. Prasser et al. 2018. Enhancing reuse of data and biological material in medical research: From FAIR to FAIR-Health. *Biopreserv Biobank* 16 (2):97–105. doi:10.1089/bio.2017.0110.

Ioannidis, J. P., and O. A. Panagiotou. 2011. Comparison of effect sizes associated with biomarkers reported in highly cited individual articles and in subsequent meta-analyses. *JAMA* 305 (21):2200–2210. doi:10.1001/jama.2011.713.

Knoppers, B. M., R. L. Chisholm, J. Kaye et al. 2013. A P3G generic access agreement for population genomic studies. *Nat Biotechnol* 31 (5):384–385. doi:10.1038/nbt.2567.

Knoppers, B. M., J. R. Harris, I. Budin-Ljosne, and E. S. Dove. 2014. A human rights approach to an international code of conduct for genomic and clinical data sharing. *Hum Genet* 133 (7):895–903. doi:10.1007/s00439-014-1432-6.

Lawler, M., D. Haussler, L. L. Siu et al. 2017. Sharing clinical and genomic data on cancer—The need for global solutions. *N Engl J Med* 376 (21):2006–2009. doi:10.1056/NEJMp1612254.

Mendy, M., Caboux, E., Lawlor, RT., Wright, J., Wild, CP. 2017. *Common Minimum Technical Standards and Protocols for Biobanks Dedicated to Cancer Research.* Geneva, Switzerland: IARC Technical Publications, WHO Press, World Health Organization.

Moore, H. M., C. C. Compton, J. Alper, and J. B. Vaught. 2011. International approaches to advancing biospecimen science. *Cancer Epidemiol Biomarkers Prev* 20 (5):729–732. doi:10.1158/1055-9965.EPI-11-0021.

Moore, H. M., A. Kelly, S. D. Jewell et al. 2011. Biospecimen reporting for improved study quality. *Biopreserv Biobank* 9 (1):57–70. doi:10.1089/bio.2010.0036.

National Research Council. 2011. *Toward Precision Medicine: Building a Knowledge Network for Biomedical Research and a New Taxonomy of Disease.* Edited by National Research Council. Washington, DC: The National Academies Press.

Newman, T. J., and J. J. Freitag. 2011. Personalized medicine development. As the popularity of personalized medicine grows the role of the CRO continues to evolve. *Appl Clin Trials* 31.

Oelmüller, U. 2017. Standardization is key in the preanalytical field. *Biobanks Europe Issue 6.*

Plebani, M. 2006. Errors in clinical laboratories or errors in laboratory medicine? *Clin Chem Lab Med* 44 (6):750–759. doi:10.1515/CCLM.2006.123.

Reports, IARC Working Group. 2007. Common minimum standards and protocols for biological resource centers dedicated to cancer research. In: Edited by World Health Organization International Agency for Research on Cancer. http://www.iarc.fr/en/publications/pdfs-online/wrk/wrk2/index.php.

Riegman, P. H., M. M. Morente, F. Betsou et al. 2008. Biobanking for better healthcare. *Mol Oncol* 2 (3):213–222. doi:10.1016/j.molonc.2008.07.004.

van Gool, A. J., F. Bietrix, E. Caldenhoven et al. 2017. Bridging the translational innovation gap through good biomarker practice. *Nat Rev Drug Discov* 16 (9):587–588. doi:10.1038/nrd.2017.72.

Vaught, J., and N. C. Lockhart. 2012. The evolution of biobanking best practices. *Clin Chim Acta* 413 (19–20):1569–1575. doi:10.1016/j.cca.2012.04.030.

Warner, A. W., H. Moore, D. Reinhard, L. A. Ball, and B. M. Knoppers. 2017. Harmonizing global biospecimen consent practices to advance translational research: A call to action. *Clin Pharmacol Ther* 101 (3):317–319. doi:10.1002/cpt.461.

Yuille, M., G. J. van Ommen, C. Brechot et al. 2008. Biobanking for Europe. *Brief Bioinform* 9 (1):14–24. doi:10.1093/bib/bbm050.

Zatloukal, K., and P. Hainaut. 2010. Human tissue biobanks as instruments for drug discovery and development: impact on personalized medicine. *Biomark Med* 4 (6):895–903. doi:10.2217/bmm.10.104.

Informatics Solutions for Biomarker Discovery and Personalized Medicine in Clinical Care

3.4

Jan-Willem Boiten, Rita Azevedo, Marinel Cavelaars, André Dekker, Remond J. A. Fijneman, Arnoud van der Maas, Jeroen A. M. Beliën, and Gerrit A. Meijer

Contents

INTRODUCTION

The central paradigm in biomarker research is that the variation in disease phenotype and, in particular, disease outcome has to be correlated to variation in the underlying disease biology, which will enable us to improve disease prevention and treatment. This translational paradigm is valid, and the process involved is standard, not only across different disease areas, but also across different types of (unmet) clinical needs such as prevention, early detection, prognosis, therapeutic intervention and response monitoring. With some nuances, terms like personalized medicine, stratified medicine and precision medicine all refer to the same principle. As a consequence, biomarker research programs share a common high-level design as illustrated in Figure 3.4.1.

Typical questions to be addressed in these research programs are "Which (out of tens of thousands) biomarkers can predict good or bad outcome?" and "Which (out of tens of thousands) biomarkers can predict whether a patient will benefit from a particular therapy?" The phenomenal size of the data sets produced in these research programs and their distribution over different research laboratories and clinics require an informatics infrastructure that allows for a seamless integration and exchange of large amounts of data as well as for complex data analysis. Still, the biggest challenge is not so much capturing, storing, processing and analyzing petabytes of data produced by NGS or high-end imaging. Most technical challenges in these domains have been or are being addressed [1]. The availability of high-quality phenotypic annotations to the omics results often proves to be the weak link in the biomarker research process chain.

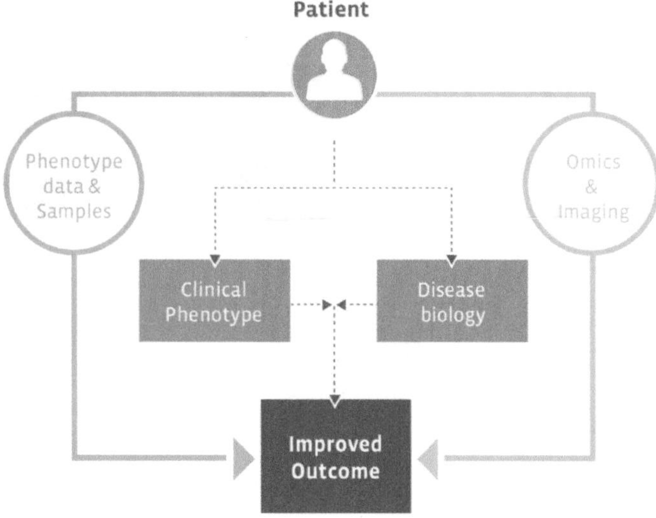

FIGURE 3.4.1 All projects in translational biomarker research share the same basic design, correlating the variation in the clinical phenotype to variations in underlying disease biology. If both sides of the workflow are available, new correlations can be made, leading to new insights that ultimately should improve the disease outcome for patients or even prevent the disease from emergence.

Practice-changing biomarker research has to adhere to the principles of evidence-based medicine. To improve outcomes, we need to change clinical practice and hence practice guidelines, fueled by solid scientific evidence. Currently, however, only a fraction of the many proof-of-concept biomarker studies is advanced to clinical trials yielding practice-changing evidence, a phenomenon often referred to as the translational "Valley of Death" [2,3]. A keyword associated with the translational valley of death is fragmentation. Fragmentation is omnipresent, not only of data but also of infrastructure initiatives, sample collections, the IT application landscape, procedures and standards. Data integration is a corner stone of the solution, but showcase initiatives in this domain have had to make tremendous and often manual efforts in interactively polishing data to make these interoperable, an underexposed activity referred to as data curation or extract-transform-load (ETL) [4]. Basically, all of this is manual labor needed to compensate for inefficient organization of the research process.

THE SOLUTION: WHAT IS NEEDED?

Fragmentation in data sources, informatics solutions, and data stewardship can be overcome by consistent application of the FAIR principles [5,6], in which FAIR stands for "findable, accessible, interoperable and reusable" data. These principles where first outlined at a Lorentz conference in Leiden (the Netherlands); the resulting FAIRport initiative (http://datafairport.org) quickly gained broad acceptance across research and research funders. To achieve the goal of FAIR data, not only should information technology (IT) systems should comply with standards, but also the researchers.

The simple association scheme between clinical phenotype and disease biology shown in Figure 3.4.1 can be further elaborated into a common workflow across biomarker projects. The entire informatics workspace can be segmented into three stages of the data life cycle (as also illustrated in Figure 3.4.2):

1. *Clinical care workflow*: Making biomarker data truly interoperable (the I from FAIR) requires consistent implementation of (meta) data standards at the source, which implies the need for tight interactions between the clinical care workflow and the research workflow.
2. *Project workspace (also called the On-line Digital Research Environment)*: A web-based suite of tools closely linked together, comparable to an "office suite" with a word processor, spreadsheet and presentation program.
3. *Data access and reuse by the general user community.* In line with the FAIR principles and the Open Science movement, biomarker research data sets should be findable and reusable. Consequently, aggregated meta data summaries of project results should be published in open data resources in order to enable the discovery of these data sets by other researchers.

Although this high-level data flow appears to be applicable across the majority of biomarker research projects, some aspects are also underexposed:

1. Direct involvement of patients and participants is needed, not only to provide input, for example, in the form of patient-reported outcome measures (PROMs), but also as a direct partner in the project providing feedback on project progress and project results. This patient interaction may require specific IT solutions that are not yet readily available.
2. Direct involvement of the data producer (e.g., hospitals) given that their information systems and standards implemented for primary use (e.g., caring for a single patient) are not designed or suited for reuse of that data in biomarker research.
3. Direct access to preliminary study data not yet officially published and validated will become more important in the near future to facilitate decision making in clinical practice through decision support tools (e.g., IBM Watson) and other machine-learning solutions. Similar IT solutions will be needed to support the shared decision-making dialog between patients and clinicians when counterweighing the risks and benefits of alternative advanced treatment options.
4. Increased sharing and usage of personal health data raises legal and ethical questions and concerns. The data infrastructure supporting biomarker research should, therefore, be supported by an efficient legal and ethical framework.

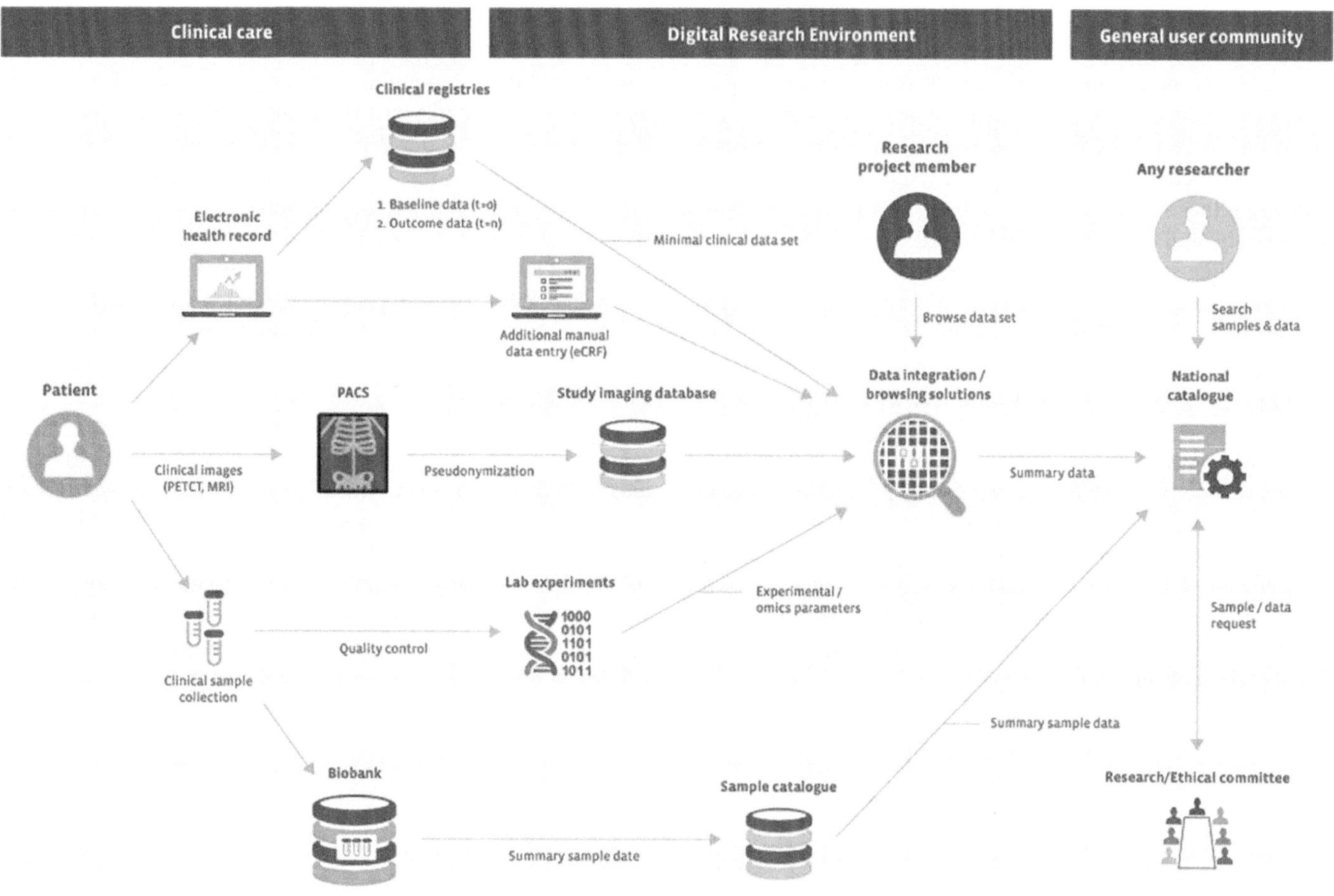

FIGURE 3.4.2 A common workflow can be drafted across biomarker discovery projects and other translational research projects. At the heart is the on-line Digital Research Environment (DRE), which is increasingly linked to the clinical care workflow. Toward the end of the project, meta data of the results will be published in the public domain (right-hand side) for findability and accessibility by other researchers for future reuse.

REAL-WORLD CLINICAL DATA

The evidence for validation of potential biomarkers is currently usually produced in prospective clinical studies. Usage of retrospective real-world clinical data would be an attractive alternative, closing feedback loops much faster and lowering costs [7,8]. Whereas a typical clinical study takes 5–7 years from design to completion, hypothesis confirmation with retrospective data could potentially be completed within 1–2 years. If routinely applied in biomarker validation, this would present an important step toward closing the innovation gap in biomarker research. Although the potential benefits are eminent, implementation in daily practice is only seen at a very limited scale. The underlying causes are multifactorial: In a 2011 review [9], the main barriers to sharing (and thus reusing) data from clinical care were summarized as "the problem is not really technical [...]. Rather, the problems are ethical, political, and administrative." Administrative barriers (i.e., unable to share) usually boils down to insufficient resources to share data, or to perform the necessary quality control. Political issues (i.e., unwilling to share) are mostly centered around a desire for personal control over the data, either to keep the preferential right to publish the data or out of concern that the data may be misinterpreted or harm the reputation of the data holder. Ethical concerns (i.e., not permitted to share) arise when the legal framework is not adequately equipped for further sharing or sharing would breach the privacy of the patient.

In addition, the control and dynamics around the information model are different between care (primary use) and research (secondary use). In clinical care, an information model is often made only once: at the introduction of an Electronic Health Record (EHR) or by a vendor. Changes after that initial model are generally kept small because they have a major impact on the system. In research, every project defines its own information model at the start, and changes during the project's lifetime are quite frequent. This analysis assumes data in EHRs are structured according to information models, whereas in fact still 80% of the hospital data are unstructured and take the form of, for example, images, scanned documents, and free text fields with prose and/or shorthand.

Much of the secondary use of clinical data for research is based on the assumption that data can generally be fully anonymized. This is not true nowadays and will become even more problematic going forward. The amount and content of the information that needs to be released for an individual to answer a research question is often potentially enough to re-identify a patient when combining these data with publicly available information (which is obviously an illegal act in most countries). For example, knowing things like sex, age, length, weight, date and location of care encounters, diagnosis, treatments and outcomes very quickly reduces the number of likely candidates. Removing the obvious identifiers (de-identified data) and replacing it with an unrelated identifier (pseudonymized data) does not change that.

A number of publications [10,11] have demonstrated that insufficient data quality and data interoperability is a roadblock for progress across the biomedical research field. One of the conditions for data interoperability is the existence of strong data and meta data standards for the underlying data. For images and genomic data, these are relatively well developed; however, for the clinical phenotype, these are often lacking. Efforts like Detailed Clinical Models are trying to fill that gap, but have not found wide acceptance in the community.

Increasing the quality of retrospective data for biomarker validation studies requires a FAIR-data IT infrastructure that supports close interactions between the research and care workflows. An important prerequisite for such a FAIR infrastructure is the decoupling between the IT architecture for primary use (i.e., clinical care) and secondary use (i.e., among others research) because of the intrinsic different nature of the underlying processes as outlined above. In the national research IT collaboration project between the Dutch University Medical Centers (UMCs), Data4lifesciences (http://data4lifesciences.nl/), an overarching information architecture was established that bridges the care and research needs and is outlined in Figure 3.4.3.

Ideally, data from multiple clinical care sources (e.g., primary care, hospital care, medication records) can be combined and linked to data from clinical registries or data specifically generated for research purposes. In most countries, this record linkage is far from trivial because data sets brought outside the hospital setting have to be de-identified in order to protect the privacy of the individuals involved. In practice, often-complicated protocols involving a trusted third party (TTP) are needed to link these data sets in a secure manner.

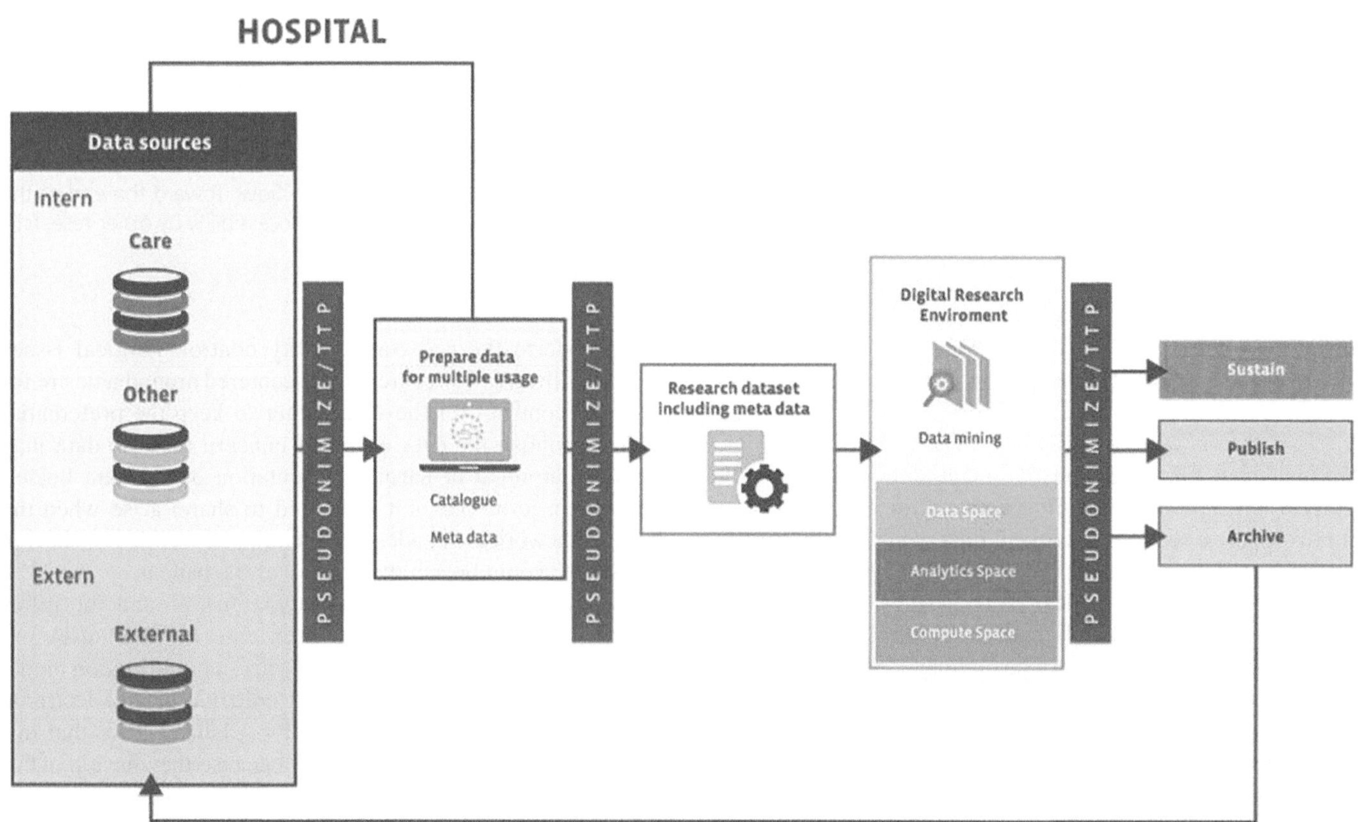

FIGURE 3.4.3 Information reference architecture as established by the Dutch UMCs collaborating in the Data4lifesciences program. The individual clinical systems in a hospital are linked to an internal data store where data sets can be "FAIRified" for external use after proper de-identification (also called pseudonymization). Findability of the data is taken care of through open publication of meta data in repositories often called a catalog or directory. The secure processing of data is organized in a research workspace (also called a Digital Research Environment).

REGISTRIES AND BIOBANKS

An attractive source of phenotype data for biomarker research could be offered by disease registries that are kept for quality assurance or health outcome evaluation purposes. Many of the data items captured for quality assurance (e.g., related to diagnosis and therapy) are also relevant phenotype characteristics for biomarker research. Such data sharing between quality registries and the translational research domain would benefit enormously from the availability of standard methods and technologies, in particular when structured reporting in the EHRs could be improved. Agreement on a joint minimal data set between all stakeholders (e.g., medical professionals, researchers, IT specialists, EHR vendors) is an important prerequisite. Internationally, ICHOM (http://www.ichom.org/) and OHDSI (https://www.ohdsi.org/) are leading collaborations working toward this overarching goal. On a local scale, more targeted efforts are under way such as the Dutch cross–UMC initiative for data registration at the source ("Registratie aan de bron").

Closely related to registries are biobanks, which are collections of samples, data, and images of individuals at different time points in their lives, either when they are healthy (population biobanks) or when they have developed disease. Biobank data may be collected without a specific research question or may be inspired by specific research questions; however, by their nature, biobanks offer materials and information to answer a wealth of new research questions in fundamental and translational biomedical research. The onset of genomics and the modern imaging era has led to an urgent need for a standardized biobank infrastructure to ensure the efficient use of research resources and avoid redundancy, to create more efficient research workflows, and to ensure effective data integration and data access. Increasingly, biobanks therefore collaborate in national and international infrastructures such as offered by BBMRI (http://www.bbmri-eric.eu/) in Europe with national nodes in the membership countries. Discovery of samples and data sets in biobanks, cohorts and registries can be served by catalogs (sometimes also called directories) [12], which are public meta data resources allowing researchers to judge the relevance of the samples and data in the collections for their research question at hand. The depth of the available data in catalogs may vary distinctly between collections: Many biobanks are only listed with their contact details (basically the catalog then serves as the "Yellow Pages" for biobanks); others offer more details like disease type, number of samples, and high-level phenotypic information, allowing potential users to make an informed decision about the usefulness of those samples. In any case, public catalogs only offer aggregated (and therefore anonymized) information: Singling out individual participants should be avoided at all times.

Many large biobanks and network organizations like BBMRI also offer request portals, facilitating the actual data and sample request from biobanks and data collections advertised in the catalog. These request services usually collect the details in an electronic fashion including a brief research proposal, which is then sent to a data/sample access committee for ethical approval before the actual samples and data can be shared.

THE DIGITAL RESEARCH ENVIRONMENT FOR BIOMARKER RESEARCH

As listed in Figure 3.4.2, a common workflow can be drafted across biomarker research projects, which should be supported with an integrated set of software solutions ideally closely integrated in an on-line digital research environment comparable to an office suite with a word processor, spreadsheet and presentation program. One example of such an integrated suite of solutions for biomarker research is offered by the Dutch TraIT (Translational Research IT; http://trait-platform.org) initiative, which actually presents the available solutions on its website in a "translational research suite"–like manner as also depicted in Figure 3.4.4. Not only are the tools directly available, but in order to facilitate user uptake, all of these tools are supported by a helpdesk, documentation, training, and a self-service portal addressing most of the obvious questions by novice users.

The need for a biomarker office suite such as TraIT is demonstrated by its adoption in biomedical research, mostly in the Netherlands but also including research institutions abroad that participate in multicenter studies. Currently, TraIT supports more than 3,000 unique registered users originating from more than 500 institutions. These users are spread out over more than 400 translational studies.

The core component in a fully operational biomarker IT suite is the Digital Research Environment (DRE), which is a website where scientists can create dedicated and secure workspaces for any kind of data project. Within these workspaces, tools may be included to import, merge, optimize, store, analyze, archive and share research data. Colleagues, both internal and external, can be invited into the workspace for collaboration, and with the same ease, access may be revoked at any time.

Each workspace should be fully scalable with regard to data quantity and computing power, thereby supporting anything from small to complex multicenter, multi-source studies. Moreover, the workspace enables scientists (and their institutional management) to perform "worry free" research, that is, security, IT infrastructure and compliance with laws and regulations are automatically addressed when using the DRE. For example, an audit trail would typically be taken care of by the DRE. Different implementations of a DRE exist, for example, the workspaces offered to analyze the large population cohort LifeLines (https://www.lifelines.nl/) in the Netherlands [13] and the DRE implementation created for the Radboudumc Nijmegen. These solutions are relatively

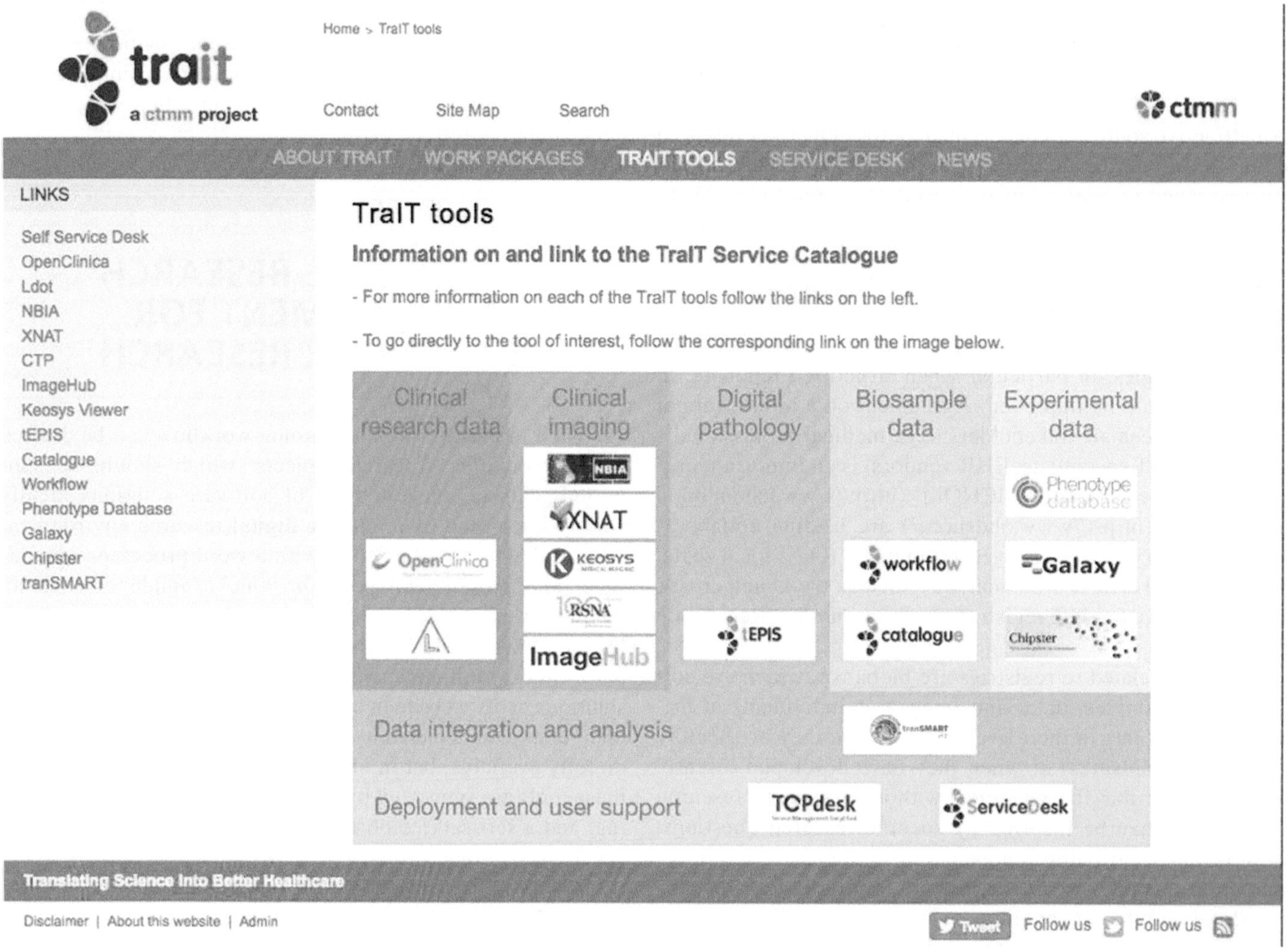

FIGURE 3.4.4 The "TraIT translational research office suite," a Dutch national implementation of a web-based digital research environment for biomarker research. The layout follows the translational research workflow with columns on top for clinical data, imaging data (including digital pathology), biobank data, and experimental molecular data. The two bottom rows provide the foundation: data integration and the link to professional-level user support. All logos are hyperlinked to the corresponding applications. See also http://www.trait-platform.org/trait-tools.

mature and self-sustaining, built from engagement in the biomedical research and innovation market over the last 5+ years. Although these workspaces offer inclusion of data capture solutions (such as OpenClinica, CASTOR, and REDcap), or data analytics tools (e.g., R, Python, Matlab), they are still quite monolithic and hard to install for individual research groups or projects without the full support of their institutions. It is foreseen that the next generation of DREs may be more modular, offering researchers the option to blend their preferred solutions in a secure environment. In addition, distributed learning will play a more prominent role, supporting the strategy of keeping the data at the source as opposed to a monolithic central data repository. The researcher will be able to acquire access to multiple remote data sources and distribute analytics to these data sources.

For the time being, individual biomarker projects in search of a data integration environment for clinical and molecular profiling data have to revert to dedicated data integration solutions. Some commercial vendors (Oracle, IDBS, SAP) have stepped into this market, but none of these solutions has yet reached a significant market penetration in biomarker research. At this time, open-source alternatives may offer a more attractive alternative, in particular for multisite biomarker projects. One solution that has gained considerable traction in biomarker research, both in academia and in the pharmaceutical industry, is tranSMART (http://transmartfoundation.org/). It enables hypothesis-free browsing and simple analytics through existing biomarker patient cohorts with on-the-fly selection of subcohorts for further analysis. A related open-source tool, particularly relevant for cancer genomics, is cBioPortal. It provides a web-based tool for exploring, visualizing, and analyzing multidimensional cancer genomics data. The portal reduces molecular profiling data from cancer tissues and cell lines into readily understandable genetic, epigenetic, gene expression, and proteomic events.

SUPPORT FOR DATA ACQUISITION WORKFLOWS

Depending on the type of biomarker under development (e.g., imaging vs. genomics-based) different data types will have to be collected in the DRE. Traditionally, biomarker researchers have to support their data acquisition workflows with homegrown software or even generic office software such as MS Excel. Nowadays, more robust and dedicated data acquisition software solutions are available for each of the data domains relevant for biomarker research: clinical, imaging, biosample, and experimental (cross-omics) data. The needs in the biosamples domain have already been discussed above in the biobanking section; the most prominent solutions for each of the other domains are briefly discussed below:

- Clinical data: Ideally, clinical data is captured from the clinical care workflow as discussed in one of the previous paragraphs. In practice, this data flow is cumbersome and incomplete for research usage. Therefore, dedicated electronic data capture solutions are required, nowadays usually implemented through a web-based electronic form, a so-called electronic case record form (eCRF). A number of solutions are popular among biomarker researchers, in particular OpenClinica (open-source: https://www.openclinica.com/), REDcap (https://www.project-redcap.org/), and CASTOR (https://www.castoredc.com/). Each of these solutions has been applied in numerous research projects and offers either open-source licenses or very liberal license conditions that are within reach of virtually every biomarker research project. Typically, they provide appropriate quality assurance features, including an audit trail, self-service form-building options, and export capabilities to the most common downstream analysis tools such as SPSS, SAS, and R, but they also export in the CDISC ODM format suited for import into data integration solutions such as tranSMART.
- Imaging data: Within the clinical workflow, the Picture Archive and Communication System (PACS) offers the central archive and workflow tool for all images (e.g., MRI, CT, PET) within the hospital with support for the widely accepted DICOM standard. Multicenter image-oriented biomarker projects are in a need for a similar solution in a research setting. Robust open-source research PACSs are available, in particular the National Biomedical Imaging Archive (NBIA: https://imaging.nci.nih.gov/), which offer a user-friendly web-shop-like interface, and the Extensible Neuroimaging Archive Toolkit (XNAT: https://www.xnat.org/) [14]. The latter is particularly suited for imaging biomarker projects ("radiomics" [15]) because it allows the access of large collections of images in a programmatic fashion to, for example, initiate external image analysis pipelines. For clinical images copied from the clinical domain to the research domain, de-identification of personal data embedded in the image files is needed [16]; the Clinical Trial Processor (CTP) originating from the Radiological Society of North America (RSNA) offers a highly configurable solution for this task. Increasingly, pathology slides are also digitized in high-resolution scanners. Sharing these slides across centers involved in common biomarker studies offers new collaboration models for, for example, remote review of those slides by foreign pathologists. Several commercial and open-source solutions are available, but this domain is still very evolving and lacking dominant solutions.
- Experimental (molecular) data: A major challenge in the experimental domain is to establish standardized pipelines, avoiding IT complexity in the transfer of these methods from the bioinformatics expert to the broader biomedical research community. The number of solutions and calculation pipelines are still diverging and are often tailor-made for the scientific problem at hand. By using Galaxy [17], established workflows for genomics and proteomics data processing can be made available to non-bioinformatician researchers. Although standardization in tooling is still completely lacking in this domain, archival and reuse of molecular profiling data are fortunately supported through a few well-established data archives, in particular, Gene Expression Omnibus [18] (GEO; https://www.ncbi.nlm.nih.gov/geo/) for expression data sets and the European Genomics-phenome Archive [19] (EGA; https://www.ebi.ac.uk/ega/) for genomics data sets.

THE FUTURE: WHAT IS NEEDED?

The broad range of promising data environments for biomarker research are mostly funded on a per-project basis. Sustainability of these efforts is therefore a critical concern. On the other hand, research-funding entities, like charities and governmental bodies have been funding research IT as an overhead on regular research projects. Much of these funds are spent in a dispersed manner with suboptimal return on investment. In particular, the fragmented storage and lack of interoperability of the study data produced virtually prohibits the reuse of the data in follow-up studies as would be needed for actual innovation in the health care system. As a first step funding bodies have been mandating the inclusion of data management plans in grant applications, but the next step would be enforcing the implementation of proper data stewardship supported by a funding model where a few percent of the research grant is not handed over in cash; instead, it can be spent only with certified data infrastructure providers, which would support the creation of an open DRE market independent of institutional funding. This variation on the fee-for-service model adequately addresses the data fragmentation and interoperability issues, while maintaining a certain level of competition in the research infrastructure domain (researchers can still choose between certified providers). Both the National Institutes of Health in the United States and the European Commission ("cloud coins" in the European Open Science Cloud) are already piloting or considering these types of funding schemes.

REFERENCES

1. Sansone, S., Rocca-Serra, P., Field, D. et al. 2012. Towards interoperable bioscience data. *Nature Gen.* **44**, 121–126.
2. Roberts, S.F., Fischhoff, M.A., Sakowski, S.A., Feldman, E.L. 2012. Transforming science into medicine: How clinician-scientists can build bridges across research's "valley of death". *Acad Med.* **87**, 266–270.
3. Van Gool, A., Bietrix, F., Caldenhoven, C. et al. 2017. Bridging the translational innovation gap through good biomarker practice. *Nat Rev Drug Discov.* **16**, 587–588. doi:10.1038/nrd.2017.72.
4. Hua, H, Correll, M., Kvecher, L. et al. 2011. DW4TR: A data warehouse for translational research. *J Biomed Inform.* **44**, 1004–1019.
5. Wilkinson, M.D., Dumontier, M., Aalbersberg, IJ.J. et al. 2016. The FAIR guiding principles for scientific data management and stewardship. *Scientific Data* **3**, 160018.
6. Mons, B., Nylon, C., Veltrop, J., Dumontier, M., Olavo Bonino da Silva Santos, L., Wilkinson, M.D. 2017. Cloudy, increasingly FAIR, revisiting the FAIR data guiding principles for the European open science cloud. *Information Services & Use* **37**, 49–56.
7. Bankin, A., 2017. The road to precision oncology. *Nature Gen.* **49**, 320–321.
8. Lau, E., Watson, K.E., Ping, P. 2016. Connecting the dots: From big data to healthy heart. *Circulation* **134**, 362–364.
9. Sullivan, R., Peppercorn, J., Sikora, K. 2011. Delivering affordable cancer care in high-income countries. *Lancet Oncol.* **12**, 933–980.
10. Prinz, F., Schlange, T., Asadullah, K. 2011. Believe it or not: How much can we rely on published data on potential drug targets? *Nat Rev Drug Discov.* **10**, 712.
11. Collins, F.S., Tabak, L.A. 2014. Policy: NIH plans to enhance reproducibility. *Nature* **505**, 612–613.
12. Holub, P., Swertz, M., Reihs, R. et al. 2016. BBMRI-ERIC directory: 515 biobanks with over 60 million biological samples. *Biopreserv Biobank.* **14**, 559–562.
13. Scholtens, S., Schmidt, N., Swertz, M.A. et al. 2015. Cohort Profile: LifeLines, a three-generation cohort study and biobank. *Int J Epidemiol.* **44**, 1172–1180.
14. Marcus, D.S., Olsen T., Ramaratnam M., Buckner, R.L. 2007. The extensible neuroimaging archive toolkit (XNAT): An informatics platform for managing, exploring, and sharing neuroimaging data. *Neuroinformatics* **5**, 11–34.
15. Gillies, R.J., Kinahan, P.E., Hricak, H. 2016. Radiomics: Images are more than pictures, they are data. *Radiology* **278**, 563–577.
16. Aryanto, K.Y.E., Oudkerk, M., van Ooijen P.M.A. 2015. Free DICOM de-identification tools in clinical research: Functioning and safety of patient privacy. *Eur Radiol.* **25**, 3685–3695.
17. Silver, A. 2017. Software simplified. Containerization technology takes the hassle out of setting up software and can boost the reproducibility of data-driven research. *Nature* **546**, 173–174.
18. Barrett, T., Troup, D.B., Wilhite, S.E. et al. 2011. NCBI GEO: Archive for functional genomics data sets—10 years on. *Nucleic Acids Res.* **39**, D1005–D1010.
19. Lappalainen, I., Almeida-King, J., Kumanduri, V. et al. 2015, European genome-phenome archive of human data consented for biomedical research. *Nature Genet.* **47**, 692–695.

Systems Biology Approaches to Identify New Biomarkers 3.5

Lars Kuepfer and Andreas Schuppert

Contents

INTRODUCTION

In 1998, the National Institutes of Health Biomarkers Definitions Working Group defined a biomarker as "a characteristic that is objectively measured and evaluated as an indicator of normal biological processes, pathogenic processes, or pharmacologic responses to a therapeutic intervention."[1]

In practice, most established biomarkers are single biological parameters, such as genetic mutations, expression of genes or physiological conditions. They are correlated to a complex biomedical process, such as disease propagation or drug action. According to their application, biomarkers should provide sensitive and specific monitoring of the process under consideration. However, often a multitude of biological mechanisms is involved in the complex processes of relevance for disease progression or therapeutic efficacy, whereas there is no one-to-one relation between the mechanisms and the respective physiological processes.[2] Hence, biomarkers that are specific for a single mechanism involved in the process of consideration, such as mutations involved in the mode of action of drugs, tend to suffer from a lack of specificity.

Systems biology, focusing on the integrated approach to describe the behavior of complex biological systems in contrast to the reductionist approach focusing on the system's entities, may be the solution toward more specific biomarkers.[3]

However, utilizing systems biology in order to develop biomarkers that are both sensitive and specific requires the identification and validation of quantitative models for the system behavior representing the complex interaction of its relevant entities. This task needs to cover all factors that have an impact on the subsystem and drive the disease, whereas the biological heterogeneity of patients needs to be taken into account—not only the genetics, but also the history of the disease, lifestyle, co-medication and many others. However, on one hand, the identification of the relevant entities and their interactions suffers from the complexity of the multi-scale architecture of biological systems requiring the integration of processes on the molecular, cellular and multicellular up to the macroscopic scale. On the other hand, the integration suffers from poorly characterized subprocesses and the lack of a mechanistic understanding of the interactions between the scales, such that quantitative models of full-scale biology are far beyond reality today. Focusing only on models of selected subsystems of lower complexity as an alternative, however, may cause a high degree of uncertainty in predictions on the macroscopic level. Hence, existing mechanistic models are limited to specific problems driven by biologically isolated subprocesses, such as pharmacokinetics (PK), or diseases that are controlled by a few key driving mechanisms like chronic myeloid leukemia. Still, the increasing availability of biomedical data repositories in combination with machine learning induces increasing interest in academia and industry in methods to tackle the biological complexity.

Because a generic systems biology approach is not yet available, we will discuss possible routes toward systematic biomarker identification using systems biology on specific examples.

UTILIZING MECHANISTIC MODELING FOR BIOMARKER IDENTIFICATION

The discovery and validation of many biomarkers is based upon targeted *in vitro* experiments with specific cell cultures. In such assays, cells are treated with a given concentration of a drug or a ligand to correlate a given exposure with the intensity of an observed cellular response, which provides the possibility of high-throughput analyses of cellular relationships across different scales of biological organization. Typical examples are pharmacodynamic (PD) or toxicodynamic (TD) measurements, where different amounts of a drug are compared with the intensity of a known biomarker to evaluate the to-be-expected efficacy or toxicity of the agent. Still, such dose–response correlations inevitably reflect the specific underlying experimental setup applied in the original test series. This might include the duration of exposure, the specific composition of the extracellular medium or simply the overall amount of a drug compared with the dose administered to a patient. Therefore, it is a challenging question how concentration–response correlations from an *in vitro* study can be translated to an *in vivo* situation in animals or patients.

One possibility for the extrapolation of PD or TD data from laboratory studies to living organisms is the model-based establishment of PK/PD relations.[4] Here, the exposure of an inducing agent in the body of an organism is estimated using PK simulations of plasma or tissue drug concentrations following the administration of a specific amount of drug. The simulated profiles may then be used to derive PK indices to specifically characterize the kinetics and effects of drug exposure. Such PK descriptors, which are routinely used in clinical practice, include, for example, the maximal concentration (C_{max}), the average concentration (C_{av}) or the area under the curve (AUC).[5] In PK/PD modeling, metrics of an *in vitro* biomarker assay and a corresponding PD or TD correlation are compared with the PK descriptors that depend on *in vivo* drug doses. Thus, the drug response is no longer limited to the original *in vitro* setup, but may be considered within the context of a living organism. Importantly, this allows translating the original concentration–response correlation (*in vitro*) to an actual dose–response correlation (*in vivo*). However, the accuracy of such correlations depends on many factors. Therefore, the experimental setup of the *in vitro* biomarker assay has to be designed in a way that it mimics the physiological conditions of a living organism as closely as possible. This may even involve the composition of the *in vitro* medium, which should ideally contain as much serum or blood as possible. However, the PK models used for contextualization also have to be carefully developed and validated against clinical data to allow an accurate translation of work-intensive *in vitro* experiments.

Previously, PK/PD concepts have been particularly used for antibacterial and antiviral applications to determine the required dose to be administered to an infected patient based on the previously identified minimum inhibitory concentration (MIC) of an *in vitro* growth experiment. However, such PK/PD concepts may also be directly used for the identification of biomarkers in any medical application. Moreover, several other complementary PK modeling concepts exist. Compartmental PK modeling, for example, is based on a few rather abstract compartments that allow the phenomenological reproduction of the kinetic behavior of an observed PK profile. Thus, by using compartmental models, general conclusions of the underlying physiological processes can be drawn. However, an extrapolation to specific clinical scenarios is rather limited. Furthermore, generally only plasma concentration profiles may be simulated with compartmental PK models, such that on- or off-target tissue concentrations may not be considered appropriately. In this regard, physiologically based pharmacokinetic (PBPK) modeling allows a significant structural extension to compartmental PK modeling.

PBPK models aim for a detailed description of physiology.[6] For example, many organs, such as stomach, spleen, pancreas, gut, liver, kidney, gonads, adipose tissue, muscle, bone, and skin, are explicitly represented in PBPK models to account for their volume of distribution as well as their specific physiological function. Thus, it is possible to specifically analyze their roles in drug absorption, distribution, excretion, and metabolism (ADME) including specific cases, such as hepatic or renal impairment. In PBPK models, the simulated organs are linked by an arterial and a venous blood pool. The organs are usually further divided into subcompartments, such as the cellular space, the interstitial space or plasma. The physiological parameters characterizing the various organs are provided from prior collections of different anthropometric and physiological databases. The passive transport processes are estimated from physicochemical properties of the drug, such as the molecular weight or the lipophilicity. The parameters are used to calculate organ–plasma partitioning coefficients, which are then used to estimate the distribution between the tissue and the supplying vascular space. Likewise, diffusion across the cellular membranes of tissue cells can be quantified through the estimation of the effective permeability surface area product. Finally, the relative contribution of active processes including drug transport or metabolism in different organs can be approximated with tissue-specific expression data.

Altogether, the detailed structural representation of the physiology combined with the possibility to quantify the underlying processes allows the simulation of concentration profiles in different tissue compartments. This is an important extension of PK/PD concepts compared with compartmental modeling because PBPK models allow an accurate match of *in vitro* experimental design with *in vivo* physiology. In particular, this is supported by equating the extracellular concentration in the supernatant of a test tube with the drug exposure profile in the interstitial space of a target organ because, in both cases, the drug concentration in the surrounding space is the driving force behind the intensity of the cellular response. Consequently, a multitude of possibilities are provided for *in vitro–in vivo* extrapolation and hence for the translation of biomarker studies to living organisms. Most importantly, this allows the direct consideration of dose–response correlations for different biomarkers within the context of an individual patient because PBPK models allow for an integration of the physiological and anthropometric information. In particular, specific patient subgroups, such as children, elderly people or diseased patients, can be easily considered in PBPK modeling. Interindividual variability can be directly taken into account by varying different physiological variables according to prior collections of corresponding parameters. In addition, PBPK modeling facilitates the evaluation of the impact of different dose levels and dosing regimes on drug PK and thus on drug exposure. This also involves potential nonlinearities in drug PK, which should implicitly be accounted for in a validated PBPK model by the structure of the underlying equations. In summary, a systematic and comprehensive assessment of the whole-body level is possible, which allows for quantitatively describing drug concentrations in the extracellular tissue environment, such that *in vitro* biomarker studies can be contextualized at the whole-body level (Figure 3.5.1).

In a recent study, PBPK modeling was used to predict clinical incidence rates of myopathy following statin administration in different genotype subgroups.[7] To this end, PBPK models were developed for simvastatin and pravastatin, which had been carefully validated with PK data of genotyped patients. In order to predict the occurrence of myopathy, the PBPK models were used to simulate statin exposure in muscle tissue, which in turn was normalized with the maximal inhibitory concentrations, which were measured *in vitro* in embryonal rhabdomyosarcoma cells[8] to obtain a toxicodynamic marker. This marker was ultimately calculated for virtual patient populations and scaled to clinical data[9] to report the documented myopathy risk in different genotype subgroups. Thus, it was possible to successfully predict myopathy incidence rates for further patient cohorts as well as for different dosing schemes.[7]

In a similar study, the impact of N-acetyltransferase type 2 (NAT2) pharmacogenomics on bactericidal activity of isoniazid was analyzed with a PBPK/PD model.[10] Although isoniazid is the standard treatment against tuberculosis, its PK shows a clear trimodal behavior that corresponds to slow, intermediate and fast NAT2 metabolizers. These three different enzymatic phenotypes differ in therapeutic efficacy as well as in drug-induced toxicity because NAT2 metabolizes both clearance of the active parent drug and formation of the toxic metabolite hydrazine. In a first step, genotype-specific PBPK models of isoniazid and several of its metabolites were developed, which were carefully validated with PK literature describing different dose levels as well as different administration routes. In order to account for the therapeutic effect, a mycobacterial growth model was developed from *in vitro* data. This PD model was subsequently coupled to the PBPK model of isoniazid using the drug concentration profile in the pulmonary interstitium, thus correlating the *in vivo* drug concentration at the specific target side, that is, in this case the side of mycobacterial infection, with a bacterial growth model. First, the PBPK/PD model was used to predict therapeutic outcomes in terms of bacterial count in different genotype subgroups. Next, the therapeutic outcomes were further correlated with liver hydrazine exposure, which was used as a surrogate marker of hepatotoxicity. Both the efficacy and the toxicity were compared for different patient subgroups in order to characterize potential safety issues and to ultimately design optimal therapeutic dosing schemes with maximum efficacy but minimal adverse side effects.

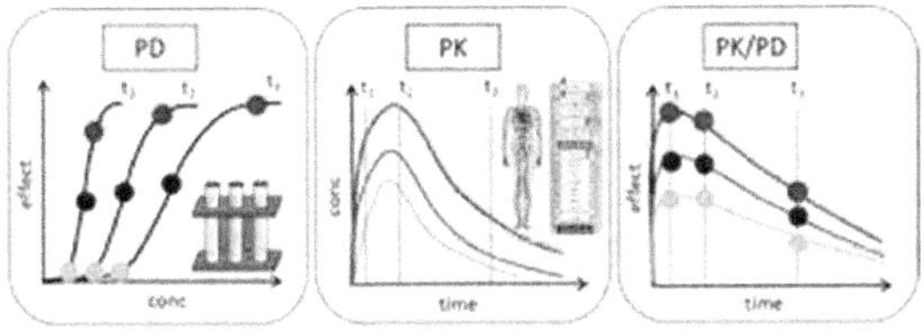

FIGURE 3.5.1 Physiologically based pharmacokinetic (PBPK)-based contextualization of *in vitro* omics data allows the derivation of dose–response correlations at the whole-body level.

The accuracy of model-based *in vitro–in vivo* extrapolation of biomarkers can be further enhanced if the underlying assay protocol complies with certain systematics in both time and concentrations. In toxicology, the Open TG-GATEs library represents a standard repository of transcriptome data after treatment with different hepatotoxicants.[11] It includes *in vitro* data from primary rat and human hepatocytes as well as *in vivo* data from rat liver biopsies treated with different drugs. For the hepatocytes, three different drug concentrations levels were considered at three different sampling times, each resulting in a 3×3 matrix of different treatment scanarios. This unqiue availability of transcriptome data for nine different sampling combinations has been used recently for refined *in vitro–in vivo* extrapolations using an approach called PBPK-based *in vivo* contextualization of *in vitro* toxicity data (PICD).[12] The application of PICD requires the development and validation of drug-specific PBPK models, which are then used to identify *in vivo* doses that are equivalent to the *in vitro* drug exposure in the TG-Gates assay. In detail, PK profiles are identified, where the *in vitro* drug exposure in the assay equals the area under the curve in the interstitial liver at each experimental time point. This allows for the establishment of model-based correlations of genes, where a cellular response in expression can be described as a consequence of drug doses administered in patients. In a study by Thiel et al., PICD was applied to identify toxicity-related pathway responses in a comparative study of 15 hepatotoxicants.[13] First, the specific gene response patterns were correlated with drug-related properties, such as its DILI-potential or physicochemical parameters. In a next step, changes in functional classes of genes involved in key cellular processes were further analyzed for a high-responsive subgroup of hepatotoxicants. Finally, the intensity of the toxic change was used to identify individual and shared biomarkers within the subset of high-responsive pathways. As a result, genes from the cytochrome P450 family, transcription regulators and drug transporters were affected most. Moreover, a comparison of individual biomarkers was used to identify potential drug–drug interactions, which were in good agreement with reported cases in the literature. Altogether, PBPK-based contextualization of *in vitro* data allows for the description of cellular responses as consequences of drug doses applied at the whole-body level. Because PBPK models can be explicitly informed using physiological and anthropometric information, the consideration of specific patient subgroups is possible. Therefore, PBPK models represent an important platform to extrapolate *in vitro* biomarker experiments into an individual *in vivo* patient context.

COMBINING MECHANISTIC MODELING AND DATA ANALYSIS FOR BIOMARKER IDENTIFICATION OF DISEASE PROGRESSION

Modeling parameters that underlie multistep carcinogenesis is the key for a better understanding of the progression of cancer and the design of individualized therapies. Reliable biomarkers indicating prognosis and disease progression are urgently needed in the clinic.

Chronic myeloid leukemia (CML) is caused by the oncogenic transformation BCR-ABL in pluripotent hematopoietic stem cells (HSCs), typically resulting in a chronic phase (CP) and a subsequent multistep disease progression toward the accelerated phase (AP) and the highly malignant blast crisis (BC). The time that the disease has been present in a patient before the diagnosis is of high relevance because patients are at a considerable risk of progression to AP and/or BC even during treatment with tyrosine kinase inhibitors (TKI). Furthermore, progression toward AP and BC mainly occurs during the first 2 years of therapy, suggesting that patients may be diagnosed in a so-called "late CP" CML. Although CML can be controlled by TKI treatment in CP, CML serves as a paradigm for model-based disease progression markers.[14,15] Systems biology based CML disease models use quantitative population dynamic models for hematopoiesis starting with both wild type (WT)–HSCs and HSCs with the BCR-ABL mutation. The respective transition rates between the cell types involved in hematopoiesis are fit to cell count data from mice and humans. Consequently, the models allow the quantitative simulation of the evolution of the hematopoietic stem cell population throughout CP (Figure 3.5.2).

The resulting disease models for CML progression can be used for the monitoring of therapeutic efficacy in individual patients.[16,17] Because of the comparably low complexity of CML in CP, monitoring cell populations is sufficient for the assessment of the disease status. Hence, CML can be used for the development of omics-based strategies for the development of biomarkers for disease progression monitoring, such as gene expression (GEX) based models. Because the distribution of the entire hematopoietic cell population shifts throughout progression within CP, many sets of genes can serve as a progression biomarker, such as CD34+, which is specifically enriched in every

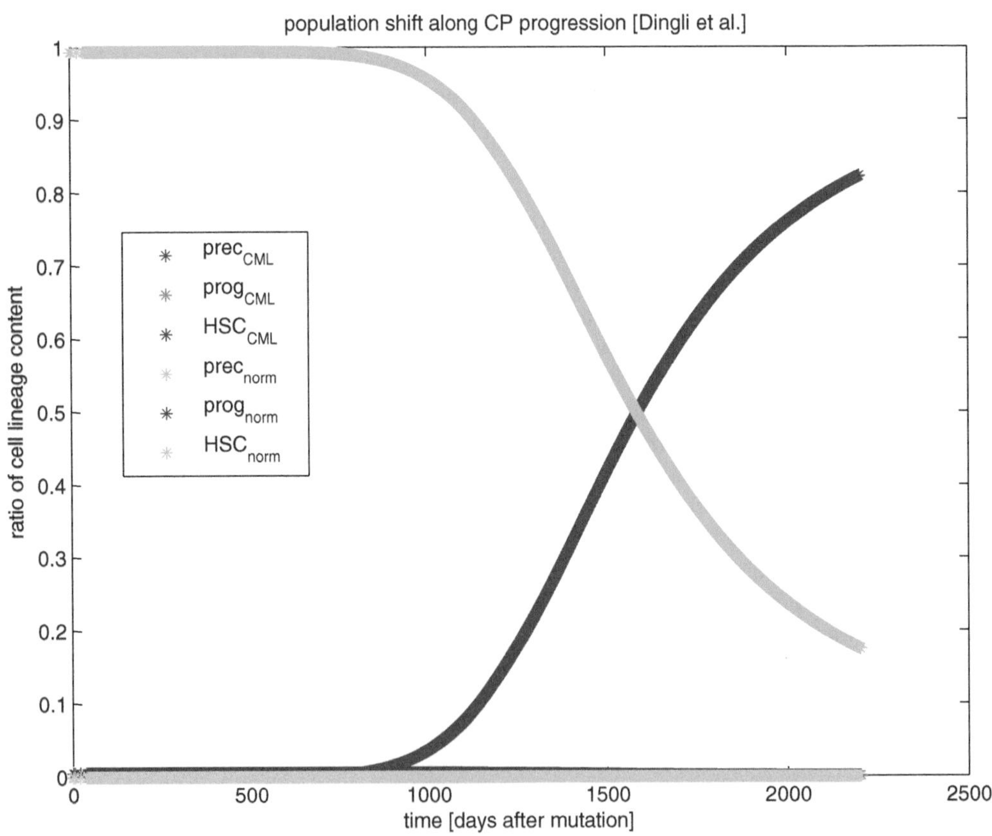

FIGURE 3.5.2 Systems biology modeling of evolution of chronic myeloid leukemia (CML) in the chronic phase enables the simulation of the hematopoietic cell population over time, starting from the first Bcr-Abl mutation. The population data can be used to simulate the evolution of cell-type specific markers.

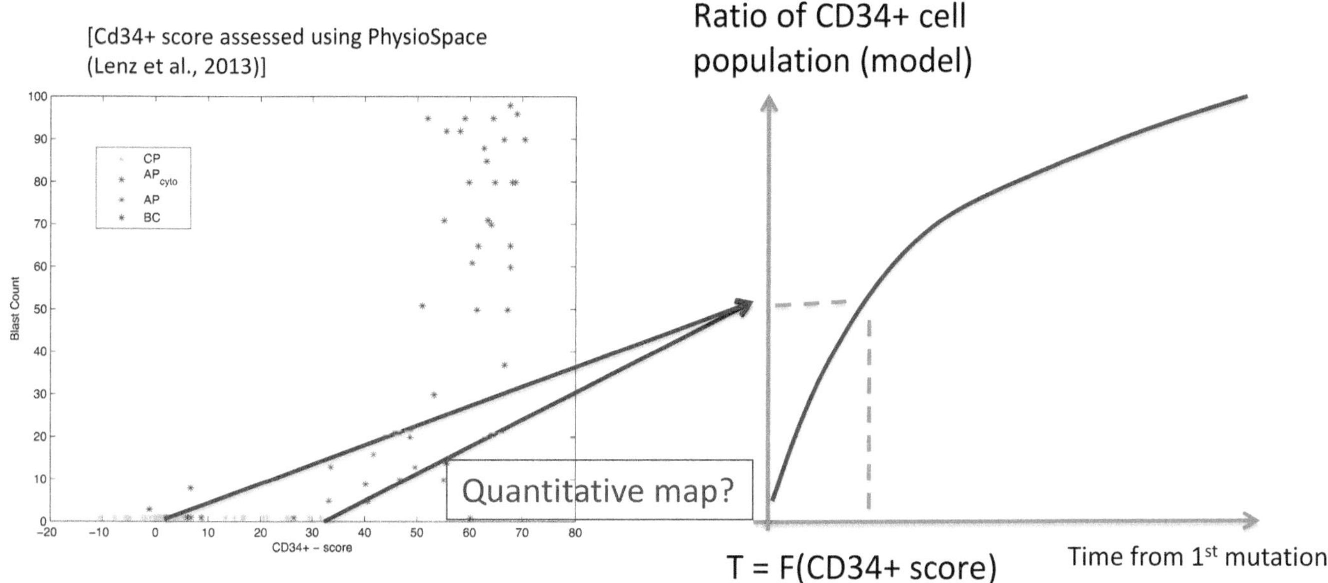

FIGURE 3.5.3 Simulations of cell-type specific markers (right) must be mapped to the observed data (left) that are affected by variations in protocols. Hence, matching of models and data requires independent additional markers for calibration.

cell type that changes concentration throughout CP.[18] However, GEX-based biomarkers are strongly affected by assay protocols and their lack of specificity, such that a direct quantitative association of a monitoring marker with a population-based mechanistic model is rarely possible (Figure 3.5.3).

Moreover, CML in CP is often diagnosed by chance only. As a consequence, the initial starting point of the model, namely the occurrence of the first BCR-ABL mutation in an HSC, is not known in an individual patient. Hence, the calibration of GEX-based monitoring markers and the mechanistic disease model requires a second GEX-based biomarker that is independent of the first one. This second "calibration marker" must not show a monotonic evolution along disease progression. Motivated by phase transitions of complex systems in statistical thermodynamics, the Shannon entropy of sample gene expression was identified as a possible calibration marker. Based on the observed dynamics of entropy along disease progression that could be monitored by the disease evolution marker, an unsupervised classification of disease states within CP as well as a quantitative association between the mechanistic population dynamic model and the GEX data were possible. The resulting classification allowed for a clear separation of the distributions in AP and in an AP with additionally occurring clonal cytogenetic changes without any increase in blast count (AP_{cyto}), which could indicate the risk of progression from CP to AP.[19]

Taken together, this approach that combines mechanistic and data-driven modeling enables an interpretation of the clinical data in terms of a generic thermodynamics phase transition paradigm. A further extension of this model to other CML disease states, such as BC and remission, and its application in assessing therapeutic efficacy can beneficially affect individual treatment. In the future, these basic mechanisms may be extended not only to other leukemias, but also to localized and metastatic solid tumors.

UTILIZING MACHINE LEARNING FOR BIOMARKER IDENTIFICATION

Machine learning from massive data, which is sometimes called the "Big Data" approach, is currently made a major issue as an alternative to mechanistic modeling. This is due to its optional cure to some problems arising in mechanistic modeling in the field of systems biology, especially the gaps in mechanistic understanding of the modeled systems.

Especially the availability of extensive omics data sets, which have been generated over the years ranging from genomics, over transcriptomics and proteomics, to metabolomics and similar data,[20] generated the hope for new, efficient routes toward new biomarkers. Distinguished examples are given by the Cancer Genome Project by the Wellcome Trust Sanger Institute,[21] the Cancer Genome Atlas TCGA Research Network[22] and the Library of Integrated Network-Based Cellular Signatures (LINCS) catalog of gene-expression data of human cells that were treated with chemical compounds and genetic reagents.[23] The variety of these programs has led to an accumulation of data, such that the bottleneck of omics approaches nowadays seems to reside in the computational methods to manage, analyze and interpret the emerging data rather than in the high-throughput techniques to generate them.

Conceptual challenges for the identification of biomarkers from complex, highly multivariate data structures with high dimensionality, such as omics databases, arise from the "curse of dimensionality," which occurs when highly multivariate data must be analyzed. It leads to counterintuitive consequences:

The distances between each pair of samples, each represented by high-dimensional data, become more and more similar: Therefore the probability of finding differences between each pair of samples by chance increases with the number of parameters used for the comparison. Hence, finding special, meaningful groups of similar samples can be more difficult in high-dimensional data sets than in low dimensional ones.

The risk for overfitting of models increases significantly with the dimensionality of data and the demand on sample size to avoid it can increase exponentially with the dimensionality of the data.

Hence, simply adding more variables to an analysis does not necessarily result in a better quality of the models. Moreover, covering all possible variations in the data requires the handling of very large data sets from heterogeneous data sources causing unmet challenges for data integration. It requires new concepts in modeling and data analysis that are adapted to the specific strengths of systems biology.

Recently, it has been shown that deep learning strategies may provide a concept to overcome the curse of dimensionality, given that the model is hierarchically organized.[24] A similar generic conceptual approach has been developed in the context of generic networks of nested functions, structured hybrid modeling (SHM),[25] and applied in learning mechanisms of drug action from data.[26,27] In addition, although the underlying computational concepts are in the exploratory stage, early applications indicate that appropriately structured combinations of partial mechanistic understanding of the underlying biology on the one hand and machine learning on the other hand can lead to hybrid models resulting in superior predictivity.

SUMMARY

Systems biology based biomarkers must satisfy a multitude of requirements:

Modeling the disease and therapy driving mechanisms must cover the relevant subprocesses at the molecular, cellular, organ and organism scale of the individual patient. It requires heterogeneous model types, ranging from logic models for the representation of intracellular signaling up to PDEs required for the simulation of diseased cell population shifts within healthy tissues affecting the overall functionality of organs.

Although the disease-driving mechanisms on the intracellular level are often sufficiently understood, their interaction, as well as the effects of environmental impact factors such as comorbidities and physiological constitution, can rarely be represented by mechanistic models. Hence, an appropriate combination of mechanistic and data-driven models must be enhanced.

On the other hand, the heterogeneity of biomedical data structures and their high dimensionality pose significant challenges for pure machine learning techniques, which prohibits their broad application in complex diseases.

Satisfying these requirements needs the integration of heterogeneous types of mechanistic models, both discrete and continuous, with data-driven models into a model network. Despite promising results in special application areas, a broad application requires the development of future computational technology. Efficient parameter- and model-identification algorithms, adapted to the variety of data types are essential. The key to establishing modeling as a technology, rather than as an art, requires the automatic selection of model types, model structures and the complexity for each submodel, such that the overall model provides the required predictivity and can be identified using the available data.

REFERENCES

1. Biomarkers Definition Working Group, Biomarkers and surrogate endpoints: Preferred definitions and conceptual framework. *Clin Pharmacol Ther.* 2001;69:89–95.
2. Schadt, E.E. et al., An integrative genomics approach to infer causal associations between gene expression and disease. *Nature Genet.* 2005;37(7):710. doi:10.1038/ng1589.
3. Noble, D. (2006). *The Music of Life: Biology Beyond the Genome.* Oxford, UK: Oxford University Press. p. 176.
4. Derendorf, H. and Meibohm, B. Modeling of pharmacokinetic/pharmacodynamic (PK/PD) relationships: Concepts and perspectives. *Pharm Res.* 1999;16(2):176–185.
5. Muller, P.Y., Milton M.N. The determination and interpretation of the therapeutic index in drug development. *Nat Rev Drug Discov.* 2012;11(10):751–761.
6. Kuepfer, L. et al. Applied concepts in PBPK modeling: How to build a PBPK/PD model. *CPT Pharmacometrics Syst Pharmacol.* 2016;5(10):516–531.
7. Lippert, J. et al. Modeling and simulation of in vivo drug effects. *Handb Exp Pharmacol.* 2016;232:313–329.
8. Kobayashi, M. et al. Association between risk of myopathy and cholesterol-lowering effect: A comparison of all statins. *Life Sci.* 2008;82:969–975.
9. Link, E. et al. SLCO1B1 variants and statin-induced myopathy—A genomewide study. *N Engl J Med.* 2008;359:789–799.
10. Cordes, H., Thiel, C., Aschmann, H.E., Baier, V., Blank, L.M., Kuepfer, L.A physiologically based pharmacokinetic model of isoniazid and its application in individualizing tuberculosis chemotherapy. *Antimicrob Agents Chemother.* 2016;60(10):6134–6145.
11. Igarashi, Y. et al. Open TG-GATEs: A large-scale toxicogenomics database. *Nucleic Acids Res* 2015;43:D921–D927.
12. Thiel, C., Cordes, H., Conde, I., Castell, J.V., Blank, L.M., Kuepfer, L. Model-based contextualization of in vitro toxicity data quantitatively predicts *in vivo* drug response in patients. *Arch Toxicol.* 2017;91(2):865–883.
13. Thiel, C. et al. A comparative analysis of drug-induced hepatotoxicity in clinically relevant situations. *PLoS Comput Biol.* 2017;13(2):e1005280.
14. Michor, F. et al. Dynamics of chronic myeloid leukaemia. *Nature* 2005;435:1267–1270.
15. Dingli, D. et al. Chronic myeloid leukemia: Origin, development, response to therapy, and relapse. *Clin Leuk.* 2008;2(2):133–139.

16. Roeder, I. et al. Dynamic modeling of imatinib-treated chronic myeloid leukemia: functional insights and clinical implications. *Nat Med.* 2006;12:1181–1184.
17. Glauche, I. et al. Therapy of chronic myeloid leukaemia can benefit from the activation of stem cells: Simulation studies of different treatment combinations. *Br J Cancer* 2012;106:1742–1752.
18. Ummanni, R. et al. Identification of clinically relevant protein targets in prostate cancer with 2D-DIGE coupled mass spectrometry and systems biology network platform. *PLoS One* 2011;6(2):e16833.
19. Brehme, M. et al. Combined population dynamics and entropy modelling supports patient stratification in chronic myeloid leukemia. *Sci Rep.* 2016;6:24057.
20. Alyass, A. et al. From big data analysis to personalized medicine for all: challenges and opportunities. *BMC Med. Genomics* 2015;8:33.
21. Garnett, M.J. et al. Systematic identification of genomic markers of drug sensitivity in cancer cells. *Nature* 2012;483(7391):570–575.
22. Weinstein, J.N. et al. The cancer genome atlas pan-cancer analysis project. *Nat Genet.* 2013;45(10):1113–1120.
23. NIH LINCS Program: Library of Integrated Network-Based Cellular Signatures. 26.08.2016; Available from: http://www.lincsproject.org/.
24. Tomaso, P. et al. Why and when can deep-but not shallow-networks avoid the curse of dimensionality: A review. *Int J Autom Comput.* 2017;14(5)503–519. doi:10.1007/s11633-017-1054-2.
25. Fiedler, B., Schuppert, A. Local identification of scalar hybrid models with tree structure. *IMA J Appl Math.* 2008;73:449–476.
26. Schuppert, A. Efficient reengineering of meso-scale topologies for functional networks in biomedical applications, *J Math Ind.* 1:6. doi:10.1186/2190-5983-1-6.
27. Balabnov, S. et al. Combination of a proteomics approach and reengineering of meso scale network models for prediction of mode-of-action for tyrosine kinase inhibitors. *PLoS One* 2013;8(1):0053668. doi:10.1371/journal.pone.

The Need for Global Collaboration in Translational Medicine

3.6

Florence Bietrix, Antoni L. Andreu, Giovanni Migliaccio, and Anton E. Ussi

Contents

Translational research in medicine, as the name suggests, is concerned with translating discoveries—whether observed in the clinic or elucidated in the research lab—into meaningful outcomes for patients and/or the community at large. It is thus the transformation of knowledge into utility, from the implementation of a small change in clinical workflow, to developing a novel drug that reverses the effects of Alzheimer's.

This broad definition thus covers the entire R&D continuum, from the moment foundational research yields an insight, to the health care and population setting in which the fruits of that early research are eventually deployed. This continuum can be broken down into "translational blocks," as can be seen in Figure 3.6.1.[1] The T0–T4 research classification describes where a given research sits on the translational continuum. Broadly speaking, the research field resides in two areas, those in the T0–T2 camp, covering discovery through to the implementation of the interventions arising from those discoveries, and those in the T3–T4 camp, whose research endeavors to understand the benefit of those interventions and other factors that influence society and its health care systems.

DRUG DEVELOPMENT IN CRISIS—THE BIRTH OF TRANSLATIONAL MEDICINE

Translational research efforts in the T0–T2 range started to gain significant momentum in the early years of the 2000s, as reported by Trochim et al.,[2] around which time several coincidental factors gave rise to the growing perception of a need for change in the way drugs are developed. First, the traditional, more linear process of chemistry-driven drug discovery was struggling to keep up in a world increasingly immersed in the biology-rich realm of the unraveled genome, a concomitant explosion of data resulting from a growing number of omics technologies, whereas in the clinic there remained a large number of highly complex diseases representing high unmet need, coupled with regulators that increasingly sought evidence of an understanding of the underlying MOA of a given therapeutic.[3] These challenges

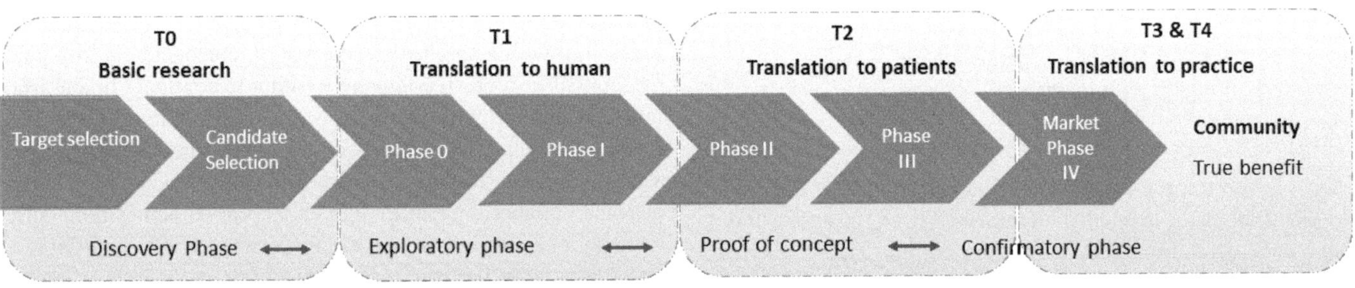

FIGURE 3.6.1 T0–T4 R&D continuum adapted from Blumberg et al. (2012). T0 applies to research ranging from basic biomedical research to preclinical studies—both *in vitro* and *in vivo*—and thus covering the identification and early development of novel biomarkers and therapeutic interventions; T1 research applies to early human trials and thereby covers proof of mechanism studies and proof of concept; T2 applies to later (phases II and III) development, assessing the clinical utility of a given intervention in a controlled study, prior to clinical introduction or roll-out of evidence-based guidelines; T3 research applies to post-marketing studies (phase IV clinical trials) as well as research covering issues around implementation of new clinical guidelines or products; T4 applies to health care system and population research, such as comparative effectiveness studies or assessing the effects of policy changes in a given system.

manifested themselves in a drug development pipeline suffering from chronically high failure rates despite rapid advancements in analytical technologies and our understanding of basic biology. Worse yet, many of those failures came at a very late stage of development, with approximately 65% of candidate drugs failing during phase II clinical trials, and a further 32%–40% of those surviving failing to prove effective during phase III trials, as measured in the period 2003–2011.[4,5]

All of the above factors conspire to perpetuate a decades-long trend of decreasing productivity in drug R&D that sadly continues today[3].

As a consequence, the cost of bringing a new drug to market is extremely high. In a 2016 publication, Di Masi et al. found that the average cost of bringing a single new prescription drug to market was calculated to be $2.5 billion, including all failures and accounting for the industry's average cost of financing their operations.[6] The causes of this crisis in productivity are no doubt manifold, nonetheless a few factors particularly stand out. First, the reproducibility of the foundational research upon which later development is based appears to be alarmingly low in biomedical research, with studies finding irreproducibility in preclinical research to range from 51%—90%.[7,8] Freedman et al. estimated that scientists in the United States spend $28 billion each year on basic research that cannot be reproduced.[9] This issue is currently the subject of intense debate in academia and policy circles. It is worth noting that a study's lack of reproducibility does not necessarily mean that the finding is incorrect: It may be attributed to other factors such as differences in study conditions.[10] However, it would be foolish not to assume that a lot of money and time is being spent trying to validate and develop interventions based on fallacious science.[9,11]

Concomitantly, the field has been impacted in the last few years by a net retreat of large pharmaceutical industry actors from discovery and early preclinical stage development leaving academia to move further along the development. A further highly significant area of concern in translational research is the lack of preclinical (*in vivo* or *in vitro*) models that reliably recapitulate the human process or pathology.[12] Results showing efficacy in *in vivo* studies form a key element of the decision to advance a candidate drug from preclinical to clinical development, and the subsequent efficacy signal in humans is typically only studied during a phase II clinical trial. Thus, a developer may spend tens of millions of U.S. dollars in the intervening period, only to find that the expected efficacy did not materialize in the human environment.[13]

A final challenge of particular note in the era of personalized medicine is that of discriminating responders from nonresponders for a given intervention. The inherent heterogeneity in populations with diseases of complex and multifactorial nature has shown that it is desirable and in some cases such as cancer, essential,[14] to be able to select *ex ante* the patients who will likely benefit from a therapy, to predict the dose that is suitable for them, and to monitor the response of the patient to the therapy.[15] The challenge, in this case, is to identify, validate and develop into a robust test the biomarker or biomarker panel in a parallel but linked development pathway such that ultimately a companion diagnostic is available and forms an intrinsic part of the therapeutic product offering.[16,17] The resulting patient-stratified drug enables the mantra of personalized medicine, namely to "provide the right treatment to the right patient, at the right dose and at the right time."[18,19]

Additionally, biomarkers are commonly sought, qualified and validated for use within the drug development process, as so-called "decision-making" biomarkers, and as such do not need to undergo the regulatory approval steps as for *in vitro* diagnostics. These biomarkers allow more accurate understanding of exposure at the disease target site and subsequent engagement, whether the drug displays activity commensurate with the MOA and dose, and potentially anticipate safety issues.[20]

SORTING THE WHEAT FROM THE CHAFF

Each year an enormous number of novel biomarker candidates are identified in the academic literature.[21] Given the above finding that a great deal of published science is not reproducible, a major challenge is, therefore, to first identify promising candidates on the basis of published data and to subsequently probe them to assess their analytical validity in the intended context. If successful, the biomarker must then survive the process of development into a robust and reliable assay that can be manufactured, distributed and utilized in a variety of conditions without excessively sacrificing performance, such as selectivity or specificity.[22] In order to influence clinical practice, the test deployed must meet regulatory requirements to ensure the performance as described above as well as show clinical utility, that is, show positive net benefit to the patient care process in the form of improved care or efficiency.[23] This simplification does not take into account U.S.-based regulations of the FDA versus EU regulations, differences in dealing with laboratory developed tests (LDT) versus *in vitro* diagnostics regulations, among others, which add an extra level of complexity.[24]

It is thus within a complex matrix requiring the latest insight into biology, cutting-edge analytics and profound clinical expertise and associated resources that characterize the environment in which translational research operates, with its myriad challenges and bottlenecks that contribute to the high cost and failure rates associated with the R&D process. Developing a personalized medicine is an undertaking that involves a great deal of interdependencies and uncertainty and should, therefore, be viewed not as a linear process progressing slowly toward marketing authorization but rather as a circular process that requires a constant and bi-directional dialog between clinic and laboratory,[25] underpinned by judicious use of the latest analytical technologies. Given the complexity and heterogeneity of many of the diseases for which personalized medicines are under development, successful translation must rely upon a multitude of disciplines, technologies and patient-related resources in order to triangulate upon the therapeutic strategy that ultimately shows a positive benefit–risk profile for the given patient.

RISK, INEFFICIENCY AND BOTTLENECKS

Looking at the above, it is a logical step to assume that an organization that successfully brings these varied disciplines and resources to bear in a goal-directed and effective manner will have the greatest chance of success. It is also fair to assume that no single organization, not even the global pharmaceutical industry behemoths of today, have within their organization all of those resources readily available. Moreover, in fact, some essential resources may be absent in entire sectors. For instance, patient cohorts are only accessible within the health care sector, consequently any pharmaceutical or biotech company wishing to conduct clinical trials must collaborate with that sector. Thus, translational research is further characterized by the need for cross-organizational, cross-sectoral and—in today's global health care market—even cross-national collaboration. Although collaboration across boundaries is no doubt an important opportunity, doing so brings with it risks and inherent inefficiencies, as anybody with academia–industry experience will be quick to point out.

Uncertainty and inefficiency in a project are drivers of risk, which in turn is a driver of cost and time. If we wish to reduce the cost and complexity of developing a personalized medicine, then we need to systematically reduce risk in the development pipeline. Given this conclusion, it is worthwhile spending some time to evaluate the types of risk that drive the high failure rates in order to develop a sense of how to handle and reduce them. By reducing or removing risk, we can reduce the aggregate cost and timelines associated with developing a novel personalized medicine.

LOOKING THROUGH THE WINDOW OF RISK

At the European Infrastructure for Translational Medicine (EATRIS), we classify risk into technical versus operational, and systemic versus specific, as explained below. By doing so, the feature of the risk can be ascertained, together with its effects, and the likely required scope of any approach to solving it can be better understood. It is worth noting that the term "risk" as we use it here is interchangeable with "bottleneck" or "challenge."

Some risks are specific to a given project and some are systemic. For instance, if Professor Smith from research institute X and Dr. Blythe from company Y develop irreconcilable differences regarding the development strategy when conducting a collaborative project and the project collapses as a result, such failure would be an example of the realization of a project-specific risk. However, a public–private partnership turning sour during a collaboration happens more than once and not only to Professor Smith and Dr. Blythe. In fact, failure with public–private partnership is inherent to an entire ecosystem and is thus a risk for all similar projects in this given system. This risk is global in nature and is thus classified as systemic risk. To elaborate further, specific risks are often the realization of a systemic risk in a given project.

Risk can by further classified into technical and operational risk. Technical risk is the potential exposure to loss or failure from design and execution of the R&D process, whereas operational risk relates to potential issues in the administration, management, human and financial processes required to perform R&D. The above example of relationship risk is an operational one, whereas the failure of a biomarker to meet a given sensitivity threshold would be a technical risk.

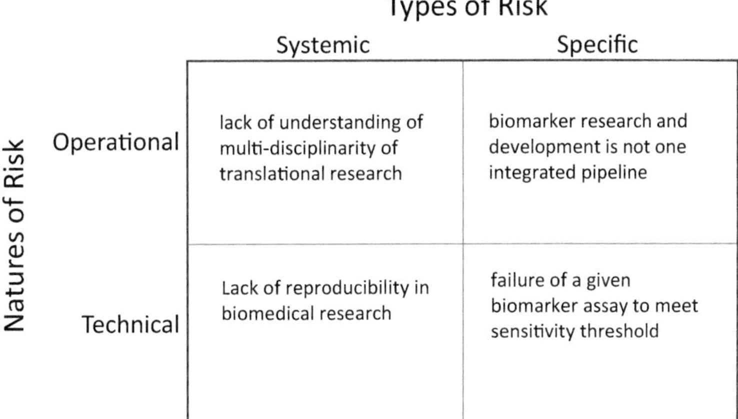

FIGURE 3.6.2 Risk matrix showing the four different types of risk that arise from the R&D field, including examples.

The matrix below shows the four different types of risk that arise from the above stratification, with examples of bottlenecks that characterize that risk (Figure 3.6.2).

ELIMINATING DIFFERENT TYPES OF RISK

By classifying types of risk, we are then able to develop a suitable solution for eliminating that risk.

The elimination of a project-specific risk requires the involvement of a small circle of parties, directly related to the project, together with any potential external parties that would be needed to technically develop a solution. In other words, the problem is limited to a small group, and the benefit of the solution, therefore, is also confined to the project. The result is a risk reduction exercise that is smaller in scope, easier in implementation and relies on fewer counterparties and, thus, is more controllable and likely to be more resource-friendly for an organization.

On the other hand, when looking at systemic issues, the whole sector or even ecosystems need to be considered globally because the solution will, in theory, need to be applicable to, and therefore endorsed by, all parties in the field. As an example, irreproducibility of an academic finding can be due to a poor experimental set-up or methodology and as such be specific to a project. However, the lack of reproducibility is not limited to a couple of (poor) studies but the result of an inadequate system, impacting the entire R&D field at all levels. As such the irreproducibility issue represents a systemic risk. The causes of irreproducibility are multiple. Possible solutions encompass, for example, reagent validation, sharing of protocols and data, and improved training.[9] However, finding a solution requires first finding consensus within the entire academic community as to the causes of irreproducibility. We subsequently would design a solution for a given cause, again seeking consensus that this is the right path, and then execute that solution. If, for example, one solution is to expand the reporting requirements of the "Methods" section for publication of research findings, we then would have to seek consensus within the publishing industry that the lack of reporting standards is indeed a driver of irreproducibility, and expanding the "Materials and Methods" section is a solution. Finally, both sectors would have to agree on the revised standards for reporting if the implementation of the solution is to be smooth. As you can see, the effort is enormous, but sadly the result is not guaranteed. As an example, the ARRIVE (Animal Research: Reporting of In Vivo Experiments) guidelines on reporting standards for publication of *in vivo* preclinical studies were endorsed by publishers in 2010, yet in practice there is little improvement in reporting standards, suggesting that the guidelines are not being followed.[26]

The difference in levels of effort required—but also the potential benefit—of tackling systemic risk versus specific risk is obvious. Working on systemic risk brings with it risks of its own that further complicate matters. For instance, the consensus approach requires the active involvement and commitment of multiple organizations in order to achieve the critical mass that can lead to consensus forming, whereas these same organizations are often in fierce competition with one another in a given sector. Thus, organizations are expected to actively collaborate with competitors to expend resources (that can otherwise be deployed to create competitive advantage over competitors, for instance, in product development or marketing efforts) that would result in net benefit for those competitors and other organizations. This is a difficult decision to make and defend, and it leads to the danger of free riders that rely on the efforts of others to solve systemic risks to which they also are exposed. Moreover, the coordination cost of highly networked collaborations is very high, and it can often be very inefficient if the governance and operational structure is not designed for effect. Highly interdependent systems are only as good as their weakest link, leaving organizations at the mercy of the efficiency of all other parties in a system.

Given the above, it is not difficult to see why systemic risk in competitive sectors is often left untouched, with organizations choosing instead to have more control by tackling only the specific risks manifested in their projects. This approach, however, has far-reaching consequences.

A PATCHWORK OF REMEDIES

As a result of the focus on solving risk within the context of a specific project in lieu of tackling the wider systemic risk, the biomedical innovation pipeline remains littered with systemic risks that subsequently delay and complicate all projects navigating the pathway. This has the effect of driving costs and development times upward to today's unsustainable levels, for several reasons.

First, by working solely within the confines of specific projects, undesired duplication of efforts will result, with multiple organizations having to tackle the same bottleneck for similar projects, each research team starting from scratch and having to deploy resources independently for the same goal. This is naturally highly inefficient, especially so when the same organizations are spending valuable resources trying to validate tools or biology that turns out to be ineffective. As we are all aware, publication of negative findings is a rare phenomenon, and so the community at large does not benefit from the finding that a certain outcome is negative. Furthermore, even if the results are positive, they are often not transferrable to other examples of the same risk, potentially due to lack of use of defined standards, technical interoperability or buy-in from other sector players, or simply because of a lack of reproducibility, as described in the previous section. Even if an organization is willing to share a positive result, the "not-invented-here" mentality can impede uptake, especially if an organization is already planning or working on their own "solution." The end result is a tragic patchwork of home-grown solutions that prevent an easy path to harmonization or standardization, prevent the possibility of combining results (such as data) to create novel pooled resources, difficulty in ascertaining and agreeing minimum quality levels, all the while increasing complexity and thereby pushing the prospect of reaching a consensus standard further and further away.

WHY COLLABORATE?

In an ideal world, only the risk inherent to the complex biology would be left, whereas operating in a framework allows the elimination of all systemic and operational risk. By doing so, the efficiency of the R&D process can be improved, leaving only the uncertainty of the science and the complexity of biological systems to confound the search for an effective medicine. The question, therefore, is whether this is possible, and if so, how would we go about reducing all these other risks and inefficiencies.

In recognition of the enormous burden that systemic risk places on the drug development pipeline, there is an increasing level of collaboration in the field, both within and across sectors, and a growing number of initiatives designed specifically to overcome systemic issues. Examples of cross-border, cross-sector collaborations are expanding in Europe and globally. The Innovative Medicines Initiative (IMI) was launched in Europe in 2008, bringing closer the key players involved in health care research, including the academia, the pharmaceutical industry, small and medium-sized enterprises, but also patient organizations, and medicines regulators with a focus on specific health issues but also on broader challenges like drug safety or knowledge management. With a budget of €2 billion euro, it was the largest public–private collaboration in the world, charged with improving the efficiency of the drug development process by working across sectors to reduce systemic risk. Its success led to the introduction of IMI2 in 2014, with a budget of €3.2 billion euro (https://www.imi.europa.eu/content/imi-2).

COORDINATED EUROPEAN ACTION IN RESEARCH INFRASTRUCTURES

Another initiative borne from Europe's increased attention to defragmentation and efficiency is the group of biomedical research infrastructures stemming from the European Strategy Forum on Research Infrastructures. These were set up with the goal of facilitating broad and efficient access to the cutting-edge research and analytical systems and facilities that are indispensable in translational medicine today. EATRIS ERIC, the European Infrastructure for Translational Medicine, was established in 2013 after an 8-year construction phase, to bring to bear academic facilities and clinical resources for the medicines and diagnostics development process. By looking at each of the risks identified above, EATRIS established a consortium of facilities and an operational process that works toward eliminating systemic risk; however, providing access to the latest technologies and know-how is needed to minimize each user's project-specific risks (http://eatris.eu). EATRIS facilitates efficient and effective access to leading analytical, biological and clinical capacity in Europe for innovative drug and diagnostic developers, and in so doing reduces the systemic risks associated with extramural collaboration. An example of a project endorsed by EATRIS is the Early Cancer Detection Europe (ECaDE) initiative. ECaDE is a multi-stakeholder approach to improve efficiency of the research and development process in early cancer detection and management by developing a biomarker research and development framework.[27] In another approach, the European biomedical research infrastructures—EATRIS-ERIC together with BBMRI-ERIC (focused on biobanking) and ELIXIR (focused on data sharing)—are working closer together to develop and share best practices for biomarker validation and bridge the translational innovation gap impacting the successful transition of biomarkers from discovery to validated tools or diagnostics.[21]

Enhancing quality in biomedical research, given the previously described reproducibility issues, is an important task for the ESFRI biomedical research infrastructures, not least of which is EATRIS. To this end, EATRIS regularly identifies and develops or joins collaborative initiatives to define standards, perform harmonization exercise, or share best practices among community practitioners.[28]

Only by combining forces can systemic risk be tackled effectively. In so doing, critical mass can be achieved for seeking consensus, resources are brought together for a common solution, and the cost of developing the solution is more evenly divided among those who will benefit. Moreover, the results are created with the purpose of wide dissemination and uptake so that all actors can benefit from the outcomes generated.

If we are to have any chance of reducing the cost and timelines of drug development in the hope of reducing the cost of health care, then we must have a systematic and coordinated approach to eliminating systemic risk. Only by tackling this form of risk can we hope to achieve the scale of change needed to have an effect. The efforts until now are commendable and noteworthy, but they must be supported and expanded so that all relevant stakeholders commit to working together to identify common solutions and put them into practice, for the sake of patients and populations.

BIBLIOGRAPHY

1. Blumberg, R. S., Dittel, B., Hafler, D., von Herrath, M. & Nestle, F. O. Unraveling the autoimmune translational research process layer by layer. *Nat Med.* **18**, 35–41 (2012).
2. Trochim, W., Kane, C., Graham, M. J. & Pincus, H. A. Evaluating translational research: A process marker model. *Clin Transl Sci.* **4**, 153–162 (2011).
3. Scannell, J. W., Blanckley, A., Boldon, H. & Warrington, B. Diagnosing the decline in pharmaceutical R&D efficiency. *Nat. Rev. Drug Discov.* **11**, 191–200 (2012).
4. Hay, M., Thomas, D. W., Craighead, J. L., Economides, C. & Rosenthal, J. Clinical development success rates for investigational drugs. *Nat Biotechnol.* **32**, 40–51 (2014).
5. Dahlin, E., Nelson, G. M., Haynes, M. & Sargeant, F. Success rates for product development strategies in new drug development. *J Clin Pharm Ther.* **41**, 198–202 (2016).
6. DiMasi, J. A., Grabowski, H. G. & Hansen, R. W. Innovation in the pharmaceutical industry: New estimates of R&D costs. *J Health Econ.* **47**, 20–33 (2016).
7. Hartshorne, J. K. & Schachner, A. Tracking replicability as a method of post-publication open evaluation. *Front Comput Neurosci.* **6**, 8 (2012).
8. Begley, C. G. & Ellis, L. M. Drug development: Raise standards for preclinical cancer research. *Nature* **483**, 531–533 (2012).
9. Freedman, L. P., Cockburn, I. M. & Simcoe, T. S. The economics of reproducibility in preclinical research. *PLOS Biol.* **13**, 1–9 (2015).
10. Flier, J. S. Irreproducibility of published bioscience research: Diagnosis, pathogenesis and therapy. *Mol Metab.* **6**, 2–9 (2017).
11. Prinz, F., Schlange, T. & Asadullah, K. Believe it or not: How much can we rely on published data on potential drug targets? *Nat Rev Drug Discov* **10**, 712 (2011).
12. Mak, I. W., Evaniew, N. & Ghert, M. Lost in translation: Animal models and clinical trials in cancer treatment. *Am J Transl Res.* **6**, 114–8 (2014).
13. Paul, S. M. How to improve R&D productivity: The pharmaceutical industry's grand challenge. *Nat Rev Drug Discov.* **9**, 203–214 (2010).
14. La Thangue, N. B. & Kerr, D. J. Predictive biomarkers: A paradigm shift towards personalized cancer medicine. *Nat Rev Clin Oncol.* **8**, 587–596 (2011).
15. Kerr, D. J. & Shi, Y. Biological markers: Tailoring treatment and trials to prognosis. *Nat Rev Clin Oncol.* **10**, 429–430 (2013).
16. Gonzalez de Castro, D., Clarke, P. A., Al-Lazikani, B. & Workman, P. Personalized cancer medicine: Molecular diagnostics, predictive biomarkers, and drug resistance. *Clin Pharmacol Ther.* **93**, 252–259 (2013).
17. Fridlyand, J. et al. Considerations for the successful co-development of targeted cancer therapies and companion diagnostics. *Nat Rev Drug Discov.* **12**, 743–755 (2013).
18. Hamburg, M. A. & Collins, F. S. The path to personalized medicine—perspective. *N Engl J Med.* **363**, 301–304 (2010).
19. Nalejska, E., Maczyska, E. & Lewandowska, M. A. Prognostic and predictive biomarkers: Tools in personalized oncology. *Mol Diagn Ther.* **18**, 273–284 (2014).
20. Anderson, D. C. & Kodukula, K. Biomarkers in pharmacology and drug discovery. *Biochem Pharmacol.* **87**, 172–188 (2014).
21. van Gool, A. J. et al. Bridging the translational innovation gap through good biomarker practice. *Nat Rev Drug Discov.* **16**, 587 (2017).
22. Guha, M. PARP inhibitors stumble in breast cancer. *Nat Biotech.* **29**, 373–374 (2011).
23. Pletcher, M. J. & Pignone, M. Evaluating the clinical utility of a biomarker: A review of methods for estimating health impact. *Circulation.* **123**, 1116–1124 (2011).
24. Pant, S., Weiner, R. & Marton, M. J. Navigating the rapids: The development of regulated next-generation sequencing-based clinical trial assays and companion diagnostics. *Front Oncol.* **4**, 78 (2014).
25. Hoelder, S., Clarke, P. A. & Workman, P. Discovery of small molecule cancer drugs: Successes, challenges and opportunities. *Mol Oncol.* **6**, 155–176 (2012).
26. Baker, D., Lidster, K., Sottomayor, A. & Amor, S. Two years later: Journals are not yet enforcing the ARRIVE guidelines on reporting standards for pre-clinical animal studies. *PLOS Biol.* **12**, e1001756 (2014).
27. Ussi, A. E. et al. Assessing opportunities for coordinated R&D in early cancer detection and management in Europe. *Int J Cancer* **140**, 1700–1701 (2017).
28. Ussi, A. E., de Kort, M., Coussens, N. P., Aittokallio, T. & Hajduch, M. In search of system-wide productivity gains—The role of global collaborations in preclinical translation. *Clin Transl Sci.* **10**, 423–425 (2017).
29. Gilliland, C. T. et al. Putting translational science on to a global stage. *Nat Rev Drug Discov.* **15**, 217–218 (2016).

4 Biomarker-Informed Clinical Trials

Overview of Biomarker Discovery and Statistical Considerations

4.1

Weidong Zhang

Contents

OVERVIEW OF OMICS BIOMARKER DISCOVERY

Biomarker discovery is critical in drug development to understand disease etiology and to evaluate drug activity. The rapid development of omics technologies over the last few decades has offered tremendous opportunities in biomarker discovery. The major omics technologies include genomics (the study of genes, mutations and their function), transcriptomics (the study of the mRNA and their expression), proteomics (the study of proteins and their expression), metabolomics (the study of molecules involved in cellular metabolism), lipomics (the study of cellular lipids and their function) and glycomics (the study of cellular carbohydrates and their function). Omics biomarkers are high dimensional in nature. Each omics technology may output thousands or millions of analytes, and the dimensionality of omics data varies from a few hundred to a few million. Advancements in omics technologies have provided us great opportunities to understand disease biology from an unbiased global landscape. The technology boom started from late twentieth century when microarray technology was first available for measuring whole transcriptomes and genomes. A DNA microarray is a solid surface with a collection of DNA fragments, known as probes or oligos, attached. A probe is a specific sequence of a section of a gene that can be used to hybridize a cDNA or cRNA from a fluorescent molecule-labeled target sample. The fluorescent intensity of a probe–target hybridization is measured and quantified to determine the abundance of DNA molecules in the target sample. As a novel alternative strategy for genomics study, next generation sequencing (NGS) technologies emerged after the completion of the Human Genome Project in 2003, and they have completely revolutionized biomedical research in the last decade. As a result, both turnaround time and cost of sequencing have been substantially reduced. According to the National Human Genome Research Institute (NHGRI), the cost of sequencing a genome dropped from $100 million in 2001 to $1,245 in 2015 (Wetterstrand, 2016), and the turnaround time was shortened from years in late 1990s to days including analysis in 2016 (Meienberg et al., 2016). Over the last decade, NGS technology has been widely applied to biomedical research in a variety of ways, including transcriptome profiling, identification of new RNA splice variants, genome-wide genetic variant identification, genome-wide epigenetic modification and DNA methylation profiling and so on. NGS technology is a particularly good fit for cancer research given the "disorder-of-genome" nature of cancers. In cancer research, NGS has significantly enhanced our ability to conduct comprehensive characterization of the cancer genome to identify novel genetic alterations, and it has significantly helped dissect tumor complexity. Coupled with sophisticated computational tools and algorithms, significant achievements have been accomplished for breast cancer, ovarian cancer, colorectal cancer, lung cancer, liver cancer, kidney cancer, head and neck cancer, melanoma, acute myeloid leukemia and others (Shyr and Liu, 2013).

Individual tumors are not uniform from cell to cell in their sequence; instead they exhibit subclonal heterogeneity and evolutionary dynamics over time, a phenomenon that could be of considerable clinical importance to the development of therapy resistance. Even rare subclones may have important clinical consequences (Beckman et al., 2012). Tracking rare subclones requires a very low noise technique, and an enhanced sequencing technique, duplex sequencing, provides several orders of magnitude improvement in accuracy compared with conventional NGS (Schmitt et al., 2012).

STATISTICAL CONSIDERATIONS IN OMICS PRECISION MEDICINE STUDY

Data Integration

Human diseases are mostly complex diseases that involve multiple biological components. The rapid rate of discovery has revealed many molecular biomarkers, including omic biomarkers that are associated with disease phenotypes. However, translating those associations into disease mechanisms and applying the discovery to the clinic remain great challenges. In genome-wide association studies (GWAS), the major issues are that either the effects of associated variants are too small or that the effects do not appear to be functionally relevant. For example, many genetic variants may be responsible for certain genetic disposition for certain diseases reside on noncoding regions of the genome (Lowe and Reddy, 2015). Using information from a single biological process, such as genetic polymorphisms, may limit our ability to reveal true biological mechanisms. On the other hand, a single data set from one experiment represents only one snapshot of the biology, which necessitates integration of data from multiple experiments or multiple biological processes. Precision medicine is a branch of systems biology that may optimally require a holistic approach to understand disease etiology. Thus, multiple data sets generated from various sources such as different labs will produce different omics data types that may be studied together to achieve maximum benefit. However, the integration of data from different technologies or platforms has posed great challenges for data analysis (Bersanelli et al., 2016). A few statistical methods have been proposed, but areas have been focused on multivariate analysis approaches such as partial least squares (PLS), principal component analysis (PCA) and network analysis (Bersanelli et al., 2016). A recent method developed by Pineda et al. seemed to work well for combining information from genetic variants, DNA methylation and gene expression data measured in bladder tumor samples, in which study penalized regression methods (LASSO and Elastic NET) were employed to explore the relationships between genetic variants, DNA methylation and gene expression measured in bladder tumor samples (Pineda et al., 2015).

Another issue with data integration is data preprocessing and normalization. For example, one may want to combine gene expression data to be derived from platforms such as PCR, microarray or NGS. Alternatively, one may want to combine gene expression data from the same technology, but the data are generated from different labs. To address these issues and to ensure valid comparisons between data sets, cross-platform normalization has been proposed before data integration (Shabalin et al., 2008; Thompson et al., 2016).

Power and Sample Size Assessment

Power and sample size estimation in precision medicine using omics technology remains statistically challenging due to the high dimensionality and uncertain effect sizes. Numerous methods have been proposed for expression-based omics data. For example, Jung and Young developed a method to take advantage of pilot data for a confirmatory experiment controlling the family-wise error rate (FWER). The family-wise error rate refers to the rate of false positive results, where "family-wise" refers to all possible situations relevant to the trial. For when pilot data are not available, a two-stage sample size recalculation was proposed using the first stage data as pilot data (Jung and Young, 2012). A false discovery rate (FDR)–based approach for RNAseq experiments was developed by Bi and Liu, by which the average power across the differentially expressed genes was first calculated, and then a sample size to achieve a desired average power while controlling FDR was followed (Bi and Liu, 2016). FDR methods use random scrambling of the data to see how often a "discovery" will falsely arise by chance. A similar FDR-based approach is also available for microarray or proteomics experiments (Liu and Hwang, 2007).

Most power and sample size calculations focus on univariate analysis. However, the need is growing for tackling this problem in multivariate analysis. Saccenti and Timmerman proposed a method for sample size estimation in a multivariate PCA (Saccenti and Timmerman, 2016). PCA is a popular dimension reduction approach that converts multiple variables to a set of variables (so-called principal components) by applying weights to the original variables such that the principal components are mutually independent. Typically, only the first or the first and second principal component may be used for further analysis. In the case of PCA, one may want to determine minimal sample size in order to obtain stable and reproducible PCA estimates of weights. Although the algorithm was developed using specific omics data, it may be used as a generalized approach to data that has a similar data type, such as gene expression and metabolomics.

Sample size estimation in GWAS requires special treatment given its unique features as compared with other omics data such as transcriptomics or proteomics data. Often GWAS is conducted in a case-control design or family-based (case–parents trio) design. Because GWAS typically evaluates hundreds of thousands of single nucleotide polymorphism (SNP) markers, a much larger sample size is expected to achieve reasonable power (Klein, 2007; Spencer et al., 2009; Wu and Zhao, 2009; Park et al., 2010). The power and sample size calculations depend on multiple factors such as effect size, the number of SNPs being tested, the distribution of minor-allele frequency (MAF), disease prevalence, linkage disequilibrium (LD), case:control ratio, and assumption of error rate in an allelic test (Hong and Park, 2012). Considering the complexity of genetic study and the data structure and objectives of the studies, numerous methods for sample size calculation have been proposed according to specific scenarios (Lee et al., 2012; Jiang and Yu, 2016).

Statistical Modeling

Conventional statistics focuses on problems with large numbers of experimental units (n) as compared with a small number of features or variables (p) measured from each unit. In drug discovery, biomarker discovery using omics data in precision medicine often deals with a "large p, small n" problem, in which hundreds of thousands of analytes are measured from a relatively much smaller number of subjects (sometimes as few as a dozen). An array of statistical methods has been developed for the analysis of high-dimensional omics data. Those methods include exploratory clustering analysis to investigate patterns and structures, and univariate or multivariate regression and classification analysis to predict disease status (Johnstone and Titterington, 2009). For omics data such as gene expression, proteomics, metabolomics, dimension reduction is considered to be the first step before subsequent analysis. Dimension-reduction techniques include descriptive statistical approaches such as coefficient of variation (CV) filtering by which analytes with low CV are removed from subsequent regression/ANOVA analysis. This approach is particularly useful when computing power is limited. Given today's high computing capacity, the CV step is typically skipped, and a univariate regression analysis is used for both dimension reduction and inference.

Although univariate single analyte analysis is still a common approach for high-dimensional data due to its simplicity and interpretation benefits, multivariate and multiple regressions considering multiple analytes in a model has become more popular due to several advantages: (1) The complexity of the disease mechanism requires integrated information from multiple biomarkers to explain more biological variations. (2) The relationships among biomarkers cannot be modeled with single biomarker analysis. Common multivariate methods include elastic net regularized regression, random forest and classification and regression trees.

High-dimensional omics data are complex in regard to not only their dimensionality but also their correlation structures. Therefore, controlling for false discovery in high-dimensional omics data may need more statistical rigor. FWER adjustment techniques such as the Bonferroni correction (i.e., requiring a proportionally lower p-value as more statistical tests are performed) is easy to implement but generally considered too conservative. Benjamini and Hochberg (BH) introduced a sequential p-value procedure that controls FDR (Benjamini and Hochberg, 1995). Compared with the FWER approach, BH is able to gain more power regarding statistical discoveries. Another FDR-related method, which is widely applied in the omics data analysis is the q-value method developed by Storey (2002). The q-value is a measure of significance in terms of the FDR. Both q-value and BH methods allow dependence of testing (i.e., the result of one test may be correlated with the results of another).

For GWAS, the selection of a genome-wide significance threshold is challenging due to the ultra-high number of statistical testing and complex genetic LD structures. Different procedures such as Bonferroni, FDR, Sidak, permutation have been proposed; however, it was suggested that a $p = 5 \times 10^{-8}$ can be used for genome-wide significance and $p = 1 \times 10^{-7}$ can be used as a suggestive threshold at a practical level (Pe'er et al., 2008; Panagiotou and Ioannidis, 2012). A recent study from Fadista et al. further updated the thresholds by investigating different scenarios. They suggested that p-value thresholds should take into account the impact of LD thresholds, MAF and ancestry characteristics. A p-value threshold of 5×10^{-8} was confirmed for a European population with MAF > 5%. However, the p-value threshold needs to be more stringent for European ancestry with low MAF (3×10^{-8} for MAF $\geq$ 1%, 2×10^{-8} for MAF = 0.5% and 1×10^{-8} for MAF $\geq$ 0.1% at LD r2 < 0.8) (Fadista et al., 2016).

DISCUSSION

Omics biomarkers have become increasingly important in understanding disease biology and providing guidance to clinical trial designs. Biomarker discovery and applying them in clinical trials are challenging due to many aspects such as the uncertainty in biology, heterogeneity of disease, quality and quantity of tissue samples, rapidly evolving technologies and high-throughput nature of the biomarker data. To ensure success, planning and execution of a biomarker project often require a multi-disciplinary team, typically consisting of clinicians, statisticians, computational biologists and precision medicine specialists.

Biomarker studies are often exploratory in nature and not prospectively designed in most situations, in which case we may not benefit much from prospective sample size and power calculation. For example, many clinical trials focus on efficacy and the studies are not powered on the biomarker, in which case, a biomarker substudy is usually conducted. Therefore, the sample size is limited by the size of the trial and patients' consents. In some circumstances a prospectively designed biomarker study, such as a proof-of-mechanism study, may be designed to evaluate biomarker effect, for example, a pharmacological effect, it is important to understand the expected effect size we hope to detect and to obtain a reliable estimate of variability around the variable of interest.

Advancement of new technologies has enabled the rapid development of new sophisticated statistical methods for the analysis of biomarkers, especially high-dimensional biomarkers.

However, it is a good practice to always understand the sources of variations because every technology used in the biomarker study has a unique method to measure signal. Failure to identify the technological artifacts may lead to biased conclusions regardless of how sophisticated the statistical model is. For example, batch, plate, equipment or protocol effect should be taken into account or at least need to be investigated in the models when possible. Overall, with thoughtful planning and careful execution of analyses, we should be able to make the best use of biomarker data and facilitate the development of modern medicines.

ACKNOWLEDGMENTS

The author is grateful to Dr. Robert A. Beckman for his thorough review and comments that significantly improved this manuscript.

REFERENCES

Beckman RA, Schemmann GS, Yeang CH. (2012). Impact of genetic dynamics and single-cell heterogeneity on development of nonstandard personalized medicine strategies for cancer. *Proc Natl Acad Sci* 109(36): 14586–14591.

Benjamini Y., Hochberg, Y. (1995). Controlling the false discovery rate: A practical and powerful approach to multiple testing. *J Royal Stat Soc* 57(1): 289–300.

Bersanelli, M., Mosca, E., Remondini, D., Giampieri, E., Sala, C., Castellani, G. Milanesi, L. (2016). Methods for the integration of multi-omics data: Mathematical aspects. *BMC Bioinformatics* 17(Suppl 2): S15.

Bi, R., Liu, P. (2016). Sample size calculation while controlling false discovery rate for differential expression analysis with RNA-sequencing experiments. *BMC Bioinformatics* 17: 146.

Fadista, J., Manning, A., Florez, J., Groop, L. (2016). The (in)famous GWAS *p*-value threshold revisited and updated for low-frequency variants. *Eur J Hum Genet* 24: 1202–1205.

Hong, E. P., Park, J. W. (2012). Sample size and statistical power calculation in genetic association studies. *Genomics Inform* 10(2): 117–122.

Jiang, W., Yu, W. (2016). Power estimation and sample size determination for replication studies of genome-wide association studies. *BMC Genomics* 17(Suppl 1): 3.

Johnstone, I., Titterington, D. (2009). Statistical challenges of high-dimensional data. *Phil Trans R Soc A* 367: 4237–4253.

Jung, S., Young, S. (2012). Power and sample size calculation for microarray studies. *J Biopharm Stat* 22(1): 30–42.

Klein, R. J. (2007). Power analysis for genome-wide association studies. *BMC Genet* 8: 58.

Lee, S., Wu, M. C., Lin, X. (2012). Optimal tests for rare variant effects in sequencing association studies. *Biostatistics* 13: 762–775.

Liu, P., Hwang, J. T. G. (2007). Quick calculation for sample size while controlling false discovery rate with application to microarray analysis. *Bioinformatics* 23(6): 739–746.

Lowe, W. L., Reddy, T. E. (2015). Genomic approaches for understanding the genetics of complex disease. *Genome Res* 25: 1432–1441.

Meienberg, J., Bruggmann, R., Oexle, K., Matyas, G. (2016). Clinical sequencing: Is WGS the better WES? *Human Genet* 135: 359–362.

Panagiotou, O. A., Ioannidis, J. P. (2012). Genome-wide significance project. What should the genome-wide significance threshold be? Empirical replication of borderline genetic associations. *Int J Epidemiol* 41(1): 273–286.

Park, J. H., Wacholder, S., Gail, M. H., Peters, U., Jacobs, K. B., Chanock, S. J., Nilanjan Chatterjee, N. (2010). Estimation of effect size distribution from genome-wide association studies and implications for future discoveries. *Nat Genet* 42: 570–575.

Pe'er, I., Yelensky, R., Altshuler, D., Daly, M. (2008). Estimation of the multiple testing burden for genomewide association studies of nearly all common variants. *Genet Epidemiol* 32: 381–385.

Pineda, S., Real, F. X., Kogevinas, M., Carrato, A., Chanock, S. J., Malats, N., Van Steen, K. (2015). Integration analysis of three omics data using penalized regression methods: An application to bladder cancer. *PLoS Genet* 11(12): e1005689.

Saccenti, E., Timmerman, M. E. (2016). Approaches to sample size determination for multivariate data: Applications to PCA and PLS-DA of omics data. *J Proteome Res* 15: 2379–2393.

Schmitt, M. W., Kennedy, S. R., Salk, J. J., Fox, E. J., Hiatt, J. B., Loeb, L. A. (2012). Detection of ultra-rare mutations by next-generation sequencing. *Proc Natl Acad Sci* 109(36): 14508–14513.

Shabalin, A., Tjelmeland, H., Fan, C., Perou, C., Nobel, A. (2008). Merging two gene-expression studies via cross-platform normalization. *Bioinformatics* 24(9): 1154–1160.

Shyr, D., Liu, Q. (2013). Next generation sequencing in cancer research and clinical application. *Biol Proced Online* 201315: 4.

Spencer, C. C., Su, Z., Donnelly, P., Marchini, J. (2009). Designing genome-wide association studies: Sample size, power, imputation, and the choice of genotyping chip. *PLoS Genet* 5: e1000477.

Storey, J. D. (2002). A direct approach to false discovery rates. *J Royal Stat Soc* 64: 479–498.

Thompson, J., Tan, J. Greene, C. (2016). Cross-platform normalization of microarray and RNA-seq data for machine learning applications. *Peer J* 4: e1621.

Wetterstrand, K. A. (2016). DNA Sequencing Costs: Data from the NHGRI Genome Sequencing Program (GSP). Available at: www.genome.gov/sequencingcostsdata. Accessed December 12, 2016.

Wu Z, Zhao H. (2009). Statistical power of model selection strategies for genome-wide association studies. *PLoS Genet* 5: e1000582.

Overview of Classical Design in Biomarker-Informed Clinical Trials

4.2

Bo Huang, Jing Wang, Robert A. Beckman, Sandeep Menon, and Weidong Zhang

Contents

RETROSPECTIVE DESIGNS

Classical designs are widely used in clinical development of personalized medicine with a predictive biomarker that does not involve any prespecified statistical adaptations based on the interim outcomes. Classical population-enrichment designs can be categorized as two types of designs: retrospective enrichment designs and prospective enrichment designs. When prospective validation and testing of a biomarker is not feasible or not assessable in time at the beginning of the trial, a retrospective enrichment design—a traditional all-comers design with retrospective validation of a biomarker, could be considered.

Retrospective validation is conducted after the completion of the study and may involve previously conducted trials in the same patient population. As stated by Mandrekar and Sargent (2009), when conducted appropriately, this design can aid in bringing forward effective treatments to marker-defined patient populations in a timely manner that might otherwise be impossible due to ethical and logistical (i.e., a large trial and a long time to complete it) considerations. For such a retrospective analysis to be valid and to minimize bias, Mandrekar and Sargent summarized a list of essential elements that are critical to retrospective validation studies.

- Data from a well-conducted randomized controlled trial
- Availability of samples on a large majority of patients to avoid selection bias
- Prospectively stated hypothesis, analysis techniques, and patient population
- Predefined and standardized assay and scoring system
- Upfront sample size and power justification

Take the development of the epidermal growth factor receptor (EGFR) inhibitors cetuximab and panitumumab in metastatic colorectal cancer (CRC) as an example. Cetuximab and panitumumab were initially marketed for the indication of EGFR + CRC, which represents 65% of advanced colorectal cancer patients. Based on a retrospective analysis of previously conducted randomized phase II and III trials (Karapetis et al., 2008; Bokemeyer et al., 2009; Van Cutsem et al., 2009), it has been demonstrated that cetuximab significantly improves the overall survival for patients with wild-type KRAS (a protein that in humans is encoded by the KRAS gene), with no survival benefit in patients harboring

KRAS-mutant status. As a result, in July 2009, the U.S. Food and Drug Administration (FDA) approved cetuximab for treatment of KRAS wild-type colon cancer. Similarly, in a prospectively specified analysis of data from a previous randomized phase III trial of panitumumab versus best supportive care (Amado et al., 2008), the hazard ratio for progression-free survival comparing panitumumab with best supportive care in the KRAS wild-type and mutant subgroups was 0.45 and 0.99, respectively, with a statistically significant treatment × KRAS status interaction ($p < 0.0001$). Given the lack of activity in the KRAS-mutant group, the label was changed to include wild-type patients only in 2009.

PROSPECTIVE DESIGNS

In contrast to retrospective enrichment designs that test and assess biomarkers of interest retrospectively, prospective enrichment designs prospectively test, assess biomarkers and select patients at the beginning of the trial. Although retrospective evaluation of predictive biomarkers could save resources and time, and make effective treatments available to patients in a much expedited time frame, it may introduce serious bias due to the nature of retrospective selection of patient subgroups and lack of controlled validation of biomarkers. Hence, in the clinical development of targeted therapies and predictive biomarkers, the prospective design is still the gold standard.

A number of prospective enrichment designs, including the classical enrichment/targeted design, biomarker stratified design, sequential testing strategy design, biomarker-analysis design, marker-based strategy design and hybrid design. These designs differ from each other by the primary hypothesis test, randomization, and multiplicity approaches (i.e., how they deal with the total false-positive rate from testing multiple hypotheses). These differences affect the operating characteristics of the design, including sample size, power and type I error rate.

Enrichment/Targeted Design

In an enrichment design or targeted design, all patients in the trial may not generally benefit from the study treatment under consideration. The goal of the enrichment designs is to study the clinical benefit in a subgroup of the patient population defined by a specific biomarker status. In this design, the patients are screened for the presence or absence of a biomarker(s) profile. After extensive screening, only patients with the presence of a certain biomarker characteristic or profile are enrolled in the clinical trial (Sargent et al., 2005; Freidlin et al., 2010). In principle, this design essentially consists of an additional criterion for patient inclusion in the trial (Figure 4.2.1).

A recent example for the enrichment design was of mutated *BRAF*-kinase (Chapman et al., 2011). Almost 50% of melanomas have an activating V600E *BRAF* mutation. This leads to the hypothesis that inhibition of mutated *BRAF* kinase will have meaningful clinical benefit. Hence, only patients who tested positive for V600E *BRAF* mutation were enrolled in the study. Patients were randomized to an inhibitor of mutated *BRAF*-kinase or control treatment. As hypothesized, a large treatment benefit was observed in the prespecified subgroup.

The following considerations should be taken into account in this design: (1) A smaller sample size is usually required, but the screening may still take the same amount of time (or even longer as explained below) as with an all-comers design given the extensive prescreen testing that will be conducted before enrollment; (2) the marketing label will be restricted; (3) even in the event of a negative study, there may still be a potential subset of patients who may benefit with the new treatment; (4) restricted enrollment does not provide data to establish that treatment is ineffective in biomarker-negative patients, and thus the clinical validity of the biomarker test remains unknown; (5) a low prevalence of the marker may be challenging operationally and financially. Operationally, the biggest challenge is in recruitment and, financially, it may not be commercially attractive. An analysis of development incentives for sponsors compared with the public showed that in the case of a low prevalence marker, sponsors may be more likely to avoid an enrichment design compared with what is optimal for the public health (Ondra et al., 2016).

Biomarker Stratified Design

The biomarker stratified design is also known as the biomarker × treatment interaction design. This design is most appropriate when there is no preliminary evidence to strongly favor restricting the trial to patients with a specific biomarker profile that would necessitate a biomarker-enrichment design. This design

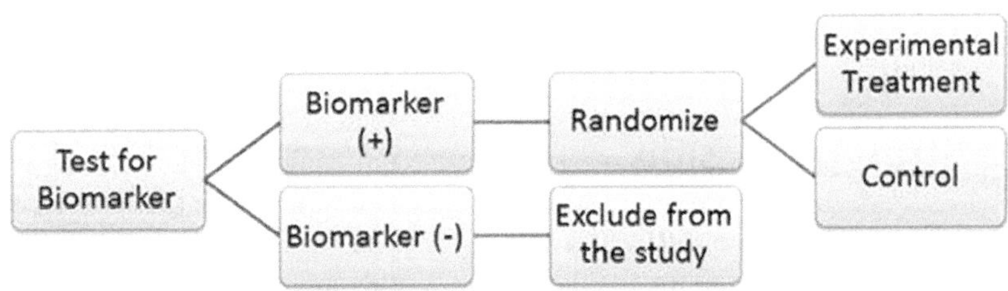

FIGURE 4.2.1 Enrichment design.

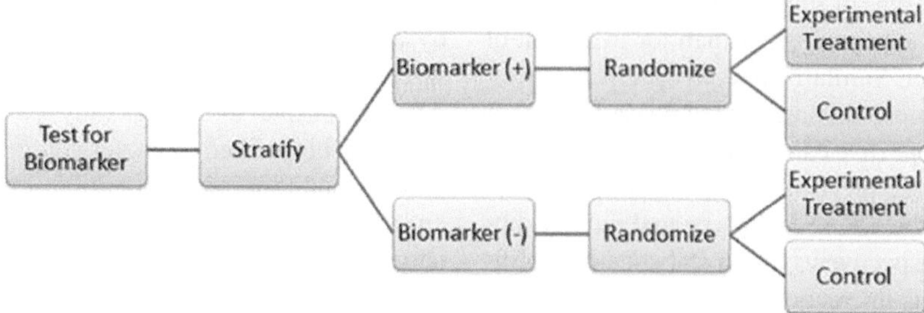

FIGURE 4.2.2 Marker × treatment interaction design.

is prospective and leads to a definitive marker validation strategy except in cases where there is a small apparent trend toward benefit in biomarker-negative patients that does not achieve statistical significance. In these latter cases, a larger second study in biomarker-negative patients may be required to clarify the situation. In this design, the patients are tested for biomarker status and then separately randomized according to their positive or negative status of the marker (Freidlin et al., 2010). Thus, the randomization is done using marker status as the stratification factor; however, only the patients with a valid measurable marker result are randomized. Patients in each marker group are then randomized to two separate treatments (Figure 4.2.2). Two separate hypotheses tests are conducted to determine the superiority of one treatment over the other separately within each marker group. The sample size is calculated separately to power the testing within each marker subgroup. Typically, one must power on larger effect sizes in such a design or the phase II program becomes very expensive (Beckman et al., 2011; Chen and Beckman, 2014). If the biomarker-negative subgroup shows numerical but not statistical benefit in a study powered for large effect sizes on the subgroups, further exploration of this subgroup as part of an adaptive phase III design may be warranted. Decision analysis can help inform development in such cases. For example, using a Bayesian approach, the optimal next step in development can be plotted on a two-dimensional grid of the results with the biomarker-positive and -negative results along the two dimensions (Beckman et al., 2011).

Another variation to the hypothesis test within the same design is to conduct a formal marker × treatment interaction test to see if the treatment effect varies within each marker status subgroup. In this case, the study is powered based on the magnitude of interaction. This design can be viewed as two stand-alone trials; however, it is different from a full population clinical trial by the considerations involved in the calculation of the sample size and the restriction of the randomization to only patients with a valid marker result, who may or may not be representative of the full population.

Sequential Testing Strategy Design

Sequential testing designs in principle can be considered as a special case of the classical randomized clinical trial for all comers or unselected patients. In all-comers trials, one may often be interested in knowing if the therapy works in the full population, but also if it works in a biomarker-positive subgroup. Each question requires a separate statistical test, and if the total false-positive rate is to be controlled at 0.05, one must give attention to the fact that each statistical test can contribute false positives. For example, in a parallel testing procedure, one could do both tests and divide up the false-positive rate between them, for example, 0.04 for the full population and 0.01 for the subgroup (this is termed "splitting alpha," where alpha is the false-positive rate). Alternatively, one could use the data from prior studies to optimize this alpha split for maximum power, which is more efficient than splitting alpha arbitrarily (Chen and Beckman, 2009).

In a sequential testing procedure, however, all patients are enrolled, and either the full population or the subgroup is tested first. In one type of sequential design, a "closed test," the first test must be positive for the second test to proceed. Although this allows the full alpha of 0.05 to be spent in the first test, it should be borne in mind that if the first test is negative in a closed testing procedure, the second test cannot be done. Thus, sequential closed testing is appropriate only when there is a high degree of confidence in the order in which the two hypotheses should be tested.

1. **Test Overall Difference Followed by Subgroup**
   Simon and Wang (2006) proposed an analysis strategy where the overall hypothesis is tested to see whether there is a difference in the response in new treatment versus the control group. If there is no difference in the response that is significant at a prespecified significance level (e.g., 0.01), then the new treatment is compared with the control group in the biomarker status–positive patients. The second comparison uses a threshold of significance, which is the proportion of the traditional 0.05 not utilized by the initial test. This is a traditional alpha-splitting approach. This approach is useful when the new treatment is believed to be effective in a wider population, and the subset analysis is supplementary and used as a fall back option.
2. **Test Subgroup Followed by the Overall Population**
   Bauer (1991) investigated multiple testing in sequential sampling. Here the hypothesis for the treatment is first tested in the biomarker-positive status patients and then tested in the overall population only if the therapy is found to be active in the biomarker-positive subgroup. In contrast to the section above, this is a so-called "closed testing procedure," that is, the second

test may be performed only if the first one is positive. This strategy is appropriate when there is a strong biological basis to believe that biomarker-positive patients will benefit more from the new drug and there is sufficient marker prevalence to appropriately power the trial on the subgroup. However, the full population will always have more patients and, therefore, potentially more power, so the design is inefficient if the biomarker hypothesis is in fact false. In this closed testing procedure, the final type I error rate is always preserved and both tests can be done at the full 0.05. That is the potential advantage of closed testing.

Biomarker-Analysis Design

The biomarker-analysis design (Baker et al., 2012) essentially has two elements. The first element is a randomized trial with the level of a continuous biomarker examined in all participants followed by the identification of a promising subgroup. This is done by using a plot of treatment benefit versus various cut points or intervals of the biomarker. It is critical in this design that the specimen be at least collected at the randomization even if it is not examined, although it is preferred that it is examined at randomization. Collection of the specimens *a priori* can mitigate the risk of noncompliance with the treatment due to knowledge of the incoming marker data (Baker and Freedman, 1995). As the data trickles in, the investigators need to assess risks and benefits, especially from an ethical perspective of concealing the new information.

This design can assist in evaluating the following hypothesis tests: (1) targeted treatment versus standard of care in the overall population; (2) targeted treatment versus standard of care in the biomarker-positive population; (3) targeted treatment versus standard of care in the biomarker-negative population; (4) marker-based treatment selection versus targeted therapy and (5) marker-based treatment selection versus standard of care. With multiple hypotheses that can be tested for (1) to (5), the significance levels and confidence intervals need to be adjusted according to the type and the number of hypotheses under consideration, that is, a scheme for allocating alpha must be provided.

The selection of a biomarker subgroup using the cut points can be done using graphics. Various graphics have been proposed in the literature that assists in better visualization and understanding of the cut points and intervals. Some of these plots present confidence intervals that adjust for multiplicity (multiple cut points create a problem of multiple hypotheses and a need for alpha allocation). Commonly used plots include (1) marker-by-treatment predictiveness curves, (2) selection impact curve, (3) tail-oriented subgroup plot, and (4) the sliding-window subgroup plot.

1. *Marker-by-treatment predictiveness curves* (Janes et al., 2011)—the probability of the response is plotted separately under targeted therapy and standard of care treatments for subjects with a marker in the interval.
2. *Selection impact curve* (Song and Pepe, 2004)—benefit of marker-based treatment selection is plotted directly as a function of marker cut points.
3. *Tail-oriented subgroup plot* (Bonetti and Gelber, 2000)—The estimated benefit of targeted therapy against the standard of care is plotted among subjects with a marker level greater than a cut point as a function of different clinically meaningful cut points. Hence, the tail of the distribution is specified when the estimated benefit is plotted for a marker level above a certain cut point.
4. *Sliding-window subgroup plot* (Bonetti and Gelber, 2004)—The estimated benefit of targeted therapy against the standard of care is plotted among subjects with a marker level within an interval as a function of marker level. Hence, the sliding window is specified when the estimated benefit is plotted for a marker level within a certain interval.

The tail-oriented subgroup plot and sliding window plots (Cai et al., 2011) give confidence intervals that account for multiple testing of several cut points or intervals.

Baker and Kramer (2005) proposed a special case of biomarker analysis design for rare events. In this design, all subjects are randomized to either the targeted therapy or the control. The specimens are collected but not examined at the time of randomization. Subjects are randomly selected at the end of the trial to test for the presence or absence of the marker. The probability of random selection is ascertained based on the positive outcome of interest. King et al. (2001) proposed testing for the marker only for the subjects with a positive outcome of interest. This type of design can be referred to as biomarker-nested case–control design. We note that this type of design risks enrolling patients with a specimen that is inadequate for analysis. This is a risk for numerous biomarkers and, in general, we recommend examining the specimens so that all enrolled patients have been assessed. In the case of missing or inadequate specimens, the patients with adequate specimens may be a nonrandom selection from the overall group.

Marker-Based Strategy Design

In this design, patients are randomly assigned to treatment dependent or independent of the marker status (Figure 4.2.3). All patients randomized to the non–biomarker-based arm receive the control treatment. In the biomarker-based arm, the patients receive the targeted or experimental therapy if the marker is positive or the control treatment if the marker is negative (Sargent et al., 2005; Freidlin et al., 2010). The outcome of all of the patients in the marker based subgroup is compared with that of all patients in the non–marker-based subgroup to investigate the predictive value of the marker. One downside of this design is that patients treated with the same regimen are included in both the marker-based and the non–marker-based subgroup, resulting in a substantial redundancy leading to many patients receiving the same treatment regimes in both subgroups. Hence, this design can reduce the treatment effect, especially if the prevalence of the marker is low, requiring

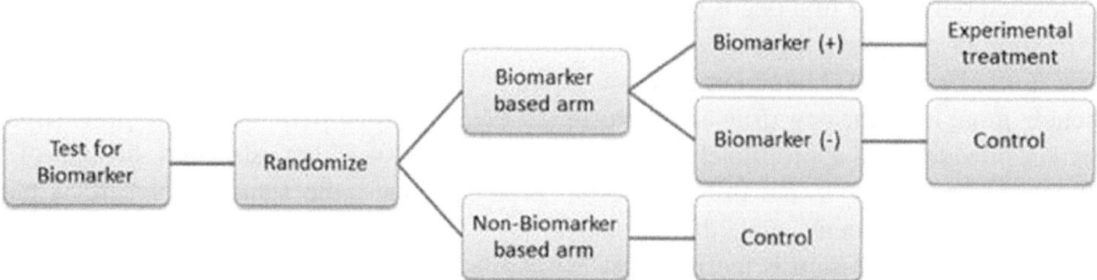

FIGURE 4.2.3 Marker-based strategy design.

a large sample size. This is illustrated in the following example. In the ERCC1 trial (Cobo et al., 2007) and presented by Freidlin et al. (2010) about 57% of the biomarker-based strategy arm patients were assigned to the same regimen of cisplatin + docetaxel as done in the standard-of-care arm. Thus, the comparison weakens the between-arm treatment effect difference and reduces the statistical power. This can lead to either getting incomplete information or possibly missing a valuable biomarker besides delaying the evaluation of the biomarker due to the increased sample size required to achieve a desired power. One other disadvantage of this design is the inability to examine the effect of targeted therapy in patients in the negative marker status group as none of these patients receive it. Even if the patients in the negative marker status group respond to the targeted therapy, this cannot be assessed. In summary, the treatment difference between the new treatment and the control treatment can be diluted by marker-based treatment selection that at times coincides with the treatment in the control arm; therefore, this design can sometimes be a poor choice as compared with the randomized design.

A modified version of this design has been proposed where the negative marker status group undergoes randomization and receives the targeted therapy or the control. Thus, the modified design allows assessment of the targeted therapy in both the biomarker-positive and -negative subgroup. This strategy helps to assess if the efficacy of the marker-positive patients to therapy is because of the marker status being positive or due to an improved treatment regardless of the marker status.

Hybrid Design

The hybrid design is very similar to the enrichment design, and all patients are examined for marker status and are randomly assigned to treatment or assigned to the standard-of-care treatment for patients with positive biomarker values. However, only a marker-positive subgroup of patients are randomly assigned to treatments, whereas patients in the marker-negative group are assigned to control or standard-of-care treatment (Figure 4.2.4). The study is powered to detect treatment difference only in the marker-positive group. This design should be considered when there is strong evidence from preclinical or prior studies that there is efficacy of some treatment(s) for the marker-based subgroup. Samples are collected from all the subjects to help testing for additional markers in the future. It should be noted that, except for the enrichment/targeted design, all classical designs covered are all-comers designs so that all patients irrespective of their biomarker status are enrolled in the study.

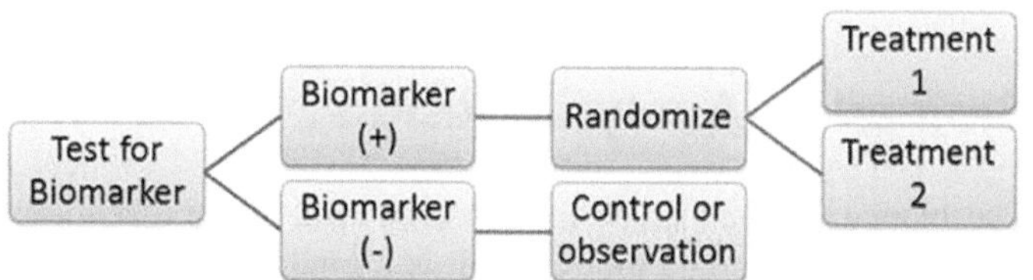

FIGURE 4.2.4 Hybrid design.

DISCUSSION

The importance of biomarkers in clinical trial design is increasing. Retrospective designs are relatively easy to conduct; however, bias may be introduced due to the post hoc nature. A further validation should be performed on the biomarkers identified in the retrospective design. Information learned from a retrospective design may be converted to a further prospective design once the biomarker is fully validated, which may improve the efficiency of the trial and overall success rate. In summary, retrospective and prospective biomarker clinical designs both have advantages and pitfalls, and the choice of a design should be considered on a case-by-case basis.

REFERENCES

Amado RG, Wolf M, Peeters M et al. (2008). Wild-type KRAS is required for panitumumab efficacy in patients with metastatic colorectal cancer. *Journal of Clinical Oncology*, 26(10): 1626–1634.

Baker SG, Freedman LS. (1995). Potential impact of genetic testing on cancer prevention trials, using breast cancer as an example. *Journal of the National Cancer Institute* 87: 1137–1144.

Baker SG, Kramer BS. (2005). Statistics for weighing benefits and harms in a proposed genetic substudy of a randomized cancer prevention trial. *Journal of Applied Statistics* 54(5): 941–954.

Baker SG, Kramer BS, Sargent DJ, Bonetti M. (2012). Biomarkers, subgroup evaluation, and clinical trial design. *Discovery Medicine* 13(70): 187–192.

Bauer P. (1991). Multiple testing in clinical trials. *Statistics in Medicine* 10: 871–890.

Beckman RA, Clark JC, Chen C. (2011). Integrating predictive biomarkers and classifiers into oncology drug development programs. *Nature Reviews Drug Discovery* 10: 735–749.

Bokemeyer C, Bondarenko I, Makhson A et al. (2009). Fluorouracil, leucovorin, and oxaliplatin with and without cetuximab in the first-line treatment of metastatic colorectal cancer. *Journal of Clinical Oncology* 27(5): 663–671.

Bonetti M, Gelber RD. (2000). A graphical method to assess treatment-covariate interactions using the Cox model on subsets of the data. *Statistics in Medicine* 19: 2595–2609.

Bonetti M, Gelber RD. (2004). Patterns of treatment effects in subsets of patients in clinical trials. *Biostatistics* 5: 465–481.

Cai T, Tian L, Wong PH, Wei LJ. (2011). Analysis of randomized comparative clinical trial data for personalized treatment selections. *Biostatistics* 12(2): 270–282.

Chapman PB, Hauschild A, Robert C et al. (2011). Improved survival with vemurafenib in melanoma with BRAF V600E mutation. *New England Journal of Medicine* 364(26): 2507–2516.

Chen C, Beckman RA. (2009). Hypothesis testing in a confirmatory phase III trial with a possible subset effect. *Statistics in Biopharmaceutical Research* 1: 431–440.

Chen C, Beckman RA. (2014). Maximizing return on socioeconomic investment in phase II proof-of-concept trials. *Clinical Cancer Research* 20: 1730–1734.

Cobo M, Isla D, Massuti B et al. (2007). Customizing cisplatin based on quantitative excision repair cross-complementing 1 mRNA expression: A phase III trial in non-small-cell lung cancer. *Journal of Clinical Oncology* 25(19): 2747–2754.

Freidlin B, McShane LM, Korn EL. (2010). Randomized clinical trials with biomarkers: Design issues *Journal of the National Cancer Institute* 102(3): 152–160.

Janes H, Pepe MS, Bossuyt PM, Barlow WE. (2011). Measuring the performance of markers for guiding treatment decisions. *Annals of Internal Medicine* 154(4): 253–259.

Karapetis CS, Khambata-Ford S, Jonker DJ et al. (2008). K-ras mutations and benefit from cetuximab in advanced colorectal cancer. *New England Journal of Medicine* 359(17): 1757–1765.

King MC, Wieand S, Hale K et al., National Surgical Adjuvant Breast and Bowel Project. (2001) Tamoxifen and breast cancer incidence among women with inherited mutations in BRCA1 and BRCA2: National Surgical Adjuvant Breast and Bowel Project (NSABP-P1) Breast Cancer Prevention Trial. *The Journal of the American Medical Association* 286(18): 2251–2256.

Mandrekar SJ, Sargent D. (2009). Clinical trial designs for predictive biomarker validation: Theoretical considerations and practical challenges. *Journal of Clinical Oncology* 27(24): 4027–4034.

Ondra T, Jobjörnsson S, Beckman RA et al. (2016). Optimizing trial designs for targeted therapies. *PLoS One* 11(9): e0163726. doi: 10.1371/journal.pone.0163726.

Sargent DJ, Conley BA, Allegra C, Collete L. (2005). Clinical trial designs for predictive marker validation in cancer treatment trials. *Journal of Clinical Oncology* 23(9): 2020–2227.

Simon R, Wang SJ. (2006). Use of genomic signatures in therapeutics development in oncology and other diseases. *The Pharmacogenomics Journal* 6(3): 166–173.

Song X, Pepe MS. (2004) Evaluating markers for selecting a patient's treatment. *Biometrics* 60(4): 874–883.

Van Cutsem E, Köhne CH, Hitre E et al. (2009). Cetuximab and chemotherapy as initial treatment for metastatic colorectal cancer. *New England Journal of Medicine* 360(14): 1408–1417.

Novel Biomarker Guided Clinical Trial Designs

A Practical Review

4.3

Bo Huang, Jing Wang, Robert A. Beckman, Sandeep Menon, and Weidong Zhang

Contents

OVERVIEW OF NOVEL DESIGNS

Adaptive Accrual Designs

If biomarker-based subgroups are predefined, but with uncertainty in the best possible population, an adaptive accrual design could be considered with an interim analysis that may lead to modification of the patient population for accrual.

Wang et al. (2007) proposed a phase III design comparing an experimental treatment with a control treatment that begins with accruing both positive- and negative-biomarker status patients. An interim futility analysis would be performed and, based on results of the interim analysis, it would be decided to either continue the study in all patients or only the biomarker-positive patients. Specifically, the trial follows the following scheme: begin with accrual to both marker-defined subgroups; an interim analysis is performed to evaluate the test treatment in the biomarker-negative patients. If the interim analysis indicates that confirming the effectiveness of the test treatment for the biomarker-negative patients is futile, then the accrual of biomarker-negative patients is halted, and the final analysis is restricted to evaluating the test treatment for the biomarker-positive patients. Otherwise, accrual of biomarker-negative and biomarker-positive patients continues to the target sample size until the end of the trial. At that time, the test treatment is compared with the standard treatment for the overall population and for biomarker-positive patients (Figure 4.3.1).

Jenkins et al. (2011) proposed a similar design but with more flexibility in the context of oncology trials. This design allows the trial to test treatment effect in the overall population, subgroup population or the co-primary populations at the final analysis based on the results from interim analysis. In addition, the decision to extend to the second stage is based on an intermediate or surrogate end point correlated to the final end point (Figure 4.3.2). Specifically, the trial has two distinct stages and follows the following scheme: (1) at the first stage, accrual in both marker-defined subgroups; an interim analysis is performed on the first-stage subjects using a short-term intermediate end point; (2) based on the interim results, the trial can (a) continue in co-primary populations; (b) continue in the marker-defined subgroup; (c) continue in the full population without an analysis in marker-defined subgroup; (d) stop for futility. Each of the above options has a prespecified, but potentially different, stage-2 sample size and length of follow up associated with it. As the trial continues to recruit new subjects for stage 2, the stage-1 subjects would remain in the trial and be monitored for the long-term end point. The final assessment for the trial is on the long-term end point for all patients from both stages.

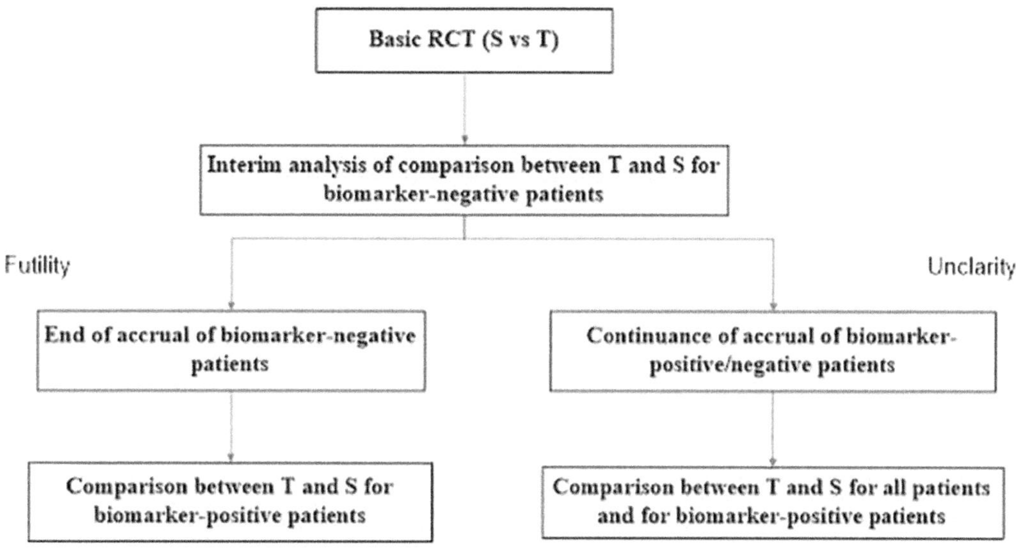

FIGURE 4.3.1 Adaptive accrual design.

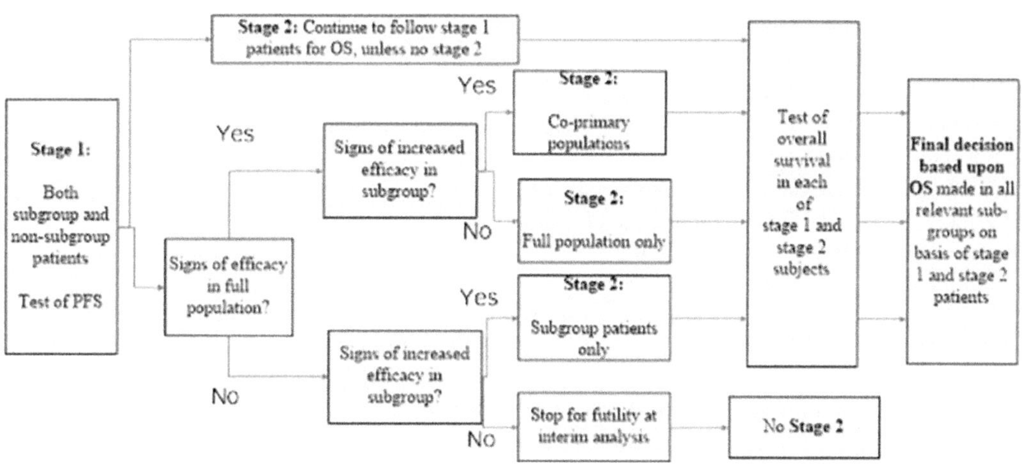

FIGURE 4.3.2 Alternative adaptive accrual design. (From Jenkins, M. et al. *Pharm. Stat.,* 10, 347–356, 2011.)

Adaptive accrual designs are very attractive due to their flexibility to change sample size and enrich the population, which greatly increase the chance of study success. However, these designs also dramatically increase the complexity of trial. From the trial management perspective, the logistics of drug supply are complicated if the sample size is increased (in particular, biologics are expensive and significant lead time is required to supply them, thus the cost for the supply) and recruitment might slow down if the population is enriched. From the statistical perspective, the type I error of the trial (false positives) would be inflated due to the potential interim adaptations and multiplicity (more than one statistical comparison). Appropriate statistical correction methods and testing procedures should be applied to preserve the type I error. Furthermore, intensive simulations should be conducted to obtain a good understanding of the various trial features such as the interim decision rule before committing to a design.

Biomarker-Adaptive Threshold Design

Biomarker development and validation is usually very expensive and time consuming. Often times by the time of the start of late phase clinical trials, a reliable biomarker, as well as its threshold, for identifying patients sensitive to an experimental treatment is not known.

When the marker is known but the threshold or the cut point for defining a positive or negative biomarker status is not clear, a biomarker-adaptive threshold design can be considered (Jiang et al., 2007). The biomarker-adaptive threshold design combines the test of overall treatment effect with the establishment and validation of a cut point for a prespecified biomarker that identifies a biomarker-based subgroup believed to be most sensitive to the experimental treatment. This design potentially provides substantial gain in efficiency.

Specifically, the main purpose of the biomarker-adaptive threshold design is to identify and validate a cutoff point for a prespecified biomarker and to compare the clinical outcome between experimental and control treatments for all patients and for the patients identified as biomarker positive in a single study. The procedure provides a prospective statistical test of the hypotheses that the experimental treatment is beneficial for the entire patient population or that the experimental treatment is beneficial for a subgroup defined by the biomarker, and it provides an estimate of the optimal biomarker cutoff point.

The statistical hypothesis test can be carried out by splitting the overall type I error rate alpha (α). First, compare the treatment response on the overall population at $\alpha 1$, and if not significant, then perform the second test at $\alpha - \alpha 1$. For example, if benefit is seen in the overall population at a desired significance level of say 0.04, then the testing is stopped. Otherwise, the testing is carried out at 0.01 to test for benefit in the identified biomarker-based subpopulation. This strategy controls overall alpha below the 0.05 level. The advantage of this procedure is its simplicity and that it explicitly separates the effect of the test treatment in the broad population from the subgroup. However, it takes a conservative approach in adjusting for multiplicity in combining the overall and subgroup analyses. Other strategies of combining the two statistical tests for overall and subgroup patients involve consideration of the correlation structure of the two test statistics. A point estimate and a confidence interval for the cutoff value could be estimated by a bootstrap resampling approach. Methods that optimize alpha allocation between full population and subgroup based on the in-trial data could also be applied (Chen and Beckman, 2009, Chen et al., 2016a).

Adaptive Signature Design

The adaptive signature design (Freidlin and Simon, 2005) is a design proposed to select the subgroup using a large number of potential biomarkers. This design is appropriate when both the potential biomarkers and the cut off are unknown; however, there is evidence that the targeted therapy may work in some of the shortlisted biomarkers. It combines a definitive test for treatment effect in the entire patient population with identification and validation of a biomarker signature for the subgroup sensitive patient population. There are three elements in this design: (1) trial powered to detect the overall treatment effect at the end of the trial; (2) identification of the subgroup of patients who are likely to benefit from the targeted therapy at the first stage of the trial; (3) statistical hypothesis test to detect the treatment difference in the sensitive patient population based only the subgroup of patients randomized in the latter half of the trial. These elements are prespecified prospectively.

Statistical tests should be conducted appropriately in this design to account for multiplicity. A proposed strategy is as follows: test for benefit in the overall population at a slightly lower significance level than the overall alpha of 0.05 (e.g., 0.04). If benefit is seen at this more stringent significance level, then the targeted therapy is declared superior to the control treatment for the overall population. The hypothesis testing and analysis is complete at this stage. If no benefit is detected in the overall population, then the signature component of the design is used to select a potentially promising biomarker subgroup. It is done by the following steps: (1) split the study population into a training subsample and a validation subsample of patients. The training subsample is used to develop a model to predict the treatment difference between targeted therapy and control as a function of baseline covariates in order to find a biomarker signature. (2) The putative signature is then applied to the validation subsample to obtain a prediction for each subject in this sample. A predicted score is calculated to classify the subject as sensitive or nonsensitive. The subgroup is selected using a prespecified cutoff for this predicted score. (3) The second hypothesis test is conducted in this sensitive subgroup to see the benefit of the targeted therapy against the control. This test is conducted at a much lower significance (e.g., 0.01). According to Freidlin and Simon (2005), this design may be ideal to use for phase II clinical trials for developing signatures to identify patients who respond better to targeted therapies. The advantage of this design is its ability to de-risk losing the label of the broader population. However, because only half of the patients are used for development or validation, and with the large number of potential biomarkers for consideration, a large sample size may be needed to adequately power the trial. In this scenario, the evidence that one of the biomarkers may identify a responsive subgroup is critical in that a large phase III level investment may be required, and therefore few would invest in it without this evidence.

Cross-validated Adaptive Signature Design

The cross-validated adaptive signature design (Freidlin et al., 2010) is an extension of the adaptive signature design and allows use of the entire study population for signature development and validation.

Similar to the adaptive signature design, the initial test of benefit of the targeted therapy against the control is conducted in the overall population and is conducted at a slightly lower significance level $\alpha 1$ than the overall alpha. The sensitive subset is determined by developing the "classifier" (i.e., the biomarker signature) using the full population. It is done in the following steps:

1. Test for treatment benefit in the overall population at $\alpha 1$, which is a slightly lower significance level than the overall α. If benefit is seen, then the targeted therapy is declared superior compared with the control treatment for the overall population and the analysis is completed. If benefit is not seen in the full population, then the next steps are carried out for signature development and validation.
2. Split study population into "k" subsamples.
3. One of the "k" subsamples is omitted to form a training subsample. Similar to the adaptive signature design, develop a model or biomarker signature to predict the treatment difference between targeted therapy and control as a function of baseline covariates using

this training subsample. Apply the developed model to each subject not in this training subsample so as to classify patients as sensitive or nonsensitive.

4. Repeat the same process leaving out a different sample from the "k" subsamples to form training subsample. After "k" iterations, every patient in the trial will be classified as sensitive or nonsensitive.
5. Compare the treatment difference within the subgroup of patients classified as sensitive using a test statistic (T). Evaluate whether the findings are due to chance by of T by permuting the two treatments and repeating the entire "k" iterations of the cross-validation process. Perform the test at $\alpha - \alpha 1$. If benefit it seen, then the superiority is claimed for the targeted therapy in the sensitive subgroup.

The cross-validation approach can considerably enhance the performance of the adaptive signature design because it permits the maximization of information contributing to the development of the signature, which is particularly useful in the high-dimensional data setting where the sample size is limited. Cross-validation also maximizes the size of the sensitive patient subset used to validate the signature. One drawback is the fact that the signature for classifying sensitive patients in each subsample might not be the same and thus can cause difficulty in interpreting the results if a significant treatment effect is identified in the sensitive subgroup.

Basket and Umbrella Trial Designs

A major issue in the clinical development of precision medicines is that genetic characterization of tumors divides common cancers such as lung or breast into a dozen or more much rarer diseases. That poses a challenge to drug companies, which in recruiting for a single-drug trial could have to screen as many as 10,000 patients to find enough patients to test a drug against a rare mutation. Screening patients for a trial involving 10 or 20 drugs instead is expected to be much more efficient and to more quickly provide patients with access to potentially beneficial treatments. To meet this challenge, the umbrella trial, in which multiple biomarker subsets and corresponding targeted agents are tested simultaneously, has been devised (Figure 4.3.3).

Further, increasing knowledge about the genetic causes of disease is prompting intense interest in the concept of precision medicine. This is particularly the case in oncology, which researchers view as the field most advanced with the strategy. The science is prompting researchers to develop treatments that target mutations or other molecular alterations regardless of where a patient's cancer is located in the body.

A key driver of the strategy is the fact that the same cancer-causing molecular traits are often found in a variety of tumor types, raising hope that a drug effective against the target in, say breast cancer, would be effective in a tumor originating in another organ with the same molecular alteration. Accordingly, the basket trial, in which patients with multiple tumor types but sharing the same molecular alteration are combined in a single study using a targeted therapy directed at that alteration, has been developed (Figure 4.3.3). Indeed, Roche Holding AG's breast-cancer drug Herceptin® (trastuzumab), which targets a receptor called Her2, turned out to be effective—and was eventually approved—for gastric tumors that have high levels of Her2. However, the drug vemurafenib, which is especially effective against melanoma skin cancer with a certain mutation in a gene called BRAF, turns out to have essentially no effect against colon cancer harboring the same mutation, raising the issue that tissue specificity of molecular effects is much more complicated than anticipated. Therefore, researchers should exercise caution when applying this approach.

Umbrella Design

Assess different molecularly targeted drugs on different mutations in one cancer type or histology

Basket Design

Assess one or more molecularly targeted drugs on one or more mutations in multiple cancer types or histologies

FIGURE 4.3.3 Umbrella trial design and basket trial design.

Both umbrella and basket trials have the potential for dramatic increases in efficiency of clinical development in an era when the cost of developing drugs is becoming increasingly unsustainable. As shown in Figure 4.3.3, an umbrella trial assesses different molecularly targeted drugs on different mutations in one cancer type of histology. Examples are Investigation of Serial Studies to Predict Your Therapeutic Response with Imaging And molecular Analysis 2 (I-SPY TRIAL 2, I-SPY 2; NCT01042379 [Barker et al., 2009]), the FOCUS4 study in advanced colorectal cancer (Kaplan et al., 2007), and the phase II adaptive randomization design Biomarker-integrated Approaches of Targeted Therapy for Lung Cancer Elimination (BATTLE [Kim et al., 2011]) in non-small cell lung cancer (NSCLC) (NCT00409968).

A basket trial assesses one or more molecularly targeted drugs on one or more mutations regardless of cancer types of histology. This design facilitates a particular targeted therapeutic strategy (i.e., inhibition of an oncogenically mutated kinase) across multiple cancer types. An example is the Molecular Profiling based Assignment of Cancer Therapeutics (MPACT; NCT01827384) trials (Conley and Doroshow, 2014).

These designs are quite powerful because they can screen and test multiple treatments and multiple biomarkers in multiple indications simultaneously.

Basket trials to date have either been exploratory phase II studies or, if confirmatory (i.e., intended to support a marketing approval), had designs that were not generally suitable for confirmatory development but were nonetheless accepted for confirmation due to exceptional levels of scientific evidence, extremely rare diseases, and very high unmet medical need. As an example of the latter, the first confirmatory basket trial involved the transformational drug imatinib in rare sarcomas and other rare cancers overexpressing known imatinib targets. Despite having a short-term surrogate end point (response rate), no control arms, and extremely small numbers (186 patients spread over 40 malignancies), in this exceptional circumstance the study was (we believe appropriately) judged sufficient for approval for a few tumor types, in one tumor type based on a single response (Demetri et al., 2011).

However, the vast majority of efficacious drugs are not transformational, and the most resource-intensive phase of development is the confirmatory phase. Accordingly, a confirmatory basket trial design was invented (Beckman et al., 2016; Chen et al., 2016b).

The confirmatory basket trial design is potentially suitable for confirmatory studies of drugs that are efficacious but not exceptional, that is, the majority of efficacious drugs. It features control arms, definitive end points, rigorous statistical powering, and control of the false-positive rate, but by pooling indications still requires far fewer patients per indication than a conventional confirmatory study. It may not be appropriate for ultra-rare indications. The design is a funnel (Figure 4.3.4) and proceeds in steps of "pruning," or removal of possibly ineffective indications based on an interim analysis and a sensitive interim end point, followed by "pooling" of the remaining high probability of success indications and evaluation of the pool according to a definitive end point for approval. Each indication

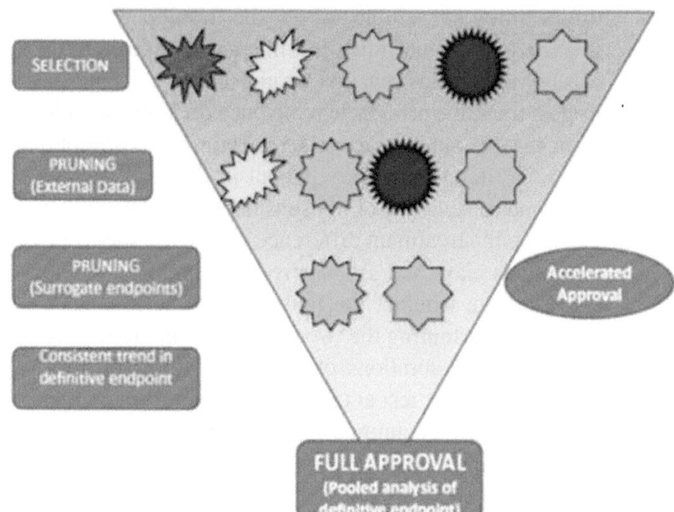

FIGURE 4.3.4 Design concept for a confirmatory basket trial. (Reprinted with permission from Beckman, R.A. et al., *Clin. Pharmacol. Ther.*, 100, 617–625, 2016.)

is powered for the interim analysis according to a sensitive interim end point, such as response rate or progression-free survival (PFS), and must achieve clinically and statistically significant results according to this end point at interim to avoid being pruned. Indications that pass this hurdle may be eligible for accelerated approval (U.S. Food and Drug Administration, 2014). Sample size adjustments for the remaining indications are required after the pruning step, and strategies for this are discussed in Chen et al. (2016b). Within the final pool, indications may be removed and not approved if they do not demonstrate at least a trend toward efficacy and a positive benefit–risk balance.

The flow chart of design concept for the confirmatory basket study is shown in Figure 4.3.4. Indications are selected and may be pruned based on external data from the class or from a maturing exploratory study. Within the study, an accepted surrogate end point is used to prune indications or to qualify them for accelerated approval. Remaining indications are then pooled for evaluation of a final approval end point. If this pooled analysis is negative, the study fails. If the pooled analysis is positive, each remaining indication is evaluated for a consistent trend in the final end point and (not shown) for a positive benefit–risk balance. Indications displaying these characteristics may be qualified for full approval, In Figure 4.3.4 indications are selected, one is pruned based on external data, two are pruned based on the interim analysis, and the remaining two proceed as a pool to full approval. Prespecified adaptations are applied for sample size adjustment after pruning and to control the false-positive rate for the case where the drug is ineffective in all indications ("global null hypothesis").

The design addresses two statistical issues. First, false negative studies: Because sharing a molecular marker does not guarantee efficacy across all indications, there is a risk that a negative indication might be included in the final pooled basket, diluting the positive result, and leading to a false negative, in the same way a single rotten fruit would affect an entire fruit basket. Careful selection of indications based on prior evidence, and pruning of indications that cannot demonstrate strong performance at interim, minimizes this possibility, giving good statistical power under most circumstances (Chen et al. 2016b). Second, random high bias (Beckman et al., 2016; Chen et al., 2016b) leading to a possible increased false-positive rate. To understand random high bias, imagine that all indications in the basket were truly negative. Nonetheless, occasionally, due to the play of chance, an indication would pass through the interim "pruning" step successfully. Such an indication or indication(s) would have a "head start," having randomly done well up to the interim and thus be somewhat biased toward a positive result. Random high bias can be controlled by working toward a lower nominal false-positive rate, which can be calculated by methods given by Chen et al. (2016b). A related issue is a slight tendency to overestimate the amount of benefit due to the therapy (Li et al., 2017).

The confirmatory basket trial design controls the false-positive rate for the case where the drug is ineffective in all indications ("global null hypothesis"). Controlling of the "family-wise" false-positive rate for any possible combination of effective and ineffective indications in a basket trial remains a challenging unsolved research problem.

DISCUSSION

As presented in this article, numerous innovative designs have been proposed in the literature and have presented tremendous opportunities for innovation in drug development. The benefits of advanced biomarker informed designs include, but are not limited to, (1) the higher probability of success with more accurate targeting of the population defined by a biomarker, (2) the flexibility to modify the trial to gain clinical benefits, (3) the possibility to shorten the development cycle, (4) the possibility to reduce the cost of development, (5) the ability to develop drugs for rare conditions and very small subgroups, and (6) the ability to leverage more data outside of the trial. However, the unknown and variable predictive power of biomarkers and the challenges in developing robust biomarker assays remain key issues to be addressed before the implementation of advanced designs involving biomarkers. The novel biomarker-informed designs should be viewed as "a" solution not "the" solution for planning a clinical trial experiment, and the use of the novel designs should be fully evaluated and applied depending on the context. Thorough statistical simulations are encouraged to be in place before any decision making. In principle, any novel design can only be implemented "as designed" and the statistical validity and integrity must be preserved. Guidelines regarding principles of how and when to use novel designs should be developed such that the risk of misuse and misinterpretation of the novel designs are kept at a minimum.

REFERENCES

Barker AD, Sigman CC, Kelloff GJ, Hylton NM, Berry DA, and Esserman LJ. (2009). ISPY2: An adaptive breast cancer trial design in the setting of neoadjuvant chemotherapy. *Clinical Pharmacology and Therapeutics* 86: 97–100.

Beckman RA, Antonijevic Z, Kalamegham R, and Chen C. (2016). Adaptive design for a confirmatory basket trial in multiple tumor types based on a putative predictive biomarker. *Clinical Pharmacology and Therapeutics* 100: 617–625.

Chen C, Beckman RA. (2009). Hypothesis testing in a confirmatory phase III trial with a possible subset effect. *Statistics in Biopharmaceutical Research* 1: 431–440.

Chen C, Li N, Shentu Y, Pang L, and Beckman RA. (2016a). Adaptive informational design of confirmatory phase III trials with an uncertain biomarker effect to improve the probability of success. *Statistics in Biopharmaceutical Research* 8: 238–247.

Chen C, Li N, Yuan S, Antonijevic Z, Kalamegham R, and Beckman RA. (2016b). Statistical design and considerations of a phase 3 basket trial for simultaneous investigation of multiple tumor types in one study. *Statistics in Biopharmaceutical Research* 8: 248–257.

Conley BA and Doroshow JH. (2014). Molecular analysis for therapy choice: NCI MATCH. *Seminars in Oncology* 41: 297–299.

Demetri, G, Becker, R, Woodcock, J, Doroshow, J, Nisen, P, and Sommer, J. (2011). Alternative trial designs based on tumor genetics/pathway characteristics instead of histology. *Issue Brief: Conference on Clinical Cancer Research*; http://www.focr.org/conference-clinical-cancer-research-2011).

Freidlin B, McShane LM, and Korn EL. (2010). Randomized clinical trials with biomarkers: Design issues. *Journal of the National Cancer Institute* 102(3): 152–160.

Freidlin B and Simon R. (2005). Adaptive signature design: An adaptive clinical trial design for generating and prospectively testing a gene expression signature for sensitive patients. *Clinical Cancer Research* 11: 7872–7878.

Jenkins M, Stone A, and Jennison C. (2011). An adaptive seamless phase II/III design for oncology trials with subpopulation selection using correlated survival endpoints. *Pharmaceutical Statistics* 10(4): 347–356.

Jiang W, Freidlin B, and Simon R. (2007). Biomarker-adaptive threshold design: A procedure for evaluating treatment with possible biomarker-defined subset effect. *Journal of the National Cancer Institute* 99(13): 1036–1043.

Kaplan R, Maughan T, Crook A, Fisher D, Wilson R, Brown L et al. (2007). Evaluating many treatments and biomarkers in oncology: A new design. *Journal of the Clinical Oncology* 31: 4562–4568.

Kim ES, Herbst RS, Wistuba II, Lee JJ, Blumenschein GR Jr, Tsao A et al. (2011). The BATTLE trial: Personalizing therapy for lung cancer. *Cancer Discovery* 1: 44–53.

Li W, Chen C, Li X, and Beckman RA. (2017). Estimation of treatment effect in two-stage confirmatory oncology trials of personalized medicines: Estimation of treatment effect in two-stage confirmatory oncology trials. *Statistics in Medicine* 36(12): 1843–1861.

U.S. Food and Drug Administration. (2014). Guidance for Industry: Expedited Programs for Serious Conditions: Drugs and Biologics, https://www.fda.gov/downloads/Drugs/Guidances/UCM358301.pdf.

Wang SJ, O'Neill RO, and Hung HM. (2007). Approaches to evaluation of treatment effect in randomized clinical trials with genomic subset. *Pharmaceutical Statistics* 6: 227–244.

Outsourcing Biomarkers in Clinical Trials

Advantages and Disadvantages

4.4

Robert Holt

Contents

The contract research organization (CRO) biomarker outsourcing market has witnessed significant growth in the past decade. In 2016 the global biomarker discovery outsourcing service market was estimated to be worth $2.7 billion[1] and the global biomarker market is expected to reach $45.5 billion by 2020.[2] The growth in outsourced biomarker studies is primarily due to the increased use of biomarkers in clinical trials as well as a rise in the amount of outsourcing being carried out by pharmaceutical companies. The use of CROs in clinical trials was originally born out of resourcing and capacity issues experienced by pharmaceutical companies' internal departments; this relationship has since matured, with CROs effectively acting as an extension of the pharmaceutical company's own laboratories.

BIOMARKERS IN CLINICAL TRIALS

Biomarkers are used in clinical trials for many purposes including patient enrolment, evaluation of safety and toxicity, as a measure of drug efficacy or for pharmacodynamic or pharmacokinetic analysis. In addition clinical trials are often used for biomarker discovery because they allow access to a relevant patient population that has been treated with a novel therapeutic compound.

INSOURCING AND OUTSOURCING MODELS

Insourcing requires the pharmaceutical company to carry out all their clinical trial biomarker work in-house. This requires specific biomarker teams and associated resources (e.g., equipment) to be established within the pharmaceutical company. This approach is uncommon, with the majority of pharmaceutical companies outsourcing some, if not all of their clinical trial biomarker work. Conversely, the outsourcing model sees CROs being paid by pharmaceutical companies to carry out biomarker work on their behalf, enabling the pharmaceutical company to focus on their key competencies.

MULTIPLE VERSUS SINGLE CROs

Some pharmaceutical companies will outsource all their biomarker work to a single company while others will outsource to multiple CROs. The use of a single vendor is advantageous because it allows the pharmaceutical company to negotiate more favourable financial terms due to the volume of work being outsourced. The use of a single vendor also decreases the workload within the pharmaceutical company, for example, procurement and project management teams have less vendor relationships to manage. However, the use of multiple vendors allows greater flexibility with respect to, for example, access to different technology platforms and allows the pharmaceutical company to tailor each of their individual biomarker programs.

FINANCIAL MODELS

There are several financial models that are used for outsourcing of biomarker activities; these fall into two major categories: transactional models and outcome based models.

Transactional models: Transactional outsourcing is the most common financial model used for biomarker analysis in clinical trials. Under this model the CRO is paid according to a pre-agreed price for each project/test/item. The CRO will determine the price for this work based on the cost of laboratory reagents and the amount of time required to carry out the work, which is generally described in a quotation, and the pharmaceutical company pays the CRO for each project. This transactional model has the advantage of allowing the pharmaceutical company to gather prices from multiple CROs for each project to obtain a competitive price. The major disadvantage of this model is that it puts pressure on procurement resources within the pharmaceutical company and business development resources within the CRO because each individual project requires a separate quotation and contract.

Once the relationship between the pharmaceutical company and CRO has become more established, it may develop such that the CRO becomes a preferred or sole supplier for certain services. The use of sole suppliers is advantageous for both pharmaceutical company and CRO because work is automatically awarded to a certain CRO, decreasing the need for procurement and business development resources. The disadvantage of this is that other CROs find it difficult to disrupt these existing relationships. These preferred/sole supplier relationships are usually managed via master service agreements rather than individual quotations, enabling a more efficient supply chain process to be established.

Outcome-based models: In outcome-based agreements, CROs are paid for the delivery of pre-agreed outcomes rather than individual specific products and services. For example, under a transactional model, a CRO tasked with developing a companion diagnostic test would price the project as individual work packages based on FTEs and the other resources required. Each work package would be billed as it were carried out, and the prices for future work packages may be amended if required. Under an outcome-based agreement, the pharmaceutical company would pay for, for example, an FDA approved companion diagnostic test rather than each individual step carried out along the way. This approach encourages the CRO to streamline their processes to deliver the outcomes using less resources than planned and, in certain cases, cost savings may also be passed onto the pharmaceutical company. The major hurdle to the implementation of outcome-based agreements is that there needs to be trust between CRO and the pharmaceutical company to allow complete transparency about each other's processes.

ADVANTAGES OF OUTSOURCING

Outsourcing biomarkers in clinical trials has advantages for both the pharmaceutical company and the CRO.

Resources: Biomarker analysis in clinical trials is not a continuous activity and will often require large numbers of resources (both people and equipment) for short periods of time. Consequently, it will be difficult for a pharmaceutical company to establish this capability in-house unless the project numbers are high enough to sustain a sufficient level of resources. CROs are able to sustain these resource levels because they are running multiple clinical trials for more than one pharmaceutical company. Despite this, resource allocation and scheduling remains one of the more challenging aspects of contract research.

Access to technology: There are hundreds of different technology platforms associated with biomarker assays and these are being continually developed. Outsourcing allows pharmaceutical companies access to novel technologies without the initial capital outlay for the equipment and training. CROs are generally able to acquire new technology platforms more easily because they have access to multiple Pharma Clients with interest in a particular platform. The downside of this from a CRO perspective is the risk associated with investing in a platform that has a low uptake or one that is rapidly replaced with a superior technology. As a consequence, a significant number of platform manufacturers are now also offering CRO services to allow pharmaceutical companies access to emerging technologies.

Expertise: Biomarker analysis requires highly trained members of staff and the training required for complex techniques can take several months. Outsourcing allows a pharmaceutical company access to this expertise when required rather than having to commit resources to training internal staff. This is of particular importance when the biomarker assay is only used in a small number of studies, in this instance, the resources required to train internal staff would far outweigh the cost of outsourcing the analysis to a CRO.

Quality: Biomarker analysis in clinical trials often requires work to be carried out in a regulatory-compliant laboratory such as Good Clinical Practice (GCP), Clinical Laboratory Improvement Amendments (CLIA) or ISO15189; these requirements often vary by country. For example, CLIA requires all facilities that perform tests on materials derived from the human body for the purposes of providing information for the diagnosis, prevention, or treatment of any disease or impairment of, or the assessment of the health of, human beings to obtain a CLIA certificate and to meet CLIA regulatory requirements. Outsourcing allows a pharmaceutical company access to multiple labs that have the relevant accreditations and regulatory approvals without having to allocate significant resources to gaining these accreditations in-house.

Turnaround: Biomarker analysis in clinical trials often requires results to be available a short time after a patient sample is collected, for example, in the case of biomarker-driven patient enrolment where a test result will lead to enrolment into a clinical trial. In this instance, a CRO will carry out an expedited analysis using assays that are kept in stock for a particular client. The CRO will generally charge an increased price for rapid turnaround work due to the additional resources that are required for analysis of this type.

Decreased costs (for Pharma): A major advantage to outsourcing is that it enables a pharmaceutical company to obtain a competitive price for the work being carried out. Pharmaceutical company procurement teams will generally identify suitable vendors through a request for information (RFI). This document determines the basic capabilities of each vendor and allows the pharmaceutical company to draw up a short list of suitable CROs. Once the shortlist has been established each CRO will generally receive a request for quote (RFQ), which is a more detailed document requiring the CRO to submit pricing and details of other relevant assessment factors such as project governance and financial terms. The pharmaceutical company will collate these data and score each vendor based on a number of criteria, including price, before awarding the work to the winning vendor.

Clinical trial failures (for Pharma): Approximately 10% of novel compounds entering Phase I clinical trials will reach the market as an approved therapeutic, this number rises significantly with the use of biomarkers. Despite this, the number of clinical trials that fail or do not reach completion is significant. The use of CROs to carry out clinical trial work allows the pharmaceutical company to manage the consequences of a failed clinical trial more easily than if these activities were carried out in-house. Contract negotiations surrounding such trial failures often form a key discussion point during the initial stages of contract planning.

Changing CRO vendors: Dependant on the terms of their agreement a pharmaceutical company can change CRO vendors if needed; if work were carried out using an in-house lab this would be more difficult.

DISADVANTAGES OF OUTSOURCING

Lack of control: The lack of control over outsourced projects is a major disadvantage for a pharmaceutical company. This can affect many areas including quality, time, project scope or even the risks associated with a CRO business failing or being acquired by another business. Communication between a CRO and pharmaceutical company is key, generally two Project Managers will form the main point of contact between each organization to ensure that core teams are kept informed of all decisions regarding the project.

Confidentiality: Pharmaceutical companies are concerned over confidentiality issues that may affect the outcome of the clinical trial. An error by a CRO may affect the ability of a drug to get to market, it is therefore essential that the pharmaceutical company ensure the CRO has all the procedures and infrastructure in place to ensure confidentiality is maintained. This is especially important where one CRO is dealing with multiple competing pharmaceutical companies. In this instance, it is common for each pharmaceutical company to be assigned completely independent project teams within a CRO.

Country specific issues: Certain countries, such as China, have regulations that do not allow patient tissue to be taken out of the country, so a separate CRO or lab will need to be found within this country. Some larger CROs have laboratory facilities in multiple global locations; in this case, it is advantageous for the pharmaceutical company to partner with such organizations so that one company can carry out all the analysis for a clinical trial.

Failed clinical trials (for CRO): The large number of failed and cancelled clinical trials is one of the largest issue facing CROs. If a biomarker study is cancelled, the resources and investment put in place by the vendor are no longer required; this in turn can have a significant effect on the CRO's business, particularly for smaller companies.

Multiple partners: The use of multiple CROs, while allowing greater flexibility in terms of technology and pricing, has a detrimental effect in terms of speed and technology transfer. This is of particular importance in the case of larger biomarker studies, for example, companion diagnostic development where the assay is passed from one company to the next, causing delays and issues with communication and technology transfer.

Changing CRO vendors: In the case where a pharmaceutical company uses a CRO as a sole supplier for a particular service, changing vendors to a different CRO can be extremely difficult, particularly if the incumbent supplier has been in place for a long period of time.

REFERENCES

1. http://www.appliedclinicaltrialsonline.com/global-biomarker-and-companion-diagnostics-outsourcing-market.
2. http://www.marketsandmarkets.com/Market-Reports/biomarkers-advanced-technologies-and-global-market-43.html?gclid=CIfu6vmH4NECFQ0R0wodjLoBmQ.

Case Studies for Biomarker-Based Clinical Trials 4.5

Bo Huang, Jing Wang, Robert A. Beckman, Sandeep Menon, and Weidong Zhang

Contents

CASE STUDY 1: DEVELOPMENT OF CRIZOTINIB IN ALK+ NSCLC

Lung cancer is currently the leading cause of cancer death in both men and women. Historically, lung cancer was categorized as two types of diseases: small cell lung cancer (SCLC) and NSCLC, with NSCLC accounting for about 85%–90% of lung cancer cases. NSCLC can also be classified according to histological type: adenocarcinoma, squamous-cell carcinoma and large-cell carcinoma. Such classification is important for determining management and predicting outcomes of the disease.

However, with the rapid advance in biological and genetic science over the last two decades, researchers have found there are various molecular alterations that are "oncogenic drivers" in that they appear to drive the growth of lung cancer. Table 4.5.1 shows potential oncogenic drivers in NSCLC based on current knowledge, such as EGFR overexpression and the anaplastic lymphoma kinase (ALK) translocation.

TABLE 4.5.1 Oncogenic drivers in lung adenocarcinoma

ONCOGENIC DRIVERS	*PREVALENCE*
KRAS	20%–25%
EGFR	13%–17%
ALK	3%–7%
MET skipping	~3%
HER2	~2%
BRAF	~2%
PIK3CA	~2%
ROS1	~1%
MET amp	~1%
NRAS	~1%
MEK	~1%
AKT	~1%
RET	~1%
NTRK1	~0.5%

The Molecularly Targeted Agent (MTA) crizotinib (XALKORI®, Pfizer Inc., New York, NY) is a potent, selective, small-molecule competitive inhibitor of Anaplastic Lymphoma Kinase (ALK), Mesenchymal Epithelial Transition (MET), and c-ros oncogene 1 (ROS-1) (Christensen et al., 2007; Shaw et al., 2014). The first-in-human phase 1 trial started in December 2015 to estimate the Maximum Tolerated Dose (MTD) opening to all-comer patients with solid tumors. The EML4-ALK translocation in NSCLC was discovered in 2007. In the same year, the study was amended to add patients with EML4-ALK translocation to the MTD cohort, and the first clinical response was observed in ALK+ tumors in early 2008. Subsequently, the clinical development program progressed rapidly, and crizotinib was approved in 2011 by the FDA for NSCLC that is ALK-positive as detected by an FDA-approved companion diagnostic test, a commercially available break-apart fluorescence *in situ* hybridization (FISH) probe for detecting the ALK gene rearrangement in NSCLC (Kwak et al., 2010). It took only 6 years from the beginning of the first-in-human study to registration.

Table 4.5.2 summarizes the clinical studies and their trial designs and end points that led to the accelerated and full approval by the global health authorities. Classical enrichment designs were used for these studies that allowed for the investigation of this novel drug in an efficient and rapid way because

TABLE 4.5.2 Clinical studies that led to accelerated approval and full approval

PROTOCOL	*SETTING*	*TRIAL DESIGN*	*PRIMARY ENDPOINTS*
A8081001 (n = 119)	All lines, solid tumors, ALK-positive NSCLC	Single-arm, open-label study of crizotinib	Safety, pharmacokinetics, response
A8081005 (n = 136)	≥2nd line ALK-positive NSCLC	Single-arm, open-label study of crizotinib	Safety, response
A8081007 (confirmatory phase 3) (n = 318)	2nd line ALK-positive NSCLC	Crizotinib versus (pemetrexed or docetaxel), randomized, open-label study	PFS

Source: clinicaltrials.gov.

patients with ALK translocation only account for approximately 5% of NSCLC population. High response rates (55%–60%) in the two single-arm enrichment studies led to accelerated approval by the FDA. Full approval was granted after positive readout of randomized confirmatory study A8081007.

In the absence of comparative data, it was unclear whether the distinct clinicopathologic characteristics of patients with ALK-positive NSCLC noted above might be contributing to the observed antitumor activity of crizotinib. Extensive retrospective statistical analyses were conducted using bootstrapping (resampling patients many times with matched covariates from the data already generated) and modeling (covariate-adjusted) to simulate outcomes of randomized controlled studies of crizotinib versus standard advanced NSCLC treatment (Selaru et al., 2016). These analyses utilized data from the control arms of three Pfizer-sponsored phase III studies evaluating first-line paclitaxel–carboplatin or gemcitabine–cisplatin and second- or later-line erlotinib regimens in patients with advanced unselected NSCLC to simulate a concurrent control. These analyses supported a clinically meaningful and statistically significant effect of crizotinib despite the lack of a concurrent active control arm.

CASE STUDY 2: BAYESIAN PREDICTIVE PROBABILITY DESIGN FOR AN ENRICHMENT PHASE II PROOF-OF-CONCEPT STUDY

Breast cancer is a common type of cancer among women. A diagnosis of triple-negative breast cancer (TNBC) means the three most common types of receptors that fuel cancer growth—estrogen receptor (ER), progesterone receptor (PR) and human epidermal growth factor receptor 2 (HER2)—are not present, which represents 15% of breast cancer patients. In TNBC with Notch genomic alternations (NA+), the inhibition of activation of the Notch pathway using single-agent Notch inhibitor therapy may induce clinical activity (Stylianou et al., 2006; Krop et al., 2012). The prevalence of Notch alteration in breast cancer is estimated to be around 10%, so Notch+ TNBC represents only 1%–2% of breast cancer, a very rare disease.

This example describes a phase 2 proof-of-concept (POC) study of an experimental Notch inhibitor, an oral drug given twice daily (b.i.d.). The hypothesis is that treatment with this drug response rate can be improved from the historical level of ≤30% to ≥60%. However, there are two main challenges in designing the trial. First, there is no prior clinical data to suggest that a high response rate of 60% can be achieved in this rare disease defined by NA+, nor is there any prior data on the analytical validity or clinical utility of the assay. As a result, it is highly desirable to stop the trial early if the observed objective response rate (ORR) is low during the trial. Second, due to the extremely low prevalence rate of 1%–2% of the target population in breast cancer, enrollment speed is expected to be slow, albeit 20–25 sites will be opened to screen hundreds of TNBC patients, and the turnaround time of the NGS assay (~2 to 3 weeks) may decrease trial acceptance.

To meet the aforementioned challenges, a Bayesian predictive probability (BPP) design was proposed with multiple interim looks (Lee and Liu, 2008), so that the trial could be stopped early if there was no or low drug effect because it is a costly study with high risk. The Bayesian approach allows greater flexibility in continuously monitoring the trial data to make a go/no-go decision.

Figure 4.5.1 illustrates the study design. Patients are tested for biomarker status. If the status is NA positive, patients are assigned to the experimental drug using the proposed BPP design. It is estimated that at least 28 patients are required to test the hypothesis controlling for the type I and type II error rates. In addition, 20 NA-negative patients would be enrolled to gather some data for an exploratory analysis but not for formal statistical testing. Because the treatment effect (if any) is expected to be much smaller (if there were any effect) in the marker-negative population, a much larger sample size would be required to formally test for it statistically. Therefore, a sample size of 20 can provide only a qualitative estimate.

The BPP design uses a statistical distribution (termed a "beta binomial conjugate") of possible values of the tumor response rate p to calculate the predictive probability, or the probability of a positive result at the end of the trial, based on the cumulative information at any given point in time. At frequent points, the predictive probability of a positive result is compared with selected upper and lower bounds. If the predictive probability is greater than the upper bound, the trial is stopped and declared successful. Conversely, if the predictive probability falls below the lower bound, the trial is stopped for futility.

It is estimated that 28 patients will be required in order to have 25 response-evaluable patients so as to control the one-sided false-positive rate at 0.05 with 90% power when

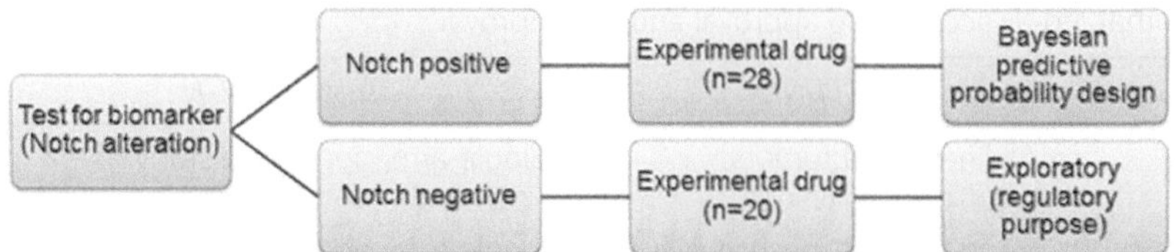

FIGURE 4.5.1 Study of an experimental notch inhibitor.

TABLE 4.5.3 Interim futility/efficacy decision rules

NUMBER OF PATIENTS	*NEGATIVE/FUTILITY*	*POSITIVE/ EFFICACY*
8	≤1	≥8
10	≤2	≥9
12	≤3	≥10
15	≤5	≥11
18	≤6	≥12

the true response rate is 60%. The design has multiple interim looks for potential early stopping, and the decision rules are provided in Table 4.5.3. At the final analysis, at least 12 responders are required of 25 evaluable patients to claim the drug efficacious.

With the proposed Bayesian design, the study has a 78% probability of early termination if the drug is innefective, and the expected sample size under these circumstance is approximately 14 patients. This design achieves the objective of terminating the trial early with minimal resources when the effect of the drug is not as high as expected.

CASE STUDY 3: ADAPTIVE GROUP SEQUENTIAL DESIGN WITH POPULATION ENRICHMENT FOR CARDIOVASCULAR RESEARCH

Developing treatments for cardiovascular disease is usually enormously expensive and time consuming due to the specifc problems such as relatively low event rates, small treatment effects, a lack of reliable surrogate end points, and diverse patient populations. Thus, the statistical methods that use accumulating data from a trial to inform and modify its design are of considerable interests in cardiovascular research. Mehta et al. (2009) proposed an adaptive multistage design with population enrichment and demostrated how to implement it in a cardiovascular trial, as noted below.

Consider a placebo-controlled randomized cardiovascular trial with a composite primary end point including death, myocardial infarction, or ischemia-driven revascularization during the first 48 hours after randomization for therapies intended to reduce the risk of acute ischemic complications in patients undergoing percutaneous coronary intervention. Assume, based on prior knowledge, that the placebo event rate is in the range of 7%–10%. The investigational drug is assumed, if effective, to reduce the event rate by 20%, but the evidence to support this assumption is limited. The actual risk reduction could be larger, but it could also easily be as low as 15%, a treatment effect that would still be of clinical interest given the severity and importance of the outcomes, but that would require a substantial increase in sample size.

Assume, besides the entire population under study G0, two subgroups of patients, G1 and G2, have also been identified by investigators as of interest. G1 is a subset of G0 and G2 is a subset of G1. In addition, it is preferred that the sample size of the study not exceed 15,000.

A classical multistage design (group sequential design) was considered first with a target sample size of 8,750 patients and interim analyses to be performed at 50% and 70% of that target (Figure 4.5.2). The first interim look would take place when the first 4,375 patients have finished the study, and the trial stops for efficacy or for futility according to specified upper and lower bounds, respectively. The second interim look would take place when 6,215 patients have finished the study, and there are corresponding efficacy and futility bounds for this second look. At the final look, when all 8,750 patients have completed the study, the final efficacy bound is exceeded in a successful study.

Such a classical multistage design could be modified by allowing an increase in sample size when the estimated power of the study ("conditional power") calculated based on interim results is below expectation. For example, assume that at the first and second look, we observed the data shown in Figure 4.5.3, which do not exceed either the efficacy or futility boundaries. Therefore, the trial can continue beyond the second look. However, based on the observed data up to the second interim look, the estimated event rate for the control arm is 8.7% and the estimated percentage risk reduction is only 15.07%. For these values, the conditional power of this study is only 67%. The sample size can then be increased to create the desired power of 80%.

Because the total sample size is preferred to not exceed 15,000, the design could be further modified as follows:

1. If the sample size reestimation for the overall population G0 produces a revised sample size >15,000, consider enrichment.
2. The enrichment strategy proceeds as follows: (a) estimate the number of additional patients needed

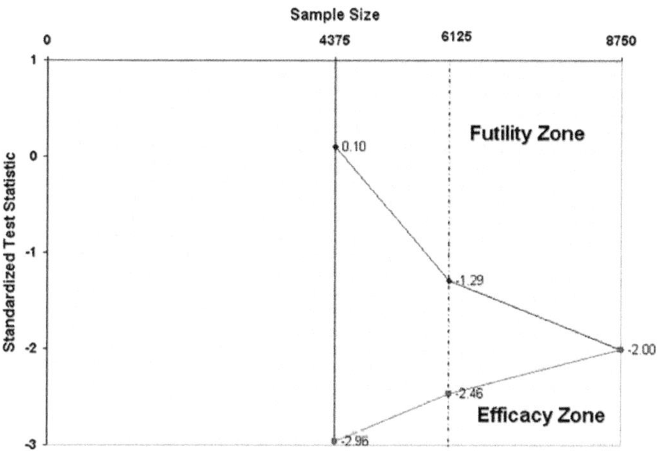

FIGURE 4.5.2 Classical multistage design.

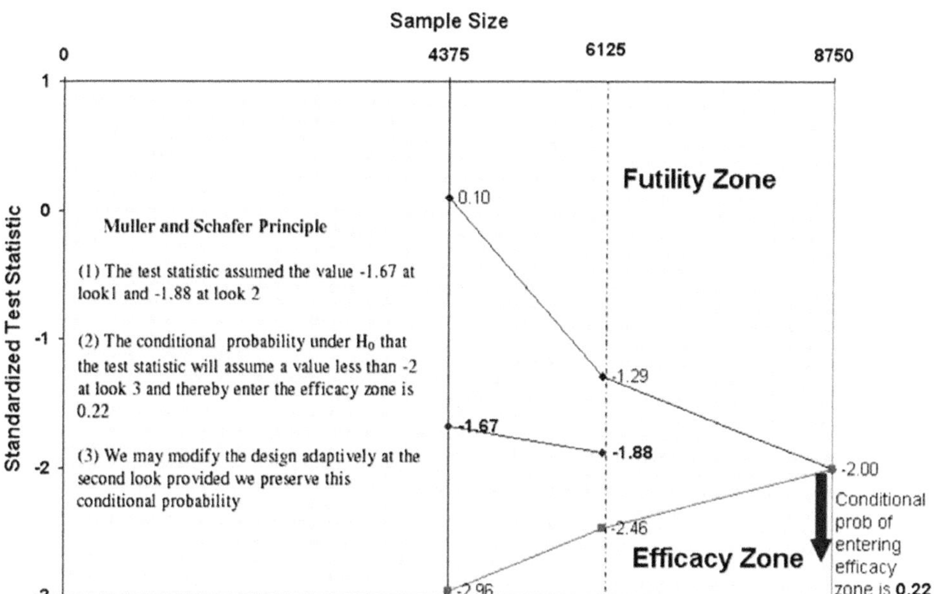

FIGURE 4.5.3 Multistage design with sample size adaption.

to for 80% conditional power in the subset G1, assuming that the observed effect size in G1 is the true effect size; (b) if that sample size plus the number of patients already enrolled is <15,000, the trial will continue until the additional number of patients is enrolled, but future eligibility will be restricted to patients belonging to subgroup G1; (c) if that enrichment strategy does not yield a sample size <15,000, the same calculation will be performed with the estimated effect size for subgroup G2 and, if that reestimation yields a sample size <15,000, the trial will continue, enrolling only members of G2; (d) if neither of the sample size calculations for the patient subgroups yields a sample size <15,000, the trial will be continued with the original eligibility criteria and sample size target, provided that the conditional power with 8,750 patients is at least 20%; (e) otherwise, the trial will be terminated for futility.

As pertains to how to test for a treatment effect at the end of the trial given the possibility of a sample size increase and population enrichment at the second interim look. To preserve the false-positive rate of the study as a whole, an option is to employ a closed testing procedure, which guarantees strong control of the type 1 error rate. Specifically, a testing procedure as noted in Figure 4.5.4 could be applied. This scheme of

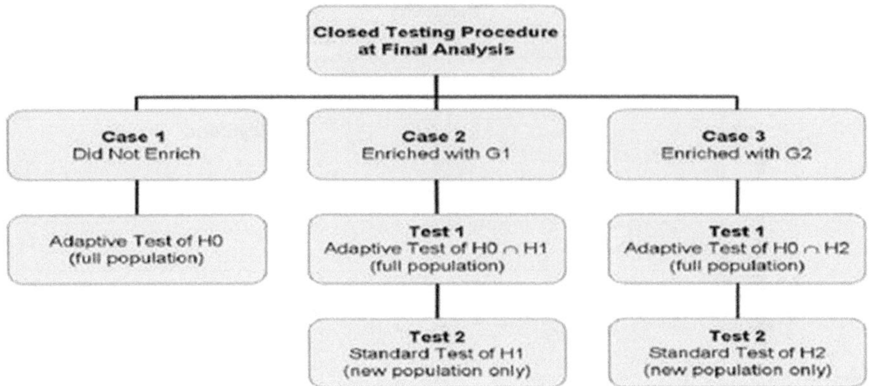

FIGURE 4.5.4 Closed testing procedure. H1 had no treatment effect in patient population G1. G, group; H, hypothesis.

statistical testing strongly controls the false-positive rate, but like all sequential tests, has potential issues if the patients in the full population dilute what would have been a positive signal in the subgroup.

SUMMARY

Novel clinical designs leveraging biomarker information provide more options for effective drug development. To address potential heterogeneity of the patient population, biomarker-driven enrichment strategies are often utilized to improve drug response and understand safety. Through the three real-world examples presented in this section, we demonstrated that great benefits—including shortened drug development cycle and reduced resource consumption—can be achieved from these novel designs. However, it is worth noting that these designs are taking advantage of fully validated biomarkers, which is a critical and difficult step to accomplish for many development programs. Therefore, the choice of an appropriate biomarker-informed trial design should be based on the confidence and evidence of strength of the predictive value of the biomarker.

REFERENCES

Christensen JG et al. (2007). Cytoreductive antitumor activity of PF-2341066, a novel inhibitor of anaplastic lymphoma kinase and c-Met, in experimental models of anaplastic large-cell lymphoma. *Mol Cancer Ther* 6, 3314–3322.

Krop I et al. (2012). Phase 1 pharmacologic and pharmacodynamic study of the gamma secretase (Notch) inhibitor MK-0752. *J Clin Oncol Rapid Commun* 30, 2307–2313.

Kwak EL et al. (2010). Anaplastic lymphoma kinase inhibition in non–small-cell lung cancer. *N Engl J Med* 363, 1693–1703.

Lee J, Liu, D. (2008). A predictive probability design for phase II cancer clinical trials. *Clin Trials* 5(2), 93–106.

Mehta C, Gao P, Bhatt DL, Harrington RA, Skerjanec S, Ware JH. (2009). Optimizing trial design: Sequential, adaptive, and enrichment strategies. *Circulation* 119(4), 597–605.

Selaru P, Tang Y., Huang B, Polli A, Wilner K, Donnelly E, Cohen D. (2016). Sufficiency of single-arm studies to support registration of targeted agents in molecularly selected patients with cancer: Lessons from the clinical development of crizotinib. *Clin Transl Sci* 9(2), 63–73.

Shaw AT et al. (2014). Crizotinib in ROS1-rearranged non-small-cell lung cancer. *N Engl J Med* 371, 1963–1971.

Stylianou S, Clarke R, Brennan K. (2006). Aberrant activation of notch signaling in human breast cancer. *Cancer Res* 66(3), 1517–1525.

Informational Designs and Potential Applications to Rare Disease

4.6

Robert A. Beckman and Cong Chen

Contents

INTRODUCTION

Rare diseases are defined by the U.S. Orphan Drug Act as conditions with an overall prevalence of 200,000 or less in the U.S. (Orphan Drug Act of 1983, 1983), or in Europe as diseases affecting less than 1 in 2,000 individuals (http://www.rarediseaseday.org/article/what-is-a-rare-disease). Estimates of the number of rare diseases range up to 7,000, and estimates of the number of affected individuals range up to 30 million in the United States alone, or 1 in 10 individuals (rarediseaseday.us/wp-content/uploads/2011/11/RDD-FAQ-2013.pdf). Fifty percent of those affected by rare diseases are children (http://www.rarediseaseday.org/article/what-is-a-rare-disease), and many rare diseases are life threatening or debilitating. Moreover, the molecular dissection of common diseases like cancer is leading to the creation of small subsets of these diseases that require individualized treatment, effectively breaking up these common diseases into multiple rare diseases (Beckman et al., 2011). Accordingly, the unmet medical need in the rare disease area is great.

Rare diseases present numerous challenges for drug development (Schwartz et al., 2015). Typically, it is not even known for certain that the disease is truly rare: The incidence and prevalence may increase with increased awareness in the medical or patient community, often triggered by new treatment options. There is typically limited information on the underlying biology or pathophysiology of the disease, making the design of therapies and the search for pharmacodynamic and predictive biomarkers especially difficult. Often little is known about the natural history of the disease, including prognostic subgroups and relevant clinical endpoints. In particular, there may be no validated endpoint for pivotal studies and, importantly, the correlation structure among candidate endpoints is unknown. There may be no previous precedents for drug development and no standard comparison therapy, and a placebo control may be unethical for severe conditions. Compounding these difficulties is the very small number of available patients, who may be widely distributed among a large number of practitioners, each with relatively limited experience in treating the rare condition.

The small numbers of patients can also equate to small markets. Indeed, when a quantitative decision analysis evaluated utility of clinical development pathways from the point of view of multiple stakeholders, it appeared that the incentive for a for-profit organization to engage in clinical development within rare subgroups of a common disease was small unless there was an extremely high probability of success, anticipated length of therapy, or anticipated price (Ondra et al., 2016).

In light of these difficulties, it is surprising that 230 drugs have been approved for orphan diseases in the last decade, and 47 percent of U.S. drug approvals in 2015 were for orphan drugs (http://www.rarediseaseday.org/article/what-is-a-rare-disease). This can largely be attributed to the flexibility the FDA has shown in evaluating scientific evidence in these conditions (U.S. Food and Drug Administration, 2015a). Indeed, the FDA draft guidance document in rare diseases states that the FDA understands the difficulties of development in rare diseases, and that it can use its judgment in applying the evidentiary standards for approval (U.S. Food and Drug Administration, 2015b).

Nonetheless, a review of the successes in the rare disease area shows that many of the successful drugs were exceptional in their efficacy (U.S. Food and Drug Administration, 2015a) and 230 drugs approved is a small number compared with the 7,000 rare diseases. Despite considerable efforts by patient advocacy groups, pharmaceutical and biotechnology companies, and national health authorities, the unmet medical need is still largely unaddressed, and arguably, a cost-effective approval pathway exists only for truly exceptional therapies, whereas benefit may also be provided by therapies with typical levels of effectiveness. Cost-effective methods to provide reliable evidence suitable for approval in the rare disease area are still urgently needed.

The recent FDA draft guidance for rare diseases suggests that a natural history study be conducted as an initial step in rare disease development to address the many uncertainties alluded to above and allow optimal design of a pivotal study (U.S. Food and Drug Administration, 2015b). Such a natural history study would be devoted to understanding the temporal course of the disease in the absence of therapy, identifying prognostic subgroups, exploring clinical endpoints and their temporal relationships and correlation structure, and validating candidate clinically meaningful and sensitive pivotal endpoint(s).

While the use of a separate dedicated natural history study is a very rational recommendation based on the reasons above, it raises the following questions:

- Will pharmaceutical and biotechnology companies likely invest in a difficult-to-enroll natural history study, where the number of available patients is small, no therapeutic benefit is offered, and follow-up may also be long, knowing that such a study is only the prelude to designing the study of potential interest and benefit to them?
- If not, can/will patient advocacy groups trigger these studies in conjunction with academic consortia and public funding? Will the absence of a prospect of therapeutic benefit in these studies dampen enthusiasm from these groups?
- Will one such study provide information about endpoints germane to therapies with different mechanisms of action, or will separate natural history studies need to be performed for sufficiently distinct therapies within the same condition?
- How long will it take to conduct adequately powered natural history studies followed by adequately powered therapeutic studies?
- Will people with debilitating or life-threatening rare diseases support the idea of waiting for sequential natural history and therapeutic studies?

An alternative to a separate natural history study is a pivotal study, which is capable of adapting its design based on what is learned about the natural history. The classical adaptive design uses early interim results to inform elements of the second half of the study according to prospective rules. However, early interim results may not provide sufficient information about the development of long-term endpoints or their relationship to earlier short-term measures. Yet, long-term endpoints are often the most clinically meaningful and most appropriate for judging whether a therapy should be approved. This is particularly true in oncology, where survival is the key endpoint and where many "rare" diseases are appearing as molecularly defined subsets of common cancers. This is equally true in progressive neurodegenerative, autoimmune and metabolic disorders, of which many qualify as rare diseases.

The informational design (Chen et al., 2016; Chen and Beckman, 2016) is an alternative to classical adaptive designs in that the adaptation is performed at the end of the trial and is based on information from a randomly selected subset of patients in the trial, essentially a phase II substudy within the pivotal study. The patients within this informational cohort may be included in the final analysis after paying an appropriate statistical penalty to control type I error, or they may be excluded without paying the penalty. The design allows the full observation of outcomes from endpoints that develop over time before the adaptation is undertaken according to prospective rules.

In the next subsection, we review the informational design within the context of other alternatives and discuss two application examples that have to do with predictive subgroup selection, presenting performance metrics for the design.

In the final brief subsection, we briefly introduce a speculative application for which research is ongoing: selection of a pivotal endpoint within the pivotal study itself.

INFORMATIONAL DESIGN

In the informational design, a random subset of patients is designated as the informational cohort, and the results from this cohort are utilized to govern a prospective adaptation at the end of the study. If the overall study is stratified, the same stratification should be applied to the informational cohort. Figure 4.6.1 contrasts the informational design with a conventional interim analysis, which is performed on all the patients in the middle of the study, using either an interim endpoint with an uncertain relationship to the final endpoint or incomplete information on the final endpoint. Figure 4.6.2 contrasts the informational design with a seamless phase II/III study, where again an interim analysis is performed at the end of the phase II portion, which informs an adaptation affecting the phase III portion (Muller et al., 2004; Bretz et al., 2006; Stallard and Todd, 2011; Mehta et al., 2014). Compared with a conventional interim analysis or seamless phase II/III, the informational analysis provides much fuller insight into long-term endpoints and their relationship to earlier measures. The disadvantage of the informational design is that the adaptation cannot occur until the end of the study. Thus, for example, if a subgroup is excluded from the final study analysis based on an informational analysis, those patients have nonetheless already been enrolled in the study.

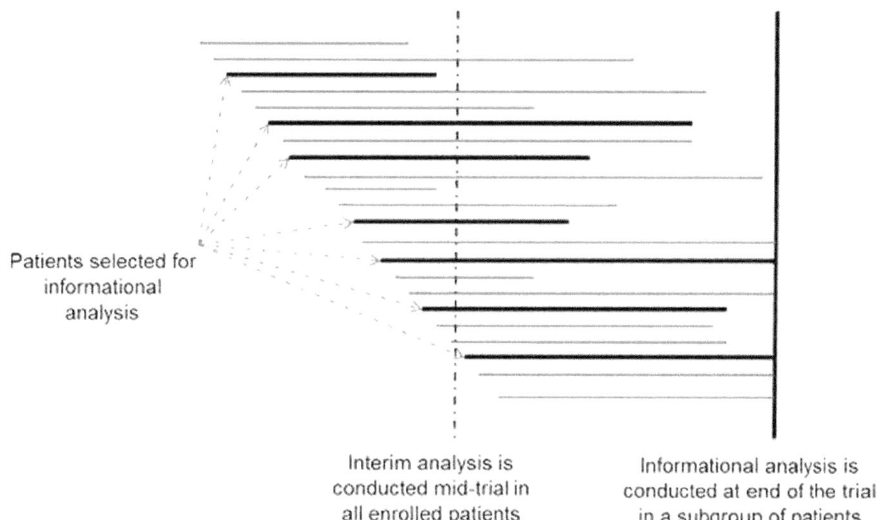

FIGURE 4.6.1 Conventional interim analysis compared with informational analysis. Patients are represented by horizontal lines. The informational analysis cohort is indicated by black horizontal lines and the informational analysis itself is indicated by a black vertical line at the end of the study. The interim analysis is indicated by the dashed vertical line in the middle of the study. (Reproduced from Chen, C. et al., *Stat. Biopharm. Res.*, 8, 238–247, 2016, with permission.)

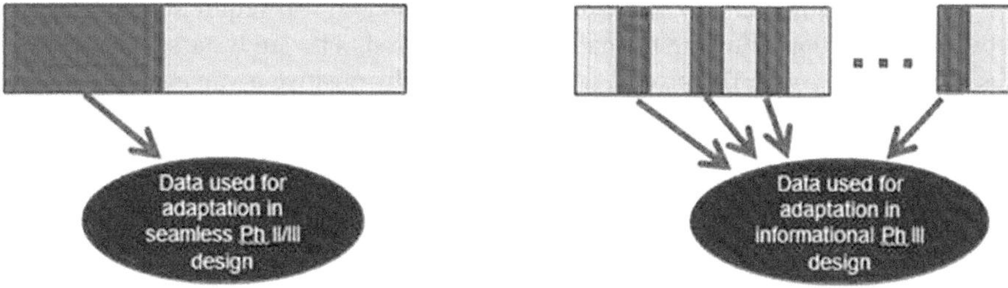

FIGURE 4.6.2 Seamless phase II/III design contrasted with the informational design. The dark blue represents data used for adaption. In a seamless phase II/III study, all the data up to a certain time point are used, whereas in the informational design, a portion of the information from all time points are used. (Reproduced from Chen, C. et al., *Stat. Biopharm. Res.*, 8, 238–247, 2016, with permission)

The information design is ideal for applications in which a decision does not have to be made until the end of the study. For example, if it is unclear whether a therapy benefits the whole population or a biomarker-defined subgroup, the whole population can be enrolled and questions of benefit in the subgroup and the overall population can both be tested at the same time in parallel. As discussed in Chapter 4.3, testing in parallel is important because when you test sequentially the second test is not allowable statistically unless the first is positive. Sequential statistical testing can produce regrettable results if the optimal order of testing is not known, as in the RADIANT trial (Kelly et al., 2015). In this study, patients with locally advanced non-small cell lung cancer received erlotinib, an agent directed against the EGFR, on the experimental arm. There was a hypothesis that the subset of patients with higher expression of EGFR would experience greater benefit. However, the statistical analysis was designed in a sequential fashion in which a benefit in the overall population was required before benefit in subgroups could be considered. The study failed in the overall population, and the nominally positive p-values in the subgroup could not be considered based on the study design.

The information design allows parallel testing of two hypotheses in a highly optimal fashion. As discussed above, such parallel testing of two questions or hypotheses means that the false-positive rate alpha must be divided between the two ("alpha allocation"). Instead of arbitrarily picking the amounts of alpha allocated to each of the two hypotheses, Chen and Beckman (2009) previously suggested using external data or an interim analysis within the study to estimate the alpha allocation that provides the maximum study power. The information design does this, but goes a step further. Because the information cohort, in contrast to an interim analysis, contains the mature time course of the time dependent final endpoint, it is a better source of data for adaptation. Similarly, compared with using external data for adaptation (which may have also been followed to maturity), the internal information cohort is from the same study and, therefore, has the same study design, study sites, and study population, again making it a superior information source for adaptation. Finally there is a further advantage with respect to alpha allocation. If two measures are correlated, they may be allowed a total false-positive rate that adds up to more than 0.05. This is because the total false-positive rate is really less than this sum for correlated endpoints. In this case, the information cohort, if stratified in the same way as the full study, provides a highly valid "unbiased" estimate of the correlative relationship between the biomarker subgroup outcome and the full population outcome.

Another important application of the information design is defining a cutoff between biomarker positive and biomarker negative for a continuous biomarker (Chen and Beckman 2016; Chen et al., 2016). The information cohort may be used to interrogate different cutoffs to find the optimum. These cutoffs are ascending, beginning with a low value, and the question is asked whether the patients with lower biomarker levels should be excluded. Thus, the study looks at different subsets.

A novel potential application of the information design that is a subject of current research is in the area of rare diseases where the proper endpoints may not be known. These rare diseases often are associated with considerable unmet medical need, creating an urgency to complete development. Yet, if the endpoints are not known, health authorities will often suggest a dedicated natural history study with no therapy to determine the best endpoints (U.S. Food and Drug Administration, 2017). Given the difficulty of enrolling patients, this study can only delay the availability of desperately needed therapies for years. The information design could clarify the relationship and correlation between multiple candidate endpoints and selection of a pivotal endpoint in the overall study, thus allowing the natural history study to be done simultaneously with the pivotal study, saving years of crucial development time (Beckman and Chen, unpublished).

For all these applications, the steps for using the information design are as follows:

- Prospectively define a fraction (i.e., 20%, 40%, 60% or 80%) of the study patients as an information cohort, what information will be sought from the information cohort, and what the rules will be for a prespecified adaptation based on this information. The higher the fraction of patients in the information cohort, the better the information, and the greater the statistical penalty that must be paid if these patients are used in the final analysis.
- Decide whether the information cohort patients will also be used in the final analysis. If so, a statistical penalty must be paid, and the study must be done at a lower nominal false-positive rate of less than 0.05. Use the procedures in Chen et al. (2016) to calculate the required lower nominal false-positive rate.
- Enroll the study with any applicable stratifications, randomly assigning the predetermined fraction of the patients to the information cohort within each stratum.
- Extract the needed information from the information cohort, and perform the study adaptation (optimal alpha allocation, continuous biomarker cutoff determination, or choice of a pivotal endpoint).
- Evaluate the whole study, using the lower nominal false-positive rate if applicable.

In the context of seamless phase II/III designs, a penalty must be paid to control the false-positive rate if the patients from the phase II portion are to be included in the final analysis. Nonetheless, in this context, it is generally worthwhile to pay this penalty. We surmise it is also worthwhile to pay the penalty to include the informational cohort patients in the final study analysis, but we have not formally compared it to a design in which these patients are excluded from the final analysis.

A CASE STUDY: INFORMATIONAL DESIGN IN A PHASE III NON-SMALL CELL LUNG CANCER TRIAL (RADIANT)

In the RADIANT study (Kelly et al., 2015), patients with locally advanced NSCLC received erlotinib, an agent directed against the EGFR, on the experimental arm. There was a hypothesis that the subset of patients with higher expression of EGFR would experience greater benefit. However, the statistical analysis was designed in a sequential fashion in which a benefit in the overall population was required before benefit in subgroups could be considered. The study failed in the overall population, and the nominally positive p-values in the subgroup could not be considered based on the study design.

In contrast, we assume the optimal order of testing is not known and seek to allocate α between the full population hypothesis (α_1) and the subpopulation hypothesis (α_2). This can be alternately be done using a conventional α spending function (Spiessens et al., 2010), applying arbitrary fixed values (often testing the subpopulation hypothesis at lower α) (Freidlin et al., 2005), or by optimizing study power using internal study data (Chen et al., 2009) or external data from a mature phase II study of a similar design, the "phase II+ method" (Beckman et al., 2011). Alternatively, in the case of the information design, the allocation will be performed based on data in the informational cohort comprising a fraction t of the patients to optimize study power (Chen and Beckman, 2016; Chen et al., 2016).

In the application example, as in RADIANT, the true (unknown) hazard ratios are 0.90 for the full population, 0.61 for the subpopulation. We looked at cases where either 17% or 34% of the events came from the subgroup. A hypothetical study with 1:1 randomization required 410 events for 83% power for a hazard ratio of 0.75 in the full population at a one-sided alpha of 0.025. A sequential procedure with the primary hypothesis being the full population hypothesis as in RADIANT would have only 19% power under these circumstances. If the subpopulation hypothesis were primary, the design would have 54% power when 17% of events are from the subpopulation and 80% power when 34% of the events are from the subpopulation. However, in the case under consideration, one does not know *a priori* which hypothesis should be primary, and if one made the biomarker hypothesis primary when the putative biomarker was not predictive, that would substantially decrease the power. The informational design provided 45% power when 17% of the events were from the subgroup and 80% power when

34% of the events were from the subgroup, nearly as good as could have been achieved by *a priori* knowledge of the hazard ratios. Indeed, a broad exploration of conditions showed that the informational design, by allocating more or less alpha to the subgroup, depending on the strength of the biomarker effect in the informational cohort, provided stable performance across the different possible versions of truth, and this performance was nearly as good as knowing the truth *a priori* (Chen and Beckman, 2016; Chen et al., 2016).

DISCUSSION

Rare diseases represent a uniquely challenging area for a variety of reasons, including lack of sufficient prior knowledge to allow design of optimal studies, as well as limited available population for study. The informational design may be a very useful design in this space, for optimally allocating α among subgroup hypotheses, for identifying optimal subgroup boundaries for continuous biomarkers, and for choosing optimal endpoints without prior knowledge and without the added delay of a separate natural history study. These applications may accelerate the availability of life-altering therapies for the many patients suffering from these conditions.

REFERENCES

Beckman RA, Clark JC, and Chen C. (2011). Integrating predictive biomarkers and classifiers into oncology drug development programs. *Nature Reviews Drug Discovery* 10: 735–749.

Bretz F, Schmidli H, König F, Racine A, and Maurer W. (2006). Confirmatory seamless phase II/III clinical trials with hypotheses selection at interim: General concepts. *Biometrical Journal* 48: 623–634.

Chen C and Beckman RA. (2009). Hypothesis testing in a confirmatory Phase III trial with a possible subset effect. *Statistics in Biopharmaceutical Research* 1: 431–440.

Chen C and Beckman RA. (2016). Informational design of confirmatory Phase III trials. *Biopharmaceutical Report* 23: 1–16.

Chen C, Li N, Shentu Y, Pang L, and Beckman RA. (2016). Adaptive informational design of confirmatory Phase III trials with an uncertain biomarker effect to improve the probability of success. *Statistics in Biopharmaceutical Research* 8: 238–247.

Freidlin B and Simon R. (2005). Adaptive signature design: An adaptive clinical trial design for generating and prospectively testing a gene expression signature for sensitive patients. *Clinical Cancer Research* 11: 7872–7878.

Kelly K, Altorki NK, Eberhardt WE, O'Brien ME, Spigal DR, Crinó L, et al. (2015). Adjuvant erlotinib versus placebo in patients with stage IB-IIIA non-small-cell lung cancer (RADIANT): A randomized, double-blind, phase III trial. *Journal of Clinical Oncology* 33: 4007–4014.

Mehta C, Schäfer H, Daniel H, and Irle S. (2014). Biomarker driven population enrichment for adaptive oncology trials with time to event endpoints. *Statistics in Medicine* 33: 4515–4531.

Muller HH and Schafer H. (2004). A general statistical principle for changing a design any time during the course of a trial. *Statistics in Medicine* 23: 2497–2508.

Ondra T, Jobjörnsson S, Beckman RA, Burman, CF, König F, Stallard N, and Posch M. (2016). Optimizing trial designs for targeted therapies. *PLoS One* 11: e0163726. http://www.fromhopetocures.org/fighting-rare-diseases.

Orphan Drug Act of 1983, Public Law 97-414, Stat. 2049. (1983). Amended by public law 98-551 (1984) to add a numeric prevalence threshold. http://www.rarediseaseday.org/article/what-is-a-rare-disease rarediseaseday.us/wp-content/uploads/2011/11/RDD-FAQ-2013.pdf. Accessed on March 5, 2017.

Schwartz J. et al. (2015). Research in rare disease: The nature and extent of evidence needed for decision. *51st Annual Meeting of the Drug Information Association*, session #318 (track 17), June 2015, Washington, DC. https://issuu.com/postscripts/docs/dia2015_finalprogram.

Spiessens B and Debois M. (2010). Adjusted significance levels for subgroup analyses in clinical trials. *Contemporary Clinical Trials* 31: 647–656.

Stallard N and Todd S. (2011). Seamless phase II/III designs. *Statistical Methods in Medical Research* 20: 623–634.

U.S. Food and Drug Administration, Novel drugs 2014 summary. (2015a). https://www.fda.gov/Drugs/DevelopmentApprovalProcess/DrugInnovation/ucm429247.htm. Accessed on March 5, 2017.

U.S. Food and Drug Administration. (2015b). Rare diseases: Common issues in drug development, guidance for industry (draft). https://www.fda.gov/downloads/Drugs/GuidanceComplianceRegulatoryInformation/Guidances/UCM458485.pdf. Accessed on March 5, 2017.

U.S. Food and Drug Administration. (2017). Multiple end points in clinical trials, guidance for industry. https://www.fda.gov/downloads/Drugs/GuidanceComplianceRegulatoryInformation/Guidances/UCM536750.pdf. Accessed on March 5, 2017.

5 The Road Ahead

Identification of the Hurdles and Solutions to the Implementation of Precision Medicine in Healthcare

Paving a New Era in Patient Centric Care through Innovations in Precision Medicine

5.1

Mark Fidock, Glen Hughes, Alain van Gool, and Claudio Carini

Contents

DEFINING THE NEED

The concept of precision medicine relies on the tailoring of medical treatments to defined patient subpopulations. The term was first used by Clayton Christensen in 2009 [1], who described it as a disruptive solution for health care. It did not gain global recognition until 2015, when U.S. President Barack Obama, in his State of the Union Address, announced details of the National Institute of Health's (NIH) Precision Medicine Initiative [2]. The program was launched in 2016 with a $215 million budget and it aimed to increase research and technology in the area, as well as provide a policy framework to help embed the approach into U.S. health care. Since then, it has entered the medical lexicon within the medical and scientific communities, regulatory authorities, and health care payers and with patients and their advocacy groups [3].

In the pharmaceutical industry, drug development runs with exceptionally high attrition rates, with only 10% of drugs tested in phase I clinical trials making it to through to drug approval [4,5]. This poor success rate is driven in part by the "all comers" dogma that maintains that all patients will benefit from a therapy. The success rate of drug discovery has been shown to be three times higher when patient selection biomarkers are used in drug development programs [6,7]. Precision medicine will alter the way medicine is delivered to patients by changing how diseases are thought of and treated, and for those running health care systems it will increase the cost effectiveness of treatment. Most importantly, for the patient, a targeted treatment approach provides a greater certainty of response to a therapy and no exposure to drugs that have a low efficacy. Tailoring medicines to patient subgroups will have clear benefits for all stakeholder groups [8].

SCIENTIFIC BASIS OF PRECISION MEDICINE

In the precision medicine model, diagnostic testing is typically used to select patients for an appropriate therapy. This usually involves measuring a biomarker with either, genomic, proteomic or radiomic technologies at the tissue or biological fluid level [9]. The identification of the patient selection biomarker is not a trivial pursuit as a detailed knowledge of disease drivers, the molecular targets and pathways are needed to establish whether there is a biological heterogeneity in the disease population and, therefore, a potential difference in drug response. Biomarker identification has been greatly influenced by the development of omic technologies and computing power [10,11]. Molecular endotype and phenotype data can now be combined and analyzed using complex bioinformatics.

Oncology is at the forefront of precision medicine due to the extensive scientific research into the molecular drivers of the disease and mechanisms of drug resistance. Additionally, the routine collection of tissue samples in oncology during diagnosis has meant their availability for biomarker studies. This is particularly true for lung cancer where detailed molecular profiling has identified drugs for patients with metastatic non-small cell lung cancer (NSCLC) and a mutated epidermal growth factor receptor (EGFRm), including osimertinib (Tagrisso®) for T790M-positive disease [12]. However, other therapeutic areas are advancing at pace driven by the establishment of consortia aimed at fast tracking research and better treatment. Notable examples include: Unbiased BIOmarkers in PREDiction of respiratory disease outcomes (U-BIOPRED) in asthma and the Alzheimer's Disease Neuroimaging Initiative (ADNI) in Neuroscience.

In respiratory, asthma was once considered a single disease. However, responses to medicines vary considerable between these patients. It is now known that asthma comprises distinct heterogenous inflammatory disorders with different phenotypes [13,14]. Approximately 5% of asthma patients have an eosinophilic phenotype, a severe uncontrolled form of the disease driven by high concentrations of eosinophils [15,16]. These patients can now be identified by their elevated blood eosinophil counts and treated with biological therapies such as benralizumab Fasenra®, a monoclonal interleukin (IL)–5R antibody that recruits Natural Killer cells to induce direct eosinophil killing, or Mepolizumab (Nucala®) which reduces eosinophils indirectly by binding with the IL-5 ligand. The field of diabetes is also developing along similar lines to respiratory. Data-driven cluster analysis has discovered that patients with newly diagnosed diabetes can be placed into five groups based on six variables (glutamate decarboxylase antibodies, age at diagnosis, body mass index, HbA1c level, and estimates of β-cell function and insulin resistance) and each of these different groups has different risks of progression and diabetic complications [17]. It now seems plausible that these new insights will permit the development of precision medicine strategies in metabolic disease clinical trials and deliver new treatments options for patients.

WHAT DOES IT TAKE TO PERFORM PRECISION MEDICINE IN CLINICAL STUDIES?

The development of a precision medicine plan often begins with and follows the translational research plan, transforming the basic science of disease processes, from cell or animal models, into a targeted therapy in humans [18]. The identification of a patient response biomarker should be a systematic and a dynamic process to increase the likelihood of discovering one. Hundreds of biomarkers may be initially screened in the biomarker discovery phase using multi-omic approaches, but these will ultimately lead to 5–10 being qualified in clinical trials [18]. Biomarker measurements may take place in many sample types including, tissue biopsies (formalin-fixed paraffin-embedded [FFPE]) or blood, as well as in urine, sputum or saliva. In transitioning to the clinical phase, the quality of the biomarker data also needs to be assured using a fit-for-purpose approach to assay validation [19], with equal thought also being placed on sample stability, how it is collected and stored.

One area of advance has been the increased inclusion of disease patients in phase I multiple ascending dose studies. This allows the early measurement of patient stratification biomarkers and an estimation of its prevalence in the intended population. The next critical step for patient selection biomarkers is their measurement in phase II clinical studies, with methods validated to bioanalytical guidelines [20,21]. The phase IIa study is used to confirm that there is a consistent association between the biomarker and the disease and that there is a heterogeneity of expression that allows the patient population to be segmented and the biomarker differences are associated with drug response. It may also be necessary to compare the biomarker measurements to normal or control populations to establish reference ranges. Given the high cost associated with diagnostic test development–positive data at this stage may be used to trigger investment into the development of a complementary or companion diagnostic (CDx) test [22].

The larger patient numbers in a phase IIb study are used to expand the clinical evidence for the predictivity and prevalence of the biomarker to select a population that benefits from the drug. Following the conclusion of the study, a statistical analysis is used to demonstrate the relationship between the level of a biomarker and clinical outcome. If a clinical association can be demonstrated, the specific threshold of the biomarker required for that clinical association is determined and is often referred to as the threshold value. In simple terms, the threshold establishes the biomarker value above or below which the patient would benefit from the therapy. It is critically important to know the variation in the test performance at the threshold, to understand how the classification of patients may change. The prototype diagnostic test, designated investigational use only (IUO) at this stage, is then readied for use in a phase III clinical trial where it is used to prospectively select subjects to be enrolled, using the phase IIb determined threshold value. If successful, the phase III data will support the contemporaneous regulatory approval of the drug and diagnostic test [23].

In therapeutic areas such as cardiovascular or respiratory disease, alternative laboratory-developed or point-of-care tests may be used, such as those developed for fractional exhaled nitric oxide (FeNo). These tests might not be directly linked to a drug but may form part of the clinical work-up of a patient with the aim to identify their phenotype for treatment. These tests are developed using noninterventional clinical trials, typically consisting of upward of 240 subjects, to establish what the normal and disease levels of the biomarker are and where the threshold between the two is.

TECHNOLOGIES THAT ENABLE PRECISION MEDICINE

Technology developments have advanced precision medicine by increasing the opportunities to identify the mechanisms of disease and development of novel biomarkers as surrogate end points and CDxs. Current abilities to detect and quantitate DNA variants, RNA levels, proteins levels and modifications and structurally distinct metabolites have greatly advanced over the past decades, which will further accelerate the availability of clinically usable biomarkers. Other biomarker read-outs based on imaging, phenotypic changes and patient-related outcomes are also generating interest in the field.

The first technology platform to deliver a CDx was immunohistochemistry (IHC) in 1998. Using FFPE tissue biopsies, clinical studies identified a subpopulation of women with breast cancer who overexpressed aberrant levels of epidermal growth factor 2 protein (HER2) and responded well to trastuzumab (Herceptin®) treatment [24]. This correlation between HER2 overexpression and clinical benefit has led to the launch of several precision medicine s including trastuzumab (Herceptin®), pertuzumab (Perjeta®) and ado-trastuzumab emtansine (Kadcyla®). IHC assays are now being used in the field of immuno-oncology to measure increases in programmed death-ligand 1 (PD-L1) in tumors for check-point inhibitors nivolumab (Opdivo®), pembrolizumab (Keytruda®) and atezolizumab (Tecentriq®) (Table 5.1.1). Currently 13 of the 32 approved complementary and CDxs utilize IHC. To overcome the subjectivity of IHC scoring, which has been long debated, digital pathology approaches such as those being developed at Definiens AG may revolutionize the way these tests are used in clinical practice [25].

In the last 30 years, there has been an increased understanding of the role of cancer susceptibility genes and gene mutations in many forms of common cancer [26]. Precision oncology has largely relied on polymerase chain reaction (PCR) technology for the qualitative detection of gene deletions and substitution mutations, typically using DNA extracted from FFPE tissues samples. Notable examples of these types of tests include the Qiagen therascreen EGFR test for gefitinib (Iressa®) in NSCLC, Roche cobas® KRAS Mutation Test cetuximab (Erbitux®) in colorectal cancer and cobas EGFR Mutation Test v2 T790M for osimertinib (Tagrisso®) in NSCLC.

In some patients it may be difficult to obtain tumor biopsies due to their disease status and an inability to tolerate further additional invasive procedures. For these patients, a blood-based circulating tumor-DNA (ctDNA) test, also known as a liquid biopsy, may be the more viable option. This has been made possible by the discovery that tumors shed DNA into the circulation and its measurement can act as a surrogate biomarker for the tumor of origin [27]. ctDNA testing represents a major advancement for patient testing and increases patient access to therapies such as osimertinib (Tagrisso®), using the cobas PCR-based T790M ctDNA test, approved by the U.S. Food and Drug Administration (FDA) in 2016. Interestingly, the liquid biopsy approach is now extending to other disease areas, including cardiovascular disease [28]. Further advances to the analytical performance of ctDNA testing to enhance its performance and compatibility with next generation sequencing (NGS) are underway by a number of companies including Guardant Health and Resolution Bioscience Inc.

The major technology development that has revolutionized precision medicine across all therapeutic areas is NGS. The advances over traditional Sanger sequencing, first cited in 1977, are threefold: (1) the low cost and speed of NGS makes large-scale sequencing affordable; (2) innovation in computational cloud-based computing now enables rapid bioinformatic analysis of big data sets, using algorithms that can detect sequence variants or rare mutations; and (3) availability of patient samples from clinical trials or from health care providers that are associated with patient phenotyping and drug response data [29]. The synergistic combination of these three components has driven large-scale genomic programs, in the biopharmaceutical industry, academia and health care systems. NGS platforms vary in their underlying chemistry and read technology with competition between the developers helping drive the science. Key players in this field are Illumina (HiSeq platform), ThermoFisher Scientific (Ion Torrent), and Pacific Bioscience (RSII). Most of these organizations offer bench-top machines with the ability to sequence DNA to a variety of read lengths, depths and sequence coverage. With this data it is now reasonable to expect a step change in our understanding of disease mechanisms, the identification of new therapeutic targets and new segments of disease with new diagnostic tests across all therapeutic areas [30]. The use of NGS as a diagnostic platform has now been recognized, with the FDA giving marketing approval for the Foundation Medicine F1CDx™ test, which can detect mutations in 324 genes and two genomic signatures. The test can identify which patients with NSCLC, melanoma, breast cancer, colorectal cancer, or ovarian cancer may benefit from 15 different FDA-approved targeted treatment options (Table 5.1.1).

Tumor mutational burden (TMB) is an emerging diagnostic approach in immuno-oncology. This methodology has shown that tumors with higher mutational burden are more immunogenic and more responsive to immunotherapies such as PD-L1 checkpoint inhibitors. TMB can be assessed using either tumor or blood ctDNA combined with NGS, with the TMB rate reported as mutations per megabase (10^6 nucleotide base pairs), where a value >20 is referred to as high [31]. In several anti-PD-L1 clinical studies for nivolumab (Opdivo), atezolizumab (Tecentriq) and pembrolizumab (Keytruda), patients with tumors that have a high TMB have been associated with a meaningful clinical benefit, such as progression-free survival.

TABLE 5.1.1 Precision medicine drugs and companion diagnostics approved by the U.S. Food and Drug Administration as of August 2018

DRUG TRADE NAME (GENERIC NAME)	*NDA/BLA*	*DEVICE TRADE NAME*	*PMA*	*DEVICE MANUFACTURER*
TIBSOVO® (ivosidenib)	NDA 211192	Abbott RealTime IDH1	P170041	Abbott
VENCLEXTA® (venetoclax)	208573	VYSIS CLL FISH PROBE KIT	P150041	Abbott Molecular, Inc.
Herceptin (trastuzumab)	BLA 103792	PATHVYSION HER2 DNA Probe Kit	P980024 S001-S010	Abbott Molecular, Inc.
Xalkori (crizotinib)	NDA 202570	VYSIS ALK Break Apart FISH Probe Kit	P110012 S001-S003	Abbott Molecular, Inc.
Idhifa® (enasidenib)	209606	Abbott RealTime IDH2	P170005	Abbott Molecular, Inc.
Herceptin (trastuzumab)	BLA 103792	INSITE HER2 NEU KIT	P040030	Biogenex Laboratories, Inc.
Mekinist (tramatenib); Tafinlar (dabrafenib)	NDA 204114; NDA 202806	THxID BRAF Kit	P120014	Biomérieux, Inc.
Herceptin (trastuzumab)	BLA 103792	HER2 CISH PharmDx Kit	P100024 S001-S004	Dako Denmark A/S
Herceptin (trastuzumab); Perjeta (pertuzumab)	BLA 103792; BLA 125409	HercepTest	P980018 S001-S017	Dako Denmark A/S
Herceptin (trastuzumab); Perjeta (pertuzumab)	BLA 103792; BLA 125409	HER2 FISH PharmDx Kit	P040005 S001-S009	Dako Denmark A/S
Erbitux (cetuximab); Vectibix (panitumumab)	BLA 125084; BLA 125147	Dako EGFR PharmDx Kit	P030044 S001-S002	Dako North America, Inc.
Gleevec/Glivec (imatinib mesylate)	NDA 021335; NDA 021588	DAKO C-KIT PharmDx	P040011 S001-S002	Dako North America, Inc.
KEYTRUDA® (pembrolizimab)	BLA 125514/S-046; sBLA125514/s24	PD-L1 IHC 22C3 PharmDx	P150013/S011; P150013/S006	Dako North America, Inc.
Tafinlar® (dabrafenib); Mekinist® (trametinib); Xalkori® (crizotinib); Iressa® (gefitinib); Rubraca™ (rucaparib); TAGRISSO™ (osimertinib); Tarceva® (erlotinib); Erbitux (cetuximab); Vectibix (panitumumab); Gilotrif (afatinib); Herceptin® (trastuzumab); Perjeta (pertuzumab); Kadcyla (ado-trastuzumab emtansine); Zelboraf (vemurafenib); Zykadia® (ceritinib); Zykadia® (ceritinib); Cotellic® (cobimetinib) in combination with Zelboraf (vemurafenib); Alecensa® (alectinib)	NDA 202806; NDA 202806/S-006; NDA 204114; NDA 204114/S-005; NDA 202570; NDA 202570/S-021; NDA 206955; NDA 209115; NDA 208065; NDA 021743; BLA 125084; BLA 125147; NDA 201292; BLA 103792; BLA 125409; BLA 125427; NDA 202429; NDA 205755/S-009; NDA 206192; NDA 208434/S-003	FoundationOne CDx	P170019	Foundation Medicine, Inc.
Vectibix® (panitumumab)	125147	Praxis Extended RAS Panel	P160033	Illumina, Inc.
Herceptin (trastuzumab)	BLA 103792	Bond Oracle Her2 IHC System	P090015	Leica Biosystems

(*Continued*)

TABLE 5.1.1 (*Continued*) Precision medicine drugs and companion diagnostics approved by the U.S. Food and Drug Administration as of August 2018

DRUG TRADE NAME (GENERIC NAME)	*NDA/BLA*	*DEVICE TRADE NAME*	*PMA*	*DEVICE MANUFACTURER*
Tafinlar® (dabrafenib); Mekinist® (trametinib); Xalkori® (crizotinib); Iressa® (gefitinib)	NDA 202806/S006; NDA 202570/S021; NDA 202570/S021; NDA 206995	Oncomine Dx Target Test	P160045	Life Technologies Corporation
Herceptin (trastuzumab)	BLA 103792	SPOT-LIGHT HER2 CISH Kit	P050040 S001-S003	Life Technologies, Inc.
TASIGNA (nilotinib)	NDA 022068, S-026	MolecularMD MRDx® BCR-ABL Test	K173492	MolecularMD Corporation
Lynparza (olaparib)	NDA 208558	BRACAnalysis CDx	P140020/S012	Myriad Genetic Laboratories, Inc.
Erbitux (cetuximab)	BLA 125084	Therascreen KRAS RGQ PCR Kit	P110030	Qiagen Manchester, Ltd.
Gilotrif (afatinib)	NDA 201292	Therascreen EGFR RGQ PCR Kit	P120022	Qiagen Manchester, Ltd.
Tarceva (erlotinib)	NDA 021743	Cobas EGFR Mutation Test	P120019	Roche Molecular Systems, Inc.
Zelboraf (vemurafenib)	NDA 202429	COBAS 4800 BRAF V600 Mutation Test	P110020 S001-S006	Roche Molecular Systems, Inc.
TAGRISSO® (osimertinib)	208065; 208065/S008; 208065/S009	Cobas® EGFR Mutation Test v2	P150044;P120019/ S016; P12 0019/S018	Roche Molecular Systems, Inc.
TECENTRIQ™ (atezolizumab)	NDA 761034/S012	PD-L1 (SP142)	P160002/S006	Ventana
Herceptin (trastuzumab)	BLA 103792	INFORM HER2 NEU	P940004 S001	Ventana Medical Systems, Inc.
Herceptin (trastuzumab)	BLA 103792	PATHWAY ANTI-HER2 NEU (4B5) Rabbit monoclonal primary antibody	P990081 S001-S016	Ventana Medical Systems, Inc.
Herceptin (trastuzumab)	BLA 103792	INFORM HER2 DUAL ISH DNA Probe Cocktail Assay	P100027 S001-S006	Ventana Medical Systems, Inc.
ZYKADIA® (ceritinib)	sDNA 205755-09; (NDA supplement)	VENTANA ALK (D5F3) CDx Assay	P140025/S005	Ventana Medical Systems, Inc.

Source: U.S. FDA list of cleared and approved Companion Diagnostic devices, https://www.fda.gov/medicaldevices/productsandmedicalprocedures/invitrodiagnostics/ucm301431.htm.

There are potentially other sources of molecular biomarkers, including circulating tumor cells (CTC), exosomes and platelets and analytes such as miRNA that may form the basis of patient selection tests. The authors refer the reader to reviews describing these in more detail [32–34].

REGULATORY PATHS FOR PRECISION MEDICINE DEVICES

Diagnostic devices can be categorized into either *in vitro* or *in vivo*. *In vitro* diagnostic (IVD) devices involve sampling the subject tissue or biofluid and measuring the biomarker outside the body. *In vivo* measurements consist of imaging modalities such as positron emission tomography (PET), computerized tomography (CT) and MRI. The regulatory approval paths vary globally, but in the major geographies they are administered by the FDA, European Medical Agency (EMA, EU), China Food and Drug Administration (CFDA, China) and Pharmaceuticals and Medical Devices Agency (PMDA, Japan). Each agency has its own evidentiary requirements, and risk assessment and review processes for test approval.

The FDA sets a high regulatory hurdle for the approval of medical devices and is noteworthy of further discussion. The Center for Devices and Radiological Health (CDRH) is the branch of the FDA that is responsible for overseeing the clearance/approval, manufacturing, performance and safety of these devices. The FDA has established three classes of device, based on the risk the test poses to the patient, namely, Class I—low risk, Class II—medium risk, Class III—high risk [35]. Class I and II tests are cleared through the 510(k)-premarket notification (PMN) route; where a predicate test does not exist to compare performance against a marketed test, the regulatory submission will need to be via the *de novo* 510(k) application route. Because a CDx is used for identifying patients for a specific therapeutic intervention, these tests are categorized as Class-III devices and require submission through the more comprehensive premarket approval (PMA) process. A PMA submission must include scientific evidence demonstrating safety and efficacy in the intended patient population through clinical studies that establish device performance and clinical utility of the biomarker [36]. A contemporaneous process of drug and test development/approval is required if the test is to be ready at drug launch (Figure 5.1.1).

Not all precision medicine strategies will require a CDx. In 2015 the FDA introduced the term "complementary diagnostic" in relation to the approval of nivolumab in NSCLC. To date, no guidelines have been developed, but it is generally understood that this type of test will follow a PMA filing route because it will be used to identify a biomarker-defined subset of patients who respond well to a drug. An important distinction can be made for this type of test because a complementary diagnostic test is recommended for the safe and efficacious use of a drug and is an aid for the treating physician. However, a CDx test result is a required before a drug can be prescribed to the patient. Alternatively, a test may be used to identify a patient's clinical characteristics such as their phenotype. These tests may be more appropriately developed through the PMN 510(k) route, for example point-of-care uric acid testing in gout patients in order to establish whether a patient is responding to drugs such as Lesinurad® [37].

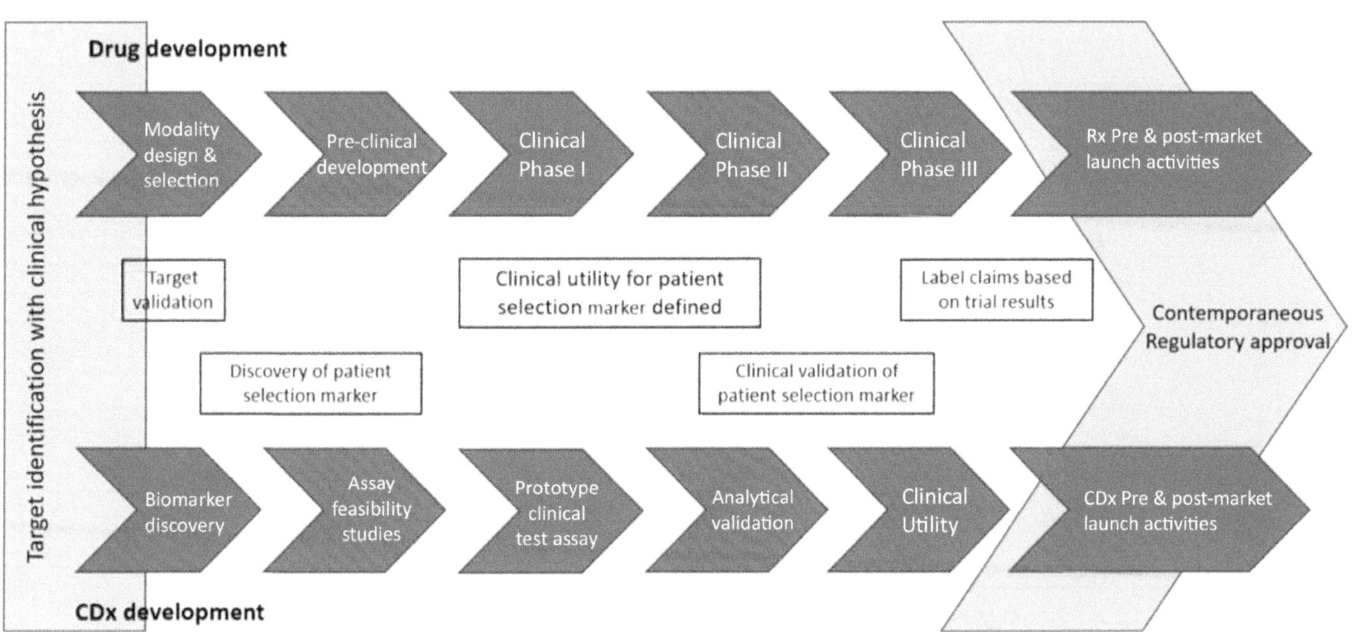

FIGURE 5.1.1 Targeted drug development linked to companion diagnostic (CDx) development for contemporaneous regulatory approval. (Adapted from Luo D. et al., *Front Genet.*, 6, 357, 2016.)

At the time of writing the European Union is introducing a new regulatory framework, moving away from a self-certified system (CE Mark) to a risk-based assessment analogous to that used by the FDA. Some current CE marked *in vitro* diagnostics, already on the market, will need to reapply to be certified under the new framework. This will be a complex process that the diagnostic test industry needs to complete by 2022 [38].

INNOVATION AND THE NEXT WAVE OF PRECISION MEDICINE

Precision medicine is a combination of innovative diagnostic testing, new targeted therapies and patient advocacy with significant progress expected to be made in all these areas over the next decade. With the rapid increase in targeted drug development across all therapeutic areas, we can expect to see a linked growth in the development of innovative, possibly disruptive, diagnostic technologies (Figure 5.1.2). This innovation being accelerated due to clear benefits to patients from early diagnosis, increased drug development success rates, pressure from health care providers for more cost-effective solutions, with the sum of this being unprecedented investment into precision medicine and the diagnostic development biotech sector [39].

Technologies underpinning precision medicine continue to develop with the advances seen in NGS having been remarkable in enabling patient treatment decisions and in transforming clinical oncology. Following this revolution, progress in mass spectrometry, noninvasive imaging and biosensors are pushing toward comparable insights in human biology of health and disease. We expect that these technologies, in combination, will further propel applications in precision medicine and provide an increasing amount of available data for mining.

These technological advances prompt us to redesign the way we analyze and use "Big Data" for precision medicine. Given its complexity the analysis of large multiparameter data sets requires the advanced computational and analytical approaches of artificial intelligence (AI), such as machine and deep learning. AI is rapidly emerging as a new technology with the potential to impact broadly across the whole of health care and beyond into the far reaches of society. A recent example of the use of AI is in enabling digital pathology to become analytically quantitative, increasing the speed, accuracy and concordance of diagnostic testing across laboratories [40]. The ultimate intent of these AI developments in health care is not to replace the experts but to allow pathologists to review only those cases where the treatment decision is borderline. Similar innovations are being developed in the analysis of imaging and omics, including the integration of clinical data [41].

The past 10 years has seen a change in the therapeutic options that deviate from classical small molecule and biological approaches. An example of a truly personalized approach, using the patient's own cells, is chimeric antigen receptor T-cell therapy (CAR-T) where chimeric antigen receptors are introduced into the patient's own isolated T-cells to target and become cytotoxic to the tumor [42]. The first two FDA CAR-T therapies are targeted against CD19, found on many types of lymphoma cells. These were approved in 2017 for relapsed/refractory diffuse large B-cell lymphoma (DLBCL) for axicabtagene ciloleucel (Yescarta®) and relapsed/refractory B-cell precursor acute lymphoblastic leukemia (ALL) for tisagenlecleucel (Kymriah®) [43].

The recently defined clustered regularly interspersed short palindromic repeat (CRISPR) method is now being developed for clinical use, in a similar manner to that for CAR-T, but also

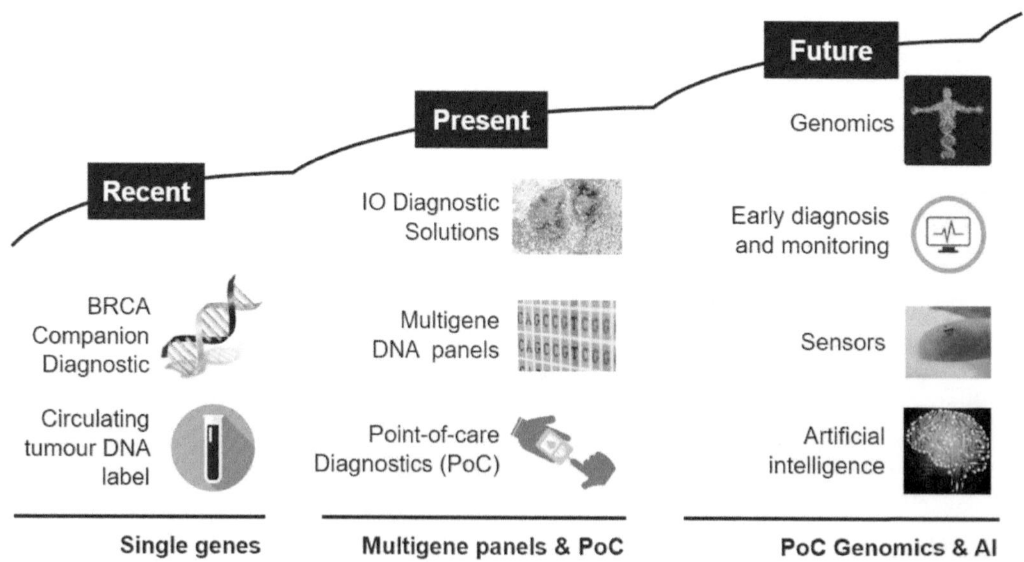

FIGURE 5.1.2 The era of precision medicine is driving technology innovation with a greater expectation of benefit to patients and health care providers. AI, artificial intelligence.

as a disruptive technology for the development diagnostic tests. With improved sensitivity compared with that available today, this will enable detection of events linked to disease etiology or progression far earlier and will lead to better patient outcomes [44,45].

In addition to the chemical, biological and molecular interventions, progress will be made in personalized therapies through optional personalized nutrition and lifestyle. Although changes in nutrition and lifestyle are the oldest therapeutic interventions, they are driven to much lesser extent by the molecular observations that are driving pharmaceutical drug development. However, they may need to be combined as an integrated pharma-nutrition approach capitalizing on the advances in understanding of the microbiome to maximize the effect of precision interventions to maintain or restore health [46].

Finally, citizen and patient advocacy is expected to be another key driver of implementing precision medicine [47]. The increased availability and use of sensors and wearables fuels potential applications as individuals gain more insight into their health/disease status through the continuous monitoring biomarker levels via their smartphones [48]. These developments are strongly pursued by start-up entrepreneurs but also by large industries such as Google, Apple and Amazon, which are investing heavily in this health care model [49]. The role of the patient and the doctor is changing, where the patient is well informed on their health status and with coaching by the doctor determines the optimal path toward a personalized health schedule. This cultural change is consistent with the modern way of living and has the potential to increase the prevention of disease through early action. A popular example is provided by the preventive surgery of Angelina Jolie given her familial history in breast and ovarian cancer [50].

In summary, there are high expectations that precision medicine approaches will become established practice in both drug development and in clinical care over the next decade. The advances in the molecular understanding of disease continue with parallel progress being made in the development of novel therapies and diagnostics. The need for regulatory oversight for medical devices that impact patient treatment decisions is clear, although having a single global framework would simplify and accelerate the approval process potentially allowing faster delivery of life changing medicines to patients. Precision medicine is and will continue to be for the foreseeable future a rapidly moving and exciting area that is shaping the delivery of clinical care to patients.

REFERENCES

1. Christensen CM, Grossman JH, Hwang J. *The Innovator's Prescription: A Disruptive Solution for Health Care*. New York: McGraw-Hill, 2009.
2. https://obamawhitehouse.archives.gov/node/333101.
3. Zhang XD. Precision medicine, personalized medicine, omics and big data: Concepts and relationships. *J Pharmacogenomics Pharmacoproteomics*. 2015; 6:e144. doi:10.4172/2153-0645.1000e144.
4. Smith P, Mytelka DS, Dunwiddie CT et al. How to improve R&D productivity: The pharmaceutical industry's grand challenge. *Nat Rev Drug Discov*. 2010; 9(3): 203–214. doi:10.1038/nrd3078.
5. Plenge R. Disciplined approach to drug discovery and early development. *Sci Transl Med*. 2016; 8(349): 349ps15. doi:10.1126/scitranslmed.aaf2608.
6. Thomas DW, Burns J, Audette J et al. *Clinical Development Success Rates 2006-2015*. 2016; San Diego, CA: Biomedtracker/Washington, DC: BIO/Bend: Amplion.
7. Morgan P, Brown DG, Lennard S et al. Impact of a five-dimensional framework on R&D productivity at AstraZeneca. *Nat Rev Drug Discov*. 2018; 17: 167–181. doi:10.1038/nrd.2017.244.
8. Mirnezami R, Nicholson J, Darzi A. Preparing for precision medicine. *N. Engl J Med*. 2012; 366(6), 489–491. doi:10.1056/NEJMp1114866.
9. Frangogiannis NG. Biomarkers: Hopes and challenges in the path from discovery to clinical practice. *Transl Res*. 2012; 159(4): 197–204. doi:10.1016/j.trsl.2012.01.023.
10. Vernon ST, Hansen T, Kott KA et al. Utilizing state-of-the-art "omics" technology and bioinformatics to identify new biological mechanisms and biomarkers for coronary artery disease. *Microcirculation*. 2018; e12488. doi:10.1111/micc.12488.
11. Mamoshina P, Vieira A, Putin E, Zhavoronkov A. Applications of deep learning in biomedicine. *Mol Pharm*. 2016; 13(5): 1445–1454. doi:10.1021/acs.molpharmaceut.5b00982.
12. Kobayashi Y, Mitsudomi T. Not all epidermal growth factor receptor mutations in lung cancer are created equal: Perspectives for individualized treatment strategy. *Cancer Sci*. 2016; 107(9): 1179–1186. doi:10.1111/cas.12996.
13. Borish L, Culp JA. Asthma: A syndrome composed of heterogeneous disease. *Ann Allergy Asthma Immunol*. 2008; 101(1): 1–8. doi:10.1016/S1081-1206(10)60826-5.
14. Ozdemir C, Kucuksezer UC, Akdis M et al. The concepts of asthma endotypes and phenotypes to guide current and novel treatment strategies. *Expert Rev Respir Med*. 2018; 12(9), 733–743. doi:10.1080/17476348.2018.1505507.
15. Chung KF, Wenzel SE, Brozek JL et al. International ERS/ATS guidelines on definition, evaluation and treatment of severe asthma. *Eur Respir J*. 2014; 43(2): 343–373. doi:10.1183/09031936.00202013.
16. Wenzel S. Severe asthma in adults. *Am J Respir Crit Care Med*. 2005; 172(2): 149–160. doi:10.1164/rccm.200409-1181PP.
17. Stefan N, Fritsche A, Schick F et al. Phenotypes of prediabetes and stratification of cardiometabolic risk. *Lancet Diabetes Endocrinol*. 2016; 4(9): 789–798. doi:10.1016/S2213-8587(16)00082-6.
18. Wheling M. (Ed) *Principles of Translational Science in Medicine: From Bench to Bedside*. Amsterdam, the Netherland: Academic Press. 2015; doi:10.1016/C2013-0-15547-X.
19. Lee JW, Devanarayan V, Barrett YC et al. Fit-for-purpose method development and validation for successful biomarker measurement. *Pharm Res*. 2006; 23(2): 312-328. doi:10.1007/s11095-005-9045-3.
20. Food and Drug Administration. Bioanalytical method validation: Guidance for industry. 2018. https://www.fda.gov/downloads/drugs/guidances/ucm070107.Pdf.
21. European Medicines Agency. Guideline on bioanalytical method validation. 2011; http://www.ema.europa.eu/docs/en_GB/document_library/Scientific_guideline/2011/08/WC500109686.pdf.
22. Fridlyand J, Simon RM, Walrath JC et al. Considerations for the successful co-development of targeted cancer therapies and companion diagnostics. *Nat Rev Drug Discov*. 2013; 12(10): 743–755. doi:10.1038/nrd4101.
23. Scheerens H, Malong A, Bassett K et al. Current status of companion and complementary diagnostics. *Clin Transl Sci*. 2017; 10**:** 84–92. doi:10.1111/cts.12455.

24. Waring PM. Matching patients with drugs: Triumphs and challenges. *Per Med.* 2006; 3(3): 335–344. doi:10.2217/17410541.3.3.335.
25. Hart SN. Will digital pathology be as disruptive as genomics? *J Pathol Inform.* 2018; 9: 27. doi:10.4103/jpi.jpi_25_18.
26. Turnbull C, Sud A, Houlston RS. Cancer genetics, precision prevention and a call to action. *Nat Genet.* 2018; 50(9): 1212–1218. doi:10.1038/s41588-018-0202-0.
27. Schwarzenbach H, Hoon DSB, Pantel K. Cell-free nucleic acids as biomarkers in cancer patients. *Nat Rev Cancer.* 2011; 11(6): 426–437. doi:10.1038/nrc3066.
28. Damani S, Bacconi A, Libiger O et al. Characterization of circulating endothelial cells in acute myocardial infarction. *Sci Transl Med.* 2012; 4(126): 126ra33. doi:10.1126/scitranslmed.3003451.
29. Leff DR, Yan GZ. Big data in precision medicine. *Engineering.* 2015; 1(3): 277–279.
30. Dhawan D. *Progress and Challenges in Precision Medicine.* Amsterdam, the Netherlands: Academic Press. 2017; Chapter 2: 35–54.
31. Chalmers ZR, Connelly CF, Fabrizio D et al. Analysis of 100,000 human cancer genomes reveals the landscape of tumor mutational burden. *Genome Med.* 2017; 9(1): 34. doi:10.1186/s13073-017-0424-2.
32. Woo D, Yu M. Circulating tumor cells as "liquid biopsies" to understand cancer metastasis. *Transl Res.* 2018; 201: 128–135. doi:10.1016/j.trsl.2018.07.003.
33. Sundararajan V, Sarkar FH, Ramasamy TS. The versatile role of exosomes in cancer progression: Diagnostic and therapeutic implications. *Cell Oncol.* 2018; 41(3): 223–252. doi:10.1007/s13402-018-0378-4.
34. Toiyama Y, Okugawa Y, Fleshman J et al. MicroRNAs as potential liquid biopsy biomarkers in colorectal cancer: A systematic review. *Biochim Biophys Acta.* 2018; doi:10.1016/j.bbcan.2018.05.006.
35. Pettitt D, Smith J, Meadows N et al. Regulatory barriers to the advancement of precision medicine. *Expert Review of Precision Medicine and Drug Development.* 2016; 1(3): 319–329. doi:10.1080/23808993.2016.1176526.
36. FDA. *In Vitro Companion Diagnostic Devices.* Washington, DC: FDA; 2014.
37. Nova Max Uric Acid Monitoring System. Uric acid test system for at home prescription use, K160990; FDA; 2017.
38. https://ec.europa.eu/growth/sectors/medical-devices/regulatory-framework_en
39. https://about.beauhurst.com/blog/precision-medicine-investment-2017/
40. Vandenberghe ME, Scott ML, Scorer PW et al. Relevance of deep learning to facilitate the diagnosis of HER2 status in breast cancer. *Sci Rep.* 2017; 7:45938. doi:10.1038/srep45938.
41. Ehteshami BB, Veta M, Johannes VDP et al. Diagnostic assessment of deep learning algorithms for detection of lymph node metastases in women with breast cancer. *JAMA.* 2017; 318(22): 2199–2210. doi:10.1001/jama.2017.14585.
42. Miliotou AN, Papadopoulou LC. CAR T-cell therapy: A new era in cancer immunotherapy. *Curr Pharm Biotechnol.* 2018; 19(1): 5–18. doi:10.2174/1389201019666180418095526.
43. Pehlivan KC, Duncan BB, Lee DW. CAR-T Cell therapy for acute lymphoblastic leukemia: Transforming the treatment of relapsed and refractory disease. *Curr Hematol Malig Rep.* 2018; 13(5): 396–406. doi:10.1007/s11899-018-0470-x.
44. Gootenberg JS, Abudayyeh OO, Kellner MJ, Joung J, Collins JJ, Zhang F. Multiplexed and portable nucleic acid detection platform with Cas13, Cas12a, and Csm6. *Science.* 2018; 360(6387): 439–444.
45. https://www.the-scientist.com/features/crispr-inches-toward-the-clinic-64535
46. Janssens Y, Nielandt J, Bronselaer A et al. Disbiome database: Linking the microbiome to disease. *BMC Microbiol.* 2018; 18(1): 50. doi:10.1186/s12866-018-1197-5.
47. https://www.npaf.org/
48. Rose SM, Perelman D, Colbert E et al. Digital health: Tracking physiomes and activity using wearable biosensors reveals useful health-related information. *PLoS Biol.* 2017; 15(1). doi:10.1371/journal.pbio.20014.
49. https:/www.economist.com/business/2018/02/03/apple-and-amazons-moves-in-health-signal-a-coming-transformation.
50. Evers C, Fischer C, Dikow N et al. Familial breast cancer: Genetic counselling over time, including patients' expectations and initiators considering the Angelina Jolie effect. *PLoS One.* 2017; 12(5). doi:10.1371/journal.pone.0177893.

Integrating Molecular Testing into Early Stage Clinical Applications

5.2

Jonathan I. Wilde

Contents

INTRODUCTION

The principle of personalized health care is to match the right treatment to the right patient. This principle sounds simple and straightforward, but the process of achieving this goal is complex as disease knowledge grows alongside the constant development of molecular technologies. In this section, I focus on the steps and considerations to utilize molecular testing in early stage clinical trials, using as an example the development of patient selection within oncology therapeutic clinical trials, where advances in sampling (e.g., ctDNA), assays (e.g., NGS, digital polymerase chain reaction [dPCR]), informatics and trial design have all been made.

DEVELOP AND DEFINE REQUIREMENTS

A clear understanding of the study's clinical, regulatory and commercial environment is needed to enable the development of the assay requirements used for partner selection.

Clinical and Technical Requirements

At the outset of the study, decisions have been made on the clinical objective, including the therapeutic under investigation, the cancer type and stage being targeted, the point of intervention, and the biomarker(s) identified from the exploratory and translational research stage that is(are) now considered ready for the prospective selection of patients in this phase I or phase IIa early stage trial. To enable the generation of clear assay requirements, the molecular testing strategy needs to be further defined. The biomarker(s) used for patient selection need to be specified not only at the level of the specific gene, but also the specific alterations which are to be used and which are to be avoided and if known, the level of detection (i.e., percentage allelic frequency) associated with clinical significance. At this early stage, a broader biomarker selection strategy can be utilized that can be refined in any future pivotal, companion diagnostic (CDx) development trial.

The next decision is on the specific sample or samples that will be used. The goal is to use the most appropriate sample for identifying the selection biomarker while minimizing the risk to the patient. It is useful to understand the patient's journey and determine the samples and collection method that is appropriate for the disease and treatment process and the facility where the patient is being treated. From the patient's perspective, it is important to try to minimize the collection of additional samples (e.g., use an archival sample) or, if required, use a minimally invasive sample (e.g., blood, plasma, urine) or minimize the size of biopsy taken. An important aspect of early stage trials is to enable further translational research and the refinement of the testing process and, therefore, the collection of additional samples for these downstream studies is often required though again the burden on the patient of these samples being collected has to be considered. The sample collection, preservation and shipping method selected needs to be validated and appropriate for use at all clinical sites in the study. The acceptable turnaround time from sample to result is also informed by the disease being investigated and the patient's condition. Many early stage oncology trials are with patients who have failed prior treatments and need to be placed on a new therapeutic as soon as possible.

To illustrate this process, two sample types are reviewed in more detail; FFPE tissue, which is by far the most common sample in oncology studies, and ctDNA, which is becoming more widely used as the field tries to use less invasive sampling methods.

Tissue biopsy samples preserved via the FFPE method are the mainstay of pathology and have therefore been utilized for biomarker identification. Variables that need to be considered include the amount of tissue available, the percent tumor present in the biopsy and percent allelic frequency of DNA alteration that is clinically significant (24). In a multicenter clinical study, each site will utilize its own FFPE protocol, and variation in tissue handling and fixation time could negatively affect the amount and quality of DNA recovered. There are a number of approaches to dealing with this: First select an assay that can cope with these variables, including extraction of DNA from over-fixed samples and an assay that works with poorer quality DNA; second, review the site's collection protocols to identify any potential problems; and third, once the study has started, monitor the assay pass rates from each site to identify site-specific problems and move quickly to rectify these. Improvements to the FFPE tissue collection method have been difficult to introduce as its use is so entrenched. Approaches such as the PreAnalytix PAXgene tissue container that preserves the integrity of the tissue as well as DNA and RNA have been developed but are currently research use only (RUO) and not widely used (25).

The use of ctDNA purified from plasma to obtain cancer related DNA alteration data is increasing, with a number of assays available, including those run within a single Clinical Laboratory Improvement Amendments (CLIA) lab, such as Guardant Health, Foundation Medicine and Inivata; and kit-based assay like the cobas assay from Roche Diagnostics. Circulating free DNA (cfDNA) is released from healthy, inflamed or cancerous tissue from cells undergoing apoptosis or necrosis. In patients with cancer, ctDNA is the fraction of cfDNA derived from tumor (33). A ctDNA test is particularly suitable when a patient is unable to provide a tissue biopsy due to ill health or the location of the tumor. However, challenges do exist. Not all tumors release ctDNA, especially early stage cancers, and mutations do not always match between tissue and plasma (21), and therefore in cases where a mutation is not found in ctDNA, it does not rule out its presence in the tumor. It is suggested that if ctDNA is used for patient selection, tissue is also collected where possible for subsequent comparison. As mentioned for FFPE tissue biopsies, as more sensitive methods are developed, it is important to understand the clinical significance and analytical reproducibility of a mutation detected at low allelic frequencies (<0.5%) and set an appropriate cut off for patient selection. There can be challenges of reproducible collection of plasma and ctDNA due to delays or variation to the centrifugation process. This has improved through the use of specific collection tubes such as the Streck tube to enable preservation of the ctDNA and remove the need for centrifugation on site (28).

Partner Requirements

In addition to the technical requirements, it is important that the assay provider and, if separate, the clinical laboratory have the necessary experience to be an assay development partner both within the early stage trial and in the future development of a CDx. As outlined in Table 5.2.1, the assay partner should be assessed to determine their prior experience in developing molecular diagnostics to the desired timeline, their experience with relevant regulatory agencies and their financial stability. In this Section, from a regulatory perspective, the focus is on the FDA; however, experience with other regions' regulatory agencies could be required. This is particularly relevant in Europe with the new EU IVD regulation and the increased requirements for CDx development (1,2,19). This assessment is not only on the company as whole but should also cover the individual people who will be directly involved in the assay development partnership, including their experience and engagement with the project.

TABLE 5.2.1 Technical and IVD partner requirements for assay selection. The requirements for each area are determined and criteria set for exceeds, meets or below acceptance. The table is completed for each assay under consideration. The decision can be made by assessing the relative colour coding or be refined by weighting each area so that high priority requirements will have a greater impact in the decision

	EXCEEDS	*MEETS*	*BELOW*
Technical Fit			
_Sample requirements	High experience with specific sample type	Some experience with specific sample type	No experience with specific sample type
_Turnaround time	Exceeds requirements, e.g. <3 days	Meets requirements, e.g. <7 days	Below requirements, e.g. >10 days
_Assay Performance	IVD, FDA approved Detects required alterations and more at > sensitivity	Analytical validation complete, CLIA assay. Detects required alterations at suitable sensitivity	RUO assay Detects some alterations or poor sensitivity
IVD Partner fit			
_Regulatory	Multiple IVD-CDx experience	Some IVD-CDx experience	No IVD-CDx experience
_Commercial	Global company	Small, public company	Start up company
_Internal experience	Extensive prior experience	Some prior experience	No prior experience

MATCHING ASSAY SYSTEM TO REQUIREMENTS

Assay Selection

Once the requirements for the diagnostic assay have been determined and agreed to the next step is assay selection. At this stage, the diagnostic scientist will draw upon their experience, ongoing horizon scanning efforts, and industry and academic contacts to prepare a list of candidate assays and companies. The generation of a selection matrix (Table 5.2.1) with clear criteria and weighting for high priority requirements and a subsequent summary table recording how each assay matches up to the requirements is a useful way select the most suitable assay in an open and unbiased manner. During this selection process, it is important to document the assays that do and do not meet expectations because this information will be useful for future selection activities.

If there is sufficient time, then pilot studies to compare a set of assays using the same samples, such as those performed by Sherwood et al. and Thress et al. are incredibly valuable (29,31). It can be challenging to obtain samples for these studies and so opportunities to obtain samples from earlier clinical trials, commercial sources or the use of cell lines or manufactured standards can all be explored. Once candidate assay or assays have been identified, the final decision will come down to in-depth assessment of the assay, a key step being the review of analytical validation data.

Analytical Validation

The aim of analytical validation is to clearly show that an assay and lab is capable of delivering a reproducible and reliable result independent of variables.

There a common set of studies that are undertaken for analytical validation of a molecular assay, and these are listed and defined in Table 5.2.2.

The extent of validation is dependent on the regulatory status of assay as set out in the FDA guidelines 42 CFR

TABLE 5.2.2 Definition of common terms used in analytical validation

TERM	*DEFINITION*
Analytical Validation	Demonstrates an assay functions successfully to defined performance characteristics, equivalent to a reference method or data
Analytical Verification	Demonstrates that a previously validated method can meet the analytical requirements (accuracy, precision etc.) in the testing lab
Accuracy	The closeness of the experimental value to the known value or truth
Precision	The closeness of a series of experimental values to each other, independent of the truth. Often a series of reproducibility studies in which the effect of varying operator, time, reagents and equipment
Analytical sensitivity	The ability of a test to detect a particular analyte, especially the minimum detectable concentration of a substance, often the input range, in some cases limit of blank is also suitable
Analytical specificity	The ability of the assay to measure a particular substance, often studies involve a study of interfering substances especially during sample collection
Reportable range or reference intervals	The functional range of an assay, over which concentrations of analytecan be measured accurately and precisely

493.1253 Standard: Establishment and verification of performance specifications (3). There are two options: First, if an assay is FDA cleared (510k), then there is only the need to establish and verify performance specifications compared with those established by the manufacturer for accuracy, precision and the reportable range of the assay fits within the labs patient population. Second, and more likely for the types of early stage studies discussed in this Section, when an assay is introduced that has either been modified from a FDA-cleared or -approved test or is an in-house developed method, such as a CLIA-approved laboratory-developed test then in addition to accuracy, precision and reportable range, studies to establish the analytical sensitivity, analytical specificity, and any other performance characteristics required for test performance such as clinical performance in relevant samples. The lab performing the verification or validation must be certified under CLIA and, ideally, accredited through additional quality schemes such as that run by College of American Pathologists (CAP) or have International Organization for Standardization (ISO) certification. If the testing lab has not been worked with previously, then a visit to the lab to review the technical facilities as well as operational processes such as sample storage, standard operating procedures (SOPs), and staff training is recommended.

The exact experiments vary by assay type (Table 5.2.3) and, most importantly, through the determination of the product requirements. It must be remembered that it is the whole process from sample collection to result that needs to be validated. For a NGS assay used for detecting mutations in DNA, this would include sample collection, DNA extraction, NGS data quality, and software identifying the disease-associated mutation (30). The availability of samples is a common challenge for these validation studies and, therefore, manufactured standards or cell line samples can be utilized for some of the studies (18), though end-to-end reproducibility with the clinical sample should be demonstrated. Documents from the FDA such as guidance, approval information, discussion papers and statements including those around the clinical use and validation of NGS assay (4–7) and 510k/PMA decisions are a good starting point for the design of these studies. Relevant Clinical Laboratory Standards Initiative (CLSI) guidelines including MM01 Molecular Methods for Clinical Genetics and Oncology Testing, 3rd Edition (8) are a useful reference. A number of analytical validation studies have been published and these provide additional guidance when designing similar studies (20,22,32).

For an assay to be considered for use in a trial, it is expected that some degree of analytical validation of the selected or closely related assay has already been undertaken and, therefore, the report will either have been published or can be requested to enable review against the product requirements. It is likely that some additional studies will need to be undertaken to address key requirements such as performance against specific study tissue types or assessing sensitivity for a specific set of study mutations.

TABLE 5.2.3 Examples of analytical validation studies performed for different assay types

VALIDATION STUDIES	*DNA MUTATION*	*RNA EXPRESSION*	*IHC*
Reproducibility (Accuracy and Precision)	Intra sample, run, inter operator, instrument, reagents & lab. From FFPE, DNA across range of variables, near cut off	Intra, sample, run, inter operator, instrument, reagents, lab. From FFPE, RNA, across range of variables, near cut off	Intra sample, run, inter operator, instrument, reagents, lab. Inter observer (pathologist) From blocks, slides across range of variables, near cut off
Analytical Sensitivity (Limit of Blank, Detection, Quantitation)	No template, range of % tumour, infiltrate, DNA input, macrodissection	No template, range of % tumour, infiltrate, RNA input, macrodissection	Range of target expression levels. Antibody specificity across multiple tissue types
Analytical Specificity (Interfering substances, Assay Carry over)	Primer probe specificity (in silico) & cross reactivity. Effect of necrotic tissue, blood, micro-organisms	Primer probes pecificity (in silico) & cross reactivity. Effect of necrotic tissue, gDNA, blood, micro-organisms	Effect of Ischemia/Fixation
Stability	Sample (block, slide, DNA), reagents (extraction, assay)	Sample (block, slide, RNA), reagents (extraction, assay)	Sample (block, cut section), reagents (antibody)
Cutoff determination and interpretation of result	Description of studies and method of determining and interpreting cutoff, e.g., % allelic freq of mutation	Description of studies and method of determining and interpreting cutoff, e.g., expression above background	Description of studies and method of determining and interpreting cutoff, e.g., target expression level
Control strategy	Positive, negative controls, sample sufficiency	Positive, negative controls, sample sufficiency	Control cell line validation, positive controls
Software validation	Confirmation of software validation	Confirmation of software validation	Not required if manual process
Independent Method	e.g., if PCR use NGS assay across clinical study cohort	e.g., if microarray use RT-PCR or NGS across study cohort	e.g., Western blot/RNA expression levels across study cohort
Clinical Validation (for PMA)	Performance in Clinical study compared to truth	Performance in Clinical study compared to truth	Performance in Clinical study compared to truth

The degree of assay validation for an early stage, non-pivotal trial should be appropriate and fit for purpose but is not as extensive as that required ahead of a pivotal CDx study.

Local and Central Testing

This topic has often been phrased as local *versus* central testing but as molecular diagnostics have become more widely available within clinical laboratories, the question of whether to enable local testing *and* central testing is becoming more common. This is especially true as new technologies, such as highly sensitive cancer mutation panels are being implemented into clinical trials and into the wider health care system. It is clear that the use of a single central facility for the biomarker testing determining patient selection is preferable to minimize variability and maintain quality. However, it would not be in a patient's interest to delay their treatment or require the collection of additional samples without hope of entry into trial. For a local testing lab to be considered it should have suitable accreditation (CLIA/CAP or similar) to perform testing that will inform a patient's treatment. Even when local testing is allowed, it is ideal that samples be collected for retrospective testing using the central lab assay to ensure consistent data on all patients is available in the trial for later data analysis.

An example where a query to use local testing could arise is reviewed here. In this clinical trial, the central test is a highly sensitive assay designed to detect low allelic frequency of a cancer-associated mutation; however, this mutation is also tested for in local clinical laboratories using a number of techniques. In this case local testing may be accepted if a patient has been tested for this specific mutation using a validated assay at either the facility's own clinical laboratory or at an external CLIA lab, such as Foundation Medicine, Personal Genome Diagnostics or similar. Patients who test negative via the local assay could be retested with the more sensitive central test. If a patient has not had any testing at time of being considered for entry into the trial, it is preferable they are tested using the central assay. An illustration of the approach to accommodate local testing is the My Pathway study (NCT02091141, 9) sponsored by Genentech, where local testing was allowed providing a CLIA-certified lab performed testing on either a blood biopsy and/or tissue. The availability of an archival or new pretreatment tissue sample is required if molecular testing was not performed by Foundation Medicine.

INCORPORATING ASSAY INTO TRIAL

The Diagnostic Testing Team

The diagnostic scientist is typically responsible for many of the activities previously discussed including developing and implementing diagnostic plans, the assessment, selection, analytical validation and deployment of an assay suitable for clinical trial use, managing the relationship with the diagnostic provider and lab, oversight of testing quality and composing and review patient identification and biomarker sections of clinical protocols and regulatory documents. However, the incorporation of a diagnostic test into a clinical trial is a complex and multistep process. The execution of this process requires a cross-functional team with each team member supplying not only their specialist knowledge but also working seamlessly together to deliver both scientific and operational success. A key aspect is the integration of team members from partner companies including the assay developer and the contract research organizations (CROs) who are often responsible for aspects of clinical operations, sample collection and data collection. With this in mind, it is vital to establish a close and open working relationship within the whole and often virtual team and important to define the roles and responsibilities within each process that is being undertaken. The exact team will vary by institution and situation, but an example is illustrated in Figure 5.2.1. This is a large group and therefore it is recommended to have an internal core team, which for the development of assays within early stage clinical trials often comprises the diagnostic scientist, translational scientist and clinical operations lead. Initially, the core team focuses on diagnostic test selection, development and testing of the assay and then transfers to a more operational function including sample management, monitoring performance of the testing and data management. The core team works with the wider group through both focused meetings, such as with the diagnostic assay partner; and ad hoc discussions such as the regulatory assessment described below. It is recommended that the core team maintain a project timeline, action log and risk log to ensure these overlapping and interdependent activities are tracked. The core team updates the main program team to gain approval for key strategic activities ensuring they are in line with overall program goals and timelines.

Regulatory Review

Throughout the process of assay evaluation, it has been important to keep in mind the overall risk–benefit to the patient. A formal assessment of this risk–benefit profile is required by the FDA prior to using an assay for patient selection. When the intended use of an assay is diagnosis of disease and, in this case, the selection of patients for treatment in a clinical trial, then the assay is considered an IVD device and is regulated by the FDA. If the assay has not been cleared for the specific intended use, it is considered an investigational device. As discussed previously, the assay must be validated within in a CLIA or equivalent laboratory. Additionally, prior to use in a trial the risks and benefits to the patient brought about from using this investigational device to make treatment decisions must be considered. The process with the FDA is to request an investigational device exemption (IDE) to permit clinical investigation of the IVD. Studies subject to the IDE regulation can be classified into either significant risk (SR) or nonsignificant risk (NSR). For non-pivotal, early stage trials, if an assay is used to determine a patient's treatment then this risk assessment needs to be undertaken. Any assay

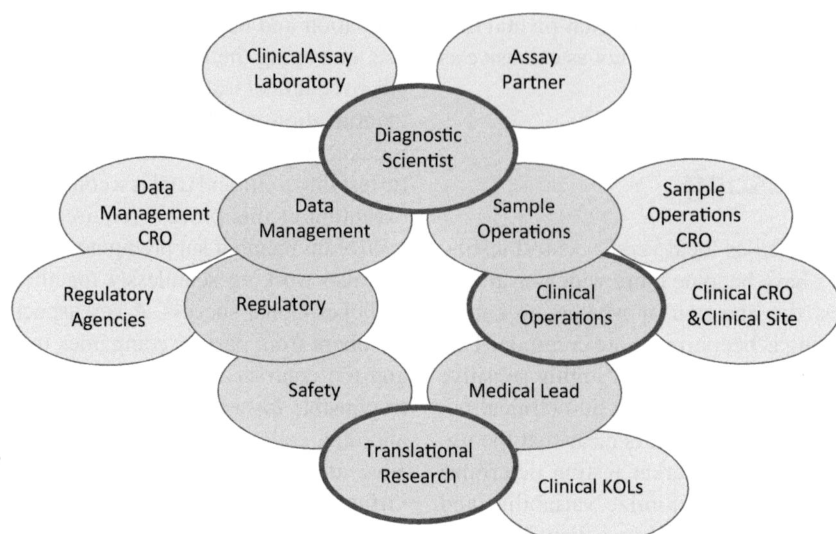

FIGURE 5.2.1 The Diagnostic Development Team. The members of a typical diagnostic development team are illustrated here. Those shown in blue are internal team members, a heavy blue outline are Core team members, those in green are external collaborators and members of selected teams.

used in a pivotal, CDx trial then it is automatically considered SR and an IDE is required.

To aid this IDE risk assessment process, a series of questions have been presented for consideration by the FDA when using diagnostics for patient selection (23). The questions are as follows: (a) Will use of the investigational test results lead to some trial subjects foregoing or delaying a treatment that is known to be effective? (b) Will use of the investigational test results expose trial subjects to safety risks (e.g., adverse events from the experimental therapy) that (in some "net" sense) exceed the risks encountered with control therapies or non-trial standard of care? (c) Is it likely, based on *a priori* information about the investigational therapy, that incorrect test results would degrade the safety or efficacy of subjects' treatment? (d) Does specimen acquisition, done for investigational testing and outside the standard of care, require an invasive sampling procedure that presents significant risk?

The assessment of risk should be undertaken and signed off by the diagnostic development team and the conclusion communicated for approval to the program team. Further comments from institutional review boards (IRBs) or ethics committees (ECs) are also incorporated. If it is determined that the use of the assay meets the regulatory criteria for SR then an IDE should be filed to support its use in the trial. If it is judged that there is a nonsignificant risk to the patient then the risk assessment is documented and one can proceed without an IDE, with appropriate summary information filed into the investigational new drug application (IND). If no comment is received during review of the IND then the study can proceed as nonsignificant risk and the abbreviated requirements followed (23). This process has now been described in a draft guidance entitled "Investigational IVDs used in clinical investigations of therapeutic products" released by the FDA in December 2017 (10).

All clinical trials on human subjects are expected to follow Good Clinical Practice (GCP) guidelines. GCP is an international quality standard for the design, conduct, recording and reporting of human trials ensuring the well-being of patients is protected and that research data are reliable (16). The guideline was developed with consideration of the current good clinical practices of the European Union, Japan, and the United States as well as those of Australia, Canada, the Nordic countries and the World Health Organization (WHO). From a molecular testing perspective, it is important to ensure that the description of the biomarker driven studies within the clinical protocol and subject informed consent provide information for the IRB/EC to assess any potential risk to the trial subjects and to ensure the informed consent accurately describes the sample collection, the data that will be generated, how it will be used and how that data will be protected.

Assay and Testing Readiness

The assay and vendor have now been identified based upon the agreed requirements. Below are reviewed activities that are recommended to be undertaken to ensure successful incorporation of the assay into the early stage clinical trial. A statement of work (SOW) should be agreed with the diagnostic testing partner, including key deliverables, timelines and costs to cover assay development, validation, testing process and data transfer. Any relevant licenses, such as CLIA licenses or CAP accreditation, should be referenced in this document. As ever, the process of agreeing a SOW takes time (months) and not every detail is known as the time to sign arrives. Therefore, the SOW can be supplemented with a roles and responsibilities document that further describe activities such as sample testing, data transfer and quality control (QC) monitoring. Additional study paperwork relating to use of assay for patient selection needs to be completed, including biomarker sections within the clinical protocol and reviewing the sample collection processes within the study manual. If the diagnostic testing method is particularly complicated or a novel technology is being utilized, consider preparing an additional document clearly describing the patient selection process. Undertaking a dry run of the testing process is a useful exercise to undercover any technical or operational challenges. The extent to which depends upon the complexity of the testing method and could range from evaluating the entire process from sending samples to the clinical lab, receiving a clinical report, and determining patient eligibility to testing only specific aspects such as data reporting. It is a requirement of the IDE to monitor performance of diagnostic testing; therefore, this monitoring process needs to be agreed and finalized with the testing lab. As an example, for an NGS assay used to detect mutations in DNA monitoring can include review of sample extraction QC pass rates, assay QC pass rates, assay turn-around time, mutation detection rates compared against predicted values. Monitoring should occur on a regular basis depending on rate of accrual, bearing in mind that, in the early rounds of monitoring, the data can be outside of expected values due to low sample numbers. Even with the best planning, the first samples can always identify a problem in the testing process; therefore, it is worth keeping a close watch on the initial set of testing runs all the way from sample collection through to the generation of the data determining the patient's eligibility.

Advances in Molecular Testing and Clinical Trial Design

A common theme throughout this section is that our increased understanding of DNA alterations present in cancer subtypes has resulted in the development of multiple targeted therapeutics, used singly and in combination, as well as a greater need for molecular testing. A challenge is how to efficiently and ethically identify and recruit patients using molecular testing into the huge number of clinical trials needed to test these therapeutics. One such approach is to design trials where several trial arms are created to study multiple therapeutics or diseases within one master protocol. Different types of master protocols exist, including umbrella, basket, platform and $n = 1$ trials (Table 5.2.4), and these are described in detail by Woodcock and LaVange (34).

This concept tackles many of the issues raised in this chapter as illustrated by the set-up of the BISCAY phase Ib umbrella trial (NCT02546661, 11, 27) The initial aim was to investigate AZD4547, an FGFR inhibitor in metastatic bladder cancer patients whose tumor DNA contained an activating FGFR mutation or gene fusion. However, these mutations were found to have a prevalence of just 10% in this bladder cancer population. This challenge of low prevalence will be faced more and more as diseases are further subdivided into smaller groups by their molecular signature. In a classical trial, patients would be screened for the specific FGFR mutations and only those testing positive would enter the trial. In this case, only 10% of patients

TABLE 5.2.4 Master protocol types

TERM	*DEFINITION*
Umbrella trial	Testing of multiple targeted therapeutics within a single disease
Basket trial	Testing a single targeted therapeutic across multiple diseases
Platform trial	A flexible design, incorporating aspects of the umbrella trial, where multiple targeted therapeutics are tested in a single disease following a master protocol enabling the updates to the trial such as entry and exit of treatment arms
N = 1 trials	Study where multiple sources are utilised in a pre-determined manner to select and refine a patient's treatment. Results are compared and analysed across multiple patients

screened will have the potential to enter the trial: This is not ideal for either the patient or the sponsor. In this case, late stage patients have their hopes raised of receiving a useful treatment, have potentially undergone additional surgery for sample collection and delayed from entering further trials for treatment they desperately need. For the sponsor, there is the delay in recruitment and the additional cost of screening 10 patients for every one that enters further screening for the trial. Therefore, a key requirement of this study was to design a trial that balanced both the needs of the patient and the sponsor in testing the safety and tolerability of AZD4547 in metastatic bladder cancer. To address these challenges the AZD4547 study was incorporated into an umbrella trial design and which became the BISCAY master protocol with the aim that every patient who undergoes biomarker testing and passes further screening can enter further screening for the trial. Clinical requirements also had to be met. Following the recent positive results with PD-1/PD-L1 therapeutics and subsequent approval of atezolizumab in bladder cancer (17), the trial was designed to have at its backbone the PD-L1 immunotherapeutic durvalumab. Study arms were designed with durvalumab given in combination with AZD4547 (also included AZD45474 monotherapy), olaparib and AZD1775. Patients were assigned to these arms based upon specific gene mutations in their tumor relevant to the molecules under investigation. The prevalence of the arm-specific mutations was expected to cover about 78% of the testing population. To ensure all patients tested have a chance to enter the trial, patients whose tumors do not contain study specific alterations entered a durvalumab-only arm. The master protocol allows for a flexible approach enabling arms to be added, delayed or stopped as existing arms are filled or new information becomes available. This approach is being utilized with two further arms added; a mutation-selected arm of durvalumab with vistusertib and an unselected arm of durvalumab with AZD9150 (11). The assay requirements were developed to address these clinical and technical needs. The assay has to utilize FFPE samples previously collected from the bladder or from metastatic sites and with variable tumor content. It needs to accurately detect all 20+ DNA alterations and have the potential to detect additional alterations required by future trial arms. The company supplying the assay and the clinical lab would have experience in processing samples from worldwide sites in its CLIA approved lab and have or develop a reporting system to simplify the process from result to decision on trial arm. The FoundationOne NGS assay from Foundation Medicine was selected to determine which if any of the study-selected mutations are present and patients assigned to modules according to tumor genomic profile utilizing a robust algorithm (Figure 5.2.2).

The wider use of NGS panels has opened up further opportunities to streamline patient testing. Following in the same principle of maximizing the use of molecular testing is the Precision Enrolment program from Foundation Medicine (13). Patients are screened at the request of their physician using Foundation Medicine assays to generate molecular profiles. Patient profiles are matched to requirements of active biomarker driven clinical trials and the patient's physician informed. The physician can then decide whether to contact the sponsor and potential entry of the patient into the trial.

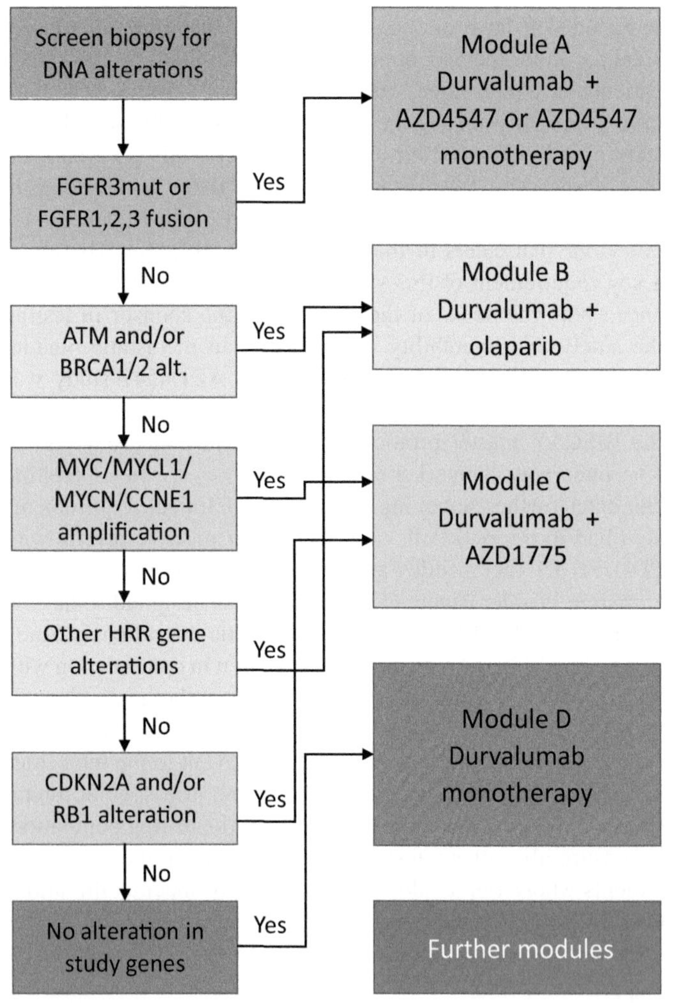

FIGURE 5.2.2 BISCAY umbrella trial patient selection algorithm. A subject's tissue biopsy is collected and analyzed using the FoundationOne multi-gene NGS assay. If alterations associated with the trial are identified the algorithm used to determine which arm of the trial a subject enters. If no trial specific alterations are found the subject is eligible for a nonselected arm. In this way, multiple therapeutics are investigated and ensuring an arm of the trial can be identified for every subject tested. (Modified from Ref. [27])

Innovation is not just occurring in the development of multi-gene panel tests. An important clinical requirement is minimizing the time taken from sample collection to result to enable patients to enter the trial and receive treatment as soon as possible. This can be accomplished through choice of sample and the assay technology. For example, in the study of AZD5363 in advanced cancer, a test system utilizing a plasma sample for ctDNA testing and BEAMing assay from Sysmex Inostics, was utilized in an attempt to avoid waiting for banked samples or a new biopsy to be tested (18). Simpler, rapid platforms are now available for use in clinical trials and being developed as CDxs such as the Idylla platform from Biocartis. Assays for KRAS and NRAS/BRAF are being developed as CDxs with Amgen and have a turnaround time of 150 min, with only 2 min of hands-on time (14). It is likely that a combination testing approach will become more common with local, rapid, targeted assessment of the most likely disease associated DNA alteration while or ahead of testing using a multi-gene NGS assay for broader screening of disease-associated alterations.

The FDA is also working to keep pace with the innovations in molecular testing in line with enabling their stated goal to "… empower patients, researchers, and providers to work together toward development of individualized care." In this respect, the FDA has had a focused effort on enabling the use of NGS. They have simplified the approval of the sequencer itself (26), undertaken a review of how to tackle the challenges of analytical validation and data analysis (4–7,12) leading to the marketing authorization of three NGS-based cancer profiling tools and the introduction of a three-tiered approach to the level of clinical and analytical evidence required depending on the claims and usage of the markers (12). Additionally, a further draft guidance document was released by the FDA in December 2017 to address the challenge of developing targeted therapies in low frequency molecular subsets of disease (15), as encountered in the BISCAY study described earlier. The guidance proposes the evidence required to create an inclusion criteria comprised of a group of low frequency molecular alterations and that, if the trial is successful, the FDA will consider approval across this broad inclusion criteria even if evidence is sparse or missing for individual alterations. This simplification and clarification of the approval process for multi-gene panel NGS test will support their expanded use in early stage trials and future development of CDxs.

CONCLUSION

There is a wealth of molecular diagnostic technology available; however, it is key that the clinical question and technical needs are understood and defined to design an appropriate, high-quality testing process that meets both the requirements of the patient and sponsor to push the boundaries and deliver on the promise of molecular targeted therapeutics.

REFERENCES

1. http://eur-lex.europa.eu/legal-content/EN/TXT/?uri=uriserv:OJ.L_.2017.117.01.0176.01.ENG&toc=OJ:L:2017:117:TOC.
2. http://www.ema.europa.eu/docs/en_GB/document_library/Scientific_guideline/2017/07/WC500232420.pdf.
3. https://www.gpo.gov/fdsys/pkg/CFR-2011-title42-vol5/pdf/CFR-2011-title42-vol5-sec493-1254.pdf.
4. https://www.fda.gov/downloads/MedicalDevices/NewsEvents/WorkshopsConferences/UCM427869.pdf.
5. https://www.fda.gov/downloads/medicaldevices/newsevents/workshopsconferences/ucm523200.pdf.
6. https://www.fda.gov/downloads/MedicalDevices/DeviceRegulationandGuidance/GuidanceDocuments/UCM509838.pdf.
7. https://www.fda.gov/downloads/medicaldevices/newsevents/workshopsconferences/ucm488271.pdf.
8. https://clsi.org/standards/products/molecular-methods/documents/mm01/
9. https://clinicaltrials.gov/ct2/show/NCT02091141.
10. https://www.fda.gov/downloads/Drugs/GuidanceCompliance RegulatoryInformation/Guidances/UCM588884.pdf.
11. https://clinicaltrials.gov/show/NCT02546661.
12. https://www.fda.gov/downloads/medicaldevices/productsand medicalprocedures/invitrodiagnostics/ucm584603.pdf.
13. https://www.foundationmedicine.com/insights-and-trials/foundation-smarttrials#enrollment.
14. https://www.fiercebiotech.com/medtech/amgen-biocartis-sign-ras-companion-diagnostic-deal.
15. https://www.fda.gov/ucm/groups/fdagov-public/@fdagov-meddev-gen/documents/document/ucm589083.pdf.
16. http://www.ema.europa.eu/docs/en_GB/document_library/Scientific_guideline/2009/09/WC500002874.pdf.
17. Bellmunt J, Powles T, Vogelzang NJ, A review on the evolution of PD-1/PD-L1 immunotherapy for bladder cancer: The future is now. *Cancer Treatment Reviews* 2017 58–67.
18. de Bruin EC, Whiteley JL, Corcoran C, Kirk PM, Fox JC, Armisen J, Lindemann JPO, Schiavon G, Ambrose HJ, Kohlmann A. Accurate detection of low prevalence AKT1 E17K mutation in tissue or plasma from advanced cancer patients. *PLOS One* 2017 12(5) e0175779.
19. Enzmann H, Meyer R, Broich K. The new EU regulation on in vitro diagnostics: Potential issues at the interface of medicines and companion diagnostics. *Future Medicine* 2016 10 (12).
20. Frampton GM, Fichtenholtz A, Otto GA, Wang K, Downing SR, He J, Schnall-Levin M et al., Development and validation of a clinical cancer genomic profiling test based on a massively parallel DNA sequencing. *Nat Biotechnol* 2013 31(11) 1023–1031.
21. Jenkins S, Yang JC, Ramalingam SS, Yu K, Patel S, Weston S, Hodge R et al., Plasma ctDNA analysis for detection of the EGFR T790M mutation in patients with advanced non-small cell lung cancer. *J Thorac Oncol* 2017 12(7) 1061–1070.
22. Lanman RB, Mortimer SA, Zill OA, Sebisanovic D, Lopez R, Collisson EA, Divers SG et al., Analytical and clinical validation of a digital sequencing panel for quantitative, highly accurate evaluation of cell free circulating tumour. *DNA PLOS One* 2015 10(10) e0140712.
23. Litwack, ED, "Investigational device exemption" (2014). 2014 Personalized Medicine? Using sequencing (and other assays) in clinical trials: FDA rules and regulations. Paper 2. http://digitalcommons.wustl.edu/hrpoconf_personalizedmed_2014/2.
24. Marton MJ, Weiner R. Practical guidance for implementing predictive biomarkers into early phase clinical studies. *Biomed Res Int* 2013 2013 891391.
25. Mathieson W, Marcon N, Antunes L, Ashford DA, Betsou F, Frasquilho SG, Kofanova OA et al., A critical evaluation of the PAXgene tissue fixation system: Morphology, immunohistochemistry, molecular biology and proteomics. *Am J Clin Pathol* 2016 146(1) 25–40.
26. Pant S, Weiner R, Marton MJ. Navigating the rapids: The development of regulated next-generation sequencing-based clinical trial assays and companion diagnostics. *Front Oncol* 2014 4 78.
27. Powles T, Kilgour E, Mather R, Galer A, Arkenau HT, Farnsworth A, Wilde J, Ratnayake J, Landers D. BISCAY, a phase 1b, biomarker-directed multidrug umbrella study in patients with metastatic bladder cancer. *J Clin Oncol* 2016 34 (suppl; abstr TPS4577).
28. Sherwood JL, Corcoran C, Brown H, Sharpe AD, Musilova M, Kohlmann A. Optimized pre-analytical methods improve KRAS mutation detection in circulating tumour DNA (ctDNA) from patients with non-small cell lung cancer (NSCLC). *PLOS One* 2016 11(2) e0150197.
29. Sherwood JL, Brown H, Rettino A, Schreieck A, Clark G, Claes B, Agrawal B et al., Key differences between 13 KRAS mutation detection technologies and their relevance for clinical practice. *ESMO Open* 2017 2(4) e000235.
30. Strom SP. Current practices and guidelines for clinical next-generation sequencing oncology testing. *Cancer Biol Med.* 2016 13(1) 3–11.
31. Thress KS, Brant R, Carr TH, Dearden S, Jenkins S, Brown H, Hammett T, Cantarini M, Barrett JC, EGFR mutation detection in ctDNA from NSCLC patient plasma: A cross platform comparison of leading technologies to support the clinical development of AZD9291. *Lung Cancer* 2015 90(3) 509–515.
32. Walsh PS, Wilde JI, Tom EY, Reynolds JD, Chen DC, Chudova DI, Pagan M et al., Analytical performance verification of a molecular diagnostic for cytology-indeterminate thyroid nodules. *J Clin Endocrinol Metab* 2012 97(12) E2297–E2306.
33. Wan JCM, Massie C, Garcia-Corbacho J, Mouliere F, Brenton JD, Caldas C, Pacey S, Baird R, Rosenfeld N. Liquid biopsies come of age: Towards implementation of circulating tumour DNA. *Nat Rev Cancer* 2017 17 223–238.
34. Woodcock J, LaVange LM. Master protocols to study multiple therapies, multiple diseases or both. *N Engl J Med* 2017 377(1) 62–70.

Ectopic Gene Deregulations and Chromosome Conformations 5.3

Integrating Novel Molecular Testing into Clinical Applications, from Leukemias to Gliomas

Francis Hector Grand, Matthew Salter, Ewan Hunter, and Alexandre Akoulitchev

Contents

AN OVERVIEW OF MOLECULAR TESTING ENTERING CLINICAL PRACTICE: MOLECULAR DIAGNOSTICS

The paradigms for molecular testing developed in the last century and today are more relevant than ever in clinical practice. Increasing volumes of sophisticated molecular diagnostic data are generated by high-throughput methods (Vogelstein et al., 2013). Diagnostic approaches to chromatin biology tend to reduce data complexity and sample processing time while adding essential clinical information not readily available with other diagnostic techniques.

The origins of somatic and constitutional molecular diagnostics are found in the identification of specific chromosomal disorders (Nowell and Hungerford, 1960). It is a sobering thought that even today, with an expanding spectrum of molecular diagnostic applications, much of clinical genetic diagnostic practice in the United Kingdom still involves a microscope and karyotype analysis (Norbury and Cresswell, 2017).

The overview of molecular diagnostics will, in the context of this chapter, be described with concise examples in leukemia and glioma and illustrate how the lessons from the development of molecular testing in the last century are now more relevant than ever in the field of cancer diagnosis.

THE PARADIGM OF CHRONIC MYELOID LEUKEMIA AND IMPLICATIONS FOR ONCOLOGY DIAGNOSTICS

The identification of the Philadelphia chromosome was a landmark in genetic diagnostics. It became the first recurrent cytogenetic diagnostic event used for the identification of a malignancy (Nowell and Hungerford, 1960). The drive to identify recurrent chromosomal abnormalities in malignancy became a research imperative through the 1980s and 1990s, and cytogenetic diagnosis of Philadelphia chromosome (Ph)-positive chronic myeloid leukemia (now classed as a myeloproliferative neoplasm [Vardiman et al., 2009, 2010]) became routine in the hematological research laboratories.

Following from the Ph chromosome's identification, the chromosomal break points were defined (Rowley et al., 1973), and the genes—*BCR* and *ABL*, disrupted in the t(9;22) translocation—were identified (Heisterkamp et al., 1985; Grosveld et al., 1987). Importantly, *BCR* and *ABL* were shown to be fused in-frame as a result of the translocation. In a revolutionary research study, the *BCR-ABL* fusion was transfected into mice and the construct was functionally demonstrated as driving the resultant leukemia (Scott et al., 1991).

In a brilliant study published by Druker et al. (1996) an ATP mimetic (STI 571) produced by Novartis was shown to selectively inhibit the BCR-ABL protein *in vitro*, causing the selective death of *BCR-ABL*–positive colony forming units granulocyte-macrophage (CFU-GM) (Druker et al., 1996).

When this study was first published, the data was considered controversial, but imatinib (Glivec) generated sustained cytogenetic responses in patients with chronic myeloid leukemia (O'Brien et al., 2003; O'Brien and Deininger, 2003) and a decade later, the survival advantages been proved beyond doubt (Hochhaus et al., 2017).

CURRENT MOLECULAR TESTING FOR EOSINOPHILIC LEUKEMIA: RT-PCR

Imatinib was also known to show activity against multiple tyrosine kinases including platelet derived growth factor receptor beta (PDFRFB) (Apperley et al., 2002) and platelet derived growth factor receptor alpha (PDGFRA) (Cools et al., 2003). The success of the clinical trials in chronic myeloid leukemia provided the impetus for the identification of a pharmacogenetic strategy (Druker, 2004). An early study of patients with eosinophillic leukemia syndromes identified patients with PDGFRA or PDGFRB translocations that responded to Glivec (Apperley et al., 2002; Cools et al., 2003; Grand et al., 2004; Deininger et al., 2005; David et al., 2007).

Cools et al. (2003), in conjunction with other teams, identified the *FIP1L1-PDGFRA* fusion gene resulting from a cryptic interstial deletion at 4q12 as a driver mutation in a proportion of patients with idiopathic hyper-eosinophilic syndrome.

As a result, FIP1L1-PDGFRA–positive eosinophilic leukemia patients were treated with imatinib, improving survival outcomes in a leukemia with poor survival prognosis (Apperley et al., 2002; Cools et al., 2003).

Imatinib remains an excellent first-line therapy based on the molecular diagnosis of the FIP1L1-PDGFRA fusion gene. Patients with imatinib-sensitive lesions are prescribed imatinib on the UK National Health Service if they were confirmed as having an imatinib-sensitive lesion.

The diagnosis of eosinophilic leukemia requires a persistent state of unexplained eosinophilia (>1,500 mm^3) lasting for longer than 6 months with evidence of organ dysfunction due to eosinophilic tissue infiltration (Valent et al., 2012). A missed diagnosis is potentially fatal and the presence of the FIP1L1-PDGFRA fusion gene predicts a favorable response to tyrosine kinase inhibitors.

A favored approach to the detection of the fusion gene include reverse transcription (RT)-PCR or fluorescence *in-situ* hybridization (FISH).

TECHNICAL CHALLENGES FOR EXISTING MOLECULAR APPROACHES

The identification of ectopic PDGFRA expression is critical for the patient's clinical management. In clinical diagnostic practice, the unambiguous detection of FIP1L1-PDGFRA is complicated by several factors.

First, eosinophils can be difficult to use as targets for FIP1L1 PDGFRA FISH, with a small, but persistent population of FIP1L1-PDGFRA–positive cells beneath the sensitivity of detection for the technique.

Second, exclusion of FIP1L1-PDGFRA is also important because some patients may benefit from alternative therapies such as mepolizumab (Roufosse et al., 2013). Application of RT-PCR is complicated by the diversity of break points within FIP1L1 and the size of genomic region deletion region (Figure 5.3.1). The situation is further complicated for RT-PCR by the alternative splicing of FIP1L1 mRNA and the variable use of cryptic splice sites in the fusion gene. This problem is also apparent with genomic breakpoints that are mostly confined to the exon 12 of PDGFRA (Score et al., 2009).

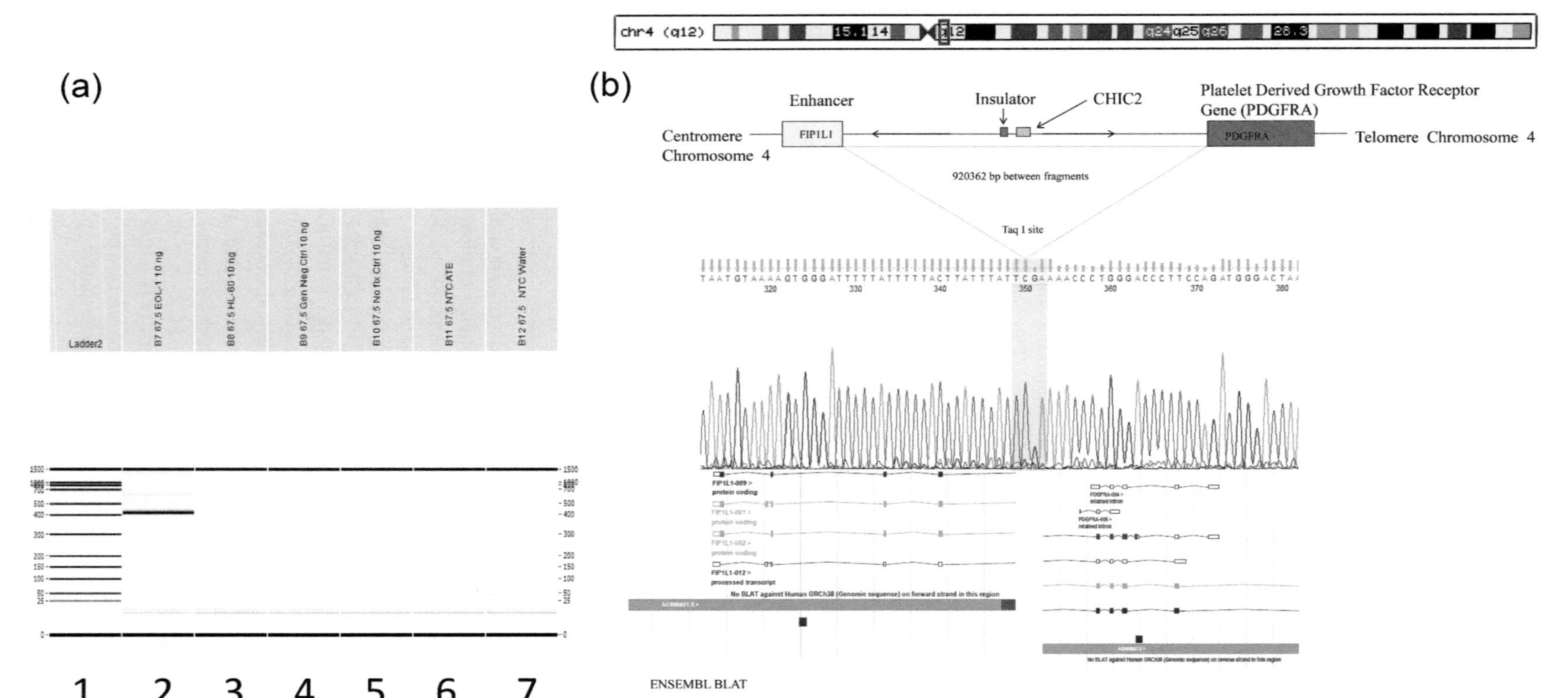

FIGURE 5.3.1 Detection of chromosome conformation with ectopic deregulation of PDGFRA oncogene by EpiSwitch™. (a) Caliper LabChip image of single step amplification. Lane marker (1); detection of chromosome conformation in the EOL-1 cell line (single step product), verified by sequencing; size—462 bp (2); Controls (3–7). (b) The insulator is hypermethylated in glioblastoma by IDH1 R132H mutations or deleted in EOL1. The chromosome 4 ideogram and interaction map the region between FIP1L1 and PDGFRA. The two 3C fragments are ligated through Taq I (TCGA). Below the sequence trace is the ENSEMBL BLAT mapping data (sequence homologies in red).

FIP1L1-PDGFRA transcripts can often only be detected by sensitive nested RT-PCR assays, even in previously untreated patients and, if the sample has been sent from distant laboratories, problems with RNA quality are likely to be a serious issue with regard to the sensitivity of detection.

Nested RT-PCR has a greatly enhanced risk of contamination resulting in false-positive results, potentially leading to misdiagnosis and inappropriate treatment.

Although real time quantitative-PCR (RQ-PCR) can detect FIP1L1-PDGFRA in many cases, the diversity of mRNA fusions makes the design of workable primer/probe sets that can detect all variants very difficult (Score et al., 2009; Walz et al., 2009).

NEW MOLECULAR APPROACHES: CHROMOSOME CONFORMATION SIGNATURES

Todays integrated data sets and public databases offer extensive, often genome-wide, exposure to readouts of various molecular regulatory modalities: from rare mutations and genetic variants for single nucleotide polymorphisms (SNPs), expression quantitative trait loci (QTLs), protein quantitative trait loci (pQTLs) and histone quantitative trait loci (hQTLs), to transcriptional profiles of gene expression and miRNAs, maps of long and short range of chromosome interactions, and continuous readouts of epigenetic patterns for DNA methylation and histone modifications, all as part of the integrated multi-omic approach and analysis.

An important level of the epigenetic regulatory framework of the three-dimensional (3D) genome architecture is a fundamental functional link between phenotypic features and the multitude of contributing regulatory modalities (Tordini et al., 2016). Individual conditional chromosome conformations represent the smallest units of integrated multi-omic regulation around the specific loci of interest. The integrative model of 3D genome architecture, based on chromosome conformation profiles, offers effective stratification modalities with functional links to dysregulated molecular mechanisms and pathological phenotypes at the same time (Crutchley et al., 2010; Tordini et al., 2016). Use of chromosome conformations as a functional readout could make a significant contribution to the omics-based molecular guidelines for disease diagnosis and treatments, improving the efficacy of decision trees currently used in clinic.

Regarding latest developments in oncology, the understanding of the epigenetics of tumors is rapidly expanding. The unfolding overview in chromatin biology implicates the epigenome, and the 3D topology of chromatin in cancer progression (Corces and Corces, 2016). The human genome is thought to organize into topological domains that represent discrete structural and regulatory units (Bickmore, 2013; Chambers et al., 2013), while high-order chromatin architecture shapes the landscape of chromosomal alterations in cancer (Fudenberg et al., 2011).

The 3D architecture of genomes has been mapped by high-throughput techniques (Dostie et al., 2006; van der Werken et al., 2012) and the genome is thought to be partitioned into contact domains that are associated with distinct patterns of histone marks. Interestingly, of all continuous epigenetic modalities, changes in H3K27ac and H3K4me1 appear to show better concordance with changes in higher-order chromatin organization (Huang et al., 2015).

The development of chromosome conformation capture (3C, Dekker et al., 2002) and other research-oriented techniques such as 4C, 5C, Hi-C and ChiA-PET, and GAM has greatly improved our perceptions of chromosome folding (Babu and Fullwood, 2015; Beagrie et al., 2017).

Chromatin loops frequently link promoters, terminators, enhancers, and genetic variants in the noncoding parts of the genome. Their formation underlies individual regulatory contributions into functional modulation of genome, expression profile, including timing and amplitudes for activation and repression of distant loci (Rao et al., 2014). Importantly, a switch in, or the formation of, conditional chromatin loops—chromatin conformation signatures—is the primary step in a cascade of genome regulation leading to changes in a phenotype (Christova et al., 2007). The chromatin signatures as a biomarker modality has several well documented advantages against both low-prevalence genetic variant–based readouts and continuous readouts of gene expression, miRNA and DNA methylation platforms. Among important features of chromosome signatures, one could mention their binary nature, high biochemical stability, early presymptomatic timing of formation and the link to the most challenging phenotypes (Crutchley et al., 2010). In principle, as a biomarker modality, chromosome conformations could offer a highly effective means of diagnostics; monitoring and prognostic analysis.

IDH MUTATIONS IN GLIOMA, LESSONS FROM THE PAST AND CHROMOSOME CONFORMATIONS

Extensive data sets of somatic mutations in chromatin regulators have been assembled for many tumor types. In general, these genetic variants are associated with dramatic changes in chromatin organization (Plass et al., 2013). The development of first chromosome conformation capture techniques led to a number of studies attempting to map in the context of mutations of such genes as IDH1, highly relevant to prognosis in gliomas. The historical diagnosis of gliomas is defined by the tumor cell type and their malignant features, while the clinical outcome of the tumor varies widely within the histologic grade. Analysis of DNA copy number aberrations and IDH status in 174 adult supratentorial gliomas of astrocytic or oligodendroglial origin by PCR-based direct sequencing and comparative genomic hybridization confirmed survival differences in the context of cytogenetic aberrations and the IDH mutation status (Hattori et al., 2016).

Gain-of-function mutations in IDH define clinical and prognostic classes of gliomas. Mutant IDH protein produces a new onco-metabolite, 2-hydroxglutarate, which interferes with the TET family of 5′-methycytosine hydroxylases (Dang et al., 2009; Figueroa et al., 2010; Xu et al., 2011; Lu et al., 2012; Cairns et al., 2013).

The CCCTC-binding factor (CTCF) insulator protein contributes to the formation of chromatin loops and boundaries that form the previously described chromatin domains. Ten-eleven translocation (TET) enzymes catalyze a key step in the removal of DNA methylation, and in the context of IDH1 mutations in glioma cells exhibit hypermethylation at cohesion and CTCF-binding sites.

CTCF is perhaps the most characterized insulator–binding protein and is a key genome organizer, thought to play a critical role in the insulation of enhancers and their interacting sequences. CTCF profiles have been determined from the genomes of several cell types, and the role of CTCF in mediating chromatin interactions has been extensively studied in several selected gene regions (Handoko et al., 2011; Katainen et al., 2015).

CTCF binding is known to be methylation-sensitive, and this led to a study of the insulation between the genes FIP1L1 and PDGFRA in the context of IDH1 R132H mutations by Flavahan et al. (2016). The Flavahan study demonstrated that a long-range chromatin interaction with PDGFRA could be identified in glioma patients and cell lines with IDH1 R132H mutations. This data has since been confirmed and evaluated at high resolution by chromosome conformation capture EpiSwitch™ assay, with a readout based on MIQE-compliant quantitative PCR (qPCR) (Grand et al. 2016). "Minimum information for the publication of real-time quantitative PCR experiments" (MIQE) guidelines provide an industry blueprint for good PCR assay design and unambiguous reporting of experimental detail and results (Bustin et al., 2009).

EpiSwitch technology, designed to the technical standards of the clinical grade assay, has now mapped and evaluated in detail the conformational juxtaposition between FIP1L1 and PDGFRA in human glioma-cell lines with and without IDH mutations.

Importantly, the original study went further and explored whether the marker of 3C interaction correlated with the response to tyrosine kinase inhibitors. The authors demonstrated the downregulation of cell growth and proliferation of glioma cells with detectable interaction between FIP1L1 and PDGFRA in response to the addition of tyrosine kinase inhibitors (Flavahan et al., (2016).

The observation of the ectopic deregulation via chromosome conformation and functional loss of insulation between FIP1L1 and PDGFRA is fascinating in the clinical story of glioma itself: In 2007, a phase II clinical trial with imatinib in glioma used direct sequencing of relevant tyrosine kinase genes but failed to identify any activating mutations in the genes (Raymond et al., 2008).

Moreover, PDGFRA is also known as a therapeutic target for tyrosine kinase inhibitors in various leukemias and myeloproliferative neoplasms. This raises the question of similar mechanisms of deregulation for PDGFRA, its detection and the prognostic stratification of patients across a number of oncological indications.

WHY DETECT CHROMOSOME CONFORMATIONS IN TYROSINE KINASE PROTO-ONCOGENES?

EpiSwitch mapping of chromosome interactions identified at high resolution several chromosome conformations as part of the genome architecture around the FIP1L1 and PDGFRA loci. All long-range chromosomal interactions could be detected with EpiSwitch MIQE-compliant qPCR (Bustin et al., 2009) in the acute myeloid leukemia (eosinophilic) cell line EOL-1 using 10 ng of processed genomic template. The results were quantified, normalized and confirmed by direct sequencing of the qPCR product as part of the EpiSwitch assay development and validation.

The relevant FIP1L1-PDGFRA chromosome conformation (Figure 5.3.1) can be detected by EpiSwitch assay reproducibly both in glioblastoma multiforme DBTRG-05MG, glioblastoma astrocytoma U-373 cells and a few biopsies from patients with oligodendroglioma; including samples with IDH1 R132H mutations, as described by Flavahan et al. (2016). 3C FIP1L1 PDGFRA is also detected in the acute myelomonocytic leukemia cell line GDM-1 (with 10 and 20 ng of template) and HL-60 cell lines (at 20 ng of template), which are imatinib-sensitive (Chase et al., 2009; Haass et al., 2012). All detection is performed with concentration-matched negative controls (Table 5.3.1).

The glioblastoma cell line DBTRG-05MG was also positive for an additional long-range interaction between FIP1L1 and PDGFRA, further implicating genome architecture in the regulation of this loci.

Interestingly, in agreement with independent analysis, positions of FIP1L1-PDGFRA chromosome conformations, as mapped at high resolution by EpiSwitch assay, aligned outside of reported chromosomal break points for the tested leukemias and aligned in concordance with peaks of H3K27ac, the chromatin marker associated with genome architecture (Huang et al., 2015). They also aligned with DNase I–hypersensitive sites specific for a glioblastoma cell line (Figure 5.3.2).

As part of EpiSwitch assay readout, all copy numbers of detected conditional chromosome conformations are normalized against an established control—a constitutive chromosome interaction near the MMP1 locus (Figure 5.3.3). Ranked normalized copy number ratio of the tested samples show no detectable FIP1L1-PDGFRA interaction in negative controls and several cell lines and clinical samples. The cohort of imatinib-sensitive

TABLE 5.3.1 Detection of chromosome conformation FIP1L1-PDGFRA in glioma and leukemia samples by EpiSwitch™. The IDH1 SNP rs11554137 is a coding single nucleotide polymorphism found adjacent to the IDH1 R132H mutation, seen at high frequency in AML

INDICATION	*CELL LINE OR CLINICAL SAMPLES*	*IDH1 SNP RS11554137 STATUS*	*IDH1 R132H STATUS*	*FIP1L1-PDGFRA 3C COPIES IN 20 NG POSITIVE/NEGATIVE*	*MMP1 3C COPIES IN 20 NG*
Acute myeloid leukemia (AML)	AML-193	Wild-type	Wild-type	**YES** (0.04)	0.37
Anaplastic oligoastrocytoma	BT-412	Wild-type	R132H[a]	Not used in this assay	—
Glioblastoma multiforme	DBTRG-05MG	Wild-type	Wild-type	**YES** (0.05)	0.35
Acute myeloid leukemia (eosinophillic)	EOL-1	Wild-type	Wild-type	**YES** (0.03)	0
Acute myelomonocytic leukemia	GDM-1	Wild-type	Wild-type	**YES** (1.99)	7.3
AML	HL-60	rs11554137[b]	Wild-type	**YES** (0.18)	1.24
AML	KG-1	rs11554137	Wild-type	NO (0)	4.2
Glioblastoma astrocytoma	U-373	Wild type	Wild-type	YES (2.01)	1.24
Glioblastoma astrocytoma	U87	Wild-type	Wild-type	Not used in this assay	—
Oligodendroglioma	47436	Wild-type	R132H	NO (0)	1.89
Oligodendroglioma	47679	Wild-type	Wild-type	**YES** (1.19)	2
Oligodendroglioma	81600	Wild-type	R132H	NO (0)	0.96
Oligodendroglioma	174694	Wild-type	R132H	**YES** (0.04)	0.61
Oligodendroglioma	184263	Wild-type	Wild-type	**YES** (0.06)	5.19
Oligoastrocytoma	30736	Wild-type	Wild-type	NO (0)	1.12
Adipose biopsy	91013A	Wild-type	Wild-type	NO (0)	6.05
Normal blood	17153	Not analyzed	Not analyzed	NO (0)	4.53
Normal blood	17335	Wild-type	Wild-type	NO (0)	2.18
Normal blood	17353	Wild type	Wild-type	NO (0)	3.09

Source: Ho, P.A. et al., *Blood*, 118, 4561–4566, 2011, and Wang, X.W. et al., *Cancer*, 119, 806–813, 2013.
[a] See also Flavahan et al., 2016. [b]See also Dodémont et al., 2011.

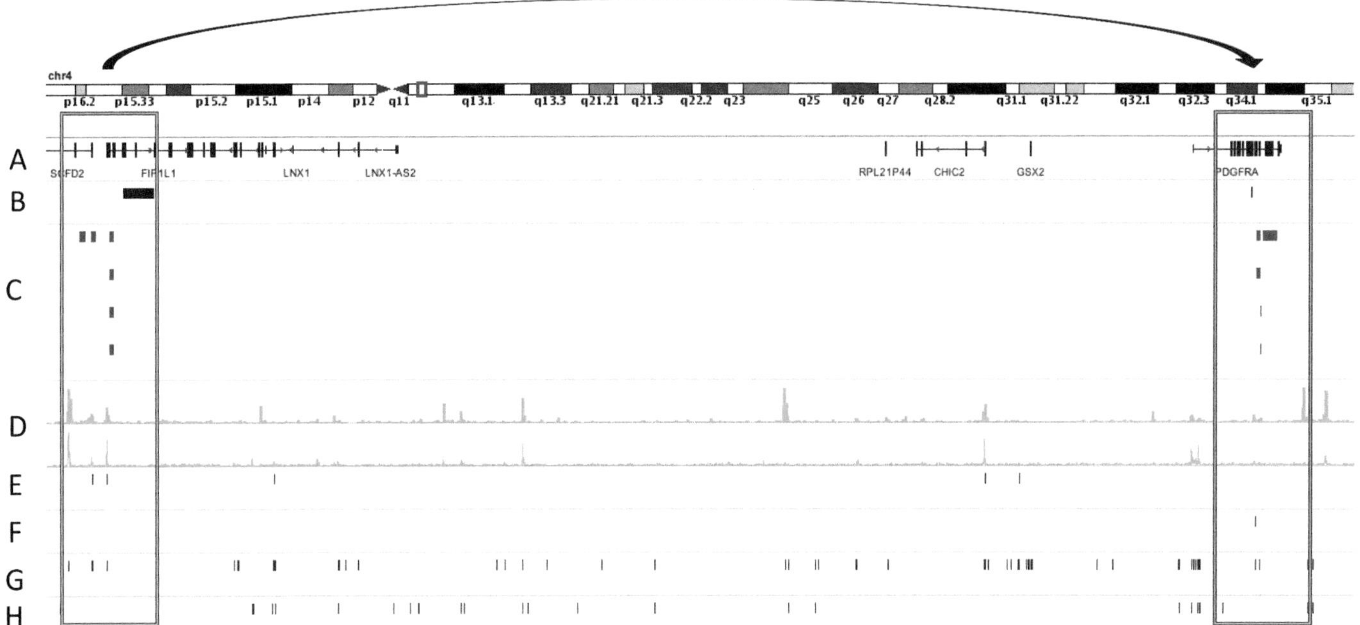

FIGURE 5.3.2 EpiSwitch™ FIP1L1-PDGFRA chromosome conformation signature and its concordance with other regulatory modalities for the PDGFRA locus. IVG genome browser image (hg19) for the concordance between publicly available genetic and epigenetic data with chromosome conformation signature for the FIP1L1 and PDFRA loci. (A) hg19 reference sequence; (B) Chromosome break point regions (FIP1L1 exons 7–10, PDGFRA exon 12); (C) EpiSwitch Chromosome conformation signature sites of interaction; (D) H3K27ac data for two Glioblastoma cell lines; (E) RNA Polymerase II peaks for glioblastoma H54 cell line; (F) site of PDGFRA p. T674I mutation–mutation gives rise to Imatinib resistance; (G) DNase-Seq (DNase I hypersensitive sites sequencing) peaks for glioblastoma H54 cell line; and (H) CTCF protein binding peaks for glioblastoma H54 cell line.

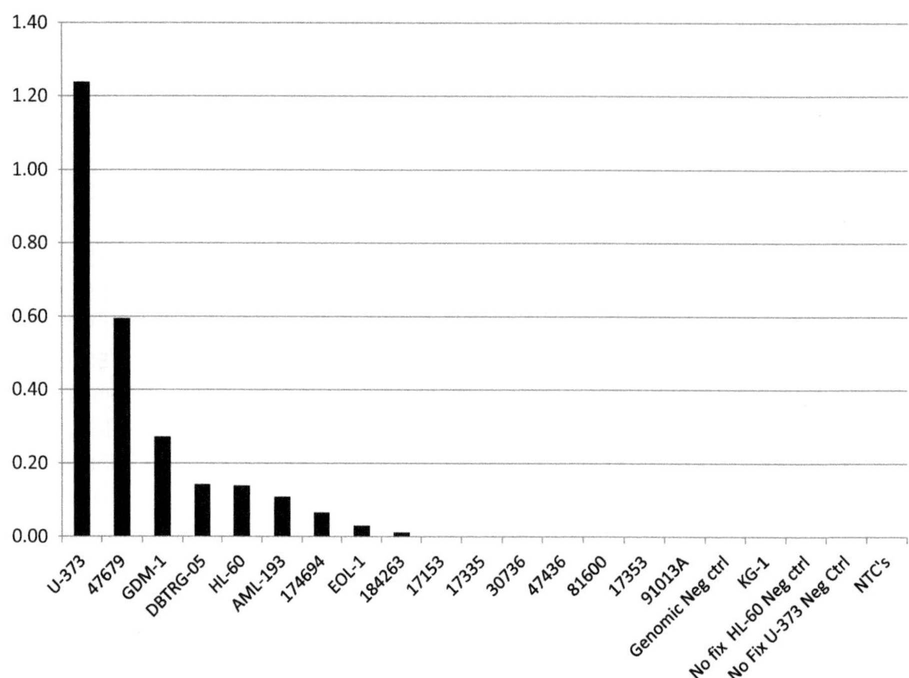

FIGURE 5.3.3 Ranked normalized copy number for the conditional FIP1L1-PDGFRA chromosome conformation quantified by EpiSwitch™ in tested cell lines and clinical samples. As part of EpiSwitch assay readout, all copy numbers of detected conditional chromosome conformations are normalized against an established control—a constitutive chromosome interaction in the vicinity of MMP1 locus (see also Table 5.3.1).

cell lines and individual patient samples showed a qualitative switch in the epigenetic profile of their genome architecture, with the detectable 3C FIP1L1-PDGFRA. The difference in copy numbers could reflect the chromosome conformation dynamics in the context of the particular cell type and tissue.

A reliable and robust detection of the conditional chromosome conformations associated with ectopic deregulation of the PDGFRA proto-oncogene (Flavahan et al., 2016) by clinical grade assay, such as EpiSwitch, potentially offers powerful prognostic tools for selection of patients for imatinib and other tyrosine kinase (TKI) treatment regimes.

Interestingly, glioma patient sample 47679 with no identified IDH1 mutation or IDH1 SNP rs11554137 (Ho et al., 2011; Wang et al., 2013) showed the FIPL1-PDGFRA conditional chromosome conformation.

The mechanism underlying the loss of topological insulation in the CTCF-enhancer regions at 4q12 in patients without IDH1 R132H mutations is unknown and suggests alternative mechanisms, including other mutational events in IDH1, IDH2 (Flavahan et al., 2016; Zadeh et al., 2016), CTCF and/or cohesin (Katainen et al., 2015).

Phase II trials (Raymond et al., 2008) of imatinib in glioma with 112 patients identified five cases with an objective partial response, no somatic mutations of the KIT or platelet-derived growth factor receptor-α or -β were identified by sequencing either gene. This observation is further supported by the study of 109 glioblastoma patients by Bleeker et al., (2014), where no mutations were identified in either KIT or PDGFRA.

As complementary to the sequencing analysis in glioma patients, and RT-PCR in leukemia patients, the ectopic deregulation of PDGFRA locus could be directly assessed by the conditional chromosome conformation signature of PDGFRA. The 3C test may assist with the predictive prognosis for TKI treatment, though this remains to be confirmed in a larger patient series.

CONCLUSIONS AND FUTURE WORK

The presented examples of identified recurrent conformational structural deregulation of FIP1L1 and PDGFRA in a proportion of glioma and leukemia cells and patient samples marks the opening of the next frontier in clinical molecular diagnostic practice. The regulatory alterations in genome architecture, associated with specified pathological conditions could not be identified diagnostically with conventional sequencing approaches. Development and use of technologies and methodologies for the detection of the structural deregulation of chromatin conformations and long-range conditional interactions promises to bring critically important insights to assist clinical decision process and patient substratifications. On the other hand, the clinical peer group could only be satisfied by molecular technologies and technical solutions that could meet diagnostic expectations for high resolution, sensitivity, robustness, fast turnover, low cost and high throughput. The pharmacogenetic lessons of the past are applicable to new chromosome conformation capture techniques.

The early definitions of macro-structural cytogenetic chromosomal changes defined the field of genetics and molecular genetic techniques. The new frontier in describing the micro-structural chromosomal changes associated with long-range interactions is defining the field of epigenetic control over the genome and the links to patient phenotype. It is opening the door for new techniques that will help stratify patients, assist in the clinical decision process and significantly improve treatment outcomes.

REFERENCES

Apperley, J. F., M. Gardembas, J. V. Melo, R. Russell-Jones, B. J. Bain, E. J. Baxter, A. Chase et al., 2002. "Response to imatinib mesylate in patients with chronic myeloproliferative diseases with rearrangements of the platelet-derived growth factor receptor beta." *N Engl J Med* 347 (7):481–487. doi:10.1056/NEJMoa020150.

Babu, D., and M. J. Fullwood. 2015. "3D genome organization in health and disease: Emerging opportunities in cancer translational medicine." *Nucleus* 6 (5):382–393. doi:10.1080/19491034.2015.1106676.

Beagrie, R. A., A. Scialdone, M. Schueler, D. C. Kraemer, M. Chotalia, S. Q. Xie, M. Barbieri et al., 2017. "Complex multi-enhancer contacts captured by genome architecture mapping." *Nature* 543 (7646):519–524. doi:10.1038/nature21411.

Bickmore, W. A. 2013. "The spatial organization of the human genome." *Annu Rev Genomics Hum Genet* 14:67–84. doi:10.1146/annurev-genom-091212-153515.

Bleeker, F. E., S. Lamba, C. Zanon, R. J. Molenaar, T. J. Hulsebos, D. Troost, A. A. van Tilborg et al., 2014. "Mutational profiling of kinases in glioblastoma." *BMC Cancer* 14:718. doi:10.1186/1471-2407-14-718.

Bustin, S. A., V. Benes, J. A. Garson, J. Hellemans, J. Huggett, M. Kubista, R. Mueller et al., 2009. "The MIQE guidelines: Minimum information for publication of quantitative real-time PCR experiments." *Clin Chem* 55 (4):611–622. doi:10.1373/clinchem.2008.112797.

Cairns, R. A., and T. W. Mak. 2013. "Oncogenic isocitrate dehydrogenase mutations: Mechanisms, models, and clinical opportunities." *Cancer Discov* 3 (7):730–741. doi:10.1158/2159-8290.CD-13-0083.

Chambers, E. V., W. A. Bickmore, and C. A. Semple. 2013. "Divergence of mammalian higher order chromatin structure is associated with developmental loci." *PLoS Comput Biol* 9 (4):e1003017. doi:10.1371/journal.pcbi.1003017.

Chase, A., Schultheis, B., Kreil, S., Baxter, J., Hidalgo-Curtis, C., Jones, A., Zhang, L., Grand, F. H., Melo, J. V., and Cross, N. C. 2009 "Imatinib sensitivity as a consequence of a CSF1R-Y571D mutation and CSF1/CSF1R signaling abnormalities in the cell line GDM1. *Leukemia* 23 (2):358–364. doi:10.1038/leu.2008.295.

Christova, R., T. Jones, P. J. Wu, A. Bolzer, A. P. Costa-Pereira, D. Watling, I. M. Kerr, and D. Sheer. 2007. "P-STAT1 mediates higher-order chromatin remodelling of the human MHC in response to IFNgamma." *J Cell Sci* 120 (Pt 18):3262–3270. doi:10.1242/jcs.012328.

Cools, J., D. J. DeAngelo, J. Gotlib, E. H. Stover, R. D. Legare, J. Cortes, J. Kutok et al., 2003. "A tyrosine kinase created by fusion of the PDGFRA and FIP1L1 genes as a therapeutic target of imatinib in idiopathic hypereosinophilic syndrome." *N Engl J Med* 348 (13):1201–1214. doi:10.1056/NEJMoa025217.

Corces, M. R., and V. G. Corces. 2016. "The three-dimensional cancer genome." *Curr Opin Genet Dev* 36:1–7. doi:10.1016/j.gde.2016.01.002.

Crutchley, J. L., X. Q. Wang, M. A. Ferraiuolo, and J. Dostie. 2010. "Chromatin conformation signatures: Ideal human disease biomarkers?" *Biomark Med* 4 (4):611–629. doi:10.2217/bmm.10.68.

Dang, L., D. W. White, S. Gross, B. D. Bennett, M. A. Bittinger, E. M. Driggers, V. R. Fantin et al., 2009. "Cancer-associated IDH1 mutations produce 2-hydroxyglutarate." *Nature* 462 (7274):739–744. doi:10.1038/nature08617.

David, M., N. C. Cross, S. Burgstaller, A. Chase, C. Curtis, R. Dang, M. Gardembas et al., 2007. "Durable responses to imatinib in patients with PDGFRB fusion gene-positive and BCR-ABL-negative chronic myeloproliferative disorders." *Blood* 109 (1):61–64. doi:10.1182/blood-2006-05-024828.

Deininger, M., E. Buchdunger, and B. J. Druker. 2005. "The development of imatinib as a therapeutic agent for chronic myeloid leukemia." *Blood* 105 (7):2640–2653. doi:10.1182/blood-2004-08-3097.

Dekker, J., K. Rippe, M. Dekker, and N. Kleckner. 2002. "Capturing chromosome conformation." *Science* 295 (5558):1306–1311. doi:10.1126/science.1067799.

Dostie, J., T. A. Richmond, R. A. Arnaout, R. R. Selzer, W. L. Lee, T. A. Honan, E. D. Rubio et al., 2006. "Chromosome conformation capture carbon copy (5C): A massively parallel solution for mapping interactions between genomic elements." *Genome Res* 16 (10):1299–1309. doi:10.1101/gr.5571506.

Druker, B. J. 2004. "Imatinib as a paradigm of targeted therapies." *Adv Cancer Res* 91:1–30. doi:10.1016/S0065-230X(04)91001-9.

Druker, B. J., S. Tamura, E. Buchdunger, S. Ohno, G. M. Segal, S. Fanning, J. Zimmermann, and N. B. Lydon. 1996. "Effects of a selective inhibitor of the Abl tyrosine kinase on the growth of Bcr-Abl positive cells." *Nat Med* 2 (5):561–566.

Figueroa, M. E., O. Abdel-Wahab, C. Lu, P. S. Ward, J. Patel, A. Shih, Y. Li et al., 2010. "Leukemic IDH1 and IDH2 mutations result in a hypermethylation phenotype, disrupt TET2 function, and impair hematopoietic differentiation." *Cancer Cell* 18 (6):553–567. doi:10.1016/j.ccr.2010.11.015.

Flavahan, W. A., Y. Drier, B. B. Liau, S. M. Gillespie, A. S. Venteicher, A. O. Stemmer-Rachamimov, M. L. Suva, and B. E. Bernstein. 2016. "Insulator dysfunction and oncogene activation in IDH mutant gliomas." *Nature* 529 (7584):110–114. doi:10.1038/nature16490.

Fudenberg, G., G. Getz, M. Meyerson, and L. A. Mirny. 2011. "High order chromatin architecture shapes the landscape of chromosomal alterations in cancer." *Nat Biotechnol* 29 (12):1109–1113. doi:10.1038/nbt.2049.

Grand, F. H., S. Burgstaller, T. Kuhr, E. J. Baxter, G. Webersinke, J. Thaler, A. J. Chase, and N. C. Cross. 2004. "p53-Binding protein 1 is fused to the platelet-derived growth factor receptor beta in a patient with a t(5;15)(q33;q22) and an imatinib-responsive eosinophilic myeloproliferative disorder." *Cancer Res* 64 (20):7216–7219. doi:10.1158/0008-5472.CAN-04-2005.

Grand, F. H., C. Bird, E. Corfield, M. Dezfouli, W. Elvidge, B. Foulkes, M. Goloschokin et al., 2016 "Chromatin conformation signatures associated with epigenetic deregulation of the FIP1L1 and PDGFRA genes." *Blood*. Publication number: 1525, Submission ID: 90884. 602, Poster I.

Grosveld, G., A. Hermans, A. De Klein, D. Bootsma, N. Heisterkamp, and J. Groffen. 1987. "The role of the Philadelphia translocation in chronic myelocytic leukemia." *Ann N Y Acad Sci* 511:262–269.

Haass, W., M. Stehle, S. Nittka, M. Giehl, P. Schrotz-King, A. Fabarius, W. K. Hofmann, and W. Seifarth. 2012. "The proteolytic activity of separase in BCR-ABL-positive cells is increased by imatinib." *PLoS One* 7 (8):e42863. doi:10.1371/journal.pone.0042863.

Handoko, L., H. Xu, G. Li, C. Y. Ngan, E. Chew, M. Schnapp, C. W. Lee et al., 2011. "CTCF-mediated functional chromatin interactome in pluripotent cells." *Nat Genet* 43 (7):630–638. doi:10.1038/ng.857.

Hattori, N., Y. Hirose, H. Sasaki, S. Nakae, S. Hayashi, S. Ohba, K. Adachi, T. Hayashi, Y. Nishiyama, M. Hasegawa, and M. Abe. 2016. "World Health Organization grade II–III astrocytomas consist of genetically distinct tumor lineages." *Cancer Sci* 107 (8):1159–1164. doi:10.1111/cas.12969.

Heisterkamp, N., K. Stam, J. Groffen, A. de Klein, and G. Grosveld. 1985. "Structural organization of the bcr gene and its role in the Ph' translocation." *Nature* 315 (6022):758–761.

Ho, P. A., K. J. Kopecky, T. A. Alonzo, R. B. Gerbing, K. L. Miller, J. Kuhn, R. Zeng et al., 2011. "Prognostic implications of the IDH1 synonymous SNP rs11554137 in pediatric and adult AML: A report from the Children's Oncology Group and SWOG." *Blood* 118 (17):4561–4566. doi:10.1182/blood-2011-04-348888.

Hochhaus, A., R. A. Larson, F. Guilhot, J. P. Radich, S. Branford, T. P. Hughes, M. Baccarani et al., 2017. "Long-term outcomes of imatinib treatment for chronic myeloid leukemia." *N Engl J Med* 376 (10):917–927. doi:10.1056/NEJMoa1609324.

Huang, J., E. Marco, L. Pinello, and G. C. Yuan. 2015. "Predicting chromatin organization using histone marks." *Genome Biol* 16:162. doi:10.1186/s13059-015-0740-z.

Katainen, R., K. Dave, E. Pitkanen, K. Palin, T. Kivioja, N. Valimaki, A. E. Gylfe et al., 2015. "CTCF/cohesin-binding sites are frequently mutated in cancer." *Nat Genet* 47 (7):818–821. doi:10.1038/ng.3335.

Lu, C., P. S. Ward, G. S. Kapoor, D. Rohle, S. Turcan, O. Abdel-Wahab, C. R. Edwards et al., 2012. "IDH mutation impairs histone demethylation and results in a block to cell differentiation." *Nature* 483 (7390):474–478. doi:10.1038/nature10860.

Norbury G and Cresswell L. ACGS Quality Subcommittee. Association of Clinical Genomic Science 2015–2016 Genetic Test Activity audit. ACGSAudit15_16VPub1.doc. 29th March 2017:1–64.

Nowell, P. C., and D. A. Hungerford. 1960. "Chromosome studies on normal and leukemic human leukocytes." *J Natl Cancer Inst* 25:85–109.

O'Brien, S. G., and M. W. Deininger. 2003. "Imatinib in patients with newly diagnosed chronic-phase chronic myeloid leukemia." *Semin Hematol* 40 (2 Suppl 2):26–30. doi:10.1053/shem.2003.50058.

O'Brien, S. G., F. Guilhot, R. A. Larson, I. Gathmann, M. Baccarani, F. Cervantes, J. J. Cornelissen et al., 2003. "Imatinib compared with interferon and low-dose cytarabine for newly diagnosed chronic-phase chronic myeloid leukemia." *N Engl J Med* 348 (11):994–1004. doi:10.1056/NEJMoa022457.

Plass, C., S. M. Pfister, A. M. Lindroth, O. Bogatyrova, R. Claus, and P. Lichter. 2013. "Mutations in regulators of the epigenome their connections to global chromatin patterns in cancer." *Nat Rev Genet* 14 (11):765–780. doi:10.1038/nrg3554.

Rao, S. S., M. H. Huntley, N. C. Durand, E. K. Stamenova, I. D. Bochkov, J. T. Robinson, A. L. Sanborn et al., 2014. "A 3D map of the human genome at kilobase resolution reveals principles of chromatin looping." *Cell* 159 (7):1665–1680. doi:10.1016/j.cell.2014.11.021.

Raymond, E., A. A. Brandes, C. Dittrich, P. Fumoleau, B. Coudert, P. M. Clement, M. Frenay et al., 2008. "Phase II study of imatinib in patients with recurrent gliomas of various histologies: A European organisation for research and treatment of cancer brain tumor group study." *J Clin Oncol* 26 (28):4659–4665. doi:10.1200/JCO.2008.16.9235.

Roufosse, F. E., J. E. Kahn, G. J. Gleich, L. B. Schwartz, A. D. Singh, L. J. Rosenwasser, J. A. Denburg et al., 2013. "Long-term safety of mepolizumab for the treatment of hypereosinophilic syndromes." *J Allergy Clin Immunol* 131 (2):461–467, e1-5. doi:10.1016/j.jaci.2012.07.055.

Rowley, J. D. 1973. "Identification of a translocation with quinacrine fluorescence in a patient with acute leukemia." *Ann Genet* 16 (2):109–112.

Score, J., C. Walz, J. V. Jovanovic, A. V. Jones, K. Waghorn, C. Hidalgo-Curtis, F. Lin et al., 2009. "Detection and molecular monitoring of FIP1L1-PDGFRA-positive disease by analysis of patient-specific genomic DNA fusion junctions." *Leukemia* 23 (2):332–339. doi:10.1038/leu.2008.309.

Scott, M. L., R. A. Van Etten, G. Q. Daley, and D. Baltimore. 1991. "v-abl causes hematopoietic disease distinct from that caused by bcr-abl." *Proc Natl Acad Sci* 88 (15):6506–6510.

Tordini, F., M. Aldinucci, L. Milanesi, P. Lio, and I. Merelli. 2016. "The genome conformation as an integrator of multi-omic data: The example of damage spreading in cancer." *Front Genet* 7:194. doi:10.3389/fgene.2016.00194.

Valent, P., G. J. Gleich, A. Reiter, F. Roufosse, P. F. Weller, A. Hellmann, G. Metzgeroth et al., 2012. "Pathogenesis and classification of eosinophil disorders: A review of recent developments in the field." *Expert Rev Hematol* 5 (2):157–176. doi:10.1586/ehm.11.81.

van de Werken, H. J., G. Landan, S. J. Holwerda, M. Hoichman, P. Klous, R. Chachik, E. Splinter et al., 2012. "Robust 4C-seq data analysis to screen for regulatory DNA interactions." *Nat Methods* 9 (10):969–972. doi:10.1038/nmeth.2173.

Vardiman, J. W. 2010. "The World Health Organization (WHO) classification of tumors of the hematopoietic and lymphoid tissues: An overview with emphasis on the myeloid neoplasms." *Chem Biol Interact* 184 (1–2):16–20. doi:10.1016/j.cbi.2009.10.009.

Vardiman, J. W., J. Thiele, D. A. Arber, R. D. Brunning, M. J. Borowitz, A. Porwit, N. L. Harris et al., 2009. "The 2008 revision of the World Health Organization (WHO) classification of myeloid neoplasms and acute leukemia: Rationale and important changes." *Blood* 114 (5):937–951. doi:10.1182/blood-2009-03-209262.

Vogelstein, B., N. Papadopoulos, V. E. Velculescu, S. Zhou, L. A. Diaz, Jr., and K. W. Kinzler. 2013. "Cancer genome landscapes." *Science* 339 (6127):1546–1558. doi:10.1126/science.1235122.

Walz, C., J. Score, J. Mix, D. Cilloni, C. Roche-Lestienne, R. F. Yeh, J. L. Wiemels et al., 2009. "The molecular anatomy of the FIP1L1-PDGFRA fusion gene." *Leukemia* 23 (2):271–278. doi:10.1038/leu.2008.310.

Wang, X. W., B. Boisselier, M. Rossetto, Y. Marie, A. Idbaih, K. Mokhtari, K. Gousias et al., 2013. "Prognostic impact of the isocitrate dehydrogenase 1 single-nucleotide polymorphism rs11554137 in malignant gliomas." *Cancer* 119 (4):806–813. doi:10.1002/cncr.27798.

Xu, W., H. Yang, Y. Liu, Y. Yang, P. Wang, S. H. Kim, S. Ito et al., 2011. "Oncometabolite 2-hydroxyglutarate is a competitive inhibitor of alpha-ketoglutarate-dependent dioxygenases." *Cancer Cell* 19 (1):17–30. doi:10.1016/j.ccr.2010.12.014.

Zadeh, G., and K. Aldape. 2016. "Bringing IDH into the Fold." *Cancer Cell* 29 (2):139–140. doi:10.1016/j.ccell.2016.01.010.

Challenges and Opportunities of Next Generation Sequencing Companion Diagnostics

5.4

Benoit Destenaves

Contents

INTRODUCTION

Recent advances in molecular analysis methodologies, namely NGS, coupled with the ever-increasing capacity for data analysis and storage have opened great possibilities for molecular diagnostics particularly in oncology (1). NGS enables rapid access to an extensive molecular profile of a tumor sample, providing oncologists with a comprehensive level of information to aid them in diagnosis, judging patient prognosis and selecting the most appropriate treatment options. When compared with running multiple individual marker tests, this results in both sparing of tissue and a rapid turnaround time for availability of results.

However, these new possibilities do not come without challenges from a diagnostic development perspective. As of August 2018, NGS-based CDxs have only been authorized in the United States and Europe and none under the new European *In Vitro* Diagnostic Regulation (IVDR), which is a far more stringent process than the *In Vitro* Diagnostic Directive (IVDD), which it replaced. Looking at published guidance and approvals (Table 5.4.1) from the FDA and other regulatory authorities, where available (e.g., PMDA) (2), allows an understanding of the requirements beyond those of previously approved molecular diagnostics, that will need to be met by NGS assays to be approved or registered as CDx and also the potential opportunities they offer.

TABLE 5.4.1 FDA-approved next generation sequencing systems with companion diagnostic claims

NAME	*COMPANY*	*DEVICE DESCRIPTION*	*SEQUENCING TECHNOLOGY*	*INTENDED USE*	*FDA PMA REFERENCE*
FoundationFocus™ CDxBRCA	Foundation Medicine	LDT. Detection of *BRCA1* and *BRCA2* alterations in FFPE ovarian tumor tissue	Illumina HiSeq™ 4000	FDA: 1 CDx claims	P160018
FoundationOne CDx™	Foundation Medicine	LDT. Detection of SNVs, detection of substitutions, insertion and deletion alterations (indels) and CNAs in 324 genes and select gene rearrangements, as well as genomic signatures including MSI and TMB from FFPE tumor tissue samples	Illumina HiSeq™ 4000	FDA: 17 CDx claims; pan-panel tumor mutation profiling to be used by qualified health care professionals in accordance with professional guidelines in oncology for cancer patients with solid malignant neoplasms	P170019
Praxis™ Extended RAS Panel	Illumina	Kit. Detection of 56 mutations in *KRAS* (exons 2, 3, and 4) and *NRAS* (exons 2, 3, and 4) from FFPE CRC tumor tissue samples	Illumina MiSeqDx®	FDA: 1 CDx claims	P160038
Oncomine™ Dx Target Test	Thermo Fisher	Kit. Detection of single nucleotide variants (SNVs) and deletions in 23 genes from DNA and fusions in *ROS1* from RNA isolated from FFPE NSCLC tumor tissue samples	Ion PGM™ Dx	FDA: 4 CDx indications; Mutation profiling (i.e., analytical validation) for 4 variants (2 *KRAS*, 1 *MET*, 1 *PIK3CA*)	P160045

A BRIEF HISTORY OF NEXT GENERATION SEQUENCING

NGS refers to high-throughput DNA and RNA sequencing technology, which emerged in the 2000s. "Next generation" is relative to "first generation" sequencing, more commonly known as Sanger sequencing, which was developed in the late 1970s (3) and improved upon until the late 1990s. NGS differs from Sanger sequencing, first in how the technologies work (3,4), but the biggest difference stems from the fact that it employs massively parallel sequencing to deliver very large numbers of sequences at a very high throughput. As a comparison, the NGS system with the current highest throughput (Illumina NovaSeq 6000) can deliver nearly 650,000 times more bases per run than the most powerful Sanger sequencer ever marketed (ThermoFisher/ABI 3730xl) over equivalent run times: 3 Tbases in one run (3 days) versus 4.65 Mbases (5,6).

NGS technologies can be separated into second generation sequencing, characterized by the need to prepare amplified sequencing libraries prior to sequencing and third generation sequencing, which have emerged more recently and employ single nucleic acid molecule sequencing, forgoing the need to create amplified libraries and enabling longer individual reads to be analyzed (7). These third-generation technologies are not as mature as second-generation NGS and are only used in research applications and will not be discussed further here; all references below to NGS refer to second-generation sequencing.

NGS SYSTEMS AND TECHNOLOGIES USED IN CLINICAL TESTING

NGS platforms employ very different methods for the actual sequencing, but all employ comparable workflows. Nucleic acid extraction is followed by library construction, where DNA fragments are ligated to adaptor sequences. Each fragment is then clonally amplified by PCR. Finally, sequencing is achieved by repeated cycles of biochemical reactions and data capture (e.g., image capture) (8). The great majority of NGS platforms used for clinical sequencing use either Illumina or Thermo Fisher Ion Torrent systems, in a number of cases using research-grade instruments validated by the clinical testing labs. The Illumina MiSeqDx and Thermo Fisher Personal Genome Machine Dx being the only IVD grade NGS systems currently available, at least in the United States and the European Union.

Illumina Sequencing Technology

Illumina sequencing technology originated from Solexa, which Illumina acquired in 2007 (9). The Illumina NGS workflow consists of the following four steps (10):

1. Library preparation: The sequencing library is prepared by random fragmentation of the DNA or cDNA sample (if needed), followed by 5′ and 3′ adapter ligation (Figure 5.4.1a). Adapter-ligated fragments are then PCR amplified and gel purified.
2. Cluster generation: the library is loaded into a flow cell where fragments are captured on a lawn of surface-bound oligos complementary to the library adapters. The fragments are then amplified into distinct, clonal clusters through bridge amplification (Figure 5.4.1b). When cluster generation is complete, the templates are ready for sequencing.
3. Sequencing: Illumina's sequencing by the synthesis method employs reversibly terminated fluorescently labeled nucleotides and detects single bases as they are incorporated into DNA template strands (Figure 5.4.1c).

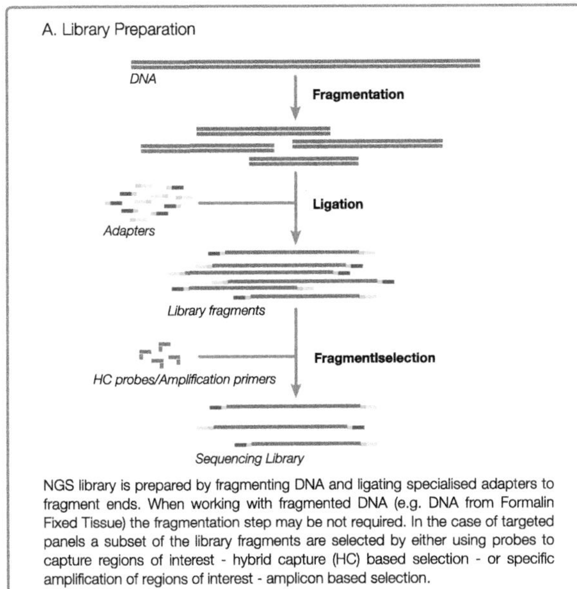

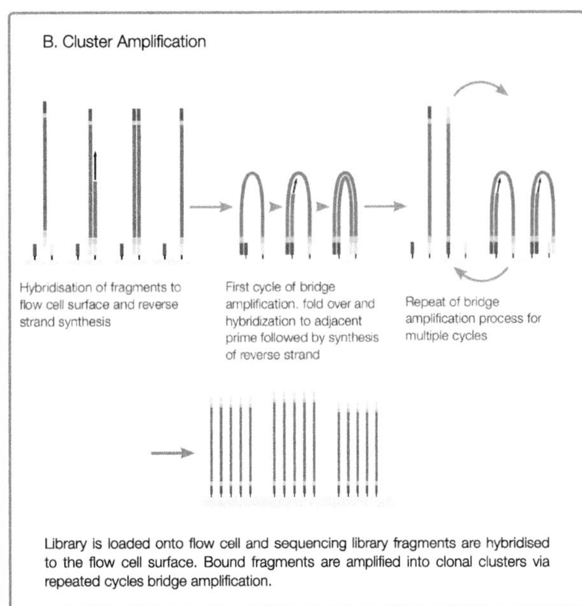

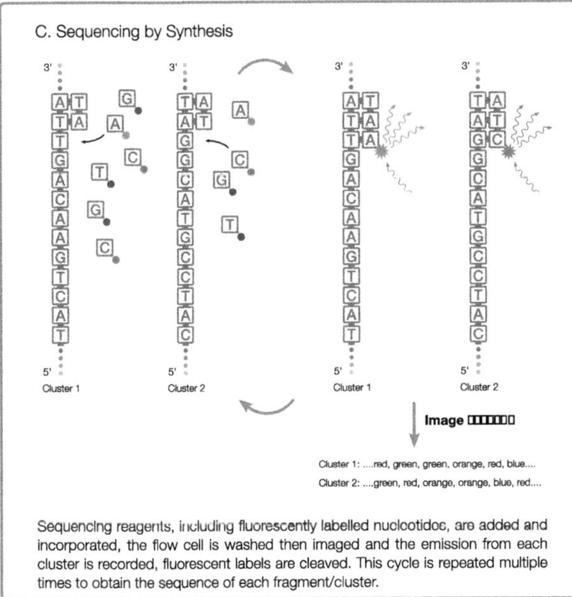

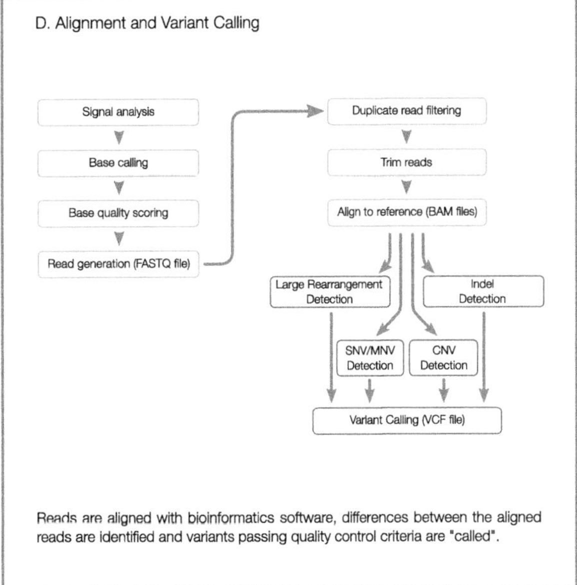

FIGURE 5.4.1 Overview of Illumina Sequencing process. (a) Library preparation, (b) cluster amplification, (c) sequencing, (d) alignment and data analysis. (www.illumina.com/documents/products/illumina_sequencing_introduction.pdf.)

4. Data analysis: sequence reads are aligned to a reference genome (Figure 5.4.1d) and variant calls are made.

Illumina Sequencing is used for the vast majority of sequencing with approximately 75% of all sequencing worldwide being done on Illumina instruments (this includes research-based applications) (11).

Ion Torrent (Thermo Fisher) Sequencing Technology

The Ion Torrent sequencing process consists for the same four basic steps as Illumina Sequencing:

1. Library preparation: The sequencing library is prepared by random fragmentation of the DNA or cDNA sample, followed by 5′ and 3′ adapter ligation adapter-ligated fragments are then PCR amplified and gel purified.
2. Cluster generation (emulsion PCR): Individual DNA fragments are isolated into emulsion droplets where PCR occurs to clonally amplify fragments of the target sequence that bind to a bead in the droplet. This enables thousands of individual PCR reactions to take place simultaneously.
3. Sequencing: the beads are loaded onto a semiconductor chip, where each single bead is deposited into a single well. The sequencing by synthesis technology employed by Ion Torrent detects the minute changes in pH caused by protons released during the incorporation of the correct nucleotide. This method has in principal, the advantage of not requiring fluorescently labeled nucleotides or optics for detection, which are expensive and fragile. Contrary to the Illumina method, in the case of homopolymers all bases will be incorporated simultaneously, causing a greater increase in the signal detected, but they are difficult to accurately quantify beyond four repeated nucleotides (12).
4. Data analysis: Sequence reads are aligned to a reference genome and variant calls are made.

NGS PANEL COMPANION DIAGNOSTICS

Over the course of the past 2 years, we have seen several NGS panels approved by the FDA as CDx. No NGS system has been registered outside the United states with CDx claims, in part due to the different regulations in some regions such as Europe where under the IVD directive (IVDD) IVD assays did not have to demonstrate clinical utility and thus were not "linked" to any particular treatment, meaning that any IVD with an analytical claim could be used. As of May 2017, the IVDD has been replaced by a far more stringent regulation (the IVDR), which requires CDx IVDs to demonstrate clinical utility (13). Although a transition period is in effect until May 2022 where IVDs can still be certified under the IVDD, these certificates will not be able to be renewed and manufacturer will have to comply with IVDR to obtain certification (13). Besides the IVDs listed in Table 5.4.1 several other companies have announced their intention to obtain approval/registration for novel NGS-based assays: Foundation Medicine ctDNA assay (Laboratory developed test), Guardant Health Guardant360 (ctDNA-based LDT); Personal Genome Diagnostics elio (ctDNA-based kit) and MSK seeking CDx claims for IMPACT (this panel is already 510k cleared by the FDA), FoundationOne CDx ex-U.S. (EU, Japan).

LESSONS LEARNED FROM RECENT APPROVALS

Analytical Validation of All Individual Variants Is Not Required

Analytical validation of NGS CDxs panels poses new challenges mainly due to the fact that instead of specifically analyzing a limited set of well-defined variants, thousands to millions of variants are analyzed at once. Even if it were possible for the diagnostic developer to identify the proper positive control samples for all possible variations, it would take years to run all required assays. Even then, the files would be so big that it would also take reviewers such as the FDA CDRH in the United States or notified bodies in the European Union, years to go through the data to enable authorization of the IVD ultimately delaying patient access to potentially life-saving therapies. Recently, NGS panels approved or cleared by the FDA have employed a strategy initially developed for the approval of the Myriad Genetics BRCAnalysis CDx test (Foundation focus BRCA, FoundationOne CDx, MSK IMPACT) (14,16). The development of Myriad Genetics is discussed in depth in chapter 6.4 of this book. The strategy is to undertake detailed analytical validation studies for each variant type, which the test aims to report rather than for each individual variant possible. Variant classes to be considered include: Single nucleotide variants (SNV), small insertion deletions (indels) typically of 15 bases or less, large rearrangements (deletions or insertions) that may be as big as whole exons or entire genes and amplifications. NGS technology currently available performs better for the detection of SNVs and small indels than for amplifications or large rearrangements as can be seen, for example, in the FoundationOne CDx summary of safety and effectiveness data published by the FDA (14); thus each of these variant types will have their own performance characteristics, mainly due to the technical limitations of the assay employed. It is also important to validate these variant types within different sequence context (e.g., normal and low complexity regions) because they may influence the ability of the assays to detect a variant.

Analytical validation studies to be carried out are broadly similar to those required for individual variant tests, but in some cases have to be adjusted to the context of NGS.

- Accuracy: For NGS assays, accuracy is the degree of concordance between the sequence obtained from the test and a reference sequence. The reference sequence can either be obtained by running the same test sample(s) on a valid comparator method or from a well known and -characterized reference standard. For *de novo* CDx claims the issue is identifying a valid comparator. Sanger sequencing was, in principal, the "gold standard" for sequencing; however, particularly within the context of testing for somatic mutations, it is far outperformed by NGS and thus results in a high level of discordance, which for validation of the first tests represented a significant challenge. As more NGS tests obtain IVD status this should be less of an issue.
- Limit of detection: For NGS assays, limit of detection is in general the lowest concentration for a variant or variant type, which has at least 95% of positive calls for a set of replicates tested at a given allele frequency (e.g., mutant allele fraction for a somatic test).
- Repeatability and reproducibility: These tests do not really differ from what is required of single assay tests and are, in principal, not impacted by variant type. Repeatability involves determining test variability when running the same set of samples multiple times within a short time frame while keeping other conditions constant (e.g., same operator, instruments, reagent lots, location). Reproducibility involves determining tests such as, but not limited to, impact of different operators, different instruments, different days and runs, different reagent lots or different locations (not applicable to single site tests).
- Analytical specificity: This relates to the ability of the test to measure uniquely the intended variant. Within the context of NGS tests, major factors that could impact specificity include chemicals used for tissue fixation and DNA extraction, cross-reactivity with homologous regions such as pseudogenes and cross contamination of patient samples from carry-over between runs of improperly multiplexed samples. Single nucleotide fingerprinting is an efficient way to control for patient sample cross contamination.

Masking on Regions Is Not Required

It has come as a surprise that the FDA did not require masking of regions of the panels not included in the CDx claims. Instead, assays have different levels of intended use claims based on the level of validation for given variants. The FoundationOne CDx test has 17 CDx claims (1 CDx claim = biomarker + drug + indication) (14). As detailed in the FoundationOne CDx summary of safety and effectiveness data each CDx claim required the most complete level of validation for variant types being claimed and in samples representative of the indication. For example, the *BRCA1* and *BRCA2* test is only valid for prescription of Rucaparib in ovarian cancer and, in principal, cannot be used to prescribe Rucaparib to a patient with breast cancer that has a *BRCA1* loss of function mutation. That being said the intended use statement of the FoundationOne CDx assay included "tumour profiling for cancer patients with solid malignant neoplasma to be used by qualified health care professionals" which can, within the context of the practice of medicine and based on professional guidelines in oncology, use this data to make treatment decisions.

Opportunities to Speed-Up Companion Diagnostic Development and Expand Access to Testing

With the approval of the FoundationOne CDx and Ion Oncomine Dx panels, the FDA has outlined a path for approval of follow-on CDxs that does not require samples for the actual clinical studies where the clinical utility of the IVD was demonstrated. This is a massive opportunity for patients because it should help broaden access to testing and identification of patients that would benefit from being given these targeted therapies.

The strategy employed by Foundation Medicine to obtain follow-on CDx claims for the FoundationOne CDx panels involved demonstrating for a set of represent clinical samples, that concordance between the FoundationOne CDx assay and the FDA-approved CDx assay (e.g., Roche Molecular Systems cobas v2 EGFR Mutation Test) was statistically non-inferior to the concordance between the same set of samples analyzed twice with the FDA-approved CDx (Figure 5.4.2) (14). In the majority of cases the CDx claims for the FoundationOne CDx panel were obtained through the analysis of representative nonclinical study samples.

Another opportunity presented by these large NGS CDx panels is that once initially approved, adding additional CDx claims, even for *de novo* CDxs, at least from the FDA's point of view, will only require a supplemental PMA (sPMA). Through this process drug and diagnostic developers will be able to leverage the existing analytical validation package for the panel and only have to add further analytical validation experiments where a gap exists in the existing panel validation studies. This greatly simplifies approval of multiple CDx for a given marker once clinical utility has been demonstrated and benefits patients by broadening access to quality testing.

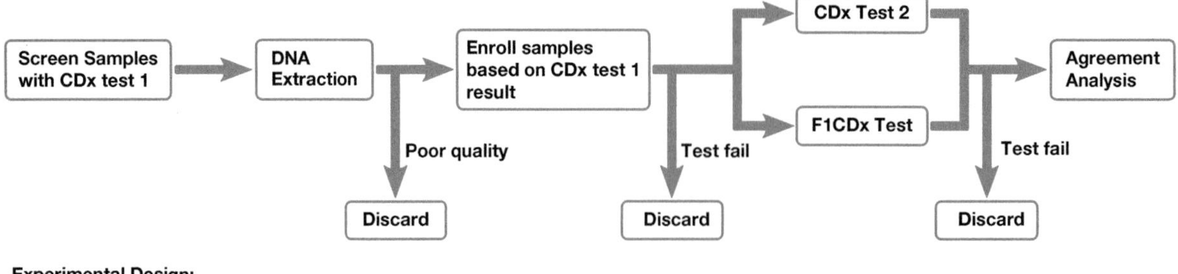

FIGURE 5.4.2 Experimental design used by Foundation Medicine for F1CDx Follow-on companion diagnostic claims. (AACR 2018, Yali Li, Clinical and analytical validation of FoundationOne CDx: The first FDA-regulated NGS-Based comprehensive genomic profiling test, regulatory science and policy session.)

CHALLENGES TO OVERCOME FOR THE DEVELOPMENT OF NGS PANELS AS COMPANION DIAGNOSTIC

Development and approval of NGS panels as CDxs is still a nascent field and a lot of unknowns and challenges remain. First and foremost, as previously touched upon the process for registration/approval of NGS-based CDx is essentially untested outside of the United States, and thus we do not know whether the same approaches taken by the FDA will be valid in other regions. In fact, if previous CDx registrations/approvals are used as an indication, then it is likely that authorities in other regions will not have identical requirements to those of the FDA.

NGS is a rapidly evolving technology both at the level of the wet lab aspects but also in terms of data analysis. Currently the FDA requires a supplemental PMA application, which can take up to 180 days, for any update to an approved CDx assays. Establishing a process for test modification by prospectively agreeing with decision-making bodies prior to registration/approval of the CDx on protocols with procedures and acceptance criteria for modification would be one possible way to enable rapid and safe update to these panels. Such a process has been outlined in the 510 k clearance of the MSK-IMPACT where seven types of changes have been anticipated (16) and could serve as a foundation to be built upon for the update of next generation-based CDxs.

The use of NGS in clinical testing is resulting in the identification of an ever-increasing number of genetic variants including rare variants. Because in a number of cases, the clinical utility claims of NGS assays will not be for individual variants but for variants meeting certain rules, it is essential to understand the clinical significance of these new variants that will not have been observed, or observed at a very low frequency, during the clinical studies. Following several workshops, the FDA issued a guidance for the development and use of public databases to support IVD diagnostics (17), but this unfortunately does not seem to be getting much traction because companies offering commercial testing are likely to see this data as a competitive advantage.

Further challenges will arise from the use of genomic signatures such as tumor mutational burden (TMB) or to a lesser extent microsatellite instability (MSI), which have recently emerged as novel markers for response to PD-1/PD-L1 inhibitors used in immune oncology. Contrary to single-variant biomarkers, where the genomic signatures are qualitative (i.e., they are either present or they are not), these genomic signatures are quantitative and thus require the determination of a cutoff to differentiate biomarker positive (e.g., TMB high) and biomarker negative. Different algorithms, often proprietary, have been developed to determine which variants are taken into account and which discarded (e.g., in most cases, hotspot mutations are eliminated). This is further complicated by how variants are initially called and filtered. In the United States, the Friends of Cancer Research (FOCR), a nonprofit organization, has convened a number of partners from public and private institutions to assess current methods of measuring tumor mutational burden with the intent to draft industry standards/guidelines for the measurement and reporting of harmonization measurement and reporting of TMB (18). Similar efforts were initiated in Europe by the German Quality Assurance Initiative Pathology (QuIP), which aims to copublish with FOCR, and by the International Quality Network for Pathology (IQN Path), which is being supported by the European Society for Medical Oncology (ESMO) (19,20).

Beyond this, implementation and uptake of NGS CDx tests in clinical testing labs also face significant challenges as these tests remain significantly more expensive than more traditional IVDs such as immunohistochemistry or single-analyte molecular diagnostics, both in terms of initial investment required but also in terms of individual testing costs. NGS assays are also perceived to have a long analysis turnaround time, relative to more traditional molecular diagnostics and remain very technical and require highly skilled personnel and large infrastructures (lab, IT, bioinformatics) to run these assays, both of which make them potentially less attractive.

Irrespective of all these challenges, NGS is poised to become the dominant testing platform for molecular diagnostics because its advantages greatly outweigh its challenges and, as technology evolves, the technical limitations of NGS testing will disappear, offering more possibilities for drug and diagnostic development and patients.

REFERENCES

1. Frampton GM, Fichtenholtz A, Otto GA, Wang K, Downing SR, He J, et al. Development and validation of a clinical cancer genomic profiling test based on massively parallel DNA sequencing. *Nat Biotechnol.* 2013;31(11):1023–1031.
2. www.pmda.go.jp/files/000214377.pdf.
3. Sanger F, Nicklen S, Coulson AR. DNA sequencing with chain-terminating inhibitors. *Proc Natl Acad Sci* 1977;74(12):5463–5467.
4. Alekseyev YO, Fazeli R, Yang S, Basran R, Maher T, Miller NS, Remick D. A Next-Generation sequencing primer-how does it work and what can it do? *Acad Pathol.* 2018;5:2374289518766521.
5. www.illumina.com/systems/sequencing-platforms/hiseq-x/specifications.html.
6. www.thermofisher.com/order/catalog/product/3730XL.
7. Ameur A, Kloosterman WP, Hestand MS. Single-Molecule sequencing: Towards clinical applications. *Trends Biotechnol.* 2018. pii: S0167-7799(18)30204-X.
8. Shendure J, Ji H. Next-generation DNA sequencing. *Nat Biotechnol.* 2008;26(10):1135–1145.
9. www.illumina.com/science/technology/next-generation-sequencing/illumina-sequencing-history.html.
10. www.illumina.com/documents/products/illumina_sequencing_introduction.pdf.
11. www.prnewswire.com/news-releases/global-next-generation-sequencing-market-assessment--forecast-2017---2021-300431518.html.
12. Khodakov D, Wang C, Zhang DY. Diagnostics based on nucleic acid sequence variant profiling: PCR, hybridization, and NGS approaches. *Adv Drug Deliv Rev.* 2016;105(Pt A):3–19.
13. EU directive (EU) 2017/745 –https://eur-lex.europa.eu/legal-content/EN/TXT/?uri=CELEX%3A32017R0745.
14. www.accessdata.fda.gov/cdrh_docs/pdf17/P170019B.pdf.
15. AACR 2018, Yali Li, Clinical and analytical validation of foundation one CDx: The first FDA-regulated NGS-Based comprehensive genomic profiling test, regulatory science and policy session – April 14 2018.
16. www.accessdata.fda.gov/cdrh_docs/reviews/DEN170058.pdf.
17. www.fda.gov/downloads/MedicalDevices/DeviceRegulationandGuidance/GuidanceDocuments/UCM509837.pdf.
18. www.focr.org/tmb.
19. quip.eu/en_GB/2018/05/14/tumor-mutational-burden-tmb-quip-organisiert-studie-und-arbeitet-mit-focr-zusammen/.
20. www.genomeweb.com/cancer/two-initiatives-seek-harmonize-tumor-mutational-burden-testing#.W5kmbUVKjwO.

Implementing Companion Diagnostics

5.5

Martina Kaufmann

Contents

INTRODUCTION

Developing personalized medicines often requires development of companion diagnostics (CDx). According to the FDA definition, a CDx is "an *in vitro* diagnostic device that provides information that is essential for the safe and effective use of the corresponding therapeutic product." This FDA definition also includes use of the CDx for monitoring (if such use this would be considered to be essential for the safe and effective use of the corresponding therapeutic product) and it could also be an imaging tool. As today CDx are mainly used in oncology for the purpose of identification of patients for efficacy reasons, this section will focus on such CDx-based on predictive biomarkers.

Several steps have to be taken for successful implementation of CDx-based patient selection approaches, starting in early research and discovery (R&D) and continuing throughout the lifecycle of the drug (Figure 5.5.1).

COMPANION DIAGNOSTICS DEVELOPMENT, REGULATORY LANDSCAPE, AND PARTNERSHIPS

Identification of the predictive biomarker(s) as basis for CDx development has to start very early in the R&D process. Biomarker hypotheses for prediction of response need to be defined and tested in preclinical models, with the aim to identify candidate predictive biomarkers and to provide a robust patient selection hypothesis early on. In development programs with a more complex biology, however, as in immune cancer therapies, this is often not feasible. In such cases, CDx development may become more challenging because a number of candidate predictive biomarkers may need to be further evaluated in early clinical studies followed by confirmation of the patient selection hypothesis in the pivotal study.

Assay development for testing of the patient selection hypothesis on clinical samples has to be already initiated at this stage to ensure to have the assays available in the required quality standards for the clinical trials.

Once formal development of a CDx is decided, a CDx assay validation and approval process has to be mapped out. This first requires an in-depth understanding of the regulatory framework for CDx in the markets of interest. Although it is beyond the scope of this section to address the CDx regulatory landscape globally in detail, it can be top-level summarized as follows:

In the United States, IVDs are under the governance of the FDA's CDRH. Consequently, the first regulatory framework for CDx was put in place in 2011 by CDRH as a draft guidance, which was replaced by the final CDx guidance titled, "*In Vitro* Companion Diagnostic Devices" in 2014 (1). In 2016, this guidance was complemented by the draft co-development guidance, titled "Principles for Co-development of an In Vitro Companion Diagnostic Device with a Therapeutic Product" (2).

In Europe, IVD regulation is currently changing. The previous IVD Directive, which did not consider CDx specifically, is about to be replaced by an IVD regulation (3) that entered into force in June 2017 with a 5-year transition period. A harmonized approach to regulatory assessment of IVD in the EU was

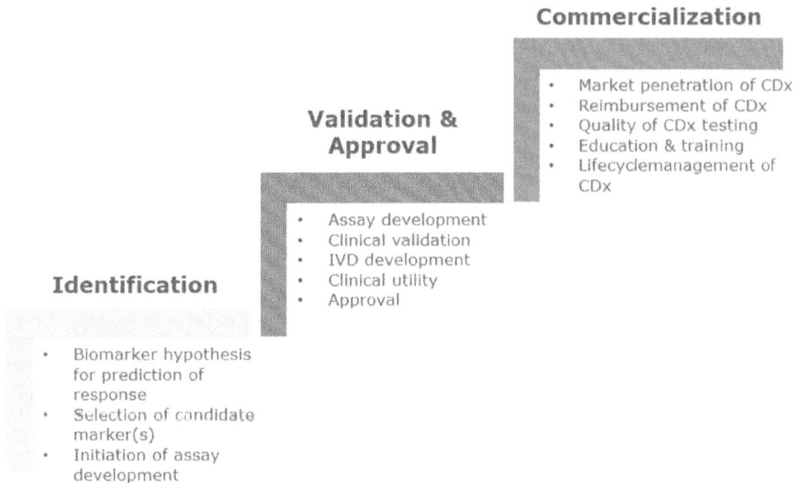

FIGURE 5.5.1 Steps in companion diagnostics implementation.

introduced. At the same time, more stringent requirements for approval of CDx will be required: Self-certification as per the previous IVD Directive will no longer be possible for CDx; instead, a so-called Notified Body will be required for conformity assessment and for consultation with the European Medicine's Agency (EMA). In July 2017, EMA has released a "Concept paper on development and lifecycle of personalized medicines and CDx" for public consultation (4), which is hoped to provide additional guidance. In Japan, the Pharmaceutical and Medical Device Agency (PMDA) released "The technical guidance for co-development of CDx and the drug" in December 2013 (5), which largely mimics the FDA guidance.

According to all these regulatory frameworks, concurrent approval of the CDx test and corresponding therapeutic product is anticipated. Therefore, co-development is needed. This is quite challenging because the development of the IVD has to become integrated into the drug development process. Because most pharmaceutical companies do not have a diagnostic division, typically a partnership with a diagnostic company or a CRO is required, which brings in another level of complexity. Identification of the CDx partner starts with choice of the CDx technology, which has to be transferable to the clinical setting. Strategic aspects and, in particular, the required capabilities such as competency in CDx development and approval, global commercial reach and a track record for successful market access of such tests, are key considerations. The CDx partner has to be involved early on to provide competent input on the CDx co-development, commercial and market access strategies, which often is achieved through a joint CDx development team.

Co-development does not necessarily require simultaneous development of drug and CDx from beginning to end. Ideally, the predictive biomarker becomes identified very early and a robust preclinically validated patient selection hypothesis allows initiation of CDx development already at this stage. However, depending on the complexity of the biology, identification of the predictive biomarker may not always be straight forward, and so in reality biomarker identification and subsequent test development can occur at any point. However, co-development should be conducted in a way that will facilitate obtaining contemporaneous marketing authorizations. Therefore, it is obvious that the later one starts, the higher the risk of delay because the CDx may not be ready for approval. To mitigate this risk, a rigorous biomarker development program has to be initiated early on (Figure 5.5.2).

With the progression of the FDA draft CDx guidance into a final guidance document in 2014, the FDA provided further clarification on the anticipation of concurrent approval. Although concurrent approval would be ideal, there might be certain cases such as a new drug for a life-threatening disease, where the FDA may decide to approve the new drug even if the

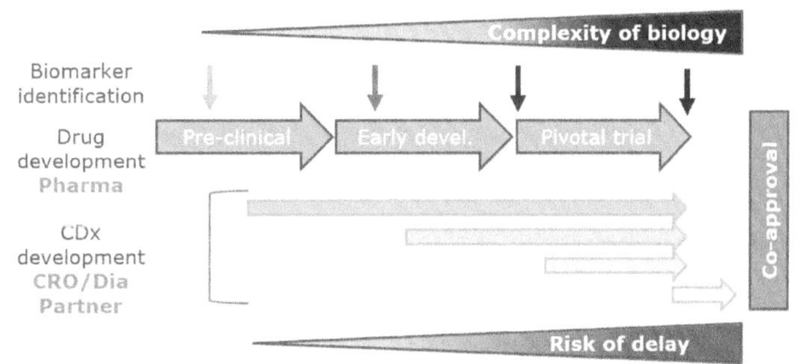

FIGURE 5.5.2 Co-development of drug and companion diagnostics.

respective CDx is not approved contemporaneously. Since then, the FDA has made such an exception twice. In March 2016, Critozinib was granted approval in ROS1-positive NSCLC. In May 2017, pembrolizumab was approved for treatment of adult and pediatric patients who have unresectable or metastatic microsatellite instability-high (MSI-H) or mismatch repair deficient (dMMR) solid tumors, in both cases with the development of the respective CDx tests as a postmarket commitment. However, these exceptions should not be interpreted as the FDA lowering the bar in general because, in the end, the strength of the clinical data and the medical need in the particular disease still matter.

The approval of pembrolizumab on the basis of molecular markers regardless of tumor type was a clear novum in drug development because in a traditional approach, new medicines have been developed based on the disease phenotype. To clarify requirements for such molecularly guided development programs, the FDA released draft guidance in December 2017 titled "Developing Targeted Therapies in Low-Frequency Molecular Subsets of Disease" (6). With this draft guidance, the agency acknowledged that the molecular markers may be particularly rare in certain disease settings and provided guidance on how sponsors can group patients with different molecular alterations if the targeted therapy may be effective across molecular subsets within a disease. The public had 60 days to comment on this draft guidance. Once final this will provide guidance on the type and quantity of evidence that can demonstrate efficacy across molecular subsets within a disease, particularly when one or more molecular subsets occur at a low frequency.

According to the current regulatory framework for CDx, the assay has to be validated in all specific indication(s) of interest. Therefore, such a pan-cancer, tissue-agnostic approval approach also has an impact on CDx development. Further guidance from FDA on this specific question, that is, which and how many tumor types have to be considered, is awaited.

Because there are differences between the general CDx regulations as outlined above for the United States, Europe, and Japan with regard to, for example, specific requirements, regulatory processes and timelines, one cannot outline a universal approach to CDx development in all detail. Therefore, the following will focus on the FDA requirements.

CDx are typically considered high-risk (class III) devices by the FDA, which requires a PMA application. However, in certain cases a CDx may become considered a moderate risk (class II) device, which would require clearance of a 510(k) premarket notification or grant of a *de novo* request.

The actual CDx development including the clinical validation of the predictive biomarker as part of the drug clinical trials and the actual IVD development has to follow the drug development timelines and registration strategies, for example, an accelerated approval strategy due to factors such as regulatory requirements like review timelines. Required validation and regulatory activities have to be anticipated, and at-risk investments will be needed in order to avoid delays in the development program of the corresponding therapeutic product. In principle, validation has to provide evidence that a method is "reliable for the intended application." For a CDx, the following performance indicators have to be considered:

- Analytical validity, that is, how accurately and robustly an assay detects the analyte of interest
- Clinical validity, that is, how well the test relates to the clinical outcome measure of interest, including determination of the clinical decision point (cut-off) in case of a biomarker measured on a continuous scale
- Clinical utility, that is, demonstrating that use of the test results in improved clinical outcome for patients, which is particularly critical from a reimbursement perspective

Analytical validation focuses on the performance characteristics of the assay. Depending on the features of the assay technology and the biomarker (quantitative vs. qualitative marker) requirements may vary and need to be agreed upon with the health authorities. A detailed analytical validation plan based on relevant regulatory documents must be developed. Assessments need to be performed on samples representative of the samples to be tested subsequently.

Clinical validation and demonstration of clinical utility needs to be addressed in the drug clinical trials. The design of such clinical trials is impacted by various factors, mainly the strength of the patient selection hypothesis, effect size and biomarker prevalence, but also operational aspects such as assay readiness, turnaround time, for example.

Typically, assay development follows a step-wise, so-called fit-for-purpose approach depending on the planned use of the assay in the drug clinical trials, also considered is the use for which the resulting data will be used (7). This also includes different levels of quality standards for the labs running these assays (Figure 5.5.3).

ICH E6 GCP requires that human clinical samples have to be analyzed in compliance with specifically defined quality standards, according to the requirements including analytical validation of the assay (8,9). This aims to ensure that the data are reliable, robust and comparable between Good Clinical Laboratory Practice (GCLP) studies. If the assay is planned to be used for patient selection (so-called "integral marker"), this is equivalent to making a medical decision, and therefore even more stringent quality standards apply. In such a case, the assay used should be the "final IVD type assay" ("design locked," as per FDA Quality system regulation and CGMP requirements) (10,11). If the final IVD-type assay is not available at this stage patient selection can be performed using an adequately validated clinical trial assay (CTA) instead. In such case bridging to the final assay or kit has to be performed but there is significant regulatory risk because, for example, at least 90% of the clinical samples need, ideally, to be available.

For patient selection, samples from U.S. patients need to be analyzed ideally in a CLIA-certified or -accredited laboratory capable of providing testing at the appropriate quality standards (12). As in most cases, no approved predecessor assay exists, the status of the assay is considered as investigational, and the assay needs to be labeled as IUO in the United States.

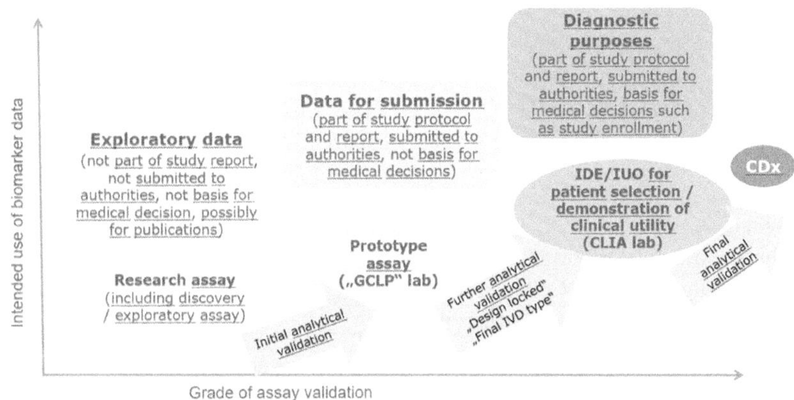

FIGURE 5.5.3 Fit-for-purpose assay validation for companion diagnostics.

On top of that, the FDA requires a risk determination procedure to be applied for use of an investigational medical device in clinical trials (i.e., an IDE) (13) for patient selection to ascertain the risk associated with such testing. This risk assessment includes consideration of the risk of the proposed clinical trial design, that is, the risk that subjects would miss out on an effective therapy, the risk of unnecessary exposure to safety risks, and the risk of severe complications due to the specimen acquisition ("biopsy issue"). Early discussion with the FDA as part of the pre-submission program (14) is advisable to clarify these aspects because the outcome of the risk determination procedure has a significant impact on development timelines and efforts. If the test use poses significant risk, approval of an IDE submission by FDA is required prior to the trial proceeding. In the case of nonsignificant risk, the test still has to comply with abbreviated requirements.

In December 2017, the FDA published a draft guidance titled "Investigational IVDs Used in Clinical Investigations of Therapeutic Products" (15), expressing concern that sponsors and investigational review boards may not appreciate that investigational IVDs used in clinical trials can pose significant risk to patients. The public had 90 days to comment on this draft guidance. Once final, this guidance will clarify the appropriate regulatory pathway for investigational IVDs used in clinical trials for therapeutic products.

In case the assay is not used for prospective testing (i.e., patient selection) but for retrospective testing only, the investigational IVD would be considered to be exempt from the requirements of the IDE regulation according to 21 CFR 812 (16) because no medical decision would be based on such testing.

Patient selection approaches are the most desired clinical trial design for targeted therapies development mainly for reasons of cost and speed but also because the inclusion of biomarker-negative patients in late stage development trials may not be considered to be ethical. However, in trials with enrollment of a biomarker-defined subgroup, the test cannot be studied for effectiveness. Therefore, such trials cannot serve as a basis for establishing the clinical utility of a CDx. As a consequence, the claim for the intended use of such a CDx is limited to "selection." "Predictive" claims for CDx rely on understanding the effect of the drug in both biomarker-positive and -negative patients and, therefore, require prospective, randomized controlled trials (RCTs) with randomization stratified by the biomarker.

Several other aspects have to be considered in the planning of clinical trials for CDx development: The initial training set for the assay needs to be distinct from the test set in the clinical trial. The cut-off(s) should be defined prior to being tested in the pivotal trial. The trials need to include sufficient subjects and allow access to appropriate samples. In addition, operational aspects, in particular sample management and reconciliation, are highly critical. Some countries may have specific regulations that also need to be acknowledged in the planning phase. Retaining samples for potential bridging studies or concordance studies should also be considered at this stage.

All these aspects have to be contemplated in the program-specific CDx strategy that has to be in line with the clinical development strategy (e.g., indications and markets of interest, registration path). This strategic document also includes the so-called target test profile (TTP), which describes the targeted characteristics of the final marketed CDx test, such as performance characteristics but also commercial aspects like market access, pricing and reimbursement.

As for the drug development program, risk assessment and a risk mitigation plan are also important for the development of the CDx test because this goes along with several risks, related to, for example, variability in specimen characteristics, specimen availability, technical issues, business risks, the clinical validation of the patient selection hypothesis, and regulatory aspects. These risks have to be identified and a risk mitigation plan has to be defined.

The regulatory submission driven by the diagnostic partner requires close alignment and coordination with and input from the pharma partner. Often complete contents of a premarket approval application are broken down into well-delineated components ("modules"), such as preclinical, clinical, and manufacturing. Each component is submitted to the FDA as soon as completed. The final clinical module has to coincide with the Biological license application (BLA)/New drug application (NDA) filing. This modular approach offers the advantage that the review may begin earlier because the FDA will review each module separately and will provide timely feedback to the applicant, offering the opportunity to resolve deficiencies earlier.

COMPANION DIAGNOSTICS COMMERCIALIZATION AND LIFECYCLE MANAGEMENT

In the case of a CDx, the use of a drug depends on the test result. Therefore, this testing is of huge strategic importance for the drug, even beyond co-approval of drug and CDx. Consequently, although these activities are driven by the diagnostic partner, the pharma partner needs to be actively involved in shaping these activities.

Understanding the demands and market needs that would result from such an approach is critical for the choice of testing technology and testing approach. As for the drug, a commercial plan and launch preparation are needed because the testing should not appear to be a bottleneck to the usage of the drug. It is critical that patients, physicians, and laboratory physicians/pathologists be aware of requirements of such a new treatment concept and that the testing is broadly available and performed with high accuracy. Training, education and awareness programs are needed; quality assurance and quality control programs are essential. Another critical element is funding. Ideally, the testing should be covered by payers. Reimbursement of diagnostic tests in general, however, is limited and highly variable. To achieve coverage by payers a body of evidence on clinical and cost effectiveness of the testing approach needs to be generated as part of a health economic strategy for the CDx. However, despite such data being available, in certain countries the pharma partner is supposed to fund the testing.

Another critical element is lifecycle management of the CDx. Less-developed countries may have different diagnostic capabilities than more advanced countries. Therefore, different technologies or testing approaches may need to be developed for these markets to complement the CDx options for a particular drug. In addition, new technologies or approaches offering certain advantages may became available, so it may be beneficial to develop a second-generation CDx. Changes to the regulatory framework may require reevaluation and recertification of already approved tests. New scientific insights may even lead to a reassessment of the patient selection hypothesis, which finally may require development of new CDx. Therefore, constant surveillance of the scientific, regulatory and technological landscape is needed.

Because CDx development is a highly complex process, a cross-functional team providing input on all the various aspects is needed. Close interactions between Research, Clinical Development, Biomarker/CDx, Regulatory and Business Development line functions are required for the CDx development and approval activities outlined in detail above. In addition, interactions are needed with Commercial and Market Access teams to consider such aspects in the development and commercialization of the CDx. Intellectual property (IP) aspects have to be taken into account and last, but not least, a dedicated CDx Project Management as well as Alliance Management functions will be required to ensure efficiency and success (Figure 5.5.4).

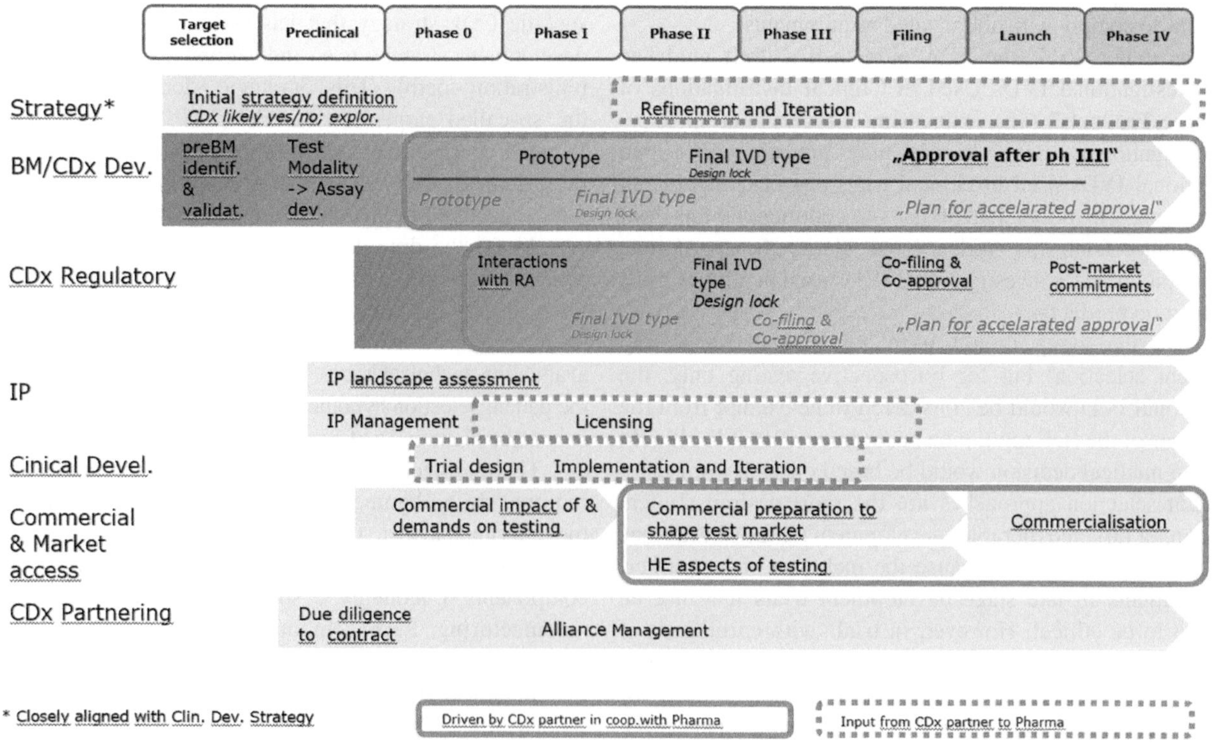

FIGURE 5.5.4 Summary of functional activities for companion diagnostics across the lifecycle.

NEW APPROACHES

The concept of CDx became introduced in 1998 with approval of trastuzumab (Herceptin®) together with the respective HER2 test for selecting patients most likely to benefit. Today, several such drug–diagnostic pairs are now established as key components of twenty-first century cancer treatment (see the FDA's "List of Cleared or Approved Companion Diagnostic Devices," (17). However, inherent challenges such as poor DNA quality from FFPE tissue, extensive intratumor heterogeneity, small sample size, ongoing clonal adaptation and limited access to samples at progression impact the preciseness of this approach. In addition, with the increasing numbers of targeted therapies, the limitations of single-marker CDx for specific drugs has become more and more obvious, calling for new approaches.

Cancer research and diagnostics have been revolutionized by high-throughput sequencing technologies, such as NGS. Significant improvements in reliability (based on advances in sample preparation, sequencing, data analysis/interpretation) and costs made the use of NGS for diagnosis, risk prediction and cancer classification in clinical routine feasible. Several studies have demonstrated the superiority of NGS panels over single assays for a wide scope of molecular oncology testing.

NGS-based multi-gene panels have probably been the most disruptive approach to CDx since this concept was established. In June 2017, the FDA approved the first multi-marker CDx to aid in the selection of NSCLC patients for treatment with specific targeted therapies. Oncomine Dx. Target Test (Thermo Fisher Scientific, currently only offered in the United States) detects and reports sequence variations in 23 genes in DNA and RNA isolated from FFPE tumor tissue, including EGFR, BRAF, and ROS1 alterations linked to the FDA-approved treatments Iressa (gefininb), the Mekinist (trametinib)/Tafinlar (dabrafenib) combo, and Xalkori (critozinib). By that, the likelihood of benefit for three different drugs can be determined by running one test, resulting in a much faster turnaround and selection of the right therapy right from the start instead of a sequential "trial-and-error–like testing." In addition, information on other markers is provided. A couple of months after approval, the test received coverage by BlueShield and BlueCross, demonstrating that payers, in principal, support such new testing concepts.

On November 30, 2017, the FDA approved FoundationOne CDx (F1CDx), another NGS-based genomic profiling test. The test is a laboratory-developed test for which the FDA has generally not enforced premarket review and other applicable requirements. However, upon request by the Foundation Medicine, the FDA worked closely with them to bring the test into the newly established Breakthrough Device Program (18), due to the test's potential consolidate multiple CDx claims for patients and health care providers in a single test. F1CDx is offering a much broader spectrum than Oncomine Dx Target Test, as the test detects genetic variants in 324 cancer genes. In addition, it includes genomic signatures to determine microsatellite instability (MSI) and TMB. By that, the test can not only help to guide selection of targeted therapies but also inform about the use of immunotherapies and last, but not least, identify opportunities for patients to get on clinical trials. Furthermore, the test was validated in five tumor types—NSCLC, colorectal cancer, melanoma, breast cancer, and ovarian cancer. All these features together allow use of this test as a CDx to identify patients with specific genetic alterations who may benefit from 15 targeted treatments approved by the FDA. In addition, reimbursement of this test followed a quite new avenue, as a parallel review program was considered by the FDA and the Centers for Medicare and Medicaid Services (CMS). This is a voluntary program aiming at reducing time between approval by the FDA and coverage by CMS. The F1CDx appeared to be the second IVD approved and covered as part of that program. Therefore, for several aspects, this test demonstrates a totally new avenue for CDx.

Liquid biopsy-based tests offer a huge opportunity for CDx because they are minimally invasive and allow a representative sampling of a patient's current whole tumor burden, thereby overcoming tumor heterogeneity and clonal adaptation issues. In addition, such tests offer the opportunity for longitudinal monitoring at regular intervals and, therefore, lower the limit of detection of recurrent disease beyond the capabilities of imaging. Through this approach, a "window of opportunity" for the initiation of salvage therapy before the actual clinical relapse can be created.

Today, a few single-analyte liquid biopsy-based CDx are approved already and a handful of liquid biopsy-based NGS panels are offered by various vendors through CLIA-certified labs. Once approved, this will result in another shift of the current CDx concept.

REFERENCES

1. FDA final guidance "In Vitro Companion Diagnostic Devices," 2014, http://www.fda.gov/downloads/medicaldevices/deviceregulationandguidance/guidancedocuments/ucm262327.pdf
2. FDA draft guidance "Principles for co-development of an in vitro companion diagnostic device with a therapeutic product," 2016, https://www.fda.gov/downloads/MedicalDevices/DeviceRegulationandGuidance/GuidanceDocuments/UCM510824.pdf
3. REGULATION (EU) 2017/746 OF THE EUROPEAN PARLIAMENT AND OF THE COUNCIL on in vitro diagnostic medical devices, 2017, http://eur-lex.europa.eu/legal-content/EN/TXT/?uri=OJ:L:2017:117:TOC
4. EMA "Concept paper on development and lifecycle of personalised medicines and companion diagnostics," 2017, http://www.ema.europa.eu/ema/index.jsp?curl=pages/news_and_events/news/2017/07/news_detail_002788.jsp&mid=WC0b01ac058004d5c1
5. PMDA "Technical guidance on development of in vitro companion diagnostics and corresponding therapeutic products," 2013, https://www.pmda.go.jp/files/000153149.pdf
6. FDA draft guidance "Developing targeted therapies in low-frequency molecular subsets of disease," 2017, https://www.fda.gov/ucm/groups/fdagov-public/@fdagov-drugs-gen/documents/document/ucm588884.pdf

7. Cummings J, Raynaud F, Jones L, Sugar R, Dive C. Fit-for-purpose biomarker method validation for application in clinical trials of anticancer drugs. *Br J Cancer* 2010; 103(9): 1313–1317, http://www.ncbi.nlm.nih.gov/pmc/articles/PMC2990602/
8. Guidance for Industry E6 Good Clinical Practice: Consolidated Guidance, The International Council for Harmonisation of Technical Requirements for Pharmaceuticals for Human Use (ICH), 1996, http://www.fda.gov/downloads/Drugs/.../Guidances/ucm073122.pdf
9. Good Clinical Laboratory Practice (GCLP): An international quality system for laboratories which undertake the analysis of samples from clinical trials, Research Quality Association, version 2, 2012, http://www.therqa.com/assets/js/tiny_mce/plugins/filemanager/files/Publications/Booklet_Downloads/Good_Clinical_Laboratory_Practice_GCLP.pdf
10. Quality System Regulation (21 CFR Part 820), U.S. Government Publishing Office, Washington, DC, http://www.accessdata.fda.gov/scripts/cdrh/cfdocs/cfcfr/CFRSearch.cfm?CFRPart=820&showFR=1
11. Good Manufacturing Processes (CGMP) m(21 CFR Part 110), U.S. Government Publishing Office, Washington, DC. http://www.fda.gov/food/guidanceregulation/cgmp/
12. Clinical Laboratory Improvement Amendments (42 USC Part 263a), https://www.gpo.gov/fdsys/pkg/USCODE-2011-title42/pdf/USCODE-2011-title42-chap6A-subchapII-partF-subpart2-sec263a.pdf
13. Investigational Device Exemptions (IDEs) for early feasibility medical device clinical studies, including certain First in Human (FIH) Studies: Guidance for Industry and Food and Drug Administration Staff, 2013, http://www.fda.gov/downloads/MedicalDevices/DeviceRegulationandGuidance/GuidanceDocuments/UCM279103.pdf
14. Requests for Feedback on Medical Device Submissions: The Pre-Submission Program and Meetings with Food and Drug Administration Staff, Guidance for Industry and Drug and Food Administration Staff, U.S. Department of Health and Human Services, U.S. Food and Drug Administration, 2014, http://www.fda.gov/downloads/medicaldevices/deviceregulationandguidance/guidancedocuments/ucm311176.pdf
15. FDA draft guidance "Investigational IVDs Used in Clinical Investigations of Therapeutic Products," 2017, https://www.fda.gov/downloads/MedicalDevices/DeviceRegulationandGuidance/GuidanceDocuments/UCM589083.pdf
16. CFR—Code of Federal Regulations Title 21, U.S. Food and Drug Administration, https://www.accessdata.fda.gov/scripts/cdrh/cfdocs/cfcfr/cfrsearch.cfm
17. FDA "List of cleared or approved companion diagnostic devices," https://www.fda.gov/MedicalDevices/ProductsandMedicalProcedures/InVitroDiagnostics/ucm301431.htm
18. FDA draft guidance "Breakthrough devices program," 2017, https://www.fda.gov/downloads/MedicalDevices/DeviceRegulationandGuidance/GuidanceDocuments/UCM581664.pdf

Publishing Biomedical Research Findings
What are the Odds?

5.6

Andreas Scherer

Contents

There is real gratification to be had from the pursuit of science, for its ideals can give purpose to life.
W. I. Beveridge
The art of scientific investigation (1953)

INTRODUCTION

Scientific progress has its foundation in the sharing of knowledge. In the biomedical field, adherence to quality standards of scientific projects is essential to increase the chances for finding cures of disease. Ideally, high-quality science for the improvement of health relies on well-designed experiments, applying standardized procedures and qualified technologies, as well as analytical methods with unbiased interpretation. Transparency and accessibility of reports and data are insufficient without acceptance of, and adherence to, standards of scientific conduct and reporting quality. All stakeholders in health science have the ethical obligation to support and conduct relevant research that is aimed to cure disease or improve health conditions. Avoiding unnecessary research is important. Such is timely publication, and unbiased and complete reporting. It has been widely recognized that, among published reports. a high percentage is of inadequate quality, making it difficult or impossible to judge scientific validity. Inappropriate reporting can occur in many ways: incompleteness, false statements, inappropriate (statistical) analysis, biased reporting on results, and hyper-claiming.

Scientific findings need to be made available to the public: raw data, materials and methods, along with the results and the interpretation. Without thorough reporting of scientific findings, it is impossible for the public to evaluate the findings and eventually accept them as new knowledge, on which again scientific progress can build. Most of these steps sound very technical, but subjectivity may cause problems or "bias" all along. Here we look at some of the obstacles toward unbiased scientific conduct and objective reporting.

SOURCES OF BIAS

In science, planning, conduct of experiments, interpretation, and reporting of data are empirical processes. Whether willingly or unwillingly, bias can be introduced into all steps of research, from the inception of an idea to publishing of results. Automation is being developed in many disciplines to minimize errors by human manipulation. However, despite their best intentions to explore the world based on facts and empirical thinking, scientists are subject to their emotions by nature and are exposed to subjective opinions. Due to the inherent nature of human irrational thinking, the two realms of empiricism and bias cannot always be segregated. Wherever data have to be generated and interpreted, objectivity may be undermined by subjective evaluation. Human nature has the tendency to develop biased opinions toward one's own data as well as data generated by others. Particularly the interpretation of results and how the results are placed within the underlying scientific context may be subject to a number of preconceptions and biases.

Various sources of bias that may influence the interpretation and reporting outcome of an experiment have been described, for example, by Kaptchuk [1]. Investigators may evaluate evidence that supports their own preconceived ideas differently from evidence that challenges these convictions, a phenomenon that is termed "confirmation bias." "Rescue bias" discounts data by finding selective faults in the experiment. Another kind of bias, which Kaptchuk terms "auxiliary hypothesis bias," occurs when modifications are implemented to imply that an unanticipated finding would have been otherwise had the experimental conditions been different [1]. Some investigators may be less skeptical toward the results of an experiment when scientific knowledge or concepts can be found that support the findings. Kaptchuk termed this "mechanism bias," although it may be worth discussing whether it is indeed farfetched to trust findings more when they can be supported by known underlying concepts.

Yet another kind of bias is of particular interest in the context of this contribution, the so-called "publication bias." Publication bias combines a number of phenomena: bias against "negative results," "editorial bias," and "funding bias." The publication strategy is influenced by all stakeholders: by the investigators whose career depend on the current system of publishing "interesting" results in high-ranked journals, on the journals whose reputation and budget depends on publishing influential findings, and on research sponsors who have financial interest in promoting the findings. The connection between publication bias and funding source, also termed "sponsorship bias," has been established in several reports [2,3]. In addition, funding or financial bias can pose pressure on researchers and journals to publish supportive and optimistic results while limiting publication of results that would otherwise not be in support of the client's hypothesis or strategy. This dependence is particularly strong in biomedical research when industry acts through either direct funding or through in-kind contribution to a study [4]. It has been found that studies sponsored by pharmaceutical companies were more likely to have outcomes favoring the goals of the sponsor than were studies with other sponsors: Selective reporting in favor of the sponsor's product or goals, selective study design to favor the product over a known weaker comparator, selective publication by journals that may regard industry-sponsored reports more trustworthy than other sources [2]. However, when Olson and colleagues analyzed controlled trials submitted to the journal *JAMA* between 1996 and 1999, they did not see evidence for such selective acceptance for publication by journal editors based on source of funding [5]. In 1998, journal editors, academics, and representatives of pharmaceutical industry convened and defined publication principles and standards to address concerns about publishing company-sponsored research [6]. The Good Publication Practice for Pharmaceutical companies (GPP1) has since been developed further and is now GPP3 [7]. GPP provides guidelines that aim to address publication bias and the public–private collaborations by encouraging companies to publish all data responsibly and to share data appropriately. Clinical trial information should be shared in a complete and timely manner; reports should be prepared according to law and guidelines, as well as author responsibilities. Based on a self-assessment of more than 500 professionals involved in publishing industry-sponsored research, the GPP2 guidelines (published in 2009) have been widely accepted and used [7].

PUBLISHING NEGATIVE RESULTS: WHAT ARE THE ODDS?

In a publishing and funding system that attributes "positive" with terms like success, trustworthiness, and quality, and "negative" to flaws and failure, positive results are still regarded higher impact than negative results. Negative results account for only 10%–30% of published articles and may have even dropped to 14% until 2007 [8].

As Ioannidis developed from a statisticians view, most published results, that is, positive findings as we have seen, are likely to be false, while negative results are more likely to be trusted (given the proper conduct of the experiments) [9]. According to Ioannidis, three factors attribute to the situation: the "statistical power" of the study as a measure of its ability to avoid false-negative (type II) errors, the unlikeliness of the hypothesis being tested, and the bias in favor of positive claims. Most studies are underpowered and do not generate data with sufficient support for a hypothesis. The chance that a result supports the null-hypothesis is about 97%, while the chance of results rejecting the null-hypothesis is only at around 0.4% [9]. Thus, the conclusion is that negative results are more trustworthy, but journals prefer to publish the positive results. Ioannou supports this finding by pointing out that published "negative results" are less subject to reporting bias, overinterpretation, and "hyper-claiming," there being a researcher's tendency of manipulating the truth in order to get published (Ioannou, A. 2009, files.eric.ed.gov/fulltext/ED504425.pdf).

THE ROLE OF THE INVESTIGATOR

The existence of the publication bias can have detrimental implications on science, ethics, and public health. Its prevention is essential for the dissemination of knowledge.

Where does this arise from?

Reports and publications are the currency of researchers. Bibliometric parameters, based on the number of publications and the journal impact factors, are used for evaluation purposes and competitive measures in the strife to get a job or academic position or to receive public or private funding. It has been well documented that reports on "positive" findings (findings that support the experimental hypothesis) are much more likely to be published than those on "negative" findings (findings that support the null-hypothesis). As Fanelli found, papers are less likely to be published and to be cited if they report "negative" results [10]. Fanelli attributes this phenomenon to the confirmation bias that favors publishing of supporting data over nonsupporting data.

Throughout a study, from the planning phase to the publication, researchers have choices as to formulating a hypothesis, following an analysis path, including or excluding data, and reporting or non-reporting, and as to which conclusions to draw. This flexibility in conducting research studies is also termed "degrees of freedom" [11]. The degrees of freedom pose a risk to scientific validity of a study: They increase the chance of "finding" false-positive (type I errors) results, and they decease the reproducibility of the findings in a reanalysis. Wicherts et al. have identified 34 researcher degrees of freedom [12], a.o., a study is conducted without prior hypothesis, a hypothesis is changed and adjusted during a study to find something "positive," manipulation of data, choosing or changing different analysis parameters, choosing or changing inclusion/exclusion parameters, misreporting results, or presenting exploratory analysis as confirmation. Flexibility in research conduct makes it more likely for a researcher to "find evidence that an effect exists rather than to correctly find evidence that it does not" [11]. According to a meta-analysis of 18 surveys among scientists, about one-third of the participants admitted to have applied questionable research or reporting practices, such as "gut feeling" data manipulation or adjusting the original study design, methodology or results "in response to pressures from a funding source" [13]. In yet another survey, John and colleagues found similar values [14]. A detailed study plan that can be "locked" before the study resumes should be useful to address these issues. An approach to solve this issue is preregistration of clinical studies with a "specific, precise, and exhaustive" research plan [12].

Besides the tool of underreporting, researchers tend to publish positive results with exaggeration and overstating the findings, trying to get attention and publicity. When Vinkers and colleagues investigated science abstracts in PubMed (https://www.ncbi.nlm.nih.gov/pubmed), published between 1974 and 2014, they found that the annual frequency of use of words with positive association, such as (quite unscientific terms as "amazing," "novel," or "excellent" increased by 880% (from 2% of abstracts in 1974 to 17.5% in 2014), while the use of negative words (e.g., again quite unscientific words such as "disappointing," "inadequate," or "weak") increased by only 257% (from 1.3% in 1974 to 3.2% in 2014). One major contributor in the increase of the positive words was the term "robust," which increased by 15,000% in the abstracts between 1974 and 2014. Vinkers and colleagues conclude that there is a likelihood that researchers may use more openly positive language to exaggerate and overstate their findings in order to have a higher chance of being published, which possibly contributes to the rise in observed positive outcome bias in scientific literature (15).

THE ROLE OF THE PUBLISHERS

In the current system, high-impact reports increase the competitive advantage and promote careers for the researchers, and at the same time secure the competitive status of journals that publish reports. Editorial bias, willingly or unwillingly, favors positive results, favors manuscripts from high-income countries, reputed first or last authors and quality of institution [10]. Editors and reviewers are often underinformed on aspects of the study and the conduct. Despite best intentions, editors and reviewers may (1) not always be able to, (2) do not intend to, or (3) are not interested in reanalyzing entire studies and may be subjected to bias [10,16]. Furthermore, a positive effect on the quality of reports that underwent peer review could not yet be determined [17]. In order to fully appreciate, for example, a clinical trial report, a reviewer would need to be expert in experimental design, statistical analysis, and clinical biology. Even for preclinical studies, hardly any reviewer is able to scrutinize the statistical analysis and look at the correctness of a software code, which ideally should be provided along with the manuscript. Despite best intentions, the lack of objective measures on the validity of a finding may, in some cases, lead reviewers and editors to follow more subjective rules in their decision-making process whether a manuscript should be accepted or rejected. In an editorial in the journal *PLoS Medicine* Editors, the authors provide suggestions to authors and editors alike to detect, address and alleviate threats of biased reporting, including paying more attention to competing (commercial) interest, making data and protocols publicly available, and to adhere to the Declaration of Helsinki (www.wma.net/en/20activities/10ethics/10helsinki), which states and demands that editors who reject "unexciting" manuscripts contribute to biased research [18].

However, not only editorial, but educational reasons may be behind the rejection of manuscripts with negative results, the so-called "publishing decision bias." Journal editors are more and more involved in creating revenue for a journal, through publishing potentially high-impact reports and thereby increasing reprints or subscription numbers. Publishers face strong competition and have strong financial interests. It has been estimated that in 2009 the number of scientific journals exceeded 26,000, publishing approximately 1.5 million peer-reviewed articles [19]. Negative results attract fewer citations, and journals publishing presumably higher impact reports will achieve better reputation and sales. It has long been assumed that published articles in scientific journals represent a biased sample of all studies carried out in laboratories [20]. Sterling and colleagues found in 1959, that 97% of published research in four psychology journals reported statistical significance [20]. Thirty-six years later, in 1995, the situation had not changed [21]. Rosenthal coined the term "file-drawer problem" as a metaphor for the assumption that file drawers of researchers are filled with results that they either believe should not be published or would not have a chance to get published [22]. Withholding negative findings or results that contradict their own or befriended colleagues' research is regarded as one means of avoiding public discredit. In addition, the mere anticipation that a report on negative results would be rejected by journals can lead to nonsubmission of such manuscripts [23,24].

It had been suggested that journals should publish more results that support the null-hypothesis. Some high-impact journals have pledged to do so. Recently, journals have been created with particular focus on publishing negative results: for instance, the *Journal of Negative Results in BioMedicine*, *Journal of Negative Results*, and the *Journal of Pharmaceutical Negative Results*. Are there any journals of "positive results in biomedical research"? Is there a "journal of positive results in preclinical biomarker studies"? The efforts to create journals that would be focused on publishing "negative" results and address publication bias, only partially solves the problem: Reports that support (a thoroughly positive term, mind you!) a null-hypothesis should be given equal chance of becoming published as those reports that reject the null-hypothesis. Publishing negative results in low-profile niche journals provide an isolated area lacking the potential of building reputation.

Some journals have already adopted a system of requiring the characterization of cell lines or antibodies used in the manuscripts, certainly a big step in the right direction. Needless to say that all these steps would affect publications with "positive" and "negative" findings alike because both types can deliver very important scientific knowledge that deserves trustworthy documentation. The current educational system; however, is set up such that a large number of publications in journals with lower quality can outweigh a small number of publications in high-impact journals. The current, highly competitive system, where the number of publications and the placement of authorship are of high relevance for a career, appears to inhibit the strife for collaborative work and for "slow publishing," which may be one way to increase the proportion of trustworthy results. There are initiatives for open source publishing, such as the F1000Research (http://f1000research.com/about), that offer immediate publishing without "reviewer-bias," promising, "all articles will benefit from transparent refereeing and the inclusion of all source data." Other web-based post-publishing review formats are PubMed Commons http://www.ncbi.nlm.nih.gov/pubmedcommons) or Scienceopen (www.scienceopen.com) that pledge bias-free acceptance of manuscripts after which an open, non-anonymous post-publication review will resume. It remains to be seen how the research community and other stakeholders will adopt such new platform.

BIOMARKER PUBLICATIONS ARE INFLATED WITH POSITIVE OUTCOME

Much of the initial insight in publication policies and strategies come from studies on social science and psychology studies. Recently, bias in biomarker studies has been investigated. Intriguingly, annually thousands of publications in reputed (and less reputed journals) suggest new biomarkers. For pancreas cancer alone, daily about five or more biomarkers are being published (www.ncbi.pubmed.org).

In 2007, Kyzas et al. examined 1,915 published articles on cancer prognostic biomarkers [25]. Articles where the abstract reported statistically significant prognostic effect for a marker and outcome received the label "positive." The authors investigated "negative" articles for the presence of statements negating the negative. Kyzas found that of the 1,915 analyzed articles on cancer prognostic biomarkers, about 93% reported positive results [25]. Only 26 articles were fully negative for the results and their interpretation.

Carvalho et al. "performed a comprehensive review of meta-analyses of peripheral nongenetic biomarkers that could discriminate individuals with major depressive disorder (MDD) from nondepressed controls [26]. Twenty of 31 meta-analyses reported statistically significant effect size estimates. The authors found evidence of selective publication of "positive studies," and that effect size estimates of meta-analyses may be inflated in this literature." As Tzoulaki and Ioannidis report, "Selective reporting biases may be common in the evidence on emerging cardiovascular biomarkers. Most of the proposed associations of these biomarkers may be inflated" [27]. When Tzoulaki and colleagues performed an analysis of meta-analyses on cardiovascular biomarkers publications in PubMed they found that 49 studies of 56 reported results at significance level, and meta-analyses were highly populated with small-study findings.

Andre et al. found, that the excess of false positive findings through multiple hypothesis testing and a concomitant failure to report negative results inflate "the potential clinical validity and utility of published biomarkers while negative results often remain hidden" [28]. A registry for biomarker studies that should contain all relevant data (e.g., raw data, hypotheses tested, material) that are necessary to evaluate proposed biomarkers and eventually lead to the development of more valuable biomarkers.

Publication bias and hidden multiple hypotheses testing distort the assessment of the true value of biomarkers.

Hidden multi-hypothesis testing arises from several biomarkers being tested by different teams using the same samples. The more hypotheses (that is, biomarker association with outcome) tested, the greater the risk of false-positive findings. These biases inflate the potential clinical validity and utility of published biomarkers while negative results often remain hidden. Trial registries have been developed where all phase II and phase III trials should be listed regardless of study outcome.

The value of publishing negative clinical biomarker results is emphasized by a publication on false-negative MRI biomarkers by Boult et al. in 2012 [29]. Boult tested the hypothesis that non-invasive imaging technology MRI could be used to investigate imaging–pathology correlation between parameters of tumor response and target inhibition in two animal studies, one using the Src inhibitor saracatinib in a murine xenograft model of prostate cancer and another study on rat mammary carcinoma with the VEGFR2 inhibitor vandetanib. Clearly, in both studies the compounds were highly effective, as shown by immunohistology and other technologies. However, the proposed MRI biomarkers failed to detect the strong biological effects of the compounds in the treated versus the control animals. The authors bring to the attention, that "reporting negative imaging biomarker responses is important, to avoid the risk of clinical trials using the same biomarkers being undertaken with a false expectation of success, and the abandonment of promising new therapeutics based on a false-negative imaging biomarker response being mistaken for a true-negative" [29]. Boult et al. specifically discussed imaging markers; however, the statements can certainly be adapted to other technologies.

REPORTING QUALITY: ACADEMIA

Besides flaws in experiment design and analysis, many research reports fall short of full description of material and methods, on details on parameters, standardized use of terms, data acquisition methods, and other details needed for evaluation and reproducing the results. As the development of scientific knowledge relies on repeatable, reproducible measurements and metadata, inaccurate and incomplete information in a publication can lead to waste of money, resources, time and the well-being of study participants in follow-up studies. Because biomarker development heavily relies on robust data, the consequences of imprecise reporting can be detrimental. Meta-analyses on the quality of research publications have recently shown that minimum standards of reporting are not met in many cases [30]. Tektonidou and Ward found in a systematic study of 170 translational research reports on exploratory biomarkers of systemic autoimmune diseases, which were published in 10 journals, less than half of the studies incorporated study-design features important for valid clinical associations [31]. Less than half had appropriate design features, such as age-, or gender matching, only 35.3% controlled for treatment. It is self-evident that the reviewing system failed dramatically in not detecting these issues in the manuscripts, and that caution as to the validity of those reports has to be applied. The consequence of poor conduct and reporting is low reproducibility rates. Poor reproducibility hurts the reputation of and trust in science. The heated discussion around the results of bexarotene-mediated reduction in brain-soluble Abeta peptide levels for the treatment of Alzheimer's disease is indicative of the current state of mind of science and its attempt to regain trust in the public eye. Researchers at Amgen have claimed that they had not been able to reproduce 67% of study results they tested in-house by trying to stay as close to the material, study design and conduct principle as the original publications. One of the studies they failed to reproduce was the bexarotene study by Cramer in 2012 [32]. That is, not until 2015, when Amgen finally made their in-house data available through the F1000 initiative (https://f1000research.com/channels/PRR), which allowed the authors of the original study to realize that Amgen had used a different mouse strain and a different formulation of the compound (comment on http://www.alzforum.org/news/community-news/truth-will-out-negative-confirmatory-orphan-data-find-new-outlets). The formulation may have a major impact on the efficacy of the compound. However, other researchers comment that the authors of the original paper have not been precise enough as to which formulation they used in which experiment. The ongoing discussion highlights the importance of reporting accuracy on one hand, as well caution toward reproducibility studies on the other.

REPORTING QUALITY: PHARMACEUTICAL INDUSTRY

Failure to publish study findings influences our knowledge and can lead to knowledge bias. Reporting over-optimistic results or withholding results can lead to a waste of time and resources at the least. All stakeholders have ethical obligations toward animal and human lives. Reporting negative results in animal studies will save lives of any animals that would otherwise be sacrificed in similar studies. Publishing (negative) clinical trial results is essential to help avoid unnecessary duplication of studies, saving money, patient distress, and possibly unsupported hope for cure. Withholding research findings may harm future patients. There are several examples where clinical treatment interventions were found to be harmful but were not published. In a 1980 clinical trial with lorcainide for the treatment of acute and recovering myocardial infarction, more deaths were observed in the treatment group than in the control group (9/48 vs. 1/47). The development of the compound was stopped for "commercial reasons" and the results were not published. Similar results were observed in other studies about 10 years later, when two other compounds related to lorcainide, encainide and flecainide, were tested in patients. Had the investigators of the later trials known about the results from 1980, the mortality of the patients in the later trials may have been avoided or reduced [33]. Another example of unnecessary research had been provided through a meta-analysis of clinical studies with aprotinin to lower perioperative blood loss.

Fergusson et al. could demonstrate that already after the twelfth of 64 trials that there would have been sufficient cumulative knowledge on the effectiveness and effect size of the compound [34]. Who is to blame for the following 52 trials, which were clearly unnecessary? As Chalmers pointed out, not only are investigators to be blamed for the lack of analysis of existing data, but also ethical boards and journals that published the results that did not contribute anything further to the field [35]. In yet another case, vanilloid receptor 1 antagonists have been tested in clinical trials for their potential pain reduction properties. Only one company has made results on its studies publicly available (www.clinicaltrials.gov); other results are only available on company websites. Obviously, the studies had not been successful, and earlier reporting on those findings would have saved the duplication of such studies.

Another prominent example of misinformation and reporting bias from industry research is the Neurontin case (http://industrydocuments.library.ucsf.edu/drug/docs/njhw0217 [36]. Neurontin (Gabapentin) is a drug developed by the pharmaceutical company Pfizer for the treatment of neuropathic and nociceptive pain, migraine prophylaxis, and treatment of bipolar disorder. On the basis of literature analysis and access to some internal Pfizer documents, Dickersin found evidence of deliberate bias in study design, reporting bias including failure to publish negative results, selective outcome reporting, selective analyses, hiding negative results in abstracts, "reframing" (i.e., making negative results appear positive), and differential citation, and time lag bias, to make Neurontin's effectiveness appear more favorable to the public [4]. Vedula et al., who also had access to company internal marketing documents, concluded that the "publication strategy was to distort the scientific literature, and thus misinform healthcare decision-makers" [36].

According to AllTrials (www. AllTrials.net), possibly as many as 50% of clinical trials have never been published as of 2015. In yet another study, of 2,028 completed or terminated cancer trials from the NIH's ClinicalTrials.gov, 17.6% were registered (status: 09/2007), only 19.5% of the 1,791 completed cancer trials and 3.4% of the 237 terminated trials had been published in peer-reviewed journals [37].

Industry has acknowledged that the demand for transparent publishing of clinical trial results and of company-sponsored research is required to improve the situation. Initiatives for the improvement of data sharing and quality reporting have been created. To increase transparency of clinical trial data, the European Federation of Pharmaceutical Industry Associations (EFPIA; www.efpia.eu) has established a portal for deposition of clinical trials data (www.clinicalsstudydatarequest.com). Since 2013, a total of 3,256 clinical trials have been listed on this site, from where access can be requested (status: December 2017).

PREREGISTRATION

A publication concept, which has recently been developed by a number of journals, is study preregistration (http://www.sciencemag.org/careers/2015/12/register-your-study-new-publication-option, [38]. Under this strategy the researcher submits the hypothesis and a research plan to the journal, where it is peer-reviewed. Upon acceptance, the researcher has ease of mind that the manuscript will be accepted, no matter what the outcome, as long the investigator adheres to the original project plan and passes a second review. The investigator will learn a lot about the scientific value of the proposed study even before a single experiment has been performed. A challenge to this format is that changes to the experiment plan that can arise for various reasons, for example, new published results, will not be easy to enforce. Only a small number of journals offer this new system, and we need to gain more experience with preregistration of studies. It will be important that not only scientists but also other stakeholders such as funders accept the system over time.

GUIDELINES AND GOOD REPORTING PRACTICE

In light of the richness of publications on, for example, biomarkers of disease, the question arises which biomarker should be chosen to move forward and develop into an applicable clinical tool? Certainly one criterion is the reproducibility of the results. As described in much more detail in the contribution by L. Freedman in this book, reproducibility of research results, particularly in biomedical sciences, is being challenged due to an increasing number of incidences of bias and non-transparent reporting. To address this issue, many journals have adopted a system that allows publishing a manuscript only when raw data are deposited in a public repository. By increasing the stringency and quality of the review system, by demanding the deposition of all descriptions of materials and protocols, and analytical tools, they can help filter high-quality manuscripts. However, in many cases descriptions of materials and protocols remain inadequately incomplete and imprecise.

To harmonize the reporting quality, several organizations have developed reporting guidelines, such as CONSORT for clinical trials (Consolidated Standards of Reporting of Trials), and EQUATOR network (Enhancing the QUAlity and Transparency Of health Research), which is an invaluable resource for reporting guidelines (Table 5.6.1). A recent survey among authors and editors showed though that a majority of authors and more than half of the editors are unaware of such community efforts [39]. However, journals increasingly endorse adherence to guidelines although specific instructions to authors and reviewers are not always in place nor do they inform reviewers or authors about the guidelines [40,41]. While many of the guidelines in Table 5.6.1 are rather specific to respective biomedical disciplines or organizations active in human health, generalization of some principles of reporting guidelines in the biomedical research field is possible.

In the following, some basic principles for adequate conduct and reporting are listed. The content is rather limited due to space restrictions of this contribution.

TABLE 5.6.1 Reporting guidelines

ORGANIZATION/GUIDELINE	*WEBSITE*	*DESCRIPTION*	
AMA Manual of Style	http://www.amamanualofstyle.com/	Manuscript style guide	A manuscript style guide for authors and editors.
ARRIVE: Animal Research: Reporting of *In Vivo* Experiments	https://www.nc3rs.org.uk/arrive-guidelines	Animal studies	ARRIVE guidelines are intended to improve the reporting of research using animals—maximizing information published and minimizing unnecessary studies.
ASSERT: A Standard for the Scientific and Ethical Review of Trials	http://www.assert-statement.org/	Review and monitoring of randomized clinical trials	Proposed standard for the review and monitoring of randomized clinical trials by research ethics committees. Checklist incorporates certain elements of CONSORT, to ensure fulfillment of the requirements for scientific validity in the ethical conduct of clinical research (18-item checklist).
Biosharing Reporting Guidelines	https://biosharing.org/	Standards in the life sciences	BioSharing works to map the landscape of community developed standards in the life sciences, broadly covering biological, natural and biomedical sciences. There are 70 reporting guidelines in BioSharing partly compiled by linking to BioPortal, MIBBI and the Equator Network.
CDE: Common Data Elements	https://www.nlm.nih.gov/cde/index.html	CDE initiatives and other tools for protocols for data collection	A common data element is an element that is common to multiple data sets across different studies. This portal provides access to NIH-supported CDE initiatives and other tools and resources that can assist investigators developing protocols for data collection.
CDISC: Clinical Data Interchange Standards Consortium	http://www.cdisc.org/standards	Standards supporting the acquisition, exchange, submission and archive of clinical research data and metadata	Standards supporting the acquisition, exchange, submission and archive of clinical research data and metadata. Develops and supports global, platform-independent data standards that enable information system interoperability to improve medical research and related areas of health care.
CoBRA: Citation of BioResources in journal Articles	https://www.ncbi.nlm.nih.gov/pubmed/?term=25855867	Citation scheme for bioresources	Members of the journal editors subgroup of the Bioresource Research Impact Factor (BRIF) initiative developed a standardized and appropriate citation scheme for bioresources. Adopting the standard citation scheme will improve the quality of bioresource reporting and will allow their traceability in scientific publications, thus increasing the recognition of bioresources' value and relevance to research.
CONSORT: Consolidated Standards of Reporting Trials	http://www.consort-statement.org/consort-2010	Evidence-based, minimum recommendations for reporting Randomized Clinical Trials (RCT)	Evidence-based, minimum recommendations for reporting RCTs. Offers a standard way for authors to prepare reports of trial findings, facilitating their complete and transparent reporting, and aiding their critical appraisal and interpretation (25-item checklist).
CONSORT Plus	http://rctbank.ucsf.edu/home/cplus	Extension of CONSORT, imposes data integrity constraints not possible in text-based reporting	Extension of CONSORT requirements that imposes data integrity constraints not possible in text-based reporting.

(*Continued*)

TABLE 5.6.1 (*Continued*) Reporting guidelines

ORGANIZATION/GUIDELINE	*WEBSITE*	*DESCRIPTION*	
COPE: Committee on Publication Ethics	http://publicationethics.org/	Forum for editors of peer-reviewed journals to discuss integrity of the scientific record	Forum for editors of peer-reviewed journals to discuss issues related to the integrity of the scientific record. Supports and encourages editors to report, catalog and instigate investigations into ethical problems in the publication process. All Elsevier journals became COPE members.
CSE: Council of Science Editors	https://www.councilscienceeditors.org/	Authoritative resource on issues in the communication of scientific information	Organization that promotes excellence in the communication of scientific information. Fosters networking, education, discussion, and exchange. Authoritative resource on current and emerging issues in the communication of scientific information.
EASE: European Association of Science Editors	http://www.ease.org.uk/	Internationally oriented community of individuals in science communication and editing	Internationally oriented community of individuals who share an interest in science communication and editing. Offers the opportunity to stay abreast of trends in the rapidly changing environment of scientific publishing, whether traditional or electronic.
EQUATOR: Enhancing the QUAlity and Transparency Of health Research	http://www.equator-network.org/	Umbrella organization that brings together developers of reporting guidelines, medical journal editors and peer reviewers, research funding bodies and other collaborators	Umbrella organization that brings together developers of reporting guidelines, medical journal editors and peer reviewers, research funding bodies and other collaborators with mutual interest in improving the quality of research publications and of research itself.
FAME Editorial Guidelines: Forum for African Medical Editors	http://www.who.int/tdr/publications/training-guideline-publications/fame-editorials/en/	68-page guidelines	68-page guidelines.
GLISC: Gray Literature International Steering Committee	http://www.glisc.info/	Guidelines for scientific and technical reports and writing/distributing gray literature	Guidelines for the production of scientific and technical reports and writing/distributing gray literature.
GNOSIS: Guidelines for Neuro-Oncology: Standards for Investigational Studies	https://www.ncbi.nlm.nih.gov/pubmed?Db=pubmed&Cmd=DetailsSearch&Term=17146595[uid]	Guidelines to standardize the reporting of surgically based phase I and phase II neuro-oncology trials	Guidelines to standardize the reporting of surgically based phase I and phase II neuro-oncology trials. The guidelines are summarized in a checklist format that can be used as a framework from which to construct a surgically based trial.
GPP2: Good Publication Practice	http://www.ismpp.org/gpp2	Guidelines for publication of the results of clinical trials sponsored by pharmaceutical companies	Guidelines that encourage responsible and ethical publication of the results of clinical trials sponsored by pharmaceutical companies.
ICMJE: International Committee of Medical Journal Editors	http://www.icmje.org/	Uniform Requirements for Manuscripts Submitted to Biomedical Journals Vancouver Group	Uniform Requirements for Manuscripts Submitted to Biomedical Journals Vancouver Group.
INANE: International Academy of Nursing Editors	https://nursingeditors.com/	Best practices in publishing and high standards in the nursing literature	International collaborative whose primary mission is to promote best practices in publishing and high standards in the nursing literature.

(*Continued*)

TABLE 5.6.1 (*Continued*) Reporting guidelines

ORGANIZATION/GUIDELINE	*WEBSITE*	*DESCRIPTION*	
Instructions to Authors in the Health Sciences: Mulford Library, University of Toledo HSL	http://mulford.utoledo.edu/instr/	Publishing guidelines for some journals	Journal titles listed in alphabetical order. Contains publishing guidelines for some journals. Indicates which journals follow CONSORT and/or other guidelines.
Mayfield Handbook of Technical & Scientific Writing	http://www.mhhe.com/mayfieldpub/tsw/home.htm	Handbook	Handbook.
MIAME: Minimum Information About a Microarray Experiment	http://fged.org/projects/miame/	Minimum information about a microarray experiment	Describes the minimum information about a microarray experiment that is needed to enable the interpretation of the results of the experiment unambiguously and potentially to reproduce the experiment.
MIBBI: Minimum Information for Biological and Biomedical Investigations		Developing guidance for the reporting of aspects of biological and biomedical science	Aims to increase the visibility of projects developing guidance for the reporting of aspects of biological and biomedical science.
MOOSE: Meta-analysis Of Observational Studies in Epidemiology	https://www.ncbi.nlm.nih.gov/pubmed?Db=pubmed&Cmd=DetailsSearch&Term=10789670[uid]	Guidelines for reporting meta-analyses of observational studies in epidemiology	Proposal for reporting meta-analyses of observational studies in epidemiology.
PLOS Editorial and Publishing Policies: Reporting Guidelines for Specific Study Designs	http://journals.plos.org/plosone/s/best-practices-in-research-reporting	Requiring that authors comply with field-specific standards for preparation and recording of data and select repositories appropriate to their field	PLOS requires that authors comply with field-specific standards for preparation and recording of data and select repositories appropriate to their field.
Practihc: Pragmatic Randomized Control Trials in Healthcare	http://www.practihc.net/	Open-access tools, training and mentoring to researchers in developing countries to design and conduct pragmatic randomized controlled trials	EU -funded converted action that provides open-access tools, training and mentoring to researchers in developing countries who are interested in designing and conducting pragmatic randomized controlled trials of health care interventions.
Principles and Guidelines for Reporting Preclinical Research	https://www.nih.gov/research-training/rigor-reproducibility/principles-guidelines-reporting-preclinical-research	Improve scientific publishing to enhance rigor and further support research that is reproducible, robust, and transparent	NIH held a joint workshop in June 2014 with the Nature Publishing Group and Science on the issue of reproducibility and rigor of research findings, with journal editors representing over 30 basic/preclinical science journals in which NIH-funded investigators have most often published. The workshop focused on the common opportunities in the scientific publishing arena to enhance rigor and further support research that is reproducible, robust, and transparent.
PRISMA: Preferred Reporting Items for Systematic Reviews and Meta-Analyses	http://www.prisma-statement.org/	Improve the reporting of systematic reviews and meta-analyses	The PRISMA Statement is to help authors improve the reporting of systematic reviews and meta-analyses. It has "focused on randomized trials, but PRISMA can also be used as a basis for reporting systematic reviews of other types of research, particularly evaluations of interventions. PRISMA may also be useful for critical appraisal of published systematic reviews, although it is not a quality assessment instrument to gauge the quality of a systematic review."

(*Continued*)

TABLE 5.6.1 (*Continued*) Reporting guidelines

ORGANIZATION/GUIDELINE	*WEBSITE*	*DESCRIPTION*	
QUOROM: QUality Of Reporting Of Meta-analyses	http://www.consort-statement.org	Guidelines on reporting meta-analyses	Checklist that describes the group's preferred way to present the abstract, introduction, methods, results, and discussion sections of a report of a meta-analysis.
RedHot: Reporting Data on Homeopathic Treatments (A Supplement to CONSORT)	https://www.ncbi.nlm.nih.gov/pubmed?term=17309373	Standard for reporting details of homeopathic treatments	Standard for reporting details of homeopathic treatments. 8-item checklist designed to be used by authors and editors when publishing reports of clinical trials.
REMARK: REporting recommendations for tumor MARKer prognostic studies	https://www.ncbi.nlm.nih.gov/pubmed?Db=pubmed&Cmd=DetailsSearch&Term=16932852[uid]	Guidelines for reporting of tumor marker studies	Guidelines for reporting of tumor marker studies.
SAGER: Sex and Gender Equity in Research	http://www.equator-network.org/reporting-guidelines/sager-guidelines/	Reporting of sex and gender information in study design, data analyses, results and interpretation of findings	Reporting of sex and gender information in study design, data analyses, results and interpretation of findings.
SMRS: Standard Metabolic Reporting Structures	https://www.ncbi.nlm.nih.gov/pubmed?Db=pubmed&Cmd=DetailsSearch&Term=16003371[uid]	Summary recommendations for standardization and reporting of metabolic analyses	Summary recommendations for standardization and reporting of metabolic analyses.
SQUIRE 2.0: Revised Standards for Quality Improvement Reporting Excellence	http://www.squire-statement.org/index.cfm?fuseaction=page.viewPage&pageID=471&nodeID=1	Quality improvement in reporting in health care	The SQUIRE Guidelines help authors write usable articles about quality improvement in health care so that findings may be easily discovered and widely disseminated. The SQUIRE website supports high quality writing about improvement through listing available resources and discussions about the writing process.
STARD: STAndards for the Reporting of Diagnostic accuracy	http://www.stard-statement.org/	Enhance quality of reporting of studies of diagnostic accuracy	Aims to improve the accuracy and completeness of reporting of studies of diagnostic accuracy, to allow readers to assess the potential for bias in the study (internal validity) and to evaluate its generalizability (25-item checklist).
STARE-HI: Statement on Reporting of Evaluation Studies in Health Informatics	https://www.ncbi.nlm.nih.gov/pmc/articles/PMC3799207/	Statement on Reporting of Evaluation Studies in Health Informatics: Explanation and Elaboration	Statement on Reporting of Evaluation Studies in Health Informatics: Explanation and Elaboration.
STRICTA (REVISED STRICTA): STandards for Reporting Interventions in Controlled Trials of Acupuncture	https://www.ncbi.nlm.nih.gov/pubmed/11890439	Designed as a supplement to CONSORT	Designed as a supplement to CONSORT, which has led to improved reporting of trial design and conduct in general. Current plans are to revise STRICTA in collaboration with the CONSORT Group, such that STRICTA becomes an "official" extension to CONSORT.
STROBE: STrengthening the Reporting of OBservational studies in Epidemiology	http://www.strobe-statement.org/index.php?id=strobe-home	Aims to establish a checklist of items that should be included in articles reporting observational research	Aims to establish a checklist of items that should be included in articles reporting observational research.

(*Continued*)

TABLE 5.6.1 (*Continued*) Reporting guidelines

ORGANIZATION/GUIDELINE	*WEBSITE*	*DESCRIPTION*	
Structured Abstracts	https://www.nlm.nih.gov/bsd/policy/structured_abstracts.html	National Library of Medicine (NLM) description of structured abstracts and how they are formatted for MEDLINE	National Library of Medicine (NLM) description of structured abstracts and how they are formatted for MEDLINE.
The REFLECT Statement: Reporting guidElines For randomized controLled trials for livEstoCk and food safeTy	http://www.reflect-statement.org/statement/	Evidence-based minimum set of items for trials reporting production, health, and food-safety outcomes	Evidence-based minimum set of items for trials reporting production, health, and food-safety outcomes (22-item checklist).
WAME: World Association of Medical Editors	http://www.wame.org/	Global association of editors of peer-reviewed medical journals to improve principles and practice of medical editing	Global association of editors of peer-reviewed medical journals who seek to foster cooperation and communication among editors, improve editorial standards, promote professionalism in medical editing through education, self-criticism, and self-regulation, and encourage research on the principles and practice of medical editing.

Source: https://www.nlm.nih.gov/services/research_report_guide.html.

HYPOTHESIS

Predefine the hypothesis and what is required to support or reject the null-hypothesis. A hypothesis should not be changed after the experiments have been performed. Because a hypothesis should be based on access to and consideration of the latest information in the field, rejection of that hypothesis based on experimental data can potentially be a very interesting addition to the scientific area. Investigators should resent the temptation of developing biased interpretations of their observations in favor of their hypothesis.

EXPERIMENT DESIGN

Experimental design is crucial to the ability to interpret the results. Sample number, effect size, batch effects, processing variation, meta-data collection, and many other factors such as the consideration of preexisting data and literature need to be taken into account to eventually be able to confirm that the experiment indeed will be able to address the hypothesis. It is advisable that interdisciplinary teams be formed to discuss and design the experiment. In the biomedical field, thorough knowledge of— for example, clinical need, ethics, surgical requirements, statistics, and ADME processes (absorption, distribution, metabolism and elimination)—may be required to design a meaningful experiment. Most of the time, a single person or group does not have all those skills. For instance, clinicians need to be able to relate the hypothesis and unmet clinical need to statisticians, while statisticians need to be able to make clear what the requirements are for a study to become statistically and scientifically sound.

Sample Size

Sample size calculations are crucial to valid research. In animal studies, using too few animals increases the risk of missing relevant differences between study groups, and is therefore unethical. Equally unethical is the use of too many animals in a study. Underpowered studies may generate false-positive results and lead to the needless use of more animals in follow-up studies; on the other hand, false-negative results may let potentially important data go undetected. "Well-powered experiments reduce the rate of false positive and false negative decisions, and consequently decrease the risk of drawing wrong conclusions" [42].

Randomization, Blocking, and Blinding

Design of experiments attempts to model all possible sources of variation that could interfere with the conduct of a study. Sources of variation can be of a technical (e.g., measurement error) or a biological nature. Technical variation can be reagent variation, temperature fluctuation, slightly different treatment conditions and so on. Biological variation, on the other hand, is inherent to living beings and accounts for the range of possible physical characteristics within a population, or within an individual when measured over a period of time (e.g. circadian rhythm). Experiment design accounts for these sources of variation and attempts to limit the variation to the factors that are relevant for the study, thereby increasing statistical power and the chance to detect true positives and true negatives. Typical design concepts are randomization (random selection of individuals from a population, and random assignment to study groups), blocking (subjects of similar characteristics are grouped together in blocks and randomly assigned to treatments), and blinding (avoiding subjectivity and expectations by nondisclosure of certain sample or patient parameters that may otherwise influence the diagnosis or other assessments). However, a review of 100 articles revealed that only 28% of papers reported that animals were randomly allocated to treatment groups, just 2% of articles reported that observers were blinded to treatment, and none stated the methods used to determine the number of animals per group, a determination required to avoid false outcomes [43].

MATERIAL

Description and identification of material used in the study, including vendor name and, lot number, for example, are of utmost importance to enable reproducibility evaluation. Failure to report on the nature of cell lines, compound formulation, antibodies, or animal strains may lead to inability to recapitulate results by other researchers and thus waste time and resources. The production of antibodies is subject to lot-to-lot variation, and some antibodies used in the literature are poorly characterized. When two research groups use antibodies from different production batches, they may experience different results. According to a review by Hughes et al., between 18% and 36% of cell lines in major cell repositories may be contaminated [44]. The effect of antibodies not binding to the desired epitope or binding to more than one epitope can be detrimental if it goes undetected. Michel and colleagues report that of 49 commercially available antibodies against 19 G-protein coupled receptors, most bound to more than one protein. Hence, they could not be trusted to distinguish between the individual proteins [45]. The overall dire situation is highlighted by Baker, who found that many of the scientists he contacted were not aware of online antibody registries and databases that have emerged, including Antibodypedia (www.antibodypedia.com/), CiteAb (www.citeab.com/), and the broader Biocompare (www.biocompare.com/) [46]. Meanwhile, journals like *Nature* have adopted a policy that requires authors to state that antibodies used in the manuscripts have been profiled for that use.

DATA HANDLING: RAW DATA AND ANALYSIS METHODS

For reviewers and scientists to be able to recapitulate the data analysis, it is essential to have access to raw data. Most journals have now policies in place that require the deposition of the raw data in repositories for reviewer and, upon acceptance, for public access. Data sharing according to the FAIR standards principle is becoming more and more adopted in the scientific community. The FAIR principle is based on sharing and reusing of data that is findable, accessible, interoperable, and reusable [47]. In recent years, data repositories have been established with the honorable intention of creating an answer to the ever-increasing amount of data from technology platforms that are developed to provide more and more data points for scientific discovery (e.g., microarray data, NGS, mass spectrometry). However, instead of developing centralized repositories with sufficient capacity, the number of repositories increased, and with that the format in which data and meta-data can be deposited. Lack of interoperability is limiting the way existing data can be accessed and utilized. The FAIR principles outline a strategy of how data should be structured in globally harmonized modus, so that they can be accessed in a digital, machine-actionable fashion. Several organizations, such as the European Life-science Infrastructure for Biological Information (ELIXIR; www.elixir-europe.org), have adopted these guidelines. A controlled data environment will help increase reproducibility and further the use of existing data.

Good reporting practice involves the clear presentation of all data handling steps. All criteria for data inclusion and exclusion, other data manipulation steps, such as how missing data will be dealt with, should be clearly noted before the experiment starts. Any adjustment has to be noted and defended. End points have to be predefined and not adjusted after the data have been collected. If an algorithm has been developed or applied, the code has to be made available [48]. Decision algorithms as they are used, for example, in clinical biomarker studies have to be defined prior to the study. Retrospective outcome adjustment is highly questionable conduct.

CONCLUSION

To be clear, there are no such things as negative findings! Each well-performed, well-analyzed, and well-documented study contributes to scientific progress and our knowledge about nature.

The terms "negative" and "positive" findings evoke subjective bias, leading to a perception of "positive" as trustworthy findings and reports, while the connotation of "negative" implies fault and failure. Despite all intention toward objective research, the human factor cannot be excluded from the scientific life. Provision of a research environment that assures promotion, support, and reward of unbiased research, reporting, interpretation, and publishing is one of the big challenges for stakeholders in the research field. While scientific fraud appears to be only an exception (about 2% of scientists have declared they have manipulated data or results in reports), insufficient quality of reporting, experimental design and conduct are problems that occur frequently and can mislead colleagues, funders, and skew the competitive landscape.

Publishers, funders, and employers are still underrating publications with "negative" results that either contradict published positive results or otherwise do not support an initial hypothesis.

Needless to stress that publishing negative results requires adherence to high scientific conduct standards, just as much as publishing positive results. Data from well-designed, well-performed experiments lead to either the acceptance or the rejection of the null-hypothesis. Hence, all results from well-conducted and thoroughly reported research have the potential to promote scientific knowledge.

Current publication strategies introduce bias into the pool of scientific publication toward publishing selected findings, causing waste of time and money for those researchers working on similar questions through following too many false positives. In a scientific environment where more and more scientists compete for public and private funding, where the number of publications is being used as a score that can decide on successful funding application and career paths, publishing strategies and reporting quality are being scrutinized. We can start though by increasing training for students on ethics and ethical scientific conduct, by offering more training on robust data analysis practices, and by increasingly introducing funders, editors and reviewers to the good scientific practices of research and reporting [49]. A reward system for completeness of reporting may be developed. Organizations like the European Infrastructure for Translational Medicine (EATRIS; www. EATRIS.eu), have begun to develop and offer services to help investigators and funders alike to improve the quality of funding applications by objectively assessing the completeness of the research plan, the clinical feasibility, and the unmet clinical need. It will be interesting to follow how the services are perceived by the individual stakeholders.

I believe that the best argument for publishing supports of the null-hypothesis is given in the Declaration of Helsinki (2004):

> Both authors and publishers have ethical obligations. In publication of the results of research, the investigators are obliged to preserve the accuracy of the results. Negative as well as positive results should be published or otherwise publicly available. Sources of funding, institutional affiliations and any possible conflicts of interest should be declared in the publication. Reports of experimentation not in accordance with the principles laid down in this Declaration should not be accepted for publication.

REFERENCES

1. Kaptchuk, T.J., Effect of interpretive bias on research evidence. *BMJ*, 2003. **326**(7404): 1453–1455.
2. Lexchin, J., et al., Pharmaceutical industry sponsorship and research outcome and quality: systematic review. *BMJ*, 2003. **326**(7400): 1167–1170.
3. Sismondo, S., Pharmaceutical company funding and its consequences: A qualitative systematic review. *Contemp Clin Trials*, 2008. **29**(2): 109–113.
4. Dickersin, K., Publication bias: recognizing the problem, understanding its origin and scope, preventing harm. In *Publication Bias in Meta-Analysis: Prevention, Assessment and Adjustments*, H.R. Rothstein, Sutton, A.J., and Borenstein, M., Editor. 2005, Chichester, UK, Wiley. 11–34.
5. Olson, C.M., et al., Publication bias in editorial decision making. *JAMA*, 2002. **287**(21): 2825–2828.
6. Wager, E., E.A. Field, and L. Grossman, Good publication practice for pharmaceutical companies. *Curr Med Res Opin*, 2003. **19**(3): 149–154.
7. Battisti, W.P., et al., Good publication practice for communicating company-sponsored medical research: GPP3. *Ann Intern Med*, 2016. **163**: 461–464.
8. Fanelli, D., Negative results are disappearing from most disciplines and countries. *Scientometrics*, 2012. **90**: 891–904.
9. Ioannidis, J.P., Why most published research findings are false. *PLoS Med*, 2005. **2**(8): e124.
10. Fanelli, D., Do pressures to publish increase scientists' bias? An empirical support from US States Data. *PLoS One*, 2010. **5**(4): e10271.
11. Simmons, J.P., L.D. Nelson, and U. Simonsohn, False-positive psychology: Undisclosed flexibility in data collection and analysis allows presenting anything as significant. *Psychol Sci*, 2011. **22**(11): 1359–1366.
12. Wicherts, J.M., et al., Degrees of freedom in planning, running, analyzing, and reporting psychological studies: A checklist to avoid p-hacking. *Front Psychol*, 2016. **7**: 1832.
13. Fanelli, D., How many scientists fabricate and falsify research? A systematic review and meta-analysis of survey data. *PLoS One*, 2009. **4**(5): e5738.
14. John, L.K., G. Loewenstein, and D. Prelec, Measuring the prevalence of questionable research practices with incentives for truth telling. *Psychol Sci*, 2012. **23**: 524–532.
15. Vinkers, C.H., J.K. Tijdink, and W.M. Otte, Use of positive and negative words in scientific PubMed abstracts between 1974 and 2014: retrospective analysis. *BMJ*, 2015. **351**: h6467.
16. Ali, P.A. and R. Watson, Peer review and the publication process. *Nurs Open*, 2016. **3**(4): 193–202.
17. Jefferson, T., et al., Editorial peer review for improving the quality of reports of biomedical studies. *Cochrane Database Syst Rev*, 2007(2): MR000016.
18. Editors, T.P.M., An unbiased scientific record should be everyone's Agenda. *PLoS Med*, 2009. **6**(2): e1000038.
19. Jinha, A., Article 50 million: an estimate of the number of scholarly articles in existence. *Learn Publ*, 2010. **23**(3): 258–263.
20. Sterling, T.D., Publication decisions and their possible effects on inferences drawn from tests of significance—Or vice versa. *J Am Stat Assoc*, 1959. **54**: 30–34.
21. Sterling, T.D., W.L. Rosenbaum, and J.J. Weinkam, Publication decision revisited: The effect of the outcome of statistical tests on the decision to publish and vice-versa. *Am Stat*, 1995. **49**(1): 108–112.
22. Rosenthal, R., The "file drawer problem" and tolerance for null results. *Psychol Bulletin*, 1979. **86**(3): 638–641.
23. Kupfersmid, J. and M. Fiala, A survey of attitudes and behaviors of authors who publish in psychology and education journals. *Am Psychol*, 1991. **46**: 249–250.
24. ter Riet, G., et al., Publication bias in laboratory animal research: A survey on magnitude, drivers, consequences and potential solutions. *PLoS One*, 2012. **7**(9): e43404.
25. Kyzas, P.A., D. Denaxa-Kyza, and J.P. Ioannidis, Almost all articles on cancer prognostic markers report statistically significant results. *Eur J Cancer*, 2007. **43**(17): 2559–2579.
26. Carvalho AF, K.C., Brunoni AR, Miskowiak KW, Herrmann N, Lanctôt KL, Hyphantis TN, Quevedo J, Fernandes BS, Berk M., Bias in peripheral depression biomarkers. *Psychother Psychosom*, 2016. **85**(2): 81–90.
27. Tzoulaki, I., et al., Bias in associations of emerging biomarkers with cardiovascular disease. *JAMA Intern Med*, 2013. **173**(8): 664–671.

28. Andre, F., et al., Biomarker studies: A call for a comprehensive biomarker study registry. *Nat Rev Clin Oncol*, 2011. **8**(3): 171–176.
29. Boult, J.K., et al., False-negative MRI biomarkers of tumour response to targeted cancer therapeutics. *Br J Cancer*, 2012. **106**(12): 1960–1966.
30. Glasziou, P., et al., Reducing waste from incomplete or unusable reports of biomedical research. *Lancet*, 2014. **383**(9913): 267–276.
31. Tektonidou, M.G. and M.M. Ward, Validity of clinical associations of biomarkers in translational research studies: The case of systemic autoimmune diseases. *Arthritis Res Ther*, 2010. **12**: 1–10.
32. Cramer, P.E., et al., ApoE-directed therapeutics rapidly clear beta-amyloid and reverse deficits in AD mouse models. *Science*, 2012. **335**(6075): 1503–1506.
33. Song, F., et al., Publication and related biases. *Health Technol Assess*, 2000. **4**(10): 1–115.
34. Fergusson, D., et al., Randomized controlled trials of aprotinin in cardiac surgery: Could clinical equipoise have stopped the bleeding? *Clin Trials*, 2005. **2**(3): 218–229; discussion 229–232.
35. Chalmers, I., The scandalous failure of science to cumulative evidence scientifically. *Clinical Trials*, 2005. **2**: 229–231.
36. Vedula, S.S., et al., Implementation of a publication strategy in the context of reporting biases. A case study based on new documents from Neurontin litigation. *Trials*, 2012. **13**: 136.
37. Ramsey, S. and J. Scoggins, commentary: Practicing on the tip of an information iceberg? Evidence of underpublication of registered clinical trials in oncology. *Oncologist*, 2008. **13**(9): 925–929.
38. Munafo, M., Open science and research reproducibility. *Ecancermedicalscience*, 2016. **10**: ed56.
39. Fuller, T., et al., What affects authors' and editors' use of reporting guidelines? Findings from an online survey and qualitative interviews. *PLoS One*, 2015. **10**(4): e0121585.
40. Hirst, A. and D.G. Altman, Are peer reviewers encouraged to use reporting guidelines? A survey of 116 health research journals. *PLoS One*, 2012. **7**(4): e35621.
41. Shamseer, L., et al., Update on the endorsement of CONSORT by high impact factor journals: A survey of journal "Instructions to Authors" in 2014. *Trials*, 2016. **17**(1): 301.
42. Grass, P., Experiment design, In Batch *E*ffects and *N*oise in *M*icroarray *E*xperiments: *S*ources and *S*olutions, A. Scherer, Editor. 2009, Chichester, UK: John Wiley and Sons. 19–32.
43. Hess, K.R., Statistical design considerations in animal studies published recently in cancer research. *Cancer Res*, 2011. **71**(2): 625.
44. Hughes, P., et al., The costs of using unauthenticated, over-passaged cell lines: How much more data do we need? *Biotechniques*, 2007. **43**(5): 575, 577–578, 581–582 passim.
45. Michel, M.C., T. Wieland, and G. Tsujimoto, How reliable are G-protein-coupled receptor antibodies? *Naunyn Schmiedebergs Arch Pharmacol*, 2009. **379**(4): 385–388.
46. Baker, M., Reproducibility crisis: Blame it on the antibodies. *Nature*, 2015. **521**(7552): 274–276.
47. Wilkinson, M.D., et al., The FAIR guiding principles for scientific data management and stewardship. *Sci Data*, 2016. **3**: 160018.
48. Shamseer, L. and J. Roberts, Disclosure of data and statistical commands should accompany completely reported studies. *J Clin Epidemiol*, 2016. **70**: 272–274.
49. Leek, J.T. and R.D. Peng, Opinion: Reproducible research can still be wrong: Adopting a prevention approach. *Proc Natl Acad Sci*, 2015. **112**(6): 1645–1646. http://www.editage.com/insights/publication-and-reporting-biases-and-how-they-impact-publication-of-research.

Computational Biology *Modeling from Molecule to Disease and Personalized Medicine*

5.7

Nina Kusch, Lisa Turnhoff, and Andreas Schuppert

Contents

INTRODUCTION

In response to the growing interest in a better understanding of the underlying biological processes driving complex diseases and in the identification of therapies that are tailored to the individual needs of each patient, extensive *omics* data sets have been generated over the years ranging from genomics, over transcriptomics and proteomics, to metabolomics and similar data [1].

With changing emphasis toward specific diseases and their treatments, an increasing number of databases and consortia also focused on clinical and preclinical data reflecting targeted perturbations of cell systems in high-throughput assessments: the Cancer Genome Project by the Wellcome Trust Sanger Institute, for example, collected, among others, mutation and gene expression data for different cancer cell lines [2], The Cancer Genome Atlas (TCGA) Research Network characterized and analyzed human tumors [3] and the Library of Integrated Network-Based Cellular Signatures (LINCS) established a catalog of gene-expression data of human cells that were treated with chemical compounds and genetic reagents [4].

The variety of these programs has led to a large accumulation of data, such that the bottleneck of *omics* approaches nowadays seems to reside in the computational methods to manage, analyze and interpret the emerging data rather than in the high-throughput techniques to generate it [1]. Bridging this gap between the generation and the interpretation of the data is the key objective in today's systems biology [5], but understanding the processes of a biological system and predicting its behavior remains complicated [6]. For the systemic understanding of drug action in patients that is needed to design targeted therapies, the processes on all system levels — from the molecular level, over the cellular, tissue and organ level up to the organism level — must be precisely analyzed and integrated [7].

Advanced modeling techniques that address these challenges therefore play a decisive role in current investigations. In this section, we provide an overview of available approaches and their applications in disease modeling for personalized medicine.

MECHANISTIC MODELING

Mechanistic models use algorithms with structures that resemble those of the system's inherent processes and model parameters that hold an explicit biological meaning [8]. They require *a priori* in-depth knowledge about the system, but can provide comprehensible insights into the model system and even allow for extrapolation to novel settings.

One of the main aims of systems biology is a holistic understanding of biological networks [9], most importantly of gene regulatory networks [10], protein-protein interaction networks [11], metabolic networks featuring biochemical reactions [12], and cellular signaling networks [13]. Two main classes of methods for biological network modeling are discerned: *stoichiometric and structural models* and *kinetic models*.

For the transition from modeling cellular networks over to modeling higher-order structures, such as tissues and organs, it is necessary to understand the behavior of all single components and their interactions at every level and to integrate this knowledge into the system context [5].

Network Modeling

Stoichiometric and structural network models constitute a set of tools that can be applied to study networks of a wide range of sizes, including genome-scale. In contrast to kinetic models, these modeling approaches manage to incorporate structural knowledge about the modeled systems without depending on critical experimental measurements [14,15].

Network-based structural models are derived from graph theory and are applied to elucidate systemic network properties, even in complex networks of considerable sizes.

Representing a network via a graph — where nodes formalize entities such as genes, proteins or metabolites and edges the relations between them [16] — constitutes a common first step in network analysis [17]. Techniques from graph theory can be used to investigate properties, such as redundancy or robustness, and to identify motifs [18,19]. Alternatively, hypergraphs may be used to model networks with edges between arbitrary numbers of nodes, such as metabolic networks [20] with reactions involving extensive protein complexes [21,22] or signaling and regulatory networks [19].

Elementary flux modes (EFMs) and extreme pathway (EP) analysis are two similar network-based modeling frameworks that are based on a uniquely defined convex vector set [23,24] calculated from the stoichiometric matrix that represents the network structure. They have been shown to be capable of determining time-invariant, topological properties of the network and the system it represents. Examples include network flexibility, structural robustness and pathway redundancy [23]. Both methods suffer from unfavorable running times when applied to networks of larger sizes, such as genome-scale networks [23], particularly the EFM approach [25,26].

Constraint-based stoichiometric models make use of the so-called pseudo-steady-state assumption, which postulates that the brief period of transient behavior in metabolic systems is negligible enough to focus on the resulting steady state behavior. They are based on a stoichiometric matrix associated with the network structure and physicochemical constraints that limit the space of possible system states, which can be interpreted as cellular phenotypes [21].

Flux balance analysis (FBA) is a very popular constraint-based modeling algorithm that determines the unique distribution of metabolite fluxes that optimizes a linear target function representing a biological objective, such as cell growth or energy production [27]. FBA is a versatile tool and applicable to large-scale networks to, for instance, study potential drug targets [28]. It is complemented by a substantial array of derived methods [29] that are designed to enable additional features, such as the integration of regulatory mechanisms or the use of additional kinetic information [30,31].

However, FBA and related methods struggle with ill-posed problems arising in systems where several distributions of fluxes in the network realize the optimal target value, for instance, due to redundant pathways [16], which are frequently found in large metabolic networks [27]. In cases like these, flux variability analysis (FVA) can be additionally applied to analyze the alternative optimal solutions and investigate the network's flexibility [32].

Kinetic modeling approaches aim to study the dynamics of the changing concentrations of biochemical entities and their interactions over time. They have been applied mainly to various moderately sized systems [26].

Networks that exclusively feature components that are assumed to exhibit a few distinct discrete states only, can be studied with *discrete logical models*: network components can be represented by logical variables and interactions between them by logical functions [33]. Notable examples for these networks are gene regulatory networks because information derived from the measurement of gene transcription levels is considered to be qualitative [34]. Other important applications of these modeling techniques have been performed on a large-scale signaling network [35], enabling the prediction of key functional network features, and on an integrated network combining metabolic and genetic regulation [36], computing the network's stable states that were linked to phenotypes.

In contrast, *continuous kinetic models* use nonlinear ordinary differential equations (ODE) and can be employed to perform quantitative network analysis, when sufficiently many detailed kinetic parameters are available [37]. Their dependency on an in-depth understanding of the mechanisms driving the system makes them computationally demanding and therefore applicable only to systems of limited size [16]. Nevertheless, they constitute a powerful tool for studying signaling networks [38] and genetic regulatory systems [18]. Moreover, they are complemented by an array of analytical tools, such as bifurcation analysis or sensitivity analysis to assess system robustness [39].

Hybrid kinetic models aim to combine the benefits of both *discrete* and *continuous kinetic modeling techniques* and may be used in the presence of partial knowledge of the system's parameters [40,41].

Multi-Scale Modeling

Advancements in measurement techniques and in the computational methodologies for knowledge extraction enable the integration of models at the cellular level into models of higher-order structures, such as tissues, organs and organ systems, or even whole-body models. These multi-scale approaches rely on the consideration of the effects of a multitude of interconnected mechanisms of multiple spatial and temporal scales [42]. They are extremely useful in the field of drug development, especially for targeted compounds [43], where they allow for the assessment of not only short-term therapeutic effects, but also for the estimation of potential long-term events, such as adverse drug reactions [44].

As a consequence of operating on several scales, multi-scale models face a few particular challenges. First, they need to be able to deal with scales of a very wide range. This holds true not only for time scales, where these models have to integrate both effects occurring within fractions of seconds and events that take months or years to fully develop, but also for spatial scales, that have to accommodate structures of widely varying sizes [45,46]. Second, multi-scale models have to capture and integrate the mechanisms of system components operating at their respective scales as well as their complex interdependencies, which in turn may exert a significant bidirectional impact across scales [47] and lead to the emergence of effects at higher and lower levels [48]. In this context, it is important to determine which interactions in particular need to be passed on while preserving a manageable model complexity [49,50].

Successful projects of multi-scale models on the scale of organs include a spatiotemporal model of liver regeneration developed by Hoehme et al. [51] with a special focus on accounting for the interconnectivity between the organ's microarchitecture and its functionality. Similarly, Schwen et al. [52] developed a multi-scale model of the liver, but focused on the first pass drug perfusion to help designing the delivery and administration of drugs.

A popular approach to build models on the whole-organism level is physiologically based pharmacokinetic (PBPK) modeling, which is capable of quantitatively describing the *ADME* processes of drugs [53]. As PBPK models feature both prior knowledge about relevant drug properties and information about the anatomy and physiology of the organism [54,55], they include only a manageable number of independent parameters that need to be fit, even though these models frequently contain a high number of ODEs [55].

In drug discovery and development, PBPK-based whole-organism models are deployed on a regular basis [56]. One emerging case of application is interspecies scaling in the preclinical phase, where these models are used to estimate an optimal dose for first-in-man trials based on *in vivo* drug pharmacokinetics in animals [57]. Another important use case for these multi-scale models is related to the clinical phase of drug development, where they are utilized during the transition from phase I to phase II. Here, PBPK-based modeling enables the extrapolation from results in healthy test subjects to not only patients in general, but also to specific subpopulations, and therefore aids in assessing risks for vulnerable subgroups of patients [58,59]. Moreover, PBPK models can be employed to systematically investigate inter-patient variability caused by physiological factors, such as age and sex, and therefore help in scaling to pediatric applications [60].

PBPK models have also been used to build multi-scale models on the whole-organism level to investigate the impact of drugs at the system level [55]. Another project notable for its vertical integration of different models across scales is the *Virtual Physiological Human*, which aims to set up an integrated framework that allows gaining detailed insight into human whole-body physiology [61].

DATA-DRIVEN MODELING

Most biological characteristics and functions, as well as disease-driving mechanisms, arise from complex interactions between cellular constituents [62]. As a consequence of this complexity, a full understanding of the processes is often difficult and the majority of current mechanistic models can only focus on specific biological subsystems and benefit from being complemented by data driven approaches.

Data-driven models analyze the system's behavior from a bird's eye perspective, describing it as one unit [14]. By using pattern recognition and machine-learning methods, they can infer knowledge from the data itself and describe the input–output relations of systems. This makes them well-suited for the simulation of systems but less capable of providing scientific insight into the mechanisms.

Based on their output type, data mining algorithms for data-driven modeling can be subdivided into two groups: *classification algorithms*, where qualitative system outputs, such as discrete classes or different states, are predicted, and *regression algorithms*, where the output is a quantitative, continuous value [63]. Moreover, data mining algorithms can be divided into two groups based on their learning behavior: *supervised learning algorithms* use a training set of data that contains known labels as examples in order to learn the function of the system, whereas *unsupervised learning algorithms'* instances are unlabeled and the function is simply inferred from the provided data structure [63,64].

Data-driven approaches usually comprise two main features of the modeling process: (1) variable selection by determining the functionally relevant variables and confining the input dimensionality, and (2) classification or regression by predicting the output value based on the predictive input variables [65,66].

Depending on the application and availability of data, the input variables can describe proteins, mutations, gene expression, copy number alterations, metabolites or other biological and chemical molecule abundances.

Extracting functionally relevant feature subsets from the enormous amount of possible inputs is a key aspect of data-driven modeling because it facilitates a faster and less computationally demanding training of the model, decreases the risk of overfitting and improves the overall performance [67,68].

In rare cases, it is possible to adopt single gene mutations as clinical biomarkers and selectively target them with anticancer therapeutics [2], as observed, for example in chronic myeloid leukemia, where the BCR-ABL fusion gene for a tyrosine kinase can be selectively targeted by imatinib [69]. Unfortunately, these one-to-one relations between single genes and diseases are exceptions [7], so commonly, molecular signatures with more than one feature are investigated for class prediction [67].

As an example, Segal et al. [70] investigated modules, which were chosen sets of genes that acted in concert to carry out a specific function, and their significant activation or repression in various conditions to classify tumors into different types. Similarly, predefined biologically meaningful gene sets—such as gene ontology groups or specific pathways that are available from public databases like the Gene Ontology Consortium [71], Kyoto Encyclopedia of Genes and Genomes (KEGG) [72] or WikiPathways [73]—can be examined for their enrichment in certain sample groups to develop informative classifiers [6]. Alternatively, Golub et al. [74] started using microarray profiling to monitor gene expression of clinical samples. They used both supervised and unsupervised algorithms to distinguish acute myeloid leukemia from acute lymphoblastic leukemia samples. Since then, a lot of different systems-based approaches emerged and found their way into clinical application [75,76].

Besides their utilization in gene expression analysis, machine learning methods can be found in a broad range of applications [77]: they have been successfully applied to genetics and genomics data, such as in the recognition of promoter sequences [78], to proteomics data, such as in the analysis of mass spectrometry data for finding biomarkers to differentiate between samples of different classes [79], or to metabolomics data, such as in finding patterns of metabolites that distinguish plasma from patients with preeclampsia from plasma of healthy ones [80]. Furthermore, they are used for drug sensitivity analyses to establish drug–disease associations for clinical application [81–83]. Different similarity measures were developed that facilitate the connection of transcriptional disease profiles to drug response profiles [84,85] to identify new drug targets, reposition already approved drugs [86] and investigate the effects of drug combinations.

The categories of data-driven approaches range from rather simple methods, such as linear regression, to more complex ones, such as regression trees, support vector machines (SVMs) or artificial neural networks (ANNs).

SVMs map the input data into a high-dimensional feature space to find an optimal hyperplane that creates a decision boundary between the different classes [66,87]. They can perform binary classification, such as in the classification of cancer tissue samples by Furey et al. [88], but also multi-categorical classification, such as the three-class classification of gene expression data from leukemia samples by Lee et al. [89].

ANNs are nonlinear regression models that consist of nodes as elementary processing units and edges as connections in between. They can learn from data by extracting linear combinations of the inputs features and mapping them to the output in a nonlinear manner using transfer functions, such as logistic or sigmoid curves [63]. After an initial guess, the parameters of the ANN are adjusted by error minimization until a specified precision is attained [68,90], which makes them a powerful tool for various applications, such as the ANN-based model to classify small, round blue cell tumors into four specific diagnostic categories by Khan et al. [90] or the ANN-based prediction of the clinical outcome of neuroblastoma patients by Wei et al. [91].

When multiple levels are included in the algorithm, such that each of them transforms the inputs of the respective level into a more integrated and more abstract output, the algorithm is referred to as *deep learning* [92]. The internal feature layers are learned from data using a general-purpose learning procedure [92], which can uncover hidden high-level features. Thereby, *deep learning* has the potential to improve the performance of biological models, increase interpretability and provide additional understanding about the structure of biological data [77]. More and more groups are already successfully applying deep learning to problems in the field of computational biology, such as Kelley et al. [93] in their investigation of DNA sequence alterations by using deep convolutional neural networks [77]. With improving software infrastructure, deep learning can expectably be applied to an even broader range of biological problems in the future, enabling the fusion of different data types and, as currently investigated in IBM's Watson project [94], even the inclusion of text mining approaches [77,95].

The training of data-driven models requires a large amount of data that exponentially increases with the number of input parameters [1]. Additionally, high-dimensional model fitting suffers from the *curse of dimensionality*—a phenomenon that describes the vastness that is generated by the exponential growth of volume for spaces of increasing dimensions [96] and the resulting sparseness of data in such a high-dimensional space [97]. As a consequence, standard statistical approaches perform poorly with high-dimensional data sets and models are caught between adequate solutions that are computationally intensive or simpler algorithms that suffer from low reliability and interpretability [1,66].

Only a few methods operate independently of data dimensionality [66]. Therefore, several mathematical strategies have been developed to address the problems caused by the high dimensionality. As described above, feature extraction algorithms, such as filtering techniques, can be used to remove irrelevant variables from the analysis [65]. Moreover, dimension reduction approaches transforming original features into new ones that can capture most of the information contained in the original data offer practical aid to encounter the challenges of high dimensionality [66]. Another means to significantly improve the performance of data-driven models for high-dimensional data is the additional use of *a priori*–known structural information with structured hybrid modeling (SHM) [98] or the incorporation of biological pathway information [83].

Dimension Reduction Approaches

A common property of high-dimensional data spaces is the existence of highly correlated data points, which can exhibit both local and global correlation structures [66]. For example, upon stimulation, cells can reliably integrate genome-wide inputs to ultimately one of a few cell fates, indicating that the regulatory network can give rise to complex dynamic patterns [99]. In order to explore this lower dimensional space of potential cell states [100], linear dimension reduction algorithms, such as singular value composition (SVD), principal component analysis (PCA), independent component analysis (ICA) or factor analysis, as well as nonlinear dimension reduction algorithms, such as nonnegative matrix factorization (NMF), are currently investigated [101].

PCA is a frequently used unsupervised multivariate approach that extracts the directions of highest variance in interdependent and intercorrelated data [102,103]. With these extracted directions, a projection space can be spanned that offers a new set of orthogonal variables to represent the information contained in the original data set.

Systematic studies suggest that only a few principal components (PCs) are sufficient to characterize most of the biological phenotypes, whereas higher-order components mainly contain irrelevant information or noise [100,101,104]. In a recent analysis however, Lenz et al. [103] investigated the information content beyond the first three PCs. In essence, they found that the underlying linear dimensionality of the gene expression space is low, but higher than previously thought and therefore needs new, refined investigations.

One method that allows a more defined analysis of the dimensionality of gene expression spaces is the *PhysioSpace* algorithm developed by Lenz et al. [105], which uses publicly available gene expression signatures representing biological phenotypes to create directions of clear biological meaning [103]. These signatures serve as a basis of comparison, when new, unknown samples are mapped into the spanned space, and offer a similarity measure between the new and the reference signatures [105]. Because the set of directions can be based on analysis-related preferences, data of different sources, such as cell line and biopsy material, can be integrated such that the *PhysioSpace* method can potentially translate knowledge from laboratory data to clinical application and provide a bigger picture of physiological processes in the gene expression space.

Hybrid Modeling Approaches

In structured hybrid modeling (SHM) approaches, *a priori* knowledge of the structure and the subprocesses of the modeled system can be incorporated in the overall model. This is achieved by creating smaller submodels that can each be characterized by their input–output behavior and connections in between that mirror the real process [98]. Consequently, the effective dimensionality of the problem is reduced and—in comparison with purely data-driven models—the amount of data required to identify the models can be reduced significantly without any loss of accuracy [7,106]. In the context of multi-scale modeling, hybrid models offer the chance to integrate different levels, for example by creating an organ-level framework and successively inserting mechanistic equations for well understood subsystems [5].

Similarly, other approaches that can systematically combine *a priori* mechanistic knowledge and data-driven algorithms can provide valuable tools for the future of integrative modeling [95,106]. During the Dialogue on Reverse Engineering Assessment and Methods (DREAM) project, for example, that compared different drug sensitivity prediction algorithms [83], approaches that incorporated additional biological pathway information in their data-driven models were among the top-performing approaches. Because both data-driven and mechanistic approaches have their benefits and drawbacks, combinations that limit the dimensionality, reduce the data demand and improve extrapolability can provide a solid foundation for the analysis of physiological systems [98].

FUTURE CHALLENGES

As demonstrated in this section, recent advances in data analysis and modeling technologies for biological systems already offer some promising advances toward the concept of personalized medicine. However, most approaches are restricted to a few specialized clinical applications only and in order to realize their full potential and push the translation of knowledge from laboratory data to clinical biomarkers, several scientific challenges remain to be resolved.

First, complex disease states, which are controlled by various intrinsic and environmental mechanisms, can rarely be conceived by only one single biological parameter. In fact, the efficacy of a therapy depends not only on genomic features, but also on a variety of other impact factors, such as comorbidities, or lifestyle and age of the patient, which can hardly be represented in laboratory experiments. Hence, computationally advanced multilevel approaches, which link variables derived from *omics* data to clinical phenotypes and disease profiles, but also real-world evidence approaches, which complement the lack of mechanistic understanding of every confounder by big data approaches, are essential for the future of personalized medicine.

Second, to enable the straightforward use of the vast amount of available data, standardization of both data structures and models will be a key contribution to further advances [107]. On the cellular level, important accomplishments have already been made by introducing the Systems Biology Markup Language (SBML) [108] to uniformly represent biochemical reaction networks and by establishing the BioModels Database [109] as a repository of computational models of biological processes. Similarly, annotation initiatives, such as the Minimal

Information Required In the Annotation of Models (MIRIAM) [110] or the Systems Biology Graphical Notation (SBGN) [111], started encouraging common development procedures.

However, to the present day, physiological and clinical monitoring data are much less standardized and can include partially unstructured formats. Hence, it is crucial to further extend the modeling standards and make them accessible for everyone. In this context, text mining and text interpretation algorithms, but also data management approaches, will further gain in importance.

Third, modeling and project strategies will have to move from individual concepts to combined efforts: crowd-sourced projects, where the workforce and expertise of a variety of researchers from different institutes and organizations is combined to find creative solutions, can contribute to an improved insight into personalized medicine [112].

Finally, methods that consider biological plasticity — the resulting intrinsic change of a biological system as a response to an external stimulus — and resulting phenomena, such as the progression of resistance, will need more attention in the future. In that context, approaches that can integrate dynamic features of disease evolution and cover phase transitions or other emerging characteristics [113,114] can offer significant contributions.

REFERENCES

1. Alyass, A., M. Turcotte, and D. Meyre, From big data analysis to personalized medicine for all: Challenges and opportunities. *BMC Medical Genomics*, 2015. **8**(1): 33.
2. Garnett, M.J. et al., Systematic identification of genomic markers of drug sensitivity in cancer cells. *Nature*, 2012. **483**(7391): 570–575.
3. Weinstein, J.N. et al., The cancer genome atlas pan-cancer analysis project. *Nature Genetics*, 2013. **45**(10): 1113–1120.
4. NIH LINCS Program: Library of Integrated Network-Based Cellular Signatures. August 26, 2016; Available from: http://www.lincsproject.org/.
5. Butcher, E.C., E.L. Berg, and E.J. Kunkel, Systems biology in drug discovery. *Nature Biotechnology*, 2004. **22**(10): 1253–1259.
6. Kristensen, V.N., et al., Principles and methods of integrative genomic analyses in cancer. *Nature Reviews Cancer*, 2014. **14**(5): 299–313.
7. Kuepfer, L. and A. Schuppert, Systems medicine in pharmaceutical research and development. *Systems Medicine*, 2016. **1386**: 87–104.
8. Liberles, D.A., et al., On the need for mechanistic models in computational genomics and metagenomics. *Genome Biology and Evolution*, 2013. **5**(10): 2008–2018.
9. Kitano, H., Systems biology: A brief overview. *Science*, 2002. **295**(5560): 1662–1664.
10. Karlebach, G. and R. Shamir, Modelling and analysis of gene regulatory networks. *Nature Reviews Molecular Cell Biology*, 2008. **9**(10): 770–780.
11. Pellegrini, M., D. Haynor, and J.M. Johnson, Protein interaction networks. *Expert Review of Proteomics*, 2004. **1**(2): 239–249.
12. Duarte, N.C., et al., Global reconstruction of the human metabolic network based on genomic and bibliomic data. *PNAS USA*, 2007. **104**(6): 1777–1782.
13. Kestler, H.A., et al., Network modeling of signal transduction: Establishing the global view. *Bioessays*, 2008. **30**(11–12): 1110–1125.
14. Bruggeman, F.J. and H.V. Westerhoff, The nature of systems biology. *Trends in Microbiology*, 2007. **15**(1): 45–50.
15. Smallbone, K., et al., Towards a genome-scale kinetic model of cellular metabolism. *BMC Systems Biology*, 2010. **4**(1): 6.
16. Tenazinha, N. and S. Vinga, A survey on methods for modeling and analyzing integrated biological networks. *IEEE/ACM Transactions on Computational Biology and Bioinformatics*, 2011. **8**(4): 943–958.
17. Llaneras, F. and J. Pico, Stoichiometric modelling of cell metabolism. *Journal of Bioscience and Bioengineering*, 2008. **105**(1): 1–11.
18. de Jong, H., Modeling and simulation of genetic regulatory systems: A literature review. *Journal of Computational Biology*, 2002. **9**(1): 67–103.
19. Klamt, S., U.U. Haus, and F. Theis, Hypergraphs and cellular networks. *PLOS Computational Biology*, 2009. **5**(5): e1000385.
20. Mithani, A., G.M. Preston, and J. Hein, Rahnuma: Hypergraph-based tool for metabolic pathway prediction and network comparison. *Bioinformatics*, 2009. **25**(14): 1831–1832.
21. Aittokallio, T. and B. Schwikowski, Graph-based methods for analysing networks in cell biology. *Briefings in Bioinformatics*, 2006. **7**(3): 243–255.
22. Mason, O. and M. Verwoerd, Graph theory and networks in Biology. *IET Systems Biology*, 2007. **1**(2): 89–119.
23. Klamt, S. and J. Stelling, Two approaches for metabolic pathway analysis? *Trends in Biotechnology*, 2003. **21**(2): 64–69.
24. Schilling, C.H., D. Letscher, and B.O. Palsson, Theory for the systemic definition of metabolic pathways and their use in interpreting metabolic function from a pathway-oriented perspective. *Journal of Theoretical Biology*, 2000. **203**(3): 229–248.
25. Samatova, N.F., A. Geist, G. Ostrouchov, A.V. Melechko, Parallel out-of-core algorithm for genome-scale enumeration of metabolic systematic pathways. in *First IEEE Workshop on High Performance Computational Biology (HiCOMB2002)*. 2002. Fort Lauderdale, FL.
26. Maarleveld, T.R., et al., Basic concepts and principles of stoichiometric modeling of metabolic networks. *Biotechnology Journal*, 2013. **8**(9): 997–1008.
27. Orth, J.D., I. Thiele, and B.O. Palsson, What is flux balance analysis? *Nature Biotechnology*, 2010. **28**(3): 245–248.
28. Raman, K., P. Rajagopalan, and N. Chandra, Flux balance analysis of mycolic acid pathway: Targets for anti-tubercular drugs. *PLOS Computional Biology*, 2005. **1**(5): 349–358.
29. Price, N.D., et al., Genome-scale microbial in silico models: The constraints-based approach. *Trends in Biotechnology*, 2003. **21**(4): 162–169.
30. Covert, M.W., C.H. Schilling, and B. Palsson, Regulation of gene expression in flux balance models of metabolism. *Journal of Theoretical Biology*, 2001. **213**(1): 73–88.
31. Covert, M.W., et al., Integrating metabolic, transcriptional regulatory and signal transduction models in Escherichia coli. *Bioinformatics*, 2008. **24**(18): 2044–2050.
32. Mahadevan, R. and C.H. Schilling, The effects of alternate optimal solutions in constraint-based genome-scale metabolic models. *Metabolic Engineering*, 2003. **5**(4): 264–276.
33. Chaves, M., R. Albert, and E.D. Sontag, Robustness and fragility of Boolean models for genetic regulatory networks. *Journal of Theoretical Biology*, 2005. **235**(3): 431–449.
34. Sanchez, L. and D. Thieffry, A logical analysis of the Drosophila gap-gene system. *Journal of Theoretical Biology*, 2001. **211**(2): 115–141.

35. Saez-Rodriguez, J., et al., A logical model provides insights into T cell receptor signaling. *PLOS Computational Biology*, 2007. **3**(8): 1580–1590.
36. Asenjo, A.J., et al., A discrete mathematical model applied to genetic regulation and metabolic networks. *Journal of Microbiology and Biotechnology*, 2007. **17**(3): 496–510.
37. Steuer, R., *Computational approaches to the topology, stability and dynamics of metabolic networks.* Phytochemistry, 2007. **68**(16–18): 2139–2151.
38. Brandman, O., et al., Interlinked fast and slow positive feedback loops drive reliable cell decisions. *Science*, 2005. **310**(5747): 496–498.
39. Saithong, T., K.J. Painter, and A.J. Millar, Consistent robustness analysis (CRA) identifies biologically relevant properties of regulatory network models. *PLoS One*, 2010. **5**(12): e15589.
40. Saadatpour, A. and R. Albert, Boolean modeling of biological regulatory networks: A methodology tutorial. *Methods*, 2013. **62**(1): 3–12.
41. Lygeros, J., et al., Dynamical properties of hybrid automata. *IEEE Transactions on Automatic Control*, 2003. **48**(1):2–17.
42. Dada, J.O. and P. Mendes, Multi-scale modelling and simulation in systems biology. *Integrative Biology*, 2011. **3**(2): 86–96.
43. Kuepfer, L., J. Lippert, and T. Eissing, Multiscale mechanistic modeling in pharmaceutical research and development, in *Advances in Systems Biology*. 2012. pp. 543–561, Springer, New York.
44. Sorger, P.K. and S.R.B. Allerheiligen, Quantitative and systems pharmacology in the post-genomic era: New approaches to discovering drugs and understanding therapeutic mechanisms, in *QSP Workshop*, R. Ward, (Ed.). 2011.
45. Hunter, P.J. and T.K. Borg, Integration from proteins to organs: The physiome project. *Nature Reviews Molecular Cell Biology*, 2003. **4**(3): 237–243.
46. Crampin, E.J., et al., Computational physiology and the physiome project. *Experimental Physiology*, 2004. **89**(1): 1–26.
47. Schnell, S., R. Grima, and P.K. Maini, Multiscale modeling in biology—New insights into cancer illustrate how mathematical tools are enhancing the understanding of life from the smallest scale to the grandest. *American Scientist*, 2007. **95**(2): 134–142.
48. Noble, D., Modeling the heart—From genes to cells to the whole organ. *Science*, 2002. **295**(5560): 1678–1682.
49. Tawhai, M.H. and J.H. Bates, Multi-scale lung modeling. *Journal of Applied Physiology*, 2011. **110**(5): 1466–1472.
50. Politi, A.Z., et al., A multiscale, spatially distributed model of asthmatic airway hyper-responsiveness. *Journal of Theoretical Biology*, 2010. **266**(4): 614–624.
51. Hoehme, S., et al., Prediction and validation of cell alignment along microvessels as order principle to restore tissue architecture in liver regeneration. *PNAS USA*, 2010. **107**(23): 10371–10376.
52. Schwen, L.O., et al., Spatio-temporal simulation of first pass drug perfusion in the liver. *PLoS Computational Biology*, 2014. **10**(3): e1003499.
53. Nestorov, I., Whole body pharmacokinetic models. *Clinical Pharmacokinetics*, 2003. **42**(10): 883–908.
54. Rodgers, T., D. Leahy, and M. Rowland, Physiologically based pharmacokinetic modeling 1: Predicting the tissue distribution of moderate-to-strong bases. *Journal of Pharmaceutical Sciences*, 2005. **94**(6): 1259–1276.
55. Krauss, M., et al., Integrating cellular metabolism into a multiscale whole-body model. *PLOS Computational Biology*, 2012. **8**(10): e1002750.
56. Jones, H. and K. Rowland-Yeo, Basic concepts in physiologically based pharmacokinetic modeling in drug discovery and development. *CPT Pharmacometrics & Systems Pharmacology*, 2013. **2**: e63.
57. Thiel, C., et al., A systematic evaluation of the use of physiologically based pharmacokinetic modeling for cross-species extrapolation. *Journal of Pharmaceutical Sciences*, 2015. **104**(1): 191–206.
58. Edginton, A.N. and S. Willmann, Physiology-based simulations of a pathological condition: Prediction of pharmacokinetics in patients with liver cirrhosis. clinical pharmacokinetics, 2008. **47**(11): 743–752.
59. Willmann, S., et al., Risk to the breast-fed neonate from codeine treatment to the mother: A quantitative mechanistic modeling study. *Clinical Pharmacology & Therapeutics*, 2009. **86**(6): 634–643.
60. Maharaj, A.R. and A.N. Edginton, Physiologically based pharmacokinetic modeling and simulation in pediatric drug development. *CPT Pharmacometrics & Systems Pharmacology*, 2014. **3**: e150.
61. Kohl, P. and D. Noble, Systems biology and the virtual physiological human. *Molecular Systems Biology*, 2009. **5**: 292.
62. Barabasi, A.-L. and Z.N. Oltvai, Network biology: Understanding the cell's functional organization. *Nature Reviews Genetics*, 2004. **5**(2): 101–113.
63. Hastie, T., R. Tibshirani, and J.H. Friedman, *The Elements of Statistical Learning: Data Mining, Inference, and Prediction.* 2nd ed. Springer series in statistics. 2009, New York: Springer. xxii, 745 p.
64. Kotsiantis, S.B., Supervised machine learning: A review of classification techniques. *Informatica*, 2007. **31**: 249–268.
65. Rodin, A.S., G. Gogoshin, and E. Boerwinkle, Systems biology data analysis methodology in pharmacogenomics. *Pharmacogenomics*, 2011. **12**(9): 1349–1360.
66. Clarke, R., et al., The properties of high-dimensional data spaces: Implications for exploring gene and protein expression data. *Nature Reviews Cancer*, 2008. **8**(1): 37–49.
67. Strunz, S., O. Wolkenhauer, and A. de la Fuente, Network-assisted disease classification and biomarker discovery, in *Systems Medicine*, U. Schmitz and O. Wolkenhauer, (Eds.). 2016. New York: Springer. pp. 353–374.
68. Larranaga, P., et al., Machine learning in bioinformatics. *Briefings in Bioinformatics*, 2006. **7**(1): 86–112.
69. Druker, B.J., et al., Five-year follow-up of patients receiving imatinib for chronic *Myeloid Leukemia. The New England Journal of Medicine*, 2006. **355**(23): 2408–2417.
70. Segal, E., et al., A module map showing conditional activity of expression modules in cancer. *Nature Genetics*, 2004. **36**(10): 1090–1098.
71. Ashburner, M., et al., Gene Ontology: tool for the unification of biology. *Nature Genetics*, 2000. **25**(1): 25–29.
72. Kanehisa, M. and S. Goto, KEGG: Kyoto encyclopedia of genes and genomes. *Nucleic Acids Research*, 2000. **28**(1): 27–30.
73. Kelder, T., et al., Mining biological pathways using WikiPathways web services. *PLoS One*, 2009. **4**(7): e6447.
74. Golub, T.R., et al., Molecular classification of cancer: Class discovery and class prediction by gene expression monitoring. *Science*, 1999. **286**(5439): 531–537.
75. Tucker-Kellogg, G., et al., Chapter 17—Systems biology in drug discovery: Using predictive biomedicine to guide development choices for novel agents in cancer, in *Systems Biomedicine*. 2010, San Diego, CA: Academic Press. pp. 399–414.
76. Mook, S., et al., Individualization of therapy using Mammaprint: From development to the MINDACT Trial. *Cancer Genomics Proteomics*, 2007. **4**(3): 147–155.
77. Angermueller, C., et al., Deep learning for computational biology. *Molecular System Biology*, 2016. **12**(7): 878.
78. Libbrecht, M.W. and W.S. Noble, Machine learning applications in genetics and genomics. *Nature Reviews Genetics*, 2015. **16**(6): 321–32.

79. Swan, A.L., et al., Application of machine learning to proteomics data: Classification and biomarker identification in postgenomics biology. *Omics-a Journal of Integrative Biology*, 2013. **17**(12): 595–610.
80. Kenny, L.C., et al., Novel biomarkers for pre-eclampsia detected using metabolomics and machine learning. *Metabolomics*, 2005. **1**(3): 227–234.
81. Menden, M.P., et al., Machine learning prediction of cancer cell sensitivity to drugs based on genomic and chemical properties. *PLoS One*, 2013. **8**(4): e61318.
82. Eduati, F., et al., Prediction of human population responses to toxic compounds by a collaborative competition. *Nature Biotechnology*, 2015. **33**(9): 933–940.
83. Costello, J.C., et al., A community effort to assess and improve drug sensitivity prediction algorithms. *Nature Biotechnology*, 2014. **32**(12): 1202–1212.
84. Guney, E., et al., Network-based in silico drug efficacy screening. *Nature Communications*, 2016. **7**: 10331.
85. Hu, G. and P. Agarwal, Human disease-drug network based on genomic expression profiles. *PLoS One*, 2009. **4**(8): e6536.
86. Iorio, F., et al., A semi-supervised approach for refining transcriptional signatures of drug response and repositioning predictions. *PLoS One*, 2015. **10**(10): e0139446.
87. Suykens, J.A.K. and J. Vandewalle, Least squares support vector machine classifiers. *Neural Processing Letters*, 1999. **9**(3): 293–300.
88. Furey, T.S., et al., Support vector machine classification and validation of cancer tissue samples using microarray expression data. *Bioinformatics*, 2000. **16**(10): 906–914.
89. Lee, Y. and C.K. Lee, Classification of multiple cancer types by multicategory support vector machines using gene expression data. *Bioinformatics*, 2003. **19**(9): 1132–1139.
90. Khan, J., et al., Classification and diagnostic prediction of cancers using gene expression profiling and artificial neural networks. *Nature Medicine*, 2001. **7**(6): 673–679.
91. Wei, J.S., et al., Prediction of clinical outcome using gene expression profiling and artificial neural networks for patients with neuroblastoma. *Cancer Research*, 2004. **64**(19): 6883–6891.
92. LeCun, Y., Y. Bengio, and G. Hinton, Deep learning. *Nature*, 2015. **521**(7553): 436–444.
93. Kelley, D.R., J. Snoek, and J.L. Rinn, Basset: Learning the regulatory code of the accessible genome with deep convolutional neural networks. *Genome Research*, 2016. **26**(7): 990–999.
94. Chen, Y., J.D. Elenee Argentinis, and G. Weber, IBM watson: How cognitive computing can be applied to big data challenges in life sciences research. *Clinical Therapeutics*, 2016. **38**(4): 688–701.
95. Dry, J.R., M. Yang, and J. Saez-Rodriguez, Looking beyond the cancer cell for effective drug combinations. *Genome Medicine*, 2016. **8**(1): 125.
96. Bellman, R.E., *Dynamic Programming*. 1957, Princeton, NJ: Princeton University Press.
97. Sammut, C. and G.I. Webb, *Encyclopedia of Machine Learning*. 1 ed. 2011. New York: Springer Science & Business Media. XXVI, 1031.
98. Fiedler, B. and A. Schuppert, Local identification of scalar hybrid models with tree structure. *IMA Journal of Applied Mathematics*, 2008. **73**(3): 449–476.
99. Huang, S., et al., Cell fates as high-dimensional attractor states of a complex gene regulatory network. *Physical Review Letters*, 2005. **94**(12): 128701.
100. Muller, F.-J. and A. Schuppert, Few inputs can reprogram biological networks. *Nature*, 2011. **478**(7369): E4–E4.
101. Schneckener, S., N.S. Arden, and A. Schuppert, Quantifying stability in gene list ranking across microarray derived clinical biomarkers. *BMC Medical Genomics*, 2011. **4**: 73.
102. Abdi, H. and L.J. Williams, Principal component analysis. *Wiley Interdisciplinary Reviews: Computational Statistics*, 2010. **2**(4): 433–459.
103. Lenz, M., et al., Principal components analysis and the reported low intrinsic dimensionality of gene expression microarray data. *Scientific Reports*, 2016. **6**: 25696.
104. Lukk, M., et al., A global map of human gene expression. *Nature Biotechnology*, 2010. **28**(4): 322–324.
105. Lenz, M., et al., PhysioSpace: Relating gene expression experiments from heterogeneous sources using shared physiological processes. *PLoS One*, 2013. **8**(10): e77627.
106. Schuppert, A., Efficient reengineering of meso-scale topologies for functional networks in biomedical applications. *Journal of Mathematics in Industry*, 2011. **1**(1): 1–20.
107. Schneider, M.V., In silico systems biology. *Methods in Molecular Biology*. 2013, New York: Humana Press; Springer. x, 313 p.
108. Hucka, M., et al., The systems biology markup language (SBML): A medium for representation and exchange of biochemical network models. *Bioinformatics*, 2003. **19**(4): 524–531.
109. Juty, N., et al., BioModels: Content, features, functionality, and use. *CPT: Pharmacometrics & Systems Pharmacology*, 2015. **4**(2): 55–68.
110. Novere, N.L., et al., Minimum information requested in the annotation of biochemical models (MIRIAM). *Nature Biotechnology*, 2005. **23**(12): 1509–1515.
111. Novere, N.L., et al., The systems biology graphical notation. *Nature Biotechnology*, 2009. **27**(8): 735–741.
112. Saez-Rodriguez, J., et al., Crowdsourcing biomedical research: Leveraging communities as innovation engines. *Nature Reviews Genetics*, 2016. **17**(8): 470–486.
113. Scheffer, M., et al., Anticipating critical transitions. *Science*, 2012. **338**(6105): 344–348.
114. Brehme, M., et al., Combined population dynamics and entropy modelling supports patient stratification in chronic Myeloid Leukemia. *Scientific Reports*, 2016. **6**: 24057.

Expectations for Quality Control

Challenges with Multiplex Assays

5.8

John L. Allinson

Contents

I will commence this section with a brief overview of Quality Control (QC) because at the time of writing there is the potential (and need) for significant change in the bioanalytical world of Biomarker measurement in drug development. Therefore, for those Bioanalytical scientists who have only thus far focused on Pharmacokinetic (PK) assays and applied standard PK acceptance rules, this will be a major change from the current criteria applied for those assays and recommended in regulatory guidance documents.

INTRODUCTION

In order to understand the expectations for QC, it is necessary to first understand what is meant by the term "Quality Control" from the perspective of the analytical laboratory. As a clinical scientist educated in clinical laboratory science from the start of my career, QC was a very important function of my work—in every single method that I used. To this end, an understanding of basic statistics was required so that method performance criteria could be used to determine how a method could be proven to be "in" or "out" of control when reviewing results and making a decision of whether to report the results or not (acceptance)—asking the basic question, "Is the data reliable or not?"

Although the criticality to individual patients of results generated in clinical laboratories is often far greater than methods used in drug development (excluding biomarkers that are used to make patient intervention decisions), nevertheless the importance of the data and the decisions being made on the results can have major implications to drug development projects. As has been mentioned by regulators many times, "Reliability of biomarker data depends on their measurement." It therefore seems sensible to incorporate the basics of QC used in clinical laboratories to the bioanalysis of biomarkers in drug development because the overall decisions are linked to the clinical interpretation of the results data and also what information can be gleaned from those results related to the clinical utility of the biomarker being studied (context of use [COU]).

QUALITY CONTROL—DEFINITION

One *general* definition of QC is as follows:

> "Quality control (QC) is a procedure or set of procedures intended to ensure that a manufactured product or performed service adheres to a defined set of quality criteria or meets the requirements of the client or customer."[1]

Most definitions that relate to laboratory analysis are in the field of clinical laboratories. A central theme in many of these amended for analytical methods *in general* could be:

> "**Quality control** in the **laboratory** is a statistical process used to monitor and evaluate the analytical process that produces analytical results. QC results are used to validate whether the analytical system (instruments and method) is operating within pre-defined specifications, inferring that test results are reliable." —amended from Ref. [2]

Biomarker methods should all go through some validation exercises so that the performance of the methods can be understood. The degree of the validation will depend upon many considerations as discussed elsewhere (Lee et al.).[3] Typically, QC is then used simply to confirm that the method stays in control and performs as it is expected to do from the benchmark of its performance during that validation exercise. Here, the "statistical process" mentioned above becomes important.

USE OF QUALITY CONTROL IN BIOANALYSIS OF BIOMARKERS

In my experience over 25 years in contract research, bioanalytical laboratories have basically followed a rule used in PK assays called the 4-6-*x* rule:

- Typically, batches of samples are processed with three levels of QC (low, medium and high), with two replicates of each level in every batch—hence "6" QC samples per batch:
    - Two-thirds (4 of 6) of all the samples must be within *x*% of the nominal.
    - Half the samples at each level must be within *x*% of the nominal.

This rule can also be applied in larger batches by increasing the number of QCs and acceptance numbers in proportion (e.g., 8-12-*x*).

These acceptance criteria are applied using the same value for "*x*" in all assays using the same type of technology for the PK method. Therefore, for small molecule drugs using liquid chromatography—tandem mass spectrometry methods, 4-6-15 is common ($x = 20$ at the lower and upper levels of quantitation [LLOQ and ULOQ, respectively]), and for large molecule drugs using ligand binding assays (LBA), 4-6-20 is common ($x = 25$ at LLOQ and ULOQ).

This rule is actually a derivation of a ±1 standard deviation (SD) rule because the ratio of "4 of 6" QCs passing statistically matches with expectations of results being within ±1 SD of the target value (±1 SD of the mean in a normally distributed population contain 68.3% of the results). Statistically, the problem that exists in applying the rule using ±15% and 20% criteria mentioned above is that these numbers do not necessarily have any valid relationship with the performance of all of the different methods being used. The criteria are ensuring that the results remain within these variable limits, but they do not ensure or monitor whether the analytical method is in control or not, as understood by expected performance.

At recent scientific meetings (Crystal City VI, September 2015; WRIB, April 2016; AAPS-NBC, May 2016; and EBF workshop, June 2016), following discussions around the subject of acceptance criteria for biomarker assays in drug development, the consensus at all meetings has been to take a more statistically driven approach such as that used in clinical laboratories due to the very different use of Biomarker data when compared with the use of PK data.

Moreover, it also takes into account the differences that exist from a physiological standpoint between different biomarkers. Physiological variation and, thus, determining clinically significant changes will differ; some biomarkers show very little variation in normal subjects and, thus, clinically significant changes may be small in terms of percentage change when compared with biomarkers that have a wide variability. Therefore, the requirements of the assays performance for different biomarkers may also vary widely.

Routinely, in simple terms, the statistics applied in clinical laboratories use multiples of the SD of the method at validation to define "warning" and "failure" criteria—typically ±2 and 3 SD respectively. These are equivalent to 95.4% and 99.7% confidence limits.

Therefore, without taking up more of this chapter on QC and the rules that should be used, suffice to say that each individual method's performance criteria—as proven at validation—should be used to calculate the acceptance criteria as mentioned above (i.e., ±2 and 3 SD, which in percentage terms is ±2 or 3 × inter-assay percentage coefficient [CV%] of the method, respectively). A single QC exceeding ±3 SD of its target value would usually be a batch failure, whereas QC results between 2 and 3 SD would be a warning to check overall performance of the methods retro- and prospectively. It is usually a good idea to review all QC data at the end of a study or on a interim basis if there are many batches. QC charts are a great help here to visualize data instead of relying just on data tables.

Additionally, if you can use the same samples for validating the assay as well as QC for the sample analysis, it is also a good idea to combine the results of those samples together and update the validation performance tables because now you will have a bigger data set and a better understanding of the long-term method performance as opposed to results from a potentially small number of batches performed during the validation exercise over a short time frame. Once the overall performance results from this exercise is available, recalculation of nominal means and SDs will allow a retrospective interrogation of the QC results data to ensure that all the batches are performing as expected.

Last, and to emphasize the point again, for multiplex assays, we must look at each analyte as a separate method. Each method will have its own performance criteria, and this must be used to calculate the expected and acceptable ranges for the QCs used. The potential is that every analyte will have different ±percentage ranges, and these percentages may be different for each QC level. The latter point is important and is just the same as the variance used for LLOQ and ULOQ in PK assays where wider criteria are allowed. However, we are basing the acceptance criteria here on data produced with those QCs, and this data reflects the precision profile of the assays (Table 5.8.1).

Figure 5.8.1 provides an example of a typical precision profile, and we can see from this why we should have variable acceptance criteria for different concentrations: simply because the method performance is not the same at different concentrations, as demonstrated by the precision profile line, which plots method CV% against BM concentration.

Table 5.8.1 gives a potential example of acceptance criteria for a multiplexed Biomarker assay using the precision data from validation to calculate the criteria for each method. The data shows that for each method the best precision is for the medium level QC, which matches expectations from the typical precision profile above. It also demonstrates why the criteria should be calculated specifically for each level because it is clearly very different; in some assays (e.g., IL-1ra) we see a near doubling of failure criteria between medium QC and high QC of ±11.1%–20.4% respectively.

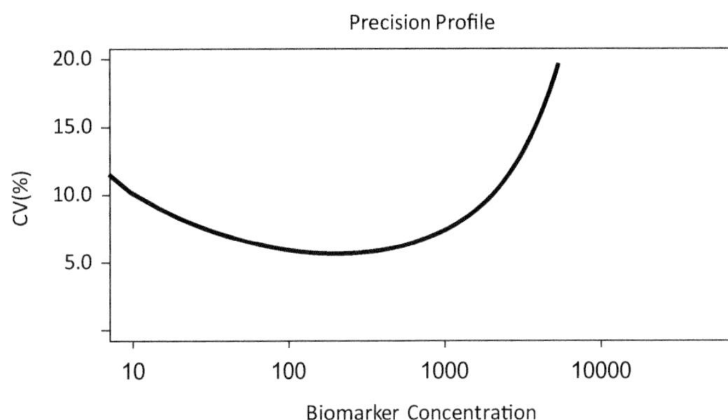

FIGURE 5.8.1 Method precision profile.

WIDENING ACCEPTANCE CRITERIA BASED ON THE CONTEXT OF USE OF THE BIOMARKER

In my experience, many multiplexed methods are used in very early drug development research and the COU should be considered when deciding how precise the individual assays need to be to allow results to be accepted and the right decisions made based on the data.

Although the description of analytical QC above is sound, there will be instances where increased variability may not have a profound effect on the results data because of study endpoints or magnitude of change, and here data may be accepted considering the factors that are related to concept of use.

Even when this is the case, I believe it is still good practice to monitor QC as explained above in terms of understanding how the method is performing in every batch. If QC results are outside the usual limits imposed based on the validation performance of the method, results may be able to be reported with caution and I would suggest in these cases that data be annotated with how those specific batches did perform in terms of QC results (e.g., with 95% confidence limits for each QC level).

TABLE 5.8.1 Example of acceptance criteria calculated from validation data in multiplexed assay

		CRITERIA	*IL-2*	*IL-6*	*IL-1B*	*IL-1ra*	*IL-15*	*IL-2ra*
Low QC	Inter-assay CV (%)		7.1	7.4	9.5	6.0	8.9	9.6
	95% CL (±2SD)	±(%)	14.2	14.8	19.0	12.0	17.8	19.2
	Failure limit (>±3SD)	>±(%)	21.3	22.2	28.5	18.0	26.7	28.8
Med QC	Inter-assay CV (%)		5.1	5.1	6.3	3.7	8.5	6.6
	95% CL (±2SD)	±(%)	10.2	10.2	12.6	7.4	17.0	13.2
	Failure limit (>±3SD)	>±(%)	15.3	15.3	18.9	11.1	25.5	19.8
High QC	Inter-assay CV (%)		6.1	8.8	7.0	6.8	8.4	7.0
	95% CL (±2SD)	±(%)	12.2	17.6	14.0	13.6	16.8	14.0
	Failure limit (>±3SD)	>±(%)	18.3	26.4	21.0	20.4	25.2	21.0

For example, an assay may be used to assess the magnitude of change of a biomarker due to drug effect, and if the expected or required change is very large (e.g., 50% or even 2- to 5-fold), then even a precise method where some QC fails may still adequately be able to predict changes such as this has occurred. It is the methods that require more precise evaluation, for instance to detect the smallest clinically significant change, where QC practices need to be stricter.

There are other QC concepts that can come into play with such biomarker assays and the reader is advised to study the concept of total allowable error (TAE), which can take into account both analytical and physiological variation to define the requirements for the performance of the assay.

A note of caution here: It is important that any method validation data be thoroughly interrogated to understand completely what the method performance criteria are—that is, to define the method characteristics. Only then should the method be evaluated against TAE requirements to ensure that the performance is adequate. Should this process be driven by the TAE, often a method will be accepted because it fulfills the criteria without understanding that there may be flaws in using the method a certain way that can be easily corrected and result in a better method performance. This can often be the case when establishing a minimum required dilution (MRD) via the parallelism experiments depending upon the techniques used. Readers are referred to the paper by Darshani et al.[4] and the C-path white paper[5] for an evaluation of parallelism/MRD.

Ultimately, it is the responsibility of the biomarker "team" of sponsor–clinical and biomarker scientists and the laboratory scientists conducting the development and validation of the assays to decide upon what are acceptable criteria to use to meet the requirements of the COU of the various biomarkers measured.

A last comment here is to remember the stage and importance of the study being conducted in the drug development timeline. In my experience, most multiplexed assays have been used in early phase development and for internal decision making by the sponsor. Here it is unlikely that the data will undergo regulatory scrutiny. However, I have also conducted several multiplexed assays in late phase trials required to meet criteria for primary and secondary end points. This data will almost certainly be audited by regulatory agencies, so bear in mind that the method validation and acceptance criteria gleaned for it will need to have robust scientific justification. Here, experience in statistically valid QC and results data interpretation is very important.

QC CHALLENGES FOR MULTIPLEXED ASSAYS

QC challenges in multiplexed assays are often overcomplicated by the choice of the actual assays themselves. It is therefore worthy to note some of the considerations to apply that may assist in simplifying the QC challenges later down the line.

1. What are the analytical ranges required for each analyte? This really comes from what the abundance of the biomarker is in the matrix of interest in the subject population that you are studying. Because most multiplexed assays that will be used will be from commercially developed sources, this becomes an important issue because the option to reoptimize methods will probably not exist. Ensuring that the analytical ranges of the methods will encompass the concentrations in your study samples will greatly assist your challenges with regard to QC material and samples.
2. What are the sensitivity requirements of each method? This alone may reduce your options for commercially available methods and will certainly impact and often challenge production of QC material to prove method performance at low concentration.
3. It is important to check each method for known matrix interferences and manufacturers will have information for some matrices but not necessarily the specific one of interest to you. The key here is if there are MRDs due to these effects and therefore it is an important investigation prior to conducting any validation experiments. Different matrices may give quite different results to those quoted by the manufacturer. The same is true for matrix from different disease populations. If so, then it is very important that all the assays require the same MRD. Multiplexing assays with different MRDs then requires that the highest MRD across all the multiplexed assays must be used, and where the MRD is different (e.g., less) for other assays, then they will be negatively impacted due to changes in sensitivity (i.e., limit of detection [LOD] and LLOQ) because they will be diluted to a higher degree than the individual methods would require. Figure 5.8.2 provides an example of this: The data of a dilutional series from two methods have been plotted. The "100%" dilution for recovery has been established to be the first dilution where subsequent dilutions show consensus in recovery. The smaller dilutions—using the Neat sample as target—show a very significant positive increase in recovery due to matrix effects that are reduced by further dilution. This is a parallelism experiment for which details can be found elsewhere (Jani et al.[4]) and C-path.[5]
4. Another factor to investigate carefully is the number of methods to include in the multiplexed assay and the platform upon which to conduct it. Some platforms will dictate limitations in terms of assay numbers (e.g., number of spots per microwell allowed due to space considerations, for example, Meso Scale Discovery [MSD], and Aushon Ciraplex); whereas others will have a wider scope, for example, xMAP (Luminex) and the Randox Bio-chip array.

The question is often answered by the COU that the biomarker results are being used for. It is common to see larger multiplex panels in the earlier phases of

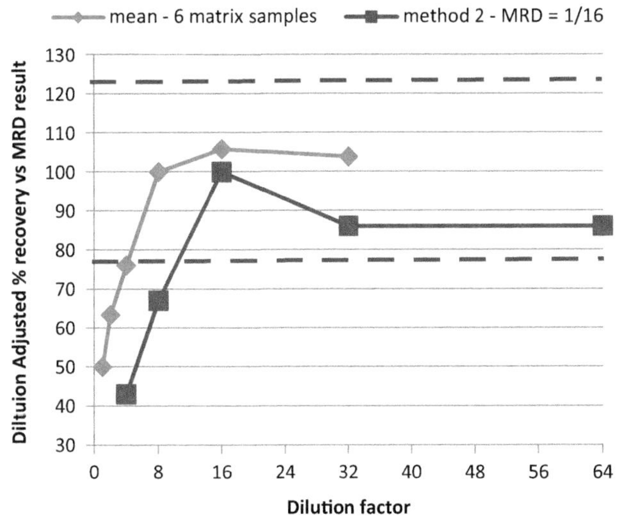

FIGURE 5.8.2 Example of two assays with different MRDs. (Red dotted line is the acceptance criteria; i.e., ± [3 × interassay CV%]) Method 1 has MRD of 1/8, Method 2 has MRD of 1/16. Here, a 1/16 dilution must be used for a multiplexed assay of these two methods and, therefore, the sensitivity on Method 1 is negatively impacted by a factor of 2.

development progressing through a refinement process in terms of both the number of biomarker methods and the degree of validation as the drug advances through different development phases. Some platforms are capable of being fully automated, which will improve on robustness (as long as the methods are sound).

As a rule of thumb, and in later phase projects, if I were using a platform that was not fully automated and for which I required an assay validated to an advanced level, I would generally not want more than four to five analytes on some platforms. Additionally, some companies offer assays with different levels of validation having been conducted by their laboratories, and in general, I stay away from "screening" or unvalidated methods and focus on those with demonstrably more robustness in terms of performance data and information that the manufacture is willing to provide.

CHOICE OF QC MATERIAL FOR USE IN MULTIPLEXED METHODS

What Is Ideal?

The ideal for QC samples is that they are fully representative of the samples to be measured. That means exactly the same matrix—serum, plasma (lithium heparin, sodium heparin, K2-EDTA, K3-EDTA, citrate, oxalate, fluoride-oxalate), urine, cerebrospinal fluid (CSF), for example—and that they are species specific. It also means that the molecules in the samples are endogenous and not artificial. Use of matrix spiked with recombinant or synthesized molecules does not really give much information about how the endogenous biomarker is behaving in the analytical method, it only gives information about how the recombinant molecule is behaving, so as a QC it is not ideal, but it may be required to be used due to the circumstances detailed below.

ENDOGENOUS QUALITY CONTROLS

Endogenous QCs (EQC) are the ideal for all biomarker analytical work. These are samples that FULLY represent the study samples being analyzed—both matrix and molecule. So, although spiked QCs described later may tell us something about the performance of the commercial assay in terms of working as manufacturer claims and rule out potential method issues in a batch, the EQCs tell us how we may expect the samples to have performed. The main problem for multiplexed assays is the ability to source samples that contain a significant concentration of ALL of the biomarkers in the multiplexed assay.

Biomarkers that have relatively high abundance in normal subjects will not be a problem, but for biomarkers that are very low or undetectable in normal subjects, it becomes a much more difficult task. Here, it is often impossible to source samples that will have all the biomarkers present and for many disease states, there is no single remedy. However, in some therapeutic areas, particularly immune-mediated and inflammatory disorders it may be worthwhile to attempt to use *ex vivo* stimulation of whole blood or Peripheral blood mononuclear cells (PBMCs) to try to generate the molecules of interest.

One possible solution is to create pools of sample from individuals who have the same biomarkers present or whereby mixing samples from these individuals does not result in one or more of the biomarkers being diluted too much so that it has no practical use. Different levels of EQCs can be created by mixing low and high concentration samples at different ratios, but again this will only result in producing EQCs for those biomarkers that are present in significant concentrations after pooling.

In my experience, it is nearly always the case that to cover all biomarkers with EQCs, more than one pool and, hence, more than one EQC may be required. Each will only be able to act as an EQC for the biomarkers they contain, and you will often have little or no control in choosing what the concentration of the biomarkers in the pool are due to the limitations of the samples you may have. Purchasing diseased-state samples from biorepositories may be expected to help in this regard, and having done this many times, I have a very mixed opinion of how successful this may be. If you decide to go this route, do so with caution and obtain as much information as possible about the collection and storage processes that the sample have undergone to help you make the decision.

It is also worth planning for the issue of obtaining biological matrix from diseased-state patients who are participating in clinical studies. This is often a very good source of relevant and useful matrices. However, it does necessitate obtaining informed consent for this use with the intention of utilizing sample residual matrices. Here, the samples with known concentrations that are the most appropriate to use to construct pools can be selected after analysis.

In consideration of all of the above possibilities and issues, the outcome, therefore, is that we need a compromise of what is achievable versus what may be desirable.

The same is true in single method biomarker assays where we are studying the biomarker in rare matrices; for example, tears, synovial fluid, vitreous humor and blister fluids. The Lee et al.[3] paper recognized this issue and acknowledged that it may not be possible to have multiple EQCs; indeed, depending upon the biomarker's abundance in the matrix, it may be difficult to produce even a single EQC. Here, the focus is on trying to produce at least one QC that is the same matrix as the samples being studied and other QCs may need to be in surrogate matrix. The compromise for QCs in multiplexed methods is the same if it is a rare matrix.

A note of caution in obtaining rare matrices from biorepositories collected from cadavers. This is less common than some years ago, but cadaverous collection of biological material with consent means that this sample material may be available. Certainly, cadaver CSF is available from a number of sources today. Consideration of suitability of this material is vital in different biomarker analytical methods. Some acute changes postmortem does occur in biological fluids and these may impact the results data that is generated from these samples. The author has experienced a number of cases where this has occurred and in terms of CSF. For instance, it is generally not suitable to use cadaver CSF for the study of biomarkers of typical interest in Alzheimer's disease.

This is for several possible reasons, some method dependent, but pH changes alone, for example, can mean that the endogenous molecule in this matrix may not behave in the same way as the calibration reference material in the calibration standards, that is, it does not demonstrate parallelism. Therefore, misleading information may be obtained from these matrix types (e.g., stability) when used to characterize and control analytical methods.

In the scenario of using other less rare matrices, however, for multiplexing, another series of compromises is required:

- It should be recognized that it may not be possible to create EQC(s) for all biomarkers in a multiplexed assay.
- To cover all biomarkers in the multiplexed assay, therefore, it will often be necessary to use more than one EQC matrix pool.
- It should be recognized that all QCs should be in the same matrix as the study samples if possible (excepting rare matrices as above).
- These QCs ideally will be EQC, but often that will not be possible.

SURROGATE AND SPIKED QUALITY CONTROLS

Note: In all cases of spiking, the materials used should, wherever possible be recognized International or WHO reference materials. This will ensure a degree of consensus between laboratories and methods in terms of concentration results if every laboratory harmonizes to using these materials as primary standards.

Where EQCs are not possible, or creation of EQC with the concentrations required to control the analytical ranges required are not possible, other alternatives are available to use in combination with any EQC(s) produced:

- *The same biological matrix as the study samples* fortified with recombinant/synthesized materials that are the same molecules as the biomarkers of interest. These QCs may be a mixture of endogenous and artificial molecules where the endogenous biomarker is present or simply artificial molecule QCs where the endogenous biomarker is undetectable.

  For Multiplexed assays, it may be necessary to spike in more than one or even all biomarkers using artificial molecules. For ease of use, it is advisable to have a single spiking solution at high concentration of all of the biomarkers that are required to be spiked into matrix to create these QCs. Where possible, it is also advisable to maintain the concentration of the original matrix at 95% or greater. Hence, a spiking solution created with at least 20 times the highest concentrations of all the biomarkers required will enable this.
- *The same biological matrix as study samples* with endogenous biomarker present diluted to produce low concentration QCs when the biomarker is typically present with a relatively high abundance, and so obtaining samples with low biomarker concentrations may be difficult. This is often required when the effect of the drug being developed is to reduce the circulating concentration of biomarker—sometimes to below clinically normal levels. Here, at least the molecules are endogenous biomarkers, but the matrix is diluted. Any parallelism studies conducted on the original pools will support the use of the dilution required to produce these QCs.
- *Different matrix to study samples*—in circumstances such as studies with rare matrices, it may not be possible to construct all QC samples with the biological matrix of interest. Here, it is important to try and have at least one QC made in the same matrix as the study samples but then use a *surrogate* matrix for the others. Although this may not be ideal, it may be all that is achievable and therefore is a practical solution, the impact of which must be considered when

interrogating the results data produced in order to evaluate and interpret those results in terms of reliability. A number of options are available.

- Using commercially produced "artificial" matrices such as artificial CSF and artificial tears. There are a number of sources and recipes for these (and other matrices) that can be found by internet searches. Best practice would dictate that the reliability and appropriateness of their use needs to be determined in the same way as other matrices. Therefore, looking at the behavior of the matrix in the method is advised. Dilutional linearity will give some information of matrix effect, and precision studies at different spiked values are essential to justify their use.
- Using protein-based buffer solutions similar to those in commercial assay kits mentioned above. Remember to use international/WHO reference materials in the preparation of QCs, if they are available.

COMMERCIAL QUALITY CONTROLS SUPPLIED BY REAGENT KIT MANUFACTURERS

Commercially manufactured QC materials (samples) may be available from reagent kit manufacturers and are usually included as a kit component. However, some manufacturers supply them either as separate items only or also as additional extra items to the kits for those users who require more material (samples) than is routinely provided in the kit.

Usually these QCs are in a surrogate matrix—most often, a protein-based buffer solution. This is often also the material that is used by manufacturers when they "validate" their assays. Therefore, when you read the literature (kit insert) provided with their kits, it is important to understand that the precision performance data quoted is nearly always produced using these surrogate matrix QCs. As such, they are optimistic and in my experience; when we test the kit performance with our own specific matrix QCs, the performance is always worse than that quoted by the manufacturer, sometimes only by a small degree but more often significantly different.

This is not necessarily a major issue, but it is the performance data of the assay produced with matrix QCs that should be used to determine target acceptance criteria.

It should be noted that some manufacturers do have available wider ranges of QC materials in various matrices, and it may be worthwhile investigating those for the range of biomarkers of interest. Note, however, that these are nearly always made with a matrix that is spiked with recombinant or synthesized molecules and hence do not truly control the methods from the endogenous biomarker's perspective.

At least two companies are big manufacturers of QC material as well as multiplexed assays, (Randox and Bio-Rad), and they offer a range of QC sets in human serum and urine that may be of use to you.

Many laboratories use the commercial or "kit" QCs primarily to demonstrate that the methods are performing according to the manufacturer's claims (because the QCs will have target values). This gives confidence that there are no method performance issues. Additional matrix QCs are usually added to these in the sample analysis batches that are more representative of the actual samples being analyzed. The results of these QCs versus the same QCs performance at validation are the really important results in terms of deciding whether or not the results are reliable and can be accepted for reporting. This is where the QC *procedures* explained earlier play the role of ensuring that the assays are *in control* using the statistically relevant boundaries of the validation performance criteria.

For multiplex assays, the difficulty is always to have QCs that cover the whole range of analytes in the assay and at levels that control the analytical range of interest. In very early stage research with large numbers of analytes being multiplexed and where there may not be a good understanding of expected results concentrations, it is not uncommon to see multiplexed assays used that have the same analytical range for all analytes. Here, it is relatively straightforward to produce matrix QCs by spiking at three different concentrations to cover the majority of the analytical range—for example, low QC 10%–20%, medium QC 40%–60% and high QC 75%–85% of the highest calibration standard. QCs for all analytes will essentially have the same concentrations at each QC level. Remember that these will not be endogenous QCs and even in early research, it will be beneficial to include at least one EQC in sample runs.

However, where more is understood about physiology and the expected concentrations in samples, we now see a number of commercial alternatives for methods where the analytical ranges for each biomarker is tailored to be more appropriate to the expected concentrations. Figure 5.8.3 gives an example of this.

In the example shown in Figure 5.8.3, where analytical ranges are different for each biomarker, then the levels of

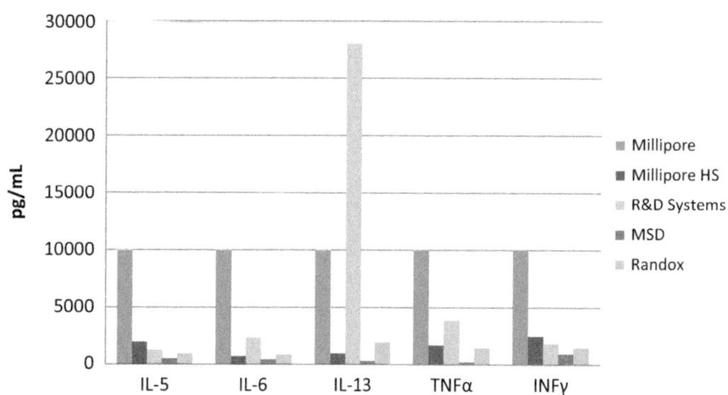

FIGURE 5.8.3 Example of commercial assays for a five-plex showing potential analytical ranges. One option has the same ranges for all analytes, whereas in others the ranges are different for each.

each QC for each biomarker will still potentially cover the same percentage of the analytical range quoted above for low, medium and high QCs, but now the concentration of the QCs will be different for each biomarker in the same level (low, med, high) of QC.

Note, that for all QC samples the target values come from the observed results in the validation experiments. Where we spike QCs, we are not necessarily concerned about recovery per se. Our main focus is method PRECISION. We need the assays to be reproducible at different observed concentrations over time. Therefore, the interassay or overall mean of results for each QC level used in the validation experiments (often called validation samples), becomes the *nominal* target value for use in sample analysis. Acceptance criteria of (± 2 and $3 \times$ interassay CV%) is applied around that nominal concentration for each QC level.

Overall, in the production of QCs for multiplexed methods, bear in mind that what is desirable may not always be achievable—especially in the early stages. Here, the practical solution may be a combination of any or all of the options above. The "gold" standard is to use endogenous molecules in their production whenever possible. The goal is to produce a range of QCs that together will be able to give assurance that all the assays are performing as required/expected to allow the laboratory to have confidence in the reporting of the results data.

SUMMARY

Multiplexed assays have been available since the late 1990s and can be powerful tools to generate biomarker results data. They have been used for many reasons including the following:

- To maximize results on multiple assays from small samples
- To generate multiple assay data quickly in a single method
- To help answer initial research questions as to which biomarkers from a large selection are of interest due to their concentration or modulation in different diseases or with different therapies
- To reduce the overall cost to the sponsor in terms of price per result compared with using individual biomarker assays

However, although they are useful tools in drug research and development, they bring with them some practical complexities. Jani et al.[4] has detailed the special attention required for many of the assay validation parameters.

QC of multiplexed assays is probably the most complex of issues in these assays, and I have often seen laboratories go to incredible lengths to validate the assays but also often without the attendant detail specifically required for QC itself.

In all cases of conducting multiplexed biomarker methods, it is good scientific practice to understand the physiological variability within- and between-subjects and to ensure a thorough and robust interrogation of all validation results data so that the method performance characteristics are fully understood. Then, in the method's application to clinical trial samples, consider the COU of each biomarker. Once these measures are conducted, a statistically and clinically valid QC procedure can be applied across all assays in the multiplexed method, ensuring the results data will be a valuable asset to decision making throughout the drug development process.

REFERENCES

1. TechTarget—WhatIs.com. http://whatis.techtarget.com/definition/quality-control-QC (accessed March 2018).
2. Bio-Rad. Basic Lessons in Laboratory Quality Control. http://www.qcnet.com/Portals/0/PDFs/QCWorkbook_Q1109_Jun08.pdf (accessed March 2018).
3. Lee, JW et al. Fit-for-purpose method development and validation for successful biomarker measurement. *Pharmaceutical Research,* 23(2): 312–328, 2006.
4. Jani, D et al. Recommendations for use and fit-for-purpose validation of biomarker multiplex ligand binding assays in drug development. *The AAPS Journal*, 18(1): 1–14, 2015.
5. Biomarker Assay Collaborative Evidentiary Considerations Writing Group, Critical Path Institute (C-Path). Points to consider document: Scientific and regulatory considerations for the analytical validation of assays used in the qualification of biomarkers in biological matrices. https://C-Path.org (accessed February 2019).

Lessons from the Past and Pioneers of the Future

Development of Maraviroc and the Companion Diagnostic HIV Tropism Assay

6.1

Elna van der Ryst, James Demarest, Jayvant Heera, Mike Westby, and Charles Knirsch

Contents

INTRODUCTION AND BACKGROUND

A review of the development of maraviroc (MVC) and companion diagnostic (CDx) tropism assays has been previously published (van der Ryst et al., 2015). This section is based on the review, but it provides a broader scope and includes more up-to-date information.

The identification of human immunodeficiency virus (HIV) infection as the causative agent of acquired immunodeficiency syndrome led to the development of laboratory assays that facilitate its diagnosis and are used as biomarkers to select optimal treatment regimens as well as to monitor treatment response. The development of antiretrovirals (ARVs) was also greatly facilitated by the use of HIV-1 RNA (a key prognostic factor for disease progression [Mellors et al., 1997]) as a surrogate marker for clinical end points (FDA guidance, 2015). Additionally, diagnostic assays to determine suitable ARV regimens for individual patients are widely used. These include phenotypic and genotypic HIV susceptibility assays (Hirsch et al., 2000) as well as assays to predict susceptibility to specific adverse reactions such as HLA-B*5701 testing for abacavir hypersensitivity (Hetherington et al., 2002; Mallal et al., 2002).

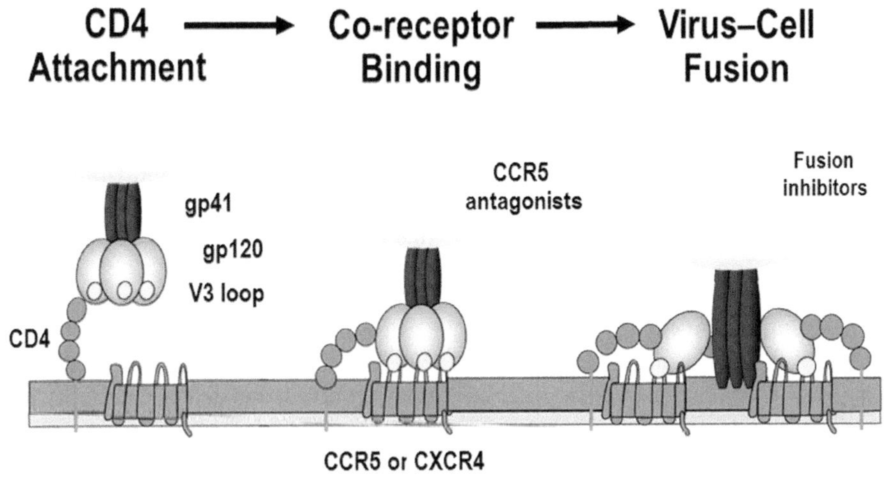

FIGURE 6.1.1 HIV-1 cell entry process and potential molecular targets for antiretroviral compounds. (Reprinted from van der Ryst, E. et al., *Ann N Y Acad Sci.*, 1346, 7–17, 2015. With permission. © 2015 New York Academy of Sciences.)

Due to the development of resistance to ARVs, combination treatment is required, preferably with drugs that target different aspects of the lifecycle of the virus, thereby resulting in a continued need for new drugs with novel mechanisms of action (d'Arminio 2000; Richman et al., 2004). The HIV entry process was identified as an attractive target. The first step in this process is binding of the virus envelope protein gp120 to CD4 on the cell surface, which leads to a conformational change in gp120 that allows binding to a chemokine receptor (CC chemokine receptor 5 [CCR5] or CXC chemokine receptor 4 [CXCR4]), which acts as a coreceptor. This is followed by fusion of the virus envelope with the host cell membrane (Figure 6.1.1) (Moore and Doms, 2003). The discoveries that individuals homozygous for the CCR5 delta 32 mutation are highly protected from infection with R5 HIV-1 strains (Liu et al., 1996; Samson et al., 1996) and that heterozygotes have reduced disease progression (Dean et al., 1996; Pasi et al., 2000) led to a number of research groups (including within Pfizer) initiating research programs to target the interaction between HIV and CCR5.

High-throughput screening of the Pfizer compound file using a chemokine radioligand binding assay to identify a lead molecule, followed by a medicinal chemistry program to optimize binding potency against the receptor, antiviral activity, absorption and pharmacokinetics, and selectivity against human cellular targets, which resulted in the identification of MVC (formerly UK-427,857) as a promising candidate for further development. Maraviroc was shown to act as a slow-offset functional antagonist of CCR5, blocking the binding of HIV gp120 to CCR5. It had no adverse effects in cell-based cytotoxicity studies, was highly selective for CCR5, safe in preclinical toxicology studies, and was predicted to have human pharmacokinetics consistent with once-daily (q.d.) or twice-daily (b.i.d.) dosing. It also demonstrated potent antiviral activity against all CCR5 tropic (R5) HIV-1 strains tested (Dorr et al., 2005, 2012).

In light of the promising preclinical data for MVC, a clinical development program was initiated. A key challenge was the identification of patients who were most likely to benefit from treatment with MVC, because the virus population from an individual patient may contain only obligate CCR5-tropic (R5) or CXCR4-tropic (X4) strains or dual-tropic strains (R5×4) that can bind both CCR5 and CXCR4. A patient sample may also contain a heterogeneous population of viruses with different tropism, which is termed *mixed tropism* (Figure 6.1.2). Dual and mixed tropic virus populations are collectively termed dual/mixed (D/M) (Westby et al., 2006). Viruses that can use the CXCR4 coreceptor for entry are inherently insensitive to drugs targeting the CCR5/HIV-1 gp120 interaction;

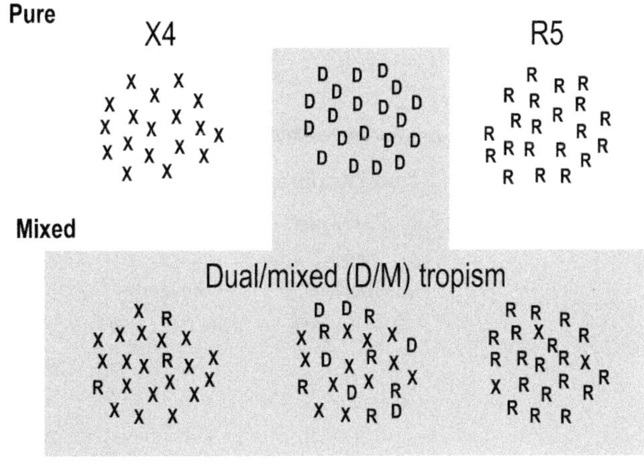

FIGURE 6.1.2 Possible tropism patterns found in people infected with HIV-1. D, dual tropic; R, CCR5 tropic (R5); X, CXCR4 tropic (X4). (Reprinted from van der Ryst, E. et al., *Ann N Y Acad Sci.*, 1346, 7–17, 2015. With permission. © 2015 New York Academy of Sciences.)

therefore, patients with a CXCR4-using virus (X4 virus, D/M virus, or both types) are unlikely to have an optimal response to these drugs.

TROFILE® HIV-1 CORECEPTOR TROPISM ASSAY

The first available assay to determine HIV-1 tropism was the Trofile® (Monogram Biosciences, South San Francisco, CA) assay. This is a phenotypic HIV-1 coreceptor tropism diagnostic assay developed to facilitate the development of coreceptor antagonists and to enable physicians to identify appropriate treatment regimens for their patients.

The development of the assay is described in detail by Whitcomb and colleagues (2007). Briefly, full-length HIV-1 envelope clones from patient samples are transferred into an expression vector and transfected into HEK-293 cells together with an envelope-deleted HIV-1 NL4-3 genomic vector expressing a luciferase reporter. The resulting luciferase reporter pseudovirus populations are then used to infect $CD4^+$ U-87 cells expressing either CCR5 or CXCR4 receptors. Infection of target cells is evaluated by the addition of a luciferase substrate and quantitation of luminescence. To demonstrate specificity, infection of target cells is also performed in the presence of CCR5 or CXCR4 antagonists (Figure 6.1.3). The assay determines whether the virus population uses CCR5, CXCR4, or both (Figures 6.1.2 and 6.1.3).

The Trofile assay was formally validated according to regulations specified by the Clinical Laboratory Improvement Amendments (CLIA). Validation experiments demonstrated that the assay was sensitive, reproducible, and able to accurately determine the coreceptor tropism of HIV-1 from patient plasma samples from all HIV-1 subtypes. It also amplified the envelope from >95% of samples that had HIV-1 RNA >1,000 copies/mL of HIV-1 RNA (with no false-positive or -negative amplifications observed) and detected minority X4 or R5 populations with 100% sensitivity when they were present in 10% of the population, whereas a population representing 5% of the total population could be detected with 85% sensitivity (Whitcomb et al., 2007).

Because treatment with CCR5 antagonists can lead to treatment failure due to preexisting minority, CXCR4-using virus populations, experiments were conducted to improve the sensitivity of the assay for detecting minor variants. From these experiments a combination of conditions designed to enhance D/M tropic virus infection of $CXCR4^+$ cells was selected and used to optimize the Trofile assay, thereby resulting in an assay with enhanced sensitivity (ES Trofile) (Reeves et al., 2009). Validation experiments demonstrated that the performance characteristics of the ES Trofile assay was similar to that of the original Trofile

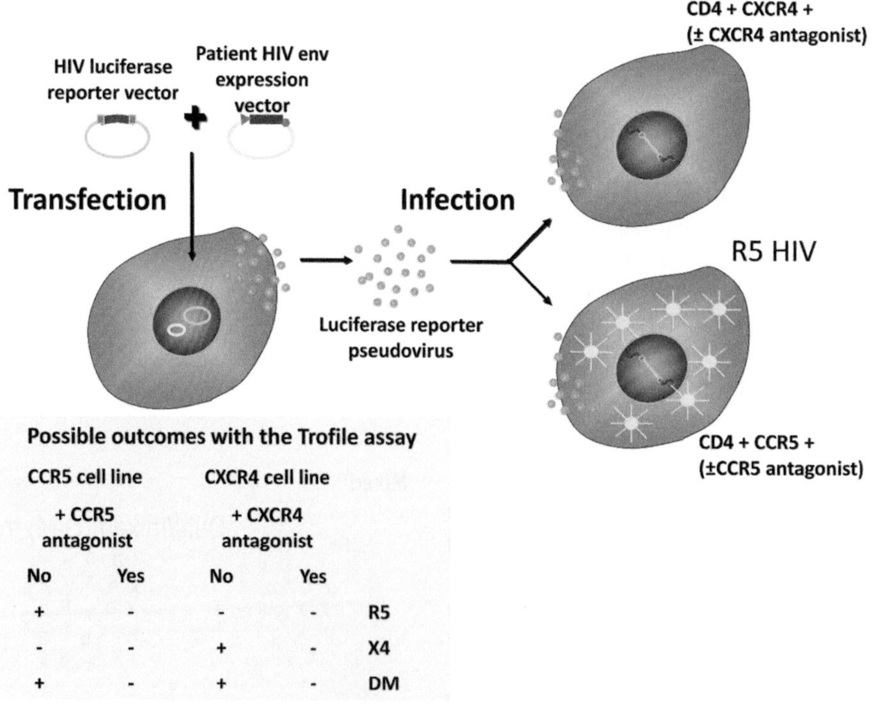

FIGURE 6.1.3 Schematic diagram of Trofile® HIV-1 tropism assay. Infection is indicated by the detection of luminescence for R5 virus. Inset summarizes possible outcomes with the Trofile assay. DM, dual/mixed tropic; R5, CCR5 tropic; X4, CXCR4 tropic. (Reprinted from van der Ryst, E. et al., *Ann N Y Acad Sci.*, 1346, 7–17, 2015. With permission. © 2015 New York Academy of Sciences.)

assay with regard to accuracy, precision, reproducibility, and viral load requirement. However, it had a significantly improved ability to detect X4 minority variants, because it was able to detect X4 *env* clones in 100% of assays when they were present in 0.3% of the population and R5 *env* clones in 100% of assays when they were present in 5% of the population (Reeves et al., 2009). The ES Trofile assay has now replaced the original assay.

DEVELOPMENT OF MARAVIROC

Data from phase I studies of healthy volunteers indicated that MVC was safe and well tolerated in dosages up to 900 mg q.d. and 300 mg b.i.d., did not influence the activity of major drug-metabolizing enzymes, and had an acceptable pharmacokinetic profile (Abel 2008). A phase IIa, proof-of-concept study was initiated using the newly developed, original Trofile assay to select appropriate patients for inclusion. Patients (n = 82) identified as having CCR5 tropic virus only were randomized to receive MVC monotherapy (dosages from 25 mg q.d. to 300 mg b.i.d.) or placebo for 10 days with follow-up until Day 40. At MVC dosages of 100 mg b.i.d. and higher, all patients with CCR5 tropic virus at baseline experienced a maximal reduction in viral load of >1.0 $\log_{10}$ copies/mL (Fätkenheuer et al., 2005). This not only demonstrated the potential utility of MVC in the treatment of HIV infection, but it also showed that the Trofile assay could be successfully used to identify patients likely to respond to a CCR5 antagonist.

Sequential analysis of virus tropism for all patients at Days 1, 11, and 40 (30 days posttreatment) were performed using the Trofile assay. Two patients with changes in virus tropism were identified. In one of these patients, transient emergence of D/M virus occurred on Day 11, with only R5 virus detected on Day 40. In the other patient, D/M virus was detected on Day 11 and remained detectable at follow-up until Day 433 when ARV therapy was initiated. To determine the origin of the CXCR4-using virus, virus envelope clones from both patients were evaluated using phylogenetic analysis, which confirmed that it emerged from a preexisting CXCR4-using population (Westby et al., 2006). These results demonstrate that the limited sensitivity of the Trofile assay for minority CXCR4-using virus can result in an outgrowth of these populations in the presence of a CCR5 antagonist. Additionally, it demonstrated the utility of the assay to evaluate changes in virus populations under selection pressure from MVC.

Following the successful demonstration of a proof of concept for MVC in monotherapy studies, a registrational, phase IIb/III clinical development program (including four large trials) evaluating MVC in dosages of 300 mg (or equivalent, depending on concomitant drugs) q.d. and b.i.d. was initiated (Table 6.1.1). The MOTIVATE 1 and 2 studies (A4001027 and A4001028) evaluated MVC q.d. or b.i.d. in combination with optimized background therapy (OBT) versus placebo plus OBT in highly treatment-experienced (TE) patients (Gulick et al., 2008). Study A4001026 (MERIT) compared MVC and efavirenz, both with zidovudine/lamivudine, in ARV-naive patients (Cooper et al., 2010). For the combined MERIT and MOTIVATE studies, 4,974 patients were screened for the presence of R5 virus only (Gulick et al., 2008; Cooper et al., 2010), using the original Trofile assay, because the samples were initiated in 2004 (prior to the availability of the ES Trofile assay).

To evaluate the safety of MVC in patients with CXCR4-using virus (because tropism assays may not detect minority CXCR4-using virus populations, and these patients may inadvertently be treated with a CCR5 antagonist), patients excluded from entering the MOTIVATE studies because they were infected with non-R5 virus (CXCR4-using virus or non-phenotypable virus) were offered the opportunity to participate in a phase IIb safety study (A4001029) (Saag et al., 2009). The designs of the phase IIb/III studies are summarized in Table 6.1.1.

TABLE 6.1.1 MVC phase IIb/III clinical development program

	R5 PATIENTS			
CHARACTERISTIC	*ARV NAIVE*	*ARV EXPERIENCED*		*NON R5 PATIENTS*
Study number/name	A4001026 MERIT	A4001027 MOTIVATE 1	A4001028 MOTIVATE 2	A4001029
Phase	IIb→III	IIb/III	IIb/III	IIb
Design	MVC vs. EFV + CBV	OBT add-on	OBT add-on	OBT add-on
Randomization	1:1:1	2:2:1	2:2:1	1:1:1
Primary end point	% HIV-1 RNA <400/<50 copies/mL at Week 48/96	ΔVL at Week 24/48	ΔVL at Week 24/48	ΔVL at Week 24/48
Randomized (*N*)	917	601	474	190

Source: Reprinted from van der Ryst, E. et al., 2015. *Ann N Y Acad Sci.*, 1346, 7–17. With permission. © 2015 New York Academy of Sciences.
ARV, antiretroviral; CBV, lamivudine/zidovudine; EFV, efavirenz; HIV, human immunodeficiency virus, MVC, maraviroc; OBT, optimized background therapy; R5, CCR5 tropic; VL, viral load.

Results of Studies in TE Patients

Of the 1,042 patients in the combined MOTIVATE studies who had an R5 tropism result at screening, 79 patients (8%) had evidence of D/M virus at baseline, reflecting the limited sensitivity for minority viral species of the original Trofile assay (Fätkenheuer et al., 2008). Nonetheless, Week 48 data from the MOTIVATE studies demonstrated a significant virologic benefit for patients receiving OBT with MVC (43% and 46% of participants receiving MVC q.d. and b.i.d. frequency, respectively, achieving HIV-1 RNA <50 copies/mL) compared with those receiving OBT only (17% achieving HIV-1 RNA <50 copies/mL [Figure 6.1.4]). Maraviroc treatment was safe and well tolerated, with a significantly higher increase in mean $CD4^+$ cell count (113 and 122 cells/mm^3 for MVC q.d. and b.i.d., respectively, compared with 54 cells/mm^3 for OBT only) (Gulick et al., 2008). In contrast, data from Week 24 of study A4001029 demonstrated no significant virologic benefit for MVC compared with placebo in patients infected with non-R5 virus (HIV-1 RNA <50 copies/mL in 27% and 21% of participants receiving MVC b.i.d. and q.d., respectively, compared with 16% for placebo; Figure 6.1.4). However, MVC was well tolerated in this population, and treatment resulted in a significantly higher increase in mean $CD4^+$ cell count from baseline (60 and 62 cells/mm^3 for MVC q.d. and b.i.d., respectively, compared with 36 cells/mm^3 for OBT only) (Saag et al., 2009).

Altogether, data from these three studies demonstrated that patients identified by the original Trofile assay as being infected with R5 HIV-1 are likely to respond to treatment with MVC, whereas those identified as being infected with non-R5 HIV-1 are unlikely to experience significant benefit. These data supported the approval from both the U.S. Food and Drug Administration (FDA) and the European Medicines Agency for the use of MVC (in combination with other ARV drugs) to treat ARV-experienced adults with R5 HIV-1 infection (ViiV Healthcare, 2018a, 2018b).

Results of Studies in Treatment-Naive Patients

The MVC q.d. treatment arm of the MERIT trial that studied treatment-naive (TN) patients with R5 HIV-1 infection was discontinued following an interim analysis of the first 205 patients at Week 16 because it did not meet prespecified, non-inferiority criteria. Data from the 48-week analysis of the MVC b.i.d. group demonstrated that MVC did not meet criteria for non-inferiority (10% level) compared with efavirenz, with 65% and 69% of patients achieving HIV-1 RNA below 50 copies/mL, respectively (Figure 6.1.5). However, significantly higher $CD4^+$ cell increases were observed in patients receiving MVC (Cooper et al., 2010). Of the 720 patients with an R5 tropism result at screening, 694 (96.4%) had evaluable baseline tropism data. Of these, 24 (3.5%) had a D/M baseline result, indicating low-level CXCR4-using virus around the detection limit of the original Trofile assay. Because the ES Trofile had become available in the interim, researchers decided to evaluate whether use of a more sensitive tropism assay could improve the outcome in the MVC group. Retesting of the MERIT screening samples revealed that 14.8% (107/721, including 1 participant randomized to receive MVC in error) of patients would have been identified as having

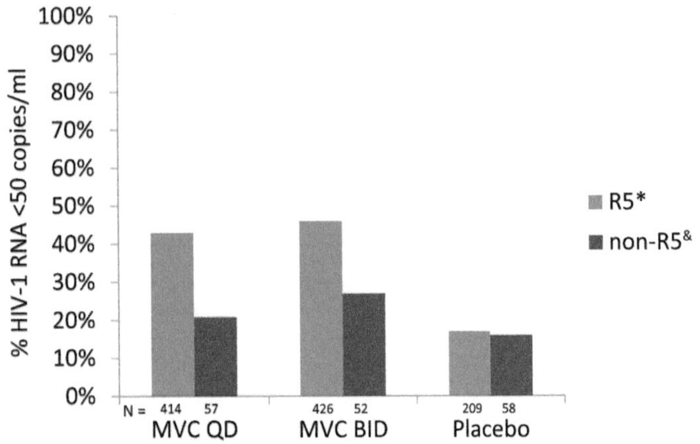

FIGURE 6.1.4 Participants with R5 HIV-1 achieving HIV-1 RNA <50 copies/mL in the combined MOTIVATE 1 and 2 studies compared with non-R5 participants from study A4001029. In both studies, participants had triple-ARV drug-class experience (± triple class resistance) and were randomized to receive MVC q.d., MVC b.i.d., or placebo, all in combination with an OBT regimen consisting of 3–6 ARVs selected by the study investigator. *$p < 0.001$ compared with placebo. &Difference from placebo not statistically significant; MVC q.d. = 7 (95% CI: –7, 20) and MVC b.i.d. = 11 (95% CI: –3, 26), respectively. ARV, antiretroviral drug; b.i.d., twice daily; CI, confidence interval; MVC, maraviroc; OBT, optimized background therapy; q.d., once daily; R5, CCR5 tropic.

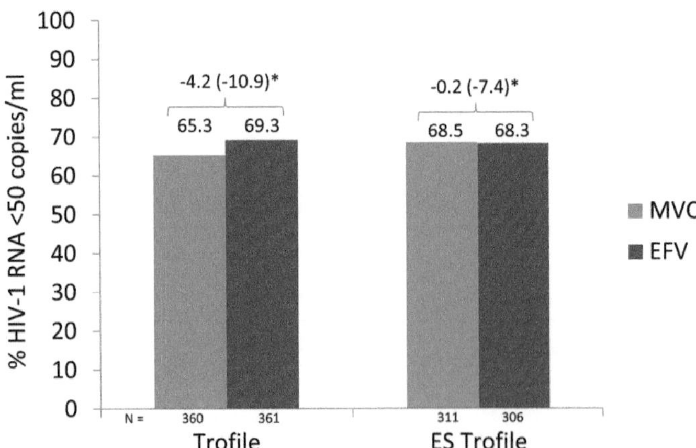

FIGURE 6.1.5 Data from MERIT. Proportions of patients achieving HIV-1 RNA <50 copies/mL at Week 48 in the primary analysis and the post hoc reanalysis using the ES Trofile® assay. Treatment-naive patients infected with HIV-1 R5 were randomized to receive MVC twice daily or EFV once daily, both in combination with lamivudine/zidovudine. *Difference adjusted for randomization strata with lower bound of the 1-sided 97.5% confidence interval. EFV, efavirenz; ES, enhanced sensitivity; MVC, maraviroc; R5, CCR5 tropic. (Reprinted from van der Ryst, E. et al., *Ann N Y Acad Sci.*, 1346, 7–17, 2015. With permission. © 2015 New York Academy of Sciences.)

CXCR4-using virus if this assay had been used for the screening process. Post hoc reanalysis of the data that excluded these patients resulted in greater response rates in the MVC arm at Week 48, which now fell within the criteria defining non-inferiority (Figure 6.1.5). Most patients (102/106) reclassified by the enhanced assay as having CXCR4-using virus at screening had tropism data collected during the study. In this group, the proportion of participants with HIV-1 RNA <50 copies/mL was lower for those treated with MVC (21/46 [46%]) than for those treated with efavirenz (42/56 [75%]) (Cooper et al., 2010). This clearly demonstrates the improved clinical utility of the ES Trofile assay for selecting patients most likely to respond to treatment with a CCR5 antagonist. These results subsequently led to the FDA approval of MVC for the treatment of ARV-naive patients with R5 HIV-1 as well as its approval in Europe (ViiV Healthcare, 2018a, 2018b).

Evaluating Reasons for Virologic Failure

Two potential pathways of virologic escape exist in patients receiving CCR5 antagonists: (1) selection of R5 virus that can use the drug-bound CCR5 receptor to enter host cells, and (2) "un-masking" of a preexisting CXCR4-using virus population through selection pressure by the CCR5 antagonist (Westby et al., 2006, 2007).

Samples collected at prespecified time points in the MVC phase IIb/III studies were evaluated using the Trofile assay (provided that HIV-1 RNA was >500 copies/mL) (Fätkenheuer et al., 2005, 2008; Cooper et al., 2010). Week 24 data demonstrated that, of the patients with R5 virus at baseline and who experienced virologic failure in the MOTIVATE studies, 57% (76/133) had evidence of CXCR4-using virus at the time of failure in the MVC arm compared with just 6% (6/95) in the placebo group (Fätkenheuer et al., 2008). This finding demonstrates that unmasking of preexisting CXCR4-using virus populations (due to selective pressure from MVC) is the major mechanism of viral escape in this population. Similar data were obtained for the MERIT study: of 644 patients with R5 tropism identified at both screening and baseline, 4.5% had emergence of CXCR4-using virus during the study (20/321 [6.2%] on MVC compared with 9/323 [2.8%] on efavirenz) (Cooper et al., 2010).

R5 virus was observed in a minority of participants who experienced treatment failure with MVC in the MOTIVATE studies. The susceptibility to MVC of these viruses were assessed using the PhenoSense® HIV entry assay (Monogram Biosciences). This is a modification of the Trofile assay in which MVC susceptibility is determined from dose-response plots of percent inhibition of viral replication versus drug concentration using the CCR5-expressing U87CD4^{+} cells. Preclinical studies have shown that MRV resistance can be identified by plateaus in the maximum percent inhibition in this assay. This phenotype is a result of the resistant virus being able to use the MVC-bound CCR5 receptor for entry and infection of CD4 T cells, with variable efficiency compared with the unbound receptor (the more efficiently it uses the bound receptor, the lower the maximum percent inhibition) (Westby et al., 2007). Using this assay, viruses with reduced susceptibility to MVC were identified in 22/58 (37.9%) and 4/29 (13.8%) patients with R5 virus infection who failed MVC treatment in the MOTIVATE (Week 48) and MERIT studies, respectively (Cooper et al., 2010; ViiV Healthcare, 2018a). In most cases, changes in the gp120 viral envelope sequence (centered around the V3 loop) were identified as conferring the resistant phenotype, but these appeared to be context dependent and no genotypes could be assigned as having "signature" MVC resistance (Jiang et al., 2015).

FURTHER DEVELOPMENT OF TROPISM ASSAYS

Although the enhanced performance of the ES Trofile assay resulted in increased clinical utility of the assay, logistical challenges continued to act as a significant barrier to the widespread use of this assay in patient care. The acquisition of Monogram Biosciences by LabCorp enabled clinicians practicing in the United States to more easily send samples to the Monogram Biosciences laboratory by using the LabCorp clinical/laboratory network. However, in the rest of the world, sample shipment logistics still resulted in long assay turnaround times, and this—together with cost—continued to limit the widespread clinical use of Trofile. This led to initiatives to develop other phenotypic assays, including Virco®Type HIV-1 (Janssen Diagnostics) (Van Baelen et al., 2007), Phenoscript® (Eurofins) (Trouplin et al., 2001; Roulet et al., 2007), PhenX-R (InPheno AG) (Braun and Wiesmann, 2007; Hamy et al., 2007), the Toulouse tropism test (Raymond et al., 2010), and the tropism coreceptor assay information test (Gonzalez-Serna et al., 2010). However, all of these assays require specialized facilities, are relatively expensive, and have relatively long turnaround times.

In an ideal situation, every assay would be evaluated in a prospective manner to demonstrate its clinical utility; however, this is not feasible in reality. Thus, a common practice (not unique to tropism assays) is to perform retrospective concordance analyses of patient samples, provided the patient has provided informed consent for such use, to characterize the performance of a new assay. In certain instances, there may be samples available from participants enrolled in a clinical trial for use in retrospective analyses of clinical outcomes; such samples were available from a limited number of studies to evaluate the positive predictive value (PPV) of alternative tropism assays with clinical outcome on MVC. Retrospective analyses of samples from the MERIT, MOTIVATE, A4001029, and other studies of MVC were used to evaluate several alternative assay approaches, including population-based Sanger sequencing, next-generation ultra-deep sequencing (UDS), heteroduplex tracking assays (HTA), and DNA-based tropism assays.

Genotypic Tropism Determination

Genotypic tropism assays that use the HIV-1 V3 loop sequence to predict tropism and could be performed in local or regional laboratories were identified as a more practical and less-expensive alternative to phenotypic tropism assays. However, there were key challenges: (1) Improvements in sequencing technology to allow for the large intra- and inter-patient variability of the HIV-1 V3 loop were required, and (2) bioinformatic algorithm interpretation is more difficult for tropism determination compared with ARV resistance, in which genotypic resistance to any given drug is often determined by signature mutations.

The high level of V3 loop diversity combined with the lack of a signature sequence that defines a V3 sequence as R5 or X4, as well as the possibility of D/M virus strains, means that complex bioinformatic algorithms are required to predict coreceptor tropism. An iterative process of combining genotypic (HIV *env*, V3 loop sequence, or both) and phenotypic (Trofile and ES Trofile and/or other methods) tropism data has been used to improve the predictive ability of the algorithms. The most widely used algorithm for predicting coreceptor tropism is the open access "geno2pheno coreceptor" algorithm (http://coreceptor.geno2pheno.org). To derive tropism, a V3 loop sequence is entered into the web-based tool that aligns and analyzes the sequence to predict the tropism. The algorithm settings can be changed to increase or decrease specificity and sensitivity rates by increasing or decreasing the false-positive rate (FPR), a threshold for classifying a sequence as non-R5. The higher the value, the more R5-like the sequence must be for it to be called R5, thereby increasing specificity. Clinical data such as CD4 and CD8 cell counts, CCR5delta32 genotype, and plasma HIV-1 RNA may be provided, if available, to further optimize the tropism prediction.

Population-based Sanger sequencing

Samples from patients enrolled in clinical trials studying MVC (MERIT and A4001078 in TN patients and MOTIVATE and A4001029 in TE patients) were retrospectively tested using a population-based Sanger sequencing method to determine concordance and predictive value for virologic response.

Data from the MERIT study demonstrated that 84% (292/351) of patients treated with MVC were concordant as R5/R5 (283 [81%]) or non-R5/non-R5 (9 [3%]) by both the ES Trofile assay and genotype. The proportion of patients achieving HIV-1 RNA levels <50 copies/mL was consistent with the assay concordance: approximately 60% over 96 weeks for R5/R5 and approximately 20% or less for non-R5/non-R5. For patients who were discordant (R5 genotype/non-R5 ES Trofile assay, n = 39; R5 ES Trofile assay/non-R5 genotype, n = 20), the response rate for either group was lower but closer to that observed in the R5/R5 patient group, suggesting that, for these patients, neither assay was fully accurate in predicting R5 tropism (McGovern et al., 2012). This finding highlights several challenges for tropism assays such as the lack of a true "gold standard," as demonstrated by the fact that the response rate of patients with discordant results was not driven by either assay; the response curves overlapped and were not separated based on being R5 by one assay over the other. Additionally, as was seen with the advances from the original Trofile assay to the current ES Trofile assay, the threshold for detecting minority species may be relevant for tropism prediction and its translation into clinical response.

Sanger sequencing methods generally have a threshold between 15% and 20% for detecting a minority species. However, optimization of both the sequencing methods to improve the detection of minority species and the algorithm settings for tropism prediction has improved the sensitivity and specificity rates of genotypic methods to assess tropism, which is illustrated by the retrospective analysis of samples from study A4001078 that used the ES Trofile assay for tropism screening (Portsmouth et al., 2013). Samples from all 199 patients screened for enrollment were retrospectively evaluated by a Sanger sequencing method that used different FPR thresholds with the geno2pheno algorithm. As expected, a higher FPR increases the rate of sensitivity for X4 variants, but it reduces the rate of specificity. The percentages of patients classified as having X4 virus increased with increasing FPR values: 5.75% FPR resulting in 15.1%, 10% FPR resulting in 23.1%, and 20% FPR resulting in 37.7% of patients being classified as having X4 virus. Of the 144 samples that were determined to be R5 by the ES Trofile assay, 132 (91.7%) and 120 (83.3%) were classified as R5 by genotype, with FPRs of 5.75% and 10%, respectively (Portsmouth et al., 2013). These data are consistent with results from other studies evaluating the correlation between genotypic and phenotypic tropism results such as the OSCAR study (Svicher et al., 2010) and the ANRS GenoTropism study (Recordon-Pinson et al., 2010).

However, MVC response is a better measure of the performance of a tropism assay than concordance with another assay because response is a direct measure of clinical utility. Therefore, the PPV of an assay, defined as the likelihood of a patient having a positive virologic response with MVC when virus tropism is defined as R5, may provide a better method for comparing assays. Results from the MODERN study, which is a prospective trial that compared genotypic (geno2pheno) with phenotypic (ES Trofile) tropism assays in predicting virologic responses to MVC, demonstrated that high concordance was present between the assays and that both assays could effectively predict MVC response. The observed response rates were 80.7% (146/181) for the genotypic assay and 74.4% (160/215) for the ES Trofile assay (stratification-adjusted difference 6.9% [95% confidence interval {CI}: 1.3, 15]). The model-based estimates of PPV (standard error) were 79.1% (92.42) and 76.3% (92.38), respectively (difference of 2.8% [95% CI: 2.1, 7.2]) (Heera et al., 2014).

Next-generation UDS

Ultra-deep sequencing has increased sensitivity for the detection of minor variants and has a threshold of <1%. Availability of these UDS platforms is relatively limited at this time, but UDS may provide an alternative genotypic tropism methodology in the future should these methodologies become more widely available.

Samples from the MOTIVATE and A4001029 studies were retrospectively analyzed to explore the sensitivity of UDS for detection as well as the clinical relevance of minority CXCR4-using variants. An initial blinded analysis using the Roche 454 UDS platform indicated that virologic outcomes based on the original Trofile assay tropism and 454 UDS method were similar (Swenson et al., 2011). Subsequently, Kagan and colleagues at Quest Diagnostics performed an analysis on 327 participants enrolled in either MOTIVATE or A4001029. Results from the ES Trofile assay were available this time and were compared with tropism calls using the Quest reflex tropism test assay in a blinded fashion. Sanger sequencing was performed as a first step. If a sample was classified as R5 via Sanger, then the 454 analysis was performed, whereas if the Sanger result was non-R5, then no further testing was conducted. In this analysis, the 454 UDS method (alone or as part of the reflex format) had a similar PPV to that seen with the ES Trofile (66% for ES Trofile vs. 65% for 454 UDS) (Kagan et al., 2012).

Heteroduplex tracking assays

Another genotypic-based approach for determining tropism is the DNA HTA. Polymerase chain reaction (PCR)-amplified V3 sequences from patient samples are hybridized to V3 sequences of known R5 or X4 viruses and subjected to electrophoresis. Differences in the V3 sequences between the probe and target sequences result in altered electrophoretic mobility due to varying degrees of mismatch in the hybridized DNA molecules (Lin and Kuritzkes, 2009). A version of the HTA (SensiTrop; Pathway Diagnostics) was found to have low sensitivity (42%) but high specificity (92.5%) for detecting CXCR4-using viruses in a study of 100 patient samples from the MVC expanded-access program (A4001050 study) previously tested by the Trofile assay (Tressler et al., 2008). An HTA from Quest Diagnostics was evaluated using 326 samples from the MOTIVATE and A4001029 studies. The data demonstrated that the HTA, either alone or with the additional support of a population genotyping method, was inferior to ES Trofile in determining tropism (Wilkin et al., 2012).

DNA-Based Tropism Assays

The standard plasma HIV-1 RNA-based tropism assays previously described are suitable for use only in viremic patients because a level of HIV-1 RNA ≥500 copies/mL is normally required. However, the ability to evaluate tropism in patients who are virologically suppressed (HIV-1 RNA <50 copies/mL) is necessary to guide the potential use of MVC when changing the treatment regimen for tolerability and/or other reasons. DNA-based tropism assays are performed using peripheral blood mononuclear cells (PBMCs) or whole blood. Cell-associated DNA, which contains the HIV-1 DNA species (integrated, unintegrated, or both), isolated from patient cells may be used in phenotypic assays, genotypic assays, or both assay types in a similar manner to the plasma HIV-1 RNA-based methods.

In the United States, two primary options exist to assess coreceptor tropism in patients with virologic suppression on combination ARV therapy: (1) Trofile DNA (Monogram Biosciences), and (2) HIV-1 Coreceptor Tropism, Proviral DNA (Quest Diagnostics). These two assays are available via a large network of sample collection/distribution across the country; other bespoke tropism assays may be available for informing treatment decisions at the local or regional level. Outside the United States, local clinical testing laboratories may offer a DNA-based tropism assay for use in patients with virologic suppression to complement their offering for use in viremic patients.

Characterization and evaluation of HIV-1 DNA-based tropism assays

Several parameters must be considered when evaluating the performance and clinical relevance of any assay. For HIV-1 DNA-based tropism assays, two key parameters were considered to assess the performance and clinical relevance of these assays; namely, concordance of cell-associated DNA- with plasma RNA-based assays and PPV (based on clinical outcomes) of DNA-based assays.

Concordance analyses attempt to address the question of whether the tropism of viral quasispecies in cell-associated DNA is a good representation of the virus(es) in circulating plasma. Two limitations of using this approach alone are that (1) it is unclear whether the use of cell-based assays in patients who are virologically suppressed is the same as when comparing samples from those who are viremic, and (2) cell-associated DNA may contain some viral *env* sequences that derive from defective, nonfunctional genomes.

Some retrospective concordance analyses performed to evaluate assay (assay 1 vs. assay 2) and sample (plasma vs. PBMC) concordance are reviewed by Poveda and colleagues (2010) and Kuritzkes (2011). A finding from these studies using genotypic tropism assays is a general tendency for a higher call rate of CXCR4-using virus in proviral DNA (i.e., PBMC) than with plasma HIV RNA as the sample template. As with comparisons of two plasma-based assays, the algorithm and associated FPR may impact the tropism readout(s) and, therefore, the level of concordance.

Baumann and colleagues (2015) performed a longitudinal analysis of 50 women living with HIV infection. The analyses included paired plasma and PBMC from time point 1 (T1) and a longitudinal sample from time point 2 (T2; median time >4 years). The study reported concordance rates of 88% with paired T1 plasma/PBMC and 80% for T1 plasma/T2 PBMC. Swenson and colleagues (2010) performed a cross-sectional analysis of tropism using population-based and deep-sequencing methodologies. Twelve sample pairs consisting of plasma RNA isolated prior to the initiation of ARV therapy and cellular DNA isolated during virologic suppression (median time 36.5 months with plasma HIV-1 RNA <50 copies/mL) were evaluated. Using deep sequencing, the levels of non-R5 virus detected were similar (within 1%) in one-third of sample pairs, whereas a somewhat greater difference was observed (ranging from 6%–72%) in the remainder. Although the sample size was relatively small, the presence of >20% non-R5 virus via deep sequencing was associated with a non-R5 tropism readout by the population-based method. Although these types of cross-sectional analyses illustrate the relative concordance between paired samples of plasma and PBMC from the same patient, they do not address questions regarding the clinical relevance of the quasispecies of the virus in the cellular compartment at the time of virologic suppression and clinical outcomes on CCR5-antagonist containing ARV therapy.

Analysis of PPV for clinical outcome of DNA-based tropism assays

In a manner similar to that previously outlined with concordance analyses, cryopreserved PBMCs were used to evaluate the performance of two DNA-based tropism assays: Trofile DNA and the genotypic HIV-1 Coreceptor Tropism, Proviral DNA assay. PBMC samples from the MOTIVATE and A4001029 phase III clinical trials in TE patients were used, thereby allowing a higher frequency of non-R5 virus to be observed than that typically seen in TN patients. The sample set included all patients enrolled in MOTIVATE, while the companion A4001029 study was open; thus, patients had equal probability of receiving MVC regardless of tropism. A positive clinical outcome or "virologic response" was defined as plasma HIV-1 RNA <50 copies/mL or a decline of $\geq 2 \log_{10}$ copies/mL at Week 8. A similar analysis was conducted using the virologic response at Week 24 (viral load <50 copies/mL). However, in this highly TE population, Week 8 was a preferred end point given the loss of virologic response commonly seen at Week 24 or later that may be impacted by other factors such as poor adherence, loss of activity for the other agents in the OBT regimen, or both.

In the first cross-sectional, retrospective analysis, 253 sample pairs were evaluated. A concordance rate of 82% was observed between the tropism determinations using Trofile DNA and the results obtained with the ES Trofile assay. The PPVs for virologic response at Week 8 were no different for tropism determinations by Trofile DNA vs ES Trofile assay: 69.6% (range 61.5%–76.9%) vs. 69.1% (range 60.7%–76.6%), respectively. Similar results with lower response rates were observed at Week 24 due to factors previously mentioned in this patient population. Thus, in this retrospective analysis using phenotypic tropism assays, the results seen with Trofile DNA were no different from those seen with the ES Trofile assay in terms of PPV for MVC-containing therapy (Chapman et al., 2012).

A second cross-sectional, retrospective analysis was performed using a similar sample pool (as far as sample availability allowed) to evaluate the Proviral Genotype assay. Paired PBMC tropism determinations ($N = 174$) showed a 74% concordance between Proviral Genotype and Trofile DNA. Interestingly, discordant results were comparable between assays, with 12% non-R5/R5 and 14% R5/non-R5 for Proviral Genotype/Trofile DNA, findings that suggest that neither assay preferentially overcalls non-R5 tropism. When determining the PPV for paired samples from the MOTIVATE and A4001029 studies, the Proviral Genotype assay had lower PPVs than the Trofile DNA (64.8% vs. 71.3% at Week 8 and 44.5% vs. 48.0% at Week 24, respectively) (Hamdan et al., 2014). The acceptable threshold for the PPV in this patient population is not clearly defined and may or may not apply to other patient populations. Changing the algorithm settings (i.e., FPR) impacted the rates of assay sensitivity and specificity, and, thus, the PPV for Proviral Genotype. For example, an FPR of 10% had sensitivity and specificity rates of 51% and 83%, respectively. Increasing the FPR to 20% resulted in increased sensitivity (61%) and decreased specificity (70%) for non-R5 detection. Reducing the FPR to 5.75% had little impact on sensitivity (51%) yet increased specificity to 90% (Hamdan et al., 2014). Thus, assay and algorithm performance characteristics may require optimization.

A small, single-arm, prospective study enrolled 74 participants with evidence of virologic suppression for the last 6 months who needed to make a change in treatment regimen due to tolerability issues with their current regimen. To inform potential MVC use, a proviral DNA-based genotypic tropism assay was performed using patient PBMC, and those with R5 tropism determination were allowed to switch to an MVC-containing regimen. A 24-week interim analysis showed that 84% of study participants maintained virologic suppression (Garcia et al., 2014). Although these data illustrate the potential use of proviral DNA-based genotypic tropism testing in patients who are virologically suppressed, additional studies with longer-term follow-up will be needed.

The retrospective and prospective data described here illustrate the challenges with the evaluation of a new sample type as well as novel assay(s). For example, although the retrospective analysis by Hamdan and colleagues (2014) suggested a less-optimal performance of a genotypic proviral DNA-based assay using PBMC from viremic patients, a similar retrospective analysis of plasma-based tropism assays on 327 samples (Kagan et al., 2012) using a reflex strategy of population-based to next-generation sequencing showed similar PPVs to the phenotypic ES Trofile assay: 40.0% vs. 42.0%, respectively. As discussed above, genotypic tropism methods using PBMC often have a higher rate of non-R5 virus reporting, and the tropism of DNA-based analyses using pre- and posttreatment samples have shown potential change in relative tropism over time with virologic suppression. Thus, in viremic TE patients, a phenotypic tropism assay or a sensitive genotypic RNA tropism assay that includes UDS may be of greater utility than genotypic proviral DNA tropism determinations for identifying patients with R5 virus who are more likely to respond to MVC. Furthermore, it is important to note that the results in viremic, heavily TE patients, such as those tested in the MOTIVATE and A4001029 trials, may or may not translate to outcomes in patients who are virologically suppressed. More data, including prospective analyses, are needed.

TREATMENT GUIDELINES FOR TROPISM TESTING

ARV treatment recommendations are continuously updated based on the rigorous assessment of emerging data and provide guidance to treating physicians. Tropism testing prior to the administration of a CCR5 antagonist is recommended by guidelines from the International AIDS Society–USA (Günthard et al., 2016), the U.S. Department of Health and Human Services (DHHS) (DHHS, 2018), and the European AIDS Clinical Society (EACS, 2018). Current guidelines do not provide specific guidance on which assay(s) should be used for assessing tropism in patients with virologic suppression; however, the DHHS includes an additional comment in its guidelines explaining that the clinical utility of Trofile DNA remains to be determined.

CONCLUSIONS

The data discussed in this section demonstrate how the codevelopment of diagnostic assays and certain therapies may have mutual benefit. The development of an assay to determine HIV-1 coreceptor tropism was critical to the development of the CCR5 antagonist MVC. Alternatively, results of clinical studies of MVC confirmed the clinical PPV and negative predictive value of the Trofile assay, as well as the improved clinical utility of the ES Trofile assay. Furthermore, retrospective evaluation of clinical samples from clinical studies of MVC have also been invaluable for the evaluation of less expensive and more convenient tropism tests with quicker turnaround times to allow for more rapid clinical decision making. The development of DNA-based tropism assays will facilitate the use of CCR5 antagonists in virologically suppressed patients, thereby increasing the patient population who might benefit from these ARVs. Altogether, this has resulted in more sensitive and accessible options for tropism testing, thereby facilitating the clinical use of CCR5 antagonists.

ACKNOWLEDGMENTS

The authors would like to thank all patients and investigators who participated in the MVC studies, as well as numerous colleagues at Pfizer and ViiV Healthcare who contributed to the program. We also thank colleagues at Monogram Biosciences.

CONFLICTS OF INTEREST

Charles Knirsch and Jayvant Heera are employees of Pfizer Inc. and are partly compensated through stock/stock options. James Demarest is an employee of ViiV Healthcare. Mike Westby was an employee of Pfizer Inc. at the time the MVC studies were conducted. Elna van der Ryst was an employee of Pfizer Inc. at the time the MVC studies were conducted and currently provides consulting services to Pfizer.

REFERENCES

Abel, S., van der Ryst, E., Rosario, M.C., Ridgway, C.E., Medhurst, C.G., Taylor-Worth, R.J. and Muirhead, G.J. (2008). Assessment of the pharmacokinetics, safety and toleration of maraviroc, a novel CCR5 antagonist, in healthy volunteers. *Br J Clin Pharmacol.*, 65, 5–18.

Baumann, R.E., Rogers, A.A., Hamdan, H.B., Burger, H., Weiser, B., Gao, W., Anastos, K. et al. (2015). Determination of HIV-1 coreceptor tropism using proviral DNA in women before and after viral suppression. *AIDS Res Ther.*, 12, 11–17.

Braun, P. and Wiesmann, F. (2007). Phenotypic assays for the determination of coreceptor tropism in HIV-1 infected individuals. *Eur J Med Res.*, 12, 463–472.

Chapman, D., Lie, Y, Paquet, A., Drews, W., Toma, J., Biswas, P., Petropoulos, C. et al. (2012). Tropism determinations derived from cellular DNA or plasma virus compartments are concordant and predict similar maraviroc treatment outcomes in an antiretroviral treatment experienced cohort. *Presented at: XIX Int AIDS Conference*; July 22–27, 2012; Washington, DC. Abstract THPE070.

Cooper, D.A., Heera, J., Goodrich, J., Tawadrous, M., Saag, M., Dejesus, E., Clumeck, N. et al. (2010). Maraviroc versus efavirenz, both in combination with zidovudine/lamivudine, for the treatment of antiretroviral-naïve subjects with CCR5-tropic HIV-1. *J Inf Dis.*, 201, 803–813.

d'Arminio Monforte, A., Lepri, A.C., Rezza, G., Pezzotti, P., Antinori, A., Phillips, A.N., Angarano, G. et al. (2000). Insights into the reasons for discontinuation of the first highly active antiretroviral therapy (HAART) regimen in a cohort of antiretroviral naïve patients. ICONA study group, Italian cohort of anti-retroviral naive patients. *AIDS*, 14, 499–507.

Dean, M., Carrington, M., Winkler, C., Huttley, G.A., Smith, M.W., Allikmets, R., Goedert, J.J. et al. (1996). Genetic restriction of HIV-1 infection and progression to AIDS by a deletion allele of the CKR5 structural gene. Hemophilia growth and development study, multicenter AIDS cohort study, multicenter hemophilia cohort study, San Francisco city cohort, ALIVE. *Science*, 274, 1856–1862.

DHHS. (2018). *Guidelines for the Use of Antiretroviral Agents in Adults and Adolescents Living with HIV.* Report from the DHHS Panel on Antiretroviral Guidelines for Adults and Adolescents, Bethesda, MD. https://aidsinfo.nih.gov/contentfiles/lvguidelines/adultandadolescentgl.pdf.

Dorr P., Stammen, B. and van der Ryst., E. (2012). Discovery and development of maraviroc, a CCR5 antagonist for the treatment of HIV infection, edited by: Aslanian, R.G. and Huang, X. *Case Studies in Modern Drug Discovery and Development.* Hoboken, NJ: John Wiley & Sons, pp. 196–226.

Dorr, P., Westby, M., Dobbs, S., Griffin, P., Irvine, B., Macartney, M., Mori, J. et al. (2005). Maraviroc (UK-427,857), a potent, orally bioavailable, and selective small molecule inhibitor of chemokine receptor CCR5 with broad-spectrum anti-human immunodeficiency virus type 1 activity. *Antimicrob Agents Chemother.*, 49, 4721–4732.

EACS. (2018). Guidelines, version 9.1. http://www.eacsociety.org/files/2018_guidelines-9.1-english.pdf.

Fätkenheuer, G., Nelson M., Lazzarin, A., Konourina, I., Hoepelman, A.I., Lampiris, H., Hirschel, B. et al. (2008). Subgroup analysis of maraviroc in previously treated R5 HIV-1 infection. *N Engl J Med.*, 359, 1442–1445.

Fätkenheuer, G., Pozniak, A.L., Johnson, M.A., Plettenberg, A., Staszewski, S., Hoepelman, A.I., Saag, M.S. et al. (2005). Efficacy of short-term monotherapy with maraviroc, a new CCR5 antagonist in HIV-1 infected patients. *Nat Med.*, 11, 1170–1172.

Garcia, F., Poveda, E., Ribas, M.A., Perez-Elias, M.J., Martinez-Madrid, O.J., Navarro, J., Ocampo, A. et al. (2014). Genotypic tropism testing of proviral DNA to guide maraviroc initiation in aviremic subjects. *Presented at: 21st Conference on Retroviruses and Opportunistic Infections*; March 2–6, 2014; Boston, MA. Abstract 607.

Gonzalez-Serna, A., Leal, M., Genebat, M., Abad, M.A., Garcia-Perganeda, A., Ferrando-Martinez, S. and Ruiz-Mateos, E. (2010). TROCAI (tropism coreceptor assay information): A new phenotypic tropism test and its correlation with Trofile enhanced sensitivity and genotypic approaches. *J Clin Microbiol.*, 48, 4453–4458.

Gulick, R.M., Lalezari, J., Goodrich, J., Clumeck, N., DeJesus, E., Horban, A., Nadler, J. et al. (2008). Maraviroc for previously treated patients with R5 HIV-1 infection. *N Engl J Med.*, 359, 1429–1441.

Günthard, H.F., Saag, M.S., Benson, C.A., del Rio, C., Eron, J.J., Gallant, J.E., Hoy, J.F. et al. (2016). Antiretroviral drugs for treatment and prevention of HIV infection in adults. *JAMA*, 316, 191–210.

Hamdan, H., Demarest, J., Jagannatha, S., Kagan, R. and Pesano, R. (2014). Proviral DNA tropism assessment predicts maraviroc treatment outcomes in an HIV-1 treatment-experienced clinical trial cohort. *Presented at: 30th Annual Clinical Virology Symposium*; April 27–30, 2014; Daytona Beach, FL. Abstract M52.

Hamy, F., Vidal, V., Hubert, S. and Klimkait, T. (2007). Hybridization-based assay and replicative phenotyping as diagnostic platform for determination of co-receptor tropism. *Presented at: 5th European HIV Drug Resistance Workshop*; March 28–30, 2007; Cascais, Portugal. Abstract 60.

Heera, J., Valluri, S., Craig, C., Fang, A., Thomas, N., Meyer, R.D. and Demarest, J. (2014). First prospective comparison of genotypic vs phenotypic tropism assays in predicting virologic responses to maraviroc (MVC) in a phase 3 study: MODERN. *J Int AIDS Soc.*, 17, 19519.

Hetherington, S., Hughes, A.R., Mosteller, M., Shortino, D., Baker, K.L., Spreen, W., Lai, E. et al. (2002). Genetic variations in HLA-B region and hypersensitivity reactions to abacavir. *Lancet*, 359, 1121–1122.

Hirsch, M., Brun-Véezinet, F., D'Acquila, F.T., Hammer, S.M., Johnson, V.A., Kuritzkes, D.R., Loveday, C. et al. (2000). Antiretroviral drug resistance testing in adult HIV-1 infection: Recommendation of an International AIDS Society-USA panel. *JAMA*, 283, 2417–2426.

Jiang, X., Feyertag, F., Meehan, C.J., McCormack, G.P., Travers, S.A., Craig, C., Westby, M., Lewis, M. and Robertson, D.L. (2015). Characterizing the diverse mutational pathways associated with R5-tropic maraviroc resistance: HIV-1 that uses the drug-bound CCR5 coreceptor. *J Virol.*, 89, 11457–11472.

Kagan, R.M., Johnson, E.P., Siaw, M., Biswas, P., Chapman, D.S., Su, Z., Platt, J.L. and Pesano, R.L. (2012). A genotypic test for HIV-1 tropism combining Sanger sequencing with ultradeep sequencing predicts virologic response in treatment-experienced patients. *PLoS One*, 7, e46334.

Kuritzkes, D.R. (2011). Genotypic tests for determining coreceptor usage of HIV-1. *J Infect Dis.*, 203, 146–148.

Lin N.H. and Kuritzkes, D.R. (2009). Tropism testing in the clinical management of HIV-1 infection. *Curr Opin HIV AIDS*, 4, 481–487.

Liu, R., Paxton, W.A., Choe, S., Ceradini, D., Martin, S.R., Horuk, R., MacDonald, M.E., Stuhlmann, H., Koup, R.A. and Landau, N.R. (1996). Homozygous defect in HIV-1 coreceptor accounts for resistance of some multiply exposed individuals to HIV-1 infection. *Cell*, 86, 367–377.

Mallal, S., Nolan, D., Witt, C., Masel, G., Martin, A.M., Moore, C., Sayer, D. et al. (2002). Association between presence of HLA-B*5701, HLA-DR7, and HLA-DQ3 and hypersensitivity to HIV-1 reverse-transcriptase inhibitor abacavir. *Lancet*, 359, 727–732.

McGovern, R.A., Thielen, A., Portsmouth, S., Mo, T., Dong, W., Woods, C.K., Zhong, X. et al. (2012). Population-based sequencing of the V3-loop can predict the virological response to maraviroc in treatment-naive patients of the MERIT trial. *J Acquir Immune Defic Syndr.*, 61, 279–286.

Mellors, J.W., Muñnoz, A., Giorgi, J.V., Margolick, J.B., Tassoni, C.J., Gupta, P., Kingsley, L.A. et al. (1997). Plasma viral load and CD4+ lymphocytes as prognostic markers of HIV-1 infection. *Ann Intern Med.*, 126, 947–954.

Moore, J.P. and Doms, R.W. (2003). The entry of entry inhibitors: A fusion of science and medicine. *Proc Nat Acad Sci.*, 100, 10598–10602.

Pasi, K.J., Sabin, C.A., Jenkins, P.V., Devereux, H.L., Ononye, C. and Lee, C.A. (2000). The effects of the 32bp CCR-5 deletion on HIV transmission and HIV disease progression in individuals with heamophilia. *Br J Hematol.*, 111, 136–142.

Portsmouth, S., Valluri, S.R., Portsmouth, S., Valluri, S.R., Däumer, M., Thiele, B., Valdez, H. et al. (2013). Correlation between genotypic (V3 population sequencing) and phenotypic (Trofile ES) methods of characterizing co-receptor usage of HIV-1 from 200 treatment-naïve HIV patients screened for study A4001078. *Antiviral Res.*, 97, 60–65.

Poveda, E., Alcami, J., Paredes, R., Córdoba, J., Gutiérrez, F., Llibre, J.M., Delgado, R. et al. (2010). Genotypic determination of HIV tropism-clinical and methodological recommendations to guide the therapeutic use of CCR5 antagonists. *AIDS Rev.*, 12, 135–148.

Raymond, S., Delobel, P., Mavigner, M., Cazabat, M., Souyris, C., Encinas, S., Bruel, P. et al. (2010). Development and performance of a new recombinant virus phenotypic entry assay to determine HIV-1 coreceptor usage. *J Clin Virol.*, 47, 126–130.

Recordon-Pinson, P., Soulié, C., Flandre, P., Descamps, D., Lazrek, M., Charpentier, C., Montes, B. et al. (2010). Evaluation of the genotypic prediction of HIV-1 coreceptor use versus a phenotypic assay and correlation with the virological response to maraviroc: The ANRS GenoTropism study. *Antimicrob Agents Chemother.*, 54, 3335–3340.

Reeves, J.D., Coakley, E., Petropoulos, C.J. and Whitcomb, J.M. (2009). An enhanced sensitivity Trofile HIV coreceptor tropism assay for selecting patients for therapy with entry inhibitors targeting CCR5: A review of analytical and clinical studies. *J Viral Entry*, 3, 94–102.

Richman, D.D., Morton, S.C., Wrin, T., Hellmann, N., Berry, S., Shapiro, M.F. and Bozzette S.A. (2004). The prevalence of antiretroviral drug resistance in the United States. *AIDS*, 18, 1393–1401.

Roulet, V., Rochas, S., Labernardiere, J.L., Mammano, F., Faudon, J.L., Raja, N., Lebel-Binay S. and Skrabal K. (2007). HIV PHENOSCRIPT ENV: A sensitive assay for the detection of HIV X4 minority species and determination of non-B subtype viral tropism. *Presented at: 14th Conference on Retroviruses and Opportunistic Infections*; February 25–28, 2007; Los Angeles, CA. Abstract 617.

Saag, M., Goodrich, J., Fätkenheuer, G., Clotet, B,. Clumeck, N., Sullivan, J., Westby, M., van der Ryst, E. and Mayer, H.; for the A4001029 Study Group. (2009). A double-blind placebo-controlled trial of maraviroc in treatment-experienced patients infected with non-CCR5 tropic HIV-1: 24-week results. *J Infect Dis.*, 11, 1638–1647.

Samson, M., Libert, F., Doranz, B., Rucker, J., Liesnard, C., Farber, C.M., Saragosti, S. et al. (1996). Resistance to HIV-1 infection in caucasian individuals bearing mutant alleles of the CCR-5 chemokine receptor gene. *Nature*, 382, 722–725.

Svicher, V., D'Arrigo, R., Alteri, C., Svicher, V., D'Arrigo, R., Alteri, C., Andreoni, M. et al. (2010). Performance of genotypic tropism testing in clinical practice using the enhanced sensitivity version of Trofile as reference assay: Results from the OSCAR Study Group. *New Microbiol.*, 33, 195–206.

Swenson, L.C., Mo, T., Dong, W.W., Zhong, X., Woods, C.K., Jensen, M.A., Thielen, A. et al. (2011). Deep sequencing to infer HIV-1 co-receptor usage: Application to three clinical trials of maraviroc in treatment-experienced patients. *J Infect Dis*, 203, 237–245.

Swenson, L.C., Moores, A., Low, A.J., Thielen, A., Dong, W., Wood, C., Jensen, M.A. et al. (2010). Improved detection of CXCR4-using HIV by V3 genotyping: Application of population-based and "deep" sequencing to plasma RNA and proviral DNA. *JAIDS*, 54, 506–510.

Tressler, R., Valdez, H., Van der Ryst, E., James, I., Lewis, M., Wheeler, J. and Than, S. (2008). Comparison of results from the SensiTrop™ vs Trofile™ assays on 100 samples from the maraviroc expanded access program. *Presented at: 15th Conference on Retrovirus and Opportunistic Infections*; February 3–6, 2008; Boston, MA. Abstract 920A.

Trouplin, V., Salvatori, F., Capello, F., Obry, V., Brelot, A., Heveker, N., Alizon, M., Scarlatti, G., Clavel, F. and Mammano, F. (2001). Determination of coreceptor usage of human immunodeficiency virus type 1 from patient plasma samples by using a recombinant phenotypic assay. *J Virol.*, 75, 251–259.

US Department of Health and Human Services and US Food and Drug Administration Center for Drug Evaluation and Research. (2015). *Guidance for Industry. Human Immunodeficiency Virus-1 Infection: Developing Antiretroviral Drugs for Treatment.* Report from DHHS and CDER. https://www.fda.gov/downloads/drugs/guidancecomplianceregulatoryinformation/guidances/ucm355128.pdf.

Van Baelen, K., Vandenbroucke, I., Rondelez, E., Van Eygen, V., Vermeiren, H. and Stuyver, L.J. (2007). HIV-1 coreceptor usage determination in clinical isolates using clonal and population-based genotypic and phenotypic assays. *J Virol Methods*, 146, 61–73.

van der Ryst, E., Heera, J., Demarest, J. and Knirsch, C. (2015). Development of maraviroc, a CCR5 antagonist for treatment of HIV, using a novel tropism assay. *Ann N Y Acad Sci.*, 1346, 7–17.

ViiV Healthcare. (2018a). Prescribing information for Selzentry. https://www.gsksource.com/pharma/content/dam/GlaxoSmithKline/US/en/Prescribing_Information/Selzentry/pdf/SELZENTRY-PI-MG-IFU.PDF.

ViiV Healthcare. (2018b). Summary of product characteristics for Celsentri. https://www.medicines.org.uk/emc/product/6159/smpc. Last update October 12, 2018.

Westby, M., Lewis, M., Whitcomb, J., Youle, M., Pozniak, A.L., James, I.T., Jenkins, T.M., Perros, M. and van der Ryst, E. (2006). Emergence of CXCR4-using human immunodeficiency virus type 1 (HIV-1) variants in a minority of HIV-1-infected patients following treatment with the CCR5 antagonist maraviroc is from a pretreatment CXCR4-using reservoir. *J Virol.*, 80, 4909–4920.

Westby, M., Smith-Burchnell, C., Mori, J., Lewis, M., Mosley, M., Stockdale, M., Dorr, P., Ciaramella, G. and Perros, M. (2007). Reduced maximal inhibition in phenotypic susceptibility assay indicates that viral strains resistant to the CCR5 antagonist maraviroc utilize inhibitor-bound receptor for entry. *J Virol.*, 81, 2359–2371.

Whitcomb, J.M., Huang, W., Fransen, S., Limoli, K., Toma, J., Wrin, T., Chappey, C., Kiss, L.D., Paxinos, E.E. and Petropoulos, C.J. (2007). Development of a novel single-cycle recombinant-virus assay to determine human immunodeficiency virus type 1 coreceptor tropism. *Antimicrob Agents Chemother.*, 51, 566–575.

Wilkin, T.J., Su, Z., Kagan, R., Heera, J. and Schapiro, J.M. (2012). Comparison of Quest Diagnostics to Trofile and Trofile-ES coreceptor tropism assays for predicting virologic response to maraviroc (MVC), a CCR5 antagonist. *Presented at: 50th Interscience Conference on Antimicrobial Agents and Chemotherapy*; September 12–15, 2010; Boston, MA. Abstract H-932a.

Case Studies *Infertility and Anti-Müllerian Hormone*

6.2

Martin Hund and Joan-Carles Arce

Contents

DEFINITION

Infertility is characterized by the failure to establish a clinical pregnancy after 12 months of regular, unprotected sexual intercourse, or is due to impairment of a person's capacity to reproduce either as an individual or with their partner (Zegers-Hochschild et al. 2017; NICE 2017). Primary infertility refers to the situation of never having achieved a clinical pregnancy, whereas secondary infertility refers to not being able to achieve a second clinical pregnancy, having previously achieved a first clinical pregnancy (Zegers-Hochschild et al. 2017).

CAUSES

Infertility can result from disorders in one or both partners and is categorized as "male factor," "female factor" or "combined." Approximately one-third of all infertility cases can be attributed solely to the female partner, one-third solely to the male partner and one-third are caused by a combination of problems in both partners or are unexplained (NHS 2017); more than one factor contributes to infertility in approximately 25% of couples with fertility problems. Non–sex-specific causes include being under- or overweight, smoking, stress, occupational or environmental hazards (e.g., exposure to

TABLE 6.2.1 Sex-specific causes of infertility

MALE-SPECIFIC CAUSES OF INFERTILITY	*FEMALE-SPECIFIC CAUSES OF INFERTILITY*
Sterilization • Vasectomy • Castration Low-quality or absent sperm • Low sperm count or mobility • Abnormal sperm • Failure of sperm to reach semen Ejaculation disorders • Premature ejaculation • Ejaculation into bladder Hypogonadism Testicular abnormalities • Congenital defects • Cancer or surgery • Infection or trauma Medication or drugs • Anti-inflammatory drugs • Anabolic steroids • Chemotherapy • Alcohol	Ovulation disorders • Premature ovarian failure • Polycystic ovary syndrome • Thyroid disorders • Chronic conditions (e.g., cancer, HIV) • Cushing's syndrome Endometriosis Tubular disorders • Sterilization • Pelvic or cervical surgery • Pelvic inflammatory disease Age • Couples in industrialized countries increasingly delay pregnancy Medicines or drugs • Nonsteroidal anti-inflammatory drugs • Chemotherapy • Neuroleptic drugs • Illicit drugs (heroin, cocaine, marijuana)

Source: Menken, J. et al., *Science* 234, 413, 1986; Janssen, N.M., and M.S. Genta. *Arch. Intern. Med.*, 160, 610–619, 2000; Kaplan, B. et al. *Eur. J. Obstet. Gynecol. Reprod. Biol.*, 123, 72–76, 2005; Homan, G.F. et al., *Hum. Reprod. Update*, 13, 209–223, 2007; Turan, V. and K. Oktay. *Expert Opin. Drug. Saf.* 13, 775–783, 2014; Chow, E.J. et al. *Lancet Oncol.*, 17, 567–576, 2016; Blumenauer, V. et al., *J. Reproduktionsmed. Endokrinol.*, 14, 272–305, 2017; National Institute for Health and Care Excellence (NICE). Clinical guideline 156: Assessment and treatment for people with fertility problems. Last updated September 2017.

Note: HIV, human immunodeficiency virus.

radiation, pesticides, heavy metals or solvents) and sexually transmitted infections such as chlamydia, hepatitis B, hepatitis C, gonorrhea or human immunodeficiency virus (Homan et al. 2007; Hassan and Killick 2004). Sex-specific causes are summarized in Table 6.2.1.

Fertility rates in women decline gradually, but significantly, from the age of 32 years and decrease more rapidly from the age of 37 years (ACOG CGP PC 2014). This is particularly relevant in Western societies where increasing age in the female partner is one of the most common explanations for infertility. Many couples choose to delay parenthood (Mills et al. 2011), which may be contributing to declining fertility rates in high-income regions. Equally, the high prevalence of sexually transmitted diseases in the developing world may contribute to higher rates of secondary infertility reported in lower income regions (Nachtigall 2006).

PREVALENCE

Infertility affects approximately 45 million couples worldwide, with one in six couples experiencing some form of infertility problem at least once during their reproductive lifetime (Mascarenhas et al. 2012; ASRM 2015). Based on 2010 data, worldwide rates of primary and secondary infertility were 1.9% (95% confidence intervals [CI]: 1.7%, 2.2%) and 10.5% (95% CI: 9.5%, 11.7%), respectively, and varied by region (Mascarenhas et al. 2012). Primary infertility in child-seeking women varies by region, from 1.0% in Latin America and the Caribbean region, to 2.6% in North Africa and the Middle East (Mascarenhas et al. 2012). Similarly, secondary infertility in women varies from 7.2% in high-income and North Africa/Middle East regions, to 18% in Central/Eastern Europe and Central Asia regions (Mascarenhas et al. 2012). Since 1990, there has been little change in the prevalence of primary and secondary infertility in most regions; however, absolute numbers of infertile couples has increased due to population growth (Mascarenhas et al. 2012).

ASSISTED REPRODUCTIVE TECHNOLOGY

Couples are increasingly seeking assisted reproductive technology (ART) for help with conceiving. Approximately 1.6 million ART cycles are performed globally each year resulting in an estimated 400,000 births (ASRM 2015). Over two-thirds of ART cycles are performed in the following high-income regions: Europe ($n = 640,144$), Japan ($n = 241,089$), United States ($n = 176,247$) and Australia/New Zealand ($n = 70,082$) (ASRM 2015).

ART can treat a range of causes of infertility (NICE 2017; Farquhar et al. 2015), and a variety of techniques are available, including *in vitro* fertilization (IVF), intracytoplasmic sperm injection (ICSI), egg or sperm donation, gamete intrafallopian transfer, artificial insemination, surrogacy and cryopreservation. IVF and ICSI are the most commonly used techniques (ASRM 2015); ICSI is indicated for male factor infertility and may be considered for patients who have previously failed to become pregnant with IVF (NICE 2017). During ART therapy, pharmacological induction of follicle development is used to obtain multiple oocytes at the time of oocyte aspiration (Zegers-Hochschild et al. 2017); this is known as ovarian stimulation. Several factors influence the success of ART, including patient characteristics (age, ovarian reserve [OR], parity, reproductive history, body mass index and prior response to ART), selection of an appropriate ART treatment protocol and embryo quality (Toner et al. 1991; Loveland et al. 2001; Lee et al. 2006; Huang and Rosenwaks 2012; La Marca and Sunkara 2014; Desai et al. 2014). Based on patient characteristics, IVF treatment protocols can be individualized to maximize the potential for a successful pregnancy (La Marca and Sunkara 2014).

ASSESSMENT OF OVARIAN RESERVE

People who are concerned about their fertility should be offered an initial assessment into lifestyle factors and sexual history. Couples seeking treatment for infertility should be informed that over 80% of couples in the general population will conceive within 1 year if the woman is aged under 40 years and if they do not use contraception and have regular sexual intercourse (every 2–3 days optimizes the chance of pregnancy) (NICE 2017). A woman of reproductive age who has not conceived within 1 year should be referred along with her partner for further clinical investigation (e.g., semen analysis, assessment for tubal and uterine abnormalities, regularity of menstrual cycle and OR) (NICE 2017). Evaluation should be expedited in women who are older than 35 years, where treatment is warranted after 6 months of failed attempts to conceive, or earlier if clinically indicated (ACOG CGP PC 2014).

The ovary contains a finite number of primordial follicles, which declines with age (Hansen et al. 2008; Iliodromiti and Nelson 2013; Dewailly et al. 2014), and the remaining pool defines true OR at a given time point (Wallace and Kelsey 2010; Grynnerup et al. 2014). An ideal test for OR should be simple to use (i.e., a single, cycle-independent measurement), noninvasive to minimize patient discomfort, and offer high precision and good reproducibility (i.e., low variability between users and laboratories). Methods for assessment of OR broadly fall into two categories: use of ultrasound imaging techniques for visualization of the ovaries, and measurement of biomarkers that reflect the biology of the aging ovary. Antral follicle count (AFC), determined by transvaginal ultrasound on Days 2–4 of the menstrual cycle, is commonly used but requires experienced sonographers and is susceptible to inter-operator variability (Broekmans et al. 2010; PC ASRM 2012; NICE 2017). Age is considered the original biomarker of OR as reproductive ageing is a continuous process that begins prior to birth and continues through to menopause (Hansen et al. 2008). However, any two individuals of the same age can have a 100-fold difference in OR (Wallace and Kelsey 2010). Endocrine markers, including follicle-stimulating hormone (FSH), estradiol (E2) or inhibin B—each measured on Day 3 of the menstrual cycle—are sometimes used for assessment of OR, but can only be performed during the early follicular phase of the menstrual cycle and are susceptible to high intra- and intercycle variability (PC ASRM 2012; Jirge 2011). Measurement of serum anti-Müllerian hormone (AMH) and AFC for assessment of OR represents the current standard of care (NICE 2017; Dewailly et al. 2014).

ANTI-MÜLLERIAN HORMONE

Introduction

Anti-Müllerian hormone, a member of the transforming growth factor-β family, is a dimeric glycoprotein produced by the ovarian granulosa cells of preantral and small antral follicles (Hansen et al. 2011). AMH regulates initial follicular recruitment, serving to inhibit transition of primordial follicles to primary follicles (Figure 6.2.1) (Durlinger et al. 1999, 2002; Carlsson et al. 2006;

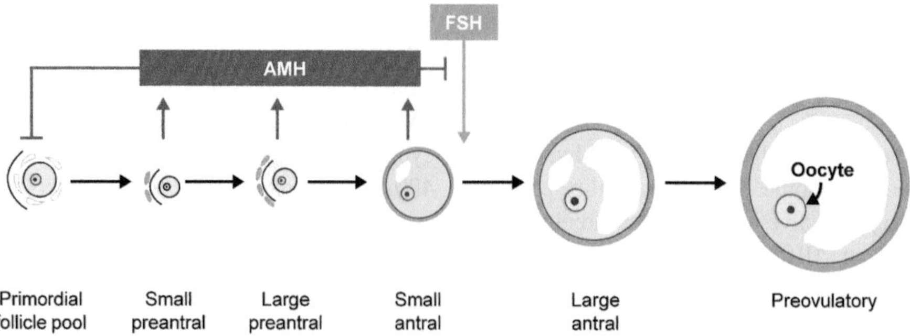

FIGURE 6.2.1 AMH production during follicular development. AMH produced in small growing follicles inhibits initial follicle recruitment and FSH-dependent growth and selection of preantral and small antral follicles. AMH, anti-Müllerian hormone; FSH, follicle-stimulating hormone. (Adapted from Dewailly, D. et al., *Hum. Reprod. Update*, 20, 370–385, 2014.)

Dewailly et al. 2014). At any age, there can be significant variation in AMH levels between individuals, with greatest variation observed in the early-to-mid 20s (Kelsey et al. 2011). Factors influencing AMH levels include race/ethnicity (Bleil et al. 2014), ovarian stimulation (Bottcher et al. 2014), use of combined hormonal contraceptives (Kallio et al. 2013), use of gonadotrophin-releasing hormone (GnRH) agonists (Drakopoulos et al. 2019) and smoking (Plante et al. 2010).

AMH as a Biomarker for Ovarian Reserve

Serum AMH levels correlate with the number of primordial follicles in the ovary (Hansen et al. 2011; Dewailly et al. 2014). Therefore, AMH can serve as a direct serum marker of functional OR (Anderson et al. 2012), which relates to the number of growing (i.e., preantral and antral) follicles that can be recruited by exogenous FSH to grow to a pre-ovulatory stage.

Serum levels of AMH can be quantified by immunoassay (Anckaert et al. 2016), offering several advantages over other methods of OR assessment. AMH levels are relatively stable throughout the menstrual cycle (intracycle variation has been observed but the magnitude of these changes is considered of limited clinical relevance), hence AMH can be measured on any day of the cycle (Hansen et al. 2011). AMH demonstrates low inter- and intracycle variation compared with other biomarkers (Tsepelidis et al. 2007; van Disseldorp et al. 2010), although fluctuations in AMH have been observed in younger women, and in women with high AMH levels (Kelsey et al. 2011; Nelson et al. 2011; Randolph et al. 2014; Lambert-Messerlian et al. 2016). The ovary is a dynamic organ and AMH may reflect this dynamic status in intercycle variation in some patients. Additionally, compared with AFC determination, experienced personnel or ultrasound equipment are not required and AMH measurement is less prone to inter-operator variability (van Disseldorp et al. 2010).

Compared with AFC, serum AMH levels are more strongly correlated with the number of oocytes retrieved during ovarian stimulation, as identified in retrospective analyses of overall data from two multicenter, randomized, active-controlled trials comparing pregnancy rates in patients undergoing IVF (Anckaert et al. 2012; Arce et al. 2013). This was subsequently confirmed at the level of the individual study centers, where AMH was more strongly correlated with oocyte yield than AFC in the GnRH agonist cohort (r = 0.56 vs. r = 0.28) and the GnRH antagonist cohort (r = 0.55 vs. r = 0.33); furthermore, AFC provided no added predictive value beyond AMH (Nelson et al. 2015a).

Automated measurement of AMH levels using the Elecsys® AMH Plus immunoassay shows good correlation with AFC. This was confirmed in a multicenter, prospective study in 451 women of reproductive age with regular menstrual cycles. AMH correlated with age (r = –0.47) and AFC (r = 0.68) (Anderson et al. 2015), and a high percentage agreement was observed between the Elecsys AMH Plus assay-based and AFC-based classification of responders (Figure 6.2.2). For example, based on AFC groupings of low (0–7), normal (8–15) and high (>15), there was classification agreement in 63.2, 56.9 and 74.5% of women in each group, respectively (Anderson et al. 2015).

Unlike AFC values that varied between sites and operators, Elecsys AMH Plus immunoassay measurements demonstrated minimal variability between sites. Importantly, automated measurement of AMH levels showed better prediction of AFC classification than FSH and E2 (Figure 6.2.3). Furthermore, combining age with AMH did not statistically significantly improve clinical performance in low and high AFC group classifications (Anderson et al. 2015).

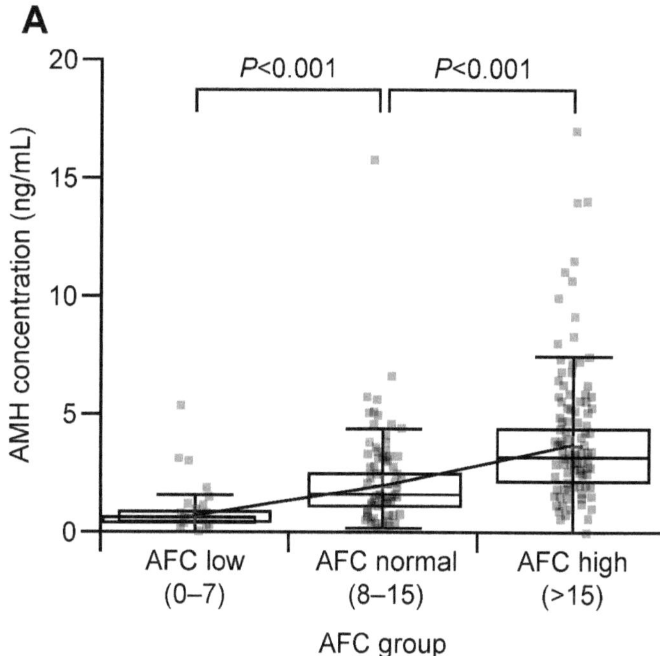

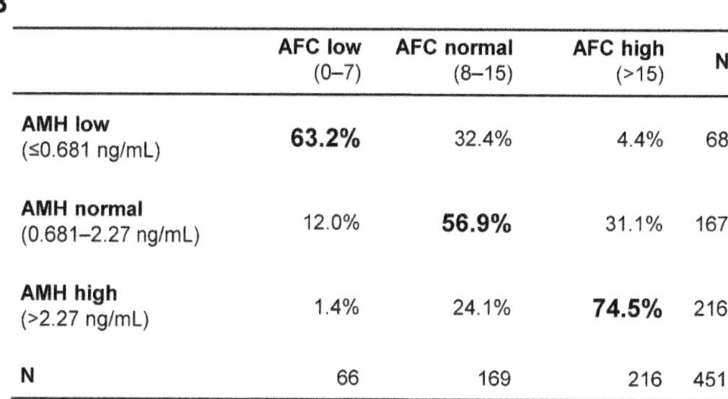
B

	AFC low (0–7)	AFC normal (8–15)	AFC high (>15)	N
AMH low (≤0.681 ng/mL)	**63.2%**	32.4%	4.4%	68
AMH normal (0.681–2.27 ng/mL)	12.0%	**56.9%**	31.1%	167
AMH high (>2.27 ng/mL)	1.4%	24.1%	**74.5%**	216
N	66	169	216	451

FIGURE 6.2.2 (A) Distribution of Elecsys® AMH Plus immunoassay-measured AMH (ng/mL) in different AFC groups in a multicenter study and (B) Elecsys AMH Plus immunoassay-derived values show high percentage agreement with AFC-based classifications of low/normal/high responder. AFC, antral follicle count; AMH, anti-Müllerian hormone. (Panel A from Anderson, R.A. et al. *Fertil. Steril.*, 103, 1074–1080, e4, 2015.)

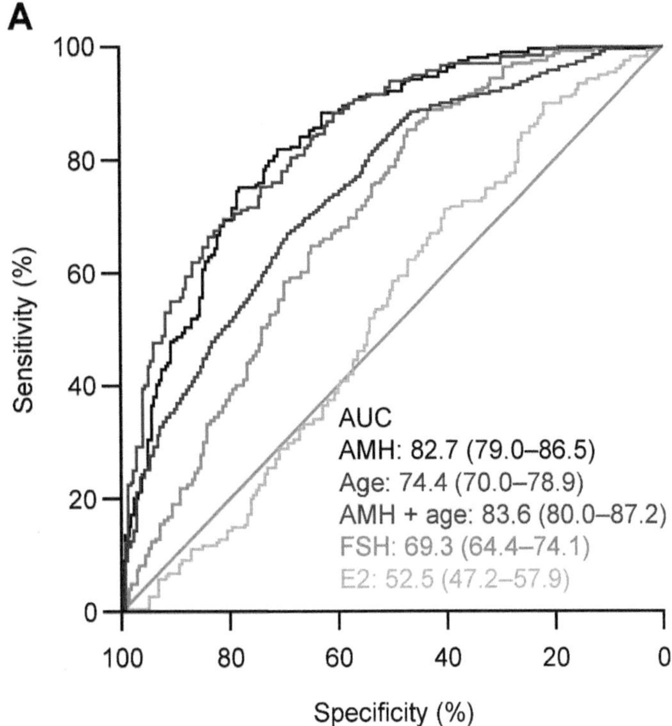

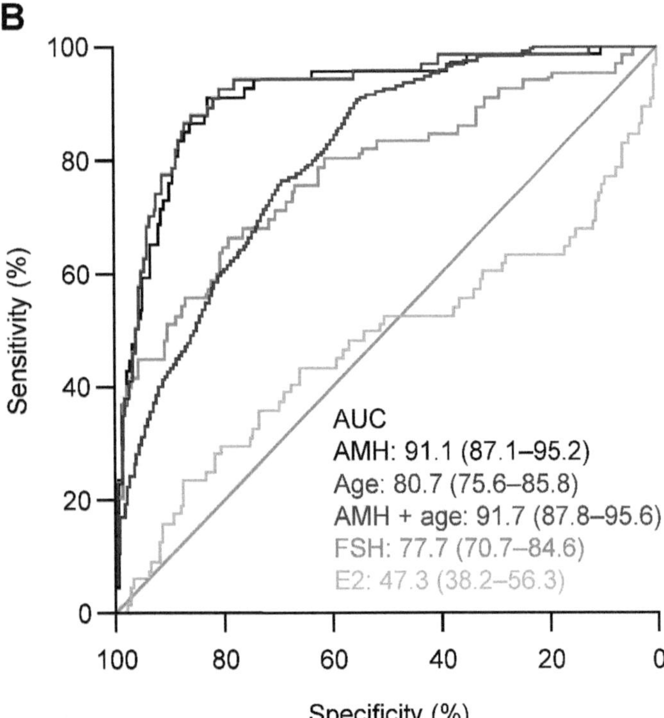

FIGURE 6.2.3 Receiver operator characteristic curves for classification of (A) high AFC and (B) low AFC, by AMH, FSH, E2 and age. Numbers in brackets represent 95% confidence intervals. AFC, antral follicle count; AMH, anti-Müllerian hormone; AUC, area under the curve; E2, estradiol; FSH, follicle-stimulating hormone. (Adapted from Anderson, R.A. et al., *Fertil. Steril.*, 103, 1074–1080, e4, 2015.)

AMH-Tailored Ovarian Stimulation

Ovarian response to ovarian stimulation, which partially determines the outcome of ART (Steward et al. 2014), varies between women due to differences in OR (Wallace and Kelsey 2010). For example, live birth rates increase with up to approximately 15 oocytes retrieved, then plateau (Figure 6.2.4A) (Sunkara et al. 2011; Steward et al. 2014). Risk of ovarian hyperstimulation syndrome (OHSS; excessive response to ovarian stimulation) also increases with the number of oocytes retrieved, particularly above 15 (Steward et al. 2014). Therefore, during ART therapy, retrieval of 15 oocytes represents an optimal yield when possible, balancing the potential to achieve a live birth against the risk of OHSS. OHSS is an iatrogenic condition characterized by ovarian enlargement and a dramatic increase in vascular permeability, causing ascites, and is potentially fatal (Nelson 2017). OHSS risk factors include young age, low body weight, symptoms of polycystic ovary syndrome (PCOS), previous history of OHSS or high E2 on the day of human chorionic gonadotropin administration (PC ASRM 2008).

There is a strong, positive correlation between basal AMH serum levels and the number of oocytes retrieved after ovarian stimulation, and AMH is a better predictor of response to ovarian stimulation than AFC (Arce et al. 2013; La Marca and Sunkara 2014; Nelson et al. 2015a). Women with an excessive ovarian response (e.g., ≥21 oocytes retrieved) have a significantly higher AMH level than normal responders (Nelson et al. 2007) and basal levels of AMH can identify women at high risk for OHSS (Ocal et al. 2011). In a multicenter clinical evaluation of the Elecsys AMH Plus assay for prediction of response to ovarian stimulation, a hyperresponse was observed in 16/149 women undergoing a GnRH antagonist treatment protocol in the course of their first cycle of ovarian stimulation for IVF (Roche Diagnostics 2018). The Elecsys AMH Plus assay provided a high sensitivity and negative predictive value for prediction of a hyperresponse, whereby 81.3% of women at risk of a hyperresponse were correctly identified, and 96.6% of women not at risk could be ruled out (Roche Diagnostics 2018).

Based on its performance as a predictor of response to ovarian stimulation, AMH can be used to help define patient-tailored stimulation protocols, with the aim of optimizing oocyte yield and minimizing the risk of an abnormal ovarian response (Figure 6.2.4B) (Nelson et al. 2009; La Marca and Sunkara 2014). In patients with a low OR, poor response to ovarian stimulation is expected. Because there is no evidence of a superior approach to treating poor responders, FSH stimulation to obtain a maximal response should be administered and a protocol associated with reduced discomfort and with the least treatment burden should be selected. In patients with an OR in the normal range, the main objective is to maximize the success rate; normal responders should receive standard treatment. In contrast, the principal objective in patients with a high OR (i.e., predicted hyperresponders) is to minimize the risk of OHSS; these patients should be counseled on the risk of OHSS

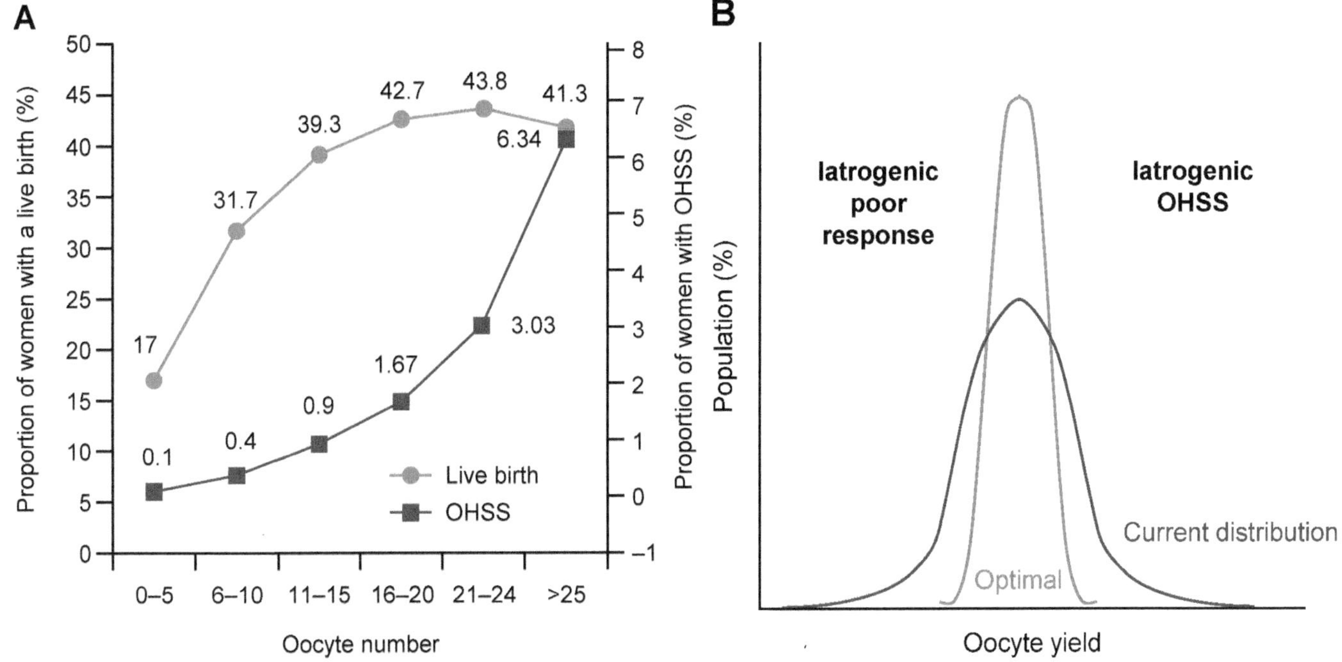

FIGURE 6.2.4 (A) Live birth rate and the risk of OHSS per number of oocytes retrieved per cycle and (B) the risk of abnormal ovarian response and optimal oocyte yield during ovarian stimulation. OHSS, ovarian hyperstimulation syndrome. (Panel A adapted from Steward, R.G. et al. *Fertil. Steril.*, 101, 967–973, 2014; Panel B adapted from Arce, J.C. *World CDx Conference*, London, UK, 2016.)

and appropriate FSH should be administered with lower gonadotropin doses (La Marca and Sunkara 2014).

Current evidence supports use of AMH-tailored ovarian stimulation protocols to increase IVF treatment success and reduce adverse outcomes and costs. For example, compared with FSH and age, AMH is a superior predictor of anticipated oocyte yield and live birth rate (Nelson et al. 2009). Moreover, recent evidence suggests that individualizing ovarian stimulation protocols based on the assessment of OR leads to improved patient outcomes (Fauser 2017). The following case study highlights the clinical benefit of individualized ovarian stimulation protocols.

CASE STUDY: INDIVIDUALIZED DOSING OF FOLLITROPIN DELTA ACCORDING TO SERUM ANTI-MÜLLERIAN HORMONE

Introduction

During ovarian stimulation, administration of exogenous recombinant FSH (rFSH) after pituitary downregulation serves to stimulate multiple follicle development. Standard dose FSH is not suitable for all women due to interindividual differences in OR and ovarian response to FSH (de Boer et al. 2003; van der Gaast et al. 2006). Several rFSHs are available for clinical use. Widely used rFSH proteins, expressed using Chinese Hamster Ovary Cells, are known as follitropin alfa or follitropin beta. More recently, a novel rFSH, follitropin delta, has been derived from a human cell line. Although follitropin alfa and follitropin delta share the same amino acid sequence, they vary in their glycosylation pattern (WIPO 2014), and follitropin delta exhibits distinct pharmacokinetic and pharmacodynamic characteristics (Olsson et al. 2014). For example, a greater exposure, lower serum clearance and a higher ovarian response (as determined by median follicle number and serum concentrations of inhibin B and E2) have been observed with follitropin delta compared with follitropin alfa when administering equal international units of FSH biological activity (Olsson et al. 2014).

Clinical data show that the biological effects of follitropin delta are influenced by serum AMH and body weight (Rose et al. 2016a). A population pharmacokinetic model predicted that average concentrations of follitropin delta at steady state are expected to reduce with increasing body weight (Rose et al. 2016b). Additionally, a prospective, randomized, controlled phase II study (NCT01426386) showed that increasing rFSH doses led to a linear increase in the number of oocytes retrieved in women undergoing IVF/ICSI, and this relationship was AMH dependent (Arce et al. 2014).

The Elecsys AMH Plus assay can be used to measure serum AMH and provides greater precision compared with other AMH assays (manual and automated) (van Helden and Weiskirchen 2015; Nelson et al. 2015b). The assay is calibrated against the Beckman Coulter AMH Gen II enzyme-linked immunosorbent assay and has been validated for prediction of a hyperresponse in women receiving a standard FSH stimulation dose of 150 IU/day during their first ovarian stimulation cycle (Roche Diagnostics 2018). An individualized dosing algorithm for follitropin delta, based on body weight and AMH levels, was evaluated in the Evidence-based Stimulation Trial with Human recombinant FSH in Europe and Rest of World-1 study (ESTHER-1) (Nyboe Andersen & Nelson et al. 2017).

Study Design

ESTHER-1 (NCT01956110) was a randomized, controlled, assessor-blinded, international, multicenter, non-inferiority trial. The study compared the efficacy and safety of a new recombinant human FSH, follitropin delta, with follitropin alfa (GONAL-F®), for ovarian stimulation in women undergoing an ART program (Nyboe Andersen & Nelson et al. 2017). Women aged 18–40 years undergoing their first IVF/ICSI cycle and diagnosed with tubal infertility, endometriosis stage I/II or unexplained infertility, or with partners diagnosed with male factor infertility, were eligible for enrollment. Key inclusion criteria were a body mass index of 17.5–32.0 kg/m², regular menstrual cycles of 24–35 days duration, presence of both ovaries and a FSH serum concentration of 1–15 IU/L during the early follicular phase. Key exclusion criteria were a history of recurrent miscarriage, endometriosis stage III/IV and the use of hormonal preparations (excluding thyroid medication) during the last menstrual cycle prior to randomization (Nyboe Andersen & Nelson et al. 2017). Dosing of follitropin delta (intervention) was individualized by AMH level obtained with the Elecsys AMH Plus immunoassay and by body weight (Figure 6.2.5), and the daily dose was maintained throughout the stimulation period. Follitropin alfa (comparator) was initiated at a daily dose of 150 IU (fixed for the first 5 stimulation days) and could subsequently be adjusted.

The primary objective was to demonstrate non-inferiority of an individualized dosing regimen of follitropin delta compared with conventional dosing of follitropin alfa, on the co-primary end points of ongoing pregnancy rate and ongoing implantation rate (assessed 10–11 weeks after embryo transfer). Prespecified secondary end points included live birth rate, targeted ovarian response (8–14 oocytes) and extreme ovarian response (<4, ≥15 or ≥20 oocytes) and embryology. Safety end points included the proportion of women with early and late OHSS and/or preventative interventions for early OHSS; adverse events were also monitored.

Results

A total of 1,329 patients from 11 countries were randomized, of whom 1,326 were exposed to study drug. Participant demographics and baseline characteristics were comparable between the two treatment arms. The women had a mean age of 33 years, the majority of participants (>90%) were white and the mean duration of infertility was 35 months. Infertility was unexplained (~40%), due to a male factor (~40%), tubal (~14%) or due to endometriosis I/II (~4%) or other reasons (<1%). At baseline in the follitropin delta and follitropin alfa groups, respectively, serum FSH concentrations were 7.5 and 7.7 IU/L, serum AMH concentrations were 16.3 and 16.0 pmol/L and AFCs were 14.7 and 14.4.

Non-inferiority of individualized follitropin delta versus conventional follitropin alfa was demonstrated for both co-primary

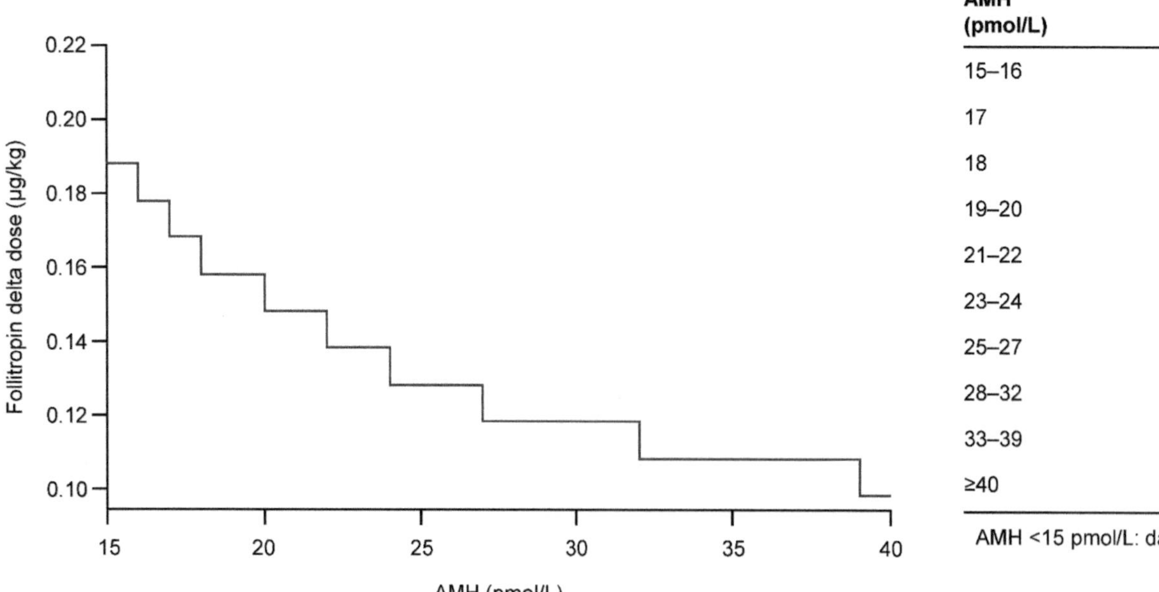

AMH (pmol/L)	Follitropin delta daily dose (µg/kg)
15–16	0.19
17	0.18
18	0.17
19–20	0.16
21–22	0.15
23–24	0.14
25–27	0.13
28–32	0.12
33–39	0.11
≥40	0.10

AMH <15 pmol/L: daily dose was 12 µg

FIGURE 6.2.5 AMH-tailored dosing algorithm for follitropin delta in the ESTHER-1 study. AMH, anti-Müllerian hormone. (Table from Nyboe Andersen & Nelson et al. 2017, *Fertil. Steril.*, 107, 387–396, 2017.)

end points: ongoing pregnancy rates were 30.7% versus 31.6%, respectively (nonsignificant difference: –0.9%), and ongoing implantation rates were 35.2% versus 35.8%, respectively (nonsignificant difference: –0.6%).

Analysis of secondary end points indicated no difference between treatment arms in terms of number of live births or number of neonates alive at 4 weeks. The rate of multiple pregnancies was also similar between groups. Importantly, individualized dosing based on AMH reduced poor and excessive responses in terms of the number of oocytes retrieved (Table 6.2.2). The number of oocytes retrieved increased markedly with rising AMH values with conventional stimulation. In contrast, the number of oocytes retrieved was more evenly distributed with more patients closer to the target response following individualized follitropin delta dosing (Figure 6.2.6a). There was no significant difference between the two treatment arms in the number of oocytes retrieved (Table 6.2.2), or in the number of good-quality blastocysts obtained (Figure 6.2.6A). Furthermore, compared with conventional dosing of follitropin alfa, fewer women receiving individualized dosing of follitropin delta had an extreme ovarian response (either <4 or ≥15 oocytes; or <4 or ≥20 oocytes; Table 6.2.2; Figure 6.2.6B); this was despite dose adjustments

TABLE 6.2.2 Ovarian response in ESTHER-1

OUTCOME VARIABLE	*INDIVIDUALIZED FOLLITROPIN DELTA (N = 665)*	*CONVENTIONAL FOLLITROPIN ALFA (N = 661)*	*P-VALUE*
Total dose (µg)	90.0 ± 25.3	103.7 ± 33.6	<0.001
Excessive response leading to triggering with GnRH agonist	10 (1.5)	23 (3.5)	0.019
Target ovarian response (8–14 oocytes retrieved)	275 (43.3)	247 (38.4)	0.019
Extreme ovarian response			
<4 or ≥15 oocytes retrieved	169 (26.6)	201 (31.3)	0.001
<4 or ≥20 oocytes retrieved	92 (14.5)	118 (18.4)	0.002
Ovarian response stratified by AMH			
Women with AMH <15 pmol/L	280	290	
Oocytes retrieved (*n*)	8.0 ± 4.3	7.0 ± 3.9	0.004
Poor responders (<4 oocytes)	33 (11.8)	52 (17.9)	0.039
Women with AMH ≥15 pmol/L	355	353	
Oocytes retrieved (*n*)	11.6 ± 5.9	13.3 ± 6.9	0.002
Excessive responders (≥15 oocytes)	99 (27.9)	124 (35.1)	0.038
Excessive responders (≥20 oocytes)	36 (10.1)	55 (15.6)	0.030

Source: Nyboe Andersen & Nelson et al. 2017, *Fertil. Steril.* 107, 387–396, 2017.
Note: Values are mean ± SD or *n* (%), unless stated otherwise. Data are for all women unless stated otherwise. AMH, anti-Müllerian hormone; GnRH, gonadotropin releasing hormone.

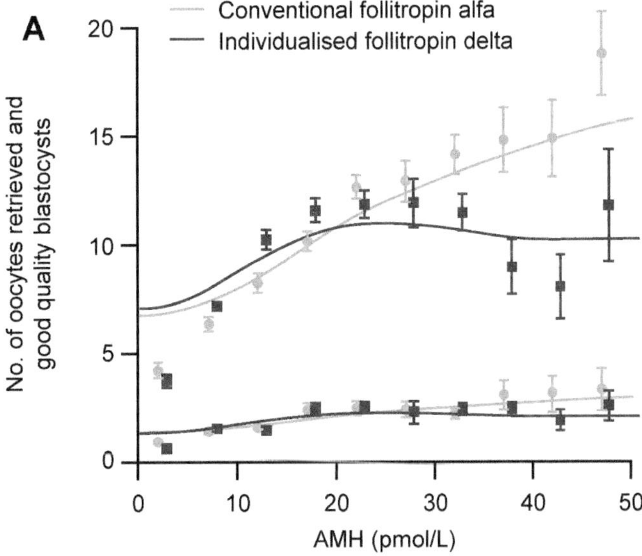

FIGURE 6.2.6 Ovarian response to AMH in ESTHER-1. (A) Number of oocytes retrieved (two upper curves) and number of good-quality blastocysts available for transfer (two lower curves) by serum AMH levels at screening for the two treatment groups. *(Continued)*

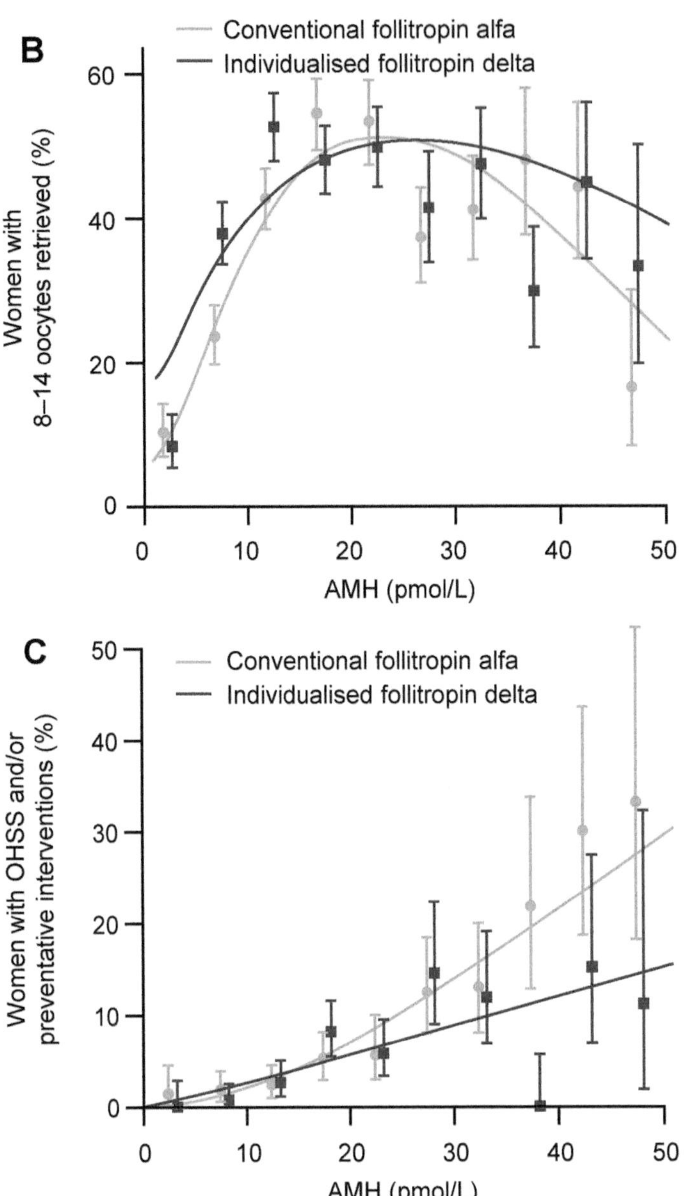

FIGURE 6.2.6 (Continued) Ovarian response to AMH in ESTHER-1. (B) Proportion of women achieving the target number of oocytes retrieved (8–14) by serum AMH levels for the two treatment groups. (C) Proportion of women requiring OHSS preventative interventions and/or experiencing OHSS by AMH levels for the two treatment groups. Data are mean ± standard error. AMH, anti-Müllerian hormone; OHSS, ovarian hyperstimulation syndrome. (Nyboe Andersen & Nelson et al. 2017, *Fertil. Steril.*, 107, 387–396, 2017.)

in 36.8% of women in the conventional follitropin alfa group versus none in the individualized follitropin delta group.

In women identified as potential hyporesponders (i.e., women with AMH <15 pmol/L), treatment with individualized follitropin delta was associated with more oocytes (8.0 vs. 7.0; $p = 0.004$) and a lower incidence of women with a poor response (i.e., <4 oocytes; 33 vs. 52; $p = 0.039$), compared with conventional follitropin alfa (Table 6.2.2). Similarly, in women identified as potential hyperresponders (i.e., women with AMH ≥15 pmol/L), treatment with individualized follitropin delta was associated with fewer oocytes retrieved (11.6 vs. 13.3; $p = 0.002$) and a lower incidence of patients with more than or equal 15 or 20 oocytes retrieved ($p = 0.038$ and $p = 0.030$, respectively), compared with follitropin alfa (Table 6.2.2).

The risk of requiring OHSS preventative interventions or experiencing OHSS increased with increasing AMH levels and differed between treatment arms (Figure 6.2.6c; Table 6.2.3). Fewer women in the individualized follitropin delta group required OHSS preventative interventions compared with the conventional follitropin alfa group ($p = 0.005$; Table 6.2.3). Two women in the individualized follitropin delta group, and six women in the conventional follitropin alfa group, were hospitalized because of OHSS (mean duration of hospitalization: 4.0 days and 8.7 days, respectively; Table 6.2.3). Apart from OHSS and OHSS preventative measures, there were no other differences in adverse events between treatment groups; the most common adverse events in both groups were headache, pelvic pain, pelvic discomfort and vomiting during pregnancy.

TABLE 6.2.3 Safety outcomes in ESTHER-1

OUTCOME VARIABLE	*INDIVIDUALIZED FOLLITROPIN DELTA (N = 665)*	*CONVENTIONAL FOLLITROPIN ALFA (N = 661)*	*P-VALUE*
Preventative interventions	15 (2.3)	30 (4.5)	0.005
Early OHSS			
Any grade	17 (2.6)	20 (3.0)	0.291
Moderate/severe	9 (1.4)	9 (1.4)	0.644
Any grade and/or preventative intervention	31 (4.7)	41 (6.2)	0.046
Moderate/severe and/or preventative intervention	24 (3.6)	34 (5.1)	0.019
All OHSS			
Any grade	23 (3.5)	32 (4.8)	0.238
Moderate/severe	14 (2.1)	19 (2.9)	0.514
Any grade and/or preventative intervention	37 (5.6)	53 (8.0)	0.037
Moderate/severe and/or preventative intervention	29 (4.4)	44 (6.7)	0.013
Hospitalization due to OHSS	2 (0.30)	6 (0.90)	0.108
Mean duration of hospitalization (days)	4.0	8.7	0.276

Source: Nyboe Andersen & Nelson et al. 2017, *Fertil. Steril.*, 107, 387–396, 2017.
Note: Values are *n* (%), unless stated otherwise. Data are for all women unless stated otherwise. OHSS, ovarian hyperstimulation syndrome.

CONCLUSIONS AND IMPLICATIONS FOR PRACTICE

Individualized follitropin delta dosing based on Elecsys AMH Plus and body weight reduces the chance of an ovarian response that is too low or too high, without the need for dose adjustments during stimulation. Compared with standard of care, tailored follitropin delta dosing based on AMH level and body weight is at least as effective in terms of pregnancy and implantation rate, is associated with significantly fewer cases of OHSS and/or preventative interventions for OHSS and requires a significantly lower total gonadotropin dose. EU marketing authorization for follitropin delta identifies the Elecsys AMH Plus from Roche Diagnostics as the assay to be used for individualized dosing (EMA 2016).

SUMMARY AND FUTURE DIRECTIONS

Infertility is relatively common, affecting one in six couples worldwide during their reproductive lifetimes, and is caused by a range of sex- and non–sex-specific factors. A number of tests are used to diagnose the cause of infertility and determine an optimal treatment strategy. Couples are increasingly seeking ART, which can help with conceiving. During ART, ovarian stimulation is used to obtain multiple oocytes with tailoring of the stimulation to obtain an optimal yield, balancing the potential to achieve a live birth against the risk of OHSS. The standard dose of FSH used for ovarian stimulation is not suitable for all women due to interindividual differences in OR and ovarian response to FSH as well as other important characteristics known to influence FSH exposure (e.g., body weight). Therefore, the success of ART relies, in part, on knowledge of an individual's OR. There are several methods available for the assessment of OR and the current standard of care is measurement of AMH, which serves as a direct serum marker of functional OR and can be quantified by immunoassay or determination of AFC by transvaginal ultrasonography. AMH demonstrates low inter- and intramenstrual cycle variation and provides better prediction of AFC classification than FSH and E2. As a strong predictor of response to ovarian stimulation, AMH can be used to define patient-tailored ovarian stimulation protocols. For example, determination of AMH levels using the Elecsys AMH Plus assay can be used in combination with body weight to establish the individual daily dose of follitropin delta, a novel recombinant FSH.

Future directions to expand the clinical utility of AMH potentially include prediction of menopause; identification of patients at risk for post-chemotherapy ovarian insufficiency; identification of testicular tissue in disorders of sex development; use as a biomarker for premature ovarian insufficiency; the diagnosis of PCOS or granulosa tumor cells; prediction of mono-follicular development in an ovulatory infertility/PCOS and reduction in OHSS risk in this population; and evaluation of hypogonadotropic hypogonadism (Nelson 2013; Leader and Baker 2014).

BIBLIOGRAPHY

American College of Obstetricians and Gynecologists Committee on Gynecologic Practice and Practice Committee (ACOG CGP PC). 2014. Female age-related fertility decline. Committee Opinion No. 589. *Fertil. Steril.* 101:633–634.

American Society for Reproductive Medicine (ASRM). Assisted reproductive technology. A guide for patients. Revised 2015. http://www.reproductivefacts.org/globalassets/rf/news-and-publications/bookletsfact-sheets/english-fact-sheets-and-info-booklets/art.pdf (accessed January 29, 2018).

Anckaert, E., J. Smitz, J. Schiettecatte, B. M. Klein, and J. C. Arce. 2012. The value of anti-Mullerian hormone measurement in the long GnRH agonist protocol: Association with ovarian response and gonadotrophin-dose adjustments. *Hum Reprod* 27:1829–1839.

Anckaert, E., M. Oktem, A. Thies et al. 2016. Multicenter analytical performance evaluation of a fully automated anti-Mullerian hormone assay and reference interval determination. *Clin Biochem* 49:260–267.

Anderson, R. A., S. M. Nelson, and W. H. Wallace. 2012. Measuring anti-Mullerian hormone for the assessment of ovarian reserve: When and for whom is it indicated? *Maturitas* 71:28–33.

Anderson, R. A., E. Anckaert, E. Bosch et al. 2015. Prospective study into the value of the automated Elecsys antimullerian hormone assay for the assessment of the ovarian growing follicle pool. *Fertil Steril* 103:1074–1080, e4.

Arce, J. C., A. La Marca, B. Mirner Klein, A. Nyboe Andersen, and R. Fleming. 2013. Antimüllerian hormone in gonadotropin releasing-hormone antagonist cycles: Prediction of ovarian response and cumulative treatment outcome in good-prognosis patients. *Feril Steril* 99:1644–1653.

Arce, J. C., A. N. Andersen, M. Fernandez-Sanchez et al. 2014. Ovarian response to recombinant human follicle-stimulating hormone: A randomized, antimullerian hormone-stratified, dose-response trial in women undergoing in vitro fertilization/intracytoplasmic sperm injection. *Fertil Steril* 102:1633–1640.

Arce, J. C. *World CDx Conference*. London, UK. March 17, 2016.

Bleil, M. E., S. E. Gregorich, N. E. Adler, B. Sternfeld, M. P. Rosen, and M. I. Cedars. 2014. Race/ethnic disparities in reproductive age: An examination of ovarian reserve estimates across four race/ethnic groups of healthy, regularly cycling women. *Fertil Steril* 101:199–207.

Blumenauer, V., U. Czeromin, D. Fehr et al. 2017. D.I.R annual 2016—The German IVF registry. *J Reproduktionsmed Endokrinol* 14:272–305.

Bottcher, B., I. Tsybulyak, T. Grubinger, L. Wildt, and B. Seeber. 2014. Dynamics of anti-mullerian hormone during controlled ovarian stimulation. *Gynecol Endocrinol* 30:121–125.

Broekmans, F. J., D. de Ziegler, C. M. Howles, A. Gougeon, G. Trew, and F. Olivennes. 2010. The antral follicle count: Practical recommendations for better standardization. *Fertil Steril* 94:1044–1051.

Carlsson, I. B., J. E. Scott, J. A. Visser, O. Ritvos, A. P. Themmen, and O. Hovatta. 2006. Anti-Mullerian hormone inhibits initiation of growth of human primordial ovarian follicles in vitro. *Hum Reprod* 21:2223–2227.

Chow, E. J., K. L. Stratton, K. C. Oeffinger et al. 2016. Pregnancy after chemotherapy in male and female survivors of childhood cancer treated between 1970 and 1999: A report from the childhood cancer survivor study cohort. *Lancet Oncol* 17:567–576.

de Boer, E. J., I. den Tonkelaar, E. R. te Velde, C. W. Burger, and F. E. van Leeuwen; OMEGA-project group. 2003. Increased risk of early menopausal transition and natural menopause after poor response at first IVF treatment. *Hum Reprod* 18:1544–1552.

Desai, N., S. Ploskonka, L. R. Goodman, C. Austin, J. Goldberg, and T. Falcone. 2014. Analysis of embryo morphokinetics, multinucleation and cleavage anomalies using continuous time-lapse monitoring in blastocyst transfer cycles. *Reprod Biol Endocrinol* 12:54.

Dewailly, D., C. Y. Andersen, A. Balen et al. 2014. The physiology and clinical utility of anti-Mullerian hormone in women. *Hum Reprod Update* 20:370–385.

Drakopoulos, P., A. van de Vijver, J. Parra et al. 2019. Serum anti-Müllerian hormone is significantly altered by downregulation with daily gonadotropin-releasing hormone agonist: A prospective cohort study. *Front. Endocrinol.* doi: 10.3389/fendo.2019.00115.

Durlinger, A. L., P. Kramer, B. Karels et al. 1999. Control of primordial follicle recruitment by anti-Mullerian hormone in the mouse ovary. *Endocrinology* 140:5789–5796.

Durlinger, A. L., M. J. Gruijters, P. Kramer et al. 2002. Anti-Mullerian hormone inhibits initiation of primordial follicle growth in the mouse ovary. *Endocrinology* 143:1076-84.

European Medicines Agency (EMA). 2019. Summary of product characteristics of Rekovelle. https://www.ema.europa.eu/documents/product-information/rekovelle-epar-product-information_en.pdf (accessed February 13, 2019).

Farquhar, C., J. R. Rishworth, J. Brown, W. L. Nelen, and J. Marjoribanks. 2015. Assisted reproductive technology: An overview of cochrane reviews. *Cochrane Database Syst Rev* CD010537. doi:10.1002/14651858.CD010537.pub4.

Fauser, B. C. J. M. 2017. Patient-tailored ovarian stimulation for in vitro fertilization. *Fertil Steril* 108:585–591.

Grynnerup, A. G., A. Lindhard, and S. Sorensen. 2014. Recent progress in the utility of anti-Mullerian hormone in female infertility. *Curr Opin Obstet Gynecol* 26:162–167.

Hansen, K. R., N. S. Knowlton, A. C. Thyer, J. S. Charleston, M. R. Soules, and N. A. Klein. 2008. A new model of reproductive aging: The decline in ovarian non-growing follicle number from birth to menopause. *Hum Reprod* 23:699–708.

Hansen, K. R., G. M. Hodnett, N. Knowlton, and L. B. Craig. 2011. Correlation of ovarian reserve tests with histologically determined primordial follicle number. *Fertil Steril* 95:170–175.

Hassan, M. A., and S. R. Killick. 2004. Negative lifestyle is associated with a significant reduction in fecundity. *Fertil Steril* 81:384–392.

Homan, G. F., M. Davies, and R. Norman. 2007. The impact of lifestyle factors on reproductive performance in the general population and those undergoing infertility treatment: A review. *Hum Reprod Update* 13:209–223.

Huang, J. Y., and Z. Rosenwaks. 2012. In vitro fertilisation treatment and factors affecting success. *Best Pract Res Clin Obstet Gynaecol* 26:777–788.

Iliodromiti, S., and S. M. Nelson. 2013. Biomarkers of ovarian reserve. *Biomark Med* 7:147–158.

Janssen, N. M., and M. S. Genta. 2000. The effects of immunosuppressive and anti-inflammatory medications on fertility, pregnancy, and lactation. *Arch Intern Med* 160:610–619.

Jirge, P. R. 2011. Ovarian reserve tests. *J Hum Reprod Sci* 4:108–113.

Kallio, S., J. Puurunen, A. Ruokonen, T. Vaskivuo, T. Piltonen, and J. S. Tapanainen. 2013. Antimullerian hormone levels decrease in women using combined contraception independently of administration route. *Fertil Steril* 99:1305–1310.

Kaplan, B., R. Nahum, Y. Yairi et al. 2005. Use of various contraceptive methods and time of conception in a community-based population. *Eur J Obstet Gynecol Reprod Biol* 123:72–76.

Kelsey, T. W., P. Wright, S. M. Nelson, R. A. Anderson, and W. H. Wallace. 2011. A validated model of serum anti-mullerian hormone from conception to menopause. *PLoS One* 6:e22024.

La Marca, A., and S. K. Sunkara. 2014. Individualization of controlled ovarian stimulation in IVF using ovarian reserve markers: From theory to practice. *Hum Reprod Update* 20:124–140.

Lambert-Messerlian, G., B, Plante, E. E. Eklund, C. Raker, and R. G. Moore. 2016. Levels of antimüllerian hormone in serum during the normal menstrual cycle. *Fertil Steril* 105:208–213.

Leader, B., and V. L. Baker. 2014. Maximizing the clinical utility of antimullerian hormone testing in women's health. *Curr Opin Obstet Gynecol* 26:226–236.

Lee, T. H., C. D. Chen, Y. Y. Tsai et al. 2006. Embryo quality is more important for younger women whereas age is more important for older women with regard to in vitro fertilization outcome and multiple pregnancy. *Fertil Steril* 86:64–69.

Loveland J. B., H. D. McClamrock, A. M. 2001. Malinow, and F. I. Sharara. Increased body mass index has a deleterious effect on in vitro fertilization outcome. *J Assist Reprod Genet* 18:382–386.

Mascarenhas, M. N., S. R. Flaxman, T. Boerma, S. Vanderpoel, and G. A. Stevens. 2012. National, regional, and global trends in infertility prevalence since 1990: A systematic analysis of 277 health surveys. *PLoS Med* 9:e1001356.

Menken, J., J. Trussell, and U. Larsen. 1986. Age and infertility. *Science* 234:413.

Mills, M., R. R. Rindfuss, P. McDonald, and E. te Velde; ESHRE Reproduction and Society Task Force. 2011. Why do people postpone parenthood? Reasons and social policy incentives. *Hum Reprod Update* 17:848–860.

Nachtigall, R. D. 2006. International disparities in access to infertility services. *Fertil Steril* 85:871–875.

National Health Service (NHS). Infertility. Last reviewed 14 February 2017. https://www.nhs.uk/conditions/infertility/causes/ (accessed February 13, 2019).

National Institute for Health and Care Excellence (NICE). Clinical guideline 156: Assessment and treatment for people with fertility problems. Last updated September 2017. https://www.nice.org.uk/guidance/cg156 (accessed February 13, 2019).

Nelson, S. M., R. W. Yates, and R. Fleming. 2007. Serum anti-Mullerian hormone and FSH: Prediction of live birth and extremes of response in stimulated cycles—implications for individualization of therapy. *Hum Reprod* 22:2414–2421.

Nelson, S. M., R. W. Yates, H. Lyall et al. 2009. Anti-Mullerian hormone-based approach to controlled ovarian stimulation for assisted conception. *Hum Reprod* 24:867–875.

Nelson, S. M., M. C. Messow, A. M. Wallace, R. Fleming, and A. McConnachie. 2011. Nomogram for the decline in serum antimullerian hormone: A population study of 9,601 infertility patients. *Fertil Steril* 95:736–741, e1–e3.

Nelson, S. M. 2013. Biomarkers of ovarian response: Current and future applications. *Fertil Steril* 99:963–969.

Nelson, S. M., B. M. Klein, and J. C. Arce. 2015a. Comparison of antimullerian hormone levels and antral follicle count as predictor of ovarian response to controlled ovarian stimulation in good-prognosis patients at individual fertility clinics in two multicenter trials. *Fertil Steril* 103:923–930, e1.

Nelson, S. M., E. Pastuszek, G. Kloss et al. 2015b. Two new automated, compared with two enzyme-linked immunosorbent, antimullerian hormone assays. *Fertil Steril* 104:1016–1021.

Nelson, S. M. 2017. Prevention and management of ovarian hyperstimulation syndrome. *Thromb Res* 151(Suppl 1):S61–S64.

Nyboe Andersen, A., S. M. Nelson, B. C. Fauser et al. 2017. Individualized versus conventional ovarian stimulation for in vitro fertilization: A multicenter, randomized, controlled, assessor-blinded, phase 3 noninferiority trial. *Fertil Steril* 107:387–396.

Ocal, P., S. Sahmay, M. Cetin, T. Irez, O. Guralp, and I. Cepni. 2011. Serum anti-Mullerian hormone and antral follicle count as predictive markers of OHSS in ART cycles. *J Assist Reprod Genet* 28:1197–1203.

Olsson, H., R. Sandstrom, and L. Grundemar. 2014. Different pharmacokinetic and pharmacodynamic properties of recombinant follicle-stimulating hormone (rFSH) derived from a human cell line compared with rFSH from a non-human cell line. *J Clin Pharmacol* 54:1299–1307.

Plante, B. J., G. S. Cooper, D. D. Baird, and A. Z. Steiner. 2010. The impact of smoking on antimullerian hormone levels in women aged 38 to 50 years. *Menopause* 17:571–576.

Practice Committee of the American Society for Reproductive Medicine (PC ASRM). 2008. Ovarian hyperstimulation syndrome. *Fertil Steril* 90:S188–S193.

Practice Committee of the American Society for Reproductive Medicine (PC ASRM). 2012. Testing and interpreting measures of ovarian reserve: A committee opinion. *Fertil Steril* 98:1407–1415.

Randolph, J. F., Jr., S. D. Harlow, M. E. Helmuth, H. Zheng, and D. S. McConnell. 2014. Updated assays for inhibin B and AMH provide evidence for regular episodic secretion of inhibin B but not AMH in the follicular phase of the normal menstrual cycle. *Hum Reprod* 29:592–600.

Roche Diagnostics, 2019. Elecsys AMH Plus Method Sheet. https://dialog1.roche.com/gb/en_gb/eLabDoc (accessed February 13, 2019).

Rose, T. H., D. Röshammar, L. Erichsen, L. Grundemar, and J. T. Ottesen. 2016a. Characterisation of population pharmacokinetics and endogenous follicle-stimulating hormone (FSH) levels after multiple dosing of a recombinant human FSH (FE 999049) in healthy women. *Drugs R D* 16:165–172.

Rose, T. H., D. Roshammar, L. Erichsen, L. Grundemar, and J. T. Ottesen. 2016b. Population pharmacokinetic modelling of FE 999049, a recombinant human follicle-stimulating hormone, in healthy women after single ascending doses. *Drugs R D* 16:173–180.

Steward, R. G., L. Lan, A. A. Shah et al. 2014. Oocyte number as a predictor for ovarian hyperstimulation syndrome and live birth: An analysis of 256,381 in vitro fertilization cycles. *Fertil Steril* 101:967–973.

Sunkara, S. K., V. Rittenberg, N. Raine-Fenning, S. Bhattacharya, J. Zamora, and A. Coomarasamy. 2011. Association between the number of eggs and live birth in IVF treatment: An analysis of 400 135 treatment cycles. *Hum Reprod* 26:1768–1774.

Toner, J. P., C. B. Philput, G. S. Jones, and S. J. Muasher. 1991. Basal follicle-stimulating hormone level is a better predictor of in vitro fertilization performance than age. *Fertil Steril* 55:784–791.

Tsepelidis, S., F. Devreker, I. Demeestere, A. Flahaut, C. Gervy, and Y. Englert. 2007. Stable serum levels of anti-Mullerian hormone during the menstrual cycle: A prospective study in normo-ovulatory women. *Hum Reprod* 22:1837–18340.

Turan, V. and K. Oktay. 2014. Sexual and fertility adverse effects associated with chemotherapy treatment in women. *Expert Opin Drug Saf* 13:775–783.

van der Gaast, M. H., M. J. Eijkemans, J. B. van der Net et al. 2006. Optimum number of oocytes for a successful first IVF treatment cycle. *Reprod Biomed Online* 13:476–480.

van Disseldorp, J., C. B. Lambalk, J. Kwee et al. 2010. Comparison of inter- and intra-cycle variability of anti-Mullerian hormone and antral follicle counts. *Hum Reprod* 25:221–227.

van Helden, J., and R. Weiskirchen. 2015. Performance of the two new fully automated anti-Mullerian hormone immunoassays compared with the clinical standard assay. *Hum Reprod* 30:1918–1926.

Wallace, W. H., and T. W. Kelsey. 2010. Human ovarian reserve from conception to the menopause. *PLoS One* 5:e8772.

World Intellectual Property Organization (WIPO; WO 2009/127826). 2019. Recombinant FSH including alpha 2,3 and alpha 2,6 sialylation. https://patentscope.wipo.int/PCT/GB2009/000978 (accessed February 13, 2019).

Zegers-Hochschild, F., G. D. Adamson, S. Dyer et al. 2017. The international glossary on infertility and fertility care, 2017. *Fertil Steril* 108:393–406.

Factor Xa
Case Study

6.3

Boris Gorovits

Contents

BIOLOGY OF FACTOR Xa AND COAGULATION CASCADE

Factor X (FX) has a critical position in the coagulation cascade and in maintaining the normal hemostasis. FX deficiency, a very rare condition caused by an insufficient amount of active FX, leads to a severe phenotype due to a substantial risk of hemorrhaging. It is a vitamin K-dependent serine endopeptidase produced in the liver. FX is activated into the Factor Xa (FXa) through the intrinsic and extrinsic pathways [1–5] (Figure 6.3.1). The FX to FXa transition is catalyzed by the tissue factor (TF)—Factor VIIa (FVIIa) complex by a limited proteolysis, a crucial step in the overall blood coagulation initiation phase. The majority of the FXa is produced in the propagation phase of coagulation cascade through the FIXa (product of Factor IX activation) catalyzed digestion of the FX while complexed with the non-proteolytic Factor VIIIa (product of FVIII activation). Upon transition, and when combined with Factor Va, Ca2+ and phospholipids on the surface of activated platelets, FXa forms the prothrombinase complex that is involved in the enzymatic transformation of prothrombin into thrombin. The thrombin polymerizes into the insoluble blood clot, thrombus. In turn, thrombus has the ability to activate platelets as well as convert fibrinogen into an insoluble fibrin clot [6,7]. Activity of FXa can be effectively inhibited by binding to the tissue factor plasma inhibitor (TFPI), a Kunitz type inhibitor found in plasma, also on the surface of vascular endothelial cells and in platelets [5]. TFPI binding to the FXa followed by binding to the TF-FVIIa complex or direct binding of TFPI to the TF-FVIIa-FXa complex leads to the formation of a quaternary TF-FVIIa-FXa-TFPI complex and abrogation of TF activity. In addition to its role in the coagulation cascade, FXa was evaluated as a cell signaling protease [8] and was proposed to have a role in pathogenesis of pro-inflammatory conditions [1].

APPLICATION OF FACTOR Xa INHIBITOR

Thrombosis remains one of the major causes of morbidity and mortality in the Western world. The pathogenesis of thrombosis lays at the basis of a number of cardiovascular conditions such as pulmonary embolism, unstable angina, myocardial infarction, sudden cardiac death, peripheral arterial occlusion, ischemic stroke and deep-vein thrombosis [9]. Postsurgical complications, such as peripheral arterial disease and venous thromboembolism (VTE), may lead to an elevated risk of coronary heart disease [10]. As a result, a significant effort has been made to identify and develop effective anti-thrombotic agents. Various anticoagulant agents exist, including vitamin K antagonists (VKA), low molecular weight heparins and warfarin. Because FXa links intrinsic and extrinsic coagulation pathways and serves as a gate keeper in the thrombin production, modulating of the FXa activity with direct and indirect FXa inhibitors allows to control thrombin generation and has been applied as an

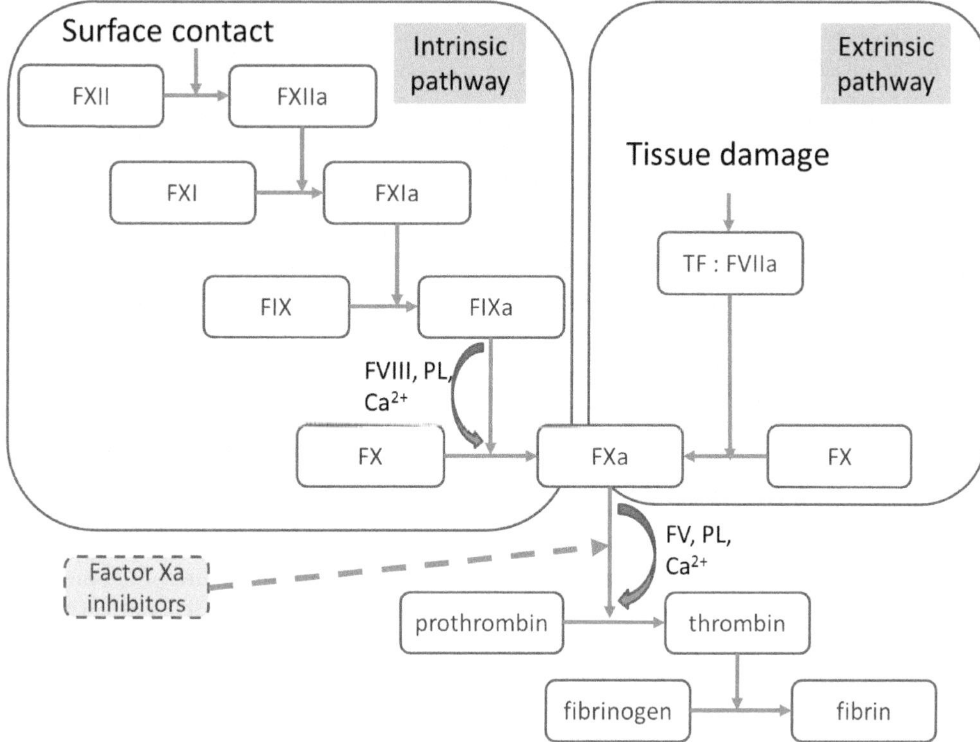

FIGURE 6.3.1 Example of coagulation pathway.

effective antithrombotic methodology. Direct and indirect FXa inhibitors have been applied for treatment of various serious thrombosis related conditions, including VTE and treatment of acute coronary syndrome [11]. The indirect FXa inhibitors such as fondaparinux and idraparinux act through interaction with antithombin, whereas the direct and specific FXa inhibitors such as rivaroxaban, directly bind to the FXa [12]. Some of these therapeutics may have a slow activity onset, require frequent monitoring, can exhibit food and drug interactions [13,14]. Based on the mechanism of thrombin activation, the anti-thrombotic inhibitors can be classified into anticoagulants, antiplatelet agents and direct thrombolytic drugs. Due to a critical role played by the FXa in the blood coagulation cascade, various FXa inhibitors have been developed as part of the effort to address thromboembolic diseases. Direct inhibition of FXa activity is expected to reduce production of additional thrombin with a minimal impact on the preexisting thrombin levels and therefore provide an improved anti-thrombotic effect with a limited effect on the primary hemostasis [7]. A significant number of compounds have been investigated as potentially specific for FXa inhibition. Several have been tested in advanced clinical studies with various results. A desired FXa inhibitor should effectively decrease thrombin generation with a limited negative effect on the levels of preexisting thrombin. Such selectivity can ensure an improved anti-thrombotic effect with a reduced impact on the overall hemostasis levels. As a result, specific FXa inhibitors have shown to possess a wider therapeutic window compare to other potential anti-thrombotic agents, including direct inhibitors of thrombin [15,16].

ASSESSMENT OF FXa INHIBITORS ACTIVITY

Assessment of FXa activity and overall impact of FXa inhibitors on the coagulation cascade is a critical element of clinical monitoring after administration of FXa or thrombin inhibitors (DTIs). Monitoring is also an important element of transition between anticoagulation therapeutic agents, an element of compliance control and during new drug development. Activity of FXa inhibitors can be evaluated by a number of coagulation cascade activity methodologies, including prothrombin time (PT), activated partial thromboplastin time (aPTT), but also by a direct monitoring of the FXa activity. During Phase I and II studies, pharmacodynamics properties of rivaroxaban were assessed by evaluating FXa activity and prolongation of the PT or aPTT [11,17,18]. The impact on the FXa activity was shown to be dose dependent and continued for several hours after the FXa inhibitor administration. FXa activity continued to be reduced even after 24 h after administration of 5 mg of the compound therefore suggesting a daily dosing regimen. Similarly, the effect of apixaban was evaluated by a number of coagulation cascade activity protocols and by a direct FXa activity assessment *in vitro* [19]. Compound effect on performance of various coagulation cascade activity protocols varied dependent on the type of the method. For example, variable results were obtained in the aPTT and PT assay kits supplied by various manufacturers. It was suggested that aPTT and PT methods produce a wide range of interindividual responses therefore complicating understanding of the FXa inhibitor (e.g., apixaban) effect on the FXa activity [20–22]. In addition to a variable impact of reagent type produced by different manufacturers, interference by various matrix components, for example by the lupus anticoagulant, was suspected as the main cause [19].

On the contrary, chromogenic FXa activity protocol produced a linear correlation with the apixaban levels [18,19,23]. The use of a chromogenic FXa activity assay was therefore recommended as a reliable approach to evaluate therapeutic concentrations, whereas aPTT and PT methods were not found to be applicable.

In addition to the evaluation of FXa inhibitor levels, monitoring for coagulation activity is critical when a reversal of anticoagulation is required. Quick reversal of anticoagulation may be applied in patients with severe bleeding or patients requiring surgery [24]. Reversal of FXa inhibitors and other anticoagulants action is achieved by administering clotting factor concentrates, including prothrombin complex or activated prothrombin complex concentrates and also FVIIa. An adequate degree of anticoagulation may be confirmed by a variety of methods including activated aPTT, thrombin time (TT) and chromogenic FXa assay. As an example, applications of monitoring methods to control reversal of the anticoagulant activity of fondaparinux and rivaroxaban in animal and clinical studies was reported [25]. The monitoring was recommended to continue after hemostasis was achieved as the clearance rates for the orally administered FXa inhibitors can be longer versus that for the clotting factor agents.

PROTOCOLS OF COAGULATION ACTIVITY MEASUREMENT

Analytical techniques intended to measure activity of anticoagulants include a number of clotting time based assays aiming to assess the entire coagulation cascade activity, also enzymatic activity assays to evaluate specific elements of the cascade and ligand binding immunosorbent (e.g., ELISA) or high performance liquid chromatography based protocols intended to analyze specific hemostatic factors [26]. A brief description of selected coagulation activity protocols is provided below.

- **Prothrombin Time and International Normalized Ratio**
  In the PT method, the time (typically in the range of 10–14 s) required for plasma clotting is measured after the addition of thromboplastin (includes a mix of TF, phospholipids and Ca2+). Presence of the extrinsic coagulation pathway inhibitors or deficiency in one of the elements of the pathway results in extended (prolonged) coagulation time measured by the PT protocol. This is a broadly available test commonly applied to evaluate extrinsic and intrinsic coagulation pathways to include activity of factors II, V, VII and X. Several commercial kits produced using various TF reagent and methods of preparation are available. Due to these differences, PT kits often vary in their sensitivity to the anticoagulation effect of therapeutics [27]. In the 1983, the World Health Organization (WHO) introduced an International Sensitivity Index (ISI) and the international normalized ratio (INR) aiming to standardize various PT tests available on the market [28]. The INR allows to normalize the individual test kit sensitivity by applying the ISI factor supplied by the kit manufacturer. The INS approach was developed for monitoring VKAs such as warfarin and may not be directly applicable to the detection of FXa inhibitors activity due to the high specificity of anti-FXa therapeutics. A dose dependent PT prolongation was reported on a reagent dependent basis in subjects treated with rivaroxaban [18,21,29]. Others reported data suggests a significant variability based on the differences in the specific nature of applied assay reagents and assay protocols utilized at various labs. Test results remained variable even after the INR conversion [17,30–32]. Similar variability and reagent nature dependency was observed in a dilute PT test where the thromboplastin reagent concentration is reduced to mimic that under the physiological conditions. A lack of PT test sensitivity to the presence of FXa inhibitors was explain by several factors, including the global clotting nature of the PT test, the transient and highly specific effect of FXa inhibitors on the PT prolongation and a relatively fast PK clearance rates of these therapeutics [21].

- **Activated Partial Thromboplastin Time**
  In the aPTT method, the time (typically 20–40 s) required for plasma clotting is measured after the addition of a partial thromboplastin. In contrast to the PT method, in the partial thromboplastin reagent the TF is not supplied with the phospholipid component. Impairments in either the intrinsic or extrinsic pathways result in prolongation of the aPTT read out time. To initiate reaction, a surface activator (e.g., silica, glass or kaolin) and phospholipid reagents are added to the plasma sample. After contact factors are activated, sample is supplied with Ca2+ ions and the clotting time (i.e., aPTT) is detected. A variation of the aPTT test, referred to as activated clotting time, is based on analysis of clotting activity in the whole blood [33]. Anti-coagulation factors, including FXa inhibitors, have a variable effect on the aPTT read out with a significant degree of reagent dependency [31,33,34]. This limits aPTT protocol application for evaluation of FXa inhibitor levels.

- **Prothrombinase Induced Clotting Time**
  In the prothrombinase-induced clotting time (PiCT) method, the time (seconds) of plasma clotting is

measured after coagulation is activated by the prothrombinase complex that contains a mix of a specific amount of bovine FXa, phospholipids, Ca2+ and a specific FV activator (Russell's viper venom-V). Because the outcome of the test depends on the residual activity of the FXa in the sample, presence of FXa inhibitors will prolong clotting time. PiCT has been evaluated as a potential protocol to detect presence of FXa inhibitors [35]. A more common variation, two-step, PiCT protocol produced an unexpected relationship between the PiCT read out and rivaroxaban concentrations where a decrease in the PiCT measured clotting time was observed at low concentrations of the FXa inhibiting therapeutic [35]. No such decrease was observed in a single step modification of the PiCT or when a human FXa reagent was applied [30]. Modified commercial PiCT protocol was suggested as a potential monitoring tool for FXa inhibitors although it was noted that therapeutics vary greatly in their ability to modulate the PiCT read out that may significantly complicate the use of PiCT as a measure of FXa inhibitor activity when switching between anti-FXa agents [30]. Although with some additional modifications PiCT test has shown a higher sensitivity to the rivaroxaban concentrations versus that for the PT protocol, it remained insufficiently sensitive to assess low concentrations of the therapeutic [36].

- **Thrombin Generation Test**
A number of thrombin generation tests are available including methods applicable for plasma or whole blood sample assessment. In the test, thrombin generation is evaluated by supplying a trigger—typically TF or collagen. Activity of generated thrombin is detected by applying a fluorogenic or chromogenic substrate reagent. The degree of standardization between the various tests is limited and the use of a thrombin calibrator is required [37,38]. Typical measured values include maximum concentration (activity) of generated thrombin, the endogenous thrombin potential (ETP, a product of the thrombin activity and the time during which it remains active), the lag time and the time required to achieve maximum thrombin concentration [39]. FXa inhibitor rivaroxaban was shown to effectively modulate thrombin generation parameters, particularly in the initial phase of the reaction. Rivaroxaban was also shown to inhibit ETP within a relatively short time after administration of therapeutic (2-h period) [36]. As a limitation, the thrombin generation test remains insufficiently sensitive to detect low concentrations of the FXa inhibitors and because of this is not commonly used in clinical setting [21].

- **HepTest**
The HepTest was developed by the American Diagnostica and is a FXa clot-based activity assay [26]. The method is based on measuring the ability of anticoagulant to modulate inhibition of exogenous FXa by plasma anti-thrombin III in the presence of naturally occurring plasma antagonists. Based on the assay principles, the rate of FXa inhibition is expected to be directly correlated with the concentration of evaluated anticoagulant. The effect is indirectly measured by determining the prolongation of the clotting time. An undiluted plasma or whole blood sample is supplemented with an exogenous FXa and after addition of a mixture containing Ca2+, phospholipids, FV and fibrinogen, the clotting time is measured. Similar to the aPTT and PT tests, HepTest protocol generated a dose dependent response when tested in the presence of apixaban although unexpected results were reported in the presence of rivaroxaban where low concentrations of the FXa inhibiting therapeutic reduced clotting time [21,31]. A slight modification of the test protocol, including a reduction in the incubation time, produced results better aligned with the expected. Due to the method potential to generate unexpected values, the HepTest is generally viewed as a nonspecific for evaluating FXa activity and is not broadly used.

- **Anti-factor FXa Chromogenic Tests**
In the FXa chromogenic assays, the enzyme activity is measured by applying a synthetic substrate that contains an enzymatic cleavage site present in the naturally occurring protein substrate. The synthetic substrate contains a chromophore moiety, the *p*-nitroaniline. Upon enzymatic release of the *p*-nitroaniline, the change is solution color is measured colorimetrically to produce final assay read out. The resulting change in the optical density is proportional to the concentration of the active FXa in the sample [26,33]. FXa chromogenic assays have been actively used to evaluate the presence and concentrations of various anticoagulants, including heparin, thrombin inhibitors and anti-FXa therapeutics [31–33,40]. Application of FXa chromogenic assays to determine anti-FXa therapeutics requires the use of a therapeutic specific standard curve that is viewed as a potential limitation due to the limited availability in clinical setting [24,41]. A chromogenic FXa activity method was evaluated to monitor samples with significant rivaroxaban concentrations [21]. Although establishment of a calibration curve is required, chromogenic test is capable of measuring a broad range of FXa inhibitor plasma concentrations. In the case of rivaroxaban, chromogenic FXa activity method was shown to detect inhibitor concentrations up to 660 ng/mL covering expected therapeutic plasma levels of the drug [42]. Compare to other coagulation protocols, including broadly available PT method, FXa activity chromogenic assays offer a sensitive and specific approach to measuring FXa inhibitor plasma concentrations.

CONCLUSION

Factor X and its activated version, FXa, are recognized as critical elements of coagulation cascade with a high potential to treatment of various thrombosis related conditions. Although a number of anticoagulants are available on the market, specific FXa activity inhibitors are emerging as highly efficient in addressing serious thrombosis related complications. Several tests have been developed to monitor activity of anticoagulation agents in the clinical setting. These include PT/INR, aPTT, HepTest, PiCT and chromogenic FXa activity assays. These tests differ based on the nature of evaluated activity and the ability to quantitatively assess activity of a specific component of the coagulation cascade. Some of the tests are viewed as more appropriate to evaluate levels of VKAs (e.g., PT), heparin (e.g., aPTT, chromogenic FXa) and FXa inhibitors (e.g., chromogenic FXa). Tests also vary in their general availability, for example, the PiCT and HepTest use is limited. New oral FXa inhibitors have an advantage of well predictable pharmacokinetics and pharmacodynamics, fewer food and drug interactions. Although FXa inhibitors promise a significant improvement in reduced degree of monitoring, the need for a predicting and quantitative evaluation of their therapeutic activity remains, specifically for patients with abnormally low or high body mass, patients with renal and hepatic impairment and patients with bleeding or thrombotic conditions [21]. Reported data suggests that only certain coagulation cascade activity tests produce reliable correlation of reported read out with the FXa inhibitor levels. Some of the routinely used assays (PT/INR, aPTT, HepTest, PiCT) cannot appropriately report the effect of FXa inhibitors on the clotting activity. The FXa inhibitors are reported to have minimal effect on the aPTT and PT and the outcome of the test greatly depends on the type of assay reagents used. The chromogenic FXa activity assay has demonstrated a good correlation with FXa inhibitor concentrations and activities. It has emerged as a potential method of choice when it comes to measuring levels of FXa inhibitors in patients.

REFERENCES

1. Ebrahimi, S. et al., Factor Xa signaling contributes to the pathogenesis of inflammatory diseases. *J Cell Physiol*, 2016.
2. Rau, J.C. et al., *Serpins in thrombosis, hemostasis and fibrinolysis. J Thromb Haemost: JTH*, 2007. **5**(Suppl 1): 102–115.
3. Borensztajn, K. and C.A. Spek, Blood coagulation factor Xa as an emerging drug target. *Expert Opin Ther Targets*, 2011. **15**(3): 341–349.
4. McVey, J.H., Tissue factor pathway. *Baillieres Clin Haematol*, 1994. **7**(3): 469–484.
5. McVey, J.H., Tissue factor pathway. *Baillieres Best Pract Res Clin Haematol*, 1999. **12**(3): 361–372.
6. Rosen, E.D., Gene targeting in hemostasis. Factor X. *Front Biosci*, 2002. **7**: d1915–d1925.
7. Guo, L. and S. Ma, Advances in inhibitors of FXa. *Curr Drug Targets*, 2015. **16**(11): 1207–1232.
8. Leadley Jr, R.J., L. Chi, and A.R. Porcari, Non-hemostatic activity of coagulation factor Xa: Potential implications for various diseases. *Curr Opin Pharmacol*, 2001. **1**(2): 169–175.
9. Agrawal, R., P. Jain, and S.N. Dikshit, Apixaban: A new player in the anticoagulant class. *Curr Drug Targets*, 2012. **13**(6): 863–875.
10. Whayne, T.F., A review of the role of anticoagulation in the treatment of peripheral arterial disease. *Int J Angiol*, 2012. **21**(4): 187–194.
11. Gulseth, M.P., J. Michaud, and E.A. Nutescu, Rivaroxaban: An oral direct inhibitor of factor Xa. *Am J Health Syst Pharm*, 2008. **65**(16): 1520–1529.
12. Turpie, A.G., Oral, direct factor Xa inhibitors in development for the prevention and treatment of thromboembolic diseases. *Arterioscler Thromb Vasc Biol*, 2007. **27**(6): 1238–1247.
13. Ansell, J. et al., Pharmacology and management of the vitamin K antagonists: American college of chest physicians evidence-based clinical practice guidelines (8th Edition). *Chest*, 2008. **133**(6 Suppl): 160s–198s.
14. Bassand, J.P., Review of atrial fibrillation outcome trials of oral anticoagulant and antiplatelet agents. *Europace*, 2012. **14**(3): 312–324.
15. Galanis, T. et al., New oral anticoagulants. *J Thromb Thrombolysis*, 2011. **31**(3): 310–320.
16. Bauer, K.A., Recent progress in anticoagulant therapy: Oral direct inhibitors of thrombin and factor Xa. *J Thromb Haemost*, 2011. **9 Suppl 1**: 12–19.
17. Kubitza, D. et al., Safety, pharmacodynamics, and pharmacokinetics of single doses of BAY 59-7939, an oral, direct factor Xa inhibitor. *Clin Pharmacol Ther*, 2005. **78**(4): 412–421.
18. Mueck, W. et al., Clinical pharmacokinetic and pharmacodynamic profile of rivaroxaban. *Clin Pharmacokinet*, 2014. **53**(1): 1–16.
19. Hillarp, A. et al., Effects of the oral, direct factor Xa inhibitor apixaban on routine coagulation assays and anti-FXa assays. *J Thromb Haemost*, 2014. **12**(9): 1545–1553.
20. Hillarp, A. et al., Effects of the oral, direct factor Xa inhibitor rivaroxaban on commonly used coagulation assays. *J Thromb Haemost*, 2011. **9**(1): 133–139.
21. Lindhoff-Last, E. et al., Assays for measuring rivaroxaban: Their suitability and limitations. *Ther Drug Monit*, 2010. **32**(6): 673–679.
22. van Ryn, J. et al., Dabigatran etexilate—A novel, reversible, oral direct thrombin inhibitor: Interpretation of coagulation assays and reversal of anticoagulant activity. *Thromb Haemost*, 2010. **103**(6): 1116–1127.
23. Barrett, Y.C. et al., Clinical laboratory measurement of direct factor Xa inhibitors: Anti-Xa assay is preferable to prothrombin time assay. *Thromb Haemost*, 2010. **104**(6): 1263–1271.
24. Mancl, E.E., A.N. Crawford, and S.A. Voils, Contemporary Anticoagulation Reversal Focus on Direct Thrombin Inhibitors and Factor Xa Inhibitors. *J Pharm Pract*, 2013. **26**(1): 43–51.
25. Desmurs-Clavel, H. et al., Reversal of the inhibitory effect of fondaparinux on thrombin generation by rFVIIa, aPCC and PCC. *Thromb Res*, 2009. **123**(5): 796–798.
26. Walenga, J.M. and D.A. Hoppensteadt, Monitoring the new antithrombotic drugs. *Semin Thromb Hemost*, 2004. **30**(6): 683–695.
27. Zucker, S. et al., Standardization of laboratory tests for controlling anticoagulent therapy. *Am J Clin Pathol*, 1970. **53**(3): 348–354.
28. Poller, L., International Normalized Ratios (INR): The first 20 years. *J Thromb Haemost*, 2004. **2**(6): 849–860.
29. Samama, M.M. et al., Assessment of laboratory assays to measure rivaroxaban—An oral, direct factor Xa inhibitor. *Thromb Haemost*, 2010. **103**(4): 815–825.

30. Harder, S., J. Parisius, and B. Picard-Willems, Monitoring direct FXa-inhibitors and fondaparinux by Prothrombinase-induced Clotting Time (PiCT): Relation to FXa-activity and influence of assay modifications. *Thromb Res*, 2008. **123**(2): 396–403.
31. Samama, M.M., Which test to use to measure the anticoagulant effect of rivaroxaban: The anti-factor Xa assay. *J Thromb Haemost*, 2013. **11**(4): 579–580.
32. Samama, M.M. et al., Evaluation of the prothrombin time for measuring rivaroxaban plasma concentrations using calibrators and controls: Results of a multicenter field trial. *Clin Appl Thromb Hemost*, 2012. **18**(2): 150–158.
33. Bates, S.M. and J.I. Weitz, Coagulation assays. *Circulation*, 2005. **112**(4): e53–e60.
34. Smogorzewska, A. et al., Effect of fondaparinux on coagulation assays: Results of College of American Pathologists proficiency testing. *Arch Pathol Lab Med*, 2006. **130**(11): 1605–1611.
35. Graff, J., B. Picard-Willems, and S. Harder, Monitoring effects of direct FXa-inhibitors with a new one-step prothrombinase-induced clotting time (PiCT) assay: Comparative in vitro investigation with heparin, enoxaparin, fondaparinux and DX 9065a. *Int J Clin Pharmacol Ther*, 2007. **45**(4): 237–243.
36. Graff, J. et al., Effects of the oral, direct Factor Xa inhibitor rivaroxaban on platelet-induced thrombin generation and prothrombinase activity. *J Clin Pharmacol*, 2007. **47**(11): 1398–1407.
37. Berntorp, E. and G.L. Salvagno, Standardization and clinical utility of thrombin-generation assays. *Semin Thromb Hemost*, 2008. **34**(7): 670–682.
38. van Veen, J.J., A. Gatt, and M. Makris, Thrombin generation testing in routine clinical practice: Are we there yet? *Br J Haematol*, 2008. **142**(6): 889–903.
39. Gerotziafas, G.T. et al., In vitro inhibition of thrombin generation, after tissue factor pathway activation, by the oral, direct factor Xa inhibitor rivaroxaban. *J Thromb Haemost*, 2007. **5**(4): 886–888.
40. Klaeffling, C. et al., Development and clinical evaluation of two chromogenic substrate methods for monitoring fondaparinux sodium. *Ther Drug Monit*, 2006. **28**(3): 375–381.
41. Wang, W. et al., Novel anthranilamide-based FXa inhibitors: Drug design, synthesis and biological evaluation. *Molecules*, 2016. **21**(4): 491.
42. Samama, M.M. et al., Evaluation of the anti-factor Xa chromogenic assay for the measurement of rivaroxaban plasma concentrations using calibrators and controls. *Thromb Haemost*, 2012. **107**(2): 379–387.

6.4 Delivering Companion Diagnostics for Complex Biomarkers

The BRCA *Companion Diagnostic Story for LYNPARZA™ (Olaparib)*

Maria C. M. Orr and Simon P. Dearden

Contents

INTRODUCTION TO LYNPARZA™ (OLAPARIB) AND MECHANISM OF ACTION

LYNPARZA™ (Olaparib, AZD2281, KU-0059436) is a potent polyadenosine 5′diphosphoribose [poly (ADPribose)] polymerization (PARP) inhibitor (PARP-1, -2 and -3). PARP enzymes, in particular PARP1 and PARP2, are essential for the repair of DNA single-strand breaks (SSBs), which occur daily, tens of thousands of times per cell (Caldecott 2008; Redon et al. 2010; Ko and Ren 2012; Beck et al. 2014).

Inhibiting PARPs leads to persistence of SSBs, which are converted to DNA double-strand breaks (DSBs) during DNA replication (Michels et al. 2013; Benafif and Hall 2015). DSBs are efficiently repaired by homologous recombination repair (HRR) (Figure 6.4.1) (Jasin and Rothstein 2013).

However, tumors with homologous recombination repair deficiencies (HRD), such as *BRCA1*, *BRCA2* or other HRR gene mutations, cannot accurately repair this DNA damage, leading to cell cycle arrest and apoptosis (Evers et al. 2010).

The mechanism by which olaparib acts is through the trapping of inactive PARP onto SSBs, thus preventing repair (Helleday 2011; Murai et al. 2012). In preclinical experiments olaparib has been shown to inhibit growth of select tumor cell lines *in vitro* and decrease tumor growth in mouse xenograft models both as monotherapy or following platinum-based chemotherapy especially in models with deficiencies in *BRCA1*, *BRCA2* or other HRR genes (Rottenberg et al. 2008; Hay et al. 2009).

The Concept of Synthetic Lethality and Its Importance in Olaparib Drug Development

Theodore Dobzhansky first described "synthetic lethality" in 1946 as the situation when mutations in one of two genes individually have no effect but combining the mutations leads to

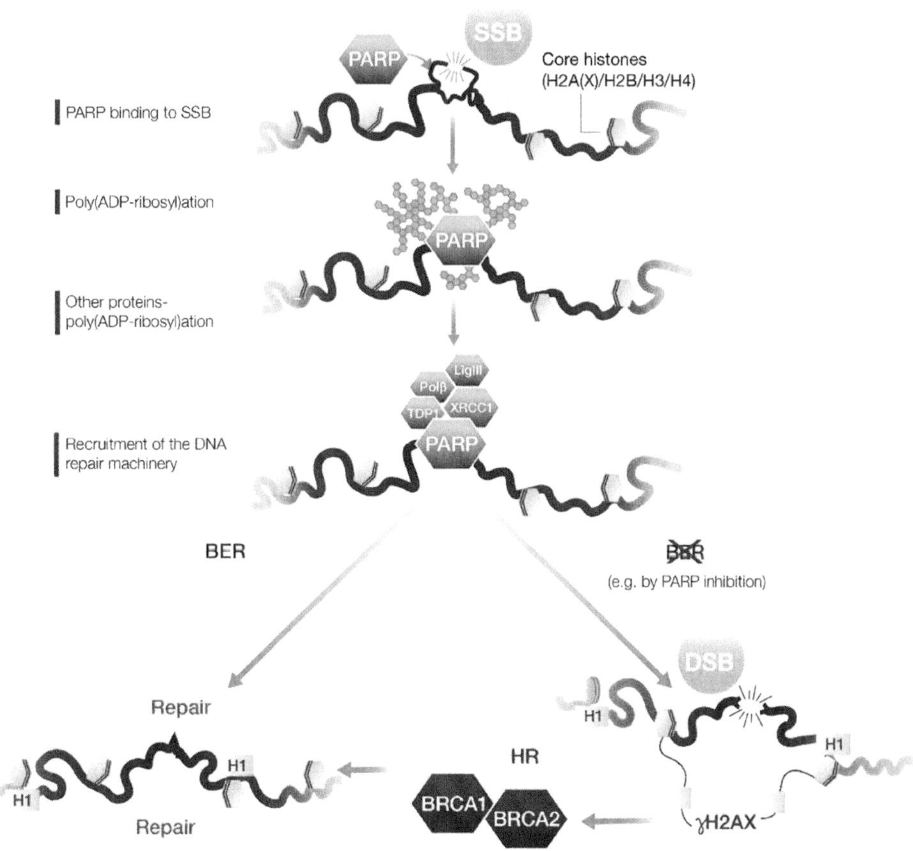

FIGURE 6.4.1 The role of PARP1 in SSB repair. PARP1 is recruited to the site of a DNA SSB once the lesion is detected whereupon it is activated. Activation of PARP1 leads to the synthesis of PAR polymers that attach to other acceptor proteins at the DNA lesion site (e.g., histones), resulting in the recruitment of proteins involved in the repair of SSB, in particular those involved in the base excision repair (XRCC1, Tdp1, Ligase III, Polβ). Loss of base excision repair, for example, through inhibition of PARP, results in SSBs being converted to DSBs and the formation of γH2AX foci. DSB repair requires *BRCA1/2* proteins that are deficient in many tumors, including breast and ovarian cancer. (Based on data from Redon, C.E. et al., *Clin. Cancer Res.,* 16, 4532–4542, 2010.)

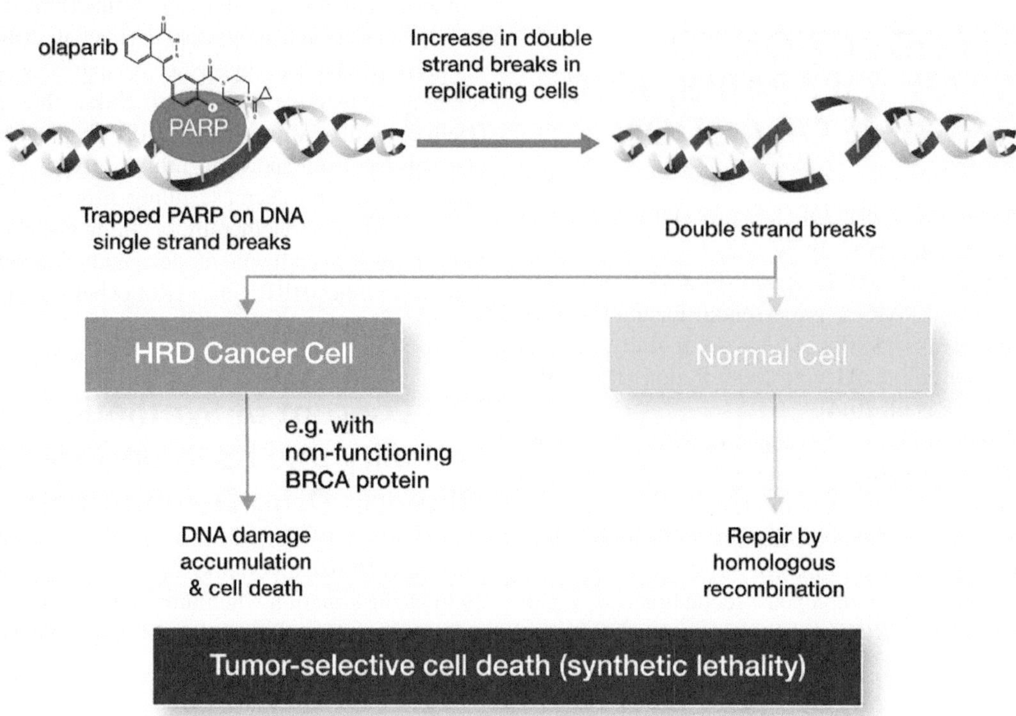

FIGURE 6.4.2 Olaparib mechanism of action. (Based on data from Tewari, K.S. et al., *Clin. Cancer Res.*, 21, 3829–3835, 2015.)

cell death (Dobzhansky 1946). The concept has been applied to anticancer treatment describing situations where a cancer mutation and a drug together cause tumor cell death. The mutation or drug alone cannot kill tumor cells but combined they have the desired effect (Figure 6.4.2) (O'Connor 2015; Rabenau and Hofstatter 2016).

The use of a PARP inhibitor in tumors with defects in HRR, for example, a *BRCA*-deficient cancer, is probably the first example of the clinical application of synthetic lethality (McCabe et al. 2006; Ellisen 2011; Curtin 2012). Thus, much of the initial development of PARP inhibitors (PARPi), such as olaparib, has focused on targeting cancers with mutations of the breast cancer–related genes, *BRCA1* and *BRCA2*, which code for proteins integral to the HRR pathway.

INTRODUCTION TO THE *BRCA* GENES, MUTATIONS IN *BRCA1/2* AND *BRCA* TESTING

Introduction to the *BRCA* Genes

BRCA1 and *BRCA2* are involved in many key cellular processes. Critically, both are involved in DNA repair and in transcriptional regulation in response to DNA damage, thus both proteins are essential to the maintenance of replication fidelity. Without functional *BRCA* proteins, cells are inefficient in homologous recombination repair of DNA damage, leading to apoptosis or cell transformation. Consequently, *BRCA1* and *BRCA2* are important tumor suppressor genes (Liu and West 2002; Yoshida and Miki 2004; Gudmundsdottir and Ashworth 2006; Roy et al. 2012).

BRCA1 and *BRCA2* are large genes located on chromosome 17q21 and 13q12-13, respectively. *BRCA1* has 22 exons encoding 1863 amino acids distributed over 125 kb of genomic DNA. *BRCA2* contains 28 exons, distributed over 85kb encoding a protein of 3418 amino acids (Shamoo 2003; Clark et al. 2012; Huret et al. 2013).

Mutations in the *BRCA* Genes and Their Clinical Relevance

Most *BRCA* mutations are predicted to truncate proteins with resultant loss of function. Mutations at splice junctions have the potential to disrupt proper mRNA processing and gene expression. Missense mutations resulting in amino acid substitutions within key functional domains can also cause loss of function in the absence of truncated proteins (Goldgar et al. 2004; Easton et al. 2007; Borg et al. 2010). In addition, single and multi-exonic deletions and duplications within *BRCA* are also predicted to disrupt protein function. A wide spectrum of mutations has been identified across the entire length of both genes, with approximately 6,000 classified as loss of function (ClinVar 2018). In certain populations, specific mutations have been reported

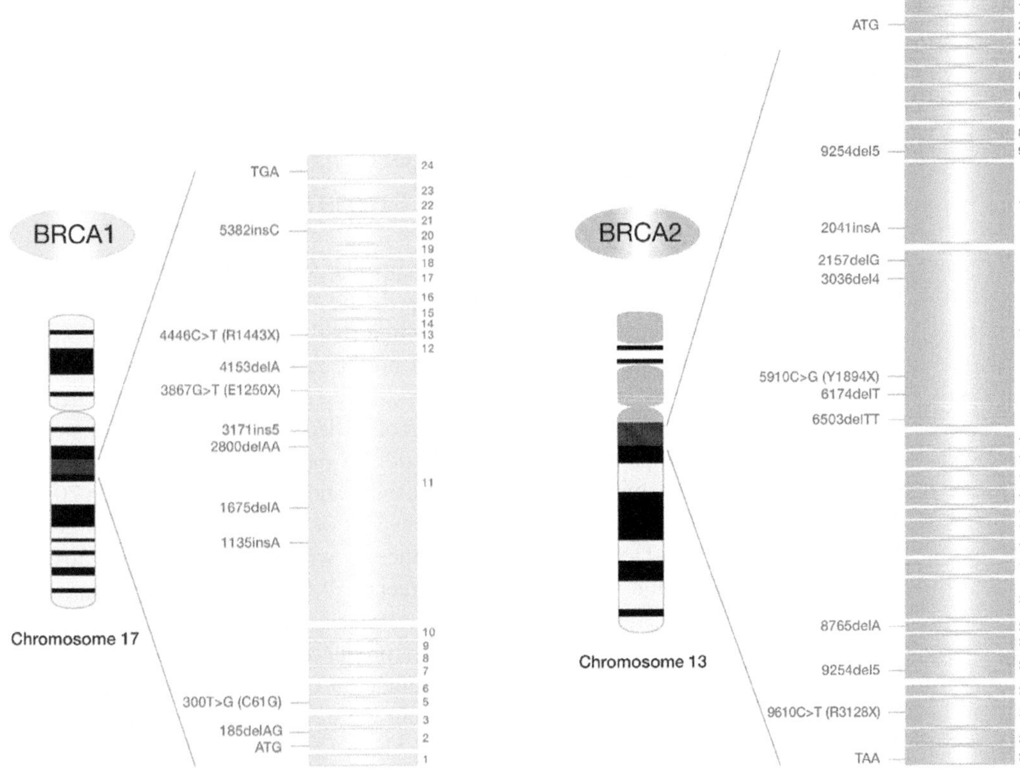

FIGURE 6.4.3 The chromosomal location and structural organization of the *BRCA1* and *BRCA2* genes including the positions of common and founder mutations. (Based on data from Fackenthal, J.D. and Olopade, O.I., *Nat. Rev. Cancer*, 7, 937–948, 2007 and BIC: Breast Cancer Information Core. http://research.nhgri.nih.gov/projects/bic/index.shtml. [Accessed in 2012].)

with high frequency due to a founder effect (Figure 6.4.3) (Ferla et al. 2007). Genetic changes in *BRCA* genes that do not lead to loss of function are referred to as *BRCA* variants.

Mutations, leading to loss of function, in the *BRCA1* and *BRCA2* genes lead to an increased risk of breast or ovarian cancer as part of hereditary breast-ovarian cancer syndrome (Tutt and Ashworth 2002). Women who are heterozygous for *BRCA1/2* pathogenic (i.e., loss of function) mutations have a lifetime risk of 40%–80% for breast cancer and 11%–40% for ovarian cancer (Petrucelli et al. 1998).

BRCA mutation carriers have lost a functioning copy of either *BRCA1* or *BRCA2*. Crucially, *BRCA* mutation carriers retain one functional copy of the affected gene and are still able to repair DSBs by HRR. When the remaining functional copy of the affected *BRCA* gene is also lost, meaning that cells are unable to undertake HRR; this "second hit" makes cells susceptible to tumorigenesis (Jasin 2002; Scully et al. 2002). PARPi selectively targets *BRCA*-deficient cancer cells while sparing tissues retaining functional *BRCA* genes that can undergo normal HRR (Helleday 2011).

Researchers have established that mutations in *BRCA* genes are also associated with the development of sporadic tumors, as a proportion of ovarian and breast cancers contain somatic (tumor only) *BRCA1/2* pathogenic mutations (Welcsh and King 2001; Janatova et al. 2005; Hennessy et al. 2010).

BRCA Testing

Current testing paradigms for determination of germline *BRCA* mutation (*gBRCAm*) status are typically based on analysis of DNA extracted from white blood cells obtained from blood samples though nucleated cells from other samples may also be used (e.g., saliva) (Meghnani et al. 2016). The exact suite of tests used depends upon the established testing service and their preferred methods. Methods typically deployed include Sanger sequencing or Next Generation Sequencing (NGS) to detect mutations and methods such as multiplex ligation-dependent probe amplification to assess exonic rearrangements (Ewald et al. 2009).

DNA sequencing is considered the most sensitive method of detecting unknown mutations in large genes such as *BRCA1/2* but a variety of techniques to screen for mutations including single-strand conformational polymorphism analysis, conformation sensitive gel electrophoresis, two-dimensional gene scanning, denaturing high performance liquid chromatography, chemical cleavage mismatch, heteroduplex analysis and protein truncation test are also used (Palma et al. 2006). In many laboratories, a combination of approaches is used to assess *gBRCAm* status. Recently, sequencing, and in particular NGS, have become the methods of choice (Michils et al. 2012; D'Argenio et al. 2015; Wallace et al. 2016; Toland et al. 2018).

Assessment of *BRCA* mutations in tumors is challenging. Analysis is typically conducted on DNA from formalin fixed paraffin embedded tumor tissues, which yield limited amounts of low quality DNA. Tumors are histologically heterogeneous and analyzed samples contain varying proportions of DNA from normal cells. Consequently, somatic mutation detection methods must be able to detect DNA changes present in a fraction of the total DNA isolated. NGS methods have the potential to detect variants at low allele frequencies due to the clonal nature of NGS and are preferred for tumor *BRCA* mutation analysis (Ellison et al. 2015; Mafficini et al. 2016).

BRCA Variant Classification

The analysis and interpretation of *BRCA1/2* mutations is complex. Although the majority of *BRCA* mutations are small insertions/deletions or point mutations, exonic insertions/deletions and rearrangements within *BRCA1/2* that are reported to disrupt function, are also seen in 6%–10% of *BRCA* mutation carriers (Judkins et al. 2012).

Given heterogeneity in *BRCA1/2* variants, it is necessary to have criteria to classify variants appropriately and ensure those likely leading to disrupted function can be discriminated from those with no functional impact. Robust, evidence-based methodology to classify *BRCA1/2* variants is used defining those leading to loss of function. Classification is usually based upon the American College of Medical Genetics (ACMG) recommendations for standards for interpretation and reporting of sequence variants (Richards et al. 2015). This process uses comprehensive literature combined with substantial genetics expertise to deliver accurate variant classification and interpretation of *BRCA* test results.

A variety of classification systems exist but all are founded on classifying the degree of likelihood of pathogenicity. Details of commonly used classifications are outlined in Table 6.4.1.

Several public databases exist to capture *BRCA* mutations and classification. These include the Breast Cancer Information Core (BIC) database, ClinVar database is maintained by the NCBI (National Center for Biotechnology Information) and the Leiden Open Variation Database (LOVD) for *BRCA1* and *BRCA2* sequence variants along with many others (Szabo et al. 2000; Fokkema et al. 2011; BIC 2012; Vallée et al. 2012; Eggington et al. 2014). Mutation classification can differ in these resources and their utility in clinical practice should be further assessed (Vail et al. 2015).

COMPANION DIAGNOSTIC DEVELOPMENT

The Concept of Targeted Therapy and Companion Diagnostics

The concept of targeting specific patient populations for treatment is not new. In the 1970s, the selective estrogen receptor modulator tamoxifen (Nolvadex®, AstraZeneca) was developed to treat advanced breast cancer. Lerner et al. (1976) reported estrogen receptor positive status correlated with treatment outcome and thus they concluded that a diagnostic test would have value in the identification of patients for tamoxifen treatment (Lerner et al. 1976; Ginsburg and Willard 2010; Jordan 2014; Jørgensen 2014).

However, it was 20 years before the first formal example of a targeted therapy with a Companion diagnostic (CDx) was delivered. Herceptin™ (trastuzumab) is a monoclonal antibody that can bind to and inactivate the HER2 receptor. The HER2 protein is overproduced in 15%–20% of all breast cancer cases thus the identification of tumors overexpressing HER2 was critical to targeting the agent to those patients. In 1998 Herceptin was approved alongside the CDx HercepTest (Slamon et al. 2001; Ross and Gray 2003; Jørgensen et al. 2010).

This example paved the way for further CDxs. As of mid-2018, the FDA had approved 33 unique CDxs for guiding treatment decisions for 25 drugs across multiple indications. Although most examples are in the oncology setting this portfolio continues to grow across all therapeutic areas (List of Cleared or Approved Companion Diagnostic Devices [In Vitro and Imaging Tools] June 2018).

However, achieving simultaneous approval of the drug and CDx remains challenging. At the time that the *BRCA* test was being developed as a CDx only a small number of examples existed which had achieved simultaneous approval, including Herceptin®(trastuzumab), Erbitux® (cetuximab), Vectibix® (panitumumab), Zelboraf® (vemurafenib), Xalkori® (crizotinib), Mekinist® (trametinib), Tafinlar® (dabrafenib), Gilotrif® (afatinib) and Gleevec/Glivec® (imatinib mesylate). Of note is the fact that all the approved CDx at that time were limited to patient selection tools based on oncogenes where the identification and validation of a limited number of activating mutations (gain of function) is relatively less complex than the effort required to

TABLE 6.4.1 Commonly used classification systems

BIOLOGICAL CLASSIFICATION	*MYRIAD CLASSIFICATION*	*BIC CLASSIFICATION*	*ACMG*
Disrupts normal gene function	Deleterious Suspected deleterious	Pathogenic Likely pathogenic	Pathogenic Likely pathogenic
Uncertain	Variants of unknown significance	Uncertain	Uncertain significance
Does not disrupt normal gene function	Variant, favor polymorphism	Likely not pathogenic/little clinical significance	Likely benign
	Benign polymorphism	Not pathogenic/low clinical significance	Benign

identify and validate all the mutations likely to lead to loss of function in tumor suppressor genes such as *BRCA1* and *BRCA2*.

Companion Diagnostics—Regulatory Situation

Definitions and requirements for the development and delivery of CDxs vary from country to country with the United States having the most established regulations (Ansari 2013; Hanamura and Aruga 2014).

U.S. companion diagnostic regulatory situation

The FDA defines an *in vitro* diagnostic (IVD) CDx device as a diagnostic device providing information that is essential for the safe and effective use of a corresponding therapeutic product. The use of an IVD CDx device with a therapeutic product is stipulated in the instructions for use of both the device and corresponding therapeutic, including any generic equivalents (Principles for Co-development of an In Vitro Companion Diagnostic Device with a Therapeutic Product: Draft Guidance for Industry and Food and Drug Administration Staff).

The United States uses a risk-based system to determine the regulatory scrutiny required for an IVD. The class of the device is based on the intended use and indications for use (Harnock 2014). This determines the type of premarketing submission required prior to marketing of the device. Class I devices require minimal scrutiny and the least regulatory control, whereas Class III devices require a detailed assessment of the device's safety and effectiveness (Figure 6.4.4).

FDA guidance on CDx development, drafted in 2011 and finalized in 2014 indicated it was likely that most IVD CDxs would be Class III devices though the agency acknowledged that future scenarios may exist where Class II submission via the premarket 510 (k) pathway might be acceptable. Further draft guidance containing details on the principles for codevelopment of an *in vitro* CDx device with a therapeutic product was released in 2016, detailing the proposed requirements to be met for the successful delivery of such a device (Principles for Co-development of an In Vitro Companion Diagnostic Device with a Therapeutic Product: Draft Guidance for Industry and Food and Drug Administration Staff).

European Union companion diagnostic regulatory situation

Until recently there were no formal CDx regulations in place for the EU, nevertheless all diagnostics, irrespective of whether they were a CDx or not, were expected to receive a certificate of conformity prior to marketing (Conformité Européene-*in vitro* diagnostic-mark [CE-IVD-mark]).

However, the regulatory landscape has changed in Europe with the introduction of the medical device and *in vitro* diagnostic regulations in May 2017 (Official Journal of the European Union, L 116 and L 117, 5 May 2017). The introduction of these new regulations will require more stringent CDx regulatory scrutiny in the future. CDx will be classified as high individual risk or moderate public risk (category C) devices requiring a conformity assessment by a notified body (NB) prior to marketing (Pignatti et al. 2014, Enzmann et al. 2016, Concept paper on predictive biomarker-based assay development in the context of drug development and lifecycle).

DELIVERING A COMPANION DIAGNOSTIC FOR OLAPARIB

The importance of defective homologous recombination to olaparib treatment response was recognized from project inception. In 2004, data suggested *BRCA* mutations were predictive of response to PARP inhibitors (Turner et al. 2004). In 2005,

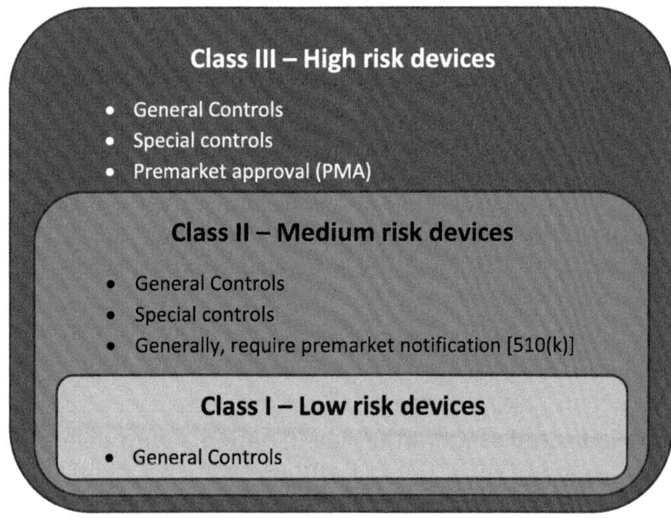

FIGURE 6.4.4 U.S. classification of medical devices for regulatory review. (From https://www.fda.gov/medicaldevices/deviceregulationandguidance/overview/classifyyourdevice/default.htm.)

Farmer et al. (2005) reported that *BRCA1/2* dysfunction sensitized cells to PARP inhibition causing cell cycle arrest and apoptosis. The underlying mechanism, mediated by PARP inhibition, led to the accumulation of DNA damage normally repaired by homologous recombination.

As a result, the development of PARP inhibitors, including olaparib, was focused initially on targeting tumors associated with mutations in *BRCA1/2*, though efforts to identify further predictive biomarkers continue.

Delivering the First Companion Diagnostic in the United States for Olaparib

In the United States, olaparib was initially indicated as a monotherapy for patients with deleterious or suspected deleterious germline *BRCA* mutated (as detected by an FDA-approved test) advanced ovarian cancer who have been treated with three or more prior lines of chemotherapy (Lynparza prescribing information in the United States). This approval was based on the results of a single-arm study in patients with *gBRCAm* advanced cancers (Domchek et al. 2016; Matulonis et al. 2016). A total of 137 ovarian cancer patients with measurable disease treated with three or more prior lines of chemotherapy, were enrolled. All patients received olaparib monotherapy (400 mg capsules b.i.d.). Objective response rate (ORR) and duration of response (DOR) were assessed by the investigator according to RECIST v1.1 (Figure 6.4.5).

To comply with U.S. CDx requirements, it was recognized that a contemporaneous premarket approval of a *BRCA1/2* mutation test, as a high risk, class III medical device, would be required. This requirement was confirmed through the FDA regulatory presubmission process.

AstraZeneca partnered with Myriad Genetics Laboratories Inc (hereafter referred to as Myriad). In 2013 to deliver the regulatory approved CDx based on Myriad's leading position in U.S. *BRCA* testing market and their extensive experience of delivering *BRCA* testing to aid hereditary breast and ovarian cancer management.

At that time, Myriad had been offering the BRAC*Analysis*® test as a laboratory developed test (LDT) run under Clinical Laboratory Improvement Amendment (CLIA)/College of American Pathologists (CAP) certification since 1996 to determine the hereditary risk of breast and ovarian cancer. Myriad chose this path due to the technical challenges of analyzing large tumor suppressor genes and the complexity of identifying and classifying the many possible loss of function mutations that arise.

Description of the Myriad BRACAnalysis CDx test

Analysis of *BRCA* mutations at Myriad involves detection and classification of variants in the *BRCA1/2* genes using genomic DNA obtained from whole blood. The test includes complete Sanger sequencing and an assessment of large rearrangements in the *BRCA1* and *BRCA2* genes. The tests are independent of each other but data are combined to give a thorough assessment of *BRCA1/2* mutation status.

For *BRCA1*, full sequence determination of approximately 5,400 base pairs (bp) comprising 22 coding exons and 700bp of adjacent intronic sequence is performed. Noncoding exons 1 and 4 are not analyzed. For *BRCA2*, approximately 10,200 bp comprising 26 coding exons and 900bp of adjacent intronic sequence is sequenced. Exon 1 is noncoding and is not analyzed. The intronic regions analyzed generally do not extend more than 20 bp proximal to the 5′ end and 10 bp distal to the 3′ end of each exon (Myriad Genetic Laboratories, Incorporated. BRACAnalysis CDx Technical Specifications).

Large rearrangement testing is a multiplex PCR assay intended to detect large genomic deletions/duplications across coding, limited flanking intronic and proximal promoter regions of the *BRCA1/2* genes. This analysis determines copy number abnormalities indicative of deletions/duplications. Myriad's proprietary software analysis normalizes the copy number of individual amplicons in *BRCA1* against *BRCA2*, plus three control genes. Any sample with potential large rearrangements is reviewed, verified and confirmed by a repeat multiplex quantitative PCR analysis (Judkins et al. 2012; Myriad Genetic Laboratories, Incorporated. BRACAnalysis CDx Technical Specifications).

Myriad variant classification

The ACMG has published standards and guidelines for clinical sequencing (Richards et al. 2015) including interpretive categories and definitions of sequence variations, based on literature and underlying scientific understanding of gene structure and function.

The process, criteria and evidence for classifying variants at Myriad have been established in accordance with ACMG guidelines and are consistently applied to all variants identified. *BRCA1/2* variants are classified into one of the five categories summarized in Table 6.4.2 (Eggington et al. 2014; Richards et al. 2015).

US indication: Lynparza is indicated as monotherapy in patients with deleterious or suspected deleterious germline *BRCA mutated (as detected by an FDA-approved test)* advanced ovarian cancer who have been treated with three or more prior lines of chemotherapy

FIGURE 6.4.5 Original U.S. indication granted for Lynparza™ (olaparib) in 2014.

TABLE 6.4.2 Myriad variant classification

VARIANT CLASSIFICATION	*DESCRIPTION*
Deleterious mutation	All mutations (nonsense, insertions, deletions) that prematurely terminate the protein product before the last documented deleterious mutation of the gene. In addition, some specific mis-sense mutations and noncoding intervening sequence mutations are recognized as deleterious on the basis of compelling scientific data derived from linkage analysis of high-risk families, functional assays, biochemical evidence and/or demonstration of abnormal mRNA transcript processing.
Genetic variant, suspected deleterious	Includes genetic variants for which available evidence indicates a strong likelihood, but not definitive proof, that the mutation is deleterious.
Genetic variant of uncertain significance (VUS)	Includes mis-sense variants and variants that occur in analyzed intronic regions whose clinical significance has not yet been determined, as well as terminating variants that truncate the gene distal to the last known deleterious mutation.
Genetic variant, favor polymorphism	Includes genetic variants for which available evidence indicates that the variant is highly unlikely to contribute substantially to cancer risk.
No mutation detected (NMD)	Includes genetic variants in the protein coding region that neither alter the amino acid sequence nor are predicted to significantly affect exon splicing and base pair alterations in the noncoding portions of the gene that have been demonstrated to have no deleterious effect on the length of stability of the mRNA transcript. These also include genetic variants for which published data demonstrate absence of clinical significance.

Source: Eggington, J.M. et al., *Clin. Genet.*, 86, 229–237, 2014.

The challenges of delivering a germline BRCA test for U.S. regulatory approval

Delivering a complex test to meet FDA premarket approval standards meant Myriad and AstraZeneca faced several challenges including the following:

- How to deliver LDTs compliant with Quality System Regulation (QSR) requirements for premarket approval?
- How to deliver analytical validation to cover thousands of possible variants that lead to loss of function of *BRCA* protein?
- How to prepare the clinical validity data set to manage missing data?

Delivering the first premarket-approved laboratory developed test and compliance with Quality System Regulation

Typically, oncology CDx are offered as complete diagnostic systems (i.e., reagents, instruments and software) which obtain FDA regulatory approval on the system (kit). These kits are made commercially available and clinical testing can be performed in any appropriate laboratory that analytically validates the assay. Due to the challenges and complexity of undertaking analysis of the *BRCA* genes, Myriad's test was offered as a test undertaken by a single laboratory only, an LDT.

The FDA does not typically review laboratory-developed tests. In the United States, LDTs are subject to the test performance standards of CLIA, which require laboratories to adhere to standards of quality control, personnel qualifications, and documentation, as well as to validate tests. However, there are no set standards for implementing such validations. The FDA requires CDx tests to be regulated according to QSR. To be QSR compliant, Myriad created a separate laboratory to deliver the CDx testing to FDA standards. This required the establishment and maintenance of a quality management system covering areas such as design, manufacture, testing, control, documentation and record maintenance including maintenance of complaint and related investigation files, and other quality assurance systems.

Delivering analytical validation of two large tumor suppressor genes

Oncology based CDx tests approved prior to BRACAnalysis CDx identified limited numbers of gain of function mutations in oncogenes such as *BRAF*, *KRAS* and *EGFR*. Delivering analytical and clinical validation for small numbers of mutations is relatively straightforward compared with the complexity of delivering such evidence for complex genes such as *BRCA1* and *BRCA2* where around 6,000 loss of function variants have been identified and new loss of function mutations continue to be identified.

It was impossible to test all variants to regulatory standards, thus Myriad performed regulatory studies using a representative set of variants and samples carrying those markers, demonstrating the CDx analytical capabilities. The performance characteristics were determined by studies using samples that represented a range of variants detected by the test, from ovarian cancer patients, breast cancer patients and unaffected individuals from families with and without a high risk for hereditary breast and ovarian cancer.

Overall, to meet the FDA requirements for premarket approval (PMA) submission, Myriad submitted 17 nonclinical analytical verification studies, 6 comparator studies, 2 extraction studies, a process validation study, equipment and software validation, the variant classification process and underwent two onsite laboratory inspections. Full details of the preclinical analytical studies are outlined in the Summary of Safety and Effectiveness Data (BRACAnalysis CDx Summary of Safety and Effectiveness Data [SSED] 2017).

Delivering the clinical validation for the companion diagnostic

Clinical validation is key in developing a CDx. The clinical benefit of BRACAnalysis CDx was demonstrated by retrospective analysis of efficacy and safety data obtained from an open-label, nonrandomized study in patients with ovarian cancer who have a deleterious or suspected deleterious *gBRCAm* previously treated with at least three lines of prior chemotherapy. All enrolled subjects had an existing *BRCA* mutation result available within their medical records. Efficacy analysis was based on Objective Response Rate (ORR) and Duration of Response (DoR) observed in 137 patients with *gBRCAm*-associated ovarian cancer. In this cohort, the ORR was 34% (95% CI: 26%, 42%) with a median DoR of 7.9 months (Domchek et al. 2016).

The effectiveness analysis for the BRACAnalysis CDx was based on a subset of 61 *gBRCAm* patients for whom specimens were available for retesting. Concordance between an existing prior test result performed locally at the clinical site, reported in the Case Report Form (CRF), and the results from the BRACAnalysis CDx was determined to be 96.7% (59/61 [95% CI: 88.7%, 99.6%]). Among the discordant results, one sample did not yield a callable result with the BRACAnalysis CDx, and another sample reported the same variant with both tests but had different classification results in the existing prior local and BRACAnalysis CDx tests (deleterious vs. variant of unknown significance, respectively). The clinical outcome data for the 59 patients with confirmed CDx *gBRCAm* status was ORR 41% (95% CI: 28%, 54%), and median DoR 8.0 months. Taken together, the results in the subset of CDx *gBRCAm* patients were comparable to those observed in the full cohort, supporting effectiveness of the device. The results are summarized in Table 6.4.3.

However, only enrolling patients with known *BRCA* mutations from existing local test results required additional robustness analyses to assess the impact of incomplete data on the intent to treat population that includes patients who will be negative on local testing but positive by the BRACAnalysis CDx test. Patients with such results are part of the intended use population of the BRACAnalysis CDx device; but these patients had been excluded from the clinical trial, which only enrolled patients whose local *BRCA* test result was positive. As a result, no efficacy data were available from such cases. To model for missing data, the efficacy of olaparib treatment in patients with positive results from the BRACAnalysis CDx was estimated assuming different combinations for the following parameters:

- The ORR among patients with positive results with both the BRACAnalysis CDx and local tests was fixed at 41%, observed from the trial
- The missing ORR among patients with positive BRACAnalysis CDx and negative local test results was modeled between 5% and 40% to exhaust all possibilities while not exceeding the ORR estimated seen in the clinical trial
- The proportion of cases with positive local test results was modeled between 5% and 30% as the prevalence of *BRCA* mutations in pretreated patients was unknown. Based on published literature, the germline *BRCA* mutation rate in untreated, unselected ovarian cancers is from 11% to 15% (Hennessy et al. 2010; Pal et al. 2010) however, it is not known if prior chemotherapy regimens could impact the *BRCA* prevalence rate in the intended indication hence a broader range of prevalence estimates was used.
- The negative percent agreement of the two tests was fixed at 0.988 (159/161), observed from multiple clinical studies and literature evidence (Kurian et al. 2014).

Combining all the above, the ORR modeled for the BRACAnalysis CDx test-positive population, including those who tested negative by local tests, was calculated. The confidence intervals were calculated based on the imputed ORR from the 137 patients in the study. The smallest ORR value modeled for the BRACAnalysis CDx test-positive population, including those who may have tested negative by local tests, was 34% (95% CI: 26, 43%), which was not significantly different from that observed for the overall subpopulation of 137 patients who had measurable disease and who had received 3 or more lines of prior chemotherapy (34% [95% CI: 26%, 42%]). The full results are listed in Table 6.4.4.

In December 2014 olaparib and BRACAnalysis CDx received simultaneous approval by the FDA marking the first LDT to be approved through the PMA process and the first LDT approved as a companion test as well as the first tumor suppressor gene mutation test approved as a CDx. Although approval was granted, post approval commitments were imposed to provide additional data related to the performance of the BRACAnalysis CDx test.

TABLE 6.4.3 Summary of clinical efficacy results

CLINICAL STUDY RESULTS SUBSET[a]	*TOTAL SUBJECTS (N)*	*SUBJECTS WITH RESPONSE (N [%])*	*ORR*	*95% CI*	*PROGRESSED (N [%])*	*MEDIAN DOR (MONTHS)*	*95% CI*
All	137	46 (33.6)	0.34	(0.26, 0.42)	30 (65.2)	7.9	(5.6, 9.6)
With BRAC*Analysis*® CDx result	59	24 (40.7)	0.41	(0.28, 0.54)	14 (58.3)	8.0	(3.8, Not calculated)
No BRAC*Analysis* CDx result	78	22 (28.2)	0.28	(0.19, 0.40)	16 (72.7)	7.9	(6.0, 9.6)

[a] Ovarian cancer patients with measurable disease who received at least three lines of prior chemotherapy.

TABLE 6.4.4 Estimated ORR for the complete BRAC*Analysis*® CDx positive population

	ESTIMATED ORR FOR CDX + (%) (95% CI)[a]	*ASSUMED OBJECTIVE RESPONSE RATE FOR BRACANALSYIS CDX - POSITIVE, LOCAL TEST - NEGATIVE*							
		5%	*10%*	*15%*	*20%*	*25%*	*30%*	*35%*	*40%*
Assumed Prevalence Local Test - Negative	**5%**	34 (26, 43)	35 (27, 44)	36 (28, 44)	37 (29, 46)	38 (30, 47)	39 (30, 47)	40 (32, 49)	41 (33, 50)
	10%	37 (29, 46)	38 (30, 47)	38 (30, 47)	39 (30, 47)	39 (30, 47)	40 (32, 49)	40 (32, 49)	41 (33, 50)
	15%	38 (30, 47)	39 (30, 47)	39 (30, 47)	39 (30, 47)	40 (32, 49)	40 (32, 49)	40 (32, 49)	41 (33, 50)
	20%	39 (30, 47)	39 (30, 47)	39 (30, 47)	40 (32, 49)	40 (32, 49)	40 (32, 49)	40 (32, 49)	41 (33, 50)
	25%	39 (30, 47)	40 (32, 49)	40 (32, 49)	40 (32, 49)	40 (32, 49)	40 (32, 49)	40 (32, 49)	41 (33, 50)
	30%	40 (32, 49)	40 (32, 49)	40 (32, 49)	40 (32, 49)	40 (32, 49)	40 (32, 49)	41 (33, 50)	41 (33, 50)

[a] 95% CI (confidence interval) assuming an $n = 137$.

BRACAnalysis CDx®

Intended Use Statement: BRACAnalysis CDx® is an in vitro diagnostic device intended for the qualitative detection and classification of variants in the protein coding regions and intron/exon boundaries of the BRCA1 and BRCA2 genes using genomic DNA obtained from whole blood specimens collected in EDTA. Single nucleotide variants and small insertions and deletions (indels) are identified by polymerase chain reaction (PCR) and Sanger sequencing. Large deletions and duplications in BRCA1 and BRCA2 are detected using multiplex PCR. Results of the test are used as an aid in identifying ovarian cancer patients with deleterious or suspected deleterious germline BRCA variants, who are or may become eligible for treatment with LYNPARZA™ (olaparib). This assay is for professional use only and is to be performed only at Myriad Genetic Laboratories, a single laboratory site located at 320 Wakara Way, Salt Lake City, UT 84108

FIGURE 6.4.6 Original U.S. Intended Use Statement for BRAC*Analysis*® CDx as granted in 2014.

Myriad successfully submitted additional analytical validation studies to cover a broader spectrum of variant types and provided updates on how often classifications change detection rates for de novo variants. In addition, AstraZeneca and Myriad submitted data from the phase III trial SOLO-2 (NCT01874353) and the pivotal trial Study 19 (NCT00753545) to convert the accelerated approval of olaparib capsules to regular approval for olaparib tablets for this indication while in addition granting approval for olaparib tablets for the maintenance treatment of adult patients with recurrent epithelial ovarian, fallopian tube, or primary peritoneal cancer, who are in a complete or partial response to platinum-based chemotherapy (Figure 6.4.6).

Delivering a Companion Diagnostic in the European Union

In December 2014 olaparib was approved in Europe as a maintenance therapy for treatment of adult patients with platinum-sensitive relapsed *BRCA*-mutated (germline and/or somatic) high-grade serous epithelial ovarian, fallopian tube, or primary peritoneal cancer who are in response (complete response or partial response) to platinum-based chemotherapy. Approval was granted on data from Study 19 (NCT00753545), a pivotal phase II study, in women with platinum-sensitive recurrent ovarian cancer who had a complete/partial response following platinum-based therapy. Patients were randomized to olaparib monotherapy (400 mg capsules b.i.d.) or placebo as maintenance therapy. The study selected patients who were platinum sensitive, thereby enriching for a HRD phenotype. Preexisting *BRCA* test results and retrospective analysis of *BRCA* status via centralized blood and tumor testing reported data on 96% of patients (254/265). This study met its primary end point of progression free survival prolongation in the overall patient population (8.4 vs. 4.8 months; PFS hazard ratio: 0.35 [95% CI: 0.25, 0.49]; $p < 0.00001$). Greater benefit was observed in patients with germline or somatic *BRCA* mutations (11.2 vs. 4.3 months; PFS HR: 0.18 [95% CI: 0.11, 0.30]; $p < 0.00001$), consistent with the biologic rationale (Figure 6.4.7).

At the time of olaparib's submission, germline *BRCA* mutation testing was established in clinical practice in Europe for the determination of risk for hereditary breast and ovarian cancer. As such, infrastructure was in place to provide germline *BRCA* testing to guide olaparib treatment decisions.

The inclusion of patients with somatic *BRCA* mutations necessitated establishment of tumor based *BRCA* testing enabling the detection of both germline and somatic mutations. Although tumor *BRCA* testing was immature at launch, Myriad obtained CE marking for tumor *BRCA* testing at their Munich laboratory in 2014 using the Tumor BRACAnalysis CDx test. In addition, a CE-IVD for detection of *BRCA* mutations from frozen tumor

EU indication: LYNPARZA is indicated as monotherapy for the maintenance treatment of adult patients with platinum-sensitive relapsed *BRCA-mutated (germline and/or* somatic) high grade serous epithelial ovarian, fallopian tube, or primary peritoneal cancer who are in response (complete response or partial response) to platinum-based chemotherapy.

FIGURE 6.4.7 Original EU indication granted for Lynparza™ (olaparib) in 2014.

tissue DNA was available via Multiplicom, the *BRCA* MASTR™ Dx test. This test was CE-IVD marked in 2015 and later updated to accommodate the assessment of *BRCA* mutations from formalin fixed paraffin embedded tumor DNA samples in 2018.

In May 2018, EMA-approved olaparib tablets for use as a maintenance therapy for patients with platinum-sensitive relapsed high-grade, epithelial ovarian, fallopian tube, or primary peritoneal cancer who are in complete response or partial response to platinum-based chemotherapy, regardless of *BRCA* status based on the data from the SOLO2 (NCT01874353) and Study 19 (NCT00753545) trials.

CONCLUSIONS AND FUTURE PERSPECTIVES

The approval of olaparib and its associated CDx marked a new era for precision medicine. Based on its mechanism of action, the early development of olaparib was targeted to the HRD population initially identified through loss of function in the *BRCA1/2* genes. Despite the many challenges associated with delivering a LDT for two large and complex tumor suppressor genes such as *BRCA1* and *BRCA2* in a trial where clinical validation data was not available for the full intent to treat (ITT) population, the PMA approval for BRACAnalysis CDx in the United States was delivered contemporaneously with the drug.

The search for further biomarkers of PARP sensitivity continues and the future will see additional complex biomarkers delivered which will aid the identification of patients most likely to benefit from the PARPi, olaparib. Identification of such biomarkers and regulatory approval will remain challenging but, as seen with the olaparib and *BRCA* exemplar, such approaches can be successfully achieved through the collaborative efforts of the scientific community, the drug and diagnostic partners and the regulatory agencies.

REFERENCES

Ansari, M. 2013. The regulation of companion diagnostics—A global perspective. *Therapeutic Innovation and Regulatory Science* 47 (4):405–415.

Beck, C., Robert, I., Reina-San-Martin, B., Schreiber, V., and Dantzer, F. 2014. Poly(ADP-ribose) polymerases in double-strand break repair: Focus on PARP1, PARP2 and PARP3. *Experimental Cell Research* 329(1):18–25.

Benafif, S., and Hall, M. 2015. An update on PARP inhibitors for the treatment of cancer. *OncoTargets and Therapy* 8:519–528.

BIC: Breast Cancer Information Core. http://research.nhgri.nih.gov/projects/bic/index.shtml (Accessed in 2012).

Borg, A., Haile, R.W., Malone, K.E., Capanu, M., Diep, A., Törngren, T., Teraoka, S. et al. 2010. Characterization of *BRCA1* and *BRCA2* deleterious mutations and variants of unknown clinical significance in unilateral and bilateral breast cancer: The WECARE study. *Human Mutation* 31(3):E1200–E1240.

BRACAnalysis CDx Summary of Safety and Effectiveness Data (SSED) https://www.accessdata.fda.gov/cdrh_docs/pdf14/P140020b.pdf (Accessed January 2017).

Caldecott, K.W. 2008. Single-strand break repair and genetic disease. *Nature Reviews Genetics* 9:619–631.

Clark, S.L., Rodriguez, A.M., Snyder, R.R., Hankins, G.D.V., and Boehning, D. 2012. Structure-function of the tumor suppressor *BRCA1*. *Computational and Structural Biotechnology Journal* 1(1):pii:e201204005.

ClinVar. https://www.ncbi.nlm.nih.gov/clinvar/ (Accessed June 2018).

Concept paper on predictive biomarker-based assay development in the context of drug development and lifecycle http://www.ema.europa.eu/docs/en_GB/document_library/Scientific_guideline/2017/07/WC500232420.pdf.

Curtin, N.J. 2012. DNA repair dysregulation from cancer driver to therapeutic target. *Nature Review Cancer* 12(12):801–817.

D'Argenio, V., Esposito, M.V., Telese, A., Precone, V., Starnone, F., Nunziato, M., Cantiello, P. et al. 2015. The molecular analysis of *BRCA1* and *BRCA2*: Next-generation sequencing supersedes conventional approaches. *Clinica Chimica Acta* 15(446):221–225.

Dobzhansky, T. 1946. Genetics of natural populations. XIII. Recombination and variability in populations of *Drosphila pseudoobscura*. *Genetics* 31:269–290.

Domchek, S.M., Aghajanian, C., Shapira-Frommer, R., Schmutzler, R.K., Audeh, M.W., Friedlander M., Balmana, J. et al. 2016. Efficacy and safety of olaparib monotherapy in germline BRCA1/2 mutation carriers with advanced ovarian cancer and three or more lines of prior therapy. *Gynecologic Oncology* 140(2):199–203.

Easton, D.F., Deffenbaugh, A.M., Pruss, D., Frye, C., Wenstrup, R.J., Allen-Brady, K., Tavtigian, S.V. et al. 2007. A systematic genetic assessment of 1,433 sequence variants of unknown clinical significance in the *BRCA1* and *BRCA2* breast cancer-predisposition genes. *American Journal Human Genetics* 81(5):873–883.

Eggington, J.M., Bowles, K.R., Moyes, K., Manley, S., Esterling, L., Sizemore, S., Rosenthal, E. et al. 2014. A comprehensive laboratory-based program for classification of variants of uncertain significance in hereditary cancer genes. *Clinical Genetics* 86(3):229–237.

Ellisen, L.W. 2011. PARP inhibitors in cancer therapy: Promise, progress and puzzles. *Cancer Cell* 19(2):165–167.

Ellison, G., Huang, S., Carr, H., Wallace, A., Ahdesmaki, M., Bhaskar, S., and Mills J. 2015. A reliable method for the detection of *BRCA1* and *BRCA2* mutations in fixed tumour tissue utilising multiplex PCR-based targeted next generation sequencing. *BMC Clinical Pathology* 15:5.

Enzmann, H., Meyer, R., and Broich, K. 2016. The new EU regulation on in vitro diagnostics: Potential issues at the interface of medicines and companion diagnostics. *Biomarkers in Medicine* 10(12):1261–1268.

Evers, B., Helleday, T., and Jonkers, J. 2010. Targeting homologous recombination repair defects in cancer. *Trends in Pharmacological Sciences* 31(8):372–380.

Ewald, I.P., Ribeiro, P.L.I., Palmero, E.I., Cossio, S.L., Giugliani, R., and Ashton-Prolla P. 2009. Genomic rearrangements in *BRCA1* and *BRCA2*: A literature review. *Genetics and Molecular Biology* 32(3):437–446.

Fackenthal, J.D., and Olopade, O.I. 2007. Breast cancer risk associated with *BRCA1* and *BRCA2* in diverse populations. *Nature Reviews Cancer* 7(12):937–948.

Farmer. H., McCabe, N., Lord, C.J., Tutt, A.N.J., Johnson, D.A., Richardson, T.B., Santarosa, M. et al. 2005. Targeting the DNA repair defect in *BRCA* mutant cells as a therapeutic strategy. *Nature* 434(7035):917–921.

Ferla, R., Calò, V., Cascio, S., Rinaldi, G., Badalamenti, G., Carreca, I., Surmacz, E., Colucci, G., Bazan, V., and Russo, A. 2007. Founder mutations in *BRCA1* and *BRCA2* genes. *Annals Oncology* 18(Suppl 6):vi93–vi98.

Fokkema, I.F., Taschner, P.E., Schaafsma, G.C., Celli, J., Laros, J.F., and den Dunnen, J.T. 2011. LOVD v.2.0: The next generation in gene variant databases. *Human Mutation* 32(5):557–563.

Ginsburg, G.S., and Willard, H.F. 2010. Chapter 1—The foundations of genomic and personalized medicine. In: Ginsburg, G.S., and Willard, H.F., editors. *Essentials of Genomic and Personalized Medicine*. Academic Press, pp. 1–10. Available from: http://www.sciencedirect.com/science/article/pii/B9780123749345000015.

Goldgar, D., Easton, D., Deffenbaugh, A., Monteiro, A., Tavtigian, S., Couch, F., and Breast Cancer Information Core (BIC) Steering Committee. 2004. Integrated evaluation of DNA sequence variants of unknown clinical significance: Application to *BRCA1* and *BRCA2*. *American Journal Human Genetics* 75(4):535–544.

Ginsburg, G.S., and Willard, H.F. 2010. Chapter 1—The foundations of genomic and personalized medicine. In: Ginsburg, G.S., and Willard, H.F., editors. *Essentials of Genomic and Personalized Medicine*. Academic Press, pp. 1–10. Available from: http://www.sciencedirect.com/science/article/pii/B9780123749345000015.

Gudmundsdottir, K., and Ashworth, A. 2006. The roles of *BRCA1* and *BRCA2* and associated proteins in the maintenance of genomic stability. *Oncogene* 25(43):5864–5874.

Hanamura, N., and Aruga, A. 2014. Global development strategy for companion diagnostics based on the usage and approval history for biomarkers in Japan, the USA and the EU. *Personalized Medicine* 11(1):27–40.

Harnack, G. 2014. *Mastering and Managing the FDA Maze: Medical Device Overview: A Training and Management Desk Reference for Manufacturers Regulated by the Food and Drug Administration*. American Society for Quality; 2nd edition. Milwaukee, WI: ASQ Quality Press.

Hay, T., Matthews, J.R., Pietzka, L., Lau, A., Cranston, A., Boulter, R., Nygren, A.O.H. et al. 2009. Poly(ADP-ribose) polymerase-1 inhibitor treatment regresses autochthonous *BRCA2*/p53-mutant mammary tumors *in vivo* and delays tumor relapse in combination with carboplatin. *Cancer Research* 69(9):3850–3855.

Helleday, T. 2011. The underlying mechanism for the PARP and *BRCA* synthetic lethality: Clearing up the misunderstandings. *Molecular Oncology* 5(4):387–393.

Hennessy, B.T., Timms, K.M., Carey, M.S., Gutin, A., Meyer, L.A., Flake, D.D. 2nd, Abkevich, V. et al. 2010. Somatic mutations in *BRCA1* and *BRCA2* could expand the number of patients that benefit from poly (ADP ribose) polymerase inhibitors in ovarian cancer. *Journal Clinical Oncology* 28(22):3570–3576.

Huret, J.L., Ahmad, M., Arsaban, M., Bernheim, A., Cigna, J., Desangles, F., Guignard, J.C. et al. 2013. Atlas of genetics and cytogenetics in oncology and haematology in 2013. *Nucleic Acids Research*. 41(Database issue):D920–D924.

In Vitro Companion Diagnostic Devices, Guidance for Industry and Food and Drug Administration Staff. http://www.fda.gov/downloads/MedicalDevices/DeviceRegulationandGuidance/GuidanceDocuments/UCM262327.pdf through http://www.fda.gov (Accessed August 2016).

Janatova, M., Zikan, M., Dundr, P., Matous, B., and Pohlreich, P. 2005. Novel somatic mutations in the *BRCA1* gene in sporadic breast tumors. *Human Mutation* 25(3):319.

Jasin, M., and Rothstein, R. 2013. Repair of strand breaks by homologous recombination. Cold Spring Harbor *Perspectives in Biology* 5(11):a012740.

Jasin, M. 2002. Homologous repair of DNA damage and tumorigenesis: The *BRCA* connection. *Oncogene* 21(58):8981–8993.

Jordan, V.C. 2014. Tamoxifen as the first targeted long-term adjuvant therapy for breast cancer. *Endocrine-related Cancer* 21(3):R235–R246.

Jørgensen, J.T., and Winther, H. The development of the HercepTest—From bench to bedside. J.T. Jørgensen, H. Winther (Eds.), *Molecular Diagnostics—The Key Driver of Personalized Cancer Medicine*. Singapore: Pan Stanford Publishing (2010), pp. 43–60.

Jørgensen, J.T. 2014. Drug-diagnostics co-development in oncology. *Frontiers in Oncology* 4:208.

Judkins, T., Rosenthal, E., Arnell, C., Burbidge, L.A., Geary, W., Barrus, T., Schoenberger, J., Trost, J., Wenstrup, R.J., and Roa, B.B. 2012. Clinical significance of large rearrangements in *BRCA1* and *BRCA2*. *Cancer* 118(21):5210–5216.

Ko, H.L., and Ren, E.C. 2012. Functional aspects of PARP1 in DNA repair and transcription. *Biomolecules* 2(4):524–548.

Kurian, A.W., Hare, E.E., Mills, M.A., Kingham, K.E., McPherson, L., Whittemore, A.S., McGuire, V. et al. 2014. Clinical evaluation of a multiple-gene sequencing panel for hereditary cancer risk assessment. *Journal Clinical Oncology* 32(19):2001–2009.

Landrum M.J., Lee J.M., Benson M., Brown G., Chao C., Chitipiralla S., Gu B. et al. 2016. ClinVar: Public archive of interpretations of clinically relevant variants. *Nucleic Acids Research* 44(D1):D862–D868.

Lerner, H.J., Band, P.R., Israel, L., and Leung, B.S. 1976. Phase II study of tamoxifen: Report of 74 patients with stage IV breast cancer. *Cancer Treatment Reports* 60:1431–1435.

List of Cleared or Approved Companion Diagnostic Devices (In Vitro and Imaging Tools). http://www.fda.gov/MedicalDevices/ProductsandMedicalProcedures/InVitroDiagnostics/ucm301431.htm (Accessed June 2018).

Liu, Y. and West, S.C. 2002. Distinct functions of *BRCA1* and *BRCA2* in double-strand break repair. *Breast Cancer Research* 4(1):9–13.

Lynparza prescribing information in the US https://www.accessdata.fda.gov/drugsatfda_docs/label/2017/208558s000lbl.pdf.

Mafficini, A., Simbolo, M., Parisi, A., Rusev, B., Luchini, C., Cataldo, I., Piazzola, E.L. et al. 2016. *BRCA* somatic and germline mutation detection in paraffin embedded ovarian cancers by next-generation sequencing. *Oncotarget* 7(2):1076–1083.

Matulonis, U.A., Harter, P., Gourley, C., Friedlander, M., Vergote, I., Rustin, G., Scott, C. et al. 2016. Olaparib maintenance therapy in patients with platinum-sensitive, relapsed serous ovarian cancer and a *BRCA* mutation: Overall survival adjusted for post-progression poly(adenosine diphosphate ribose) polymerase inhibitor therapy. *Cancer* 122(12):1844–1852.

McCabe, N., Turner, N.C., Lord, C.J., Kluzek, K., Bialkowska, A., Swift, S., Giavara, S. et al. 2006. Deficiency in the repair of DNA damage by homologous recombination and sensitivity to poly(ADP-Ribose) polymerase inhibition. *Cancer Research* 66(16):8109–8115.

Meghnani, V., Mohammed, N., Giauque, C., Nahire, R., and David, T. 2016. Performance characterization and validation of saliva as an alternative specimen source for detecting hereditary breast cancer mutations by next generation sequencing. *International Journal of Genomics*: 2059041.

Michels, J., Vitale, I., Saparbaev, M., Castedo, M., and Kroemer, G. 2013. Redictive biomarkers for cancer therapy with PARP inhibitors. *Oncogene* 33(30):3894–3907.

Michils, G., Hollants, S., Dehaspe, L., Van Houdt, J., Bidet, Y., Uhrhammer, N., Bignon, Y-J., Vermeesch, J.R., Cuppens, H., and Matthijs, G. 2012. Molecular analysis of the breast cancer genes *BRCA1* and *BRCA2* using amplicon-based massive parallel pyrosequencing. *Journal Molecular Diagnostics* 14(6):623–630.

Murai, J., Huang, S.Y., Das, B.B., Renaud, A., Zhang, Y., Doroshow, J.H., Ji, J., Takeda, S., and Pommier, Y. 2012. Trapping of PARP1 and PARP2 by clinical PARP inhibitors. *Cancer Research* 72(21):5588–5599.

Myriad Genetic Laboratories, Incorporated. BRACAnalysis® CDx Technical Specifications. https://myriad-web.s3.amazonaws.com/myriadpro.com/Test%20Request%20Forms/CTRL%20

0538%20rev2%20BRACAnalysis%20CDx%20Technical%20 Information_FINAL%20PMRC%20APPROVED.pdf (Accessed January 2017).

O'Connor, M.J. 2015. Targeting the DNA damage response in cancer. *Molecular Cell* 60(4):547–560.

Official Journal of the European Union, L 116 and L 117, 5 May 2017.

Osborne, C.K. 1998. Tamoxifen in the treatment of breast cancer. *New England Journal Medicine* 339(22):1609–1618.

Osborne, C.K., Yochmowitz, M.G., Knight, W.A. III, and McGuire, W.L. 1980. The value of estrogen and progesterone receptors in the treatment of breast cancer. *Cancer* 46(12 Suppl):2884–2888.

Palma, M., Ristori, E., Ricevuto, E., Giannini, G., and Gulino, A. 2006. *BRCA1* and *BRCA2*: The genetic testing and the current management options for mutation carriers. *Critical Reviews in Oncology/Hematology* 57(1):1–23.

Pal, T., Permuth-Wey, J., Betts, J.A., Krischer, J.P., Fiorica, J., Arango, H., LaPolla, J. et al. 2010. *BRCA1* and *BRCA2* mutations account for a large proportion of ovarian carcinoma cases. *Cancer* 104(12):2807–2816.

Petrucelli, N., Daly, M.B., and Feldman, G.L. *BRCA1* and *BRCA2* hereditary breast and ovarian cancer. 1998. In: Pagon R.A., Adam, M.P., Ardinger, H.H., et al., editors. GeneReviews® [Internet]. Seattle, WA: University of Washington; 1993-2016. Available from: http://www.ncbi.nlm.nih.gov/books/NBK1247/.

Pignatti, F., Ehmann, F., and Hemmings, R. 2014. Cancer drug development and the evolving regulatory framework for companion diagnostics in the European Union. *Clinical Cancer Research* 20(6):1458–1468.

Principles for Codevelopment of an *In Vitro* Companion Diagnostic Device with a Therapeutic Product: Draft Guidance for Industry and Food and Drug Administration Staff https://www.fda.gov/downloads/MedicalDevices/DeviceRegulationandGuidance/GuidanceDocuments/UCM510824.pdf (Accessed June 2018).

Rabenau, K., and Hofstatter, E. 2016. DNA damage repair and the emerging role of Poly(ADP-ribose) polymerase inhibition in cancer therapeutics. *Clinical Therapeutics* 38(7):1577–1588.

Redon, C.E., Nakamura, A.J., Zhang, Y.-W., Ji, J., Bonner, W.M., Kinders, R.J., Parchment, R.E., Doroshow, J.H., and Pommier, Y. 2010. Histone γH2AX and Poly(ADP-Ribose) as clinical pharmacodynamic biomarkers. *Clinical Cancer Research* 16(18):4532–4542.

Regulation (EU) 2017/745 of the European Parliament and of the Council of 5 April 2017 on medical devices, amending Directive 2001/83/EC, Regulation (EC) No 178/2002 and Regulation (EC) No 1223/2009 and repealing Council Directives 90/385/EEC and 93/42/EEC (1). 2017. *Official Journal of the European Union*, L 116 and L 117:1–175.

Regulation (EU) 2017/746 of the European Parliament and of the Council of 5 April 2017 on in vitro diagnostic medical devices and repealing Directive 98/79/EC and Commission Decision 2010/227/EU (1). 2017. *Official Journal of the European Union*, L 116 and L 117:176–336.

Richards, C.S., Bale, S., Bellissimo, D.B., Das, S., Grody, W.W., Hegde, M.R., Lyon, E., and Ward, B.E., Molecular subcommittee of the ACMG laboratory quality assurance committee. 2008. ACMG recommendations for standards for interpretation and reporting of sequence variations: Revisions 2007. *Genetics in Medicine* 10(4):294–300.

Richards, S., Aziz, N., Bale, S., Bick, D., Das, S., Gastier-Foster, J., Grody, W.W. et al., and ACMG Laboratory Quality Assurance Committee. 2015. Standards and guidelines for the interpretation of sequence variants: A joint consensus recommendation of the American College of Medical Genetics and Genomics and the Association for Molecular Pathology. *Genetics in Medicine* 17(5):405–423.

Ross, J.S., and Gray, G.S. 2003. Targeted therapy for cancer: The HER-2/neu and Herceptin story. *Clinical Leadership and Management Review* 17(6):333–340.

Rottenberg, S., Jaspers, J.E., Kersbergen, A., van der Burg, E., Nygren, A.O., Zander, S.A., Derksen, P.W. et al. 2008. High sensitivity of *BRCA1*-deficient mammary tumors to the PARP inhibitor AZD2281 alone and in combination with platinum drugs. *Proceedings National Academy Sciences U S A* 105(44):17079–17084.

Roy, R., Chun, J., and Powell, S.N. 2012. *BRCA1* and *BRCA2*: Different roles in a common pathway of genome protection. *Nature Reviews Cancer* 12(1):68–78.

Scully, R. 2002. Role of *BRCA* gene dysfunction in breast and ovarian cancer predisposition. *Breast Cancer Research* 2(5):324–330.

Shamoo, Y. 2003. Structural insights into *BRCA2* function. *Current Opinion in Structural Biology* 13(2):206–211.

Slamon, D.J., Leyland-Jones, B., Shak, S., Fuchs, H., Paton, V., Bajamonde, A., Fleming, T. et al. 2001. Use of chemotherapy plus a monoclonal antibody against HER2 for metastatic breast cancer that overexpresses HER2. *New England Journal Medicine* 344(11):783–792.

Szabo, C., Masiello, A., Ryan, J.F., and Brody, L.C. 2000. The breast cancer information core: Database design, structure, and scope. *Human Mutation* 16(2):123–131.

Tewari, K.S., Eskander, R.N., and Monk, B.J. 2015. Development of olaparib for BRCA-deficient recurrent epithelial ovarian cancer. *Clinical Cancer Research* 21(17):3829–3835.

Toland, A.E., Forman, A., Couch, F.J., Culver, J.O., Eccles, D.M., Foulkes, W.D., Hogervorst, F.B.L. et al. On behalf of the BIC Steering Committee. 2018. Clinical testing of *BRCA1* and *BRCA2*: A worldwide snapshot of technological practices. *Genomic Medicine* 3:7.

Turner, N., Tutt, A., and Ashworth, A. 2004. Hallmarks of BRCAness in sporadic cancers. *Nature Reviews Cancer* 4(10): 814–819.

Tutt, A., and Ashworth, A. 2002. The relationship between the roles of *BRCA* genes in DNA repair and cancer predisposition. *Trends in Molecular Medicine* 8(12):571–576.

Vail, P.J., Morris, B., Van-Kan, A., Burdett, B.C., Moyes, K., Theisen, A., Kerr, I.D., Wenstrup, R.J., and Eggington, J.M. 2015. Comparison of locus-specific databases for *BRCA1* and *BRCA2* variants reveals disparity in variant classification within and among databases. *Journal of Community Genetics* 6(4):351–359.

Vallée, M.P., Francy, T.C., Judkins, M.K., Babikyan, D., Lesueur, F., Gammon, A., Goldgar, D.E., Couch, F.J., and Tavtigian, S.V. 2012. Classification of missense substitutions in the *BRCA* genes: A database dedicated to Ex-UVs. *Human Mutation* 33(1):22–28.

Wallace, A.J. 2016. New challenges for *BRCA* testing: A view from the diagnostic laboratory. *European Journal of Human Genetics* 24:S10–S18.

Welcsh, P.L., and King, M.C. 2001. *BRCA1* and *BRCA2* and the genetics of breast and ovarian cancer. *Human Molecular Genetics* 10(7):705–713.

Yoshida, K., and Miki, Y. 2004. Role of *BRCA1* and *BRCA2* as regulators of DNA repair, transcription, and cell cycle in response to DNA damage. *Cancer Science* 95(11):866–871.

6.5 Delivering the Personalized Medicine Promise

Best Practice Recommendations for Designing Biomarker Development Processes

Karen K. Y. Lam, Mary Zacour, Ji-Young V. Kim, Zsuzsanna Hollander, Rhonda Wideman, Raymond T. Ng, Scott J. Tebbutt, and Bruce M. McManus

Contents

BIOMARKERS: VALUABLE TOOLS FOR PREVENTIVE AND PERSONALIZED MEDICINE

Biomarkers, defined as "characteristic[s] that [are] objectively measured and evaluated as an indicator of normal processes, pathological processes, or pharmacologic responses to a therapeutic intervention" (Colburn et al. 2001, 89–95), are now widely recognized as valuable tools for personalized medicine with applications spanning the health science and drug development spectra. In particular, predictive and/or early diagnostic applications of biomarkers are being increasing recognized as key means for transforming the current "reactive damage-control" approach into proactive care and the maintenance of wellness. Here, we discuss some of the crucial decisions and steps involved in building a process map of biomarker development from its earliest stages to implementation, as well as the critical partnerships needed to ensure success.

BARRIERS TO BIOMARKERS

Recent years have brought impressive advances in both knowledge and technology, accelerating the discovery of biomarkers with potential diagnostic and/or therapeutic applications. Unfortunately, this burgeoning progress in biomarker discovery has not had translational success and failed to advance in their development to realize their clinical benefits (termed, "the biomarker barrier"). The creation of clinically useful biomarker-based tests that overcomes the biomarker barrier demands not just large-scale discovery efforts, but also biomarker validation and development or exploitation of assay platform technologies that are accurate, reliable, sensitive and specific in the clinically-appropriate measurement ranges, easy to use and interpret, rapid, and cost-effective. However, all too often, basic researchers focus on biomarkers, nominally to improve medicine through personalization and disease prevention, yet without any consideration of issues such as the clinical relevance of their projects, how the new knowledge might be applied, regulatory hurdles, clinical utility or resource-use implications of their tests, and other practicalities critical to implementation of their discoveries. For example, commercialization and implementation of new biomarker-based diagnostics require significant capital, acceptance by regulatory authorities and health-service payers, and ultimately, technology adoption by clinicians, clinical laboratories, and patients and therefore, require a carefully crafted strategic plan needs to be in place even before starting the biomarker development process. Lack of consideration for these crucial elements effectively prevents discoveries from being translated into a form capable of achieving their intended goals. Together with a general imbalance in the support for translational research (as compared with that for innovation, for example) these issues all contribute to the current biomarker barrier.

The primary objective of the current article is to facilitate the advancement of this important field by sharing a number of valuable lessons our group at the Centre of Excellence for the Prevention of Organ Failure (PROOF Centre), and others have learned over the past decade or more. Learnings related to the pre-discovery planning that involves defining clinical need and utility, developing economic justification, and initiating the involvement of relevant partners and stakeholders very early on.

BEFORE BEGINNING: DESIGNING THE DEVELOPMENT PROCESS

Clearly Identifying the Driving Forces: Clinical Need and Utility

The biomarker development path should begin and end with the patients our discoveries are intended to benefit. Indeed, for biomarkers to fulfill their promise of improving health care, the driving forces for their development need to be founded in

- Clinical need (i.e., what do physicians and patients need to improve care?)
- Clinical utility (i.e., will discovering/developing markers to address the clinical need significantly improve patient care and outcomes, compared with current practice?)

It is imperative to define a scenario in which a biomarker panel would provide clinical benefit. There may be little incentive, for example, to adopt diagnostic/prognostic tests in areas where there are no differential therapies or drug dosing options available to follow up on after test results. In contrast, with clinical need and utility established, the development team can build strategies to support the development process, to demonstrate sufficient clinical efficacy end points to regulators, to convince health-payers to reimburse for the tests, and to drive clinical uptake. Thus, early engagement of stakeholders, including patients, clinicians, and regulators, is critical.

Economic Considerations as Key Drivers of Biomarker Development

Regardless of the burden of the underlying disease, the amount of investment during the development process, and the clinical performance of the biomarker, the ultimate value of the biomarker must *provide acceptable value for the resources it consumes, better than the standard care.* Addressing this fundamental question requires understanding the projected costs and health outcomes associated with biomarker implementation in the target subgroups, using appropriate clinical algorithms.

Collaborations and partnerships with health economics and outcomes experts should occur early in the life cycle of biomarker development.

During early stages of biomarker development planning, the potential of biomarkers to reduce the burden of the target disease should be explored by understanding the changes in outcomes and costs by the biomarker being developed (e.g., how many person-years will be gained and how much cost will be saved if early clinical event rates are reduced by 20, 30, or 40%?). Quantitative projections can provide broad benchmarks for the investigative team in terms of the minimal performance characteristics of the biomarker (e.g., combination of sensitivity, specificity, and costs) that will define whether there are favorable cost-effectiveness ratios. These projections can also help in convincing the investors and granting agencies of the potential return on investment for the proposed biomarker discovery and development project.

Mid-cycle decision analysis can also play a critical role in guiding the development process. An important step is to determine the cutoff on an assay's value for clinical decision making, which requires a thoughtful, discussion-based tradeoff between the sensitivity and specificity of the biomarker. It is all about having the assay being functionally "fit-for-purpose." The tradeoff between false positive and false negative rates can be objectively informed by considering long-term costs and health outcomes. In fact, from a decision-analytic perspective, there will be one optimal cutoff that will result in maximal net monetary benefit (health gains relative to net of costs). Similarly, economic considerations will be informative in determining which subgroups of individuals will benefit the most from the biomarker-derived test. Decision-analytic methodology can be used for risk stratification of patients such that individuals above a minimal risk cutoff are considered for the application of the biomarker (Weinstein et al. 2003, 9–17).

Many aspects of the biomarker are not finalized until late stages, and many unknowns will remain even after development with regard to the clinical uptake in the population, appropriate pricing, and the test's long-term impact on health care cost and benefit. A typical position of biomarkers on the "upstream" disease pathway, which refers to the determinants of health that are not necessarily directly connected to the biological and behavioral bases for disease, often requires modeling long-term outcomes well beyond the availability of data (e.g., social relations, communities, and policies) (Smedley and Syme 2000). Thus, it is of utmost importance that the decision-analytic framework follows the best practice standards and recommendations for model-based evaluations in order to best leverage the currently available data, and allow one to understand that the economic evaluations of biomarkers are likely to face context-specific challenges.

Check the Validity and Value of the Test

Once appropriate targets are identified for development, the validity and value of the biomarker tests to be developed must be established. To assess these issues the research and development team needs to answer these questions:

- How accurately and reliably does the test measure the biomarker of interest? (i.e., analytical validity)
- How consistently and accurately does the test detect or predict the intermediate outcome of interest, in patients? (i.e., clinical validity)
- Does the new test influence care such that it represents good value for money spent as compared with current practice? (i.e., value or efficiency)

Early in the research process, the information necessary to answer the foregoing questions is not usually available; however, simply considering them in advance with the appropriate partners and end-users can enable good strategic direction for targeting resources to the development of the most promising laboratory tools. One exemplary approach is to identify current "gold standards" in care and the metrics a new test needs to achieve in order to match or outperform such standards. Such questions will require formal or informal surveys of knowledge opinion leaders in the field, particularly involving the clinicians and clinical laboratorians who will use or run the test, and an early partnership can help developers gain this insight. Furthermore, the exercise can help introduce forward-thinking and iterative improvements early in the research process that allows the developers to clearly understand the purpose of the test being developed.

Formulate the Path Forward by Imagining the Work Backwards

For successful and meaningful biomarker development, the importance of starting with considerations of final phases at the very beginning of the process cannot be over-emphasized. If biomarkers are to fulfill their promise in clinical practice, then the current norm of focusing efforts primarily on new discoveries must be updated to one of a focus on commercialization issues and the perspective of the health system and health care payers, right from the earliest stages of development. Typically, many aspects important to commercialization or implementation of the tool are overlooked during development, and the lack of experience with later stage translation results in incorrect assumptions being made during early design. It is just too late to start considering commercialization issues once lead candidates are already prioritized and laboratory tests are already developed.

The Circuitry: In Parallel, Not in Series

In addition to considering end-stage issues early, it is critically important to develop pathways for navigating later development phases in parallel with development of the biomarker test. These pathways should be considered iterative, with feedback funneled into advancing development of the biomarkers. The success of this process depends on early engagement and active involvement of stakeholders in the development of intellectual property, the regulatory path, cost-effectiveness modeling, engaging health care payers, educating physicians regarding the novel tests, and so forth. For example, developing a novel multiplex biomarker test is expensive to implement but such an approach can be more effective in disease screening and for assessing multiple physiological pathways that contribute to disease activity and prognosis (Rifai et al. 2006, 971–983). Each of these factors has direct impacts on reducing health care cost. Building a business case from the start sets the stage for the reimbursement by demonstrating to governments and health insurers the cost savings to be realized, not only by avoiding the health care burdens associated with unfavorable clinical outcomes, but also by keeping the target population healthy, able to work, and contributing to the economy. Cost-effectiveness and the case for reimbursement are also critical drivers for investors and commercial partners, who must see the potential for recouping their investments before they will buy in at the earliest stages of research. It is far too late to wait until biomarker validation is done before starting to develop commercialization strategies.

BIOMARKER DISCOVERY, DEVELOPMENT AND VALIDATION

As outlined above, once clinical need and utility have been used to pinpoint clinical target areas, discovery work can be designed. Discovery of multiplex biomarker panels involves analyzing vast numbers of compounds contained in relevant samples collected from target and control cohorts, and establishing statistically significant links between sets of compounds and clinical criteria.

Cohorts

Access to rigorously phenotyped patient populations, with in depth expert adjudication of phenotypes, is essential for discovery of clinically relevant biomarkers. In our experience, establishing clinical criteria for patient cohorts and accruing sufficient quantities of appropriately collected and banked biospecimens to support discovery of clinically valid biomarkers have been significant challenges.

Accrual of patients willing to consent to the use of their samples and information depends upon a level of public trust that the research will benefit the study subjects themselves, loved ones, or society at large, and that the research will be conducted with a high level of ethical ideals and respect of participant rights. One of the positive lessons learned in this regard has been the value of organizing deliberative democracy consultations between the public and the scientific enterprise, as a mechanism of engaging the public as partners in translational research. In these forums, we have gathered advice and different perspectives from a variety of citizens and experts with different backgrounds and needs. We have used the forums to make decisions and build processes that reflect social realities, while at the same time educating and enhancing the public trust in the endeavor. This approach, combined with the use of a very simple and clear communication style when engaging the public has proven effective in our experience, improving the percentage of patients agreeing to provide research biosamples to at least 60%, for example. A how-to model for deliberative democracy forum organization is provided in a publication by Button and Mattson (Button and Mattson 1999, 609–637).

Biobanking

Biobanking—the collection, cataloguing, and storage of human biological samples and associated clinical data, for research purposes—is another key enabler of biomarker discovery and development. Specimen handling, preparation, storage, and transport (i.e., "pre-analytical") procedures are critical determinants of sample quality and the results produced in biomarker discovery projects; for reproducible results, these procedures must be stringently controlled. Often, however, sample preparation protocols vary between biobanks or even between specimens deposited by different investigators within the same biobank. This has constituted a significant challenge for biomarker science as a whole. Even seemingly minor variations such as the number of minutes between collection of the samples and their cryopreservation can have important effects on the results obtained.

Standardization of pre-analytical conditions and working procedures between individuals and biobanks that collect and house the same types of specimens, as well as strict adherence to standard operating procedures (SOPs) is recommended. In the United States much work to reduce the problematic issues has been accomplished in recent years by the Office of Biorepositories and Biospecimen Research (OBBR) and in Canada, biobanks can standardize practices and become certified through the Canadian Tumour Repository Network (www.ctrnet.ca). Although both of OBBR and Canadian Tumour Repository Network focus on samples for cancer research, they also offer general best practice recommendations for biobanking, SOPs, and many other useful resources regarding standardization of biobanked samples.

Technological Platform

There are many potential platforms to use for biomarker work, each with its own technical caveats and best practices. The relative merits of these platforms are beyond the scope of this paper but a review by Jain et al. (2010, 23–72) can serve as a good introduction to popular platforms currently being used for biomarker discovery. Some general recommendations may nonetheless be offered. The successful advancement of our biomarker panels to later phase development has followed on the PROOF Centre's use of non-targeted approaches for discovery activities. Targeted approaches are useful later in the process, such as to refine the information gathered during discovery trials. However, applying targeted approaches for the discovery work itself may result in undermining the chance of discovering informative targets because such an approach would not permit discovery of unexpected biomarkers. Moreover, results from non-targeted trials help fill the knowledge gaps not only by providing potential new biomarkers of utility, but also by shedding light on entire pathways whose role in disease may previously have been underappreciated.

Of further consideration for diagnostic biomarkers are the needs for measurability in easily accessed body fluids (e.g., whole blood, plasma, urine) and for capturing the complexity of disease processes, including considerations of the time-course of disease advancement. Most diseases involve perturbations to multiple biological pathways that are interrelated, and such complexity can be captured by measuring samples from multiple compartments to provide systems-level understanding of disease progression. PROOF Centre biomarker tests, for example, detect and measure multiple RNAs, proteins, and metabolites, from different body compartments (such as genes expressed in blood cells, proteins secreted into blood plasma and metabolites excreted in urine). One of the challenges inherent in this type of multiplex biomarker development is the need to balance the tradeoff between better tests and the ease/costs of tests when deciding the numbers and types of biomarkers to include on a given panel. Multiplex panels are likely to provide biomarkers that are more sensitive, specific, and robust to biological variability between individuals (Rifai et al. 2006, 971–983), yet as the number and type of biomarkers in a panel increase, so do the costs, complexities, and challenges associated with their regulatory approval and implementation in clinical laboratories. Because such practical challenges could stifle advancement of a biomarker panel into clinical use, the benefits associated with different sized candidate panels must be carefully evaluated prior to decisions on which panels are taken forward in the development process.

As a further "lesson learned" we underline the importance of pre-analytical triage (i.e., "cleaning" and pre-filtering) of data before high-end analysis. In our experience, for example, despite manufacturer claims of strict quality assurance, some microarrays include spots that produce spurious results. Such quality control (QC) problems seem under-discussed in the peer-reviewed literature; in-house QC testing should be applied, such that aberrant results can be removed. Appropriate normalization of data must also be applied (e.g., such as normalizing whole-blood-derived data to differential cell counts). Pre-filtering is another important step, given the vast amounts of data typically produced from each sample in genomic/proteomic discovery. Without noise subtraction the data would simply be meaningless. Indeed, many seemingly exciting biomarker findings of early "omics" years (before the complexities and pitfalls of the data analysis were fully appreciated) were later discounted as having been merely artifacts of "overfitting the noise" (Kern 2012, 6097–6101). Although there is certainly a much greater general awareness of the importance of noise reduction currently, the *caveat emptor* principle should still be applied in evaluating manufacturer claims regarding the noise reduction functions of their software. These may rely on blind algorithm-based subtraction without consideration of important biological knowledge specific to the context.

Assay Development

Once the core biomarker candidates are identified, refined, and validated, the method that was used to measure them in the discovery and refinement stages must be migrated/adapted/changed to a platform that is feasible for use in the intended clinical setting. Cross-validation may have initiated this process; if not, this may entail developing a new laboratory analysis method for quantifying the biomarkers. No universally accepted roadmap exists for developing and validating multiplex assays for measurement of gene and/or protein expression. Instead assays must be tailor-made, taking into account their intended use, conditions of their application, and the needs and capabilities of the end-users (i.e., to be fit for their intended purpose [Lee et al. 2006b, 312–328]). This again underlines the importance of partnerships where end-users contribute to earlier development. Method development strategies are beyond the scope of this paper, however, details can be found in Lee et al. (2006a, 269–298) and Rifai et al. (2006, 971–983).

Once the method has been developed and optimized, pre-validation results are usually generated and used to establish assay acceptance or rejection criteria according to the fit-for-purpose strategy (Lee et al. 2006b, 312–328), SOPs must also be prepared and rigorously applied, including those for pre-analytical sample processes, if this has not already been done (e.g., collection, handling, preservation, means of correcting any concentration variabilities, etc.). Pre-analytical variability can greatly compromise data utility. One of the lessons we have learned is the importance of establishing early on that assay performance is up to regulatory standards, such that the exploratory data produced by these methods are sufficiently reliable to support advancement of the biomarker test along the regulatory path.

Analytical validation of the assay must be performed next. Sometimes the basic research laboratories do not have the experience or the resources to validate assays with the level of rigor that the regulatory agencies require. Therefore utilizing the service from third party research firms that specialize in developing biomarker assays can be advantageous. This can be an attractive option to consider if the service provider has past experience with the translating biomarker candidates and has established proprietary materials, and thus translation from candidates to assay can be expedited, saving time and money.

Engage Regulators and End-users from Early Development

It is of critical importance to engage expert regulatory advisors (and regulatory agencies themselves) early in the biomarker development process to help guide strategic choices that will prepare and smooth the path to regulatory acceptance. Without such acceptance, novel diagnostics cannot reach the patients they are designed to help. Developers who have formerly relied on the

LDT pathway to avoid more onerous regulatory processes should be aware that LDTs are also expected to come under FDA regulation in the near future, following on concerns that increasing numbers of improperly validated LDTs may be putting patients at risk for missed and/or wrong diagnosis that have, in turn, led to patients' exposure to unnecessary harmful medical procedures and failure to receive appropriate treatment (Food and Drug Administration 2014).

The two largest and most commercially significant regulatory agencies are the FDA and European Medicines Evaluation Agency (EMEA), the U.S. and EU bodies, respectively. Which regulatory approval to pursue will depend on where the biomarker test is to be tested clinically and made available. Because Health Canada's regulations follow closely on those of FDA and commercialization plans typically focus on the large market-size jurisdiction of the United States, it may be advantageous for Canadian groups to also involve the FDA in its biomarker development plans. Investigators who plan to perform trials and provide their biomarker tests outside of North America should consult the regulators for the appropriate countries or groups of countries. In general, the FDA standards are more stringent than those for EMEA, and as such a product designed for both markets can sometimes be marketed sooner in Europe; however, the developer should be aware that products that meet evidentiary requirements of EMEA may not be acceptable to the FDA. Consideration of the more stringent standards during development is thus advised.

Regulatory pathways are complex and expert knowledge of them is a must to chart the best path. IVDs such as blood tests for biomarkers are considered "medical devices" by the FDA and getting approval to test them in clinical trials to assess their efficacy and safety is subject to the Investigational Device Exemptions (IDE) regulation. Certain investigational IVD device studies are exempt from most of the provisions. The further requirements for approval of IVDs depends on their class, which is determined based on the level of risk of the intended use and not based upon the class of technology involved. Risk level for IVDs is primarily established based on harmful possible outcomes were the IVD to falsely diagnose patients, either positively or negatively. A 510(k) evaluation is generally acceptable for lower risk tests whereas a Premarket Approval Application is required for tests with higher risk. These and many other points are clarified in *Guidance for Industry and FDA Staff-In Vitro Diagnostic (IVD) Device Studies* (UCM 262327, August 6, 2014) and the other guidance documents to which it refers.

As a final consideration, multiplex biomarker panels for diagnostic, predictive and other applications represent novel developments for which regulatory agencies have few precedents or guidance policies to follow (Regnier et al. 2010, 165–171). Hence, the processes to obtain approval can be fraught with difficulties, which can slow down or even prevent novel tools from being brought into clinical practice. The FDA is struggling to resolve issues and keep pace with biomedical innovation and is thus providing many opportunities to participate and influence how regulatory agencies will proceed in this area. These include public discourse and open invitations to submit comments regarding draft versions of guidance documents that the agency will later enforce (Highly Multiplexed Microbiological/Medical Countermeasure In Vitro Nucleic Acid Based Diagnostic Devices [UCM 327294, August 27, 2017]). Participation in these opportunities is a constructive way for clinicians, biomarker developers, and other stakeholders to help move this field forward along the best path, and is highly encouraged.

THE PROOF CENTRE: BREAKING THE BARRIER

The PROOF Centre formed with competitively garnered support from Canada's Networks of Centres of Excellence (NCE), to address aforementioned biomarker barriers. PROOF Centre's mandate is not only to discover, but also to develop, commercialize, and implement biomarkers to better prevent, predict, diagnose and treat heart, lung, and kidney diseases. Over the recent years, PROOF Centre has advanced a number of promising multiplex biomarker panels for diagnostic and predictive uses in organ disease, failure, and transplant monitoring. Diagnostic biomarkers offer a variety of potential uses in clinical practice; for instance, blood tests could be used to diagnose or identify risk for disease early enough that preventive action can be taken, to stratify patients so that they receive the best treatment choice for their personal needs, to assess severity/disease progression, to predict prognosis, and/or to assess responses to treatment, for optimal disease management.

The development path for multiplex biomarkers, such as the PROOF Centre's proteomic and genomic panels, in many ways represents "uncharted waters" (i.e., is lacking in guidelines and precedents for development, regulatory acceptance, and clinical use alike). By example, the PROOF Centre has been charting a path through these waters over recent years, with its now well-advanced transplantation-related biomarkers program (Biomarkers in Transplantation [BiT] program) (Lin et al. 2009, 927–935; Hollander et al. 2010, 1388–1393, 2013, 259–265; Günther et al. 2012, 1; Freue et al. 2013, e1002963; Lin et al. 2013, 723–733). This program is aimed at developing a multiplex biomarker panel, used for diagnosing acute immune rejection after heart transplant. BiT has received support from various government partners including Genome Canada, Genome British Columbia, and the Canadian Institutes of Health Research, and industry partners such as Pfizer, Astellas, and IBM. The Centre works closely with business, scientific and technical teams from various platform technology companies to ensure new assays are robust, reproducible and deployable in clinical settings. Key collaborators for aiding implementation have included clinicians, regional and provincial health authorities, health economists, and clinical laboratorians. The large repository of properly recorded clinical and genomics data now serves as a resource for new collaborations and projects. Building on this consortial model established through BiT, PROOF Centre has successfully launched other biomarker development programs in chronic obstructive pulmonary disease (COPD), chronic kidney disease (CKD), heart failure, as well as other acute and chronic conditions.

Lessons Learned and Future Outlook

In recognition of the need for a broad spectrum of advisors, meetings involving all relevant stakeholders should be held for biomarker development efforts. Drawing on the expertise of industry executives, government and health authority representatives, clinicians, computational scientists, patient advocates, and statisticians, both locally and worldwide, such "best practices" meetings should charge participants with helping to define a systematic, effective approach to the discovery, development, and clinical deployment of robust and clinically useful biomarkers for personalized medicine. These sessions and our past experience strongly suggest that the key path to breaking the biomarker barrier and to create valid tools that will have an impact on health care is that we must work together—recognize that the resources and perspective required are beyond the capabilities of any one single group. Developers should instead participate in partnerships, ideally those that span the entire spectrum from biomarker developer to health system implementation. Broad partnerships or consortia offer the best chance of directing discoveries that both improve outcomes and provide value, along the critical paths that can lead them into clinical use.

REFERENCES

Button, Mark and Kevin Mattson. 1999. Deliberative democracy in practice: Challenges and prospects for civic deliberation. *Polity* 31 (4): 609–637.

Colburn, WA, Victor G. DeGruttola, David L. DeMets, Gregory J. Downing, Daniel F. Hoth, John A. Oates, Carl C. Peck et al. 2001. Biomarkers and surrogate endpoints: preferred definitions and conceptual framework. Biomarkers definitions working group. *Clinical Pharmacol and Therapeutics* 69: 89–95.

Food and Drug Administration. 2014. *Framework for Regulatory Oversight of Laboratory Developed Tests (LDTS)*. U.S. Department of Health and Human Services, Food and Drug Administration, Rockville, MD.

Freue, Gabriela V. Cohen, Anna Meredith, Derek Smith, Axel Bergman, Mayu Sasaki, Karen KY Lam, Zsuzsanna Hollander, Nina Opushneva, Mandeep Takhar, and David Lin. 2013. Computational biomarker pipeline from discovery to clinical implementation: Plasma proteomic biomarkers for cardiac transplantation. *PLoS Computational Biology* 9 (4): e1002963.

Günther, Oliver P., Virginia Chen, Gabriela Cohen Freue, Robert F. Balshaw, Scott J. Tebbutt, Zsuzsanna Hollander, Mandeep Takhar, W. Robert McMaster, Bruce M. McManus, and Paul A. Keown. 2012. A computational pipeline for the development of multi-marker bio-signature panels and ensemble classifiers. *BMC Bioinformatics* 13 (1): 1.

Hollander, Zsuzsanna, David Lin, Virginia Chen, Raymond Ng, Janet Wilson-McManus, Andrew Ignaszewski, Gabriela Cohen Freue et al. 2010. Whole blood biomarkers of acute cardiac allograft rejection: Double-crossing the biopsy. *Transplantation* 90 (12): 1388–1393.

Hollander, Zsuzsanna, Virginia Chen, Keerat Sidhu, David Lin, Raymond T. Ng, Robert Balshaw, Gabriela V. Cohen-Freue, Andrew Ignaszewski, Carol Imai, and Annemarie Kaan. 2013. Predicting acute cardiac rejection from donor heart and pre-transplant recipient blood gene expression. *The Journal of Heart and Lung Transplantation* 32 (2): 259–265.

Jain, Kewal K. 2010. Technologies for discovery of biomarkers. In *The Handbook of Biomarkers*, pp. 23–72. Totowa, NJ: Springer.

Kern, Scott E. 2012. Why your new cancer biomarker may never work: Recurrent patterns and remarkable diversity in biomarker failures. *Cancer Research* 72 (23): 6097–6101.

Lee, Jean W., Daniel Figeys, and Julian Vasilescu. 2006a. Biomarker assay translation from discovery to clinical studies in cancer drug development: Quantification of emerging protein biomarkers. *Advances in Cancer Research* 96: 269–298.

Lee, Jean W., Viswanath Devanarayan, Yu Chen Barrett, Russell Weiner, John Allinson, Scott Fountain, Stephen Keller, Ira Weinryb, Marie Green, and Larry Duan. 2006b. Fit-for-purpose method development and validation for successful biomarker measurement. *Pharmaceutical Research* 23 (2): 312–328.

Lin, David, Gabriela Cohen Freue, Zsuzsanna Hollander, GB John Mancini, Mayu Sasaki, Alice Mui, Janet Wilson-McManus, Andrew Ignaszewski, Carol Imai, and Anna Meredith. 2013. Plasma protein biosignatures for detection of cardiac allograft vasculopathy. *The Journal of Heart and Lung Transplantation* 32 (7): 723–733.

Lin, David, Zsuzsanna Hollander, Raymond T. Ng, Carol Imai, Andrew Ignaszewski, Robert Balshaw, Gabriela Cohen Freue, Janet E. Wilson-McManus, Pooran Qasimi, and Anna Meredith. 2009. Whole blood genomic biomarkers of acute cardiac allograft rejection. *The Journal of Heart and Lung Transplantation* 28 (9): 927–935.

Regnier, Fred E., Steven J. Skates, Mehdi Mesri, Henry Rodriguez, Živana Težak, Marina V. Kondratovich, Michail A. Alterman, et al. 2010. Protein-based multiplex assays: Mock presubmissions to the US food and drug administration. *Clinical Chemistry* 56 (2): 165–171.

Rifai, Nader, Michael A. Gillette, and Steven A. Carr. 2006. Protein biomarker discovery and validation: The long and uncertain path to clinical utility. *Nature Biotechnology* 24 (8): 971–983.

Smedley, BD and SL Syme. 2000. *Promoting Health: Intervention Strategies from Social and Behavioral Sciences*. National Academy Press, Washington, DC.

Weinstein, Milton C., Bernie O'Brien, John Hornberger, Joseph Jackson, Magnus Johannesson, Chris McCabe, and Bryan R. Luce. 2003. Principles of good practice for decision analytic modeling in health-care evaluation: Report of the ISPOR Task Force on Good Research Practices–Modeling Studies. *Value in Health* 6 (1): 9–17.

Development, Launch, and Adoption of a Pharmacodiagnostic

6.6

Planning versus Reacting for Pharmaceutical Companies

Michael C. Montalto and George A. Green

Contents

INTRODUCTION

Research in immuno-oncology has unlocked great potential in the fight against cancer. Several different therapeutic antibodies selective for checkpoint receptors and ligands, such as cytotoxic T lymphocyte antigen-4 (CTLA-4), programmed cell death protein 1 (PD-1), and programmed cell death ligand 1 (PD-L1), have demonstrated durable responses in a growing number of tumor types.[1–5] However, such responses are not observed in all patients, and it is becoming increasingly clear that strategies to identify likely responders are critical to development and commercialization of the next wave of immuno-oncology agents and combinations.[6] Biomarkers that are predictive of response, such as those based on biological pathways, presence of drug targets, or immune status of the tumor, have a key role in directing treatment decisions for oncology patients.[7]

Biomarkers are used in all phases of drug discovery and development, including hypothesis generation, clinical mechanisms of action and patient selection, target identification, preclinical research, pharmacodynamics, drug safety, and toxicology.[8] This chapter focuses on biomarkers that guide patient selection, particularly those that may enter the clinical setting as a companion diagnostic (CDx) or complementary diagnostic.[9]

CDx refers to a diagnostic product that is used to inform selection of patients for treatment with a specific drug.[9,10] The drug and diagnostic are referred to as "codeveloped products." Pharmacodiagnostics (PDx) refers to the science and processes used in CDx development. When patients are selected as candidates for treatment by a drug using a CDx, both the drug and CDx are subject to simultaneous regulatory review, and will refer to each other in their labeling. Typically, the drug label references the diagnostic test in the Indication(s) section. In the United States, current practice is to list the biomarker used for patient selection in the Indication(s) section, but not on product labels.[11] The diagnostic product will commonly reference the drug in the Intended Use section. If the test can be used for several drugs of similar mechanism, they are listed individually, not as a class. These examples are based on U.S. regulatory requirements; worldwide requirements vary, and are typically similar for the drug. Lesser requirements for the diagnostic are common.

Several published documents provide guidance for the codevelopment of a diagnostic and therapeutic. However, many focus on the scientific and clinical strategies, and ultimate performance of the biomarker test, as well as the general clinical and regulatory timeline coordination for the drug and diagnostic (reviews by Fridlyand et al.[12] and Scheerens et al.[9]). Beyond such reviews and guidelines, many other practical factors must be considered for the timely delivery of a commercially viable PDx. The practical development of a PDx is a complicated process that involves the coordination of multiple functional areas within the pharmaceutical company, the diagnostic partner company, and other external organizations. Codevelopment requires alignment across preclinical research, biomarker technology platforms, clinical development programs, medical education and awareness, commercial organizations, business development, standards and proficiency bodies, regulatory functions and agencies, contract research laboratories, diagnostic reference laboratories, and diagnostic companies. These functions, most of which cannot be directly controlled by the pharmaceutical company, can have a direct impact on the timely access to the drug by patients.

This chapter will first examine the economic drivers that may differ between pharmaceutical and diagnostic industries to highlight the reasons that pharmaceutical companies invest in all areas of PDx development, regardless of which functions are housed within the pharmaceutical company. Subsequently, we will discuss the key practical considerations beyond technical performance, such as choice of diagnostic platforms, manufacturers, and market distribution models, bridging from exploratory to diagnostic validations, business development considerations, and the growing role of medical organizations in pre- and post-PDx launch activities.

ECONOMIC MISALIGNMENTS

Two major industries, pharmaceutical and in vitro diagnostic (IVD), have economic stakes in the successful commercialization of a PDx. Although it may be tempting to think these two sectors are equally incentivized to develop and commercialize a PDx, this is hardly true and the economic misalignment between these industries has repercussions on planning activities for both industries. This is a major driver with respect to how pharmaceutical companies should organize themselves to maximize the chances of a successful PDx program. It also requires diagnostic companies to adapt to different plans, goals, and expectations for their development program, compared with internally developed products.

The economic misalignment between diagnostic and pharmaceutical companies may arise for several reasons. The long lead-time for a drug to be approved coupled with the high risk of not gaining approval in early phases of development make it challenging for a diagnostic company to invest early.[13,14] There are also data that suggest the value of a PDx is not tied to the value of a therapy post-launch, which is counterintuitive. Correlative analysis has shown that drug prices (presumably a surrogate for value) do not have a significant statistical relationship to PDx test prices; further, the drug effectiveness or population served also did not correlate to PDx prices.[15] These discrepancies make it difficult to justify a revenue-sharing model between the diagnostic company and pharmaceutical company and thus far such agreements have not been announced publicly. In addition to reimbursement challenges, in most regions, diagnostic products do not have as extensive intellectual property protections as therapeutics. Also, although requirements for regulatory approval of PDx products can be extensive and time-consuming, most regulatory environments allow for the use of locally developed tests, commonly known as "lab-developed tests."[16] These tests allow individual labs to develop and commercialize biomarker tests following regulatory approval of the commercial PDx, sometimes significantly taking market share and profit from the approved PDx.

A final challenge to diagnostic company commitment to PDx development is market size. Moore et al. highlighted the fact that in 2011, only 1% of marketed therapeutics had a CDx, which was smaller than expected given the interest in this class of diagnostics.[14] With rare exceptions, PDx products target ever-narrowing treatment indications and small patient populations. Commercial PDx products require volume to enable manufacturing-cycle management, distribution, and lifecycle management. This can be difficult to justify with low-volume products having inadequate reimbursement and low selling prices. Thus, for the reasons stated, it is not surprising that most development and commercialization support in diagnostic companies is for non-PDx tests with more rapid product-development cycles, lower risk, and higher profitability.

In pharmaceutical companies, most R&D and commercialization resources are dedicated to the drug development, launch, and post-market activity. There is a current trend in the industry recognizing the need for PDx development to be established as a separate function, or as a part of the translational medicine activities in the drug development program. However, development and clinical strategies for drugs still rarely fully account for the investment, timing, and management of PDx development. Thus, there is a large gap in aligning PDx activity and investment between the diagnostic and pharmaceutical industries.[17] Given that most of the economic incentive to develop a PDx is tied to the drug adoption, pharmaceutical companies typically assume a larger than expected burden of development and commercialization of a PDx. It is therefore mandatory for the pharmaceutical company to prepare early and resource the technology development, business development, assay transfer, regulatory guidelines, commercialization, and post-launch education through medical training. It is also important for the pharmaceutical company to anticipate and accommodate the requirements of the diagnostic company to successfully undertake the PDx program. However, the ultimate responsibility for the success of the diagnostic product is with the diagnostic company. Both companies must work collaboratively toward aligned goals for a successful development and launch effort.

CHOICE OF BIOMARKER PLATFORMS FOR EXPLORATORY STRATIFICATION RESEARCH

For clinical development programs, there are many different types of biomarker platforms that can measure endpoints at a range of physical resolutions and in various matrices.[18,19] For practical purposes, we can think of biomarker platforms in two tiers. The first-tier platforms are those that are used commonly by most drug development programs and include noninvasive imaging, immunohistochemistry (IHC), polymerase chain reaction (PCR), flow cytometry, immunoassays, mass spectroscopy, and more recently next-generation sequencing (NGS). The second-tier platforms are used for specific use cases, but perhaps not universally. Second-tier platforms may include RNA/DNA hybridization and DNA sequencing, fluorescent *in situ* hybridization (FISH), high-level multiplexed assays (of varying types), pathology image analysis, liquid chromatography, surface plasmon resonance, and laser capture microdissection.[20] Second-tier platforms may also include those that are emerging and are setting new standards for sensitivity, specificity, and throughput. Emerging technologies are often only used in limited pilot studies to demonstrate feasibility of applications toward drug development.

Both common and less common platforms are employed widely in biomarker research along the entire continuum of discovery through development and provide essential information to drive drug programs forward. However, when choosing platforms to potentially select patients for possible prospective enrollment during registrational clinical trials, the choices become significantly reduced. Any patient-selection strategy, if successful in early development, may ultimately find PDx utility as a marketed IVD product. The designation of IVD requires conformity with an entirely new set of criteria on the platform beyond technical performance, including the need to develop under medical device controls (FDA[21]; e-CFR21:820[22]; EUdir98/79[23]; and/or ISO:13485[24]), and approval by regulatory agencies and market access efforts in relation to reimbursement.[25]

Of those listed above, there are relatively few platforms that have been approved for IVD use and even fewer that are marketed as a CDx.[26,27] These include IHC, FISH, PCR, and NGS.[28] To date, only these four platforms have been developed as PDx platforms under stringent medical device regulations. Although it is very likely that other platforms could achieve these requirements, the first-time establishment of a medical device program as well as review by regulatory agencies is often fraught with unforeseen complications that manifest in redesigns, intense documentation, and continued requests for additional data. These factors increase costs and can extend timelines by months or even years. Thus, when critical timelines must be met for drug approvals, selecting diagnostic platforms that have been previously approved as a PDx is often desired as a first choice. If a platform has not been previously approved, the next best option is to choose a company that has relevant experience in developing IVD medical devices, ideally as a PDx, which can reduce, but not eliminate, the risk of timeline delays mentioned above.

There are often situations in which a technology platform is employed in exploratory "signal finding" phases but has not been previously developed as an IVD. In such instances, it is important to develop and plan for a clear strategy to "bridge" the results from the non-IVD platform to one that has been approved.[19] One example is the use of multiplexed quantitative IHC.[29] This technology is currently being used in exploratory clinical research to understand mechanisms of action, as well as signal seeking for potential enrichment strategies.[30] If a stratification marker is observed using this technology that can identify responding patients in early studies, it is likely that an expansion for prospective enrollment based on this platform will be desired.[31] Currently multiplexed IHC is not an approved or cleared IVD medical device, thus it would be important to establish a standard IHC test and validate outcome results using the original multiplexed test in comparison studies.

MOVING FROM EXPLORATORY BIOMARKER ASSAYS TO PHARMACODIAGNOSTIC DEVELOPMENT: PLANNING EARLY AND EXECUTION

The codevelopment paradigm of a drug/diagnostic combination is well understood by the industry. However, some complex aspects of codevelopment and the role of the pharmaceutical company in diagnostic development are less appreciated. Early consideration of a diagnostic assay in the pharmaceutical development process is recommended, and conceptual planning should begin as early as preclinical development (Figure 6.6.1).[32]

In practice, this is rarely the case. Since biomarker discovery and evidence supporting correlation with outcomes is highly dependent on human studies, assay development schedules do not coordinate well with drug clinical studies. While assays can be developed analytically in preclinical development, knowing exactly which biomarkers and analysis methods are most relevant in relation to response outcomes is challenging.[33,34] Thus, it is very challenging to arrive at the optimal assay for selecting patients prior to having clinical outcomes data from studies large enough to test the patient-selection hypothesis. Moreover, clinical development teams often change stratification strategies based on data from phase I or early phase II studies, and there can be changing requirements related to the type and utility of biomarker data. Further, defining clinically relevant values for assays based on data from phase II studies is difficult due to small sample sizes or study design.[19,35] Frequently, clinical studies large enough to objectively establish clinically relevant values are only found in phase III, and further studies that would clinically validate the utility of the diagnostic are not planned or possible (Figure 6.6.2).

Thus, diagnostic assay development is often significantly compressed starting at phase I/II, with analytically validated

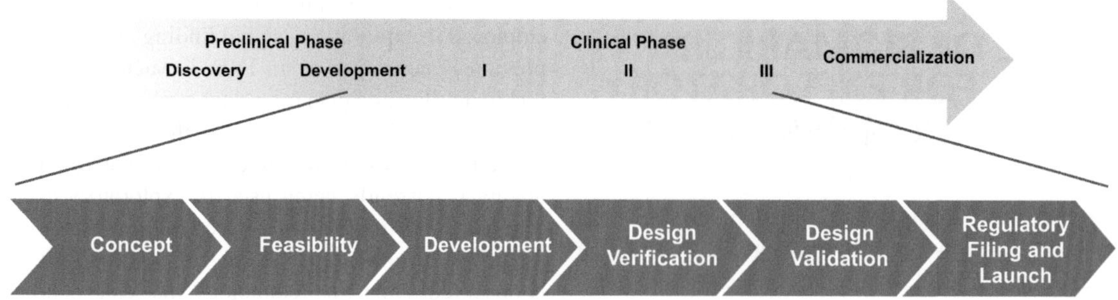

FIGURE 6.6.1 Phases of the drug development process.

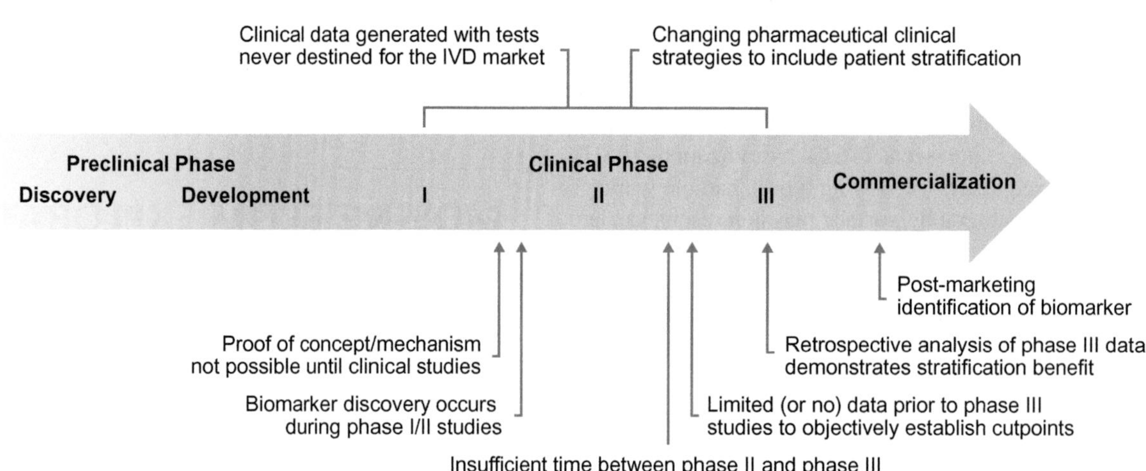

FIGURE 6.6.2 Timeline of factors that impact the drug development process.

assays and clinically relevant values based on limited data that need to be delivered by the end of expanded early phase development or phase III.[36]

As with all complex and compressed product development paradigms, being prepared, planning early, having cross-functional teams, good communication mechanisms, and working in a flexible and forward-thinking fashion can facilitate timely delivery. For the pharmaceutical company, creation and engagement of a multidisciplinary PDx team composed of key functions from the drug development team as well as other key functions is critical to maintaining aggressive codevelopment timelines. The PDx team should have members from business development, clinical development, technical assay leaders, bioinformatics, medical, regulatory, and commercial among others (Table 6.6.1).

Although establishing and implementing a codevelopment strategy is the main function of the team, the operating framework of the team itself can serve as an important platform to manage expectations of the clinical development team and facilitate an agile environment critical to maintaining schedule.[37] The team serves as a forum to educate about the diagnostic development process, specimen planning, regulatory considerations, progress, and risks, as well as preparing the company for changes to strategy, partnerships, development programs, and other unforeseen complications. Pharmaceutical companies are well advised to engage veterans from the diagnostic industry to serve as key leaders of these cross-functional teams to facilitate alignment of the dynamic tensions between the industries.

While the biomarker plan and strategy to incorporate a diagnostic into the drug development scheme originates from the clinical development team, most members of the pharmaceutical clinical development team do not have direct experience with the development, manufacture, and registration of a diagnostic product. Therefore, it is important to educate drug development teams about all the aspects of IVD development that may impact drug development strategies and timelines.[38] For device development, this includes engineering paradigms of design controls such as requirements, risk assessment, verification, and validation. As patient-selection approaches become more refined and specific, the PDx development strategy will increasingly influence drug development and clinical and regulatory strategies. Some of the factors that may impact the drug development timeline are listed in Table 6.6.2.

Another complication of the codevelopment paradigm is the conflicting priorities and focus between pharmaceutical and diagnostic companies. For example, pharmaceutical companies emphasize "no delays on drug trials" and "no limitations on drug indications" and they will be intently focused on the drug development schedule as the governing schedule for a PDx

TABLE 6.6.1 Pharmaceutical PDx codevelopment team members

FUNCTION	*ROLE*
PDx matrix team leader	• Set the strategy and execution timeline for the codevelopment team. Main point of contact for diagnostic company development team
Business development	• PDx contracting requires unique skills and knowledge. Multiple diagnostic partners are a frequent possibility
Clinical lead	• Set the clinical strategy and biomarker plan to accomplish patient enrichment schemes and accommodate the diagnostic verification and validation plans
Technical	• Provide technical input into assay design and assess results
Medical	• Input on design goals and assess product design
Regulatory	• Help develop PDx submission strategy and incorporate into drug submissions
Commercial	• Input on design goals and launch planning
Intellectual property	• Ensure freedom to operate, and protection of novel intellectual property
Sample acquisition	• Develop plan to meet development, verification, and validation needs. Interface with academic, consortia, and commercial sources to acquire specimens
Supply chain and procurement	• Develop and validate critical raw material manufacturing and establish raw material supply plan for prototypes, assays, and transfer

TABLE 6.6.2 Factors of PDx development that can impact drug development

FACTORS	*COMMENTS*
Schedule	• Typical PDx development programs take 6–18 months from proof of concept to a test ready for use in clinical studies • An additional 6–18 months is required to progress to a submission-ready product
Cost	• Can be substantial, depending on the technology • Frequently must be underwritten by the pharmaceutical company as the business case is rarely strongly positive for the diagnostic company
Device development requirements	• The PDx test has separate verification and validation requirements, but they likely must be integrated with the drug clinical studies and may impact on schedule and design
Regulatory	• In the United States, separate submissions are required; review schedules differ and must be coordinated through the U.S. FDA Center for Drug Evaluation and Research and Center for Devices and Radiological Health. Requirements are less complicated outside of the United States

launch.[14,39] Diagnostic companies will emphasize limited portfolio risk exposure, profitability of the diagnostic product, and conformity to quality system requirements and regulations.[40,41] Thus, codevelopment programs can be stressed by different, and at times conflicting, needs and strategies of the two industries.

TEST ADOPTION: IT CANNOT BE AN AFTERTHOUGHT

A critical aspect of successful PDx development in terms of commercial and medical requirements for successful pre- and post-launch activities is test adoption. It is perhaps the most overlooked and underestimated component of codevelopment. Areas of test adoption that require coordination are shown in Figure 6.6.3.

Pharmaceutical companies must take an active role in PDx test adoption, especially with tests that are complex to interpret. An interesting case study discussed in the periodical of the College of American Pathologists, *Cap Today*, reviews the role Genentech played in the adoption of the HER2 biomarker IHC test, which is the CDx for the breast cancer drug trastuzumab.[42] Genentech placed field representatives in pathology laboratories across the United States to educate about the safe and effective use of the CDx. It also developed very clear guidelines that helped to answer questions regarding sample preparation and potential confusion related to interpretation. A more recent example is that of PD-L1 testing, for which there are four different tests for different companion and complementary drug/diagnostic combinations.[43] PD-L1 testing is further complicated by differing expression-level values that are considered as clinically relevant and differing interpretation guidelines from each manufacturer.[44–47] In response to confusion in the market, several pharmaceutical companies and diagnostic companies, with

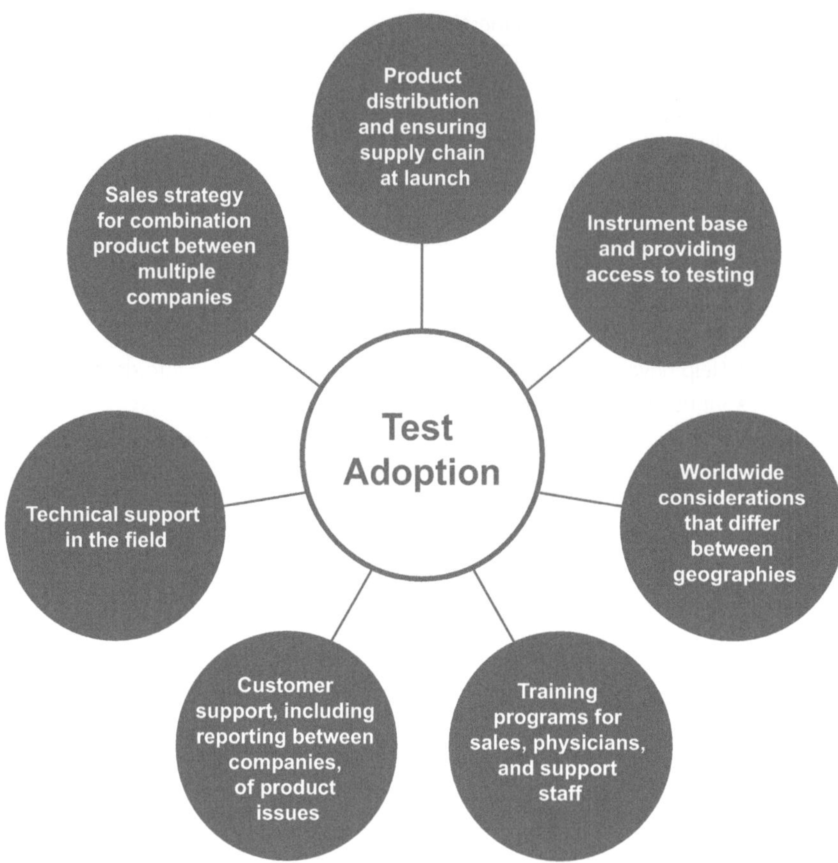

FIGURE 6.6.3 Areas of test adoption that require coordination.

input from the U.S. FDA, the American Association for Cancer Research, and American Society of Clinical Oncology, designed a series of studies to understand the differences and similarities of these assays with respect to analytical and clinical performance and provide unbiased data to the scientific and clinical communities.[48] Standardization activities such as these would optimally be planned prior to launch, and ideally in a pre-competitive setting between pharmaceutical, diagnostic companies, regulatory agencies, advocacy organizations, and payers.

Other factors that can impact adoption and require thoughtful planning by both the pharmaceutical and diagnostic company include post-marketing commitments that are commonly required to answer questions not addressed by initial clinical studies, rather than to monitor performance. Supporting clinical evaluations beyond validation are also often needed because adoption is best supported by a body of evidence, not just a study performed for regulatory approval. Lastly, working toward incorporating testing into treatment/diagnosis recommendations and demonstrating health economic value are critical to reimbursement, coding, and payment considerations. All factors must be addressed to get physicians to adopt and routinely utilize the PDx assay.

SUMMARY

For the successful development, launch, and adoption of a PDx, thorough planning, resourcing, and operational coordination within a pharmaceutical company is critical. Understanding the strategic misalignments between pharmaceutical and diagnostic companies is important to create the necessary stakeholder buy-in for organizing the ideal infrastructure to execute a successful PDx program. Regardless of strategic misalignments, the pharmaceutical company and diagnostic company must align at multiple functional areas and cooperate during all phases of drug development to achieve collaborative goals and successfully launch a diagnostic product to market. Development and adoption of a diagnostic assay requires coordination of several factors, including economic, schedule, and market targets; adherence to regulations and guidelines; the choice of diagnostic platform; early consideration of the design alongside the drug development process; planning and strategy development by multifunctional team members; and test adoption processes. The field of personalized medicine is expanding and the use of predictive biomarkers to identify patient populations that may respond to therapies is becoming more significant for guiding clinical treatment of patients. A limited number of diagnostics become FDA-approved; therefore, it is essential that pharmaceutical companies work effectively and efficiently with diagnostic companies to successfully develop, launch, and encourage adoption of PDx assays.

REFERENCES

1. Nishino M, Ramaiya NH, Hatabu H, Hodi FS. Monitoring immune-checkpoint blockade: Response evaluation and biomarker development. *Nat Rev Clin Oncol.* 2017: doi:10.1038/nrclinonc.2017.1088.
2. Zou W, Wolchok JD, Chen L. PD-L1 (B7-H1) and PD-1 pathway blockade for cancer therapy: Mechanisms, response biomarkers, and combinations. *Sci Transl Med.* 2016;8(328):328rv324.
3. Brahmer JR, Tykodi SS, Chow LQM, et al. Safety and activity of anti–PD-L1 antibody in patients with advanced cancer. *N Engl J Med.* 2012;366(26):2455–2465.
4. Reck M, Rodríguez-Abreu D, Robinson AG, et al. Pembrolizumab versus chemotherapy for PD-L1–positive non–small-cell lung cancer. *N Engl J Med.* 2016;375(19):1823–1833.
5. Snyder A, Makarov V, Merghoub T, et al. Genetic basis for clinical response to CTLA-4 blockade in melanoma. *N Engl J Med.* 2014;371(23):2189–2199.
6. Mehnert JM, Monjazeb AM, Beerthuijzen JMT, et al. The challenge for development of valuable immuno-oncology biomarkers. *Clin Cancer Res.* 2017;23(17):4970–4979.
7. Topalian SL, Taube JM, Anders RA, Pardoll DM. Mechanism-driven biomarkers to guide immune checkpoint blockade in cancer therapy. *Nat Rev Cancer.* 2016;16(5):275–287.
8. Hodgson DR, Whittaker RD, Herath A, Amakye D, Clack G. Biomarkers in oncology drug development. *Mol Oncol.* 2009;3(1):24–32.
9. Scheerens H, Malong A, Bassett K, et al. Current status of companion and complementary diagnostics: Strategic considerations for development and launch. *Clin Transl Sci.* 2017;10(2):84–92.
10. Jørgensen JT, Hersom M. Companion diagnostics—A tool to improve pharmacotherapy. *Ann Transl Med.* 2016;4(24):482.
11. e-CFR21:201-B. Requirements on content and format of labeling for human prescription drug and biological products. *Electronic Code of Federal Regulations.* Washington, DC: US Government Publishing Office; 2017.
12. Fridlyand J, Simon RM, Walrath JC, et al. Considerations for the successful co-development of targeted cancer therapies and companion diagnostics. *Nat Rev Drug Discov.* 2013;12(10):743–755.
13. Taube SE, Clark GM, Dancey JE, et al. A perspective on challenges and issues in biomarker development and drug and biomarker codevelopment. *J Natl Cancer Inst.* 2009;101(21):1453–1463.
14. Moore MW, Babu D, Cotter PD. Challenges in the codevelopment of companion diagnostics. *J Pers Med.* 2012;9(5):485–496.
15. Luo D, Smith JA, Meadows NA, et al. A quantitative assessment of factors affecting the technological development and adoption of companion diagnostics. *Front Genet.* 2015;6:357.
16. FDA. Laboratory developed tests. https://www.fda.gov/MedicalDevices/ProductsandMedicalProcedures/InVitroDiagnostics/LaboratoryDevelopedTests/default.htm. 2017. Accessed September 26, 2017.
17. Kling J. Diagnosis or drug? Will pharmaceutical companies or diagnostics manufacturers earn more from personalized medicine? *EMBO Rep.* 2007;8(10):903–906.
18. Yuan J, Hegde PS, Clynes R, et al. Novel technologies and emerging biomarkers for personalized cancer immunotherapy. *J Immunother Cancer.* 2016;4:3.
19. Marton MJ, Weiner R. Practical guidance for implementing predictive biomarkers into early phase clinical studies. *Biomed Res Int.* 2013;2013:9.
20. AdvaMed. An overview of diagnostic test technologies. https://www.advamed.org/sites/default/files/resource/overviewofdiagnostictestingtechnologies.pdf. 2016. Accessed September 26, 2017.
21. FDA. Principles for codevelopment of an in vitro companion diagnostic device with a therapeutic product. https://www.fda.gov/downloads/MedicalDevices/DeviceRegulationandGuidance/GuidanceDocuments/UCM510824.pdf. 2016. Accessed September 26, 2017.
22. e-CFR21:820. Part 820—Quality System Regulation. *Electronic Code of Federal Regulations.* Washington, DC: US Government Publishing Office; 2017.
23. EUdir98/79. Directive 98/79/EC of the European Parliament and of the Council of 27 October 1998 on in vitro diagnostic medical devices. 2017; http://eur-lex.europa.eu/legal-content/EN/TXT/?uri=celex:31998L0079. Accessed September 26, 2017.
24. ISO:13485. ISO 13485:2016—medical devices—quality management systems—requirements for regulatory purposes. https://www.iso.org/obp/ui/#iso:std:iso:13485:ed-3:v1:en. 2016. Accessed September 26, 2017.
25. IOM, NCPF. *Developing Biomarker-Based Tools for Cancer Screening, Diagnosis, and Treatment: The State of the Science, Evaluation, Implementation, and Economics: Workshop Summary.* Washington, DC: National Academies Press; 2006.
26. Dracopoli NC, Boguski MS. The evolution of oncology companion diagnostics from signal transduction to immuno-oncology. *Trends Pharmacol Sci.* 2017;38(1):41–54.
27. Twomey JD, Brahme NN, Zhang B. Drug-biomarker co-development in oncology—20 years and counting. *Drug Resist Updat.* 2017; 30:48–62.
28. FDA. List of cleared or approved companion diagnostic devices (in vitro and imaging tools). https://www.fda.gov/MedicalDevices/ProductsandMedicalProcedures/InVitroDiagnostics/ucm301431.htm. 2017. Accessed September 26, 2017.
29. Taylor CR. Predictive biomarkers and companion diagnostics. The future of immunohistochemistry—"in situ proteomics," or just a "stain?" *Appl Immunohistochem Mol Morphol.* 2014;22(8):555–561.
30. Stack EC, Wang C, Roman KA, Hoyt CC. Multiplexed immunohistochemistry, imaging, and quantitation: A review, with an assessment of tyramide signal amplification, multispectral imaging and multiplex analysis. *Methods.* 2014;70(1):46–58.
31. Fryburg DA, Song DH, de Graaf D. Early patient stratification is critical to enable effective and personalised drug discovery and development. http://www.ddw-online.com/personalised-medicine/p149258-early-patient-stratification-is-critical-to-enable-effective-and-personalised-drug-discovery-and-development.html. 2011. Accessed September 26, 2017.

32. Strovel J, Sittampalam S, Coussens NP, et al. Early drug discovery and development guidelines: For academic researchers, collaborators, and start-up companies. Assay guidance manual. 2012. https://www.ncbi.nlm.nih.gov/books/NBK92015/. Accessed September 26, 2017.
33. EPR. Biomarkers in drug discovery and development. https://www.europeanpharmaceuticalreview.com/article/4357/biomarkers-drug-discovery-development/. 2010. Accessed September 26, 2017.
34. Frank R, Hargreaves R. Clinical biomarkers in drug discovery and development. *Nat Rev Drug Discov.* 2003;2(7):566–580.
35. Schilsky RL, Doroshow JH, LeBlanc M, Conley BA. Development and use of integral assays in clinical trials. *Clin Cancer Res.* 2012;18(6):1540–1546.
36. Xu L. Biomarker considerations for early phase clinical trials. http://www.clinicaltrialsarena.com/news/operations/biomarker-considerations-in-early-phase-clinical-trials-5789431. 2017. Accessed September 26, 2017.
37. Pisani J, Lee M. A critical makeover for pharmaceutical companies: Overcoming industry obstacles with a crossfunctional strategy. https://www.strategyand.pwc.com/media/file/A-critical-makeover-for-pharmaceutical-companies.pdf. 2017. Accessed September 26, 2017.
38. Cross RL, Singer J, Colella S, Thomas RJ, Silverstone Y. *The Organizational Network Fieldbook: Best Practices, Techniques and Exercises to Drive Organizational Innovation and Performance.* New York: Wiley; 2010.
39. Agarwal A, Ressler D, Snyder G. The current and future state of companion diagnostics. *Pharmgenomics Pers Med.* 2015;8:99–110.
40. OECD. Policy issues for the development and use of biomarkers in health. https://www.oecd.org/health/biotech/49023036.pdf. 2011. Accessed September 26, 2017.
41. WHO. A risk based approach for the assessment of in vitro diagnostics. http://www.who.int/diagnostics_laboratory/evaluations/140513_risk_based_assessment_approach_buffet.pdf. 2014. Accessed September 26, 2017.
42. Ford A. Pharma strives to aid companion diagnostics. http://www.captodayonline.com/pharma-strives-aid-companion-diagnostics/. 2015. Accessed September 26, 2017.
43. Liu D, Wang S, Bindeman W. Clinical applications of PD-L1 bioassays for cancer immunotherapy. *J Hematol Oncol.* 2017;10(1):110.
44. Dako. PD-L1 IHC 22C3 pharmDx. http://www.accessdata.fda.gov/cdrh_docs/pdf15/P150013c.pdf. 2016. Accessed September 26, 2017.
45. Ventana. VENTANA PD-L1 (SP142) assay. http://www.accessdata.fda.gov/cdrh_docs/pdf16/P160002c.pdf. 2016. Accessed September 26, 2017.
46. Dako. PD-L1 IHC 28-8 pharmDx. https://www.accessdata.fda.gov/cdrh_docs/pdf15/P150025c.pdf. 2017. Accessed September 26, 2017.
47. Ventana. VENTANA PD-L1 (SP263) assay. https://usinfo.roche.com/rs/975-FPO-828/images/PD-L1SP263IG.pdf. 2017. Accessed September 26, 2017.
48. Hirsch FR, McElhinny A, Stanforth D, et al. PD-L1 immunohistochemistry assays for lung cancer: Results from phase 1 of the Blueprint PD-L1 IHC assay comparison project. *J Thorac Oncol.* 2017;12(2):208–222.

Biomarker Consortia to Drive BM Discovery and Clinical Applications in the UK 6.7

Lessons Learned from RA-MAP

John D. Isaacs, Sarah Brockbank, Ayako Pedersen, and Andrew P. Cope

Contents

INTRODUCTION

Multiple and diverse facilities and skills are required to identify, and subsequently verify, a clinically useful biomarker. For example, access to patients and clinical material is required as the substrate for biomarker discovery. The development of robust assays and, ultimately, their conversion into clinically useful tests is also critical, and if the biomarker is derived from a systems analysis of complex data sets then bioinformatics and advanced statistical modeling skills may be required. It is highly unlikely that all of these requirements exist in a single academic or commercial entity, hence the advantage of consortia.

HISTORY OF THE RA-MAP CONSORTIUM

The RA-MAP consortium is an academic–industry collaboration comprising 10 academic and 11 industry partners. It originated from a desire of the UK Government in the mid-2000s to accelerate discovery in human immunology, which was lagging behind the murine science.

The consequent Medical Research Council's (MRC) Strategic Review of Human Immunology [1] considered knowledge gaps alongside emerging technologies and tools, and developed a roadmap to support delivery of transformative research.

Key requirements included the creation of an interdisciplinary environment and increasing connectivity between institutions and sectors. Biomarkers and stratified medicine were not an explicit target of the Strategy at the outset but, to catalyze progression, the MRC joined forces with the Association of the British Pharmaceutical Industry (ABPI) to develop the MRC/ABPI Immunology and Inflammation Initiative. This was the first time MRC and industry partners had worked together in this way, catalyzed by a residential workshop in July 2009, to discuss a number of issues highlighted in a "view from industry" discussion paper. Themes of the workshop included the following:

- Develop awareness of academic and industry research and training priorities and activities.
- Encourage alignment, both at a strategic and practical level, in areas where there are good prospects for added value through joint working.
- Foster collaboration with the help of appropriate funding mechanisms.

Through the discussions, a number of key issues were raised that were perceived to form obstacles to industry/academic collaborations, including a need for improved networking, a requirement for improved access to human tissues, and a deep desire for improved support for clinical research in the UK. The workshop also identified several ways in which industry and academia could work together to tackle these issues. There was a strong steer from delegates that the best way to proceed and tackle these problems was to take a disease focused approach. This led to two disease-focused workshops, in 2010, one of which focused on rheumatoid arthritis (RA). The scope of each workshop was to include discussion of novel biomarkers, stratified medicine approaches, sharing of chemical tools, development and refinement of animal models and *in vitro* approaches. The workshops coincided with a funding call from the MRC to support a major initiative in each disease area.

RA-MAP

Informed by the outputs of the workshop a group of rheumatologists and scientific colleagues, representing the major academic rheumatology centers in the UK, developed a funding application. The ultimate focus was to develop a deeper understanding of the immune dysregulation that underpinned RA development and perpetuation. The hypothesis was that this would identify key pathways in RA pathogenesis and lead to the identification of prognostic and theragnostic biomarkers, as well as a better understanding of the remission state (Figure 6.7.1). The eventual funding application contained two work packages. The first, "Towards a Cure for Rheumatoid Arthritis" (TACERA) would provide the considerable infrastructure required for an observational clinical study of a large cohort of early, treatment-naïve, RA patients. The second, "The Immune Toolkit" would capitalize on TACERA

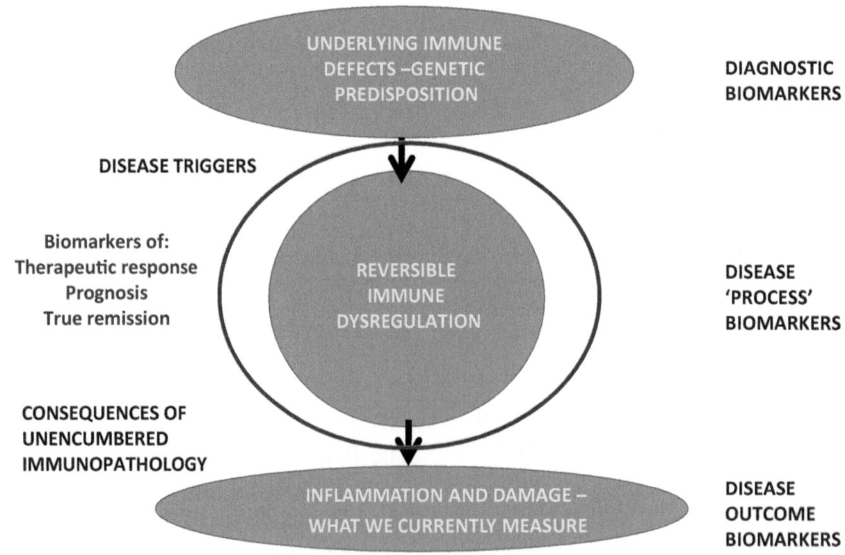

FIGURE 6.7.1 The RA-MAP hypothesis. Although understanding a considerable amount about rheumatoid arthritis predisposition, and possessing reasonable diagnostic biomarkers (top ellipse), what we currently measure in terms of outcomes are inflammation and damage—the consequences of unencumbered immunopathology (bottom ellipse). The "missing link" between these is a deep understanding of the dysregulated RA immune system (disease process, circle). The RA-MAP hypothesis is that understanding the latter will provide prognostic and theragnostic biomarkers, as well as a better understanding of disease remission.

and collect a variety of clinical samples, longitudinally, to provide the substrate for laboratory studies. RA-MAP was also designed to identify clinical and demographic predictors of prognosis, and a further aspect (PREVENT-RA) would collect a cohort of first-degree relatives to enable study of RA predisposition.

WORKING AT THE ACADEMIA–INDUSTRY INTERFACE

RA-MAP was clearly not the first example of academia and industry working together on a discovery project. For example, the EU Innovative Medicines Initiatives (IMI) already existed but RA-MAP was fundamentally different. Because the MRC distributed UK Government, and ultimately UK taxpayer, funds, industrial partners could not directly benefit from the grant awarded to the RA-MAP consortium (approximately $3.5 m). Furthermore, unlike IMI, there was no expectation for industrial partners to contribute funding up-front. The aspiration, however, was that industry partners would be prepared to commit "in-kind" contributions at various stages of the consortium's life, underpinning a true synergistic and collaborative effort to solve an important translational question. Consequently, establishing and maintaining the engagement of all partners was a significant early task.

The panel that reviewed and guided both outline and full funding applications comprised academic and industry representation, ensuring the focus had relevance to both parties. Furthermore, via the ABPI and the UK Bio Industry Association (BIA), the consortium was widely advertised and membership open to all. Nonetheless, a lot more was needed to ensure the meaningful engagement of all partners. Ultimately, this was achieved by developing a consortium agreement, which spoke to the interests of all partners. This took considerable time, effort, and negotiation, and required the imagination and skills of a RA-MAP project manager (PM) with experience of working in both academic and industrial sectors. Although a major driver for all partners was the potential for impactful discovery in the pre-competitive space, the key to driving partner engagement was ultimately access to data and samples, and ownership of eventual Intellectual Property (IP). Thus, it was determined that anonymized patient data would be equally accessible to all consortium partners. In terms of biological samples, the aspiration was that industry partners may, where relevant, assist with their analysis via access to technological platforms unavailable in academia.

The results generated by the consortium would be owned, in equal share, by the university partners. The universities would enter into an exploitation agreement in relation to a collaboration patent, subject to approval by each party's technology transfer offices, with patent costs shared equally. Industry partners agreed to support registration and protection of IP rights arising from the data, including filing and prosecuting patent applications, or assisting in actions relating to infringement of IP. To support subsequent exploitation, the university partners agreed to grant a worldwide, nonexclusive license to any industry partner to use the IP for commercial purposes, taking into account the relative contribution to the consortium of that partner.

CATALYSING EARLY ENGAGEMENT AND INTERACTION

Although the eventual consortium agreement proved attractive to all partners, there was to be an inevitable lag phase before patient-level data, and samples, arrived. A number of strategies were adopted to maintain engagement during this lengthy period.

One of the overarching aims of RA-MAP was to define prognostic markers for patients with RA. This could include clinical and demographic features as well as biomarkers. All partners were engaged early in the life of the consortium by developing a project focused on participants that received placebo in clinical trials. It is well recognized that some individuals "respond" despite receiving placebo and we wanted to explore whether specific features could be identified that were associated with a good outcome. There was enthusiasm for this idea, but it was not without its challenges. For example, industrial partners were asked to release (anonymized) data from phase II and phase III trials of marketed products, to pool with similar data from other companies. In addition to the associated legal challenges, it was not always obvious who owned the data, particularly when the decision had been taken not to further develop an asset that had been returned to the original developer. In some cases, the legal processes involved with releasing data were sufficiently lengthy to preclude participation but, ultimately, data were amassed and analyzed from more than 3,500 placebo/control participants from 19 trials, involving more than 30,000 individual trial visits.

Workshops were also convened during this period focused on biosample collection, handling and storage, including discussions around priorities for different types of analysis. One option was for all samples to be shipped to a central site for processing and storage, conceivably a contract research organization (CRO). However, recruitment sites were disseminated around the UK and, conscious of the potential importance of rapid sample processing, the consortium decided on a model of regional hub sites (academic centers) where samples would arrive within 4 h of blood draw. From these sites, processed samples could be further distributed to partners with expertise in particular analytic technologies (e.g., transcriptional profiling, metabolomics) (Figure 6.7.2). With regard to flow cytometry, frozen PBMCs were shipped to several sites, each of which had an interest and expertise in particular panels (e.g., T cells, B cells, myeloid

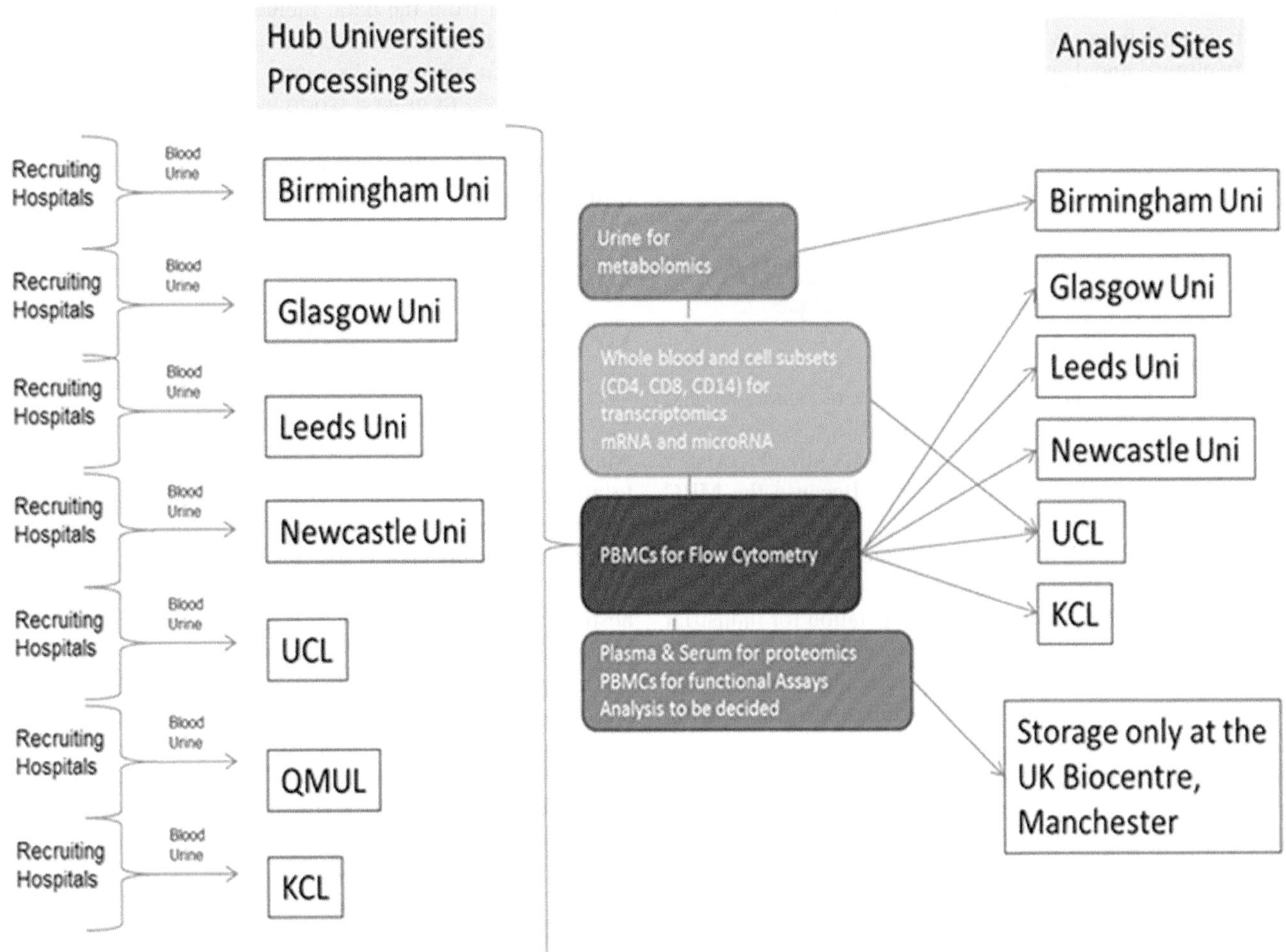

FIGURE 6.7.2 RA-MAP biosample distribution, processing and analysis. Samples collected at recruiting hospitals were rapidly transferred to academic hub sites, where they arrived within 4 h of collection. Following initial processing at the hub sites, samples were batched and forwarded to analysis sites, reflecting the specific expertise of academic partners. A number of flow cytometry panels were studied at different academic partner sites, again reflecting specific academic Interests.

cells). Processing sites also isolated three leucocyte subsets from PBMC (CD4+ and CD8+ T cells and monocytes) for transcriptional analysis.

To ensure homogeneity of processing, SOPs were developed and distributed to processing sites, and face-to-face training provided to RA-MAP technicians. A laboratory information management system was commissioned and installed in all processing sites to document sample arrival, processing and storage. Once sample collection commenced, regular and frequent teleconferences (TCs) were held between the PM and all RA-MAP technicians. These provided the opportunity to highlight any difficulties and discuss SOPs and such and were supplemented with face-to-face technical workshops twice a year. The latter were essential for streamlining work processes; however, more importantly, they provided a shared sense of "belonging" for the consortium technicians who were spread across the UK (Figure 6.7.2). Ultimately, samples that were not required for immediate analysis were batched and shipped to the UK Biocentre for secure storage [2].

In these ways, early engagement was encouraged and sustained. These early interactions also built the "RA-MAP team," which served us well in later years, breaking down barriers between institutions as well as between academia and industry. They also engendered our ethos of sharing.

CLINICAL STUDIES IN THE ACADEMIC SETTING—THE TACERA STUDY

The TACERA study was the underpinning resource for RA-MAP data and biosample collection. The aspiration was to collect a cohort of patients with seropositive RA and collect longitudinal clinical data (deep clinical phenotyping) and biosamples (for systems immunology analysis), starting on the day of diagnosis prior to antirheumatic drug prescription.

TACERA was a significant undertaking for an academic consortium with relatively limited resources. In addition to ethical review, over 30 recruiting hospitals had to independently sign a site agreement before recruitment could start at that hospital. This required individualized paperwork for each site, agreement over funding, and a site visit to explain the study and train key personnel. A study steering and data monitoring committee (SSDMC) was established, fulfilling the roles of a trial steering committee and data monitoring committee, and incorporating independent rheumatologists as well as patients. After a slow start to recruitment, advice from the SSDMC and encouragement from the MRC Steering Group (see below) provided strategies and guidance to accelerate accruals.

As a consequence of guidance, the consortium developed an agreement with a CRO. To conserve limited resources, the agreement was bespoke, focused purely on site setup, which at that point was the rate-limiting step in patient recruitment. This greatly accelerated site set-up, reducing the average time for site set-up from an average of 78 (range 49–139) to 32 (range 22–41) days, and provided a huge boost to recruitment. Recruitment was encouraged by regular newsletters and monthly "prizes" for high recruiting sites, as well as investigator meetings. Once all sites were established, recruitment proceeded expeditiously. It is worth mentioning that a key learning from RA-MAP, at least for the academic partners, was that outsourcing, when selected carefully, can pay big dividends.

Notably, TACERA did not dictate a particular treatment regime because this would have required additional regulatory approvals, as well as dis-incentivizing investigators. The advantage was that TACERA was a real-life study. As such, investigators were asked to manage their patients according to National Institute for Health and Care Excellence (NICE) guidance—which essentially meant that the vast majority received methotrexate as their first disease-modifying drug, either as monotherapy or in combination with another synthetic disease-modifying anti-rheumatic drug (DMARD).

Alongside TACERA, the consortium also performed a small vaccination study, in which healthy volunteers were vaccinated with a neoantigen (hepatitis B). Similar samples for immune phenotyping were collected as from the TACERA cohort; the vaccine study served as a small (n = 49) control cohort, representing the evolution of a healthy immune response to compare with the dysregulated immune response in TACERA.

CONSORTIUM MANAGEMENT

Over the lifetime of RA-MAP, more than 300 individuals contributed to the project. At any one time, there were around 60 individuals actively engaged with the consortium (excluding recruitment sites), reflecting the distinct skillsets required at different stages of the project. The principle investigators comprised a significant proportion of the UK rheumatology community. Consequently, the management of RA-MAP over its 6-year lifetime required excellent communication between partners, and strong management skills. This was achieved with a relatively streamlined and therefore nimble management structure that involved all partners. The consortium management board (CMB) was chaired by a representative of one of the industry partners and contained the two academic work package leads, the PM, one other industry and one other academic representative. The CMB met face to face every 3 months, immediately prior to the project steering group (PSG); it was responsible for coordinating all activities and for reporting progress to the funder. The PSG was chaired by the academic consortium lead and acted as the decision making body for the consortium. The PSG included a representative from each of the RA-MAP partners and, again, met once a quarter immediately after the consortium management board. As an example of the strong partnership engendered within RA-MAP, these meetings were well attended and the vast majority of partners represented on each occasion, sometimes via a telephone link. These two bodies were supplemented by a number of small and *ad hoc* working groups, which met by teleconference and ensured that nimble decisions could be made in between quarterly CMB and PSG. For example, the data analysis committee was key to translating the primary clinical questions into experimental design and analysis, ultimately producing the data analysis plan. Similarly, the PM held regular teleconferences with the hub technicians to ensure sample processing was proceeding smoothly.

The key individual(s) for ensuring the consortium hit its milestone targets were the PM. There was always one full-time PM, which is not always the case for an academic consortium of this kind, and we were fortunate to hire two successive individuals with both academic and industry experience. They were familiar with the key skills of project and risk management, and ensured the project flowed smoothly from start to finish. As well as organizing meetings and writing reports, they took responsibility for financial management and liaised with external vendors and suppliers. They visited all partners and were responsible for trouble shooting. Only very rarely did problems need to be escalated to the PSG and chief investigator. Essentially, they took ownership of the project, taking considerable daily pressure away from the academic and industry leads, who were able to focus on more strategic aspects of the project.

An unusual feature of RA-MAP was ongoing funder involvement. For the first three years of the consortium the academic and industry leads, as well as the PM, were interviewed twice a year by a panel convened by the MRC. The focus was on project progress, including data snap shots. The panel included academic and industry representatives, including some members of the original funding committee, ensuring knowledge of the objectives of RA-MAP was not forgotten. Prior to the meetings milestone and financial reports were prepared and presented to the funders. At times these were tough meetings, for example in the early stages when site set-up and recruitment lagged far behind projections. However, their aim was to be constructive and they provided very useful interactions, and an ability to discuss bottlenecks. As highlighted earlier, RA-MAP was pioneering and so learning was two-way, with the funders learning what was possible for such an academic–industry consortium. These face-to-face meetings were far more effective than the annual reports usually required for consortia such as RA-MAP. The outcomes of these meetings with the funder were relayed to the RA-MAP team a few weeks later, and cascaded to RA-MAP membership.

As a result of these structures, as well as investigator meetings at which progress to date was discussed, there was strong engagement of all partners throughout the program of work. This was greatly catalyzed by the nature of the consortium agreement, which ensured that all members were equal partners in the project.

RA-MAP—THE SAMPLES

A major challenge for the consortium was the collection, processing and storage of precious clinical samples. As discussed earlier, these were transported to hub sites where they arrived within 4 h of collection, and were immediately processed according to RA-MAP SOPS. Some were frozen and batched for future onward transport to analysis sites whereas others were analyzed locally.

A number of challenges were resolved during the course of the project. Some were relatively minor, such as a batch effect with flow cytometry samples. Others were more challenging. For example, the sheer number of samples needing simultaneous processing required the consortium to outsource one particular analysis to a CRO at relatively short notice. This was a good example of the PM's skills in scoping a number of potential vendors and ultimately choosing one, in discussion with the academic lead, that provided the optimal balance between quality and affordability.

There were also some analyses that were not incorporated in the original budget, which was less than the initial request. Proteomic analysis of serum was one of these, which subsequently provided an excellent example of industrial partner engagement. Industry partners were very keen that proteomic analysis was performed on at least a subset of RA-MAP serum samples. As there was no resource within the budget to perform this expensive analysis, one option would have been for the academic lead to request funding from one or more industry partners but this was likely to be time consuming and not necessarily successful. Instead, one of the industry partners developed a "crowd-sourcing" arrangement whereby each industry partner contributed a small but significant sum that ultimately allowed the analysis to proceed expeditiously. As per the consortium agreement, the results were owned collectively by consortium members and will ultimately be published with RA-MAP as the sole author.

RA-MAP—THE DATA

RA-MAP generated a complex immune phenotyping data set from a range of "omic" platforms, as well as linked clinical data. Following early discussions, the consortium decided to utilize a customized TranSMART database for data storage, hosted on the MRC-eMedLab cloud computing facility [3,4]. TranSMART is a data warehouse that provides for data access, visualization, exploration and download to all the members of the consortium. Data available on TranSMART include TACERA and vaccine data, alongside 19 publicly available RA clinical trials. TACERA data include clinical data, X-ray scores, flow cytometry, metabolomics, proteomics and gene expression profiles including small RNA (all at multiple time points). The vaccine study collected information on 49 patients, including demographics, medical history and vaccine response serology. In addition, the study served as a control population for TACERA flow cytometry, metabolomics, proteomics and gene expression profiles. The RA-MAP local platform of TranSMART currently provides service to 82 users from 23 organizations and stores 37 GB of data in order to offer high performance computing capacity, a long-term data sustainability solution and an appropriate environment for future meta-analyses by the rheumatology and wider immune-mediated inflammatory disease research community.

RA-MAP—SYSTEMS IMMUNOLOGY

The most exciting aspect of RA-MAP was the data analysis. This was again a multi-partner effort, involving biostatisticians, bioinformaticians, data modelers and clinicians from several academic and industry partners. Although face-to-face meetings were essential to kick-start data analysis, much of the interaction took place in "MAP-LAB" teleconferences, which occurred every 2 weeks. At each TC, specific aspects of the data were discussed. Manuscript progress was also discussed, and MAP-LAB served as an excellent way to coordinate and focus efforts as the consortium reached its climax.

MAP-LAB was again coordinated by the PM, who ensured that data and PowerPoint presentations were prepared and distributed ahead of the TCs, in concert with the teams presenting the data. MAP-LAB was a huge success and timed to suit partners on both sides of the Atlantic. It was an aspect of RA-MAP where our overarching partnerships shone through, in large part due to the variety of skills and expertise available—as well as genuine and understandable interest in the outputs of the consortium. Indeed, its legacy continues beyond RA-MAP, with continued biweekly engagement of partners, particularly around manuscript preparation.

RA-MAP—THE LEGACY

RA-MAP funding has now ended. MAP-LAB meetings continue, but RA-MAP has also left a much greater legacy. Some excellent science is emerging, as expected. In addition, however, we have provided a new model for academia–industry collaboration in the pre-competitive space, in line with the original aspirations of the MRC/ABPI Immunology and Inflammation Initiative. RA-MAP evolved into a true partnership: By the close of the consortium, it was difficult to recall the affiliations of partners when in a room together. The equitable data sharing and single authorship publications undoubtedly contributed to this ethos, underpinned by a very novel partnership contract and agreement. In addition, there are the data and samples. We are obliged to make these available to others, because of our funding agreement, but inevitably this would have happened anyway. Our RA-MAP patient partners dedicated a lot of their time and donated significant amounts of blood, with the aim of advancing our understanding of RA. They have done so, but it remains essential that opportunities to catalyze further advances are not lost, and a data and sample sharing committee has now been established to examine and expedite reasonable requests. The RA-MAP TranSMART/eTRIKs infrastructure has additionally served as a role model for many related MRC-stratified medicine consortia, such as MATURA (RA), PSORT (psoriasis), and MASTERPLANS (systemic lupus erythematosus). All projects use a similar infrastructure in a common eMedLab hosting environment. Collectively these projects have curated and integrated a substantial public domain data corpus for use as a reference for pan-inflammatory disease data analysis.

CONCLUSION

RA-MAP was conceived to catalyze the development of novel academic–industry partnerships in the pre-competitive space, and to serve as a template and pioneer for future consortia, particularly around stratified medicine. Eight years later, we feel we have achieved this, providing novel ideas around contractual agreements, data sharing and publication. Furthermore, "in-kind" contributions from our industrial partners totaled well over $750,000; we also received cash contributions of $150,000 plus the priceless patient-level data for the Predictors of Remission study [5]. We have additionally provided ideas for the organization of multi-partner, trans-continental consortia, to retain interest and engagement over a prolonged time period. The MRC has now invested more than £70 million in 12 stratified medicine consortia and related infrastructure. The disease focus for these consortia ranges from schizophrenia to hypertension, with further funding to be announced at the time of writing [6]. These also form part of the RA-MAP legacy, several using the RA-MAP consortium agreement as a model contractual template.

REFERENCES

1. MRC strategic review of human immunology. https://www.mrc.ac.uk/publications/browse/strategic-review-of-human-immunology/, Accessed October 17, 2017.
2. Biocentreuk. The centre for biomedical services. http://www.ukbiocentre.com, Accessed October 24, 2017.
3. i2b2 tranSMART Foundation. http://transmartfoundation.org/, Accessed October 24, 2017.
4. Medical Research Council eMedLab. http://www.emedlab.ac.uk/, Accessed October 24, 2017.
5. Medical Research Council Research: MRC/ABPI Inflammation and Immunity Initiative. https://www.mrc.ac.uk/research/initiatives/stratified-medicine/research/, Accessed October 24, 2017.

Measuring the Success of U.S. Biomarker Consortia

6.8

Joseph P. Menetski

Contents

SUMMARY AND OVERVIEW

It has been more than 10 years since the FDA released its initial report on the "Challenge and Opportunity on the Critical Path to New Medical Products" (2004). The report included an extensive description of the critical need for drug development tools (Link to FDA report, 2004), including biological markers (biomarkers). In 2014, the FDA published a guidance document that described the process for qualifying these drug development tools in order to accelerate drug approvals. They also directly encouraged the formation of collaborative groups that could share the burden of biomarker development and stated that several consortia had previously shown the usefulness of this approach.

Since 2004, the number of consortia carrying out biomarker development has indeed grown, but there have been only a few attempts to assess their impact, and these are mostly limited to analysis of individual efforts. Among these, Salter and Holland (2014) have highlighted several examples of emerging biomarkers that can be linked to progress in the field. The Quantitative Imaging Biomarker Alliance (QIBA), sponsored by the Radiological Society of North America, has reviewed the impact of standardization of imaging methods and discussed specific examples (Mulshine et al. 2015). The Chronic Kidney Disease Biomarker Consortium has assessed the lessons learned for that effort over a 5-year period (Hsu et al., 2015). Wagner et al. (2010) provided a description of lessons learned from a biomarker project that was completed as part of the Foundation for the National Institutes of Health (FNIH) Biomarkers Consortium. Although these publications focus on a single consortium at a time, a more general indication of biomarker consortia activity in general can be found in the high percentage of consortia-based projects currently in the FDA qualification pipeline. This seems to provide the strongest evidence that consortia-based efforts have had clear benefits for drug development in terms of actual drug approvals and related regulatory decision making.

In addition to providing specific decision-making tools, consortia have in fact begun to address critical issues in the overall biomarker development process. These partnerships have been able to provide a unique mechanism to gather data and build consensus to support inquiry into areas of biomarker development that have been difficult for any single scientific entity to address on its own. As the regulatory use of biomarkers has matured, the unique characteristics of consortia have proved increasingly useful in addressing the fundamental processes that the FDA uses to evaluate and qualify biomarkers. This chapter will describe and categorize specific biomarker development activities of U.S.-based consortia and provide some examples of semiquantitative metrics of their impact on the field, as well as highlight efforts to clarify the qualification process and provide more efficient approaches to future biomarker development.

PARTNERSHIPS THAT BENEFIT THE ENTIRE FIELD

The development of clinically relevant biomarkers is a time-consuming, data intensive and costly process (Amur et al., 2015; Lavezzari and Womack, 2016). The overall biomarker development process has been compared by some to that of developing a drug in terms of the amount of time and the breadth of resources that are required to generate enough confidence in a given biomarker to allow critical decision making. Those organizations that would benefit most from the use of a biomarker (primarily drug companies) generally do not plan to sell the biomarker as a product on its own, and so the incentive to commit the time and resources to generate the required information is low. Diagnostic companies that commercialize a biomarker generally do not have the resources needed to generate the amount of data required to support regulatory qualification of the biomarker. The basic scientists that are closest to the molecular characteristics of a disease generally do not engage in regulatory development of biomarkers as part of their mission and normally lack the expertise, access to data, and level of financial resources required to support such development. Finally, confidence in the use of a biomarker in drug development often requires a broad consensus about what clinical outcomes the biomarker is correlated with and what the limits are to its use. Consortia are uniquely organized to provide the necessary sharing of resources and expertise and a forum to drive consensus in the field: the major reasons why the FDA recommended consortia as vehicles for biomarker development and qualification in its original 2014 Drug Development Tools Guidance.

The results of any partnership should be judged in light of the stated goals and expectations of the initiating group in order to appropriately characterize the success of the consortium effort. These goals vary among the many different biomedical research partnerships that have been initiated over the past 10 years. However, Altshuler et al. (2010) provided an overarching set of four goals that describe the activities of most partnerships (developing standards and infrastructure, data generation and aggregation, knowledge creation, and product development). This analysis has been extended and refined in the FasterCures Consortia-pedia Framework report (2013) to examine the public landscape and to highlight differences in goals and expectations for active consortia. The Consortia-pedia project has analyzed the characteristics of each partnership based on several categories (e.g., sponsor, main focus of output, disease area). The FasterCures Framework Report (2013) shows that the number of new consortia increased every year between 2004 and 2012, leading to a final total of 250 individual consortia worldwide. Although the report did not specifically separate U.S. and international consortia, it did identify several interesting characteristics of these organizations that influence the type of results the partnership may generate. Most important for this discussion, it found that 26% of all partnerships state a main focus of their work as "broadly used biomarkers." Thus, the majority of these groups fit into the "knowledge creation" category as described by Altshuler et al. (2010). The largest number of partnerships are affiliated with the areas of cancer, rare diseases, Alzheimer's disease, diabetes and tuberculosis. Thus, to evaluate their overall viability in this growing complex research environment, one needs to generate and analyze metrics with which to measure the success of knowledge generation and, more importantly, if and how that knowledge is being used.

HOW SUCCESS IN BIOMARKER CONSORTIA MAY BE MEASURED

Although a comprehensive evaluation of the relative success of biomarker consortia has been lacking, Altshuler et al. (2010) have cited "knowledge generation" as useful yardstick of success. The type of information that is generated by a biomarker group will be dependent on the stage of development of the biomarker collaboration project. Biomarker development can be separated roughly into three stages: (1) biomarker identification, (2) biomarker development, and (3) biomarker utilization. Biomarker identification groups are designed to find new molecular characteristics that may relate to a disease or symptom of a disease. Consortia that focus on the next stage—biomarker development—are designed to confirm the correlation between the biomarker and disease (or physiological process), generally by reproducing the original result(s) in an unrelated data set. Finally, a biomarker utilization stage consortium shows that the potential marker can be used with some degree of confidence in a particular clinical setting for decision making. Each stage involves very different metrics of success. A successful biomarker discovery program is very different from obtaining a successful result from a utilization program, which might generate enough data for regulatory qualification of a biomarker.

In previous publications on the utility of biomarker consortia, authors have focused on the promise and progress that individual partnerships have demonstrated (Hsu et al., 2015; Mulshine et al., 2015; Salter and Holland, 2014; Wholley, 2014; Woolsey et al., 2010). Although interesting, this focus does not provide a high level view of what areas of consortium work provide value and which types of work should be improved or avoided. An overall description of the environment should address concerns around "consortium fatigue" that have been described recently (Papadaki and Hirsch, 2013), allowing stakeholders to identify "what works and what doesn't" and to feel confident in the choices of consortium projects in which they wish to participate.

Figure 6.8.1 links the goals of biomarker projects to the stage of biomarker development and a reasonable estimation of expected outcomes. The discovery of biomarkers typically generates publications of potential markers, which may list several that will transition to the next stage of development but are most often focused on describing the hypothesis that led to the discovery. Development

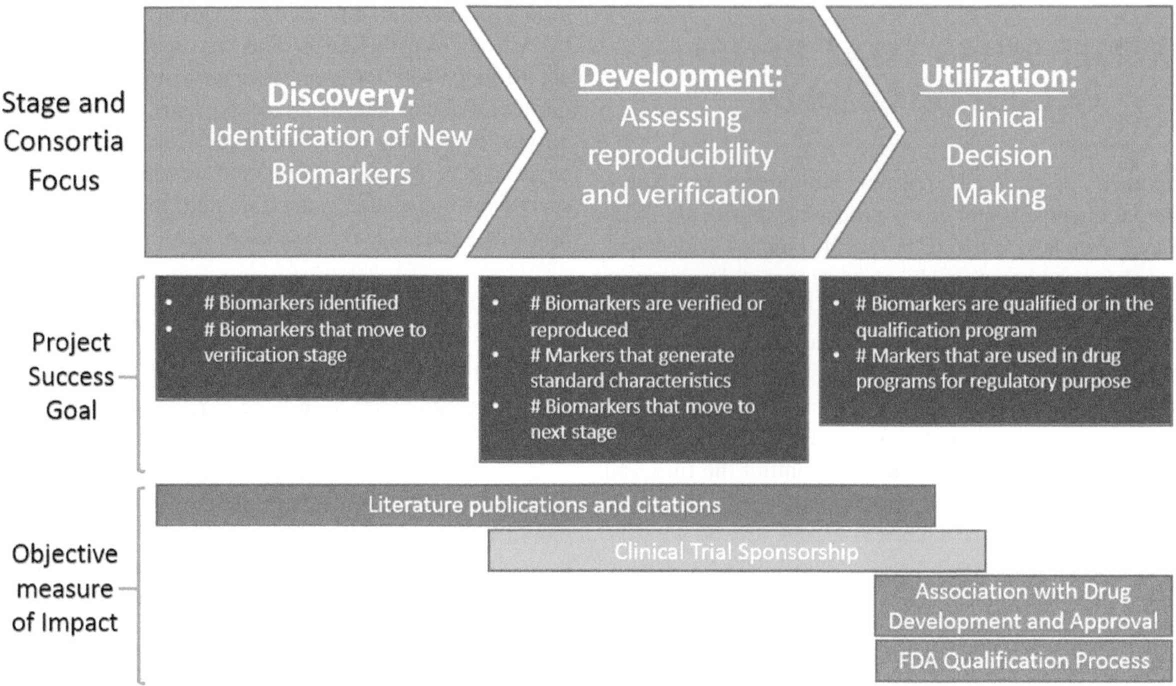

FIGURE 6.8.1 Stages of biomarker development and associated impact measures.

stage projects (clinical characterization and reproducibility) also mostly generate publications, although these are often accompanied by more extensive references to supporting data. As mentioned previously, clinical study projects, in the development stage, that measure or use markers will also often be described in public clinical study registries or databases such as clinicaltrials.gov. In the final biomarker utilization stage, the markers are used clinically and enough information may be generated to allow regulatory qualification and/or decision making in drug development, requiring submission of a formal briefing document to FDA or the European Medicines Agency. Evidence for results from these stages can often be found from the prominent public disclosure of the information into several database repositories.

DISCOVERY OF NEW MARKERS

The initial stage of biomarker development involves the identification of new biomarkers linked to a specific disease or disease trait. These projects are exploratory efforts that normally generate many candidate markers that require further verification by reproduction in additional studies. With the adoption of omics techniques, the ability to find potential RNA, protein or metabolic markers has increased substantially, and omics techniques are likely to continue to generate many such candidate markers. According to the FasterCures Consortia-pedia, 65% of all consortia-based biomarker research is focused on genomic or genetic approaches. The criteria for success in this stage of biomarker development are generally (1) publication of the initial findings and (2) demonstration of sufficient evidence to continue to the next stage of biomarker development. Although the most definitive sign of success in this stage is moving to the next stage, that information is not reported in a consistent manner. Thus, the number of this type of publication provides a more accessible, objective and broad assessment of biomarker discovery. A list of publications on biomarker identification can be obtained as a simple literature search and provides a quantitative assessment of activity. Another quantitative approach that is available using publication records is assessing the number of citations each publications has generated, and this has been used as an estimate of the impact that a given research effort has had on the field (Mingers and Yang, 2017). A literature search for papers associated with biomarker identification showed an increase from less than 900 publications in 2006 to almost 4,500 in 2016.[1] As expected, the number of these publications that reference the involvement of consortia for scientific or funding input also increases and, in 2016, represented 2% of the total number of biomarker identification papers. Interestingly, the biomarker identification publications that reported an association with consortium efforts between 2006 and 2016 were cited about 50% more often than the average citation rate of all publications (for the same search criteria) during this same period. These figures therefore suggest both that the activity of

[1] Publications associated with biomarker discovery were identified using Thompson Reuters, Web of Science searching topics for biomarker and (discovery or identif*). The search was further refined to assess consortia association by searching for consortium in topic, group author, author or funding organization. The search was done in January 2017. The search is not expected to be a comprehensive identification of every biomarker discover project, but to provide a consistent comparison in the literature.

consortia in the biomarker identification stage has increased in recent years and that this activity appears to have had an above-average impact on the biomarker research field.

BIOMARKER VERIFICATION AND REPRODUCIBILITY IN THE CLINICAL SETTING

The second stage of biomarker development is the verification and reproduction of the marker in a sample set other than the identification cohort. At this stage of development, the marker is shown to have potential clinical validity, and the analytical characteristics of the marker measurement and their relationship to the correlation of the marker to clinical outcomes are clearly defined. This is, by most accounts, the most labor- and resource-intensive part of the development process. Generating the appropriate clinical evidence and providing confident analytical limits for decision making requires access to multiple sample sets in all patient types where the marker will be used. As shown in Figure 6.8.1, the public release of information of this type can also be found in the literature record as well as in clinical trial databases.

A literature search for biomarker use in clinical trials shows a consistent increase in articles each year from 2006 to 2016.[2] Over that time period, the number of publications, referencing biomarker use in a clinical trial, increased from less than 1,000 to more than 6,000. Initially, very few of these papers were associated with the work of a consortium. However, the percentage of publications that were associated with consortia has increased to 2% in 2015 and 2016. As with biomarker identification papers, the average number of citations for clinical biomarker study publications from consortia is consistently higher than the average, occurring at 1.2- to 2.9-fold the average, as determined for each year. These results suggest that the use of consortia is increasing slowly and that the publications generated by consortia are having slightly more impact on the field than those not associated with consortia.

ClinicalTrials.gov is a registry and results database of publicly and privately supported clinical studies of human participants conducted around the world and offers additional evidence of the impact of consortia on biomarker development. As of January 2017, ClinicalTrials.gov listed 19,999 trials that described a biomarker as a key measure used in part of the trial.[3] Three hundred of these acknowledged a consortium as involved in or as a sponsor of the trial. The number of trials that involve consortia each year has increased over time; approximately 1.5% of trials are associated with a consortium. Interestingly, between 25% and 34% of these trials report an industry association. Thus, a significant percentage of the consortium-associated trials include industry members, and this result suggests that this stage of research may be particularly interesting to industry partners.

BIOMARKERS FOR BROAD USE DECISION MAKING FOR CLINICAL TRIALS

The last stage of biomarker development (i.e., utilization) is characterized by the actual use of the biomarker in clinical settings for decisions on therapeutic treatment or to support milestone (so-called "go/no-go") decisions in the drug development process. For the purposes of this review, the discussion will be limited to those markers that are being used in a general decision making process (e.g., safety markers) and not those that are developed by a company during a drug development program. Although the number of general use programs in the later stages of biomarker development are few, several very definitive sources of information can be used to assess the relative level of success in this important and final step of the biomarker development process. The FDA's Center for Drug Evaluation and Research (CDER) has instituted a process to qualify biomarkers and drug discovery tools through a dedicated Biomarkers Qualification Review Team (Link to Biomarker Guidance 2014). A qualified biomarker allows the FDA to approve the use of the biomarker for general use in drug development outside the context of a specific drug approval, though within a specific clinical context of use, and represents a definitive metric of success in the clinical use stage of biomarker development. The FDA maintains a list on its website (Link to FDA Qualified Markers) of biomarker programs that are in some stage of qualification. As of January 2017, the FDA had qualified six biomarkers through the Biomarker Qualification Program, and three of the markers qualified were for human clinical decision making. These qualified markers are biofluid and imaging markers and are intended to be used for patient selection in clinical trials. The FDA also provides a public description of the current projects in each stage of the qualification process that illuminates the degree to which consortia are involved.

The FDA Biomarker Qualification program reports the status of qualification programs at several stages prior to final qualification. As of January 2017, there were 23 biomarker programs in some stage of FDA qualification review (Link to Drug Development Tool (DDT) Qualification Projects at CDER, FDA). The 21 programs that have moved beyond initiation stage can be found on the FDA website (Drug Developmental Tool

[2] Publications associated with biomarkers in clinical trials were identified using Thompson Reuters, Web of Science searching topics for (biomarker and trial and clinical) or (biomarker and study and clinical). The search was further refined to assess consortia association by searching for consortium in topic, group author, author or funding organization. The search was done in January 2017. The search is not expected to be a comprehensive identification of every clinical trial that measured a biomarker, but to provide a consistent comparison in the literature.

[3] Clinical trials associated with biomarkers were identified using the clincaltrials.gov database searching for biomarker. The search was further refined to assess consortia association by searching for consortium or consortia. The search was done in January 2017. The search is not expected to be a comprehensive identification of every clinical trial that mentioned a biomarker, but to provide a consistent comparison in the database.

Qualification projects link). It is very important to note that 90% of the 21 programs that are currently in the Biomarker Qualification Program are directly attributed to consortia activities. Thus, from this objective point of view, consortia activity is having a significant positive impact in the process of qualification.

Two organizations account for the majority of consortium-associated programs currently in the qualification review process. The Critical Path Institute was established in 2005 in response to the FDA's 2004 Critical Path Initiative report, and has established a strong relationship with the FDA to support its mission "to foster development of new evaluation tools and standards for therapy trials, which accelerates regulatory qualification and medical product approval and adoption (Woosley et al., 2010)." Similarly, the FNIH Biomarkers Consortium was launched in 2006 with the support and backing of the founding members of the National Institutes of Health (NIH), FDA and the Pharmaceutical Research and Manufacturers of America (PhRMA). The mission of the FNIH Biomarkers Consortium is to discover, develop, and seek regulatory approval for biomarkers to support and accelerate development of new drugs, preventive medicine and medical diagnostics by combining the forces of the public and private sectors (Wholley, 2014). A handful of different consortia led by the Critical Path Institute and the FNIH's Biomarkers Consortium have submitted a total of 15 of the 18 qualification efforts currently under review.

A detailed analysis of the projects currently in the qualification program provides insight into the focus of the FDA qualification effort. Because the majority of these programs are associated with consortium efforts, this also provides insight into the success of these organizations. The qualification of safety markers is well represented in the list of programs (10/21). The FDA has an extensive and long history of understanding safety issues in drug discovery, and these markers are used in multiple therapeutic development programs. The next most prevalent marker types are prognostic (5/21) and predictive (4/21) marker types that are used in patient selection. (A prognostic biomarker is used to identify the likelihood of a clinical event, disease recurrence or progression in patients who have the disease or medical condition of interest; a predictive biomarker is used to identify individuals who are more likely than similar individuals without the biomarker to experience a favorable or unfavorable effect from exposure to a medical product or an environmental agent.) The definition of these biomarker categories has been standardized in a joint collaboration between the FDA and the NIH and are referred to as the BEST (Biomarkers, EndpointS, and other Tools) Resource definitions (Link to BEST Resource: FDA-NIH Biomarker Working Group, 2016). Comparison of the BEST biomarker types shows the majority of the markers in the FDA qualification program fall into three of the seven biomarker types defined by the BEST resource. Thus, the FDA qualification pipeline and, by extension, the regulatory success of consortium activity show a focus on biomarkers that can be used in multiple therapeutic development efforts and on safety markers and patient selection markers.

The qualification process is a clear and measurable indicator of consortium success. However, there are other aspects of biomarker use that are useful to assess the productivity of consortia: for instance, if a marker or tool that was developed by a consortium can be linked to the regulatory approval of a particular drug or to clinical decisions made in a specific drug development program. Although a clear indicator of success, the ability to analyze this type of data is unfortunately limited because it is not generally released publically due to program sponsors' concerns about confidentiality and intellectual property. To evaluate success in this context, therefore, one must rely on more anecdotal evidence.

The FNIH Biomarkers Consortium offers several such examples. The development of six drugs has been accelerated or enabled using tools that were developed by projects in the consortium (four antibiotic [Talbot et al., 2016] and two oncology therapeutics [Printz, 2013]). In addition, multiple project team member companies are currently using tools in their own drug development trials that were developed in Biomarkers Consortium projects (e.g., Robertson et al., 2014). These tools may or may not be submitted for formal regulatory qualification using the FDA process, but their impact in drug development and clinical patient care is significant. In the future, it would be useful to generate a standard approach to capturing information about adoption across all consortia that are working at this stage to document the utility and successes of the active consortia in this space.

PARTNERING TO DEVELOP CONSENSUS

As described, the involvement and success of consortia in specific biomarker development has increased markedly over the last decade. During the past several years in particular, the biomarker community has recognized that the biomarker development and qualification process itself requires additional clarity to allow continued progress (Lavezzari and Womack, 2016). Once again, the unique ability of consortia to drive and codify consensus among diverse stakeholders has been critical in helping define the qualification process. The most recent focus has been on establishing the evidentiary criteria necessary to qualify a biomarker, the appropriate levels of analytical validation needed to describe a biomarker measurement platform and the most appropriate statistical approaches to use in establishing the correlation of a marker to the clinical outcomes or traits of interest. Each of these areas has generated a distinct interest group led by consortia to help provide guiding principles for addressing the needs in these areas.

In August 2015, a symposium cosponsored by the University of Maryland Center for Excellence for Regulatory Science and Innovation, the FDA and the Critical Path Institute began to explore the requirements for important parts of the qualification process (Link to M-CERSI Symposium on Biomarkers in Drug Development, 2015). This meeting set the stage for the generation of working groups on analytical validation and statistical approaches. In addition, discussion about the type and amount of evidence needed for biomarker qualification within the biomedical research community lead to several consortium-driven efforts to define an approach to assessing evidentiary criteria needs in biomarker development. The National Biomarker Development Alliance (Link to A National Biomarker Development Alliance (NBDA) workshop) held a workshop titled "Collaboratively Building a Foundation for FDA Biomarker Qualification." The FNIH Biomarkers Consortium and the FDA cosponsored a workshop in April 2016, which generated a *Framework for Defining Evidentiary Criteria in Biomarker Development.* This framework document provides specific guidance to biomarker developers on the steps that need to be addressed in generating a successful qualification project (Link to FNIH Biomarkers Consortium—Workshop: Developing an Evidentiary Criteria Framework for Safety Biomarkers Qualification). In addition, the framework will inform future FDA efforts to develop relevant guidance for evidentiary criteria in biomarker qualification.

The work to drive consensus in these areas of development is expected to continue as the issues initially identified are completed and as new gaps in the development path are found. For example, the Duke Margolis Center for Health Policy, along with the Critical Path Institute and the FDA, is organizing a workshop for the spring of 2017 that will provide a community consensus on details of analytical validation. The working group on statistical considerations that began at the 2015 Maryland Center of Excellence in Regulatory Science and Innovation (MCERSI) conference is generating a consensus document that is expected to be released in early 2017. Consortia are thus continuing to play a major role in addressing the basic processes of biomarker development and qualification on behalf of the entire biomedical research community.

SUMMARY AND FUTURE PROMISE

The "team science" approach to biomarker development enabled by the emergence of public–private consortia has become what is likely to prove to be a permanent part of the biomedical research and development landscape. The number of consortia and their demonstrable impact have clearly increased over the past decade. It is important to recognize that consortia have had a positive influence in the all stages of biomarker development but particularly in the later (utilization) stage, when the need to marshal sufficient resources and to achieve consensus to confirm the value of markers is most critical. However, in order for the consortium approach to continue to be supported by industry and nonprofit groups, it will be essential for consortia to find clear and demonstrable metrics of success in areas such as enhancing decision making in clinical trials for new therapeutics or choosing the right therapies for the right patients. Finally, as consortia move potential biomarkers from the identification stage to the decision-making stage, we can expect to see an explosion of new approaches to clinical trial measurements with ensuing increase in trial efficiency. There is no doubt that biomarkers are the key to efficient targeted therapeutic development, and there is substantial evidence that the scientific community must work together on biomarker development in order to deliver the promise of safe, effective and affordable medicines.

REFERENCES

A National Biomarker Development Alliance (NBDA) workshop titled "Collaboratively Building a Foundation for FDA Biomarker Qualification." (December 2015). http://www.nbdabiomarkers.org/nbda-events/nbda-workshop-vii-"collaboratively-building-foundation-fda-biomarker-qualification."

Altshuler, JS, Balogh, E, Barker, AD, Eck, SL, Friend, SH, Ginsburg, GS, Herbst, RS, Nass, SJ Streeter, CM, Wagner, JA. 2010. Opening up to precompetitive collaboration. *Sci Transl Med* 2(52): 52cm26.

Amur S, LaVange L, Zineh I, Buckman-Garner S, Woodcock J. 2015. Biomarker qualification: Toward a multiple stakeholder framework for biomarker development, regulatory scceptance, and utilization. *Clin Pharmacol Ther* 98(1): 34–46.

BEST (Biomarkers, EndpointS, and other Tools) Resource: FDA-NIH Biomarker Working Group. 2016. https://www.ncbi.nlm.nih.gov/books/NBK338448/.

Biomarkers Consortium—Workshop: Developing an Evidentiary Criteria Framework for Safety Biomarkers Qualification. http://fnih.org/sites/default/files/final/pdf/Evidentiary%20Criteria%20Framework%20Final%20Version%20Oct%2020%202016.pdf.

Center for Drug Evaluation and Research. Guidance for Industry and FDA Staff Qualification Process for Drug Development Tools. 2014. http://www.fda.gov/Drugs/DevelopmentApprovalProcess/DrugDevelopmentToolsQualificationProgram/BiomarkerQualificationProgram/default.htm.

Challenge and Opportunity on the Critical Path to New Medical Products. 2004. FDA critical path opportunities report. http://www.fda.gov/downloads/scienceresearch/specialtopics/criticalpathinitiative/criticalpathopportunitiesreports/ucm077254.pdf.

Drug Development Tool (DDT) Qualification Projects at CDER, FDA. http://www.fda.gov/Drugs/DevelopmentApprovalProcess/DrugDevelopmentToolsQualificationProgram/ucm409960.htm.

FasterCures Consortia-pedia report developed by the Milken Institute to catalogue and characterize the consortia landscape and activity. 2013. http://consortiapedia.fastercures.org/assets/PDF/45700-ConsortiaReport.pdf.

FDA Biomarker Qualification Program. http://www.fda.gov/Drugs/DevelopmentApprovalProcess/DrugDevelopmentToolsQualificationProgram/ucm284076.htm.

FDA List of Qualified Biomarkers. http://www.fda.gov/Drugs/DevelopmentApprovalProcess/DrugDevelopmentToolsQualificationProgram/BiomarkerQualificationProgram/ucm535383.htm.

Hsu C-Y, Ballard S, Batlle D, Bonventre JV, Bottinger EP, Feldman HI, Klein JB et al. and Nelson RG for the CKD Biomarkers. 2015. Consortium cross-disciplinary biomarkers research: Lessons learned by the CKD biomarkers consortium. *Clin J Am Soc Nephrol* 10: 894–902.

Lavezzari G, Womack AW. 2016. Industry perspectives on biomarker qualification. *Clin Pharmacol Ther* 99: 208–213.

M-CERSI Symposium on Biomarkers in Drug Development. 2015. http://www.pharmacy.umaryland.edu/centers/cersievents/biomarkers/.

Mingers, J and Yang, L. 2017. Evaluating journal quality: A review of journal citation indicators and ranking in business and management. *Eur J Oper Res* 257(1): 323–337.

Mulshine, JL, Gierada, DS, Armato III, SG, Rick S. Avila, RS, Yankelevitz, DF, Kazerooni, EA, McNitt-Gray, MF, Buckler, AJ, Sullivan, DC. 2015. Role of the quantitative imaging biomarker alliance in optimizing CT for the evaluation of lung cancer screen detected nodules. *J Am Coll Radiol* 12:390–395.

Papadaki, M and Hirsch, G. 2013. Curing consortium fatigue. *Sci Transl Med* 5: 200fs35.

Printz, C. 2013. I-SPY 2 may change how clinical trials are conducted: Researchers aim to accelerate approvals of cancer drugs. *Cancer* 119(11): 1925–1927.

Robertson, RP, Raymond, RH, Lee, DS, Calle, RA, Ghosh A, Savage, PJ, Shankar, SS, Vassileva, MT, Weir GC, Fryburg, DA. 2014. Arginine is preferred to glucagon for stimulation testing of B-cell function. *Am J Physiol Endocrinol Metab* 307(8): E720–E727.

Salter, H and Holland, R. 2014. Biomarkers: Refining diagnosis and expediting drug development—Reality, aspiration and the role of open innovation, *J Intern Med* 276: 215–228.

Talbot, GH, Powers, JH, Hoffmann, SC. 2016. Developing outcomes assessments as endpoints for registrational clinical trials of antibacterial drugs: 2015 update from the biomarkers consortium of the foundation for the National Institutes of Health. Biomarkers consortium of the foundation for the National Institutes of Health CABP-ABSSSI and HABP-VABP project teams. *Clin Infect Dis* 62(5): 603–607.

Wagner JA, Prince M, Wright EC, Ennis MM, Kochan J, Nunez DJR, Schneider B et al. 2010 The biomarkers consortium: Practice and pitfalls of open-source precompetitive collaboration. *Clin Pharm Thera* 87(5): 539–542.

Wholley, D. 2014. The biomarkers consortium. *Nat Rev Drug Discov* 13: 791–792.

Woosley, RL, Myers, RT, Goodsaid, F. 2010. The Critical Path Institute's approach to precompetitive sharing and advancing regulatory science *Clin Pharmacol Ther* 87(5): 530–533.

The Value of Public–Private Partnerships in the Netherlands

6.9

Hanna L. Groen and Nico L. U. van Meeteren

Contents

INTRODUCTION: MISSION-ORIENTED PUBLIC–PRIVATE COLLABORATION

In the light of global societal challenges, the research, development and innovation (RD&I) landscape in the European Union is shifting toward a more mission-oriented approach (1). By defining and following a clear directive mission, this landscape is expected to achieve more societal impact and make far better use of the economic opportunities in the coming years. Directive missions can be powerful tools to obtain a programmatic focus in RD&I and improve pay-off of collective investments—public as well as private. This approach should especially help in the challenge of reducing waste of investments in biomedical RD&I, which was shown to be quite evident in the life sciences and health (LSH) sector (2). As a consequence, mission-oriented policies enable orchestration of the complex and dynamic RD&I landscape and the role of its many public and private stakeholders. The Dutch LSH sector especially considered close collaborations between citizens, researchers, government, and entrepreneurs and between disciplines, institutes, and sectors essential by its governance to deal with the societal challenge of "Health and Care."[1] This *quadruple helix* collaboration is considered able to bridge between *societal pull* and *technological push*. As a consequence, such collaboration in parallel provokes the necessary combination of societal and technological innovation (Figure 6.9.1), thereby rendering investments in RD&I via public–private partnership (PPP).

MISSION-ORIENTED PUBLIC–PRIVATE PARTNERSHIPS: THE WAY FORWARD

Societal Challenge "Health and Care"

The mission-oriented approach will gradually be incorporated into the EU RD&I-policies when it receives part of the next European Framework Programme for Research and Innovation

[1] The Dutch societal challenge "Health and Care" is an equivalent of EU's "Health, Demographic Change and Wellbeing."

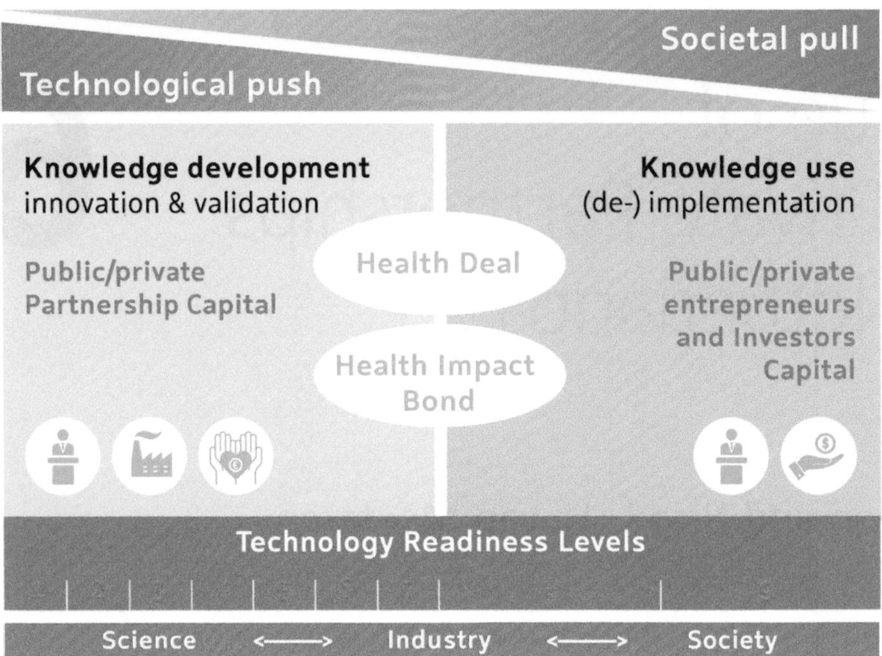

FIGURE 6.9.1 Overview of the process that can guide an idea (TRL 1) via innovation and validation to the health care market (TRL 9). (Adjusted from Knowledge and Innovation Agenda 2018–2021, Grow~Motion. https://www.health-holland.com/public/downloads/kia-kic/knowledge-and-innovation-agenda-2018–2021.pdf.)

(Horizon Europe, 3). Next to societal challenges on climate change and energy resources, citizens' health and well-being is a clear grand challenge that is faced in the EU and worldwide. Within the EU, member states and regions can act on the EU-wide missions within this grand challenge, as did the Netherlands, top-down via its sequential governments as well as bottom-up from out of its 11 regions (Figure 6.9.2).

Urged by the already anticipated stepwise loss of RD&I investments from the national gas-resources, the Dutch government initiated a nationwide "industry policy" in 2012. Hereto nine industrial sectors of main societal interest were identified and labeled as "Top Sectors" (4). Distinguishing marks of these sectors were high productivity of labor combined with (potential) high export capacities and holding an outstanding scientific infrastructure and track record. In these Top Sectors, that mostly coincide with the EU grand societal challenges, public RD&I-investments were stepwise increased as a *quid pro quo* of gradual growth of private investments in RD&I-infrastructure. This growth is due both to more joint investments in projects and to an increasing number of long-lasting strategic PPPs. At the start, coming from an economic and financial crisis, this industry policy focused almost exclusively on each one of the sectors individually, trying to counteract the years of built-up arrears of maintenance. A successful approach, given the facts and figures on several indices nationally (5) and internationally (6) that demonstrate the increase of national and foreign RD&I-investments, installment of new PPPs and companies (both startups and scale-ups), the number of patents issued, and the increase of exports from most of these Top Sectors. Having these sectors ultimately up and running again, the industry policy and its revitalized Top Sectors thereupon shifted their ambitions to the societal challenges, first nationally, but clearly with the outlook to scientifically, socially and economically contribute to the EU as well as globally. As one of the nine, the Dutch Top Sector LSH formed no exception here, given that it was revitalized and successful during these past years. Additionally, the Top Sector LSH—with its widely recognized brand name Health~Holland (7)—succeeded in receiving outstanding figures with this industry policy for its sector (8) and was consequently asked to adopt the societal challenge "Health and Care" by the government as one of the focus points in its future RD&I. A coalition of public and private stakeholders in the Dutch LSH sector therefore joined forces in its mission to invest in and harvest from evidence-based innovations for the *vital functioning of citizens in a healthy economy*.

Within this shift of focus—from building up Top Sectors toward Top Sectors in the front line of the societal challenges—several mission-oriented national PPPs were set up in recent years for fundamental, experimental and industrial LSH RD&I. PPPs are long-lasting multi-stakeholder partnerships within a specific field of expertise. In the LSH sector, PPPs focus on prevention, cure and care related to, for instance, people with cardiovascular diseases, dementia, and cancer. Preferably, PPPs consist of teams of experts and both private and public institutes that combine interdisciplinary visionary and conceptual knowledge bases, creativity and in kind and cash resources in robust, long-term, intra- and inter-sector collaborations. With efficient RD&I in mind, PPPs work on a precompetitive mission-led program with excellent science toward outcomes and deliverables with both societal and economic impact. Health and Care–related PPPs work closely together with Dutch health foundations and

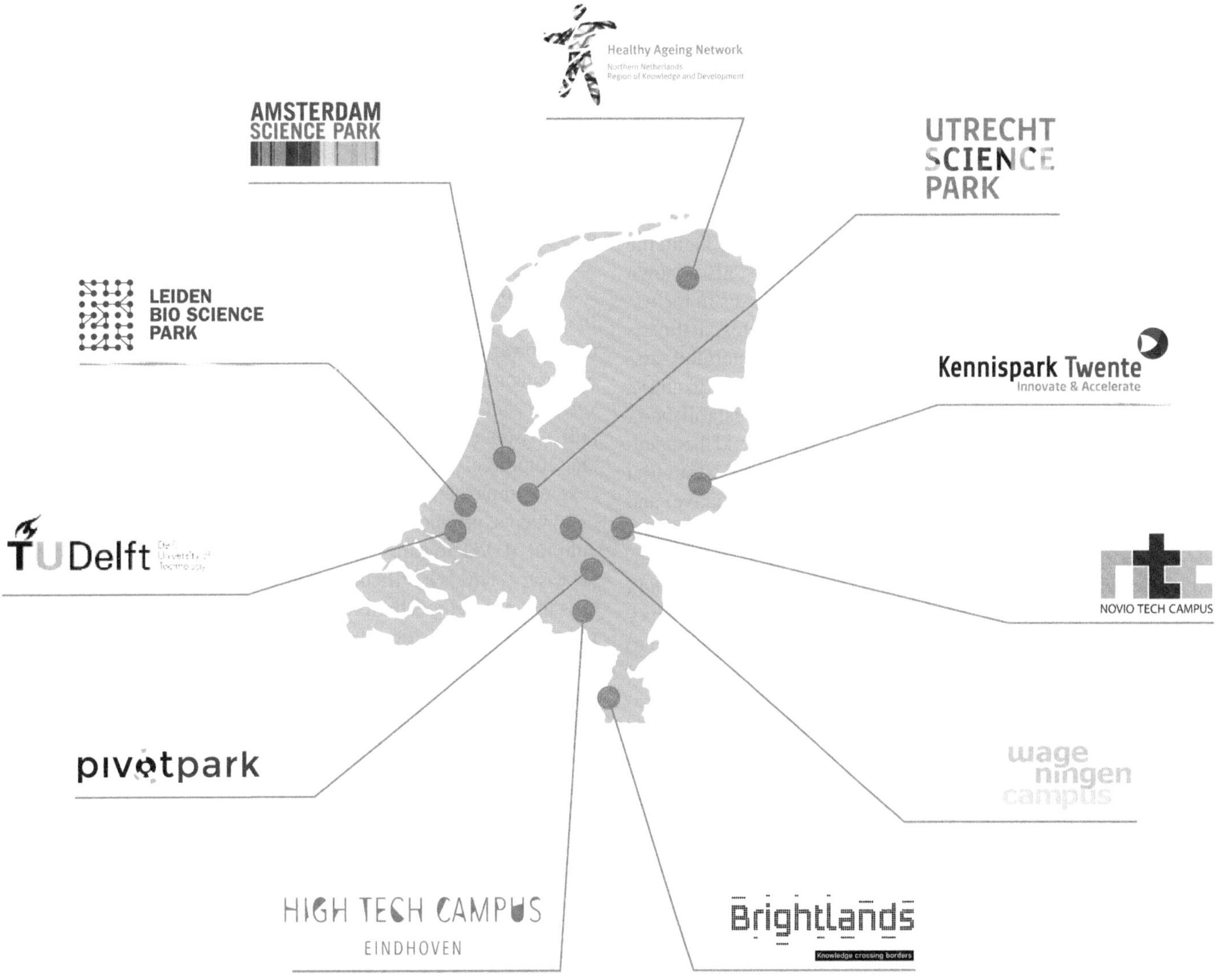

FIGURE 6.9.2 Overview of Dutch LSH campuses. (Based on Knowledge and Innovation Agenda 2018–2021.)

patients and patient representatives, health care professionals, and regulatory organizations at their start, next to academic and industrial parties as well as national and regional governments. By doing so, societal and patient-engagement is guaranteed from scratch in their RD&I strategy. On top of this multi-stakeholder engagement, the government, more specifically the Ministry of Economic Affairs and Climate Policy, encourages partnerships like these as it stimulates such public–private collaboration by provision of 30% superposed investment of each private euro spent in the PPP (PPP Allowance regulation, 9). Over the years, the amount of private cash spent in these collaborations has grown toward more than 500 million euros annually in the Top Sectors. Within the Top Sector LSH, growth of PPP allowance of 4 million euros in 2013 to almost 60 million euros in 2018 was established, representing the enormous influx of private investments in the LSH-knowledge and innovation infrastructure from private companies. In return, this infrastructure serves as a motor for evidence-based innovation in the sector assisting the systems transformation that is needed in the societal challenge Health and Care. In conclusion, by sustainable and intensive facilitation of PPPs in Health and Care–related innovation, the public–private strategy of the Top Sector contributes to institutional renewal of the knowledge and innovation landscape and, via its innovative and validated concepts, products and services, to the growth potential of this sector.

Generic vs. Disease-Specific Public–Private Partnerships

In the Netherlands strategic PPPs have been initiated via a top-down approach by the Top Sector's governance or bottom-up via one or more of the many stakeholders in the sector, especially via the representatives of health foundations and patient organizations. In addition to these two routes, some strategic PPPs evolve from a conventional PPP project funded by the Top Sector LSH. The Health and Care coalition led by the Top Sector LSH and its governance will, for the coming years, primarily facilitate already existing strategic PPPs, helping them to become viable and successful. Moreover, disease-specific PPPs that were installed during recent years receive intensive guidance: initiatives for and with patients with cardiovascular diseases (Dutch CardioVascular Association), oncology (Oncode Institute), infectious diseases (Netherlands Centre for One Health, including the antimicrobial resistance-targeted initiative of the Netherlands Antibiotic Development Platform) and, dementia (Delta Plan on Dementia). Likewise, the Dutch-Flanders regenerative medicine initiative (Regenerative Medicine Crossing Borders, also known as RegMed XB) is facilitated. This large PPP focusses in its first phase on a real organic cure for people with diabetes, arthrosis or kidney failure via the principle of restoring degenerated, diseased or damaged cells, tissues and organs, a typical example of mission-oriented RD&I, before the concept was established.

In addition, generic PPPs focusing on diagnostics and therapeutics are facilitated, namely those that develop intra- and extramural medical technologies (Innovative Medical Devices Initiative) and eHealth and ICT infrastructure (Health Research Infrastructure and *GO-FAIR*), critically evaluate success and implementation potential of medical innovations (Health innovation initiative Holland), and, pharmacology and drug development (European Lead Factory and PharmInvestHolland). Finally, special attention will be paid to PPPs working on personalized prevention. National initiatives of the Prevention coalition (providing novel social, mental and physical approaches in the contexts of the neighborhoods, in schools and workplaces and in the intramural care facilities), will be aligned with international initiatives in which the Netherlands is already a partner (that are EU Joint Program Initiatives as *More Years, Better Lives* and *A Healthy Diet for a Health life*).

To complete this knowledge and innovation infrastructure, a substantial part of the public and private LSH organizations have invested in highly fundamental biotechnical RD&I (*Building Blocks of Life*) and, on the other side of the spectrum, in practice-based research by the universities of applied sciences. These applied research universities—also known under the name of Polytechnics—and their integrated mission-oriented strategic Health Agenda harbor interesting potentials for local and regional Small and medium-sized enterprises (SMEs) and their academic staff and students, thereby including the workforce of the future. They combine RD&I efforts with implementation and dissemination efforts via "living labs" or so-called "learning communities" that are guided by embedded innovators and scientists from mixed–educational, societal and local governmental, entrepreneurial and academic–network structures. All of these are built up out of the conviction that progress on societal challenges is best made by interdisciplinary local RD&I initiatives, with the Dutch national and province governments to help invest in long-lasting connections. Moreover the hereto relevant sectors, like LSH, and also in key enabling technologies (i.e., Bioinformatics and Biotechnology), strengthen these more regional and local ecosystems. All in all, this connection of national and local RD&I PPPs provides instrumental means that facilitates getting innovative and validated concepts, products and services out of the numerous biosciences parks and into the capillaries of regional LSH facilities and its (future) workforce, all the way "from lab to life."[2]

Push and Pull: Technological and Societal Innovation

Getting from lab to life evidentially is not just a process of technological push. Given that the *technological push* of knowledge and innovation toward Technology Readiness Level (TRL, 10) 7 is traditionally organized well, something seems to be still missing. To make this push socially and economically successful, the system requires private and public stakeholders to intermingle with pivotal societal and economic partners of the societal challenges at hand. In order to ensure that innovative concepts, products and services move toward TRL 8 and 9 and are thus applicable for societal use and economic earnings for private companies and entrepreneurs a *societal pull* will have to be initiated (Figure 6.9.1). This societal pull—also known as deployment, or implementation, valorization, and commercialization—should, therefore, be organized: a process of implementation and, simultaneously, de-implementation of concepts, products and services through evidence-based sociocultural RD&I toward Societal Readiness Levels (11) is highly necessary. This is a difficult and time- and energy-consuming process that normally is hard to get funding for. In this respect, much is expected of specific instruments that should help this process thrive, namely the combination of a *Health Deal*[3] and a *Health Impact Bond*,[4] which should help guide innovations to the health care market and in the lives of patients in order to improve their health as well as their daily functioning, both in the Netherlands and, for export and extra revenue, abroad. Without an emphasis on this market pull, end-users (public or private) will not profit fully from the combined efforts out of the *quadruple helix*.

CONDITIONS FOR SUCCESSFUL PUBLIC–PRIVATE PARTNERSHIPS

As stated earlier, both the existing and new strategic PPPs are key to the Dutch strategy to tackle the Health and Care societal challenge. To be successful, PPPs require initial investments (i.e., capital), exposure via public relations and communication and, eventually, valorization, implementation and commercialization of their validated concepts, products and services via a health care human capital agenda, as well as via national and international entrepreneurs through worldwide diplomacy guided export. The latter should preferably be based on a solid internationalization strategy.

Inspired by the successes of LSH-PPPs, many new strategic PPP initiatives targeting Health and Care challenges are being planned, constructed and/or explored, mostly initiated bottom-up by LSH stakeholders. With respect to the LSH coalition's prominent

[2] A nowadays more appropriate considered semantic formula than "from bench to bedside."

[3] *Health Deal*: a regulatory and legislator set of agreements between public and private partners signed by the involved ministries.

[4] *Health Impact Bond*: a financial set of agreements on "profits and losses" and "social and economic return of investments" between investors and financiers, innovators and societal customer and provider organizations.

emphasis on the success of PPPs, public and private parties are very much willing to consider facilitating new strategic PPPs. These PPPs are fitted into the evolving knowledge and innovation infrastructure, as *stand-alone* PPPs are no longer an option. PPPs that can fulfil these conditions and really add to the ecosystem are reviewed and selected based on the following three main factors:

1. Organization
   - End-user involvement (from patients to professionals and entrepreneurs)
   - Energy, diligence and efficiency
   - Continuity potential
   - Congruence of the operational, logistic and personal prerequisites of the PPP with those of the founding organizations
2. Relevance
   - For the societal challenge on Health and Care and based on the Dutch National Research Agenda and LSH Roadmaps
   - In synergy with strategy of Departmental Agendas of the Ministries of Health, Welfare and Sport, Economic Affairs and Climate Policy, Education, Culture and Science, and, when and where necessary, of the Ministries of Social Affairs and Employment, and Agriculture, Nature and Food Quality
3. Impact
   - Economic potential
   - Scientific excellence
   - Technology transfer potential to establish valorization and (de-)implementation of products and services
   - Interdisciplinary and interinstitutional cooperation

Examples of PPPs that may fit these criteria in the future are—mature and new—disease-specific options like those for respiratory and neurological diseases. Generic PPP options that will be reviewed in the near future are those targeting RD&I on aging, physical functioning (as meant by the WHO's International Classification of Human Functioning, 12), reducing animals used in research, and unravelling pathology pathways. Each PPP will be reviewed and, according to its foreseen potentials, facilitated, first in collective sessions to inspire and learn from each other by best practices and to avoid duplicated efforts and redundancy. Selectively, some PPPs will also receive intensive guidance to boost their development and position in the LSH ecosystem. To qualify for this guidance, long-term strategy, theme and program, as well as economic aspects, are crucial. When beneficial the PPP will be stimulated to integrate with other up and running PPPs to enable mutual reinforcement. A strategy that no longer is restricted to the public and private institutions and institutes within the LSH sector, but all the more so focuses, with respect to the societal challenge Health Care, on cross-sector involvements and. The Top Sector LSH is thus seeking for top notch and, preferably, already validated innovations in order to have these *assembled, integrated* and *ecologically*—that is: in the appropriate real-life context(s)—*validated* as complete and solid end-solutions that make societal and economic profits as a whole.

SUSTAINABLE NATIONAL AND EUROPEAN FUNDING OF PUBLIC–PRIVATE PARTNERSHIPS

A large part of the coalition's success in the societal challenge Health and Care relies on synergy, cooperation and RD&I inventiveness. Adequate financial support is, however, a prerequisite for success. Yet, in recent years, it has become increasingly difficult to gain funding for RD&I. This has had a wide range of negative effects on the whole process, from developing innovative ideas to product market entry. The PPP Allowance regulation, as established by the Dutch Ministry of Economic Affairs and Climate Policy, enabled LSH and its partners to lubricate the process for the PPP initiatives in TRL 1–7. For those initiatives focusing on valorization and commercialization (in TRL 8–9), alternative funding options and new financiers will have to be found.

With respect to the future, the challenge of Health and Care will require extra public and private investments in an even more robust coalition of Top Sectors and ministries that will cooperate and further strengthen the existing ecosystem. The coalition encompasses public and private players that are more and more acquainted with the industry policy frameworks and inspired by the relevant routes and game changers in the Dutch National Research Agenda of the Ministry of Education, Sciences and Culture. The coalition will also be assisted in the future by a government that facilitates, stimulates and is involved in RD&I as well as in implementing, valorizing and commercializing its revenues and the results. This guarantees social innovation based on missions formulated in multi-stakeholder approaches based on the societal challenges.

Besides more investments in mission-guided projects and strategic PPPs, financial support comes from many others, for example, health foundations, the Dutch Ministries of Economic Affairs and Climate Policy, Health, Welfare and Sports, and Education, Research and Culture, and from the provincial administrations. These investments permit initiation of strategic PPPs for fundamental, industrial and experimental RD&I in the LSH sector and cross sectoral. Various processes are available to allocate the financial means to deploy the PPP funding as efficiently as possible. Several health foundations use their resources as collected from their many private donors, mostly the public at large, to support and partner with strategic PPPs such as RegMed XB (Diabetes Foundation, Rheumatism foundation and Kidney foundation) and Oncode Institute (via Dutch Cancer Foundation).

LSH STAKEHOLDERS AND MARKET DEVELOPMENT IN THE LSH ECOSYSTEM

The LSH coalition also collaborates with regional RD&I-clusters to facilitate their regional and interregional cooperation. These regional clusters are installed to create synergy and to interconnect public and private partners via the consolidation of a critical mass in the form of startups and scaleups out of an economic interest of preservation of existing jobs and create new ones. As such, the regions will be supported to further expand their collaborations with regional citizen cooperations and initiatives, research institutes, higher education partners and institutions, SMEs, public health and primary, secondary and tertiary care facilities including university medical centers, universities of applied sciences, regional training centers and, where applicable, the Ministry of Health, Welfare and Sport's Test Beds. These test beds will correspondingly serve as showcases for national upscaling and international export by enabling entrepreneurs with means to demonstrate the technological supremacy and societal values of their innovations for the public at large and investors. This will mostly be done under the guidance of regional and local economic boards that integrate the regional, societal and economic challenges in profitable RD&I coalitions.

In this way science parks and Test Beds form inspiring best practices that facilitate and stimulate economic and societal value of innovations. To continue this fruitful market development, other important options are to be addressed in the coming years:

- Regulatory and legal affairs that coincide with barriers, but taking into account safety and privacy based on the EU framework
- Financial instruments that provide profitable innovation
- Clear mixture of national and European funding, credits and tax regulation that reward desired developments and acts
- Transparent and end user-included
- Deals such as *Health Deals*, *Health Export Deals* and *Educational Health Deals*, in which public and private parties remove system barriers in implementation of validated products

CONCLUSIONS AND FUTURE AMBITIONS

The public and private coalition involved in the societal challenge of Health and Care envisages a thriving Dutch LSH and partnering sectors and key enabling technologies that, through research excellence, will develop health and health-related technological and sociocultural innovations focused on the real needs of citizens as well as on affordability and productivity in the prevention, cure and care cycle. Such RD&I will be achieved with serious and continuous end-user involvement and will, most importantly, create propositions that help citizens participate in society and at the same time create business value for public and private entrepreneurs in the Netherlands as well as abroad. Strategic mission-oriented public–private partnerships are key factors in this movement and have to be largely supported in the coming decades, nationally and internationally. In order to do so, the LSH-led coalition and especially its citizens and entrepreneurs will profit from this rebuilt and aligned national ecosystem. A good societal, economic and scientific position has been established in the recent strategic periods; this successful ecosystem can be further expanded in the future.

REFERENCES

1. Mazzucato, M. 2018. *Mission-Oriented Research & Innovation in the European Union: A Problem-Solving Approach to Fuel Innovation-led Growth*. Luxembourg, UK: Publications Office of the European Union.
2. Moher, D. et al. 2018. Increasing value and reducing waste in biomedical research, who's listening? *Lancet*, 387, 1573–1586.
3. Horizon Europe. Adopted by European Commission. 2018. (https://ec.europa.eu/info/designing-next-research-and-innovation-framework-programme/what-shapes-next-framework-programme_en).
4. Dutch Top Sectors. Appointed by Dutch Parliament. 2012. (https://www.topsectoren.nl/).
5. Progress report. 2017. Top Sectors (https://www.topsectoren.nl/publicaties/publicaties/rapporten-2017/oktober/09-10-17/vgr-2017).
6. Euro Health Consumer Index. 2017. (https://healthpowerhouse.com/ehci-2017/).
7. Health~Holland. Top Sector LSH. 2014. (https://www.health-holland.com).
8. Top Sectors responsible for 25% of Dutch GDP. 2017. (https://www.cbs.nl/nl-nl/achtergrond/2017/41/topsectoren-goed-voor-een-kwart-van-het-bbp).
9. PPP Allowance regulation of Top Sector LSH. 2012. (https://www.health-holland.com/calls/tki-match).
10. Technology Readiness Levels according to H2020 programme. 2014. (https://ec.europa.eu/research/participants/data/ref/h2020/wp/2014_2015/annexes/h2020-wp1415-annex-g-trl_en.pdf).
11. Societal Readiness Levels defined according to Innovation Fund Denmark. 2018. (https://innovationsfonden.dk/sites/default/files/2018-08/societal_readiness_levels_-_srl.pdf).
12. International Classification of Functioning. May 22, 2001. (http://www.who.int/classifications/icf/en/).

Paving a New Era of Personalized Medicine

Emerging Technologies

7

A New Era of Personalized Medicine **7.1**

Emerging Technologies—Functional Proteomics, Companion Diagnostics, and Precision Cancer Medicine

Michela Perani and Peter J. Parker

Contents

INTRODUCTION

In developed countries, about 50% of people diagnosed with cancer survive their disease for 10 years or more (Cancer Research UK; http://www.cancerresearchuk.org/health-professional/cancer-statistics/survival). This represents a doubling of survival over the last 40 years and reflects significant progress in the fight against cancer. Nevertheless, cancer remains one of the leading causes of mortality worldwide and, as a disease more prevalent in middle and old age, the chances of getting cancer increase as the longevity of the population increases. How will we get ahead in the race to improve outcomes?

For the last decade or so, we have endorsed the belief that the answer is "personal tumor medicine," defined as "a clinical approach that attempts to select the most appropriate therapeutic strategy for each patient on the basis of individual variability" (Srivastava 2012, Jackson and Chester 2015). This entails generating and using information derived from the breadth of omic approaches to patient profiling (genomics, transcriptomics, metabolomics and proteomics) in order to better define an individual's disease processes and so direct clinical decision making toward the most effective, nontoxic, targeted therapy for that individual. Our ability to generate this sort of information has itself been revolutionized by the development of new assays, tools and technologies that offer better precision, sensitivity and depth of molecular data (Duffy and Crown 2014, Rubin et al. 2014).

All the above has without doubt revolutionized medicine and the way we apply scientific endeavor to its practice; however, we have a long way to go. Despite the new omic technologies impacting in recent years, better biomarkers are urgently needed in both research and clinical application. We need robust biomarkers and reliable assays to improve (early) diagnosis, screening and prognosis and to monitor therapeutic responses (pharmacodynamic biomarkers), detect minimal residual disease, and guide molecularly targeted therapy to provide a dynamic and powerful approach across a wide spectrum of diseases (Lonergan et al. 2017).

Despite the fact that standard and high-throughput omics methodologies are rapidly developing, it is becoming clear that, to understand the complexity of cancer biology, current omics approaches are not sufficient to provide the insight into tumor behavior. Even with the range of omic approaches, there are limitations. For example, a genome sequence describes the repertoire of opportunities; however, a sequence as such does not reveal whether a given protein is expressed nor does it inform on its location or activation status. Similarly, the (coding) transcriptome is a useful descriptor that can be deconvoluted to predict transcription factor function, but again, it does not give insight into the protein(s) produced. To state the obvious, in cells, many levels of regulation take place after genes have been transcribed; this is most evident in the posttranslational modification of proteins (PTMs), but of course embraces all forms of biochemical control, including allosteric regulation, complex assembly influencing localization, and so forth.

The importance of PTMs of proteins is immense, encompassing many of the critical signaling events that are triggered during neoplastic transformation. This embraces many forms of modification, including the isopeptide linkages associated with ubiquitylation and sumoylation as well as sidechain modifications associated with methylation, acetylation, glycosylation and the prevalent phosphorylation of proteins (Markiv et al. 2012). This last modification is rampant in eukaryotes and exemplifies the need for more detailed, functional proteomics. The complex phosphorylation network governs nearly all cellular processes and offers a plethora of biological targets. This intricate network, is regulated at several levels to achieve signaling specificity, including protein expression, substrate recognition, and spatiotemporal modulation (Newman et al. 2014). Thus, inside the cell, protein kinases, phosphatases, and their cellular targets govern nearly all cellular processes and interplay between kinases and phosphatases make phosphorylation a highly dynamic process. These spatiotemporal and dynamic equilibria influence the function, stability, structure and subcellular localization of proteins. Proteins with their spectrum of PTMs define cellular physiology: They are the real players of this very complex game, such that when dysregulated, they have a profound impact on neoplastic transformation. As a consequence, regulatory kinases and phosphatases are key nodes that can be manipulated to influence the evolution of cancer. Much effort has been devoted to elucidate how these processes are controlled and coordinated in cells (Guha et al. 2008, Deschenes-Simard et al. 2014).

Well-known examples of proteins where aberrant regulation of phosphorylation contributes to neoplastic transformation include the receptor tyrosine kinases, the PI3-kinase/Akt/mTOR and the Ras/Raf/MEK/ERK signaling pathways (Steelman et al. 2011, McCubrey et al. 2012, Zhao et al. 2017). Examples in the cell cycle control pathways include p16/pRB/cyclin D1 and p19/p53/MDM2 (Zhang et al. 2014, Inoue and Fry 2015). It is believed that about 40% of proteins are phosphorylated at some point during their lifetime, however, proteins are often phosphorylated at multiple sites that may be present on the same substrate at the same time, adding combinatorial behavior as an extra layer of complexity (Zhang et al. 2002, Dephoure et al. 2013). To study such a complex network, even without considering its dynamic regulation, is a very intimidating mission. Moreover, this heterogeneous PTM landscape may not represent the majority of a protein species but might represent only a very small part making demands on the specificity and sensitivity of analyses (Newman et al. 2014).

Clearly, in this scenario, the analysis of only the genome and the transcriptome is not enough to untangle the extraordinarily complicated functional and regulatory mechanisms that exist inside a cell and are pertinent to disease. To address these limitations, new techniques and clinical proteomics applications are rapidly developing to bridge these clinical and research necessities.

Here we highlight some examples of established technologies and arising new ones, together with the efforts that led to platforms that integrate omics approaches.

MASS SPECTROMETRY-BASED TECHNIQUES

Clinical proteomics procedures are progressively relying more on mass spectrometry (MS) due to its high-throughput screening ability to analyze diverse biological samples. MS is used for the detection and measurement of proteins and peptides as well as to identify PTMs in proteins, although the PTMs are more challenging (Crutchfield et al. 2016, Szasz et al. 2016). MS also affords biomedical applications in tissue biopsies and cell lines and also in human bodily fluids such as plasma, serum, blood and urine. Human plasma and serum contain a wide range of proteins and circulating free DNA (cfDNA), and many are below the nanogram range. With the introduction of multiple reaction monitoring (MRM)–based MS, also known as selected reaction monitoring (SRM), it is now possible to identify and quantify low abundance peptides with very high sensitivity in a complex mixture. Thus, MRM results in greater selectivity and sensitivity at least one to two orders of magnitude higher than standard MS techniques (Mitchell 2010, Himmelsbach 2012). Two other developing MS-based techniques are the parallel reaction monitoring (PRM) and data-independent acquisition (DIA), with targeted data extraction on fast scanning high-resolution accurate-mass (HR/AM) instruments (Himmelsbach 2012, Shi et al. 2016). MS has also entered clinical laboratories in combination with either gas chromatography (GC) or liquid chromatography (LC) (Himmelsbach 2012, van den Ouweland and Kema 2012). In particular, the use of LC–tandem MS (LC-MS/MS) has grown enormously in clinical laboratories during the last decade, bringing increased specificity compared with conventional high-performance LC (HPLC), especially for low-molecular-weight analytes (Grebe and Singh 2011). Notwithstanding the fact that MS-based techniques are considered very valuable, there are also several limitations (Mitchell 2010, Solari et al. 2015, Crutchfield et al. 2016, Panis et al. 2016). For example, for the detection of low abundant proteins, enrichment steps are required that complicate the workflow, leading to loss of information and decreasing accuracy. In addition, MS applications typically yield no information on spatial localization or compartmentalization inside the cell.

IMMUNO-CAPTURE PROTEOMICS STRATEGIES

Forward and Reverse Phase Protein Arrays

Protein microarrays provide versatile platforms for characterization of hundreds of thousands of proteins in a high-throughput way. There are two different types of protein microarrays, which can be classified into forward phase protein arrays (FPPAs) and reverse phase protein arrays (RPPAs). The concept is based on printing small volumes of antibodies or tissue/cell lysate (pL scale), on distinct microspots (in an ordered pattern) on solid microscope slides, such as glass, plastic or silicon chips (Sutandy et al. 2013, Atak et al. 2016).

In the FPPA, antibodies or other capture reagents are immobilized onto solid array surfaces and incubated with complex samples such as tissue or cell lysates. The slides are incubated with microliter quantities of lysates and the arrayed antibodies will recognize the epitopes of the targeted proteins. The captured proteins can now be detected directly if the analyte is labeled (e.g., biotinylated lysate) or a second labeled antibody can be used for signal readout. The microarray images are then converted into protein expression profiles informative on the protein composition of the sample. The RPPA is based on the same principle, but in this case the lysate is spotted directly on the slide and then probed with antibodies against the targeted proteins of interest. This reverse array will allow a higher throughput compared with the forward one where the number of high-performing antibodies that can be generated is still relatively low. A limiting factor of these technologies is the dispensing of very small volumes of complex samples, which means that the number of molecules absorbed per spot can be a limiting factor because low abundance proteins fail to be observed. In addition, when arraying proteins, batch-to-batch variability can be problematic (Mitchell 2010). In both types of array, signal readouts are generally chemiluminescent, fluorescent or colorimetric although there are emerging alternative readouts, including label-free systems.

An alternative format of protein array is the suspension bead array (SBA) (Atak et al. 2016). Here antibodies are covalently coupled to color-coded magnetic beads labeled with different fluorescent identifiers. Protein lysates are labeled with biotin prior to incubation with the bead array. Streptavidin–phycoerythrin (SAPE) recognizes then the antibody–protein binding event and flow cytometry detects bead ID and SAPE with dual laser readouts.

These protein array techniques give interesting high throughput data sets, and without doubt, they provide valuable information. However, they are not clearly linked to the specific functional proteomic questions that many want to answer. As for MS, these applications also yield no information on subcellular localization, and to extract functional insight into proteins there is the need to go beyond the determination of concentration—there is the need to place them in a biological space and context. This demands orthogonal approaches that would replace or complement MS and protein array information (Szasz et al. 2016).

Besides the above exemplifications, there are other emerging approaches using immune-capture or alternative-capture techniques in proteomics. Some examples are listed in Table 7.1.1 and have been reviewed recently (Gold et al. 2010, Kodama and Hu 2012, Lee et al. 2013, Tighe et al. 2013, Lake and Aboagye 2014, Zhu et al. 2015, Atak et al. 2016, Fredolini et al. 2016).

Imaging Mass Cytometry

An interesting approach to proteomics that is gaining traction is imaging mass cytometry wherein tissue sections can be "visualized" with complex mixtures of metal-labeled antibodies (Chang et al. 2017). These are subsequently detected by systematic laser ablation of the sample exploiting the metal label to determine antigen presence through coupled, inductive MS. This platform technology is just becoming commercially available through Fluidigm (https://www.fluidigm.com/press/early-access-program-for-new-imaging-mass-cytometry-platform). Typically this technology, like flow cytometry, has been exploited to date to reveal cell phenotypes from the complexity of the cell types present in tissue samples, using cell surface protein markers. Imaging mass cytometry (IMC) in the future may offer opportunities to obtain spatially resolved information on protein location with a significant degree of complexity (perhaps multiplexing 50 antigens per tissue section).

TABLE 7.1.1 Examples of emerged and emerging immune-capture and alternative-capture techniques in proteomics.

TECHNOLOGY PLATFORM	*WEBSITE*
Luminex xMAP technology	https://www.luminexcorp.com/research/our-technology/xmap-technology/
Mesoscale discovery	http://www.mesoscale.com
Q-Plex technology	http://www.quansysbio.com/
BD cytometric bead array (CBA)	http://www.bdbiosciences.com/eu/applications/research/bead-based-immunoassays/m/745474/overview
Full Moon BioSystems Inc. antibody arrays	http://www.fullmoonbio.com/
Panorama antibody arrays	http://www.sigmaaldrich.com/life-science/cell-biology/protein-arrays/antibody-arrays.html
RayBiotech Inc. antibody arrays	http://www.raybiotech.com/antibody-array.html
SMC Erenna immunoassay	http://www.merckmillipore.com/GB/en/life-science-research/protein-detection-quantification/Immuno assay-Platform-Solutions/smc-erenna-immunoassay-platform/Cleb.qB.KJUAAAFOhkA1lU_l,nav?Referrer URL=https%3A%2F%2Fwww.google.co.uk%2F&bd=1
Quanterix-SIMOA (SIngle MOlecule Array)	http://www.quanterix.com/node/214
Arrayed immuno-multiplexing (AIM)	http://www.intuitivebio.com/resources/technology/
Somalogic-SOMAmer	http://www.somalogic.com/technology
Bimolecular fluorescence complementation (BiFC)	http://ruo.mbl.co.jp/bio/e/product/flprotein/fluo-chase.html
Enzyme fragment complementation	https://www.discoverx.com/technologies-platforms/enzyme-fragment-complementation-technology
Nucleic acid programmable protein arrays (NAPPA)	http://nappaproteinarray.org/technology.html

Note: Technology platforms for the exploitation of proteomics are listed, alongside sites through which access to the underlying technologies can be obtained.

Antibody and Antibody Detection–Based Assays

Antibody-based assays still remain the most common assays used in proteomics research and clinical applications, and they are essential approaches alongside MS analysis. Immunohistochemistry (IHC), ELISA, western blots and flow cytometry are still considered the gold standard methods for these affinity-based assays (Bordeaux et al. 2010). The number of biomarkers in medical research and in the clinical setting is growing rapidly, and so is the accompanying market for immunoassay kits where the community relies heavily on antibody performance to deliver accurate results. However, research and clinical laboratories recognize that widely variant results exist if the same test is performed in different laboratories, using different batches of reagents, and different platforms or process methods. It may seem trivial to state it, but the specificity of an antibody cannot be taken for granted (Bordeaux et al. 2010, Begley and Ellis 2012, Baker 2015). It is now widely acknowledged that the main concerns regarding antibodies are the antibody-limited specificity and cross-reactivity. For example, antibodies can be application-specific, meaning that it is important to validate them for their specific application, and for polyclonal antibodies there is the problem of batch-to-batch variability even from a single source. Specificity concerns have also been reported within the Human Protein Atlas (HPA) project. The HPA is a unique gene-centric initiative, aiming to raise antibodies to all human proteins (to every protein-coding gene) (http://www.proteinatlas.org/). The intention is to provide a map of the human proteome based on IHC tissue microarrays (both for normal and cancer tissues) and IF imaging on cell lines for spatial distribution of proteins at a subcellular level. The open access database contains images together with application-specific validation for each antibody. Within the HPA, with more than 9,000 antibodies tested, approximately half (49%) failed to show specificity, with a staining pattern that was not consistent with what was reported in literature or by bioinformatics data (Berglund et al. 2008). This exemplifies very clearly the scale of the problem and the challenge we face ahead.

Furthermore, besides the efforts required to validate our commonly used reagents, there is also a determination to establish new affinity reagents for which, some examples already on the market include single-domain antibodies, designed ankyrin repeat proteins (DARPins), affibody molecules, peptide and aptamer libraries (Stumpp et al. 2008, Lofblom et al. 2010, Hoon et al. 2011, Muyldermans 2013). However, even with these new tools, we still need to drive specificity in a way that a single-recognition assay cannot guarantee. In this context, a well-valued solution is to use multiple antibodies against the same target (or target complex), with the key strategy of driving specificity with two site-assays. In this configuration, only when both antibodies (or alternative affinity reagents) are bound to a specific target can the signal be generated. Specificity is assured by the fact that the probability that unspecific binding will occur in close proximity on different epitopes is very rare and the margin of error is therefore tolerable.

An example of a well-established dual-target, antibody-based recognition is based on Förster resonance energy transfer (FRET). However, other emerging bench assays also visualize proteins, protein modifications or protein interactions and examples are described below.

DUAL-TARGET RECOGNITION TECHNOLOGIES

FRET-Based Imaging Techniques

FRET-based imaging techniques can be used to detect events in formalin-fixed paraffin-embedded (FFPE) tissue, fresh tissues, fixed and live cells. Here two epitopes in close proximity are recognized by fluorescently labeled antibodies, leading to the resonant transfer energy from the excited fluorophore donor to a fluorophore acceptor, emitting a signal that is detected under fluorescence microscopy or monitored as a lifetime change in the donor (a more precise approach). This resonant transfer is dependent on the distance (≤10 nm) between the two fluorophores, which drives the specificity. FRET-based imaging techniques have been reviewed recently (Fruhwirth et al. 2010, Larijani et al. 2015, Shrestha et al. 2015). A useful advance in this approach is aFRET, where amplification is associated with the acceptor fluorophore, enabling generic Fab secondary reagents to be employed with a range of primary antisera (Veeriah et al. 2014). This provides good flexibility and avoids the limitations imposed by the need to directly and stoichiometrically label each primary antiserum (with the potential for loss of function). FRET approaches tackle the issue of specificity and distribution, providing information on location. However, like MS, FRET is challenging in terms of high-end technology and expertise to extract the data. As a result, they are less affordable platforms and less practical in a routine hospital setting (Larijani et al. 2015). Therefore, these technologies need to be complemented and sustained with simpler assays, cheap to run, specific and amenable to routine use in a hospital setting.

Emerging Bench-Top Assays

Proximity ligation assay

One alternative approach that exploits two-site detection, with a bench-top assay that moves away from sophisticated equipment, is the DuoLink assay (Fredriksson et al. 2002, Greenwood et al. 2015). DuoLink utilizes a proximity ligation assay (PLA) approach to assess the interaction of two proteins or the presence of two epitopes in close proximity along with their subcellular localization in fixed cells and tissue. This assay is based on the use of two primary antibodies of choice (from different species) that bind to two antigens in close proximity followed by the so called PLUS and MINUS PLA probes, which consist of secondary antibodies (species specific), each conjugated to an oligonucleotide. The assay requires a ligation step and a rolling circle amplification (RCA) reaction in conjunction with fluorescently labeled oligonucleotides for signal readouts. The advantages and limitations of this emerging technology have been discussed recently (Larijani et al. 2015). From a utility perspective, this assay has engaged researchers worldwide and has been particularly effective in permitting analysis of protein complex formation/colocalization in a relatively facile manner.

Coincidence biodetection technology

A novel assay termed coincidence biodetection technology is under development in our laboratory. This involves the use of dimer-dependent readouts where the individual components of the dimer are associated with the two recombinant detection reagents (antisera or other device) that can only dimerize and generate a signal when brought together in close proximity by the recognition of target antigens. Coincidence biodetection allows the detection and quantification of protein–protein interactions and PTMs on a target protein or protein complex; the assay has been commented on recently (Larijani et al. 2015). An illustration of the coincidence biodetection assay is depicted in Figure 7.1.1. This technology, like FRET, also retains subcellular

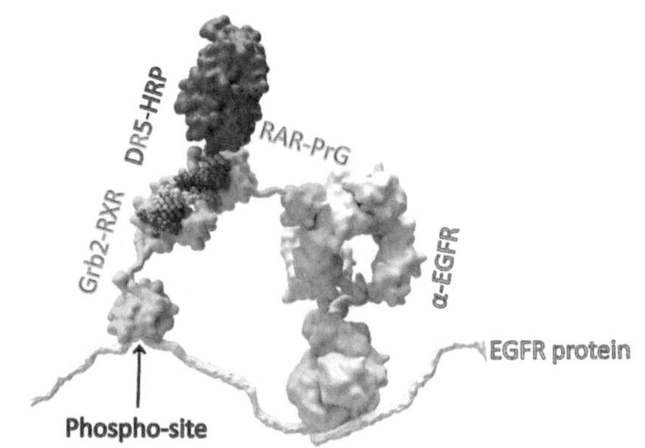

FIGURE 7.1.1 Schematic diagram of the coincidence biodetection technology exemplified by phospho-EGFR dual recognition. An antibody against the epidermal growth factor receptor backbone (α-EGFR) and Grb2 phosphotyrosine-binding domain SH2 linked to RXR (Grab2-RXR) are first added; followed by the addition of RAR linked to proteinG (RAR-PrG), where ProteinG binds to the Fc region of the antibody α-EGFR. The DR5 oligonucleotide linked to horse radish peroxidase (HRP) is then added, and only when both RXR and RAR are present in close proximity can a heterotrimeric complex be formed with direct repeat consensus sequence for RXR/RAR heterodimer formation. HRP will generate the signal readout when HRP substrates are added, such as DAB or fluorescently labeled tyramide (TSA). Signal detection under brightfield or confocal microscopy can then be quantified. (Image adapted from Phospho Biomedical Animation [phospho.co.uk].)

localization information, being amenable for use in fixed cells and tissues (including tissue micro-arrays). It has the advantage of being a simple bench-top assay that can be easily quantified and has the potential to be multiplexed.

AN INTEGRATED OMICS APPROACH—THE HUMAN PROTEOME PROJECT

Overall, with the diversified spectrum of omics approaches, we have generated an enormous amount of biological data sets. However, the integration of these data sets has become a real challenge. The Human Proteome Organization (HUPO) is an international scientific organization, aiming to develop an integrated omics approach that promotes international collaborations, educational training and technology breakthroughs in proteomics, with the aim of better understanding human disease (https://hupo.org/human-proteome-project). HUPO is the main organization of global human proteomics and has two major programs: the chromosome-based Human Proteome Project (C-HPP) and the Biology/Disease Human Proteome Project (B/D-HPP) (Kim et al. 2014, Lam et al. 2016, Szasz et al. 2016). The C-HPP aims to map the individual coding regions of each protein in each of the 23 human chromosomes. The goal is to classify at least one protein encoded by each of the approximately 20,000 human genes. The target is to elucidate the function of every gene on each chromosome. The proteins will be categorized based on tissue expression and major isoforms together with their different PTMs. The B/D-HPP provides a structure for the organization of biology- and diseased-based research groups. The aim is to assemble and link proteins to the relevant diseases and biological systems and make them openly available. These programs will be using technologies such as quantitative MS and antibody capture techniques supported by strong bioinformatics data analysis. This attempt to incorporate proteomics data into the genomic framework is of fundamental importance to elucidate the intricate differences and variations of protein expression patterns during development in healthy individuals and/or in disease development.

CONCLUSIONS

Given all the multidisciplinary and versatile approaches and efforts in developing new strategies to improve the outcomes of cancer treatments, it is considered essential that we tackle the problem by both (1) developing an integrated stratification approach that goes beyond the individual investigations of DNA, RNA, epigenomics, proteomics and PTMs analysis; and (2) developing reliable and reproducible routine benchtop assays that are inexpensive, sensitive, and selective and which do not require teams of highly trained personnel. Both approaches are of vital importance if we want to tackle the molecular pathology of human disease with its dynamic regulation and, at the same time, bring fast screening solutions to the bedside.

What are the challenges that lie ahead? To overcome all the difficulties, omics approaches in cancer studies should (1) reduce data variation by producing standard reliable protocols and platforms, (2) standardize and regulate protocols to obtain comparable results, (3) improve on databases and bioinformatic infrastructures for integrated omics analysis, (4) improve on data sharing and analysis by computational and visualization tools, (5) be able to assess tumor heterogeneity, and (6) improve on serum proteomic strategies for minimally-invasive monitoring. Studies of solid tumors are generally performed in FFPE tumor tissue samples or fresh biopsies of the primary or metastatic tumors. However, sampling of tumor tissue from primary or metastatic lesions is an invasive procedure for patients; thus, there is a rising interest in blood-based biomarkers, such as exosomes, circulating tumor cells (CTCs) or circulating tumor DNA (ctDNA), for their accessibility and thus for their potential clinical impact for screening and monitoring disease progression.

Where do we want to arrive? We still do not have the capability to be able to predict emerging properties from a transcriptional/translational profile that tells you what the state of a cell is. Techniques and information must all be linked with an ability to create an integrated genotype–phenotype report and a predictive model of behavior and drug response. The ultimate goal will be to take a comprehensive proteome live view of cell functions by connecting the genome and the proteome and to provide a real-time view of the dynamic pathology: a comprehensive view that will be able to predict emerging properties and to deliver precise and informed answers leading, ultimately, to accessible and affordable personalized treatments.

REFERENCES

Atak, A., S. Mukherjee, R. Jain, S. Gupta, V. A. Singh, N. Gahoi, P. M. K, and S. Srivastava. 2016. Protein microarray applications: Autoantibody detection and posttranslational modification. *Proteomics* 16 (19):2557–2569. doi:10.1002/pmic.201600104.

Baker, M. 2015. Reproducibility crisis: Blame it on the antibodies. *Nature* 521 (7552):274–276. doi:10.1038/521274a.

Begley, C. G., and L. M. Ellis. 2012. Drug development: Raise standards for preclinical cancer research. *Nature* 483 (7391):531–533. doi:10.1038/483531a.

Berglund, L., E. Bjorling, P. Oksvold, L. Fagerberg, A. Asplund, C. A. Szigyarto, A. Persson et al. 2008. A genecentric human protein atlas for expression profiles based on antibodies. *Mol Cell Proteomics* 7 (10):2019–2027. doi:10.1074/mcp.R800013-MCP200.

Bordeaux, J., A. Welsh, S. Agarwal, E. Killiam, M. Baquero, J. Hanna, V. Anagnostou, and D. Rimm. 2010. Antibody validation. *Biotechniques* 48 (3):197–209. doi:10.2144/000113382.

Chang, Q., O. I. Ornatsky, I. Siddiqui, A. Loboda, V. I. Baranov, and D. W. Hedley. 2017. Imaging mass cytometry. *Cytometry A* 91 (2):160–169. doi:10.1002/cyto.a.23053.

Crutchfield, C. A., S. N. Thomas, L. J. Sokoll, and D. W. Chan. 2016. Advances in mass spectrometry-based clinical biomarker discovery. *Clin Proteomics* 13:1. doi:10.1186/s12014-015-9102-9.

Dephoure, N., K. L. Gould, S. P. Gygi, and D. R. Kellogg. 2013. Mapping and analysis of phosphorylation sites: A quick guide for cell biologists. *Mol Biol Cell* 24 (5):535–542. doi:10.1091/mbc.E12-09-0677.

Deschenes-Simard, X., F. Kottakis, S. Meloche, and G. Ferbeyre. 2014. ERKs in cancer: Friends or foes? *Cancer Res* 74 (2):412–419. doi:10.1158/0008-5472.CAN-13-2381.

Duffy, M. J., and J. Crown. 2014. Precision treatment for cancer: Role of prognostic and predictive markers. *Crit Rev Clin Lab Sci* 51 (1):30–45. doi:10.3109/10408363.2013.865700.

Fredolini, C., S. Bystrom, E. Pin, F. Edfors, D. Tamburro, M. J. Iglesias, A. Haggmark et al. 2016. Immunocapture strategies in translational proteomics. *Expert Rev Proteomics* 13 (1):83–98. doi:10.1586/14789450.2016.1111141.

Fredriksson, S., M. Gullberg, J. Jarvius, C. Olsson, K. Pietras, S. M. Gustafsdottir, A. Ostman, and U. Landegren. 2002. Protein detection using proximity-dependent DNA ligation assays. *Nat Biotechnol* 20 (5):473–477. doi:10.1038/nbt0502-473.

Fruhwirth, G. O., S. Ameer-Beg, R. Cook, T. Watson, T. Ng, and F. Festy. 2010. Fluorescence lifetime endoscopy using TCSPC for the measurement of FRET in live cells. *Opt Express* 18 (11):11148–11158. doi:0.1364/OE.18.011148.

Gold, L., D. Ayers, J. Bertino, C. Bock, A. Bock, E. N. Brody, J. Carter et al. 2010. Aptamer-based multiplexed proteomic technology for biomarker discovery. *PLoS One* 5 (12):e15004. doi:10.1371/journal.pone.0015004.

Grebe, S. K., and R. J. Singh. 2011. LC-MS/MS in the clinical laboratory—Where to from here? *Clin Biochem Rev* 32 (1):5–31.

Greenwood, C., D. Ruff, S. Kirvell, G. Johnson, H. S. Dhillon, and S. A. Bustin. 2015. Proximity assays for sensitive quantification of proteins. *Biomol Detect Quantif* 4:10–16. doi:10.1016/j.bdq.2015.04.002.

Guha, U., R. Chaerkady, A. Marimuthu, A. S. Patterson, M. K. Kashyap, H. C. Harsha, M. Sato et al. 2008. Comparisons of tyrosine phosphorylated proteins in cells expressing lung cancer-specific alleles of EGFR and KRAS. *Proc Natl Acad Sci USA* 105 (37):14112–14117. doi:10.1073/pnas.0806158105.

Himmelsbach, M. 2012. 10 years of MS instrumental developments—Impact on LC-MS/MS in clinical chemistry. *J Chromatogr B Analyt Technol Biomed Life Sci* 883–884:3–17. doi:10.1016/j.jchromb.2011.11.038.

Hoon, S., B. Zhou, K. D. Janda, S. Brenner, and J. Scolnick. 2011. Aptamer selection by high-throughput sequencing and informatic analysis. *Biotechniques* 51 (6):413–416. doi:10.2144/000113786.

Inoue, K., and E. A. Fry. 2015. Aberrant expression of cyclin D1 in cancer. *Sign Transduct Insights* 4:1–13. doi:10.4137/STI.S30306.

Jackson, S. E., and J. D. Chester. 2015. Personalised cancer medicine. *Int J Cancer* 137 (2):262–266. doi:10.1002/ijc.28940.

Kim, M. S., S. M. Pinto, D. Getnet, R. S. Nirujogi, S. S. Manda, R. Chaerkady, A. K. Madugundu et al. 2014. A draft map of the human proteome. *Nature* 509 (7502):575–581. doi:10.1038/nature13302.

Kodama, Y., and C. D. Hu. 2012. Bimolecular fluorescence complementation (BiFC): A 5-year update and future perspectives. *Biotechniques* 53 (5):285–298. doi:10.2144/000113943.

Lake, M. C., and E. O. Aboagye. 2014. Luciferase fragment complementation imaging in preclinical cancer studies. *Oncoscience* 1 (5):310–325. doi:10.18632/oncoscience.45.

Lam, M. P., V. Venkatraman, Y. Xing, E. Lau, Q. Cao, D. C. Ng, A. I. Su, J. Ge, J. E. Van Eyk, and P. Ping. 2016. Data-driven approach to determine popular proteins for targeted proteomics translation of six organ systems. *J Proteome Res* 15 (11):4126–4134. doi:10.1021/acs.jproteome.6b00095.

Larijani, B., M. Perani, K. Alburai'si, and P. J. Parker. 2015. Functional proteomic biomarkers in cancer. *Ann N Y Acad Sci* 1346 (1):1–6. doi:10.1111/nyas.12749.

Lee, J. R., D. M. Magee, R. S. Gaster, J. LaBaer, and S. X. Wang. 2013. Emerging protein array technologies for proteomics. *Expert Rev Proteomics* 10 (1):65–75. doi:10.1586/epr.12.67.

Lofblom, J., J. Feldwisch, V. Tolmachev, J. Carlsson, S. Stahl, and F. Y. Frejd. 2010. Affibody molecules: Engineered proteins for therapeutic, diagnostic and biotechnological applications. *FEBS Lett* 584 (12):2670–2680. doi:10.1016/j.febslet.2010.04.014.

Lonergan, M., S. J. Senn, C. McNamee, A. K. Daly, R. Sutton, A. Hattersley, E. Pearson, and M. Pirmohamed. 2017. Defining drug response for stratified medicine. *Drug Discov Today* 22 (1):173–179. doi:10.1016/j.drudis.2016.10.016.

Markiv, A., N. D. Rambaruth, and M. V. Dwek. 2012. Beyond the genome and proteome: Targeting protein modifications in cancer. *Curr Opin Pharmacol* 12 (4):408–413. doi:10.1016/j.coph.2012.04.003.

McCubrey, J. A., L. S. Steelman, W. H. Chappell, S. L. Abrams, G. Montalto, M. Cervello, F. Nicoletti et al. 2012. Mutations and deregulation of Ras/Raf/MEK/ERK and PI3K/PTEN/Akt/mTOR cascades which alter therapy response. *Oncotarget* 3 (9):954–987. doi:10.18632/oncotarget.652.

Mitchell, P. 2010. Proteomics retrenches. *Nat Biotechnol* 28 (7):665–670. doi:10.1038/nbt0710-665.

Muyldermans, S. 2013. Nanobodies: Natural single-domain antibodies. *Annu Rev Biochem* 82:775–797. doi:10.1146/annurev-biochem-063011-092449.

Newman, R. H., J. Zhang, and H. Zhu. 2014. Toward a systems-level view of dynamic phosphorylation networks. *Front Genet* 5:263. doi:10.3389/fgene.2014.00263.

Panis, C., L. Pizzatti, G. F. Souza, and E. Abdelhay. 2016. Clinical proteomics in cancer: Where we are. *Cancer Lett* 382 (2):231–239. doi:10.1016/j.canlet.2016.08.014.

Rubin, E. H., J. D. Allen, J. A. Nowak, and S. E. Bates. 2014. Developing precision medicine in a global world. *Clin Cancer Res* 20 (6):1419–1427. doi:10.1158/1078-0432.CCR-14-0091.

Shi, T., E. Song, S. Nie, K. D. Rodland, T. Liu, W. J. Qian, and R. D. Smith. 2016. Advances in targeted proteomics and applications to biomedical research. *Proteomics* 16 (15–16):2160–2182. doi:10.1002/pmic.201500449.

Shrestha, D., A. Jenei, P. Nagy, G. Vereb, and J. Szollosi. 2015. Understanding FRET as a research tool for cellular studies. *Int J Mol Sci* 16 (4):6718–6756. doi:10.3390/ijms16046718.

Solari, F. A., M. Dell'Aica, A. Sickmann, and R. P. Zahedi. 2015. Why phosphoproteomics is still a challenge. *Mol Biosyst* 11 (6):1487–1493. doi:10.1039/c5mb00024f.

Srivastava, S. C. 2012. Paving the way to personalized medicine: Production of some promising theragnostic radionuclides at Brookhaven National Laboratory. *Semin Nucl Med* 42 (3):151–163. doi:10.1053/j.semnuclmed.2011.12.004.

Steelman, L. S., W. H. Chappell, S. L. Abrams, R. C. Kempf, J. Long, P. Laidler, S. Mijatovic et al. 2011. Roles of the Raf/MEK/ERK and PI3K/PTEN/Akt/mTOR pathways in controlling growth and sensitivity to therapy—Implications for cancer and aging. *Aging (Albany NY)* 3 (3):192–222. doi:10.18632/aging.100296.

Stumpp, M. T., H. K. Binz, and P. Amstutz. 2008. DARPins: A new generation of protein therapeutics. *Drug Discov Today* 13 (15–16):695–701. doi:10.1016/j.drudis.2008.04.013.

Sutandy, F. X., J. Qian, C. S. Chen, and H. Zhu. 2013. Overview of protein microarrays. *Curr Protoc Protein Sci* Chapter 27:Unit 27 1. doi:10.1002/0471140864.ps2701s72.

Szasz, A. M., B. Gyorffy, and G. Marko-Varga. 2016. Cancer heterogeneity determined by functional proteomics. *Semin Cell Dev Biol.* doi:10.1016/j.semcdb.2016.08.026.

Tighe, P., O. Negm, I. Todd, and L. Fairclough. 2013. Utility, reliability and reproducibility of immunoassay multiplex kits. *Methods* 61 (1):23–29. doi:10.1016/j.ymeth.2013.01.003.

van den Ouweland, J. M., and I. P. Kema. 2012. The role of liquid chromatography-tandem mass spectrometry in the clinical laboratory. *J Chromatogr B Analyt Technol Biomed Life Sci* 883–884:18–32. doi:10.1016/j.jchromb.2011.11.044.

Veeriah, S., P. Leboucher, J. de Naurois, N. Jethwa, E. Nye, T. Bunting, R. Stone et al. 2014. High-throughput time-resolved FRET reveals Akt/PKB activation as a poor prognostic marker in breast cancer. *Cancer Res* 74 (18):4983–4995. doi:10.1158/0008-5472.CAN-13-3382.

Zhang, H., X. Zha, Y. Tan, P. V. Hornbeck, A. J. Mastrangelo, D. R. Alessi, R. D. Polakiewicz, and M. J. Comb. 2002. Phosphoprotein analysis using antibodies broadly reactive against phosphorylated motifs. *J Biol Chem* 277 (42):39379–39387. doi:10.1074/jbc.M206399200.

Zhang, Q., S. X. Zeng, and H. Lu. 2014. Targeting p53-MDM2-MDMX loop for cancer therapy. *Subcell Biochem* 85:281–319. doi:10.1007/978-94-017-9211-0_16.

Zhao, W., Y. Qiu, and D. Kong. 2017. Class I phosphatidylinositol 3-kinase inhibitors for cancer therapy. *Acta Pharm Sin B* 7 (1):27–37. doi:10.1016/j.apsb.2016.07.006.

Zhu, Q., Y. Chai, Y. Zhuo, and R. Yuan. 2015. Ultrasensitive simultaneous detection of four biomarkers based on hybridization chain reaction and biotin-streptavidin signal amplification strategy. *Biosens Bioelectron* 68:42–48. doi:10.1016/j.bios.2014.12.023.

The Impact of Epigenetics on the Future of Personalized Medicine

7.2

Jane Mellor

Contents

INTRODUCTION

The science of epigenetics reveals how environmental and lifestyle factors influence the form and function of our genomes, often reflecting altered patterns of gene expression. As changes in gene expression often underpin the development of disease, there is a great deal of interest in assessing epigenetic states as biomarkers for predicting and diagnosing disease and for monitoring individual patients' responses to particular drug regimes. Despite the rapid advances in and lowering cost of high-throughput technologies such as next generation sequencing (NGS), it remains challenging to assess many of the well-known epigenetic markers and relate these to particular states (Stricker et al., 2017). This is particularly the case where an epigenetic change occurs a long distance from the gene or genes whose expression they influence. These problems are exacerbated when examining individuals because the composition of the epigenome is so influenced by the individuals' ethnicity, age, and lifestyle as well as the heterogeneity between cells within the same person. Put simply, it is hard to predict the consequences of one particular epigenetic change in an individual. Instead, an *unbiased* integrated approach is needed. This could involve, for example, relating phenotypic or clinical readout characteristics shared by a number of individuals to a spectrum of different epigenetic states. This can be done by assessing *higher-order structures in chromatin*, also known as the chromosome conformation, which integrate a range of epigenetic markers and states into *one* simple readout (Figure 7.2.1). Taking this strategy provides a robust and powerful approach to using epigenetics as a diagnostic and prognostic tool in personalized medicine.

This review covers two major areas. First is an introduction to the major components of the epigenome—DNA, nucleosomal histones, and noncoding RNA (ncRNA)—and their covalent modification. This section also includes a discussion of the impact of environmental factors such as diet and lifestyle on the epigenome via differences in cellular metabolism that influence the covalent modifications. Second is an illustration of how using chromosome conformation analysis in chromatin isolated from blood, which captures the sum of the contribution of the main epigenetic markers, rapidly and accurately reports the epigenetic state of an individual.

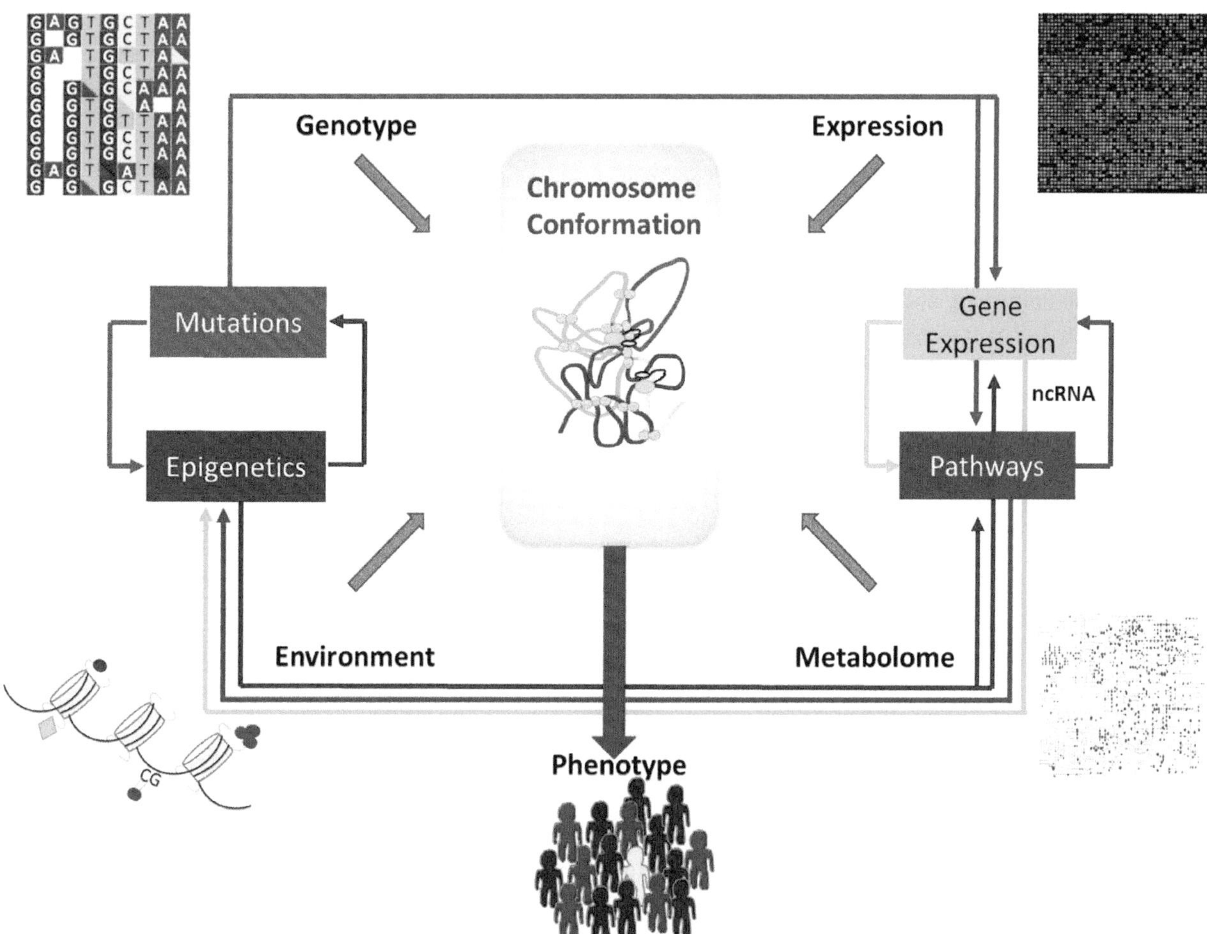

FIGURE 7.2.1 Integration of genetic, epigenetic, metabolomic and gene expression profiles at chromosome conformations that influence phenotype. In the schematic of a higher order structure in the chromatin, or chromosome conformation, CTCF dimers mediating long and short-range interactions between different regions are shown as yellow spheres, RNA polymerase II as a cyan sphere, and nascent transcripts as black lines. Direction of cross talk between the mutations (genotype), epigenetics (environment), gene expression and the metabolome (metabolic pathways) are represented by color-coded lines and arrows. These profiles are dynamic and will change over time as a disease progresses or during aging. The combination of changes to the metabolome, gene expression, the epigenome and the genotype are reflected in the chromosome conformation, which, in turn, influences the phenotype or clinical readout. (Adapted from Tordini, F. et al., *Front Genet.*, 7, 194, 2016.)

WHAT IS EPIGENETICS?

Epigenetics describes the factors that influence the phenotype of the cell or organism in a way that is inheritable, but without altering the underlying DNA sequence. In the context of personalized medicine, the concept of the somatic epitype is more useful. Like conventional epigenetics, a somatic epitype does not involve changes to the primary DNA sequence, but differs in that it is nonheritable. In conventional epigenetics, the phenotype that is inherited is determined by epigenetic changes in the grandparents' generation, although demonstrating rigorous epigenetic inheritance in humans is extremely difficult. Somatic epitypes are interesting in that they can arise *in utero* or after birth in response to environmental stresses (e.g., maternal care, nutrition toxins) and their effects on phenotype—particularly diseases that are strongly associated with epigenetic changes such as diabetes, cardiovascular disease, neurodegeneration and cancer—can be modulated by environmental factors such as diet. Diet provides the essential components for the maintenance of a healthy epigenome.

There are three main features of the epigenome. The first is covalent modification of DNA, the second is posttranslational modification of histone proteins that form the nucleosome, the basic building block of chromatin, and the third component is the ncRNA, tightly associated with the chromatin, which is also covalently modified. The enzymes that modify DNA, chromatin and ncRNA are known as "writers" because they write or deposit the epigenetic code. The code, in turn, is "read" by specific domains on proteins. These proteins help to alter the epigenetic state of the chromatin by, for example, contributing to the repression or activation of genes, facilitating repair of mutations in the DNA, or ensuring the stability of centromeric or telomeric regions of chromosomes. The code is reversible and enzymes known as "erasers" function to remove modifications.

COVALENT MODIFICATIONS TO DNA

DNA can be methylated at position 5 of the cytosine base (5mC) when followed by guanine (CpG dinucleotide) by "writers" known as DNA methyltransferases (DNMTs). DNA methylation is widespread throughout genomes except at certain promoters, which are rich in the CpG dinucleotide but remain unmethylated (these regions are known as CpG islands). 5mC is generally considered to be a repressive modification that functions by recruiting 5mC binding proteins in complexes that promote protein lysine deacetylation, often on the histone proteins in the surrounding chromatin (See "Posttranslational Covalent Modifications to Histones"). As shown in Figure 7.2.2, when found at gene promoters, 5mC is generally considered to be associated with their repression. However, 5mC can be associated with gene activity by preventing the binding of factors such as CTCF to CG-rich DNA at highly specific sites known as insulators (Bell and Felsenfeld, 2000; Maurano et al., 2015). CTCF, when functioning as an insulator binding protein, interferes with communication between enhancers and promoters and is associated with repressed gene expression (Cubenas-Potts and Corces, 2015; Kim et al., 2015). CTCF can also promote oxidation of 5mC by the ten-eleven translocation (TET) enzymes (methylcytosine dioxygenase) to produce 5-hydroxymethylcytosine (5hmC) and thus alter expression state because 5hmC is associated with more active chromatin (Cimmino and Aifantis, 2016). Indeed, the "reader" proteins that bind 5hmC specifically include DNA repair factors, splicing mediators and transcription regulators. 5hmC is a stable modification but can be thought of as an intermediate in the demethylation of DNA. TET dioxygenases require α-ketoglutarate and are inhibited by the oncometabolites 2-hydroxyglutarate, fumarate and succinate (Chowdhury et al., 2011; Xiao et al., 2012; Laukka et al., 2016; Sciacovelli et al., 2016). Thus, levels of DNA methylation are very

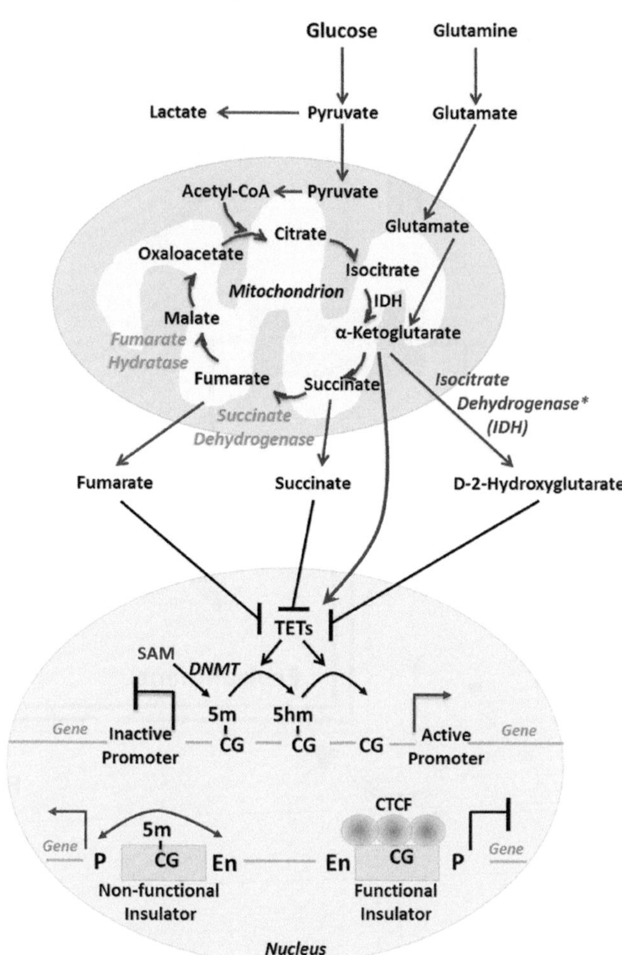

FIGURE 7.2.2 Mutations in genes encoding metabolic enzymes influence methylation of DNA and the formation of higher-order structures in chromatin (chromosome conformations). Metabolic intermediates resulting from the metabolism of glucose in the mitochondrion. Loss of function (green) and gain of function (blue*) mutations in genes encoding enzymes of the TCA cycle lead to the accumulation of fumarate, succinate or 2-hydroxyglutarate (2HG), respectively, that inhibit (block blocked lines) the activity of the α-ketoglutarate-dependent (red arrow) ten-eleven translocation (TET) family of DNA demethylases that function in the nucleus. Methylation of cytosine (5mC) by DNA methyltransferases (DNMTs) is generally considered to be a repressing modification (blocked line) although this is not uniformly observed. 5mC can be converted to hydroxymethyl cytosine (5hmC), generally considered to be an activating modification, as an intermediate in full demethylation of DNA by TETs. Blue arrows represent the initiation site of transcription. Increased genome-wide DNA methylation resulting from reduced TET activity also alters the higher-order structures in the chromatin (chromosome conformation) by reducing CTCF (pink spheres) binding to DNA. DNA-bound CTCF can create functional insulators that block or reduce gene activation (blocked black line) by preventing long-range interactions (blue double-headed arrow) between an enhancer (En) and a promoter (P). In the text, increased DNA methylation in *IDH1*-mutant gliomas leads to loss of insulator function that normally protects the *PDGFRA* oncogene from high-level activation by the *FIP1L1* enhancer (see Figure 7.2.5). In *IDH1*-mutant gliomas, flux through the TCA cycle is maintained by increased uptake of glutamine that, via glutamate, maintains levels of α-ketoglutarate for conversion to succinate. (Adapted from Fuhler, G.M. et al., *Trends Mol. Med.*, 23, 3–5, 2017.)

sensitive to the metabolic state of a cell and, indeed, mutations in the genes encoding the tricarboxylic acid (TCA)-cycle metabolic enzymes such as isocitrate dehydrogenase (IDH), fumarate hydratase (FH) or succinate dehydrogenase (SDH) that lead to accumulation of fumarate and 2-hydroxyglutarate are associated with a wide range of cancers (Fuhler et al., 2017) (Figure 7.2.2). Changes to DNA methylation patterns, particularly around promoters, are widely associated with aging and disease, and promoted as a reliable diagnostic tool. However, promoter hypermethylation is not correlated with changes in gene expression (Noushmehr et al., 2010), suggesting no causality in the relationship between DNA methylation and gene expression. Furthermore, both in normal and cancerous tissues, and during the evolution of a cancer phenotype, DNA methylation patterns change over time, indicating a stochastic component to the writing and erasing of DNA methylation that leads eventually to deterministic methylation profiles in the population of cells associated with disease (Landan et al., 2012). Thus, although DNA methylation can be an informative readout, the stochastic component to the changes in DNA methylation limits its application in personalized medicine, where accurate stratification of an individual's epitype is required.

POSTTRANSLATIONAL COVALENT MODIFICATIONS TO HISTONES

Histones are highly abundant positively charged proteins that serve to compact the negatively charged DNA so that it fits into the nucleus and can be segregated effectively to daughter cells at mitosis and meiosis. The discovery that the *N*- and *C*-terminal regions of histones extend beyond the compact nucleosomal core (Luger et al., 1997) and are the sites of extensive post-translation modifications (Bannister and Kouzarides, 2011) led to development of tools to study where and when modifications are found. Much work focused on gene expression, although we now have a greater understanding of how histone modifications are associated with processes such as DNA replication, repair and recombination and with *higher-order structures in chromatin*. For example, active or repressed genes are associated with distinctive types and distributions of histone modifications and recruit distinct reader proteins that contribute to each particular state. The marked consistency from the simplest yeast to humans in the type and distribution of histone modifications, and the massive dysregulation of histone modifications in cancer (Chi et al., 2010), has led to a widespread idea that histone modifications play an instructive role in the processes with which they are associated. More recently, this view has been challenged (Howe et al., 2017). Most data on histone modifications are obtained from cells in which the "writer" enzymes are ablated, resulting in the loss of a particular modification. This approach is problematic because these enzymes modify a wide range of proteins, not just histones, and the consequential effects of ablation on a molecular process such as gene expression could be indirect. Strong correlations should not lead to inferences of causality and new techniques such as (Clustered Regularly Interspaced Short Palindromic Repeats)-dCAS9 (dead CRISPR associated protein 9)-mediated targeting of specific histone modifications to particular regions of chromatin are needed to address this rigorously (Mendenhall et al., 2013; Stricker et al., 2017). Nevertheless, particular combinations of histone modifications are generally predictive for particular regions of the genome. For example, acetylation of lysine 27 on histone H3 (H3K27ac) and trimethylation at H3K4 (H3K4me3) and lysine 36 (H3K36me3) predict genes that have recently been transcribed. In unbiased data analysis, H3K27ac and H3K4me1 (with little or no H3K4me3) show a strong correlation with higher-order chromatin organization, although additional features are required for high-resolution analysis and data deconvolution (Dogan et al., 2015; Tordini et al., 2016; Whalen et al., 2016). Thus, although some single modifications predict features, such as H3K4me3 and the beginning of active genes, others such as H3K27ac can only discriminate regions likely to form higher-order structures from active gene promoters when the context of H3K4 methylation is included. Other modifications, such as H3K9me2/3 or H3K27me3, tend to be associated with repressed chromatin (at genes and chromosomal architectural features such as centromeres and telomeres) often by virtue of the nature of the reader proteins recruited to the modifications. As shown in Figure 7.2.3, all modifications to histone proteins, like DNA methylation, can be erased and are highly sensitive to the metabolic state of the cell because the cofactors for the writers and erasers are key central metabolites. Regulation of some classes of protein lysine demethylases, such as α-ketoglutarate-dependent lysine demethylases, is similar to the TET dioxygenases and highly sensitive to changes in metabolic state. This is not limited to demethylases because control of all classes of histone-modifying enzymes—including acetyltransferases, methyltransferases, other demethylases and deacetylases—are also influenced by levels of metabolic intermediates and the quality of an individual's diet. Over- or undernutrition, coupled to chaotic lifestyles and disrupted circadian rhythms, can disturb metabolism and influence DNA and histone modifications with consequential effects on an individual's epitypes and thus their disease susceptibility. Dietary fiber is essential for a healthy epigenome. These complex carbohydrates are fermented in the large intestine by the gut microbiome to produce short chain fatty acids such as butyrate, which in turn act to modify host chromatin by inhibiting histone deacetylation (Krautkramer et al., 2017). During aging, NAD^+ levels drop, reducing sirtuin-dependent lysine deacetylation, leading to epigenetic deregulation that may be associated with the increased likelihood of developing age-related conditions (Li et al., 2017). Given the interrelated nature of central metabolites shown in Figure 7.2.3, there are likely to be many confounding events in individuals, making it very hard to link a change at a single modification to a clinical readout.

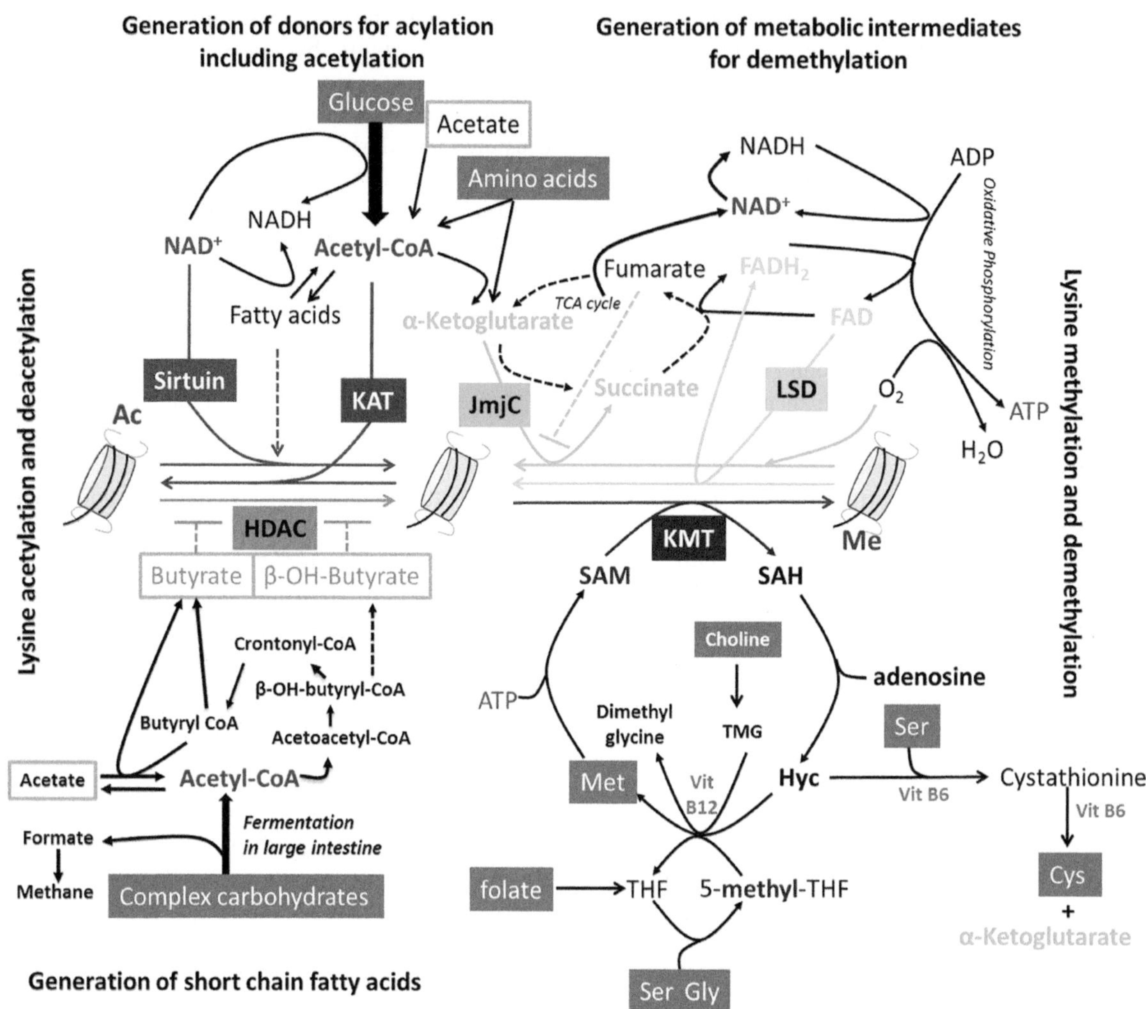

FIGURE 7.2.3 Integration of metabolic pathways that influence post-translational modifications to proteins, and modifications to DNA and RNA. **Left**: pathways controlling the acetylation and deacetylation of histone in nucleosomes (gray cylinders of histone proteins) wrapped in DNA (black line). Note acetylation is just one of a number of lysine acylations (crotonylation, butryrylation) that occur on histones and other proteins. Top: acetyl-CoA is the cofactor for lysine acetyltransferases, whereas NAD+ is the cofactor for the sirtuin class of lysine deacetylases. Bottom: Short chain fatty acids, such as butyrate, produced by the fermentation of fiber by microbes in the large intestine, inhibit lysine deacetylases. **Right**: pathways controlling the methylation and demethylation of lysines. Top: mitochondrial metabolites act as substrates or cofactors for two classes of lysine demethylases (JmjC family and Lsd family). Note regulation of the JmjC family is similar to the ten-eleven translocation (TET) dioxygenases that demethylate DNA. Bottom: pathways that generate methyl donors (blue) for methylation of DNA, proteins and RNA. Each pathway is color-coded. Key intermediates from the diet are shown in orange filled boxes, vitamins in orange text and short chain fatty acids from the gut in gray boxes. (From Fan, J. et al., *ACS Chem. Biol., 10*, 95–108, 2015; Kasubuchi, M. et al., *Nutrients*, 7, 2839–2849, 2015; Pryde, S.E. et al., *FEMS Microbiol. Lett.*, 217, 133–139, 2002.)

NONCODING RNA

Noncoding transcripts are key regulators of events on chromatin, including gene expression and genome stability. Epigenetic-related ncRNAs vary in size and function and include small (microRNA [miRNA], small interfering RNA [siRNA], piwi-interacting RNA [piRNA]) and long transcripts (long noncoding RNA [lncRNA]; >200 nts). One function for both the small and the long ncRNAs is to act as scaffolds to recruit chromatin-modifying complexes, to modify DNA and histones and to complement the role of transcription factors or epigenetic modifications and specific reader domains in recruiting histone-modifying enzymes to specific sites as illustrated in Figure 7.2.4. Because many of these events occur co-transcriptionally, using nascent RNA as the scaffold, directly or indirectly, helps provide a memory of these events via self-reinforcing epigenetic loops, which are often associated with maintaining silenced chromatin (Holoch and Moazed, 2015). However, lncRNAs are also associated with activation as well as repression of genes via their role in the formation of higher-order structures in chromatin (Lai et al., 2013) (Figure 7.2.4).

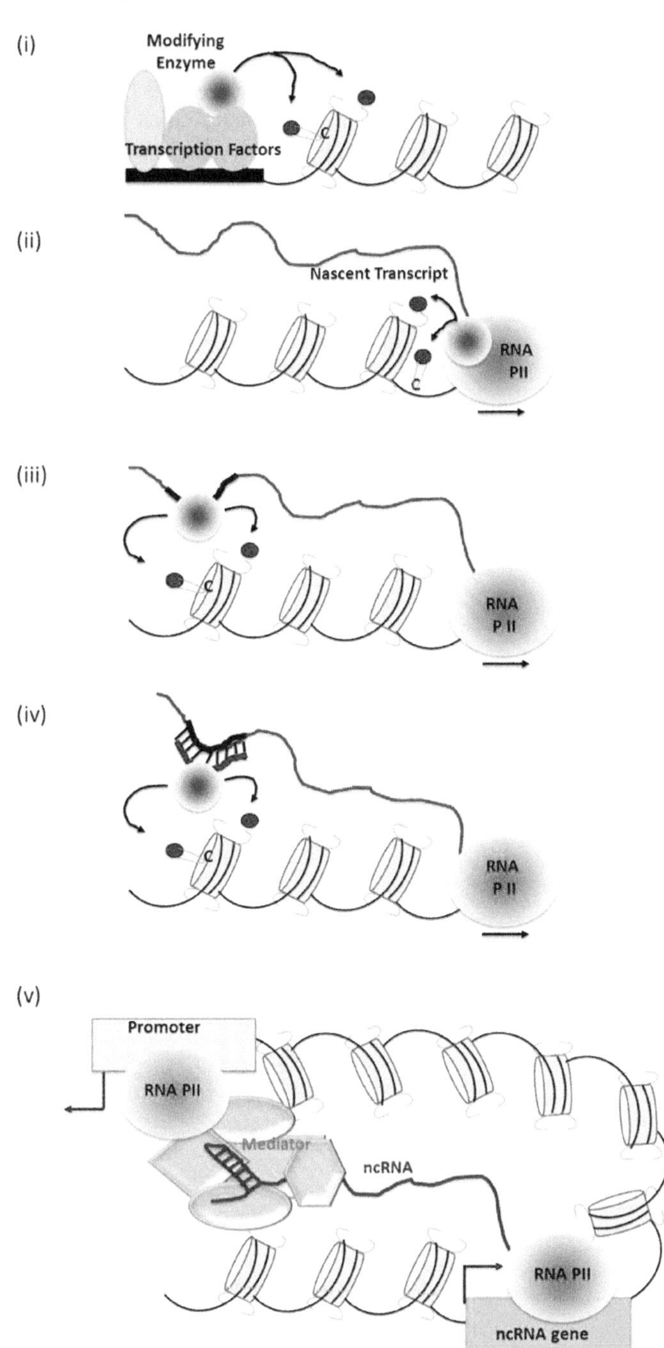

FIGURE 7.2.4 The role of non **coding RNAs in recruitment of epigenetic enzymes and in the formation of chromosome conformations related to gene expression**. Enzymes that modify histones or DNA are shown as red spheres and can be recruited by (*i*) transcription factors, (*ii*) to RNA polymerase II (RNA PII) (*iii*) directly to a nascent transcript (red line) (*iv*) or to nascent transcripts by small RNAs with homology to the RNA (blue line). (*v*) Nascent transcripts also help form higher-order structures in chromatin, shown here resulting in gene activation via recruitment of the mediator complex to a noncoding transcript, leading to activation at a distal promoter. (From Holoch, D. and Moazed, D., *Nat. Rev. Genet.*, 16, 71–84.; Lai, F. et al., *Nature*, 494, 497–501, 2013; Lai, F. and Shiekhattar, R., *Curr. Opin. Genet. Dev.*, 25, 38–42, 2014.)

This reinforces the idea that genome conformations integrate information provided by the local chromatin environment, including lncRNAs, DNA methylation and histone modifications (Tordini et al., 2016). Specificity in the local function of nascent lncRNAs may be provided by specific base sequences and by their methylation (Cao et al., 2016; Gilbert et al., 2016), modulating the type of RNA:protein interactions that can form and linking the functionality of lncRNAs to metabolic state and nutrition. Finally, circulating RNA molecules may allow the spreading of epitypes from one tissue to another, directing changes to the chromatin (Suraj et al., 2017).

HIGHER-ORDER STRUCTURES IN CHROMATIN

Because much current thinking in molecular biology is reductionist, we tend to view genomes not as a whole, but as a collection of individual loci on which molecular machines act to reproduce genetic information during cell division, DNA repair or gene expression. Although this thinking helps provide mechanistic insights into these machines, it is not very useful in trying to understand how epigenetics influences phenotype, how changes to the epigenome influence diseases or how signals are transmitted from generation to generation during cell division. Indeed, it is now recognized that the stable and inheritable epigenetic information, and the ability to correctly express this information, is *not* carried at the level of the individual gene, but requires the formation and regulation of higher-order chromatin structures (Dixon et al., 2012). High-order structures are proposed to be the prime coordinators of gene expression (Nora et al., 2012). The nucleus is a dense and complex nanoenvironment that partitions chemical reactions, and this partitioning is required to form chromatin and its structural hierarchies and for the functioning of the molecular machines that replicate the genetic information (Zhu and Brangwynne, 2015; Hnisz et al., 2017). Indeed a change in the fractal dimension of the chromatin, corresponding to changes in both chromatin accessibility and compaction heterogeneity, leads to enhancement or repression of transcription. This appears to be coordinated genome-wide because the changes are not observed just at single loci, but at multiple genes leading to alterations in whole pathways or networks (Almassalha et al., 2017).

It is also recognized that higher-order structures in chromatin form as a result of the integration of events such as DNA methylation, histone modification and ncRNA with the actions of architectural proteins, such as cohesin and CTCF (Tark-Dame et al., 2014; Merelli et al., 2015; Tordini et al., 2016). Folding appears to occur on at least two levels—long-range and shorter range—and the balance between the two states allows switching between open and more compact chromatin states (Dekker and Mirny, 2016). Interestingly, both long-range and short-range interactions show unbiased associations with acetylation at lysine 27 on histone H3, whereas at least some short-range interactions are associated with the architectural proteins cohesin and CTCF (Tark-Dame et al., 2014). Interactions of CTCF molecules with one another and with locus-specific enhancer and promoter binding transcription factors often mediate the interactions required for these higher-order structures (Dekker and Mirny, 2016; Narendra et al., 2016; Jerkovic et al., 2017). Dynamic physical contacts between promoters and enhancers are one of the elements required for cell type-specific transcription. However, a single promoter can form interactions with a number of different elements, dependent on tissue type or disease status, that may be located various distances both upstream and downstream from the promoter, as illustrated in Figure 7.2.5. For example, in normal cells the platelet-derived growth factor receptor A (*PDGFRA*) promoter forms at least four long-range interactions upstream from, and in addition to, those with the intragenic *PDGFRA* enhancer and upstream insulator, forming a potentially complex higher-order structure (Flavahan et al., 2016; Ing-Simmons and Merkenschlager, 2016). It is not known whether these interactions represent an ensemble view of a series of separate or temporally controlled events in a population of cells or whether all potential interactions occur in one cell at one moment in time. The first option lends itself to the idea of a dynamic chromatin environment, with some element of sampling of the local environment by the *PDGFRA* promoter, looking for information about how often to initiate transcription (Wang et al., 2011). In addition, the type of interactions that occur with the wild-type (WT) *PDGFRA* promoter raise interesting questions about the selective nature of the insulator because it appears to protect *PDGFRA* from ectopic activation by the *FIP1L1* enhancer but does not stop other interactions in the vicinity. In *IDH1* mutant cells, new interactions with the *FIP1L1* enhancer and the *PDGFRA* promoter are observed, coupled with much reduced interactions between the *PDGFRA* promoter and insulator, suggesting a large-scale reorganization of the chromosome conformation as a result of the increased DNA methylation (Flavahan et al., 2016). Thus, short-range interactions can be complex and are key to maintaining an open chromatin state because selective loss of factors such as CTCF leads to chromatin compaction (Tark-Dame et al., 2014; Narendra et al., 2015) and altered gene expression (Lupianez et al., 2015; Narendra et al., 2016). The highly dynamic nature of loop formation is poorly understood, but early insights support a loop-extrusion model (Dekker and Mirny, 2016). By this model, topological machines set up the three-dimensional (3D) genome organization allowing the promotion and inhibition of communications along and between chromosomes (Doyle et al., 2014).

To illustrate how assessment of higher-order structures in chromatin stratifies epitypes in personalized medicine, examples of patients presenting with gliomas or eosinophilic leukemias will be used. Both can be caused by overexpression of *PDGFRA*, encoding the tyrosine kinase platelet-derived growth factor receptor α (PDFGRA), but overexpression may result from distinct molecular mechanisms. The first mechanism has a metabolic epigenetic basis resulting from activating mutations in *IDH1* commonly associated with glioma (Flavahan et al., 2016). Glioma cells maintain normal levels of key TCA metabolites in the presence of *IDH1* mutations by increasing the relative flux of glutamine and glutamate into the

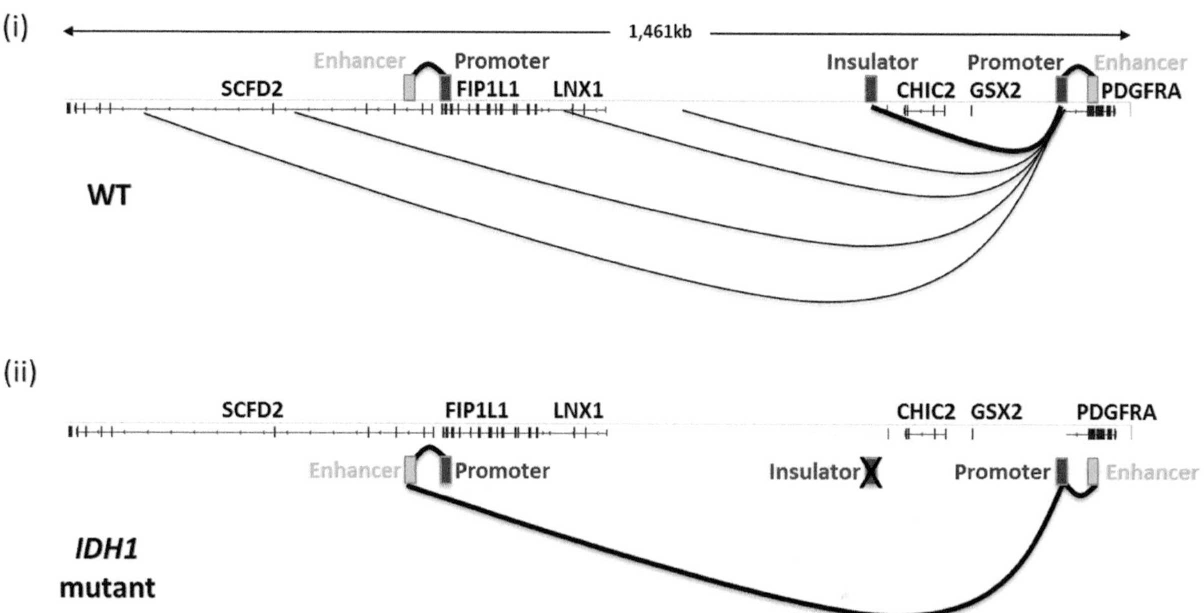

FIGURE 7.2.5 Chromosome conformations detectable in the vicinity of *PDGFRA*. A schematic of a 1.5-Mb region of chromosome 4 containing the genes indicated in blue (boxes exons, lines introns) upstream of *PDGFRA*. The promoter and enhancer for *FIP1L1* and *PDGFRA* are shown as blue- and green-filled boxes and the CTCF-dependent insulator that in wild-type (WT) cells protects ectopic activation of *PDGFRA* by the *FIP1L1* enhancer is red filled. Black lines represent known short-range chromosome conformations anchored to the *PDGFRA* promoter in (*i*) WT cells detected by HiC. (From Rao, S.S. et al., *Cell*, 159, 1665–1680, 2014) or (*ii*) *IDH1*-mutant gliomas in which insulator function is compromised (black X) detected by local 3C (chromosome conformation capture). (From Flavahan, W.A. et al., *Nature*, 529, 110–114, 2016). It is not known whether interactions in addition to those shown also exist.

TCA cycle (Waitkus et al., 2016) while also accumulating the oncometabolite 2-hydroxyglutarate (2HG) (Guo et al., 2011) (Figure 7.2.2). This leads to inhibition of the TET family of DNA demethylases, with genome-wide consequences resulting from global hypermethylation of the DNA (Xiao et al., 2012). Remarkably, and contrary to expectations, hypermethylation of promoter DNA is not generally associated with changes in gene expression (Noushmehr et al., 2010). Instead, activation of the *PDGFRA* oncogene is associated with changes in the higher-order structure of the chromatin, consistent with the underlying chromosome conformation being the primary determinant of gene expression. In these *IDH1* gain-of-function mutant glioma cells, increased DNA methylation at an insulator upstream of *PDGFRA* reduces CTCF binding, allowing a long-range interaction with the strong *FIP1L1* enhancer 900 kb upstream. This results in fivefold higher *PDGFRA* expression and increased cell growth compared with nonmutant cells (Flavahan et al., 2016). The high levels of 2HG in the *IDH1* mutant glioma cells result in the loss of about 600 CTCF-binding events and the gain of about 300 sites, and so perturbing the 3D organization of the genome. Patients with glioblastoma respond to Imatinib (Hassler et al., 2014), a costly, potent and specific inhibitor of protein-tyrosine kinases including PDGFRA. It is not known how many other tyrosine kinases are ectopically activated as a result of the *IDH1* gain-of-function mutations and increased global methylation. Nevertheless, the robust test for the novel chromosome conformations involving *FIP1L1* and *PDGFRA* identifies patients suitable for tyrosine kinase inhibitor (TKI) therapy. Currently, DNA sequencing is used to detect mutations in the patients' DNA, but this test suffers from a high level of false-positive or false-negative results. Detecting the novel higher order structure between the *FIP1L1* enhancer and the *PDGFRA* promoter, with the associated ectopic activation of *PDGFRA*, offers a cost-effective and potentially robust personalized test for glioma patients suitable for Imatinib treatment.

The second molecular mechanism is observed in some, but not all, eosinophilic leukemias where large deletions leading to fusions between *FIP1L1* and *PDGFRA* are associated with activation of the tyrosine kinase and sensitivity to Imatinib therapy (Cools et al., 2003; Griffin et al., 2003). Many patients are not given effective Imatinib therapy because the polymerase chain reaction (PCR)–based assays available (Score et al., 2009) are not robust enough to detect their deletions and thus to identify mis-regulated *PDGFRA* in their disease. Due to infiltration of the major organs by the eosinophilic leukemia cells, this misdiagnosis can be fatal with, for example, a high risk of cardiac arrest (reviewed by Valent et al., 2012). Accurate stratification of patients can be done most robustly by detecting the unique chromosome conformations at the *FIP1L1-PDGFRA* locus (Flavahan et al., 2016), which will be present in patients with local 4q12 deletions.

In summary, and as illustrated in Figure 7.2.1, chromatin conformations are the smallest compartmentalized unit in the 3D genome that accurately reflect phenotype (clinical readout) by integrating features such as gene expression, genetics, the metabolome and the epigenome (Merelli et al., 2015; Tordini et al., 2016; Whalen et al., 2016; Stricker et al., 2017). As an inevitable conclusion, it also means that any reductionist approach to trying to stratify phenotypes and clinical readouts on the basis of only one of this group of regulatory features, without taking into account the chromosome conformation, will not be successful.

INTEGRATED EPIGENETICS: ASSESSING CHROMOSOME CONFORMATION

The assessment of higher-order structures in chromatin is challenging because of the current status of the biochemical protocols used in the field. Without functional filters and a robust detection base, the most common detection methodologies capture significant numbers of short-lived, high off-rate interactions, generating stochastic artefacts, which affect the reproducibility and sensitivity of the assay. However, when these conditions are adequately controlled, the techniques for assessing chromosome conformations demonstrate the necessary phenotypic relevance and biochemical properties to support their use as unique biomarkers. Indeed, conditional chromosome conformations have been proposed as "ideal human biomarkers" (Crutchley et al., 2010). As a biomarker, chromosome conformation analysis outperforms other approaches for detecting genetic variants such as single nucleotide polymorphisms (SNPs) expression quantitative trait loci (eQTL), protein quantitative trait loci (pQTL), histone quantitative trait loci (hQTL), and the classical profiling of changes in transcript levels, proteins, DNA methylation or histone modifications to detect associations with clinical readouts (Dube et al., 2016; Wang and Dostie, 2016; Beagrie et al., 2017). It is becoming clear that development of robust and reproducible detection technologies for chromosome conformations is an essential step toward their routine adoption in diagnostic and prognostic medicine (Bastonini et al., 2014; Mukhopadhyay et al., 2014; Jakub et al., 2015; Flavahan et al., 2016; Beagrie et al., 2017). Indeed rapid and reliable assays for chromosome conformations are being developed commercially and adopted in clinical settings. When combined with standard regression analysis and machine learning algorithms, chromosome conformation analysis provides the "gold standard" for stratification of patients to meet the future challenges of personalized medicine.

REFERENCES

Almassalha, L.M., Tiwari, A., Ruhoff, P.T., Stypula-Cyrus, Y., Cherkezyan, L., Matsuda, H., Dela Cruz, M.A., Chandler, J.E., White, C., Maneval, C., *et al.* (2017). The global relationship between chromatin physical topology, fractal structure, and gene expression. *Sci Rep 7*, 41061.

Bannister, A.J., and Kouzarides, T. (2011). Regulation of chromatin by histone modifications. *Cell Res 21*, 381–395.

Bastonini, E., Jeznach, M., Field, M., Juszczyk, K., Corfield, E., Dezfouli, M., Ahmat, N., Smith, A., Womersley, H., Jordan, P., *et al.* (2014). Chromatin barcodes as biomarkers for melanoma. *Pigment Cell & Melanoma Res 27*, 788–800.

Beagrie, R.A., Scialdone, A., Schueler, M., Kraemer, D.C., Chotalia, M., Xie, S.Q., Barbieri, M., de Santiago, I., Lavitas, L.M., Branco, M.R., *et al.* (2017). Complex multi-enhancer contacts captured by genome architecture mapping. *Nature 543*, 519–524.

Bell, A.C., and Felsenfeld, G. (2000). Methylation of a CTCF-dependent boundary controls imprinted expression of the Igf2 gene. *Nature 405*, 482–485.

Cao, G., Li, H.B., Yin, Z., and Flavell, R.A. (2016). Recent advances in dynamic m6A RNA modification. *Open Biol 6*, 160003.

Chi, P., Allis, C.D., and Wang, G.G. (2010). Covalent histone modifications–miswritten, misinterpreted and mis-erased in human cancers. *Nat Rev Cancer 10*, 457–469.

Chowdhury, R., Yeoh, K.K., Tian, Y.M., Hillringhaus, L., Bagg, E.A., Rose, N.R., Leung, I.K., Li, X.S., Woon, E.C., Yang, M., *et al.* (2011). The oncometabolite 2-hydroxyglutarate inhibits histone lysine demethylases. *EMBO Rep 12*, 463–469.

Cimmino, L., and Aifantis, I. (2016). Alternative roles for oxidized mCs and TETs. *Curr Opin Genet Dev 42*, 1–7.

Cools, J., DeAngelo, D.J., Gotlib, J., Stover, E.H., Legare, R.D., Cortes, J., Kutok, J., Clark, J., Galinsky, I., Griffin, J.D., *et al.* (2003). A tyrosine kinase created by fusion of the PDGFRA and FIP1L1 genes as a therapeutic target of imatinib in idiopathic hypereosinophilic syndrome. *N Engl J Med 348*, 1201–1214.

Crutchley, J.L., Wang, X.Q., Ferraiuolo, M.A., and Dostie, J. (2010). Chromatin conformation signatures: Ideal human disease biomarkers? *Biomark Med 4*, 611–629.

Cubenas-Potts, C., and Corces, V.G. (2015). Architectural proteins, transcription, and the three-dimensional organization of the genome. *FEBS Lett 589*, 2923–2930.

Dekker, J., and Mirny, L. (2016). The 3D genome as moderator of chromosomal communication. *Cell 164*, 1110–1121.

Dixon, J.R., Selvaraj, S., Yue, F., Kim, A., Li, Y., Shen, Y., Hu, M., Liu, J.S., and Ren, B. (2012). Topological domains in mammalian genomes identified by analysis of chromatin interactions. *Nature 485*, 376–380.

Dogan, N., Wu, W., Morrissey, C.S., Chen, K.B., Stonestrom, A., Long, M., Keller, C.A., Cheng, Y., Jain, D., Visel, A., *et al.* (2015). Occupancy by key transcription factors is a more accurate predictor of enhancer activity than histone modifications or chromatin accessibility. *Epigenet Chromatin 8*, 16.

Doyle, B., Fudenberg, G., Imakaev, M., and Mirny, L.A. (2014). Chromatin loops as allosteric modulators of enhancer-promoter interactions. *PLoS Comput Biol 10*, e1003867.

Dube, J.C., Wang, X.Q., and Dostie, J. (2016). Spatial organization of epigenomes. *Curr Mol Biol Rep 2*, 1–9.

Fan, J., Krautkramer, K.A., Feldman, J.L., and Denu, J.M. (2015). Metabolic regulation of histone post-translational modifications. *ACS Chem Biol 10*, 95–108.

Flavahan, W.A., Drier, Y., Liau, B.B., Gillespie, S.M., Venteicher, A.S., Stemmer-Rachamimov, A.O., Suva, M.L., and Bernstein, B.E. (2016). Insulator dysfunction and oncogene activation in IDH mutant gliomas. *Nature 529*, 110–114.

Fuhler, G.M., Eppinga, H., and Peppelenbosch, M.P. (2017). Fumarates and cancer. *Trends Mol Med 23*, 3–5.

Gilbert, W.V., Bell, T.A., and Schaening, C. (2016). Messenger RNA modifications: Form, distribution, and function. *Science 352*, 1408–1412.

Griffin, J.H., Leung, J., Bruner, R.J., Caligiuri, M.A., and Briesewitz, R. (2003). Discovery of a fusion kinase in EOL-1 cells and idiopathic hypereosinophilic syndrome. *Proc Natl Acad Sci USA 100*, 7830–7835.

Guo, C., Pirozzi, C.J., Lopez, G.Y., and Yan, H. (2011). Isocitrate dehydrogenase mutations in gliomas: Mechanisms, biomarkers and therapeutic target. *Curr OpinNeurol 24*, 648–652.

Hassler, M.R., Vedadinejad, M., Flechl, B., Haberler, C., Preusser, M., Hainfellner, J.A., Wohrer, A., Dieckmann, K.U., Rossler, K., Kast, R., *et al.* (2014). Response to imatinib as a function of target kinase expression in recurrent glioblastoma. *SpringerPlus 3*, 111.

Hnisz, D., Shrinivas, K., Young, R.A., Chakraborty, A.K., and Sharp, P.A. (2017). A phase separation model for transcriptional control. *Cell 169*, 13–23.

Holoch, D., and Moazed, D. (2015). RNA-mediated epigenetic regulation of gene expression. *Nat Rev Genet 16*, 71–84.

Howe, F.S., Fischl, H., Murray, S.C., and Mellor, J. (2017). Is H3K4me3 instructive for transcription activation? *Bioessays 39*, 1–12.

Ing-Simmons, E., and Merkenschlager, M. (2016). Oncometabolite tinkers with genome folding, boosting oncogene expression. *Trends Mole Med 22*, 185–187.

Jakub, J.W., Grotz, T.E., Jordan, P., Hunter, E., Pittelkow, M., Ramadass, A., Akoulitchev, A., and Markovic, S. (2015). A pilot study of chromosomal aberrations and epigenetic changes in peripheral blood samples to identify patients with melanoma. *Melanoma Res 25*, 406–411.

Jerkovic, I., Ibrahim, D.M., Andrey, G., Haas, S., Hansen, P., Janetzki, C., Gonzalez Navarrete, I., Robinson, P.N., Hecht, J. et al. (2017). Genome-wide binding of posterior HOXA/D transcription factors reveals subgrouping and association with CTCF. *PLoS Genet 13*, e1006567.

Kasubuchi, M., Hasegawa, S., Hiramatsu, T., Ichimura, A., and Kimura, I. (2015). Dietary gut microbial metabolites, short-chain fatty acids, and host metabolic regulation. *Nutrients 7*, 2839–2849.

Kim, S., Yu, N.K., and Kaang, B.K. (2015). CTCF as a multifunctional protein in genome regulation and gene expression. *Exp Mol Med 47*, e166.

Krautkramer, K.A., Rey, F.E., and Denu, J.D. (2017). Chemical signaling between gut microbiota and host chromatin: What is your gut really saying? *J Biol Chem 292* (21), 8582–8593.

Lai, F., and Shiekhattar, R. (2014). Enhancer RNAs: The new molecules of transcription. *Curr Opin Genet Dev 25*, 38–42.

Lai, F., Orom, U.A., Cesaroni, M., Beringer, M., Taatjes, D.J., Blobel, G.A., and Shiekhattar, R. (2013). Activating RNAs associate with Mediator to enhance chromatin architecture and transcription. *Nature 494*, 497–501.

Landan, G., Cohen, N.M., Mukamel, Z., Bar, A., Molchadsky, A., Brosh, R., Horn-Saban, S., Zalcenstein, D.A., Goldfinger, N., Zundelevich, A., *et al.* (2012). Epigenetic polymorphism and the stochastic formation of differentially methylated regions in normal and cancerous tissues. *Nat Genet 44*, 1207–1214.

Laukka, T., Mariani, C.J., Ihantola, T., Cao, J.Z., Hokkanen, J., Kaelin, W.G., Jr., Godley, L.A., and Koivunen, P. (2016). Fumarate and succinate regulate expression of hypoxia-inducible genes via TET enzymes. *J Biol Chem 291*, 4256–4265.

Li, J., Bonkowski, M.S., Moniot, S., Zhang, D., Hubbard, B.P., Ling, A.J., Rajman, L.A., Qin, B., Lou, Z., Gorbunova, V., *et al.* (2017). A conserved NAD+ binding pocket that regulates protein-protein interactions during aging. *Science 355*, 1312–1317.

Luger, K., Mader, A.W., Richmond, R.K., Sargent, D.F., and Richmond, T.J. (1997). Crystal structure of the nucleosome core particle at 2.8 A resolution. *Nature 389*, 251–260.

Lupianez, D.G., Kraft, K., Heinrich, V., Krawitz, P., Brancati, F., Klopocki, E., Horn, D., Kayserili, H., Opitz, J.M., Laxova, R., *et al.* (2015). Disruptions of topological chromatin domains cause pathogenic rewiring of gene-enhancer interactions. *Cell 161*, 1012–1025.

Maurano, M.T., Wang, H., John, S., Shafer, A., Canfield, T., Lee, K., and Stamatoyannopoulos, J.A. (2015). Role of DNA methylation in modulating transcription factor occupancy. *Cell Rep 12*, 1184–1195.

Mendenhall, E.M., Williamson, K.E., Reyon, D., Zou, J.Y., Ram, O., Joung, J.K., and Bernstein, B.E. (2013). Locus-specific editing of histone modifications at endogenous enhancers. *Nat Biotechnol 31*, 1133–1136.

Merelli, I., Tordini, F., Drocco, M., Aldinucci, M., Lio, P., and Milanesi, L. (2015). Integrating multi-omic features exploiting chromosome conformation capture data. *Front Genet 6*, 40.

Mukhopadhyay, S., Ramadass, A.S., Akoulitchev, A., and Gordon, S. (2014). Formation of distinct chromatin conformation signatures epigenetically regulate macrophage activation. *Int Immunopharmacol 18*, 7–11.

Narendra, V., Bulajic, M., Dekker, J., Mazzoni, E.O., and Reinberg, D. (2016). CTCF-mediated topological boundaries during development foster appropriate gene regulation. *Genes Dev 30*, 2657–2662.

Narendra, V., Rocha, P.P., An, D., Raviram, R., Skok, J.A., Mazzoni, E.O., and Reinberg, D. (2015). CTCF establishes discrete functional chromatin domains at the Hox clusters during differentiation. *Science 347*, 1017–1021.

Nora, E.P., Lajoie, B.R., Schulz, E.G., Giorgetti, L., Okamoto, I., Servant, N., Piolot, T., van Berkum, N.L., Meisig, J., Sedat, J., *et al.* (2012). Spatial partitioning of the regulatory landscape of the X-inactivation centre. *Nature 485*, 381–385.

Noushmehr, H., Weisenberger, D.J., Diefes, K., Phillips, H.S., Pujara, K., Berman, B.P., Pan, F., Pelloski, C.E., Sulman, E.P., Bhat, K.P., *et al.* (2010). Identification of a CpG island methylator phenotype that defines a distinct subgroup of glioma. *Cancer Cell 17*, 510–522.

Pryde, S.E., Duncan, S.H., Hold, G.L., Stewart, C.S., and Flint, H.J. (2002). The microbiology of butyrate formation in the human colon. *FEMS Microbiol Lett 217*, 133–139.

Rao, S.S., Huntley, M.H., Durand, N.C., Stamenova, E.K., Bochkov, I.D., Robinson, J.T., Sanborn, A.L., Machol, I., Omer, A.D., Lander, E.S., *et al.* (2014). A 3D map of the human genome at kilobase resolution reveals principles of chromatin looping. *Cell 159*, 1665–1680.

Sciacovelli, M., Goncalves, E., Johnson, T.I., Zecchini, V.R., da Costa, A.S., Gaude, E., Drubbel, A.V., Theobald, S.J., Abbo, S.R., Tran, M.G., *et al.* (2016). Fumarate is an epigenetic modifier that elicits epithelial-to-mesenchymal transition. *Nature 537*, 544–547.

Score, J., Walz, C., Jovanovic, J.V., Jones, A.V., Waghorn, K., Hidalgo-Curtis, C., Lin, F., Grimwade, D., Grand, F., Reiter, A., *et al.* (2009). Detection and molecular monitoring of FIP1L1-PDGFRA-positive disease by analysis of patient-specific genomic DNA fusion junctions. *Leukemia 23*, 332–339.

Stricker, S.H., Koferle, A., and Beck, S. (2017). From profiles to function in epigenomics. *Nat Rev Genet 18*, 51–66.

Suraj, S., Dhar, C., and Srivastava, S. (2017). Circulating nucleic acids: An analysis of their occurrence in malignancies. *Biomed Rep 6*, 8–14.

Tark-Dame, M., Jerabek, H., Manders, E.M., van der Wateren, I.M., Heermann, D.W., and van Driel, R. (2014). Depletion of the chromatin looping proteins CTCF and cohesin causes chromatin compaction: Insight into chromatin folding by polymer modelling. *PLoS Comput Biol 10*, e1003877.

Tordini, F., Aldinucci, M., Milanesi, L., Lio, P., and Merelli, I. (2016). The genome conformation as an integrator of multi-omic data: The example of damage spreading in cancer. *Front Genet 7*, 194.

Valent, P., Gleich, G.J., Reiter, A., Roufosse, F., Weller, P.F., Hellmann, A., Metzgeroth, G., Leiferman, K.M., Arock, M., Sotlar, K., *et al.* (2012). Pathogenesis and classification of eosinophil disorders: A review of recent developments in the field. *Expert Rev Hematol 5*, 157–176.

Waitkus, M.S., Diplas, B.H., and Yan, H. (2016). Isocitrate dehydrogenase mutations in gliomas. *Neuro-oncology 18*, 16–26.

Wang, X.Q., and Dostie, J. (2016). Chromosome folding and its regulation in health and disease. *Curr Opin Genet Dev 43*, 23–30.

Wang, X.Q., Crutchley, J.L., and Dostie, J. (2011). Shaping the genome with non-coding RNAs. *Curr Genomics 12*, 307–321.

Whalen, S., Truty, R.M., and Pollard, K.S. (2016). Enhancer-promoter interactions are encoded by complex genomic signatures on looping chromatin. *Nat Genet 48*, 488–496.

Xiao, M., Yang, H., Xu, W., Ma, S., Lin, H., Zhu, H., Liu, L., Liu, Y., Yang, C., Xu, Y., *et al.* (2012). Inhibition of alpha-KG-dependent histone and DNA demethylases by fumarate and succinate that are accumulated in mutations of FH and SDH tumor suppressors. *Genes Dev 26*, 1326–1338.

Zhu, L., and Brangwynne, C.P. (2015). Nuclear bodies: The emerging biophysics of nucleoplasmic phases. *Curr Opin Cell Biol 34*, 23–30.

7.3

Incorporating Predictive Imaging Biomarkers in Clinical Trials for Personalised Healthcare

John C. Waterton

Contents

IMAGING BIOMARKERS

According to the 2016 Biomarkers, EndpointS, and other Tools (BEST) Glossary from the U.S. Food and Drug Administration–National Institutes of Health (FDA-NIH) Biomarker Working Group,[1] a biomarker is "a defined characteristic that is measured as an indicator of normal biological processes, pathogenic processes, or responses to an exposure or intervention, including therapeutic interventions...molecular, histologic, radiographic, or physiologic characteristics are types of biomarkers." Other chapters in this volume concern molecular and histologic (i.e., biospecimen) biomarkers, which require recovery of a specimen, such as blood, biopsy, urine, exhalate, or cerebrospinal fluid (CSF), from the patient, followed by remote analysis by trained personnel, using a dedicated *in vitro* diagnostic device approved for that specific analysis. Imaging (radiographic) biomarkers

(IBs), however, collect only a signal, that is, they are biosignal biomarkers. In stark contrast to biospecimen biomarkers, the quality and usability of biosignal biomarker data depends overwhelmingly on operations and events at the very moment the patient is coupled to the scanner, which is a device not usually designed, approved, marketed, or operated for quantitative biomarker work. Because of these differences, the validation of IBs requires a very different approach and roadmap[2] from the validation of biospecimen biomarkers: similarly, their risk-management in practice[3] involves very different considerations.

Imaging is sometimes referred to as a new or emerging biomarker modality, but the opposite is the case. The origin of modern biomarker thinking, the 1999 FDA/NIH workshop,[4] cites as examples both a predictive IB (rapid clearance of ^{99m}Tc-sestamibi. predictive of lack-of-benefit from adjuvant chemotherapy[5] in some forms of breast cancer) and a prognostic IB (tumor shrinkage: objective response). Indeed, the latter biomarker, tumor shrinkage, was already established by the 1950s,[6] when molecular biology was in its infancy. Its current manifestation (RECIST1.1[7]) is used daily in clinical trials and clinical practice, and is used frequently by regulators as a surrogate endpoint in new drug approvals.[8] Indeed, IBs were used by FDA as surrogate endpoints for 30 separate drug approvals between 2010 and 2014.[8]

IMAGING MODALITIES

Medical imaging depends on distinguishing the biosignals emanating from different locations in the body. Any medical imaging modality must meet three requirements:

- Some physics (such as magnetic resonance or Compton scattering) that couples a detector, usually located outside the body, to molecules or structures inside the body
- A source of contrast to distinguish between these molecules or structures
- Penetration of the biosignal to and from the depth at which the molecules or structures reside in the body

Wilhelm Röntgen established this medical imaging paradigm in 1895 with X-radiography, leading swiftly to the Nobel Prize in 1901. Since then, a dozen or more medical imaging modalities have been invented (Figure 7.3.1), of which six are generally of interest to the drug developer (Box 7.3.1).

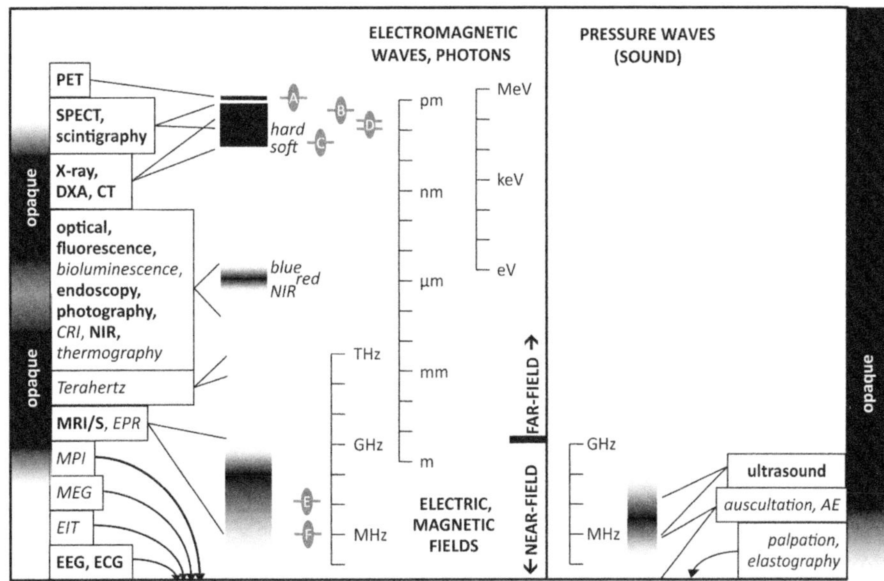

FIGURE 7.3.1 Imaging and related biosignal modalities located in the sound or pressure wave spectrum. Windows where tissue is not opaque are identified. Imaging modalities of most interest to the drug developer are labeled in **bold**. AE, acoustic emission; CRI, Cerenkovradiation imaging; CT, X-ray computed tomography; DXA, dual-energy X-ray absorptiometry; ECG, electrocardiography; EEG, electroencephaloeraphy; EIT, electrical impedance tomography; EPR, electron paramagnetic resonance; MEG, magnetoencephalography; MPI, magnetic particle imaging; MRI/S, magnetic resonance imaging/in vivo magnetic resonance spectroscopy; NIR, near infrared imaging; PET, positron emission tomography; SPECT, single photon emission computed tomography. *A–F*, see Box 7.3.1.

BOX 7.3.1 MEDICAL IMAGING MODALITIES OF INTEREST TO THE DRUG DEVELOPER

Positron-Emission Tomography. Tracer substances incorporating positron-emitting isotopes are administered. Annihilation of the emitted positron with a tissue electron produces a pair of 511 keV photons (Figure 7.3.1A) that are detected. Common radiotracers for positron-emission tomography (PET) incorporate ^{18}F, although there are clinical applications for some other positron-emitting isotopes including ^{82}Rb, ^{13}N and ^{11}C, and investigational uses for ^{15}O, ^{64}Cu, ^{68}Ga, ^{124}I, and ^{89}Zr. Although many PET tracers have been used in a research setting, fewer than 10 have FDA new drug application (NDA) approval.*

Single-Photon Emission Computed Tomography (incorporating Scintigraphy, its planar equivalent). Gamma-emitting radiotracers are employed. Gamma photons are detected following collimation. Common radiotracers for single-photon emission computed tomography (SPECT) incorporate ^{99m}Tc whose decay produces a 140-keV photon (Figure 7.3.1B), although there are clinical applications for some other gamma-emitting isotopes including ^{67}Ga, ^{111}In, ^{123}I, ^{131}I, ^{201}Tl and ^{133}Xe. Most SPECT radiotracers emit gamma photons with energies between 80 keV and 200 keV, although ^{131}I produces higher energy photons at 364 keV. Approximately 30 SPECT tracers have FDA NDA approval.

X-ray Computed Tomography, or X-radiography, its planar equivalent. "Soft" (low energy, down to 15 keV) X-rays used, for example, in mammography tend more to photoelectric absorption (Figure 7.3.1C), which depends on the cube of the atomic number. "Hard" (high energy, up to 150 keV) X-rays used in whole-body imaging tend more to Compton scattering. dual-energy X-ray absorptiometry (DXA) uses two energies, typically 40 and 70 keV (Figure 7.3.1D), for tissue characterization. Some applications require intravenous contrast agents containing high atomic-number elements, usually iodine (atomic number 53). Approximately 10 such X-ray CT (computed tomography) contrast agents have FDA NDA approval.

Magnetic Resonance Imaging. Nuclear magnetic resonance is induced, and the resulting free induction decay is detected using a radiofrequency receiver. Most applications of magnetic resonance in humans image the ^{1}H isotope in water molecules, mostly using 1.5 T (63 MHz) or 3 T (126 MHz; Figure 7.3.1E) magnets, although systems as high as 7 T (300 MHz) and as low as 0.25 T (10 MHz; Figure 7.3.1F) are currently marketed. Some applications require intravenous contrast agents designed to accelerate return (relaxation) of the enthalpy and entropy of the nuclear spins back to equilibrium: these usually incorporate gadolinium or iron. Approximately 10 such MRI contrast agents have FDA NDA approval. There are some, mainly investigational, applications of MR to detect other substances, using isotopes such as ^{19}F, ^{3}He, ^{129}Xe, ^{23}Na, ^{31}P, ^{13}C, or ^{7}Li, typically using frequencies in the 10–100 MHz range. The related technique of electron paramagnetic resonance (EPR) detects unpaired electrons, employing frequencies typically between 200 MHz and 2.2 GHz.

Optical and near-infrared. The absorption, scattering or fluorescence of colored molecules or structures is detected, particularly in the red or near infrared part of the spectrum, where background absorption is lower. Some applications require colored intravenous contrast agents: two** such agents have FDA NDA approval.

Ultrasound. Scattering of ultrasound by structural boundaries is detected. Some applications require intravenous microbubble contrast agents: two† such agents have FDA NDA approval.

* These are fludeoxyglucose F-18; fluoride F-18; the amyloid tracers florbetapir F-18, florbetaben F-18 and flutemetamol F-18; fluciclovine F-18; choline C-11; ammonia N-13; and rubidium Rb-82.

** Fluorescein and indocyanine green.

† "Definity" perflutren lipid microspheres and "Optison" perflutren protein-type A microspheres.

EXAMPLES OF IMAGING BIOMARKERS

BEST recognizes several categories of biomarker, including susceptibility/risk biomarker, diagnostic biomarker, monitoring biomarker, prognostic biomarker, predictive biomarker, pharmacodynamic/response biomarker, and safety biomarker. All these roles can be filled by IBs, and any of the six modalities in Figure 7.3.1 can be used, as illustrated below.

Some IBs are already very familiar to the drug developer, having been widely used for many years. Some illustrative examples are

- Bone mineral density (BMD) measured by DXA in osteoporosis[9] and used as a risk, diagnostic, monitoring, prognostic, predictive, response or safety IB, also used in regulatory drug approvals
- Left ventricular ejection fraction (LVEF) measured by ultrasound[10] (or by SPECT or MRI) in cardiology and used as a diagnostic, monitoring, prognostic, predictive, pharmacodynamic, response or safety IB, also in regulatory drug approvals

- Objective tumor response,[7] TNM (tumor–node–metastasis (TNM) stage,[11] time to progression, and progression-free survival in solid tumors: these biomarkers rely on imaging measurements, may use CT, MRI, ultrasound, SPECT and/or PET, and are used as diagnostic, monitoring, prognostic, predictive, or response biomarkers, also in regulatory drug approvals
- Carotid intima-media thickness[12] measured by ultrasound in atherosclerosis and used as a risk, prognostic, or response IB
- Sharp van der Heijde erosion score[13] measured by X-radiography in rheumatoid arthritis and used as a response IB

Other IBs may be less familiar because they are newly emerging, or used only in limited settings, for example:

- Change in [^{18}F] fludeoxyglucose (FDG) standardized uptake value (SUV) in PET used[14,15,16] as a pharmacodynamic IB to measure target and pathway inhibition in the PI3K-AKT-mToR* pathway
- Macular hole diameter measured by optical coherence tomography[17] used as a response IB and surrogate endpoint in regulatory drug approval
- Change in the transfer constant K^{trans} in dynamic contrast-enhanced MRI used[18,19] as a pharmacodynamic IB of drug efficacy in synovitis in addition to the more familiar use in oncology[20]
- Low-attenuating area in lung CT used[21] as a prognostic IB in emphysema
- Inhibition of liver organic anion transporters measured using dynamic gadoxetate-enhanced MRI as an IB for risk of drug-induced liver injury[22,23]
- The proprietary FerriScan MRI transverse relaxation rate (R_2) IB used by FDA as a companion diagnostic (CDx)[24] in the approval of Exjade (deferasirox)
- Dopamine receptor occupancy using [^{11}C]-raclopride PET used[25] as a pharmacodynamic IB of antipsychotic drug efficacy

A distinction is sometimes drawn between "structural" and "molecular" imaging modalities. As illustrated above, this distinction is inaccurate and unhelpful given that all imaging modalities can yield both between "structural" and "molecular" insights: the important questions for the drug developer are the inferences made from biomarker data, and the level of confidence in those inferences.

PREDICTIVE IMAGING BIOMARKERS IN THE DRUG DEVELOPMENT PATHWAY

As exemplified above, IBs have been widely and successfully used in drug development for many years, and numerous standards and guidelines have been established. Recently there has been a surge in interest in stratified medicine and personalised health care (PHC), leading to much-increased interest by drug developers in predictive biomarkers. These pose unusual and unfamiliar challenges to the drug developer.

WHAT IS A PREDICTIVE BIOMARKER?

Consider[26] observations made on a patient in, for example, a clinical trial, at times $t = b$ (baseline) and $t = f$ (follow-up).

> C_t is some clinical measurement of a patient's quality of life: how he or she feels or functions, or perhaps has survived, made at time t.
> B_b is a biomarker measurement made at baseline.
> M is a specified medical intervention (e.g., drug therapy, surgery, radiotherapy) initiated at $t = b$.
> $F|(p)$ is a forecast of C_f, made at $t = b$ using prior information p.
> E is the absolute forecast error: $E|(p) = \text{modulus}(C_f - F|(p))$.

For a prognostic biomarker:

$$E|(C_b, B_b) < E|(C_b) \tag{7.3.1}$$

in other words, by measuring the biomarker we can better forecast the patient's ultimate clinical outcome than had we not measured the biomarker. The same is generally true for susceptibility/risk and diagnostic biomarkers because the concepts of susceptibility, risk and diagnosis almost always incorporate a prognostic implication. A purely prognostic biomarker does not help forecast whether one treatment will provide a better clinical outcome than an alternative treatment.

For a predictive biomarker,

$$E|(C_b, B_b, M) < E|(C_b, M) \tag{7.3.2}$$

and

$$E|(C_b, B_b, M) < E|(C_b, B_b) \tag{7.3.3}$$

in other words, by measuring the biomarker can we better forecast† whether a particular patient will respond to the specific treatment M. The forecast may be of benefit (improved outcome) or of harm (worse outcome). A purely predictive biomarker does not improve our forecast of the patient's ultimate clinical outcome without specifying treatment.

* phosphoinositide 3-kinase / protein kinase B / mammalian target of rapamycin.

† Note that both Eqs [7.3.2] and [7.3.3] must be satisfied. In the case where B is purely prognostic, then [7.3.2] but not [7.3.3] would be satisfied; in the case where M is effective and B is uninformative, then [7.3.3] but not [7.3.2] would be satisfied. A biomarker may be simultaneously prognostic and predictive if [7.3.1] and [7.3.3] are satisfied.

A regulatory authority who holds evidence that a particular biomarker is predictive for a particular therapy may require the biomarker to appear in the drug labeling either under "indications" (if the biomarker forecasts benefit from treatment) or under "contra-indications" (if the biomarker forecasts harm from treatment). Such use of the biomarker in patient management requires a cut point: if the biomarker is numerical, then the cut point will be a number, whereas if the biomarker is an ordered categorical, then the cut point will be the boundary between two categories or scores.

Predictive biomarkers are often nonproprietary, existing before the invention of the treatment for which they are predictive, for example, from the list above, BMD, LVEF or TNM. However, in some cases, a drug developer may invent the biomarker (or adapt an existing biomarker) and codevelop it with the investigational drug. This combination of a proprietary biomarker with a proprietary investigational therapeutic is often called a CDx as, for example, with the FerriScan/Exjade pairing.

EXAMPLES OF PREDICTIVE IMAGING BIOMARKERS

Many studies have reported that a certain IB predicts a favorable response to a particular treatment.[27] Unfortunately, investigators often fail to demonstrate that the results are specific to that treatment, that is, Eqs [7.3.1] and [7.3.2] are satisfied but not [7.3.3] (the biomarker is prognostic but not necessarily predictive). Tables 7.3.1 and 7.3.2 collect illustrative examples of IBs that appear to be truly predictive in human studies. Table 7.3.1 includes examples where the regulatory labeling implies that the biomarker is predictive. These can be broadly divided into predictive IBs of benefit (e.g., FerriScan/Exjade, Quadramet/bone SPECT, bicalutamide/TNM, Fosamax/BMD), predictive IBs of benefit or harm associated with biodistribution (e.g., Zevalin, Bexxar), and predictive IBs of harm (e.g., Activase/hemorrhage, Tykerb/LVEF). Beyond Table 7.3.1, BMD can also be a predictive IB of harm. Many drugs, including PPARγ inhibitors,[28] aromatase inhibitors[29] and many corticosteroids[30] are associated with osteoporosis and fracture risk: BMD is a predictive IB for harm (fractures) from the drug. However, although regulatory labeling and clinical guidelines may recommend monitoring of BMD before and during treatment, BMD below cutoff is rarely an absolute contraindication. In the case of anastrozole, phase III trial data[31] showed that BMD predicted treatment-associated harm, and anti-osteoporotic therapy[32] can be introduced in patients with low or declining BMD (i.e., BMD is a predictive IB for lack-of-harm of the combination therapy).

Table 7.3.2 shows some investigational predictive IBs. Predictive biomarkers have transformed cancer treatment: of 41 cleared or approved CDx devices listed[33] by FDA, 38 are for oncology products, using modalities such as *in situ* hybridization, polymerase chain reaction, or immunohistochemistry, typically to detect tumor mutations. These of course are not imaging biomarkers. However, as indicated in Table 7.3.2, there are many investigational imaging CDxs in cancer, of which the vintafolide/etarfolatide folate combination has advanced the most. In addition, large-molecule anticancer therapeutics such as monoclonal antibodies or nanomedicine formulations may need CDxs if they exhibit unpredictable pharmacokinetics in human tumors. These agents may exhibit enhanced permeability and retention, or they may fail to access tumor cells at all. Tumor vasculature, endothelial permeability and lymphatic drainage are heterogeneous within and between tumors,[34] so IBs of heterogeneous pharmacokinetics, or heterogeneous microenvironment[35] may have potential[36] as predictive biomarkers. Outside oncology, the important approval by FDA of three PET amyloid tracers has allowed clinical trials of patients with mild cognitive impairment to be selected for amyloid pathology, as exemplified by solanezumab.

IMPLICATIONS FOR THE DRUG DEVELOPER

If, at the outset of drug development, it is suspected that a particular IB may be predictive for the investigational drug, the drug developer faces both opportunities and risks. The opportunities are common to all predictive biomarkers, not just predictive imaging biomarkers. Incorporation of the predictive biomarker in the label will help assure the regulatory authority that patients who will benefit can be distinguished from those who do not benefit, or who may be harmed. For the drug developer, this may mean a better chance of regulatory approval, faster approval, or approval with a smaller dossier. It may also strengthen the health economic case because payers will pay for treatment of those who will benefit, but avoid the cost of treating patients who do not benefit or are harmed. In addition, the larger effect size will make trials in a selected population smaller and faster.

Some of the risks apply generally to all predictive biomarkers, whereas some are particularly germane to predictive imaging biomarkers. Firstly, if the indications or contraindications depend on the predictive IB, then that IB must be measured before the drug can be prescribed. If the imaging test is only available in specialized academic centers, it is unlikely to be suitable for a primary care drug, or for a drug to be widely used in less-developed countries (unlike many biospecimen biomarkers, where a sample can be taken locally and shipped for central analysis). If the imaging test is expensive, then the health economic case is weakened. Although it is easy to mandate that a biospecimen biomarker be analyzed using equipment designed and operated for that specific quantitative assessment, an imaging biomarker usually relies on the scanners that happen to be available in the local hospital, so the imaging biomarker must be proactively standardized and reproducible across many makes and models of scanners, already installed and yet-to-be introduced. Finally, the IB must be standardized and widely available even before the start of clinical trials, in order that phase IIb and phase III can be designed to support the intended personalised health care-based regulatory submission.

TABLE 7.3.1 Examples of predictive imaging biomarkers associated with regulatory approvals

DRUG APPROVAL	*IMAGING BIOMARKER*	*WHAT THE IMAGING BIOMARKER PREDICTS*	*RELEVANT INDICATION/ CONTRAINDICATION TEXT*	*COMMENTS*
Bicalutamide 150 mg, UK PL 17901/0006, revision of December 3, 2015	TNM stage (typically determined using MRI, CT and/or SPECT)	Lack-of-benefit if biomarker is below lower cutoff (i.e., T1-T2, N0, M0); and Lack-of-benefit if biomarker is above upper cutoff (i.e., any T, any N, M1).	Management of patients with locally advanced, nonmetastatic (either T3-T4, any N, M0; or T1-T2, N+, M0) prostate cancer for whom surgical castration or other medical intervention is not appropriate or acceptable.	Studies failed to show benefit either in patients with localized disease (T1-T2, N0/Nx, M0) or in metastatic patients (any T, any N, M1).[55]
Quadramet (Samarium [153Sm]-lexidronam pentasodium), marketing authorization EU/1/97/057/001, renewal of December 12, 2007	Number of skeletal metastases that take up [99mTc]-labeled bisphosphonates (SPECT)	Lack-of-benefit if biomarker below cutoff (i.e., without multiple painful skeletal metastases with [99mTc]-labeled bisphosphonate uptake).	Relief of bone pain in patients with multiple painful osteoblastic skeletal metastases that take up technetium [99mTc]-labeled bisphosphonates on bone scan. The presence of osteoblastic metastases that take up technetium [99mTc]-labeled bisphosphonates should be confirmed prior to therapy.	Patients should have undergone recent bone scintigraphy documenting increased osteoblastic activity at painful sites to help exclude other causes of chronic pain, which would be unlikely to respond.[56]
Zevalin ([90Y]-ibritumomab tiuxetan) original FDA BLA approval 125019 of February 19, 2002	Biodistribution of [111In]-ibritumomab tiuxetan ("expected" or "altered") (SPECT). Expected biodistribution: easily detectable uptake in blood pool on D1, with less activity in blood pool on D2/3; moderately high to high uptake in normal liver and spleen D1/D2/D3; moderately low or very low uptake in normal kidneys, urinary bladder, D1/D2/D3. Altered biodistribution diffuse uptake in normal lung more intense than cardiac blood pool on D1 or more intense than the liver on D2/3; kidneys with greater intensity than the liver D2/3; intense uptake throughout the normal bowel comparable to uptake by the liver D2/3	Altered biodistribution thought to predict lack-of-benefit.	ZEVALIN, as part of the ZEVALIN therapeutic regimen...is indicated for the treatment of patients with relapsed or refractory low-grade, follicular, or transformed B-cell non-Hodgkin's lymphoma... Y-90 ZEVALIN should not be administered to patients with altered biodistribution of In-111 ZEVALIN.	Zevalin was approved by 2002 for treatment only with the predictive imaging biomarker showing "expected biodistribution" of [111In]-ibritumomab tiuxetan. Mandatory requirement for this predictive IB was relaxed by FDA in 2011, *inter alia* because of the low incidence of "altered biodistribution" in clinical practice.
Exjade (deferasirox) FDA labeling of August 12, 2016	FerriScan R2-MRI Analysis System, FDA approved liver iron concentration (LIC) imaging companion diagnostic for deferasirox, which relies on MRI transverse relaxation rate R_2	LIC ≥ 5 mg Fe/g dw predicts benefit.	Exjade is indicated for the treatment of chronic iron overload in patients with...non- transfusion-dependent thalassemia (NTDT) syndromes and with a liver iron concentration (LIC) of at least 5 milligrams of iron per gram of liver dry weight (mg Fe/g dw)... Prior to starting therapy, obtain LIC by liver biopsy or by an FDA-cleared or approved method for identifying patients for treatment with deferasirox therapy.	The proprietary FerriScan R2-MRI Analysis System incorporates a specific MRI protocol; a Phantom (test object) to verify the MRI protocol set up; analysis software to derive R_2 values; and a calibration curve relating R_2 to LIC.[22]

(*Continued*)

TABLE 7.3.1 (*Continued*) Examples of predictive imaging biomarkers associated with regulatory approvals

DRUG APPROVAL	*IMAGING BIOMARKER*	*WHAT THE IMAGING BIOMARKER PREDICTS*	*RELEVANT INDICATION/ CONTRAINDICATION TEXT*	*COMMENTS*
Tykerb (lapatinib) tablets FDA NDA approval 022059 of March 13, 2007	Left ventricular ejection fraction (ultrasound or SPECT)	LVEF below normal limits predicts harm.	TYKERB, a kinase inhibitor, is indicated...in combination...for the treatment of patients with advanced or metastatic breast cancer whose tumors overexpress HER2...LVEF should be evaluated in all patients prior to initiation of treatment with TYKERB to ensure that the patient has a baseline LVEF that is within the institution's normal limits. LVEF should continue to be evaluated during treatment with TYKERB to ensure that LVEF does not decline below the institution's normal limits.	Many other anticancer therapies, including doxorubicin, radiotherapy, trastuzumab, and sunitinib carry significant risk of cardiotoxicity. Guidelines for LVEF in cancer patient care have been defined, with consensus opinion defining cancer therapeutics-related cardiac dysfunction as a decrease in LVEF of >10%, confirmed with repeated imaging.[57]
Fosamax (alendronate sodium) NDA 020560 FDA labeling of June 8, 1999	Bone mineral density (DXA)	Predicted benefit if biomarker below cutoff.	Treatment of osteoporosis [which] may be confirmed by the finding of low bone mass (for example, at least 2 standard deviations below the premenopausal mean) or by the presence or history of osteoporotic fracture.	Initial NDA approval was based on[58] reduction in fracture risk in premenopausal white women with bone mineral density of the lumbar spine that was at least 2.5 SD below the mean value. Initial labeling employed BMD as a predictive imaging biomarker. Subsequent studies[59] also found benefit in patients with normal bone mineral density, removing the requirement to measure the biomarker.
Activase (alteplase) FDA BLA 103172. product license supplement approved June 18, 1996	Evidence of intracranial hemorrhage (yes/no)	Predicts harm if biomarker above cutoff (i.e., "yes").	Management of acute ischemic stroke in adults for improving neurological recovery and reducing the incidence of disability. Treatment should only be initiated within 3 h of the onset of stroke and after the exclusion of extracranial hemorrhage by a cranial computerized tomography (CT) scan or other diagnostic imaging method similarly sensitive for the presence of hemorrhage.	Because of a strong rationale for alteplase harming hemorrhagic stroke patients, such patients with were excluded in phase III using the IB.[60,61]

TABLE 7.3.2 Example investigational uses of predictive imaging biomarkers

DRUG, INDICATION, AND STAGE IN DEVELOPMENT	*IMAGING BIOMARKER*	*WHAT THE IMAGING BIOMARKER WAS OR IS EXPECTED TO PREDICT*	*COMMENTS*
Vintafolide, platinum-resistant ovarian cancer, phase III	FR+ folate receptor-positive status assessed using SPECT with [^{99m}Tc]-etarfolatide, a proprietary investigational tracer codeveloped with vintafolide by the sponsor, Endocyte.	FR+ status predicted benefit from vintafolide[62] in phase II and was expected to do so in phase III.	EMA CHMP[63] recommended this predictive imaging biomarker for approval as a companion imaging diagnostic in patients with platinum-resistant ovarian cancer treated with vintafolide therapy. This recommendation was conditional on the outcome of the PROCEED trial of vintafolide, whose findings were reportedly[64] negative. [^{99m}Tc]-etarfolatide was not used for patient selection in PROCEED: all participants were planned[65] to undergo imaging during the screening period.
Solanezumab, mild Alzheimer disease, phase III	PET measurement of [^{18}F]-florbetapir brain uptake (SUV ratio for anterior and posterior cingulate, frontal cortex, temporal cortex, parietal cortex, and precuneus normalized to cerebellum[66]). This provides an imaging biomarker of brain amyloid burden. [^{18}F]-florbetapir is proprietary tracer previously NDA approved by FDA and EMA as a diagnostic.	Florbetapir uptake was expected to predict benefit of solanezumab.	[^{18}F]-florbetapir was used in several phase III[67] studies of solanezumab. [^{18}F]-florbetapir was not used prospectively for patient selection in the EXPEDITION-1 and 2 studies, although a retrospective analysis with a cut point of SUV ratio≥1.1 was performed.[68] In EXPEDITION-3, only patients exhibiting [^{18}F]-florbetapir uptake or CSF evidence of amyloid pathology were eligible for inclusion.[69] In the A4 study,[70] only patients exhibiting [^{18}F]-florbetapir uptake were eligible for inclusion.[71]
Onivyde (nanoliposomal irinotecan), solid tumors, post-marketing	Ferumoxytol tumor uptake derived from MRI T_2*. This provides an imaging biomarker of nanoparticle permeability and retention. Ferumoxytol is nanoparticle with diagnostic potential previously NDA approved by FDA as a therapeutic in an unrelated indication.	Ferumoxytol uptake is expected to predict benefit of nanoliposomal irinotecan.	Biodistribution of nanomedicines may be variable, reflecting enhanced permeability and retention. In a study[72] of nanoliposomal irinotecan in solid tumors, patients undergo prior ferumoxytol MRI as a potential companion diagnostic.[39] In 13 patients, lesions with ferumoxytol uptake above median exhibited greater reductions in tumor size following treatment with nanoliposomal irinotecan.[73,74]
HER2-targeted PEGylated liposomal doxorubicin (MM-302) antibody-liposomal drug conjugate, solid tumors overexpressing HER2, phase I	Uptake of [^{64}Cu]-MM-302, an investigational PET tracer, assessed by combined PET-MRI.	[^{64}Cu]-MM-302 uptake is expected to predict benefit of MM-302.	Biodistribution of nanomedicines may be variable, reflecting enhanced permeability and retention. In a study[75] of MM-302 in solid tumors, patients undergo prior [^{64}Cu]-MM-302 PET. In an exploratory analysis, patients whose lesions exhibited high [^{64}Cu]-MM-302 uptake tended to have better progression-free survival.[76]

(*Continued*)

TABLE 7.3.2 (*Continued*) Example investigational uses of predictive imaging biomarkers

DRUG, INDICATION, AND STAGE IN DEVELOPMENT	*IMAGING BIOMARKER*	*WHAT THE IMAGING BIOMARKER WAS OR IS EXPECTED TO PREDICT*	*COMMENTS*
Tirapazamine in combination with chemoradiotherapy, phase III	1. Hypoxic tumor volume derived from maps of voxelwise uptake of [^{18}F]-fluoroazomycin arabinoside (FAZA) above histogram-derived threshold. 2. Uptake of [^{18}F]-misonidazole ([^{18}F]-MISO) above background. [^{18}F]-FAZA and [^{18}F]-MISO are nonproprietary investigational PET hypoxia tracers.	PET-detected hypoxia was expected to predict benefit of tirapazamine in combination with chemoradiotherapy.	Neither [^{18}F]-FAZA nor [^{18}F]-MISO was used prospectively for patient selection. Retrospective analysis reported that PET-detected hypoxia by either tracer was predictive of benefit of tirapazamine,[77,78]
Neoadjuvant chemotherapy in solid tumors, investigational.	Rate of clearance of [^{99m}Tc]-sestamibi, a proprietary SPECT tracer previously NDA approved by FDA as a diagnostic.	Rapid clearance was expected to predict lack-of-benefit of neoadjuvant chemotherapy.	In an initial study[5], a time-to-half-clearance of $\leq$204 min predicted lack-of-benefit. Other PET and SPECT tracers[79] have also been investigated as potentially predictive of lack-of-efficacy in oncology and epilepsy arising from over-expression of ATP-binding cassette transporters, such as P-glycoprotein, breast cancer resistance protein and multidrug resistance-associated proteins.
Tariquidar, pharmacoresistant temporal lobe epilepsy, investigational.	(R)-[^{11}C]-verapamil plasma-to-brain transport rate constant, K_1 (mL/min/cm^3). [^{11}C]-verapamil is a nonproprietary investigational PET tracer.	Low baseline K_1, corresponding to higher baseline P-glycoprotein activity, was expected to predict benefit of P-glycoprotein-inhibitors such as tariquidar.	Pharmacoresistant patients had lower baseline K_1, corresponding to higher baseline P-glycoprotein activity, than seizure-free patients, and K_1 was less responsive to tariquidar than in healthy controls, suggesting that K_1 could act as a predictive Imaging biomarker for P-glycoprotein-inhibitors such as tariquidar.[80]

CHOOSING THE IMAGING BIOMARKER: BIOLOGICAL AND CLINICAL VALIDATION

The decision to incorporate a predictive IB in a PHC drug development incurs complexity, cost and risks, yet necessarily is based on incomplete information. Compelling evidence in support of Eqs. [7.3.2] and [7.3.3] will be lacking at the outset because the object of the planned development program will be to gather that very evidence. If an IB has previously been well standardized, informative, and satisfactory to regulatory authorities (as with BMD, TNM or LVEF), there are fewer risks in a new PHC development than with a completely novel IB. Adoption of an imaging-based PHC strategy requires a platform of supporting evidence that can be built using some of the Bradford Hill criteria.[2,37,38] It is important to show that the IB faithfully reports the underlying pathology it purports to measure. Using examples from Tables 7.3.1 and 7.3.2, the florbetapir IB, measured antemortem, reflected amyloid plaque in human brains at autopsy,[39] whereas the MRI R_2 IB was shown[40] to reflect liver iron in man. Ideally, imaging-pathology correlation should be made in the target patient population; however, where it is practically and ethically difficult to obtain tissue from patients, well-designed animal experiments are enormously valuable. For example, with the [^{64}Cu]-MM-302 PET biomarker, studies in HER2-overexpressing xenografts verified that PET faithfully reports the biodistribution of the liposomes,[73] whereas the MRI T_2* biomarker reported colocalization of ferumoxytol, MM-398 liposomes and tumor-associated macrophages,[41] and in a mouse model, deferasirox-induced changes in the MRI R_2 biomarker faithfully reported drug-induced amelioration of iron overload.[42] Even more helpful is evidence that the IB is predictive in animal models: the fluoroazomycin arabinoside (FAZA) IB was shown to be predictive for benefit of tirapazamine in combination with radiotherapy in EMT6 tumor-bearing nude mice: tirapazamine only delayed growth of the tumors taking up FAZA.[43]

PROVIDING THE ASSAY: CAPABILITY, TECHNICAL VALIDATION, QA/QC AND STANDARDISATION

The Tracer

Not all IBs require a tracer or contrast agent. However, if the tracer is novel (e.g., [^{99m}Tc]-etarfolatide or [^{64}Cu]-MM-302), it will require approval for investigational use in human subjects (e.g., investigational medicinal product/investigational new drug [IMP/IND][44]), before trials in combination with the investigational therapeutic can start. Even if the tracer has IND or NDA approval, it may not be available in all jurisdictions (as is currently the case for ferumoxytol). The costs incurred in winning IND and NDA approval for the tracer are high,[45] and there may be little or no return on this investment if the associated therapeutic (e.g., vintafolide in the case of [^{99m}Tc]-etarfolatide) does not achieve NDA approval. Tracers must be manufactured to ICH GMP: for ^{11}C-containing tracers (half-life 20 min) this manufacturing requires an on-site cyclotron and radiopharmacy. Although ^{18}F-containing tracers (half-life 110 min) can be shipped, the trial site must still be within around 4 h travel time from the cyclotron. ^{99m}Tc-containg tracers require a ^{99}Mo generator, the supply of which has been threatened[46] in recent years.

The drug developer may explore a PHC strategy based on several similar tracers, for example, the three different NDA-approved amyloid tracers in Alzheimer's trials, or two different hypoxia tracers as in the case of tirapazamine. This increases the statistical and regulatory complexity during drug development but mitigates the risk that a particular tracer is unavailable at some sites after the drug is marketed.

The Device

Most IBs use devices such as PET, SPECT, MRI and CT scanners. These are commonly available in radiology and nuclear medicine departments, but not always: adoption of BMD as a predictive IB for alendronate was slow until concerted efforts made low-cost DXA instruments widely available outside specialist care settings.[47] In addition, there may be restrictions on the choice of scanner. The FerriScan R2-MRI CDx uses 1.5T magnets[48]: it relies on the relaxation time T_2, which is strongly field dependent. Most commonly, however, the drug developer will want the IB to be available from as many different makes and model of device as possible so that the drug with its predictive IB can be marketed globally. Manufacturers have little interest in demonstrating the equivalence of their equipment to competitors, so the task of cross-manufacturer standardization often falls on academic societies and public–private partnerships such as the Innovative Medicines Initiative.

Acquisition Protocol

Unlike typical biospecimen biomarkers, optimization of the IB acquisition protocol often requires patient participation in iterative methodology trials. Such trials may involve inconvenience, investigational tracers and ionizing radiation, without benefit to patient volunteers. These methodology trials must be concluded before development of the therapeutic is very far advanced: slow patient recruitment in methodology trials may delay the entire program. Protocols should be locked down before phase IIb to ensure that inferences made about Predictivity in phase IIb and III are consistent between regulatory submissions and subsequent labeling and use. Ideally, an acquisition protocol must be devised and optimized for every make and model of scanner in the trial. If aspects of the acquisition protocol are proprietary to one manufacturer, then equivalents must be devised for other makes. For example, the NIH Osteoarthritis Initiative[49] aims to discover and validate imaging biomarkers in osteoarthritis, with an acquisition protocol implemented using proprietary pulse sequences on the Siemens 3T Trio platform. Attempts to replicate the protocol on other vendors' scanners revealed important systematic differences in imaging biomarker values[50,51] when the same patients were imaged on different scanners.

Analysis Protocol

Unlike the data acquisition, there is some scope for iterating the IB analysis protocol while clinical trials of the therapeutic are underway, providing blinding is maintained. In the trial setting, central reading helps ensure consistency. It may be possible to maintain central reading even after the product and its CDx are launched, as with FerriScan. Alternatively, analysis could be performed locally following guidelines specific for the drug-biomarker combination as for Zevalin, or following general international guidelines, established and updated by expert bodies such as the American Joint Committee on Cancer for TNM[52] or European Society of Cardiology for LVEF.[10]

Quality Assurance/Quality Control

Radiologic imaging devices are designed and marketed for user interface and picture quality, not for quantitative use. When used for quantitation, different devices may yield different data, and meticulous standardization is essential. In the trial setting,[3] significant effort is needed to ensure compliance with the protocol, which may be unfamiliar to the operator, because a mistake in the acquisition usually means the IB will be unreliable or unavailable from that patient. Regulatory authorities have created specific guidelines[53] to help ensure that IBs are obtained in a manner that complies with the trial's protocol, and that quality is maintained over time and between trial sites. They also expect verifiable records of the acquisition and analysis process. The "imaging charter," ancillary to the trial protocol, defines the process in exhaustive detail. Phantoms, such as the European Spine Phantom for BMD[54] or the proprietary FerriScan Phantom Pack[45] for liver R_2, are often central to quality control (QC), although adequate phantoms do not exist for all IBs. The drug developer will often engage specialist imaging clinical research organizations to perform site qualification and training, QC, analysis, data management and quality assurance (QA).

FUTURE PROOFING

Drug labeling often relies heavily on phase III data, even for products that were first approved many years previously. It is therefore undesirable if technological developments cause drift in the predictive IB so that values are no longer comparable with those used to derive the phase III outcome. Scoring systems such as TNM are particularly vulnerable to biomarker drift because every new generation of scanner will tend to upstage some patients. It is inconceivable that some patients found to be T2 M0 using 1990s CT, scintigraphy or 0.5T MRI would not now be found T3 or M1 with current high-resolution CT, SPECT or 3T MRI. Equally, an IB based on "uptake above background" may drift, as more-sensitive scanners push down the noise floor and exposing weaker signals.

More generally, new makes and models of scanner will be introduced, and tracers may become unavailable. Periodic revalidation, often using standard phantoms, and leading to new guidelines, is almost inevitable and must be built into the PHC strategy.

CONCLUSION

Predictive imaging biomarkers have been successfully incorporated in clinical trials for personalised health care in several therapeutic areas, but with a distinct roadmap[2] for validation, development and use.

ENDNOTES

1. FDA-NIH Biomarker Working Group (2016). BEST (Biomarkers, EndpointS, and other Tools) Resource. http://www.ncbi.nlm.nih.gov/books/NBK326791/ (accessed 2018-02-23).
2. O'Connor JPB, Aboagye EO, Adams JE, Aerts HJ, Barrington SF, Beer AJ, Boellaard R et al. (2017). Imaging biomarker roadmap for cancer studies. *Nature Reviews Clinical Oncology* **14**:169–186.
3. Liu Y, deSouza NM, Shankar LK, Kauczor H-U, Trattnig S, Collette S, Chiti A (2015). A risk management approach for imaging biomarker-driven clinical trials in oncology. *Lancet Oncology* **16**:e622–e628.
4. Atkinson Jr AJ, Magnuson WG, Colburn WA, DeGruttola VG, DeMets DL, Downing GJ, Hoth DF et al. (2001). Biomarkers and surrogate endpoints: Preferred definitions and conceptual framework. *Clinical Pharmacology and Therapeutics* **69**:89–95.
5. Ciarmiello A, Vecchio S, Silvestro P, Potena MI, Carriero MV, Thomas R, Botti G, D'Aiuto G, Salvatore M (1998). Tumor clearance of technetium 99m-sestamibi as a predictor of response to neoadjuvant chemotherapy for locally advanced breast cancer. *Journal of Clinical Oncology* **16**:1677–1683.
6. Brindley CO, Markoff E, Schneiderman MA (1959). Direct observation of lesion size and number as a method of following the growth of human tumors. *Cancer* **12**:139–146.
7. Eisenhauer EA, Therasse P, Bogaerts J, Schwartz LH, Sargent D, Ford R, Dancey J et al. (2009). New response evaluation criteria in solid tumours: Revised RECIST guideline (version 1.1). *European Journal of Cancer* **45**:228–247.
8. FDA (2015). Novel drugs approved using surrogate endpoints. http://www.fda.gov/downloads/newsevents/testimony/ucm445375.pdf (accessed 2018-02-25).
9. Schousboe JT, Shepherd JA, Bilezikian JP, Baim S (2013). Position development conference on bone densitometry. Executive summary of the 2013 international society for clinical densitometry position development conference on bone densitometry. *Journal of Clinical Densitometry* **16**:455–466.

10. Lang RM, Badano LP, Mor-Avi V, Afilalo J, Armstrong A, Ernande L, Flachskampf FA et al. (2015). Recommendations for cardiac chamber quantification by echocardiography in adults: An update from the American Society of Echocardiography and the European Association of Cardiovascular Imaging. *Journal of the American Society of Echocardiography* **28**:1–39.e14.
11. American Joint Committee on Cancer (2016). Cancer staging manual. https://cancerstaging.org/references-tools/deskreferences/Pages/default.aspx (accessed 2018-02-25).
12. Nezu T, Hosomi N, Aoki S, Matsumoto M (2016). Carotid intima-media thickness for atherosclerosis. *Journal of Atherosclerosis & Thrombosis* **23**:18–31.
13. van der Heijde D, Dankert T, Nieman F, Rau R, Boers M (1999). Reliability and sensitivity to change of a simplification of the Sharp/van der Heijde radiological assessment in rheumatoid arthritis. *Rheumatology* **38**:941–947.
14. Sarker D, Ang JE, Baird R, Kristeleit R, Shah K, Moreno V, Clarke PA et al. (2015). First-in-human phase I study of pictilisib (GDC-0941), a potent pan-class I phosphatidylinositol-3-kinase (PI3K) inhibitor, in patients with advanced solid tumors. *Clinical Cancer Research* **21**:77–86.
15. Bendell JC, Rodon J, Burris HA, de Jonge M, Verweij J, Birle D, Demanse D, De Buck SS, Ru QC, Peters M, Goldbrunner M, Baselga J (2012). Phase I, dose-escalation study of BKM120, an oral pan-Class I PI3K inhibitor, in patients with advanced solid tumors. *Journal of Clinical Oncology* **30**:282–290.
16. Maynard J, Ricketts SA, Gendrin C, Dudley P, Davies BR (2013). 2-Deoxy-2-[18.F]fluoro-D-glucose positron emission tomography demonstrates target inhibition with the potential to predict anti-tumour activity following treatment with the AKT inhibitor AZD5363. *Molecular Imaging & Biology* **15**: 476–485.
17. Dugel PU, Regillo C, Eliotte D (2015). Characterization of anatomic and visual function outcomes in patients with full-thickness macular hole in ocriplasmin phase 3 trials. *American Journal of Ophthalmology* **160**:94–99.e1.
18. Hodgson RJ, Connolly S, Barnes T, Eyes B, Campbell RS, Moots R (2007). Pharmacokinetic modeling of dynamic contrast-enhanced MRI of the hand and wrist in rheumatoid arthritis and the response to anti-tumor necrosis factor-alpha therapy. *Magnetic Resonance in Medicine* **58**:482–489.
19. Waterton JC, Ho M, Nordenmark LH, Jenkins M, diCarlo J, Guillard G, Roberts C et al. (2017). Repeatability and response to therapy of dynamic contrast enhanced magnetic resonance imaging biomarkers in rheumatoid arthritis in a large multicentre trial setting. *European Radiology* **27**:3662–3668.
20. O'Connor JPB, Jackson A, Parker GJM, Roberts C, Jayson GC (2012). Dynamic contrast-enhanced MRI in clinical trials of antivascular therapies. *Nature Reviews Clinical Oncology* **9**:167–177.
21. Johannessen A, Skorge TD, Bottai M, Grydeland TB, Nilsen RM, Coxson H, Dirksen A, Omenaas E, Gulsvik A, Bakke P (2013). Mortality by level of emphysema and airway wall thickness. *American Journal of Respiratory & Critical Care Medicine* **187**:602–608.
22. Ulloa JL, Stahl S, Yates J, Woodhouse N, Kenna JG, Jones HB, Waterton JC, Hockings PD (2013). Assessment of gadoxetate DCE-MRI as a biomarker of hepatobiliary transporter inhibition. *NMR in Biomedicine* **26**:1258–1270.
23. Sourbron S, Sommer WH, Reiser MF, Zech CJ (2012). Combined quantification of liver perfusion and function with dynamic gadoxetic acid–enhanced MR imaging. *Radiology* **263**:874–883.
24. FDA. FerriScan R2-MRI Analysis System Liver Iron Concentration Imaging Companion Diagnostic For Deferasirox http://www.accessdata.fda.gov/cdrh_docs/reviews/K124065.pdf (accessed 2018-02-28).
25. Nord M, Farde L (2011). Antipsychotic occupancy of dopamine receptors in schizophrenia. *CNS Neuroscience & Therapeutics* **17**:97–103.
26. Waterton JC, Pylkkanen L (2012). Qualification of imaging biomarkers for oncology drug development. *European Journal of Cancer* **48**:409–415.
27. European Society of Radiology (2011). Medical imaging in personalised medicine: A white paper of the research committee of the European Society of Radiology (ESR). *Insights into Imaging* **2**:621–630.
28. FDA AVANDIA (rosiglitazone maleate) tablets highlights of prescribing information. http://www.accessdata.fda.gov/drugsatfda_docs/label/2007/021071s031lbl.pdf (accessed 2016-11-24).
29. FDA ARIMIDEX (anastrozole) tablet for oral use highlights of prescribing information. http://www.accessdata.fda.gov/drugsatfda_docs/label/2014/020541s029lbl.pdf (accessed 2016-11-24).
30. FDA SYMBICORT (budesonide 80/160 mcg and formoterol fumarate dihydrate 4.5 mcg) Inhalation Aerosol highlights of prescribing information. http://www.accessdata.fda.gov/drugsatfda_docs/label/2010/021929s021lbl.pdf (accessed 2016-11-24).
31. Eastell R, Adams JE, Coleman RE, Howell A, Hannon RA, Cuzick J, Mackey JR, Beckmann MW, Clack G (2008). Effect of anastrozole on bone mineral density: 5-year results from the anastrozole, tamoxifen, alone or in combination trial. *Journal of Clinical Oncology* **26**:1051–1057.
32. Markopoulos C, Tzoracoleftherakis E, Polychronis A, Venizelos B, Dafni U, Xepapadakis G, Papadiamantis J et al. (2010). Management of anastrozole-induced bone loss in breast cancer patients with oral risedronate: Results from the ARBI prospective clinical trial. *Breast Cancer Research* **12**:R24.
33. FDA. List of Cleared or Approved Companion Diagnostic Devices (In Vitro and Imaging Tools). http://www.fda.gov/MedicalDevices/ProductsandMedicalProcedures/InVitroDiagnostics/ucm301431.htm (accessed 2018-02-28).
34. Ekdawi SN, Jaffray DA, Allena C (2016). Nanomedicine and tumor heterogeneity: Concept and complex reality. *Nano Today* **11**:402–414.
35. O'Connor JPB, Rose CJ, Waterton JC, Carano RAD, Parker GJM, Jackson A (2015). Imaging intratumor heterogeneity: Role in therapy response, resistance, and clinical outcome. *Clinical Cancer Research* **21**:249–257.
36. Stapleton S, Allen C, Pintilie M, Jaffray DA (2013). Tumor perfusion imaging predicts the intra-tumoral accumulation of liposomes. *Journal of Controlled Release* **172**:351–357.
37. Bradford Hill, A (1965). The environment and disease: Association or causation? *Proceedings of the Royal Society of Medicine* **58**:295–300.
38. Chetty RK, Ozer JS, Lanevschi A, Schuppe-Koistinen I, McHale D, Pears JS, Vonderscher J, Sistare FD, Dieterle F (2010). A systematic approach to preclinical and clinical safety biomarker qualification incorporating Bradford Hill's principles of causality association. *Clinical Pharmacology and Therapeutics* **88**:260–262.
39. Clark CM, Pontecorvo MJ, Beach TG, Bedell BJ, Coleman RE, Doraiswamy PM, Fleisher AS et al. (2012). Cerebral PET with florbetapir compared with neuropathology at autopsy for detection of neuritic amyloid-beta plaques: A prospective cohort study. *Lancet Neurology* **11**:669–678.
40. St Pierre TG, Clark PR, Chua-anusorn W, Fleming AJ, Jeffrey GP, Olynyk JK, Pootrakul P, Robins E, Lindeman R (2005). Noninvasive measurement and imaging of liver iron concentrations using proton magnetic resonance. *Blood* **105**:855–861.
41. Kalra AV, Spernyak J, Kim J, Sengooba A, Klinz S, Paz N, Cain J et al. (2014). Magnetic resonance imaging with an iron oxide nanoparticle demonstrates the preclinical feasibility of predicting intratumoral uptake and activity of MM-398, a nanoliposomal irinotecan (nal-IRI). *Cancer Research* **74**(19 Supplement):2065.

42. Nick H, Allegrini PR, Fozard L, Junker U, Rojkjaer L, Salie R, Niederkofler V, O'Reilly T (2009). Deferasirox reduces iron overload in a murine model of juvenile hemochromatosis. *Experimental Biology & Medicine* **234**:492–503.
43. Beck R, Roper B, Carlsen JM, Huisman MC, Lebschi JA, Andratschke N, Picchio M, Souvatzoglou M, Machulla HJ, Piert M (2007). Pretreatment ^{18}F-FAZA PET predicts success of hypoxia-directed radiochemotherapy using tirapazamine. *Journal of Nuclear Medicine* **48**:973–980.
44. Todde S, Windhorst AD, Behe M, Bormans G, Decristoforo C, Faivre-Chauvet A, Ferrari V et al. (2014). EANM guideline for the preparation of an Investigational Medicinal Product Dossier (IMPD). *European Journal of Nuclear Medicine and Molecular Imaging* **41**:2175–2185.
45. Nunn AD (2007). Molecular imaging and personalized medicine: An uncertain future. *Cancer Biotherapy and Radiopharmceuticals* **22**:722–739.
46. Ballinger JR (2010). Short- and long-term responses to molybdenum-99 shortages in nuclear medicine. *British Journal of Radiology* **83**:899–901.
47. Blake GM, Fogelman I (2007). The role of DXA bone density scans in the diagnosis and treatment of osteoporosis. *Postgraduate Medical Journal* **83**:509–517.
48. Resonance Health Ltd. Ferriscan fact sheet. http://www.resonancehealth.com (accessed 2016-11-23).
49. Eckstein F, Wirth W, Nevitt MC (2012). Recent advances in osteoarthritis imaging—the Osteoarthritis Initiative. *Nature Reviews Rheumatology* **8**:622–630.
50. Balamoody S, Williams TG, Wolstenholme C, Waterton JC, Bowes M, Hodgson R, Zhao S, Scott M, Taylor CJ, Hutchinson CE (2013). Magnetic resonance transverse relaxation time T_2 of knee cartilage in osteoarthritis at 3-T: A cross-sectional multicentre, multivendor reproducibility study. *Skeletal Radiology* **42**:511–520.
51. Balamoody S, Williams TG, Waterton JC, Bowes M, Hodgson R, Taylor CJ, Hutchinson CE (2010). Comparison of 3T MR scanners in regional cartilage-thickness analysis in osteoarthritis: A cross-sectional multicenter, multivendor study. *Arthritis Research and Therapy* **12**:R202.
52. Amin MB, Edge S, Greene F, Byrd DR, Brookland RK, Washington MK, Gershenwald JE et al. (2017). *AJCC Cancer Staging Manual*. New York: Springer.
53. FDA. Clinical Trial Imaging Endpoint Process Standards. Draft Guidance for Industry. http://www.fda.gov/downloads/drugs/guidancecomplianceregulatoryinformation/guidances/ucm268555.pdf (accessed 2016-11-23).
54. Kolta S, Ravaud P, Fechtenbaum J, Dougados M, Roux C (1999). Accuracy and precision of 62 bone densitometers using a European Spine Phantom. *Osteoporosis International* **10**:14–19.
55. Iversen P, McLeod DG, See WA, Morris T, Armstrong J, Wirth MP, Casodex Early Prostate Cancer Trialists' Group (2010). Antiandrogen monotherapy in patients with localized or locally advanced prostate cancer: final results from the bicalutamide Early Prostate Cancer programme at a median follow-up of 9.7 years. *BJU International* **105**:1074–1081.
56. Bodei L, Lam M, Chiesa C, Flux G, Brans B, Chiti A, Giammarile F (2008). EANM procedure guideline for treatment of refractory metastatic bone pain. *European Journal of Nuclear Medicine and Molecular Imaging* **35**:1934–1940.
57. Plana JC, Galderisi M, Barac A, Ewer MS, Ky B, Scherrer-Crosbie M, Ganame J et al. (2014). Expert consensus for multimodality imaging evaluation of adult patients during and after cancer therapy: A report from the American Society of Echocardiography and the European Association of Cardiovascular Imaging. *European Heart Journal Cardiovascular Imaging* **10**:1063–1093.
58. Liberman UA, Weiss SR, Bröll J, Minne HW, Quan H, Bell NH, Rodriguez-Portales J, Downs RW Jr, Dequeker J, Favus M (1995). Effect of oral alendronate on bone mineral density and the incidence of fractures in postmenopausal osteoporosis. The Alendronate phase III Osteoporosis Treatment Study Group. *New England Journal of Medicine* **333**:1437–1443.
59. Wells GA, Cranney A, Peterson J, Boucher M, Shea B, Robinson V, Coyle D, Tugwell P (2008). Alendronate for the primary and secondary prevention of osteoporotic fractures in postmenopausal women. *Cochrane Database of Systematic Reviews* CD001155.
60. National Institute of Neurological Disorders and Stroke rt-PA Stroke Study Group (1995). Tissue plasminogen activator for acute ischemic stroke. *New England Journal of Medicine* **333**:1581–1587.
61. Hacke W, Kaste M, Fieschi C, Toni D, Lesaffre E, von Kummer R, Boysen G, Bluhmki E, Höxter G, Mahagne MH, Hennerici, M (1995). Intravenous thrombolysis with recombinant tissue plasminogen activator for acute hemispheric stroke. The European Cooperative Acute Stroke Study (ECASS) *JAMA* **274**:1017–1025.
62. Morris RT, Joyrich RN, Naumann RW, Shah NP, Maurer AH, Strauss HW, Uszler JM, Symanowski JT, Ellis PR, Harb WA (2014). Phase II study of treatment of advanced ovarian cancer with folate-receptor-targeted therapeutic (vintafolide) and companion SPECT-based imaging agent (^{99m}Tc-etarfolatide). *Annals of Oncology* **25**:852–858.
63. European Medicines Agency Committee for Medicinal Products for Human Use. Summary of opinion. http://www.ema.europa.eu (accessed 2016-11-30).
64. Endocyte, Inc. Merck and Endocyte Announce Independent DSMB Recommends Vintafolide Proceed Phase 3 Trial Be Stopped For Futility Following Interim Analysis Press release May 2, 2014. http://investor.endocyte.com/releasedetail.cfm?ReleaseID=844838 (accessed 2016-11-30).
65. Endocyte Inc. Study for Women With Platinum Resistant Ovarian Cancer Evaluating EC145 in Combination With Doxil (PROCEED). https://clinicaltrials.gov/ct2/show/NCT01170650 (accessed 2016-11-30).
66. Siemers ER, Sundell KL, Carlson C, Case M, Sethuraman G, Liu-Seifert H, Dowsett SA, Pontecorvo MJ, Dean RA, Demattos R (2016). Phase 3 solanezumab trials: Secondary outcomes in mild Alzheimer's disease patients. *Alzheimer's & Dementia: The Journal of the Alzheimer's Association* **12**:110–120.
67. Doody RS, Thomas RG, Farlow M, Iwatsubo T, Vellas B, Joffe S, Kieburtz K et al. (2014). Phase 3 trials of solanezumab for mild-to-moderate Alzheimer's disease. *New England Journal of Medicine* **370**:311–321.
68. Dean R, Shaw LM, Waligorska TW, Korecka M, Figurski M, Trojanowski JQ, Sundell K et al. (2014). Inclusion of patients with Alzheimer's disease pathology in solanezumab EXPEDITION 3 using florbetapir PET imaging or INNO-BIA ALZBIO3 CSF Aβ1-42. *Alzheimer's & Dementia: The Journal of the Alzheimer's Association* **10**(4 supplement): 811 P4–076.
69. Eli Lilly and Company. Progress of Mild Alzheimer's Disease in Participants on Solanezumab Versus Placebo (EXPEDITION 3). https://clinicaltrials.gov/ct2/show/record/NCT01900665 (accessed 2016-11-30).
70. Sperling RA, Rentz DM, Johnson KA, Karlawish J, Donohue M, Salmon DP, Aisen P (2014). The A4 study: Stopping AD before symptoms begin? *Science Translational Medicine* **6**:228fs13.
71. Eli Lilly and Company. Clinical Trial of Solanezumab for Older Individuals Who May be at Risk for Memory Loss (A4) https://clinicaltrials.gov/ct2/show/NCT02008357 (accessed 2016-11-30).
72. Merrimack Pharmaceuticals. Pilot Study to Determine Biodistribution of MM-398 and Feasibility of Ferumoxytol as a Tumor Imaging Agent. https://clinicaltrials.gov/ct2/show/record/NCT01770353 (accessed 2018-02-28).

73. Sachdev JC, Ramanathan RK, Raghunand N, Anders C, Munster P, Minton S, Northfelt D et al. (2016). A phase 1 study in patients with metastatic breast cancer to evaluate the feasibility of magnetic resonance imaging with ferrumoxytol as a potential biomarker for response to treatment with nanoliposomal irinotecan (nal-IRI, MM-398). *Cancer Research* **76** (4 Suppl. 1): OT3-02-14.
74. Ramanathan RK, Korn RL, Sachdev JC, Fetterly GJ, Jameson, G Marceau K, Marsh V et al. (2014). Lesion characterization with ferumoxytol MRI in patients with advanced solid tumors and correlation with treatment response to MM-398, nanoliposomal irinotecan (nal-IRI). *European Journal of Cancer.* **50**(Suppl. 6): 87.
75. Merrimack Pharmaceuticals. 64-Cu Labeled Brain PET/MRI for MM-302 in Advanced HER2+ Cancers With Brain Mets https://clinicaltrials.gov/ct2/show/NCT02735798 (accessed 2016-11-30).
76. Lee H, Shields AF, Siegel BA, Miller KD, Krop I, Ma CX, LoRusso PM et al. (2017). ^{64}Cu-MM-302 positron emission tomography quantifies variability of enhanced permeability and retention of nanoparticles in relation to treatment response in patients with metastatic breast cancer. *Clinical Cancer Research* **23**:4190–4202.
77. Graves EE, Hicks RJ, Binns D, Bressel M, Le Q-T, Peters L, Young RJ, Rischin D (2016). Quantitative and qualitative analysis of [^{18}F]FDG and [^{18}F]FAZA positron emission tomography of head and neck cancers and associations with HPV status and treatment outcome. *European Journal of Nuclear Medicine and Molecular Imaging* **43**:617–625.
78. Rischin D, Hicks RJ, Fisher R, Binns D, Corry J, Porceddu S, Peters LJ (2006). Prognostic significance of [^{18}F]-misonidazole positron emission tomography-detected tumor hypoxia in patients with advanced head and neck cancer randomly Assigned to chemoradiation with or without tirapazamine: A substudy of Trans-Tasman Radiation Oncology Group Study 98.02. *Journal of Clinical Oncology* **24**:2098–2104.
79. Mairinger S, Erker T, Muller M, Langer O (2011). PET and SPECT radiotracers to assess function and expression of ABC transporters in vivo. *Current Drug Metabolism* **12**:774–792.
80. Feldmann M, Asselin MC, Liu J, Wang S, McMahon A, Anton-Rodriguez J, Walker M et al. (2013). P-glycoprotein expression and function in patients with temporal lobe epilepsy: A case-control study. *Lancet Neurolology* **12**:777–785.

Innovation in Metabonomics to Improve Personalized Health Care

7.4

Nelma Pértega-Gomes and Massimo Loda

Contents

METABOLOMICS

Many fields of medicine have a growing interest in characterizing diseases at the molecular level in order to develop personalized and tailored diagnostic and therapeutic approaches [1]. Metabolomics is emerging as a promising area to characterize various disease phenotypes and identify individual metabolic features that can be used to predict response to therapy [2–6]. Metabolomics is defined as the high-throughput detection and quantification of metabolites in a biological sample. Metabolites are the intermediate end products of metabolism and are involved in various cellular functions such as providing fuel or structural components or have stimulatory and inhibitory signaling effects. Metabolomics covers a broad range of small molecules that include, but are not limited to, amino acids, lipids, carbohydrates, vitamins, steroids, and ketones [3,7–9]. Although genes and proteins are subject to epigenetic and pretranslational modulations, as intermediate products metabolites provide an integrated and additional insight into the biological system [9–11]. Metabolic profiling can, therefore, contribute to the molecular characterization of various disease phenotypes and to the identification of distinguishable features that can improve the prediction for therapy response in patients [12].

In this section, we will discuss how to approach metabolomics and how innovation in metabolomics is being used to improve personalized health care.

Measuring the Human Metabolome

The total metabolite content of a biological sample is called the metabolome. Although other omics techniques focus on sets of relatively chemically similar biopolymers composed of a limited number of building blocks, that is, 4 nucleotides or 22 amino acids, estimating the number of metabolites in human metabolome is a more challenging task. The Human Metabolome Database (HMDB) [13] is the most comprehensive database available to date and includes more than 40,000 metabolite entries. Metabolomics covers structurally heterogeneous and physicochemical diverse molecules, whereas the dynamic range of metabolite concentrations exists across nine orders of magnitude [2,14,15]. Although highly informative, the analysis of metabolomics data is not easy and relies on interactions with multiple scientific areas such as, biology, bioinformatics, biochemistry, epidemiology and clinical research [16,17].

The approach

Various sophisticated methods and techniques, often complementary to one another, are used to identify and quantify metabolites. Different analytical tools can be used alone, in parallel or in combination. The choice is driven by the goals of individual studies and by the strength and limitations of the different techniques [18,19]. Two different approaches are often used in metabolomics studies: targeted analysis and untargeted analysis [17,20].

Targeted analysis

The targeted metabolomics approach is driven by a specific hypothesis to interrogate a particular biochemical pathway. It refers to the exact quantification of known and expected metabolites. Here, predefined metabolite-specific signals are used to quantify concentrations of a limited number of metabolites. The sample preparation and instrument adjustment are crucial for an accurate quantification of the metabolites of interest [17,20]. The major limitations of the targeted analysis for metabolomics are that it requires the compounds of interest to be previously known and that these compounds must be available in a purified form. For many of the identifiable metabolites, purified standards are not yet available and setting up the experimental conditions for novel metabolites is generally an expensive endeavor.

Untargeted analysis

In contrast to targeted metabolomics, the untargeted approach aims to measure simultaneously as many metabolites as possible. No prior knowledge of the nature and identity of assessed metabolites is needed, and minimal pretreatment has to be applied so as to prevent loss of metabolites. The data sets obtained are particularly complex and a number of metabolites remain uncharacterized. However, this approach enables discoveries and the generation of new hypotheses [17]. Both metabolic profiling and metabolite fingerprinting are approaches in which the metabolites expected to be altered are not known. Metabolic profiling is used for diagnosis, to help as an extension of genomics and to better define the function of genes [21]. Metabolic fingerprinting does not aim to identify or quantify all the metabolites in a sample, but rather considers a total profile (fingerprint) as a unique pattern characterizing a snapshot of the metabolism in a particular sample just looking at patterns. Here, pattern recognition tools are used to classify the fingerprints and identify the features of the profile characteristic of each pattern. Fingerprinting is usually performed with spectroscopic techniques such as nuclear magnetic resonance (NMR) [22–24] or mass spectrometry (MS) [25] by directly acquiring physical spectra without prior separation techniques (chromatography or electrophoresis) (described in more detail below).

The technology

Sample preparation

Metabolite extraction is an important initial step prior to analysis. The analysis of a large number of metabolites with very diverse properties makes extraction a very challenging step. No single extraction methodology exists for all metabolites within a cell or tissue. Protocols are therefore adapted according to the metabolites of interest and the type of specimen. Extraction conditions that favor preservation of one metabolite class can destroy other metabolite species, making it difficult to find a balance between comprehensiveness and optimal measurement for different classes of metabolites. Because the metabolite levels represent the ultimate response of a biological system to environmental or genetic changes, a rapid freeze of the cell metabolism and any enzymatic activity is the first step in obtaining reliable results. The rapid stopping of the cell metabolism is termed quenching. Quenching aims at preventing the turnover of metabolites. Arrest of metabolic activities can be achieved through fast changes in temperature or pH. A commonly used method is based on the use of a cold aqueous methanol solution into which the sample is submerged. The cells are separated from the quenching mixture by centrifugation at low temperatures.

Table 7.4.1 shows six commonly utilized extraction methods and the respective recovery of spiked metabolite standards:

TABLE 7.4.1 Overview of the different extraction methods based on the recovery of spiked metabolite standards

	CLASSES OF METABOLITES							
EXTRACTION METHOD	AMINO ACIDS (%)	ORGANIC ACIDS (%)	FATTY ACIDS (%)	NUCLEOTIDES (%)	PEPTIDES (%)	SUGARS (%)	SUGAR ALCOHOLS (%)	SUGAR PHOSPHATES
Chloroform:methanol:buffer	>80	>60	NR	>40	>40	>40	>60	>20
Boiling ethanol	>60	>60	>40	0	>20	0	>40	NR
Perchloric acid	>20	0	0	0	>40	>40	>60	NR
Potassium hydroxide	>60	>20	>60	>20	>40	>20	>60	NR
Methanol:water	>60	>60	>40	>60	>20	>20	>60	NR
Pure methanol	>80	>80	>60	>80	>40	>20	>60	NR

Source: Villas-Boas, S.G. et al., *Yeast.*, 22, 1155–1169, 2005.
Note: NR, not recovered.

chloroform:methanol:buffer [26], boiling ethanol [27], perchloric acid [28], potassium hydroxide [28], methanol:water [29] and pure methanol (an adaptation of the pure methanol:water extraction method using pure cold methanol [−40°C]) extraction methods.

Methods of analysis

Numerous analytical platforms offer multifaceted and powerful approaches to describe parts of the metabolome [18]. NMR, gas chromatography (GC) coupled to MS or high-resolution MS (HRMS) are recognized as the most powerful techniques used for the high-throughput investigation of the metabolome. These high-throughput technologies have enabled the identification of biomarker profiles that characterize disease phenotypes, providing the rationale for the development of new, targeted drugs, and potentially enabling the right therapy for every patient and identifying groups of biomarkers characteristic of a particular disease [30,31]. Each technique has associated advantages and disadvantages, thus, different analytical technologies should be used depending on the query because no single methodology is ideal for all of the many metabolites within a biological sample.

Mass spectrometry MS-based approaches are the most sensitive. The process involves the conversion of the sample into gaseous ions, with or without fragmentation, which are then characterized by their mass-to-charge ratios and relative abundance. This technique basically studies the effect of ionizing energy on molecules. It depends upon chemical reactions in the gas phase in which sample molecules are consumed during the formation of ionic and neutral species.

Gas chromatography–mass spectrometry MS is usually coupled with GC or LC to separate different classes of metabolites. GC-MS is the most robust technique in MS-based metabolomics, widely used to identify and quantify volatile, thermally stable, low-molecular-weight metabolites such as organic acids, amino acids, nucleic acids, sugars, amines, and alcohols. Volatile metabolites can be separated and quantified by GC-MS directly. A list of available GC-MS libraries can be found in [32]. Unfortunately, most mass spectral libraries are tailored to suit the chemical industry, or drug studies, and therefore do not represent a large number of naturally occurring metabolites and their intermediates, limiting their applicability to metabolomics studies.

Liquid chromatography–mass spectrometry LC offers some advantages over GC, including the possibility of separating compounds at room temperature. LC-MS is being increasingly used in metabolomics applications because of its high sensitivity and a range in polarity and molecular mass that is wider than that of GC-MS. It can be tailored for the analysis of a specific metabolite or class of compounds. Thus, LC-MS techniques are typically more sensitive and show a higher accuracy over a larger size range than GC-MS techniques. GC-MS and LC-MS techniques can thus be considered complementary. Non-Steroidal Anti-Inflammatory Drugs (NSAIDs) can be used to elucidate the structure of unknown compounds.

Capillary electrophoresis–mass spectrometry Capillary electrophoresis–MS (CE-MS) is a very recent technique that is becoming one of the major analytical tools for metabolomics. CE-MS provides several important advantages over other separation techniques and possesses very high resolving power with very small sample requirements. One of the most significant advantages of this technique is the ability to separate cations, anions and uncharged molecules in one single analysis, making it useful for simultaneous profiling of many different metabolite classes [33,34]. Human pluripotent stem cell (hPSC) has been used for both targeted and untargeted approaches [35,36], including analysis of inorganic ions, organic acids, amino acids, vitamins, carbohydrates and peptides.

Nuclear magnetic resonance NMR is a physical phenomenon in which nuclei in a magnetic field absorb and reemit electromagnetic radiation. NMR spectroscopy is particularly useful in the detection of compounds that are less tractable by GC-MS and LC-MS (such as amines, sugars, and volatile and nonreactive compounds). NMR is nondestructive, so samples can be used for further analyses. Sample preparation for NMR is simple and largely automated. NMR-based metabolomics is a particularly powerful approach when applied to the high-throughput analysis of biofluids such as blood. The drawback of NMR is that its sensitivity is orders of magnitude lower than MS, making it inappropriate for the analysis of large number of low-abundant metabolites. One common approach for increasing the sensitivity is the use of a higher magnetic field. Hyperpolarization, which offers a potential different strategy to overcome the sensitivity limitation, allows the measurement of chemical reactions in real time (e.g., the use of 13C isotopes). The hyperpolarization of selected molecules (e.g., pyruvate) and subsequent injection in a living organism can be used in imaging techniques *in vivo*. This approach allows the monitoring of tumor metabolism *in vivo* without radioactive isotopes. NMR has been extensively used for metabolite fingerprinting, profiling and metabolic flux analysis [23,37–39].

Other technologies *Fourier-transform infrared and Raman spectroscopy* are spectroscopic techniques that rely on vibrational frequencies of metabolites to provide a fingerprint of metabolism. Although selectivity and sensitivity are not as high as in NMR and MS, these techniques are able to profile carbohydrates, amino acids, lipids, and fatty acids as well as proteins simultaneously. They have been recognized as valuable tools for metabolomics fingerprinting. Although the main drawback of FT-IT is the limitation of using only dried samples, Raman spectroscopy can detect metabolites directly from tissues even *in vivo*.

TABLE 7.4.2 Principles, advantages and limitations for NMR and MS

	BRIEF DESCRIPTION	*METHOD*	*ADVANTAGE*	*DISADVANTAGE*
Mass spectrometry–based metabolomics	Low analytical reproducibility	GC-MS	Sensitive	Slow
	Extensive sample preparation		Robust	
	High cost per sample		Large linear range	
	In vivo measurements are not possible		Large commercial and public libraries	
		LC-MS	Large sample capacity	Limited commercial libraries
			Many modes of separation available	Slow
		CE-MS	High separation power	Limited commercial libraries
			Small sample requirements	
			Rapid analysis	
NMR-based metabolomics	Very high analytical sample reproducibility	NMR	Rapid analysis	Low sensitivity
	Minimal sample preparation		High resolution	Libraries are of limited use due to complex matrix
	Low cost per sample		Nondestructive	
	In vivo measurements can be made			

Table 7.4.2 shows a summary of the different techniques used to perform metabolomics measurements in biological samples.

Data Analysis

As stated previously, the dynamic range of metabolite concentrations exists across nine orders of magnitude, making the analysis very challenging. The handling, processing and analysis of metabolomics data represents a clear challenge and requires unique mathematical, statistical and bioinformatics tools in addition to those in common with proteomics or microarray data. Metabolomics raw data involves noise reduction, peak detection and integration, compound identification and quantification; in addition, implementation of low values is often required. Metabolomics data have been analyzed using a wide range of statistical and machine-learning algorithms mostly divided in two major classes: supervised and unsupervised algorithms [40,41].

Examples of supervised methods include ANOVA [42], partial least squares (PLS) [43] and discriminant function analysis (DFA) [44]. Unsupervised methods that have been routinely used in analyzing metabolomics data are hierarchical clustering [45], principal component analysis, and self-organizing maps [46]. Database management systems for metabolomics are required to collect both metadata, raw, and processed experimental data. Storing metadata, covering experimental design, the nature of the samples and their treatment prior to the analysis, and information about the analytical technique and data-processing details are important to be able to reproduce the experimental conditions and compare results obtained in different research facilities. A number of databases, data management, analysis and visualization tools are currently publicly available. These include metabolic pathway databases and pathway viewers such as KEGG (http://www.genome.ad.jp/kegg/), MetaCyc (http://metacyc.org/), MapMan (http://gabi.rzpd.de/projects/MapMan/), KaPPA-View (http://kpv.kazusa.or.jp/kappa-view/), MetaboAnalyst (http://www.metaboanalyst.ca/) and the Human Metabolome Database (http://www.hmdb.ca/).

Metabolic Profiling in the Clinic: The Contribution of Metabolomics to Personalized Health Care

Many diseases have a homogeneous clinical presentation; however, from a molecular standpoint, diseases can be molecularly heterogeneous and diagnostic methods are still lagging behind in redefining disease classification. Personalized medicine aims to design therapeutic interventions based on the molecular alterations of individual patients, including metabolic reprogramming secondary to the disease state. Associations of specific alterations with disease has relied on different sets of "high-throughput" data. Most efforts have been focused on the identification of genetic changes—germline or somatic in the case of tumors—as well as expression profiling. However, these genetic variations only account for small percentages of the occurrence of common diseases [47,48]. The gap between genomic alterations and transcriptional profiling has been increasingly recognized. Metabolomics can play an important role in filling this gap. Recently the integration of metabolites concentration using LC and GC coupled with MS, and genetic mutations was used to explain disease risk in healthy volunteers [49]. The study included 80 adults of normal health and showed that the comprehensive metabolic profiles obtained provided a functional readout to assess the penetrance of gene mutations identified by whole-exome sequencing on these individuals. In addition, metabolic abnormalities identified by statistical analysis uncovered potential damaging mutations that were previously discarded. Additionally, metabolic signatures consistent with early signs of disease and drug effects were found, demonstrating that metabolomics represents a powerful and effective tool in precision medicine to help with risk stratification of disease and customized drug therapy.

The example of metabolomics as applied to cancer diagnosis and therapy is one of the most popular of all. Specific genetic alterations appear to drive unique metabolic programs in cancer cells. These can be used as biomarkers of genetic subtypes of cancer or as discovery tools for therapeutic targeting of metabolic enzymes. Given the example of prostate cancer, the detection of specific metabolites in blood and urine has been suggested as reliable biomarkers of tumor state, including inferring driving oncogenes [50–53]. *In vivo*, the use of metabolic tracers such as 18F-fluorodeoxyglucose-positron emission tomography (18F-FDG-PET) allow a direct and dynamic analysis. Tissues that are characterized by significant glucose metabolism, such as malignant tumors and brain tissue, can be imaged radiologically. An important application of this technique is diagnosing and staging malignancies as well as measuring responses to therapies, but these tools can also be extended to other pathologies such as inflammation [54,55]. Nonradioactive, stable isotope-labeled tracers are also powerful metabolic probes, offering good safety profiles for *in vivo* dosing and the opportunity to measure the metabolism of just about any biological molecule that can be synthesized. Using techniques such as 13C hyperpolarization magnetic resonance, metabolism can be monitored *in vivo* [56]. Following administration of stable isotope-labeled substrates, it can also be informative to collect biological fluids or tissues, via biopsy or surgical resection, and apply analytical methods to samples *ex vivo* to finely trace metabolism of the substrate. For example, recently reported results from a 13C6-glucose tracer study in human subjects with non-small cell lung cancer (NSCLC) revealed enhanced pyruvate carboxylase pathway activity, as read out by increased levels of the downstream metabolite 13C3-aspartate, occurred in cancer bearing lung tissue. This effort serves as an example of how metabolic tracer studies reveal potentially novel therapeutic targets in personalized medicine [57].

More recently, there has been increased interest in finding early metabolic indicators of disease in longitudinal cohorts, years before symptoms are clinically apparent. For example, in pancreatic cancer [58] the elevation of circulating branched-chain amino acids was shown to be an early event in human pancreatic adenocarcinoma development. Type 2 diabetes lipid profiling identified a triacylglycerol signature associated with insulin resistance and was able to improve diabetes prediction in humans [59,60]. Another interesting study describes the detecting preclinical Alzheimer's disease in a group of cognitively normal older adults. The authors discovered and validated a set of lipids in peripheral blood that predicted phenoconversion to either amnestic mild cognitive impairment or Alzheimer's disease within a 2- to 3-year time frame with over 90% accuracy [61].

Examples of metabolomics for biomarker study in other diseases include diabetes [62,63], macular degeneration [64], asthma [65], Parkinson's disease [66], nonalcoholic fatty liver disease [67], and tuberculosis [68] and inborn metabolic diseases [69].

Pharmacometabolomics represents another big potential for clinical application of metabolomics studies. It consists of the identification of individual metabolomics characteristics able to predict the effectiveness and/or toxicity of a particular drug or combination of drugs. This concept was first introduced by a study performed in mice where the authors predicted the hepatotoxic effects of paracetamol using metabolomics urinary profiles [70]. Pharmacometabolomics has shown significant potential in its ability to contribute to personalized health care.

Table 7.4.3 shows the most relevant examples of clinical applications of metabolomics in various diseases, discriminates the most interesting metabolites for each case and shows the variety of biological samples that can be used to apply metabolomics to clinics.

TABLE 7.4.3 Clinical applications of metabolomics

DISORDERS	*METABOLITE(S)*	*BIOLOGICAL MATERIAL*	*APPLICABILITY*	*REFERENCES*
Inflammatory bowel diseases (IBD)	*N*-methylhistamine	Urine	Indicator of disease activity in patients; enhanced in IBD	[71]
Ulcerative colitis (UC) and Crohn's disease (CD)	*N*-methylhistamine	Urine	Use to diagnose UC and CD and monitor pathological progression	[71]
Ulcerative colitis (UC)	*N*-methylhistamine	Urine	Differentiate active and quiescent UC	[72]
UC	Metabolic profiles	Fecal extracts	Differentiate UC from controls	[73]
UC	Metabolic profiles	Blood plasma	Insight into the molecular processes associated to the development of UC	[74]

(Continued)

TABLE 7.4.3 (*Continued*) Clinical applications of metabolomics

DISORDERS	*METABOLITE(S)*	*BIOLOGICAL MATERIAL*	*APPLICABILITY*	*REFERENCES*
UC	Metabolic profiles, methylamine metabolism	Urine	Insight into the molecular processes associated to the development of UC	[75]
CD and UC	Metabolic profiles	Urine	Distinguish CD from UC; critical for disease management	[76]
IBD	Metabolic profiling	Fecal extracts	Diagnostic value for IBD	[77]
CD and UC	Reduced concentrations of short chain fatty acids and methylamines	Fecal extracts	Diagnostic value	[78]
Insulin resistance	Branched chain amino acid (BCAA) related signature	Blood	Associated with insulin resistance	[79–81]
Diabetes	Isoleucine, tyrosine and phenylalanine	Blood plasma	Predictive of diabetes	[60]
Nonalcoholic fatty liver disease (NAFLD)	Reduced concentrations of creatine and increased bile acids and eicosanoids	Serum	Diagnostic value	[82]
Nonalcoholic steatohepatitis (NASH)	9- and 13-HODEs and 9 and 13-oxoODEs	Plasma	Circulating biomarkers of NASH	[83]
Blood pressure	Low levels of formate	Urine	Indicative of high blood pressure	[84]
Chronic heart failure	Increased levels of Pseudouridine and 2-oxoglutarate	Serum	Diagnostic value	[85]
Ovarian	Glycerolipid, pyrimidine, purine, amino acid, propanoate, free fatty acid metabolism	Fresh-frozen tumor samples	Significantly different between borderline tumors and carcinomas	[86]
Prostate	Sarcosine	Urine and plasma	Distinguish benign prostate, clinically localized Pca and metastatic disease	[52]
Bladder	Urinary metabolomics	Urine	Staging, grading and diagnostic capabilities	[87]
Lung, gastric, colorectal, breast and prostate	Plasma free amino acid profiling, tryptophan identified as a key amino acid associated with cancer progression	Plasma	Diagnosis of lung, gastric, colorectal, breast and prostate cancer	[88]
Breast	Elevated concentrations of taurine and choline-containing compound in cancer samples	Tissue biopsies	Diagnosis	[89]
Breast	Metabolic profiling	Serum samples	Clinical prediction for early detection of recurrent breast cancer	[90]
Autism	Increased concentrations of *N*-methyl-2-pyridone-5-carboxamide, *N*-methyl nicotinic acid, *N*-methyl nicotinamide, taurine and lower concentration of glutamate	Urine	Diagnosis	[91]
Alzheimer	Reductions of sphingomyelin and significant increases in two ceramide species (N16:0 and N21:0)	Plasma	Early detection, risk assessment and therapeutic monitoring	[92]
Schizophrenia and risperidone treatment	Citrate, palmitic acid, myoinositol and allantoin used to classify schizophrenia and myo-inositol, uric acid and tryptophan to distinguish between pretreatment and pretreatment patients	Plasma	Biomarkers	[93]
Three main psychotic disorders (schizophrenia, other nonaffective psychosis and affective psychosis)	Increased levels of glutamic acid in all disorders when compared with the healthy controls	Serum	Diagnosis	[94]

CONCLUDING REMARKS

Historically, metabolites have been used in simple tests to diagnose complex diseases, for example, the development of blood glucose test strips to test for diabetes and the measurement of phenylalanine in newborns to screen for phenylketonuria. Metabolomics has evolved from inborn errors of metabolism to an omics discipline that now covers virtually all diseases.

Being a relatively young omics discipline, metabolomics has already developed as an important integrative part of biology research. Metabolomics represents a novel approach, holding the promise to enable the detection of states of disease, monitor disease progression, and orient the choice of therapy by identifying individual responders and predicting toxicity. Despite the proven potential applications of metabolomics to diagnostic and prognostic screenings, the use of metabolomics in personalized medicine is still at an early stage. The current metabolomics technologies go beyond the scope of standard clinical chemistry techniques and are capable of precise analysis of thousands of metabolites. Metabolic profiling can be used in the discovery of biologically active molecules, diagnostic, prognostic and predictive markers and as an adjunct to other high-throughput technologies as an integration platform, among other applications. In addition, metabolic profiling can provide the biological and chemical basis for the identification of new radio chemicals for *in vivo* imaging.

One of the major actual drawbacks is the fact that the clinical community is largely unfamiliar with the field of metabolomics and its clinical uses. Yet, the challenge in the bioinformatics analysis of metabolomics data can bring about novel approaches to complex, high-throughput techniques.

To meet the growing challenges in metabolomics, a community-wide effort is required. The recently created Metabolomics Society represents a major step toward the progress of the discipline and its applicability in the medical environment. All this can only positively affect the increasingly complex era of molecular medicine.

REFERENCES

1. Rosenblum D, Peer D. Omics-based nanomedicine: The future of personalized oncology. *Cancer Lett* 2014, **352**(1):126–136.
2. Cacciatore S, Loda M. Innovation in metabolomics to improve personalized healthcare. *Ann N Y Acad Sci* 2015, **1346**(1):57–62.
3. Collino S, Martin FP, Rezzi S. Clinical metabolomics paves the way towards future healthcare strategies. *Br J Clin Pharmacol* 2013, **75**(3):619–629.
4. Johnson CH, Ivanisevic J, Siuzdak G. Metabolomics: Beyond biomarkers and towards mechanisms. *Nat Rev Mol Cell Biol* 2016, **17**(7):451.
5. Li S, Todor A, Luo R. Blood transcriptomics and metabolomics for personalized medicine. *Comput Struct Biotechnol J* 2016, **14**:1–7.
6. Zhang GF, Sadhukhan S, Tochtrop GP, Brunengraber H. Metabolomics, pathway regulation, and pathway discovery. *J Biol Chem* 2011, **286**(27):23631–23635.
7. Aboud OA, Weiss RH. New opportunities from the cancer metabolome. *Clin Chem* 2013, **59**(1):138–146.
8. Junot C, Fenaille F, Colsch B, Becher F. High resolution mass spectrometry based techniques at the crossroads of metabolic pathways. *Mass Spectrom Rev* 2014, **33**(6):471–500.
9. Zhao YY, Lin RC. UPLC-MS(E) application in disease biomarker discovery: The discoveries in proteomics to metabolomics. *Chem Biol Interact* 2014, **215**:7–16.
10. Kordalewska M, Markuszewski MJ. Metabolomics in cardiovascular diseases. *J Pharm Biomed Anal* 2015, **113**:121–136.
11. Fuhrer T, Zamboni N. High-throughput discovery metabolomics. *Curr Opin Biotechnol* 2015, **31**:73–78.
12. Nowicki S, Gottlieb E. Oncometabolites: Tailoring our genes. *The FEBS Journal* 2015, **282**(15):2796–2805.
13. Wishart DS, Tzur D, Knox C, Eisner R, Guo AC, Young N, Cheng D, Jewell K, Arndt D, Sawhney S et al. HMDB: The Human Metabolome Database. *Nucleic Acids Res* 2007, **35**(Database issue):D521–D526.
14. Villas-Boas SG, Mas S, Akesson M, Smedsgaard J, Nielsen J. Mass spectrometry in metabolome analysis. *Mass Spectrom Rev* 2005, **24**(5):613–646.
15. Roberts LD, Gerszten RE. Toward new biomarkers of cardiometabolic diseases. *Cell Metabolism* 2013, **18**(1):43–50.
16. Liesenfeld DB, Habermann N, Owen RW, Scalbert A, Ulrich CM. Review of mass spectrometry-based metabolomics in cancer research. *Cancer Epidemiol Biomarkers Prev* 2013, **22**(12):2182–2201.
17. Naz S, Vallejo M, Garcia A, Barbas C. Method validation strategies involved in non-targeted metabolomics. *J Chromatogr A* 2014, **1353**:99–105.
18. Gowda GA, Djukovic D. Overview of mass spectrometry-based metabolomics: Opportunities and challenges. *Methods Mol Biol* 2014, **1198**:3–12.
19. Lei Z, Huhman DV, Sumner LW. Mass spectrometry strategies in metabolomics. *J Biol Chem* 2011, **286**(29):25435–25442.
20. Griffiths WJ, Koal T, Wang Y, Kohl M, Enot DP, Deigner HP. Targeted metabolomics for biomarker discovery. *Angew Chem Int Ed Engl* 2010, **49**(32):5426–5445.
21. Clarke CJ, Haselden JN. Metabolic profiling as a tool for understanding mechanisms of toxicity. *Toxicol Pathol* 2008, **36**(1):140–147.
22. Kim HK, Choi YH, Erkelens C, Lefeber AW, Verpoorte R. Metabolic fingerprinting of Ephedra species using 1H-NMR spectroscopy and principal component analysis. *Chem Pharm Bull (Tokyo)* 2005, **53**(1):105–109.
23. Choi HK, Choi YH, Verberne M, Lefeber AW, Erkelens C, Verpoorte R. Metabolic fingerprinting of wild type and transgenic tobacco plants by 1H NMR and multivariate analysis technique. *Phytochemistry* 2004, **65**(7):857–864.
24. Lamers RJ, Wessels EC, van de Sandt JJ, Venema K, Schaafsma G, van der Greef J, van Nesselrooij JH. A pilot study to investigate effects of inulin on Caco-2 cells through in vitro metabolic fingerprinting. *J Nutr* 2003, **133**(10):3080–3084.
25. Allen J, Davey HM, Broadhurst D, Heald JK, Rowland JJ, Oliver SG, Kell DB. High-throughput classification of yeast mutants for functional genomics using metabolic footprinting. *Nat Biotechnol* 2003, **21**(6):692–696.
26. de Koning W, van Dam K. A method for the determination of changes of glycolytic metabolites in yeast on a subsecond time scale using extraction at neutral pH. *Anal Biochem* 1992, **204**(1):118–123.

27. Gonzalez B, Francois J, Renaud M. A rapid and reliable method for metabolite extraction in yeast using boiling buffered ethanol. *Yeast* 1997, **13**(14):1347–1355.
28. Villas-Boas SG, Hojer-Pedersen J, Akesson M, Smedsgaard J, Nielsen J. Global metabolite analysis of yeast: Evaluation of sample preparation methods. *Yeast* 2005, **22**(14):1155–1169.
29. Maharjan RP, Ferenci T. Global metabolite analysis: The influence of extraction methodology on metabolome profiles of Escherichia coli. *Anal Biochem* 2003, **313**(1):145–154.
30. Nicholson JK, Wilson ID. Opinion: Understanding 'global' systems biology: Metabonomics and the continuum of metabolism. *Nat Rev Drug Discov* 2003, **2**(8):668–676.
31. Sauer U, Heinemann M, Zamboni N. Genetics. Getting closer to the whole picture. *Science* 2007, **316**(5824):550–551.
32. Halket JM, Waterman D, Przyborowska AM, Patel RK, Fraser PD, Bramley PM. Chemical derivatization and mass spectral libraries in metabolic profiling by GC/MS and LC/MS/MS. *J Exp Bot*. 2005, **56**(410):219–243.
33. Soga T, Imaizumi M. Capillary electrophoresis method for the analysis of inorganic anions, organic acids, amino acids, nucleotides, carbohydrates and other anionic compounds. *Electrophoresis* 2001, **22**(16):3418–3425.
34. Soga T, Ueno Y, Naraoka H, Ohashi Y, Tomita M, Nishioka T. Simultaneous determination of anionic intermediates for Bacillus subtilis metabolic pathways by capillary electrophoresis electrospray ionization mass spectrometry. *Anal Chem* 2002, **74**(10):2233–2239.
35. Perrett D, Alfrzema L, Hows M, Gibbons J. Capillary electrophoresis for small molecules and metabolites. *Biochem Soc Trans* 1997, **25**(1):273–278.
36. Perrett D. Capillary electrophoresis in clinical chemistry. *Ann Clin Biochem* 1999, **36 (Pt 2)**:133–150.
37. Viant MR, Rosenblum ES, Tieerdema RS. NMR-based metabolomics: A powerful approach for characterizing the effects of environmental stressors on organism health. *Environ Sci Technol* 2003, **37**(21):4982–4989.
38. Krishnan P, Kruger NJ, Ratcliffe RG. Metabolite fingerprinting and profiling in plants using NMR. *J Exp Bot*. 2005, **56**(410):255–265.
39. Griffin JL. Metabonomics: NMR spectroscopy and pattern recognition analysis of body fluids and tissues for characterisation of xenobiotic toxicity and disease diagnosis. *Curr Opin Chem Biol* 2003, **7**(5):648–654.
40. Mendes P. Emerging bioinformatics for the metabolome. *Brief Bioinform* 2002, **3**(2):134–145.
41. Sumner LW, Mendes P, Dixon RA. Plant metabolomics: Large-scale phytochemistry in the functional genomics era. *Phytochemistry* 2003, **62**(6):817–836.
42. Churchill GA. Using ANOVA to analyze microarray data. *Biotechniques* 2004, **37**(2):173–175, 177.
43. Musumarra G, Barresi V, Condorelli DF, Scire S. A bioinformatic approach to the identification of candidate genes for the development of new cancer diagnostics. *Biol Chem* 2003, **384**(2):321–327.
44. Raamsdonk LM, Teusink B, Broadhurst D, Zhang N, Hayes A, Walsh MC, Berden JA, Brindle KM, Kell DB, Rowland JJ et al. A functional genomics strategy that uses metabolome data to reveal the phenotype of silent mutations. *Nat Biotechnol* 2001, **19**(1):45–50.
45. Eisen MB, Spellman PT, Brown PO, Botstein D. Cluster analysis and display of genome-wide expression patterns. *Proc Natl Acad Sci USA* 1998, **95**(25):14863–14868.
46. Tamayo P, Slonim D, Mesirov J, Zhu Q, Kitareewan S, Dmitrovsky E, Lander ES, Golub TR. Interpreting patterns of gene expression with self-organizing maps: Methods and application to hematopoietic differentiation. *Proc Natl Acad Sci USA* 1999, **96**(6):2907–2912.
47. Lander ES. Initial impact of the sequencing of the human genome. *Nature* 2011, **470**(7333):187–197.
48. Manolio TA, Collins FS, Cox NJ, Goldstein DB, Hindorff LA, Hunter DJ, McCarthy MI, Ramos EM, Cardon LR, Chakravarti A et al. Finding the missing heritability of complex diseases. *Nature* 2009, **461**(7265):747–753.
49. Guo L, Milburn MV, Ryals JA, Lonergan SC, Mitchell MW, Wulff JE, Alexander DC, Evans AM, Bridgewater B, Miller L et al. Plasma metabolomic profiles enhance precision medicine for volunteers of normal health. *Proc Natl Acad Sci USA* 2015, **112**(35):E4901–4910.
50. Lima AR, Bastos Mde L, Carvalho M, Guedes de Pinho P. Biomarker Discovery in Human Prostate Cancer: An Update in Metabolomics Studies. *Transl Oncol* 2016, **9**(4):357–370.
51. Sroka WD, Boughton BA, Reddy P, Roessner U, Slupski P, Jarzemski P, Dabrowska A, Markuszewski MJ, Marszall MP. Determination of amino acids in urine of patients with prostate cancer and benign prostate growth. *Eur J Cancer Prev* 2016.
52. Sreekumar A, Poisson LM, Rajendiran TM, Khan AP, Cao Q, Yu J, Laxman B, Mehra R, Lonigro RJ, Li Y et al. Metabolomic profiles delineate potential role for sarcosine in prostate cancer progression. *Nature* 2009, **457**(7231):910–914.
53. Priolo C, Pyne S, Rose J, Regan ER, Zadra G, Photopoulos C, Cacciatore S, Schultz D, Scaglia N, McDunn J et al. AKT1 and MYC induce distinctive metabolic fingerprints in human prostate cancer. *Cancer Res* 2014, **74**(24):7198–7204.
54. Almuhaideb A, Papathanasiou N, Bomanji J. 18F-FDG PET/CT imaging in oncology. *Ann Saudi Med* 2011, **31**(1):3–13.
55. Hess S, Blomberg BA, Zhu HJ, Hoilund-Carlsen PF, Alavi A. The pivotal role of FDG-PET/CT in modern medicine. *Acad Radiol* 2014, **21**(2):232–249.
56. Brindle KM. Imaging metabolism with hyperpolarized (13) C-labeled cell substrates. *J Am Chem Soc* 2015, **137**(20):6418–6427.
57. Sellers K, Fox MP, Bousamra M, 2nd, Slone SP, Higashi RM, Miller DM, Wang Y, Yan J, Yuneva MO, Deshpande R et al. Pyruvate carboxylase is critical for non-small-cell lung cancer proliferation. *J Clin Invest* 2015, **125**(2):687–698.
58. Mayers JR, Wu C, Clish CB, Kraft P, Torrence ME, Fiske BP, Yuan C, Bao Y, Townsend MK, Tworoger SS et al. Elevation of circulating branched-chain amino acids is an early event in human pancreatic adenocarcinoma development. *Nat Med* 2014, **20**(10):1193–1198.
59. Rhee EP, Cheng S, Larson MG, Walford GA, Lewis GD, McCabe E, Yang E, Farrell L, Fox CS, O'Donnell CJ et al. Lipid profiling identifies a triacylglycerol signature of insulin resistance and improves diabetes prediction in humans. *J Clin Invest* 2011, **121**(4):1402–1411.
60. Wang TJ, Larson MG, Vasan RS, Cheng S, Rhee EP, McCabe E, Lewis GD, Fox CS, Jacques PF, Fernandez C et al. Metabolite profiles and the risk of developing diabetes. *Nat Med* 2011, **17**(4):448–453.
61. Mapstone M, Cheema AK, Fiandaca MS, Zhong X, Mhyre TR, MacArthur LH, Hall WJ, Fisher SG, Peterson DR, Haley JM et al. Plasma phospholipids identify antecedent memory impairment in older adults. *Nat Med* 2014, **20**(4):415–418.
62. Suhre K, Meisinger C, Doring A, Altmaier E, Belcredi P, Gieger C, Chang D, Milburn MV, Gall WE, Weinberger KM et al. Metabolic footprint of diabetes: A multiplatform metabolomics study in an epidemiological setting. *PLoS One* 2010, **5**(11):e13953.
63. Roberts LD, Koulman A, Griffin JL. Towards metabolic biomarkers of insulin resistance and type 2 diabetes: Progress from the metabolome. *Lancet Diabetes Endo* 2014, **2**(1):65–75.
64. Osborn MP, Park Y, Parks MB, Burgess LG, Uppal K, Lee K, Jones DP, Brantley MA, Jr. Metabolome-wide association study of neovascular age-related macular degeneration. *PLoS One* 2013, **8**(8):e72737.

65. Fitzpatrick AM, Park Y, Brown LA, Jones DP. Children with severe asthma have unique oxidative stress-associated metabolomic profiles. *J Allergy Clin Immunol* 2014, **133**(1):258–261, e251–e258.
66. Roede JR, Uppal K, Park Y, Lee K, Tran V, Walker D, Strobel FH, Rhodes SL, Ritz B, Jones DP. Serum metabolomics of slow vs. rapid motor progression Parkinson's disease: A pilot study. *PLoS One* 2013, **8**(10):e77629.
67. Kalhan SC, Guo L, Edmison J, Dasarathy S, McCullough AJ, Hanson RW, Milburn M. Plasma metabolomic profile in nonalcoholic fatty liver disease. *Metabolism* 2011, **60**(3):404–413.
68. Weiner J, 3rd, Parida SK, Maertzdorf J, Black GF, Repsilber D, Telaar A, Mohney RP, Arndt-Sullivan C, Ganoza CA, Fae KC et al. Biomarkers of inflammation, immunosuppression and stress with active disease are revealed by metabolomic profiling of tuberculosis patients. *PLoS One* 2012, **7**(7):e40221.
69. Miller MJ, Kennedy AD, Eckhart AD, Burrage LC, Wulff JE, Miller LA, Milburn MV, Ryals JA, Beaudet AL, Sun Q et al. Untargeted metabolomic analysis for the clinical screening of inborn errors of metabolism. *J Inherit Metab Dis* 2015, **38**(6):1029–1039.
70. Clayton TA, Lindon JC, Cloarec O, Antti H, Charuel C, Hanton G, Provost JP, Le Net JL, Baker D, Walley RJ et al. Pharmaco-metabonomic phenotyping and personalized drug treatment. *Nature* 2006, **440**(7087):1073–1077.
71. Winterkamp S, Weidenhiller M, Otte P, Stolper J, Schwab D, Hahn EG, Raithel M. Urinary excretion of N-methylhistamine as a marker of disease activity in inflammatory bowel disease. *Am J Gastroenterol* 2002, **97**(12):3071–3077.
72. Bjerrum JT, Nielsen OH, Hao F, Tang H, Nicholson JK, Wang Y, Olsen J. Metabonomics in ulcerative colitis: Diagnostics, biomarker identification, and insight into the pathophysiology. *J Proteome Res* 2010, **9**(2):954–962.
73. Le Gall G, Noor SO, Ridgway K, Scovell L, Jamieson C, Johnson IT, Colquhoun IJ, Kemsley EK, Narbad A. Metabolomics of fecal extracts detects altered metabolic activity of gut microbiota in ulcerative colitis and irritable bowel syndrome. *J Proteome Res* 2011, **10**(9):4208–4218.
74. Martin FP, Rezzi S, Philippe D, Tornier L, Messlik A, Holzlwimmer G, Baur P, Quintanilla-Fend L, Loh G, Blaut M et al. Metabolic assessment of gradual development of moderate experimental colitis in IL-10 deficient mice. *J Proteome Res* 2009, **8**(5):2376–2387.
75. Murdoch TB, Fu H, MacFarlane S, Sydora BC, Fedorak RN, Slupsky CM. Urinary metabolic profiles of inflammatory bowel disease in interleukin-10 gene-deficient mice. *Anal Chem* 2008, **80**(14):5524–5531.
76. Williams HR, Cox IJ, Walker DG, North BV, Patel VM, Marshall SE, Jewell DP, Ghosh S, Thomas HJ, Teare JP et al. Characterization of inflammatory bowel disease with urinary metabolic profiling. *Am J Gastroenterol* 2009, **104**(6):1435–1444.
77. Marchesi JR, Holmes E, Khan F, Kochhar S, Scanlan P, Shanahan F, Wilson ID, Wang Y. Rapid and noninvasive metabonomic characterization of inflammatory bowel disease. *J Proteome Res* 2007, **6**(2):546–551.
78. Jansson J, Willing B, Lucio M, Fekete A, Dicksved J, Halfvarson J, Tysk C, Schmitt-Kopplin P. Metabolomics reveals metabolic biomarkers of Crohn's disease. *PLoS One* 2009, **4**(7):e6386.
79. Newgard CB, An J, Bain JR, Muehlbauer MJ, Stevens RD, Lien LF, Haqq AM, Shah SH, Arlotto M, Slentz CA et al. A branched-chain amino acid-related metabolic signature that differentiates obese and lean humans and contributes to insulin resistance. *Cell Metabolism* 2009, **9**(4):311–326.
80. Huffman KM, Shah SH, Stevens RD, Bain JR, Muehlbauer M, Slentz CA, Tanner CJ, Kuchibhatla M, Houmard JA, Newgard CB et al. Relationships between circulating metabolic intermediates and insulin action in overweight to obese, inactive men and women. *Diabetes Care* 2009, **32**(9):1678–1683.
81. Fiehn O, Garvey WT, Newman JW, Lok KH, Hoppel CL, Adams SH. Plasma metabolomic profiles reflective of glucose homeostasis in non-diabetic and type 2 diabetic obese African-American women. *PLoS One* 2010, **5**(12):e15234.
82. Barr J, Vazquez-Chantada M, Alonso C, Perez-Cormenzana M, Mayo R, Galan A, Caballeria J, Martin-Duce A, Tran A, Wagner C et al. Liquid chromatography-mass spectrometry-based parallel metabolic profiling of human and mouse model serum reveals putative biomarkers associated with the progression of nonalcoholic fatty liver disease. *J Proteome Res* 2010, **9**(9):4501–4512.
83. Feldstein AE, Lopez R, Tamimi TA, Yerian L, Chung YM, Berk M, Zhang R, McIntyre TM, Hazen SL. Mass spectrometric profiling of oxidized lipid products in human nonalcoholic fatty liver disease and nonalcoholic steatohepatitis. *J Lipid Res* 2010, **51**(10):3046–3054.
84. Makinen VP, Soininen P, Forsblom C, Parkkonen M, Ingman P, Kaski K, Groop PH, FinnDiane Study G, Ala-Korpela M. 1H NMR metabonomics approach to the disease continuum of diabetic complications and premature death. *Mol Syst Biol* 2008, **4**:167.
85. Kell DB. Metabolomic biomarkers: Search, discovery and validation. *Expert Rev Mol Diagn* 2007, **7**(4):329–333.
86. Denkert C, Budczies J, Kind T, Weichert W, Tablack P, Sehouli J, Niesporek S, Konsgen D, Dietel M, Fiehn O. Mass spectrometry-based metabolic profiling reveals different metabolite patterns in invasive ovarian carcinomas and ovarian borderline tumors. *Cancer Res* 2006, **66**(22):10795–10804.
87. Pasikanti KK, Esuvaranathan K, Ho PC, Mahendran R, Kamaraj R, Wu QH, Chiong E, Chan EC. Noninvasive urinary metabonomic diagnosis of human bladder cancer. *J Proteome Res* 2010, **9**(6):2988–2995.
88. Miyagi Y, Higashiyama M, Gochi A, Akaike M, Ishikawa T, Miura T, Saruki N, Bando E, Kimura H, Imamura F et al. Plasma free amino acid profiling of five types of cancer patients and its application for early detection. *PLoS One* 2011, **6**(9):e24143.
89. Li M, Song Y, Cho N, Chang JM, Koo HR, Yi A, Kim H, Park S, Moon WK. An HR-MAS MR metabolomics study on breast tissues obtained with core needle biopsy. *PLoS One* 2011, **6**(10):e25563.
90. Gu H, Pan Z, Xi B, Asiago V, Musselman B, Raftery D. Principal component directed partial least squares analysis for combining nuclear magnetic resonance and mass spectrometry data in metabolomics: Application to the detection of breast cancer. *Anal Chim Acta* 2011, **686**(1–2):57–63.
91. Yap IK, Angley M, Veselkov KA, Holmes E, Lindon JC, Nicholson JK. Urinary metabolic phenotyping differentiates children with autism from their unaffected siblings and age-matched controls. *J Proteome Res* 2010, **9**(6):2996–3004.
92. Han X, Rozen S, Boyle SH, Hellegers C, Cheng H, Burke JR, Welsh-Bohmer KA, Doraiswamy PM, Kaddurah-Daouk R. Metabolomics in early Alzheimer's disease: Identification of altered plasma sphingolipidome using shotgun lipidomics. *PLoS One* 2011, **6**(7):e21643.
93. Xuan J, Pan G, Qiu Y, Yang L, Su M, Liu Y, Chen J, Feng G, Fang Y, Jia W et al. Metabolomic profiling to identify potential serum biomarkers for schizophrenia and risperidone action. *J Proteome Res* 2011, **10**(12):5433–5443.
94. Oresic M, Tang J, Seppanen-Laakso T, Mattila I, Saarni SE, Saarni SI, Lonnqvist J, Sysi-Aho M, Hyotylainen T, Perala J et al. Metabolome in schizophrenia and other psychotic disorders: A general population-based study. *Genome Med* 2011, **3**(3):19.

Predictive Cancer Biomarkers *Current Practice and Future Challenges*

7.5

Vladimira Koudelakova and Marian Hajduch

Contents

INTRODUCTION

Personalized cancer treatment became a standard approach in the treatment of human cancers in twenty-first century. Among predictive cancer biomarkers, the estrogen and progesterone receptors (ER, PR) have the longest history: they have been used in the clinic to predict responses to endocrine therapy in breast cancer for over 40 years.[1] Similarly, the HER2 receptor status measurements have been used clinically to predict responsiveness to trastuzumab (Herceptin) in breast cancers for over 15 years. The success of the personalized approach in oncology has multiple determinants. Among others, most prominent is the nature of the disease, which is multifactorial in principle and requires patient stratification in order to select the most appropriate treatment to achieve meaningful therapeutic benefit and to substantiate the relative high cost of therapy. Thus, biomarker driven drug discovery is a widely accepted paradigm in biopharmaceutical industry and molecular tumor examination and became standard prerequisite for indication of targeted biological therapy in routine clinical practice. During the last decade, many new validated predictive biomarkers for several cancer types have been introduced into clinical practice, and several promising biomarkers are undergoing clinical evaluation. This review is to recapitulate the current predictive biomarkers used in clinical practice (Figure 7.5.1, Table 7.5.1).

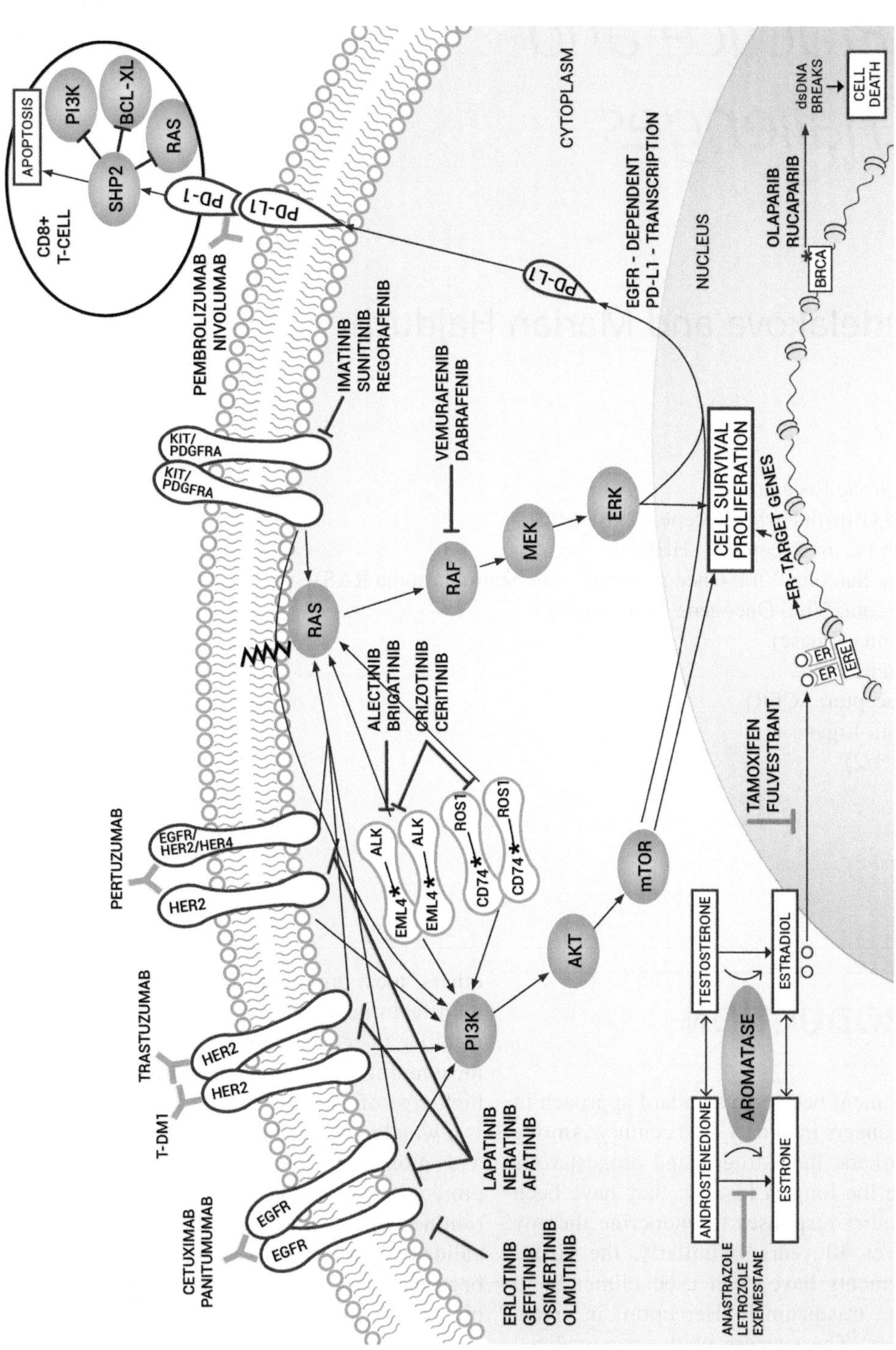

FIGURE 7.5.1 Targeting predictive biomarkers using monoclonal antibodies (green), tyrosine kinase inhibitors (red lines), hormone therapy (orange lines), and PARP inhibitors (red arrow). BRCA mutation is represented by a red asterisk; black asterisks symbolize the possible fusion partners of *ALK* (*KIF5B, TFG, HIP1, KLC1, TPR, STRN, DCTN1, SQSTM1* and *BIRC6*) and *ROS1* (*SLC34A2, TPM3, SDC4, EZR, FIG, LRIG3, KDELR2* and *CCDC6*).

TABLE 7.5.1 Predictive markers currently in use for FDA-approved therapies

BIOMARKER	*TESTING*	*ABERRATION TYPE*	*DRUG*	*DRUG ACTION*	*CANCER TYPE*	*SETTING*	*LINE OF THERAPY*
ER/PR	IHC	Expression	Tamoxifen	ER modulator	Breast cancer (pre/postmenopausal)	Neoadjuvant, adjuvant, palliative	First-line
			Anastrozole	Aromatase inhibitor	Breast cancer (postmenopausal)	Neoadjuvant, adjuvant, palliative	First-line
			Letrozole	Aromatase inhibitor	Breast cancer (postmenopausal)	Neoadjuvant, adjuvant, palliative	First-line
			Exemestane	Aromatase inhibitor	Breast cancer (postmenopausal)	Neoadjuvant, adjuvant, palliative	First-line
			Leuprolide	LH-RH analog	Breast cancer (premenopausal)	Adjuvant, palliative	First-line
			Goserelin	LH-RH analog	Breast cancer (premenopausal)	Adjuvant, palliative	First-line
			Fulvestrant	ER degrader	Breast cancer (postmenopausal)	Palliative	Beyond-line
HER2	IHC, FISH	Amplification/overexpression	Trastuzumab	MoAb (HER2)	Breast cancer	Neoadjuvant, adjuvant, palliative	First-line
					GEA	Palliative	First-line
			T-DM1	MoAb-drug conjugate (HER2)	Breast cancer	Palliative	Second-line
			Pertuzumab	MoAb (anti-HER2 dimerization domain)	Breast cancer	Neoadjuvant, palliative	First-line
			Lapatinib	Reversible pan-HER TKI	Breast cancer	Palliative	Second-line
			Neratinib	Irreversible pan-HER TKI	Breast cancer	Adjuvant	Second-line
EGFR	PCR, NGS	Exon 19del, L858R, G719X, L861X	Erlotinib	Reversible EGFR TKI (1st generation)	NSCLC	Palliative	First-line
			Gefitinib	Reversible EGFR TKI (1st generation)	NSCLC	Palliative	First-line
			Afatinib	Irreversible pan-HER TKI (2nd generation)	NSCLC	Palliative	First-line
		T790M	Osimertinib	Irreversible EGFR TKI (3rd generation)	NSCLC	Palliative	Second-line
			Olmutinib	Irreversible EGFR TKI (3rd generation)	NSCLC	Palliative	Second-line

(Continued)

TABLE 7.5.1 (*Continued*) Predictive markers currently in use for FDA-approved therapies

BIOMARKER	*TESTING*	*ABERRATION TYPE*	*DRUG*	*DRUG ACTION*	*CANCER TYPE*	*SETTING*	*LINE OF THERAPY*
ALK	FISH, IHC	ALK rearrangement/ expression	Crizotinib	Reversible ALK/ROS1/MET TKI (1st generation)	NSCLC	Palliative	First-line
		ALK rearrangement/ expression, including L1196M	Ceritinib	reversible ALK/IGF1/ ROS1 TKI (2nd generation)	NSCLC	Palliative	First-line, second-line
			Alectinib	Reversible ALK/RET TKI (2nd generation)	NSCLC	Palliative	First-line, second-line
			Brigatinib	Reversible ALK/EGFR TKI (2nd generation)	NSCLC	Palliative	Second-line
KRAS/NRAS	PCR, NGS	Exon 2 codon 12/13, exon 13 codon 59/61, exon 4 codon 117/146	Cetuximab	MoAb (EGFR)	CC	Palliative	First-line
			Panitumumab	MoAb (EGFR)	CC	Palliative	First-line
BRAF	PCR, NGS	V600E	Vemurafenib	Reversible BRAF TKI	Melanoma	Palliative	First-line
			Dabrafenib	Reversible BRAF TKI	Melanoma	Adjuvant, palliative	First-line
					NSCLC	Palliative	First-line (+ trametinib)
ROS1	FISH	ROS1 rearrangement	Crizotinib	Reversible ALK/ROS1/MET TKI (1st generation)	NSCLC	Palliative	First-line
			Ceritinib	reversible ALK/IGF1/ ROS1 TKI (2nd generation)	NSCLC	Palliative	First-line
PD-L1	IHC	Expression ≥50%	Pembrolizumab	MoAb (PD-1 inhibitor)	NSCLC	Palliative	First-line, second-line
			Nivolumab	MoAb (PD-1 inhibitor)	NSCLC	Palliative	Second-line
MSI-H/dMMR	PCR, IHC	BAT25, BAT26, D2S123, D5S346, D17S250/MLH1, MSH2, MSH6, PMS2	Pembrolizumab	MoAb (PD-1 inhibitor)	Solid tumors	Palliative	Beyond-line
KIT/PDGFRA	IHC, PCR, NGS	KIT expression, exon 11, 9,13, 17 (KIT); exon 18 (except D842V), 12, 14	Imatinib	Multi-targeted receptor TKI	GIST	Adjuvant, palliative	First-line
			Sunitinib	Multi-targeted receptor TKI	GIST	Palliative	Second-line
			Regorafenib	Multi-targeted receptor TKI	GIST	Palliative	Third-line
BRCA1/2	PCR, NGS	Germline BRCA mutation	Olaparib	PARP1 inhibitor	Ovarian cancer	Palliative	Beyond-line
					Breast cancer (HER2-)	Palliative	Beyond-line
			Rucaparib	PARP1/2/3 inhibitor	Ovarian cancer	Palliative	Beyond-line

Abbreviations: LH-RH, Luteinizing hormone-releasing hormone; ER, estrogen receptor; PR, progesterone receptor, IHC, immunohistochemistry; FISH, fluorescence in situ hybridization; PCR, polymerase chain reaction; NGS, next generation sequencing; dMMR, mismatch repair deficiency; MSI-H, microsatellite instability-high; MoAb, monoclonal antibody; TKI, tyrosinkinase inhibitor; GEA, gastroesophageal adenocarcinoma; NSCLC, non-small cell lung cancer; GIST, gastrointestinal stromal tumor; CC, colorectal cancer.

ER/PR (ESTROGEN/ PROGESTERONE RECEPTOR)

ER, i.e. ERα, is a transcription factor activated by estrogen (estradiol) binding that enhances the transcription of genes associated with cell proliferation such as MYC and CCND1 (cyclin D1).[2,3] ER surrogate stimulates the expression of PR, which is therefore used as a marker of ER function. However, in the presence of progestogens, PR associates with ERα and changes its chromatin binding location, resulting in the expression of genes associated with cell cycle arrest, apoptosis, and differentiation. Almost three-quarters of all breast cancer cases are ER/PR-positive.[1,2] Patients with ER+/PR+ tumors respond better to endocrine therapy than those with ER+/PR- tumors.[4–6] Although the wild-type (wt) ER is a predictor of responsiveness, mutations in the sequence encoding the ligand-binding domain of the ER gene (*ESR1*) are associated with acquired resistance. These mutations occur in 10%–40% of breast cancer recurrences and are particularly common after long-term treatment with aromatase inhibitors. It may be possible to overcome this resistance using next-generation selective estrogen receptor downregulators that are currently undergoing clinical evaluation.[7]

Immunohistochemical (IHC) analysis of ER/PR expression is mandatory for all newly diagnosed breast cancer cases and recurrences. The American Society for Clinical Oncology/College of American Pathologists (ASCO/CAP) has introduced a testing algorithm and validated methods for this purpose and recommends a threshold of at least 1% of ER/PR positive tumor nuclei for indication.[1,8] As predictive markers, ER and PR are used in neoadjuvant, adjuvant, and palliative settings (Table 7.5.1).

HER2 (HUMAN EPIDERMAL GROWTH FACTOR RECEPTOR 2; ERBB2)

The *HER2* gene belongs to the epidermal growth factor receptor family (EGFR/ERBB/HER), is localized at 17q12, and encodes the transmembrane tyrosine kinase p185. It is amplified and overexpressed in 15%–20% of all invasive breast cancer cases (Figures 7.5.2 and 7.5.3). When overexpressed, p185 forms homo- and heterodimers with EGFR and HER3, leading to constitutive activation of the downstream MAPK and PI3K/AKT pathways and, thus, enhancement of cell proliferation, invasion, and metastasis.[9,10] Compared to their HER2-negative counterparts, HER2-positive tumors are more aggressive in terms of nodal and distant metastasis, have lower ER levels, and are less responsive to chemotherapy. Despite this poor prognosis, HER2 positivity predicts responsiveness to anti-HER2 therapy. Several anti-HER2 therapeutics with different mechanisms of action are currently approved for use in neoadjuvant, adjuvant, and palliative settings (Table 7.5.1), and combinations of these agents could be used to overcome the serious problem of acquired resistance.[9]

In addition to its role in breast cancer, *HER2* amplification and/or overexpression has been described in 7%–38% cases of gastroesophageal adenocarcinoma (GEA), with a higher frequency in tumors localized at the gastroesophageal junction. The combination of trastuzumab therapy and chemotherapy has been approved for the treatment of HER2-positive GEA. Testing for HER2 positivity is recommended by the IHC as a screening method followed FISH (or other ISH) in IHC-equivocal (2+) cases of breast cancer as well as GEA.[11,12]

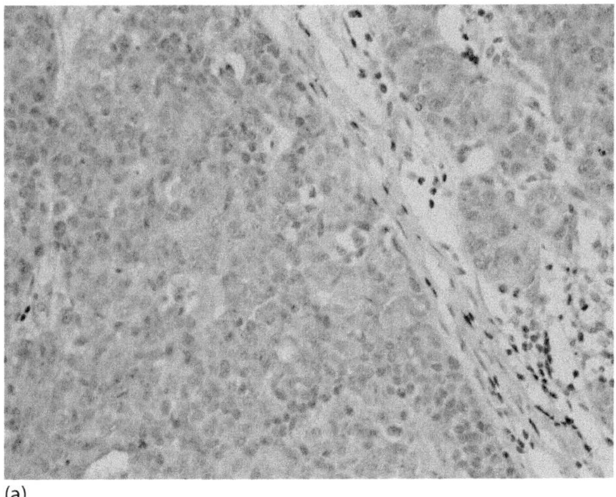
(a)

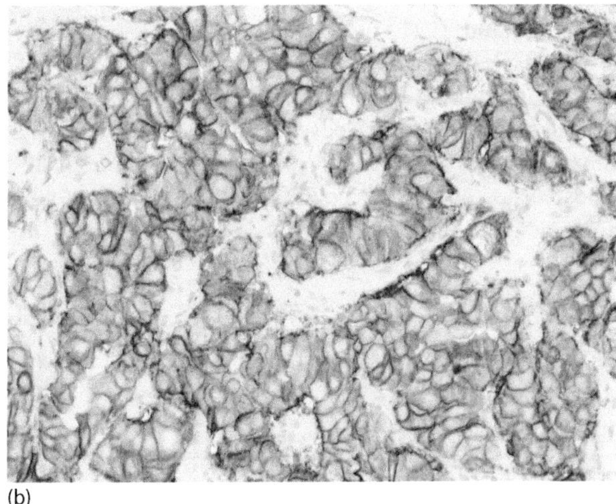
(b)

FIGURE 7.5.2 Example of immunohistochemical HER2 protein expression analyses (HercepTest) for a HER2-negative breast cancer patient with no membrane staining and an IHC score of 0 (a) and a HER2-positive patient with complete intense membrane staining and an IHC score of 3+ (b).

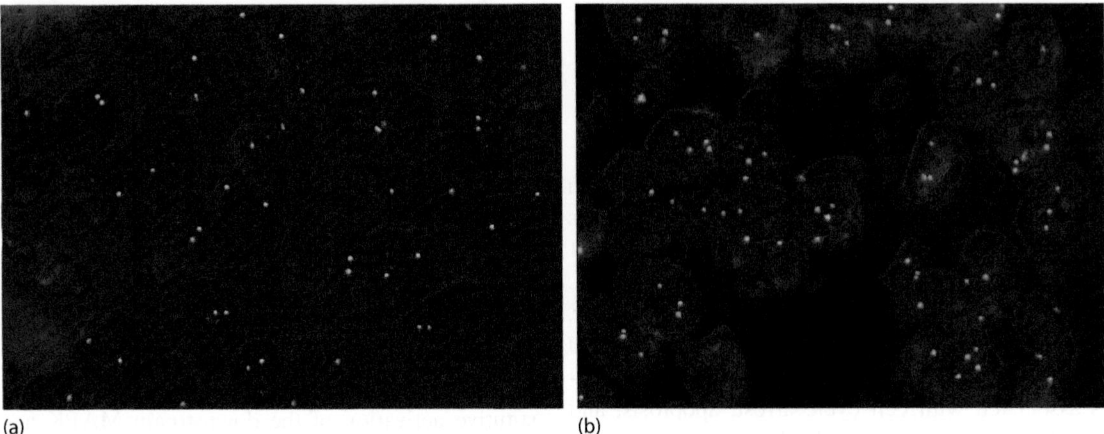

FIGURE 7.5.3 Example of HER2 copy number assessments using fluorescent *in situ* hybridization for a breast cancer patient with a normal HER2 copy number (a) and one with HER2 amplification (b).

EGFR (EPIDERMAL GROWTH FACTOR RECEPTOR 1; HER1)

The epidermal growth factor receptor (*EGFR*) gene is located on 7p11 and encodes a tyrosine-kinase receptor from the HER family that is involved in the development, progression, angiogenesis, and metastasis of various cancer types.[13] Three mechanisms of EGFR activation in tumor cells have been described: mutation, amplification/gene copy number gain (CNG), and overexpression. However, only mutation has been validated as a predictive biomarker in non-small cell lung cancer (NSCLC).

Activating mutations causing independent *EGFR* activation have been detected in approximately 30%–40% of Asian and 10%–15% of Caucasian NSCLC cases and are most common in non/former smokers, females, and patients with adenocarcinoma. In-frame deletions in exon 19 (bases 746–753) and the L858R substitution in exon 21 are the most common alterations, being found in over 90% of all cases, followed by mutations in exons 18 (G719X) and 21 (L861X).[14,15] These *EGFR* mutations predict responsiveness to the first and second generation of tyrosine-kinase inhibitors (TKIs) (Table 7.5.1). A serious problem is acquired resistance, which develops in almost all initially responsive patients and is caused by a secondary T790M substitution in most cases (60%). This could potentially be overcome by the third generation of TKIs (Table 7.5.1).[16,17]

KRAS/NRAS (KIRSTEN RAT SARCOMA VIRAL ONCOGENE HOMOLOG/ NEUROBLASTOMA RAS)

The *KRAS* gene, localized on 12p12, encodes a membrane-bound GTPase protein that, alongside other members of the RAS protein family (NRAS and HRAS), plays an important role in EGFR-mediated signal transduction (RAF/MAPK) and, thus, regulates cellular processes including proliferation, differentiation, and survival. KRAS mutations have been detected in 35%–45% of patients with colorectal cancer (CRC). The most common mutations are G12V and G12D in codon 12 of exon 2, and G13D in exon 13 of codon 2. Less commonly mutated positions are codons 59 and 61 in exon 3 and codons 117 and 146 in exon 4. Mutated KRAS proteins accumulate in the GTP-bound active form that constitutively activates the downstream signaling cascade irrespective of upstream EGFR signalling.[18]

Similarly, mutations in codons 12, 13, and 61 of *NRAS* (localized on 1p13) have been detected in 3%–5% of all colorectal cancer cases, with the most common mutations being Q61K and Q61R.[19] CRC patients with *KRAS* and *NRAS* mutations are resistant to anti-EGFR antibodies (cetuximab/panitumumab) and, thus, are excluded from therapy.[20] Although *KRAS* mutations are detected in about 30% of lung adenocarcinoma cases, they are mutually exclusive with EGFR activating mutations and ALK rearrangements[21] and have been reported to associate with unresponsiveness to synthetic EGFR inhibitors (gefitininb, erlotinib). However, *KRAS* testing is not recommended in NSCLC. *KRAS* mutations are also common in other cancer types, including pancreatic and thyroid cancers, with no recognized predictive value in clinical practice.

BRAF (V-RAF MURINE SARCOMA VIRAL ONCOGENE HOMOLOG B)

The *BRAF* gene, localized on 7q34, encodes a serine/threonine kinase directly downstream of KRAS in the MAP kinase signaling cascade. *BRAF* mutations have been reported in approximately 60% of malignant melanoma cases, with the V600E substitution being observed in the majority. As with *RAS*, *BRAF* mutation leads to constitutive activation of the proliferative

MAPK pathway. Targeted BRAF inhibitors (vemurafenib and dabrafenib) are approved for the treatment of *BRAF*-mutated advanced malignant melanoma.[22] Interestingly, BRAF inhibitors selectively inhibit melanoma cells with *BRAF* mutations and can activate BRAF in wt-*BRAF* cells, thus stimulating their growth.[23] A vertical inhibition strategy targeting different steps in the same pathway by combining inhibitors of BRAF and MEK (trametinib/cobimetinib) has recently become the standard of care for advanced *BRAF*-mutated malignant melanoma.[24] Similarly, the dabrafenib/trametinib combination has been approved for treatment of advanced *BRAF*-mutated NSCLC.

BRAF mutations have also been detected (at lower rates) in colorectal, lung, and ovarian cancer, as well as in gastrointestinal stromal tumors. Their prevalence in colorectal cancer patients is 5%–10%, and they are probably associated with resistance to anti-EGFR antibodies. However, the predictive capacity of such mutations is unclear due to the rarity of the V600E mutation.[25] Moreover, the V600E *BRAF* mutation is present in approximately two-thirds of microsatellite-unstable CRC caused by MLH1 hypermethylation/protein loss, but not in CC with Lynch syndrome. BRAF testing is therefore recommended for Lynch syndrome risk evaluation in CRC.[20]

ALK (ANAPLASTIC LYMPHOMA KINASE)

The *ALK* gene belongs to the transmembrane insulin receptor tyrosine kinase family, is localized on 2p23, and is active in the development of the nervous system. Its native expression in adults is negligible. However, aberrant cytoplasmic expression occurs in 3%–5% of all NSCLC cases, usually as a result of an inversion on the 2p chromosome resulting in a fusion with echinoderm microtubule associated protein like-4 (*EML4*). ALK kinase activity can also be activated by fusion with *KIF5B*, *TFG*, *HIP1*, *KLC1*, *TPR*, *STRN*, *DCTN1*, *SQSTM1*, or *BIRC6*.[26,27] These rearrangements lead to constitutive ligand-independent activity, resulting in activation of the RAS/MAPK, PI3K/AKT, and JAK/STAT pathways. *ALK* rearrangement is usually mutually exclusive (i.e. it occurs without parallel mutations in *EGFR* or *KRAS*) and is most common in younger patients, non/former smokers, and patients with adenocarcinoma histology.[26,28] *ALK* rearrangement has also been identified in anaplastic large cell lymphoma, renal cell, breast, esophageal, and colorectal cancers.

Aberrant ALK activity can be blocked by crizotinib in advanced NSCLC. Despite its high efficacy in ALK-positive NSCLC, most patients develop resistance to this TKI by various mechanisms, e.g. secondary mutations in the *ALK* kinase domain (usually L1196M), *ALK* CNG/amplification, or bypass signaling through aberrant activation of an alternative kinase (*EGFR* or *KIT*). This resistance can be partially overcome by next generation ALK inhibitors, which are more potent and have superior blood-brain barrier penetration (Table 7.5.1).[28,29] *ALK* rearrangement should be tested for together with EGFR mutation in all advanced-stage lung adenocarcinoma or lung tumors with an adenocarcinoma component.[21]

ROS1 (ROS PROTO-ONCOGENE 1)

ROS1, localized on 6q22, encodes a transmembrane tyrosine-kinase receptor from the insulin receptor family that is structurally similar to the *ALK* oncogene product. Like those of *ALK*, *ROS1* rearrangements result in fusion followed by receptor dimerization that causes constitutive kinase activity and oncogenic signaling. Such rearrangements have been observed in several cancer types, including NSCLC, colorectal, ovarian, and gastric cancer, glioblastoma multiforme, and cholangiocarcinomas. *ROS1* fusions occur in about 2% of all NSCLC cases, and 9 fusion partners have been identified in NSCLC to date. The most common fusion partner is *CD74* (>30%), followed by *SLC34A2* (17%), *TPM3* (15%), *SDC4* (11%), *EZR* (6%), and more rarely, *FIG*, *LRIG3*, *KDELR2*, and *CCDC6*. *ROS1* fusions are typically found in adenocarcinoma and are most common in younger female patients and non/former smokers.[30,31]

NSCLC patients with *ROS1* rearrangement are sensitive to crizotinib therapy, which has been approved for advanced ROS1-positive NSCLC. Most treated patients develop acquired resistance, which hopefully will be overcome by next-generation TKIs that have yielded promising results in ongoing clinical trials (e.g. entrectinib).[32]

KIT (STEM CELL FACTOR RECEPTOR; SCFR)

The *KIT* gene, localized on 4q11, encodes a transmembrane tyrosine-kinase SCF receptor from the platelet-derived growth factor receptor family, whose members regulate proliferation, differentiation, and survival. Activating *KIT* mutations are associated with gastrointestinal stromal tumors (GIST) in which their prevalence is above 90%. These mutations cause ligand-independent dimerization and constitutive activation of KIT signaling. KIT is expressed in almost all GIST cases and is therefore used for GIST diagnosis. Activation *KIT* mutations in exon 11 (65%) are most common, followed by mutations in exons 9 (10%), 13 (2%), and 17 (1%). Mutations in alpha-type platelet-derived growth factor receptor (*PDGFRA*), localized on 4q12, occurs in another 5% of GIST cases. *PDGFRA* mutations occur mainly in exon 18 and, more rarely, in exons 12 and 14. *KIT* and *PDGFRA* mutations seem to be mutually exclusive.[33,34]

KIT expression serves as both a diagnostic marker and a predictive marker of the response to KIT TKI therapies (Table 7.5.1), which are approved for treatment of KIT/PDGFR-positive GIST in palliative settings; imatinib is also approved for use in adjuvant settings. The effectiveness of TKI therapy depends on the mutation type: tumors with exon 11 mutations are most sensitive to imatinib, whereas those with exon 9 mutations respond to higher imatinib doses and to sunitinib therapy. Tumors with *PDGFRA* mutations are sensitive to imatinib except those with the D842V substitution (exon 18), which do not respond to imatinib. There is currently no approved treatment for patients with the D842V mutation, but clinical studies with the novel PDGFR inhibitor crenolanib are underway.[34,35]

PD-L1 (PROGRAMMED DEATH-LIGAND 1)

PD-L1 is a transmembrane protein that binds to the PD1 (Programmed death 1) receptor, which is expressed on activated cytotoxic T-cells. This interaction activates a signaling pathway leading to reduced T-cell proliferation and apoptosis induction. This mechanism was found to be the main mechanism by which NSCLC, but also other solid tumors evades detection and elimination by the immune system.[36]

Immune cell infiltration is common; it has been known to occur for a long time in many tumor types and could serve as prognostic factor. Immune checkpoint inhibition is an effective treatment strategy for some cancers (e.g. melanoma), but there is currently no clear predictor of tumor responsiveness. Several clinical studies have shown that NSCLC cases in which at least 50% of tumor cells express PD-L1 have twice the reference response rate to anti-PD1 agents (pembrolizumab, nivolumab). PD-L1 IHC testing is therefore recommended for advanced NSCLC patients with negative or unknown test results for *EGFR* mutations, *BRAF* mutations, and *ALK* and *ROS1* rearrangements. More recently, pembrolizumab was approved for treatment of unresectable or metastatic solid tumors with mismatch repair deficiency or microsatellite instability. Interestingly, this is the first FDA approval based on a specific genetic feature without tumor origin specification.[37] Recent studies also indicate positive predictive value of higher tumor mutation burden in response to the check-point inhibitors.

BRCA1/2 (BREAST CANCER 1/2)

The breast cancer 1 and 2 (*BRCA1/2*) genes located on 17q21 and 13q12 are tumor suppressor genes that play essential roles in DNA double-strand break repair by homologous recombination and the maintenance of DNA stability. Mutations in *BRCA* genes lead to defects in the double-strand break repair system, accumulation of DNA alterations, and substantial genomic instability. Inherited *BRCA1/2* mutations may cause hereditary breast and ovarian cancer syndrome with elevated risks of breast (50%–70%) and ovarian cancer (15%–55%) and other malignancies (e.g. prostate and pancreatic cancer). The population frequency of pathogenic *BRCA1/2* mutations is 1:400. *BRCA1/2* mutations are observed in about 5%–10% of breast cancer cases and account for a particularly high proportion of triple-negative breast cancers (TNBC). They are also found in 15% of ovarian cancer cases.[38,39]

DNA in cancer cells with damaged *BRCA* genes is repaired by alternative mechanisms. Since the poly(ADP-ribose) polymerase (PARP) plays an important role in base excision repair as well as double-strand repair mechanisms, PARP inhibition increases the occurrence of irreparable toxic DNA double-strand breaks resulting in cell death. PARP inhibitors therefore effectively kill the tumor cells through the principle of synthetic lethality. The PARP inhibitors olaparib and rucaparib have been approved for the treatment of advanced previously treated ovarian cancer with germline *BRCA* mutation. More recently, olaparib was also approved for treatment of metastatic HER2-negative breast cancer with *BRCA* mutation previously treated by chemotherapy.[39]

TESTING

Testing of predictive biomarkers is usually required by the summary of product characteristics (SPC) for indication of targeted therapy. SPC is approved by regulatory authorities and is an essential part of publicly available drug documentation. Practical recommendations and standards are given by learned societies such as the ASCO, the CAP, the Association for Molecular Pathology (AMP), the National Comprehensive Cancer Network (NCCN), and (for lung cancer) the International Association for the Study of Lung Cancer (IASLC). The process of validating the test, diagnostic criteria, and interpretation differ between genes and tumor types, and tests must be performed by pathologists or molecular biologist with subspecialty expertise, integrated knowledge, and sufficient experience. Therefore, to maximize the accuracy and reliability of test results, predictive biomarkers should be tested in central laboratories rather than local laboratories. High discrepancies between test results obtained in local and central laboratories have been reported in HER2 testing.[40,41] Moreover, molecular testing should be conducted as soon as possible to avoid delays in allocating patients to appropriate therapies.

Immunohistochemistry is recommended as a method of choice for several markers (e.g. ER, PR, PD-L1, KIT) because it is easy, fast, inexpensive, and can be performed in most laboratories. Conversely, the FISH and NGS techniques are more expensive and time consuming, requiring special equipment and personnel training. In HER2 testing, IHC is used as a screening method, with ISH being used to clarify equivocal results. Similarly for ALK detection, both IHC and FISH could be used, with ISH being used to clarify weak positive results. For ROS1 detection, the recommendations are non-uniform; FISH remains the gold standard for the moment pending clinical validation of IHC methods via clinical trials that will allow test results to be correlated with clinical responses to ROS1 inhibitors and survival analysis.[42] The main advantage of PCR and NGS techniques is that they require minimal quantities of sample, can be used to test unusual samples (e.g. plasma, blood for circulating DNA based biomarkers), and allow several genes to be tested simultaneously. Therefore, they are becoming increasingly widely used.

The most common sample type for testing is tissue preserved in a paraffin block, which is usually used for pathologic classification. Samples should be retested after tumor progression because of the possibility of essential changes in tumor genome. However, repeated sampling is often impossible in advanced cancer patients. Consequently, there is an urgent need to validate new sample types for testing (e.g. circulating tumor cells or cell-free circulating DNA), with some examples already implemented in clinical practice (e.g. EGFR mutations analysis from circulating tumor DNA).

FUTURE CHALLENGES

Over the last few decades, intensive research efforts have been focused on tumor biology and on identifying and clinically validating prognostic, predictive, and diagnostic biomarkers. High-throughput analysis tools such as sequencing and microarray technologies have the potential to elucidate the nature of several cancer types, but well-defined and extensive datasets will be needed to make this possible.

The discovery of genetic changes affecting *EGFR*, *ALK*, *ROS1*, and *BRAF* significantly improves the outcomes of small but important subsets of NSCLC patients. Almost one-third of lung adenocarcinoma patients harbor the *KRAS* mutation, but despite decades of intensive research on various strategies, KRAS is not yet targetable. However, new oncogenes, such as *MET* and *HER2* mutants, and *RET* (rearranged during transfection) and *NTRK1* (neurotrophic tyrosine receptor kinase) rearrangements have joined the list of potential targetable drivers.

The *RET* rearrangement is reported in 1%–2% of NSCLC cases and is mutually exclusive with other NSCLC gene alterations, including *EGFR*, *ALK*, and *ROS1*. *RET* rearrangement is targetable by several multi-kinase inhibitors, the most promising being alectinib. The estimated prevalence of *NTRK1* rearrangement in NSCLC patients is 0.1% and it is targetable by several TKIs (e.g. entrectinib and larotrectinib) that are currently undergoing clinical evaluation. Exon splicing mutations in the *MET* gene are found in 4% of NSCLC patients (exon 14), while *MET* copy number gains are seen in 2%–5% of newly diagnosed lung adenocarcinoma cases and almost 20% of NSCLC patients treated with EGFR TKI. Patients with *MET* alterations could benefit from MET TKI (e.g. crizotinib, cabozantinib) or monoclonal antibody (e.g. emibetuzumab, rilotuzumab) therapies. *HER2* mutations in exon 20 occur in 1%–3% of lung adenocarcinomas and are targetable by dual EGFR/HER2 TKI (afatinib) and pan-HER inhibitors (neratinib, dacomitinib).[43–45]

Current predictive biomarkers have significantly improved the prognosis of many breast cancer patients, but two major outstanding problems are therapy resistance and the high heterogeneity of triple-negative breast cancer (TNBC); i.e. breast cancers that are ER-, PR-, and HER2-negative. Despite intensive research, there is a lack of validated biomarkers for TNBC, although several promising predictive biomarkers are being evaluated in clinical studies. The PI3K-AKT-mTOR pathway is the predominant and most commonly altered oncogenic pathway in BC including TNBC, with *PTEN* and *INPP4B* losses and *PIK3CA* mutations being particularly common. These alterations are frequently found in other tumor types (e.g. colorectal, prostate, ovarian cancers, melanoma) and are targetable by mTOR, PI3K, AKT, and mTOR/PI3K inhibitors (e.g. everolimus, temsirolimus, or BKM120).[46] Androgen receptor positivity is another promising biomarker present in 13%–37% of TNBC cases; it is targetable using antiandrogens (e.g. bicalutamide or enzalutamide) in monotherapies or in combination with CDK and/or PI3K inhibitors or chemotherapy.[47]

The number of predictive biomarkers transferred to clinical practice will continue to increase in parallel with the number of new drugs in the coming years. Major outstanding challenges in cancer research include extremely heterogeneous cancer subtypes (e.g. TNBC), cancers with treatment limitations (e.g. brain tumors), resistance to gold standard treatments and ways of overcoming it, the toxicity of some combination therapies, and changes in tumor biology caused by initial therapy. Well-advised clinical studies with stringent criteria and logical endpoints will be essential in overcoming these problems. However, despite these issues, there has been immense progress in cancer research over the last few years, and targeted therapeutic approaches have saved huge numbers of human lives.

ACKNOWLEDGEMENTS

The research on prognostic, predicitive and diagnostic cancer biomarkers was supported by the Czech Ministry of School and Education within the National Suistainability Program I (grant No. LO 1304).

REFERENCES

1. Davies C, Godwin J, Gray R, Clarke M, Cutter D, Darby S, McGale P, et al.: Relevance of breast cancer hormone receptors and other factors to the efficacy of adjuvant tamoxifen: Patient-level meta-analysis of randomised trials. *Lancet* 2011, 378: 771–784.
2. Carroll JS: Mechanisms of oestrogen receptor (ER) gene regulation in breast cancer. *Eur J Endocrinol* 2016, 175:R41–R49.
3. Ikeda K, Horie-Inoue K, Inoue S: Identification of estrogen-responsive genes based on the DNA binding properties of estrogen receptors using high-throughput sequencing technology. *Acta Pharmacol Sin* 2015, 36: 24–31.
4. Nordenskjold A, Fohlin H, Fornander T, Lofdahl B, Skoog L, Stal O: Progesterone receptor positivity is a predictor of long-term benefit from adjuvant tamoxifen treatment of estrogen receptor positive breast cancer. *Breast Cancer Res Treat* 2016, 160: 313–322.
5. Mohammed H, Russell IA, Stark R, Rueda OM, Hickey TE, Tarulli GA, Serandour AA, et al.: Progesterone receptor modulates ERalpha action in breast cancer. *Nature* 2015, 523: 313–317.
6. Carroll JS, Hickey TE, Tarulli GA, Williams M, Tilley WD: Deciphering the divergent roles of progestogens in breast cancer. *Nat Rev Cancer* 2017, 17: 54–64.
7. Nicolini A, Ferrari P, Duffy MJ: Prognostic and predictive biomarkers in breast cancer: Past, present and future. *Semin Cancer Biol* 2017.

8. Hammond ME, Hayes DF, Wolff AC, Mangu PB, Temin S: American society of clinical oncology/college of American pathologists guideline recommendations for immunohistochemical testing of estrogen and progesterone receptors in breast cancer. *J Oncol Pract* 2010, 6: 195–197.
9. Rimawi MF, Schiff R, Osborne CK: Targeting HER2 for the treatment of breast cancer. *Annu Rev Med* 2015, 66: 111–128.
10. Gutierrez C, Schiff R: HER2: Biology, detection, and clinical implications. *Arch Pathol Lab Med* 2011, 135: 55–62.
11. Bartley AN, Washington MK, Colasacco C, Ventura CB, Ismaila N, Benson AB, III, Carrato A, et al.: HER2 testing and clinical decision making in gastroesophageal adenocarcinoma: Guideline from the College of American Pathologists, American Society for Clinical Pathology, and the American Society of Clinical Oncology. *J Clin Oncol* 2017, 35: 446–464.
12. Wolff AC, Hammond ME, Hicks DG, Dowsett M, McShane LM, Allison KH, Allred DC, et al.: Recommendations for human epidermal growth factor receptor 2 testing in breast cancer: American Society of Clinical Oncology/College of American Pathologists clinical practice guideline update. *J Clin Oncol* 2013, 31: 3997–4013.
13. Huang SM, Harari PM: Epidermal growth factor receptor inhibition in cancer therapy: Biology, rationale and preliminary clinical results. *Invest New Drugs* 1999, 17: 259–269.
14. Haspinger ER, Agustoni F, Torri V, Gelsomino F, Platania M, Zilembo N, Gallucci R, Garassino MC, Cinquini M: Is there evidence for different effects among EGFR-TKIs? Systematic review and meta-analysis of EGFR tyrosine kinase inhibitors (TKIs) versus chemotherapy as first-line treatment for patients harboring EGFR mutations. *Crit Rev Oncol Hematol* 2015, 94: 213–227.
15. Koudelakova V, Kneblova M, Trojanec R, Drabek J, Hajduch M: Non-small cell lung cancer—genetic predictors. *Biomed Pap Med Fac Univ Palacky Olomouc Czech Repub* 2013, 157: 125–136.
16. Kuiper JL, Heideman DA, Thunnissen E, Paul MA, van Wijk AW, Postmus PE, Smit EF: Incidence of T790M mutation in (sequential) rebiopsies in EGFR-mutated NSCLC-patients. *Lung Cancer* 2014, 85: 19–24.
17. Russo A, Franchina T, Ricciardi GRR, Smiroldo V, Picciotto M, Zanghi M, Rolfo C, Adamo V: Third generation EGFR TKIs in EGFR-mutated NSCLC: Where are we now and where are we going. *Crit Rev Oncol Hematol* 2017, 117: 38–47.
18. Tan C, Du X: KRAS mutation testing in metastatic colorectal cancer. *World J Gastroenterol* 2012, 18: 5171–5180.
19. Okada Y, Miyamoto H, Goji T, Takayama T: Biomarkers for predicting the efficacy of anti-epidermal growth factor receptor antibody in the treatment of colorectal cancer. *Digestion* 2014, 89: 18–23.
20. Sepulveda AR, Hamilton SR, Allegra CJ, Grody W, Cushman-Vokoun AM, Funkhouser WK, Kopetz SE, et al.: Molecular biomarkers for the evaluation of colorectal cancer: Guideline from the American Society for clinical pathology, college of American pathologists, association for molecular pathology, and the American society of clinical oncology. *J Clin Oncol* 2017, 35: 1453–1486.
21. Leighl NB, Rekhtman N, Biermann WA, Huang J, Mino-Kenudson M, Ramalingam SS, West H, Whitlock S, Somerfield MR: Molecular testing for selection of patients with lung cancer for epidermal growth factor receptor and anaplastic lymphoma kinase tyrosine kinase inhibitors: American Society of Clinical Oncology endorsement of the College of American Pathologists/International Association for the study of lung cancer/association for molecular pathology guideline. *J Clin Oncol* 2014, 32: 3673–3679.
22. Sosman JA, Kim KB, Schuchter L, Gonzalez R, Pavlick AC, Weber JS, McArthur GA, et al.: Survival in BRAF V600-mutant advanced melanoma treated with vemurafenib. *N Engl J Med* 2012, 366: 707–714.
23. Halaban R, Zhang W, Bacchiocchi A, Cheng E, Parisi F, Ariyan S, Krauthammer M, McCusker JP, Kluger Y, Sznol M: PLX4032, a selective BRAF(V600E) kinase inhibitor, activates the ERK pathway and enhances cell migration and proliferation of BRAF melanoma cells. *Pigment Cell Melanoma Res* 2010, 23: 190–200.
24. Tolcher AW, Peng W, Calvo E: Rational approaches for combination therapy strategies targeting the MAP kinase pathway in solid tumors. *Mol Cancer Ther* 2018, 17: 3–16.
25. Lee HS, Kim WH, Kwak Y, Koh J, Bae JM, Kim KM, Chang MS, et al.: Molecular testing for gastrointestinal cancer. *J Pathol Transl Med* 2017, 51:103–121.
26. Toyokawa G, Seto T: Anaplastic lymphoma kinase rearrangement in lung cancer: Its biological and clinical significance. *Respir Investig* 2014, 52: 330–338.
27. Iyevleva AG, Raskin GA, Tiurin VI, Sokolenko AP, Mitiushkina NV, Aleksakhina SN, Garifullina AR, et al.: Novel ALK fusion partners in lung cancer. *Cancer Lett* 2015, 362: 116–121.
28. Thai AA, Solomon BJ: Treatment of ALK-positive nonsmall cell lung cancer: Recent advances. *Curr Opin Oncol* 2017.
29. Dagogo-Jack I, Shaw AT: Crizotinib resistance: Implications for therapeutic strategies. *Ann Oncol* 2016, 27 Suppl 3:iii42–iii50.
30. Zhu Q, Zhan P, Zhang X, Lv T, Song Y: Clinicopathologic characteristics of patients with ROS1 fusion gene in non-small cell lung cancer: A meta-analysis. *Transl Lung Cancer Res* 2015, 4: 300–309.
31. Pal P, Khan Z: ROS1-1. *J Clin Pathol* 2017, 70: 1001–1009.
32. Drilon A, Siena S, Ou SI, Patel M, Ahn MJ, Lee J, Bauer TM, et al.: Safety and antitumor activity of the multitargeted Pan-TRK, ROS1, and ALK inhibitor entrectinib: Combined results from Two Phase I trials (ALKA-372-001 and STARTRK-1). *Cancer Discov* 2017, 7: 400–409.
33. Iorio N, Sawaya RA, Friedenberg FK: Review article: The biology, diagnosis and management of gastrointestinal stromal tumors. *Aliment Pharmacol Ther* 2014, 39: 1376–1386.
34. Balachandran VP, DeMatteo RP: Gastrointestinal stromal tumors: Who should get imatinib and for how long? *Adv Surg* 2014, 48: 165–183.
35. Hemmings C, Yip D: The changing face of GIST: Implications for pathologists. *Pathology* 2014, 46: 141–148.
36. Grigg C, Rizvi NA: PD-L1 biomarker testing for non-small cell lung cancer: Truth or fiction? *J Immunother Cancer* 2016, 4:48.
37. Syn NL, Teng MWL, Mok TSK, Soo RA: De-novo and acquired resistance to immune checkpoint targeting. *Lancet Oncol* 2017, 18:e731–e741.
38. Neff RT, Senter L, Salani R: BRCA mutation in ovarian cancer: Testing, implications and treatment considerations. *Ther Adv Med Oncol* 2017, 9: 519–531.
39. Okuma HS, Yonemori K: BRCA gene mutations and poly(ADP-Ribose) polymerase inhibitors in triple-negative breast cancer. *Adv Exp Med Biol* 2017, 1026: 271–286.
40. Kaufman PA, Bloom KJ, Burris H, Gralow JR, Mayer M, Pegram M, Rugo HS, et al.: Assessing the discordance rate between local and central HER2 testing in women with locally determined HER2-negative breast cancer. *Cancer* 2014, 120: 2657–2664.
41. Perez EA, Suman VJ, Davidson NE, Martino S, Kaufman PA, Lingle WL, Flynn PJ, Ingle JN, Visscher D, Jenkins RB: HER2 testing by local, central, and reference laboratories in specimens from the North Central Cancer Treatment Group N9831 intergroup adjuvant trial. *J Clin Oncol* 2006, 24: 3032–3038.
42. Niu X, Chuang JC, Berry GJ, Wakelee HA: Anaplastic Lymphoma Kinase Testing: IHC vs. FISH vs. NGS. *Curr Treat Options Oncol* 2017, 18:71.

43. Farago AF, Azzoli CG: Beyond ALK and ROS1: RET, NTRK, EGFR and BRAF gene rearrangements in non-small cell lung cancer. *Transl Lung Cancer Res* 2017, 6: 550–559.
44. Salgia R: MET in Lung Cancer: Biomarker Selection Based on Scientific Rationale. *Mol Cancer Ther* 2017, 16: 555–565.
45. Pillai RN, Behera M, Berry LD, Rossi MR, Kris MG, Johnson BE, Bunn PA, Ramalingam SS, Khuri FR: HER2 mutations in lung adenocarcinomas: A report from the Lung Cancer Mutation Consortium. *Cancer* 2017, 123: 4099–4105.
46. Dey N, De P, Leyland-Jones B: PI3K-AKT-mTOR inhibitors in breast cancers: From tumor cell signaling to clinical trials. *Pharmacol Ther* 2017, 175: 91–106.
47. Mina A, Yoder R, Sharma P: Targeting the androgen receptor in triple-negative breast cancer: Current perspectives. *Onco Targets Ther* 2017, 10: 4675–4685.

Glycomics as an Innovative Approach for Personalized Medicine

7.6

Stephanie Holst, Nurulamin Abu Bakar, Monique van Scherpenzeel, Manfred Wuhrer, and Dirk J. Lefeber

Contents

INTRODUCTION INTO PROTEIN GLYCOSYLATION AND GLYCOMICS

Protein Glycosylation in Disease

Approximately 50% of all human proteins are glycosylated (Apweiler et al. 1999). Because glycosylation takes place in the secretory pathway of the cell, the vast majority (>90%) of proteins on the cell surface and secreted in body fluids is modified by glycosylation. Aberrant glycosylation of plasma proteins has been reported in many human diseases ranging from monogenetic inherited disorders to common diseases such as cancer (Walt et al. 2012). The majority of tumor biomarkers that are currently approved by the FDA are glycoproteins (Fuzery et al. 2013). However, the glycan part of these markers has hitherto been largely neglected.

Glycans have important functions in many biological processes such as the interaction of cells with their extracellular environment to mediate cell adhesion, macromolecular interactions (e.g., binding of antibodies to their receptors), and pathogen invasion (e.g., toxin binding, virus attachment; Figure 7.6.1) (Defaus et al. 2014). Glycans can also modulate protein function like signaling and influence the half-life of cell surface receptors (e.g., via multivalent lectin-glycan complexes) (Varki 2017). Furthermore, glycans play a role in several other important processes, such as protein secretion, quality control for protein folding in the endoplasmic reticulum (ER), and selective protein targeting (Moremen et al. 2012).

Unlike the biosynthesis of DNA and proteins, glycosylation is a non–template-driven multi-enzymatic process that is localized

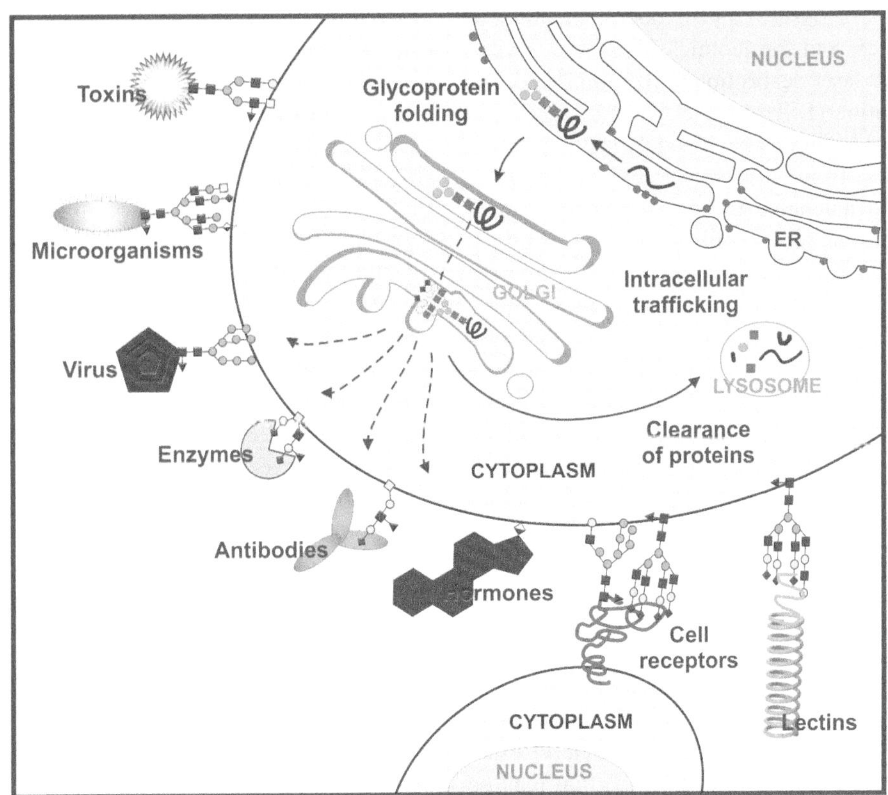

FIGURE 7.6.1 Scheme visualizing the diverse roles of glycans in biological processes. (Adapted from Defaus, S. et al., *Analyst,* 139, 2944–2967, 2014, and used with permission.)

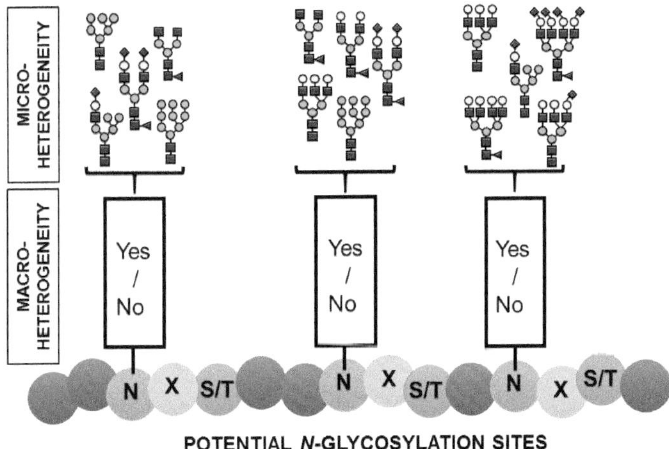

FIGURE 7.6.2 Glycan micro- and macro-heterogeneity. Glycosylation micro-heterogeneity describes the diversity of glycan structures that can occupy a certain glycosylation site, whereas macro-heterogeneity is the diversity of occupation of glycosylation sites. The concept of micro- and macro-heterogeneity is exemplified in this scheme using different N-glycan motifs. N, asparagine; S/T, serine/threonine; X, any amino acid except proline.

to the secretory pathway of the cell. Because no strict quality control system is known for glycan modifications, most proteins contain an enormous diversity of glycan structures, differing in attachment sites, composition, branching and linkage types (Figure 7.6.2). In addition to the glycosyltransferases that add monosaccharides during the biosynthetic process, many other factors influence protein glycosylation, including cytosolic metabolites (Jaeken 2013), lipid biosynthesis (Cantagrel et al. 2010), metal ions (Potelle et al. 2016), pH (Rivinoja et al. 2012, Jaeken 2013, Jansen et al. 2016a, Van Damme et al. 2017,), Golgi homeostasis (Jansen et al. 2016b, 2016c) and vesicular transport (Zeevaert et al. 2008). Abnormal glycosylation in human disease can therefore reflect abnormalities in a large number of biological processes and offers unique opportunities to develop biomarkers for early diagnosis as well as monitoring therapy and disease progression.

Glycomics and Personalized Medicine

To fully exploit the opportunities of protein glycosylation for personalized medicine, accurate analysis of protein glycosylation is a prerequisite. The emergence of mass spectrometry (MS) and bioinformatics techniques have significantly improved glycomics profiling, especially for the discovery of disease glycomarkers with potential in diagnosis and treatment monitoring. The technology allows the positioning of glycomics profiling in systems biology approaches next to genomics and other MS-based omics technologies as another, complementary layer of personalized information. In addition, targeted analysis of individual glycoproteins is possible for refined glycomic biomarker analysis.

Glycomics comprises the analyses of glycans (the glycome), released from proteins and/or lipids in any biological sample (e.g., blood, urine, CSF, or tissues) and is complementary to genomics, metabolomics and proteomics. For MS glycan analysis, matrix-assisted laser desorption ionization (MALDI) and electrospray ionization (ESI) are the two most common ionization techniques. MALDI has been widely used to obtain total N-glycan profiles from diverse biological samples. MALDI-MS is fast and amenable to automation, making it well suited for high-throughput profiling of, for example, large sample cohorts. However, the detection of acidic glycans (e.g., highly sialylated or sulfated glycans) is challenging. Derivatization methods for glycans such as permethylation (Kang et al., 2008) or linkage-specific sialic acid modification such as ethyl esterification (Wheeler et al. 2009, Reiding et al. 2014) overcome the ionization biases by neutralizing the negatively charged residues and largely prevent in- and post-source decay. The soft ionization technique ESI-MS allows analysis of native glycans and often allows discrimination of isomeric species when coupled with liquid chromatography (LC) separation. In addition, it is amenable for the analysis of intact glycoproteins (Wuhrer 2013, van Scherpenzeel et al. 2015a). The analysis of protein glycosylation is particularly challenging because the high macro-heterogeneity (i.e., glycosylated site variation, differences in site occupancy) and micro-heterogeneity (i.e., glycan structure variety) often result in a multitude of glycoforms (Figure 7.6.2) that may be combined with other post translational modifications (PTM) on a specific protein, resulting in a large set of protein variants, the so-called proteoforms. At the same time, this offers vast possibilities for highly specific and sensitive markers for personalized medicine (Smith and Kelleher 2013).

In this section, we will highlight developments in glycobiomarker analysis as exemplified by two well-studied groups of disorders, the monogenic congenital disorders of glycosylation (CDG) and cancer. In addition, the added value of glycomics for other common diseases will be discussed, based on recent population-based studies and an outlook is provided on what is needed to fully implement glycomics in all aspects of personalized medicine.

GLYCOMICS APPROACHES IN THE WARD

CDG: Clinical Consequences of Abnormal Glycosylation

CDGs are a group of genetic defects with abnormal glycosylation of proteins and/or lipids. Since their first description in 1980, more than 100 different defects have been identified. Genetic defects have been found in all major glycosylation pathways, including protein N-glycosylation, protein O-glycosylation, lipid and glycosylphosphatidylinositol (GPI) anchor glycosylation, and glycosaminoglycan biosynthesis (Jaeken 2011). In addition, an increasing number of genetic defects are being identified that result in abnormalities in multiple pathways. For example, defects in sugar metabolism or metal ions affect multiple enzymes. The majority of at least 50 defects are known to occur in the N-linked glycosylation pathway (Freeze et al. 2014), which will be further described below.

Generally, clinical presentations of CDGs are extremely heterogeneous and a challenge for the clinician to diagnose early. The majority of CDGs with N-glycosylation defects are multiorgan diseases with neurological involvement. For example, the classical clinical features of CDG as displayed in the most prevalent type of CDG, PMM2-CDG (CDG-Ia), are abnormal fat pads, inverted nipples, feeding problems (anorexia, vomiting and diarrhea), severe failure to thrive, dysmorphic feature (large hypoplastic/dysplastic ears, abnormal subcutaneous adipose tissue distribution), hepatomegaly, skeletal abnormalities and hypogonadism. However, there are some CDGs that do not produce neurological symptoms like MPI-CDG (CDG-Ib) with thrombosis and enteropathy, and DPM3-CDG (CDG-Io), which exhibited mostly muscular dystrophy and dilated cardiomyopathy (Jaeken 2011, 2013).

CDG: Classical Screening

Transferrin is one of the most abundant glycoproteins in human plasma and has been widely used as a rapid biomarker for N-glycosylation defects. Serum transferrin profiling by means of isoelectric focusing (IEF) is used as routine screening test for CDG. Two types of CDG can be distinguished based on serum transferrin IEF: CDG-type I (CDG-I) and CDG-type II (CDG-II). The CDG-I pattern is characterized by a (partial) lack of N-glycans and points to a defect in the ER glycosylation pathway. The CDG-II pattern is characterized by truncated N-glycans, indicating a processing defect in the Golgi. Fructosemia, galactosemia, alcohol abuse, hepatopathy and bacterial sialidases are known as secondary causes of abnormal transferrin glycosylation (Lefeber et al. 2011). Normal transferrin IEF profiles have been observed in some CDG cases, such as ALG14-CDG and ALG11-CDG (CDG-Ip) (Al Teneiji et al. 2017) for CDG-I subtypes, GCS1-CDG (CDG-IIb), SLC35A3-CDG and SLC35C1-CDG (CDG-IIc) for CDG-II subtypes, as well as defects in sugar metabolism like GNE-CDG, PGM3-CDG and GFPT1-CDG and in a Golgi homeostasis defect named Cohen Syndrome (Lefeber 2016, Al Teneiji et al. 2017). Because of this limitation, there is a need to expand the panel of glycomarkers to other types of glycoproteins such as those having O-glycosylation, or proteins that are synthesized by specific cell types, especially for covering tissue- or organ-specific clinical symptoms. Moreover, the combination of whole exome sequencing and glycomics will be essential for an optimal diagnostic result. In the next section, we will describe the application of glycomics methodologies for CDG diagnostics.

CDG: Application of Glycomics

Traditional CDG screening by serum transferrin IEF was found to fail in the detection of several CDG subtypes. In addition, although this approach is used to discriminate CDG-I from CDG-II, it can neither provide evidence for the exact genetic defect, nor does it reveal the glycan structures. Hence, advanced glycoprofiling through MS is vital to develop potential biomarkers for disease characterization. Mainly two glycomics approaches have been applied for glycomarker discovery for CDG (Figure 7.6.3), namely global glycoprofiling of total serum proteins and protein-specific glycoprofiling of mainly intact transferrin (van Scherpenzeel et al. 2015a).

Classical analysis of permethylated N-glycans of total serum proteins by MALDI-MS has successfully been applied in CDG-II patients to determine diagnostic glycomarkers. Accumulation of single, specific N-glycans as a result of a defective enzyme was observed in some glycosyltranferase deficiencies such as MGAT2-CDG (CDG-IIa; *N*-acetylglucosaminyltransferase defect) and B4GALT1-CDG (CDG-IId; galactosyltransferase defect). In the case of CDG-IIc (GDP-fucose transporter defect), presenting with normal transferrin IEF pattern, a significant decrease of fucosylated N-glycans on serum proteins was observed for direct diagnosis. Other, more general glycomic alterations such as loss of tri-antennary N-glycans and appearance of truncated N-glycans can be seen in serum profiles of COG7-CDG and ATP6V0A2-CDG patients (Guillard et al. 2011). Recently, the application of total serum glycoprofiling in CDG-I patients led to the discovery of a specific "N-tetrasaccharide" as a novel small N-glycomarker for the diagnosis of ALG1-CDG (Zhang et al. 2016), reflecting a novel disease mechanism leading to specifically altered glycosylation.

Protein-specific, HRMS of intact serum transferrin has significantly improved CDG diagnostics due to the robustness, speed and accuracy with which glycomarkers for CDG can be obtained for several genetic defects. Intact transferrin glycoprofiling using ESI-MS was first introduced in 2001 at the Mayo Clinic (Rochester, Minnesota) as a routine diagnostic test for the identification of CDG-I defects and alcohol abuse (Lacey et al. 2001). Recently, the introduction of high-resolution quadrupole time-of-flight (QTOF) MS for intact transferrin glycoprofiling has significantly improved the mass resolution for a more accurate annotation of the glycan structures and has further improved the diagnosis of CDG. Moreover, only one-fifth of the material is needed because of the sensitive nanoLC-chip chromatography. Intact transferrin QTOF glycoprofiling has contributed to the identification of novel CDGs such as PGM1-CDG (phosphoglucomutase-1 deficiency) (Tegtmeyer et al. 2014) and Man1B1-CDG (1-2-α-mannosidase defect) (Van Scherpenzeel et al. 2014) and also allowed the fast identification of CDG-II subtypes of SLC35A1-CDG (CMP-sialic acid transporter defect) and SLC35A2-CDG (UDP-galactose transporter defect). In addition, abnormal transferrin glycoprofiles were identified for several golgi homeostasis defects such as COG1-CDG, ATP6V0A2-CDG, TMEM165-CDG, TMEM199-CDG, and CCDC115-CDG, however, they were not directly diagnostic for the specific gene defect (van Scherpenzeel et al. 2015a, Jansen et al. 2016a, 2016b, 2016c, Van Damme et al. 2017).

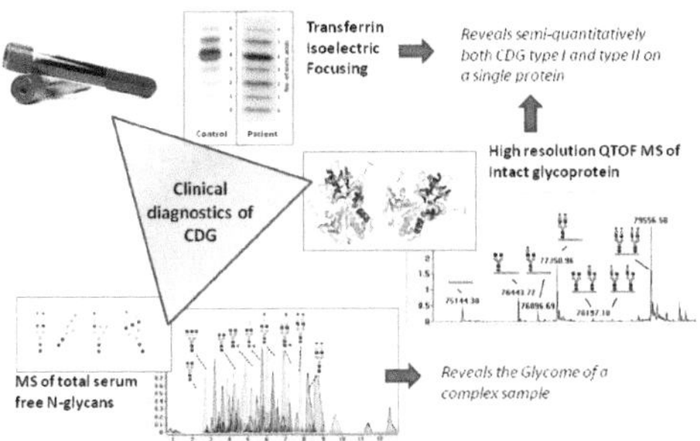

FIGURE 7.6.3 Clinical glycomics to diagnose congenital disorders of glycosylation. The three most commonly applied techniques used to analyze biological fluid were presented here. Mass spectrometry (MS) provides sensitivity and structural insights compared with isoelectric focusing.

Protein-specific glycoprofiling has a couple of advantages over the more generally applied total serum glycomics. First, total serum glycomics always require transferring IEF or intact transferrin glycoprofiling as initial step because the lack of whole glycans as is observed in CDG-I cannot be detected. In case that a CDG-II pattern is obtained, total serum glycan profiling can be performed for glycan structural analysis. Based on our experience, a characteristic profile was observed with specific truncated glycans for some glycosyltransferase deficiencies such as MGAT2-CDG or B4GALT1-CDG (Guillard et al. 2011), but most of the subtypes show rather nonspecific profiles with an increase of fucosylation or decrease of sialylation, which often falls within the broad control reference range. This broad reference range became evident with the recent publication of Hennig and coworkers, where they show that healthy individuals present with large variations in their glycosylation phenotype (Hennig et al. 2016). Likely, the glycome reflects the genetic, epigenetic and metabolic system, influenced by individual lifestyles and environmental factors, leading to large interindividual glycome heterogeneity. Importantly, within one individual, changes in glycosylation are rather small over time (Hennig et al. 2016).

In clinical diagnostics, commonly a one-time sampling is performed, which is not always adequate to show the genetic defect. For precision medicine, however, the glycome might be a highly sensitive marker given that changes within one individual are small. Initial studies are highly promising to mediate novel early diagnosis and disease stratification markers, subsequently resulting in improved patient well-being and reduced treatment costs (Almeida and Kolarich 2016). In addition, for therapy monitoring, which will be discussed in the next paragraph, glycomics is a highly sensitive and fast approach to optimize the treatment of the individual patient.

CDG: Treatment and Monitoring

Most of the CDG subtypes can only be treated symptomatically. However, there are new developments in the field of CDG research in understanding the mechanisms of glycosylation and how these mechanisms can be interfered with to overcome the genetic defect. There are a few approaches that are currently considered: chaperone therapy, enzyme replacement therapy and monosaccharide supplementation therapy. Three monosaccharide supplementation therapies play a role as promising treatments for specific CDG subtypes:

1. Mannose therapy (1 g/kg body weight per day, divided in 4–6 doses) for MPI-CDG (CDG-IIb; phosphomannose isomerase deficiency): This therapy can be explained biochemically by phosphorylation of oral mannose to mannose-6-phosphate by hexokinases, which bypasses the defect in MPI to restore the GDP-mannose pool for N-glycosylation (Jaeken 2010). The therapy could be sensitively monitored by intact transferrin MS (Janssen et al. 2014).
2. Fucose therapy (depending on nature of the mutation) for SLC35C1-CDG (GDP-fucose transporter defect): The treatment is reported to be efficient in some patients with regard to the typical recurrent infections with hyperleukocytosis (Marquardt et al. 1999, Jaeken 2010).
3. Galactose therapy (0.5–1 g/kg per day) for PGM1-CDG (phosphoglucomutase-1 deficiency): Oral D-galactose supplementation in PGM1-CDG could increase the production of intracellular UDP-galactose to restore N-glycosylation and has improved hepatomegaly and liver function tests as well as prevented hypoglycemic episodes. Intact transferrin glycoprofiling has been applied successfully to monitor the biochemical improvement during galactose supplementation: a good example of how glycoprofiling can be used to monitor disease severity and treatment efficacy (Tegtmeyer et al. 2014).

GLYCOMIC APPROACHES FOR CANCER DIAGNOSTICS AND THERAPY

Cancer (Glyco-)Biology

Cancer development is a multistep process in which cells acquire the capability to evade the immune system, to proliferate in a rather unregulated manner, and to invade surrounding tissues and disseminate to distant organs. Tumor initiation is thought to be caused by genetic alterations of a single cell that will proliferate and develop to a cell population (Marusyk and Polyak 2010). During tumor progression, more mutations accumulate that influence cellular processes giving advantage to the cancer cells. However, genetic alterations are not the only players in cancer development and other nongenetic factors contribute to a large intra-tumor heterogeneity and complex tumor biology (Caiado et al. 2016). Several glycosylation changes have been reported and reviewed (Pinho and Reis 2015, Kailemia et al. 2016, Tarbell and Cancel 2016) and include increases in sialylation, fucosylation, N-glycan branching, and truncation of O-glycans, which impact important cancer-associated processes like proliferation, invasion, metastasis, and angiogenesis. Changes in glycosylation can influence cancer progression (Pinho and Reis 2015, Taniguchi and Kizuka 2015) and treatment response in multiple ways (Park et al. 2012, Croci et al. 2014, Feng et al. 2016), including modification of the stability, solubility, or activity of the protein, as well as interactions with glycan-binding proteins and the extracellular matrix. Identifying the glycosylation changes and understanding underlying mechanisms has, therefore, vast potential in the development of novel biomarkers for diagnosis, stratification, and prognosis. Likewise, new therapies are emerging that intervene with cancer-related, glycan-mediated processes, including novel immunotherapeutics (Hudak and Bertozzi 2014, Vankemmelbeke et al. 2016).

Glycomics for Improved Cancer Diagnostics

As mentioned in the introduction, various glycoproteins and glycolipids have already been routinely used and measured in clinics for a long period. Examples include the approved cancer biomarkers CA 125 for ovarian cancer, CA 19-9 for pancreatic cancer and prostate-specific antigen (PSA) in prostate cancer (Lauc et al. 2016). Recent studies have shown that including information on the glycosylation of the glycoconjugates provide important insights and could increase sensitivity and/or specificity for diagnosis and prognosis. For example, detection of aberrant glycosylation on PSA, in particular increased α2,3-sialylation, has been shown to improve specificity and sensitivity of the screening for prostate cancer (Yoneyama et al. 2014, Pihikova et al. 2016). Another study found up to hexa-fucosylated N-glycans on haptoglobin in hepatocellular carcinoma, whereas N-glycans on haptoglobin from controls contained only one or none fucose (Pompach et al. 2013). In breast cancer, autoantibodies against MUC1 were elevated in sera from patients as compared with controls and specific glycoforms (core3-MUC1: GlcNAcβ1-3GalNAc-MUC1 and sialylTn-MUC1: NeuAcα2,6GalNAc-MUC1) were found to be associated with better prognosis (Blixt et al. 2011).

Glycosylation in the Context of Cancer Progression and Metastasis

Various cancer-associated processes, including invasion and metastasis, are dependent on receptors. It has been proposed that GnT-V-mediated β1,6GlcNAc-branching of N-glycans on different receptors such as epidermal growth factor receptor (EGFR), the transforming growth factor-β receptor (TGFβR) and the vascular endothelial growth factor receptors (VEGFR) induces the formation of molecular lattices via interaction with galectin 3, thereby leading to delayed internalization of these receptors (Partridge et al. 2004, Guo et al. 2009, Markowska et al. 2011). Consequently, the response to their ligands is prolonged and can influence tumor progression and angiogenesis, making the glycosylation status of the receptors a potential prognostic marker (Figure 7.6.4A). Furthermore, β1,6GlcNAc-branching of N-glycans on E-cadherin leads to loose adherens junctions, thereby weakening cell–cell adhesion and altering downstream signaling in cancer progression, whereas bisecting GlcNAc, mediated through Gnt-III, had the reverse effect and suppressed invasion and metastasis (Figure 7.6.4B) (Pinho et al. 2013, Pinho and Reis 2015). Strikingly, in gastric cancer the specific glycosylation site at Asn-554 seems to be the major player in this progress (Carvalho et al. 2016), providing an interesting treatment target as well as a prognostic indicator.

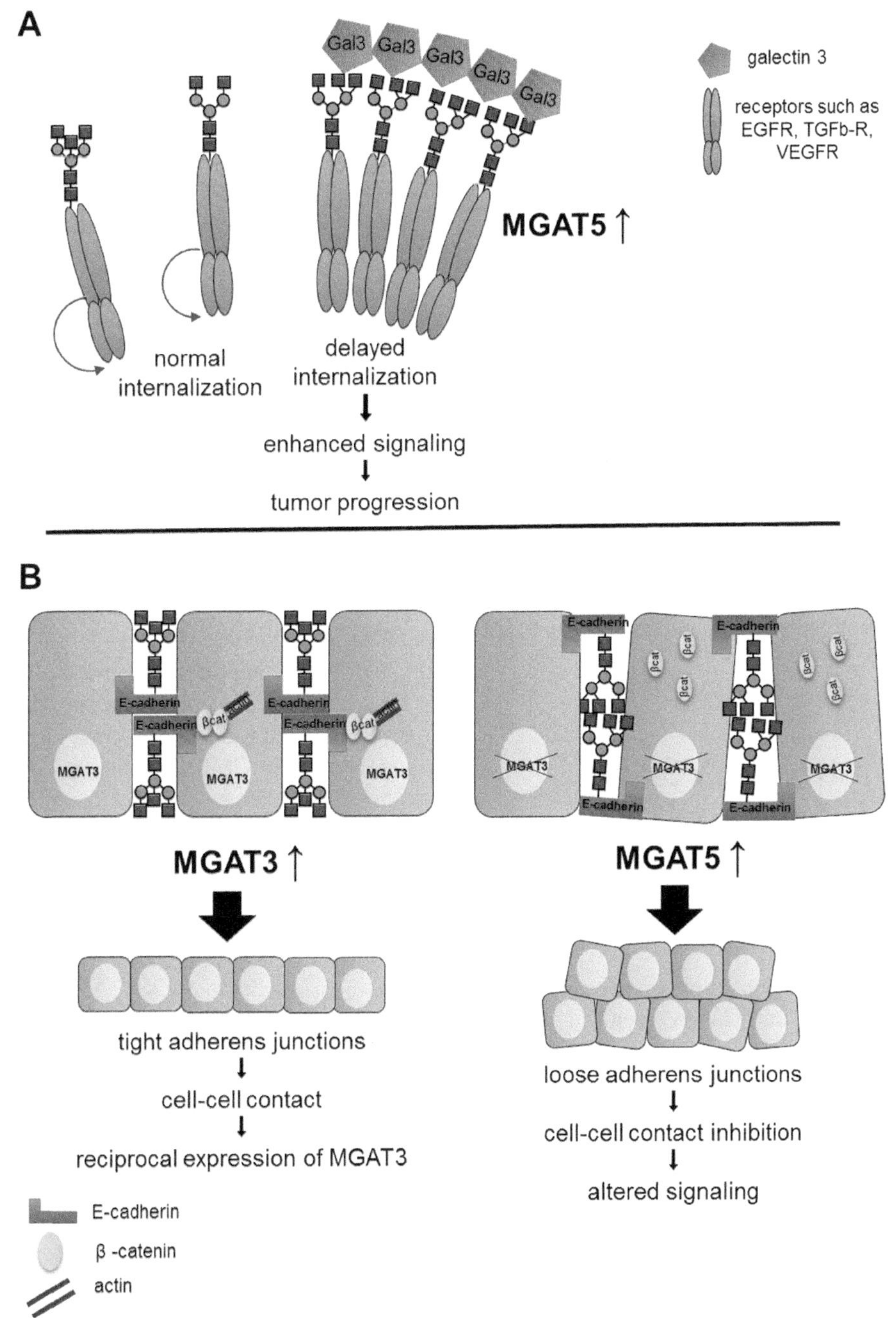

FIGURE 7.6.4 Schematic representation of modulated cell behavior through GnT-V-mediated β1,6GlcNAc-branching of N-glycans. (*A*) The epidermal growth factor receptor (EGFR), the transforming growth factor-β receptor (TGFβR) and the vascular endothelial growth factor receptors (VEGFR) are growth-, arrest- and angiogenesis-promoting receptors, respectively. The β1,6GlcNAc-branching of N-glycans on these receptors can lead to delayed internalization of the receptors through lattice formation via galectin 3 and thereby enhance signaling, which can alter cell behavior and promote tumor progression. (From Shankar, J. et al., *Essays Biochem.*, 57, 189–120, 2015.) (*B*) The expression of bisecting *N*-acetylglucosamine (MGAT3 gene) on E-cadherin promotes stable cell-cell adhesion and stimulates reciprocal expression of MGAT 3, whereas β1,6GlcNAc-branched N-glycans destabilize the tight junctions and inhibit MGAT3 expression through free cytoplasmic β-catenin. (From de Freitas Junior, J.C. and J. A. Morgado-Díaz, *Oncotarget,* 7, 19395–19413, 2016.)

Glycosylation in the Context of Treatment Responses

Eight of the 10 best selling drugs in Europe are glycoproteins, and many drug mechanisms involve receptors whose function or half-life can be largely influenced by glycosylation (Lauc et al. 2016). Polymorphisms in the glycosylation pathway can alter the receptor glycosylation and, therefore, result in different drug responses and efficacies between patients (Lauc et al. 2016). In this context, α2,6-sialylation appears to have different consequences on different types of receptors and in different types of cancer with respect to the efficacy of anticancer treatments: Although the presence of α2,6-sialylation on EGFR decreased the effect of the anticancer drug gefitinib, an EGFR kinase inhibitor, in colon cancer cells (Park et al. 2012), expression of α2,6-sialylation is beneficial for the anti-VEGF treatment because it results in anti-VEGF-sensitive tumors (Croci et al. 2014). Taking these results into account, levels of sialylation might be a good personalized indicator for the prediction of treatment responses.

Strikingly, many of the proteins related to multidrug resistance are glycoproteins that often support resistance by pumping drugs out of the cell, including P-glycoprotein (P-gp), multidrug resistance-associated protein 1 (MRP1), and breast cancer resistance protein (BCRP) (Li et al. 2013). Li et al. identified different glycosylation sites of these and other aberrantly glycosylated proteins that correlated with drug resistance in gastric cancer and should be further explored for personalized prediction of treatment responses (Li et al. 2013).

PROMISING GLYCAN EPITOPES AND GLYCAN-RELATED MUTATIONS FOR PERSONALIZED MEDICINE

As shown from the two disease groups discussed above, glycans have a large potential for patient stratification given that glycomic differences in part reflect differences in disease susceptibilities and progression—also in a personalized manner. Histo-blood group antigens, for example, are glycan epitopes and result from individual genetic polymorphisms in glycosyltransferase genes. Their potential in precision medicine has been recently reviewed by Dotz and Wuhrer (2015). Setting secretor status and Lewis genotype–dependent cutoff values for CA19-9 has been found promising in the early detection of pancreatic cancer in a recent study (Luo et al. 2016), whereas ABO glycosyltransferase activity has been associated with higher risk of pancreatic cancer (Wolpin et al. 2010).

Another study revealed that, in colon cancer, high binding of C-type lectin macrophage galactose-type lectin (MGL) to tissues from stage III patients is associated with poor survival. It was further shown that the expression of MGL ligands is induced by BRAFV600E mutation, which could be reduced *in vitro* by specific BRAFV600E inhibitors in colon cancer cell lines (Lenos et al. 2015) and could be explored in colon cancer treatment for patients with this specific mutation. Another mutation in colon cancer was identified in the GDP-mannose-4,6-dehydratase gene (GMDS), leading to significantly lower fucosylation levels. GMDS mutations have been detected with higher frequency in metastatic lesions as compared with the original tumor tissues, whereas no mutations were observed in control tissue (Nakayama et al. 2013), presenting another candidate for tailored treatments. Lauc et al. identified in a genome-wide association study (GWAS) combined with high-throughput glycomics analysis associations between hepatocyte nuclear factor 1α (HNF1α) and fucosylation on N-glycans on human plasma proteins and subsequently showed that HNF1α and HNF4α regulated the expression of several fucosyltransferases (Lauc et al. 2010). They used these findings further to explore glycan alterations as biomarkers for the diagnosis of HNF1α dysfunction, as in the case of maturity-onset diabetes of the young (MODY), which is caused by HNF1α mutations (Thanabalasingham et al. 2013). Here, the ratio of fucosylated to non-fucosylated triantennary glycans was successful in discriminating HNF1α-MODY from controls and other diabetes types.

Furthermore, the nonhuman sialic acid, *N*-glycolylneuraminic acid (NeuGc), has gained a lot of attention, especially for cancer vaccines (Vazquez et al. 2012). It can, however, also be a promising candidate for personalized medicine strategies because it is taken up through food like red meat and milk products and therefore relates to a person's diet (Padler-Karavani 2014). For patients with high intake of NeuGc-containing dietary products, novel anti-NeuGc antibodies might be a promising tailored treatment or drug delivery approach because NeuGc is especially enriched in cancer tissues. In different studies, anti-NeuGc-Sialyl-Tn IgG is being explored for early detection of carcinomas, whereas gangliosides (NeuGc)GM2 and (NeuGc)GM3 have been suggested as therapeutic targets (Padler-Karavani 2014). Figure 7.6.5

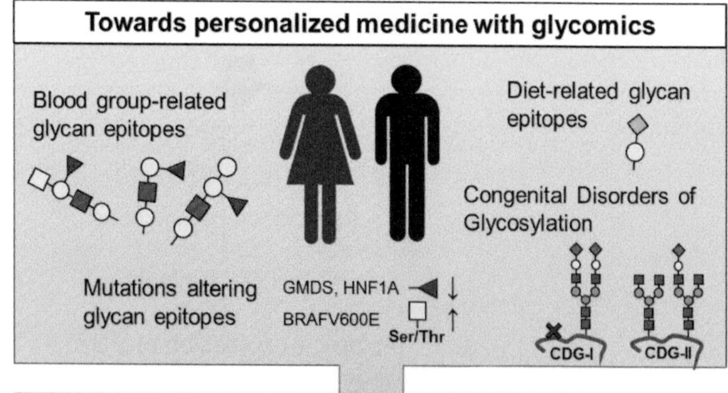

FIGURE 7.6.5 Scheme summarizing the described glycomic approaches toward personalized medicine.

summarizes glycomic approaches toward personalized medicine as described in this section.

Glycomics is also a promising approach for personalized medicine in other disease settings and clinical applications beyond cancer and CDG. The risk for viral and bacterial infections, for example, may depend on the secretor status (Underwood et al. 2015, Andreas et al. 2016). Moreover, antibody glycosylation associates with disease activity and progression in rheumatoid arthritis (Mesko et al. 2012), and antiviral activity of HIV-specific antibodies appears to be modulated by glycosylation (Ackerman et al. 2013). A recent publication from Fortune et al. highlights the potential of protein glycosylation in clinical diagnostics because they show that latent tuberculosis can be discriminated from an active infection on the basis of antibody glycosylation profiles (Lu et al. 2016). Disease control has always been hampered by a lack of tools, including rapid, point-of-care diagnostics, but in their study, they showed that Fc-effector functions are distinct between these two groups of tuberculosis patients and highlighted that antibodies from latent patients can activate antimicrobial responses more effectively than antibodies from patients with active infections.

FUTURE DIRECTIONS AND CHALLENGES

In order to advance clinical glycomics further in the context of individualized patient care, two major challenges need to be tackled: (1) further improvement of methodology to adequately analyze glycans and glycopeptides in clinical diagnostics, and (2) understanding the glycans' complexity in an individual patient via systems biology approaches.

Large technological progress in MS-based glycomics has been made over the recent years, covering different approaches of glycan analysis ranging from released glycans, over protein-specific glycosylation to site-specific glycosylation. Robust and sensitive high-throughput screenings of large sample sets enable population-based cross-sectional as well as longitudinal studies and have already resulted in the characterization of cancer- and other disease-associated glycans changes, revealing potential novel biomarkers (Lauc et al. 2010, Almeida and Kolarich 2016, Hennig et al. 2016). A recent longitudinal study has shown that the serum glycan profile of an individual is rather stable over a longer period of time. It changes, however, significantly with inflammation and various diseases (Hennig et al. 2016) and is dependent on age and sex (Ding et al. 2011). It might well be that the individual glycome reflects the risk for certain diseases, disease course and the response to therapy. To prove these hypotheses, large cohort studies need a layer of biological information to explain the large variations between individuals (Lauc et al. 2016). Nevertheless, regular monitoring of a person's glycome offers a promising tool for an individualized approach to detect pathologic events, which might be missed in cross-sectional population-based studies where small disease-associated effects may be diluted due to interindividual variation.

Although these high-throughput screening methods are mainly applied on released glycans often from serum, plasma or other biofluids, protein-specific or site-specific glycosylation changes are less amenable to high-throughput analysis, but instead unravel important biological information and can improve diagnostic and prognostic performance through better sensitivity and specificity, as discussed earlier. The advantage of site-specific analysis is the detection of subtle glycosylation differences, but it remains the analytically most challenging approach. Advanced chip-based LC separations in combination with glycan libraries increase robustness and facilitate handling, while providing a deep level of structural information, thereby forming a promising tool of translation to the clinics (Ruhaak et al. 2013, Song et al. 2015). For example, high-resolution chip-based glycoprofiling of intact transferrin has been shown to be a fast and reliable method for the diagnosis of several subtypes of CDGs (van Scherpenzeel et al. 2015b). In cases where the detection is performed with Multiple reaction monitoring (MRM) could add the sensitivity and specificity as well as the precision that is often desirable in clinical quantitative glyco-proteomics applications (Huang et al. 2016, Ruhaak et al. 2017). Likewise, MRM quantitative proteomic workflows are already in use in clinical chemistry settings, which should facilitate the translation of other MS glycomics analysis into clinical diagnostic applications.

In summary, further developments for the measurement of extremely small sample amounts as well as absolute quantification and robust platforms that easily transfer to the clinics are the subject of ongoing research.

With regard to the complexity of glycosylation as a cellular process, the different factors involved in this process as well as the non–template-driven biosynthesis, it remains challenging to identify regulatory and causal events as well as the direct functional relevance for the disease (Almeida and Kolarich 2016). Combining different omic approaches (e.g., genomics, epigenomics, transcriptomics, metabolomics, glycomics and proteomics) to gain a broader insight in the interplay between genes, transcription factors, metabolic changes and glycosylation can aid in unraveling disease mechanisms. Consequently, novel bioinformatics and systems biology tools are urgently needed to integrate different data sets, considering the potential relationships between the different players (Bennun et al. 2016).

To conclude, glycomics provides a novel discipline with unprecedented density of biological information. This is exactly why this approach is so promising for clinical research, especially in the context of personalized medicine. Ongoing technological advances facilitate the analysis of complex glycosylation and need to make their way into routine clinical applications.

ACKNOWLEDGMENTS

We acknowledge support by the European Union Seventh Framework Programme (HighGlycan project, grant number 278535, and IBD-BIOM project, grant number 305479) and Horizon 2020 Programme (GlyCoCan project, grant number

676421, and GlySign project, grant number 722095). In addition, support is acknowledged of the Dutch Organization for Scientific Research (ZONMW medium investment grant 40-00506-98-9001 and VIDI grant 91713359 to DJL; VENI grant 722015012 to MvS) and the Malaysian government (to HLP: KKM510-4/4/4/1 Jld. 4 [42]).

REFERENCES

Ackerman, M. E., M. Crispin, X. Yu, K. Baruah, A. W. Boesch, D. J. Harvey, A. S. Dugast et al., 2013. Natural variation in Fc glycosylation of HIV-specific antibodies impacts antiviral activity. *J Clin Invest* 123 (5):2183–2192. doi:10.1172/JCI65708.

Al Teneiji, A., T. U. Bruun, S. Sidky, D. Cordeiro, R. D. Cohn, R. Mendoza-Londono, M. Moharir et al., 2017. Phenotypic and genotypic spectrum of congenital disorders of glycosylation type I and type II. *Mol Genet Metab* 120 (3):235–242. doi:10.1016/j.ymgme.2016.12.014.

Almeida, A., and D. Kolarich. 2016. The promise of protein glycosylation for personalised medicine. *Biochim Biophys Acta* 1860 (8):1583–1595. doi:10.1016/j.bbagen.2016.03.012.

Andreas, N. J., A. Al-Khalidi, M. Jaiteh, E. Clarke, M. J. Hyde, N. Modi, E. Holmes, B. Kampmann, and K. Mehring Le Doare. 2016. Role of human milk oligosaccharides in Group B streptococcus colonisation. *Clin Transl Immunology* 5 (8):e99. doi:10.1038/cti.2016.43.

Apweiler, R., H. Hermjakob, and N. Sharon. 1999. On the frequency of protein glycosylation, as deduced from analysis of the SWISS-PROT database. *Biochim Biophys Acta* 1473 (1):4–8.

Bennun, S. V., D. B. Hizal, K. Heffner, O. Can, H. Zhang, and M. J. Betenbaugh. 2016. Systems glycobiology: Integrating glycogenomics, glycoproteomics, glycomics, and other omics data sets to characterize cellular glycosylation processes. *J Mol Biol* 428 (16):3337–3352. doi:10.1016/j.jmb.2016.07.005.

Blixt, O., D. Bueti, B. Burford, D. Allen, S. Julien, M. Hollingsworth, A. Gammerman, I. Fentiman, J. Taylor-Papadimitriou, and J. M. Burchell. 2011. Autoantibodies to aberrantly glycosylated MUC1 in early stage breast cancer are associated with a better prognosis. *Breast Cancer Res* 13 (2):R25. doi:10.1186/bcr2841.

Caiado, F., B. Silva-Santos, and H. Norell. 2016. Intra-tumour heterogeneity—going beyond genetics. *FEBS J* 283 (12):2245–2258. doi:10.1111/febs.13705.

Cantagrel, V., D. J. Lefeber, B. G. Ng, Z. Guan, J. L. Silhavy, S. L. Bielas, L. Lehle, H. et al., 2010. SRD5A3 is required for converting polyprenol to dolichol and is mutated in a congenital glycosylation disorder. *Cell* 142 (2):203–217. doi:10.1016/j.cell.2010.06.001.

Carvalho, S., T. A. Catarino, A. M. Dias, M. Kato, A. Almeida, B. Hessling, J. Figueiredo et al., 2016. Preventing E-cadherin aberrant N-glycosylation at Asn-554 improves its critical function in gastric cancer. *Oncogene* 35 (13):1619–1631. doi:10.1038/onc.2015.225.

Croci, D. O., J. P. Cerliani, T. Dalotto-Moreno, S. P. Mendez-Huergo, I. D. Mascanfroni, S. Dergan-Dylon, M. A. Toscano et al., 2014. Glycosylation-dependent lectin-receptor interactions preserve angiogenesis in anti-VEGF refractory tumors. *Cell* 156 (4):744–758. doi:10.1016/j.cell.2014.01.043.

de Freitas Jr. J. C., and J. A. Morgado-Díaz. 2016. The role of N-glycans in colorectal cancer progression: Potential biomarkers and therapeutic applications. *Oncotarget* 7 (15):19395.

Defaus, S., P. Gupta, D. Andreu, and R. Gutierrez-Gallego. 2014. Mammalian protein glycosylation—structure versus function. *Analyst* 139 (12):2944–2967. doi:10.1039/c3an02245e.

Ding, N., H. Nie, X. Sun, W. Sun, Y. Qu, X. Liu, Y. Yao, X. Liang, C. C. Chen, and Y. Li. 2011. Human serum N-glycan profiles are age and sex dependent. *Age Ageing* 40 (5):568–575. doi:10.1093/ageing/afr084.

Dotz, V., and M. Wuhrer. 2015. Histo-blood group glycans in the context of personalized medicine. *Biochim Biophys Acta.* 1860 (8):1596–1607. doi:10.1016/j.bbagen.2015.12.026.

Feng, X., L. Zhao, S. Gao, X. Song, W. Dong, Y. Zhao, H. Zhou, L. Cheng, X. Miao, and L. Jia. 2016. Increased fucosylation has a pivotal role in multidrug resistance of breast cancer cells through miR-224-3p targeting FUT4. *Gene* 578 (2):232–241. doi:10.1016/j.gene.2015.12.028.

Freeze, H. H., J. X. Chong, M. J. Bamshad, and B. G. Ng. 2014. Solving glycosylation disorders: Fundamental approaches reveal complicated pathways. *Am J Hum Genet* 94 (2):161–175. doi:10.1016/j.ajhg.2013.10.024.

Fuzery, A. K., J. Levin, M. M. Chan, and D. W. Chan. 2013. Translation of proteomic biomarkers into FDA approved cancer diagnostics: issues and challenges. *Clin Proteomics* 10 (1):13. doi:10.1186/1559-0275-10-13.

Guillard, M., E. Morava, F. L. van Delft, R. Hague, C. Korner, M. Adamowicz, R. A. Wevers, and D. J. Lefeber. 2011. Plasma N-glycan profiling by mass spectrometry for congenital disorders of glycosylation type II. *Clin Chem* 57 (4):593–602. doi:10.1373/clinchem.2010.153635.

Guo, H. B., H. Johnson, M. Randolph, I. Lee, and M. Pierce. 2009. Knockdown of GnT-Va expression inhibits ligand-induced down-regulation of the epidermal growth factor receptor and intracellular signaling by inhibiting receptor endocytosis. *Glycobiology* 19 (5):547–559. doi:10.1093/glycob/cwp023.

Hennig, R., S. Cajic, M. Borowiak, M. Hoffmann, R. Kottler, U. Reichl, and E. Rapp. 2016. Towards personalized diagnostics via longitudinal study of the human plasma N-glycome. *Biochim Biophys Acta* 1860 (8):1728–1738. doi:10.1016/j.bbagen.2016.03.035.

Huang, J., M. J. Kailemia, E. Goonatilleke, E. A. Parker, Q. Hong, R. Sabia, J. T. Smilowitz, J. B. German, and C. B. Lebrilla. 2016. Quantitation of human milk proteins and their glycoforms using multiple reaction monitoring (MRM). *Anal Bioanal Chem.* 409 (2):589–606. doi:10.1007/s00216-016-0029-4.

Hudak, J. E., and C. R. Bertozzi. 2014. Glycotherapy: New advances inspire a reemergence of glycans in medicine. *Chem Biol* 21 (1):16–37. doi:10.1016/j.chembiol.2013.09.010.

Jaeken, J. 2010. Congenital disorders of glycosylation. *Ann N Y Acad Sci* 1214:190–198. doi:10.1111/j.1749-6632.2010.05840.x.

Jaeken, J. 2011. Congenital disorders of glycosylation (CDG): It's (nearly) all in it! *J Inherit Metab Dis* 34 (4):853–858. doi:10.1007/s10545-011-9299-3.

Jaeken, J. 2013. Congenital disorders of glycosylation. *Handb Clin Neurol* 113:1737–1743. doi:10.1016/b978-0-444-59565-2.00044-7.

Jansen, E. J., S. Timal, M. Ryan, A. Ashikov, M. van Scherpenzeel, L. A. Graham, H. Mandel et al., 2016a. ATP6AP1 deficiency causes an immunodeficiency with hepatopathy, cognitive impairment and abnormal protein glycosylation. *Nat Commun* 7:11600. doi:10.1038/ncomms11600.

Jansen, J. C., S. Cirak, M. van Scherpenzeel, S. Timal, J. Reunert, S. Rust, B. Perez, D. et al., 2016b. CCDC115 Deficiency causes a disorder of golgi homeostasis with abnormal protein glycosylation. *Am J Hum Genet* 98 (2):310–321. doi:10.1016/j.ajhg.2015.12.010.

Jansen, J. C., S. Timal, M. van Scherpenzeel, H. Michelakakis, D. Vicogne, A. Ashikov, M. Moraitou et al., 2016c. TMEM199 deficiency is a disorder of golgi homeostasis characterized by elevated aminotransferases, alkaline phosphatase, and cholesterol and abnormal glycosylation. *Am J Hum Genet* 98 (2):322–330. doi:10.1016/j.ajhg.2015.12.011.

Janssen, M. C., R. H. de Kleine, A. P. van den Berg, Y. Heijdra, M. van Scherpenzeel, D. J. Lefeber, and E. Morava. 2014. Successful liver transplantation and long-term follow-up in a patient with MPI-CDG. *Pediatrics* 134 (1):e279-e283. doi:10.1542/peds.2013-2732.

Kailemia, M. J., D. Park, and C. B. Lebrilla. 2016. Glycans and glycoproteins as specific biomarkers for cancer. *Anal Bioanal Chem.* 409 (2):395–410. doi:10.1007/s00216-016-9880-6.

Kang, P., Y. Mechref, and M. V. Novotny. 2008. High-throughput solid-phase permethylation of glycans prior to mass spectrometry. *Rapid Commun Mass Spectrom* 22 (5):721–734. doi:10.1002/rcm.3395.

Lacey, J. M., H. R. Bergen, M. J. Magera, S. Naylor, and J. F. O'Brien. 2001. Rapid determination of transferrin isoforms by immunoaffinity liquid chromatography and electrospray mass spectrometry. *Clin Chem* 47 (3):513–518.

Lauc, G., A. Essafi, J. E. Huffman, C. Hayward, A. Knezevic, J. J. Kattla, O. Polasek et al., 2010. Genomics meets glycomics-the first GWAS study of human N-Glycome identifies HNF1alpha as a master regulator of plasma protein fucosylation. *PLoS Genet* 6 (12):e1001256. doi:10.1371/journal.pgen.1001256.

Lauc, G., M. Pezer, I. Rudan, and H. Campbell. 2016. Mechanisms of disease: The human N-glycome. *Biochim Biophys Acta* 1860 (8):1574–1582. doi:10.1016/j.bbagen.2015.10.016.

Lefeber, D. J. 2016. Protein-Specific glycoprofiling for patient diagnostics. Clin Chem 62 (1):9–11. doi:10.1373/clinchem.2015.248518.

Lefeber, D. J., E. Morava, and J. Jaeken. 2011. How to find and diagnose a CDG due to defective N-glycosylation. *J Inherit Metab Dis* 34 (4):849–852. doi:10.1007/s10545-011-9370-0.

Lenos, K., J. A. Goos, I. M. Vuist, S. H. den Uil, P. M. Delis-van Diemen, E. J. Belt, H. B. Stockmann et al., 2015. MGL ligand expression is correlated to BRAF mutation and associated with poor survival of stage III colon cancer patients. *Oncotarget* 6 (28):26278–26290. doi:10.18632/oncotarget.4495.

Li, K., Z. Sun, J. Zheng, Y. Lu, Y. Bian, M. Ye, X. Wang, Y. Nie, H. Zou, and D. Fan. 2013. In-depth research of multidrug resistance related cell surface glycoproteome in gastric cancer. *J Proteomics* 82:130–140. doi:10.1016/j.jprot.2013.02.021.

Lu, L. L., A. W. Chung, T. R. Rosebrock, M. Ghebremichael, W. H. Yu, P. S. Grace, M. K. Schoen et al., 2016. A functional role for antibodies in tuberculosis. *Cell* 167 (2):433–443 e14. doi:10.1016/j.cell.2016.08.072.

Luo, G., M. Guo, K. Jin, Z. Liu, C. Liu, H. Cheng, Y. Lu, J. Long, L. Liu, J. Xu, Q. Ni, and X. Yu. 2016. Optimize CA19-9 in detecting pancreatic cancer by lewis and secretor genotyping. *Pancreatology.* doi:10.1016/j.pan.2016.09.013.

Markowska, A. I., K. C. Jefferies, and N. Panjwani. 2011. Galectin-3 protein modulates cell surface expression and activation of vascular endothelial growth factor receptor 2 in human endothelial cells. *J Biol Chem* 286 (34):29913–29921. doi:10.1074/jbc.M111.226423.

Marquardt, T., K. Luhn, G. Srikrishna, H. H. Freeze, E. Harms, and D. Vestweber. 1999. Correction of leukocyte adhesion deficiency type II with oral fucose. *Blood* 94 (12):3976–3985.

Marusyk, A., and K. Polyak. 2010. Tumor heterogeneity: causes and consequences. *Biochim Biophys Acta* 1805 (1):105–117. doi:10.1016/j.bbcan.2009.11.002.

Mesko, B., S. Poliska, S. Szamosi, Z. Szekanecz, J. Podani, C. Varadi, A. Guttman, and L. Nagy. 2012. Peripheral blood gene expression and IgG glycosylation profiles as markers of tocilizumab treatment in rheumatoid arthritis. *J Rheumatol* 39 (5):916–928. doi:10.3899/jrheum.110961.

Moremen, K. W., M. Tiemeyer, and A. V. Nairn. 2012. Vertebrate protein glycosylation: Diversity, synthesis and function. *Nat Rev Mol Cell Biol* 13 (7):448–462. doi:10.1038/nrm3383.

Nakayama, K., K. Moriwaki, T. Imai, S. Shinzaki, Y. Kamada, K. Murata, and E. Miyoshi. 2013. Mutation of GDP-mannose-4,6-dehydratase in colorectal cancer metastasis. *PLoS One* 8 (7):e70298. doi:10.1371/journal.pone.0070298.

Padler-Karavani, V. 2014. Aiming at the sweet side of cancer: aberrant glycosylation as possible target for personalized-medicine. *Cancer Lett* 352 (1):102–112. doi:10.1016/j.canlet.2013.10.005.

Park, J. J., J. Y. Yi, Y. B. Jin, Y. J. Lee, J. S. Lee, Y. S. Lee, Y. G. Ko, and M. Lee. 2012. Sialylation of epidermal growth factor receptor regulates receptor activity and chemosensitivity to gefitinib in colon cancer cells. *Biochem Pharmacol* 83 (7):849–857. doi:10.1016/j.bcp.2012.01.007.

Partridge, E. A., C. Le Roy, G. M. Di Guglielmo, J. Pawling, P. Cheung, M. Granovsky, I. R. Nabi, J. L. Wrana, and J. W. Dennis. 2004. Regulation of cytokine receptors by Golgi N-glycan processing and endocytosis. *Science* 306 (5693):120–124. doi:10.1126/science.1102109.

Pihikova, D., P. Kasak, P. Kubanikova, R. Sokol, and J. Tkac. 2016. Aberrant sialylation of a prostate-specific antigen: Electrochemical label-free glycoprofiling in prostate cancer serum samples. *Anal Chim Acta* 934:72–79. doi:10.1016/j.aca.2016.06.043.

Pinho, S. S., and C. A. Reis. 2015. Glycosylation in cancer: mechanisms and clinical implications. *Nat Rev Cancer* 15 (9):540–555. doi:10.1038/nrc3982.

Pinho, S. S., J. Figueiredo, J. Cabral, S. Carvalho, J. Dourado, A. Magalhaes, F. Gartner et al., 2013. E-cadherin and adherens-junctions stability in gastric carcinoma: functional implications of glycosyltransferases involving N-glycan branching biosynthesis, N-acetylglucosaminyltransferases III and V. *Biochim Biophys Acta* 1830 (3):2690–2700. doi:10.1016/j.bbagen.2012.10.021.

Pompach, P., Z. Brnakova, M. Sanda, J. Wu, N. Edwards, and R. Goldman. 2013. Site-specific glycoforms of haptoglobin in liver cirrhosis and hepatocellular carcinoma. *Mol Cell Proteomics* 12 (5):1281–1293. doi:10.1074/mcp. M112.023259.

Potelle, S., W. Morelle, E. Dulary, S. Duvet, D. Vicogne, C. Spriet, M. A. Krzewinski-Recchi et al., 2016. Glycosylation abnormalities in Gdt1p/TMEM165 deficient cells result from a defect in Golgi manganese homeostasis. *Hum Mol Genet* 25 (8):1489–1500. doi:10.1093/hmg/ddw026.

Reiding, K. R., D. Blank, D. M. Kuijper, A. M. Deelder, and M. Wuhrer. 2014. High-throughput profiling of protein N-glycosylation by MALDI-TOF-MS employing linkage-specific sialic acid esterification. *Anal Chem* 86 (12):5784–5793. doi:10.1021/ac500335t.

Rivinoja, A., F. M. Pujol, A. Hassinen, and S. Kellokumpu. 2012. Golgi pH, its regulation and roles in human disease. *Ann Med* 44 (6):542–554. doi:10.3109/07853890.2011.579150.

Ruhaak, L. R. 2017. The use of multiple reaction monitoring on QQQ-MS for the analysis of protein- and site-specific glycosylation patterns in serum. *Methods Mol Biol* 1503:63–82. doi:10.1007/978-1-4939-6493-2_6.

Ruhaak, L. R., S. L. Taylor, S. Miyamoto, K. Kelly, G. S. Leiserowitz, D. Gandara, C. B. Lebrilla, and K. Kim. 2013. Chip-based nLC-TOF-MS is a highly stable technology for large-scale high-throughput analyses. *Anal Bioanal Chem* 405 (14):4953–4958. doi:10.1007/s00216-013-6908-z.

Shankar, J., C. Boscher, and I. R. Nabi. 2015. Caveolin-1, galectin-3 and lipid raft domains in cancer cell signalling. *Essays in Biochemistry* 57 (2015):189–201.

Smith, L. M., and N. L. Kelleher. 2013. Proteoform: A single term describing protein complexity. *Nat Methods* 10 (3):186–187. doi:10.1038/nmeth.2369.

Song, T., D. Aldredge, and C. B. Lebrilla. 2015. A method for In-Depth structural annotation of human serum glycans that yields biological variations. *Anal Chem* 87 (15):7754–7762. doi:10.1021/acs.analchem.5b01340.

Taniguchi, N., and Y. Kizuka. 2015. Glycans and cancer: role of N-glycans in cancer biomarker, progression and metastasis, and therapeutics. *Adv Cancer Res* 126:11–51. doi:10.1016/bs.acr.2014.11.001.

Tarbell, J. M., and L. M. Cancel. 2016. The glycocalyx and its significance in human medicine. *J Intern Med* 280(1):91–113. doi:10.1111/joim.12465.

Tegtmeyer, L. C., S. Rust, M. van Scherpenzeel, B. G. Ng, M. E. Losfeld, S. Timal, K. Raymond et al., 2014. Multiple phenotypes in phosphoglucomutase 1 deficiency. *N Engl J Med* 370 (6):533–542. doi:10.1056/NEJMoa1206605.

Thanabalasingham, G., J. E. Huffman, J. J. Kattla, M. Novokmet, I. Rudan, A. L. Gloyn, C. Hayward et al., 2013. Mutations in HNF1A result in marked alterations of plasma glycan profile. *Diabetes* 62 (4):1329–1337. doi:10.2337/db12-0880.

Underwood, M. A., S. Gaerlan, M. L. De Leoz, L. Dimapasoc, K. M. Kalanetra, D. G. Lemay, J. B. German, D. A. Mills, and C. B. Lebrilla. 2015. Human milk oligosaccharides in premature infants: Absorption, excretion, and influence on the intestinal microbiota. *Pediatr Res* 78 (6):670–677. doi:10.1038/pr.2015.162.

Van Damme, T., T. Gardeitchik, M. Mohamed, S. Guerrero-Castillo, P. Freisinger, B. Guillemyn, A. Kariminejad et al., 2017. Mutations in ATP6V1E1 or ATP6V1A cause autosomal-recessive cutis laxa. *Am J Hum Genet* 100 (2):216–227. doi:10.1016/j.ajhg.2016.12.010.

van Scherpenzeel, M., G. Steenbergen, E. Morava, R. A. Wevers, and D. J. Lefeber. 2015a. High-resolution mass spectrometry glycoprofiling of intact transferrin for diagnosis and subtype identification in the congenital disorders of glycosylation. *Transl Res* 166 (6):639–649.e1. doi:10.1016/j.trsl.2015.07.005.

van Scherpenzeel, M., G. Steenbergen, E. Morava, R. A. Wevers, and D. J. Lefeber. 2015b. High-resolution mass spectrometry glycoprofiling of intact transferrin for diagnosis and subtype identification in the congenital disorders of glycosylation. *Transl Res* 166 (6):639–649 e1. doi:10.1016/j.trsl.2015.07.005.

Van Scherpenzeel, M., S. Timal, D. Rymen, A. Hoischen, M. Wuhrer, A. Hipgrave-Ederveen, S. Grunewald et al., 2014. Diagnostic serum glycosylation profile in patients with intellectual disability as a result of MAN1B1 deficiency. *Brain* 137 (Pt 4):1030–1038. doi:10.1093/brain/awu019.

Vankemmelbeke, M., J. X. Chua, and L. G. Durrant. 2016. Cancer cell associated glycans as targets for immunotherapy. *Oncoimmunology* 5 (1):e1061177. doi:10.1080/2162402X.2015.1061177.

Varki, A. 2017. Biological roles of glycans. *Glycobiology* 27 (1):3–49. doi:10.1093/glycob/cww086.

Vazquez, A. M., A. M. Hernandez, A. Macias, E. Montero, D. E. Gomez, D. F. Alonso, M. R. Gabri, and R. E. Gomez. 2012. Racotumomab: an anti-idiotype vaccine related to N-glycolyl-containing gangliosides—preclinical and clinical data. *Front Oncol* 2:150. doi:10.3389/fonc.2012.00150.

Walt, D. A., Aoki-Kinoshita, A. F., Bendiak, B., Bertozzi, C. R., Boons, G. J., Darvill, A., Hart, G. et al. 2012. *Transforming Glycoscience: A Roadmap for the Future.* Washington, DC: The National Academies Press.

Wheeler, S. F., P. Domann, and D. J. Harvey. 2009. Derivatization of sialic acids for stabilization in matrix-assisted laser desorption/ionization mass spectrometry and concomitant differentiation of alpha(2 --> 3)- and alpha(2 --> 6)-isomers. *Rapid Commun Mass Spectrom* 23 (2):303–312. doi:10.1002/rcm.3867.

Wolpin, B. M., P. Kraft, M. Xu, E. Steplowski, M. L. Olsson, A. A. Arslan, H. B. Bueno-de-Mesquita et al., 2010. Variant ABO blood group alleles, secretor status, and risk of pancreatic cancer: results from the pancreatic cancer cohort consortium. *Cancer Epidemiol Biomarkers Prev* 19 (12):3140–3149. doi:10.1158/1055-9965.EPI-10-0751.

Wuhrer, M. 2013. Glycomics using mass spectrometry. *Glycoconj J* 30 (1):11–22. doi:10.1007/s10719-012-9376-3.

Yoneyama, T., C. Ohyama, S. Hatakeyama, S. Narita, T. Habuchi, T. Koie, K. Mori et al., 2014. Measurement of aberrant glycosylation of prostate specific antigen can improve specificity in early detection of prostate cancer. *Biochem Biophys Res Commun* 448 (4):390–396. doi:10.1016/j.bbrc.2014.04.107.

Zeevaert, R., F. Foulquier, J. Jaeken, and G. Matthijs. 2008. Deficiencies in subunits of the Conserved Oligomeric Golgi (COG) complex define a novel group of congenital disorders of glycosylation. *Mol Genet Metab* 93 (1):15–21. doi:10.1016/j.ymgme.2007.08.118.

Zhang, W., P. M. James, B. G. Ng, X. Li, B. Xia, J. Rong, G. Asif et al., 2016. A novel N-tetrasaccharide in patients with congenital disorders of glycosylation, including asparagine-linked glycosylation protein 1, Phosphomannomutase 2, and Mannose Phosphate Isomerase Deficiencies. *Clin Chem* 62 (1):208–217. doi:10.1373/clinchem.2015.243279.

miRNAs as Novel Biomarkers for Health and Disease

7.7

María Laura García Bermejo

Contents

miRNAs GENERAL FEATURES

MicroRNAs (miRNAs) are small (20–25 nucleotides) noncoding RNAs and key players in gene posttranscriptional regulation. More than 80% of the genes in mammals are under their control. Moreover, almost every cellular function is tightly regulated by miRNAs. Indeed, miRNAs show very specific expression patterns among tissues and cell types.

Around 2,000 miRNAs have been already identified in the human genome. Their mechanism of action is based on the recognition of small sequences (6–8 nt) in their target mRNAs. Due to the small size of the recognition site, one miRNA could regulate hundreds of target mRNAs and one mRNA can be regulated by several miRNAs. This dynamic regulation has unveiled them as critical regulators of a wide range of cellular events. Therefore, their deregulation is frequently associated with disease triggering and development.

miRNA BIOGENESIS

miRNAs are mainly transcribed by RNA polymerase II as longer primary transcripts called pri-miRNAs. miRNAs genes are often located in noncoding DNA regions and they are frequently organized in clusters. Clustered miRNAs are transcribed as a single, longer pri-miRNA that generates several functional miRNAs by subsequent processing. miRNA genes can also be found in protein-coding genes, often located in introns. In these cases, splicing of the coding mRNAs leads to generation of the functional miRNA. As other transcripts produced by RNA polymerase II, pri-miRNAs present a 5′cap and a 3′poly-A tail.

Pri-miRNAs molecules are imperfect stem–loop structures that can be recognized by a processing complex formed by the RNAse III enzyme Drosha and the RNA binding protein (RBP) DGCR8. Stem–loop double-stranded structures of pri-miRNAs are recognized by DGCR8, which guides the positioning of Drosha. This catalytic site cleaves pri-miRNAs, liberating a hairpin RNA molecule of 70–100 nucleotides known as pre-miRNA.

Pre-miRNAs are exported to the cytoplasm by the nuclear export receptor Exportin 5 in a Ran-GTP dependent manner. Once in the cytoplasm pre-miRNAs are further processed by another RNAse III enzyme called Dicer. A new cleavage produces a double-stranded RNA molecule of 22 nucleotides. One of the strands (the mature miRNA) is transferred to the Argonaute protein in order to build the RNA-induced silencing complex (RISC). The other strand, called minor, passenger or strand, is frequently degraded. Strand selection mechanism is still under investigation but it has been proposed that the less stable base-pairing in its 5′end is often chosen as the guide strand to be loaded into the RISC complex. The RISC complex is the key effector of miRNA regulation: It is responsible for driving mRNA degradation or translation repression.

miRNA TARGET RECOGNITION, FUNCTION AND TURNOVER

Once loaded into the RISC complex, miRNAs recognize their target mRNAs by base-pair complementarity. Target sequences are mainly located in the 3′ untranslated region (UTR) of mRNAs. However, functional miRNA binding sites can also be found in the 5′ UTR and open reading frame regions.

Nucleotides in positions 2–8, called the seed sequence, are essential for pairing with the target mRNA and miRNA function. In the case of perfect complementarity of the seed sequence of the miRNA with the target sequence, miRNAs act as a short interfering RNA (siRNA), promoting mRNA by cleaving the RISC complex. When pairing with target sequences is partially complementary, which is the most frequent mechanism in mammals, miRNA regulation takes place by mRNA translation repression or degradation. However, this degradation process is different and involves recruitment of deadenylase complexes that remove or shorten the poly-A tail of the target transcript. Poly-A tail shortening induces decapping of the 5′ extreme of the transcript and uncapped mRNAs are rapidly degraded by 5′ to 3′ exoribonucleases.

Although the contribution rate of mRNA decay and translational repression to miRNA action is a controversial topic, it seems clear that target degradation provides a major contribution to silencing in mammal cells. In this regard, it has been estimated that mRNA decay is present in 85% of the miRNA regulation process, whereas 15% corresponds to translational repression.

miRNAs exhibit inherent half-lives that could be determined by their sequence or after maturation, by posttranscriptional mechanisms such as uracile and adenosine addition to their 3′end. Half-lives of some miRNAs could reach hours or even days in organs such the liver or heart. However, this slow turnover is not suitable for their capacity to regulate rapid cell responses to environmental signals in other contexts. Indeed, some miRNAs expressed in the retina and involved in darkness adaptation present a half-life of approximately 1 h. Therefore, miRNA half-life and miRNA decay regulation appear critical for miRNA function, and they are considered big challenges in miRNA biology research for the near future.

miRNA SECRETION: CIRCULATING miRNAs

Initial studies proposed that miRNAs were only present inside the cell, but it has been widely demonstrated that miRNAs can be secreted to the extracellular environment, with relevant functional consequences. Because of secretion, miRNAs can be detected in a wide range of cell-free body fluids such as urine, serum or saliva.

miRNA secretion is a highly regulated process and selection of miRNAs that can be secreted is not a random process. As mentioned before, miRNA deregulation has been associated with the development of a wide range of pathologies. Due to both features, physiological or pathological regulation of intracellular miRNAs may also modify the panel of secreted miRNAs. Indeed, changes in serum miRNAs profiles have been unveiled as useful markers of a wide range of diseases including cancer, cardiovascular disease, neurodegeneration or nephropathies as well as altered physiological states such as pregnancy.

Moreover, serum or plasma miRNAs have demonstrated great stability and resistance to aggressive conditions such as RNAse treatment or drastic pH changes. The mechanism underlying this unexpected miRNA stability includes that circulating microRNAs are released from cells in membrane vesicles (exosomes or microvesicles) that protect them from the environment. Recent studies have demonstrated that serum circulating miRNAs can also be carried by Argonaute2 proteins or combined with high-density lipoprotein cholesterol (HDL-C).

miRNAs IN LIQUID BIOPSY AS USEFUL BIOMARKERS

Serum or plasma miRNAs achieve nearly all the required characteristics for an ideal biomarker. Their presence in a peripheral fluid allows diagnosis by minimum invasive methods because samples can be easily and routinely obtained in clinical practice. In addition to their high stability in fresh samples, several studies have demonstrated that miRNAs maintain stability and reliability in long-term stored serum/plasma samples, even if the samples have been held at room temperature for hours or upon freeze–thaw cycles. Not all the blood markers are so stable, thus affecting the tests reproducibility.

Additionally, circulating miRNAs can be easily quantified by quantitative real-time PCR (qRT-PCR), a very reliable and affordable technique. Moreover, several biosensors involving nanoparticles and electrochemistry methods are being currently developed in order to generate point-of-care devices for miRNAs detection in any clinical context. These characteristics, joined to their tissue and cell type specificity, point out circulating miRNAs as promising biomarkers for more accurate diagnosis and monitoring of diseases.

Remarkably, circulating miRNAs belong to a new concept of biomarkers in contrast to the previous ones. Secreted miRNAs are not the result of terminal tissue damage or a final cell response. Most of them are secreted at the initial steps of the pathophysiological mechanisms responsible for diseases. Moreover, because they are fine-tuned regulators and orchestrate continuous cell response to stimuli, they bring a dynamic knowledge of the pathology evolution. Therefore, miRNAs could provide to the clinician very valuable additional information that is currently not accessible.

REFERENCES

Arroyo JD et al. (2011). Argonaute2 complexes carry a population of circulating microRNAs independent of vesicles in human plasma. *Proc Natl Acad Sci USA*. 108(12):5003–5008.

Bartel DP. (2009). MicroRNAs: Target recognition and regulatory functions. *Cell* 136(2):215–233.

Baskerville S, Bartel DP (2005). Microarray profiling of microRNAs reveals frequent coexpression with neighboring miRNAs and host genes. *RNA* 11(3):241–247.

Chang TC, Mendell JT (2007). microRNAs in vertebrate physiology and human disease. *Annu Rev Genomics Hum Genet.* 8:215–239.

Huntzinger E, Izaurralde E. (2011). Gene silencing by microRNAs: Contributions of translational repression and mRNA decay. *Nat Rev Genet.* 12(2):99–110.

Kozomara A et al. (2011). miRBase: Integrating microRNA annotation and deep-sequencing data. *Nucleic Acids Res.* 39(Database issue):D152–D157.

Krol J et al. (2010). Characterizing light-regulated retinal microRNAs reveals rapid turnover as a common property of neuronal microRNAs. *Cell.* 141(4):618–631.

Krol J, Loedige I, Filipowicz W. (2010). The widespread regulation of microRNA biogenesis, function and decay. *Nat Rev Genet.* 11(9):597–610.

Krützfeldt J et al. (2006). Strategies to determine the biological function of microRNAs. *Nat Genet.* 38 Suppl: S14–S19.

Landgraf P et al. (2007). A mammalian microRNA expression atlas based on small RNA library sequencing. *Cell.* 129(7):1401–1404.

Leni Moldovan et al. (2014). Methodological challenges in utilizing miRNAs as circulating biomarkers. *J Cell Mol Med.* 18(3): 371–390.

Rana TM. (2007). Illuminating the silence: Understanding the structure and function of small RNAs. *Nat Rev Mol Cell Biol.* 8(1):23–36.

Treiber T et al. (2012). Regulation of microRNA biogenesis and function. *Thromb Haemost.* 107(4):605–610.

Valadi H et al. (2007). Exosome-mediated transfer of mRNAs and microRNAs is a novel mechanism of genetic exchange between cells. *Nat Cell Biol.* 9(6):654–659.

Targeted Proteomics for Absolute Quantification of Protein Biomarkers in Serum and Tissues

7.8

Coşkun Güzel, Christoph Stingl, Frank Klont, Roel Tans, Esther Willems, Rainer Bischoff, Alain van Gool, Theo M. Luider, and the Biomarker Development Center Consortium

Contents

INTRODUCTION

In every living cell, a large number of biological processes take place that are mainly regulated by proteins. In these biological processes, molecular and chemical interactions between proteins form the basis of health and disease [1]. Moreover, proteins by themselves are complex biomolecules that do not exist mostly as single structures. Different modifications such as phosphorylation can exist but also, for instance, differences in the primary, secondary structure of proteins. The different individual protein structures can be well studied regarding their functions; however, the high diversity of modifications of a protein makes it hard to implement the full understanding of how it relates to molecular mechanisms in health and disease. In other words, a certain protein biomarker in a body fluid does not depict its importance per se given that it is likely present in different isoforms (e.g., PTMs such as glycosylation, methylation, and phosphorylation) [2].

Ligand binding assays (LBAs) and notably ELISAs are routinely used for protein bioanalysis. Advantages of the immunoassay technology are the high selectivity and sensitivity (e.g., plasma detection limits in the low picograms-per-milliliter range) [3] and the ease with which they can be performed in a high-throughput format. However, immunoassays have limitations such as the high development cost for sensitive and well-characterized antibodies as well as cross-reactivity with other proteins or interference from other ligands bound to the target protein [4]. Mostly, the primary structure of the antibody is not available and, therefore, the antibody delivered is a kind of black box [5]. Although multiplexing is possible with immunoassays (e.g., those based on flow cytometry), analytical quality generally suffers from applicability issues because the analytical conditions need to be a compromise between the reagents of all components of the multiplex set [6]. Most importantly, the often-limited specificity of the antibodies that are key components of immunoassays limit their use in biomarker analysis, which requires optimal specificity in addition to sensitivity. As a consequence, in many cases differential behavior of protein biomarkers cannot be confirmed in follow-up studies [7] and require a more robust analytical method to quantitate proteins.

MS has emerged as an alternative analytical method to quantify proteins and protein isoforms. Protein analysis favors the use of LC coupled to MS based on well-established ionization principles. The various proteomics methods can roughly be classified in bottom-up proteomics (focusing on identification of protein fragments following proteolytic digestion), top-down proteomics (focusing on intact proteins) and targeted proteomics (focusing on quantifying preselected peptides or proteins). The latter has received much attention for biomarker analysis, validation and further evaluation.

SELECTIVE AND PARALLEL REACTION MONITORING MASS SPECTROMETRY

Of targeted proteomics, selected reaction monitoring (SRM) MS emerged as the most widely used experimental approach to quantify peptides in biological samples by MS and thereby infer the corresponding protein levels [8–13, 23] (for further review, see Vidova and Spacil 2017) [14]. However, interferences in complex biological samples often limit sensitivity in comparison with immunoassays unless appropriate sample preparation is performed [12–17]. Co-eluting peptides with a precursor ion mass close to the peptide of interest may result in fragment ions that overlap with the targeted transitions, resulting in considerable chemical noise. Such noise limits detection sensitivity and contributes to diminished accuracy and precision. In addition, it is challenging to quantify low levels of proteins in biological samples like serum or tissue due to the limited sensitivity and dynamic range of MS detectors and finite loading capacity of LC columns, as well as insufficient resolution for separating interfering compounds.

Parallel reaction monitoring (PRM) using high-resolution MS [18] goes beyond SRM and provides data with higher mass accuracy (part-per-million to sub–parts-per-million levels), thus reducing interferences caused by co-eluting compounds with similar but not identical mass transitions [18,19] (Figure 7.8.1). The higher selectivity allows covering a wider dynamic concentration range [18]. Moreover, PRM methods for individual peptides are easier to set up because all transitions are monitored and optimal transitions in terms of sensitivity and specificity can be retrieved and combined *in silico* after the analysis [20]. Literature on PRM shows the feasibility of the approach for quantification of proteins in complex biological samples after proteolytic digestion [19,21,22]. Notably, Domon and coworkers published on the use of PRM in large-scale experiments [23–27]. However, reaching the nanograms-per-milliliter level in body fluids without using affinity binders (e.g., immunoglobulins) remains a challenge also for PRM approaches. Measuring low protein levels (ng/mL) in trypsin-digested and fractionated serum in a reproducible manner is possible [28]. As an example, we targeted HSP90α, a protein that is upregulated in various cancers and is thus pursued as a target for anticancer therapy [29]. We compared the concentration of HSP90α in 43 sera from healthy subjects measured by SRM, by PRM and by a commercially available ELISA with respect to comparability, repeatability and sensitivity [28]. We demonstrated a reproducible, robust and sensitive PRM assay to determine HSP90α concentrations in strong cation exchange (SCX)-fractionated sera at low nanograms-per-milliliter levels. Sensitivity of the PRM assay was in agreement with data obtained by ELISA although the coefficient of variance (CV) was lower in PRM. In SRM and especially PRM, the quality of measurement can easily be

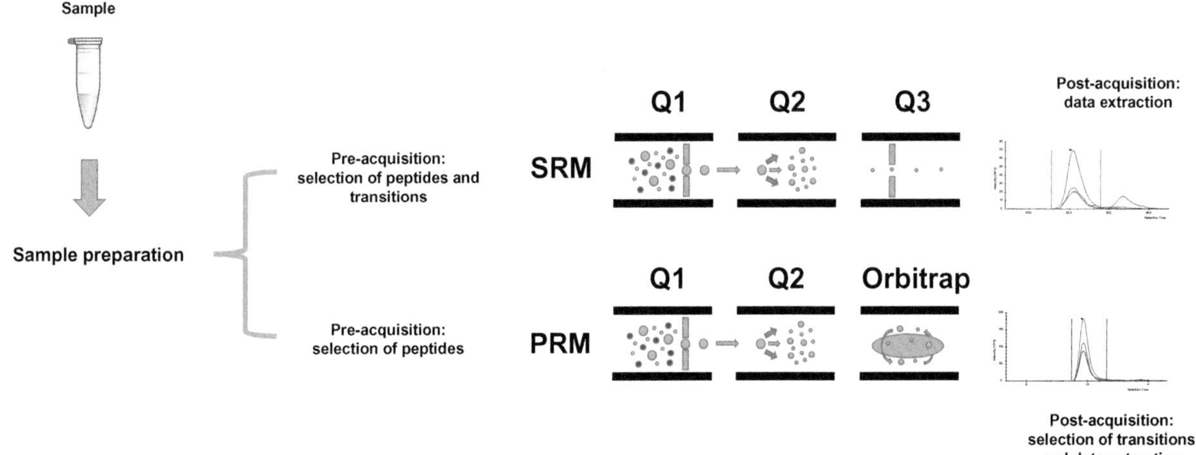

FIGURE 7.8.1 Selected reaction monitoring (SRM) and parallel reaction monitoring (PRM).

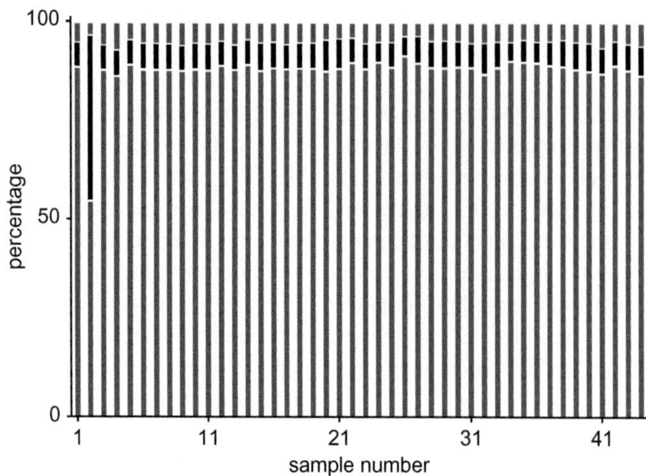

FIGURE 7.8.2 Distribution of endogenous HSP90α transitions for peptide YIDQEELNK measured by PRM. Bars represent normalized peak areas (percentage of total) of the transitions y5 (red), y6 (black) and y7 (blue) determined in 43 SCX-fractionated serum samples. Sample number 44 corresponds to 1 fmol of the pure (0.1% aqueous FA) SIL HSP90α peptide. It can be seen that one outlier (sample number 2) was observed by PRM. (From Güzel, C. et al., *Proteomics Clin. Appl.*, 12, 2018, doi: 10.1002/prca.201700107.)

assessed by an aberrant ratio between transitions (Figure 7.8.2), whereas in ELISA, results caused by aberrations in the assay are much more difficult to recognize as outliers. If fractionation of biological samples is technically feasible, PRM can be used as an attractive alternative to immunoassays to quantify multiple proteins at the ng/mL level in complex protein mixtures, including serum. The major analytical advantages of MS include the more specific detection and the excellent technical reproducibility that can be reached (CV lower than 5%).

To use SRM and PRM in the most optimal way, the peptides of interest ought to be chosen thoughtfully. The peptides chosen ought to be unique for the protein targeted (proteotypic peptides), and the peptide must not be too short (less than 7 amino acids) or too long (more than 20 amino acids) to prevent loss of specificity and sensitivity, respectively. One should avoid ragged ends that could lead to partial enzymatic digestion and, ideally, they should contain no amino acids that are prone to chemical modifications such as oxidation (e.g., methionine, histidine, tryptophan), deamidation (e.g., combination of N-G/P or Q-G/P), or N-terminal cyclization of glutamine and glutamate and N-terminal carboxymethylation of cysteine. Moreover, N- and C-terminal peptides are in general more prone to degradation and should be avoided if possible. Additionally, one should be aware of protein-specific amino acid polymorphisms, posttranslational modifications (e.g., methylation, phosphorylation, glycosylation) and other natural variants resulting in different proteoforms. Publicly available databases such as the SRMAtlas compendium (http://www.srmatlas.org), Uniprot, ENSEMBLE or dbsnip are imperative in that respect. Moreover, most software programs (e.g., Skyline) incorporate build-in libraries and filters to exclude peptides with specific amino acids features as outlined above to facilitate target selection. Nonetheless, unexpected modification or loss of peptides due to biological cleavage or degradation of protein subunits can introduce aberrant ratios between peptides within one protein. One is, therefore, encouraged to select at least two peptides per protein for adequate quantification. (For an excellent review on this topic, see [12,30].)

SAMPLE PREPARATION

Sample preparation for analysis in both SRM and PRM remains a point of specific concern. Most often, sample preparation is performed stepwise: first, lysis and denaturation; followed by reduction and alkylation of sulfhydryl groups; and finally, proteolytic (predominately tryptic) digestion. Because the peptides targeted by PRM and SRM are mostly selected in such a way that sulfhydryl groups are not present in these peptides, it is not necessary to use reducing and alkylation reagents that may produce adverse effects because undesired side reactions with an alkylating reagent (e.g., iodoacetamide) can occur (Figure 7.8.3). Additionally, without enrichment, both methodologies generally remain less sensitive as compared with immunoassays. Optimally, biomarker analysis methods possess optimal selectivity and specificity. The combination of the highly selective MS with an affinity purification to obtain optimal sensitivity would be a golden combination, and several possibilities exist using various binders, for example, antibody (including Stable Isotope Standards and Capture by Anti-Peptide Antibodies [SISCAPA]) and affimers. In general, without sample fractionation, SRM and PRM can reach micrograms protein of interest per milliliter serum. With sample fractionation, or affinity enrichment by antibodies or other affinity separation (ion exchange columns or metal bound chromatography [e.g., Ni^{2+}-IMAC, TiO_2]) sensitivities of nanograms per milliliter serum [28,31,32] or nanograms per grams total protein in tissue can be reached [33,34].

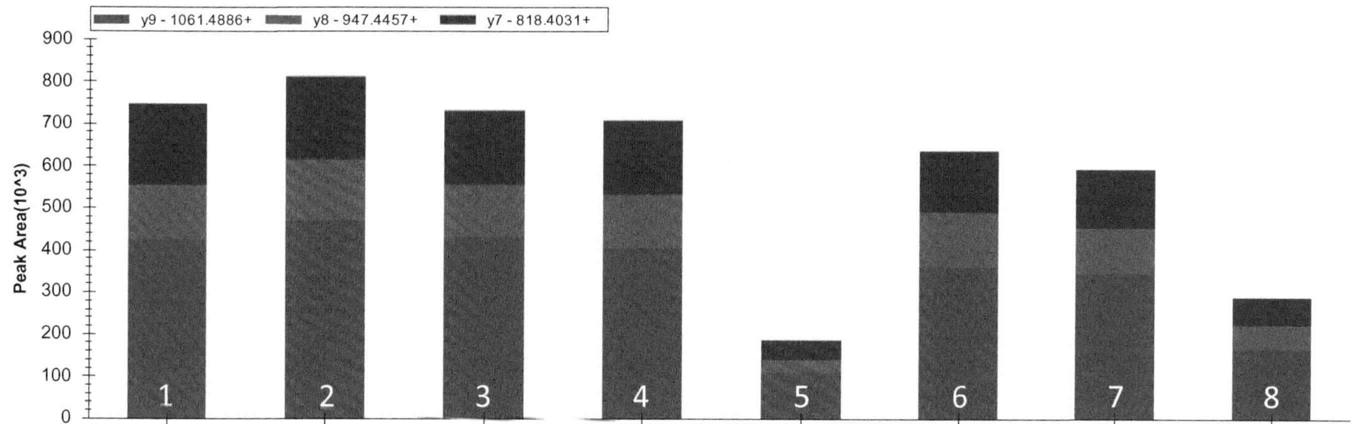

FIGURE 7.8.3 Repetitive (n = 4) PRM measurements of a tissue digest with an alkylation and reduction treatment (bars 5–8) and without treatment (bars 1–4). In this example, the targeted stable isotope-labeled peptide SDLVNEEATGQFR related to CEACAM5 showed high variances of intensities with the alkylation and reduction procedure, while without treatment remained stable.

ABSOLUTE QUANTITATION

Stable isotope-labeled peptides are essential for targeted MS in order to obtain absolute quantification and reliability of the measurements performed. We can divide the function of these stable isotope-labeled peptides in two directions: (1) correction for variation introduced, for instance, by the matrix of a patient sample; and (2) calibrants. The latter is covered largely by FDA (https://www.fda.gov) and EMI guidelines (http://www.ema.europa.eu); thus, they are not further discussed here. The reference peptides give information on the analytical procedure for, for example, recovery and reproducibility as indicators for performance of the sample preparation [35,36]. Stable isotope-labeled peptides can be added in the various steps of sample preparation during method development to provide detailed information for each step and the potential problems that can occur. Ideally, the stable isotope-labeled peptide is spiked at the start of the sample preparation. In this way, nonspecific cleavage during digestion can be assessed to a certain level by adding peptides with specific enzyme cleavage sites. However, there might be biological reasons that this spiking in the starting material can be difficult to perform, for instance, due to the presence of enzymatic activity affecting the spiked peptide or a sample preparation step aiming to remove small molecules (e.g., precipitation). Most ideally, the protein of interest is spiked as a stable isotope-labeled protein; in most cases, this is difficult and expensive to realize and even recombinant proteins are chemically not exactly comparable to the endogenous protein in a complex sample environment (e.g., where there are other secondary, tertiary or quaternary structures or alternative modifications). This problem, obtaining the most ideal reference standard, is comparable to immunoassay techniques and it is not specific for MS. MS gives the possibility to standardize in a specific way because more than one peptide of a protein can be taken and a thorough assessment for correct measurement among the different peptides can be performed. The use of pure endogenous proteins or corresponding recombinant proteins can help considerably in that respect. In complex samples such as serum and tissue, one observes that disagreements between peptides can exist; however, after sample preparation, these measurements often tend to agree much better, but not always. The use of a quantifying peptide and a qualifier peptide cannot solve this problem of disagreement between peptides from the same protein because the ultimate problem is that if sample preparation cannot reduce the complexity of the peptide or protein mixture, disagreeing quantitative results will remain and only a biased semiquantitative comparison can be made, most often for the peptide with the highest intensity.

The signal-to-noise ratio of SRM and PRM measurements depends on the quadrupole characteristics. In quadrupoles of different vendors, mass windows may be adjusted to different mass widths (ranging from 0.2 to 2 m/z) that have a pronounced effect on sensitivity and selectivity. One can imagine that, if the mass window is too wide, many interfering compounds will pass the quadrupole, generating possibly interfering fragment ions. However, if the window of the quadrupole is too small, fewer ions will pass and sensitivity will suffer. Because the quadrupole window has a large effect on the signal-to-noise characteristics of the measurement, the combination of a high-resolution mass analyzer for the fragment ions with a quadrupole mass filter for precursor ion selection can significantly improve selectivity by reducing chemical noise without necessarily decreasing sensitivity. An optimal width of the quadrupole settings with high-resolution fragment ion analysis will decrease the lower limit of detection significantly.

The analysis of SRM and PRM data can be readily performed in dedicated programs such as the Skyline software [37]. For correct annotation and integration of the peaks manual inspection is still recommended after automatic data processing. For assessment of larger numbers of samples, specific algorithms can be used to streamline this process.

The Skyline software [37] is supported by all MS vendors, and the software is maintained and kept at a continuously high level and the company has a strong interaction with the users to improve the software.

TOP-DOWN PROTEOMICS

The proteome is a highly complex and variable collection of biological entities. The combination of allelic variation, alternative splicing and numerous PTMs yields numerous "protein variants" from a single gene, designated proteoforms [38]. Bottom-up proteomics has prevailed in studying proteins at the peptide level in proteomics discovery studies and yields a detailed reflection of protein content and relative abundance in a biological sample. However, this approach loses a lot of information contained within full-length proteins and associated structural information due to the tryptic cleavage of proteins [2,39]. Indeed, one cannot determine the presence of multiple pre- and/or posttranslational modifications on the same intact protein. This can be of vital importance as an activity or the stability of a protein is determined by multiple modifications so measuring single modifications in a protein biomarker may not yield useful mechanistic or diagnostic information.

Top-down proteomics circumvents this issue by measuring intact proteins or proteoforms. Once intact proteoforms are prepared for ionization, an MS1 scan provides the mass of the intact proteoform including its PTMs. When precursor ions are selected, they are fragmented and analyzed in an MS2 scan (also called MS/MS or tandem MS). These fragments can then be analyzed to characterize the proteoform, protein isoforms, the amino acid sequence and PTMs in the context of the intact protein [2,39].

Omitting the tryptic cleavage step in protein analysis seems more simple and straightforward. However, top-down proteomics is highly challenging because it relies even more strongly on sample preparation, MS instrumentation and data-analysis as compared with shotgun proteomics [39]. First, top-down proteomics requires at least some purification of the protein of interest. These purifications are often conducted with antibodies, a process which strongly depends on the quality of the binding agent and the accessibility of the binding epitope in the intact native protein. Second, intact proteins yield multiple charge states after electrospray ionization (4–20 rather than 2–3 for peptides) where each ion with a unique charge is a specific proteoform. This requires a high resolution in LC separation of intact proteins and in MS analysis (often using ultra high-resolution quadrupole time-of-flight [QTOF] MS or Orbitrap MS). The complexity of proteoform MS and MS/MS spectra requires novel ways for MS data handling and interpretation and intact protein MS databases, both of which are only emerging at the moment [2,40].

As part of the biomarker analysis workflow, we recommend that proteins identified by bottom-up proteomics are to be further characterized through top-down proteomics. If possible, measuring intact proteins as the diagnostic biomarker is preferred. We have made such a step using intact glycotransferin as a clinically applied biomarker for specific classes of CDG [41]. The feasibility of this approach is a great promise to future perspectives.

FUTURE PERSPECTIVES IN MULTIPLEXING IN TARGETED PROTEOMICS

SRM and PRM have the potential to measure and quantify multiple proteins of interest (based on up to ~100 peptides per run) including specific mutations and modifications without using antibodies. As such, it has matured to a powerful method for specific analysis of protein and peptide biomarkers. For serum proteins in the micrograms-per-milliliter range, this has been nicely illustrated by the Borchers group who developed a 30-min MRM method to quantify 67 plasma proteins in several clinical studies [42].

Further innovation is needed to apply SRM and PRM to large numbers of samples in routine diagnostic laboratories and population studies. Preferably, in a bioassay the measurement of a sample must, on average, not take more than 5 min per sample. SRM or PRM protein measurements may be performed in 5 min in exceptional cases, but longer time frames of up to 1 h exist, which is much longer as compared with immunoassays with low minute scale per analysis and that can use automated, high-throughput technology. Technically, MS can measure on a second–minute timescale. However, in the multiplex analysis of multiple peptides, chromatography remains the bottleneck with respect to increasing throughput, whereas in SRM mode, the dwell time (the duration in which each *m/z* ion signal is collected) can become a limiting factor. However, technically and theoretically this improvement in throughput might be achievable based on robust, fast and parallelized chromatography approaches [43].

Recently, profiling of mass peaks by data-independent acquisition (DIA) MS received quite some interest [14,44]. Using this approach, biological samples are profiled in an unbiased manner, usually using ultra-high resolution QTOF mass spectrometers, yielding a data-rich profile of tryptic peptides. Through a subsequent data-dependent acquisition (DDA), one can focus on a predefined set of peptide biomarkers. This avoids the time- and labor-intensive step of peptide-specific assay development as discussed above in SRM and PRM because the same laboratory workflow is used for analysis of samples from the same matrix. For comparative studies, this is a potentially powerful approach; however, DIA-DDA MS will only yield semiquantitative data at best, so it is currently unsuitable for the absolute quantitation of biomarker peptides. Further innovation is needed through combination of DIA-DDA MS with multiplex labeling techniques to fully realize the potential of MS as a high-throughput, multiplex analytical technique. Although this may come at the cost of increased complexity, decreased sensitivity and higher costs, it might open alternative avenues for clinically applicable quantification of multiple proteins in complex clinical samples.

REFERENCES

1. Aebersold, R., and Mann, M. (2016). Mass-spectrometric exploration of proteome structure and function. *Nature* 537, 347.
2. Schmit, P.O. et al. (2017). Towards a routine application of Top-Down approaches for label-free discovery workflows. *J Prot* 175, 12–26.
3. Rusling, J. F., Kumar, C. V., Gutkind, J. S., and Patel, V. (2010). Measurement of biomarker proteins for point-of-care early detection and monitoring of cancer. *Analyst* 135, 2496–2511.
4. Bults, P., van de Merbel, N. C., and Bischoff, R. (2015). Quantification of biopharmaceuticals and biomarkers in complex biological matrices: A comparison of liquid chromatography coupled to tandem mass spectrometry and ligand binding assays. *Expert Rev Proteomics* 12, 355–374.
5. Bradbury, A., Plückthun, A. and 110 co-signatories. (2015). Standardize antibodies in research. *Nature* 518, 27–29.
6. Tighe, P. J., Ryder, R. R., Todd, I., and Fairclough, L. C. (2015). ELISA in the multiplex era: Potentials and pitfalls. *Proteomics Clin Appl* 9, 406–422.
7. Freedman, L.P., Cockburn, I.M., and Simcoe, T.S. (2015). The economics of reproducibility in preclinical research. *PLoS Biol* 13: e1002165.
8. Bereman, M. S., MacLean, B., Tomazela, D. M., Liebler, D. C., and MacCoss, M. J. (2012). The development of selected reaction monitoring methods for targeted proteomics via empirical refinement. *Proteomics* 12, 1134–1141.
9. Keshishian, H., Addona, T., Burgess, M., Kuhn, E., and Carr, S. A. (2007). Quantitative, multiplexed assays for low abundance proteins in plasma by targeted mass spectrometry and stable isotope dilution. *Mol Cell Proteomics* 6, 2212–2229.
10. Kim, K. H., Ahn, Y. H., Ji, E. S., Lee, J. Y., Kim, J. Y., An, H. J., and Yoo, J. S. (2015). Quantitative analysis of low-abundance serological proteins with peptide affinity-based enrichment and pseudo-multiple reaction monitoring by hybrid quadrupole time-of-flight mass spectrometry. *Anal Chim Acta* 882, 38–48.
11. Guzel, C., Ursem, N. T., Dekker, L. J., Derkx, P., Joore, J., van Dijk, E., Ligtvoet, G., Steegers, E. A., and Luider, T. M. (2011). Multiple reaction monitoring assay for pre-eclampsia related calcyclin peptides in formalin fixed paraffin embedded placenta. *J Proteome Res* 10, 3274–3282.
12. Lange, V., Picotti, P., Domon, B., and Aebersold, R. (2008). Selected reaction monitoring for quantitative proteomics: A tutorial. *Mol Syst Biol* 4, 222.
13. Boichenko, A. P., Govorukhina, N., Klip, H. G., van der Zee, A. G., Guzel, C., Luider, T. M., and Bischoff, R. (2014). A panel of regulated proteins in serum from patients with cervical intraepithelial neoplasia and cervical cancer. *J Proteome Res* 13, 4995–5007.
14. Vidova, V., and Spacil, Z. (2017). A review on mass spectrometry-based quantitative proteomics: Targeted and data independent acquisition. *Anal Chim Acta* 964, 7–23.
15. Hembrough, T. et al. (2012). Selected reaction monitoring (SRM) analysis of epidermal growth factor receptor (EGFR) in formalin fixed tumor tissue. *Clin Proteomics* 9, 5.
16. Shi, T., Su, D., Liu, T., Tang, K., Camp, D. G., 2nd, Qian, W. J., and Smith, R. D. (2012). Advancing the sensitivity of selected reaction monitoring-based targeted quantitative proteomics. *Proteomics* 12, 1074–1092.
17. Zhi, W., Wang, M., and She, J. X. (2011). Selected reaction monitoring (SRM) mass spectrometry without isotope labeling can be used for rapid protein quantification. *Rapid Commun Mass Spectrom* 25, 1583–1588.
18. Peterson, A. C., Russell, J. D., Bailey, D. J., Westphall, M. S., and Coon, J. J. (2012). Parallel reaction monitoring for high resolution and high mass accuracy quantitative, targeted proteomics. *Mol Cell Prot* 11, 1475–1488.
19. Kim, Y. J., Gallien, S., El-Khoury, V., Goswami, P., Sertamo, K., Schlesser, M., Berchem, G., and Domon, B. (2015). Quantification of SAA1 and SAA2 in lung cancer plasma using the isotype-specific PRM assays. *Proteomics* 15, 3116–3125.
20. Ronsein, G. E., Pamir, N., von Haller, P. D., Kim, D. S., Oda, M. N., Jarvik, G. P., Vaisar, T., and Heinecke, J. W. (2015). Parallel reaction monitoring (PRM) and selected reaction monitoring (SRM) exhibit comparable linearity, dynamic range and precision for targeted quantitative HDL proteomics. *J Proteomics* 113, 388–399.
21. Sowers, J. L., Mirfattah, B., Xu, P., Tang, H., Park, I. Y., Walker, C., Wu, P., Laezza, F., Sowers, L. C., and Zhang, K. (2015). Quantification of histone modifications by parallel-reaction monitoring: A method validation. *Anal Chem* 87, 10006–10014.
22. Yu, Q., Liu, B., Ruan, D., Niu, C., Shen, J., Ni, M., Cong, W., Lu, X., and Jin, L. (2014). A novel targeted proteomics method for identification and relative quantitation of difference in nitration degree of OGDH between healthy and diabetic mouse. *Proteomics* 14, 2417–2426.
23. Gallien, S., Peterman, S., Kiyonami, R., Souady, J., Duriez, E., Schoen, A., and Domon, B. (2012). Highly multiplexed targeted proteomics using precise control of peptide retention time. *Proteomics* 12, 1122–1133.
24. Kim, Y. J., Gallien, S., van Oostrum, J., and Domon, B. (2013). Targeted proteomics strategy applied to biomarker evaluation. *Proteom Clin Appl* 7, 739–747.
25. Lesur, A., and Domon, B. (2015). Advances in high-resolution accurate mass spectrometry application to targeted proteomics. *Proteomics* 15, 880–890.
26. Gallien, S., Kim, S. Y., and Domon, B. (2015). Large-scale targeted proteomics using internal standard triggered-parallel reaction monitoring (IS-PRM). *Mol Cell Prot* 14, 1630–1644.
27. Gallien, S., Bourmaud, A., Kim, S. Y., and Domon, B. (2014). Technical considerations for large-scale parallel reaction monitoring analysis. *J Proteomics* 100, 147–159.
28. Güzel, C., Govorukhina, N.I., Stingl, C., Dekker, L.J.M., Boichenko, A., van der Zee, A.G.J., Bischoff, R.P.H., and Luider, T.M (2018). Comparison of targeted mass spectrometry techniques with an immunoassay: A case study for HSP90α. *Proteomics Clin Appl* 12, doi:10.1002/prca.201700107.
29. Haque, A., Alam, Q., Alam, M. Z., Azhar, E. I., Sait, K. H., Anfinan, N., Mushtaq, G., Kamal, M. A., and Rasool, M. (2016). Current understanding of HSP90 as a novel therapeutic target: An emerging approach for the treatment of cancer. *Curr Pharm Des* 22, 2947–2959.
30. Calderón-Celis, F., Encinar, J.R., and Sanz-Medel, A. (2017). Standardization approaches in absolute quantitative proteomics with mass spectrometry. *Mass Spectrom Rev.* doi:10.1002/mas.21542.
31. Wilffert, D., Reis, C.R., Hermans, J., Govorukhina, N.,Tomar, T., de Jong, S., Quax, W.J., van de Merbel, N.C., and Bischoff, R. (2013). Antibody-free LC-MS/MS quantification of rhTRAIL in human and mouse serum. *Anal Chem* 85, 10754–10760.
32. Wilffert, D., Bischoff, R., and van de Merbel, N. C. (2015). Antibody-free workflows for protein quantification by LC–MS/MS. *Bioanalysis* 7, 763–779.
33. Guo, L., Wang, Q., Weng, L., Hauser, L.A., Strawser, C.J., Rocha, A.G., Dancis, A., Mesaros, C.A., Lynch, D.R., and Blair, I.A. (2017). Analytical chemistry. doi:10.1021/acs.analchem.7b04590.

34. Güzel, C. et al. (2018). Proteomic alterations in early stage cervical cancer. *Oncotarget* 9, 18128–18147.
35. Bronsema, K.J., Bischoff, R., and van de Merbel, N.C. (2013). High-sensitivity LC-MS/MS quantification of peptides and proteins in complex biological samples: The impact of enzymatic digestion and internal standard selection on method performance. *Anal Chem* 85, 9528–9535.
36. Bronsema, K.J., Bischoff, R., and van de Merbel, N.C. (2012). Internal standards in the quantitative determination of protein biopharmaceuticals using liquid chromatography coupled to mass spectrometry. *J Chrom B* 893–894, 1–14.
37. MacLean, B., Tomazela, D. M., Shulman, N., Chambers, M., Finney, G. L., Frewen, B., Kern, R., Tabb, D. L., Liebler, D. C., and MacCoss, M. J. (2010). Skyline: An open source document editor for creating and analyzing targeted proteomics experiments. *Bioinformatics* 26, 966–968.
38. Smith, L.M., Kelleher, N.L., and Consortium for top down proteomics. (2013). Proteoform: A single term describing protein complexity. *Nat Methods* 10, 186–187.
39. Toby, T.K., Fornelli, L., and Kelleher, N.L. (2016). Progress in top-down proteomics and the analysis of proteoforms. *Ann Rev Anal Chem* (Palo Alto, Calif). 9, 499–519.
40. Catherman, A.D., Skinner, O.S., and Kelleher, N.L. (2014). Top down proteomics: Facts and perspectives. *Biochem Biophys Res Commun* 445, 683–693.
41. Tegtmeyer, L.C. et al. (2014). Multiple phenotypes in phosphoglucomutase 1 deficiency. *N Engl J Med* 370, 533–543.
42. Domanski, D., Percy, A.J., Yang J., Chambers, A.G., Hill, J.S., Freue, G.V., and Borchers, C.H. (2012). MRM-based multiplexed quantitation of 67 putative cardiovascular disease biomarkers in human plasma. *Proteomics* 12, 1222–1243.
43. Meier, F., Beck S., Grassl N., Lubeck, M., Park, M.A., Raether, O., Mann, M (2015). Parallel accumulation-serial fragmentation (PASEF): Multiplying sequencing speed and sensitivity by synchronized scans in a trapped ion mobility device. *J Proteome Res* 14, 5378–5387.
44. Bruderer, R., Bernhardt, O.M., Gandhi, T., Xuan, Y., Sondermann, J., Schmidt, M., Gomez-Varela, and D., Reiter, L. (2017). Optimization of experimental parameters in data-independent mass spectrometry significantly increases depth and reproducibility of results. *Mol Cell Proteomics* 16, 2296–2309.

Integrated Omics for Tissue-Based Mapping of the Human Proteome

7.9

Cecilia Lindskog

Contents

INTRODUCTION

In the advancement of medical diagnostics and treatment from traditional health care to individualized therapy, the evolving era of "Big Data" provides the basis for such a transition. The first step was taken by the completion of the Human Genome Project in 2001, and since then, Big Data for science and medicine have received increased attention based on the growing data sets from different omics technologies such as genomics, transcriptomics and proteomics. Although genomic medicine has lead to a plethora of new treatments, the genome sequence cannot predict the dynamics of protein expression patterns, posttranslational modifications, or protein–protein interactions that control an individual's response to treatment. As a result, an integration of genomics research in combination with other omics technologies paves the way for further progress in molecular medicine and targeted treatment.

Proteins are critical for life, ensuring the daily function of every living cell and organism. Proteome-wide research offers a basis for biomarker discovery in order to understand biological function and complexity in human health and disease, and several recent efforts have been established generating a map of all human proteins. These efforts include the Human Proteome Map [1] and the Proteomics DB [2] based on MS of human tissues, as well as the Human Proteome Project (HPP), launched in 2010 by the Human Proteome Organization (HUPO) [3]. The HPP project is a major, comprehensive initiative aiming at identifying and characterizing an accurate map of at least one proteoform of each of the approximately 20,000 human protein-coding genes, as well as their co- and posttranslational modifications [4]. Over 80% of the human proteins predicted by the genome have already been identified using either MS or antibody-based techniques [5]. Another crucial task in mapping the human proteome is resolving the spatial resolution of all human proteins in organs, tissues, cells and organelles. One project underlying the HPP project with focus on spatial proteomics is the Human Protein Atlas project, with the ultimate goal to generate a complete map of the human proteome using integration of transcriptomics and antibody-based proteomics.

THE HUMAN PROTEIN ATLAS PROJECT

The Human Protein Atlas project was initiated in 2003 with the aim to reveal the spatial resolution in different human tissues, cancer types and cell lines. The analysis is based on immunohistochemistry and immunocytochemistry, showing the protein expression at a single cell level. All primary data, including more than 10 million high-resolution images are publicly available at www.proteinatlas.org, constituting a valuable tool for in-depth studies of protein localization and expression in human tissues and cells. The analysis is combined with transcriptomics using mRNA sequencing (RNA-Seq). This integrated omics strategy allows not only for studies looking at single proteins, but also for global efforts analyzing different functional groups of proteins in various tissues or organelles based on expression level and distribution. The specific proteomes, summarizing for example tissue-elevated proteins or housekeeping proteins, are described in different knowledge chapters in the Human Protein Atlas, providing the basis for biomarker discovery efforts. The Human Protein Atlas database constitutes the largest and most comprehensive knowledge resource for spatial distribution of proteins and has several potential implications for use in personalized medicine. The gene-centric data is updated on a yearly basis, and the most recent version (version 18) based on Ensembl version 88.38 contains antibody-based protein data using >26,000 antibodies, covering 87% of all human protein-coding genes [6]. Both in-house–generated antibodies and commercial antibodies are used, and only antibody data that have passed rigorous quality tests for antigen specificity and validation are added to the Human Protein Atlas. Version 18 is divided into three sub-atlases: the Tissue Atlas, the Pathology Atlas and the Cell Atlas, each one with different implications for human biology and precision medicine, described more in detail below. The three parts of the Human Protein Atlas are interconnected and complement each other, enabling the user to explore the distribution of each protein in tissues and organs, studying the subcellular localization and determining the protein's relation to cancer.

THE TISSUE ATLAS

The Tissue Atlas allows for exploration of the protein expression in the context of neighboring cells. This gives a possibility to study the protein expression in different cell types within the same tissue (such as proximal vs. distal tubules of the kidney) to determine single cell variation within the same cell type such as differences in staining intensity and to yield an overview of the main subcellular localization in different cell types. In the Tissue Atlas, each gene page on the Human Protein Atlas contains a comprehensive summary of expression across tissues both on the mRNA and protein level [7,8]. The protein expression data currently covers 15,317 human genes (78%), and is based on immunohistochemistry using tissue microarrays. Histology-based manual annotation of protein expression levels shows the localization across 76 cell types corresponding to 44 different non-diseased human tissue types, covering all major parts of the human body (Figure 7.9.1a). By integrating the analysis with RNA-Seq, each gene is associated with a specific RNA tissue category. The RNA tissue categories group all human protein-coding genes based on pattern of expression, including expressed in all, tissue enriched, group enriched, tissue enhanced, mixed or not detected (Table 7.9.1) [9]. The classification is based on both internally generated RNA-Seq data using 37 different tissues, as well as two different external RNA expression data sets, allowing for direct comparison; RNA-Seq data from the Genotype-Tissue Expression (GTEx) project [10], including 31 tissues, and the FANTOM5 consortium [11], including 35 tissues (Figure 7.9.1b). Altogether, the three different RNA expression data sets cover 40 of the 44 tissues analyzed with immunohistochemistry, providing an important complement for analysis of tissue-specific expression. In a comparison between the independent data sets, consistent results were shown to support the distribution in which almost half of the proteins are expressed in all tissues and relatively few are unique for one or a few tissues [12]. Information on the distribution of protein and RNA expression levels, previous gene/protein characterization data and bioinformatics predictions are taken into consideration in the interpretation of each gene, which is manually summarized in descriptive sentences and data reliability scores. This knowledge-based information provides a comprehensive overview of the data, complemented with representative images that give the user a visual summary of the protein expression pattern, which can be analyzed more in detail in the primary immunohistochemistry images. For each antibody, non-diseased samples from 144 different individuals are available, and all images are clickable for an enlarged high-resolution view (Figure 7.9.1c).

Numerous comprehensive knowledge chapters describe the proteome and transcriptome of each organ, as well as sub-proteomes corresponding to particular functional groups of genes [7,13]. The tissue and organ proteome chapters include catalogs of proteins expressed in a tissue-specific manner, based on RNA tissue categories. About one-third of the human genes (Figure 7.9.1d) show some level of elevated expression in a certain tissue compared with all other analyzed tissues (tissue enriched, group enriched or tissue enhanced). Such tissue-specific proteomes—for example, the testis-specific proteome [14], lung-specific proteome [15], and the liver-specific proteome [16]—are believed to play a major role in the organ physiology and constitute interesting starting points for organ-specific research. Knowledge of the molecular repertoire of each non-diseased tissue may have important implications for medicine and aid in identifying and stratifying high-risk individuals, contributing to further understanding of underlying disease mechanisms and guiding treatment modalities. Tissue-elevated proteins could also be searched for in serum given that this group of proteins constitutes an interesting starting point for identifying candidates for serum-based diagnostic tests. Each tissue and organ proteome chapter contains complete lists of expression of all genes in a certain organ, clickable with direct links to search results or Tissue Atlas gene summary pages, where proteins of interest can be explored further. Another feature is a network plot for each organ or tissue (Figure 7.9.1e), highlighting the group-enriched genes that are simultaneously elevated in a group of two to seven tissues, compared with all other analyzed tissues. These plots aid in finding common features between different organs. In addition to the tissue and organ-specific proteome chapters, other chapters found in the Tissue Atlas are the sub-proteomes, summarizing different functional groups of genes. Such proteomes include "the housekeeping proteome," "the druggable

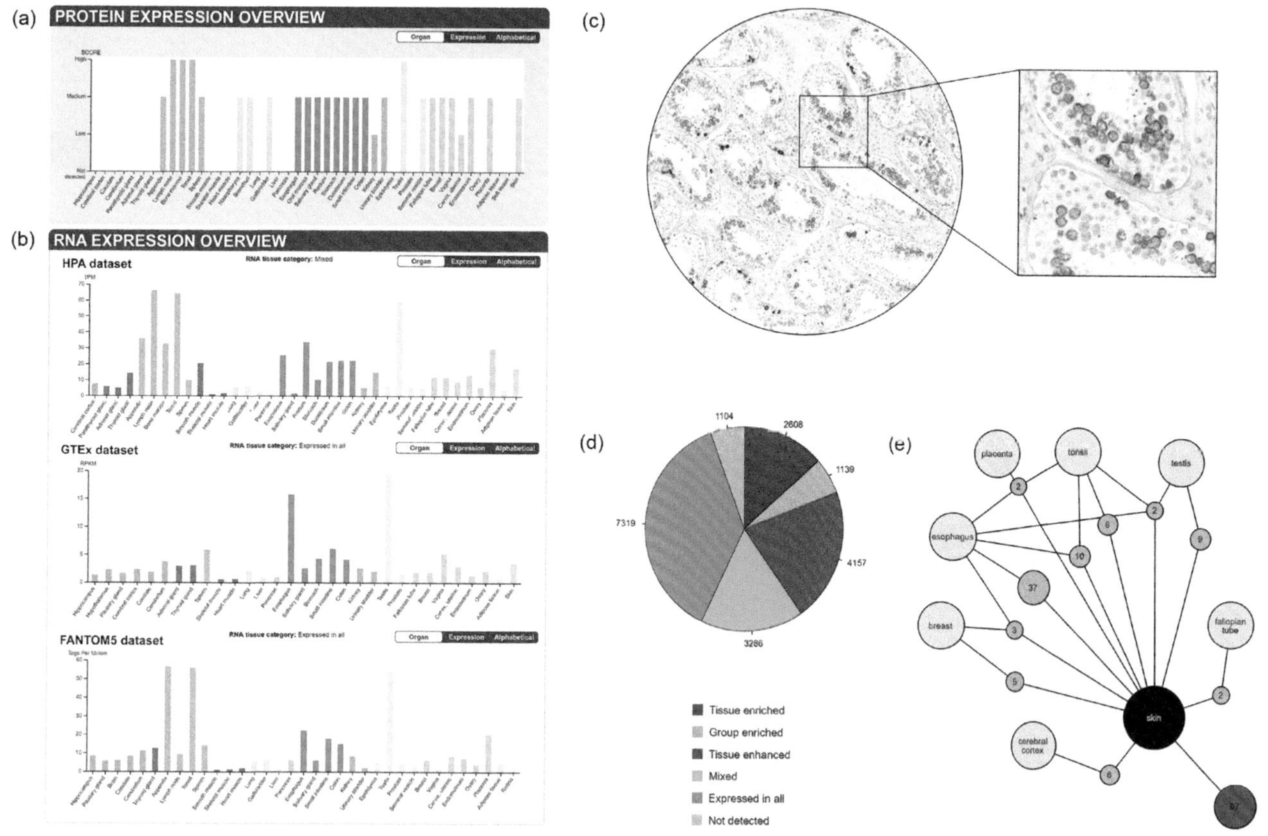

FIGURE 7.9.1 Overview of data in the Tissue Atlas. (a) Protein expression levels based on immunohistochemistry toward Cyclin B1 (CCNB1) in 44 human non-diseased tissue types. (b) RNA expression levels based on three independent data sets: Human Protein Atlas, GTEx consortium and FANTOM5 consortium. (c) Immunohistochemical staining pattern of CCNB1 in human testis, showing the protein localization in the context of neighboring cells. (d) RNA tissue categories based on classification of all human genes according to RNA expression levels in the Human Protein Atlas RNA-Seq data set. (e) Example of a network plot, showing genes enriched in skin (red nodes), as well as group-enriched genes that share an enriched expression between skin and other tissues (orange nodes).

TABLE 7.9.1 Categorization of all human protein–coding genes according to expression

RNA CATEGORY	*DESCRIPTION*	*NUMBER OF GENES*
Tissue enriched	Expression in one tissue at least fivefold higher than all other tissues	2,608
Group enriched	Fivefold higher average Transcript per million (TPM) in a group of two to seven tissues compared with all other tissues	1,139
Tissue enhanced	Fivefold higher average TPM in one or more tissues compared with the mean TPM of all tissues	4,157
Expressed in all	≥1 TPM in all tissues	7,319
Not detected	<1 TPM in all tissues	1,104
Mixed	Detected in at least one tissue and in none of the above categories	3,286

proteome," "the secretome and membrane proteome," "the cancer proteome," "the regulatory proteome," and "the isoform proteome." Each of these proteomes provides various implications for human biology and precision medicine, and gives the users of the database the possibility of identifying proteins expressed in a certain manner belonging to a specific functional category of genes. Especially interesting is "the druggable proteome," focusing on 620 proteins targeted by the FDA-approved drugs. Surprisingly, almost one-third of these proteins were shown to be expressed in all tissues and organs, highlighting the need for further studies on potential unwanted side effects in the therapeutical administration of drugs [7].

THE PATHOLOGY ATLAS

With the launch of the Pathology Atlas [17], showing the consequence of all human genes on overall patient survival in all major cancers, the dream of personalized treatment took a major step

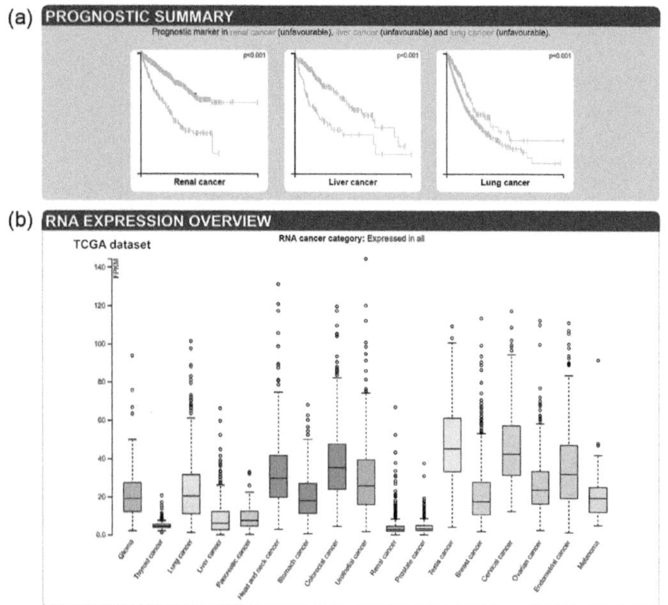

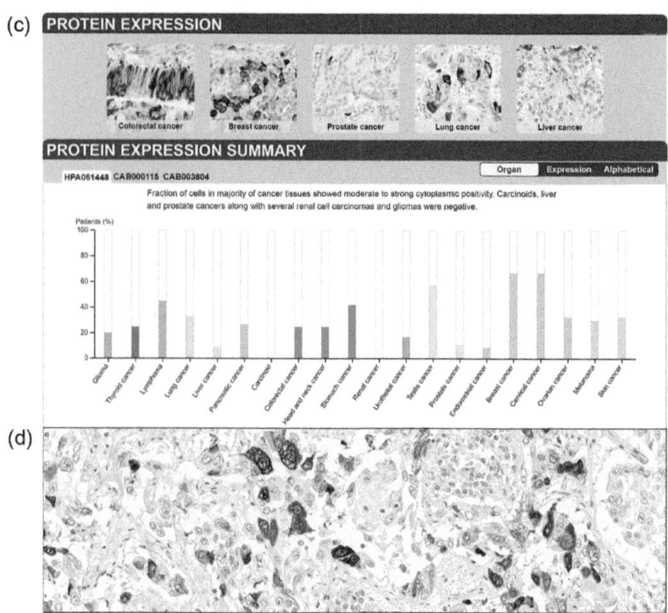

FIGURE 7.9.2 Overview of data in the Pathology Atlas. (a) Summary of cancer types where RNA expression levels of Cyclin B1 (CCNB1) were associated with prognosis, illustrated with Kaplan-Meier plots. (b) RNA expression levels based on The Cancer Genome Atlas data in 17 different cancer types. (c) Examples of immunohistochemical staining pattern and protein expression levels in 20 different cancer types. (d) Immunohistochemical staining pattern of CCNB1 in lung cancer, showing the protein localization in the context of neighboring cells.

forward. The Pathology Atlas is an example demonstrating the power of Big Data, based on a systems-level approach analyzing 17 major cancer types using data from 8,000 patients derived from publicly available data from The Cancer Genome Atlas (TCGA). In contrast to earlier cancer investigations, the analysis was not focused on mutations in cancers, but the downstream effects of such mutations across all protein-coding genes. More than 500,000 Kaplan-Meier plots were generated describing the consequence of RNA levels on clinical survival, identifying potential prognostic genes (Figure 7.9.2a). These genes were further divided into favorable genes, where high RNA expression correlated with longer survival time, and unfavorable genes, where high RNA expression correlated with shorter survival time. Shorter patient survival was generally associated with up-regulation of genes involved in mitosis and cell growth, and down-regulation of genes involved in differentiation. The RNA-Seq data from TCGA is also used for classification of all genes in the same manner as in the Tissue Atlas, allowing for exploration of genes elevated in a certain cancer type in comparison to all other analyzed cancers (Figure 7.9.2b). The transcriptomic analysis is combined with 5 million high-resolution pathology-based immunohistochemistry images from 20 cancer types, corresponding to up to 12 patients per cancer type (Figure 7.9.2c). These publicly available images are generated in the same manner as the non-diseased images in the Tissue Atlas and allow for analysis of the protein expression pattern at a single cell resolution, determining the expression level in the context of neighboring cells, such as the relation to tumor stroma (Figure 7.9.2d). Difference in expression levels of individual cancer patients strongly highlights the need for personalized cancer treatment based on precision medicine, and the Pathology Atlas constitutes an important resource for identification of potential cancer biomarkers.

Similar to the Tissue Atlas, the Pathology Atlas also provides knowledge chapters for each cancer type, serving as comprehensive overviews of the transcriptomic and proteomic landscape of each cancer. This allows access to cancer tissue-elevated genes and prognostic genes, linked with the corresponding protein expression data and provides the basis for cancer type-specific research.

THE CELL ATLAS

The Cell Atlas visualizes the location of over 12,000 proteins (61% of the human protein-coding genes) in high-resolution, multicolor images of immunofluorescently labeled cells [18]. Fixed cells are imaged using a confocal laser scanning microscope, and each image consists of four channels that can be toggled on and off, protein of interest, and markers for nucleus, microtubules and ER. The proteins are mapped to 32 different subcellular structures, which is a first essential step toward novel insights into protein function (Figure 7.9.3a). Surprisingly, the cellular architecture was shown to be complex, with more than half of all analyzed proteins localized to multiple compartments. A significant proportion (15%) of the proteins also displayed single cell variation in abundance or localization, many of which are linked to cell cycle dependency.

In the same manner as in the Tissue Atlas and Pathology Atlas, RNA-Seq is used for classifying genes according to expression level in 56 different cell lines (Figure 7.9.3b). The transcriptomics data serve as a powerful tool for selecting appropriate

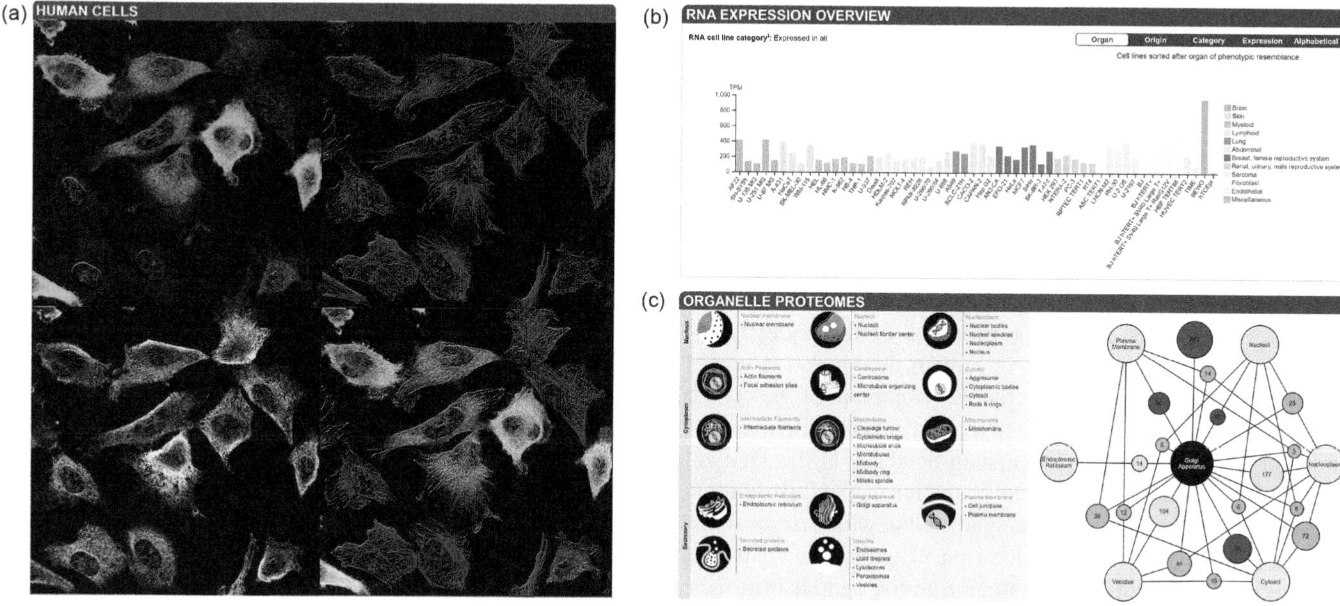

FIGURE 7.9.3 Overview of data in the Cell Atlas. (a) Immunofluorescence staining of Cyclin B1 (CCNB1) in U-251 MG cells, showing DAPI in blue, CCNB1 in green, microtubules in red and endoplasmic reticulum (ER) in yellow. (b) RNA expression levels of CCNB1 in 56 different cell lines. (c) Overview of different organelle proteomes, summarized in separate knowledge chapters, as well as example of a network plot showing the colocalization of proteins expressed in the Golgi apparatus and other subcellular structures.

cell lines for studying certain proteins or pathways. Knowledge chapters of different organellar proteomes provide an insight into each organelle and provide images for different morphologies and substructures, as well as network plots showing proteins localized to multiple organelles (Figure 7.9.3c).

QUALITY ASSURANCE AND ANTIBODY VALIDATION

The centerpiece of the Human Protein Atlas is its unique collection of high-resolution images stained using antibody-based proteomics in non-diseased and cancer tissues as well as various cell lines, displaying protein expression patterns at a single cell resolution. In all three sub-atlases, the immunohistochemistry and immunofluorescence stainings are combined with transcriptomics analyses using RNA-Seq, allowing for a quantitative measurement that can be used for classification of human genes according to expression patterns across tissues and cells. Several previous studies suggest that RNA levels cannot be used for predicting protein levels [19]; however, recent efforts suggest that RNA is a good proxy for protein levels if a gene-specific RNA-to-protein factor is used [20]. Based on this assumption, the Human Protein Atlas uses RNA levels for validation, as well as a starting point for further exploration of the distribution of protein expression across tissue samples. Reliability of the antibody staining pattern must, however, be determined for every specific protein, taking into consideration also previously published gene/protein characterization data and bioinformatics predictions. For some proteins, RNA is not a good proxy for protein levels, and other methods must be used for validation. Nevertheless, the combination of transcriptomics with antibody-based proteomics constitutes a powerful tool for further exploration of the human proteome in health and disease.

Antibodies are among the most frequently used tools in basic research and clinical assays, and require high sensitivity and specificity in order to provide the best estimate of protein expression across tissues and cells. Research antibodies have been shown to lack generally accepted guidelines for quality control, which led to a proposal for antibody validation generated by an international consortium of researchers called the International Working Group on Antibody Validation (IWGAV) [21]. The IWGAV suggested the use of five "pillars" for validation of antibodies, with the aim to establish standards for evaluation and quality control that can be implemented by both users and producers of antibodies. The Human Protein Atlas adheres to the principles suggested by IWGAV, and in the most recent version of the database more than 10,540 antibodies directed to 6,787 human protein targets have passed the criteria of enhanced validation, validated using at least one of the five validation strategies. For proteins that do not meet the criteria for enhanced antibody validation, off-target binding and false annotation is higher, but the effect on global proteomic analyses is small [18]. Nevertheless, users can filter search results for antibody reliability, and the number of enhanced validated antibodies will increase in upcoming releases of the Human Protein Atlas.

DISCUSSION

In the era of Big Data and the transition toward personalized health care, there is an urgent need for better understanding of target protein functions in human health and disease. Comprehensive databases and the development of powerful *in silico* techniques enable effective data mining for identification of novel biomarkers. One such database is the Human Protein Atlas, aiming to generate a spatial map of all human protein-coding genes in tissues and cells, which represents an invaluable resource for exploration of expression patterns at a single cell resolution. By integrating the spatial information based on antibody-based proteomics with transcriptomic strategies, the database constitutes a knowledge resource for in-depth understanding of biology, molecular repertoire and architecture of every human cell. In the Tissue and Pathology Atlas, the cell type-specific protein expression in the context of neighboring cells is provided, allowing for identification of proteins exclusively expressed in a subset of cells, differences in expression levels between organs, as well as identification of proteins up- or down-regulated in cancer as compared with normal tissue. The Pathology Atlas demonstrates the power of Big Data, taking the dream of personalized treatment a major step forward with information on how all human genes are related to patient survival. This opens up for identification of potential biomarkers in future diagnostic schemes and personalized cancer treatments. The Cell Atlas complements the other sub-atlases, providing important high-resolution information on the exact localization of every protein inside the cell, in different subcellular compartments. Such information constitutes a first essential step toward novel insights into understanding the function of each protein.

The protein data in the Human Protein Atlas currently covers 87% of all human protein-coding genes, and both the coverage as well as the details provided in the different sub-atlases will increase in upcoming releases. All data is publicly available with unrestricted access and can be downloaded in different formats, allowing for large-scale bioinformatics analyses. The database also contains a search function, with filters that can be combined into complex queries, generating a specific search for a list of genes that matches selected characteristics. On a monthly basis, the Human Protein Atlas database is used by an estimated 200,000 researchers worldwide. In 1–2 yearly updates, new data and functionalities are added, and the data is synchronized according to the most recent Ensembl version.

Another recent initiative with the aim to generate a reference atlas of all cells in the human body as a resource for studies in health and disease is the Human Cell Atlas, an international consortium initiated by the Chan Zuckerberg Initiative. The initiative collaborates with researchers from the Human Protein Atlas, as well as groups focusing on single cell analysis, including the Allen Institute for Cell Science, Wellcome Trust, European Bioinformatics Institute (EBI), Chan Zuckerberg Biohub, Karolinska Institute, Broad Institute, Sanger Institute and UC Santa Cruz. Integration of technologies from these groups is likely to accelerate research in precision medicine and aid in generating a more detailed map of the human building blocks of life.

In summary, the Human Protein Atlas constitutes a comprehensive stand-alone open-access resource available for researchers worldwide and is believed to help accelerating efforts to find biomarkers meeting future needs in personalized health care, and leading to products that will benefit humanity.

REFERENCES

1. Kim, M.S. et al., A draft map of the human proteome. *Nature*, 2014. **509**(7502): 575–581.
2. Wilhelm, M. et al., Mass-spectrometry-based draft of the human proteome. *Nature*, 2014. **509**(7502): 582–587.
3. Omenn, G.S. et al., Metrics for the human proteome project 2016: Progress on identifying and characterizing the human proteome, including post-Translational modifications. *J Proteome Res*, 2016. **15**(11): 3951–3960.
4. Omenn, G.S., The strategy, organization, and progress of the HUPO Human Proteome Project. *J Proteomics*, 2014. **100**: 3–7.
5. Omenn, G.S. et al., Progress on the HUPO draft human proteome: 2017 metrics of the human proteome project. *J Proteome Res*, 2017. **16**(12): 4281–4287.
6. Aken, B.L. et al., Ensembl 2017. *Nucleic Acids Res*, 2017. **45**(D1): D635–D642.
7. Uhlen, M. et al., Proteomics. Tissue-based map of the human proteome. *Science*, 2015. **347**(6220): 1260419.
8. Thul, P.J. and C. Lindskog, The human protein atlas: A spatial map of the human proteome. *Protein Sci*, 2018. **27**(1): 233–244.
9. Fagerberg, L. et al., Analysis of the human tissue-specific expression by genome-wide integration of transcriptomics and antibody-based proteomics. *Mol Cell Proteomics*, 2014. **13**(2): 397–406.
10. Consortium, G.T., Human genomics. The Genotype-Tissue Expression (GTEx) pilot analysis: Multitissue gene regulation in humans. *Science*, 2015. **348**(6235): 648–660.
11. Yu, N.Y. et al., Complementing tissue characterization by integrating transcriptome profiling from the human protein atlas and from the FANTOM5 consortium. *Nucleic Acids Res*, 2015. **43**(14): 6787–6798.
12. Uhlen, M. et al., Transcriptomics resources of human tissues and organs. *Mol Syst Biol*, 2016. **12**(4): 862.
13. Lindskog, C., The potential clinical impact of the tissue-based map of the human proteome. *Expert Rev Proteomics*, 2015. **12**(3): 213–215.
14. Djureinovic, D. et al., The human testis-specific proteome defined by transcriptomics and antibody-based profiling. *Mol Hum Reprod*, 2014. **20**(6): 476–488.
15. Lindskog, C. et al., The lung-specific proteome defined by integration of transcriptomics and antibody-based profiling. *FASEB J*, 2014. **28**(12): 5184–5196.

16. Kampf, C. et al., The human liver-specific proteome defined by transcriptomics and antibody-based profiling. *FASEB J*, 2014. **28**(7): 2901–2914.
17. Uhlen, M. et al., A pathology atlas of the human cancer transcriptome. *Science*, 2017. **357**(6352).
18. Thul, P.J. et al., A subcellular map of the human proteome. *Science*, 2017. **356**(6340).
19. Fortelny, N. et al., Can we predict protein from mRNA levels? *Nature*, 2017. **547**(7664): E19–E20.
20. Edfors, F. et al., Gene-specific correlation of RNA and protein levels in human cells and tissues. *Mol Syst Biol*, 2016. **12**(10): 883.
21. Uhlen, M. et al., A proposal for validation of antibodies. *Nat Methods*, 2016. **13**(10): 823–827.

Proteogenomics and Multi-omics Data Integration for Personalized Medicine

7.10

Peter Horvatovich, Corry-Anke Brandsma, Frank Suits, Rainer Bischoff, and Victor Guryev

Contents

INTRODUCTION

The recent decade demonstrated an explosion of high-throughput molecular profiling platforms such as genomics, transcriptomics, proteomics and metabolomics, which allow for the collection of individual- and sample-specific genome information and molecular profiles. These new profiling technologies allow for the obtaining of personalized and precision (patient subgroup)–specific information, which opens an unprecedented prospect for personalized and precision health management, diagnostics and medicine. The different omics layers provide complementary information, and adequate integration of this data provides an additional new insight into molecular mechanism of aging and disease development and opens the possibility to identify novel drug targets and biomarkers. This section discusses the different approaches to integrate genomics, transcriptomics and proteomics data, which is referred as proteogenomics data integration in the literature, but will touch in limited extent the metabolomics profiles, considered to represent the overall biological status of a biological system. The different omics integration approaches are discussed in two separate sections. The first part presents the identification of sample-specific peptides and protein sequences from protein sequence predicted from genomics/transcriptomics data. The second part deals with data integration approach using quantitative aspect of data such as correlation between molecular layers, identification of copy number alteration (CNA), genome mutations/variants regulating transcripts, proteins, metabolites and quantifying phenome levels with quantitative trait loci analysis and correlation analysis and finally exploring shared and molecular layer-specific interaction network of transcriptome, proteome and metabolome. Proteogenomics data integration is discussed in the context of personalized and precision medicine, highlighting the strengths and weaknesses for this application.

IDENTIFICATION OF PROTEIN SEQUENCE VARIANTS

Efficient proteogenomics data integration requires the understanding of the data collected at genomics and proteomics level (Figure 7.10.1). Genomics data is obtained mainly through next-generation massive parallel sequencing of four nucleotide bases of DNA fragments of 250–500 base pairs (Figure 7.10.1a). Sequencing is performed by cyclic rebuilding of complementary DNA strands using nucleotide bases labeled with fluorescence and DNA building process terminating chemical groups. These groups are cleaved and removed by washing upon reading the fluorescence signal, and the cyclic process rebuilding of complementary DNA continues. When it comes to mRNA, the molecules are first fragmented and reverse transcribed to cDNA (Figure 7.10.1a), which is subjected to the DNA cyclic sequencing. Poly-adenylated mRNAs are mature transcripts after the splicing event, therefore their sequences can be used to predict the sequence of possible translated proteins. In order to isolate poly-adenylated transcripts, ribosomal RNA (rRNA) are depleted including all other non-translated RNA such as miRNA or long noncoding RNA. Poly-adenylated transcripts are broken into fragments of approximately 500 bases length before reversed cDNA transcription and are then ligated with a molecular barcode and sequence adaptor. These fragments following PCR amplification are then sequenced at single or both ends with 50–150 base reads. This results in fragmented pieces of sequence, which after filtering for sequencing quality, can be either aligned to a reference genome or processed with *de novo* transcriptome assembly. The aligned sequence is annotated with genome annotation or protein-coding part is extracted by searching open reading frames, followed by the translation of the mRNA sequence to the protein amino acid sequence based on nucleotide base codon triplets.

The most commonly applied proteomics profiling is based on shotgun bottom-up LC-MS/MS proteomics approach, where proteins are cleaved first to smaller peptides using protease with specific cleavage rules such as trypsin, which enzymatically cleaves proteins at lysine or arginine except if the sequence has

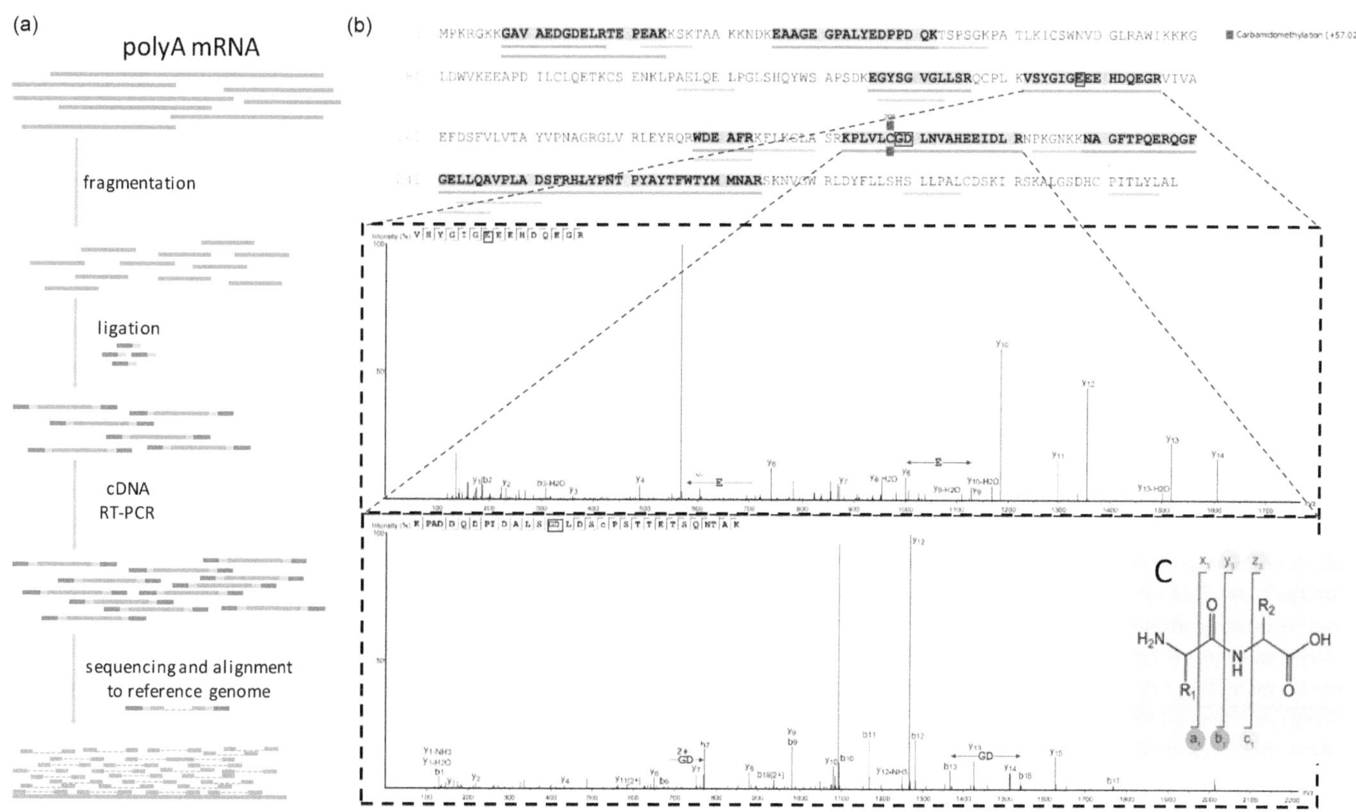

FIGURE 7.10.1 Main analysis steps and data structure of poly-adenylated transcripts and LC-MS/MS protein molecular profile used in proteogenomics data integration. (a) Poly-adenylated transcripts are sequenced with multi-parallel next generation sequencing, whereas (b) peptides generated from proteins using protease are analyzed with LC-MS/MS, and (c) the analyzed peptides are fragmented with collision-induced dissociation (CID) and electron transfer dissociation (ETD) fragmentation approaches. Sequencing and amplification of the transcript provides complete genome coverage of the expressed transcriptome, whereas proteomics data is highly fragmented with respect to sequence coverage, resulting in large gaps in the identified protein sequence. The sequence of a protein of apurinic/apyrimidinic endodeoxyribonuclease 1 (APEX1, ENST00000398030) is shown in plot *B* with a sequence coverage of 41%, which holds one single amino acid variant of D148E (i.e., changing aspartic acid to glutamic acid in position 148). The MS/MS spectra of the peptide VSYGIGEEEHDQEGR mapping to this location and holding the D148E amino acid change (highlighted in red in the sequence and in the MS/MS spectra) is shown below the protein sequence. Another MS/MS spectra of the peptide KPLVLCGDLNVAHEEIDLR is shown below the MS/MS spectra with a single amino acid variant peptide. This MS/MS spectrum shows an incomplete ion series, with four gaps (PAD, PI, GD, NT) where there is no fragment ion between the amino acids forming the gap, which means there is no possibility for *de novo* spectra reconstruction because the exact order of amino acids cannot be determined in the gaps.

proline at the C terminal side of these amino acids. The generated peptides are separated by LC and are analyzed by MS online. It is important to note that there is no peptides and proteins amplification method, therefore proteomics profiling is limited to the dynamic concentration range of LC-MS/MS instrument, which span typically 4–5 orders of magnitude, while tissue and cellular proteome is estimated to span 7–9, and blood proteome to span 11 orders of magnitude [1,2]. In addition, genome sequencing is based on the four nucleotide bases of DNA with 16 known modifications, which have close physicochemical properties for which step-wise degradation and synthesis enzyme systems exist. In contrary, proteins consist of chemically highly diverse 20 amino acids with more than 300 known chemical modifications spanning the full chemical space of organic chemistry, for which there is no known enzyme system that enables step-wise sequencing. Therefore, the quantification and identification of proteins represent more important challenges for an analytical/bioinformatics system. The most widely used approach to analyze protein-derived peptides is the data-dependent acquisition mode (DDA). DDA measure peptides by cycles constituting different types of mass spectra, in which the first scan is a non-fragmented scan, which is used to select the most *n* abundant ions for top *n* fragmentation by collision-induced dissociation (CID) or electron transfer dissociation (ETD), resulting in *n* number of MS/MS fragment spectra for each cycle if at least *n* fragment ions pass an intensity threshold and are not excluded to avoid repeats within one chromatographic peak. In CID, peptide ions are collided with neutral gas introduced into the mass analyzer, whereas ETD performs fragmentation by transferring an electron from a negatively ionized polycyclic aromatic hydrocarbon such as anthracene or fluoranthene. CID provides efficient fragmentation for tryptic peptides of length up to 30–40 amino acids, whereas ETD more efficiently fragments larger than shorter peptides. Fragmentation of peptides is not random, but follows probabilistic rules (Figure 7.10.1c). CID for tryptic peptides produces *y* ions as most abundant and *b* ions as the second most abundant ion series, whereas *a* ions are rarely observed with this fragmentation approach. ETD provides mainly intensive *c* and *z* ion series which is complementary to CID fragmentation.

Because fragmentation of peptides is probabilistic, there is no assurance that fragments between all possible amino acids are present in the MS/MS spectrum. However, in the LC-MS/MS data set, the large fraction of MS/MS spectra has gaps in the fragment ion series, preventing *de novo* MS/MS spectrum identification. In *de novo* identification, the peptide primary amino acid sequence is reconstructed by using the consecutive mass difference between fragment pairs corresponding to intact or PTM-modified amino acids such as methionine oxidation. This strategy does not work for MS/MS spectra with gaps, due to missing consecutive sequence information. Thus, the targeted identification called database search (DS) is more efficient and widely used to identify spectrum by spectrum-peptide-match (PSM) procedure [3]. For MS/MS spectrum identification, DS uses protein sequences supposed to be present in the sample, which is generally made up of high-quality curated so-called canonical protein sequences available in public databases such as Uniprot [4] and Ensembl [5]. Canonical sequence is the most prevalent and longest sequence variant containing all known exomes; it is most similar to orthologous sequences in other species, which allows for the clearest description of protein domains, isoforms and polymorphism. PSM is performed by *in silico*–generated peptides from protein sequences and scoring the mass of fragments in MS/MS spectra against the mass of theoretical fragments calculated from the *in silico* peptides, using peptides that have a molecular mass falling into the precursor selection mass window of the mass spectrometer. The score reflects the similarity of the two fragment mass lists and provides a ranked list of best peptide matches for each MS/MS spectra. Because the similarity score between the two mass lists is not zero even for incorrect matches, several statistical procedures have been developed to devise a false-discovery rate (FDR)–controlled peptide identification list. The most popular and simple approach to estimate incorrect PSM content is to use decoy sequences obtained by reversing protein- or protease-derived *in silico* peptide sequences. In this case, the FDR can be estimated by making a sorted list of scores for all best PSMs in a data set. Then the threshold for PSM score is set at a value that includes the fraction of double the number of decoy PSM as best hit with respect to all PSM having higher scores than the threshold corresponding to the user-defined FDR level. The reason to set the number of decoy PSMs to double is because it is assumed that the PSM identified with the original peptide/protein sequence will contain the same number of incorrect identifications corresponding to the number of decoy best PSMs.

The identification procedure rarely leads to complete coverage of a protein sequence, and the median protein sequence coverage of a typical proteomics experience is around 20%. In fact, a protein is considered to be present if at least two or more peptides uniquely mapping to the protein sequence in question has been identified. This has the consequence that LC-MS/MS proteomics data is highly fragmented and does not contain peptide evidence for large part of the protein sequence for proteins identified in the sample. This can be particularly problematic when it comes to identifying sequence variants or mutations affecting only one or more amino acids at a particular position of the protein sequences. It is well known that protease, with clear cleavage rules, introduces bias into peptide distributions determined by the distribution of the amino acid or sequence motifs of the protease cleavage sites [6]. The bias is further substantiated, since not all of a compound is submitted to MS/MS fragmentation in a complex sample and that not all compound-related MS/MS spectra can be identified by *de novo* or DS approaches in LC-MS/MS data due to presence of gaps in MS/MS spectra or due to the large number of possible PTMs, which are not possible to search efficiently or due to sequence variants not present in the protein sequence database used in DS [7,8]. Additionally, the analysis of single protein without PTMs will rarely allow identification of all peptides theoretically possible to obtain with the applied protease due to ionization competition and the loss of peptides during sample preparation and LC-MS/MS analysis [9]. In consequence of these biases, the proteomics LC-MS/MS data set provides highly fragmented peptide identifications and results in the previously mentioned low protein sequence coverage, making it challenging to target a particular part of a protein sequence holding a mutation and variants such as single amino variants, splice junction peptides or indels. In fact, the number of reported variants differing from canonical sequences in the public database detected in proteogenomics analysis is not high; it represents 0.1%–5% of the total number of identified peptides and includes only a small fraction of variants that are predicted from genomics and transcriptomics data [10–13].

Identification efficiency depends on the resolution of the MS and MS/MS spectra, on the protein sequence search space and on the parameters used for PSM mapping. In many proteogenomics studies, the endeavor to use the latest generation of high-resolution fast scanning instruments such as Orbitrap or QTOF machines and optimized PSM search parameters leaves the choice of which protein sequence search space should be used. The most economic way to select the protein sequence variant database for peptide and protein identification is to use variant databases such as dbSNP [14], dbVar [15], COSMIC cancer mutation [16] or public genomics databases [17] for this purpose. However, these databases contain a large number of variants, which do not reflect the most accurate protein constitution of the analyzed sample therefore providing a large and unnecessary search space for DS that leads to worsening the statistics of the PSM process and lowering the peptide identification performance. The best protein sequence database can be obtained primarily from poly-adenylated mRNA, and if this is not available, then the secondary choice is from exome sequencing data [18,19]. This approach provides the most accurate prediction of protein sequence variants present in the analyzed samples and provides the most accurate peptide and protein identification. However, it should be noted that translation of proteins takes time and that half-life of transcript is shorter than the half-life of proteins [20], therefore proteins and mRNA taken from the same sample may differ in terms of protein species distribution. For this reason, it is advised to perform searches with the protein sequences database predicted from multiple samples of the same patient or from combined sequences from multiple patents' samples of the same patient group.

The FDR calculation in DS depends on the used search space. The canonical sequence database holding one sequence for each human protein-coding gene, such as the manually curated Swissprot, which provides the most dissimilar protein and peptide sequence set. When this sequence set is diluted with isoforms and sequence variants, then peptide sequence distribution will change and the FDR calculation may be compromised to a certain extent. It was also noticed that peptide mapping to variants generally provide low quality scores, and to remedy this problem, the cascade DS search strategy has been introduced, where in the first stage MS/MS spectra are filtered out by spectra quality and the high-quality spectra are searched with the canonical sequence of the public database (Swissprot, Uniprot or Ensembl). Then the non-identified high-quality spectra are submitted for search using sequence variants, which may be followed by several search rounds that include various classes of PTMs. At each DS stage, the FDR is calculated using the protein/peptide sequence used for that particular search stage, removing a substantial part of the FDR estimation bias and increasing identification reliability [3,21]. The experts at the Human Proteome Organization (HUPO) compiled a MS data interpretation guideline to announce identification of missing proteins and new previously unknown peptide sequences [22]. This guideline has been adapted by several journals such as *Journal of Proteome Research* to provide a checklist to avoid FDR error accumulation, to avoid false peptide identifications by checking the completeness of *y* ion series and the quality of MS/MS spectra, for example. The reader is invited to read further details on genomics and proteomics data generation and properties used in proteogenomics data integration in a book chapter [23] and a review wrote by the chapter's authors [19].

From the definition of the canonical sequence, it is evident that regular proteomics pipeline using canonical sequences is not suitable for personalized or precision medicine approaches given that it does not enable identification of patient- and sample-specific protein variants. To the contrary, the proteogenomics approach may provide patient- or sample group-specific protein sequence variants. However, it is evident that efforts should be made to increase the low sequence coverage of proteins in proteogenomics studies to improve the relatively low number of identified variant-specific peptides compared with the number of variants detected at the genomics and transcriptome levels. Attempts to improve this situation such as using multiple proteases [24] or using multidimensional LC for proteomics deep mining [18] come at the cost of increasing the time required to analyze one sample and thus decrease sample analysis throughput. For this reason, these approaches cannot be applied in the clinical setting or large population analysis. Another possible approach is the use of combined fragmentation approaches, combining, for example, HCD (high energy form of CID) and ETD fragmentation in Orbitrap Fusion or Lumos called EThcD [25,26] or middle-down proteomics [27,28], approaches which generate longer peptides to improve sequence coverage. Additionally, middle-down proteomics is reducing protein peptide mapping or inference problem by increasing the length of the analyzed peptides, which makes them more specific to a particular protein form.

To provide more accurate prediction of the translated protein sequences, ribo-Seq representing mRNA at the ribosome under translation process in the sample can be used. In ribo-Seq analysis, an enzyme is used to degrade mRNA outside of the ribosome providing approximately 30 nucleotide base length mRNA fragments. This information can be used to further filter the protein-coding transcript sequences and provide more accurate prediction of the translated protein sequences, leading ultimately to better peptide and protein identification [29].

One important goal of proteogenomics data integration is to determine how the different variants effect phenotype and change the activity of proteins. This is not a trivial task and the most precise method is to perform protein activity studies with the wild-type and mutant variants, for example, in a cell line with gene editing or gene silencing experiment. This is, however, a time- and resource-consuming process and the effects measured in cell line experiments cannot always be translated to tissues in living organism; therefore, proteogenomics data integration can help to prioritize variants, where the proteomics data serve as filter. Because the detection of a non-synonymous missense variant at the protein level can be missed due to the large probability to miss a peptides present in the sample, this filtering process may be completed by *in silico* protein structure prediction tools [30,31], which allow for assessing alteration of protein activities due to sequence changes or by assessing the clinical importance using genetic mutation clinical databases [32].

QUANTITATIVE RELATIONSHIPS BETWEEN OMICS LAYERS

The correlation between the abundance of mRNA and proteins is known to be mediocre, around 0.4–0.5 (Pearson correlation) in cell line and tissue, meaning that only part of quantitative effects of the mRNA can be translated at the protein level [20,33,34]. Therefore, quantitative aspects such as differential expression and correlation between transcripts/proteins of the same molecular type should be considered as separate complement information. mRNA has a shorter half-life and is changing more dynamically compared with proteins, therefore the protein's signature provides a more stable signal than mRNA. In addition, proteins are actively participating in biological processes such as in metabolisms, cell signaling and receptor/ligand interactions, therefore their differential levels better reflect the activity of pathways than mRNA. It should also be noted that protein levels are not directly reflecting the activity of proteins responsible for molecular changes behind biological/clinical events. For example, high concentration of matrix metalloproteases (MMPs) may not necessarily mean high activity of these proteases in the samples because the activity of proteases is dependent on the presence of natural inhibitors such as T-cell immunoglobulin and mucin domain (TIM) proteins [35,36] as well the presence of other proteases that can activate by cleavage of the pro form of MMPs. To measure activity of proteins or identify interaction partners, protein activity-specific sample preparation methods should be applied that provide readout for activity and not for concentration. These approaches are called activity-based profiling and use immobilized inhibitor/affinity ligands to perform activity-based enrichment of interactive partners, which allows for the identification and quantification of the amount of active interacting protein partners [37,38].

One often-used aspect of quantitative molecular profile is differential expression between pre-classified sample groups. Many genetics studies use differential analysis of mRNA to make inference on protein abundance and activity changes and to relate mRNA changes to specific biological events and particular phenotype. Because proteins better reflect the biology, it would be logical to use solely proteomics profiles for such a type of analysis. However, the proteomics profile is never comprehensive due to a lack of signal amplification and because many peptides of proteins are not identified even if their abundance is sufficiently high for their detection [39]. Therefore, proteomics cannot provide a quantitative value for all protein species, and the availability of quantitative mRNA profile provides a useful complementary information. Additionally, to fully understand the molecular mechanism of biological events such as aging, disease onset and progression, it is necessary to obtain a picture of the complete molecular regulatory events, including molecular events happening at the mRNA level before translation.

Understanding the molecular mechanism of disease is required to identify the genomics elements that promote disease development. To this end, large-scale GWASs are performed to determine which single nucleotide polymorphism (SNP) or genetic variants are enriched in the genome holding the disease or certain phenotype. However, GWAS studies do not provide information on how genetic variability influences molecular profiles and quantifiable phenome parameters. On the other hand, quantitative trait loci analysis (QTL) is widely used to determine the quantitative relationship between SNPs or sequence structure variants and the molecular profile. For example, expression QTL (eQTL) investigates how SNPs or structure variants influence mRNA expression [40–43]. This analysis can be also performed for other molecular profiles or any quantifiable phenotype parameters, leading to proteins (pQTL), metabolites (mQTL) and quantifiable clinical phenotype (cQTL) quantitative trait loci analysis. QTL studies in general can identify how a genomic variant affects the molecular profile and can be used to assess the genomics regulation on the quantitative variability of expressed transcripts, proteins or metabolites. Schröder et al. performed eQTL analysis to study the effect of >300,000 SNPs on the expression of >48,000 mRNA using Illumina microarray profiling of 149 human liver samples to identify which *cis* and *trans* SNPs (*cis* eQTL influence the mRNA of the gene located close to SNP, whereas *trans* eQTL influence the mRNA far away) influences genes involved in the absorption, distribution, metabolism and excretion (ADME) of drugs [43]. This study therefore provides important information how the expression of genes involved in drug ADME can be influenced by genetic variability, which is important information for understanding the patient-specific ADME behavior of drugs. Extension of QTL analysis to proteome and metabolome provides more precise information on phenotype and genetic regulations of disease. Williams et al. [34] performed genomics, transcriptomics (25,136 mRNA), proteomics (2,622 proteins), metabolomics (98 metabolites) and phenomics study in liver of 386 individuals from 80 cohorts of the BXD mouse genetic reference population to reveal the genetic regulation of phenotypes related to metabolism of low- and high-fat diets and as a consequence of mitochondrial and cardiovascular functions. This data integration revealed multiple genotype/molecular profile/phenotype links, which were only possible to identify by using all the studied omics layers, but not by using an individual omic layer. This was substantiated by the fact that 85% of QTL influenced either transcript or protein levels but not simultaneously both, showing the complementarity of transcriptomic and proteomic (and metabolomics) layers required to understand complex phenotypes. For example, Bckdha gene, which encodes E1 α polypeptide of the BCKD complex has *trans* pQTL on chromosome 9 in low- and high-fat diets, but does not have the same *trans* eQTL at the transcript level. Additionally, transcripts of BCKD complex do not show tight co-regulation (correlation), but it is tightly co-regulated at protein levels. Figure 7.10.2 shows the relationship between GWAS and xQTL analysis (where *x* can be e for mRNA

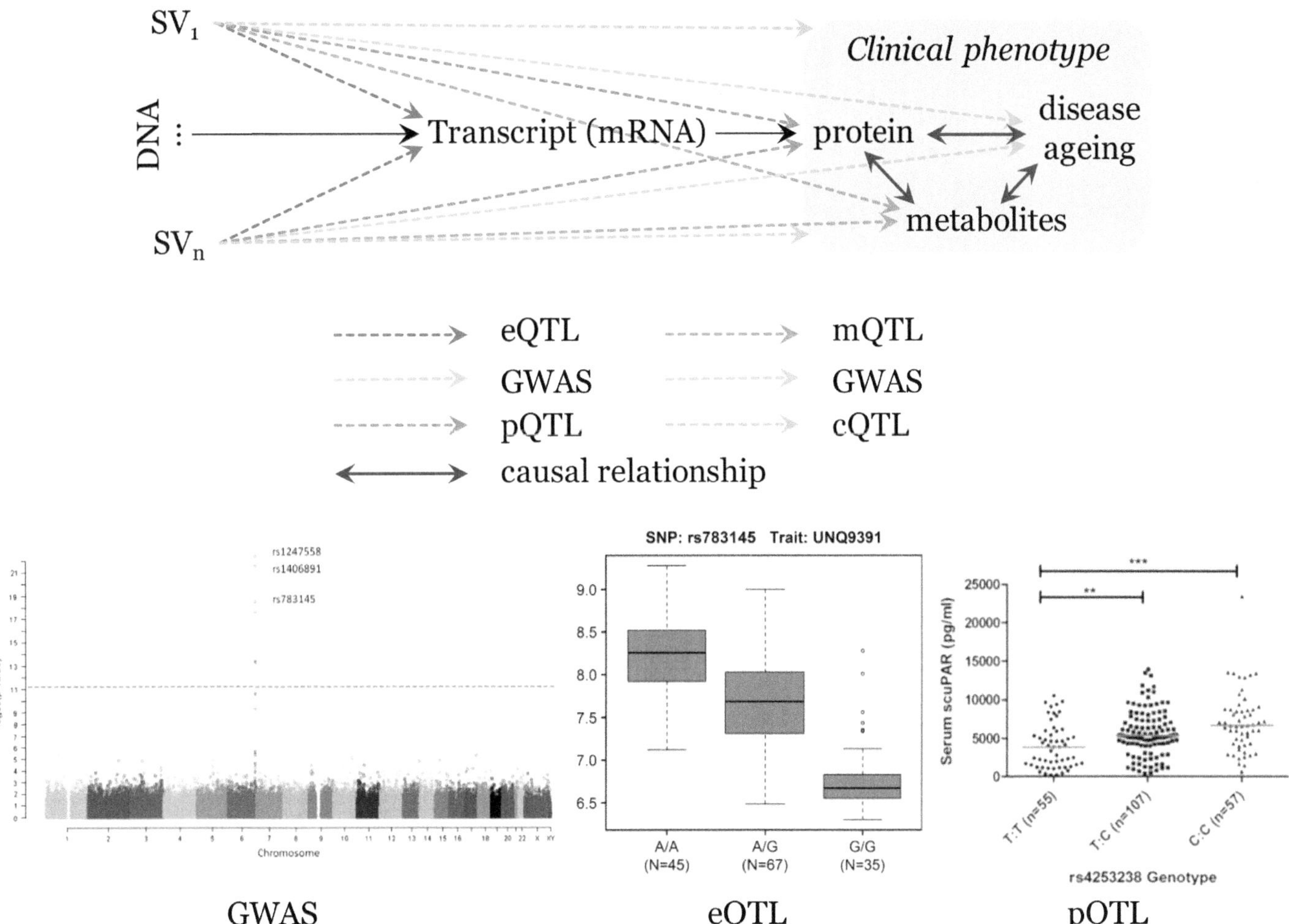

FIGURE 7.10.2 Statistical test to assess the effect of genome variants on human disease through genome-wide association studies (GWAS) and quantitative trait loci analysis of transcript expression (mRNA, eQTL), proteins (pQTL), metabolites (mQTL) and clinical phenotype (cQTL) to identify the genetic contribution for disease development. Proteins and metabolites are active compounds in disease and are closer to disease development than transcripts, making them as the causal effect of disease onset and development. GWAS and eQTL plots originate from Schröder, A. et al., *Pharmacogenomics J.* 13, 12–20, 2013, studying the genomic influence on drug ADME properties, whereas the pQTL example of scuPAR levels in serum of COPD patients is obtained from Portelli, M.A. et al., *FASEB J.* 28, 923–934, 2014.)

expression, *p* for proteins, *m* for metabolomics and *c* for clinical phenotype data) performed between different molecular layers and clinical phenotypes and present examples of significant GWAS and xQTL interactions.

Another manner how genomic variation can influence the abundance molecular profile in individuals is through copy number variation (CNV, germline origin) or copy number alteration (CNA, somatic mutation origin). In a deep proteogenomics study aiming to reveal the effects of somatic mutation in breast cancer by CPTAC consortium members, the effect of CNA on transcripts, proteins and phosphosites was investigated [45]. This analysis showed that not all CNA/transcript regulation is present at CNA/proteins and at CNA/phosphosite levels for corresponding gene, but CNA/proteins and CNA/phosphosite seem to be the subset of CAN/transcript for *cis* regulation (i.e., abundance of compounds is effected by copy number of a chromosomal location close to the location of the studied compounds) and that genes involved in joint CNA/protein and CNA/mRNA regulation show high correlation between mRNA and protein levels. These co-regulated genes with CNA/mRNA and CNA/proteins interaction was more enriched in oncogenes and tumor suppressors compared with genes with CNA/mRNA interactions only. *Trans* CNA regulations show a similar subsetting effect in the direction of mRNA→proteins→phosphosite than *cis* CNA/mRNA and CNA/proteins regulators. *Trans* CNA regulators are an indirect effect of losses or gains of DNA, and therefore the driver gene identification remain challenging. In this study, the authors used the Library of Integrated Network-based Cellular Signatures (LINCS) database (http://www.lincsproject.org/) [46–48] to identify driver genes, which database includes the functional knockdown effects of genes in cell lines. Figure 7.10.3 shows CNA/proteins and CNA/mRNA correlation

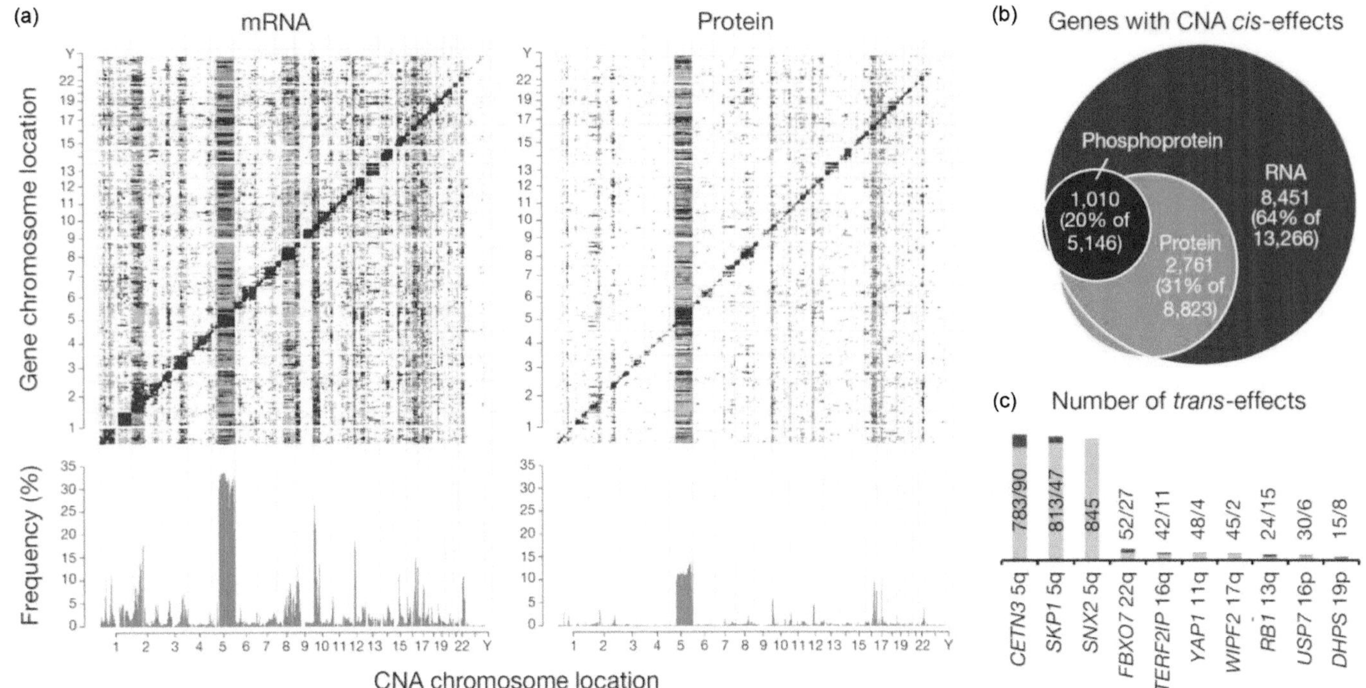

FIGURE 7.10.3 (a) Correlation matrix of copy number alteration (CNA; *x* axes) to RNA and protein levels (*y* axes) highlight new CNA *cis* (diagonal) and *trans* (off-diagonal) effects. Positive significant correlations (false-discovery rate [FDR] < 0.05) are shown in red, while significant negative correlations are highlighted in green. Histograms at the bottom show the percentage of significant *trans* CNA effects according to chromosomal location. (b) Overlap of significant (FDR < 0.05) CNA *cis* effects for mRNA, protein, and phosphoprotein levels. (c) Total numbers of significant *trans* CNA/protein effects (bars height, FDR < 0.05) showing the overlap with genes in LINCS knockdown profiles (red bars obtained from four cell lines, *t*-test FDR < 0.1). (Reproduced from Mertins, P. et al., *Nature* 534, 55–62, 2016.)

plots and the number of *cis* and *trans* CNA interactions identified in this study. This analysis identified 10 genes as potential regulatory candidates of *trans* CNA/protein regulation such as showing the ERBB2 gene was only functionally connected to CNA gain as *trans*-effect. Another example is the E3 ligase SKP1 and the ribonucleoprotein export factor CETN3, which genes are both located on chromosome arm 5q with less frequent gains in luminal B breast cancer and frequent losses in basal-like breast cancer. These genes were identified as potential regulators influencing the expression of the tyrosine kinase and EGFR, which is a main target for treatment in breast cancer.

Co-expression network analysis using correlation (relevance network), partial correlation (graphical Gaussian models) [49] or joint random forest [50] is another approach to integrate multi-omics data. For example, co-expression networks constructed from mRNA data obtained under multiple experimental conditions are often used to identify genes with similar functionality. Co-expression networks constructed from proteomics data are more stable and allow more accurate prediction of a gene's molecular functions. The reason behind that mRNA co-expression depends on mixed effects of molecular functions and chromosomal location, whereas the later effect is less emphasized in the co-expression network of proteins. Wang et al. [51] demonstrated a higher accuracy in predicting molecular function of genes using proteomics data over mRNA by the higher agreement between co-expression clusters of proteins and gene ontology biological processes compared with co-expression clusters of mRNA. Another study reported the opposite trend in the co-expression cluster of 73 mouse genes forming the cytosolic ribosome complex and reported higher connectivity between mRNA than corresponding proteins' data [34]. The results of these studies reinforce that an additional regulatory mechanism exists between transcriptomic and proteomics, and therefore the two molecular profiles provide different complementary information and different networks of molecular interactions based on co-expressions. In the study by Wang et al. [51], it was also demonstrated that protein co-expression networks can identify new molecular functions of genes. For example, a new function of the ERBB2 gene was identified as the gene contributing to lipid biosynthetic processes in breast cancer and for PLG gene, which is involved in complement activation.

The difference between mRNA and proteomics co-expression networks is further highlighted in a number of other studies such as Petralia et al. [50] who constructed a co-expression network from mRNA and proteins from the CPTAC/TCGA breast cancer study and concluded that a network constructed from proteins has higher connectivity in immune response and MMP-specific genes compared with clusters constructed from mRNA data (Figure 7.10.4).

FIGURE 7.10.4 Protein (a) and mRNA (b) co-expression network obtained with the joint random forest approach. Green edges (687) are shared between proteomic and mRNA data, red edges (502) are unique to protein, while blue edges (382) are unique to mRNA data. Gene names with at least 10 edges are labeled with gene symbols. (Taken from Petralia, F. et al., *J. Proteome. Res.*, 15, 743–754, 2016.)

CONCLUSION

Proteogenomics data integration starts with accurate preprocessing of transcriptomic and proteomics data and includes identification of sample-specific protein sequence variants by predicting sample-specific protein sequences from polyadenylated mRNA or exome sequencing data. Further multi-omics data integration address quantitative aspects, such as the correlation between molecular layers, xQTL analysis studying the effect of genomics variants on the quantitative molecular profiles, associations between CNAs and quantitative molecular profiles and co-expression analysis to identify molecular functions of genes.

In current multi-omics studies, elements of genome, transcriptome, proteome and phospho-proteome are collected and integrated. In the future, it is expected that multi-omics studies will include additional molecular layers such as riboSeq profiling (mRNA standing in translation phase on the ribosome), histone methylation and acetylation, DNA methylation and dynamic time course profile will be included in order to gain deeper understanding of molecular mechanism of disease and healthy aging biological events. These data completed with other information sources such as clinical imaging, clinical metadata, life-style monitoring sensors such as heart rate and physical activity trackers will bring clinical Big Data diagnostics that support precision health and disease management of individuals during the entire lifetime.

REFERENCES

1. R. Schiess, B. Wollscheid, R. Aebersold, Targeted proteomic strategy for clinical biomarker discovery, *Mol. Oncol.* 3 (2009) 33–44. doi:10.1016/J.MOLONC.2008.12.001.
2. P.E. Geyer, L.M. Holdt, D. Teupser, M. Mann, Revisiting biomarker discovery by plasma proteomics, *Mol. Syst. Biol.* 13 (2017) 942. doi:10.15252/MSB.20156297.
3. A.I. Nesvizhskii, A survey of computational methods and error rate estimation procedures for peptide and protein identification in shotgun proteomics, *J. Proteomics.* 73 (2010) 2092–2123. doi:10.1016/J.JPROT.2010.08.009.
4. A. Bateman, M.J. Martin, C. O'Donovan, M. Magrane, E. Alpi, R. Antunes, B. Bely et al., UniProt: The universal protein knowledge base, *Nucleic Acids Res.* 45 (2017) D158–D169. doi:10.1093/nar/gkw1099.
5. D.R. Zerbino, P. Achuthan, W. Akanni, M.R. Amode, D. Barrell, J. Bhai, K. Billis et al., Ensembl 2018, *Nucleic Acids Res.* 46 (2018) D754–D761. doi:10.1093/nar/gkx1098.
6. M. Peng, N. Taouatas, S. Cappadona, B. van Breukelen, S. Mohammed, A. Scholten, A.J.R. Heck, Protease bias in absolute protein quantitation, *Nat. Methods.* 9 (2012) 524–525. doi:10.1038/nmeth.2031.
7. R.A. Scheltema, J.P. Hauschild, O. Lange, D. Hornburg, E. Denisov, E. Damoc, A. Kuehn, A. Makarov, M. Mann, The Q Exactive HF, a Benchtop mass spectrometer with a pre-filter, high-performance quadrupole and an ultra-high-field Orbitrap analyzer, *Mol. Cell. Proteomics.* 13 (2014) 3698–708. doi:10.1074/mcp. M114.043489.
8. A. Michalski, J. Cox, M. Mann, More than 100,000 detectable peptide species elute in single shotgun proteomics runs but the majority is inaccessible to data-dependent LC-MS/MS, *J. Proteome Res.* 10 (2011) 1785–1793. doi:10.1021/pr101060v.
9. P.M. van Midwoud, L. Rieux, R. Bischoff, and E. Verpoorte, H.A.G. Niederländer, Improvement of recovery and repeatability in liquid chromatography–mass spectrometry analysis of peptides, *J Proteome Res.* 6.2 (2007) 781–791. doi:10.1021/PR0604099.
10. J.M. Proffitt, J. Glenn, A.J. Cesnik, A. Jadhav, M.R. Shortreed, L.M. Smith, K. Kavanagh, L.A. Cox, M. Olivier, Proteomics in non-human primates: Utilizing RNA-Seq data to improve protein identification by mass spectrometry in vervet monkeys, *BMC Genomics.* 18 (2017) 877. doi:10.1186/s12864-017-4279-0.
11. S. Ma, R. Menon, R.C. Poulos, J.W.H. Wong, S. Ma, R. Menon, R.C. Poulos, J.W.H. Wong, Proteogenomic analysis prioritises functional single nucleotide variants in cancer samples, *Oncotarget.* 8 (2017) 95841–95852. doi:10.18632/oncotarget.21339.
12. B. Zhang, J. Wang, X. Wang, J. Zhu, Q. Liu, Z. Shi, M.C. Chambers et al., Proteogenomic characterization of human colon and rectal cancer, *Nature* 513 (2014) 382–387. doi:10.1038/nature13438.
13. S. Woo, S.W. Cha, G. Merrihew, Y. He, N. Castellana, C. Guest, M. MacCoss, V. Bafna, Proteogenomic database construction driven from large scale RNA-seq data, *J. Proteome Res.* 13 (2014) 21–28. doi:10.1021/pr400294c.
14. S.T. Sherry, M.H. Ward, M. Kholodov, J. Baker, L. Phan, E.M. Smigielski, K. Sirotkin, dbSNP: The NCBI database of genetic variation, *Nucleic Acids Res.* 29 (2001) 308–311.
15. I. Lappalainen, J. Lopez, L. Skipper, T. Hefferon, J.D. Spalding, J. Garner, C. Chen, et al., DbVar and DGVa: Public archives for genomic structural variation, *Nucleic Acids Res.* 41 (2013) D936–D941. doi:10.1093/nar/gks1213.
16. S.A. Forbes, D. Beare, H. Boutselakis, S. Bamford, N. Bindal, J. Tate, C.G. Cole et al., Campbell, COSMIC: Somatic cancer genetics at high-resolution, *Nucleic Acids Res.* 45 (2017) D777–D783. doi:10.1093/nar/gkw1121.
17. M.S. Kim, S.M. Pinto, D. Getnet, R.S. Nirujogi, S.S. Manda, R. Chaerkady, A.K. Madugundu et al., A draft map of the human proteome, *Nature* 509 (2014) 575–581. doi:10.1038/nature13302.
18. A.A. Lobas, D.S. Karpov, A.T. Kopylov, E.M. Solovyeva, M. V. Ivanov, I.Y. Ilina, V.N. Lazarev et al., Exome-based proteogenomics of HEK-293 human cell line: Coding genomic variants identified at the level of shotgun proteome, *Proteomics.* 16 (2016) 1980–1991. doi:10.1002/pmic.201500349.
19. R. Bischoff, H. Permentier, V. Guryev, P. Horvatovich, Genomic variability and protein species—Improving sequence coverage for proteogenomics, *J. Proteomics.* 134 (2016) 25–36. doi:10.1016/j.jprot.2015.09.021.
20. B. Schwanhäusser, D. Busse, N. Li, G. Dittmar, J. Schuchhardt, J. Wolf, W. Chen, M. Selbach, Global quantification of mammalian gene expression control, *Nature.* 473 (2011) 337–342. doi:10.1038/nature10098.
21. A. Kertesz-Farkas, U. Keich, W.S. Noble, Tandem mass spectrum identification via cascaded search, *J. Proteome Res.* 14 (2015) 3027–3038. doi:10.1021/pr501173s.
22. G.S. Omenn, L. Lane, E.K. Lundberg, R.C. Beavis, C.M. Overall, E.W. Deutsch, Metrics for the Human Proteome Project 2016: Progress on identifying and characterizing the Human Proteome, including post-translational modifications, *J. Proteome Res.* 15 (2016) 3951–3960. doi:10.1021/acs.jproteome.6b00511.

23. R. Barbieri, V. Guryev, C.-A. Brandsma, F. Suits, R. Bischoff, P. Horvatovich, Proteogenomics: Key driver for clinical discovery and personalized medicine, *Proteogenomics*. (2016) 21–47. doi:10.1007/978-3-319-42316-6_3.
24. S. Trevisiol, D. Ayoub, A. Lesur, L. Ancheva, S. Gallien, B. Domon, The use of proteases complementary to trypsin to probe isoforms and modifications, *Proteomics*. 16 (2016) 715–728. doi:10.1002/pmic.201500379.
25. C.K. Frese, A.F.M. Altelaar, H. van den Toorn, D. Nolting, J. Griep-Raming, A.J.R. Heck, S. Mohammed, toward full peptide sequence coverage by dual fragmentation combining electron-transfer and higher-energy collision dissociation tandem mass spectrometry, *Anal. Chem.* 84 (2012) 9668–9673. doi:10.1021/ac3025366.
26. G.P.M. Mommen, C.K. Frese, H.D. Meiring, J. van Gaans-van den Brink, A.P.J.M. de Jong, C.A.C.M. van Els, A.J.R. Heck, Expanding the detectable HLA peptide repertoire using electron-transfer/higher-energy collision dissociation (EThcD), *Proc. Natl. Acad. Sci.* 111 (2014) 4507–4512. doi:10.1073/pnas.1321458111.
27. A. Cristobal, F. Marino, H. Post, H.W.P. van den Toorn, S. Mohammed, A.J.R. Heck, toward an optimized workflow for middle-down Proteomics, *Anal. Chem.* 89 (2017) 3318–3325. doi:10.1021/acs.analchem.6b03756.
28. J.D. Sanders, S.M. Greer, J.S. Brodbelt, Integrating carbamylation and ultraviolet photodissociation mass spectrometry for middle-down proteomics, *Anal. Chem.* 89 (2017) 11772–11778. doi:10.1021/acs.analchem.7b03396.
29. G.M. Sheynkman, M.R. Shortreed, A.J. Cesnik, L.M. Smith, Proteogenomics: Integrating next-generation sequencing and mass spectrometry to characterize human proteomic variation, *Annu. Rev. Anal. Chem.* (Palo Alto. Calif). 9 (2016) 521–545. doi:10.1146/annurev-anchem-071015-041722.
30. M.F. Rogers, H.A. Shihab, T.R. Gaunt, C. Campbell, CScape: A tool for predicting oncogenic single-point mutations in the cancer genome, *Sci. Rep.* 7 (2017) 11597. doi:10.1038/s41598-017-11746-4.
31. S. Castellana, C. Fusilli, G. Mazzoccoli, T. Biagini, D. Capocefalo, M. Carella, A.L. Vescovi, T. Mazza, High-confidence assessment of functional impact of human mitochondrial non-synonymous genome variations by APOGEE, *PLOS Comput. Biol.* 13 (2017) e1005628. doi:10.1371/journal.pcbi.1005628.
32. J. Li, L. Shi, K. Zhang, Y. Zhang, S. Hu, T. Zhao, H. Teng et al., VarCards: An integrated genetic and clinical database for coding variants in the human genome, *Nucleic Acids Res.* 46 (2018) D1039–D1048. doi:10.1093/nar/gkx1039.
33. T.Y. Low, S. Van Heesch, H. Van den Toorn, P. Giansanti, A. Cristobal, P. Toonen, S. Schafer et al., Quantitative and qualitative proteome characteristics extracted from In-Depth integrated genomics and proteomics analysis, *Cell Rep.* 5 (2013) 1469–1478. doi:10.1016/J.CELREP.2013.10.041.
34. E.G. Williams, Y. Wu, P. Jha, S. Dubuis, P. Blattmann, C.A. Argmann, S.M. Houten et al., , Systems proteomics of liver mitochondria function, *Science*. 352 (2016) aad0189. doi:10.1126/science.aad0189.
35. T. Klein, R. Bischoff, Physiology and pathophysiology of matrix metalloproteases, *Amino Acids*. 41 (2011) 271–290. doi:10.1007/s00726-010-0689-x.
36. M.G. Masciantonio, C.K.S. Lee, V. Arpino, S. Mehta, S.E. Gill, The balance between metalloproteinases and TIMPs: Critical regulator of microvascular endothelial cell function in health and disease, *Prog. Mol. Biol. Transl. Sci.* 147 (2017) 101–131. doi:10.1016/BS.PMBTS.2017.01.001.
37. L. Prely, T. Klein, P.P. Geurink, K. Paal, H.S. Overkleeft, R. Bischoff, *Activity-Dependent Photoaffinity Labeling of Metalloproteases*, Humana Press, New York, 2017: pp. 103–111.
38. E.J. van Rooden, B.I. Florea, H. Deng, M.P. Baggelaar, A.C.M. van Esbroeck, J. Zhou, H.S. Overkleeft, M. van der Stelt, Mapping in vivo target interaction profiles of covalent inhibitors using chemical proteomics with label-free quantification, *Nat. Protoc.* 13 (2018) 752–767. doi:10.1038/nprot.2017.159.
39. F. Klont, L. Bras, J.C. Wolters, S. Ongay, R. Bischoff, G.B. Halmos, P. Horvatovich, assessment of sample preparation bias in mass spectrometry-based proteomics, *Anal. Chem.* 90 (2018) 5405–5413. doi:10.1021/acs.analchem.8b00600.
40. L. Franke, R.C. Jansen, *eQTL Analysis in Humans*, Humana Press, Totowa, NJ2009: pp. 311–328.
41. D.C. Jeffares, C. Jolly, M. Hoti, D. Speed, L. Shaw, C. Rallis, F. Balloux, C. Dessimoz, J. Bähler, F.J. Sedlazeck, Transient structural variations have strong effects on quantitative traits and reproductive isolation in fission yeast, *Nat. Commun.* 8 (2017) 14061. doi:10.1038/ncomms14061.
42. M. Imprialou, A. Kahles, J.G. Steffen, E.J. Osborne, X. Gan, J. Lempe, A. Bhomra et al., Genomic rearrangements in arabidopsis considered as quantitative traits, *Genetics*. 205 (2017) 1425–1441. doi:10.1534/genetics.116.192823.
43. A. Schröder, K. Klein, S. Winter, M. Schwab, M. Bonin, A. Zell, U.M. Zanger, Genomics of ADME gene expression: Mapping expression quantitative trait loci relevant for absorption, distribution, metabolism and excretion of drugs in human liver, *Pharmacogenomics J.* 13 (2013) 12–20. doi:10.1038/tpj.2011.44.
44. M.A. Portelli, M. Siedlinski, C.E. Stewart, D.S. Postma, M.A. Nieuwenhuis, J.M. Vonk, P. Nurnberg et al., Genome-wide protein QTL mapping identifies human plasma kallikrein as a post-translational regulator of serum uPAR levels, *FASEB J.* 28 (2014) 923–934. doi:10.1096/fj.13-240879.
45. P. Mertins, D.R. Mani, K. V. Ruggles, M.A. Gillette, K.R. Clauser, P. Wang, X. Wang et al., Proteogenomics connects somatic mutations to signalling in breast cancer, *Nature*. 534 (2016) 55–62. doi:10.1038/nature18003.
46. A.B. Keenan, S.L. Jenkins, K.M. Jagodnik, S. Koplev, E. He, D. Torre, Z. Wang, et al., The library of integrated network-based cellular signatures NIH program: System-Level cataloging of human cells response to perturbations, *Cell Syst.* 6 (2018) 13–24. doi:10.1016/j.cels.2017.11.001.
47. A. Koleti, R. Terryn, V. Stathias, C. Chung, D.J. Cooper, J.P. Turner, D. Vidović et al., Data portal for the library of Integrated Network-based Cellular Signatures (LINCS) program: Integrated access to diverse large-scale cellular perturbation response data, *Nucleic Acids Res.* 46 (2018) D558–D566. doi:10.1093/nar/gkx1063.
48. Q. Duan, C. Flynn, M. Niepel, M. Hafner, J.L. Muhlich, N.F. Fernandez, A.D. Rouillard et al., LINCS Canvas Browser: Interactive web app to query, browse and interrogate LINCS L1000 gene expression signatures, *Nucleic Acids Res.* 42 (2014) W449–W460. doi:10.1093/nar/gku476.
49. R. Opgen-Rhein, K. Strimmer, From correlation to causation networks: A simple approximate learning algorithm and its application to high-dimensional plant gene expression data, *BMC Syst. Biol.* 1 (2007) 37. doi:10.1186/1752-0509-1-37.
50. F. Petralia, W.M. Song, Z. Tu, P. Wang, New method for joint network analysis reveals common and different coexpression patterns among genes and proteins in breast cancer, *J. Proteome Res.* 15 (2016) 743–754. doi:10.1021/acs.jproteome.5b00925.
51. J. Wang, Z. Ma, S.A. Carr, P. Mertins, H. Zhang, Z. Zhang, D.W. Chan et al., Proteome profiling outperforms transcriptome profiling for coexpression based gene function prediction, *Mol. Cell. Proteomics*. 16 (2017) 121–134. doi:10.1074/mcp.M116.060301.

Sequencing Approaches for Personalized Cancer Therapy Selection in Pathology 7.11

Wendy de Leng, Manon Huibers, John Hinrichs, Edwin Cuppen, and Stefan Willems

Contents

INTRODUCTION

The best treatment of each individual cancer patient starts with the best molecular pathological diagnosis. This does not only include appropriate tissue-based classification of a specific disease, it also includes reporting relevant histopathological tumor features (e.g., size, grade, vascular/perineural invasion). Pathological analysis also increasingly includes predictive and prognostic biomarkers. These biomarkers can be at the DNA level (e.g., BRAF mutations in melanoma), RNA level (e.g., Anaplastic Lymphoma Kinase (ALK) translocation in NSCLC) or protein level (overexpression of ALK) or even apply at the epigenetic level (e.g., MethylGuanine DNA MethylTransferase (MGMT) hypermethylation in brain tumors) [1]. For some tumor types (e.g., gastrointestinal stroma tumor (GIST), melanoma, NSCLC, CorloRectalCarcinoma (CRC)), genetic analysis has become recommended in advanced disease stages. The awareness that the basis of cancer (including biological course and drug response) is determined by the genetic make-up of the cancer cells led to the development and application of broader sequencing assays, shifting from single hotspot mutation analysis (e.g., Sanger sequencing for KRAS mutations) to multiplex (next generation) gene sequencing. In an ideal world, the more DNA information (i.e., sequence data obtained), the better the disease prediction can be made. In reality, despite progress in biomarker exploration, especially though fundamental research, there is still an unmet need to detect more selective biomarkers in precision/personalized cancer treatment. In this section we briefly describe the most commonly used genetic/genomic sequencing approaches currently used in pathology practice and speculate on the future of pathology sequencing.

TARGETED NEXT GENERATION SEQUENCING

In routine pathology diagnostics the available tissue is most frequently FFPE, resulting in suboptimal fragmented DNA. Furthermore, tissue samples can be very small, for instance, in the case of small needle biopsies. To allow for a reliable sequencing approach, the technique should be able to work on small amounts of fragmented DNA. Therefore targeted NGS is an excellent option [2]. It is a targeted approach because only a selection of genes and genomic regions of interest will be analyzed. This allows for a fast and cost effective way of sequencing on as little as 10 ng DNA.

Technique

Targeted NGS is based on a multiplex PCR reaction to allow amplification of the different regions of interest (Figure 7.11.1). After several purification steps and the ligation of barcodes and sequencing adaptors, the amplified DNA pool is sequenced. For FFPE material, the length of the amplicons should not be longer than 140 bp [2]. Because targeted NGS is a PCR based assay with the risk of introducing PCR errors, which can later, by mistake, be interpreted as variants. To prevent the identification of these false positive variants, both the positive and negative DNA strand can be sequenced after which the data is bioinformatically compared [3]. An alternative is the use of molecular barcodes where a unique sequence of several nucleotides is added to all initial DNA molecules that are amplified during the multiplex PCR. The benefit of this approach is that PCR artefacts will be present for only a few unique molecular barcodes, whereas a somatic mutation will be identified for multiple unique molecular barcodes [4]. Finally, depending on the size of the gene panel, a PCR amplification of areas of interest may be replaced by hybridization capture, which is most frequently used for very large gene panels.

Application in Pathology

Because the genes of interest for several solid tumor types overlap, many labs use one dedicated panel to detect mutations in lung cancer, colon cancer, melanoma and GIST [2,5–9]. However, when, for instance, hematological disorders or gliomas need to be analyzed, completely different genes are of interest. Therefore, the use of dedicated NGS panels is necessary when clinically actionable variants in a broad range of tumor types need to be identified.

Besides information on mutations, this targeted NGS approach can provide information on the presence of amplifications based on the coverage of the different amplicons. High-level amplifications, for instance EGFR amplifications in glioblastoma or HER2 amplifications in breast cancer, can reliably be detected when sufficient amplicons are analyzed for the gene of interest [10].

Next to coverage-based analysis, SNPs can provide information on copy number variations (CNVs). In the case of CNVs, the

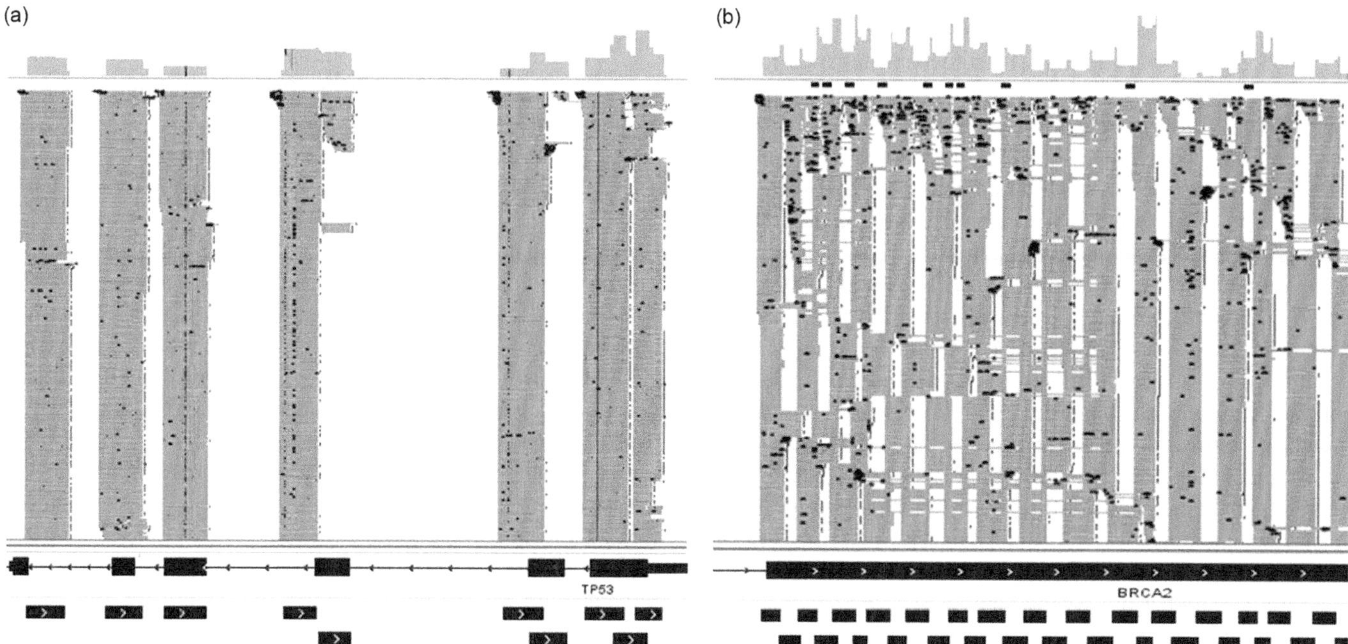

FIGURE 7.11.1 Next generation sequencing (NGS)-targeted panel design. (a) TP53 amplicon design to detect small exons, one or two amplicons per exon suffice to cover the entire coding region; (b) BRCA2 amplicon design to detect large amplicons where multiple exons overlap to cover the entire exon.

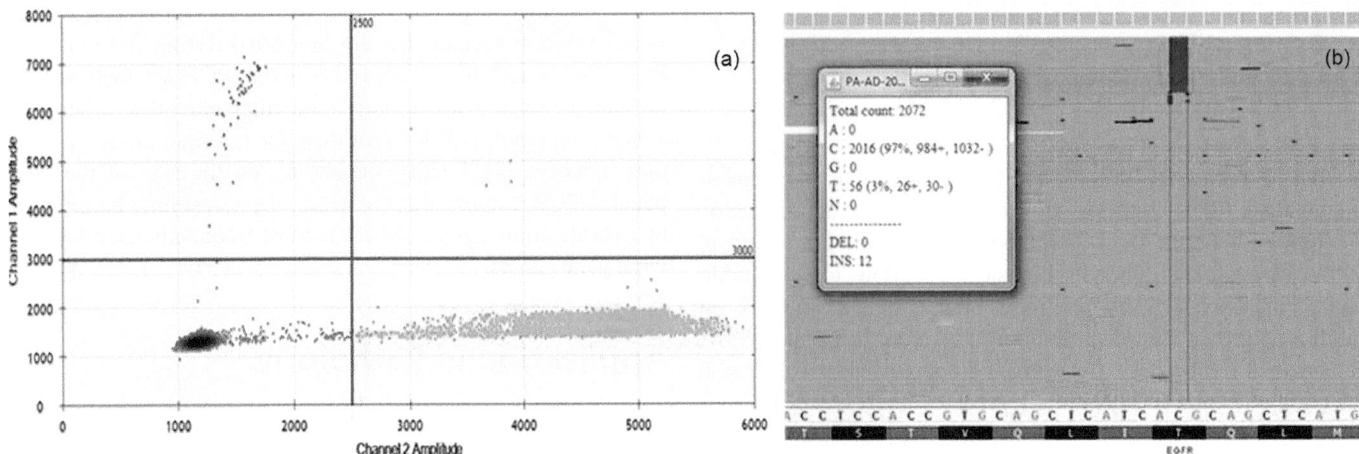

FIGURE 7.11.2 Detection of EGFR T790M mutation. (a) ddPCR results, 61 mutant droplets and 4,463 wild-type droplets were detected resulting in 1.4% mutant; (b) next generation sequencing (NGS) results, 56 of the 2,072 (3%) of the reads contain the T790M mutation.

B-allele frequency (BAF) will change as the ratio between both alleles of a heterozygous SNP will change [11].

In addition, RNA analysis for the detection of translocations is possible using targeted NGS and this analysis can be combined with a DNA analysis for simultaneous mutation detection [7,8].

The most recent application of targeted NGS is the analysis of liquid biopsies where a targeted NGS approach could aid in tumor follow-up and the very rapid detection of therapy resistance by the identification of resistance mutations (Figure 7.11.2) [12].

Pros and Cons

Currently many laboratories use targeted NGS for their routine diagnostics. The assay can easily be automated and therefore performed in a standardized way. Furthermore, the smaller bench top sequencers allow for a fast throughput as fewer samples need to be sequenced in one sequencing run and the sequencing and analysis time is much shorter compared with whole exome-sequencing (WES) and whole genome sequencing (WGS). However, still a certain number of analyses need to be performed on a regular basis to maintain this fast throughput and to maintain a sufficient knowledge base within the lab for appropriate troubleshooting in case of technical difficulties. A major benefit of targeted NGS is that only regions of interest can be analyzed and that the size of the panel can be up to 15,000 amplicons. The drawback, however, is that the genes of interest change in time with the identification of new actionable variants, resulting in the need to update the gene panel regularly. Furthermore, structural variants cannot be detected and most panels are of limited value for reliable detection of CNVs.

Finally, small gene panels make very deep sequencing of small gene sets affordable, thereby providing the opportunity to follow tumor mutations in the blood as the tumor is shedding its DNA. This may allow the early detection of resistance mutations, resulting in quick changes in treatment strategies.

New Developments

Within pathology diagnostics, two different trends occur simultaneously where, on one hand, larger gene panels are used to allow the analysis of different tumor types using one standardized gene panel. On the other hand, the use of smaller tumor type-specific gene panels is of interest because this will also allow more cost-effective extremely deep sequencing for liquid biopsy. Furthermore, the number of actionable mutations is suggested to increase over time, as does the number of known resistance mechanism of targeted treatments, resulting in the need of analyzing more genes per tumor type.

WHOLE-EXOME SEQUENCING

Technique

In contrast to targeted NGS, WES will gather information regarding the complete protein-coding region of the genome (all exons). The human genome consists of >200,000 exons containing approximately 30 million base pairs [13]. Compared with targeted NGS, this method contains much more information regarding the genomic landscape; however, compared with WGS, it can still be regarded as a "targeted" approach that is scaled up to the whole exome, which is still "only" 1%–3% of our whole genome.

WES is currently used in research setting and can be performed on human samples; however, this technique is slowly approaching the diagnostic field [14]. If WES would be widely implemented in molecular pathology diagnostics this would change the diagnostic workflow dramatically [1]. Before implementation in diagnostics some hurdles should be overcome; for example, DNA yield, DNA quality, turnaround time, costs, bioinformatics and reporting complexity [15].

Application in Pathology

WES is slowly being implemented in clinical genetic departments, with promising results thus far [16]. If the use of WES is widely implemented in routine genetic diagnostics [17], it may not be long before also pathology departments start to introduce this technique. However, there are important differences between implementation in genetics versus pathology.

First, the expected variant frequency of mutational findings in genetics and pathology is different: Germline alterations are found in 50% or 100% frequency, whereas somatic alterations can range from <5% to 99%. For somatic sequencing, the expected variant frequency greatly depends on tumor percentage and tumor heterogeneity. This should be taken into account when introducing a mutation analysis assay because in somatic sequencing, sensitivity of <5% should be reached.

Second, the origin of material between the two disciplines requires a different approach. Medical genetics uses DNA derived from blood cells, with good quality. The pathology department works mainly with FFPE material, resulting in poor-quality fragmented DNA. This demands a forgiving sequence strategy that handles fragmented DNA well. Ideally, the material used should be improved by freezing biopsy material or molecular fixation solutions.

If it is not clear whether a variant found is disease causing (pathogenic) or not disease causing (benign), in these cases the term variant of unknown significance (VUS) is used. Unfortunately enlarging the genomic regions one sequences, makes data analysis more complicated; more variants will be found using WES compared to targeted NGS. To identify pure somatic variants, a paired analysis of tumor and normal tissue (e.g., buccal swab or a blood sample) can be used [18]. If fresh biopsy tissue is included to filter germline variants for WES (200 ng input DNA, 95× mean coverage), on average 16 somatic variants will be found per tumor sample. It was shown that from all mutations detected a majority of somatic findings were classified as VUS (mutations found in 97 patients; 16× "targeted therapy available," 98× "biologically relevant" and 1,474× "VUS") [18]. These data show that the wider we look in the genome, the more we will find which we cannot explain. This makes WES an elegant technique to explore new options for future targeted therapy.

A recent comparison between WES and targeted NGS [19] showed that WES has a high specificity (96%) and precision (99%) but a low sensitivity (73%) and accuracy (75%); in other words, there is a greater chance of false-negative findings in WES compared with currently used methods.

A recent study validated a pipeline for WES for cancer care [20] on FFPE material; the EXaCT-1 test. This method uses HaploPlex (Agilent) target enrichment and NGS (Illumina) to detect somatic variants. They stated that this whole exome approach has comparable results between FFPE and fresh-frozen tumor samples and concluded that this method is accurate and sensitive for clinical-relevant mutation analysis with reasonable costs and turnaround time. This study shows that it is possible to apply WES in diagnostics, but some major pitfalls remain.

Pros and Cons

The major advantage of WES over targeted NGS is that all coding regions are covered, containing approximately 85% of all disease-related mutations [21]. It therefore requires only one technical validation of the technique to allow diagnostic use for all tumor types. The large validation will be costly, but will in the end be more efficient than validating numerous smaller targeted gene panels. A second advantage is that one can trace back mutations, insertions, deletions and CNVs across the genome (based on BAF of SNPs or coverage analyses) [11]. Finally, WES is a great method for detecting novel aberrations in research setting. If WES is used in a diagnostic setting, a novel variant will first be annotated as VUS, whereas if more knowledge is gained, the classification can change into novel pathogenic findings, which would not have been discovered only if targeted sequencing is performed.

As a disadvantage, many of the currently described WES techniques require an input of large amounts of good quality DNA (>50 to 200 ng [15]), although the first publication on FFPE WES shows promising results [20]. The input threshold of 200 ng DNA might result in a number of diagnostic samples that cannot be sequenced. In cancer diagnostics, it is not desirable that many samples dropout due to technical constraints; for every patient a test result is requested, and a new biopsy means more burden for the patient. A second disadvantage is the large amount of data gathered by WES, which makes filtering of variants (e.g., technically incorrect, artefacts, SNPs) complicated. Besides, if variants are filtered correctly, a moderate list will remain to be truly somatic, which need to be classified and interpreted for clinical use. This makes reporting of the WES findings a more labor-intensive and subjective matter compared to targeted sequencing and should be standardized in diagnostic use. A third drawback is that detection of CNVs and structural variants is not reliable. A fourth drawback is the high costs that come with implementation of WES; although they have dropped dramatically over the last few years they are still higher compared to targeted NGS. In a recent publication from the Association for Molecular Pathology, costs of targeted NGS were compared with WES [22]. The costs for a 5- to 50-gene targeted NGS panel is approximately $700 (range $578–908), compared with exome sequencing which is approximately $2,500 (range $1,499–$3,388). Another pitfall is that turnaround time should be fast enough, where for targeted NGS this is 5 days, for WES the workup from sample to report is currently 2–5 weeks (assuming no complicated reporting issues [20]).

New Developments

With the wide use of NGS in diagnostics (genetics or pathology), a fierce debate should be conducted regarding ethical issues [23]. One of the ethical aspects should be informed consent, which ideally should be consistent among different clinical centers and understandable for patient and clinician [16,24]. Before wide implementation in clinical practice, we should sort out the ethical issues concerning WES.

To implement WES in the diagnostic workflow not only technical, ethical and logistic issues should be solved, also health care costs should be taken into account. Although WES costs have dropped dramatically over the past years [21], they are still higher than current sequencing approaches. When the switch was made from single exon Sanger sequencing to targeted NGS for multiple genes, this was mainly driven by more effective hands-on time per sample and costs. To introduce WES in a diagnostic workflow without making health care more expensive, these costs have to drop. Targeted NGS already showed to be cost effective for NSCLC when taking into account a decrease in treatment costs (–$2.7 million; more targeted therapy, less nontargeted therapy, and more treatment in clinical trials) [22]. How this cost effectiveness is translated in WES is not yet known and should be investigated when implemented in pathology diagnostics.

WHOLE GENOME SEQUENCING

With WGS, we mean sequencing the complete DNA sequence of an organism's genome. This includes sequencing chromosomal DNA as well as mitochondrial DNA. In practice, only 95% of a human genome will be sequenced, because most sequenced genomes contain "gaps" owing to repetitive DNA-related assembly difficulties [25]. This is still an enormous amount of sequence data compared with WES (approximately 1% of the genome) or SNP genotyping (approximately 0.1% of the genome). WGS has largely been used as a research tool and resulted in an enormous boost in knowledge about cancer including new targeted therapies. While personalized medicine is developing further, WGS is slowly making its entrance in the clinic to guide therapeutic intervention.

Technique

Most of WGS today is based on "shotgun" sequencing in which the DNA is first broken up randomly into numerous small segments (100–1,000 bp) [26]. The reason for this is that most sequencing techniques today can only sequence DNA fragments up to 1,000 bp. All generated segments are sequenced separately. Multiple overlapping reads for the target DNA are obtained by performing several rounds of this fragmentation and sequencing. Computer software uses the overlapping ends of different reads to assemble them into a continuous sequence (Figure 7.11.3). Although capillary sequencing (Sanger) was the first approach to successfully sequence a nearly full human genome [27], it is still too expensive and takes too long for commercial purposes. Since 2005, capillary sequencing has been progressively displaced by high-throughput ("next-generation," massive parallel) sequencing technologies such as dye sequencing, pyrosequencing, and semiconductor sequencing [28]. These technologies continue to employ the basic shotgun strategy, namely, parallelization and template generation via genome fragmentation. In practice enzymes called transposomes randomly cut the DNA into short segments ("tags"). Adapters are added at either side of the introduced breaks (ligation). These adapters consist of terminal sequences, for attachment of the

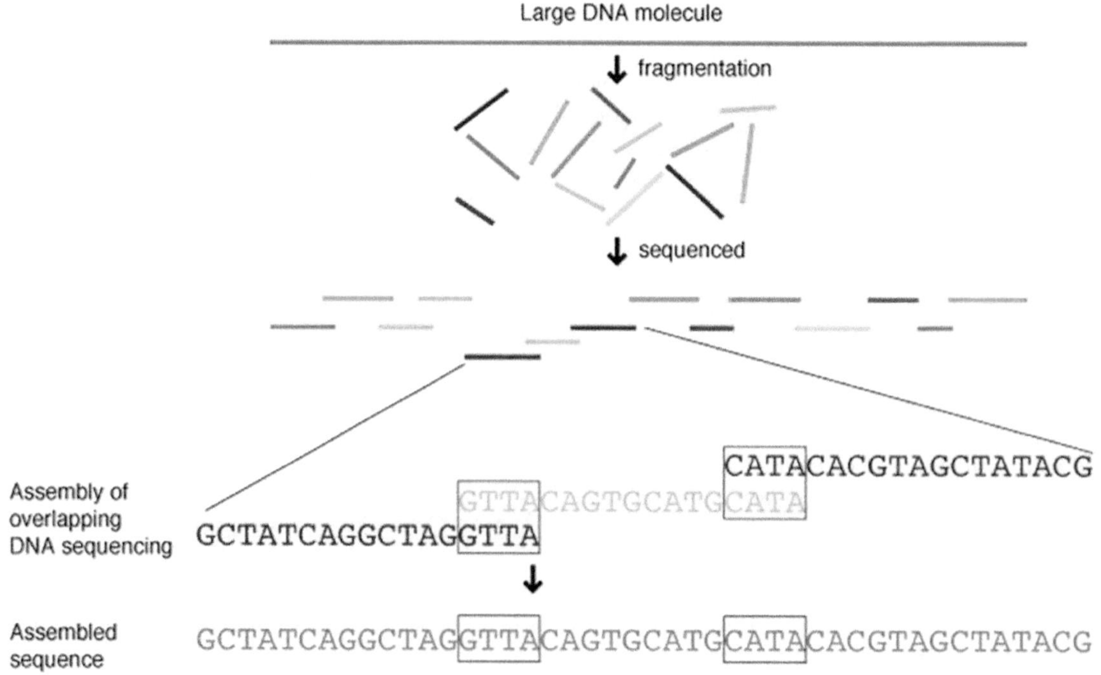

FIGURE 7.11.3 Shotgun sequencing. Shotgun sequencing is a laboratory technique for determining the DNA sequence of an organism's genome. The method involves breaking the genome into a collection of small DNA fragments that are sequenced individually. A computer program looks for overlaps in the DNA sequences and uses them to place the individual fragments in their correct order to reconstitute the genome (National Human Genome Research Institute, Bethesda, MD.)

fragment, primer binding sites, as starting point of polymerase based sequencing, and indices as identifier when more genomes are sequenced simultaneously. The "tagged" segments are attached to either a solid phase (glass plate) or a liquid phase (beads in oil/water solution) and subsequently clonally amplified before they are sequenced [28]. The primary sequencing is still based on the classical Sanger sequencing involving DNA polymerase and single nucleotides. The main difference is the massive parallel sequencing in which millions of different sequencing reactions occur simultaneously. The main challenge is the reassembly of all this separate sequences into one continuous sequence.

Application in Pathology Practice

Although in medical genetics practice WES is sometimes already replaced by WGS, pathology mainly uses targeted sequencing. This is largely caused by the difference in diagnostic questionnaire. Genetics is often searching for unique (unknown) DNA mutations associated with rare diseases. Therefore, a sequence approach in which many genes are sequenced is important. Pathology, on the other hand, focuses on a limited number of known cancer mutations that are predictive in targeted therapy. This makes a targeted sequencing approach with a focus on sensitivity more important.

Pros and Cons

The main advantage of WGS sequencing is that you will not miss any clinical relevant information. Furthermore, it allows for reliable detection of CNV and structural variation. On the other hand, WGS will reveal hundreds or thousands of mutations of which we do not know the clinical relevance and treatment solutions. Because of the complexity, results will only become available after 3 or 4 weeks, including information that directly influences treatment. The diagnostic use of WGS may have ethical implications [29]. WGS is not only limited to tumor DNA, it will also give information on a person's genetic predisposition and possible genetic diseases. The same holds true for WES (as described above). Another aspect is the sensitivity of sequencing. If you consider the trade-off between sensitivity (sequence depth) and information (sequence width), WGS and targeted sequencing are both on the outer limit of the spectrum. When analyzing a small solid biopsy or even a liquid biopsy, with minor amounts of tumor DNA, targeted sequencing is the best option. However, when no activating mutations can be found in this way, a broader search with the use of WES or WGS is preferred.

New Developments

In future cancer treatment we will need an extensive genetic makeup of the tumor to optimal predict, treat and monitor cancer. For this we need the information of WES and WGS and the sensitivity of targeted sequencing. With current NGS techniques, we rely on a two- or three-step approach, which uses all three techniques combined. Emerging nanopore sequencing could in the future eliminate this dilemma. This technique holds the promise of sequencing the whole genome at flexible sequence depth without preamplification [30].

FUTURE PERSPECTIVE

With decreasing sequencing costs and increased computational power, it is likely that WES and WGS will become current practice in pathology. Appropriate bioinformatics pipelines will allow to size down all information to clinical relevant output. Meanwhile, potentially clinical relevant genetic data obtained can be used to (1) direct eligibility for trial enrollment, or (2) guide research. This will mean that interpretation and integration of complex data in the right clinical and pathological context becomes the future endeavor for (molecular) pathologists/biologists. (Table 7.11.1 lists the pros and cons of WGS, WES, and targeted NGS.)

If the whole exome or genome is analyzed, there is an increased chance of encountering "incidental findings" of mutations (most often germline) that were not the initial reason for sequencing. Therefore, an ethical and legal framework for dealing with these "incidental" or "unsolicited" findings is necessary as well as guidelines on how to inform patients and their families [31,32].

DNA/RNA sequencing will not only stay a leading assay for predicting response to targeted therapy. Computing mutational load might also become an important clinical biomarker for directing immunotherapy, next to programmed death ligand 1 (PD-L1) protein expression and characterization of the immune infiltrate. Next to tissue-based analysis, blood-based sequencing assays (liquid biopsies) will become increasingly important. DNA/RNA analysis of liquid biopsies allow for minimally invasive monitoring of disease and early disease detection.

REFERENCES

1. Dietel, M., Molecular pathology: A requirement for precision medicine in cancer. *Oncol Res Treat*, 2016. **39**(12): 804–810.
2. de Leng, W.W. et al., Targeted next generation sequencing as a reliable diagnostic assay for the detection of somatic mutations in tumours using minimal DNA amounts from formalin fixed paraffin embedded material. *PLoS One*, 2016. **11**(2): e0149405.
3. Fisher, K.E. et al., Clinical validation and implementation of a targeted next-generation sequencing assay to detect somatic variants in non-small cell lung, melanoma, and gastrointestinal malignancies. *J Mol Diagn*, 2016. **18**(2): 299–315.
4. Stahlberg, A. et al., Simple multiplexed PCR-based barcoding of DNA for ultrasensitive mutation detection by next-generation sequencing. *Nat Protoc*, 2017. **12**(4): 664–682.

TABLE 7.11.1 Pros and cons of whole-genome sequencing (WGS), whole-exome sequencing (WES), and targeted next generation sequencing (NGS).

	TARGETED NGS	*WES*	*WGS*
Input material	FFPE/fresh material 10–20 ng	Fresh (FFPE?) 50–200 ng	Fresh 100–350 ng
Depth of sequencing (cost dependent)	500–1,000×	<100×	30–60×
Width of sequencing	Multiple genes and/or hotspots	All exons	The whole genome (exons and introns)
Novel findings; interesting for research	Minimal	High	Very high
Possibility of finding mutations, insertions, deletions, copy number variations	Targeted mut, insert and del. Small areas for SNP analysis	All exome mut, insert, del, SNPs	All genome mut, insert, del, SNPs
Complexity of data analysis	Fairly easy	Complex	Highly complex
Chance for "unsolicited findings"	Small	High	High
Validation	For every panel separately	Once	Once
Turnaround time	5 days	2–5 weeks	4–8 weeks
Costs	$700	$2,500	$3,000–5,000

5. Frampton, G.M. et al., Development and validation of a clinical cancer genomic profiling test based on massively parallel DNA sequencing. *Nat Biotechnol*, 2013. **31**(11): 1023–1031.
6. Hamblin, A. et al., Clinical applicability and cost of a 46-gene panel for genomic analysis of solid tumours: Retrospective validation and prospective audit in the UK National Health Service. *PLoS Med*, 2017. **14**(2): e1002230.
7. Hovelson, D.H. et al., Development and validation of a scalable next-generation sequencing system for assessing relevant somatic variants in solid tumors. *Neoplasia*, 2015. **17**(4): 385–399.
8. Luthra, R. et al., A Targeted high-throughput next-generation sequencing panel for clinical screening of mutations, gene amplifications, and fusions in solid tumors. *J Mol Diagn*, 2017. **19**(2): 255–264.
9. Vendrell, J.A. et al., High-throughput detection of clinically targetable alterations using next-generation sequencing. *Oncotarget*, 2017. **8**(25): 40345–40358.
10. Hoogstraat, M. et al., Simultaneous detection of clinically relevant mutations and amplifications for routine cancer pathology. *J Mol Diagn*, 2015. **17**(1): 10–18.
11. Dubbink, H.J. et al., Diagnostic detection of allelic losses and imbalances by next-generation sequencing: 1p/19q Co-Deletion analysis of gliomas. *J Mol Diagn*, 2016. **18**(5): 775–786.
12. Weerts, M.J.A. et al., Somatic tumor mutations detected by targeted next generation sequencing in minute amounts of serum-derived cell-free DNA. *Sci Rep*, 2017. **7**(1): 2136.
13. Ng, S.B. et al., Targeted capture and massively parallel sequencing of 12 human exomes. *Nature*, 2009. **461**(7261): 272–276.
14. Lapin, V. et al., Regulating whole exome sequencing as a diagnostic test. *Hum Genet*, 2016. **135**(6): 655–673.
15. Ballester, L.Y. et al., Advances in clinical next-generation sequencing: Target enrichment and sequencing technologies. *Expert Rev Mol Diagn*, 2016. **16**(3): 357–372.
16. Vrijenhoek, T. et al., Next-generation sequencing-based genome diagnostics across clinical genetics centers: Implementation choices and their effects. *Eur J Hum Genet*, 2015. **23**(9): 1270.
17. O'Donnell-Luria, A.H. and D.T. Miller, A Clinician's perspective on clinical exome sequencing. *Hum Genet*, 2016. **135**(6): 643–654.
18. Beltran, H. et al., Whole-exome sequencing of metastatic cancer and biomarkers of treatment response. *JAMA Oncol*, 2015. **1**(4): 466–474.
19. Chang, Y.S. et al., Evaluation of whole exome sequencing by targeted gene sequencing and sanger sequencing. *Clin Chim Acta*, 2017. **471**: 222–232.
20. Rennert, H. et al., Development and validation of a whole-exome sequencing test for simultaneous detection of point mutations, indels and copy-number alterations for precision cancer care. *NPJ Genom Med*, 2016. **1**.
21. van Dijk, E.L. et al., Ten years of next-generation sequencing technology. *Trends Genet*, 2014. **30**(9): 418–426.
22. Sabatini, L.M. et al., Genomic sequencing procedure microcosting analysis and health economic cost-impact analysis: A report of the association for molecular pathology. *J Mol Diagn*, 2016. **18**(3): 319–328.
23. Bredenoord, A.L., M.C. de Vries, and J.J. van Delden, Next-generation sequencing: Does the next generation still have a right to an open future? *Nat Rev Genet*, 2013. **14**(5): 306.
24. Fowler, S.A., C.J. Saunders, and M.A. Hoffman, Variation among consent forms for clinical whole exome sequencing. *J Genet Couns*, 2017. 27(1): 104–114.
25. De Bustos, A., A. Cuadrado, and N. Jouve, Sequencing of long stretches of repetitive DNA. *Sci Rep*, 2016. **6**: 36665.
26. Favello, A., L. Hillier, and R.K. Wilson, Genomic DNA sequencing methods. *Methods Cell Biol*, 1995. **48**: 551–569.
27. Lander, E.S. et al., Initial sequencing and analysis of the human genome. *Nature*, 2001. **409**(6822): 860–921.
28. Metzker, M.L., Sequencing technologies—The next generation. *Nat Rev Genet*, 2010. **11**(1): 31–46.
29. Davey, S., Next generation sequencing: Considering the ethics. *Int J Immunogenet*, 2014. **41**(6): 457–462.
30. Schmidt, J., Membrane platforms for biological nanopore sensing and sequencing. *Curr Opin Biotechnol*, 2016. **39**: 17–27.
31. Lolkema, M.P. et al., Ethical, legal, and counseling challenges surrounding the return of genetic results in oncology. *J Clin Oncol*, 2013. **31**(15): 1842–1848.
32. Bijlsma, R.M. et al., Unsolicited findings of next-generation sequencing for tumor analysis within a Dutch consortium: Clinical daily practice reconsidered. *Eur J Hum Genet*, 2016. **24**(10): 1496–500.

Mass Spectrometry Imaging 7.12

Enabling Comprehensive Local Analysis of Molecular Biomarkers in Tissue for Personalized Medicine

Tiffany Porta, Steven W. M. Olde Damink, and Ron M. A. Heeren

Contents

INTRODUCTION

With approximately 14 million new cases and more than 8 million deaths each year worldwide, cancer is a major global health issue [1,2]. The vast majority of cancers develop as a consequence of inborn or acquired genetic mutations. Each tumor type possesses its own molecular specificity and changes. Despite recent advances with clinico-genomics and chemotherapeutic treatments, most patients still receive non-personalized treatment and many cancers remain incurable. Precision oncology requires novel molecular descriptors of cancer, and attempts have been made to redefine common cancer phenotypes according to their genomic features [2,3]. However, these strategies are slow and costly, and do not account for tumor heterogeneity and evolution. This clearly points to a need for rapid, point-of-care tissue-based diagnostics that can assist clinical decision making. Early screening, preoperative diagnostics, surgical pathology, intraoperative diagnostics and postoperative follow-up are all key to the routine implementation of P4-medicine [3]. This system medicine approach critically depends on the availability of smart, speedy, specific, selective and sensitive diagnostics (S5-diagnostics).

Of the diagnosed cancer cases, 80% require surgery as the only curative treatment possibility, with a caveat that the diagnosis was made early enough. Early and precise detection of cancer is a key strategy for significantly reducing the risk of mortality and thus improving prognosis. Preoperative histological examination from biopsies or fine needle aspiration (FNA) can be very challenging due to inaccessibility of the lesion or presence of the lack of disease specific material in the initial biopsy/cytology specimen. As a result, histology based on light microscopy is not always reliable or even possible.

Surgical resection with microscopically negative, thus tumor-free margins are left under the responsibility of the surgeon who often faces challenging tumor delineation during the operation. Intraoperative pathological assessment of the resected specimen using frozen sections has limited sensitivity and specificity due to time constraints that preclude use of targeted immune staining. This is where innovative personalized medicine based on novel, broad biomarker profiles can have a major impact on health care. The tumor recurrence rates in patients can be reduced through

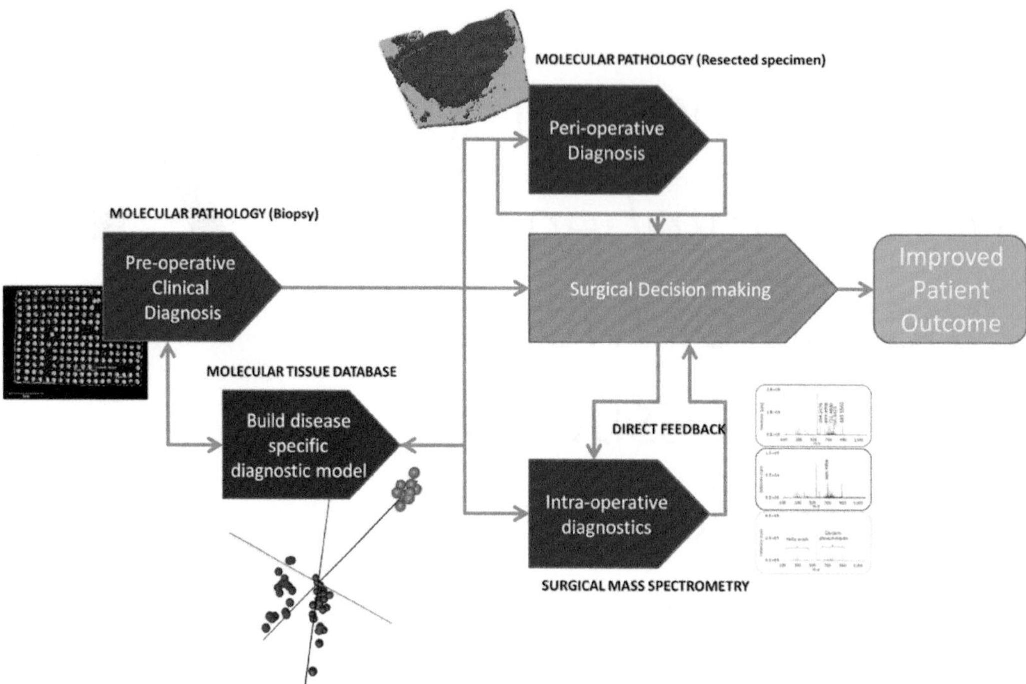

FIGURE 7.12.1 An overview of the different time points in the framework of a surgical intervention where mass spectrometry (in brown) can improve the quality of information that is fed into the clinical decision-making process. The dark blue workflow element is using the breadth of molecular information combined with the pathological assessment to use modern bioinformatics approaches to build a disease specific model. This model can subsequently be evaluated during perioperative mass spectrometry imaging and intraoperative surgical mass spectrometry. The connection lines indicate the flow of personalized information that ultimately result in an improved patient outcome.

more accurate and on-site determination of the resection margin during surgery, using personalized diagnostic profiles. This minimizes removal and/or damage of healthy tissues as an undesirable effect for the patient and thus potentially improves patient postoperative organ function, oncological outcome and survival rates. Breast cancer surgery, as an example, would benefit from real-time tissue-type information to guide and optimize surgical resection within the anatomic boundaries, reducing the incidence of incomplete tumor resection. Pathological assessment of the resected specimen using a common inking approach is often compromised by ink "running" into the tissue [4]. The current incidence of incomplete tumor resection is estimated to be 4.6% in lumpectomies performed for invasive breast cancer and 19% in lumpectomies performed for ductal carcinoma [5].

Clearly, tissue evaluation, be it during surgery or postoperatively, will benefit from comprehensive spatial analysis of molecular biomarkers. In the current clinical routine, a resected tissue/biopsy is sent to the pathologist for histopathological evaluation of frozen sections. Morphology-based information on grade and tumor type and is usually returned within 30 min. In Figure 7.12.1, the different diagnostic time points for tissue evaluation have been summarized. The limited accuracy of the conventional pathological methods available for evaluation in this time frame can pose a challenge to the pathologist (and consequently the surgeon), depending on the tumor and tissue type. In addition, only a limited number of resected specimens can be analyzed during the operation. On the other hand, one would prefer to obtain instant diagnostics, at the time the surgeon is resecting the tumor. This exemplifies the need to develop innovative, smart, speedy, specific and sensitive diagnostic molecular approaches that result in more personalized information in support of surgical decision making *in situ* and in real time.

Over the last two decades, MS-based analytical techniques have revolutionized clinical research. Evolutions in speed, sensitivity, selectivity and specificity of MS instrumentation have accelerated the introduction in many clinical (diagnostic) laboratories. There, MS-based assays are routinely employed for the diagnostic evaluation of various liquid biopsies in clinical care. Direct tissue analysis with MSI is making its way into molecular pathology where it improves the accuracy of cellular phenotyping using molecular fingerprints on the proteome and the metabolome level at clinically relevant speeds [6].

MASS SPECTROMETRY IMAGING AND CLINICAL DIAGNOSTICS

In the past 20 years, MS imaging (MSI)—a label-free imaging modality—has demonstrated enormous potential for biomedical diagnostics and research [7–10]. Unlike immunohistochemistry, which targets specific protein markers, MSI does not require any previous knowledge of the samples and can be employed in an untargeted approach. MSI combines the molecular analytic properties of MS with the morphological, detailed spatial information obtained by traditional histology. MS-based molecular pathology in that way enables to bridge the gap between molecular information and tissue morphology. This correlation enables a deeper understanding of tissue complexity and pathophysiological processes underlying disease [11,12]. MSI has been demonstrated to be particularly powerful for disease-specific tissue classification, evaluation of therapy response and prognosis prediction [13]. Recent work has proven the power of such approach for discrimination of cancer and delineation of tumor margins [14]. Ambient ionization MS (imaging) requires minimal sample preparation and therefore provides unprecedented throughput compatible with clinical needs and especially with the time restrictions required for intraoperative tumor diagnosis and margin delineation [15,16]. A rapid molecular profiling approach can be used to provide real-time feedback to the surgeon. Currently, the majority of the tissue-typing research is based on lipid and metabolite profiling. The fact that the diagnostic procedure is conducted immediately at the source of the clinical problem, the tissue itself, and not on plasma, serum or urine samples offers a clear advantage. No remote dilution, processing, metabolism or degradation of relevant molecules has occurred yet and the metabolic profiles offer a direct insight in the molecular state of health and disease. Evidently, point-of-care diagnostics are a condition sine qua non for personalized and precise clinical treatments on many levels.

Here, we provide a concise review of recent innovative MS-based diagnostic approaches for intraoperative diagnostics and molecular pathology during a surgical procedure. This combination, the molecular operating room (OR), is foreseen to have a tremendous impact to improve surgical decision making in the very near future. We report examples of techniques being employed either *ex vivo* (i.e., perioperatively) and/or with a potential of being employed *in vivo* (i.e., intraoperatively). Each example highlighted in this section is illustrative of successful translational research.

High-Throughput Molecular Pathology for Perioperative Tissue Assessment

Molecular pathology is a subdiscipline in pathology that emphasizes the study and analysis of the state of a tissue section based on the examination of the molecular composition of cells, tissues and organs. It brings together in an interdisciplinary manner the study of the morphology and anatomy found in clinical samples with genomics, proteomics and metabolomics. This unique combination enables molecular pathologists to evaluate the molecular state of health and disease more accurately and more precise. In fact, it is a key concept of personalized medicine and the required diagnostics. Molecular pathology [17] was enabled and accelerated when technological innovations in rapid DNA sequencing and genome wide screening became commonly available. It has resulted in a paradigm change in clinical treatment and diagnosis. Now, molecular pathologists approach a disease differently in a true systems biology approach. A comprehensive understanding of molecular patterns of health and disease is needed to pave the way for personalized medicine and tissue regeneration. One barrier to predictive, personalized medicine is the lack of a comprehensive molecular understanding at the tissue level during the entire perioperative (preoperative, intraoperative and postoperative combined) process. As we grasp the astonishing complexity of biological systems (whether single cells or whole organisms), it becomes more and more evident that within this complexity lies the information needed to provide insight in the origin, progression and treatment of various diseases.

Ideally, a diagnosis is based on several, complementary information levels utilizing several classes of molecules, in a sensitive, specific and accurate manner. The introduction of molecular pathology as an intraoperative diagnostic and surgical guidance tool needs one other element: speed. Ideally, a resected tissue specimen, tissue biopsy or liquid biopsy is evaluated immediately during a surgical procedure on various molecular levels with such a systems approach. There is ample evidence that more precise intraoperative diagnostics improves patient outcome, reduces the number of recurrent surgical procedures and assists in the decision-making processes that determine optimal postoperative treatment. Unfortunately, many of the molecular pathological methods require more time than most surgical procedures. Most intraoperative diagnostic information is still obtained from cryopreserved specimens followed by H&E staining and morphological/anatomical assessment of the tissue sections. The introduction of MSI in molecular pathology rapidly changes this limitation, and shifts paradigms once again.

The best way to capture disease complexity is to chart and connect multilevel molecular information within a tissue using MS and correlative tissue classification algorithms. Charting this territory through the generation of molecular atlases from cells and tissue has become reality through the clinical implementation of imaging MS complemented with high-throughput omics approaches. These atlases now allow the unsupervised classification of the anatomy of a tissue section, as illustrated in Figure 7.12.2.

MSI [11] has matured since its inception in the 1960s [19] and is revolutionizing the concept of comprehensive tissue atlases. Direct tissue analysis on various molecular levels has become reality at analytical speeds compatible with clinical needs. Recent innovations have now enabled direct tissue phenotyping based on thousands of molecules in parallel on a surgically relevant timescale [6,20]. It is evident that a single analytical technology merely yields a subset of the molecular information needed to obtain an in-depth understanding of a clinical problem. Multimodal approaches enable the study of clinical samples at a variety of molecular and spatial scales. The distribution of several hundreds of molecules on the surface of complex (biological) surfaces can be determined directly in complementary MSI experiment with different desorption and ionization strategies. High-throughput, high-resolution MALDI techniques offer three-dimensional molecular information on the tissue level. The combination with tools from structural biology makes it possible to build diagnostic and prognostic models that allow for direct classification of disease. Several approaches using MSI have already been demonstrated to allow for tumor classification and staging [21–25] based on locally correlated molecular profiles. The classifiers and models developed in these studies are predominantly based on proteome, lipidome and metabolome profiles obtained from

FIGURE 7.12.2 Unsupervised classification (using principal component analysis) of three different tissue types in a frozen mouse brain section based on lipid mass spectrometry imaging. The different colors (red, green and blue) indicate the contributions of the individual distinctive principal components using a high-performance MALDI-MSI method. (From Belov, M.E. et al. *Anal. Chem.*, 89, 7493–7501, 2017.)

tissue microarrays or larger patient cohorts samples taken from biobanked material. They enable the analysis of disease subtypes, treatment response and cellular heterogeneity. These studies also demonstrate the potential to build databases and models of molecular tissue types that can be deployed in the preoperative pathological examination of needle biopsy material and intra- or postoperative examination of resected tissue. The feasibility of tumor delineation by classifying cancerous and not cancerous tissues based on molecular signatures in perioperative settings by DESI-MSI (Desorption ElectroSpray Ionisation) also was demonstrated for breast cancer where fatty acids, including oleic acid were more abundant in cancer tissue [25] and in gastric cancer [26].

Santagata et al. [27] demonstrated in 2004 that DESI-MS could guide brain tumor removal. A rapid detection of the tumor metabolite 2-hydroxyglutarate (2-HG) from tissue sections of surgically resected gliomas was readily performed under ambient conditions and without complex or time-consuming preparation. Tumor delineation was achieved—using DESI-MS—by monitoring a specific onco-metabolite (2-hydroxybutyrate) generated exclusively by isocitrate dehydrogenase 1-mutant gliomas. In their approach, frozen sections of resected gliomas performed to control the free margin analyzed by DESI was correlated with MRI reconstruction of tumor to improve decision making, showing how ambient MSI techniques can bring a real improvement of intraoperative decisions.

The ability to reveal molecular changes in heterogeneous tissues prior to visible morphological changes is a clear advantage of label-free molecular imaging techniques in perioperative molecular pathology. This same type of information has recently been applied during surgical procedures and enabled direct tissue typing when surgical instruments are combined with MS analysis.

Intraoperative and *In Situ* Margin Assessment Techniques Based on Molecular Tissue Profiling

Recently, ambient ionization MS has emerged for cancer diagnosis and surgical margin evaluation [14] and is foreseen as revolutionary approach to provide molecular information in real-time from tissues. Analysis of tissue can be done either from sectioned tissue or directly *in situ* without processing of the tissue. Rapid evaporative ionization MS (REIMS) was developed in 2010 by the group of Zoltan Takats for rapid and *in vivo* classification of human tissue through MS analysis of aerosols released during electrosurgical dissection by mean of electrical scalpel or forceps [28,29]. This is the first ambient technique introduced with a direct potential of being introduced in surgical setting. The beauty of the REIMS technique is the possible combination to surgical devices without any modification of the later. The surgical procedure remains the same, and surgeons do not need extra training to perform their operation. During electrosurgery, an electrical current is used to heat the tissue very rapidly. The tissue is vaporized and a smoke is generated. Commercial standard electrosurgical handpieces incorporate a tubing (generally between 2- and 5-m long) connected to an extraction system (usually HEPA-filtered, in the operating room) to aspirate the smoke generated. For on-line chemical analysis of the smoke, the tubing of the handpiece is simply derived and connected a Venturi gas jet pump mounted into the source interface of the mass analyzer [30]. MS measures the mass-to-charge ratio of molecular ions generated at the site of dissection. The smoke aspirated is rich in molecular information and contains tissue-specific profiles discriminating between tumor and surround tissue, which is typically not obvious to the naked eye during surgery. Through this rapid and on-line analysis of electrosurgical aerosol, one can dream toward real-time margin assessment. The data collected in real-time is compared with a reference library to determine the type of tissue being cut (i.e., tumor vs. non-tumor) and the feedback is provided to the surgeon within seconds, at the time of the resection. So far, the reference library is built using an *ex vivo* analysis of tissue samples, which permits the creation of a spectral database for prospective use. Validation of the database is done by histological examination of expert pathologists [31].

REIMS has been further developed by utilizing surface induced dissociation and was integrated with an endoscopic polypectomy snare to allow *in vivo* analysis of the gastrointestinal tract. The classification performance of this novel endoscopic REIMS method was tested *in vivo* and Balog et al. demonstrated the device to be capable of differentiating between healthy layers of the intestinal wall, cancer, and adenomatous polyps based on the REIMS fingerprint of each tissue type [32]. Additional investigations with REIMS demonstrated high diagnostic accuracy for tumor type and for established histological features of poor prognostic outcome in CRC based on a multivariate analysis of the mucosal lipidome [31].

Instead of using a cutting electrode, thermal ablation of surgical tissues can also be achieved by means of a laser. Laser desorption using lasers both in the ultraviolet and far-infrared wavelength regimes [33]. Resonant infrared laser ablation allows the production of ions by excitation of O-H bonds present in water molecules, naturally present in most biological tissues. It was demonstrated that the generated molecular patterns are specific to the cell phenotypes and benign versus cancer regions of patient biopsies can be easily differentiated. One of the strength of laser surgery is the possibility of performing direct sampling *in vivo* and in real time under minimally invasive conditions [34]. This opens large possibilities for application in a wide variety of clinical area, such as in dermatology because skin can be sampled with minimal damage and pain [34]. Commercial and surgical CO_2 lasers have been employed for combination with MS and found to yield sufficiently high ion current during normal use. The principal component analysis–based real-time data analysis method was developed for the quasi real-time identification of mass spectra. Performance of the system was demonstrated in the case of various malignant tumors of the gastrointestinal tract [32].

Real-time analysis of brain tissue by direct combination of cavitron ultrasonic surgical aspiration (CUSA) and sonic spray MS was also described [33,35]. CUSA is widely used for brain and liver cancer surgery. In its normal operation mode, the surgical device employs ultrasound to disintegrate tissues. The resulting tissue debris is continuously transported from the surgical site to the mass analyzer by a continuous drain of an aqueous solution. The liquefied tissue debris is aspirated and nebulized for sonic spray ionization-type gas-phase ion production using a Venturi air jet pump following existing technology [33]. This approach enables direct real-time MS analysis. The mass spectra generated—also based on lipid signature—were found to be highly specific to the different tissue types analyzed. Multivariate data analysis was developed for real-time tissue identification in a surgical environment and was successful for differentiating of *ex vivo* human tissues such as astrocytomas, meningiomas, metastatic brain tumors, and healthy brain tissue could be achieved [33]. Recently, another approach that utilizes liquid sampling and direct analysis during surgery, the MasSpec Pen [36], was employed to *ex vivo* screen 20 human cancer thin tissue sections and 253 human patient tissue samples including normal and cancerous tissues from breast, lung, thyroid, and ovary. The approach delivers a discrete water droplet to a tissue surface to extract biomolecules prior to rapid MS analysis. This approach shows promise for direct profiling of tumors and margins in resected tissue during surgery prior to the routine pathology.

THE FUTURE: IMAGING DATA AND PERSONALIZED HEALTH CARE

It is evident that personalized medicine based on molecular imaging modalities has entered the era of Big Data. The multidisciplinary nature of translational molecular imaging results in different data streams contributed by different imaging experts. All of this data can contribute to an increased precision of a treatment, provided systems exist that combine the molecular and clinical data to underpin clinical decision making. Radiology images can already be employed for patient classification in the field of radiomics [37]. Numerous groups in MSI are deploying innovative patient classification and stratification strategies to find novel diagnostic or prognostic markers [38]. Active learning strategies are used to segment MS-based molecular images and can already automatically annotate and classify different cellular phenotypes within a single tissue section taking advantage of available information in other domains [39,40]. All of these tools require input from clinical diagnostics disciplines, such as pathology and clinical diagnostics, to define, validate and cross-reference the cellular phenotypes. Similar classification strategies are rolled out to facilitate rapid intraoperative diagnostics. It is becoming clear that de mathematical models and the corresponding databases are the economic commodities of the future. In the near future, we will see an increase of multicenter studies in which these broad panel-based molecular patterns are validated to accurately define a personal state of health and disease. This is the translational future for molecular imaging in personalized health care.

REFERENCES

1. Torre, L.A., Bray, F., Siegel, R.L., Ferlay, J., Lortet-Tieulent, J., Jemal, A. Global cancer statistics, 2012. *CA Cancer J Clin.* **65**, 87–108 (2015).
2. Torre, L.A., Siegel, R.L., Ward, E.M., Jemal, A. Global cancer incidence and mortality rates and trends—an update. *Cancer Epidemiol Biomarkers Prev.* **25**, 16–27 (2016).
3. Weston, A.D., Hood, L. Systems biology, proteomics, and the future of health care: Toward predictive, preventative, and personalized medicine. *J Proteome Res.* **3**, 179–196 (2004).
4. Barrio, A.V., Morrow, M. Appropriate margin for lumpectomy excision of invasive breast cancer. *Chin Clin Oncol.* **5**, 35 (2016).
5. St John, E.R., Al-Khudairi, R., Ashrafian, H., Athanasiou, T., Takats, Z., Hadjiminas, D.J. et al. Diagnostic accuracy of intraoperative techniques for margin assessment in breast cancer surgery a meta-analysis. *Ann Surg.* **265**, 300–310 (2017).
6. Potocnik, N.O., Porta, T., Becker, M., Heeren, R.M.A., Ellis, S.R. Use of advantageous, volatile matrices enabled by next-generation high-speed matrix-assisted laser desorption/ionization time-of-flight imaging employing a scanning laser beam. *Rapid Commun Mass Sp.* **29**, 2195–2203 (2015).
7. Norris, J.L., Caprioli, R.M. Analysis of tissue specimens by matrix-assisted laser desorption/ionization imaging mass spectrometry in biological and clinical research. *Chem Rev.* **113**, 2309–2342 (2013).
8. Ko, K.H., Han, N.Y., Kwon, C.I., Lee, H.K., Park, J.M., Kim, E.H. et al. Recent advances in molecular imaging of premalignant gastrointestinal lesions and future application for early detection of barrett esophagus. *Clin Endosc.* **47**, 7–14 (2014).
9. Rodrigo, M.A., Zitka, O., Krizkova, S., Moulick, A., Adam, V., Kizek, R. MALDI-TOF MS as evolving cancer diagnostic tool: A review. *J Pharm Biomed Ana.* **95**, 245–255 (2014).

10. Vaysse, P.M., Heeren, R.M.A., Porta, T., Balluff, B. Mass spectrometry imaging for clinical research—Latest developments, applications, and current limitations. *Analyst.* **142**, 2690–2712 (2017).
11. Chughtai, K., Heeren, R.M. Mass spectrometric imaging for biomedical tissue analysis. *Chem Rev.* **110**, 3237–3277 (2010).
12. Aichler, M., Walch, A. MALDI Imaging mass spectrometry: Current frontiers and perspectives in pathology research and practice. *Lab Invest*; A journal of technical methods and pathology. **95**, 422–431 (2015).
13. Mascini, N.E., Eijkel, G.B., ter Brugge, P., Jonkers, J., Wesseling, J., Heeren, R.M.A. The use of mass spectrometry imaging to predict treatment response of patient-derived xenograft models of triple-negative breast cancer. *J Proteome Res.* **14**, 1069–1075 (2015).
14. IFA, D.R., Eberlin, L.S. Ambient ionization mass spectrometry for cancer diagnosis and surgical margin evaluation. *Clin Chem.* **62**, 111–123 (2016).
15. Calligaris, D., Norton, I., Feldman, D.R., Ide, J.L., Dunn, I.F., Eberlin, L.S. et al. Mass spectrometry imaging as a tool for surgical decision-making. *J Mass Spectrom.* **48**, 1178–1187 (2013).
16. Eberlin, L.S., Norton, I., Dill, A.L., Golby, A.J., Ligon, K.L., Santagata, S. et al. Classifying human brain tumors by lipid imaging with mass spectrometry. *Cancer Res.* **72**, 645–654 (2012).
17. Harris, T.J.R., McCormick, F. The molecular pathology of cancer. *Nat Rev Clin. Oncol.* **7**, 251–265 (2010).
18. Belov, M.E., Ellis, S.R., Dilillo, M., Paine, M.R.L., Danielson, W.F., Anderson, G.A. et al. Design and performance of a novel interface for combined matrix-assisted laser desorption ionization at elevated pressure and electrospray lonization with orbitrap mass spectrometry. *Anal Chem.* **89**, 7493–7501 (2017).
19. Heeren, R.M.A. Getting the picture: The coming of age of imaging MS. *Int J Mass Spectrom.* **377**, 672–680 (2015).
20. Norris, J.L., Caprioli, R.M. Imaging mass spectrometry: A new tool for pathology in a molecular age. *Proteomics Clin Appl.* **7**, 733–738 (2013).
21. Djidja, M.-C., Claude, E., Snel, M.F., Francese, S., Scriven, P., Carolan, V. et al. Novel molecular tumour classification using MALDI–mass spectrometry imaging of tissue micro-array. *Anal Bioanal Chem.* **397**, 587–601 (2010).
22. Hinsch, A., Buchholz, M., Odinga, S., Borkowski, C., Koop, C., Izbicki, J.R. et al. MALDI imaging mass spectrometry reveals multiple clinically relevant masses in colorectal cancer using large-scale tissue microarrays. *J Mass Spectrom.* **52**, 165–173 (2017).
23. Quaas, A., Bahar, A.S., von Loga, K., Seddiqi, A.S., Singer, J.M., Omidi, M. et al. MALDI imaging on large-scale tissue microarrays identifies molecular features associated with tumour phenotype in oesophageal cancer. *Histopathology.* **63**, 455–462 (2013).
24. Abbassi-Ghadi, N., Veselkov, K., Kumar, S., Huang, J., Jones, E., Strittmatter, N. et al. Discrimination of lymph node metastases using desorption electrospray ionisation-mass spectrometry imaging. *Chem Commun.* **50**, 3661–3664 (2014).
25. Calligaris, D., Caragacianu, D., Liu, X., Norton, I., Thompson, C.J., Richardson, A.L. et al. Application of desorption electrospray ionization mass spectrometry imaging in breast cancer margin analysis. *Proc Natl Acad Sci.* **111**, 15184–15189 (2014).
26. Eberlin, L.S., Tibshirani, R.J., Zhang, J., Longacre, T.A., Berry, G.J., Bingham, D.B. et al. Molecular assessment of surgical-resection margins of gastric cancer by mass-spectrometric imaging. *Proc Natl Acad Sci.* **111**, 2436–2441 (2014).
27. Santagata, S., Eberlin, L., Norton, I., Ide, J., Feldman, D., Liu, X.H. et al. Metabolite-imaging mass spectrometry to guide brain surgery. *J Neuropath Exp Neur.* **72**, 575–576 (2013).
28. Balog, J., Szaniszlo, T., Schaefer, K.C., Denes, J., Lopata, A., Godorhazy, L. et al. Identification of biological tissues by rapid evaporative ionization mass spectrometry. *Anal Chem.* **82**, 7343–7350 (2010).
29. Balog, J., Sasi-Szabo, L., Kinross, J., Lewis, M.R., Muirhead, L.J., Veselkov, K. et al. Intraoperative tissue identification using rapid evaporative ionization mass spectrometry. *Sci Transl Med.* **5**, (2013).
30. Szabo, L.S., Balog, J., Szaniszlo, T., Szalay, D., Godorhazy, L., Toth, M. et al. Rapid evaporative ionization mass spectrometry (Reims) as a new method for intraoperative in-situ, real time identification of gastrointestinal malignancies. *Ann Oncol.* **21**, 102–102 (2010).
31. St John, E.R., Balog, J., McKenzie, J.S., Rossi, M., Covington, A., Muirhead, L. et al. Rapid evaporative ionisation mass spectrometry of electrosurgical vapours for the identification of breast pathology: Towards an intelligent knife for breast cancer surgery. *Breast Cancer Res.* **19**, (2017).
32. Balog, J., Kumar, S., Alexander, J., Golf, O., Huang, J.Z., Wiggins, T. et al. In vivo endoscopic tissue identification by Rapid Evaporative Ionization Mass Spectrometry (REIMS). *Angew Chem Int Edit.* **54**, 11059–11062 (2015).
33. Schafer, K.C., Szaniszlo, T., Gunther, S., Balog, J., Denes, J., Keseru, M. et al. In situ, real-time identification of biological tissues by ultraviolet and infrared laser desorption ionization mass spectrometry. *Anal Chem.* **83**, 1632–1640 (2011).
34. Fatou, B., Saudemont, P., Leblanc, E., Vinatier, D., Mesdag, V., Wisztorski, M. et al. In vivo real-time mass spectrometry for guided surgery application. *Sci Rep-UK.* **6**, (2016).
35. Huang, J., Gao, Y., Zhuo, H., Zhang, J., Ma, X. Can ionization mass spectrometry coupled with ultrasonic scalpel a fine detection method for intraoperative pathological analysis? *Med Hypotheses.* **84**, 509–510 (2015).
36. Zhang, J.L., Rector, J., Lin, J.Q., Young, J.H., Sans, M., Katta, N. et al. Nondestructive tissue analysis for ex vivo and in vivo cancer diagnosis using a handheld mass spectrometry system. *Sci Transl Med.* **9**, (2017).
37. Lambin, P., Rios-Velazquez, E., Leijenaar, R., Carvalho, S., van Stiphout, R.G.P.M., Granton, P. et al. Radiomics: Extracting more information from medical images using advanced feature analysis. *Eur J Cancer.* **48**, 441–446 (2012).
38. Balluff, B. MALDI imaging mass spectrometry for proteomic segmentation of tumor heterogeneity in gastric cancer tissues. *Virchows Arch.* **463**, 106–106 (2013).
39. Hanselmann, M., Kirchner, M., Renard, B.Y., Amstalden, E.R., Glunde, K., Heeren, R.M.A. et al. Concise representation of mass spectrometry images by probabilistic latent semantic analysis. *Anal Chem.* **80**, 9649–9658 (2008).
40. Hanselmann, M., Roder, J., Kothe, U., Renard, B.Y., Heeren, R.M.A., Hamprecht, F.A. Active learning for convenient annotation and classification of secondary ion mass spectrometry images. *Anal Chem.* **85**, 147–155 (2013).

Insights of Personalized Medicine from the Gut Microbiome

7.13

Lessons Learned from Celiac Disease

Gloria Serena, Anna Sapone, and Alessio Fasano

The human microbiome is composed of the collection of microorganisms (10^{14}) that coexist with their host and it colonizes virtually every surface of the human body exposed to the external environment [1]. The largest and most complex microbial ecosystem is located in the gastrointestinal (GI) tract. These intestinal commensal microorganisms contribute to nutrients absorption, food fermentation, stimulation of the host immune system, and competition against enteric pathogens [2]. The studies regarding the GI microbiome composition have been for long time limited by the lack of proper techniques to fully appreciate its complexity. The majority of microorganisms, and specifically the bacteria colonizing the intestine, are in fact, arduous to culture because of their strictly anaerobic properties. The development of NGS techniques such as 454 pyrosequencing and Illumina sequencing have permitted the detection of bacteria that were not culturable through the analysis of their hypervariable regions on the ribosomal 16s gene [3]. The closer examination of the intestinal microflora has highlighted its important role in modulating the host health and the importance of how its composition can be influenced by external factors. It has been shown that the health of an adult subject is closely dependent the gut microbiome composition starting from infancy and that it can be influenced by elements such as delivery mode (vaginal vs. cesarean section), feeding regimen (breast feeding vs. formula), maternal prenatal diet, psychological factors during pregnancy and use of antibiotics [4]. Given its role in modulating the host health it is not surprising that intestinal dysbiosis (alterations in the microbiome composition) has been associated with several conditions and diseases such as obesity [5], inflammatory bowel diseases (IBD) [6], and type 1 and 2 diabetes [7], just to name a few.

Studies on obese animals have highlighted the important role that the diet plays in shaping the intestinal microbiome composition. Walker et al. have reported that a fat and protein rich diet triggers a microflora enriched in *Bacteroidetes* enterotypes, whereas *Prevotella* species are abundant when a carbohydrate-rich diet is introduced [8]. Furthermore, *in vivo* studies have shown that the microbiome of obese animals is characterized by specific bacteria taxa [9,10] as compared with lean animals: Obese mice showed a high abundance of *Firmicutes* correlated with a reduction of *Bacteroides* and *Clostridium perfingens* [11,12].

Studies on patients with Crohn's disease have reported an increase in *Bacteroidetes* and *Firmicutes* in Crohn's subjects as compared with healthy controls [13]. Furthermore, the severity of the disease has been associated with a decreased diversity of *Bacteroidetes* [14]. Interestingly these studies have also highlighted the contribution that the sampling location and the site of disease inflammation have on the microbiome composition [15].

It is now confirmed that even diseases that are not strictly associated with the GI tract such as type 1 diabetes can be correlated with intestinal dysbiosis. Human studies have shown that the fecal microbiome of type 1 diabetes patients is characterized by a reduction in the *Actinobacteria:Firmicutes/Bactreroidetes* ratio and short-chain fatty acids producer bacteria such as *Clostridia* [16,17]. Additionally, alterations in the microbiome composition have been reported also in the small intestine of type 1 diabetes patients with specific bacteria taxa associated with expression of pro-inflammatory cytokines [18].

The fundamental contribution that the gut microbiome has on intestinal mucosal homeostasis suggests that its modulation may represent an optimal tool to prevent, diagnose and treat conditions associated with dysbiosis. Each individual is characterized by a unique intestinal microflora that is differently shaped depending on the host genomic material and external factors such as stress, diet and use of antibiotics. Several studies have already shown the impact that microbiome modulation can have on human health through probiotics use [5,19,20]. In addition to the therapeutic properties that a healthy microbiome has, detection of specific bacteria species and their bioproducts could be used as biomarkers for diseases characterized by dysbiosis. Following the Human Microbiome Project (HMP), the protocols and guidelines to study and analyze commensal bacteria are increasingly precise, standardized and less expensive, therefore making these tasks more accessible [21]. Bacteria belonging to the intestinal microflora population can be easily detected not only in biopsies specimens, but also in fecal and blood samples [21,22]. Although the research studies exploring the microbiome has been rapidly expanding in the last decades, the exact role that dysbiosis plays in the onset of different diseases is still under debate. Although the current research is focused on describing the differences in microbiome compositions, it does not evaluate the mechanistic and causative links between dysbiosis and different disorders. Prospective mechanistic studies are necessary to better understand the role that the microbiome plays in the development of diseases and therefore apply this mechanistic understanding to useful intervention and prevention.

Given its unique characteristics, celiac disease (CD) represents a good example of a condition in which the microbiome analysis may be used for causative studies in order to develop personalized medicine tools.

CD is an immune-mediated enteropathy occurring in genetically predisposed individuals upon ingestion of gluten [23]. The HLA haplotypes DQ2 and/or DQ8 have been shown to be necessary, but not sufficient, for the development of the disease with 95% of patients carrying an HLA DQ2 and 5% an HLA DQ8 [24].

The cascade of events that lead to the development of CD initiates when gliadin peptides, one of the proteins component of gluten, interact with the intestinal epithelium and induce the release of zonulin, a regulator of intercellular tight junctions [25,26]. The consequent increase in intestinal permeability contributes to the translocation of gliadin into the lamina propria therefore triggering both an innate and adaptive immune response [27]. Gliadin has a neutrophilic chemoattractant property and induces the release of IL-8 and IL-15 [28–31]. Another important element of CD pathogenesis is represented by $CD8^+$ intraepithelial lymphocytes (IELs) that contribute to the intestinal damage through their cytotoxic activity [32]. Following innate immune-mediated cellular insult with subsequent release of intracellular tissue transglutaminase (tTG), gliadin is deamidated and then presented by antigen presenting cells to CD4 lymphocytes [33,34]. In turn, this activation leads to both type 1 T-helper cells (Th1) and Th17 immune activation with subsequent production of pro-inflammatory cytokines such as IL-12, interferon-gamma (IFNγ) and IL-17A [35,36]. Finally, the activation of B cells leads to their maturation in response to gluten intake in a fluctuating pattern and triggers the production of anti transglutaminase 2 and anti-gliadin antibodies.

It has been suggested that environmental factors, other than gluten, may contribute to the loss of tolerance in genetically predisposed individuals [37]. This hypothesis is based on data showing a dramatic increase of CD incidence in the last decades [38,39]. Given its rapidity over time, this increment cannot be attributed exclusively to genetic changes in the population. Furthermore, it is now well recognized that CD can develop later in life independently of the timing of gluten introduction in the diet. Combined, these epidemiological data strongly suggest that genetic predisposition and exposure to gluten are necessary but not sufficient to lead to the break of tolerance and that additional environmental factors are probably necessary as modulators of oral tolerance [40].

Among the different elements that may contribute to the onset of CD, alterations in the microbiome composition appear to be one of the most probable candidates. Several groups have reported that patients with CD are characterized by a certain level of intestinal dysbiosis. Studies on pediatric CD patients, for example, have shown respectively a decrease in the total *Bifidobacterium* species [41], an increase in *Clostridium* and *Prevotella* [42] and a lower ratio between *Lactobacillus* and *Bifidobacterium* to *Bacteroides* [43]. Lower microbial diversity has been described among adult CD patients [44] and two studies have reported an increased production of short chain fatty acids such as butyrate and propionate in fecal samples from active CD patients [45,46]. Interestingly, changes in the microbiota composition have been associated with clinical manifestations and severity of the disease [44].

The exact role of the microbiome in CD's onset has not been fully established yet, and it is still unclear if the dysbiosis that characterizes active CD patients has to be considered cause or a consequence for the loss of tolerance to gluten. *In vitro* studies have shown that pro-inflammatory cytokines abundantly produced in CD patients upon ingestion of gluten, such as IL-15, are able to alter the microbial composition in the intestinal compartment [47]. Conversely, the microbiome and its derived metabolites are able to modulate the immune response, therefore suggesting a causative role of the microbiome in the onset of CD [48].

To this day, a life-long elimination of gluten from the diet represents the only treatment available for CD [40]. This diet interrupts the immune response activation and allows the complete recovery from the clinical manifestations. In a small group of patients, however, the gluten free diet is not sufficient to restore the normal intestine architecture. Subjects that do not respond to gluten free diet are defined as affected by nonresponsive CD, and a subgroup of them are true refractory celiac disease (RCD) cases. Two main types of RCD have been described: RCDI patients have IELs with normal phenotype and no clonal T cell receptor (TCR) rearrangement. RCDII patients present IELs with abnormal phenotype and exhibit severe epithelial lesions and a high risk to develop malignancy. Although pediatric CD patients usually respond well to the gluten free diet, RCD appears to be more common in adult subjects over 50 years old [49]. Given the higher number of CD diagnosis among the adult population, RCD is becoming a worrisome issue. Studies have reported that delay in diagnosis or lack of recovery in CD patients may be associated to increased mortality, primarily because of malignancy [27] and untreated CD has been correlated to the onset of several comorbidities [50–52].

The challenges concerning the treatment of CD patients that have recently emerged have highlighted the need of new tools in the field.

Given the role that the microbiome has in maintaining the intestinal immune homeostasis and the dysbiosis that characterize CD subjects, regulating its composition could represent a valid therapy to be added to the gluten free diet for patients with RCD, at least for RCDI. Some groups have already reported encouraging results. *In vitro* experiments have shown that the intestinal inflammatory response induced by gliadin is attenuated after treatment with *Bifidobacteria* [53,54]. Furthermore, a recent study has reported a full recovery of a patient with RCDII after a fecal transplant [55].

In addition to their therapeutic potential, bacteria belonging to the intestinal microflora could be used as possible biomarkers to detect the initial phase of the disease in order to prevent the complete loss of tolerance. Data describing dysbiosis in genetically predisposed infants support the hypothesis that alterations in the intestinal microflora may actively contribute to the development of CD [56]. These infants showed a decreased abundance of *Bacteroides* as compared with the control group. Furthermore, the genetically predisposed kids that developed autoimmune diseases presented an increased production of lactate that correlated with enrichment in *Lactobacilli* spp. during the preclinical phase of the disease, followed, during the active phase, by a drop of lactate, an increase in butyrate production and a general reduction in microbial diversity. Given the changes in the microbiome composition prior the onset of the disease and the effect that metabolites such as lactate and butyrate have in shaping the immune response [57–60], it is plausible to hypothesize that intestinal dysbiosis may have a causative role in the loss of tolerance to gluten.

One mechanistic link between the microbiome/immune system crosstalk and the development of CD has been recently described for the first time [48]. T regulatory cells (Treg) represent a crucial component of the immune system machinery by preventing onset of autoimmune diseases and chronic inflammation [61,62]. Furthermore, they have been shown to play an important role in the onset of oral tolerance [63]. Their function and differentiation is regulated by the expression of transcription factor FOXP3 [64,65]. Two main functionally different isoforms of FOXP3 have been described: the full-length isoform (FL) is fundamental for the suppressive activity of Treg cells and their differentiation, whereas the FOXP3 isoform Delta 2 (Δ2) appears to be less efficient in down regulating the immune response against pro-inflammatory Th17 cells [66–68]. The expression of two different isoforms is the direct product of alternative splicing, a strictly regulated posttranscriptional modification that occurs under specific circumstances [69]. Despite its "reduced efficiency," FOXP3 Δ2 is expressed in similar quantity to its counterpart FL in human Treg cells and its exact function is still unknown [66]. Conversely to other autoimmune diseases, CD is characterized by a higher percentage of intestinal Treg cells as compared with healthy controls [63,70]. Interestingly it has been shown that celiac patients also show an increased expression of FOXP3 Δ2 and that the FL/Δ2 ratio in these patients is skewed toward a less functional Δ2 phenotype [48]. Given the inability of FOXP3 Δ2 in suppressing a Th17 pro-inflammatory response, these findings suggest that epigenetic changes on Treg cells may play a role in the loss of tolerance to gluten.

It has been recently reported that a pro-inflammatory microenvironment, characterized by high concentration of IFNγ, associated with specific microbiome derived metabolites such as butyrate, is able to modulate the expression of FOXP3 isoforms in celiac patients by leading to an over expression of FOXP3 Δ2. The same effect was not seen in healthy controls, therefore highlighting the presence of innate differences between healthy and celiac individuals in the way immune cells respond to the same microenvironment. Given the presence of high concentration of butyrate in the active phase of CD [45,56], these findings give new insights on how the microbiome may modulate the immune response and therefore contribute to the pathogenesis of CD. The higher expression of FOXP3 Δ2 triggered by the short chain fatty acid could, in fact, negatively affect the capability of Treg cells of suppressing Th17 driven immune response therefore leading to chronic inflammation.

Lactate, another well-described microbial metabolite, has been shown to trigger higher expression of both FOXP3 isoforms in celiac patients as compared with controls [48]. This effect appears to be mediated by antigen presenting cells such as macrophage. The knowledge of the effect that lactate has on Treg cells in celiac patients combined with its described abundance in genetically predisposed infants during the preclinical phase of CD, followed by a dramatic drop during the active stage [56], suggest that changes in the microbiome composition preceding the onset of the disease may trigger the recruitment of Treg cells in the small intestine, prime the immune system and eventually lead to the loss of tolerance to gluten (Figure 7.13.1).

These findings mechanistically link for the first time dysbiosis with specific epigenetic changes in the immune system. Based on these results, it is tempting to infer that monitoring and modulating the microbiome composition in infants that are genetically susceptible to CD may represent a possible way to treat and prevent the disease. The knowledge of the environmental trigger (gluten), of the genetic predisposition (HLA DQ2/DQ8) and of the effect that microbial products have on the regulation of the immune system makes CD a unique example of autoimmune disease for which personalized medicine may become a reality.

The numerous *in vivo* and clinical studies showing that probiotics can be used to successfully prevent infections [71] suggest that targeting specific changes in the microbiome composition may be sufficient to put a brake on the cascade of events that lead to the loss of tolerance. Furthermore, given that the infants' microbiome depends on the delivery mode, the mother's microbiome and the diet [37], its final composition could be modulated at early age. Retrospective studies looking at how these factors are able to contribute to the final microbiome phenotype are needed to better understand the changes that lead to the disease.

Finally, adopting bacteria from the microflora as a biomarker in CD patients could represent a valid tool to confirm diagnosis and to follow the efficacy of the treatment in follow up patients.

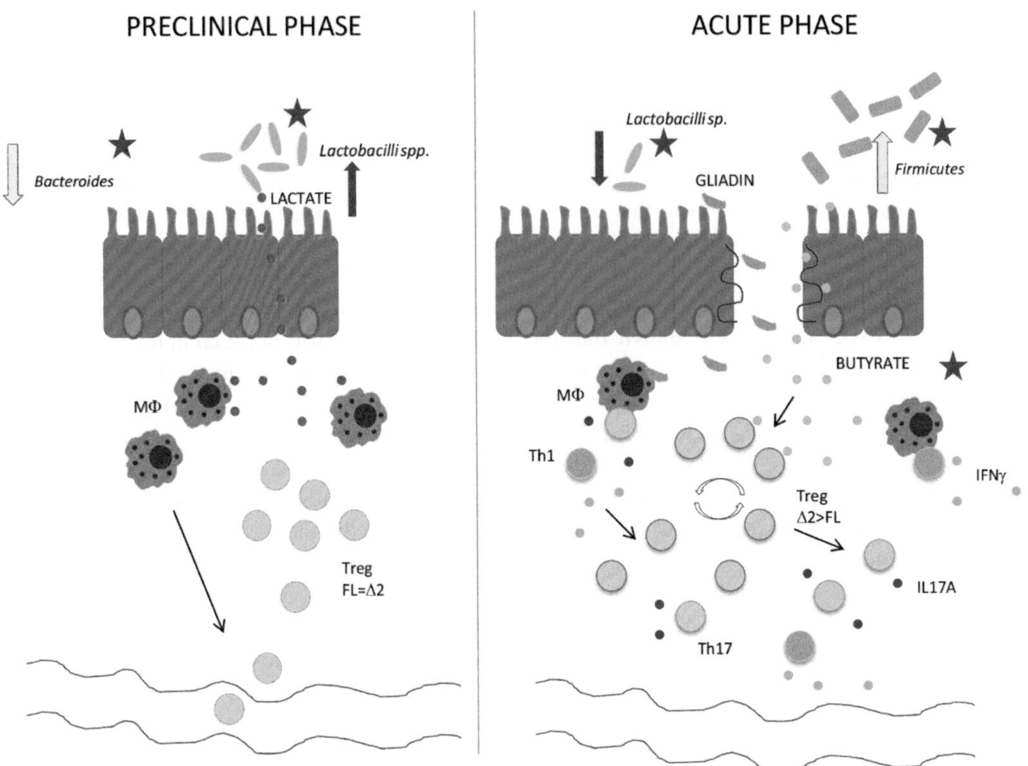

FIGURE 7.13.1 Proposed contribution of dysbiosis to celiac disease (CD). Graphic representation of how changes in microbial composition and bioproducts may play a role in the loss of tolerance to gluten and where specifically targeted approaches may help to prevent the onset of CD (red stars). In genetically predisposed individuals, abundance of the *Bacteroides phylum* has been show to be dramatically reduced (56). Furthermore, the individuals that developed CD were characterized, during the preclinical phase of the disease, by a high production of lactate associated with an increased presence of *Lactobacilli* spp. Through interaction with immune cells (probably antigen presenting cells), lactate is able to indirectly recruit Treg cell from the blood stream to the lamina propria of the intestine, therefore triggering an increased expression of FOXP3 (48). When gluten is introduced in the diet, the interaction between gliadin peptides and epithelial cells lead to disruption of tight junctions and increased intestinal permeability (25,26) that will initiate the inflammatory process. During the acute phase of the CD, the production of lactate is dramatically reduced, while the short chain fatty acids producer bacteria belonging to the *Firmicutes* phylum increase (42,56), with subsequent high production of butyrate (45,46). This short chain fatty acid, in combination with the pro-inflammatory cytokine interferon-gamma (IFNγ), has been shown to directly trigger the expression of FOXP3 D2 on the intestinal resident Treg cells, therefore reducing their suppressive function (48). The highly pro-inflammatory environment creates a positive feedback effect that will lead to high numbers of defectiveT regulatory (Treg) cells and to chronic inflammation.

The gold standard practice for diagnosing CD is represented by a duodenal biopsy with increased number of IELs, elongated crypts and villous atrophy. Additionally, the serologic analysis represents another useful tool used to test for CD. The serological testing is commonly used not only at time of diagnosis, but also during follow up visits to confirm the adherence and the efficacy of the gluten free diet. The array of "celiac" antibodies is represented by IgA and IgG anti-tissue transglutaminases, endomysial antibodies and IgG anti-deamidated gliadin peptide antibodies [40]. Although several studies have demonstrated the high specificity and sensitivity of these tests, the lack of standardization in regard to the values attributable to a positive tests and the differences emerged among tests used by different institutions have raised some concerns about their reliability to diagnose CD without being associated to a positive endoscopy. Furthermore, in rare cases patients with active CD have been described to have negative serology [72].

The microbiome is unique to each individual and can be considered as an additional "fingerprint" of the host. Its bioproducts represent a dynamic and fundamental component for the intestinal homeostasis and changes in its bacteria composition have been associated with several diseases. Given the unique properties that CD has in terms of autoimmune disease (known external trigger and genetic predisposition), it represents the perfect tool to better investigate how the analysis of the microbiome may play a role in the field of personalized medicine. Furthermore, a better understanding of the players that contribute to the pathogenesis of CD would also give new insights on how to prevent and treat other autoimmune diseases in which dysbiosis is involved.

REFERENCES

1. Eckburg PB, Bik EM, Bernstein CN, Purdom E, Dethlefsen L, Sargent M, Gill SR, Nelson KE, Relman DA. Diversity of the human intestinal microbial flora. *Science* 2005; **308**:1635–1638.
2. Mariat D, Firmesse O, Levenez F, Guimaraes V, Sokol H, Dore J, Corthier G, Furet JP. The firmicutes/bacteroidetes ratio of the human microbiota changes with age. *BMC Microbiol* 2009; **9**:123.
3. Deurenberg RH, Bathoorn E, Chlebowicz MA, Couto N, Ferdous M, Garcia-Cobos S, Kooistra-Smid AM et al., Application of next generation sequencing in clinical microbiology and infection prevention. *J Biotechnol* 2017; **243**:16–24.
4. Moya-Perez A, Luczynski P, Renes IB, Wang S, Borre Y, Anthony Ryan C, Knol J, Stanton C, Dinan TG, Cryan JF. Intervention strategies for cesarean section-induced alterations in the microbiota-gut-brain axis. *Nutr Rev* 2017; **75**:225–240.
5. Dahiya DK, Renuka, Puniya M, Shandilya UK, Dhewa T, Kumar N, Kumar S, Puniya AK, Shukla P. Gut microbiota modulation and its relationship with obesity using prebiotic fibers and probiotics: A review. *Front Microbiol* 2017; **8**:563.
6. Round JL, Mazmanian SK. The gut microbiota shapes intestinal immune responses during health and disease. *Nat Rev Immunol* 2009; **9**:313–323.
7. de Goffau MC, Luopajarvi K, Knip M, Ilonen J, Ruohtula T, Harkonen T, Orivuori L et al., Fecal microbiota composition differs between children with beta-cell autoimmunity and those without. *Diabetes* 2013; **62**:1238–1244.
8. Walker AW, Ince J, Duncan SH, Webster LM, Holtrop G, Ze X, Brown D et al., Dominant and diet-responsive groups of bacteria within the human colonic microbiota. *ISME J* 2011; **5**:220–230.
9. Goodrich JK, Waters JL, Poole AC, Sutter JL, Koren O, Blekhman R, Beaumont M et al., Human genetics shape the gut microbiome. *Cell* 2014; **159**:789–799.
10. Conterno L, Fava F, Viola R, Tuohy KM. Obesity and the gut microbiota: Does up-regulating colonic fermentation protect against obesity and metabolic disease? *Genes Nutr* 2011; **6**:241–260.
11. Zuo HJ, Xie ZM, Zhang WW, Li YR, Wang W, Ding XB, Pei XF. Gut bacteria alteration in obese people and its relationship with gene polymorphism. *World J Gastroenterol* 2011; **17**:1076–1081.
12. Turnbaugh PJ, Ley RE, Mahowald MA, Magrini V, Mardis ER, Gordon JI. An obesity-associated gut microbiome with increased capacity for energy harvest. *Nature* 2006; **444**:1027–1031.
13. Manichanh C, Rigottier-Gois L, Bonnaud E, Gloux K, Pelletier E, Frangeul L, Nalin R et al., Reduced diversity of faecal microbiota in Crohn's disease revealed by a metagenomic approach. *Gut* 2006; **55**:205–211.
14. Willing BP, Dicksved J, Halfvarson J, Andersson AF, Lucio M, Zheng Z, Jarnerot G, Tysk C, Jansson JK, Engstrand L. A pyrosequencing study in twins shows that gastrointestinal microbial profiles vary with inflammatory bowel disease phenotypes. *Gastroenterology* 2010; **139**:1844–1854 e1.
15. Swidsinski A, Ladhoff A, Pernthaler A, Swidsinski S, Loening-Baucke V, Ortner M, Weber J et al., Mucosal flora in inflammatory bowel disease. *Gastroenterology* 2002; **122**:44–54.
16. Brown CT, Davis-Richardson AG, Giongo A, Gano KA, Crabb DB, Mukherjee N, Casella G et al., Gut microbiome metagenomics analysis suggests a functional model for the development of autoimmunity for type 1 diabetes. *PLoS One* 2011; **6**:e25792.
17. Giongo A, Gano KA, Crabb DB, Mukherjee N, Novelo LL, Casella G, Drew JC et al., Toward defining the autoimmune microbiome for type 1 diabetes. *ISME J* 2011; **5**:82–91.
18. Pellegrini S, Sordi V, Bolla AM, Saita D, Ferrarese R, Canducci F, Clementi M et al., Duodenal mucosa of patients with type 1 diabetes shows distinctive inflammatory profile and microbiota. *J Clin Endocrinol Metab* 2017; **102**:1468–1477.
19. Homayouni-Rad A, Soroush AR, Khalili L, Norouzi-Panahi L, Kasaie Z, Ejtahed HS. Diabetes management by probiotics: Current knowledge and future perspective. *Int J Vitam Nutr Res* 2017:1–13.
20. Brunkwall L, Orho-Melander M. The gut microbiome as a target for prevention and treatment of hyperglycaemia in type 2 diabetes: From current human evidence to future possibilities. *Diabetologia* 2017; **60**:943–951.
21. Kumar R, Eipers P, Little RB, Crowley M, Crossman DK, Lefkowitz EJ, Morrow CD. Getting started with microbiome analysis: sample acquisition to bioinformatics. *Curr Protoc Hum Genet* 2014; **82**:18 8 1–29.
22. Paisse S, Valle C, Servant F, Courtney M, Burcelin R, Amar J, Lelouvier B. Comprehensive description of blood microbiome from healthy donors assessed by 16S targeted metagenomic sequencing. *Transfusion* 2016; **56**:1138–1147.
23. Fasano A, Catassi C. Clinical practice. Celiac disease. *N Engl J Med* 2012; **367**:2419–2426.
24. Garrote JA, Gomez-Gonzalez E, Bernardo D, Arranz E, Chirdo F. Celiac disease pathogenesis: The proinflammatory cytokine network. *J Pediatr Gastroenterol Nutr* 2008; **47**:S27–S32.
25. Lammers KM, Lu R, Brownley J, Lu B, Gerard C, Thomas K, Rallabhandi P, Shea-Donohue T et al., Gliadin induces an increase in intestinal permeability and zonulin release by binding to the chemokine receptor CXCR3. *Gastroenterology* 2008; **135**:194–204.e3.
26. Fasano A. Zonulin and its regulation of intestinal barrier function: the biological door to inflammation, autoimmunity, and cancer. *Physiol Rev* 2011; **91**:151–175.
27. Rubio-Tapia A, Murray JA. Celiac disease. *Curr Opin Gastroenterol* 2010; **26**:116–122.
28. Lammers KM, Khandelwal S, Chaudhry F, Kryszak D, Puppa EL, Casolaro V, Fasano A. Identification of a novel immunomodulatory gliadin peptide that causes interleukin-8 release in a chemokine receptor CXCR3-dependent manner only in patients with coeliac disease. *Immunology* 2011; **132**:432–440.
29. Monteleone G, Pender SL, Alstead E, Hauer AC, Lionetti P, McKenzie C, MacDonald TT. Role of interferon alpha in promoting T helper cell type 1 responses in the small intestine in coeliac disease. *Gut* 2001; **48**:425–429.
30. Maiuri L, Ciacci C, Ricciardelli I, Vacca L, Raia V, Auricchio S, Picard J, Osman M, Quaratino S, Londei M. Association between innate response to gliadin and activation of pathogenic T cells in coeliac disease. *Lancet* 2003; **362**:30–37.
31. Lammers KM, Chieppa M, Liu L, Liu S, Omatsu T, Janka-Junttila M, Casolaro V, Reinecker HC, Parent CA, Fasano A. Gliadin induces neutrophil migration via engagement of the formyl peptide receptor, FPR1. *PLoS One* 2015; **10**:e0138338.
32. Sanchez-Castanon M, Castro BG, Toca M, Santacruz C, Arias-Loste M, Iruzubieta P, Crespo J, Lopez-Hoyos M. Intraepithelial lymphocytes subsets in different forms of celiac disease. *Auto Immun Highlights* 2016; **7**:14.
33. Palova-Jelinkova L, Rozkova D, Pecharova B, Bartova J, Sediva A, Tlaskalova-Hogenova H, Spisek R, Tuckova L. Gliadin fragments induce phenotypic and functional maturation of human dendritic cells. *J Immunol* 2005; **175**:7038–7045.

34. Thomas KE, Sapone A, Fasano A, Vogel SN. Gliadin stimulation of murine macrophage inflammatory gene expression and intestinal permeability are MyD88-dependent: role of the innate immune response in Celiac disease. *J Immunol* 2006; **176**:2512–2521.
35. Sapone A, Lammers KM, Mazzarella G, Mikhailenko I, Carteni M, Casolaro V, Fasano A. Differential mucosal IL-17 expression in two gliadin-induced disorders: Gluten sensitivity and the autoimmune enteropathy celiac disease. *Int Arch Allergy Immunol* 2010; **152**:75–80.
36. Castellanos-Rubio A, Santin I, Irastorza I, Castano L, Carlos Vitoria J, Ramon Bilbao J. TH17 (and TH1) signatures of intestinal biopsies of CD patients in response to gliadin. *Autoimmunity* 2009; **42**:69–73.
37. Martin VJ, Leonard MM, Fiechtner L, Fasano A. Transitioning from descriptive to mechanistic understanding of the microbiome: The need for a prospective longitudinal approach to predicting disease. *J Pediatr* 2016; **179**:240–248.
38. Rubio-Tapia A, Kyle RA, Kaplan EL, Johnson DR, Page W, Erdtmann F, Brantner TL et al., Increased prevalence and mortality in undiagnosed celiac disease. *Gastroenterology* 2009; **137**:88–93.
39. Vilppula A, Kaukinen K, Luostarinen L, Krekela I, Patrikainen H, Valve R, Maki M, Collin P. Increasing prevalence and high incidence of celiac disease in elderly people: A population-based study. *BMC Gastroenterol* 2009; **9**:49.
40. Serena G, Camhi S, Sturgeon C, Yan S, Fasano A. The role of gluten in celiac disease and type 1 diabetes. *Nutrients* 2015; **7**:7143–7162.
41. Collado MC, Donat E, Ribes-Koninckx C, Calabuig M, Sanz Y. Specific duodenal and faecal bacterial groups associated with paediatric coeliac disease. *J Clin Pathol* 2009; **62**:264–269.
42. Ou G, Hedberg M, Horstedt P, Baranov V, Forsberg G, Drobni M, Sandstrom O et al., Proximal small intestinal microbiota and identification of rod-shaped bacteria associated with childhood celiac disease. *Am J Gastroenterol* 2009; **104**:3058–3067.
43. Nadal I, Donat E, Ribes-Koninckx C, Calabuig M, Sanz Y. Imbalance in the composition of the duodenal microbiota of children with coeliac disease. *J Med Microbiol* 2007; **56**:1669–1674.
44. Wacklin P, Kaukinen K, Tuovinen E, Collin P, Lindfors K, Partanen J, Maki M, Matto J. The duodenal microbiota composition of adult celiac disease patients is associated with the clinical manifestation of the disease. *Inflamm Bowel Dis* 2013; **19**:934–941.
45. Tjellstrom B, Hogberg L, Stenhammar L, Falth-Magnusson K, Magnusson KE, Norin E, Sundqvist T, Midtvedt T. Faecal short-chain fatty acid pattern in childhood coeliac disease is normalised after more than one year's gluten-free diet. *Microb Ecol Health Dis* 2013; **24**.
46. Nistal E, Caminero A, Herran AR, Arias L, Vivas S, de Morales JM, Calleja S, de Miera LE, Arroyo P, Casqueiro J. Differences of small intestinal bacteria populations in adults and children with/without celiac disease: effect of age, gluten diet, and disease. *Inflamm Bowel Dis* 2012; **18**:649–656.
47. Meisel M, Mayassi T, Fehlner-Peach H, Koval JC, O'Brien SL, Hinterleitner R, Lesko K et al., Interleukin-15 promotes intestinal dysbiosis with butyrate deficiency associated with increased susceptibility to colitis. *ISME J* 2017; **11**:15–30.
48. Serena G, Yan S, Camhi S, Patel S, Lima RS, Sapone A, Leonard MM, Mukherjee R, Nath BJ, Lammers KM, Fasano A. Proinflammatory cytokine interferon-gamma and microbiome-derived metabolites dictate epigenetic switch between forkhead box protein 3 isoforms in coeliac disease. *Clin Exp Immunol* 2017; **187**:490–506.
49. Malamut G, Murray JA, Cellier C. Refractory celiac disease. *Gastrointestinal endoscopy clinics of North America* 2012; **22**:759–772.
50. Casella G, Bordo BM, Schalling R, Villanacci V, Salemme M, Di Bella C, Baldini V, Bassotti G. Neurological disorders and celiac disease. *Minerva Gastroenterol Dietol* 2016; **62**:197–206.
51. Casella G, Orfanotti G, Giacomantonio L, Bella CD, Crisafulli V, Villanacci V, Baldini V, Bassotti G. Celiac disease and obstetrical-gynecological contribution. *Gastroenterol Hepatol Bed Bench* 2016; **9**:241–249.
52. Farnetti S, Zocco MA, Garcovich M, Gasbarrini A, Capristo E. Functional and metabolic disorders in celiac disease: New implications for nutritional treatment. *J Med Food* 2014; **17**:1159–1164.
53. Laparra JM, Sanz Y. Bifidobacteria inhibit the inflammatory response induced by gliadins in intestinal epithelial cells via modifications of toxic peptide generation during digestion. *J Cell Biochem* 2010; **109**:801–807.
54. Lindfors K, Blomqvist T, Juuti-Uusitalo K, Stenman S, Venalainen J, Maki M, Kaukinen K. Live probiotic Bifidobacterium lactis bacteria inhibit the toxic effects induced by wheat gliadin in epithelial cell culture. *Clin Exp Immunol* 2008; **152**:552–558.
55. van Beurden YH, van Gils T, van Gils NA, Kassam Z, Mulder CJ, Aparicio-Pages N. Serendipity in refractory celiac disease: Full recovery of duodenal villi and clinical symptoms after fecal microbiota transfer. *J Gastrointestin Liver Dis* 2016; **25**:385–388.
56. Sellitto M, Bai G, Serena G, Fricke WF, Sturgeon C, Gajer P, White JR et al., Proof of concept of microbiome-metabolome analysis and delayed gluten exposure on celiac disease autoimmunity in genetically at-risk infants. *PLoS One* 2012; **7**:e33387.
57. Iraporda C, Errea A, Romanin DE, Cayet D, Pereyra E, Pignataro O, Sirard JC, Garrote GL, Abraham AG, Rumbo M. Lactate and short chain fatty acids produced by microbial fermentation downregulate proinflammatory responses in intestinal epithelial cells and myeloid cells. *Immunobiology* 2015; **220**:1161–1169.
58. Samuvel DJ, Sundararaj KP, Nareika A, Lopes-Virella MF, Huang Y. Lactate boosts TLR4 signaling and NF-kappaB pathway-mediated gene transcription in macrophages via monocarboxylate transporters and MD-2 up-regulation. *J Immunol* 2009; **182**:2476–2484.
59. Fontenelle B, Gilbert KM. n-Butyrate anergized effector CD4+ T cells independent of regulatory T cell generation or activity. *Scand J Immunol* 2012; **76**:457–463.
60. Furusawa Y, Obata Y, Fukuda S, Endo TA, Nakato G, Takahashi D, Nakanishi Y et al., Commensal microbe-derived butyrate induces the differentiation of colonic regulatory T cells. *Nature* 2013; **504**:446–450.
61. Chang X, Gao JX, Jiang Q, Wen J, Seifers N, Su L, Godfrey VL, Zuo T, Zheng P, Liu Y. The Scurfy mutation of FoxP3 in the thymus stroma leads to defective thymopoiesis. *J Exp Med* 2005; **202**:1141–1151.
62. Chang X, Zheng P, Liu Y. FoxP3: A genetic link between immunodeficiency and autoimmune diseases. *Autoimmun Rev* 2006; **5**:399–402.
63. Hmida NB, Ben Ahmed M, Moussa A, Rejeb MB, Said Y, Kourda N, Meresse B, Abdeladhim M, Louzir H, Cerf-Bensussan N. Impaired control of effector T cells by regulatory T cells: A clue to loss of oral tolerance and autoimmunity in celiac disease? *Am J Gastroenterol* 2012; **107**:604–611.
64. Albert MH, Liu Y, Anasetti C, Yu XZ. Antigen-dependent suppression of alloresponses by Foxp3-induced regulatory T cells in transplantation. *Eur J Immunol* 2005; **35**:2598–2607.
65. Devaud C, Darcy PK, Kershaw MH. Foxp3 expression in T regulatory cells and other cell lineages. *Cancer Immunol Immunother* 2014; **63**:869–876.

66. Allan SE, Passerini L, Bacchetta R, Crellin N, Dai M, Orban PC, Ziegler SF, Roncarolo MG, Levings MK. The role of 2 FOXP3 isoforms in the generation of human CD4+ tregs. *J Clin Invest* 2005; **115**:3276–3284.
67. Du J, Huang C, Zhou B, Ziegler SF. Isoform-specific inhibition of ROR alpha-mediated transcriptional activation by human FOXP3. *J Immunol* 2008; **180**:4785–4792.
68. Joly AL, Liu S, Dahlberg CI, Mailer RK, Westerberg LS, Andersson J. Foxp3 lacking exons 2 and 7 is unable to confer suppressive ability to regulatory T cells in vivo. *J Autoimmun* 2015; **63**:23–30.
69. Lynch KW. Consequences of regulated pre-mRNA splicing in the immune system. *Nat Rev Immunol* 2004; **4**:931–940.
70. Granzotto M, dal Bo S, Quaglia S, Tommasini A, Piscianz E, Valencic E, Ferrara F, Martelossi S, Ventura A, Not T. Regulatory T-cell function is impaired in celiac disease. *Dig Dis Sci* 2009; **54**:1513–1519.
71. Uberos J, Aguilera-Rodriguez E, Jerez-Calero A, Molina-Oya M, Molina-Carballo A, Narbona-Lopez E. Probiotics to prevent necrotising enterocolitis and nosocomial infection in very low birth weight preterm infants. *Br J Nutr* 2017; **117**:994–1000.
72. Malamut G, Meresse B, Cellier C, Cerf-Bensussan N. Refractory celiac disease: From bench to bedside. *Semin Immunopathol* 2012; **34**:601–613.

8 The Next Frontiers of Therapeutic Target Areas

Autoimmunity, Inflammation, Respiratory, Metabolic, Cardiovascular, and Neurological Diseases

Precision Mucosal Sampling and Biomarkers in Allergic Rhinitis and Asthma

8.1

Akhilesh Jha, Ryan S. Thwaites, Nehmat Singh, and Trevor T. Hansel

Contents

INTRODUCTION

From birth onward, the human immune system is modulated by interactions with viruses, bacteria, and allergens (Hansel, Johnston, and Openshaw 2013). Most respiratory diseases cause an inflammatory response with abnormal airway mucosal responses being a particular feature of asthma. A variety of asthma phenotypes are now recognized at a clinical, pathological and molecular level, which results in intermittent and recurrent bronchospasm (Martinez and Vercelli 2013; Wenzel 2016). Allergen-specific IgE and eosinophilic inflammation are key features of allergic rhinitis (AR) (European Academy of Allergy and Clinical Immunology 2015) and viral infections and allergy have a synergistic effect on upper airway symptoms (Cirillo et al. 2007).

There have been important insights into mechanisms of disease in allergy and asthma obtained using precision sampling methods. However, there is an urgent need to utilize and further develop non-invasive methods of taking samples from the respiratory mucosa and other compartments; because combined with novel "omics"-based assay platforms, these techniques may identify key biomarkers in viral infections, AR and asthma. We discuss and compare the progress made in assessing respiratory mucosal inflammation using upper and lower airway samples together with systemic measures of airway inflammation.

UPPER AIRWAY SAMPLING

The upper airway provides a readily accessible route to investigate respiratory mucosal inflammation, permitting sampling at serial time points and in key populations such as children. The nose is a highly specialized organ with multiple roles ranging from olfactory sensing, humidification and filtration of large particles. At the entrance of the nares, the mucosa transitions from a keratinized, stratified, squamous epithelium to one that is composed of pseudostratified ciliated epithelial cells interspersed with goblet cells. The latter lines the nasal inferior turbinates and is where precision nasal mucosal sampling techniques are focused. It represents the beginning of the respiratory epithelium that extends to the bronchioles and provides the basis for the concept of the one-airway hypothesis (Licari et al. 2017). The similarity between nasal and bronchial mucosa has been demonstrated using histological techniques (Bourdin et al. 2009), transcriptomic analysis (Poole et al. 2014) and in clinical settings as evidenced by improvement in asthma symptoms with treatment for AR (J Bousquet et al. 2012).

Nasal Lavage

The nasal lavage (NL) method has been a longstanding tool to measure nasal inflammation, having been used to measure inflammatory mediators for a variety of clinical contexts (Nicola et al. 2014; Greiff et al. 2015), viral load (Piedra et al. 2017; Sobel Leonard et al. 2016), neutrophil subsets (Arebro et al. 2017) and mucosal IgA (Habibi et al. 2015; Salk et al. 2016). It is also used as a therapy for the alleviation of symptoms associated with AR and upper respiratory tract infection (Choo et al. 2017; King et al. 2015). The technique involves instilling saline from a syringe (usually 5–10 mL) either directly or via a 'nasal olive' into the nasal mucosa through each nostril. NL fluid is then aspirated back in to the syringe or collected in a nasal pool device with subsequent measurement of soluble mediators or immune cells (Figure 8.1.1a,b).

The main disadvantage of nasal lavage is the dilution of secretions that can lead to a significant reduction in the levels of mediators being detected compared with direct sampling of the nasal mucosa (Freedman et al. 2004; Gelardi et al. 2014; Gelardi et al. 2012; Jochems et al. 2017). NL has been found to be reproducible (Howarth et al. 2005), but it has been found that diagnostic NL can only be performed once daily to get comparable results (Hentschel et al. 2014).

Nasal Curettage

Nasal curettage using the Rhinoprobe curette (Arlington Scientific) is a non-invasive way of obtaining mucosal cells that does not require local anesthesia, and is a useful alternative to nasal biopsy (Howarth et al. 2005) (Figure 8.1.2a,b). The obtained cellular sample can be used for viral diagnosis, cytology, histology, transcriptomics or flow cytometry. This technique was found to be safe and easy to use in studying 1257 infants and children with respiratory syncytial virus (RSV) infection (Jalowayski et al. 1990), and has also been successfully used to look at gene expression after experimental human rhinovirus (HRV) infection (Proud, Sanders, and Wiehler 2004; Proud et al. 2008) and to detect respiratory viruses in chronic rhinosinusitis (Cho et al. 2013). Nasal transcriptional signatures have also been assessed in serial samples during asthma exacerbations (McErlean et al. 2014). Nasal cytology has been employed in AR (Jirapongsananuruk and Vichyanond 1998; Lin et al. 2001; Gelardi et al. 2014) and nonallergic rhinitis (NAR) (Gelardi et al. 2012) (33), as well as in cystic fibrosis (Freedman et al. 2004; Harris et al. 2004).

Nasal Brush

Nasal epithelial samples can also be collected non-invasively using an interdental or cytology brush (Harris et al. 2004) and have been used to study epithelial cell markers and function in cystic fibrosis (van Meegen et al. 2011; Mosler et al. 2008) and childhood asthma (Poole et al. 2014) as well as to investigate changes in DNA methylation (Bergougnoux, Claustres, and De Sario 2015). Comparison of different nasal sampling techniques suggest that brushings are more suited to obtaining

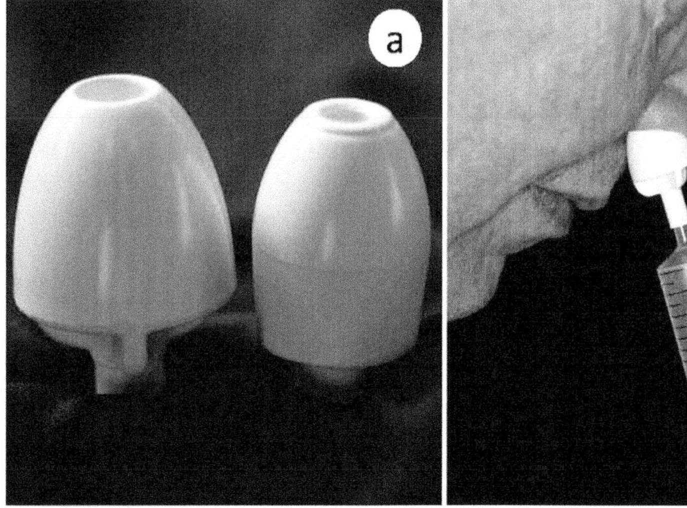

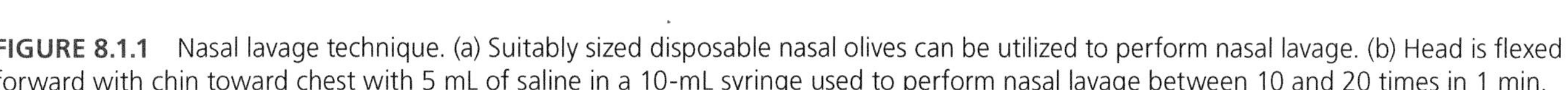
FIGURE 8.1.1 Nasal lavage technique. (a) Suitably sized disposable nasal olives can be utilized to perform nasal lavage. (b) Head is flexed forward with chin toward chest with 5 mL of saline in a 10-mL syringe used to perform nasal lavage between 10 and 20 times in 1 min.

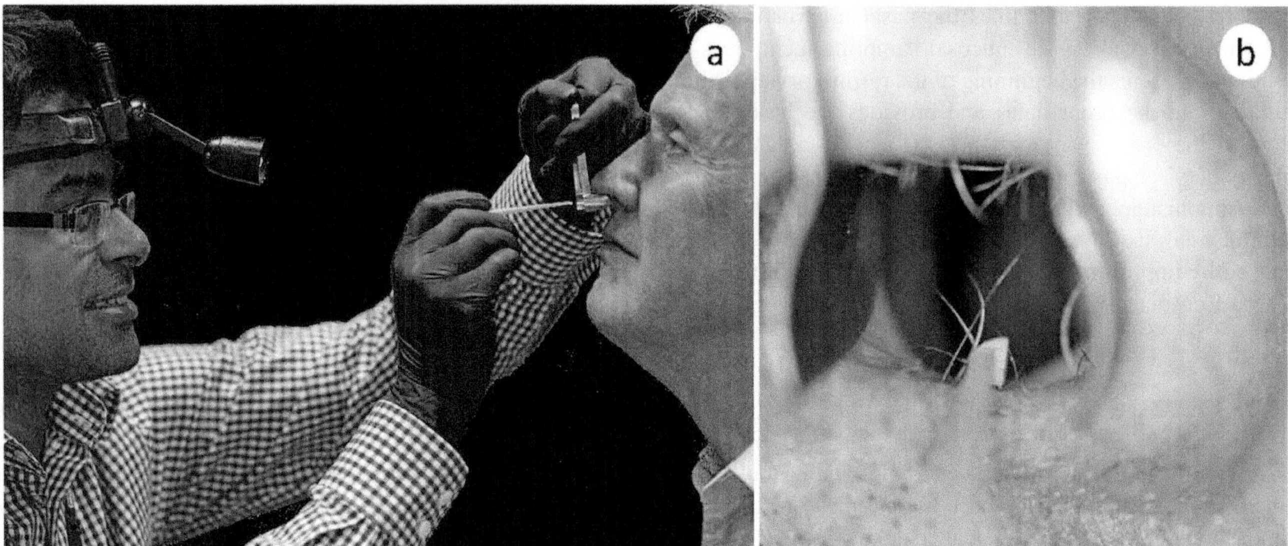

FIGURE 8.1.2 Nasal curettage technique. (a) Nasal curettage should be performed with adequate lighting (e.g., using a headlamp) and use of a Thudicum's speculum (b) for appropriate visualization of the inferior turbinate. This area is chosen because it is lined with pseudostratified ciliated columnar epithelium, which extends to the respiratory bronchioles.

epithelial cells for culture (Stokes et al. 2014), whereas curettage has superior ability to quantify leukocytes and eosinophils (Lin et al. 2001).

Nasosorption

A variety of natural origin materials have been used to absorb mucosal lining fluid (MLF) from mucosal surfaces: including paper, rayon (wood pulp) and sponge. Filter paper consisting of natural cellulose from the cotton plant has been widely used to absorb nasal secretions since the classical paper of Alam and colleagues in 1992 (Alam et al. 1992; Sim et al. 1994; Sim et al. 1995; Weido et al. 1996; Linden et al. 2000; Bensch, Nelson, and Borish 2002; Riechelmann et al. 2003). Ophthalmic Weck-Cel sponges composed of natural cellulose have been used to sample saliva, cervical and vaginal secretions (Rohan et al. 2000). Following nasal allergen challenge (NAC), levels of interleukin (IL)-5 and IL-13 are higher in nasosorption eluates than in nasal lavage (Erin et al. 2005). In addition, filter paper discs have been produced from Shandon filter cards, and have been utilized to measure histamines and cytokines after NAC and with natural allergen exposure (Wagenmann et al. 1997; Wagenmann, Schumacher, and Bachert 2005; Baumann et al. 2013). However, it has been found that different batches of filter paper vary in their degree of protein binding, some failing to release cytokines (internal data).

Therefore, improved methods of sampling mucosal lining fluid have been developed using various synthetic absorptive matrices (SAM). These are absorbent materials that are well tolerated and have the ability to obtain MLF even from inflamed noses at frequent intervals over extended periods of time.

Polyurethane foam has been used to obtain nasal samples, and a pressure system suggested for elution of MLF (Riechelmann et al. 2003; Lü and Esch 2010). Guy Scadding and colleagues have used strips of polyurethane foam after NAC, eluting nasal MLF by spin filtration through a cellulose acetate filter (Scadding et al. 2012; Steveling et al. 2015; Scadding et al. 2015).

Accuwik Ultra (Pall Life Sciences) has been utilized for nasosorption in children with AR (Chawes et al. 2010), and adults following NAC (Nicholson et al. 2011; Leaker et al. 2016). The Bisgaard group in Copenhagen has studied infants with maternal atopy (Følsgaard et al. 2012), nasal bacterial and viral infection (Følsgaard et al. 2013; Wolsk et al. 2016), and type 1/17 immune responses (Wolsk et al. 2016). However, Accuwik Ultra is no longer produced because it became poorly absorbent on storage.

Synthetic sponges made from polyvinyl alcohol (PVA) and hydroxylated polyvinyl acetate (HOPVA) have been utilized to sample uterine cervical secretions (Castle et al. 2004). Seven different absorptive materials have been compared for sampling oral fluid prior to measuring antibodies (Chang, Cohen, and Bienek 2009), whereas polyurethane mini-sponges have been used to collect human tears (López-Cisternas et al. 2006).

Leukosorb (Pall Life Sciences) has been used in a nasal Lipopolysaccharide (LPS) challenge study to show IL-1β and IL-6 responses (Dhariwal et al. 2015). In addition, following HRV challenge in asthma patients, nasosorption and bronchosorption were performed to measure IL-15 (Jayaraman et al. 2014), IL-25 (Beale et al. 2014), IL-33 (Jackson et al. 2014) and IL-18 (Jackson et al. 2015). More recently, leukosorb has been used for nasosorption to study influenza, RSV bronchiolitis of infancy, TB, vasculitis (chronic granulomatous angiitis,

CGA) and following nasal challenge with resiquimod (unpublished data). The recovery of cytokines is unaffected in the first 24 h by storage at room temperature or freezer storage (Rebuli et al. 2017) and an extensive validation study has been carried out comparing performance of nasosorption with nasal lavage (Jochems et al. 2017).

Nasosorption is performed by maneuvering the strip up the lumen of the nostril, avoiding rubbing or rotation against the nasal mucosa as is required for some swabs. The outside of the nose is then pressed with a finger to cause apposition of the SAM against the mucosa (Figure 8.1.3a–c). The procedure may tickle slightly but is painless, and MLF can be obtained even from non-inflamed noses at frequent intervals, without the need for local anesthetic. There is minimal protein binding to the SAM strip, and fluid can be eluted by spin filtration. High levels of mediators of inflammation can then be measured in the MLF: higher than detectable by nasal lavage.

LOWER AIRWAY SAMPLING

Sputum

The significance of sputum eosinophilia was noted by the late Morrow Brown in his original study showing the efficacy of oral prednisolone in asthma (Brown 1958), although sputum has been of interest to clinicians since the time of Hippocrates (Finlayson 1958). The clinical application of quantitation of levels of eosinophils in induced sputum was pioneered by the late Freddy Hargreave (Pin et al. 1992; Djukanovic et al. 2002). As an extension of this work, normalization of sputum eosinophil counts has been shown to be effective in the reduction of asthma exacerbations (Petsky et al. 2012). In addition, adult asthma phenotypes have been defined by sputum eosinophil and

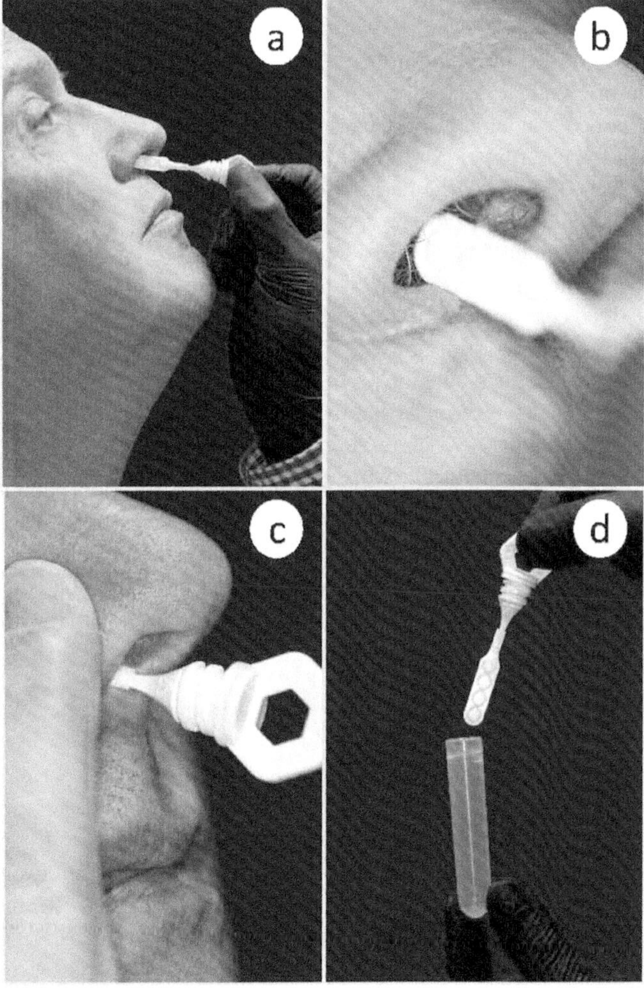

FIGURE 8.1.3 Nasosorption technique. (a) Head is extended backward prior to insertion of nasosorption (NS) probe. (b) This should be manipulated up the lumen until alongside the inferior turbinate. (c) Nostril with nasosorption probe is gently pressed with a finger for 1 min. (d) NS probe is removed and placed in a sealed tube for subsequent elution or freezing.

neutrophil percentages (Green et al. 2002; Fahy 2009). There are reports that blood eosinophil counts are a poor surrogate for sputum eosinophil counts (Hastie et al. 2013; Schleich et al. 2014), whereas other studies have found blood eosinophil counts to be predictive of sputum eosinophil counts (Schleich et al. 2013; Wagener et al. 2015). The analysis of fluid-phase mediators derived from sputum samples has a large number of technical problems (Keatings et al. 2002): these range from degradation by proteases and bacteria, loss of protein secondary structure due to reduction by dithiothreitol (DTT), binding to mucus, contamination with saliva and oropharyngeal contents, and variable leakage of mediators from dead and dying cells. Elegant attempts have been made to validate the measurement of fluid phase levels of IL-5 in sputum (Pizzichini et al. 1996), and this has highlighted the effects of proteases (Kelly et al. 2001).

Bronchoscopic Methods: Biopsy, Brushings and Bronchoalveolar Lavage

Conventional flexible bronchoscopy is normally performed on patients with respiratory disease in order to carry out bronchial mucosal biopsy, bronchial brushings and bronchoalveolar lavage (BAL). This has given important mechanistic insights into the evaluation of inflammation in asthma (Busse et al. 2005). A European Society Task Force has issued guidelines for measurements of cellular components and standardization of BAL (Haslam and Baughman 1999). However, BAL sampling has the major problems of unknown dilution and failure to retrieve lavage fluid, which makes accurate detection and quantification of soluble mediators difficult (Kavuru, Dweik, and Thomassen 1999) and fever and malaise is an acknowledged complication after the procedure (Krause et al. 1997).

Prescott Woodruff, John Fahy and colleagues have helped to define the molecular basis for heterogeneity seen in asthma patients by assessing IL-13 driven gene signatures in bronchial epithelial brushings and correlating these with cytokine expression in bronchial biopsies and responsiveness to inhaled corticosteroids (Woodruff et al. 2009). They have performed genome-wide profiling of bronchial epithelial cells to identify important biomarkers such as periostin (POSTN), chloride channel calcium-activated family member 1 (CLCA1) and serine peptidase inhibitor clade B (ovalbumin) member 2 (SERPINB2), which are upregulated in asthma and show good response to corticosteroids (Woodruff et al. 2007).

Bronchoscopic Microsampling and Bronchosorption

Bronchosorption is a novel sampling method, that has the advantage of being less invasive and causing less dilution of mucosal fluids than BAL. A straw of SAM or synthetic sponge is attached to a leading plastic wire, and placed down a sheath within the operating portal of a bronchoscope. Under direct bronchoscopic vision the SAM is advanced against the mucosa of a main bronchus or segmental bronchus for up to 30 secs, under direct vision the SAM is then withdrawn up the sheath, and MLF eluted from the detached SAM. The eluted MLF may then be analyzed for levels of inflammatory mediators.

Bronchoscopic microsampling (BMS) is a procedure for bronchosorption that has been developed by Olympus (Ishizaka et al. 2001; Ishizaka et al. 2004; Komaki et al. 2005) but is currently only licensed in Japan. Olympus have 2 different BMS systems: BC402C with a fibrous hydroxylated polyester (FHPE) probe (Yamazaki et al. 2003; Kikuchi et al. 2009; Kodama et al. 2009), and BC401C with a cotton probe (Komaki et al. 2005; Sasabayashi et al. 2007; Kipnis et al. 2008; Kanazawa et al. 2010; Sugasawa et al. 2011; Sugasawa et al. 2012; Cohen et al. 2008). BMS has been used for sampling in various respiratory conditions; acute respiratory distress syndrome (ARDS) for IL-6 (Ishizaka et al. 2001), KL-6 (Ishizaka et al. 2004) and endothelin-1 (Nakano et al. 2007); bacterial infections (Sasabayashi et al. 2007; Yanagi et al. 2007) including *Mycobacterium avium* (Nishimura et al. 2008); pharmacokinetics of antibiotics (Yamazaki et al. 2003; Kikuchi et al. 2009; Kikuchi et al. 2007) and lung cancer (Watanabe et al. 2003; Yasuda et al. 2011). A major stumbling block has been that the BMS probe (BC401C) used for bronchosorption in patients with asthma caused mucosal contact bleeding: half of all samples were contaminated with blood. The authors concluded that it was not feasible to sample MLF using this BMS system from peripheral airways in asthma patients (Cohen et al. 2008).

The Respiratory Medicine Group in Imperial College have developed a bronchosorption device (in conjunction with Hunt Developments (UK) Ltd) that employs a different proprietary medical grade fibrous SAM at the tip. A HRV infection model used in volunteers with allergic asthma and healthy controls, utilized both nasosorption and bronchosorption to measure IL-15 (Jayaraman et al. 2014), IL-25 (Beale et al. 2014), IL-33 (Jackson et al. 2014), IL-18 (Jackson et al. 2015) and interferons (Hansel et al. 2017). The department now has a total experience of over 1000 bronchosorption samples being successfully taken. The procedure is well tolerated and up to 40μl of fluid is absorbed on a single occasion. The bronchosorption kit is supplied in sterile packs, and has undergone durability and functional testing, and is supplied gamma-irradiated.

Exhaled Breath

Exhaled breath has also been extensively studied as a non-invasive means to assess airway inflammation, by the measurement of mediators in exhaled breath condensate (EBC) (Horváth et al. 2017). Richard Effros and colleagues have elegantly highlighted the issues of salivary contamination and dilution in condensed water vapor that occurs during collection of EBC (Effros et al. 2012; Horváth et al. 2005); and this is likely to be a serious obstacle to measuring EBC pH (Effros et al. 2006) and levels of inflammatory mediators that are in breath droplets. However, assessment of volatile organic compounds (VOCs) and metabolomics on EBC looks to be more promising (Wagener et al. 2013; Bikov et al. 2013; van der Schee et al. 2015).

Nitric oxide production in the bronchial mucosa is increased in asthma patients with eosinophilic inflammation and the measurement of fractional exhaled nitric oxide (FeNO) offers a quick, non-invasive and reproducible surrogate of airway inflammation (Högman, Lehtimäki, and Dinh-Xuan 2017; Pavord et al. 2017; American Thoracic Society and European Respiratory Society 2005). Although utilizing FeNO may be of benefit in monitoring inflammation, systematic reviews investigating the role of FeNO guided management on the use of inhaled corticosteroid therapy have been hampered by the heterogeneity of study design characteristics (Essat et al. 2016; Gomersal et al. 2016). FeNO levels can also be influenced by age, medication use and smoking, and disagreements have arisen in national and international guidelines for its use as a biomarker for diagnosis and monitoring of therapy in asthma patients (National Institute for Health and Care Excellence [NICE] 2014; Global Initiative for Asthma [GINA] 2017).

Table 8.1.1 compares upper and lower airway sampling techniques to measure airway inflammation.

TABLE 8.1.1 Comparison of upper and lower airway sampling techniques to measure inflammation

SAMPLING METHOD	*ADVANTAGES*	*DISADVANTAGES*	*BIOMARKERS*	*KEY REFERENCES*
Upper Airway				
Nasal lavage	Good tolerability Non-invasive	Variable dilution Limited repeatability (>24 h)	Inflammatory mediators RSV and influenza viral load Secretory IgA	Nicola et al. (2014) Piedra et al. (2017) Sobel Leonard et al. (2016) Habibi et al. (2015)
Nasal curettage	Obtains epithelial and immune cells Ability to perform gene expression, epithelial cell culture and flow cytometry	Nasolacrimal reaction Limited repeatability (but can alternate nostrils) Fragile small sample	Detection of RSV PCR ILC2s on flow cytometry Cell-specific (e.g., eosinophil) gene signatures after HRV and allergen	Jalowayski et al. (1990) Proud et al. (2008) Leaker et al. (2016) Dhariwal et al. (2017) Jochems et al. (2017)
Nasal biopsy	Full thickness mucosal and lamina propria Preserved architecture	Requires an expert clinician and a local anesthetic Bleeding	Immunohistology	European Academy of Allergy and Clinical Immunology (2015)
Nasal brush	Non-invasive Ability to perform gene expression and epithelial cell culture	Obtains mainly epithelial cells	Epithelial cell markers (e.g.,EPCAM1, E-cadherin)	Stokes et al. (2014) Poole et al. (2014) Mosler et al. (2008)
Nasosorption	Direct mucosal sampling Excellent tolerability Non-invasive Repeatable	Requires validation for viral load and mRNA detection	IFN response to HRV in asthma IL-5, IL-13 after allergen and HRV challenge	Hansel et al. (2017) Leaker et al. (2016) Jackson et al. (2014) Jochems et al. (2017) Lin et al. (2001)
Lower Airway				
Sputum	Non-invasive and easy to obtain Standardized techniques	Loss of protein structure with use of DTT Salivary contamination Limited repeatability (>24 h)	Utilized for asthma phenotyping (eosinophilic, neutrophilic) Correlation between blood and sputum eosinophils has conflicting evidence	Brown (1958) Djukanovic et al. (2002) Hastie et al. (2013) Wagener et al. (2015) Keatings et al. (2002)
Bronchoalveolar lavage (BAL)	Good for ascertaining differential BAL cell counts Extensive literature base on range of respiratory conditions	Variability in dwell time and volume of aspirated lavage fluid Variable dilution making measurement of soluble mediators difficult	Asthma phenotyping (Th1/Th2) Cellular and cytokine response to segmental and whole lung allergen challenge	Busse et al. (2005) Haslam and Baughman (1999) Kavuru et al. (1999)

(Continued)

TABLE 8.1.1 (*Continued*) Comparison of upper and lower airway sampling techniques to measure inflammation

SAMPLING METHOD	*ADVANTAGES*	*DISADVANTAGES*	*BIOMARKERS*	*KEY REFERENCES*
Bronchial brush	Genome-wide profiling Safe and well tolerated	Mainly obtains epithelial cells Occasional bleeding	IL-13 induced genes Role of POSTN, CLCA1, SERPINB2 genes in asthma and response to steroids	Busse et al. (2005) Woodruff et al. (2007) Woodruff et al. (2009)
Bronchosorption	Direct mucosal sampling Detection of cytokines, interferons and DNA methylation analysis Well tolerated	Reduced sample volume	Increased IFN-γ and type 2 cytokines in asthma after HRV SerpinB3 and Uteroglobin increased in COPD after smoking	Hansel et al. (2017) Yasuda et al. (2011) Franciosi et al. (2014)
Exhaled breath	FeNO is non-invasive, quick and can be used to monitor eosinophilic inflammation Measuring VOCs and metabolomics in EBC is promising	FeNO can be affected by smoking, age, medications and can be raised in other eosinophilic diseases Salivary contamination and dilution in condensed water vapor during collection of EBC	FeNO to guide ICS therapy is the subject of debate VOCs: Ability to differentiate asthma phenotypes and children with and without wheeze	Högman et al. (2017) Global Initiative for Asthma GINA (2017) Horváth et al. (2017) Wagener et al. (2013) van der Schee et al. (2015)

Abbreviations: RSV Respiratory Syncytial Virus; IFN Interferon; HRV Human Rhinovirus; DTT Dithiothreitol; VOC Volatile Organic Compounds; EBC Exhaled Breath Condensate; ICS Inhaled Corticosteroids.

MEASURING AIRWAY INFLAMMATION IN BLOOD

The peripheral blood represents an easily accessible site for sampling and biomarker discovery, but has the major disadvantage of being remote from the site of airway disease. Blood-associated biomarkers of asthma, like those of the airway, have mainly focused on type 2 inflammation (Chung et al. 2014; Wenzel 2012). The eosinophil is regarded to play a central role in the pathogenesis of asthma and comprehensive reviews have been published on this granulocyte that migrates from the bloodstream into inflamed respiratory and gut sites (Hogan et al. 2008; Rosenberg, Dyer, and Foster 2012). Historically, the humble blood eosinophil count has been extensively used in the management of asthma (Horn et al. 1975; Jean Bousquet et al. 1990; Tefferi 2005). Blood eosinophilia has been associated with asthma and found to correlate with sputum eosinophilia (Wagener et al. 2015) with blood eosinophil counts >400/μL associated with higher asthma exacerbation rates (Price et al. 2015; Zeiger et al. 2017). A mathematical algorithm has been used to predict elevated sputum eosinophils: the ratio of blood Eosinophil to Lymphocyte and the ratio of blood eosinophil to neutrophil (ELEN) index (Zhang et al. 2014), and has been used to assess response to treatment (Castro et al. 2014).

Recently, there has been renewed interest in using blood eosinophil counts to select asthmatic patients for monoclonal antibody therapy. Blood eosinophilia is decreased by anti-IL-5 therapies (Menzies-Gow et al. 2003; Leckie et al. 2000), which is associated with therapeutic effect (Pavord et al. 2012; Ortega et al. 2014; Lugogo et al. 2016). However, blood eosinophil counts undergo diurnal variation in healthy volunteers and patients with moderate asthma (Sennels et al. 2011; Spector and Tan 2012). Exercise also has the capacity to increase the eosinophil count (Sand et al. 2013), which does not make it an ideal biomarker for clinical stratification.

The blood eosinophil count is favored by recent American Thoracic Society/European Respiratory Society international guidelines on severe asthma, and suggests that the utility of other biomarkers in identifying asthma phenotypes needs further validation (Chung et al. 2014).

T-cell and group 2 innate lymphoid cell (ILC2) counts have been suggested as alternative blood cell biomarkers. CD4 T-cell numbers are equivalent between moderate and severe asthmatics; however, ILC2 numbers and activation are higher in severe asthma (Smith et al. 2016) and distinguish asthma from AR (Bartemes et al. 2014).

Blood proteins have also been investigated as biomarkers of asthma. Total IgE levels are elevated in asthmatics (Burrows et al. 1989), particularly in paediatric asthma (Borish et al. 2005), whereas IgE blockade with Omalizumab is effective in a subgroup of severe asthmatics (Busse et al. 2001). However, definitions of elevated IgE levels depend on age, sex, genetics and environmental factors, restricting the use of IgE level alone as a biomarker for initiation of anti-IgE therapy (Borish et al. 2005). Elevated eosinophil cationic protein (ECP) has been associated with lower forced expiratory volume in 1 second (FEV1) (Meyer et al. 2014) but the finding that ECP levels are equivalent between allergic asthmatics and rhinitic controls and unrelated to eosinophil activation status or disease severity (Ferguson et al. 1995), limits its role as a blood

TABLE 8.1.2 Blood biomarkers for assessment of airway inflammation in allergy and asthma

CELL/PROTEIN	*ADVANTAGES*	*DISADVANTAGES*	*BIOMARKERS*	*KEY REFERENCES*
Eosinophil count	Ease of collection and interpretation Associated with airway eosinophilia	Diurnal and exercise mediated variability	Asthmatics with >400/µL have higher exacerbation rate Decreased by anti-IL-5 therapy, especially in patients with blood eosinophilia	Wagener et al. (2015) Sennels et al. (2011) Sand et al. (2013) Price et al. (2015) Zeiger et al. (2017) Menzies-Gow et al. (2003) Leckie et al. (2000) Pavord et al. (2012) Ortega et al. (2014) Lugogo et al. (2016)
T cells	Type 2 mediators contribute to disease, one source of which are Th2 cells	CD4+ T cell counts and activation status equivalent between mild and severe asthma		Smith et al. (2016)
Group 2 innate lymphoid cells (ILC2)	Severe asthma associated with higher blood ILC2 counts and activation than mild asthma ILC2 count may distinguish allergic asthma from rhinitis	Rapidly induced following allergen exposure	Elevated count associated with asthma and disease activity	Smith et al. (2016) Bartemes et al. (2014) Doherty et al. (2014)
IgE and blocking antibodies in immunotherapy	Elevated in asthma Blockade by Omalizumab effective in severe asthma	Age, genetics, sex and smoking influence serum IgE levels Elevated in severe asthma in infants but not adults	Generally high in asthma and may be associated with disease severity	Burrows et al. (1989) Busse et al. (2001) Borish et al. (2005)
Eosinophilic cationic protein (ECP)	Part of phenotypic cluster of lower FEV1, elevated ECP and greater response to therapy	No relation to disease severity, eosinophil activation status, does not distinguish asthma from atopy	Mixed evidence for relation to disease severity and response to therapy	Meyer et al. (2014) Ferguson et al. (1995)
Periostin	Some reports show association with airway eosinophilia	Distinction of eosinophilic airway inflammation contended	Persistently high serum periostin associated with greater response to Lebrikizumab (anti-IL-13)	Wagener et al. (2015) Jia et al. (2012) Corren et al. (2011)

biomarker. Periostin, which is induced by type 2 mediators, has been reported to be associated with airway eosinophilia (Jia et al. 2012), but this finding has not been confirmed by others (Wagener et al. 2015). However, persistently high serum periostin levels have been associated with an enhanced response to IL-13 blockade (Corren et al. 2011), suggesting that periostin has a role in helping define type 2 mediated asthma.

Many exciting advances toward visualizing the asthmatic airway for biomarker development have occurred in recent years. Though these techniques are beyond the scope of this section, the authors encourage reference to a recent review (Trivedi et al. 2017), and also highlight recent advances using advanced fiber optic methodology to perform molecular alveomics (Craven et al. 2016; Krstajic et al. 2016).

Table 8.1.2 summarizes blood cellular and protein components as well as biomarkers for the assessment of airway inflammation in allergy and asthma.

CONCLUSION AND FUTURE PERSPECTIVE

Direct mucosal sampling methods from the respiratory tract such as nasosorption and bronchosorption have yielded important insights in identifying molecular signatures of respiratory disease and are promising techniques in the era of precision medicine. This will refine the diagnosis, stratification and monitoring of a variety of airway diseases, and may provide targets and biomarkers for new therapy. However, related methodologies can be applied to sampling from any moist mucosal surface to identify biomarkers of disease. Modern analytic techniques are progressing rapidly and metabolomics and immunoassay by electrical impedance could be usefully applied to the mucosa. Hence, advances in the development of point-of-care assays may enable immediate clinical decisions for the rational administration of therapeutics. It is hoped that precision sampling with advanced bioanalytics will assist the diagnosis and management of the complex range of phenotypes that comprise asthma, as well as other diseases of mucosal surfaces.

REFERENCES

Alam, R, Tommy C Sim, Kimberley Hilsmeier, and J Andrew Grant. 1992. "Development of a New Technique for Recovery of Cytokines from Inflammatory Sites *In Situ*." *Journal of Immunological Methods* 155 (1): 25–29.

American Thoracic Society, and European Respiratory Society. 2005. "ATS/ERS Recommendations for Standardized Procedures for the Online and Offline Measurement of Exhaled Lower Respiratory Nitric Oxide and Nasal Nitric Oxide, 2005." *American Journal of Respiratory and Critical Care Medicine* 171 (8): 912–930. doi:10.1164/rccm.200406-710ST.

Arebro, Julia, Sandra Ekstedt, Eric Hjalmarsson, Ola Winqvist, Susanna Kumlien Georén, and Lars-Olaf Cardell. 2017. "A Possible Role for Neutrophils in Allergic Rhinitis Revealed after Cellular Subclassification." *Scientific Reports* 7 : 43568. doi:10.1038/srep43568.

Bartemes, Kathleen R., Gail M. Kephart, Stephanie J. Fox, and Hirohito Kita. 2014. "Enhanced Innate Type 2 Immune Response in Peripheral Blood from Patients with Asthma." *The Journal of Allergy and Clinical Immunology* 134 (3): 671–678.e4. doi:10.1016/j.jaci.2014.06.024.

Baumann, R., M. Rabaszowski, I. Stenin, L. Tilgner, M. Gaertner-Akerboom, K. Scheckenbach, J. Wiltfang, A. Chaker, J. Schipper, and M. Wagenmann. 2013. "Nasal Levels of Soluble IL-33R ST2 and IL-16 in Allergic Rhinitis: Inverse Correlation Trends with Disease Severity." *Clinical and Experimental Allergy: Journal of the British Society for Allergy and Clinical Immunology* 43 (10): 1134–1143. doi:10.1111/cea.12148.

Beale, Janine, Annabelle Jayaraman, David J Jackson, Jonathan D R Macintyre, Michael R Edwards, Ross P Walton, Jie Zhu et al. 2014. "Rhinovirus-Induced IL-25 in Asthma Exacerbation Drives Type 2 Immunity and Allergic Pulmonary Inflammation." *Science Translational Medicine* 6 (256): 256ra134. doi:10.1126/scitranslmed.3009124.

Bensch, Greg W., Harold S. Nelson, and Larry C. Borish. 2002. "Evaluation of Cytokines in Nasal Secretions after Nasal Antigen Challenge: Lack of Influence of Antihistamines." *Annals of Allergy, Asthma & Immunology: Official Publication of the American College of Allergy, Asthma, & Immunology* 88 (5): 457–462. doi:10.1016/S1081-1206(10)62382-4.

Bergougnoux, Anne, Mireille Claustres, and Albertina De Sario. 2015. "Nasal Epithelial Cells: A Tool to Study DNA Methylation in Airway Diseases." *Epigenomics* 7 (1): 119–126. doi:10.2217/epi.14.65.

Bikov, Andras, Koralia Paschalaki, Ron Logan-Sinclair, Ildiko Horváth, Sergei A Kharitonov, Peter J Barnes, Omar S Usmani, and Paolo Paredi. 2013. "Standardised Exhaled Breath Collection for the Measurement of Exhaled Volatile Organic Compounds by Proton Transfer Reaction Mass Spectrometry." *BMC Pulmonary Medicine* 13 (4). *BioMed Central*: 43. doi:10.1186/1471-2466-13-43.

Borish, Larry, Bradley Chipps, Yamo Deniz, Sheila Gujrathi, Beiyao Zheng, and Chantal M. Dolan. 2005. "Total Serum IgE Levels in a Large Cohort of Patients with Severe or Difficult-to-Treat Asthma." *Annals of Allergy, Asthma & Immunology* 95 (3): 247–253. doi:10.1016/S1081-1206(10)61221-5.

Bourdin, Arnaud, D. Gras, Isabelle Vachier, and Pascal Chanez. 2009. "Upper Airway 1: Allergic Rhinitis and Asthma: United Disease through Epithelial Cells." *Thorax* 64 (11): 999–1004. doi:10.1136/thx.2008.112862.

Bousquet, Jean, Holger J. Schünemann, Boleslaw Samolinski, Pascal Demoly, Carlos E Baena-Cagnani, Claus Bachert, Sergio Bonini et al. 2012. "Allergic Rhinitis and its Impact on Asthma (ARIA): Achievements in 10 Years and Future Needs." *The Journal of Allergy and Clinical Immunology* 130 (5): 1049–1062. doi:10.1016/j.jaci.2012.07.053.

Bousquet, Jean, Pascal Chanez, Jean Yves Lacoste, Gilbert Barnéon, Nouchine Ghavanian, Ingrid Enander, Per Venge et al. 1990. "Eosinophilic Inflammation in Asthma." *New England Journal of Medicine* 323 (15): 1033–1039. doi:10.1056/NEJM199010113231505.

Brown, H M. 1958. "Treatment of Chronic Asthma with Prednisolone; Significance of Eosinophils in the Sputum." *Lancet* 2 (7059): 1245–1247.

Burrows, Benjamin, Fernando D. Martinez, Marilyn Halonen, Robert A. Barbee, and Martha G. Cline. 1989. "Association of Asthma with Serum IgE Levels and Skin-Test Reactivity to Allergens." *New England Journal of Medicine* 320 (5): 271–277. doi:10.1056/NEJM198902023200502.

Busse, William, Jonathan Corren, Bobby Quentin Lanier, Margaret McAlary, Angel Fowler-Taylor, Giovanni Della Cioppa, Andre van As, and Niroo Gupta. 2001. "Omalizumab, Anti-IgE Recombinant Humanized Monoclonal Antibody, for the Treatment of Severe Allergic Asthma." *The Journal of Allergy and Clinical Immunology* 108 (2): : 184–190. doi:10.1067/mai.2001.117880.

Busse, William W., Adam Wanner, Kenneth Adams, Herbert Y. Reynolds, Mario Castro, Badrul Chowdhury, Monica Kraft, Robert J. Levine, Stephen P. Peters, and Eugene J. Sullivan. 2005. "Investigative Bronchoprovocation and Bronchoscopy in Airway Diseases." *American Journal of Respiratory and Critical Care Medicine* 172 (7): 807–816. doi:10.1164/rccm.200407-966WS.

Castle, Philip E, Ana-Cecilia Rodriguez, Frederick P Bowman, Rolando Herrero, Mark Schiffman, M Concepcion Bratti, Lidia Ana Morera, Danny Schust, Peggy Crowley-Nowick, and Allan Hildesheim. 2004. "Comparison of Ophthalmic Sponges for Measurements of Immune Markers from Cervical Secretions." *Clinical and Diagnostic Laboratory Immunology* 11 (2): 399–405.

Castro, Mario, Sally E Wenzel, Eugene R Bleecker, Emilio Pizzichini, Piotr Kuna, William W Busse, David L Gossage et al. 2014. "Benralizumab, an Anti-Interleukin 5 Receptor α Monoclonal Antibody, versus Placebo for Uncontrolled Eosinophilic Asthma: A Phase 2b Randomised Dose-Ranging Study." *The Lancet Respiratory Medicine* 2 (11): 879–890. doi:10.1016/S2213-2600(14)70201-2.

Chang, C. K., M. E. Cohen, and D. R. Bienek. 2009. "Efficiency of Oral Fluid Collection Devices in Extracting Antibodies." *Oral Microbiology and Immunology* 24 (3): 231–235. doi:10.1111/j.1399-302X.2008.00500.x.

Chawes, Bo L K, Matthew J Edwards, Betty Shamji, Christoph Walker, Grant C Nicholson, Andrew J Tan, Nilofar V Følsgaard, Klaus Bønnelykke, Hans Bisgaard, and Trevor T Hansel. 2010. "A Novel Method for Assessing Unchallenged Levels of Mediators in Nasal Epithelial Lining Fluid." *The Journal of Allergy and Clinical Immunology* 125 (6): 1387–1389.e3. doi:10.1016/j.jaci.2010.01.039.

Cho, Gye Song, Byung-Jae Moon, Bong-Jae Lee, Chang-Hoon Gong, Nam Hee Kim, You-Sun Kim, Hun Sik Kim, and Yong Ju Jang. 2013. "High Rates of Detection of Respiratory Viruses in the Nasal Washes and Mucosae of Patients with Chronic Rhinosinusitis." *Journal of Clinical Microbiology* 51 (3): 979–984. doi:10.1128/JCM.02806-12.

Choo, Min-Kyung, Yasuyo Sano, Changhoon Kim, Kei Yasuda, Xiao-Dong Li, Xin Lin, Mary Stenzel-Poore et al. 2017. "TLR Sensing of Bacterial Spore-Associated RNA Triggers Host Immune Responses with Detrimental Effects." *The Journal of Experimental Medicine*. doi:10.1084/jem.20161141.

Chung, Kian Fan, Sally E Wenzel, Jan L Brozek, Andrew Bush, Mario Castro, Peter J Sterk, Ian M Adcock et al. 2014. "International ERS/ATS Guidelines on Definition, Evaluation and Treatment of Severe Asthma." *The European Respiratory Journal* 43 (2): 343–373. doi:10.1183/09031936.00202013.

Cirillo, Ignazio, Gianluigi Marseglia, Catherine Klersy, and Giorgio Ciprandi. 2007. "Allergic Patients Have More Numerous and Prolonged Respiratory Infections than Nonallergic Subjects." *Allergy* 62 (9): 1087–1090. doi:10.1111/j.1398-9995.2007.01401.x.

Cohen, J., W. Rob Douma, Nick ten Hacken, Judith M Vonk, Matthijs Oudkerk, and Dirkje S Postma. 2008. "Ciclesonide Improves Measures of Small Airway Involvement in Asthma." *European Respiratory Journal* 31 (6): 1213–1220. doi:10.1183/09031936.00082407.

Corren, Jonathan, Robert F Lemanske, Nicola A Hanania, Phillip E Korenblat, Merdad V Parsey, Joseph R Arron, Jeffrey M Harris et al. 2011. "Lebrikizumab Treatment in Adults with Asthma." *The New England Journal of Medicine* 365 (12): 1088–1098. doi:10.1056/NEJMoa1106469.

Craven, Thomas, Tashfeen Walton, Ahsan Akram, Neil McDonald, Emma Scholefield, Tim Walsh, Christopher Haslett, Mark Bradley, and Kevin Dhaliwal. 2016. "In-Situ Imaging of Neutrophil Activation in the Human Alveolar Space with Neutrophil Activation Probe and Pulmonary Optical Endomicroscopy." *The Lancet* 387: S31. doi:10.1016/S0140-6736(16)00418-9.

Dhariwal, Jaideep, Aoife Cameron, Maria-Belen Trujillo-Torralbo, Ajerico Del Rosario, Eteri Bakhsoliani, Malte Paulsen, David J Jackson et al. 2017. "Mucosal Type 2 Innate Lymphoid Cells are a Key Component of the Allergic Response to Aeroallergen." *American Journal of Respiratory and Critical Care Medicine* 193 (3): 273–280. doi:10.1164/rccm.201609-1846OC.

Dhariwal, Jaideep, Jeremy Kitson, Reema E. Jones, Grant Nicholson, Tanushree Tunstall, Ross P. Walton, Grace Francombe et al. 2015. "Nasal Lipopolysaccharide Challenge and Cytokine Measurement Reflects Innate Mucosal Immune Responsiveness." *Plos One* 10 (9): e0135363. doi:10.1371/journal.pone.0135363.

Djukanovic, Ratko, Peter J. Sterk, John V. Fahy, and Frederick E. Hargreave. 2002. "Standardised Methodology of Sputum Induction and Processing." *European Respiratory Journal* 20 (Supplement 37): 1S–2s. doi:10.1183/09031936.02.00000102.

Doherty, Taylor A., David Scott, Hannah H. Walford, Naseem Khorram, Sean Lund, Rachel Baum, Jinny Chang et al. 2014. "Allergen Challenge in Allergic Rhinitis Rapidly Induces Increased Peripheral Blood Type 2 Innate Lymphoid Cells That Express CD84." *Journal of Allergy and Clinical Immunology* 133 (4): 1203–1205.e7. doi:10.1016/j.jaci.2013.12.1086.

Effros, Richard M, Richard Casaburi, Janos Porszasz, Edith M Morales, and Virender Rehan. 2012. "Exhaled Breath Condensates: Analyzing the Expiratory Plume." *American Journal of Respiratory and Critical Care Medicine* 185 (8): 803–804. doi:10.1164/rccm.201109-1702ED.

Effros, Richard M., Richard Casaburi, Jennifer Su, Marshall Dunning, John Torday, Julie Biller, and Reza Shaker. 2006. "The Effects of Volatile Salivary Acids and Bases on Exhaled Breath Condensate pH." *American Journal of Respiratory and Critical Care Medicine* 173 (4): 386–392. doi:10.1164/rccm.200507-1059OC.

Erin, Edward M., Angela S. Zacharasiewicz, Grant C. Nicholson, Andrew J. Tan, L. A. Higgins, T. J. Williams, R. D. Murdoch, Stephen. R. Durham, Peter J. Barnes, and Trevor T. Hansel. 2005. "Topical Corticosteroid Inhibits Interleukin-4, -5 and -13 in Nasal Secretions Following Allergen Challenge." *Clinical and Experimental Allergy: Journal of the British Society for Allergy and Clinical Immunology* 35 (12): 1608–1614. doi:10.1111/j.1365-2222.2005.02381.x.

Essat, Munira, Sue Harnan, Tim Gomersall, Paul Tappenden, Ruth Wong, Ian Pavord, Rod Lawson, and Mark L. Everard. 2016. "Fractional Exhaled Nitric Oxide for the Management of Asthma in Adults: A Systematic Review." *The European Respiratory Journal* 47 (3): 751–768. doi:10.1183/13993003.01882-2015.

European Academy of Allergy and Clinical Immunology. 2015. *Global Atlas of Allergic Rhinitis and Chronic Rhinosinusitis*. Edited by CA Akdis, PW Hellings, and I Agache. *EAACI*. Zurich, Switzerland: EAACI.

Fahy, J. V. 2009. "Eosinophilic and Neutrophilic Inflammation in Asthma: Insights from Clinical Studies." *Proceedings of the American Thoracic Society* 6 (3): 256–259. doi:10.1513/pats.200808-087RM.

Ferguson, Alexander C., Radana Vaughan, Huguette Brown, and Carol Curtis. 1995. "Evaluation of Serum Eosinophilic Cationic Protein as a Marker of Disease Activity in Chronic Asthma." *Journal of Allergy and Clinical Immunology* 95 (1): 23–28. doi:10.1016/S0091-6749(95)70148-6.

Finlayson, R. 1958. "The Vicissitudes of Sputum Cytology." *Medical History* 2 (1): 24–35.

Følsgaard, Nilofar V., Bo L. Chawes, Morten a. Rasmussen, Anne L. Bischoff, Charlotte G. Carson, Jakob Stokholm, Louise Pedersen et al. 2012. "Neonatal Cytokine Profile in the Airway Mucosal Lining Fluid Is Skewed by Maternal Atopy." *American Journal of Respiratory and Critical Care Medicine* 185 (3): 275–280. doi:10.1164/rccm.201108-1471OC.

Følsgaard, Nilofar V., Susanne Schjørring, Bo L. Chawes, Morten A. Rasmussen, Karen A. Krogfelt, Susanne Brix, and Hans Bisgaard. 2013. "Pathogenic Bacteria Colonizing the Airways in Asymptomatic Neonates Stimulates Topical Inflammatory Mediator Release." *American Journal of Respiratory and Critical Care Medicine* 187 (6): 589–595. doi:10.1164/rccm.201207-1297OC.

Franciosi, Lorenza, Dirkje S. Postma, Maarten van den Berge, Natalia Govorukhina, Peter L. Horvatovich, Fabrizia Fusetti, Bert Poolman et al. 2014. "Susceptibility to COPD: Differential Proteomic Profiling after Acute Smoking." *PLoS ONE* 9 (7): e102037. doi:10.1371/journal.pone.0102037.

Freedman, Steven D., Paola G. Blanco, Munir M. Zaman, Julie C. Shea, Mario Ollero, Isabel K. Hopper, Deborah A. Weed et al. 2004. "Association of Cystic Fibrosis with Abnormalities in Fatty Acid Metabolism." *The New England Journal of Medicine* 350 (6): 560–569. doi:10.1056/NEJMoa021218.

Gelardi, Matteo, Diego G Peroni, Cristoforo Incorvaia, Nicola Quaranta, Concetta De Luca, Salvatore Barberi, Ilaria Dell'albani, Massimo Landi, Franco Frati, and Olivier de Beaumont. 2014. "Seasonal Changes in Nasal Cytology in Mite-Allergic Patients." *Journal of Inflammation Research* 7: 39–44. doi:10.2147/JIR.S54581.

Gelardi, Matteo, Gian Luigi Marseglia, Amelia Licari, Massimo Landi, Ilaria Dell'Albani, Cristoforo Incorvaia, Franco Frati, and Nicola Quaranta. 2012. "Nasal Cytology in Children: Recent Advances." *Italian Journal of Pediatrics* 38 (1): 51. doi:10.1186/1824-7288-38-51.

Global Initiative for Asthma (GINA). 2017. "Global Strategy for Asthma Management and Prevention." http://ginasthma.org/2017-gina-report-global-strategy-for-asthma-management-and-prevention/ Accessed on May 1, 2018.

Gomersal, Tim, Sue Harnan, Munira Essat, Paul Tappenden, Ruth Wong, Rod Lawson, Ian Pavord, and Mark Lloyd Everard. 2016. "A Systematic Review of Fractional Exhaled Nitric Oxide in the Routine Management of Childhood Asthma." *Pediatric Pulmonology* 51 (3): 316–328. doi:10.1002/ppul.23371.

Green, Ruth H., Christopher E. Brightling, Gerrit Woltmann, Debbie Parker, Andrew J. Wardlaw, and Ian D. Pavord. 2002. "Analysis of Induced Sputum in Adults with Asthma: Identification of Subgroup with Isolated Sputum Neutrophilia and Poor Response to Inhaled Corticosteroids." *Thorax* 57 (10): 875–879. doi:10.1136/thorax.57.10.875.

Greiff, Lennart, Cecilia Ahlström-Emanuelsson, Mikaela Alenäs, Gun Almqvist, Morgan Andersson, Anders Cervin, Jan Dolata et al. 2015. "Biological Effects and Clinical Efficacy of a Topical Toll-like Receptor 7 Agonist in Seasonal Allergic Rhinitis: A Parallel Group Controlled Phase IIa Study." *Inflammation Research* 64 (11): 903–915. doi:10.1007/s00011-015-0873-2.

Habibi, Maximillian S, Agnieszka Jozwik, Spyridon Makris, Jake Dunning, Allan Paras, John P DeVincenzo, Cornelis a M de Haan et al. 2015. "Impaired Antibody-Mediated Protection and Defective IgA B-Cell Memory in Experimental Infection of Adults with Respiratory Syncytial Virus." *American Journal of Respiratory and Critical Care Medicine* 191 (9): 1040–1049. doi:10.1164/rccm.201412-2256OC.

Hansel, Trevor T, Sebastian L Johnston, and Peter J Openshaw. 2013. "Microbes and Mucosal Immune Responses in Asthma." *The Lancet* 381 (9869): 861–873. doi:10.1016/S0140-6736(12)62202-8.

Hansel, Trevor T., Tanushree Tunstall, Maria-Belen Trujillo-Torralbo, Betty Shamji, Ajerico Del-Rosario, Jaideep Dhariwal, Paul D.W. Kirk et al. 2017. "A Comprehensive Evaluation of Nasal and Bronchial Cytokines and Chemokines Following Experimental Rhinovirus Infection in Allergic Asthma: Increased Interferons (IFN-γ and IFN-λ) and Type 2 Inflammation (IL-5 and IL-13)." *EBioMedicine* 1–11. doi:10.1016/j.ebiom.2017.03.033.

Harris, Ceinwen M., Filipa Mendes, Anca Dragomir, Iolo J.M. Doull, I. Carvalho-Oliveira, Zsuzsanna Bebok, John P. Clancy et al. 2004. "Assessment of CFTR Localisation in Native Airway Epithelial Cells Obtained by Nasal Brushing." *Journal of Cystic Fibrosis: Official Journal of the European Cystic Fibrosis Society* 3 (Suppl 2): 43–48. doi:10.1016/j.jcf.2004.05.009.

Haslam, Patricia L., and Robert P. Baughman. 1999. "Report of ERS Task Force: Guidelines for Measurement of Acellular Components and Standardization of BAL." *The European Respiratory Journal* 14 (2): 245–248.

Hastie, Annette T, Wendy C Moore, Huashi Li, Brian M Rector, Victor E Ortega, Rodolfo M Pascual, Stephen P Peters, Deborah A Meyers, Eugene R Bleecker, and Blood Institute's Severe Asthma Research Program National Heart, Lung. 2013. "Biomarker Surrogates Do Not Accurately Predict Sputum Eosinophil and Neutrophil Percentages in Asthmatic Subjects." *The Journal of Allergy and Clinical Immunology* 132 (1): 72–80. doi:10.1016/j.jaci.2013.03.044.

Hentschel, Julia, Ulrike Müller, Franziska Doht, Nele Fischer, Klas Böer, Jürgen Sonnemann, Christina Hipler et al. 2014. "Influences of Nasal Lavage Collection-, Processing- and Storage Methods on Inflammatory Markers--Evaluation of a Method for Non-Invasive Sampling of Epithelial Lining Fluid in Cystic Fibrosis and Other Respiratory Diseases." *Journal of Immunological Methods* 404 (1): 41–51. doi:10.1016/j.jim.2013.12.003.

Hogan, Simon P, Helene F Rosenberg, Redwan Moqbel, Simon Phipps, Paul S Foster, Paige Lacy, A Barry Kay, and Marc E Rothenberg. 2008. "Eosinophils: Biological Properties and Role in Health and Disease." *Clinical and Experimental Allergy: Journal of the British Society for Allergy and Clinical Immunology* 38 (5): 709–750. doi:10.1111/j.1365-2222.2008.02958.x.

Högman, Marieann, Lauri Lehtimäki, and Anh Tuan Dinh-Xuan. 2017. "Utilising Exhaled Nitric Oxide Information to Enhance Diagnosis and Therapy of Respiratory Disease - Current Evidence for Clinical Practice and Proposals to Improve the Methodology." *Expert Review of Respiratory Medicine* 11 (2): 101–109. doi:10.1080/17476348.2017.1281746.

Horn, Barry R., Eugene D. Robin, James Theodore, and Antonius Van Kessel. 1975. "Total Eosinophil Counts in the Management of Bronchial Asthma." *New England Journal of Medicine* 292 (22): 1152–1155. doi:10.1056/NEJM197505292922204.

Horváth, Ildiko, John Hunt, and Peter J. Barnes 2005. "Exhaled Breath Condensate: Methodological Recommendations and Unresolved Questions." *The European Respiratory Journal* 26 (3): 523–48. doi:10.1183/09031936.05.00029705.

Horváth, Ildiko, Peter J Barnes, Stelios Loukides, Peter J Sterk, Marieann Högman, Anna-carin Olin, Anton Amann et al. 2017. "A European Respiratory Society Technical Standard: Exhaled Biomarkers in Lung Disease." *The European Respiratory Journal* 49 (4). doi:10.1183/13993003.00965-2016.

Howarth, Peter H., Carl G A Persson, Eli O. Meltzer, Mikila R. Jacobson, Stephen R. Durham, and Philip E. Silkoff. 2005. "Objective Monitoring of Nasal Airway Inflammation in Rhinitis." *The Journal of Allergy and Clinical Immunology* 115 (3 Suppl 1): S414–S441. doi:10.1016/j.jaci.2004.12.1134.

Ishizaka, Akitoshi, Masazumi Watanabe, Tetsuji Yamashita, Yasuyo Ogawa, Hidefumi Koh, Naoki Hasegawa, Hidetoshi Nakamura et al. 2001. "New Bronchoscopic Microsample Probe to Measure the Biochemical Constituents in Epithelial Lining Fluid of Patients with Acute Respiratory Distress Syndrome." *Critical Care Medicine* 29 (4): 896–898.

Ishizaka, Akitoshi, Tomoyuki Matsuda, Kurt H Albertine, Hidefumi Koh, Sadatomo Tasaka, Naoki Hasegawa, Nobuoki Kohno et al. 2004. "Elevation of KL-6, a Lung Epithelial Cell Marker, in Plasma and Epithelial Lining Fluid in Acute Respiratory Distress Syndrome." *American Journal of Physiology. Lung Cellular and Molecular Physiology* 286 (6): L1088–L1094. doi:10.1152/ajplung.00420.2002.

Jackson, David J., Heidi Makrinioti, Batika M. J. Rana, Betty W. H. Shamji, Maria-Belen Trujillo-Torralbo, Joseph Footitt, Jerico del-Rosario et al. 2014. "IL-33–Dependent Type 2 Inflammation during Rhinovirus-Induced Asthma Exacerbations *In Vivo*." *American Journal of Respiratory and Critical Care Medicine* 190 (12): 1373–1382. doi:10.1164/rccm.201406-1039OC.

Jackson, David J., Nicholas Glanville, Maria-Belen Trujillo-Torralbo, Betty WH Shamji, Jerico del-Rosario, Patrick Mallia, Matthew J. Edwards, Ross P. Walton, Michael R. Edwards, and Sebastian L. Johnston. 2015. "IL-18 is Associated with Protection against Rhinovirus-Induced Colds and Asthma Exacerbations." *Clinical Infectious Diseases* 60 (10): 1528–1531. doi:10.1093/cid/civ062.

Jalowayski, Alfredo A, Pramila Walpita, Barbara A. Puryear, and James D. Connor. 1990. "Rapid Detection of Respiratory Syncytial Virus in Nasopharyngeal Specimens Obtained with the Rhinoprobe Scraper." *Journal of Clinical Microbiology* 28 (4): 738–41.

Jayaraman, Annabelle, David J. Jackson, Simon D. Message, Rebecca M. Pearson, Julia Aniscenko, Gaetano Caramori, Patrick MalliaJayaraman, Annabelle, David J. Jackson, Simon D. Message, Rebecca M. Pearson, Julia Aniscenko, Gaetano Caramori, Patrick Mallia et al. 2014. "IL-15 Complexes Induce NK- and T-Cell Responses Independent of Type I IFN Signaling during Rhinovirus Infection." *Mucosal Immunology* 7 (5): 1151–1164. doi:10.1038/mi.2014.2.

Jia, Guiquan, Richard W. Erickson, David F. Choy, Sofia Mosesova, Lawren C. Wu, Owen D. Solberg, Aarti Shikotra et al. 2012. "Periostin Is a Systemic Biomarker of Eosinophilic Airway Inflammation in Asthmatic Patients." *The Journal of Allergy and Clinical Immunology* 130 (3): 647–654.e10. doi:10.1016/j.jaci.2012.06.025.

Jirapongsananuruk, Orathai, and Pakit Vichyanond. 1998. "Nasal Cytology in the Diagnosis of Allergic Rhinitis in Children." *Annals of Allergy, Asthma & Immunology: Official Publication of the American College of Allergy, Asthma, & Immunology* 80 (2): 165–170. doi:10.1016/S1081-1206(10)62950-X.

Jochems, Simon P, Katherine Piddock, Jamie Rylance, Hugh Adler, Beatriz F. Carniel, Andrea Collins, Jenna F Gritzfeld et al. 2017. "Novel Analysis of Immune Cells from Nasal Microbiopsy Demonstrates Reliable, Reproducible Data for Immune Populations, and Superior Cytokine Detection Compared to Nasal Wash." *PLOS ONE* 12 (1): e0169805. doi:10.1371/journal.pone.0169805.

Kanazawa, Hiroshi, Toyoki Kodama, Kazuhisa Asai, Saeko Matsumura, and Kazuto Hirata. 2010. "Increased Levels of N ε -(Carboxymethyl)lysine in Epithelial Lining Fluid from Peripheral Airways in Patients with Chronic Obstructive Pulmonary Disease: A Pilot Study." *Clinical Science* 119 (3): 143–149. doi:10.1042/CS20100096.

Kavuru, Mani S., Raed A. Dweik, and Mary Jane Thomassen. 1999. "Role of Bronchoscopy in Asthma Research." *Clinics in Chest Medicine* 20 (1): 153–189. doi:10.1016/S0272-5231(05)70133-7.

Keatings, Vera, Richard Leigh, C. Peterson, Jan Shute, Per Venge, and Ratko Djukanović. 2002. "Standardsed Methodology of Sputum Induction and Processing (ERS Task Force): Analysis of Fluid Phase Mediators." *European Respiratory Journal* 20 (Supplement 37): 24S–39S. doi:10.1183/09031936.02.00002402.

Kelly, Margaret. M., Richard Leigh, S. Carruthers, P. Horsewood, G. J. Gleich, Frederick E. Hargreave, and Gerard Cox. 2001. "Increased Detection of Interleukin-5 in Sputum by Addition of Protease Inhibitors." *The European Respiratory Journal* 18 (4): 685–691.

Kikuchi, Eiki, Junko Kikuchi, Yasuyuki Nasuhara, Satoshi Oizumi, Akitoshi Ishizaka, and Masaharu Nishimura. 2009. "Comparison of the Pharmacodynamics of Biapenem in Bronchial Epithelial Lining Fluid in Healthy Volunteers given Half-Hour and Three-Hour Intravenous Infusions." *Antimicrobial Agents and Chemotherapy* 53 (7): 2799–2803. doi:10.1128/AAC.01578-08.

Kikuchi, Junko, Koichi Yamazaki, Eiki Kikuchi, Akitoshi Ishizaka, and Masaharu Nishimura. 2007. "Pharmacokinetics of Gatifloxacin after a Single Oral Dose in Healthy Young Adult Subjects and Adult Patients with Chronic Bronchitis, with a Comparison of Drug Concentrations Obtained by Bronchoscopic Microsampling and Bronchoalveolar Lavage." *Clinical Therapeutics* 29 (1): 123–130. doi:10.1016/j.clinthera.2007.01.005.

King, David, Ben Mitchell, Christopher P Williams, and Geoffrey K P Spurling. 2015. "Saline Nasal Irrigation for Acute Upper Respiratory Tract Infections."*The Cochrane Database of Systematic Reviews* 4: CD006821. doi:10.1002/14651858.CD006821.pub3.

Kipnis, Eric, Kirk Hansen, Teiji Sawa, Kiyoshi Moriyama, Ashley Zurawel, Akitoshi Ishizaka, and Jeanine Wiener-Kronish. 2008. "Proteomic Analysis of Undiluted Lung Epithelial Lining Fluid." *Chest* 134 (2): 338–345. doi:10.1378/chest.07-1643.

Kodama, Toyoki, Hiroshi Kanazawa, Yoshihiro Tochino, Shigenori Kyoh, Kazuhisa Asai, and Kazuto Hirata. 2009. "A Technological Advance Comparing Epithelial Lining Fluid from Different Regions of the Lung in Smokers." *Respiratory Medicine* 103 (1): 35–40. doi:10.1016/j.rmed.2008.09.004.

Komaki, Yuichi, Hisatoshi Sugiura, Akira Koarai, Masafumi Tomaki, Hiromasa Ogawa, Takefumi Akita, Toshio Hattori, and Masakazu Ichinose. 2005. "Cytokine-Mediated Xanthine Oxidase Upregulation in Chronic Obstructive Pulmonary Disease's Airways." *Pulmonary Pharmacology & Therapeutics* 18 (4): 297–302. doi:10.1016/j.pupt.2005.01.002.

Krause, Andreas, Beate Hohberg, Felicitas Heine, Matthias John, Gerd R. Burmester, and Christian Witt. 1997. "Cytokines Derived from Alveolar Macrophages Induce Fever after Bronchoscopy and Bronchoalveolar Lavage." *American Journal of Respiratory and Critical Care Medicine* 155 (5): 1793–1797. doi:10.1164/ajrccm.155.5.9154894.

Krstajic, Nikola, Ahsan R. Akram, Tushar R. Choudhary, Neil McDonald, Michael G. Tanner, Ettore Pedretti, Paul A. Dalgarno et al. 2016. "Two-Color Widefield Fluorescence Microendoscopy Enables Multiplexed Molecular Imaging in the Alveolar Space of Human Lung Tissue." *Journal of Biomedical Optics* 21 (4): 46009. doi:10.1117/1.JBO.21.4.046009.

Leaker, Brian R, V. A. Malkov, R. Mogg, M. K. Ruddy, Grant C. Nicholson, Andrew J. Tan, C. Tribouley et al. 2016. "The Nasal Mucosal Late Allergic Reaction to Grass Pollen Involves Type 2 Inflammation (IL-5 and IL-13), the Inflammasome (IL-1β), and Complement." *Mucosal Immunology*: 1–13. doi:10.1038/mi.2016.74.

Leckie, Margaret J., Anneke ten Brinke, Jamey Khan, Zuzana Diamant, Brian J. O'connor, Christine M. Walls, Ashwini K. Mathur et al. 2000. "Effects of an Interleukin-5 Blocking Monoclonal Antibody on Eosinophils, Airway Hyper-Responsiveness, and the Late Asthmatic Response." *Lancet (London, England)* 356 (9248): 2144–2148. doi:10.1016/S0140-6736(00)03496-6.

Licari, Amelia, Riccardo Castagnoli, Chiara Francesca Denicolò, Linda Rossini, Alessia Marseglia, and Gian Luigi Marseglia. 2017. "The Nose and the Lung: United Airway Disease?" *Frontiers in Pediatrics* 5 : 44. doi:10.3389/fped.2017.00044.

Lin, Robert Y., Ayoub Nahal, Moon Lee, and Howard Menikoff. 2001. "Cytologic Distinctions between Clinical Groups Using Curette-Probe Compared to Cytology Brush." *Annals of Allergy, Asthma & Immunology: Official Publication of the American College of Allergy, Asthma, & Immunology* 86 (2): 226–231. doi:10.1016/S1081-1206(10)62696-8.

Linden, Margareta, Christer Svensson, Eva Andersson, Morgan Andersson, Lennart Greiff, and Carl GA Persson. 2000. "Immediate Effect of Topical Budesonide on Allergen Challenge-Induced Nasal Mucosal Fluid Levels of Granulocyte-Macrophage Colony-Stimulating Factor and Interleukin-5." *American Journal of Respiratory and Critical Care Medicine* 162 (5): 1705–1708. doi:10.1164/ajrccm.162.5.9910094.

López-Cisternas, Juan, Jessica Castillo-Díaz, Leonidas Traipe-Castro, and Remigio O López-Solís. 2006. "Use of Polyurethane Minisponges to Collect Human Tear Fluid." *Cornea* 25 (3): 312–318. doi:10.1097/01.ico.0000183531.25201.0d.

Lü, Fabien X., and Robert E. Esch. 2010. "Novel Nasal Secretion Collection Method for the Analysis of Allergen Specific Antibodies and Inflammatory Biomarkers." *Journal of Immunological Methods* 356 (1–2): 6–17. doi:10.1016/j.jim.2010.03.004.

Lugogo, Njira, Christian Domingo, Pascal Chanez, Richard Leigh, Martyn J. Gilson, Robert G. Price, Steven W. Yancey, and Hector G. Ortega. 2016. "Long-Term Efficacy and Safety of Mepolizumab in Patients With Severe Eosinophilic

Asthma: A Multi-Center, Open-Label, Phase IIIb Study." *Clinical Therapeutics* 38 (9): 2058–2070.e1. doi:10.1016/j.clinthera.2016.07.010.

Martinez, Fernando D., and Donata Vercelli. 2013. "Asthma." *The Lancet* 382 (9901): 1360–1372. doi:10.1016/S0140-6736(13)61536-6.

McErlean, Peter, Sergejs Berdnikovs, Silvio Favoreto, Junqing Shen, Assel Biyasheva, Rebecca Barbeau, Chris Eisley et al. 2014. "Asthmatics with Exacerbation during Acute Respiratory Illness Exhibit Unique Transcriptional Signatures within the Nasal Mucosa." *Genome Medicine* 6 (1): 1. doi:10.1186/gm520.

Menzies-Gow, Andrew, Patrick Flood-Page, Roma Sehmi, John Burman, Qutayba Hamid, Douglas S Robinson, A. Barry Kay, and Judah Denburg. 2003. "Anti–IL-5 (Mepolizumab) Therapy Induces Bone Marrow Eosinophil Maturational Arrest and Decreases Eosinophil Progenitors in the Bronchial Mucosa of Atopic Asthmatics." *Journal of Allergy and Clinical Immunology* 111 (4): 714–719. doi:10.1067/mai.2003.1382.

Meyer, Norbert, Sarah Janine Nuss, Alexander Siebenhüner, Cezmi A. Akdis, Günter Menz, and Thomas Rothe. 2014. "Differential Serum Protein Markers and the Clinical Severity of Asthma." *Journal of Asthma and Allergy* 7 : 67. doi:10.2147/JAA.S53920.

Mosler, Katharina, Christelle Coraux, Konstantina Fragaki, Jean-Marie Zahm, Odile Bajolet, Katia Bessaci-Kabouya, Edith Puchelle, Michel Abély, and Pierre Mauran. 2008. "Feasibility of Nasal Epithelial Brushing for the Study of Airway Epithelial Functions in CF Infants." *Journal of Cystic Fibrosis: Official Journal of the European Cystic Fibrosis Society* 7 (1): 44–53. doi:10.1016/j.jcf.2007.04.005.

Nakano, Yasushi, Sadatomo Tasaka, Fumitake Saito, Wakako Yamada, Yoshiki Shiraishi, Yuko Ogawa, Hidefumi Koh et al. 2007. "Endothelin-1 Level in Epithelial Lining Fluid of Patients with Acute Respiratory Distress Syndrome." *Respirology (Carlton, Vic.)* 12 (5): 740–743. doi:10.1111/j.1440-1843.2007.01115.x.

National Institute for Health and Care Excellence (NICE). 2014. "Diagnostic Guidance 12. Measuring Fractional Exhaled Nitric Oxide Concentration in Asthma: NIOX MINO, NIOX VERO and NObreath." https://www.nice.org.uk/guidance/dg12. Accessed on May 1, 2018.

Nicholson, Grant C, Harsha H Kariyawasam, Andrew J Tan, Jens M Hohlfeld, Deborah Quinn, Christoph Walker, David Rodman et al. 2011. "The Effects of an Anti-IL-13 mAb on Cytokine Levels and Nasal Symptoms Following Nasal Allergen Challenge." *The Journal of Allergy and Clinical Immunology* 128 (4): 800–807.e9. doi:10.1016/j.jaci.2011.05.013.

Nicola, Marina Lazzari, Heráclito Barbosa de Carvalho, Carolina Tieko Yoshida, Fabyana Maria dos Anjos, Mayumi Nakao, Ubiratan de Paula Santos, Karina Helena Morais Cardozo et al. 2014. "Young 'Healthy' Smokers Have Functional and Inflammatory Changes in the Nasal and the Lower Airways." *Chest* 145 (5): 998–1005. doi:10.1378/chest.13-1355.

Nishimura, Tomoyasu, Naoki Hasegawa, Masazumi Watanabe, Toru Takebayashi, Sadatomo Tasaka, and Akitoshi Ishizaka. 2008. "Bronchoscopic Microsampling to Analyze the Epithelial Lining Fluid of Patients with Pulmonary Mycobacterium Avium Complex Disease." *Respiration; International Review of Thoracic Diseases* 76 (3): 338–343. doi:10.1159/000148063.

Ortega, Hector G., Mark C. Liu, Ian D. Pavord, Guy G. Brusselle, J. Mark FitzGerald, Alfredo Chetta, Marc Humbert et al. 2014. "Mepolizumab Treatment in Patients with Severe Eosinophilic Asthma." *New England Journal of Medicine* 371 (13): 1198–1207. doi:10.1056/NEJMoa1403290.

Pavord, I. D., S. Afzalnia, A. Menzies-Gow, and L. G. Heaney. 2017. "The Current and Future Role of Biomarkers in Type 2 Cytokine-Mediated Asthma Management." *Clinical & Experimental Allergy* 47 (2): 148–160. doi:10.1111/cea.12881.

Pavord, Ian D, Stephanie Korn, Peter Howarth, Eugene R Bleecker, Roland Buhl, Oliver N Keene, Hector Ortega, and Pascal Chanez. 2012. "Mepolizumab for Severe Eosinophilic Asthma (DREAM): A Multicentre, Double-Blind, Placebo-Controlled Trial." *The Lancet* 380 (9842): 651–659. doi:10.1016/S0140-6736(12)60988-X.

Petsky, Helen L., Christopher J. Cates, Toby J. Lasserson, A. M. Li, C. Turner, Jennifer A. Kynaston, and A. B. Chang. 2012. "A Systematic Review and Meta-Analysis: Tailoring Asthma Treatment on Eosinophilic Markers (Exhaled Nitric Oxide or Sputum Eosinophils)." *Thorax* 67 (3): 199–208. doi:10.1136/thx.2010.135574.

Piedra, Felipe-Andrés, Minghua Mei, Vasanthi Avadhanula, Reena Mehta, Letisha Aideyan, Roberto P. Garofalo, and Pedro A. Piedra. 2017. "The Interdependencies of Viral Load, the Innate Immune Response, and Clinical Outcome in Children Presenting to the Emergency Department with Respiratory Syncytial Virus-Associated Bronchiolitis." *PLOS ONE* 12 (3): e0172953. doi:10.1371/journal.pone.0172953.

Pin, I, P G Gibson, R Kolendowicz, A Girgis-Gabardo, J A Denburg, F E Hargreave, and J Dolovich. 1992. "Use of Induced Sputum Cell Counts to Investigate Airway Inflammation in Asthma." *Thorax* 47 (1): 25–29. doi:10.1136/THX.47.1.25.

Pizzichini, Emilio, M. M. Pizzichini, Ann Efthimiadis, Susan Evans, Marilyn M. Morris, Diane Squillace, Gerald J. Gleich, Jerry Dolovich, and Frederick E. Hargreave. 1996. "Indices of Airway Inflammation in Induced Sputum: Reproducibility and Validity of Cell and Fluid-Phase Measurements." *American Journal of Respiratory and Critical Care Medicine* 154 (2 Pt 1): 308–317. doi:10.1164/ajrccm.154.2.8756799.

Poole, Alex, Cydney Urbanek, Celeste Eng, Jeoffrey Schageman, Sean Jacobson, Brian P O'Connor, Joshua M Galanter et al. 2014. "Dissecting Childhood Asthma with Nasal Transcriptomics Distinguishes Subphenotypes of Disease." *The Journal of Allergy and Clinical Immunology* 133 (3): 670–678.e12. doi:10.1016/j.jaci.2013.11.025.

Price, David B, Anna Rigazio, Jonathan D Campbell, Eugene R Bleecker, Christopher J Corrigan, Mike Thomas, Sally E Wenzel et al. 2015. "Blood Eosinophil Count and Prospective Annual Asthma Disease Burden: A UK Cohort Study." *The Lancet Respiratory Medicine* 3 (11): 849–858. doi:10.1016/S2213-2600(15)00367-7.

Proud, David, Ronald B Turner, Birgit Winther, Shahina Wiehler, Jay P Tiesman, Tim D Reichling, Kenton D Juhlin et al. 2008. "Gene Expression Profiles during in Vivo Human Rhinovirus Infection: Insights into the Host Response." *American Journal of Respiratory and Critical Care Medicine* 178 (9): 962–968. doi:10.1164/rccm.200805-670OC.

Proud, David, Scherer P. Sanders, and Shahina Wiehler. 2004. "Human Rhinovirus Infection Induces Airway Epithelial Cell Production of Human -Defensin 2 Both *In Vitro* and *In Vivo*." *The Journal of Immunology* 172 (7): 4637–4645. doi:10.4049/jimmunol.172.7.4637.

Rebuli, Meghan E., Adam M. Speen, Phillip W. Clapp, and Ilona Jaspers. 2017. "Novel Applications for a Non-invasive Sampling Method of the Nasal Mucosa." *American Journal of Physiology—Lung Cellular and Molecular Physiology* 312 (2): L288–L296. doi:10.1152/ajplung.00476.2016.

Riechelmann, H., T. Deutschle, E. Friemel, H-J. J. Gross, and M. Bachem. 2003. "Biological Markers in Nasal Secretions." *The European Respiratory Journal* 21 (4): 600–605. doi:10.1183/09031936.03.00072003.

Rohan, Lisa Cencia, Robert P. Edwards, Lori A. Kelly, Kelly A. Colenello, Frederick P. Bowman, and Peggy A. Crowley-Nowick. 2000. "Optimization of the Weck-Cel Collection Method for

Quantitation of Cytokines in Mucosal Secretions." *Clinical and Diagnostic Laboratory Immunology* 7 (1): 45–48. doi:10.1128/CDLI.7.1.45-48.2000.

Rosenberg, Helene F., Kimberly D. Dyer, and Paul S. Foster. 2012. "Eosinophils: Changing Perspectives in Health and Disease." *Nature Reviews Immunology* 13 (1): 9–22. doi:10.1038/nri3341.

Salk, Hannah M., Whitney L. Simon, Nathaniel D. Lambert, Richard B. Kennedy, Diane E. Grill, Brian F. Kabat, and Gregory A. Poland. 2016. "Taxa of the Nasal Microbiome Are Associated with Influenza-Specific IgA Response to Live Attenuated Influenza Vaccine." *PLOS ONE* 11 (9): e0162803. doi:10.1371/journal.pone.0162803.

Sand, Kristin L, Torun Flatebo, Marian Berge Andersen, and Azzam A Maghazachi. 2013. "Effects of Exercise on Leukocytosis and Blood Hemostasis in 800 Healthy Young Females and Males." *World Journal of Experimental Medicine* 3 (1): 11–20. doi:10.5493/wjem.v3.i1.11.

Sasabayashi, Mari, Yoshitaka Yamazaki, Kenji Tsushima, Orie Hatayama, and Tadashi Okabe. 2007. "Usefulness of Bronchoscopic Microsampling To Detect the Pathogenic Bacteria of Respiratory Infection." *Chest* 131 (2): 474–479. doi:10.1378/chest.06-0989.

Scadding, Guy W., Aarif O. Eifan, Martin Penagos, Florentina A. Dumitru, A. Switzer, O. McMahon, D. Phippard, A. Togias, Stephen R. Durham, and Mohamed H. Shamji. 2015. "Local and Systemic Effects of Cat Allergen Nasal Provocation." *Clinical and Experimental Allergy: Journal of the British Society for Allergy and Clinical Immunology* 45 (3): 613–623. doi:10.1111/cea.12434.

Scadding, Guy W, Moises a Calderon, Virginia Bellido, Gitte Konsgaard Koed, Niels-Christian Nielsen, Kaare Lund, Alkis Togias et al. 2012. "Optimisation of Grass Pollen Nasal Allergen Challenge for Assessment of Clinical and Immunological Outcomes." *Journal of Immunological Methods* 384 (1–2): 25–32. doi:10.1016/j.jim.2012.06.013.

Schleich, Florence N, Maité Manise, Jocelyne Sele, Monique Henket, Laurence Seidel, and Renaud Louis. 2013. "Distribution of Sputum Cellular Phenotype in a Large Asthma Cohort: Predicting Factors for Eosinophilic vs Neutrophilic Inflammation." *BMC Pulmonary Medicine* 13 (1): 11. doi:10.1186/1471-2466-13-11.

Schleich, Florence Nicole, Anne Chevremont, Virginie Paulus, Monique Henket, Maité Manise, Laurence Seidel, and Renaud Louis. 2014. "Importance of Concomitant Local and Systemic Eosinophilia in Uncontrolled Asthma." *European Respiratory Journal* 44 (1): 97–108. doi:10.1183/09031936.00201813.

Sennels, Henriette P, Henrik L Jørgensen, Anne-Louise S Hansen, Jens P Goetze, and Jan Fahrenkrug. 2011. "Diurnal Variation of Hematology Parameters in Healthy Young Males: The Bispebjerg Study of Diurnal Variations." *Scandinavian Journal of Clinical and Laboratory Investigation* 71 (7): 532–541. doi:10.3109/00365513.2011.602422.

Sim, Tommy C., J. Andrew Grant, Kimberly A. Hilsmeier, Yoshiaki Fukuda, and Rafeul Alam. 1994. "Proinflammatory Cytokines in Nasal Secretions of Allergic Subjects after Antigen Challenge." *American Journal of Respiratory and Critical Care Medicine* 149 (2 Pt 1): 339–344. doi:10.1164/ajrccm.149.2.8306027.

Sim, Tommy C., Lisa M. Reece, Kimberly A. Hilsmeier, J. Andrew Grant, and Rafeul Alam. 1995. "Secretion of Chemokines and Other Cytokines in Allergen-Induced Nasal Responses: Inhibition by Topical Steroid Treatment." *American Journal of Respiratory and Critical Care Medicine* 152 (3): 927–933. doi:10.1164/ajrccm.152.3.7545059.

Smith, Steven G., Ruchong Chen, Melanie Kjarsgaard, Chynna Huang, John-Paul Oliveria, Paul M. O'Byrne, Gail M. Gauvreau et al. 2016. "Increased Numbers of Activated Group 2 Innate Lymphoid Cells in the Airways of Patients with Severe Asthma and Persistent Airway Eosinophilia." *The Journal of Allergy and Clinical Immunology* 137 (1): 75–86.e8. doi:10.1016/j.jaci.2015.05.037.

Sobel Leonard, Ashley, Micah T McClain, Gavin J D Smith, David E Wentworth, Rebecca A Halpin, Xudong Lin, Amy Ransier et al. 2016. "Deep Sequencing of Influenza A Virus from a Human Challenge Study Reveals a Selective Bottleneck and Only Limited Intrahost Genetic Diversification." *Journal of Virology* 90 (24): 11247–11258. doi:10.1128/JVI.01657-16.

Spector, Sheldon Laurence, and Ricardo Antonio Tan. 2012. "Is a Single Blood Eosinophil Count a Reliable Marker For 'eosinophilic Asthma?'." *The Journal of Asthma: Official Journal of the Association for the Care of Asthma* 49 (8): 807–810. doi:10.3109/02770903.2012.713428.

Steveling, Esther Helen, Mongkol Lao-Araya, Christopher Koulias, Guy Scadding, Aarif Eifan, Louisa K. James, Alina Dumitru et al. 2015. "Protocol for a Randomised, Double-Blind, Placebo-Controlled Study of Grass Allergen Immunotherapy Tablet for Seasonal Allergic Rhinitis: Time Course of Nasal, Cutaneous and Immunological Outcomes." *Clinical and Translational Allergy* 5 (1): 43. doi:10.1186/s13601-015-0087-2.

Stokes, Andrea B, Elisabeth Kieninger, Aline Schögler, Brigitte S Kopf, Carmen Casaulta, Thomas Geiser, Nicolas Regamey, and Marco P Alves. 2014. "Comparison of Three Different Brushing Techniques to Isolate and Culture Primary Nasal Epithelial Cells from Human Subjects." *Experimental Lung Research* 40 (7): 327–332. doi:10.3109/01902148.2014.925987.

Sugasawa, Yusuke, Keisuke Yamaguchi, Seiichiro Kumakura, Taisuke Murakami, Toyoki Kugimiya, Kenji Suzuki, Isao Nagaoka, and Eiichi Inada. 2011. "The Effect of One-Lung Ventilation upon Pulmonary Inflammatory Responses during Lung Resection." *Journal of Anesthesia* 25 (2): 170–177. doi:10.1007/s00540-011-1100-0.

Sugasawa, Yusuke, Keisuke Yamaguchi, Seiichiro Kumakura, Taisuke Murakami, Kenji Suzuki, Isao Nagaoka, and Eiichi Inada. 2012. "Effects of Sevoflurane and Propofol on Pulmonary Inflammatory Responses during Lung Resection." *Journal of Anesthesia* 26 (1): 62–69. doi:10.1007/s00540-011-1244-y.

Tefferi, Ayalew. 2005. "Blood Eosinophilia: A New Paradigm in Disease Classification, Diagnosis, and Treatment." *Mayo Clinic Proceedings* 80 (1): 75–83. doi:10.4065/80.1.75.

Trivedi, Abhaya, Chase Hall, Eric A. Hoffman, Jason C. Woods, David S. Gierada, and Mario Castro. 2017. "Using Imaging as a Biomarker for Asthma." *Journal of Allergy and Clinical Immunology* 139 (1): 1–10. doi:10.1016/j.jaci.2016.11.009.

van der Schee, Marc P., Simone Hashimoto, Annemarie C. Schuurman, Janine S. Repelaer van Driel, Nora Adriaens, Romy M. van Amelsfoort, Tessa Snoeren et al. 2015. "Altered Exhaled Biomarker Profiles in Children during and after Rhinovirus-Induced Wheeze." *European Respiratory Journal* 45 (2): 440–448. doi:10.1183/09031936.00044414.

van der Schee, Marc Philippe, Tamara Paff, Paul Brinkman, Willem Marinus Christiaan van Aalderen, Eric Gerardus Haarman, and Peter Jan Sterk. 2015. "Breathomics in Lung Disease." *Chest* 147 (1): 224–231. doi:10.1378/chest.14-0781.

van Meegen, Marit A., Suzanne W. J. Terheggen-Lagro, Cornelis K. van der Ent, and Jeffrey M. Beekman. 2011. "CFTR Expression Analysis in Human Nasal Epithelial Cells by Flow Cytometry." *PLoS ONE* 6 (12): e27658. doi:10.1371/journal.pone.0027658.

Wagener, Ariane H., Ching Yong Yick, Paul Brinkman, Marc P. van der Schee, Niki Fens, and Peter J. Sterk. 2013. "Toward Composite Molecular Signatures in the Phenotyping of Asthma." *Annals of the American Thoracic Society* 10 (Supplement): S197–S205. doi:10.1513/AnnalsATS.201302-035AW.

Wagener, Ariane H., Selma B. de Nijs, Rene Lutter, Ana R. Sousa, Els JM Weersink, Elisabeth H. Bel, and Peter J. Sterk. 2015. "External Validation of Blood Eosinophils, FE NO and Serum Periostin as Surrogates for Sputum Eosinophils in Asthma." *Thorax* 70 (2): 115–120. doi:10.1136/thoraxjnl-2014-205634.

Wagenmann, M., L. Schumacher, and C. Bachert. 2005. "The Time Course of the Bilateral Release of Cytokines and Mediators after Unilateral Nasal Allergen Challenge." *Allergy* 60 (9): 1132–1138. doi:10.1111/j.1398-9995.2005.00867.x.

Wagenmann, Martin, Fuad M. Baroody, Cheng-Chou Cheng, Anne Kagey-Sobotka, Lawrence M. Lichtenstein, and Robert M. Naclerio. 1997. "Bilateral Increases in Histamine after Unilateral Nasal Allergen Challenge." *American Journal of Respiratory and Critical Care Medicine* 155 (2): 426–431. doi:10.1164/ajrccm.155.2.9032173.

Watanabe, Masazumi, Akitoshi Ishizaka, Eiji Ikeda, Akira Ohashi, and Koichi Kobayashi. 2003. "Contributions of Bronchoscopic Microsampling in the Supplemental Diagnosis of Small Peripheral Lung Carcinoma." *The Annals of Thoracic Surgery* 76 (5): 1668–1672. doi:10.1016/S0003-4975(03)01015-4.

Weido, Anthony J, Lisa M Reece, Rafeul Alam, Cindy K Cook, and Tommy C Sim. 1996. "Intranasal Fluticasone Propionate Inhibits Recovery of Chemokines and Other Cytokines in Nasal Secretions in Allergen-Induced Rhinitis." *Annals of Allergy, Asthma & Immunology: Official Publication of the American College of Allergy, Asthma, & Immunology* 77 (5): 407–415. doi:10.1016/S1081-1206(10)63340-6.

Wenzel, Sally E. 2012. "Asthma Phenotypes: The Evolution from Clinical to Molecular Approaches." *Nature Medicine* 18 (5): 716–725. doi:10.1038/nm.2678.

Wenzel, Sally E. 2016. "Emergence of Biomolecular Pathways to Define Novel Asthma Phenotypes. Type-2 Immunity and Beyond." *American Journal of Respiratory Cell and Molecular Biology* 55 (1): 1–4. doi:10.1165/rcmb.2016-0141PS.

Wolsk, Helene Mygind, Bo L. Chawes, Nilofar V. Følsgaard, Morten A. Rasmussen, S. Brix, and H. Bisgaard. 2016. "Siblings Promote a Type 1/Type 17-Oriented Immune Response in the Airways of Asymptomatic Neonates." *Allergy* 71 (6): 820–828. doi:10.1111/all.12847.

Woodruff, Prescott G, Homer A Boushey, Gregory M Dolganov, Chris S Barker, Yee Hwa Yang, Samantha Donnelly, Almut Ellwanger et al. 2007. "Genome-Wide Profiling Identifies Epithelial Cell Genes Associated with Asthma and with Treatment Response to Corticosteroids." *Proceedings of the National Academy of Sciences of the United States of America* 104 (40): 15858–15863. doi:10.1073/pnas.0707413104.

Woodruff, Prescott G., Barmak Modrek, David F. Choy, Guiquan Jia, Alexander R. Abbas, Almut Ellwanger, Laura L. Koth, Joseph R. Arron, and John V. Fahy. 2009. "T-Helper Type 2-Driven Inflammation Defines Major Subphenotypes of Asthma." *American Journal of Respiratory and Critical Care Medicine* 180 (5): 388–395. doi:10.1164/rccm.200903-0392OC.

Yamazaki, Koichi, Shigeaki Ogura, Akitoshi Ishizaka, Toshinari Oh-hara, and Masaharu Nishimura. 2003. "Bronchoscopic Microsampling Method for Measuring Drug Concentration in Epithelial Lining Fluid." *American Journal of Respiratory and Critical Care Medicine* 168 (11): 1304–1307. doi:10.1164/rccm.200301-111OC.

Yanagi, S, J Ashitani, K Imai, Y Kyoraku, A Sano, N Matsumoto, and M Nakazato. 2007. "Significance of Human Beta-Defensins in the Epithelial Lining Fluid of Patients with Chronic Lower Respiratory Tract Infections." *Clinical Microbiology and Infection: The Official Publication of the European Society of Clinical Microbiology and Infectious Diseases* 13 (1): 63–69. doi:10.1111/j.1469-0691.2006.01574.x.

Yasuda, Hiroyuki, Kenzo Soejima, Sohei Nakayama, Ichiro Kawada, Ichiro Nakachi, Satoshi Yoda, Ryosuke Satomi et al. 2011. "Bronchoscopic Microsampling Is a Useful Complementary Diagnostic Tool for Detecting Lung Cancer." *Lung Cancer* 72 (1): 32–38. doi:10.1016/j.lungcan.2010.07.016.

Zeiger, Robert S., Michael Schatz, Anand A. Dalal, Wansu Chen, Ekaterina Sadikova, Robert. Y. Suruki, Aniket A. Kawatkar, and Lei Qian. 2017. "Blood Eosinophil Count and Outcomes in Severe Uncontrolled Asthma: A Prospective Study." *The Journal of Allergy and Clinical Immunology: In Practice* 5 (1): 144–153. e8. doi:10.1016/j.jaip.2016.07.015.

Zhang, X.-Y., J. L. Simpson, H. Powell, I. A. Yang, J. W. Upham, P. N. Reynolds, S. Hodge et al. 2014. "Full Blood Count Parameters for the Detection of Asthma Inflammatory Phenotypes." *Clinical and Experimental Allergy: Journal of the British Society for Allergy and Clinical Immunology* 44 (9): 1137–1145. doi:10.1111/cea.12345.

Stratified Medicine in Autoimmune Diseases

8.2

John A. Reynolds and Ian N. Bruce

Contents

INTRODUCTION

The past two decades has witnessed an unprecedented and rapid expansion in the understanding of the pathogenesis of autoimmune diseases. As a result, we have also witnessed the development of a large number of novel therapeutic agents (both biologics and small molecules) across a broad spectrum of autoimmune conditions. Despite this huge effort, not all diseases have had the same amount of "success" when it comes to drugs making it through pivotal phase III trials and on to becoming approved medications for widespread use.

Rheumatoid arthritis (RA) is a condition where a number of new targeted therapies have been licensed and approved for use. Indeed, some of the drugs now used to treat RA are among the largest selling pharmaceutical agents globally (Lindsley 2013). In contrast, drug development in many other autoimmune conditions, most notably the connective tissue diseases (CTDs)—systemic lupus erythematosus (SLE), Sjögren's syndrome and systemic sclerosis—has been much more challenging. As a result, across the spectrum of CTDs only one drug has been licensed by the U.S. Food and Drug Administration (FDA) and other regulators in the past 50 years (Horowitz and Furie 2012) (Figure 8.2.1 and Table 8.2.1.)

When we consider stratified medicine in autoimmune diseases, it is useful to consider the contrasting challenges posed in these groups of disorders. In RA, for example, there are now soaring costs of biological agents and an increasing drug bill for health care payers for this condition. Although many drugs have been successful, there is very little evidence to demonstrate clear superiority of one drug over another. At an individual patient level it remains unclear which drug is most likely to elicit an excellent response/remission. Current therapeutic decision making is therefore driven largely by order of entry to the market, drug pricing and national approval processes. There is a pressing need for a better evidence base to inform therapeutic choices (Isaacs and Ferraccioli 2011).

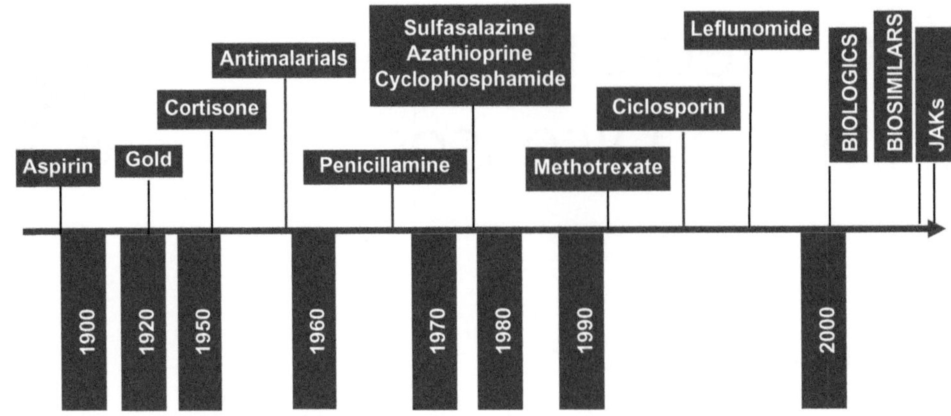

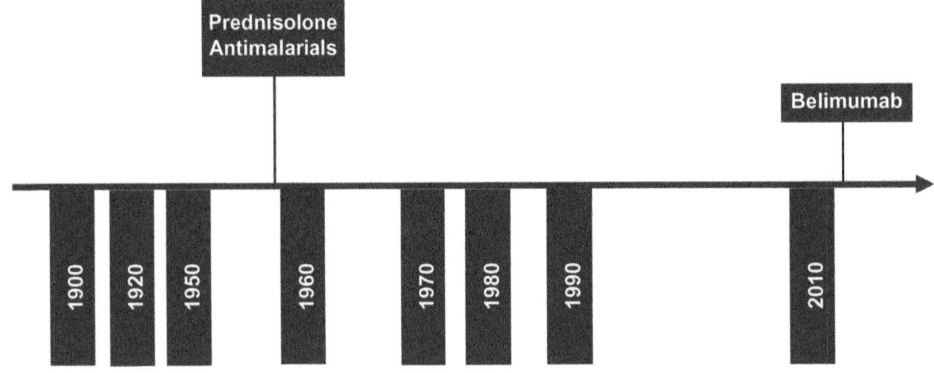

FIGURE 8.2.1 Historic schematic summarizing medications approved and licensed for use in (1A) rheumatoid arthritis (RA) and (1B) systemic lupus erythematosus (SLE). In RA, a wide range of different MOA agents have been approved even since 2000. In SLE, only one drug has been licensed and approved in the past 50 years.

TABLE 8.2.1 Summary of potential biomarkers to predict drug response in RA and SLE

BIOMARKER	*DRUG*	*DETAILS*	*REFERENCES*
RA			
RF+ ACPA+	RTX	Increased response at 6 months	Chatzidionysiou et al. (2011)
Shared epitope (SE)	DMARD	Increased response to "triple therapy" compared with MTX alone	O'Dell et al. (1998)
	Anti-TNF	VKA haplotype associated with large reduction in DAS-28 score and increased overall response	Ling et al. (2016)
TNFα promoter region	Anti-TNF	Increased frequency of A allele in responders compared with nonresponders	O'Rielly et al. (2009)
FCGR3A	RTX	Increased likelihood of response to RTX in patients with V allele at position 159	Ruyssen-Witrand et al. (2012)
	RTX	Improved response at 6 in patients with VV genotype compared with other genotypes	Quartuccio et al. (2014)
Plasmablasts	RTX	Fewer CD19+/IgD–/CD27– associated with reduced response in patients with seronegative RA	Tony et al. (2017)
	RTX	Increased plasmablasts and baseline lymphocyte count associated with reduced response	Stradner et al. (2016)

(Continued)

TABLE 8.2.1 (*Continued*) Summary of potential biomarkers to predict drug response in RA and SLE

BIOMARKER	*DRUG*	*DETAILS*	*REFERENCES*
Type 1 interferon gene signature	RTX	Increased expression of interferon genes at 3 months in responders	Vosslamber et al. (2011)
	RTX	Patients with increased interferon score less likely to respond Effect most pronounced if not taking corticosteroids	de Jong et al. (2015), (2017)
	Anti-TNF	Poor response to treatment in patients with increased interferon gene expression at 1 month posttreatment	van Baarsen et al. (2010)
IL-33	RTX	Increased response to RTX in seropositive patients with high IL-33 and total IgG	Sellam et al. (2016)
MRP8/14 (calprotectin)	Anti-TNF/RTX	Higher baseline MRP8/14 in responders compared with nonresponders	Choi et al. (2013)
CXCL-10, CXCL-13	Anti-TNF	Increased baseline levels in responders compared with nonresponders	Han et al. (2016)
SLE			
Anti-ENA antibodies	RTX	Associated with a reduced time to flare compared with antibody negative patients	Ng et al. (2007)
	RTX	Reduced likelihood in improvement in skin in patients with positive anti-Ro or anti-RNP antibodies	Vital et al. (2015)
Raised anti-dsDNA antibodies, low serum complement	RTX	Serological response in patients with raised dsDNA and/or low complement	Merrill et al. (2010)
	Belimumab	Increased response compared with placabo	van Vollenhoven et al. (2012)
Plasmablasts	RTX	More rapid repopulation of plasmablasts (CD19+/CD27++/CD38++) and memory B cells (CD19++/CD27+),)in patients with earlier relapse	Vital et al. (2011)
Type 1 interferon gene signature	Sifalimumab	Greater improvement in patients with higher interferon expression at baseline	Khamashta et al. (2016)
	Anifrolumab	Increase response in patients with increased interferon expression at baseline	Furie et al. (2017)

RTX = rituximab, DMARD = Non-biological disease-modifity anti-rheumatic drug, MMF = mycophenolate mofetil, MTX = methotrexate, CXCL = chemokine C-X-C motif ligand.

In CTDs, in contrast, the challenge is that, with only a limited number of agents showing any success in early phase trials, how do we better understand the heterogeneity, and indeed the actual true taxonomy of lupus and other CTDs, in order to better match pathogenesis with targeted therapy (Teruel et al. 2017).

This section will seek to address some of these issues using RA and SLE as two key exemplars of the current status of stratified medicine in autoimmune diseases.

RHEUMATOID ARTHRITIS

RA is a chronic inflammatory disease affecting synovial joints and which presents as a symmetrical polyarthritis affecting the small joints of the hands, wrists, feet and ankles in particular. The therapeutic choices for RA have expanded greatly over the past 20 years and a better understanding of the disease processes have led to much earlier introduction of disease-modifying antirheumatic drugs (DMARDs). Until 20 years ago, there was a so called "therapeutic pyramid" where symptomatic relief and less potent DMARDs were used early and more potent immunomodulatory therapies were reserved for more advanced cases (Fries 2000). A large number of clinical trials and treatment strategy studies over the past 15–20 years have completely inverted this pyramid such that national and international guidelines now promote the early use of effective DMARD therapy. A mainstay of therapy in RA is methotrexate (MTX), which is used alone or in combination with other classic DMARDs once the diagnosis is established. MTX can be considered an "anchor" drug onto which other DMARDs or biological therapies are added (Pincus et al. 2017).

The therapeutic success of targeted anti-tumor necrosis factor (TNF) agents paved the way for the widespread use of biological agents in RA (Maini et al. 1999). Following on from the development of anti-TNF drugs, multiple additional targeted therapies have since been approved for use including rituximab (RTX), abatacept, tocilizumab and more recently, JAK-kinase inhibitors (Chatzidionysiou et al. 2017; Nam et al. 2017). Biosimilars are now entering the marketplace meaning that the choice of individual agent for any patient at any stage of their disease is much greater than previously (Nam et al. 2017).

In spite of the wide and proven efficacy of many of these agents most biological therapies in RA, whatever the target, broadly conform to the "60-40-20" rule, that is, approximately 60% of patients will have an American College of Rheumatology

(ACR) 20 improvement, 40% will have an ACR 50 improvement and only 20% will have an ACR 70 improvement, which represents a major therapeutic response (Weinblatt et al. 2003; Emery et al. 2008). There is also very little evidence of any clear superiority in efficacy of one drug over the other in "all-comer" trials of these agents (Weinblatt et al. 2013). There have therefore been significant endeavors over the past 5–10 years to better understand biomarkers related to RA and in particular those that may help stratify patients according to prognostic subsets, as well as stratifying patient according to prediction of response and/or adverse events to therapy.

Seropositive Disease

The presence of rheumatoid factor (RF) and in particular anti-cyclic citrullinated peptide antibodies (ACPA) identifies patients with a poorer long-term prognosis including mortality, cardiovascular risk and disability (Farragher et al. 2008; Naz et al. 2008; Humphreys et al. 2014). Seropositivity has also been studied in regard to whether it predicts responses to particular RA therapies. A number of studies have addressed this with anti-TNF agents with conflicting results. In terms of RF, a large real-world observational cohort based in the UK Biologics Registry studied 2879 RA patients found no difference in likelihood of having a clinically meaningful response to either infliximab or etanercept in RF positive patients (Hyrich et al. 2006). In an additional analysis of 642 UK patients with RA, Potter et al., however, found that RF or ACPA positive patients had a smaller improvement in DAS28 compared with RF/ACPA negative patients when treated with infliximab, etanercept or adalimumab (Potter et al. 2009). However, a recent meta-analysis that included data from 14 studies and 5561 patients concluded that RF or ACPA status did not significantly affect response to anti-TNF (Lv et al. 2014).

Response to RTX seems more clearly influenced by RF/ACPA status. In a study of pooled data from 10 European registries, RF and ACPA positive patients had increased response rates to RTX (Chatzidionysiou et al. 2011). The presence of ACPA resulted in an almost three times increased likelihood of having a European League Against Rheumatism (EULAR) good response at 6 months (Chatzidionysiou et al. 2011).

Genetic Markers

A number of studies have also examined whether genetic polymorphisms associated with disease susceptibility or severity may also act as predictors of response to therapeutic agents. One major susceptibility factor for RA is the presence of particular HLADRB1 genetic variants the so-called shared epitope (SE) (Viatte et al. 2016). In a secondary analysis of the original trial of triple DMARD therapy in RA, O'Dell at al noted that in patients negative for the SE, the response to MTX monotherapy was similar to that of triple DMARD therapy; in contrast in SE positive patients, triple therapy was superior to MTX monotherapy (O'Dell et al. 1998). More recent work on SE has found that three key amino acid positions (11, 71, and 74) within HLA-DRB1 are the major drivers of the association between HLADRB1 and RA (Ling et al. 2016; Viatte et al. 2016). Using this more detailed analysis, Viatte et al. also showed that the VKA haplotype was associated with a better response to TNF inhibitor drugs (odds ratio [OR] = 1.23 [95% confidence interval {CI}, 1.06, 1.43]) (Viatte et al. 2016).

A number of other genetic markers have also been studied. For anti-TNF therapy, no consistent findings have emerged across a number of studies. Potter et al. found no association between SE or PTPN22 polymorphisms and response to anti-TNF (Potter et al. 2009). A few studies have found an association between the TNFα promoter 308 region and response to anti-TNF. In a meta-analysis of studies involving 692 patients the odds of having the A allele was lower in anti-TNF responders versus nonresponders (OR = 0.43 [95% CI: 0.28, 0.68]) (O'Rielly et al. 2009). A subsequent systematic review suggested, however, that despite a number of studies showing similar results, the effect size was small and may therefore not have any clinical utility in decision making (Cuppen et al. 2016).

In respect to Rituximab (RTX) use, a number of studies have shown associations between response and the FCGR3A variants with carriage of a V allele at position 158 being associated with better responses. Ruyssen-Witrand et al. found in 111 patients treated with RTX, the V allele was associated with a higher odds of response (OR = 3.8 [95% CI: 1.2, 11.7]) (Ruyssen-Witrand et al. 2012). Similarly, Quartuccio et al. showed that of 212 RA patients treated with RTX significantly more patients with the 158 VV genotype responded at 6 months (86.5%) compared with the VF or FF genotypes (66%) (Quartuccio et al. 2014).

Overall limitations of these genetic studies to date have been small numbers in many cohorts and lack of a confirmation set. In addition, a number of findings have not been assessed in other drugs exposures, for example, the work by Viatte et al. has not yet been examined for other biologics or for DMARD therapy (Viatte et al. 2016). In addition, for example, one small study of 63 patients treated with RTX found that the TNF 308 A allele occurred in 15% of nonresponders and 32% of responders (Daïen et al. 2012). A larger study will be needed to determine whether this many also be a predictor of RTX response.

Synovial Tissue

The synovium is an important end-organ in the pathogenesis of RA. Although synovial biopsies can only be performed by trained operators, the synovium offers an attractive ad highly relevant tissue to sample for the purpose of identifying predictive biomarkers. In 2007, a small study of 18 RA patients (12 responders and 6 nonresponders) used gene expression profiling to uncover overrepresented molecular pathways in patients with good response to infliximab (van der Pouw Kraan et al. 2008). Infliximab responders had significant up-regulation of immune-related pathways including those involved in cytokine signaling and cell adhesion. These findings are supported by the observations that the synovium of infliximab responders had increased infiltration with macrophages and T cells, and increased cell adhesion molecule and cytokine expression (notably TNFα) (Wijbrandts et al. 2008). In addition to being a biomarker of those more likely to respond, lymphocyte aggregates in the synovium of RA patients resolve during disease treatment, suggesting an important pathogenic role (Klaasen et al. 2009).

Understandably, the use of synovial biopsy has not yet entered routine clinical practice due to at least, in part, it being an invasive procedure requiring expert operators. The natural progression is therefore to try to extrapolate biopsy findings to other more readily available biofluids (e.g., blood). Although increased levels of TNFα in synovium may be a good indicator of likelihood to response to anti-TNF treatment, the utility of measuring circulating TNFα levels is less convincing (discussed below).

Cellular and Cytokine Biomarkers

Cellular biomarkers may also offer some predictive value and modify the effects of some of the serum biomarkers although results have been conflicting. Plasmablasts are a distinct subtype of plasma cells that secrete large quantities of immunoglobulin. In a study by Tony et al. fewer CD19+, IgD–, CD27– cells resulted in a lower response to RTX in seronegative patients (Tony et al. 2017). Interestingly, although seronegative patients have reduced response to RTX in general, higher numbers of plasmablasts may identify a subgroup with increased response rates. Conversely, first treatment with RTX was less likely to result in low-disease activity (LDA) in patients with increased plasmablasts or lymphocytes at baseline (Stradner et al. 2016). The sensitivity of this combination for failure to achieve LDA at Week 24 was high (93.3%); however, the specificity was relatively poor (44.9%). Furthermore, this observation may not be specific to RTX as it was also associated with higher pretreatment disease activity, which would need a greater relative reduction in order to achieve LDA.

Although much effort has been made to identify serum cytokine biomarkers to predict response, there has been little success. The interferon signature is a collective term for a number of type 1 interferon stimulated genes (ISGs). In patients with RA, expression of this signature is associated with a reduced response to RTX (Vosslamber et al. 2011) (de Jong et al. 2017) and an increase post anti-TNF therapy was also associated with poorer responses (van Baarsen et al. 2010). Importantly, this signature can be modified by the use of corticosteroids (de Jong et al. 2015). This has the undesirable effect of reducing the predictive value of using the signature to predict response in patients also taking steroids. More recently, a small observational study has proposed that elevated serum IL-33 levels when considered alongside total IgG level and seropositivity was significantly associated with response to RTX (Sellam et al. 2016). The immunopathological basis for this observation is not known. However, IL-33 is unlikely to have clinical utility as a predictive biomarker as it was only detected in 33.5% of the 185 patients. The number of patients with elevated IL-33, combined with seropositivity and raised IgG was even smaller (<10%) (Sellam et al. 2016).

Myeloid-related proteins (MRP) 8 and 14 are endogenous TLR-4 ligands that have profound pro-inflammatory effects. MRP8 and 14 have been shown to promote joint inflammation in animal models and serum and faecal levels are used to monitor a number of inflammatory conditions (the MRP8/14 complex is also known as calprotectin). Choi et al. recently found that baseline MRP8/14 was higher in patients with RA who subsequently responded to adalimumab, infliximab or RTX with odds ratios of 3.30 (95% CI 1.14, 9.60), 9.75 (1.93, 49.33) and 55 (4.30, 703.43) respectively (Choi et al. 2013). Adding in MRP8/14 levels may therefore improve the ability to stratify patients for biological therapies and may add to the ability to decide the likelihood of response at an individual level (Nair et al. 2016).

TNF and Response to Anti-TNF

It seems logical perhaps that those patients with predominantly TNF-driven disease may have an increased response to anti-TNF agents. Conversely, however, in a small study of 55 RA patients, whereas pretreatment TNF levels were increased in those patients with more active disease, a posttreatment fall was only observed in the low disease activity group (Edrees et al. 2017). Although baseline TNF levels may not predict response, there is some evidence that patients with higher baseline levels need increased anti-TNF doses (Takeuchi et al. 2011).

Interestingly, in a small pilot study, patients with increased baseline levels of Chemokine C-X-C motif ligand (CXCL)-10 and CXCL-13, both of which are downstream of TNF, appear to have increased response to the anti-TNF agents adalimumab and etanacept (Han et al. 2016). Similar to the resolution of synovial abnormalities with treatment (described above), CXCL-10 and CXCL-13 levels resolved with treatment (Han et al. 2016).

SYSTEMIC LUPUS ERYTHEMATOSUS

SLE is a multisystem autoimmune disease that affects approximately 40–50 per 100,000 of the adult population. Clinical features include a wide spectrum of manifestations including joint pain, rashes, photosensitivity and alopecia but also deeper organ involvement including serositis, renal involvement (lupus nephritis), hematological disorders including hemolytic anemia, and central nervous system involvement. The therapeutic paradigm for lupus broadly addresses the extent and severity of disease manifestations. Early therapy will include symptomatic therapy and antimalarial drugs particularly hydroxychloroquine. Low dose steroids are often used to control symptoms in patients with mild/moderate disease. A wide range of immunosuppressive agents is also used, including azathioprine, methotrexate, mycophenolate mofetil (MMF) and cyclophosphamide. The latter two are often reserved for more moderate to severe disease including nephritis and central nervous system (CNS) involvement. Although there is clinical trial evidence to support the efficacy of a number of these drugs, it is important to note that apart from antimalarial drugs and corticosteroids (broadly approved for inflammatory disorders) none of the usual classic immunosuppressant therapies used in lupus have are licensed for this condition. Regarding biological therapy, there is wide use of RTX in a number of countries particularly in Europe and it is usually targeted toward refractory disease in particular lupus nephritis. To date, only one agent, belimumab, has shown success in two pivotal phase III trials and thus has been licensed by the FDA and European Medicines Agency (EMA) for use in SLE (Furie et al. 2011; Navarra et al. 2011). There has, however, been a wide spectrum of drugs trialed in lupus and many of these have had early success in phase I or phase II trials (Bruce et al. 2010). However, confirmatory phase III trials have not succeeded and the drugs have not gone forward for approval. Such agents include RTX where the clinical trial program did not reach its primary end points. Other agents include abatacept, atacicept, LJP 511, ocrelizumab, rontalizumab, anifrolumab and epratuzumab (Wallace 2010).

A major hurdle in SLE trials is disease heterogeneity. Classically, SLE trials have been split into lupus nephritis (LN) and so called "extra-renal" trials, where severe active LN is an exclusion criterion. For LN, a key outcome is the ability of an agent to prevent the development of chronic renal impairment after 3–5 years rather than simply affecting short-term changes in proteinuria etc. (Austin et al. 1986). For extra-renal disease a wide variety of potential clinically active manifestations may contribute to the disease activity score that gets the patient in to the trial. Therefore, although the majority enter a trial with mucocutaneous and/or articular disease (Navarra et al. 2011) (Wallace et al. 2014), large trial populations will also include serological and hematological changes as well as serositis, vasculitis and some neurological manifestations (Manzi et al. 2012). The overall assumption has been that all extra-renal SLE can be studied as a single trial population. Whether this classic separation, or stratification, of patient groups for clinical trials is the correct approach for the future is a matter of considerable debate.

Evidence of Heterogeneity in SLE and CTD

There is very clear evidence of basic and immunopathogenic heterogeneity within the various CTDs. When one considers the spectrum of autoantibodies associated with lupus and other CTD these autoantibody spectra do suggest that there are immunopathogenic subsets driving certain manifestations. For example, recent cross-sectional studies from patients with lupus have shown that although over 95% of patients have a positive anti-nuclear antibody, a lower proportion will have other more "specific" antibodies such as anti-double stranded DNA (57%) and anti-Smith (anti-Sm) (26%) (Ippolito et al. 2011). Other autoantibodies that seem to relate to specific manifestations such as lupus nephritis; for example, anti-C1q antibodies occur in approximately a quarter of patients with lupus (Orbai et al. 2015). This heterogeneity is also observed in other CTDs where clinical trials have proved challenging. In inflammatory myopathies, a wide spectrum of myositis specific antibodies has been described with very little overlap but which seem to track with particular clinical manifestations (Betteridge and McHugh 2016). Similarly, in systemic sclerosis there are major subsets understood based on the presence of certain antibodies such as topoisomerase-1, centromere and RNA III polymerase (Choi and Fritzler 2016). At a genetic level there has been enormous increase in our understanding of the genetic susceptibility of lupus and related conditions. It should be noted that there is a wide degree of overlap in genetic susceptibility markers between a number of these related conditions. Also of relevance within SLE, a large number of genetic polymorphisms have been associated with lupus in genome-wide association studies (GWAS), many of these genes cluster around several distinct pathways such as immune signal transduction, the Toll-like receptor and type 1 interferon pathway and immune complex processing pathways (Bentham and Vyse 2013). Although these point to common immunological pathways of relevance in the pathogenesis of lupus, it also suggests subsets of patients where particular targeted therapies may be more likely to succeed. This heterogeneity might of course also explain why trials to date have been unsuccessful. Recent elucidation of monogenic forms of lupus also suggest that certain distinct pathways may have a particular strong drive toward the pathogenesis and development of lupus (Crow and Manel 2015; An et al. 2017).

Transcription profiling has also shed some further light on the pathogenic mechanisms underlying SLE and in particular studies of the interferon (IFN) gene signature have suggested a prominent role of the IFN pathway in driving SLE. Kalunian et al. (Kalunian et al. 2016) identified patients with SLE from recent early phase clinical trials. In patients with moderate to severe lupus approximately 70%–80% of active patients had a high interferon gene signature. Even in this group, however, there was a bimodal pattern where 20%–30% did not have a high signature. In milder disease the bimodal pattern was also demonstrated and the population was approximately 50:50 with regard to elevated interferon gene transcripts (Kalunian et al. 2016). More recently, in a longitudinal analysis of transcript only profiling of whole blood of patients with paediatric onset lupus, within this population there were seven subset or "strata" of patients identified (Banchereau et al. 2016). In addition to a prevalent interferon signature, there was strong evidence of a plasmablast signature being associated with disease activity. In addition, neutrophil transcripts appeared associated with progression to active lupus nephritis and several other signatures appeared to correlate with response to specific therapy (Banchereau et al. 2016).

Evidence of Stratified Responses in SLE and CTD Studies

With regard to typical clinically available testing, routine monitoring patients with lupus will include assessment of levels of double stranded DNA antibodies and complement. In addition, the autoantibody profile of most patients will have been known to the physician at the time of deciding on new therapies. It is these immunological characteristics therefore that have been most widely studied in terms of stratified medicine in SLE and CTD.

MMF has emerged as a "cornerstone" drug for LN and extrarenal disease. In the Aspreva Lupus Maintenance Study (ALMS) trial, the 6-month response rate was 56.2% (Appel et al. 2009), however higher response rates were noted in patients of non-white race/ethnicity (Isenberg et al. 2010). With RTX, clinical trials suggested no overall benefit from standard of care (SOC) (Merrill et al. 2010; Rovin et al. 2012). However, Hispanic and black race/ethnicity patients also showed a difference between active treatment and SOC, albeit mainly due to a lower placebo "SOC" response in these subsets (Merrill et al. 2010).

Several studies have sought to address how these biomarkers relate to response to therapy. Ng et al. noted in a prospective cohort of SLE patients treated with RTX that positive extractable nuclear antigen (ENA) antibodies were associated with an earlier time to flare compared with patients who were negative for these antibodies (Ng et al. 2007). Vital et al. also noted that patients with chronic cutaneous lupus particularly those with positive antibodies to Ro or RNP were less likely to have a cutaneous response post RTX compared with patients without these antibodies (Vital et al. 2015). Belimumab (anti-BLyS) has a European license and showed superiority to SOC in two pivotal phase III trials (Furie et al. 2011; Navarra et al. 2011). Pooled analysis of patients with high anti-dsDNA antibodies and low complement found that the treatment difference between active treatment and placebo was greater (19.8%) compared with the overall differences seen in the individual trial populations (van Vollenhoven et al. 2012). It should be pointed out however that this was mainly due to a lower placebo-SOC response in this subset rather than a higher overall response to belimumab. In the EXPLORER trial of RTX, patients with low complement or high anti-dsDNA antibodies also had sustained improvements in serology following RTX (Merrill et al. 2010).

Genetic Factors

Evidence is beginning to emerge that in certain situations genetic factors (mainly associated with drug specific pathways) may have an influence on efficacy and/or adverse events to certain treatments used in SLE. Shu et al. (2016) studied Chinese patients with LN treated with cyclophosphamide. Cyclophosphamide is a prodrug requiring 4-hydroxylation. Several genes within the cytochrome P450 pathway were associated with wide variability in the CPA4 hydroxylation (4-OH-CPA) levels achieved and 49% of the variability was explained by the genetic changes. "Extensive metabolizers" had higher 4-OH-CPA plasma concentrations following treatment and a shorter time to achieving complete remission. In addition, this population also had a higher risk of leukopenia and gastrointestinal adverse (Shu et al. 2016) events. With regard to MMF several enzymes are involved in MMF metabolism including uridine glucuronosyltransferase metabolizing enzymes (UGTs) ABCC2 gene, which encodes the multidrug resistance-associated protein MRP2 and MDR1 MDR1, IMPDH and cytochrome CYP-2C8. In a number of studies with patients undergoing renal transplant these enzymes have relevance in levels of immunosuppression and adverse events seen. To date, one study that included 19 patients with LN examined these genetic polymorphisms and suggested that UGT2B7 C802T heterozygosity predicted increased renal clearance of Mycophenolic acid (MPA) (Joy et al. 2010). In patients treated for hematological malignancy with RTX, the FCGR3A polymorphisms are associated with treatment response to RTX-based chemotherapeutic regimes (Liu et al. 2016). A previous study of 12 SLE patients treated with RTX also suggested that FC gamma R2A and 3A genotypes were associated with the level of B-cell depletion the patient achieved (Anolik et al. 2003). A more recent larger study of 214 rheumatology patients (including RA, SLE and vasculitis) demonstrated that the FCGR3A V allele was associated with increased flare-free survival, but additionally and increased risk of late-onset neutropenia (Ajeganova et al. 2017).

Pharmacodynamic Changes

A dynamic immunological environment exists following B-cell targeted therapies (Vital et al. 2011; Lazarus et al. 2012). Several groups have identified that the degree of B cells depletion in patients following treatment with RTX has an influence on overall responses. RTX efficacy is closely correlated with efficiency of B-cell depletion and speed of repopulation, for instance rapid return of plasmablasts is associated with relapse (Vital et al. 2011). Post-RTX, flares associated with high anti-dsDNA antibody levels were associated with low total B cell numbers and an increased percentage of IgD-CD27^{hi} plasmablasts. In contrast, relapses with low anti-dsDNA antibody levels were accompanied by a return to near normal total B cell numbers and an increased percentage of IgD-CD27- B cells and (Lazarus et al. 2012).

Transcriptomic Biomarkers

Several clinical trials target the type-I IFN pathway have recently been reported. Several agents have been developed to specifically target this pathway including rontalizumab, sifalimumab and anafrolimab. Sifalimumab is an anti-IFN alpha monoclonal antibody and has been tested in early phase clinical trials. Pharmacodynamically, this drug inhibits the type 1 interferon gene signature (Higgs et al. 2013; Petri et al. 2013). In a phase II trial ($n = 431$) patients treated with sifalimumab were more likely to achieve an SRI4[1] response compared with placebo treated patients (Khamashta et al. 2016). A subgroup analysis examined the responses according to the baseline IFN gene signature and found that major improvements were particularly noted in those with a high baseline IFN signature. Only 19% of patients had a low IFN signature at baseline and this group had a high placebo response rate (60%), which was comparable to the responses seen in the actively treated arms (Khamashta et al. 2016). Similarly, in a phase Ib trial of sifalimumab in patients with inflammatory myopathies, patients who had a more profound suppression of the IFN gene signature posttreatment were also more likely to have a meaningful improvement in their manual muscle testing (Higgs et al. 2013). Anifrolumab is a fully human IgG1 monoclonal

[1] Defined as a 4-point improvement in the SLEDAI-2K score, no new BILAG-2004 "A" or ">1 B" score and no significant worsening in the physician global assessment score (Khamashta et al. 2016).

antibody that binds the interferon alpha receptor and therefore prevents signaling of all type 1 interferons (Furie et al. 2015). A recent phase II clinical trial randomized 305 patients to two doses of anifrolumab or placebo over a 48-week period. There was a higher rate of achieving the primary end point in patients taking anifrolumab compared with placebo. When patients were stratified by their interferon signature at baseline patients with a high baseline interferon signature had a greater effect size on drug compared with placebo (again this was because of a lower placebo response in this population 13.2%, compared with the overall placebo response of 17.6%) (Furie et al. 2017).

The results of IFN signature testing in anti-IFN targeted therapies have not, however, been entirely consistent. In the phase II trial of rontalizumab a humanized IgG1 anti-IFNα monoclonal antibody, 238 patients were randomized to receive active drug or placebo. At baseline 75.6% of patients had a high IFN gene signature. Overall, this trial showed similar response rates using the British Isles Lupus Assessment Group (BILAG) and Systemic Lupus Erythematosus Responder Index (SRI) scoring systems between active treatment and placebo. Notably however it was the patients with the low IFN gene signature where the Systemic Lupus Erythematosus Responder Index (SRI) response was higher and steroid use over the period of trial was lower in patients treated with rontalizumab. This group also had less clinical flares over the period of follow up (Kalunian et al. 2016).

LIMITATIONS IN CURRENT STUDIES

The small number of studies and inconsistent findings to date limits our ability to draw conclusions on how best to stratify lupus and other CTDs. Recent systematic reviews have focused on an analysis of the quality of data to support any stratification of patients being treated with MMF or RTX for SLE (Mendoza-Pinto et al. 2017) (Pirone et al. In Press) In both cases only a limited number of studies could be assessed and many of these were post hoc and secondary analyses and as such carried a high risk of bias based on the lack of adjustment for additional confounding variables etc. There was also evidence that some of the factors that were associated with response in general rather than being drug specific (Mendoza-Pinto et al. 2017) (Pirone et al. In Press).

Another major limitation in SLE and related conditions is the lack of real consensus on the right outcome measures to use. The metric properties of the clinical outcome measures can restrict their ability to fully address the biology of disease particularly in situations where a differential effect or a substrata of patients is being studied. For example, the SLEDAI-2K and related instruments score the presence or absence of particular clinical manifestations for example, skin rashes. The scoring system demands that the skin rash has to be completely cleared before any reduction in the score for that item changing. As such, a patient with an extensive skin rash may experience a large improvement to only a mild rash but will actually still be deemed a "nonresponder" for the purposes of any end point analysis (Yee et al. 2011). Similarly, outcome measures used in other conditions such as skin thickening scores in systemic sclerosis (Clements et al. 1993) and outcomes measures in Sjögren's syndrome may not be sensitive enough to demonstrate high responsiveness to change over short periods of time (Bowman et al. 2017). In these settings, it may be difficult therefore to distinguish true clinically meaningful biological effect.

SUMMARY

Autoimmune rheumatic diseases are an important group of conditions in which drug development to date has had a mixture of great success (e.g., RA) and many disappointments (e.g., SLE). In both contexts however we are beginning to learn important lessons about disease heterogeneity and potential biomarker that may help to better stratify populations for more targetted therapies. In RA, seropositivity and the VKA haplotype within the SE may be important stratifiers. The predicitive value of circulating markers studied to date have been limited. The role of synovial tissue biomarkers also remains unclear and ultimately, because this is not routinely collected in RA clinics, any such biomakers will need rigourous evaluation for their added predictive value in rountine management and decision making.

In SLE and CTDs, disease heterogeneity is incresasingly clear and may be one of the major confounders that has limited the success of trials to date. Stratification of populations according to anti-dsDNA antibody and complement status has already influenced prescribing policies for belimumab. The value of the IFN gene signature to predict responses to anti-IFN therapies seems intuitive and may lead to specific populations receiving such agents in the future. A much clearer understanding of the taxonomy of CTDs may also point the way to new approches to studying and treating these conditions and pave the way to succesful trials of targetting therapies to the key immunopathogenic mechanism in individual patients.

REFERENCES

Ajeganova, S., T. Daniel, H. Hägglund, B. Fadeel, I. Vedin, A. L. Zignego, and J. Palmblad. 2017. Effect of FCGR Polymorphism on the Occurrence of Late-Onset Neutropenia and Flare-Free Survival in Rheumatic Patients Treated with Rituximab. *Arthritis Research & Therapy* 19 (1): 44. doi:10.1186/s13075-017-1241-0.

An, J., T. A. Briggs, A. Dumax-Vorzet, M. E. Alarcón-Riquelme, A. Belot, M. Beresford, I. N. Bruce et al. 2017. Tartrate-Resistant Acid Phosphatase Deficiency in the Predisposition to Systemic Lupus Erythematosus. *Arthritis & Rheumatology (Hoboken, N.J.)* 69 (1): 131–142. doi:10.1002/art.39810.

Anolik, J. H., D. Campbell, R. E. Felgar, F. Young, I. Sanz, J. Rosenblatt, and R. J. Looney. 2003. The Relationship of Fc? RIIIa Genotype to Degree of B Cell Depletion by Rituximab in the Treatment of Systemic Lupus Erythematosus. *Arthritis & Rheumatism* 48 (2): 455–459. doi:10.1002/art.10764.

Appel, G. B., G. Contreras, M. A. Dooley, E. M. Ginzler, D. Isenberg, D. Jayne, L.-S. Li et al. 2009. Mycophenolate Mofetil versus Cyclophosphamide for Induction Treatment of Lupus Nephritis. *Journal of the American Society of Nephrology: JASN* 20 (5): 1103–1112. doi:10.1681/ASN.2008101028.

Austin, H. A., J. H. Klippel, J. E. Balow, N. G.H. Le Riche, A. D. Steinberg, P. H. Plotz, and J. L. Decker. 1986. Therapy of Lupus Nephritis. *New England Journal of Medicine* 314 (10): 614–619. doi:10.1056/NEJM198603063141004.

Banchereau, R., S. Hong, B. Cantarel, N. Baldwin, J. Baisch, M. Edens, A.-M. Cepika et al. 2016. Personalized Immunomonitoring Uncovers Molecular Networks That Stratify Lupus Patients. *Cell* 165 (3): 551–565. doi:10.1016/j.cell.2016.03.008.

Bentham, J. and T. J. Vyse. 2013. The Development of Genome-Wide Association Studies and Their Application to Complex Diseases, Including Lupus. *Lupus* 22 (12): 1205–1213. doi:10.1177/0961203313492870.

Betteridge, Z. and N. McHugh. 2016. Myositis-Specific Autoantibodies: An Important Tool to Support Diagnosis of Myositis. *Journal of Internal Medicine* 280 (1): 8–23. doi:10.1111/joim.12451.

Bowman, S. J., C. C. Everett, J. L. O'Dwyer, P. Emery, C. Pitzalis, W.-F. Ng, C. T. Pease et al. 2017. Randomized Controlled Trial of Rituximab and Cost-Effectiveness Analysis in Treating Fatigue and Oral Dryness in Primary Sjogren's Syndrome. *Arthritis & Rheumatology*. doi:10.1002/art.40093.

Bruce, I. N., C. Gordon, J. T. Merrill, and D. Isenberg. 2010. Clinical Trials in Lupus: What Have We Learned so Far? *Rheumatology (Oxford, England)* 49 (6): 1025–1027. doi:10.1093/rheumatology/kep462.

Chatzidionysiou, K., E. Lie, E. Nasonov, G. Lukina, M. L. Hetland, U. Tarp, C. Gabay et al. 2011. Highest Clinical Effectiveness of Rituximab in Autoantibody-Positive Patients with Rheumatoid Arthritis and in Those for Whom No More than One Previous TNF Antagonist Has Failed: Pooled Data from 10 European Registries. *Annals of the Rheumatic Diseases* 70 (9): 1575–1580. doi:10.1136/ard.2010.148759.

Chatzidionysiou, K., S. Emamikia, J. Nam, S. Ramiro, J. Smolen, D. van der Heijde, M. Dougados et al. 2017. Efficacy of Glucocorticoids, Conventional and Targeted Synthetic Disease-Modifying Antirheumatic Drugs: A Systematic Literature Review Informing the 2016 Update of the EULAR Recommendations for the Management of Rheumatoid Arthritis. *Annals of the Rheumatic Diseases*. doi:10.1136/annrheumdis-2016-210711.

Choi, I. Y., D. M. Gerlag, M. J. Herenius, R. M. Thurlings, C. A. Wijbrandts, D. Foell, T. Vogl, J. Roth, P. P. Tak, and D. Holzinger. 2013. MRP8/14 Serum Levels as a Strong Predictor of Response to Biological Treatments in Patients with Rheumatoid Arthritis. *Annals of the Rheumatic Diseases* 1–9. doi:10.1136/annrheumdis-2013-203923.

Choi, M. Y. and M. J. Fritzler. 2016. Progress in Understanding the Diagnostic and Pathogenic Role of Autoantibodies Associated with Systemic Sclerosis. *Current Opinion in Rheumatology* 28 (6): 586–594. doi:10.1097/BOR.0000000000000325.

Clements, P. J., P. A. Lachenbruch, J. R. Seibold, B. Zee, V. D. Steen, P. Brennan, A. J. Silman, N. Allegar, J. Varga, and M. Massa. 1993. Skin Thickness Score in Systemic Sclerosis: An Assessment of Interobserver Variability in 3 Independent Studies. *The Journal of Rheumatology* 20 (11): 1892–1896.

Crow, Y. J. and N. Manel. 2015. Aicardi–Goutières Syndrome and the Type I Interferonopathies. *Nature Reviews Immunology* 15 (7): 429–440. doi:10.1038/nri3850.

Cuppen, B. V. J., P. M. J. Welsing, J. J. Sprengers, J. W. J. Bijlsma, A. C. A. Marijnissen, J. M. van Laar, F. P. J. G. Lafeber, and S. C. Nair. 2016. Personalized Biological Treatment for Rheumatoid Arthritis: A Systematic Review with a Focus on Clinical Applicability. *Rheumatology (Oxford, England)* 55 (5): 826–839. doi:10.1093/rheumatology/kev421.

Daïen, C. I., S. Fabre, C. Rittore, S. Soler, V. Daïen, G. Tejedor, D. Cadart et al. 2012. TGF beta1 Polymorphisms Are Candidate Predictors of the Clinical Response to Rituximab in Rheumatoid Arthritis. *Joint Bone Spine* 79 (5): 471–475. doi:10.1016/j.jbspin.2011.10.007.

de Jong, T. D., J. Sellam, R. Agca, S. Vosslamber, B. I. Witte, M. Tsang-A-Sjoe, E. Mantel et al. 2017. A Multi-Parameter Response Prediction Model for Rituximab in Rheumatoid Arthritis. *Joint, Bone, Spine: Revue Du Rhumatisme*. doi:10.1016/j.jbspin.2017.02.015.

de Jong, T. D., S. Vosslamber, M. Blits, G. Wolbink, M. T. Nurmohamed, C. J. van der Laken, G. Jansen, A. E. Voskuyl, and C. L. Verweij. 2015. Effect of Prednisone on Type I Interferon Signature in Rheumatoid Arthritis: Consequences for Response Prediction to Rituximab. *Arthritis Research & Therapy* 17 (1): 78. doi:10.1186/s13075-015-0564-y.

Edrees, A. F., S. N. Misra, and N. I. Abdou. 2017. Anti-Tumor Necrosis Factor (TNF) Therapy in Rheumatoid Arthritis: Correlation of TNF-Alpha Serum Level with Clinical Response and Benefit from Changing Dose or Frequency of Infliximab Infusions. *Clinical and Experimental Rheumatology* 23 (4): 469–474.

Emery, P., E. Keystone, H. P. Tony, A. Cantagrel, R. van Vollenhoven, A. Sanchez, E. Alecock, J. Lee, and J. Kremer. 2008. IL-6 Receptor Inhibition with Tocilizumab Improves Treatment Outcomes in Patients with Rheumatoid Arthritis Refractory to Anti-Tumour Necrosis Factor Biologicals: Results from a 24-Week Multicentre Randomised Placebo-Controlled Trial. *Annals of the Rheumatic Diseases* 67 (11): 1516–1523. doi:10.1136/ard.2008.092932.

Farragher, T. M., N. J. Goodson, H. Naseem, A. J. Silman, W. Thomson, D. Symmons, and A. Barton. 2008. Association of the HLA-DRB1 Gene with Premature Death, Particularly from Cardiovascular Disease, in Patients with Rheumatoid Arthritis and Inflammatory Polyarthritis. *Arthritis and Rheumatism* 58 (2): 359–369. doi:10.1002/art.23149.

Fries, J. F. 2000. Current Treatment Paradigms in Rheumatoid Arthritis. *Rheumatology (Oxford, England)* 39 (Suppl 1): 30–35.

Furie, R., J. Merrill, V. Werth, M. Khamashta, K. Kalunian, P. Brohawn, G. Illei, J. Drappa, L. Wang, and S. Yoo. 2015. Anifrolumab, an Anti-Interferon Alpha Receptor Monoclonal Antibody, in Moderate to Severe Systemic Lupus Erythematosus (SLE). *Arthritis Rheumatol* 67 (2): suppl 10 abstract number 3223. doi:10.1002/art.39962.

Furie, R., M. Petri, O. Zamani, R. Cervera, D. J. Wallace, D. Tegzová, J. Sanchez-Guerrero et al. 2011. A Phase III, Randomized, Placebo-Controlled Study of Belimumab, a Monoclonal Antibody That Inhibits B Lymphocyte Stimulator, in Patients with Systemic Lupus Erythematosus. *Arthritis & Rheumatism* 63 (12): 3918–3930. doi:10.1002/art.30613.

Furie, R., M. Khamashta, J. T. Merrill, V. P. Werth, K. Kalunian, P. Brohawn, G. G. Illei et al. 2017. Anifrolumab, an Anti-Interferon-α Receptor Monoclonal Antibody, in Moderate-to-Severe Systemic Lupus Erythematosus. *Arthritis & Rheumatology (Hoboken, N.J.)* 69 (2): 376–386. doi:10.1002/art.39962.

Han, B. K., I. Kuzin, J. P. Gaughan, N. J. Olsen, and A. Bottaro. 2016. Baseline CXCL10 and CXCL13 Levels Are Predictive Biomarkers for Tumor Necrosis Factor Inhibitor Therapy in Patients with Moderate to Severe Rheumatoid Arthritis: A Pilot, Prospective Study. *Arthritis Research & Therapy* 18 (1): 93. doi:10.1186/s13075-016-0995-0.

Higgs, B. W., W. Zhu, C. Morehouse, W. I. White, P. Brohawn, X. Guo, M. Rebelatto et al. 2013. A Phase 1b Clinical Trial Evaluating Sifalimumab, an Anti-IFN- Monoclonal Antibody, Shows Target Neutralisation of a Type I IFN Signature in Blood of Dermatomyositis and Polymyositis Patients. *Annals of the Rheumatic Diseases* 73 (1): 256–262. doi:10.1136/annrheumdis-2012-202794.

Horowitz, D. L. and R. Furie. 2012. Belimumab Is Approved by the FDA: What More Do We Need to Know to Optimize Decision Making? *Current Rheumatology Reports* 14 (4): 318–323. doi:10.1007/s11926-012-0256-4.

Humphreys, J. H., A. Warner, J. Chipping, T. Marshall, M. Lunt, D. P. M. Symmons, and S. M. M. Verstappen. 2014. Mortality Trends in Patients with Early Rheumatoid Arthritis over 20 Years: Results from the Norfolk Arthritis Register. *Arthritis Care & Research* 66 (9): 1296–1301. doi:10.1002/acr.22296.

Hyrich, K. L., K. D. Watson, A. J. Silman, D. P. M. Symmons, and British Society for Rheumatology Biologics Register. 2006. Predictors of Response to Anti-TNF-Alpha Therapy among Patients with Rheumatoid Arthritis: Results from the British Society for Rheumatology Biologics Register. *Rheumatology (Oxford, England)* 45 (12): 1558–1565. doi:10.1093/rheumatology/kel149.

Ippolito, A., D. J. Wallace, D. Gladman, P. R. Fortin, M. Urowitz, V. Werth, M. Costner et al. 2011. Autoantibodies in Systemic Lupus Erythematosus: Comparison of Historical and Current Assessment of Seropositivity. *Lupus* 20 (3): 250–255. doi:10.1177/0961203310385738.

Isaacs, J. D. and G. Ferraccioli. 2011. The Need for Personalised Medicine for Rheumatoid Arthritis. *Annals of the Rheumatic Diseases* 70 (1): 4–7. doi:10.1136/ard.2010.135376.

Isenberg, D., G. B. Appel, G. Contreras, M. A. Dooley, E. M. Ginzler, D. Jayne, J. Sánchez-Guerrero, D. Wofsy, X. Yu, and N. Solomons. 2010. Influence of Race/ethnicity on Response to Lupus Nephritis Treatment: The ALMS Study. *Rheumatology (Oxford, England)* 49 (1): 128–140. doi:10.1093/rheumatology/kep346.

Joy, M. S., T. Boyette, Y. Hu, J. Wang, M. La, S. L. Hogan, P. W. Stewart, R. J. Falk, M. A. Dooley, and P. C. Smith. 2010. Effects of Uridine Diphosphate Glucuronosyltransferase 2B7 and 1A7 Pharmacogenomics and Patient Clinical Parameters on Steady-State Mycophenolic Acid Pharmacokinetics in Glomerulonephritis. *European Journal of Clinical Pharmacology* 66 (11): 1119–1130. doi:10.1007/s00228-010-0846-x.

Kalunian, K. C., J. T. Merrill, R. Maciuca, J. M. McBride, M. J. Townsend, X. Wei, J. C. Davis, and W. P. Kennedy. 2016. A Phase II Study of the Efficacy and Safety of Rontalizumab (rhuMAb Interferon-α) in Patients with Systemic Lupus Erythematosus (ROSE). *Annals of the Rheumatic Diseases* 75 (1): 196–202. doi:10.1136/annrheumdis-2014-206090.

Khamashta, M., J. T. Merrill, V. P Werth, R. Furie, K. Kalunian, G. G. Illei, J. Drappa, L. Wang, W. Greth, and CD1067 study investigators. 2016. Sifalimumab, an Anti-Interferon- α Monoclonal Antibody, in Moderate to Severe Systemic Lupus Erythematosus: A Randomised, Double-Blind, Placebo-Controlled Study. *Annals of the Rheumatic Diseases* 75 (11): 1–8. doi:10.1136/annrheumdis-2015-208562.

Klaasen, R., R. M. Thurlings, C. A. Wijbrandts, A. W. van Kuijk, D. Baeten, D. M. Gerlag, and P. P. Tak. 2009. The Relationship between Synovial Lymphocyte Aggregates and the Clinical Response to Infliximab in Rheumatoid Arthritis: A Prospective Study. *Arthritis and Rheumatism* 60 (11): 3217–3224. doi:10.1002/art.24913.

Lazarus, M. N., T. Turner-Stokes, K.-M. Chavele, D. A. Isenberg, and M. R. Ehrenstein. 2012. B-Cell Numbers and Phenotype at Clinical Relapse Following Rituximab Therapy Differ in SLE Patients according to Anti-dsDNA Antibody Levels. *Rheumatology (Oxford, England)* 51 (7): 1208–1215. doi:10.1093/rheumatology/ker526.

Lindsley, C. W. 2013. The Top Prescription Drugs of 2012 Globally: Biologics Dominate, But Small Molecule CNS Drugs Hold on to Top Spots. *ACS Chemical Neuroscience* 4 (6): 905–907. doi:10.1021/cn400107y.

Ling, S. F., S. Viatte, M. Lunt, A. M. Van Sijl, L. Silva-Fernandez, D. P. M. Symmons, A. Young, A. J. Macgregor, and A. Barton. 2016. HLA-DRB1 Amino Acid Positions 11/13, 71, and 74 Are Associated With Inflammation Level, Disease Activity, and the Health Assessment Questionnaire Score in Patients With Inflammatory Polyarthritis. *Arthritis & Rheumatology (Hoboken, N.J.)* 68 (11): 2618–2628. doi:10.1002/art.39780.

Liu, D., Y. Tian, D. Sun, H. Sun, Y. Jin, and M. Dong. 2016. The FCGR3A Polymorphism Predicts the Response to Rituximab-Based Therapy in Patients with Non-Hodgkin Lymphoma: A Meta-Analysis. *Annals of Hematology* 95 (9): 1483–1490. doi:10.1007/s00277-016-2723-x.

Lv, Q., Y. Yin, X. Li, G. Shan, X. Wu, D. Liang, Y. Li, and X. Zhang. 2014. The Status of Rheumatoid Factor and Anti-Cyclic Citrullinated Peptide Antibody Are Not Associated with the Effect of Anti-TNF a Agent Treatment in Patients with Rheumatoid Arthritis: A Meta-Analysis. *PLoS ONE* 9 (2). doi:10.1371/journal.pone.0089442.

Maini, R., E. W. St Clair, F. Breedveld, D. Furst, J. Kalden, M. Weisman, J. Smolen et al. 1999. Infliximab (Chimeric Anti-Tumour Necrosis Factor Alpha Monoclonal Antibody) versus Placebo in Rheumatoid Arthritis Patients Receiving Concomitant Methotrexate: A Randomised Phase III Trial. ATTRACT Study Group. *Lancet (London, England)* 354 (9194): 1932–1939. http://www.ncbi.nlm.nih.gov/pubmed/10622295.

Manzi, S., J. Sánchez-Guerrero, J. T. Merrill, R. Furie, D. Gladman, S. V. Navarra, E. M. Ginzler et al. 2012. Effects of Belimumab, a B Lymphocyte Stimulator-Specific Inhibitor, on Disease Activity across Multiple Organ Domains in Patients with Systemic Lupus Erythematosus: Combined Results from Two Phase III Trials. *Annals of the Rheumatic Diseases* 71 (11): 1833–1838. doi:10.1136/annrheumdis-2011-200831.

Mendoza-Pinto, C., C. Pirone, D. A. van der Windt, B. Parker, and I. N. Bruce. 2017. Can We Identify Who Gets Benefit or Harm from Mycophenolate Mofetil in Systemic Lupus Erythematosus? A Systematic Review. *Seminars in Arthritis and Rheumatism*. doi:10.1016/j.semarthrit.2017.01.009.

Merrill, J. T., C. Michael Neuwelt, D. J. Wallace, J. C. Shanahan, K. M. Latinis, J. C. Oates, T. O. Utset et al. 2010. Efficacy and Safety of Rituximab in Moderately-to-Severely Active Systemic Lupus Erythematosus: The Randomized, Double-Blind, Phase II/III Systemic Lupus Erythematosus Evaluation of Rituximab Trial. *Arthritis and Rheumatism* 62 (1): 222–233. doi:10.1002/art.27233.

Nair, S. C., P. M. J. Welsing, I. Y. K. Choi, J. Roth, D. Holzinger, J. W. J. Bijlsma, J. M. Van Laar, D. M. Gerlag, F. P. J. G. Lafeber, and P. P. Tak. 2016. A Personalized Approach to Biological Therapy Using Prediction of Clinical Response Based on MRP8/14 Serum Complex Levels in Rheumatoid Arthritis Patients. *PLoS ONE* 11 (3): 1–12. doi:10.1371/journal.pone.0152362.

Nam, J. L., K. Takase-Minegishi, S. Ramiro, K. Chatzidionysiou, J. S. Smolen, D. van der Heijde, J. W. Bijlsma et al. 2017. Efficacy of Biological Disease-Modifying Antirheumatic Drugs: A Systematic Literature Review Informing the 2016 Update of the EULAR Recommendations for the Management of Rheumatoid Arthritis. *Annals of the Rheumatic Diseases*. doi:10.1136/annrheumdis-2016-210713.

Navarra, S. V., R. M. Guzmán, A. E. Gallacher, S. Hall, R. A. Levy, R. E. Jimenez, E. K.-M. Li et al. 2011. Efficacy and Safety of Belimumab in Patients with Active Systemic Lupus Erythematosus: A Randomised, Placebo-Controlled, Phase 3 Trial. *The Lancet* 377 (9767): 721–731. doi:10.1016/S0140-6736(10)61354-2.

Naz, S. M., T. M. Farragher, D. K. Bunn, D. P. M. Symmons, and I. N. Bruce. 2008. The Influence of Age at Symptom Onset and Length of Followup on Mortality in Patients with Recent-Onset Inflammatory Polyarthritis. *Arthritis and Rheumatism* 58 (4): 985–989. doi:10.1002/art.23402.

Ng, K. P., G. Cambridge, M. J. Leandro, J. C. W. Edwards, M. Ehrenstein, and D. A. Isenberg. 2007. B Cell Depletion Therapy in Systemic Lupus Erythematosus: Long-Term Follow-up and Predictors of Response. *Annals of the Rheumatic Diseases* 66 (9): 1259–1262. doi:10.1136/ard.2006.067124.

O'Dell, J. R., B. S. Nepom, C. Haire, V. H. Gersuk, L. Gaur, G. F. Moore, W. Drymalski et al. 1998. HLA-DRB1 Typing in Rheumatoid Arthritis: Predicting Response to Specific Treatments. *Annals of the Rheumatic Diseases* 57 (4): 209–213. doi:10.1136/ard.57.4.209.

O'Rielly, D. D., N. M. Roslin, J. Beyene, A. Pope, and P. Rahman. 2009. TNF-Alpha-308 G/A Polymorphism and Responsiveness to TNF-Alpha Blockade Therapy in Moderate to Severe Rheumatoid Arthritis: A Systematic Review and Meta-Analysis. *The Pharmacogenomics Journal* 9 (3): 161–167. doi:10.1038/tpj.2009.7.

Orbai, A.-M., L. Truedsson, G. Sturfelt, O. Nived, H. Fang, G. S. Alarcón, C. Gordon et al. 2015. Anti-C1q Antibodies in Systemic Lupus Erythematosus. *Lupus* 24 (1): 42–49. doi:10.1177/0961203314547791.

Petri, M., D. J. Wallace, A. Spindler, V. Chindalore, K. Kalunian, E. Mysler, C. M. Neuwelt et al. 2013. Sifalimumab, a Human Anti-Interferon? Monoclonal Antibody, in Systemic Lupus Erythematosus: A Phase I Randomized, Controlled, Dose-Escalation Study. *Arthritis and Rheumatism* 65 (4): 1011–1021. doi:10.1002/art.37824.

Pincus, T., Y. Yazici, T. Sokka, D. Aletaha, and J. S. Smolen. 2017. Methotrexate as the "Anchor drug" for the Treatment of Early Rheumatoid Arthritis. *Clinical and Experimental Rheumatology* 21 (5 Suppl 31): S179–S185. http://www.ncbi.nlm.nih.gov/pubmed/14969073.

Pirone, C, C. Mendoza-Pinto, D. A. van der Windt, B. Parker, M. O'Sullivan, and I. N. Bruce. 2017. Predictive and prognostic factors influencing outcomes of rituximab therapy in systemic lupus erythematosus. Seminars in Arthritis and Rheumatism 47 (3): 384–396. doi:10.1016/j.semarthrit.2017.04.010.

Potter, C., K. L. Hyrich, A. Tracey, M. Lunt, D. Plant, D. P. M. Symmons, W. Thomson et al. 2009. Association of Rheumatoid Factor and Anti-Cyclic Citrullinated Peptide Positivity, but Not Carriage of Shared Epitope or PTPN22 Susceptibility Variants, with Anti-Tumour Necrosis Factor Response in Rheumatoid Arthritis. *Annals of the Rheumatic Diseases* 68 (1): 69–74. doi:10.1136/ard.2007.084715.

Quartuccio, L., M. Fabris, E. Pontarini, S. Salvin, A. Zabotti, M. Benucci, M. Manfredi et al. 2014. The 158VV Fcgamma Receptor 3A Genotype Is Associated with Response to Rituximab in Rheumatoid Arthritis: Results of an Italian Multicentre Study. *Annals of the Rheumatic Diseases* 73 (4): 716–721. doi:10.1136/annrheumdis-2012-202435.

Rovin, B. H., R. Furie, K. Latinis, R. J. Looney, F. C. Fervenza, J. Sanchez-Guerrero, R. Maciuca et al. 2012. Efficacy and Safety of Rituximab in Patients with Active Proliferative Lupus Nephritis: The Lupus Nephritis Assessment with Rituximab Study. *Arthritis and Rheumatism* 64 (4): 1215–1226. doi:10.1002/art.34359.

Ruyssen-Witrand, A., S. Rouanet, B. Combe, M. Dougados, X. L. Loet, J. Sibilia, J. Tebib, X. Mariette, and A. Constantin. 2012. Fc Receptor Type IIIA Polymorphism Influences Treatment Outcomes in Patients with Rheumatoid Arthritis Treated with Rituximab. *Annals of the Rheumatic Diseases* 71: 875–877. doi:10.1136/annrheumdis-2011-200337.

Sellam, J., E. Rivière, A. Courties, P.-O. Rouzaire, B. Tolusso, E. M. Vital, P. Emery et al. 2016. Serum IL-33, a New Marker Predicting Response to Rituximab in Rheumatoid Arthritis. *Arthritis Research & Therapy* 18 (1): 294. doi:10.1186/s13075-016-1190-z.

Shu, W., S. Guan, X. Yang, L. Liang, J. Li, Z. Chen, Y. Zhang, L. Chen, X. Wang, and M. Huang. 2016. Genetic Markers in CYP2C19 and CYP2B6 for Prediction of Cyclophosphamide's 4-Hydroxylation, Efficacy and Side Effects in Chinese Patients with Systemic Lupus Erythematosus. *British Journal of Clinical Pharmacology* 81 (2): 327–340. doi:10.1111/bcp.12800.

Stradner, M. H., C. Dejaco, K. Brickmann, W. B. Graninger, and H. P. Brezinschek. 2016. A Combination of Cellular Biomarkers Predicts Failure to Respond to Rituximab in Rheumatoid Arthritis: A 24-Week Observational Study. *Arthritis Research & Therapy* 18 (1): 190. doi:10.1186/s13075-016-1091-1.

Takeuchi, T., N. Miyasaka, Y. Tatsuki, T. Yano, T. Yoshinari, T. Abe, and T. Koike. 2011. Baseline Tumour Necrosis Factor Alpha Levels Predict the Necessity for Dose Escalation of Infliximab Therapy in Patients with Rheumatoid Arthritis. *Annals of the Rheumatic Diseases* 70 (7): 1208–1215. doi:10.1136/ard.2011.153023.

Teruel, M., C. Chamberlain, and M. E. Alarcón-Riquelme. 2017. Omics Studies: Their Use in Diagnosis and Reclassification of SLE and Other Systemic Autoimmune Diseases. *Rheumatology (Oxford, England)* 56 (suppl_1): i78–i87. doi:10.1093/rheumatology/kew339.

Tony, H.-P., P. Roll, H. E. Mei, E. Blümner, A. Straka, L. Gnuegge, T. Dörner, and FIRST/ReFIRST study teams. 2017. Combination of B Cell Biomarkers as Independent Predictors of Response in Patients with Rheumatoid Arthritis Treated with Rituximab. *Clinical and Experimental Rheumatology* 33 (6): 887–894. http://www.ncbi.nlm.nih.gov/pubmed/26517829.

van Baarsen, L. G., C. A. Wijbrandts, F. Rustenburg, T. Cantaert, T. C. van der Pouw Kraan, D. L. Baeten, B. A. Dijkmans, P. P. Tak, and C. L. Verweij. 2010. Regulation of IFN Response Gene Activity during Infliximab Treatment in Rheumatoid Arthritis Is Associated with Clinical Response to Treatment. *Arthritis Research & Therapy* 12 (1): R11. doi:10.1186/ar2912.

van der Pouw Kraan, T. C., C. A. Wijbrandts, L. G. van Baarsen, F. Rustenburg, J. M. Baggen, C. L. Verweij, and P. P. Tak. 2008. Responsiveness to Anti-Tumour Necrosis Factor Alpha Therapy Is Related to Pre-Treatment Tissue Inflammation Levels in Rheumatoid Arthritis Patients. *Annals of the Rheumatic Diseases* 67 (4): 563–566. doi:10.1136/ard.2007.081950.

van Vollenhoven, R. F., M. A. Petri, R. Cervera, D. A. Roth, B. N. Ji, C. S. Kleoudis, Z. J. Zhong, and W. Freimuth. 2012. Belimumab in the Treatment of Systemic Lupus Erythematosus: High Disease Activity Predictors of Response. *Annals of the Rheumatic Diseases* 71 (8): 1343–1349. doi:10.1136/annrheumdis-2011-200937.

Viatte, S., D. Plant, B. Han, B. Fu, A. Young, K. L. Hyrich, and W. Ann. 2016. HHS Public Access 313 (16): 1645–1656. doi:10.1001/jama.2015.3435.Association.

Vital, E. M., M. Wittmann, S. Edward, M. Y. Md Yusof, H. MacIver, C. T. Pease, M. Goodfield, and P. Emery. 2015. Brief Report: Responses to Rituximab Suggest B Cell-Independent Inflammation in Cutaneous Systemic Lupus Erythematosus. *Arthritis & Rheumatology* (*Hoboken, N.J.*) 67 (6): 1586–1591. doi:10.1002/art.39085.

Vital, E. M., S. Dass, M. H. Buch, K. Henshaw, C. T. Pease, M. F. Martin, F. Ponchel, A. C. Rawstron, and P. Emery. 2011. B Cell Biomarkers of Rituximab Responses in Systemic Lupus Erythematosus. *Arthritis and Rheumatism* 63 (10): 3038–3047. doi:10.1002/art.30466.

Vosslamber, S., H. G. Raterman, T. C. T. M. van der Pouw Kraan, M. W. J. Schreurs, B. M. E. von Blomberg, M. T. Nurmohamed, W. F. Lems, B. A. C. Dijkmans, A. E. Voskuyl, and C. L. Verweij. 2011. Pharmacological Induction of Interferon Type I Activity Following Treatment with Rituximab Determines Clinical Response in Rheumatoid Arthritis. *Annals of the Rheumatic Diseases* 70 (6): 1153–1159. doi:10.1136/ard.2010.147199.

Wallace, D. J. 2010. Advances in Drug Therapy for Systemic Lupus Erythematosus. *BMC Medicine* 8 (1): 77. doi:10.1186/1741-7015-8-77.

Wallace, D. J., K. Kalunian, M. A. Petri, V. Strand, F. A. Houssiau, M. Pike, B. Kilgallen et al. 2014. Efficacy and Safety of Epratuzumab in Patients with Moderate/severe Active Systemic Lupus Erythematosus: Results from EMBLEM, a Phase IIb, Randomised, Double-Blind, Placebo-Controlled, Multicentre Study. *Annals of the Rheumatic Diseases* 73 (1): 183–190. doi:10.1136/annrheumdis-2012-202760.

Weinblatt, M. E., E. C. Keystone, D. E. Furst, L. W. Moreland, M. H. Weisman, C. A. Birbara, L. A. Teoh, S. A. Fischkoff, and E. K. Chartash. 2003. Adalimumab, a Fully Human Anti-Tumor Necrosis Factor Alpha Monoclonal Antibody, for the Treatment of Rheumatoid Arthritis in Patients Taking Concomitant Methotrexate: The ARMADA Trial. *Arthritis and Rheumatism* 48 (1): 35–45. doi:10.1002/art.10697.

Weinblatt, M. E., M. Schiff, R. Valente, D. van der Heijde, G. Citera, C. Zhao, M. Maldonado, and R. Fleischmann. 2013. Head-to-Head Comparison of Subcutaneous Abatacept versus Adalimumab for Rheumatoid Arthritis: Findings of a Phase IIIb, Multinational, Prospective, Randomized Study. *Arthritis and Rheumatism* 65 (1): 28–38. doi:10.1002/art.37711.

Wijbrandts, C. A., M. G. W. Dijkgraaf, M. C. Kraan, M. Vinkenoog, T. J. Smeets, H. Dinant, K. Vos et al. 2008. The Clinical Response to Infliximab in Rheumatoid Arthritis Is in Part Dependent on Pretreatment Tumour Necrosis Factor Alpha Expression in the Synovium. *Annals of the Rheumatic Diseases* 67 (8): 1139–1144. doi:10.1136/ard.2007.080440.

Yee, C.-S., V. T. Farewell, D. A. Isenberg, B. Griffiths, L.-S. Teh, I. N. Bruce, Y. Ahmad et al. 2011. The Use of Systemic Lupus Erythematosus Disease Activity Index-2000 to Define Active Disease and Minimal Clinically Meaningful Change. Based on Data from a Large Cohort of Systemic Lupus Erythematosus Patients. *Rheumatology* (*Oxford, England*) 50 (5): 982–988. doi:10.1093/rheumatology/keq376.

Biomarkers of Systemic Lupus Erythematosus

8.3

Hiroshi Kato and Andras Perl

Contents

INTRODUCTION

Systemic lupus erythematosus (SLE) is a systemic autoimmune disease characterized by production of autoantibodies against DNA, and immune-complex deposition culminating in visceral organ damage and vasculopathy (1). Although SLE is one of the most commonly encountered clinical problems in rheumatology clinic, the definition of disease has still been evolving (2,3), reflecting its overwhelming degree of diversity of clinical manifestations and outcomes. As such, the lack of "gold standard" poses a major diagnostic challenge in the care for SLE patients. It cannot be overemphasized that the lupus diagnostic criteria were developed to identify patients suitable for clinical trials, and that strict adherence to the criteria may not allow us to thoroughly capture patients who would benefit from lupus-directed interventions. Furthermore, it is important to keep in mind that the reference employed to define the diagnostic power of criteria were experts' opinions because there are no clinical, serological, or histological findings that definitively define SLE. When it comes to lupus treatment, the currently available drugs do not directly address the underlying mechanisms of the disease. As such, a substantial portion of patients succumb to their illness or toxicities from global immunosuppression. In light of these challenges in the diagnosis and management of SLE, it is abundantly clear that it is critical to define disease mechanisms and develop target-specific therapeutics in order to improve the quality of care for patients with SLE. In this section, we first review biomarkers commonly used in the management of SLE in the current rheumatology practice along with their limitations. This is followed by discussions of emerging biomarkers that are expected to improve the accuracy of our diagnostic assessment and help develop target-selective therapeutics.

Biomarkers Currently Used in the Diagnosis and Management of SLE

Antinuclear antibody and anti-double-stranded DNA antibodies

Despite the lack of gold standard in lupus diagnosis, antinuclear antibodies (ANA) assay remains a critical component of laboratory assessment whenever SLE or systemic rheumatic disease is suspected. ANA is widely ordered by a number of practitioners; however, this test does not facilitate diagnostic assessment unless it is ordered and interpreted in the pertinent clinical context. Indirect immunofluorescence assay (IIF) employs cell lines that express a wide range of clinically important nuclear autoantigens, most commonly human epidermoid carcinoma (HEp2) cell line, and remains the reference method of choice for the detection of ANA (4,5). Immunofluorescence-based methods using HEp2 cells are approximately twofold more sensitive than flow cytometry or ELISA for detection of ANA in patients with SLE (6), and considered the gold standard for ANA testing by the American College of Rheumatology (https://www.rheumatology.org/Portals/0/Files/Methodology%20of%20Testing%20Antinuclear%20Antibodies%20Position%20Statement.pdf). HEp2 cells are first incubated with diluted patient serum. This is followed by incubation with a fluorescein-conjugated secondary antibody directed against human immunoglobulin. If fluorescence is detected at one or more screening dilutions (often 1:40 and 1:160), the serum is serially diluted and retested. An endpoint is reached when fewer than half of the cells on the slide show detectable fluorescence. The ANA titer is reported as the dilution prior to this endpoint. When the serum contains immunoglobulins that bind any nuclear antigens expressed by HEp2 cells, it will be detected by the fluorescence of secondary antibody. Accordingly, this method allows for identification of any immunoglobulins that have reactivity against nuclear autoantigens expressed in HEp2 cells, which, by definition, are ANA. The staining pattern may help predict the target nuclear antigens, although such pattern recognition is subject to arbitrary factors, and does not provide definitive specification of the ANA. In contrast, a solid-phase assay employs a panel of purified native or recombinant autoantigens, such as Ro, La, Sm, U1 RNP, Scl-70, PM-Scl, Jo-1, centromere, histone, ribosomal P, and DNA, immobilized on a solid surface (microtiter plate, fluorescent microsphere, or membrane). The already known antigens are incubated with diluted patient serum, and autoantibody is detected in the analogous manner to that of IIF. As such, this method enables the identification of specific ANA. The pros and cons of each method are summarized in the Table 8.3.1.

Accordingly, when the pretest probability of SLE or related systemic autoimmune disease is substantially low on clinical grounds, it is sensible to order IIF assay only because a negative result will exclude these diagnoses. Alternatively, a decent pretest probability would warrant the orders of both of these assays, which is expected to better delineate the entity of autoimmune disease because many autoantibodies have specific clinical associations as summarized in the Table 8.3.2.

TABLE 8.3.1 Comparison of immunofluorescence and solid phase assays in ANA detection

	IIF ASSAY	*SOLID PHASE ASSAY*
Labor	More laborious	Less laborious
Sensitivity	High (may not detect SSA, Jo-1, and ribosomal P antibodies (7,8))	Low (9, 10)
Specificity	Low (5%–30% false positive) (4)	High
Scientific rigor	Low	High

TABLE 8.3.2 Clinical implications of various antinuclear antibodies

AUTOANTIBODY	*CLINICAL ASSOCIATION*
SSA (Ro)	Sjögren's syndrome, subacute cutaneous lupus erythematosus, congenital heart block
SSB (La)	Sjögren's syndrome, subacute cutaneous lupus erythematosus, congenital heart block
Smith	SLE specific
dsDNA	SLE specific, correlation with activity of lupus nephritis
RNP	Mixed connective tissue disease (MCTD)
Phospholipid	Antiphospholipid syndrome, CNS lupus
Ribosomal P	? Neuropsychiatric lupus (11)
Jo-1	Anti-synthetase syndrome (fever + polymyositis + arthritis + ILD + mechanic's hand)
Centromere	Limited systemic sclerosis
Scl-70	Diffuse systemic sclerosis

Source: Sieper, J., Management of ankylosing spondylitis, in *Rheumatology*, 4th edition, M.C. Hochberg, et al. (Eds.), Mosby, London, UK, West, S.G., *Rheumatology Secrets*, Hanley and Belfus, Philadelphia, PA, pp. 57–58, 2002.

In humans, clinical expression of SLE is preceded by the progressive accumulation of specific autoantibodies over 9 years prior to diagnosis. Ro, La, and phospholipid antibodies appear first, followed by double-stranded DNA (dsDNA) and then Smith and nuclear ribonucleoprotein antibodies (12). Anti-dsDNA Ab is one of the lupus-specific antibodies along with Smith antibody, which helps distinguish patients with SLE from other systemic rheumatic diseases (13). In principle, a lupus diagnosis requires the identification of ANA by IIF assay; however, positive anti-dsDNA in a patient with histopathological evidence of lupus nephritis (LN) permits a SLE diagnosis even in the absence of ANA (3), exemplifying the importance of anti-dsDNA Ab as a biomarker in SLE. Because the titers of anti-dsDNA often fluctuate with SLE disease activity, when anti-dsDNA levels are integrated with other measures of disease activity, they are useful in the clinical management of SLE patients (14,15). In addition, there is a well-recognized association between high titer IgG anti-dsDNA and active glomerulonephritis (16,17). However, anti-dsDNA does not always reflect the disease activity or correlate with glomerulonephritis; that is, the pathogenicity of anti-dsDNA may vary from individual to individual or over time in the same individual, which may be attributed to the divergence of properties of anti-dsDNA, including avidity for antigen, isoelectric point, isotype, ability to fix complement, and idiotype. With regard to the pathogenicity of anti-dsDNA in glomerulonephritis, the location of immune complex (IC) deposition within the kidney and the amount of IgG anti-dsDNA antibody deposition may affect the degree of renal damage (18–20). These observations might explain why there are patients who have elevated levels of anti-dsDNA in the setting of inactive or minimally active lupus (21), whereas a minority of patients have active nephritis without elevations of anti-dsDNA titer. In general, monitoring anti-dsDNA is most informative for patients in whom changes in titer correlate with clinical status before or during flares (22). Conversely, there is no relationship between antibody titer and disease activity for neuropsychiatric manifestations of SLE (23).

Complements

Along with anti-dsDNA, hypocomplementemia (CH50, C3, and C4) is the most useful laboratory test to predict a SLE flare, in particular lupus nephritis (22,24–26), yet these markers do not accurately mirror the clinical activity of SLE in a large proportion of patients, and the roles of complements in lupus pathogenesis remain to be defined. We will review data from complement deficient patients and animal models, and discuss how such information will facilitate the understanding of disease mechanisms, and help develop complement-targeted therapy below.

There are three distinct pathways of complement activation that are crucial components of innate immune response, serving to clear ICs, apoptotic debris, and pathogens. Each pathway triggers an enzyme cascade involving more than 25 proteins that converge on the common downstream effector molecules. The complement proteins employ three different mechanisms to protect against infection by (1) promoting phagocytosis of complement-coated particles, (2) promoting chemotaxis from complement split products, and (3) direct pathogen lysis via the formation of membrane attack complex (27,28). The classical complement pathway is triggered by C1 complex (C1q, C1r, and C1s) interacting with ICs, or by C1q binding to the surface of pathogens, whereas the lectin pathway is initiated when mannose-binding lectin or ficolins bind to certain carbohydrate residues on pathogen surfaces (29,30). The alternative pathway is elicited by spontaneous hydrolysis of C3 or via properdin (31). In addition, the components of the alternative pathway help amplify C3 activation triggered by the classical or lectin initiators.

Compelling evidence has emerged that homozygous deficiency of any of the early components of the classical pathway of complement activation (C1q, C1r, C1s, C4, and C2) predisposes to SLE. These deficiencies are the strongest susceptibility factors for the development of SLE (32). Overall, there appears to be a hierarchy of association of both disease prevalence and severity within the classical pathway, with patients deficient in one of C1-complex proteins (33–35) or C4 (33,36) exhibiting the highest prevalence (>80%) and the most severe disease, whereas the strength of the association significantly decreases in C2-deficient patients (33,37). Inherited C1q deficiency is exceedingly rare, although many SLE patients without C1q gene mutations have reduced serum C1q activity due to increased consumption, decreased production, and/or functional neutralization by anti-C1q. 53%–92% of SLE patients have low serum levels of C1q during active disease especially glomerulonephritis (38). C2 deficiency is associated with female-to-male ratio of 7:1, which is similar to that among patients without complement deficiencies. This is in striking contrast to the female-to-male ratio in patients with C1 or C4 deficiency, which is approximately 1:1, illustrating the profound effect of these deficiencies that overcome normal female preponderance seen in SLE (39). C3 deficiency, descried only in 23 subjects, has only been associated with lupus-like disease in three patients in two families (33,40,41). These clinical observations collectively suggest that a physiological activity of the early part of the classical pathway of complements, independent of C3 activation, plays a protective role against lupus development. In addition to the inherited complement deficiencies, acquired low levels of complements predispose to SLE. There is an increased prevalence of SLE among patients with hereditary angioedema, in which autosomal dominant deficiency of C1 inhibitor fails to regulate classical complement pathway activation and results in chronically low levels of C4 and C2. Hereditary angioedema is more common in women and associated with a high incidence of ANA, skin lesions, and photosensitivity, features reminiscent of SLE (33,42). Complements are strongly activated in SLE. Deposits of C3, C4, and associated complement proteins are readily detected in inflamed tissue specimens from patients with SLE (39). Complement activity and classical pathway protein levels are generally reduced in relation to disease activity and increase following treatment. SLE is thought to be initiated by high levels of ICs activating classical complement pathway. However, following the disease onset, there are a number of factors that influence the degree of reduction of serum complements. These include disease activity per se, the rate of production versus catabolism, and importantly, the presence of autoantibodies directed against complement proteins, such as anti-C1q. As such, even though hypocomplementemia constitutes one of the diagnostic criteria and SLE disease activity index (SLEDAI) score (2,3,43), studies on cohorts of SLE patients have demonstrated that complement levels in the circulation provide only a rough guide to disease activity (44,45).

The paradox in SLE is that complete activation of complement pathway promotes tissue injury, yet inherited deficiency of classical complement components predisposes to SLE. This has been reconciled by studies demonstrating the protective role of classical complement components (C1, C2, and C4) in facilitating the clearance of SLE ICs (27,46) as well as autoantigens, such as apoptotic debris. C1q-coated apoptotic cells suppress macrophage inflammation by inducing IL-10, IL-27, IL-33, and IL-37 and inhibit inflammasome activation (47). C1q-deficient mice develop glomerulonephritis associated with reduced clearance of apoptotic cells in the glomeruli (48). The outcome of C1q interaction with putative receptors is predominantly immunosuppressive. As an example, monocyte-derived dendritic cells (DCs) differentiated in the presence of C1q produce less IL-12 and IL-23, but greater amount of IL-10 and have an impaired ability to stimulate T cells *in vitro* (49,50). Although C1q binding to apoptotic cells is a critical mechanism that mitigates inflammatory response, it is not sufficient to prevent the development of SLE as exemplified by the mutation in an SLE patient with neurological manifestations affecting the collageneous arms of C1q, which allows C1q to bind to apoptotic cells and immunoglobulins, but disables downstream complement activation due to the inability to form a C1 complex (51). Besides direct binding, C1q promotes clearance of apoptotic cell through the induction of additional opsonins, growth arrest-specific 6 (Gas6) and protein S, as well as the receptor that recognizes these ligands, Mer tyrosine kinase (52). Gas6 binds to phosphatidylserine (PS) exposed on the surface of apoptotic cells and, and activates signaling pathways leading to phagocytosis of the apoptotic cells upon interaction with Mer or related receptors Axl or Tyro3 (53). These observations are likely clinically relevant as demonstrated when serum levels of soluble forms of Mer (sMer) correlated positively with SLE disease activity and negatively with serum C1q levels (54).

Although the aforementioned model linking the C1q deficiency and SLE development sounds plausible, it still does not explain why C4 or C2 deficiency also predisposes to SLE. In this respect, it is conceivable that C1q activation on apoptotic cells, likely initiated by natural IgM binding to neoepitopes (55,56), promotes apoptotic cell clearance by activating the classical pathway (C1q, C2 and C4) and leading to deposition of C3b on the cells (57). Note that C3b is the most abundant complement protein deposited on particles following activation of the classical or other complement pathways (58). In support of this model, C1q did not enhance ingestion of apoptotic cells in the absence of fresh serum (59). Along this line, anti-C3 antibodies were detected in the serum of three different mouse strains of lupus and suppressed apoptotic cell clearance (60). Moreover, serum from SLE patients contains elevated anti-C3b IgG that blocks detection of C3 on apoptotic cells (60).

Because IC formation is a hallmark of lupus pathogenesis, incorporation of ICs into the aforementioned models might offer a unifying paradigm on this subject. Indeed, C1q binding to ICs diverts ICs away from plasmacytoid DCs (pDCs) and promotes binding to and clearance by monocytes (61). This observation led to a model in which C1q-sufficient individuals, monocytes and macrophages rapidly clear small amounts of ICs containing self-antigens, but in the absence of C1q, ICs would be inefficiently cleared by these cells and would be more likely to engage pDCs (58). Once pDCs release interferon-alpha (IFNα), this cytokine would not only prime other cells of the immune system, but would also render pDCs more resistant to C1q-mediated inhibition (61). Consistent with this *in vitro* finding, increased levels of IFNα were detected in serum samples from a family with 4 siblings with C1q deficiency but not in the non-affected sibling or healthy controls (62). When peripheral blood mononuclear cells (PBMCs) and monocytes were cultured together with SLE-ICs in the presence or absence of C1q, C1q suppressed the expression of SLE-IC-induced IFN-stimulated genes, such as B-cell activation factor of the TNF family (BAFF) and TNF-related apoptosis-inducing ligand (TRAIL) (38). Furthermore, C1q-containing ICs were less potent in stimulating pDCs to produce IFNα than ICs without C1q (63). Increased IFNα production associated with diminished C1q is not restricted to rare mutations in the C1q gene, but is highly relevant to polygenic SLE patients because C1q levels are frequently reduced due to various mechanisms as discussed earlier. Anti-C1q antibody correlates with hypocomplementemia and lupus nephritis (64). Anti-C1q in combination with anti-dsDNA and hypocomplementemia was shown to be the strongest serological association with renal involvement (65). Mechanistically, anti-C1q antibody elicits pro-inflammatory cytokine secretion by macrophages from SLE patients in an FcγRII-dependent manner, which is accompanied by upregulation of CD80, CD274m and MHC class II (66).

How do a series of these observations inform the development of complement-targeted therapies in SLE? Because the majority of patients with SLE do not have inherited complement deficiency, the current clinical investigation focuses on targeting complements as a pro-inflammatory factor rather than replacing deficient complement components. The targeting of natural complement inhibitors (such as Crry and Factor H) to C3 degradation products binding to CR2 attenuates the activation of the late complement components via inhibition of C3 convertase. This strategy resulted in a significant decrease in anti-dsDNA and renal disease, and an increase in survival in lupus-prone mice (67,68). The complement components downstream of C3 promote inflammation and tissue injury through the generation of anaphylatoxins (C3a and C5a) that elicit the recruitment of neutrophils and by the formation of membrane attack complex (27,28). Eculizumab is a monoclonal antibody against C5 and thereby interferes with C5a/C5a receptor interaction. It is an FDA approved treatment for complement-mediated Hemolytic Uremic Syndrome (HUS) (69), in which a large proportion of patients exhibit mutations in regulatory proteins of alternative complement pathway (70). Interestingly, however, eculizumab appeared to be therapeutic in a series of cases of thrombotic microangiopathy (TMA) secondary to SLE, in which most patients did *not* have mutations in complement-regulatory proteins (71). This observation implies a distinct mechanism of lupus-related TMA in which complement genotyping might *not* inform a decision to pursue treatment with eculizumab.

Urine studies

LN is one of the most serious manifestations of SLE with a cumulative incidence of 54% (72). Proteinuria is the most common manifestation of LN and has been reported in nearly 100% of patents, followed by granular casts, cellular casts, hematuria, and reduced renal function (72). In biopsy-proven LN, the reduction of casts or proteinuria correlated with GFR 6 months after treatment (73). In addition, cellular casts are associated with anti-dsDNA (74) and anti-Smith (75), and were observed before or at the onset of renal relapses in >80% of patients (76). Hematuria and sterile pyuria appear to reflect the disease activity in close temporal proximity to the occurrence (77,78). Along this line, isolated pyuria is associated with previous ACR criteria for renal involvement and higher nonrenal disease activity (79). Furthermore, sterile pyuria and cellular casts are associated with class III or IV LN (80). A retrospective cohort study of Puerto Ricans revealed low C4 and proteinuria greater than 0.5 g/dL as predictors of earlier decline of GFR (81). In the Euro-Lupus

Nephritis Trial, a proteinuria value of <0.8 gm/day at 12 months after randomization was the single best predictor of good long-term renal function (sensitivity 81% and specificity 78%) (82). Likewise, normalization of C3 and/or C4 or reduction of proteinuria was predictive of renal response in patients with class III or IV LN randomized to mycophenolate or cyclophosphamide (26). Based on these findings, urinary casts (heme-granular or red blood cell casts), hematuria (>5 red blood cell [RBC]s per high power field [hpf]), proteinuria (>0.5 gm/24 h), and sterile pyuria (>5 white blood cell [WBC]s/hpf) are all incorporated into the SLEDAI score (43), among which proteinuria is used as a screening test for LN (83), to monitor a treatment response (83,84), and to monitor kidney disease progression given its excellent sensitivity close to 100%.

Although 24-h urine collection is the gold standard of quantification of proteinuria, this test can be cumbersome, and specimen is sometimes under-collected (85). Accordingly, a spot urine creatinine:protein ratio (PCR) was proposed as an alternative means to quantify proteinuria, reasoning that if the protein excretion remained stable, then PCR would reflect the cumulative protein excreted during a day (86). This was followed by a number of researchers reporting moderate correlation between PCR and 24-h urine protein in different diseases including diabetes (87), LN (88,89), and CKD (90). These studies established the basis for American College of Rheumatology to recommend using PCR in clinical trials of LN (91). However, the results of aforementioned studies needed to be interpreted with caution because correlation analysis would not be sufficient to evaluate the utility of a new test against the gold standard test (92). In fact, a systemic review revealed poor agreement between these two tests, signifying that PCR should not replace the gold standard 24-h urine collection (93). Along this line, PCR estimates of 24-h proteinuria were shown to be more unreliable in LN than other forms of chronic glomerular disease (94). Table 8.3.3 summarizes the biomarkers commonly used in the current rheumatology practice and their limitations.

TABLE 8.3.3 Commonly used biomarkers an their clinical implications

BIOMARKERS	*DIAGNOSTIC IMPLICATIONS*	*LIMITATIONS*
Anti-dsDNA	SLE specific (13) Utility in monitoring disease activity (14,15) Correlation with active LN (16,17)	Does not reflect the disease activity in a substantial portion of patients (21)
C3, C4, and CH50	Herald a SLE flare especially LN (22,24–26)	Do not always provide an accurate guide of disease activity (44,45)
Urinary casts	Their reduction predicts a renal response (73) Herald a renal relapse (76)	Not specific to LN
Hematuria (>5 RBCs/hpf)	Reflection of active LN (77,78)	Not specific to LN
Proteinuria (>0.5 gm/24 h)	Serves as an excellent screening test for LN (83) Its reduction predicts a renal response (82–84)	Not specific to LN
Sterile pyuria (>5 WBCs/hpf)	Reflection of active renal (77,78), and nonrenal (79) SLE	Not specific to LN

Emerging Biomarkers

SLE is a complex disease that involves a wide range of immune and nonimmune cells, in which dissecting the disease mechanisms is indeed a daunting task. In fact, there are overwhelming numbers of immunological, genetic, and epigenetic factors implicated in the lupus pathogenesis. Because it is not possible to sift through all of the available literature in this section, we will focus on biomarkers whose roles in SLE are more clearly discerned and are close to clinical translation in the near future.

Anti-nucleosome antibody

The structural unit of chromatin is the nucleosome, which consists of segments of dsDNA coiled around a histone core. Antichromatin antibodies are segregated into antibodies directed against the individual components of chromatin such as DNA, and antibodies with a high affinity to the intact nucleosome. There is evidence that the nucleosome is the primary antigen in SLE (95) and that anti-dsDNA represents a subset of this antibody population. Plasma from lupus patients contain microparticles that display DNA and nucleosomal molecules in an antigenic form, which could lead to IC formation (96). Clinically, anti-nucleosome is associated with lupus nephritis (97,98). Furthermore, the combination of anti-nucleosome and C3 exhibited stronger correlation with disease activity than anti-dsDNA and C3 (99). A meta-analysis showed that anti-nucleosome had greater sensitivity (59.9% vs. 52.4%) and a slightly higher specificity (94.9% vs. 94.2%) than anti-dsDNA in the diagnosis of SLE (100). Roles of anti-nucleosome in the SLE diagnosis deserve to be further investigated.

Type I interferon and plasmacytoid dendritic cells

Type I IFNs; IFNα/β, are signature cytokines in SLE. There is a large body of evidence implicating the essential roles of these cytokines in lupus pathogenesis. IFNα is increased in patients with active SLE. PBMCs from patients with SLE manifest an IFN signature (101,102), which correlates better with disease activity than anti-dsDNA (101). Treatment with steroid abrogates the IFNαβ signature in PBMCs (101). Furthermore, baseline type I IFN activity serves as a marker to predict future disease activity, in particular kidney involvement (103). Animal models offer further mechanistic insights into the roles of type I IFN in lupus pathogenesis. *In vivo* delivery of IFNαβ to NZB/W F1 mice leads to development of severe lupus nephritis associated with anti-dsDNA (104), whereas type I IFN receptor deficiency protects against lupus development (105–107). pDCs are a distinct lineage of DCs specialized in type I IFN production in response to viral nucleic acids sensed through Toll-like receptor 7 (TLR7) and TLR9 (108–110). Accumulating evidence suggests that aberrant recognition of self-nucleic acids through TLR7 and TLR9 (specific for single-stranded RNA and unmethylated CpG DNA, respectively) is central to autoantibody production and SLE pathogenesis (111,112). Duplication of mouse *Tlr7* gene accelerates SLE development, whereas multiple copies of *Tlr7* are sufficient to cause SLE-like disease (113–115). Along this line, chromatin- and snRNP-containing ICs are internalized by pDCs via FcγRIIa and reach the endosomal compartment where they activate TLR9 and TLR7, respectively, leading to secretion of cytokines including IFNα/β (116120). Conversely, macrophages and conventional DCs produce IFNα/β in a TLR-independent manner, but through RIG-I and MDA5 pathways that recognize viral RNAs and poly I:C, respectively (121,122).

Type I IFNs have diverse effects on a wide variety of immune cells in SLE. Interferon-α induces the differentiation of normal monocytes to mature DCs (123). Type I IFNs activate and promote the differentiation of immature myeloid DCs (mDCs) (123), which directly act on growth and differentiation of B cells (124,125). IFNα/β induces DCs to produce BAFF and a proliferation-inducing ligand (APRIL) that contributes to the survival, differentiation, and class switching of B cells including those expressing autoreactive B-cell receptors (126). In addition, IFNα/β promotes the differentiation of activated B cells into plasmablasts, and IL-6 permits plasmablasts to become antibody-secreting plasma cells (127). *In vitro*, chromatin-containing ICs activate transgenic autoreactive B cells via sequential engagement of the B-cell-antigen receptor and TLR9 (128). *In vivo*, TLR9 contributes to anti-dsDNA production in MRL-B6-129 mouse. In a B6 mouse lacking FcγRIIb, TLR9 signaling is required for anti-DNA polyreactive IgM$^+$ B cells that have escaped central tolerance to switch to pathogenic isotypes (129). TLR7 signaling plays an important role in SLE pathogenesis as well. ICs containing RNA and RNA-associated autoantigens activate autoreactive B cells *in vitro* (130). Injection of the TLR7 ligand, Imiquimod, increases IL-12p70, IFNα, and IL-6 and aggravates lupus nephritis in MRL/lpr mice (131).

IFNα/β might elicit mobilization of neutrophils by inducing G-CSF secretion (132). G-CSF-stimulated neutrophils secrete BAFF as efficiently as activated monocytes and DCs (133). In turn, DNA complexes released from activated neutrophils induce pDCs to secrete type I IFN (134,135), creating a vicious cycle of myeloid cell activation in SLE. Neutrophils may also contribute to DC activation via production of pro-inflammatory cytokines, chemokines, and defensins (136), and interactions between the integrin Mac-1 and DC-SIGN (137). Administration of IFN induces accumulation of neutrophils in the spleen of B6.yaa. Sle1 mice (138).

In SLE, there are several potential mechanisms that operate to perpetuate IFN production. Under normal circumstances, IFN secretion by healthy pDCs upon viral infection is abrogated by subsequent secretion of other cytokines such as TNF. Genetic alterations in SLE compromises the shutting down of IFN as illustrated by SLE-like syndrome in SOCS-deficient mice (139). Increased amounts of soluble TNF-receptors in SLE serum (140) may contribute to sustained IFNα/β production. Alternatively, ICs containing RNA or DNA may sustain IFN production by pDC via triggering of TLR7 and TLR9, respectively (116,130,141). In addition, IFNα/β elicits TLR7 expression on monocytes and DCs, which produce IFNα/β in response to TLR7 agonists (142); that is, IFNαβ creates a self-amplification loop via TLR7 induction.

Building on these findings, interferon-blocking therapy in SLE has gained much enthusiasm over the past decade. Monoclonal antibody against IFNα, sifalimumab (MEDI-545), led to partial inhibition of interferon signature in blood and skin in some patients (143). In a recent phase IIb study, sifalimumab met the primary efficacy end point based on SLE Response Index-4 (SRI4) in patients with moderate-severe SLE (144). Although a phase II trial of another monoclonal antibody against IFNα, rontalizumab, did not reach primary or secondary end points, rontalizumab treatment improved disease activity and was steroid sparing (145). IFN-kinoid, a complex of recombinant IFNα with keyhole limpet hemocyanin, has been developed as a means to elicit endogenous polyclonal anti-IFNα with the goal of blocking pathogenic type I IFN. Injection of IFN-kinoid resulted in T-cell dependent production of IFNα-specific neutralizing antibodies that decreased the expression of interferon-stimulated genes in blood (146). It is important to note that none of the aforementioned drugs blocks IFNβ, which might also contribute to lupus pathogenesis. In this respect, blockade of IFNAR led to nearly complete inhibition of interferon signature gene expression in peripheral blood and skin of patients with systemic sclerosis (147). Along this line, treatment with anifrolumab, a type I IFN receptor antagonist, yielded a significantly greater proportion of patients who achieved SRI4 response (148).

Neutrophils, neutrophil extracellular traps

Neutrophil extracellular traps (NETs) formation, a cell death pathway characterized by extrusion of chromatin bound to cytosolic and granular contents, has been implicated in autoimmunity (134,135,149). NETosis is induced by pathogens and sterile stimuli including cytokines, ICs, and antoantibodies (150), in which reactive oxygen species appear to play an important role. In fact, RNP ICs, which induce NETosis by type I IFN-primed neutrophils from patients with SLE (134), elicit extracellular release of oxidized mitochondrial DNA in a mitochondrial reactive oxygen species (ROS)-dependent manner (151). In turn, oxidized mitochondrial DNA elicits inflammatory response in a Stimulator of interferon genes (STING)-dependent manner (151). Low density granulocytes (LDGs), a distinct subset of pro-inflammatory neutrophils found in patients with SLE, exhibit spontaneous NETosis (152). These NETs contain higher level of autoantigens and immunostimulatory molecules such has LL-37, MMP9, and dsDNA (134,135,153,154), and are more immunogenic than NETs generated by healthy control neutrophils (154). LDGs from SLE patients release oxidized mitochondrial DNA in a superoxide-dependent manner that leads to an inflammatory response (151). Conversely, scavenger of mitochondrial ROS reduces NETosis and ameliorates glomerulonephritis by suppressing type I IFN response and anti-dsDNA production in

MRL/lpr lupus-prone mice (151). NETs also promote vasculopathy and inflammasome activation (134,135,152,155–158). NETs can further trigger an autoimmune response in SLE by exposing cathelicidin-DNA complexes, which, through activation of endosomal TLRs, promote the synthesis of type I IFN by pDCs (134,135). Cathelicidin and other NET proteins can activate the NLRP3 inflammasome in lipopolysaccharide-primed macrophages by interacting with their P2 × 7 receptor and inducing potassium efflux, which leads to release of IL-1 and IL-18 (155). These cytokines in turn promote NETosis, thereby amplifying inflammatory pathways. Neutrophils in skin and kidneys in patients with SLE undergo NETosis and externalize nucleic acids in these tissues, which, in turn, is associated with increased anti-dsDNA, implicating a direct contribution of NETosis to visceral organ disease in SLE (152). In addition to its pro-inflammatory potential of NET formation per se, NETs are not properly cleared from the circulation in patients with SLE. This defect is attributed to either complement activation within NETs, oxidation of nucleic acids externalized by NETs, or the presence of inhibitors of DNase-1 (149,151,153,159). The inability to degrade NETs is associated with higher incidence of nephritis in patients with SLE (149). Building on these findings implicating prominent roles for NETosis in SLE, neutrophils have emerged as a therapeutic target. *N*-acetyl cysteine (NAC) reveres glutathione depletion and thereby serves as a ROS scavenger, which renders NAC an appealing strategy to mitigate NETosis in SLE. Indeed, NAC appears to be therapeutic in patients with SLE (160). Because efficient NET induction requires mobilization of intracellular and extracellular calcium pools, inhibition of calcineurin might serves as an alternative means to suppress or modulate NET formation. In fact, cyclosporine A and tacrolimus proved to be therapeutic (161,162). Blocking C5a interaction with its receptor on neutrophils by eculizumab may also open a new avenue in NET-targeted therapy because complement activation plays a role in the defective NET clearance in SLE (153). Complement blockade is particularly appealing for lupus nephritis given the implication of NETosis in the renal pathology (152). In this respect, it is noteworthy that C5a and its neutrophil receptor C5aR create an amplification loop in Anti-neutrophil cytoplasmic antibody (ANCA)-associated vasculitis in which NETs activate the complement cascade culminating in production of C5a, which, in turn, stimulates neutrophils (163).

BAFF/APRIL/TACI

One of the ground-breaking discoveries in lupus research is that of an important B-cell survival factor, TNF-ligand superfamily member 13B (also known as BAFF) or B lymphocyte stimulator [BLyS]), which plays a critical role in autoimmunity, in particular in SLE (164,165). In 2011, belimumab, a monoclonal antibody targeting BAFF, proved to be therapeutic in a subset of SLE patients and has now become the first approved targeted therapy in SLE (166,167). Two ligands, BAFF and APRIL, and three receptors, TNF receptor super family member 13C (also known as BAFF receptor [BAFF-R] or BLyS receptor 3 [BR3]), TNF receptor superfamily member 17 (also known as B-cell maturation antigen [BCMA]) and TNF receptor superfamily member 13B (also known as transmembrane activator and cyclophilin ligand interactor [TACI) form the backbone of the BAFF/APRIL system. BAFF and APRIL can both interact with BCMA and TACI, whereas BAFF is the sole ligand for BAFF-R (168). BAFF-R is essential for both survival and maturation of immature B cells, whereas TACI is critical for T-cell-independent responses of B cells to type I and type II antigens, negative regulation of the B-cell compartment, and class-switch recombination of B cells (169). BCMA is expressed by plasmablasts and plasma cells, and promotes plasma cell survival. BAFF and APRIL are produced by myeloid cells, predominantly by macrophages, neutrophils, and dendritic cells, and also by radiation-resistant stromal cells. Lymphoid cells, including B cells and activated T cells, can also produce BAFF and APRIL. TLR9-activated pDCs and IL-2-activated natural killer cells produce BAFF.

BAFF is essential for B-cell maturation and survival (164,165). It is overexpressed in SLE patients, and its level correlates with anti-dsDNA titers (170). Mice genetically deficient in BAFF lack mature B cells and are immunodeficient, whereas mice that overexpress BAFF have high numbers of mature B cells and antibodies and develop a SLE-like autoimmune disease. BAFF directly stimulates T-cell proliferation and cytokine production (171,172). BAFF blockade in BCMA-deficient Nba2 mice resulted in markedly reduced number of follicular helper T (Tfh) cells in the spleens (173). In turn, Tfh cells promote the survival of high-affinity B cells in the germinal centers (174). Administration of recombinant BAFF to BCMA-deficient mice increased the numbers of total and germinal center Tfh cells, germinal center B cells, and plasma cells, suggesting that, in the absence of BCMA, excessive BAFF signaling via BAFF-R and/or TACI could drive the proliferation of Tfh cells, germinal center B cells, and splenic plasma cells. Along this line, Tfh cells from patients with SLE, which did not express BCMA, exhibited higher frequency of cells that express BAFF-R (173).

APRIL is important for antibody class-switching and plasma-cell survival (169). APRIL-overexpressing transgenic mice develop B1 B-cell neoplasia, but do not develop SLE-like pathology (175). Selective APRIL blockade can delay the development of disease in a lupus-prone murine model (NZB/W F1 mice) (176). Conversely, genetic ablation of APRIL did not protect against lupus development, and combined blockade of BAFF and APRIL led to comparable renal immunopathology to that of BAFF blockade, raising a question about the essentiality of APRIL in SLE pathogenesis (177). Further, a phase II/III clinical trial of the chimeric recombinant fusion protein atacicept, that neutralizes BAFF, APRIL, and BAFF-APRIL heterodimers, in patients with moderate-severe lupus did not prove to be effective (178).

Belimumab is a fully humanized monoclonal IgG targeting soluble human BAFF (179). The addition of belimumab to standard therapy has been studied in two multicenter, double-blind, placebo-controlled, randomized phase III trials examining 1684 patients with SLE for up to 76 weeks (166,167), in which patients with positive lupus-associated autoantibodies benefited from belimumab compared with placebo with respect to disease activity and health-related quality of life with a satisfactory drug safety profile (166,167,180,181). Belimumab treatment reduced autoantibody level and normalized hypocomplementemia (182). Patients with high disease activity, high anti-dsDNA, and hypocomplementemia had superior treatment responses to belimumab (183). Belimumab reduced the number of circulating naïve B cells, activated B cells, and plasma cells, but did not reduce the number of circulating memory B cells and T cells (182). Responses to previous immunizations for pneumococcus, tetanus, and influenza were not affected after 1 year of treatment (184).

Although serum concentrations of BAFF and APRIL were increased in patients with SLE, Sjögren's syndrome, and rheumatoid arthritis, whether BAFF serves as a biomarker for SLE has been a matter of debate. There are studies reporting correlation between serum BAFF and APRIL concentrations and overall disease activity (185–187), whereas others found not such correlations (170,182,188). Elevated baseline serum BAFF concentration was predictive of moderate-severe SLE flares in patients receiving standard therapy; that is, prednisone with antimalarial or immunosuppressant (methotrexate, mycophenolate mofetil, or azathioprine) (25), which appears to hold true for patients status post rituximab treatment (189). Because survival of autoreactive B cells requires higher levels of BAFF than does maintenance of self-tolerant population (126), the presence of elevated levels of BAFF lowers the stringency of B-cell selection and thus rescues autoreactive clones that would otherwise be deleted (190–192). BAFF supports the survival of plasmablasts (193), whereas BAFF inhibition blocks the differentiation of marginal zone B cells into plasmablasts (194). Collectively, rising levels of BAFF following rituximab therapy coupled with reduced B-cell competition could culminate in a SLE flare. In this regard, a combination of B-cell depletion and BAFF inhibition removes B cells in the marginal zone and follicular compartments more effectively than either treatment alone (195). The combination was also superior to B-cell depletion alone with respect to reducing the numbers of plasmablasts and plasma cells, as well as ameliorating disease severity in three different murine models of SLE (196). These findings established the foundation of two clinical trials of rituximab followed by belimumab therapy; CALIBRATE study and SYNBIoSe study. On the other hand, baseline serum BAFF concentration was not found to be predictive of outcome in either anti-BAFF or control-treated patients (182). These discrepancies might be attributed to differences in assay sensitivity, disease activity scores, or study population. Serum BAFF and APRIL concentrations might not accurately reflect total production of BAFF or, in the case of APRIL, which binds proteoglycans, the concentration of active cytokine in tissues. In this regard, it is noteworthy that an analysis of a cohort of 36 SLE patients showed a nonsignificant trend toward a positive correlation between serum BAFF-APRIL heterotrimer concentration and SLEDAI (197). A better understanding of the relative proportions and activity of BAFF-APRIL homotrimers and heterotrimers might redirect therapeutic targeting of the BAFF/APRIL system. In addition to the molecular biological aspects, it is important to take into account the expression of clinical phenotypes when it comes to the question of whether BAFF or APRIL serves as a biomarker of SLE. Because a post hoc analysis of BLISS trials showed that BAFF blockade was most effective in patients with active disease and high serum autoantibody titers (183), this subset of SLE patients might be best suited for analysis of clinical association of BAFF and APRIL expression. Likewise, serum BAFF is increased in patients with CNS or renal involvement (198), raising a possibility that BAFF is associated with specific clinical phenotypes of SLE.

We have reviewed a series of observations underscoring the indispensable roles of type I IFN in SLE earlier in this section. In this regard, accumulating evidence suggests that IFNs are the upstream regulators of BAFF expression. IFNα elicits BAFF expression by mouse macrophages (199). IFNα and -γ upregulate BAFF and APRIL expression by human DCs, whereas IFNα and -γ induce BAFF expression by human monocytes (200). In response to IFNγ, monocytes from patients with SLE produced a greater amount of BAFF than those from healthy subjects (201). IFNα blockade reduced BAFF expression in whole blood of patients with SLE (143). In NZM lupus-prone mice, BAFF deficiency prevented the development of glomerulonephritis in response to IFNα treatment (202). There was a positive correlation between the serum activity of IFNα and BAFF concentration in SLE patients (203).

Plasmablasts

Increased frequency of $CD27^{hi}CD38^{+}CD19^{dim}$Surface $Ig^{low}CD20^{-}CD138^{+}$ plasmablasts in peripheral blood is associated with active disease (204). Further, post-rituximab flare in patients with high-titer anti-dsDNA is characterized by increased proportion of $CD19^{+}IgD^{-}CD27^{hi}$ plasmablasts within the repopulating B-cell pool (205). Those who have high proportion of plasmablasts have few transitional B cells, a compartment rich in regulatory B cells, which could contribute to an earlier relapse (206). The number of $CD27^{hi}$ plasmablasts in patients with SLE is closely correlated with SLEDAI, and anti-dsDNA level (207). Human peripheral blood plasmablasts promote the differentiation of Tfh cells in part via IL-6 (208,209). In turn, Tfh cells stimulate plasmablast formation, thereby creating a positive amplification loop and perpetuating the disease. Along this line, Tfh-cell differentiation in mouse model of infection is dependent on B-cell production of IL-6 (210). The frequency of Tfh cells correlates with plasmablast numbers, anti-dsDNA levels, and disease activity in SLE (211–213). Because a series of these observations shed light on IL-6 secretion as a cardinal mechanism by which plasmablasts drive Tfh cell differentiation and thereby auto-amplify an inflammatory response, roles of IL-6 blockade benefit from investigation. Tocilizumab, a monoclonal antibody against IL-6 receptor, improved the disease activity, associated with the reduction of plasma cells and anti-dsDNA although the treatment was limited by dose-dependent neutropenia (214). Importantly, tocilizumab

TABLE 8.3.4 Emerging non-T-cell markers and their diagnostic and therapeutic implications

BIOMARKERS	*MECHANISMS OF CONTRIBUTION*	*DIAGNOSTIC APPLICATION*	*THERAPEUTIC INTERVENTION*
Anti-nucleosome Ab	Forms ICs with microparticles (96)	Correlation with nephritis (97,98) Allows more accurate assessment of disease activity when combined with C3 (99)	N/A
Type I IFN	Elicits maturation of DCs (123), differentiation of B cells to plasmablasts (127), and mobilization of neutrophils (132) Self-amplifies its own expression by inducing TLR7 on DCs (142) Induces BAFF expression (199,200)	Correlation with disease activity (101) Heralds a disease flare particularly nephritis (103)	Sifalimumab (phase II) (144) Rontalizumab (phase II) (145) Interferon α-kinoid (146) Anifrolumab (phase II) (148)
NETs	Induce a type I IFN response in a STING-dependent manner (151) Activate NLRP3 inflammasome in macrophages (155) Netting neutrophils infiltrate kidneys in association with anti-dsDNA (152)	Association with cutaneous and renal lupus (152)	NAC (160) Cyclosporine A (161) Tacrolimus (162) Eculizumab (71)
BAFF	Critical for B-cell survival and maturation (164,165) and plasmablast survival (193) High levels of BAFF allows expansion of autoreactive B cells (190–192) Stimulates T-cell proliferation and cytokine production, especially those of Tfh cells (171–173)	Heralds a disease flare especially after rituximab treatment (25,189). BAFF-APRIL heterotrimer better correlates with SLEDAI (197)	Belimumab (166,167) Rituximab + Belimumab (195,196) Blisibimod (217)
$CD19^+IgD^-CD27^{hi}$ or $CD27^{hi}CD38^+CD19^{dim}$surface $Ig^{low}CD20^-CD138^+$ plasmablasts	Promote Tfh cell differentiation via IL-6 (208,210)	Herald a disease flare especially after rituximab treatment (204–207)	Tocilizumab (214,215) Sirukumab (216)

decreased the frequency of $CD27^{hi}CD38^{hi}IgD^-$ plasmablasts/plasma cells, IgD^-CD27^+ post-switched memory B cells, and IgG^+ memory B cells (214,215). Likewise, treatment with another IL-6 receptor antagonist, sirukumab, was well tolerated and reduced inflammatory markers (216). Table 8.3.4 summarizes B cell and innate immune system-related biomarkers and their potential roles in diagnosis and treatment of SLE.

Mechanistic Target of Rapamycin

Mechanistic target of rapamycin (mTOR) is an evolutionally conserved serine-threonine kinase, which translates a variety of environmental cues to signals that dictate cell growth, proliferation, and differentiation (218), and has recently emerged as a central regulator of T cell proliferation and differentiation (219–223). In particular, mTOR complex 1 (mTORC1) is essential for Th1 and Th17 differentiation, whereas mTOR complex 2 (mTORC2) is indispensable for Th2 differentiation in mice (224). Constitutively active mTOR and Akt abrogate Treg differentiation (225). Along this line, genetic ablation of rheb and rictor in mice indicates that a dual blockade of mTORC1 and mTORC2 is required to allow Treg differentiation (224). Importantly, mTOR is increased in SLE T cells (226), and administration of rapamycin improves the clinical outcome in murine models of lupus (227,228) and patients with SLE (229,230). Our study documented mTORC2 deficiency in SLE T cells, and highlighted mTORC1 activation as an important mechanism that drives expansion of Th17 cells and IL-4-expressing $CD4^-CD8^-$ double-negative T cells in SLE (229,231). mTOR is a suppressor of autophagy, which plays a critical role in linage stability and function of regulatory T (Treg) cells (232). Consistently, SLE Treg cells exhibit increased mTORC1 and mTORC2 activities and diminished autophagy, where a dual blockade of mTORC1 and mTORC2 through a 4-week rapamycin treatment expands SLE Treg cells and restores their diminished function in association with induction of autophagy (233). Besides rapamycin, NAC blocks mTOR via reversal of antioxidant depletion, and improves lupus pathology in association with correction of T cell defects and suppression of anti-dsDNA production (160).

$CD3^+CD4^-CD8^-$ T Cells

A subset of TCR $\alpha\beta^+$ T cells that express neither CD4 nor CD8—known as $CD4^-CD8^-$ double-negative (DN) T cells—constitute at most 5% of T cells in human and murine peripheral blood. Of note, DN T cells are increased in SLE patients (234,235) and exhibited prominently increased mTORC1 activity (231). DN T cells produce greater amount of IL-4 than $CD4^+$ or $CD8^+$ T cells in an mTORC1-dependent manner (231), and assist B cells to produce anti-dsDNA (234,236). Lupus DN T cells secrete both IFNγ and IL-4, whereas healthy control DN T cells secrete IFNγ only (237). DN T cells from SLE patients expand significantly following anti-CD3 stimulation and produce significant amount of IFNγ and IL-17 (238). Given the prominently elevated mTORC1 in DN T cells, and their mTORC1-dependent production of IL-4, mTOR blockade merits further consideration to deplete this potentially pathogenic T-cell population.

Th1 vs. Th2 Cells

As to the roles of helper T (Th) cell subsets in SLE, it has been controversial whether SLE is driven by Th1 or Th2 immunity given the various animal models showing discrepant findings. In humans, some studies showed increased IL-4, but decreased IFNγ in lupus patients (239,240), whereas others indicate the importance of IFNγ in diffuse proliferative lupus nephritis (241,242). SLE patients with higher SLEDAI score have lower IFNγ but higher IL-4 expression than those with lower SLEDAI score (243). Frequency of polymorphism of IFNγ receptor gene was more frequent in lupus patients and was associated with skewing toward Th2 response (244). Our study showed that IL-4 expression was increased in SLE T cells, in particular in $CD8^+$ T cells, whereas IFNγ expression was downregulated in $CD4^+$, $CD8^+$, and $CD4^-CD8^-$ double-negative (DN) T cells. Interestingly, treatment with rapamycin expanded IFNγ-producing DN T cells (231). These findings would lead us to propose IFNγ as a potentially protective cytokine in SLE.

Th17 Cells

There is a growing body of evidence highlighting the importance of IL-17 in SLE. SLE patients have increased serum IL-17 and frequency of Th17 cells (245–248). There is a positive correlation between plasma IL-17 or Th17 cell frequency and SLEDAI score (245–247,249). IL-17 expressing DN T cells were increased in patients with lupus nephritis (238). The increased IL-17 production in SLE has been linked to the ability of IL-23 to promote Th17-cell differentiation, and genetic ablation of IL-23 receptor or treatment with anti-IL-23 ameliorates lupus pathology (250,251). IL-17 signaling is critical for the development of lupus nephritis in FcγRIIb deficient mice (252). Along this line, IL-17 deficient mice are protected from development of lupus-associated autoantibodies and glomerulonephritis, which is associated with reduction of DN T cells and expansion of $CD4^+CD25^+FoxP3^+$ Treg cells (253). Clinically, ustekinumab inhibits pathogenic Th17 differentiation via IL-12 and -23 blockades, and appears to be therapeutic (2017 ACR abstract #6L).

IL-21 and Follicular Helper T Cells

IL-21 has emerged as one of the pro-inflammatory cytokines in SLE. IL-21 and IL-21 receptor polymorphisms confer risk for SLE (254,255), and IL-21 expression is increased in SLE T cells (256). IL-21 induces Tfh cell development in a signal transducer and activator of transcription 3 (STAT3)-dependent manner (257–259). In turn, Tfh cells facilitate germinal center formation, memory B- and plasma-cell development, and class-switching of B cells via IL-21 secretion (260–262). Mice homozygous for Rc3h1san (the san allele encoding the sanroque form of roquin-1, a Really Interesting New Gene [RING]-type ubiquitin ligase) develop germinal centers, excessive Tfh-cell numbers, and glomerulonephritis (263). Adoptive transfer of sanroque-expressing Tfh cells into wild-type mice resulted in spontaneous germinal center formation (263). In BXD2 mice, Tfh cells drive the increased autoantibody production and enhanced germinal center B-cell responses through an IL-21-dependent mechanism (264). Importantly, $CXCR5^{high}$ $ICOS^{high}$ $PD\text{-}1^{high}$ Tfh-like cells are expanded in SLE patients, and the PD-1 expression in Tfh-like cells correlates with the disease activity and anti-dsDNA positivity (213), corroborating the relevance of Tfh-cell-driven humoral immunity to lupus pathogenesis. Likewise, Tfh cells are positively correlated with renal involvement, venous thrombosis, and thrombocytopenia, as well as autoantibodies including anti-dsDNA and anticardiolipin antibody (212). Although aforementioned studies highlight IL-21 as a promising therapeutic target, IL-21 blockade is yet far from clinical application. Nonetheless, it is encouraging that IL-21 blockade ameliorates a murine model of lupus nephritis in association with diminished renal infiltration of Tfh cells and reduction of autoantibodies, germinal center B cells, and $CD138^{hi}$ plasmablasts (265). In addition to directly promoting inflammatory response, IL-21 mitigates Treg-cell differentiation and function via mTOR-dependent suppression of autophagy, which involves the blockade of GATA-3 and CTLA-4 expression in $CD4^+CD25^+FOXP3^+$ Treg cells (233).

Treg Cells

Treg cells play indispensable roles in maintaining peripheral tolerance. Although it is an appealing notion that Treg-cell defect contributes to systemic autoimmunity, there have been contradictory observations concerning this notion. In SLE patients, the number of Treg cells was shown to be reduced (266–271), unchanged (272,273), or increased (274,275). Likewise, the suppressive function of Treg cells was shown to be decreased in active SLE (270,276,277), decreased only in a portion of patients (272), or unimpaired (268,273,278). It is important to note that various methods have been used to phenotypically define Treg cells, which may in part account for these discrepant findings. Other lines of evidence indicate negative correlation between Treg-cell frequency or suppressive function and SLEDAI score (246,268–270). Rapamycin blocks mTORC1 within Treg cells, and expands $CD4^+CD25^+FOXP3^+$ Treg-cell population among $CD3^+$ T cells stimulated *in vitro* (231) and *in vivo* (229). However, it is unknown what pro-inflammatory cues either upstream or downstream of mTOR drive Treg-cell depletion and dysfunction.

TABLE 8.3.5 Emerging T-cell related markers implicating novel therapeutic targets

BIOMARKER	*MECHANISMS OF CONTRIBUTION TO SLE*	*THERAPEUTIC INTERVENTION*
mTOR	Expands IL-4⁺ DN T cells and Th17 cells (229,231) Blocks Treg cell differentiation and function (233)	Rapamycin (230,231,233) NAC (160)
DN T cells	Produce IL-4 (229,231) Produce IL-17 (238) Stimulate B cells to produce anti-dsDNA (234,236)	Rapamycin (230,231)
Th17 cells	Directly drive nephritis (238,252) by recruiting neutrophils to renal tubules (283)	Ustekinumab (2017 ACR abstract #6L)
Tfh cells	Facilitate germinal center formation, memory B- and plasma-cell development, class-switching of B cells, and autoantibody production (260–264)	IL-21 blockade (265)
$CD4^+CD25^{hi}$, $CD4^+CD25^{hi}CD127^{low}$, or $CD4^+CD25^+FOXP3^+$ Treg cells	Their depletion or dysfunction leads to breakdown of peripheral tolerance (270,276)	IL-2 (280) Rapamycin (229,231,233)

In this regard, neutralization of IL-17 expands Treg-cells comparably to what is seen with rapamycin treatment *in vitro* (231), suggesting that suppression of IL-17 expression may be one of the mechanisms by which rapamycin promotes Treg cell expansion. As noted above, we have recently identified IL-21 as a stimulator of mTORC1 and mTORC2 activation in SLE Treg cells that abrogates the differentiation and suppressor function via inhibition of autophagy and expression of GATA-3 and CTLA-4 (233). Conversely, a dual blockade mTORC1 and mTORC2 by a prolonged rapamycin treatment restores autophagy and expression of GATA-3 and CTLA-4, and normalizes suppressor function of SLE Treg cells (233). Despite these encouraging results suggesting the critical roles of quantitative and qualitative Treg cell deficiency in SLE, Treg-cell targeted therapy is a long way from clinical application. Nonetheless, it is promising that administration of low dose IL-2, a cytokine critical for Treg-cell differentiation (279), reduced the disease activity along with restoration of Treg cell frequency and contraction of Tfh- and Th17-cells (280). Because SLE Treg cells exhibit diminished CTLA-4 expression (233), its replenishment may open a new avenue to restore Treg cell function. In this respect, abatacept has been investigated as a potentially promising therapy both in renal and nonrenal lupus even though primary efficacy end points were not met in the previous trials (281,282). Table 8.3.5 summarizes a growing list of T-cell related biomarkers and their implications for therapeutic interventions.

CONCLUDING REMARKS

The foregoing review of available and emerging biomarkers of SLE helps us realize the overwhelming degree of diversity of immune cells and molecular signaling pathways involved in the lupus pathogenesis. This seemingly daunting paradigm presents numerous potential opportunities for therapeutic interventions. Although the importance of further mechanistic studies cannot be overemphasized, we expect that future clinical studies address the feasibility of a combination of multiple drugs with different mechanisms of action, as exemplified by the combination of rituximab and belimumab, in the event when one drug turns out to be partially effective or loses efficacy over time. Harnessing a combination of nonredundant therapeutic mechanisms may not only allow for synergistic benefits, but would also minimize toxicities associated with individual drugs.

FUNDING

This work was supported by grants R01-AI072648, R01-AI122176, R34 AR068052, and R34 AI141304 from the National Institutes of Health and the Department of Medicine at the SUNY Upstate Medical University.

REFERENCES

1. Tsokos GC. Systemic lupus erythematosus. *N Engl J Med.* 2011;365(22):2110–2121.
2. Tan EM, Cohen AS, Fries JF, Masi AT, McShane DJ, Rothfield NF, et al. The 1982 revised criteria for the classification of systemic lupus erythematosus. *Arthritis Rheum.* 1982;25(11):1271–1277.
3. Petri M, Orbai AM, Alarcon GS, Gordon C, Merrill JT, Fortin PR, et al. Derivation and validation of the Systemic Lupus International Collaborating Clinics classification criteria for systemic lupus erythematosus. *Arthritis Rheum.* 2012;64(8):2677–2686.
4. Tan EM, Feltkamp TE, Smolen JS, Butcher B, Dawkins R, Fritzler MJ, et al. Range of antinuclear antibodies in "healthy" individuals. *Arthritis Rheum.* 1997;40(9):1601–1611.
5. Meroni PL, Schur PH. ANA screening: An old test with new recommendations. *Ann Rheum Dis.* 2010;69(8):1420–1422.
6. Bonilla E, Francis L, Allam F, Ogrinc M, Neupane H, Phillips PE, et al. Immunofluorescence microscopy is superior to fluorescent beads for detection of antinuclear antibody reactivity in systemic lupus erythematosus patients. *Clin Immunol.* 2007;124(1):18–21.

7. Pollock W, Toh BH. Routine immunofluorescence detection of Ro/SS-A autoantibody using HEp-2 cells transfected with human 60 kDa Ro/SS-A. *J Clin Pathol.* 1999;52(9):684–687.
8. Mahler M, Ngo JT, Schulte-Pelkum J, Luettich T, Fritzler MJ. Limited reliability of the indirect immunofluorescence technique for the detection of anti-Rib-P antibodies. *Arthritis Res Ther.* 2008;10(6):R131.
9. Bruner BF, Guthridge JM, Lu R, Vidal G, Kelly JA, Robertson JM, et al. Comparison of autoantibody specificities between traditional and bead-based assays in a large, diverse collection of patients with systemic lupus erythematosus and family members. *Arthritis Rheum.* 2012;64(11):3677–3686.
10. Op De Beeck K, Vermeersch P, Verschueren P, Westhovens R, Marien G, Blockmans D, et al. Antinuclear antibody detection by automated multiplex immunoassay in untreated patients at the time of diagnosis. *Autoimmun Rev.* 2012;12(2):137–143.
11. Bonfa E, Golombek SJ, Kaufman LD, Skelly S, Weissbach H, Brot N, et al. Association between lupus psychosis and anti-ribosomal P protein antibodies. *N Engl J Med.* 1987;317(5):265–271.
12. Arbuckle MR, McClain MT, Rubertone MV, Scofield RH, Dennis GJ, James JA, et al. Development of autoantibodies before the clinical onset of systemic lupus erythematosus. *N Engl J Med.* 2003;349(16):1526–1533.
13. Kavanaugh AF, Solomon DH. Guidelines for immunologic laboratory testing in the rheumatic diseases: Anti-DNA antibody tests. *Arthritis Rheum.* 2002;47(5):546–555.
14. ter Borg EJ, Horst G, Hummel EJ, Limburg PC, Kallenberg CG. Measurement of increases in anti-double-stranded DNA antibody levels as a predictor of disease exacerbation in systemic lupus erythematosus. A long-term, prospective study. *Arthritis Rheum.* 1990;33(5):634–643.
15. Bootsma H, Spronk P, Derksen R, de Boer G, Wolters-Dicke H, Hermans J, et al. Prevention of relapses in systemic lupus erythematosus. *Lancet.* 1995;345(8965):1595–1599.
16. Cortes-Hernandez J, Ordi-Ros J, Labrador M, Bujan S, Balada E, Segarra A, et al. Antihistone and anti-double-stranded deoxyribonucleic acid antibodies are associated with renal disease in systemic lupus erythematosus. *Am J Med.* 2004;116(3):165–173.
17. Schur PH, Sandson J. Immunologic factors and clinical activity in systemic lupus erythematosus. *N Engl J Med.* 1968;278(10):533–538.
18. Vlahakos DV, Foster MH, Adams S, Katz M, Ucci AA, Barrett KJ, et al. Anti-DNA antibodies form immune deposits at distinct glomerular and vascular sites. *Kidney Int.* 1992;41(6):1690–1700.
19. D'Andrea DM, Coupaye-Gerard B, Kleyman TR, Foster MH, Madaio MP. Lupus autoantibodies interact directly with distinct glomerular and vascular cell surface antigens. *Kidney Int.* 1996;49(5):1214–1221.
20. Chubick A, Sontheimer RD, Gilliam JN, Ziff M. An appraisal of tests for native DNA antibodies in connective tissue diseases. Clinical usefulness of Crithidia luciliae assay. *Ann Intern Med.* 1978;89(2):186–192.
21. Gladman DD, Urowitz MB, Keystone EC. Serologically active clinically quiescent systemic lupus erythematosus: A discordance between clinical and serologic features. *Am J Med.* 1979;66(2):210–215.
22. Ho A, Magder LS, Barr SG, Petri M. Decreases in anti-double-stranded DNA levels are associated with concurrent flares in patients with systemic lupus erythematosus. *Arthritis Rheum.* 2001;44(10):2342–2349.
23. Winfield JB, Brunner CM, Koffler D. Serologic studies in patients with systemic lupus erythematosus and central nervous system dysfunction. *Arthritis Rheum.* 1978;21(3):289–294.
24. Lloyd W, Schur PH. Immune complexes, complement, and anti-DNA in exacerbations of systemic lupus erythematosus (SLE). *Medicine.* 1981;60(3):208–217.
25. Petri MA, van Vollenhoven RF, Buyon J, Levy RA, Navarra SV, Cervera R, et al. Baseline predictors of systemic lupus erythematosus flares: Data from the combined placebo groups in the phase III belimumab trials. *Arthritis Rheum.* 2013;65(8):2143–2153.
26. Dall'Era M, Stone D, Levesque V, Cisternas M, Wofsy D. Identification of biomarkers that predict response to treatment of lupus nephritis with mycophenolate mofetil or pulse cyclophosphamide. *Arthritis Care Res.* 2011;63(3):351–357.
27. Walport MJ. Complement. Second of two parts. *N Engl J Med.* 2001;344(15):1140–1144.
28. Walport MJ. Complement. First of two parts. *N Engl J Med.* 2001;344(14):1058–1066.
29. Fraser DA, Tenner AJ. Directing an appropriate immune response: The role of defense collagens and other soluble pattern recognition molecules. *Curr Drug Targets.* 2008;9(2):113–122.
30. Galvan MD, Greenlee-Wacker MC, Bohlson SS. C1q and phagocytosis: The perfect complement to a good meal. *J Leukoc Biol.* 2012;92(3):489–497.
31. Spitzer D, Mitchell LM, Atkinson JP, Hourcade DE. Properdin can initiate complement activation by binding specific target surfaces and providing a platform for de novo convertase assembly. *J Immunol.* 2007;179(4):2600–2608.
32. Harley JB, Kelly JA, Kaufman KM. Unraveling the genetics of systemic lupus erythematosus. *Springer Semin Immunopathol.* 2006;28(2):119–130.
33. Pickering MC, Botto M, Taylor PR, Lachmann PJ, Walport MJ. Systemic lupus erythematosus, complement deficiency, and apoptosis. *Adv Immunol.* 2000;76:227–324.
34. Stone NM, Williams A, Wilkinson JD, Bird G. Systemic lupus erythematosus with C1q deficiency. *Br J Dermatol.* 2000;142(3):521–524.
35. Dragon-Durey MA, Quartier P, Fremeaux-Bacchi V, Blouin J, de Barace C, Prieur AM, et al. Molecular basis of a selective C1s deficiency associated with early onset multiple autoimmune diseases. *J Immunol.* 2001;166(12):7612–7616.
36. Rupert KL, Moulds JM, Yang Y, Arnett FC, Warren RW, Reveille JD, et al. The molecular basis of complete complement C4A and C4B deficiencies in a systemic lupus erythematosus patient with homozygous C4A and C4B mutant genes. *J Immunol.* 2002;169(3):1570–1578.
37. Agnello V. Association of systemic lupus erythematosus and SLE-like syndromes with hereditary and acquired complement deficiency states. *Arthritis Rheum.* 1978;21(5 Suppl):S146–S152.
38. Santer DM, Wiedeman AE, Teal TH, Ghosh P, Elkon KB. Plasmacytoid dendritic cells and C1q differentially regulate inflammatory gene induction by lupus immune complexes. *J Immunol.* 2012;188(2):902–915.
39. Manderson AP, Botto M, Walport MJ. The role of complement in the development of systemic lupus erythematosus. *Annu Rev Immunol.* 2004;22:431–456.
40. Imai K, Nakajima K, Eguchi K, Miyazaki M, Endoh M, Tomino Y, et al. Homozygous C3 deficiency associated with IgA nephropathy. *Nephron.* 1991;59(1):148–152.
41. Nilsson UR, Nilsson B, Storm KE, Sjolin-Forsberg G, Hallgren R. Hereditary dysfunction of the third component of complement associated with a systemic lupus erythematosus-like syndrome and meningococcal meningitis. *Arthritis Rheum.* 1992;35(5):580–586.
42. Koide M, Shirahama S, Tokura Y, Takigawa M, Hayakawa M, Furukawa F. Lupus erythematosus associated with C1 inhibitor deficiency. *J Dermatol.* 2002;29(8):503–507.

43. Bombardier C, Gladman DD, Urowitz MB, Caron D, Chang CH. Derivation of the SLEDAI. A disease activity index for lupus patients. The Committee on Prognosis Studies in SLE. *Arthritis Rheum.* 1992;35(6):630–640.
44. Cameron JS, Lessof MH, Ogg CS, Williams BD, Williams DG. Disease activity in the nephritis of systemic lupus erythematosus in relation to serum complement concentrations. DNA-binding capacity and precipitating anti-DNA antibody. *Clin Exp Immunol.* 1976;25(3):418–427.
45. Valentijn RM, van Overhagen H, Hazevoet HM, Hermans J, Cats A, Daha MR, et al. The value of complement and immune complex determinations in monitoring disease activity in patients with systemic lupus erythematosus. *Arthritis Rheum.* 1985;28(8):904–913.
46. Webster SD, Galvan MD, Ferran F, Garzon-Rodriguez W, Glabe CG, Tenner AJ. Antibody-mediated phagocytosis of the amyloid beta-peptide in microglia is differentially modulated by C1q. *J Immunol.* 2001;166(12):7496–7503.
47. Benoit ME, Clarke EV, Morgado P, Fraser DA, Tenner AJ. Complement protein C1q directs macrophage polarization and limits inflammasome activity during the uptake of apoptotic cells. *J Immunol.* 2012;188(11):5682–5693.
48. Botto M, Dell'Agnola C, Bygrave AE, Thompson EM, Cook HT, Petry F, et al. Homozygous C1q deficiency causes glomerulonephritis associated with multiple apoptotic bodies. *Nat Genet.* 1998;19(1):56–59.
49. Teh BK, Yeo JG, Chern LM, Lu J. C1q regulation of dendritic cell development from monocytes with distinct cytokine production and T cell stimulation. *Mol Immunol.* 2011;48(9–10):1128–11238.
50. Castellano G, Woltman AM, Schlagwein N, Xu W, Schena FP, Daha MR, et al. Immune modulation of human dendritic cells by complement. *Eur J Immunol.* 2007;37(10):2803–2811.
51. Roumenina LT, Sene D, Radanova M, Blouin J, Halbwachs-Mecarelli L, Dragon-Durey MA, et al. Functional complement C1q abnormality leads to impaired immune complexes and apoptotic cell clearance. *J Immunol.* 2011;187(8):4369–4373.
52. Galvan MD, Foreman DB, Zeng E, Tan JC, Bohlson SS. Complement component C1q regulates macrophage expression of Mer tyrosine kinase to promote clearance of apoptotic cells. *J Immunol.* 2012;188(8):3716–3723.
53. Rothlin CV, Lemke G. TAM receptor signaling and autoimmune disease. *Curr Opin Immunol.* 2010;22(6):740–746.
54. Wu J, Ekman C, Jonsen A, Sturfelt G, Bengtsson AA, Gottsater A, et al. Increased plasma levels of the soluble Mer tyrosine kinase receptor in systemic lupus erythematosus relate to disease activity and nephritis. *Arthritis Res Ther.* 2011;13(2):R62.
55. Kim SJ, Gershov D, Ma X, Brot N, Elkon KB. I-PLA(2) activation during apoptosis promotes the exposure of membrane lysophosphatidylcholine leading to binding by natural immunoglobulin M antibodies and complement activation. *J Exp Med.* 2002;196(5):655–665.
56. Chen Y, Park YB, Patel E, Silverman GJ. IgM antibodies to apoptosis-associated determinants recruit C1q and enhance dendritic cell phagocytosis of apoptotic cells. *J Immunol.* 2009;182(10):6031–6043.
57. Mevorach D, Mascarenhas JO, Gershov D, Elkon KB. Complement-dependent clearance of apoptotic cells by human macrophages. *J Exp Med.* 1998;188(12):2313–2320.
58. Elkon KB, Santer DM. Complement, interferon and lupus. *Curr Opin Immunol.* 2012;24(6):665–670.
59. Fraser DA, Laust AK, Nelson EL, Tenner AJ. C1q differentially modulates phagocytosis and cytokine responses during ingestion of apoptotic cells by human monocytes, macrophages, and dendritic cells. *J Immunol.* 2009;183(10):6175–6185.
60. Kenyon KD, Cole C, Crawford F, Kappler JW, Thurman JM, Bratton DL, et al. IgG autoantibodies against deposited C3 inhibit macrophage-mediated apoptotic cell engulfment in systemic autoimmunity. *J Immunol.* 2011;187(5):2101–2111.
61. Santer DM, Hall BE, George TC, Tangsombatvisit S, Liu CL, Arkwright PD, et al. C1q deficiency leads to the defective suppression of IFN-alpha in response to nucleoprotein containing immune complexes. *J Immunol.* 2010;185(8):4738–4749.
62. Kirou KA, Lee C, George S, Louca K, Peterson MG, Crow MK. Activation of the interferon-alpha pathway identifies a subgroup of systemic lupus erythematosus patients with distinct serologic features and active disease. *Arthritis Rheum.* 2005;52(5):1491–1503.
63. Lood C, Gullstrand B, Truedsson L, Olin AI, Alm GV, Ronnblom L, et al. C1q inhibits immune complex-induced interferon-alpha production in plasmacytoid dendritic cells: A novel link between C1q deficiency and systemic lupus erythematosus pathogenesis. *Arthritis Rheum.* 2009;60(10):3081–3090.
64. Marto N, Bertolaccini ML, Calabuig E, Hughes GR, Khamashta MA. Anti-C1q antibodies in nephritis: Correlation between titres and renal disease activity and positive predictive value in systemic lupus erythematosus. *Ann Rheum Dis.* 2005;64(3):444–448.
65. Orbai AM, Truedsson L, Sturfelt G, Nived O, Fang H, Alarcon GS, et al. Anti-C1q antibodies in systemic lupus erythematosus. Lupus. 2015;24(1):42–49.
66. Thanei S, Trendelenburg M. Anti-C1q autoantibodies from systemic lupus erythematosus patients induce a proinflammatory phenotype in macrophages. *J Immunol.* 2016;196(5):2063–2074.
67. Sekine H, Kinser TT, Qiao F, Martinez E, Paulling E, Ruiz P, et al. The benefit of targeted and selective inhibition of the alternative complement pathway for modulating autoimmunity and renal disease in MRL/lpr mice. *Arthritis Rheum.* 2011;63(4):1076–1085.
68. Sekine H, Ruiz P, Gilkeson GS, Tomlinson S. The dual role of complement in the progression of renal disease in NZB/W F(1) mice and alternative pathway inhibition. *Mol Immunol.* 2011;49(1–2):317–323.
69. Legendre CM, Licht C, Muus P, Greenbaum LA, Babu S, Bedrosian C, et al. Terminal complement inhibitor eculizumab in atypical hemolytic-uremic syndrome. *N Engl J Med.* 2013;368(23):2169–2181.
70. Noris M, Remuzzi G. Atypical hemolytic-uremic syndrome. *N Engl J Med.* 2009;361(17):1676–1687.
71. de Holanda MI, Porto LC, Wagner T, Christiani LF, Palma LMP. Use of eculizumab in a systemic lupus erythemathosus patient presenting thrombotic microangiopathy and heterozygous deletion in CFHR1-CFHR3. A case report and systematic review. *Clin Rheumatol.* 2017;36(12):2859–2867.
72. Bastian HM, Roseman JM, McGwin G, Jr., Alarcon GS, Friedman AW, Fessler BJ, et al. Systemic lupus erythematosus in three ethnic groups. XII. Risk factors for lupus nephritis after diagnosis. *Lupus.* 2002;11(3):152–160.
73. Nived O, Hallengren CS, Alm P, Jonsen A, Sturfelt G, Bengtsson AA. An observational study of outcome in SLE patients with biopsy-verified glomerulonephritis between 1986 and 2004 in a defined area of southern Sweden: The clinical utility of the ACR renal response criteria and predictors for renal outcome. *Scand J Rheumatol.* 2013;42(5):383–389.
74. Hoffman IE, Peene I, Meheus L, Huizinga TW, Cebecauer L, Isenberg D, et al. Specific antinuclear antibodies are associated with clinical features in systemic lupus erythematosus. *Ann Rheum Dis.* 2004;63(9):1155–1158.
75. Lu R, Robertson JM, Bruner BF, Guthridge JM, Neas BR, Nath SK, et al. Multiple autoantibodies display association with lymphopenia, proteinuria, and cellular casts in a large, ethnically diverse SLE patient cohort. *Autoimmune Dis.* 2012;2012:819634.

76. Hebert LA, Dillon JJ, Middendorf DF, Lewis EJ, Peter JB. Relationship between appearance of urinary red blood cell/white blood cell casts and the onset of renal relapse in systemic lupus erythematosus. *Am J Kidney Dis.* 1995;26(3):432–438.
77. Ding JY, Ibanez D, Gladman DD, Urowitz MB. Isolated hematuria and sterile pyuria may indicate systemic lupus erythematosus activity. *J Rheumatol.* 2015;42(3):437–440.
78. Rahman P, Gladman DD, Ibanez D, Urowitz MB. Significance of isolated hematuria and isolated pyuria in systemic lupus erythematosus. *Lupus.* 2001;10(6):418–423.
79. Appenzeller S, Clark A, Pineau C, Vasilevsky M, Bernatsky S. Isolated pyuria in systemic lupus erythematosus. *Lupus.* 2010;19(7):793–796.
80. Mavragani CP, Fragoulis GE, Somarakis G, Drosos A, Tzioufas AG, Moutsopoulos HM. Clinical and laboratory predictors of distinct histopathogical features of lupus nephritis. *Medicine.* 2015;94(21):e829.
81. Nieves-Plaza M, Ortiz AP, Colon M, Molina MJ, Castro-Santana LE, Rodriguez VE, et al. Outcome and predictors of kidney disease progression in Puerto Ricans with systemic lupus erythematosus initially presenting with mild renal involvement. *J Clin Rheumatol.* 2011;17(4):179–184.
82. Dall'Era M, Cisternas MG, Smilek DE, Straub L, Houssiau FA, Cervera R, et al. Predictors of long-term renal outcome in lupus nephritis trials: Lessons learned from the Euro-Lupus Nephritis cohort. *Arthritis Rheumatol.* 2015;67(5):1305–1313.
83. Balow JE. Clinical presentation and monitoring of lupus nephritis. *Lupus.* 2005;14(1):25–30.
84. Touma Z, Urowitz MB, Ibanez D, Gladman DD. Time to recovery from proteinuria in patients with lupus nephritis receiving standard treatment. *J Rheumatol.* 2014;41(4):688–697.
85. Mitchell SC, Sheldon TA, Shaw AB. Quantification of proteinuria: A re-evaluation of the protein/creatinine ratio for elderly subjects. *Age Ageing.* 1993;22(6):443–449.
86. Ginsberg JM, Chang BS, Matarese RA, Garella S. Use of single voided urine samples to estimate quantitative proteinuria. *N Engl J Med.* 1983;309(25):1543–1546.
87. Rodby RA, Rohde RD, Sharon Z, Pohl MA, Bain RP, Lewis EJ. The urine protein to creatinine ratio as a predictor of 24-hour urine protein excretion in type 1 diabetic patients with nephropathy. The Collaborative Study Group. *Am J Kidney Dis.* 1995;26(6):904–909.
88. Choi IA, Park JK, Lee EY, Song YW, Lee EB. Random spot urine protein to creatinine ratio is a reliable measure of proteinuria in lupus nephritis in Koreans. *Clin Exp Rheumatol.* 2013;31(4):584–588.
89. Leung YY, Szeto CC, Tam LS, Lam CW, Li EK, Wong KC, et al. Urine protein-to-creatinine ratio in an untimed urine collection is a reliable measure of proteinuria in lupus nephritis. *Rheumatology.* 2007;46(4):649–652.
90. Wahbeh AM, Ewais MH, Elsharif ME. Comparison of 24-hour urinary protein and protein-to-creatinine ratio in the assessment of proteinuria. *Saudi J Kidney Dis Transpl.* 2009;20(3):443–447.
91. The American College of Rheumatology response criteria for proliferative and membranous renal disease in systemic lupus erythematosus clinical trials. *Arthritis Rheum.* 2006;54(2):421–432.
92. Chitalia VC, Kothari J, Wells EJ, Livesey JH, Robson RA, Searle M, et al. Cost-benefit analysis and prediction of 24-hour proteinuria from the spot urine protein-creatinine ratio. *Clin Nephrol.* 2001;55(6):436–447.
93. Medina-Rosas J, Yap KS, Anderson M, Su J, Touma Z. Utility of urinary protein-creatinine ratio and protein content in a 24-hour urine collection in systemic lupus erythematosus: A systematic review and meta-analysis. *Arthritis Care Res.* 2016;68(9):1310–1319.
94. Birmingham DJ, Shidham G, Perna A, Fine DM, Bissell M, Rodby R, et al. Spot PC ratio estimates of 24-hour proteinuria are more unreliable in lupus nephritis than in other forms of chronic glomerular disease. *Ann Rheum Dis.* 2014;73(2):475–476.
95. Mohan C, Adams S, Stanik V, Datta SK. Nucleosome: A major immunogen for pathogenic autoantibody-inducing T cells of lupus. *J Exp Med.* 1993;177(5):1367–1381.
96. Ullal AJ, Reich CF, 3rd, Clowse M, Criscione-Schreiber LG, Tochacek M, Monestier M, et al. Microparticles as antigenic targets of antibodies to DNA and nucleosomes in systemic lupus erythematosus. *J Autoimmun.* 2011;36(3–4):173–180.
97. Simon JA, Cabiedes J, Ortiz E, Alcocer-Varela J, Sanchez-Guerrero J. Anti-nucleosome antibodies in patients with systemic lupus erythematosus of recent onset. Potential utility as a diagnostic tool and disease activity marker. *Rheumatology.* 2004;43(2):220–224.
98. Bigler C, Lopez-Trascasa M, Potlukova E, Moll S, Danner D, Schaller M, et al. Antinucleosome antibodies as a marker of active proliferative lupus nephritis. *Am J Kidney Dis.* 2008;51(4):624–629.
99. Li T, Prokopec SD, Morrison S, Lou W, Reich H, Gladman D, et al. Anti-nucleosome antibodies outperform traditional biomarkers as longitudinal indicators of disease activity in systemic lupus erythematosus. *Rheumatology.* 2015;54(3):449–457.
100. Bizzaro N, Villalta D, Giavarina D, Tozzoli R. Are anti-nucleosome antibodies a better diagnostic marker than anti-dsDNA antibodies for systemic lupus erythematosus? A systematic review and a study of metanalysis. *Autoimmun Rev.* 2012;12(2):97–106.
101. Bennett L, Palucka AK, Arce E, Cantrell V, Borvak J, Banchereau J, et al. Interferon and granulopoiesis signatures in systemic lupus erythematosus blood. *J Exp Med.* 2003;197(6):711–723.
102. Baechler EC, Batliwalla FM, Karypis G, Gaffney PM, Ortmann WA, Espe KJ, et al. Interferon-inducible gene expression signature in peripheral blood cells of patients with severe lupus. *Proc Natl Acad Sci U S A.* 2003;100(5):2610–2615.
103. Crow MK, Olferiev M, Kirou KA. Targeting of type I interferon in systemic autoimmune diseases. *Transl Res.* 2015;165(2):296–305.
104. Mathian A, Weinberg A, Gallegos M, Banchereau J, Koutouzov S. IFN-alpha induces early lethal lupus in preautoimmune (New Zealand Black x New Zealand White) F1 but not in BALB/c mice. *J Immunol.* 2005;174(5):2499–2506.
105. Braun D, Geraldes P, Demengeot J. Type I Interferon controls the onset and severity of autoimmune manifestations in lpr mice. *J Autoimmun.* 2003;20(1):15–25.
106. Santiago-Raber ML, Baccala R, Haraldsson KM, Choubey D, Stewart TA, Kono DH, et al. Type-I interferon receptor deficiency reduces lupus-like disease in NZB mice. *J Exp Med.* 2003;197(6):777–788.
107. Buechler MB, Teal TH, Elkon KB, Hamerman JA. Cutting edge: Type I IFN drives emergency myelopoiesis and peripheral myeloid expansion during chronic TLR7 signaling. *J Immunol.* 2013;190(3):886–891.
108. Liu YJ. IPC: Professional type 1 interferon-producing cells and plasmacytoid dendritic cell precursors. *Annu Rev Immunol.* 2005;23:275–306.
109. Gilliet M, Cao W, Liu YJ. Plasmacytoid dendritic cells: Sensing nucleic acids in viral infection and autoimmune diseases. *Nat Rev Immunol.* 2008;8(8):594–606.
110. Reizis B, Bunin A, Ghosh HS, Lewis KL, Sisirak V. Plasmacytoid dendritic cells: recent progress and open questions. *Annu Rev Immunol.* 2011;29:163–183.
111. Marshak-Rothstein A, Rifkin IR. Immunologically active autoantigens: The role of toll-like receptors in the development of chronic inflammatory disease. *Annu Rev Immunol.* 2007;25:419–441.

112. Shlomchik MJ. Activating systemic autoimmunity: B's, T's, and tolls. *Curr Opin Immunol.* 2009;21(6):626–233.
113. Pisitkun P, Deane JA, Difilippantonio MJ, Tarasenko T, Satterthwaite AB, Bolland S. Autoreactive B cell responses to RNA-related antigens due to TLR7 gene duplication. *Science.* 2006;312(5780):1669–1672.
114. Deane JA, Pisitkun P, Barrett RS, Feigenbaum L, Town T, Ward JM, et al. Control of toll-like receptor 7 expression is essential to restrict autoimmunity and dendritic cell proliferation. *Immunity.* 2007;27(5):801–810.
115. Fairhurst AM, Hwang SH, Wang A, Tian XH, Boudreaux C, Zhou XJ, et al. Yaa autoimmune phenotypes are conferred by overexpression of TLR7. *Eur J Immunol.* 2008;38(7):1971–1978.
116. Barrat FJ, Meeker T, Gregorio J, Chan JH, Uematsu S, Akira S, et al. Nucleic acids of mammalian origin can act as endogenous ligands for Toll-like receptors and may promote systemic lupus erythematosus. *J Exp Med.* 2005;202(8):1131–1139.
117. Bave U, Magnusson M, Eloranta ML, Perers A, Alm GV, Ronnblom L. Fc gamma RIIa is expressed on natural IFN-alpha-producing cells (plasmacytoid dendritic cells) and is required for the IFN-alpha production induced by apoptotic cells combined with lupus IgG. *J Immunol.* 2003;171(6):3296–3302.
118. Boule MW, Broughton C, Mackay F, Akira S, Marshak-Rothstein A, Rifkin IR. Toll-like receptor 9-dependent and -independent dendritic cell activation by chromatin-immunoglobulin G complexes. *J Exp Med.* 2004;199(12):1631–1640.
119. Honda K, Yanai H, Negishi H, Asagiri M, Sato M, Mizutani T, et al. IRF-7 is the master regulator of type-I interferon-dependent immune responses. *Nature.* 2005;434(7034):772–777.
120. Means TK, Latz E, Hayashi F, Murali MR, Golenbock DT, Luster AD. Human lupus autoantibody-DNA complexes activate DCs through cooperation of CD32 and TLR9. *J Clin Invest.* 2005;115(2):407–417.
121. Kawai T, Akira S. TLR signaling. *Cell Death Differ.* 2006;13(5):816–825.
122. Meylan E, Tschopp J. Toll-like receptors and RNA helicases: two parallel ways to trigger antiviral responses. *Mol Cell.* 2006;22(5):561–569.
123. Blanco P, Palucka AK, Gill M, Pascual V, Banchereau J. Induction of dendritic cell differentiation by IFN-alpha in systemic lupus erythematosus. *Science.* 2001;294(5546):1540–1543.
124. Dubois B, Vanbervliet B, Fayette J, Massacrier C, Van Kooten C, Briere F, et al. Dendritic cells enhance growth and differentiation of CD40-activated B lymphocytes. *J Exp Med.* 1997;185(5):941–951.
125. Garcia De Vinuesa C, Gulbranson-Judge A, Khan M, O'Leary P, Cascalho M, Wabl M, et al. Dendritic cells associated with plasmablast survival. *Eur J Immunol.* 1999;29(11):3712–3721.
126. Thien M, Phan TG, Gardam S, Amesbury M, Basten A, Mackay F, et al. Excess BAFF rescues self-reactive B cells from peripheral deletion and allows them to enter forbidden follicular and marginal zone niches. *Immunity.* 2004;20(6):785–798.
127. Jego G, Palucka AK, Blanck JP, Chalouni C, Pascual V, Banchereau J. Plasmacytoid dendritic cells induce plasma cell differentiation through type I interferon and interleukin 6. *Immunity.* 2003;19(2):225–234.
128. Leadbetter EA, Rifkin IR, Hohlbaum AM, Beaudette BC, Shlomchik MJ, Marshak-Rothstein A. Chromatin-IgG complexes activate B cells by dual engagement of IgM and Toll-like receptors. *Nature.* 2002;416(6881):603–607.
129. Ehlers M, Fukuyama H, McGaha TL, Aderem A, Ravetch JV. TLR9/MyD88 signaling is required for class switching to pathogenic IgG2a and 2b autoantibodies in SLE. *J Exp Med.* 2006;203(3):553–561.
130. Lau CM, Broughton C, Tabor AS, Akira S, Flavell RA, Mamula MJ, et al. RNA-associated autoantigens activate B cells by combined B cell antigen receptor/Toll-like receptor 7 engagement. *J Exp Med.* 2005;202(9):1171–1177.
131. Pawar RD, Patole PS, Zecher D, Segerer S, Kretzler M, Schlondorff D, et al. Toll-like receptor-7 modulates immune complex glomerulonephritis. *J Am Soc Nephrol.* 2006;17(1):141–149.
132. Fukuda A, Kobayashi H, Teramura K, Yoshimoto S, Ohsawa N. Effects of interferon-alpha on peripheral neutrophil counts and serum granulocyte colony-stimulating factor levels in chronic hepatitis C patients. *Cytokines Cell Mol Ther.* 2000;6(3):149–154.
133. Scapini P, Nardelli B, Nadali G, Calzetti F, Pizzolo G, Montecucco C, et al. G-CSF-stimulated neutrophils are a prominent source of functional BLyS. *J Exp Med.* 2003;197(3):297–302.
134. Garcia-Romo GS, Caielli S, Vega B, Connolly J, Allantaz F, Xu Z, et al. Netting neutrophils are major inducers of type I IFN production in pediatric systemic lupus erythematosus. *Sci Transl Med.* 2011;3(73):73ra20.
135. Lande R, Ganguly D, Facchinetti V, Frasca L, Conrad C, Gregorio J, et al. Neutrophils activate plasmacytoid dendritic cells by releasing self-DNA-peptide complexes in systemic lupus erythematosus. *Sci Transl Med.* 2011;3(73):73ra19.
136. Yang D, Chertov O, Bykovskaia SN, Chen Q, Buffo MJ, Shogan J, et al. Beta-defensins: Linking innate and adaptive immunity through dendritic and T cell CCR6. *Science.* 1999;286(5439):525–528.
137. van Gisbergen KP, Ludwig IS, Geijtenbeek TB, van Kooyk Y. Interactions of DC-SIGN with Mac-1 and CEACAM1 regulate contact between dendritic cells and neutrophils. *FEBS Lett.* 2005;579(27):6159–6168.
138. Subramanian S, Tus K, Li QZ, Wang A, Tian XH, Zhou J, et al. A Tlr7 translocation accelerates systemic autoimmunity in murine lupus. *Proc Natl Acad Sci U S A.* 2006;103(26):9970–9975.
139. Hanada T, Yoshida H, Kato S, Tanaka K, Masutani K, Tsukada J, et al. Suppressor of cytokine signaling-1 is essential for suppressing dendritic cell activation and systemic autoimmunity. *Immunity.* 2003;19(3):437–450.
140. Gill MA, Blanco P, Arce E, Pascual V, Banchereau J, Palucka AK. Blood dendritic cells and DC-poietins in systemic lupus erythematosus. *Hum Immunol.* 2002;63(12):1172–1180.
141. Vollmer J, Tluk S, Schmitz C, Hamm S, Jurk M, Forsbach A, et al. Immune stimulation mediated by autoantigen binding sites within small nuclear RNAs involves Toll-like receptors 7 and 8. *J Exp Med.* 2005;202(11):1575–1585.
142. Mohty M, Vialle-Castellano A, Nunes JA, Isnardon D, Olive D, Gaugler B. IFN-alpha skews monocyte differentiation into Toll-like receptor 7-expressing dendritic cells with potent functional activities. *J Immunol.* 2003;171(7):3385–3393.
143. Yao Y, Richman L, Higgs BW, Morehouse CA, de los Reyes M, Brohawn P, et al. Neutralization of interferon-alpha/beta-inducible genes and downstream effect in a phase I trial of an anti-interferon-alpha monoclonal antibody in systemic lupus erythematosus. *Arthritis Rheum.* 2009;60(6):1785–1796.
144. Khamashta M, Merrill JT, Werth VP, Furie R, Kalunian K, Illei GG, et al. Sifalimumab, an anti-interferon-alpha monoclonal antibody, in moderate to severe systemic lupus erythematosus: A randomised, double-blind, placebo-controlled study. *Ann Rheum Dis.* 2016;75(11):1909–1916.
145. Kalunian KC, Merrill JT, Maciuca R, McBride JM, Townsend MJ, Wei X, et al. A Phase II study of the efficacy and safety of rontalizumab (rhuMAb interferon-alpha) in patients with systemic lupus erythematosus (ROSE). *Ann Rheum Dis.* 2016;75(1):196–202.
146. Lauwerys BR, Hachulla E, Spertini F, Lazaro E, Jorgensen C, Mariette X, et al. Down-regulation of interferon signature in systemic lupus erythematosus patients by active immunization with interferon alpha-kinoid. *Arthritis Rheum.* 2013;65(2):447–456.

147. Wang B, Higgs BW, Chang L, Vainshtein I, Liu Z, Streicher K, et al. Pharmacogenomics and translational simulations to bridge indications for an anti-interferon-alpha receptor antibody. *Clin Pharmacol Ther.* 2013;93(6):483–492.
148. Furie R, Khamashta M, Merrill JT, Werth VP, Kalunian K, Brohawn P, et al. Anifrolumab, an anti-interferon-alpha receptor monoclonal antibody, in moderate-to-severe systemic lupus erythematosus. *Arthritis Rheumatol.* 2017;69(2):376–386.
149. Hakkim A, Furnrohr BG, Amann K, Laube B, Abed UA, Brinkmann V, et al. Impairment of neutrophil extracellular trap degradation is associated with lupus nephritis. *Proc Natl Acad Sci U S A.* 2010;107(21):9813–9818.
150. Brinkmann V, Zychlinsky A. Neutrophil extracellular traps: Is immunity the second function of chromatin? *J Cell Biol.* 2012;198(5):773–783.
151. Lood C, Blanco LP, Purmalek MM, Carmona-Rivera C, De Ravin SS, Smith CK, et al. Neutrophil extracellular traps enriched in oxidized mitochondrial DNA are interferogenic and contribute to lupus-like disease. *Nat Med.* 2016;22(2):146–153.
152. Villanueva E, Yalavarthi S, Berthier CC, Hodgin JB, Khandpur R, Lin AM, et al. Netting neutrophils induce endothelial damage, infiltrate tissues, and expose immunostimulatory molecules in systemic lupus erythematosus. *J Immunol.* 2011;187(1):538–552.
153. Leffler J, Martin M, Gullstrand B, Tyden H, Lood C, Truedsson L, et al. Neutrophil extracellular traps that are not degraded in systemic lupus erythematosus activate complement exacerbating the disease. *J Immunol.* 2012;188(7):3522–3531.
154. Carmona-Rivera C, Zhao W, Yalavarthi S, Kaplan MJ. Neutrophil extracellular traps induce endothelial dysfunction in systemic lupus erythematosus through the activation of matrix metalloproteinase-2. *Ann Rheum Dis.* 2015;74(7):1417–1424.
155. Kahlenberg JM, Carmona-Rivera C, Smith CK, Kaplan MJ. Neutrophil extracellular trap-associated protein activation of the NLRP3 inflammasome is enhanced in lupus macrophages. *J Immunol.* 2013;190(3):1217–1226.
156. Denny MF, Yalavarthi S, Zhao W, Thacker SG, Anderson M, Sandy AR, et al. A distinct subset of proinflammatory neutrophils isolated from patients with systemic lupus erythematosus induces vascular damage and synthesizes type I IFNs. *J Immunol.* 2010;184(6):3284–3297.
157. Knight JS, Zhao W, Luo W, Subramanian V, O'Dell AA, Yalavarthi S, et al. Peptidylarginine deiminase inhibition is immunomodulatory and vasculoprotective in murine lupus. *J Clin Invest.* 2013;123(7):2981–2993.
158. Smith CK, Vivekanandan-Giri A, Tang C, Knight JS, Mathew A, Padilla RL, et al. Neutrophil extracellular trap-derived enzymes oxidize high-density lipoprotein: an additional proatherogenic mechanism in systemic lupus erythematosus. *Arthritis Rheumatol.* 2014;66(9):2532–2544.
159. Gehrke N, Mertens C, Zillinger T, Wenzel J, Bald T, Zahn S, et al. Oxidative damage of DNA confers resistance to cytosolic nuclease TREX1 degradation and potentiates STING-dependent immune sensing. *Immunity.* 2013;39(3):482–495.
160. Lai ZW, Hanczko R, Bonilla E, Caza TN, Clair B, Bartos A, et al. N-acetylcysteine reduces disease activity by blocking mammalian target of rapamycin in T cells from systemic lupus erythematosus patients: A randomized, double-blind, placebo-controlled trial. *Arthritis Rheum.* 2012;64(9):2937–2946.
161. Zavada J, Pesickova S, Rysava R, Olejarova M, Horak P, Hrncir Z, et al. Cyclosporine A or intravenous cyclophosphamide for lupus nephritis: The Cyclofa-Lune study. *Lupus.* 2010;19(11):1281–1289.
162. Lee YH, Lee HS, Choi SJ, Dai Ji J, Song GG. Efficacy and safety of tacrolimus therapy for lupus nephritis: a systematic review of clinical trials. *Lupus.* 2011;20(6):636–640.
163. Schreiber A, Xiao H, Jennette JC, Schneider W, Luft FC, Kettritz R. C5a receptor mediates neutrophil activation and ANCA-induced glomerulonephritis. *J Am Soc Nephrol.* 2009;20(2):289–298.
164. Schneider P, MacKay F, Steiner V, Hofmann K, Bodmer JL, Holler N, et al. BAFF, a novel ligand of the tumor necrosis factor family, stimulates B cell growth. *J Exp Med.* 1999;189(11):1747–1756.
165. Batten M, Groom J, Cachero TG, Qian F, Schneider P, Tschopp J, et al. BAFF mediates survival of peripheral immature B lymphocytes. *J Exp Med.* 2000;192(10):1453–1466.
166. Navarra SV, Guzman RM, Gallacher AE, Hall S, Levy RA, Jimenez RE, et al. Efficacy and safety of belimumab in patients with active systemic lupus erythematosus: A randomised, placebo-controlled, phase 3 trial. *Lancet.* 2011;377(9767):721–731.
167. Furie R, Petri M, Zamani O, Cervera R, Wallace DJ, Tegzova D, et al. A phase III, randomized, placebo-controlled study of belimumab, a monoclonal antibody that inhibits B lymphocyte stimulator, in patients with systemic lupus erythematosus. *Arthritis Rheum.* 2011;63(12):3918–3930.
168. Vincent FB, Morand EF, Mackay F. BAFF and innate immunity: New therapeutic targets for systemic lupus erythematosus. *Immunol Cell Biol.* 2012;90(3):293–303.
169. Vincent FB, Saulep-Easton D, Figgett WA, Fairfax KA, Mackay F. The BAFF/APRIL system: Emerging functions beyond B cell biology and autoimmunity. *Cytokine Growth Factor Rev.* 2013;24(3):203–215.
170. Stohl W, Metyas S, Tan SM, Cheema GS, Oamar B, Xu D, et al. B lymphocyte stimulator overexpression in patients with systemic lupus erythematosus: Longitudinal observations. *Arthritis Rheum.* 2003;48(12):3475–3486.
171. Huard B, Schneider P, Mauri D, Tschopp J, French LE. T cell costimulation by the TNF ligand BAFF. *J Immunol.* 2001;167(11):6225–6231.
172. Ng LG, Sutherland AP, Newton R, Qian F, Cachero TG, Scott ML, et al. B cell-activating factor belonging to the TNF family (BAFF)-R is the principal BAFF receptor facilitating BAFF costimulation of circulating T and B cells. *J Immunol.* 2004;173(2):807–817.
173. Coquery CM, Loo WM, Wade NS, Bederman AG, Tung KS, Lewis JE, et al. BAFF regulates follicular helper t cells and affects their accumulation and interferon-gamma production in autoimmunity. *Arthritis Rheumatol.* 2015;67(3):773–784.
174. Goenka R, Matthews AH, Zhang B, O'Neill PJ, Scholz JL, Migone TS, et al. Local BLyS production by T follicular cells mediates retention of high affinity B cells during affinity maturation. *J Exp Med.* 2014;211(1):45–56.
175. Mackay F, Schneider P. Cracking the BAFF code. *Nat Rev Immunol.* 2009;9(7):491–502.
176. Huard B, Tran NL, Benkhoucha M, Manzin-Lorenzi C, Santiago-Raber ML. Selective APRIL blockade delays systemic lupus erythematosus in mouse. *PLoS One.* 2012;7(2):e31837.
177. Jacob CO, Guo S, Jacob N, Pawar RD, Putterman C, Quinn WJ, 3rd, et al. Dispensability of APRIL to the development of systemic lupus erythematosus in NZM 2328 mice. *Arthritis Rheum.* 2012;64(5):1610–1619.
178. Isenberg D, Gordon C, Licu D, Copt S, Rossi CP, Wofsy D. Efficacy and safety of atacicept for prevention of flares in patients with moderate-to-severe systemic lupus erythematosus (SLE): 52-week data (APRIL-SLE randomised trial). *Ann Rheum Dis.* 2015;74(11):2006–2015.
179. Baker KP, Edwards BM, Main SH, Choi GH, Wager RE, Halpern WG, et al. Generation and characterization of LymphoStat-B, a human monoclonal antibody that antagonizes the bioactivities of B lymphocyte stimulator. *Arthritis Rheum.* 2003;48(11):3253–3265.

180. Strand V, Levy RA, Cervera R, Petri MA, Birch H, Freimuth WW, et al. Improvements in health-related quality of life with belimumab, a B-lymphocyte stimulator-specific inhibitor, in patients with autoantibody-positive systemic lupus erythematosus from the randomised controlled BLISS trials. *Ann Rheum Dis.* 2014;73(5):838–844.
181. Wallace DJ, Navarra S, Petri MA, Gallacher A, Thomas M, Furie R, et al. Safety profile of belimumab: Pooled data from placebo-controlled phase 2 and 3 studies in patients with systemic lupus erythematosus. *Lupus.* 2013;22(2):144–154.
182. Stohl W, Hiepe F, Latinis KM, Thomas M, Scheinberg MA, Clarke A, et al. Belimumab reduces autoantibodies, normalizes low complement levels, and reduces select B cell populations in patients with systemic lupus erythematosus. *Arthritis Rheum.* 2012;64(7):2328–2337.
183. van Vollenhoven RF, Petri MA, Cervera R, Roth DA, Ji BN, Kleoudis CS, et al. Belimumab in the treatment of systemic lupus erythematosus: High disease activity predictors of response. *Ann Rheum Dis.* 2012;71(8):1343–1349.
184. Chatham WW, Wallace DJ, Stohl W, Latinis KM, Manzi S, McCune WJ, et al. Effect of belimumab on vaccine antigen antibodies to influenza, pneumococcal, and tetanus vaccines in patients with systemic lupus erythematosus in the BLISS-76 trial. *J Rheumatol.* 2012;39(8):1632–1640.
185. Petri M, Stohl W, Chatham W, McCune WJ, Chevrier M, Ryel J, et al. Association of plasma B lymphocyte stimulator levels and disease activity in systemic lupus erythematosus. *Arthritis Rheum.* 2008;58(8):2453–2459.
186. Stohl W, Metyas S, Tan SM, Cheema GS, Oamar B, Roschke V, et al. Inverse association between circulating APRIL levels and serological and clinical disease activity in patients with systemic lupus erythematosus. *Ann Rheum Dis.* 2004;63(9):1096–1103.
187. Morel J, Roubille C, Planelles L, Rocha C, Fernandez L, Lukas C, et al. Serum levels of tumour necrosis factor family members a proliferation-inducing ligand (APRIL) and B lymphocyte stimulator (BLyS) are inversely correlated in systemic lupus erythematosus. *Ann Rheum Dis.* 2009;68(6):997–1002.
188. Zhang J, Roschke V, Baker KP, Wang Z, Alarcon GS, Fessler BJ, et al. Cutting edge: A role for B lymphocyte stimulator in systemic lupus erythematosus. *J Immunol.* 2001;166(1):6–10.
189. Carter LM, Isenberg DA, Ehrenstein MR. Elevated serum BAFF levels are associated with rising anti-double-stranded DNA antibody levels and disease flare following B cell depletion therapy in systemic lupus erythematosus. *Arthritis Rheum.* 2013;65(10):2672–2679.
190. Lesley R, Xu Y, Kalled SL, Hess DM, Schwab SR, Shu HB, et al. Reduced competitiveness of autoantigen-engaged B cells due to increased dependence on BAFF. *Immunity.* 2004;20(4):441–453.
191. Thorn M, Lewis RH, Mumbey-Wafula A, Kantrowitz S, Spatz LA. BAFF overexpression promotes anti-dsDNA B-cell maturation and antibody secretion. *Cell Immunol.* 2010;261(1):9–22.
192. Liu Z, Davidson A. BAFF and selection of autoreactive B cells. *Trends Immunol.* 2011;32(8):388–394.
193. Avery DT, Kalled SL, Ellyard JI, Ambrose C, Bixler SA, Thien M, et al. BAFF selectively enhances the survival of plasmablasts generated from human memory B cells. *J Clin Invest.* 2003;112(2):286-97.
194. Balazs M, Martin F, Zhou T, Kearney J. Blood dendritic cells interact with splenic marginal zone B cells to initiate T-independent immune responses. *Immunity.* 2002;17(3):341–352.
195. Gong Q, Ou Q, Ye S, Lee WP, Cornelius J, Diehl L, et al. Importance of cellular microenvironment and circulatory dynamics in B cell immunotherapy. *J Immunol.* 2005;174(2):817–826.
196. Lin W, Seshasayee D, Lee WP, Caplazi P, McVay S, Suto E, et al. Dual B cell immunotherapy is superior to individual anti-CD20 depletion or BAFF blockade in murine models of spontaneous or accelerated lupus. *Arthritis Rheumatol.* 2015;67(1):215–224.
197. Dillon SR, Harder B, Lewis KB, Moore MD, Liu H, Bukowski TR, et al. B-lymphocyte stimulator/a proliferation-inducing ligand heterotrimers are elevated in the sera of patients with autoimmune disease and are neutralized by atacicept and B-cell maturation antigen-immunoglobulin. *Arthritis Res Ther.* 2010;12(2):R48.
198. Vincent FB, Northcott M, Hoi A, Mackay F, Morand EF. Association of serum B cell activating factor from the tumour necrosis factor family (BAFF) and a proliferation-inducing ligand (APRIL) with central nervous system and renal disease in systemic lupus erythematosus. *Lupus.* 2013;22(9):873–884.
199. Panchanathan R, Choubey D. Murine BAFF expression is up-regulated by estrogen and interferons: Implications for sex bias in the development of autoimmunity. *Mol Immunol.* 2013;53(1–2):15–23.
200. Litinskiy MB, Nardelli B, Hilbert DM, He B, Schaffer A, Casali P, et al. DCs induce CD40-independent immunoglobulin class switching through BLyS and APRIL. *Nat Immunol.* 2002;3(9):822–829.
201. Harigai M, Kawamoto M, Hara M, Kubota T, Kamatani N, Miyasaka N. Excessive production of IFN-gamma in patients with systemic lupus erythematosus and its contribution to induction of B lymphocyte stimulator/B cell-activating factor/TNF ligand superfamily-13B. *J Immunol.* 2008;181(3):2211–2219.
202. Jacob N, Guo S, Mathian A, Koss MN, Gindea S, Putterman C, et al. B Cell and BAFF dependence of IFN-alpha-exaggerated disease in systemic lupus erythematosus-prone NZM 2328 mice. *J Immunol.* 2011;186(8):4984–4993.
203. Ritterhouse LL, Crowe SR, Niewold TB, Merrill JT, Roberts VC, Dedeke AB, et al. B lymphocyte stimulator levels in systemic lupus erythematosus: Higher circulating levels in African American patients and increased production after influenza vaccination in patients with low baseline levels. *Arthritis Rheum.* 2011;63(12):3931–3941.
204. Odendahl M, Jacobi A, Hansen A, Feist E, Hiepe F, Burmester GR, et al. Disturbed peripheral B lymphocyte homeostasis in systemic lupus erythematosus. *J Immunol.* 2000;165(10):5970–5979.
205. Lazarus MN, Turner-Stokes T, Chavele KM, Isenberg DA, Ehrenstein MR. B-cell numbers and phenotype at clinical relapse following rituximab therapy differ in SLE patients according to anti-dsDNA antibody levels. *Rheumatology.* 2012;51(7):1208–1215.
206. Bosma A, Abdel-Gadir A, Isenberg DA, Jury EC, Mauri C. Lipid-antigen presentation by CD1d(+) B cells is essential for the maintenance of invariant natural killer T cells. *Immunity.* 2012;36(3):477–490.
207. Jacobi AM, Odendahl M, Reiter K, Bruns A, Burmester GR, Radbruch A, et al. Correlation between circulating CD27high plasma cells and disease activity in patients with systemic lupus erythematosus. *Arthritis Rheum.* 2003;48(5):1332–1342.
208. Chavele KM, Merry E, Ehrenstein MR. Cutting edge: Circulating plasmablasts induce the differentiation of human T follicular helper cells via IL-6 production. *J Immunol.* 2015;194(6):2482–2485.
209. de Wit J, Jorritsma T, Makuch M, Remmerswaal EB, Klaasse Bos H, Souwer Y, et al. Human B cells promote T-cell plasticity to optimize antibody response by inducing coexpression of T(H)1/T(FH) signatures. *J Allergy Clin Immunol.* 2015;135(4):1053–1060.
210. Karnowski A, Chevrier S, Belz GT, Mount A, Emslie D, D'Costa K, et al. B and T cells collaborate in antiviral responses via IL-6, IL-21, and transcriptional activator and coactivator, Oct2 and OBF-1. *J Exp Med.* 2012;209(11):2049–2064.

211. Feng X, Wang D, Chen J, Lu L, Hua B, Li X, et al. Inhibition of aberrant circulating Tfh cell proportions by corticosteroids in patients with systemic lupus erythematosus. *PLoS One.* 2012;7(12):e51982.
212. Simpson N, Gatenby PA, Wilson A, Malik S, Fulcher DA, Tangye SG, et al. Expansion of circulating T cells resembling follicular helper T cells is a fixed phenotype that identifies a subset of severe systemic lupus erythematosus. *Arthritis Rheum.* 2010;62(1):234–244.
213. Choi JY, Ho JH, Pasoto SG, Bunin V, Kim ST, Carrasco S, et al. Circulating follicular helper-like T cells in systemic lupus erythematosus: Association with disease activity. *Arthritis Rheumatol.* 2015;67(4):988–999.
214. Illei GG, Shirota Y, Yarboro CH, Daruwalla J, Tackey E, Takada K, et al. Tocilizumab in systemic lupus erythematosus: Data on safety, preliminary efficacy, and impact on circulating plasma cells from an open-label phase I dosage-escalation study. *Arthritis Rheum.* 2010;62(2):542–552.
215. Shirota Y, Yarboro C, Fischer R, Pham TH, Lipsky P, Illei GG. Impact of anti-interleukin-6 receptor blockade on circulating T and B cell subsets in patients with systemic lupus erythematosus. *Ann Rheum Dis.* 2013;72(1):118–128.
216. Szepietowski JC, Nilganuwong S, Wozniacka A, Kuhn A, Nyberg F, van Vollenhoven RF, et al. Phase I, randomized, double-blind, placebo-controlled, multiple intravenous, dose-ascending study of sirukumab in cutaneous or systemic lupus erythematosus. *Arthritis Rheum.* 2013;65(10):2661–2671.
217. Furie RA, Leon G, Thomas M, Petri MA, Chu AD, Hislop C, et al. A phase 2, randomised, placebo-controlled clinical trial of blisibimod, an inhibitor of B cell activating factor, in patients with moderate-to-severe systemic lupus erythematosus, the PEARL-SC study. *Ann Rheum Dis.* 2015;74(9):1667–1675.
218. Bhaskar PT, Hay N. The two TORCs and Akt. *Dev Cell.* 2007;12(4):487–502.
219. Li Q, Rao RR, Araki K, Pollizzi K, Odunsi K, Powell JD, et al. A central role for mTOR kinase in homeostatic proliferation induced CD8+ T cell memory and tumor immunity. *Immunity.* 2011;34(4):541–553.
220. Jones RG, Thompson CB. Revving the engine: Signal transduction fuels T cell activation. *Immunity.* 2007;27(2):173–178.
221. Fox CJ, Hammerman PS, Thompson CB. Fuel feeds function: energy metabolism and the T-cell response. *Nat Rev Immunol.* 2005;5(11):844–852.
222. Powell JD, Delgoffe GM. The mammalian target of rapamycin: linking T cell differentiation, function, and metabolism. *Immunity.* 2010;33(3):301–311.
223. Zheng Y, Collins SL, Lutz MA, Allen AN, Kole TP, Zarek PE, et al. A role for mammalian target of rapamycin in regulating T cell activation versus anergy. *J Immunol.* 2007;178(4):2163–2170.
224. Delgoffe GM, Pollizzi KN, Waickman AT, Heikamp E, Meyers DJ, Horton MR, et al. The kinase mTOR regulates the differentiation of helper T cells through the selective activation of signaling by mTORC1 and mTORC2. *Nat Immunol.* 2011;12(4):295–303.
225. Haxhinasto S, Mathis D, Benoist C. The AKT-mTOR axis regulates de novo differentiation of CD4+Foxp3+ cells. *J Exp Med.* 2008;205(3):565–574.
226. Fernandez DR, Telarico T, Bonilla E, Li Q, Banerjee S, Middleton FA, et al. Activation of mammalian target of rapamycin controls the loss of TCRzeta in lupus T cells through HRES-1/Rab4-regulated lysosomal degradation. *J Immunol.* 2009;182(4):2063–2073.
227. Caza TN, Fernandez DR, Talaber G, Oaks Z, Haas M, Madaio MP, et al. HRES-1/Rab4-mediated depletion of Drp1 impairs mitochondrial homeostasis and represents a target for treatment in SLE. *Ann Rheum Dis.* 2014;73(10):1888–1897.
228. Warner LM, Adams LM, Sehgal SN. Rapamycin prolongs survival and arrests pathophysiologic changes in murine systemic lupus erythematosus. *Arthritis Rheum.* 1994;37(2):289–297.
229. Lai ZW, Borsuk R, Shadakshari A, Yu J, Dawood M, Garcia R, et al. Mechanistic target of rapamycin activation triggers IL-4 production and necrotic death of double-negative T cells in patients with systemic lupus erythematosus. *J Immunol.* 2013;191(5):2236–2246.
230. Fernandez D, Bonilla E, Mirza N, Niland B, Perl A. Rapamycin reduces disease activity and normalizes T cell activation-induced calcium fluxing in patients with systemic lupus erythematosus. *Arthritis Rheum.* 2006;54(9):2983–2938.
231. Kato H, Perl A. Mechanistic target of rapamycin complex 1 expands Th17 and IL-4+ CD4-CD8- double-negative T cells and contracts regulatory T cells in systemic lupus erythematosus. *J Immunol.* 2014;192(9):4134–4144.
232. Wei J, Long L, Yang K, Guy C, Shrestha S, Chen Z, et al. Autophagy enforces functional integrity of regulatory T cells by coupling environmental cues and metabolic homeostasis. *Nat Immunol.* 2016;17(3):277–285.
233. Kato H, Perl A. The IL-21-mTOR axis blocks differentiation and function of regulatory T cells of systemic lupus erythematosus patients by suppression of autophagy. *Arthritis Rheumatol.* 70:427–438.
234. Shivakumar S, Tsokos GC, Datta SK. T cell receptor alpha/beta expressing double-negative (CD4-/CD8-) and CD4+ T helper cells in humans augment the production of pathogenic anti-DNA autoantibodies associated with lupus nephritis. *J Immunol.* 1989;143(1):103–112.
235. Anand A, Dean GS, Quereshi K, Isenberg DA, Lydyard PM. Characterization of CD3+ CD4- CD8- (double negative) T cells in patients with systemic lupus erythematosus: activation markers. *Lupus.* 2002;11(8):493–500.
236. Rajagopalan S, Zordan T, Tsokos GC, Datta SK. Pathogenic anti-DNA autoantibody-inducing T helper cell lines from patients with active lupus nephritis: Isolation of CD4-8- T helper cell lines that express the gamma delta T-cell antigen receptor. *Proc Natl Acad Sci U S A.* 1990;87(18):7020–7024.
237. Sieling PA, Porcelli SA, Duong BT, Spada F, Bloom BR, Diamond B, et al. Human double-negative T cells in systemic lupus erythematosus provide help for IgG and are restricted by CD1c. *J Immunol.* 2000;165(9):5338–5344.
238. Crispin JC, Oukka M, Bayliss G, Cohen RA, Van Beek CA, Stillman IE, et al. Expanded double negative T cells in patients with systemic lupus erythematosus produce IL-17 and infiltrate the kidneys. *J Immunol.* 2008;181(12):8761–8766.
239. Amel-Kashipaz MR, Huggins ML, Lanyon P, Robins A, Todd I, Powell RJ. Quantitative and qualitative analysis of the balance between type 1 and type 2 cytokine-producing CD8(-) and CD8(+) T cells in systemic lupus erythematosus. *J Autoimmun.* 2001;17(2):155–163.
240. Horwitz DA, Gray JD, Behrendsen SC, Kubin M, Rengaraju M, Ohtsuka K, et al. Decreased production of interleukin-12 and other Th1-type cytokines in patients with recent-onset systemic lupus erythematosus. *Arthritis Rheum.* 1998;41(5):838–844.
241. Akahoshi M, Nakashima H, Tanaka Y, Kohsaka T, Nagano S, Ohgami E, et al. Th1/Th2 balance of peripheral T helper cells in systemic lupus erythematosus. *Arthritis Rheum.* 1999;42(8):1644–1648.
242. Tucci M, Lombardi L, Richards HB, Dammacco F, Silvestris F. Overexpression of interleukin-12 and T helper 1 predominance in lupus nephritis. *Clin Exp Immunol.* 2008;154(2):247–254.
243. Min DJ, Cho ML, Cho CS, Min SY, Kim WU, Yang SY, et al. Decreased production of interleukin-12 and interferon-gamma is associated with renal involvement in systemic lupus erythematosus. *Scand J Rheumatol.* 2001;30(3):159–163.

244. Tanaka Y, Nakashima H, Hisano C, Kohsaka T, Nemoto Y, Niiro H, et al. Association of the interferon-gamma receptor variant (Val14Met) with systemic lupus erythematosus. *Immunogenetics.* 1999;49(4):266–271.
245. Shah K, Lee WW, Lee SH, Kim SH, Kang SW, Craft J, et al. Dysregulated balance of Th17 and Th1 cells in systemic lupus erythematosus. *Arthritis Res Ther.* 2010;12(2):R53.
246. Yang J, Chu Y, Yang X, Gao D, Zhu L, Wan L, et al. Th17 and natural Treg cell population dynamics in systemic lupus erythematosus. *Arthritis Rheum.* 2009;60(5):1472–1483.
247. Wong CK, Lit LC, Tam LS, Li EK, Wong PT, Lam CW. Hyperproduction of IL-23 and IL-17 in patients with systemic lupus erythematosus: Implications for Th17-mediated inflammation in auto-immunity. *Clin Immunol.* 2008;127(3):385–393.
248. Mok MY, Wu HJ, Lo Y, Lau CS. The relation of interleukin 17 (IL-17) and IL-23 to Th1/Th2 cytokines and disease activity in systemic lupus erythematosus. *J Rheumatol.* 2010;37(10):2046–2052.
249. Doreau A, Belot A, Bastid J, Riche B, Trescol-Biemont MC, Ranchin B, et al. Interleukin 17 acts in synergy with B cell-activating factor to influence B cell biology and the pathophysiology of systemic lupus erythematosus. *Nat Immunol.* 2009;10(7):778–785.
250. Kyttaris VC, Zhang Z, Kuchroo VK, Oukka M, Tsokos GC. Cutting edge: IL-23 receptor deficiency prevents the development of lupus nephritis in C57BL/6-lpr/lpr mice. *J Immunol.* 2010;184(9):4605–4609.
251. Dai H, He F, Tsokos GC, Kyttaris VC. IL-23 Limits the Production of IL-2 and Promotes Autoimmunity in Lupus. *J Immunol.* 2017;199(3):903–910.
252. Pisitkun P, Ha HL, Wang H, Claudio E, Tivy CC, Zhou H, et al. Interleukin-17 cytokines are critical in development of fatal lupus glomerulonephritis. *Immunity.* 2012;37(6):1104–1115.
253. Amarilyo G, Lourenco EV, Shi FD, La Cava A. IL-17 promotes murine lupus. *J Immunol.* 2014;193(2):540–543.
254. Sawalha AH, Kaufman KM, Kelly JA, Adler AJ, Aberle T, Kilpatrick J, et al. Genetic association of interleukin-21 polymorphisms with systemic lupus erythematosus. *Ann Rheum Dis.* 2008;67(4):458–461.
255. Webb R, Merrill JT, Kelly JA, Sestak A, Kaufman KM, Langefeld CD, et al. A polymorphism within IL21R confers risk for systemic lupus erythematosus. *Arthritis Rheum.* 2009;60(8):2402–2407.
256. Dolff S, Abdulahad WH, Westra J, Doornbos-van der Meer B, Limburg PC, Kallenberg CG, et al. Increase in IL-21 producing T-cells in patients with systemic lupus erythematosus. *Arthritis Res Ther.* 2011;13(5):R157.
257. Nurieva RI, Chung Y, Hwang D, Yang XO, Kang HS, Ma L, et al. Generation of T follicular helper cells is mediated by interleukin-21 but independent of T helper 1, 2, or 17 cell lineages. *Immunity.* 2008;29(1):138–149.
258. Nurieva RI, Chung Y, Martinez GJ, Yang XO, Tanaka S, Matskevitch TD, et al. Bcl6 mediates the development of T follicular helper cells. *Science.* 2009;325(5943):1001–1005.
259. Ray JP, Marshall HD, Laidlaw BJ, Staron MM, Kaech SM, Craft J. Transcription factor STAT3 and type I interferons are corepressive insulators for differentiation of follicular helper and T helper 1 cells. *Immunity.* 2014;40(3):367–377.
260. Linterman MA, Beaton L, Yu D, Ramiscal RR, Srivastava M, Hogan JJ, et al. IL-21 acts directly on B cells to regulate Bcl-6 expression and germinal center responses. *J Exp Med.* 2010;207(2):353–363.
261. Zotos D, Coquet JM, Zhang Y, Light A, D'Costa K, Kallies A, et al. IL-21 regulates germinal center B cell differentiation and proliferation through a B cell-intrinsic mechanism. *J Exp Med.* 2010;207(2):365–378.
262. Ettinger R, Sims GP, Fairhurst AM, Robbins R, da Silva YS, Spolski R, et al. IL-21 induces differentiation of human naive and memory B cells into antibody-secreting plasma cells. *J Immunol.* 2005;175(12):7867–7879.
263. Linterman MA, Rigby RJ, Wong RK, Yu D, Brink R, Cannons JL, et al. Follicular helper T cells are required for systemic autoimmunity. *J Exp Med.* 2009;206(3):561–576.
264. Kim YU, Lim H, Jung HE, Wetsel RA, Chung Y. Regulation of autoimmune germinal center reactions in lupus-prone BXD2 mice by follicular helper T cells. *PLoS One.* 2015;10(3):e0120294.
265. Choi JY, Seth A, Kashgarian M, Terrillon S, Fung E, Huang L, et al. Disruption of pathogenic cellular networks by IL-21 blockade leads to disease amelioration in murine lupus. *J Immunol.* 2017;198(7):2578–2588.
266. Crispin JC, Martinez A, Alcocer-Varela J. Quantification of regulatory T cells in patients with systemic lupus erythematosus. *J Autoimmun.* 2003;21(3):273–276.
267. Liu MF, Wang CR, Fung LL, Wu CR. Decreased CD4+CD25+ T cells in peripheral blood of patients with systemic lupus erythematosus. *Scand J Immunol.* 2004;59(2):198–202.
268. Miyara M, Amoura Z, Parizot C, Badoual C, Dorgham K, Trad S, et al. Global natural regulatory T cell depletion in active systemic lupus erythematosus. *J Immunol.* 2005;175(12):8392–8400.
269. Mellor-Pita S, Citores MJ, Castejon R, Tutor-Ureta P, Yebra-Bango M, Andreu JL, et al. Decrease of regulatory T cells in patients with systemic lupus erythematosus. *Ann Rheum Dis.* 2006;65(4):553–534.
270. Bonelli M, Savitskaya A, von Dalwigk K, Steiner CW, Aletaha D, Smolen JS, et al. Quantitative and qualitative deficiencies of regulatory T cells in patients with systemic lupus erythematosus (SLE). *Int Immunol.* 2008;20(7):861–868.
271. Lee HY, Hong YK, Yun HJ, Kim YM, Kim JR, Yoo WH. Altered frequency and migration capacity of CD4+CD25+ regulatory T cells in systemic lupus erythematosus. *Rheumatology.* 2008;47(6):789–794.
272. Alvarado-Sanchez B, Hernandez-Castro B, Portales-Perez D, Baranda L, Layseca-Espinosa E, Abud-Mendoza C, et al. Regulatory T cells in patients with systemic lupus erythematosus. *J Autoimmun.* 2006;27(2):110–118.
273. Zhang B, Zhang X, Tang FL, Zhu LP, Liu Y, Lipsky PE. Clinical significance of increased CD4+CD25-Foxp3+ T cells in patients with new-onset systemic lupus erythematosus. *Ann Rheum Dis.* 2008;67(7):1037–1040.
274. Suarez A, Lopez P, Gomez J, Gutierrez C. Enrichment of CD4+ CD25high T cell population in patients with systemic lupus erythematosus treated with glucocorticoids. *Ann Rheum Dis.* 2006;65(11):1512–1517.
275. Yan B, Ye S, Chen G, Kuang M, Shen N, Chen S. Dysfunctional CD4+,CD25+ regulatory T cells in untreated active systemic lupus erythematosus secondary to interferon-alpha-producing antigen-presenting cells. *Arthritis Rheum.* 2008;58(3):801–12.
276. Valencia X, Yarboro C, Illei G, Lipsky PE. Deficient CD4+CD25high T regulatory cell function in patients with active systemic lupus erythematosus. *J Immunol.* 2007;178(4):2579–2588.
277. Vargas-Rojas MI, Crispin JC, Richaud-Patin Y, Alcocer-Varela J. Quantitative and qualitative normal regulatory T cells are not capable of inducing suppression in SLE patients due to T-cell resistance. *Lupus.* 2008;17(4):289–294.
278. Yates J, Whittington A, Mitchell P, Lechler RI, Lightstone L, Lombardi G. Natural regulatory T cells: Number and function are normal in the majority of patients with lupus nephritis. *Clin Exp Immunol.* 2008;153(1):44–55.

279. Burchill MA, Yang J, Vogtenhuber C, Blazar BR, Farrar MA. IL-2 receptor beta-dependent STAT5 activation is required for the development of Foxp3+ regulatory T cells. *J Immunol.* 2007;178(1):280–290.
280. He J, Zhang X, Wei Y, Sun X, Chen Y, Deng J, et al. Low-dose interleukin-2 treatment selectively modulates CD4(+) T cell subsets in patients with systemic lupus erythematosus. *Nat Med.* 2016;22(9):991–993.
281. Merrill JT, Burgos-Vargas R, Westhovens R, Chalmers A, D'Cruz D, Wallace DJ, et al. The efficacy and safety of abatacept in patients with non-life-threatening manifestations of systemic lupus erythematosus: Results of a twelve-month, multicenter, exploratory, phase IIb, randomized, double-blind, placebo-controlled trial. *Arthritis Rheum.* 2010;62(10):3077–3087.
282. Furie R, Nicholls K, Cheng TT, Houssiau F, Burgos-Vargas R, Chen SL, et al. Efficacy and safety of abatacept in lupus nephritis: A twelve-month, randomized, double-blind study. *Arthritis Rheumatol.* 2014;66(2):379–389.
283. Disteldorf EM, Krebs CF, Paust HJ, Turner JE, Nouailles G, Tittel A, et al. CXCL5 drives neutrophil recruitment in TH17-mediated GN. *J Am Soc Nephrol.* 2015;26(1):55–66.
284. Sieper, J., Management of ankylosing spondylitis, in *Rheumatology*, 4th edition, M.C. Hochberg, et al. (Eds.), Mosby, London, UK.
285. West SG. *Rheumatology Secrets* 2002;57–58. Hanley & Belfus, Philadelphia, PA.

Advances in Cardiovascular Diagnostics

8.4

Use of Cardiac Biomarkers in Clinical Research and Care of the Patient

Mark B. Effron and Zahoor Khan

Contents

BRIEF DESCRIPTION OF BIOMARKERS IN THE CARDIOVASCULAR ENVIRONMENT

National Institutes of Health (NIH) defines a biomarker as "a characteristic that is objectively measured and evaluated as an indicator of normal biological processes, pathogenic processes, or pharmacologic responses to a therapeutic intervention" (Biomarkers Definitions Working Group 2001). Biomarkers in cardiovascular (CV) environment can be characterized whether they are measurable, contain new information outside of routine testing, and help the clinician to manage patients (Morrow 2007). Biomarkers have been put forward as surrogate end points for clinical trials as they are used as diagnostic and prognostic markers of myocardial disease.

In 2008, Braunwald proposed that cardiovascular biomarkers be divided into seven categories: inflammation, oxidative stress, extracellular matrix remodeling, neurohormones, myocyte injury, myocyte stress, and new biomarkers (Braunwald 2008). Although he based these categories primarily on effects in heart failure, they can apply to almost any type of cardiac disease. In this section, we will discuss principally two of the most commonly used clinical biomarkers one from the myocyte injury category used predominantly in ischemic heart disease (troponin) and one in the myocyte stress category used predominantly in heart failure patients (BNP and its inactive precursor NTproBNP).

BIOMARKERS OF CARDIAC INJURY—ACUTE CORONARY SYNDROME (ACS)

When injury to cardiac tissue results in disruption of the plasma membrane of myocyte, it results in the breakdown of intracellular components and their release into extracellular space. These components include structural proteins such as myoglobin, heart-type fatty acid binding protein, lactate dehydrogenase-1 and 2 (LDH-1 and 2), creatine kinase and troponin (Mueller 2004, O'Brien 2008). Acute coronary syndrome (ACS) is the clinical manifestation of injury to cardiac muscle cell in the setting of ischemia or infarction, which occurs for the most part due to atherosclerosis (Vasan 2006). Research has shown that every stage of the above process is associated with release of detectable plasma/serum biomarkers (Vasan 2006) (Figure 8.4.1).

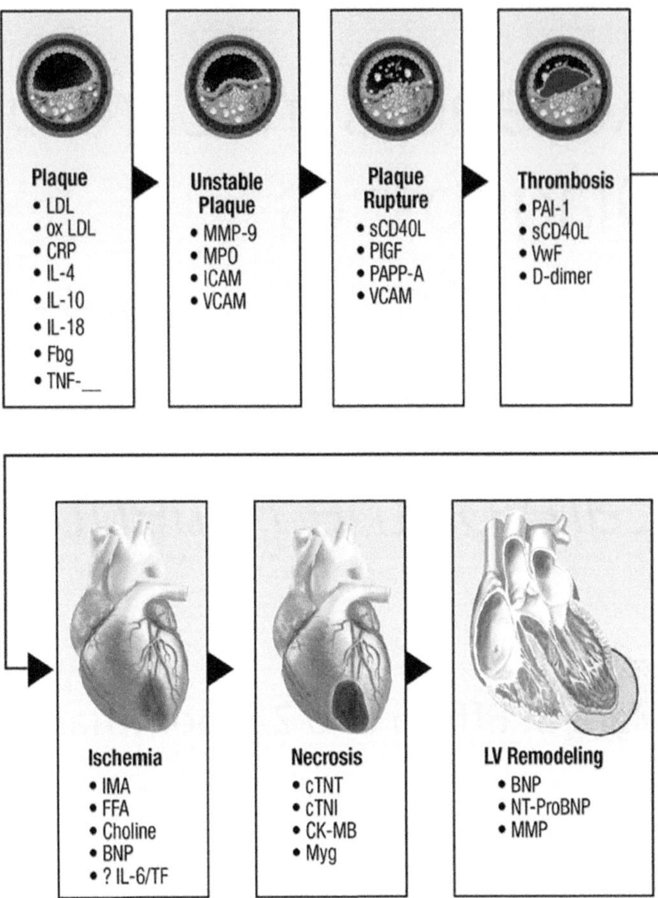

FIGURE 8.4.1 Biomarkers of acute coronary syndromes (ACS). The arrows indicate the order of events during an acute coronary syndrome. Biomarkers that are elevated and may be involved at each phase of ACS are displayed. sCD40L = soluble CD40 ligand; Fbg = fibrinogen; FFA = free fatty acid; ICAM = intercellular adhesion molecule; IL = interleukin; IMA = ischemia modified albumin; MMP = matrix metalloproteinases; MPO = myeloperoxidase; Myg = myoglobin; NT-proBNP = N-terminal proBNP; Ox-LDL = oxidized low-density lipoprotein; PAI-1 = plasminogen activator inhibitor; PAPP-A = pregnancy-associated plasma protein-A; PIGF = placental growth factor; TF = tissue factor; TNF = tumor necrosis factor; TNI = troponin I; TNT = troponin T; VCAM = vascular cell adhesion molecule; VWF = von Willebrand factor. Source: Vasan RS. Circulation 2006; 113: 2335–2362.

Creatine Kinase and CK-MB

For many years, the release of muscle related enzymes (creatine kinase, ALT, AST LDH) was used to define myocardial necrosis in the setting of chest pain. In the 1970s and 1980s, the ability to measure the MB fraction of creatine kinase (CK-MB) lead to improved differentiation of myocardial infarction from other syndromes resulting in muscle necrosis and became the biomarker of choice to diagnose MI because it was more specifically related to myocardial necrosis than other biomarkers (Jaffe 1991). CK-MB is one of the three isoenzymes of CK and has a higher concentration inside myocardial cells than the other isoenzymes of CK: BB and MM (Bessman 1985, Roberts 1975).

Measurement of CK-MB has been used in most studies of periprocedural (percutaneous coronary intervention [PCI] or coronary bypass graft surgery [CABG]) myocardial infarction (MI) and for the detection of early reinfarction (The EPIC Investigators 1994, The EPILOG Investigators 1997, The EPISTENT Investigators 1998, The ESPRIT Investigators 2000, Wiviott 2007, Wallentin 2009; Steinhubl 2002). The early rise and fall of CK-MB (Figure 8.4.2) allowed this marker to be useful in the periprocedural environment. (Puleo 1990, Puleo 1994, Wu 1999). CK-MB elevations usually return to baseline within 36 to 48 h, in contrast to elevations in troponin, which can persist for as long as 10–14 days (Jaffe 1996).

Although the use of CK-MB has been the standard for diagnosing MI in clinical trials for many years, the analysis is usually run in a central laboratory and rapid analysis is not available. In research studies, the use of CK-MB has been the principal way to diagnose a periprocedural myocardial infarction. Many studies have utilized different definitions of CK-MB usually in the range of 2–3 times the upper limit of normal (ULN) (The EPIC Investigators 1994, The EPILOG Investigators 1997, The EPISTENT Investigators 1998, The ESPRIT Investigators 2000, Wiviott 2007, Wallentin 2009, Steinhubl 2002). However, there has been controversy as to the relevance of periprocedural CK-MB elevations. Although some studies report small elevations of CK-MB to have clinical significance resulting in adverse outcomes, others report that there should be an increase of at least 5–10 times the ULN for clinical relevance to be seen (Saucedo 2000).

Recently, two different recommendations for definition of myocardial have been published and used for newer clinical trials. The first was the Universal Definition of Myocardial Infarction (Thygesen 2007), which has subsequently been revised three times, the last in 2018 (Thygesen 2018). However, the Society of Cardiac Angiography and Intervention (SCAI) published their guidelines for defining myocardial infarction in the setting of a PCI because SCAI believed the Universal Definition lacked appropriate specificity for periprocedural infarction (Type 4) (Moussa 2014).

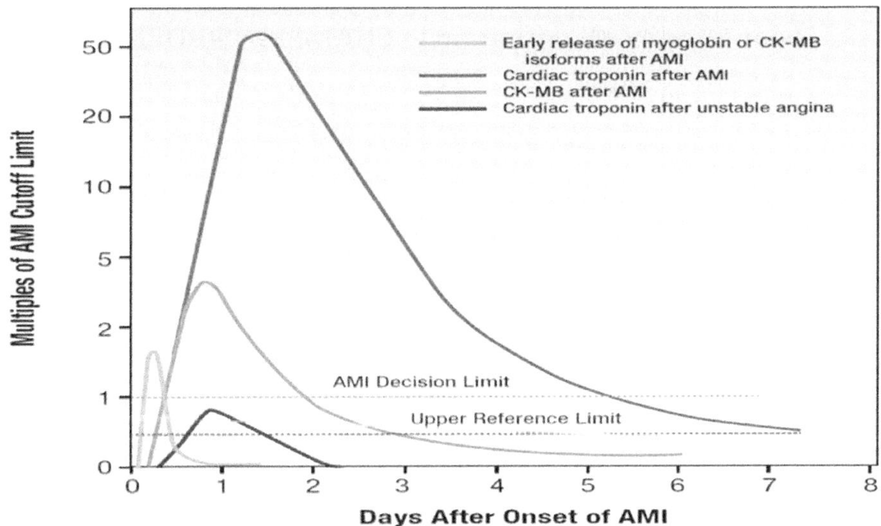

FIGURE 8.4.2 Plot of the appearance of cardiac markers in blood versus time after onset of symptoms. *Peak A*, early release of myoglobin or CK-MB isoforms after AMI; *peak B*, cardiac troponin after AMI; *peak C*, CK-MB after AMI; *peak D*, cardiac troponin after unstable angina. Data are plotted on a relative scale, where 1.0 is set at the AMI cutoff concentration. (*Source*: Wu. A.H.B. et al., *Clin Chem*, 45, 1104–1121, 1999.)

A recent example of both criteria being applied is the CHAMPION PHOENIX trial (Cavender 2016). The CHAMPION PHOENIX studied the use of cangrelor, an intravenous short acting $P2Y_{12}$ ADP receptor inhibitor, versus clopidogrel in preventing the occurrence of peri-PCI infarction. One of the stratification factors for randomization was "baseline status" (normal or abnormal, based on a combination of biomarker levels (predominantly cardiac troponin [discussed below]), electrocardiographic changes, and symptoms. However, for end point adjudication, the study defined criteria using CK-MB, not cardiac troponin, which was analyzed at a central laboratory. However, after pushback from the FDA on the end point definition, the CHAMPION PHOENIX investigators reanalyzed the data using both the Third Universal Definition of Myocardial Infarction (Thygesen 2012) and the SCAI definition of Myocardial Infarction (Moussa 2014), because almost all events were periprocedural occurring within the first 2 h following randomization. The different analyses produced similar results, a reduction of MI in the first 2 h post randomization with cangrelor compared with clopidogrel (Cavender 2016) (Figure 8.4.3). It is interesting to note, that even though cardiac troponin had been available for use in clinical studies, this important study used CK-MB to define its end point of MI (Bhatt 2013).

Cardiac Troponin

Cardiac troponin I (cTnI) and T (cTnT) are cardiac regulatory proteins that control the calcium-mediated interaction of actin and myosin (Adams 1993). Both have early releasable and structural pools, with most troponin in the structural pool (Katus 1991, Adams 1994). These proteins are protein products of specific genes, with amino acid sequences that are cardiospecific and, therefore, have the potential to be unique for the heart. Studies performed with cTnI have failed to find any cTnI outside of the heart at any stage of neonatal development (McManus 1993). Rapid analysis for cardiac troponins has been developed and decreases the time to diagnosis in patients with chest pain (Ilva 2005, Kavsak 2007). As seen in Figure 8.4.2, cardiac troponin increases faster than CK-MB in acute MI, and with the more rapid analysis, can lead to an earlier diagnosis (or exclusion) of MI in patients presenting with chest pain. However, with today's assays, there is a false positive rate of approximately 1% due to cross reacting heterophilic antibodies and to conditions other than coronary artery disease that are also associated with elevations in troponins (Lum 2006).

cTnI and cTnT can be measured in serum using immunoassays; such assays use two or three antibodies that are specific for the cardiac forms of cTnT and cTnI. Because of differences in the antibodies and other reagents used in the assays, numbers cannot be compared between assays and differences in numbers cannot be presumed to define sensitivity (Apple 2009).

In the late 1980s, investigators developed immunoassays for troponin I and troponin T (Katus 1991, Cummins 1987). Refinements in the antibodies, reagents, and automation have made the current commercial troponin assays exquisitely sensitive and precise (Apple 2011, Bhardwaj 2011). The newest, most sensitive assays can detect troponin in the bloodstream of patients without myocardial damage, perhaps due to normal myocardial cell turnover or formation of exosomes that release small amounts of free troponin into the bloodstream (Jarolim 2015).

Older assays are less sensitive than newer assays. The former is referred to as "conventional" or "sensitive" assays and the latter "high sensitivity" (hs) assays (Ilva 2005, Wu 2006, Giannitsis 2010, Venge 2009, Wilson 2009, Bhardwaj 2011). One criterion for calling an assay "hs" is the proportion of normal subjects in whom the assay is capable of detecting cTn (Apple 2012). It is now clear with hs-cTn assays that all individuals have small amounts of measurable cTn in their blood (Apple 2012). This may lead to an over diagnosis of MI in patients with heart disease, Indeed, the Third Universal Definition of Myocardial Infarction clearly states enzymatic confirmation of an MI should clearly

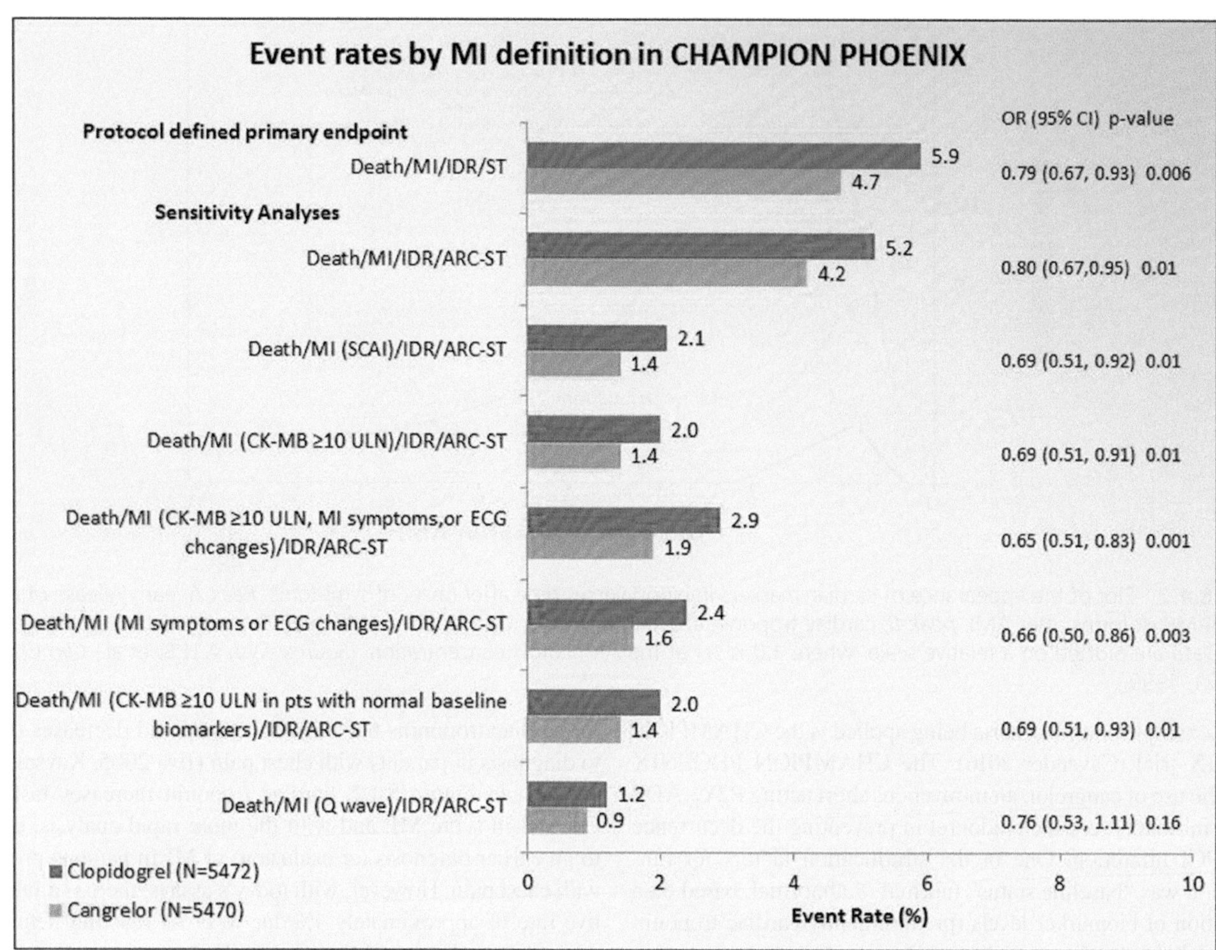

FIGURE 8.4.3 Protocol-defined and sensitivity analyses of primary efficacy end points at 48 h using different definitions of myocardial infarction (MI). ARC indicates Academic Research Consortium; CI, confidence interval; CK-MB, creatinine kinase-MB; IDR, ischemia-driven revascularization; OR, odds ratio; SCAI, Society of Coronary Angiography and Intervention; ST, stent thrombosis; and ULN, upper limit of normal. (*Source*: Cavender, M.A. et al., *Circulation,* 134, 723–733, 2016.)

be associated with a typical rise and fall in troponin (Thygesen 2012). The 4th Universal Definition of Myocardial Infarction states that a stable troponin elevation pattern should be referred to as "myocardial injury" (Thysgssen 2018).

Diagnosis of non-ST elevation acute coronary syndrome/myocardial infarction (NSTE ACS/MI).

The Third Universal Definition of Myocardial Infarction give a class I recommendation to the use of cardiac troponin I and T as the primary diagnostic biomarkers of myocardial necrosis and specifies obtaining cardiac specific troponin at presentation, 3 and 6 h, in the clinical context of suspected NSTE-ACS (Thygesen 2012, Kavsak 2007). In the Fourth Universal Definition of Myocardial Infarction, the presence of a typical rise and/or fall in biomarkers with ≥1 value above the 99th percentile of the upper reference level is the key differentiating factor of myocardial infarction (i.e., myocardial necrosis) from unstable angina, where there is only myocardial ischemia (injury) without necrosis (Thygesen 2018). The 2014 AHA/ACC Guideline for the Management of Patients With Non-ST-Elevation Acute Coronary Syndromes give a Class I, Level of evidence A, recommendation of using cardiac-specific troponin (I or T) at presentation and 3–6 h lar for a biomarker diagnosis of myocardial infarction (Amsterdam 2014). The European Society of Cardiology (ESC) 2015 Task Force for the Management of Acute Coronary Syndromes in Patients Presenting without Persistent ST-Segment Elevation use high-sensitivity cardiac troponin at presentation and 3 h later to further stratify a clinical context of chest pain for their "rule-in" "rule-out" algorithm (Figure 8.4.4) (Roffi 2015).

The problem seen in the clinical use of this highly diagnostic test is that cardiac troponin elevation can be present in other clinical syndromes where a myocardial infarction has not occurred. These include nonischemic congestive heart failure, sepsis, renal failure, post cardiac arrest, and non-penetrating chest trauma (O'Brien 2008). To establish the appropriate diagnosis of an acute myocardial infarction, there should be a change in the level of troponin and not just a result that is higher the ULN.

Numerous studies have shown that an increase in troponin leads to a worse prognosis (Kontos 2004, Ndrepepa 2011, Scirica 2011). This is important because it will identify a population in which more aggressive therapeutic intervention will demonstrate a risk greater than a benefit. This has been applied because of stratification from clinical trials, particularly in the use of PCI

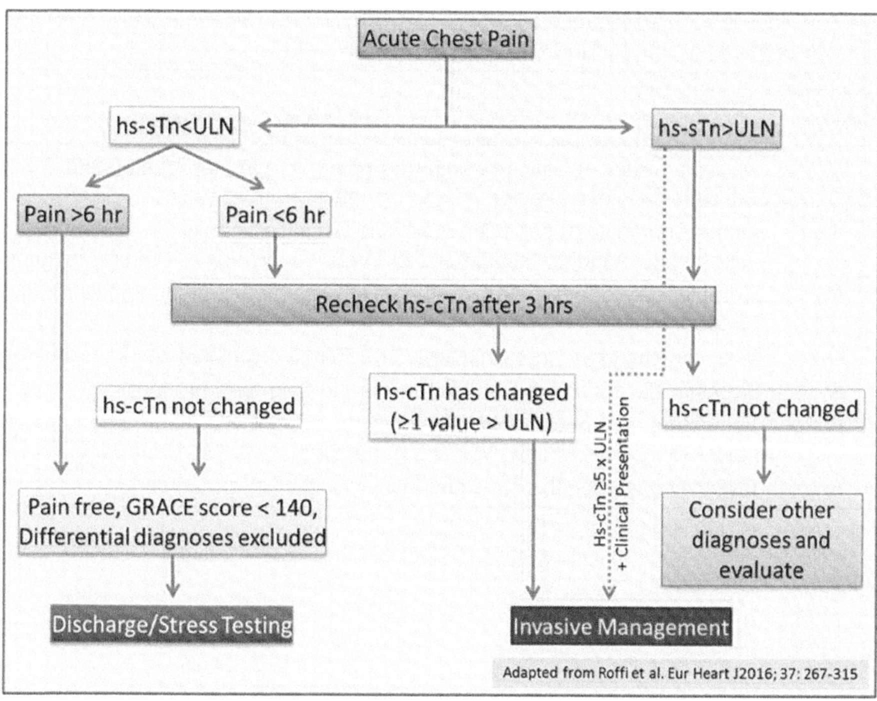

FIGURE 8.4.4 0 h/3 h rule-out algorithm of non-ST-elevation acute coronary syndromes using high-sensitivity cardiac troponin assays. GRACE, Global Registry of Acute Coronary Events score; hs-cTn, high sensitivity cardiac troponin; ULN, upper limit of normal. (*Source:* Roffi M et al. Eur Heart J. 2016; 37: 267–315.)

and potent antiplatelet agents such as glycoprotein IIb/IIIa antagonists. Subsequently, clinical trials evaluating interventional therapy have been designed using an elevation in troponin in the setting of typical signs and symptoms of ischemia for entry criteria to ensure a higher risk population (Cannon 2001).

However, the Society of Cardiac Angiology and Intervention (SCAI) published their Guidelines on defining clinically relevant MIs (Moussa 2014). In this document they noted a marked increase in periprocedural myocardial infarction, despite major improvements in performance of PCI. They reviewed the available literature and noted that cardiac troponin was too sensitive for diagnosing periprocedural MI as outlined in the Third Universal Definition. They found that periprocedural elevations of cardiac troponin, unlike in studies not dependent on periprocedural infarctions, was poorly predictive of long term, and in many cases, short term outcomes. Therefore, they proposed using CK-MB with elevations of ≥5x with Q waves or ≥10x in the absence of Q waves as the definition of periprocedural myocardial infarction. The SCAI definition of MI was used in the reanalysis of the CHAMPION-PCI study of cangrelor to help win its approval for marketing by the FDA (Cavender 2016) as previously discussed.

Beyond a diagnostic aid, elevated troponin levels are also shown to guide the type of therapeutic intervention such patients need. Early revascularization and intensive management benefits such patients both in short and long term outcomes (Hamm 1999, Valeur 2005, Mahmarian 2006, Desideri 2005). Troponin determination also has been shown to be associated with reduced hospital stay and charges from an administration point of view (Zarich 2001).

Clinical outcome trials in the cardiovascular world usually involve a composite end point that almost always includes myocardial infarction as one of the composite end points. Interventional trials have used troponin elevation as part of the end point definition, but the rate of myocardial infarction in these studies have been fairly high (Roger 2006). The use of cardiac troponin as a biomarker of infarction following a procedure has not be validated to the same extent as CK-MB, with studies showing relatively little correlation with elevated cardiac troponins post PCI and mortality (Lansky 2010). Many studies, as well as the Third Universal Definition of Myocardial Infarction, have required a fall in enzyme activity associated with a subsequent rise in enzyme leakage to signify myocardial necrosis. This is harder to see with troponins than with CK-MB, and if there is renal dysfunction present, the troponin may stay elevated without other evidence for myocardial infarction (Hamm 2002).

Therefore, troponin appears to play its role in clinical trials for the initial diagnosis of ACS and for the subsequent diagnosis of type I MI. Because the studies state that MI is defined per Universal Definition of MI and the UD now establishes the rise and/or fall of troponin as the biomarker to use, it is safe to assume that all MI patients in those clinical trials were due to an elevation of troponin. In 2012, the FDA Biomarker Qualification Review Team (BQRT) approved the use of plasma/serum cardiac troponin in safety assessment of drugs in dogs and rats (Hausner 2013). Finally, in the 2014 AHA/ACC Guideline for the Management of Patients With Non-ST-Elevation Acute Coronary Syndromes, cardiac troponin has a class I recommendation for use, whereas other biomarkers for myocardial infarction, including CK-MB, have a class II: no benefit designation (Table 8.4.1) (Amsterdam 2014).

TABLE 8.4.1 2014 AHA/ACC Guideline for the Management of Patients with Non-ST-Elevation Acute Coronary Syndromes—Summary of Recommendations for Cardiac Biomarkers and the Universal Definition of MI

CLASS OF RECOMMENDATION	*LEVEL OF EVIDENCE*	*RECOMMENDATION*
I	A	Measure cardiac-specific troponin (troponin I or T) at presentation and 3–6 h after symptom onset in all patients with suspected ACS to identify pattern of values
I	A	Obtain additional troponin levels beyond 6 h in patients with initial normal serial troponins with electrocardiographic changes and/or intermediate/high risk clinical features
I	A	Consider time of presentation the time of onset with ambiguous symptom onset for assessing troponin values
III: No Benefit	A	With contemporary troponin assays, CK-MB and myoglobin are not useful for diagnosis of ACS
I	B	Troponin elevations are useful for short- and long-term prognosis
IIb	B	Remeasurement of troponin value once on d 3 or 4 in patients with MI may be reasonable as an index of infarct size and dynamics of necrosis
IIb	B	BNP may be reasonable for additional prognostic information

Source: Amsterdam, E.A. et al., *Circulation*, 130, e344–e426, 2014.
ACS indicates acute coronary syndromes; BNP, B-type natriuretic peptide; CK-MB, creatine kinase myocardial isoenzyme; COR, Class of Recommendation; LOE, Level of Evidence; and MI, myocardial infarction.

Other Biomarkers for Diagnosing Myocardial Ischemia or Infarction

Myoglobin

Myoglobin is a small sized heme protein inside the cardiomyocytes and therefore it is released faster after injury occurs, with a plasma half-life of around 9 min (Isakov 1988, Klocke 1982). However, cardiac troponin is released earlier in blood than myoglobin and therefore the advantage of assessing early stage of cardiac injury obtained from myoglobin is very little. Additionally, comparative studies have shown both cardiac troponin and CK-MB to be more sensitive at diagnosing MI essentially stopping the use of myoglobin in the clinical diagnosis of MI (Eggers 2004, Scirica 2011).

Copeptin

Copeptin is the C-terminal portion of the arginine vasopressin precursor peptide. In the setting of acute MI, it is secreted from the pituitary gland. Its role is in the evaluation of MI is being investigated and is not recommended for clinical use at this point in time (Khan 2007, Nickel 2012).

Heart-type fatty acid binding protein (H-FABP)

The physicochemical characteristics, including low molecular weight and comparable release mechanism and kinetics to myoglobin, make H-FABP a reasonable marker for early injury. The diagnostic and prognostic value of H-FABP has shown some promise; however, the standards used in such studies were non-sensitive troponin assays (Seino 2003, Sabatine 2002). A major problem is the rapid clearance of the molecule rendering it a poor clinical marker in patients presenting more than 6 h from onset of infarction (Haltern 2010).

BIOMARKERS IN THE SETTING OF HEART FAILURE

Natriuretic Peptides (BNP, Pro-BNP)

Heart failure, as defined in the 2013 ACC-AHA Guidelines, is a "complex clinical syndrome that results from any structural or functional impairment of ventricular filling or ejection of blood" (Yancy 2013). Heart failure is usually manifested by dyspnea and early fatigue and/or edema, but volume overload need not be present. Hence, the change from the use of "congestive heart failure" to just heart failure (HF) failure with the 2013 Guidelines (Yancy 2013). Because dyspnea, fatigue, and edema are the major symptoms and signs of heart failure, heart failure is just one of many diagnoses that can result in these manifestations of HF. Therefore, it was recognized that it would be important to have a biomarker to help with diagnosis and prognosis of the disease.

Although many biomarkers in the arena of heart failure have been identified, the natriuretic peptides (NP) have been the most useful. The first naturetic peptide, atrial natriuretic peptide (ANP), was elucidated by Adolfo de Bold in 1981 (de Bold, 1981) in rat atria. After being demonstrated to be elevated in heart failure in rats, ANP was found to be elevated in humans with heart failure. With further characterization of the naturetic peptides, a similar peptide was found in the brain of rats and labeled "B-type" naturetic peptide (BNP) (Hosoda 1991). However, BNP was found to be made by human cardiac myocytes in the ventricles to a much higher degree than in the brain of humans. BNP is released from ventricular myocytes in response to myocytes stretch due to increase cavity size or hypertrophy of the ventricles, as opposed to ANP, which is released predominantly from the atria. BNP and its inactive by-product, NTproBNP, are maintained in the circulation longer than ANP allowing for better detection with testing.

The third member of the naturetic peptide family is CNP, discovered in 1990 (Sudoh 1990). CNP is low in cardiac tissue and is mostly found in vascular tissue. There is a fourth naturetic peptide that has been isolated (DNP) from the venom of the green mamba snake (Stein 1998, Lisy 1999) (Table 8.4.2).

The natriuretic peptides are hormones of cardiac and vascular endothelial origin (Potter 2011). They act in endocrine, acrine and paracrine way to exert their physiologic functions (de Lemos 2003). As patients develop heart failure, there is significant fluid retention and subsequent increase in peripheral resistance. These mechanisms are stimulated through the increased activity of the renin-angiotensin-aldosterone system (RAAS). The naturetic peptides increase diuresis by increasing glomerular filtration rate (GFR) and filtration fraction, producing natriuresis and diuresis (de Lemos 2003, Vanderheyden 2004). Naturetic peptides will interact with the adrenal glomerulosa to reduce aldosterone production and release. Natriuretic peptides essentially act in an antagonistic fashion to RAAS (Richards 1988). (Figure 8.4.5)

TABLE 8.4.2 Naturetic peptide physical characteristics

	ANP	*BNP*	*CNP*
Precursor with signal peptide	peproANP	peproBNP	peproCNP
Prohormone	proANP	proBNP	proCNP
Circulating fragments	NT-proANP	NTproBNP	NT-proANP
	ProANP	ProBNP	ProCNP
	ANP	BNP (3 different fragments of proBNP)	CNP (2 different fragments of proCNP)
Clearance	NPR-C, NEP	NPR-C, NEP	NPR-C, NEP
Circulating half-life	3 min	21 min	3 min

Source: Omland, T. et al., *Heart Failure Clin.*, 5, 471–487, 2009.
NEP, neutral endopeptidase; NPR-C, natriuretic peptide receptor-C.

Natriuretic peptides in general act on a family of transmembrane receptors known as natriuretic peptide receptors (NPR). There are three receptors in this family: NPR-A, NPR-B and NPR-C. ANP and BNP bind with high affinity to NPR-A, whereas CNP binds primarily to NPR-B. NPR-C is a clearance receptor and removes NPs from the circulation. NPR-A receptors are abundant in larger blood vessels, NPR-B in pituitary gland and NPR-C receptors is abundant in the brain, heart, kidney, adrenal, mesentery, and vascular smooth muscle tissue (Vanderheyden 2004).

Naturetic peptides have become valuable in the diagnosis, prognosis, and possible therapeutic response of heart failure to treatment. Because of their longer half-lives, BNP and NT-proBNP have become the biomarkers of choice in heart failure. With the onset of ventricular dilatation, ventricular hypertrophy, or increased wall tension leading to heart failure, the 134 amino acid pre-prohormone BNP is cleaved to the 108 amino-acid prohomone BNP. Prohormone BNP is subsequently cleaved by corin, an endopepitadase, into the 32 amino acid active hormone BNP and the inactive 76 amino acid N-terminal (NT)-proBNP (NTproBNP). Clearance of BNP and NTproBNP is via the kidney, so levels may be higher in patients with renal failure, which is common in heart failure patients (McCullough 2003). Increased body mass will have the opposite effect and will result in obtaining lower levels of BNP in patients with heart failure (Krauser 2005).

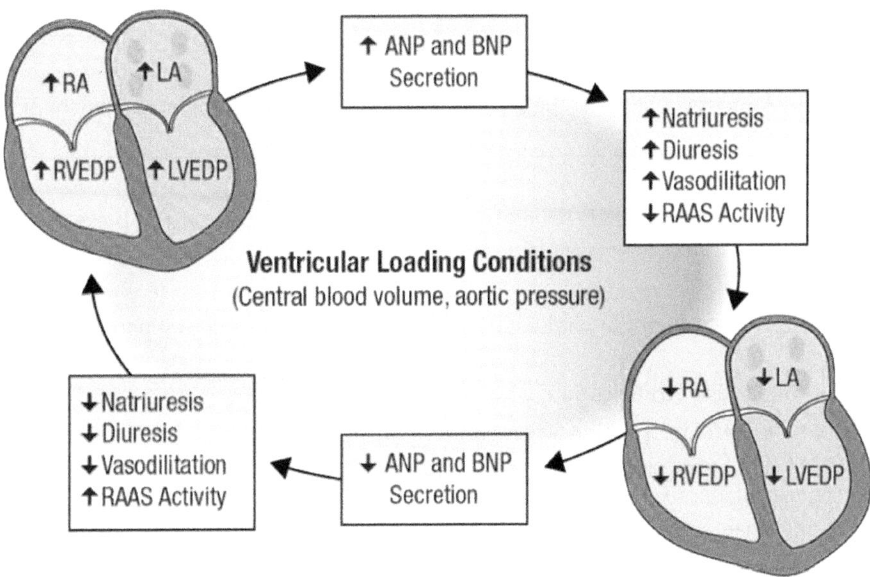

FIGURE 8.4.5 Schematic diagram depicting the mechanism by which cardiac secretion of natriuretic peptides serves to maintain intravascular volume homeostasis. ANP atrial natriuretic peptide; BNP B-type natriuretic peptide; LA left atrium; RA right atrium; LVEDP left ventricular end-diastolic pressure; RAAS renin-angiotensin-aldosterone system; RVEDP right ventricular end-diastolic pressure. (*Source:* Rodeheffer, R.J. *J Am Coll Cardiol*, 44, 740–749, 2004.)

Often shortness of breath (dyspnea) can be of unknown etiology in a clinical scenario. Although history and physical exam can help in differentiating between the two, BNP has been instrumental in aiding in diagnosing dyspnea of cardiac origin (Rodeheffer 2004). The Breathing Not Properly (BNP) study prospectively evaluated the use of BNP as a diagnostic tool for heart failure in the ambulatory and emergency settings (Maisel 2002). The study included 1586 patients with symptoms of shortness of breath with unclear etiology. In that study, the diagnostic accuracy of B-type natriuretic peptide at a cutoff of 100 pg/mL was 83.4%, and the negative predictive value at levels of less than 50 pg/mL was 96% (Maisel 2002). Subsequently, the ProBNP Investigation of Dyspnea in the Emergency Department (PRIDE) study evaluated the accuracy of NTproBNP in diagnosing heart failure (Januzzi 2005). The study showed that elevated NTproBNP at cut points of >450 pg/mL for patients <50 years of age and >900 pg/mL for patients >50 years of age were highly sensitive and specific for the diagnosis of acute heart failure (Maisel 2002). An NTproBNP level <300 pg/mL was optimal for excluding acute heart failure as a cause of dyspnea, with a negative predictive value of 99% (Maisel 2002, Januzzi 2005). Additionally, BNP has been found to be elevated in HFpEF as well as HFrEF, where a high BNP level in HFpEF may indicate high filling pressure and may indicate myocardial dysfunction in HFrEF (Ishigaki 2015). The 2017 ACC/AHA/HFSA Focused Update of the 2013 ACCF/AHA Guideline for the Management of Heart Failure (Yancy 2017) give the highest recommendation (Class I) supported by the highest level of evidence (LOE A) to obtaining BNP to aid in the diagnosis of heart failure (Table 8.4.3).

BNP and NTproBNP have also been evaluated as a prognostic tool in the treatment of heart failure and have showed a strong correlation between the levels (or changes in levels) of BNP and future outcomes. In the TACTICS-TIMI 18 trial, an elevated BNP was associated with increased risk of mortality at 7 days and 6 months, as well as the development of heart failure by 30 days (Morrow 2003). The ValHeft trial demonstrated that the relative change between baseline and 4-month BNP levels had an inverse correlation to all-cause mortality (Masson 2008). In the OPTIMIZE-HF trial, the discharge BNP was a stronger predictor of death and death or rehospitalization at 1 year than the admission BNP, admission to discharge BNP ration or clinical predictors alone in patients admitted for heart failure (Kociol 2011). NTproBNP has also been valuable to determine prognosis in patients (Jernberg 2003, Richards 2003), particularly when coupled with ejection fraction post MI (Richards 2003).

Finally, BNP and NTproBNP are being used to monitor therapy in patients with heart failure, although few studies have been published comparing the effectiveness of monitoring BNP and NTproBNP on outcomes with standard clinical therapy. Two studies, PONTIAC (NTproBNP Selected PreventiOn of cardiac events in a population of diabetic patients without A history of Cardiac disease) and STOP-HF (St Vincent's Screening to Prevent Heart Failure) provide evidence that aggressive therapy guided by initial elevation in BNP/NTproBNP can reduce clinical outcomes (Huelsmann 2013, Ledwidge 2013). Most studies comparing actual reduction of BNP to usual therapy were small and a meta-analysis of these studies were published to increase the numbers of patient in the evaluation (Pascual-Figal 2008, Savarese 2013). Twelve randomized trials were included in the analysis, which showed a clear reduction in heart failure-related hospitalizations. However, the effect of any individual trial on mortality was small and not significant, but the combined analysis showed an overall 26% reduction in all-cause mortality with BNP/NTproBNP guided therapy that was statistically significant (p=0.005) (Savarese 2013) (Figure 8.4.6). The NIH was conducting a large randomized trial, GUIDE-IT (Guiding Evidence Based Therapy Using Biomarker Intensified Treatment in

TABLE 8.4.3 2017 ACC/AHA/HFSA Focused update of the 2013 ACCF/AHA guideline for the management of heart failure - biomarker recommendations

BIOMARKERS FOR PREVENTION
For patients at risk of developing HF, natriuretic peptide biomarker–based screening can be useful to prevent the development of left ventricular dysfunction (systolic or diastolic) or new-onset HF (*Class IIa, LOE B-R*)
Biomarker for Diagnosis
In patients presenting with dyspnea, measurement of natriuretic peptide biomarkers is useful to support a diagnosis or exclusion of HF (*Class I, LOE A*)
Biomarkers for Prognosis or Added Risk Stratification
Measurement of BNP or NTproBNP is useful for establishing prognosis or disease severity in chronic HF (*Class I, LOE A*)
Measurement of baseline levels of natriuretic peptide biomarkers and/or cardiac troponin on admission to the hospital is useful to establish a prognosis in acutely decompensated HF (*Class I, LOE A*)
During a HF hospitalization, a predischarge natriuretic peptide level can be useful to establish a post discharge prognosis (*Class IIa, LOE B-NR*)
In patients with chronic HF, measurement of other clinically available tests, such as biomarkers of myocardial injury or fibrosis, may be considered for additive risk stratification (*Class IIb, LOE B-NR*)

Source: Yancy, C.W. et al., *JACC*, 70, 776–803, 2017.
BNP = B-type natriuretic peptide; HF = heart failure; LOE = level of evidence; NR = Non-randomized; R = Randomised; NT-proBNP = N-terminal pro-B-type natriuretic peptide

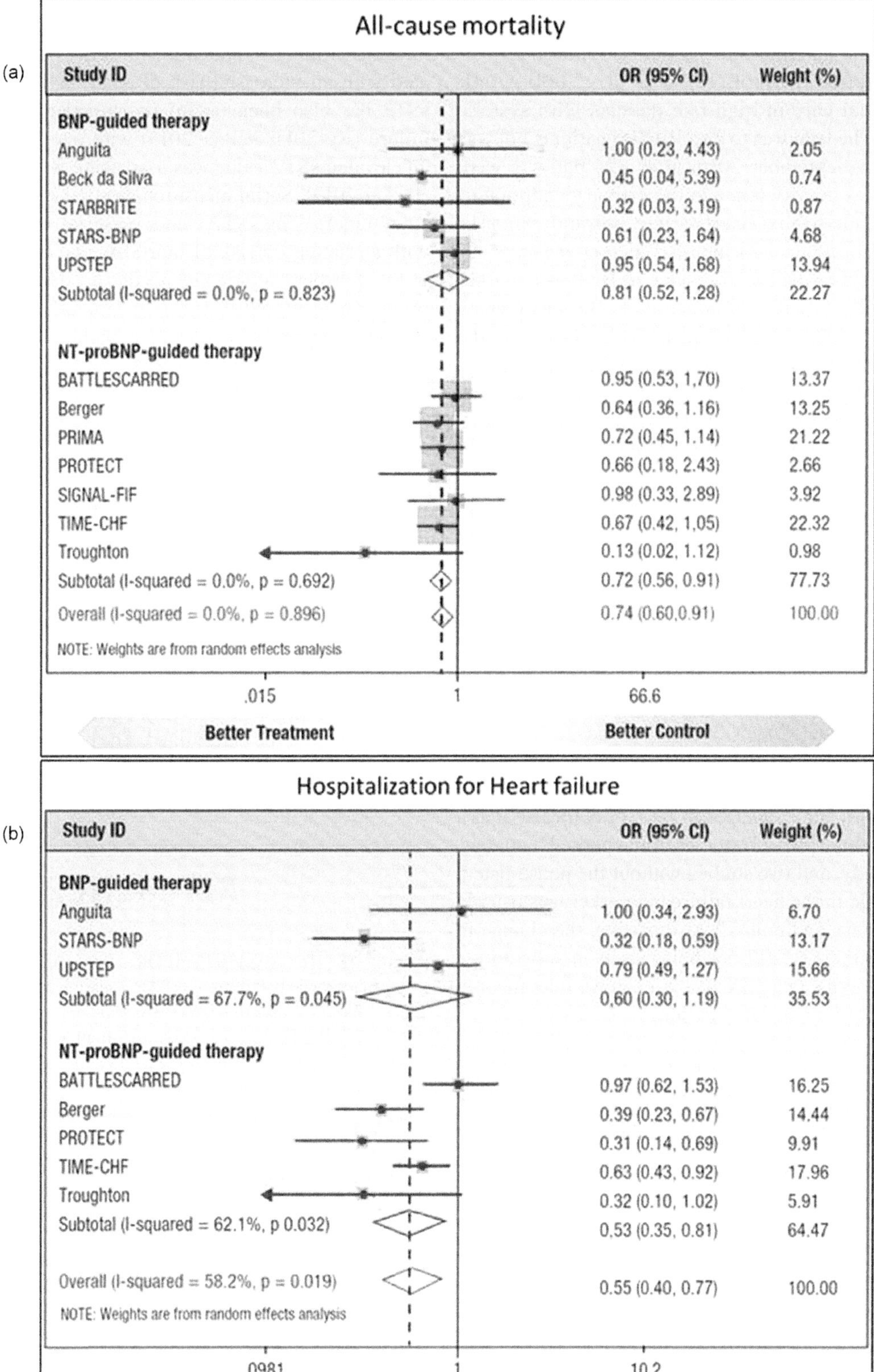

FIGURE 8.4.6 Meta-analysis of available natriuretic peptide-guided therapy trials. Solid squares represent odds ratios in trials and have a size proportional to the number of events. The 95% confidence intervals for individual trials are denoted by lines and those for the pooled odd ratios by empty diamonds. (a) All-cause mortality; (b) Heart failure hospitalizations. (*Source:* Savarese, G. et al., *PLoS One*, 8, e58287, 2013.)

Heart Failure), designed to determine the safety, efficacy, and cost-effectiveness of a strategy of adjusting therapy with the goal of achieving and maintaining a target N-terminal pro–B-type natriuretic peptide (NTproBNP) level of <1,000 pg/mL compared with usual care in high-risk patients with systolic HF (Felker 2014). The trial was to enroll 1,100 patients but was terminated early in September 2016, after 894 patients were enrolled as there was no difference in the primary endpoint of the composite of time-to-first heart failure hospitalization or cardiovascular mortality over 24 months. (Felker 2017).

A recent problem in the use of change in BNP as a marker for a therapeutic effect is the introduction of the angiotensin receptor/neprilysin inhibitor (ARNI) sacubutril/valsartan in the armamentarium of heart failure therapy. Neprilysin uses BNP as a substrate, so inhibition of neprilysin by sacubitril will cause an increase in BNP. In the PARADIGM-HF trial, patients treated with sacubitril/valsartan, which lowered the event rate compared with enalapril, had higher levels of BNP but lower levels of NTproBNP than patients treated with enalapril (Packer 2015). In patients treated with sacubutril/valsartan, NTproBNP should be monitored and not BNP, as BNP will reflect the action of the drug, whereas levels of NTproBNP will reflect the effects of the drug on the heart (Packer 2015). Currently, the most recent (2017) ACC/AHA/HFSA Guidelines for treatment of heart failure give a class I recommendations for use of naturetic peptides in the diagnosis and prognosis of disease, but not for monitoring the adequacy of treatment (Yancy 2017). As far as using reduction in naturetic peptides as a primary end point for clinical trials, as of August 2017 there were nine clinical trials listed on clinicaltrials.gov with either changes in BNP or NTproBNP as a primary end point: three phase II studies, three phase IV studies, one phase II/III study, and two studies without the phase listed. However, the change in the heart failure biomarker was usually one of many "primary end points." At this time, the change in heart failure biomarkers is still not being used in substitution for functional end points (e.g., CV death, exercise tolerance) in proving the efficacy of a new intervention.

Other Biomarkers in Heart Failure

The most promising other biomarker in heart failure is soluble ST2 (sST2). ST2 is an interleukin (IL)-1 receptor-like protein expressed on fibroblasts and cardiomyocytes in response to mechanical stress (Weinberg 2002, Daniels 2014). It is expressed in a transmembrane form (ST2L) and a soluble circulating form (sST2) and is associated with adverse outcomes in heart failure (Weinberg 2003). sST2 levels show less variability over the long term compared with NPs so that the molecule can function as a marker of duration of heart failure the way hemoglobin A1C acts as a marker for average glucose levels in patients with diabetes. sST2 performs well in both heart failure with preserved ejection fraction and heart failure with reduced ejection fraction. (Manzano-Fernandez 2011), It has worked synergistically with NPs, enhancing mortality prediction in acute and chronic heart failure. (Manzano-Fernandez 2011, Rehman 2008, Bayes-Genis 2012, Daniels 2014). In the PRIDE study, sST2 values were significantly higher in patients with heart failure and were associated with an increased risk of death at 1 year (Januzzi 2007). sST2 has also been useful in outpatients with chronic heart failure (Ky 2011, Felker 2013) with one study confirming that an elevated sST2 value was associated with increased mortality (Felker 2013). Serial measurements sST2 in outpatients demonstrated increasing sST2 values on serial samples was associated with an increased risk of mortality and hospitalization (Anand 2014). Further studies are needed to determine if serial sampling is helpful and if a similar biomarker-guided strategy can be applied to sST2 as it has been with naturetic peptides.

CONCLUSION

Biomarkers are becoming useful in the confirming diagnoses and prognosticating outcomes in cardiovascular disease. However, there are many potential pitfalls in relying only on the biomarkers and in not understanding the limitations of their use. Although change in biomarkers have been proposed as end points in trials to decrease the size of future cardiovascular outcomes trials, we are still not at the point where that can happen. Whether biomarkers are predicative enough to truly represent the actual clinical outcome remains to be determined. The benefit and risk of new therapies will still need to be correlated with actual clinical outcomes.

REFERENCES

Adams JE III, Abendschein DR, Jaffe AS. Biochemical markers of myocardial injury. Is MB creatine kinase the choice for the 1990s? *Circulation* 1993; 88: 750–763.

Adams JE III, Bodor GS, Dávila-Román VG et al. Cardiac troponin I: A marker with high specificity for cardiac injury. *Circulation* 1993; 88: 101–106.

Adams JE III, Schechtman KB, Landt Y, Ladenson JH, Jaffe AS. Comparable detection of acute myocardial infarction by creatine kinase MB isoenzyme and cardiac troponin I. *Clin Chem* 1994; 40: 1291–1295.

Amsterdam EA, Wenger NK, Brindis RG et al. 2014 AHA/ACC guideline for the management of patients with non–ST-elevation acute coronary syndromes: A report of the American College of Cardiology/American Heart Association Task Force on Practice Guidelines. *Circulation* 2014; 130: e344–e426.

Anand IS, Rector TS, Kuskowski M, Snider J, Cohn JN. Prognostic value of soluble ST2 in the valsartan heart failure trial. *Circ Heart Fail* 2014; 7: 418–426.

Apple FS, Collinson PO. For the IFCC task force on clinical applications of cardiac biomarkers. Analytical characteristics of high-sensitivity cardiac troponin assays. *Clin Chem* 2012; 58: 54–61.

Apple FS, Ler R, Murakami MM. Determination of 19 cardiac troponin I and T assay 99th percentile values from a common presumably healthy population. *Clin Chem* 2012; 58: 1574–1581.

Apple FS. A new season for cardiac troponin assays: It's time to keep a scorecard. *Clin Chem* 2009; 55: 1303–1306.

Bayes-Genis A, de Antonio M, Galan A et al. Combined use of high sensitivity ST2 and NTproBNP to improve the prediction of death in heart failure. *Eur J Heart Fail.* 2012; 14: 32–38.

Bessman SP, Carpenter CL. The creatine-creatine phosphate energy shuttle. *Annu Rev Biochem* 1985; 54: 831–862.

Bhardwaj A, Truong QA, Peacock WF et al. A multicenter comparison of established and emerging cardiac biomarkers for the diagnostic evaluation of chest pain in the emergency department. *Am Heart J* 2011; 162: 276–282, e1.

Bhatt DL, Stone GW, Mahaffey KW et al. Effect of platelet inhibition with cangrelor during PCI on ischemic events. *N Engl J Med* 2013; 368: 1303–1313.

Biomarkers Definitions Working Group. Biomarkers and surrogate endpoints: Preferred definitions and conceptual framework. *Clin Pharmacol Ther* 2001; 69: 89–95.

Braunwald E. Biomarkers in heart failure. *N Engl J Med* 2008; 358: 2148–2159.

Cannon CP, Weintraub WS, Demopoulos LA et al. Comparison of early invasive and conservative strategies in patients with unstable coronary syndromes treated with the glycoprotein IIb/IIIa inhibitor tirofiban. *N Engl J Med* 2001; 344: 1879–1887.

Cavender MA, Bhatt DL, Stone GW et al. Consistent reduction in periprocedural myocardial infarction with cangrelor as assessed by multiple definitions: Findings from CHAMPION PHOENIX (cangrelor versus standard therapy to achieve optimal management of platelet inhibition). *Circulation* 2016; 134: 723–733.

Cummins B, Auckland ML, Cummins P et al. Cardiac-specific troponin-I radioimmunoassay in the diagnosis of acute myocardial infarction. *Am Heart J* 1987; 113: 1333–1344.

Daniels LB, Bayes-Genis A. Using ST2 in cardiovascular patients: A review. *Future Cardiol* 2014; 10: 525–539.

de Bold AJ, Borenstein HB, Veress AT, Sonnenberg H. A rapid and potent natriuretic response to intravenous injection of atrial myocardial extract in rats. *Life Sci.* 1981; 28: 89–94.

de Lemos JA, McGuire DK, Drazner MH. B-type natriuretic peptide in cardiovascular disease. *Lancet* 2003; 362: 316–322.

Desideri A, Fioretti PM, Cortigiani L et al. Pre-discharge stress echocardiography and exercise ECG for risk stratification after uncomplicated acute myocardial infarction: Results of the COSTAMI-II (cost of strategies after myocardial infarction) trial. *Heart* 2005; 91: 146–151.

Eggers KM, Oldgren J, Nordenskjöld A, Lindahl B. Diagnostic value of serial measurement of cardiac markers in patients with chest pain: Limited value of adding myoglobin to troponin I for exclusion of myocardial infarction. *Am Heart J* 2004; 148: 574–581.

EPILOG Investigators. Platelet glycoprotein IIb/IIIa receptor blockade and low-dose heparin during percutaneous coronary revascularization. *N Engl J Med* 1997; 336: 1689–1697.

Felker GM, Ahmad T, Anstrom KJ et al. Rationale and design of the GUIDE-IT study: Guiding evidence based therapy using biomarker intensified treatment in heart failure. *JACC Heart Fail* 2014; 2: 457–465.

Felker GM, Fiuzat M, Thompson V et al. Soluble ST2 in ambulatory patients with heart failure: Association with functional capacity and long-term outcomes. *Circ Heart Fail* 2013; 6: 1172–1179.

Felker GM, Anstrom KJ, Adams KF et al. Effect of natriuretic peptide–guided therapy on hospitalization or cardiovascular mortality in high-risk patients with heart failure and reduced ejection fraction: A randomized clinical trial. *JAMA*. 2017; 318: 713–720.

Giannitsis E, Kurz K, Hallermayer K, Jarausch J, Jaffe AS, Katus HA. Analytical validation of a high-sensitivity cardiac troponin T assay. *Clin Chem* 2010; 56: 254–261.

Haltern G, Peiniger S, Bufe A, Reiss G, Gülker H, Scheffold T. Comparison of usefulness of heart-type fatty acid binding protein versus cardiac troponin T for diagnosis of acute myocardial infarction. *Am J Cardiol* 2010; 105: 1–9.

Hamm CW, Giannitsis E, Katus HA. Cardiac troponin elevations in patients without acute coronary syndrome (editorial). *Circulation.* 2002; 106: 2871–2872.

Hamm CW, Heeschen C, Goldmann B et al. Benefit of abciximab in patients with refractory unstable angina in relation to serum troponin T levels. c7E3 Fab Antiplatelet Therapy in Unstable Refractory Angina (CAPTURE) Study Investigators. *N Engl J Med* 1999; 340: 1623–1629.

Hausner EA, Hicks KA, Leighton JK, Szarfman A, Thompson AM, Harlow P. Qualification of cardiac troponins for nonclinical use: A regulatory perspective. *Regul Toxicol Pharmacol* 2013; 67: 108–114.

Hosoda K, Nakao K, Mukoyama M et al. Expression of brain natriuretic peptide gene in human heart: Production in the ventricle. *Hypertension* 1991; 17: 1152–1156.

Huelsmann M, Neuhold S, Resl M et al. PONTIAC (NT-proBNP Selected PreventiOn of cardiac events in a population of dIabetic patients without A history of Cardiac disease): A prospective randomized controlled trial. *J Am Coll Cardiol* 2013; 62: 1365–1372.

Ilva T, Eriksson S, Lund J et al. Improved early risk stratification and diagnosis of myocardial infarction, using a novel troponin I assay concept. *Eur J Clin Invest* 2005; 35: 112–116.

Isakov A, Shapira I, Burke M, Almog C. Serum myoglobin levels in patients with ischemic myocardial insult. *Arch Intern Med* 1988; 148: 1762–1765.

Ishigaki M, Yoshida T, Izumi H, Fujisawa Y, Shimizu S, Masuda K, Asanuma T, Okabe H, Nakatani S. Different implication of elevated B-Type natriuretic peptide level in patients with heart failure with preserved ejection fraction and in those with reduced ejection fraction. *Echocardiography* 2015; 32: 623–629.

Jaffe AS, Landt Y, Parvin CA, Abendschein DR, Geltman EM, Ladenson JH. Comparative sensitivity of cardiac troponin I and lactate dehydrogenase isoenzymes for diagnosing acute myocardial infarction. *Clin Chem* 1996; 42: 1770–1776.

Jaffe AS. Biochemical detection of acute myocardial infarction. In: Gersh B, Rahimtoola S, eds. *Acute Myocardial Infarction.* New York: Elsevier; 1991, pp. 110–127.

Januzzi JL, Camargo CA, Anwaruddin S et al. The N-terminal Pro-BNP investigation of dyspnea in the emergency department (PRIDE) study. *Am J Cardiol* 2005; 95: 948–954.

Januzzi JL Jr, Peacock WF, Maisel AS et al. Measurement of the interleukin family member ST2 in patients with acute dyspnea: Results from the PRIDE (Pro-Brain Natriuretic Peptide Investigation of Dyspnea in the Emergency Department) study. *J Am Coll Cardiol* 2007; 50: 607–613.

Jarolim P. High sensitivity cardiac troponin assays in the clinical laboratories. *Clin Chem Lab Med* 2015; 53: 635–652.

Jernberg T, Lindahl B, Siegbahn A et al. N-terminal pro-brain natriuretic peptide in relation to inflammation, myocardiac necrosis, and the effect of an invasive strategy in unstable coronary artery disease. *J Am Coll Cardiol* 2003; 42: 1909–1916.

Karjalainen PP, Vikman S, Niemelä M et al. Safety of percutaneous coronary intervention during uninterrupted oral anticoagulant treatment. *Eur Hear J* 2008; 29: 1001–1010.

Katus HA, Remppis A, Neumann FJ et al. Diagnostic efficiency of troponin T measurements in acute myocardial infarction. *Circulation* 1991; 83: 902–912.

Katus HA, Remppis A, Scheffold T, Diederich KW, Kuebler W. Intracellular compartmentation of cardiac troponin T and its release kinetics in patients with reperfused and nonreperfused myocardial infarction. *Am J Cardiol* 1991; 67: 1360–1367.

Kavsak PA, MacRae AR, Newman AM et al. Effects of contemporary troponin assay sensitivity on the utility of the early markers myoglobin and CKMB isoforms in evaluating patients with possible acute myocardial infarction. *Clin Chim Acta* 2007; 380: 213–216.

Khan SQ, Dhillon OS, O'Brien RJ et al. C-terminal provasopressin (copeptin) as a novel and prognostic marker in acute myocardial infarction: Leicester Acute Myocardial Infarction Peptide (LAMP) Study. *Circulation* 2007; 115: 2103–2110.

King M, Kingery J, Casey B. Diagnosis and evaluation of heart failure. *Am Fam Physician* 2012; 85: 1161–1168.

Klocke FJ, Copley DP, Krawczyk JA, Reichlin M. Rapid renal clearance of immunoreactive canine plasma myoglobin. *Circulation* 1982; 65: 1522–1528.

Kociol RD, Horton JR, Fonarow GC et al. Admission, discharge, or change in B-type natriuretic peptide and long-term outcomes: Data from organized program to initiate lifesaving treatment in hospitalized patients with heart failure (OPTIMIZE-HF) linked to medicare claims. *Circ Hear Fail* 2011; 4: 628–636.

Kontos MC, Shah R, Fritz LM et al. Implication of different cardiac troponin I levels for clinical outcomes and prognosis of acute chest pain patients. *J Am Coll Cardiol* 2004; 43: 958–965.

Krauser DG, Lloyd-Jones DM, Chae CU, Cameron R, Anwaruddin S, Baggish AL, Chen A, Tung R, Januzzi JL Jr. Effect of body mass index on natriuretic peptide levels in patients with acute congestive heart failure: A ProBNP Investigation of Dyspnea in the Emergency Department (PRIDE) substudy. *Am Heart J.* 2005; 149: 744–750.

Ky B, French B, McCloskey K et al. High-sensitivity ST2 for prediction of adverse outcomes in chronic heart failure. *Circ Heart Fail.* 2011; 4: 180–187.

Lansky AJ, Stone GW. Periprocedural myocardial infarction: Prevalence, prognosis, and prevention. *Circ Cardiovasc Inters* 2010; 3: 602–610.

Ledwidge M, Gallagher J, Conlon C et al. Natriuretic peptide-based screening and collaborative care for heart failure: The STOP-HF randomized trial. *JAMA* 2013; 310: 66–74.

Lisy O, Jougasaki M, Heublein DM et al. Renal actions of synthetic dendroaspis natriuretic peptide. *Kidney Int* 1999; 56: 502–508.

Lum G, Solarz DE, Farney L,. False positive cardiac troponin results in patients without acute myocardial infarction. *Lab Med* 2006; 37: 546–550.

Mahmarian JJ, Shaw LJ, Filipchuk NG et al. A multinational study to establish the value of early adenosine technetium-99m sestamibi myocardial perfusion imaging in identifying a low-risk group for early hospital discharge after acute myocardial infarction. *J Am Coll Cardiol* 2006; 48: 2448–2457.

Maisel AS, Krishnaswamy P, Nowak RM et al. Rapid measurement of B-type natriuretic peptide in the emergency diagnosis of heart failure. *N Engl J Med* 2002; 347: 161–167.

Manzano-Fernandez S, Mueller T, Pascual-Figal D, Truong QA, Januzzi JL. Usefulness of soluble concentrations of interleukin family member ST2 as predictor of mortality in patients with acutely decompensated heart failure relative to left ventricular ejection fraction. *Am J Cardiol* 2011; 107(2): 259–267.

Masson S, Latini R, Anand IS et al. Prognostic value of changes in N-terminal pro-brain natriuretic peptide in Val-HeFT (Valsartan Heart Failure Trial). *J Am Coll Cardiol* 2008; 52: 997–1003.

McCullough PA, Duc P, Omland T et al. B-type natriuretic peptide and renal function in the diagnosis of heart failure: An analysis from the breathing not properly multinational study. *Am J Kidney Dis* 2003; 41: 571–579.

McManus JL. Cardiac troponin I: A marker with high specificity for cardiac injury. *Ann Emerg Med.* 1993; 22: 1930–1931.

Morrow DA, de Lemos JA, Sabatine MS et al. Evaluation of B-type natriuretic peptide for risk assessment in unstable angina/non–ST elevation myocardial infarction: B-type natriuretic peptide and prognosis in TACTICS–TIMI-18. *J Am Coll Cardiol* 2003; 41: 1264–1272.

Morrow DA, de Lemos JA. Benchmarks for the assessment of novel cardiovascular biomarkers. *Circulation* 2007; 115: 949–952.

Moussa ID, Klein LW, Shah B et al. Consideration of a new definition of clinically relevant myocardial infarction after coronary revascularization: An expert consensus document from the society for cardiovascular angiography and interventions (SCAI). *Catheter Cardiovasc Interv* 2014; 83: 27–36.

Mueller C. Biomarkers and acute coronary syndromes: an update. *Eur Hear J* 2014; 35: 552–556.

Ndrepepa G, Braun S, Mehilli J et al. Prognostic value of sensitive troponin T in patients with stable and unstable angina and undetectable conventional troponin. *Am Heart J* 2011; 161: 68–75.

Nickel CH, Bingisser R, Morgenthaler NG. The role of copeptin as a diagnostic and prognostic biomarker for risk stratification in the emergency department. *BMC Med* 2012; 10:7.

O'Brien PJ. Cardiac troponin is the most effective translational safety biomarker for myocardial injury in cardiotoxicity. *Toxicology* 2008; 245: 206–218.

Omland T, Hagve TA. Natriuretic peptides: Physiologic and analytic considerations. *Heart Failure Clin* 2009; 5: 471–487.

Packer M, McMurray JJV, Desai AS et al. For the PARADIGM-HF Investigators and Coordinators. Angiotensin receptor neprilysin inhibition compared with enalapril on the risk of clinical progression in surviving patients with heart failure. *Circulation* 2015; 131: 54–61.

Potter LR, Yoder AR, Flora DR, Antos LK, Dickey DM. Natriuretic peptides: Their structures, receptors, physiologic functions and therapeutic applications. *Handb Exp Pharmacol* 2009; (191): 341–366.

Potter LR. Natriuretic peptide metabolism, clearance and degradation. *FEBS Journal* 2011; 278: 1808–1817.

Puleo PR, Guadagno PA, Roberts R et al. Early diagnosis of acute myocardial infarction based on assay for subforms of creatine kinase-MB. *Circulation* 1990; 82: 759–764.

Puleo PR, Meyer D, Wathen C et al. Use of a rapid assay of subforms of creatine kinase MB to diagnose or rule out acute myocardial infarction. *N Engl J Med* 1994; 331: 561–566.

Rehman SU, Mueller T, Januzzi JL Jr. Characteristics of the novel interleukin family biomarker ST2 in patients with acute heart failure. *J Am Coll Cardiol.* 2008; 52: 1458–1465.

Richards AM, McDonald D, Fitzpatrick MA, et al. Atrial natriuretic hormone has biological effects in man at physiological plasma concentrations. *J Clin Endocrinol Metab* 1988; 67: 1134–1139.

Richards AM, Nicholls MG, Espiner EA, Lainchbury JG, Troughton RW, Elliott J, Frampton C, Turner J, Crozier IG, Yandle TG. B-Type natriuretic peptides and ejection fraction for prognosis after myocardial infarction. *Circulation* 2003; 107: 2786–2792.

Roberts R, Gowda KS, Ludbrook PA, Sobel BE. Specificity of elevated serum MB creatine phosphokinase activity in the diagnosis of acute myocardial infarction. *Am J Cardiol* 1975; 36: 433–437.

Rodeheffer RJ. Measuring plasma B-type natriuretic peptide in heart failure: Good to go in 2004. *J Am Coll Cardiol* 2004; 44: 740–749.

Roffi M, Patrono C, Collet JP et al. 2015 ESC Guidelines for the management of acute coronary syndromes in patients presenting without persistent ST-segment elevation task force for the management of acute coronary syndromes in patients presenting without persistent ST-segment elevation of the European Society of Cardiology (ESC). *Eur Heart J* 2016; 37: 267–315.

Roger VL, Killian JM, Weston SA, Jaffe AS, Kors J, Santrach PJ, Tunstall-Pedoe H, Jacobsen SJ. Redefinition of myocardial infarction: Prospective evaluation in the community. *Circulation* 2006; 114: 790–797.

Sabatine MS, Morrow DA, de Lemos JA et al. Multimarker approach to risk stratification in non-ST elevation acute coronary syndromes: Simultaneous assessment of troponin I, C-reactive protein, and B-type natriuretic peptide. *Circulation* 2002; 105: 1760–1763.

Saucedo JF, Mehran R, Dangas G et al. Long-term clinical events following creatine kinase–myocardial band isoenzyme elevation after successful coronary stenting. *J Am Coll Cardiol* 2000; 35: 1134–1141.

Savarese G, Trimarco B, Dellegrottaglie S, Prastaro M, Gambardella F, Rengo G, Leosco D, Perrone-Filardi P. Natriuretic peptide-guided therapy in chronic heart failure: A meta-analysis of 2,686 patients in 12 randomized trials. *PLoS One* 2013; 8: e58287.

Scirica BM, Sabatine MS, Jarolim P et al. Assessment of multiple cardiac biomarkers in non-ST-segment elevation acute coronary syndromes: Observations from the MERLIN-TIMI 36 Trial. *Eur Heart J* 2011; 32: 697–705.

Seino Y, Ogata K, Takano T et al. Use of a whole blood rapid panel test for heart-type fatty acid-binding protein in patients with acute chest pain: Comparison with rapid troponin T and myoglobin tests. *Am J Med* 2003; 115: 185–190.

Stein BC, Levin RI. Natriuretic peptides: Physiology, therapeutic potential, and risk stratification in ischemic heart disease. *Am Heart J* 1998; 135: 914–923

Steinhubl SR, Berger PB, Mann JT III et al. Early and sustained dual oral antiplatelet therapy following percutaneous coronary intervention: A randomized controlled trial. *JAMA* 2002; 288: 2411–2420.

Sudoh T, Minamino N, Kangawa K, Matsuo H. C-type natriuretic peptide (CNP): A new member of the natriuretic peptide family identified in porcine brain. *Biochem Biophys Res Commun* 1990; 168: 863–870.

The EPIC Investigators. Use of a monoclonal antibody directed against the platelet glycoprotein IIb/IIIa receptor in high-risk coronary angioplasty. *N Engl J Med* 1994; 330: 956–961.

The EPISTENT Investigators. Randomised placebo-controlled and balloon-angioplasty-controlled trial to assess safety of coronary stenting with use of platelet glycoprotein-IIb/IIIa blockade. *Lancet* 1998; 352: 87–92.

The ESPRIT Investigators. Novel dosing regimen of eptifibatide in planned coronary stent implantation (ESPRIT): A randomised, placebo-controlled trial. *Lancet* 2000; 356: 2037–2044.

Thygesen K, Alpert JS, White HD, et al. Universal definition of myocardial infarction. *Eur Heart J* 2007; 28: 2525–2538.

Thygesen K, Alpert JS, Jaffe AS, Simoons ML, Chaitman BR, White HD. Joint ESC/ACCF/AHA/WHF task force for the universal definition of myocardial infarction. Third universal definition of myocardial infarction. *Circulation* 2012; 126: 2020–2203.

Thygesen K, Alpert JS, Jaffe AS, et al. Fourth universal definition of myocardial infarction. Circulation 2018; 126: 2020–2203 *Circulation*. 2018;138:e618–e651.

Valeur N, Clemmensen P, Saunamäki K, Grande P; DANAMI-2 investigators. The prognostic value of pre-discharge exercise testing after myocardial infarction treated with either primary PCI or fibrinolysis: A DANAMI-2 sub-study. *Eur Heart J* 2005; 26: 119–127.

Vanderheyden M, Bartunek J, Goethals M. Brain and other natriuretic peptides: Molecular aspects. *Eur J Heart Fail* 2004; 6: 261–268.

Vasan RS. Biomarkers of cardiovascular disease: Molecular basis and practical considerations. *Circulation* 2006; 113: 2335–2362.

Venge P, Johnston N, Lindahl B, James S. Normal plasma levels of cardiac troponin i measured by the high-sensitivity cardiac troponin i access prototype assay and the impact on the diagnosis of myocardial ischemia. *J Am Coll Cardiol* 2009; 54: 1165–1172.

Wallentin L, Becker RC, Budaj A et al; PLATO Investigators. Ticagrelor versus clopidogrel in patients with acute coronary syndromes. *N Engl J Med* 2009; 361: 1045–1057.

Weinberg EO, Shimpo M, De Keulenaer GW et al. Expression and regulation of ST2, an interleukin-1 receptor family member, in cardiomyocytes and myocardial infarction. *Circulation* 2002; 106: 2961–2966.

Weinberg EO, Shimpo M, Hurwitz S, Tominaga S, Rouleau JL, Lee RT. Identification of serum soluble ST2 receptor as a novel heart failure biomarker. *Circulation* 2003; 107: 721–726.

Wilson SR, Sabatine MS, Braunwald E, Sloan S, Murphy SA, Morrow DA. Detection of myocardial injury in patients with unstable angina using a novel nanoparticle cardiac troponin I assay: Observations from the PROTECT-TIMI 30 Trial. *Am Heart J* 2009; 158: 386–391.

Wiviott SD, Braunwald E, McCabe CH et al; TRITON-TIMI 38 Investigators. Prasugrel versus clopidogrel in patients with acute coronary syndromes. *N Engl J Med* 2007; 357: 2001–2015.

Wu AHB, Apple FS, Gibler WB, Jesse RL, Warshaw MM, Valdes R. National academy of clinical biochemistry standards of laboratory practice: Recommendations for the use of cardiac markers in coronary artery diseases. *Clinical Chemistry* 1999; 45: 1104–1121.

Wu AHB, Fukushima N, Puskas R, Todd J, Goix P. Development and preliminary clinical validation of a high sensitivity assay for cardiac troponin using a capillary flow (single molecule) fluorescence detector. *Clin Chem* 2006; 52: 2157–2159.

Wu AHB, Jaffe AS. The clinical need for high-sensitivity cardiac troponin assays for acute coronary syndromes and the role for serial testing. *Am Heart J* 2008; 155: 208–214.

Yancy CW, Jessup M, Bozkurt B et al. 2017 ACC/AHA/HFSA Focused update of the 2013 ACCF/AHA Guideline for the management of heart failure. *JACC* 2017. doi:10.1016/j.jacc.2017.04.025.

Zarich S, Bradley K, Seymour J et al. Impact of troponin T determinations on hospital resource utilization and costs in the evaluation of patients with suspected myocardial ischemia. *Am J Cardiol* 2001; 88: 732–736.

8.5 Biomarkers for Psychiatric Disease

An Overview

Roy H. Perlis

Contents

INTRODUCTION

The past decade has witnessed an explosion in studies of biomarkers in psychiatry, paralleling development of high-dimensionality and high-throughput methods for genomic and transcriptomic study. However, to date, the vast majority of reports of biomarkers for psychiatric disorders remain unreplicated, reflecting small sample sizes for discovery, disease heterogeneity, and (most likely) publication bias that precludes publication of failed replication efforts. The notable exception here is genomic studies of disease association, which have begun to yield consistent results, although the application of such results remains a work in progress.

Here, we selectively review efforts to develop psychiatric biomarkers spanning multiple modalities, including genomics, transcriptomics and proteomics, and cellular models. We also more briefly address other biomarker strategies drawn from imaging and mobile health, and consider the emerging role for machine learning. In each case, the goal is to understand broadly speaking progress toward actionable biomarkers, rather than to catalog every result in what is a rapidly moving field. We conclude by discussing strategies for accelerating biomarker discovery and development for neuropsychiatric diseases.

GENOMICS

The pursuit of genomic biomarkers for psychiatric disease has transitioned from a decade of futility to an embarrassment of riches for investigators, but not yet for clinicians or translational researchers. Candidate gene studies examining a slowly widening pool, beginning with monoaminergic genes and later exploring more exotic hypotheses, typically yielded unreplicated findings reflecting the limited understanding of the neurobiology of psychiatric diseases. In light of subsequent studies, it is easy to forget that the architecture of these diseases could have been oligogenic, as early studies assumed—it just proved to be otherwise [1].

The transition to unbiased, genome-wide association studies (GWASs), coupled with an awareness of the need for large cohorts to generate sufficient statistical power to reliably identify associations, has now led to numerous loci being associated with psychiatric disorders. For schizophrenia, a landmark publication described 108 loci associated at a genome-wide threshold [2]. These genomic regions typically did not point at functional variants, and often included either intergenic regions, or regions spanning multiple genes. Indeed, the strongest association identified in that study spanned the major histocompatibility complex locus. Subsequent work demonstrated that the primary signal associated with schizophrenia reflected variation in the complement factor 4A (C4A) locus, the first time a psychiatric risk locus was linked to a functional variant [3]. Still, given the modest magnitude of this risk locus, it seems unlikely to be a useful biomarker by itself.

A second aspect of the initial schizophrenia meta-analysis has proven to be at least as important because it helped introduce analysis of polygenicity. That is, beyond looking at a handful of risk variants, Purcell and colleagues examined even modest associations across thousands of variants, demonstrating that examining many loci improved ability to predict case status [2]. That report suggested that ~3% of variance could be explained by the polygenic risk score; more recent work suggests this number may approach 15%. Another way of looking at polygenic risk is in terms of odds of disease in the greatest risk quartiles or deciles—that is, how much greater is disease risk among individuals with a large common-variant genetic load.

Beyond spawning a host of efforts at human reverse genetics—that is, understanding the phenotypic impact of polygenic loading—the notion of aggregated risk may enable the application of genomics as biomarkers in psychiatry. For example, clinical trials may be analyzed post hoc to examine whether there are differential effects in highly loaded individuals, or even stratified *a priori* to allow examination of such effects [4].

Outside of schizophrenia, efforts to identify common variant associations have proceeded more slowly, primarily because of a lack of availability of large enough cohorts. In 2017, the author and colleagues utilized consumer genomics data with self-reported depression diagnoses, in combination with other major depressive disorder cohorts, to identify the first 15 loci associated with depression in individuals of Northern European ancestry [5]. This list has since been expanded to more than 40, and includes loci containing genes known to be expressed in the brain but no compelling evidence of pathophysiology. Although polygenic risk scores explain less variance in disease risk than schizophrenia, they do provide a parallel opportunity to begin to apply them as biomarkers for clinical investigation.

Investigations of common genetic variation in other psychiatric diseases have also yielded some success, although generally with fewer variants identified to date. Recent progress has been reported for bipolar disorder, PTSD, OCD [6], and several other diseases. Notably, the development of new methods to understand the relationship between these diseases at a genetic level also suggests the extent to which risk is shared [7,8]. As a result, even for psychiatric disorders lacking genome-wide association, or with only a few variants identified, it may be valuable to consider polygenic risk for better-understood diseases such as schizophrenia or MDD.

Efforts also continue to identify rare-variant associations with psychiatric disease, motivated largely by the recognition that such variants—particularly copy number variations—may be more likely to be functional, deleting or duplicating an entire gene, for example. As such, these variants could be more readily interpreted in terms of their impact. A handful have been identified to date, primarily in autism and schizophrenia [9] but also in other childhood-onset disorders including Tourette syndrome [10].

Finally, an emerging area of biomarker investigation seeks to characterize DNA methylation, particularly as a means of trying to integrate genomic and environmental risk [11]. Although efforts are continuing to identify regions of genome preferentially methylated in disease—for example, in small-scale case-control studies [12]—no larger-scale studies or replications have yet been reported.

PHARMACOGENOMICS

The relationship between variation in cytochrome P450 genes and blood levels of psychotropic medications has been understood for two decades, although the relationship between blood levels and clinical response is less clear for most medications [13]. Still, FDA labels for more than 20 medications commonly applied in the treatment of psychiatric illness include information about testing for genetic variation [14], and guidelines exist for the clinical application of CYP450 variation for some antidepressants [15].

Identifying pharmacodynamic variants, rather than pharmacokinetic variants, has proven to be more challenging: meta-analysis of GWASs of antidepressant response failed to identify any loci significant at a traditional genome-wide threshold [16,17]. In bipolar disorder, a single locus has been identified that associates with lithium response [18]. Additional studies have identified associations between rare genomic variation and adverse outcomes with psychotropic treatment. In particular, two human leukocyte antigen (HLA) alleles have been associated with risk for Stevens-Johnson syndrome with carbamazepine [19]. Similarly, rare HLA alleles have been associated with agranulocytosis among clozapine-treated patients [20,21].

Among all biomarker modalities in psychiatry, pharmacogenomics has come closest to clinical application, with multiple small studies investigating pharmacogenomic-guided treatment in mood and anxiety disorders. One typical study among 316 individuals with major depressive disorder did not show significant benefit in sustained response, the primary outcome measure, but did suggest improvement on secondary measures [22]. A challenge in these types of trials likely to impact other biomarker studies in psychiatry has been preserving the blind: although participants and raters are generally blinded, the clinicians themselves cannot be blinded without some means of providing "dummy" or masked reports.

TRANSCRIPTOMICS AND PROTEOMICS

Numerous investigations of psychiatric disease focused on endocrinologic features, generally individual hormones sampled randomly or following a chemical "challenge." Perhaps the most prominent of these was the dexamethasone suppression test (DST), which provides a valuable cautionary tale for biomarker development. Building on the initial finding that (some) individuals with depression exhibited abnormal change in cortisol following dexamethasone infusion as a probe for hypothalamic-pituitary-adrenal axis function, multiple studies sought to establish this test as a clinically useful strategy to diagnose depression, distinguish depression subtypes, or identify optimal treatment [23]. Despite a lack of validation studies, DST was routinely applied in some clinical settings to diagnose and guide treatment choice.

This lack of validation—and in particular, a lack of understanding about how DST results vary across clinical populations, as well as of its specificity—led to substantial disappointment about biomarkers in psychiatry. Ironically, the original finding that some individuals with depression have disrupted HPA axis function has not been disproven. Other endocrinologic challenge paradigms also exhibited intriguing results, such as the fenfluramine challenge, a measure of serotonergic response that was associated with aggression and irritability [24]. Unfortunately, the failure of the DST led to a general move away from endocrine investigation.

Although perhaps not practical as a biomarker, postmortem studies of specific brain regions have suggested transcriptomic abnormalities, primarily in schizophrenia. For example, a study of dorsolateral prefrontal cortex from 258 individuals with schizophrenia and an equivalent number of healthy controls [25] identified nearly 700 differentially expressed genes, although most exhibited modest effects.

CELLULAR MODELS

Intensive study of peripheral blood cells from neuropsychiatric disease patients, generally using smaller-scale approaches that characterized a feature of such cells, failed to identify consistent markers of disease. As one example, red blood cell lithium dynamics were investigated for their relationship to lithium response, similar to other studies of monoamine oxidase activity or platelet binding to tricyclic antidepressants, all with mixed results [26]. Multiple studies also examined lymphocyte function, focusing on the hypothesis that some psychiatric disorders might reflect immune-mediated disease [27].

Investigation of cellular biomarkers was revitalized with the Nobel Prize-winning discovery by Yamanaka and colleagues in 2007 [28] that induced pluripotent stem (iPS) cells could be generated from differentiated cells such as dermal fibroblasts, initially in rodent and then in human. With subsequent progress in manipulating iPS cells to differentiate into neural cells, it became possible for the first time to generate and study neurons from psychiatric patients in sufficient numbers to allow case-control analyses.

A report by Gage and colleagues [29] was the first to compare schizophrenia-derived neurons derived from iPS cells to those derived from healthy control individuals. That study suggested a range of overt abnormalities, including decreased synapse formation and prominent changes in transcriptome. In bipolar disorder, one of the first papers reporting neural models used neurons transdifferentiated from fibroblasts, omitting the iPS generating step [30]. This study used live cellular imaging to suggest differences in neural adhesion among lithium-responsive bipolar patients. Until recently, it also represented one of the largest cellular biomarker studies in psychiatry, examining 18 patient-derived lines. A subsequent study that did use iPS-derived neurons reported mitochondrial dysfunction in neurons from individuals with bipolar disorder, along with electrophysiologic evidence suggesting hyperexcitability in these neurons. Notably, this hyperexcitability could be "rescued" with lithium—but only among the subset of lines from patients reported to be lithium-responsive [31]. Although all of these studies are notable for representing "firsts," none of these results has been convincingly replicated, perhaps unsurprising given the modest sample sizes and range of potential phenotypes to be explored. As the prior discussion of candidate gene studies in psychiatry notes, the risk for type I error in all of these small samples is substantial; the highly polygenic nature of psychiatric disease, and the fact that common psychiatric disorders are associated with at most subtle changes in gross morphology or histopathology, makes it somewhat less plausible that cellular phenotypes could track so closely with disease.

Still, with ongoing work enabling iPS generation from blood rather than fibroblasts, and more efficient generation of more specific neuron types from iPS, the feasibility of cellular models as biomarkers in clinical investigation has increased. The notion of parallel *in vitro*/*in vivo* studies—that is, modeling drug response in cells in parallel with patients—may be an appealing strategy for future proof-of-concept studies, especially with increased recognition of the importance of early demonstration of target engagement in drug discovery [32].

Importantly, neural cells other than neurons may also play a role in neuropsychiatric disease, and methods to generate astrocytes and microglial cells are also in active development. The author's lab reported a method to generate patient-derived microglia-like cells from peripheral blood, rather than iPS [33]; multiple methods for iPS-derived microglia-like cells have also been reported [34]. Similar progress with astrocytes offers increasingly efficient means of generating these cells from human iPS precursors [35].

With the broadening range of cell types available, investigation of cellular biomarkers has moved beyond cell-autonomous (i.e., seen in individual cell) phenotypes to allow investigation of coculture models. As an example, the author and colleagues described a model of synaptic pruning employing microglia-like cells and processed synaptic structures called synaptosomes [33], amenable to incorporation in high-throughput screens or clinical investigations.

In addition, as a model of the developing brain, 3 dimensional organoids can be characterized, and may capture features of neurodevelopment requiring multiple cell types to be grown in proximity over time [36]. Further innovation will be required to allow this method to be applied at scale as a potential disease biomarker; for example, neurospheroids grown in array format may be more amenable to higher-throughput investigation [37].

IMAGING AND ELECTROENCEPHALOGRAPHY

The availability of high-resolution brain imaging via magnetic resonance imaging has yielded a wealth of data regarding group differences between psychiatric illness and healthy control participants, beginning with structural (morphometric) data and progressing to functional imaging of blood flow differences, spectroscopy to look at certain metabolites, and specialized means of imaging white matter tract integrity, cortical thickness, and other measures.

In general, the cost of obtaining such images, and the modest between-group differences, has limited their application as biomarkers. A review of structural imaging in first-episode psychosis found an absence of consistent results [38]. Similarly, meta-analysis of 537 fMRI studies suggested that, whereas region-specific effects were observed, they did not appear to be disease-specific [39].

Similar to imaging studies, electroencephalography (EEG) studies have yielded promising results with insufficiently large group effects to warrant application as biomarkers. A recent review suggested oscillatory effects associated with bipolar disorder [40]. Another promising direction of work has examined disruption of sleep spindle activity in schizophrenia, positing that these changes may be associated with cognitive dysfunction [41].

Because EEG data are more straightforward to gather in the outpatient setting, EEGs (and particularly reduced-lead EEG arrays) have also been incorporated into efforts to achieve more personalized antidepressant prescribing. A particularly notable element of these studies has been the ability to gather pre/post treatment measures—that is, to move toward using EEG as a marker of treatment response. A number of small studies suggest the potential utility of EEG-informed prescribing [42,43].

PASSIVE MEASURES AS BIOMARKERS

As the cost of passive monitors like accelerometers and GPS devices has fallen, interest has increased in use of such measures as biomarkers, particularly to complement or supplement traditional rating scale-based measures. For example, rather than asking about psychomotor slowing or agitation, wearable accelerometers allow direct measurement of such symptoms; rather than notoriously unreliable self-reports of sleep duration, accelerometers can help characterize such sleep. Importantly, even where these measures are inferior to gold standards such as polysomnography, their low cost and portability makes them practical for use in far larger studies. Moreover, they allow for the first time the collection of massive amounts of time-series data, rather than standard weekly or monthly measures.

An underappreciated aspect of passive measures is the amount of computation required to render them in an analyzable form, which is substantially greater than, for example, array-based genomics and transcriptomics. There are challenges in handling missing data and imputing phenotypes, for example, and development of such methods is still in progress.

An early example of passive measurement used a wearable patch with an accelerometer among 28 individuals with schizophrenia or bipolar disorder [44], yielding data on sleep duration and activity. That study, focused on adherence, found generally high acceptability but faced numerous technical challenges (e.g., in uploading data), underscoring the extent to which emerging technology often faces challenges in implementation in clinical investigations.

More recently, smart phones have provided an opportunity to take advantage of built-in sensors in a more scalable fashion. A typical study provided a smartphone app to 17 individuals with schizophrenia, and followed them for up to 3 months [45], detecting changes in behavior prior to symptomatic relapse. With standardized platforms for health data collection such as Apple's HealthKit/ResearchKit, it is likely that passive measures will increasingly be employed as biomarkers for quantifying a range of behaviors. Another frontier with this form of data collection includes processing of affective data from voice and vision, for example, as markers of depression [46].

NEW DIRECTIONS: COMPUTATIONAL APPROACHES TO PSYCHIATRIC BIOMARKERS

What is apparent for nearly all the modalities considered in this section is that the detectable effects of disease or treatment response are quite modest: when they replicate at all, they likely exemplify winner's curse, in which the initial report overstates the magnitude of effect [47]. The use of polygenic risk scores illustrates one potential response to this problem, aggregating over very large numbers of modest effects to approach clinically actionable effect sizes (e.g., explaining 15% of schizophrenia risk). More generally, machine learning may provide a systematic means of integrating small effects, and even effects drawn from different modalities that interact in a nonlinear fashion. The challenge in applying these complex models as biomarkers is a lack of interpretability: that is, they may yield a biomarker with high discrimination between disease and health, or response and nonresponse, but the rationale behind this prediction may not be readily understandable. Methods that balance discrimination and interpretability are in active development, and may be particularly relevant for derivation of these computed biomarkers [48].

SUMMARY

Efforts to identify biomarkers in psychiatric illness have left few stones unturned, with endocrinologic and blood cell studies progressing to imaging and electroencephalography, genomics, and transcriptomics. Multiple clinical trials have begun to investigate the genomics of psychotropic treatment response; studies of quantitative EEG to optimize antidepressant treatment are also ongoing. Emerging work in cellular models has yielded promising but unreplicated results, and passive measurement of behaviors like activity and sleep are also becoming more common. Overall, the lesson from psychiatric biomarker discovery so far has been consistent with medical genomics more generally: disease effects are likely to be quite small, requiring very large cohorts for discovery and replication, and methods to aggregate over multiple small differences. Still, after many decades of futility, as these challenges are overcome, biomarkers are poised for application in psychiatric clinical and translational investigation.

REFERENCES

1. Committee PGCC, Cichon S, Craddock N, Daly M, Faraone SV, Gejman PV et al. Genomewide association studies: History, rationale, and prospects for psychiatric disorders. *Am J Psychiatry.* 2009;166(5):540–556.
2. Purcell SM, Wray NR, Stone JL, Visscher PM, O'Donovan MC, Sullivan PF, et al. Common polygenic variation contributes to risk of schizophrenia and bipolar disorder. *Nature.* 2009;460(7256):748–752.
3. Sekar A, Bialas AR, de Rivera H, Davis A, Hammond TR, Kamitaki N et al. Schizophrenia risk from complex variation of complement component 4. *Nature.* 2016;530(7589):177–183.
4. Perlis R. Translating biomarkers to clinical practice. *Mol Psychiatry.* 2011;16(11):1076–1087.
5. Hyde CL, Nagle MW, Tian C, Chen X, Paciga SA, Wendland JR et al. Identification of 15 genetic loci associated with risk of major depression in individuals of European descent. *Nat Genet.* 2016;48(9):1031–1036.
6. International Obsessive Compulsive Disorder Foundation Genetics Collaborative (IOCDF-GC) and OCD Collaborative Genetics Association Studies (OCGAS). Revealing the complex genetic architecture of obsessive-compulsive disorder using meta-analysis. *Mol Psychiatry.* 2018; 23(5):1181–1188.
7. Cross-Disorder Group of the Psychiatric Genomics Consortium, Lee SH, Ripke S, Neale BM, Faraone SV, Purcell SM et al. Genetic relationship between five psychiatric disorders estimated from genome-wide SNPs. *Nat Genet.* 2013;45(9):984–994.
8. Bulik-Sullivan B, Finucane HK, Anttila V, Gusev A, Day FR, Loh PR et al. An atlas of genetic correlations across human diseases and traits. *Nat Genet.* 2015;47(11):1236–1241.
9. International Schizophrenia Consortium. Rare chromosomal deletions and duplications increase risk of schizophrenia. *Nature.* 2008;455(7210):237–241.
10. Huang AY, Yu D, Davis LK, Sul JH, Tsetsos F, Ramensky V et al. Rare copy number variants in NRXN1 and CNTN6 increase risk for tourette syndrome. *Neuron.* 2017;94(6):1101–1111.e7.
11. Burns SB, Szyszkowicz JK, Luheshi GN, Lutz P-E, Turecki G. Plasticity of the epigenome during early-life stress. *Semin Cell Dev Biol.* 2018;77:115–132.
12. Shimada M, Otowa T, Miyagawa T, Umekage T, Kawamura Y, Bundo M et al. An epigenome-wide methylation study of healthy individuals with or without depressive symptoms. *J Hum Genet.* 2018;63(3):319–326.
13. Perlis RH. Pharmacogenomic testing and personalized treatment of depression. *Clin Chem.* 2014;60(1):53–59.
14. Administration FD. Table of Pharmacogenomic Biomarkers in Drug Labeling [Internet]. November 11, 2018. Available from: http://www.fda.gov/Drugs/ScienceResearch/ResearchAreas/Pharmacogenetics/ucm083378.htm
15. Dean L. Imipramine Therapy and CYP2D6 and CYP2C19 Genotype. In: Pratt V, McLeod H, Dean L, Malheiro A, Rubinstein W, editors. *Medical Genetics Summaries [Internet].* Bethesda, MD: National Center for Biotechnology Information (US); 2012 [cited 2018 April 16]. Available from: http://www.ncbi.nlm.nih.gov/books/NBK425164/
16. GENDEP Investigators, MARS Investigators, STAR*D Investigators. Common genetic variation and antidepressant efficacy in major depressive disorder: A meta-analysis of three genome-wide pharmacogenetic studies. *Am J Psychiatry.* 2013;170(2):207–217.
17. Fabbri C, Tansey KE, Perlis RH, Hauser J, Henigsberg N, Maier W et al. New insights into the pharmacogenomics of antidepressant response from the GENDEP and STAR*D studies: Rare variant analysis and high-density imputation. *Pharmacogenomics J.* 2017;18:413–421.
18. Hou L, Heilbronner U, Degenhardt F, Adli M, Akiyama K, Akula N et al. Genetic variants associated with response to lithium treatment in bipolar disorder: A genome-wide association study. *Lancet Lond Engl.* 2016;387(10023):1085–1093.
19. Phillips EJ, Sukasem C, Whirl-Carrillo M, Müller DJ, Dunnenberger HM, Chantratita W et al. Clinical pharmacogenetics implementation consortium guideline for HLA genotype and use of carbamazepine and oxcarbazepine: 2017 update. *Clin Pharmacol Ther.* 2018;103(4):574–581.
20. Goldstein JI, Jarskog LF, Hilliard C, Alfirevic A, Duncan L, Fourches D et al. Clozapine-induced agranulocytosis is associated with rare HLA-DQB1 and HLA-B alleles. *Nat Commun.* 2014;5:4757.
21. Legge SE, Hamshere ML, Ripke S, Pardinas AF, Goldstein JI, Rees E et al. Genome-wide common and rare variant analysis provides novel insights into clozapine-associated neutropenia. *Mol Psychiatry.* 2017;22(10):1502–1508.
22. Pérez V, Salavert A, Espadaler J, Tuson M, Saiz-Ruiz J, Sáez-Navarro C et al. Efficacy of prospective pharmacogenetic testing in the treatment of major depressive disorder: Results of a randomized, double-blind clinical trial. *BMC Psychiatry.* 2017 14;17(1):250.
23. Nierenberg AA, Feinstein AR. How to evaluate a diagnostic marker test. Lessons from the rise and fall of dexamethasone suppression test. *JAMA.* 1988;259(11):1699–1702.
24. Fava M, Vuolo RD, Wright EC, Nierenberg AA, Alpert JE, Rosenbaum JF. Fenfluramine challenge in unipolar depression with and without anger attacks. *Psychiatry Res.* 2000;94(1):9–18.

25. Fromer M, Roussos P, Sieberts SK, Johnson JS, Kavanagh DH, Perumal TM et al. Gene expression elucidates functional impact of polygenic risk for schizophrenia. *Nat Neurosci*. 2016;19(11):1442–1453.
26. Muscettola G, Di Lauro A, Giannini CP. Blood cells as biological trait markers in affective disorders. *J Psychiatr Res*. 1984;18(4):447–456.
27. Barbosa IG, Rocha NP, Assis F, Vieira ÉLM, Soares JC, Bauer ME et al. Monocyte and lymphocyte activation in bipolar disorder: A new piece in the puzzle of immune dysfunction in mood disorders. *Int J Neuropsychopharmacol*. 2014;18(1) pyu 021.
28. Takahashi K, Tanabe K, Ohnuki M, Narita M, Ichisaka T, Tomoda K et al. Induction of pluripotent stem cells from adult human fibroblasts by defined factors. *Cell*. 2007;131(5):861–872.
29. Brennand KJ, Simone A, Jou J, Gelboin-Burkhart C, Tran N, Sangar S et al. Modelling schizophrenia using human induced pluripotent stem cells. *Nature*. 2011;473(7346):221–225.
30. Wang JL, Shamah SM, Sun AX, Waldman ID, Haggarty SJ, Perlis RH. Label-free, live optical imaging of reprogrammed bipolar disorder patient-derived cells reveals a functional correlate of lithium responsiveness. *Transl Psychiatry*. 2014;4:e428.
31. Mertens J, Wang Q-W, Kim Y, Yu DX, Pham S, Yang B et al. Differential responses to lithium in hyperexcitable neurons from patients with bipolar disorder. *Nature*. 2015;527(7576):95–99.
32. Haggarty SJ, Silva MC, Cross A, Brandon NJ, Perlis RH. Advancing drug discovery for neuropsychiatric disorders using patient-specific stem cell models. *Mol Cell Neurosci*. 2016;73:104–115.
33. Sellgren C, Sheridan S, Gracias J, Xuan D, Fu T, Perlis R. Patient-specific models of microglia-mediated engulfment of synapses and neural progenitors. *Mol Psychiatry*. 2017;22:170–177.
34. Muffat J, Li Y, Yuan B, Mitalipova M, Omer A, Corcoran S et al. Efficient derivation of microglia-like cells from human pluripotent stem cells. *Nat Med*. 2016;22(11):1358–1367.
35. Lundin A, Delsing L, Clausen M, Ricchiuto P, Sanchez J, Sabirsh A et al. Human iPS-Derived astroglia from a stable neural precursor state show improved functionality compared with conventional astrocytic models. *Stem Cell Rep*. 2018;10(3):1030–1045.
36. Arlotta P. Organoids required! A new path to understanding human brain development and disease. *Nat Methods*. 2018;15(1):27–29.
37. Jorfi M, D'Avanzo C, Tanzi RE, Kim DY, Irimia D. Human neurospheroid arrays for *in vitro* studies of Alzheimer's Disease. *Sci Rep*. 2018;8(1):2450.
38. Fusar-Poli P, Meyer-Lindenberg A. Forty years of structural imaging in psychosis: Promises and truth. *Acta Psychiatr Scand*. 2016;134(3):207–224.
39. Sprooten E, Rasgon A, Goodman M, Carlin A, Leibu E, Lee WH et al. Addressing reverse inference in psychiatric neuroimaging: Meta-analyses of task-related brain activation in common mental disorders. *Hum Brain Mapp*. 2017;38(4):1846–1864.
40. Maggioni E, Bianchi AM, Altamura AC, Soares JC, Brambilla P. The putative role of neuronal network synchronization as a potential biomarker for bipolar disorder: A review of EEG studies. *J Affect Disord*. 2017;212:167–170.
41. Manoach DS, Pan JQ, Purcell SM, Stickgold R. Reduced sleep spindles in Schizophrenia: A treatable endophenotype that links risk genes to impaired cognition? *Biol Psychiatry*. 2016;80(8):599–608.
42. Iosifescu DV, Neborsky RJ, Valuck RJ. The use of the psychiatric electroencephalography evaluation registry (PEER) to personalize pharmacotherapy. *Neuropsychiatr Dis Treat*. 2016;12:2131–2142.
43. Cook IA, Hunter AM, Gilmer WS, Iosifescu DV, Zisook S, Burgoyne KS et al. Quantitative electroencephalogram biomarkers for predicting likelihood and speed of achieving sustained remission in major depression: A report from the biomarkers for rapid identification of treatment effectiveness in major depression (BRITE-MD) trial. *J Clin Psychiatry*. 2013;74(1):51–56.
44. Kane JM, Perlis RH, DiCarlo LA, Au-Yeung K, Duong J, Petrides G. First experience with a wireless system incorporating physiologic assessments and direct confirmation of digital tablet ingestions in ambulatory patients with schizophrenia or bipolar disorder. *J Clin Psychiatry*. 2013;74(6):e533–e540.
45. Barnett I, Torous J, Staples P, Sandoval L, Keshavan M, Onnela J-P. Relapse prediction in schizophrenia through digital phenotyping: A pilot study. *Neuropsychopharmacology*. 2018; 43(8):1660–1666.
46. Taguchi T, Tachikawa H, Nemoto K, Suzuki M, Nagano T, Tachibana R et al. Major depressive disorder discrimination using vocal acoustic features. *J Affect Disord*. 2018;225:214–220.
47. Young NS, Ioannidis JPA, Al-Ubaydli O. Why current publication practices may distort science. *PLoS Med*. 2008;5(10):e201.
48. Hughes MC, Elibol HM, McCoy T, Perlis R, Doshi-Velez F. Supervised topic models for clinical interpretability. ArXiv Prepr ArXiv161201678. 2016.

Progressing Towards Precision Psychiatry

Current Challenges in Applying Biomarkers in Psychiatry

8.6

Victor L. Kallen and Brisa S. Fernandes

Contents

After decades of fundamental research into the neurobiological and genetic origins of psychiatric disorders, the field has progressed into a thrilling new era, being driven by the concept of precision psychiatry (see Figure 8.6.1). Although precision medicine takes interindividual differences in genetic, environmental, and life style factors into account (e.g., in prevention, diagnostics, and treatment: National Research Council, 2011), psychiatry seems not to have harvested the potential of advanced, personalized, diagnostics and therapeutic technologies to the extent other fields of medicine have done (Fernandes et al., 2017). Due to an immense body of scientific findings, the number of *potential* biomarkers in psychiatry is nevertheless considerable, though diverse, divergent, and unstructured. Ideally, a biomarker should be present before the onset of symptoms; accurately discriminate non-risk individuals from individuals at risk (sensitivity); and should be specific for a well-defined disorder (Yerys & Pennington, 2011). In psychiatry, most candidates unfortunately fail however, generally on the last criterion (disorder specificity). This implies that, notwithstanding considerable efforts, nearly 40 years of Diagnostic and Statistical Manual of Mental Disorders (DSMs) seem to have failed to convincingly connect mental disorders to biology (e.g., Yee et al., 2015). As a result, in the field of psychiatry remarkably few claimed "biomarkers" have matured into more or less standardized methods supporting clinical decision-making, like diagnostics[1] and outcome monitoring. A conclusion that arguably inflates the reach of the biomarker concept, at least in the context of mental health because "biomarker" does not seem to have a consistent meaning beyond "correlate," although it is often used as if it would carry more significance than that (Miller et al., 2016).

In psychiatry, multiple classification methods are currently employed (e.g., ICD-11; DSM-5). And although our insights in the epigenetic and biobehavioral phenomena underlying psychopathology evolved considerably, these methodologies typically classify psychiatric disorders by means of clustering phenotypical behavioral and/or (neuro) psychological observations and assessments. For example, in autism behavioral observations, neuropsychological assessments (e.g., the well-known Rey's complex Figure test, see Figure 8.6.2), and standardized parent reports still guide diagnostics and clinical practice (McPartland, 2017). The only exceptions to this are mental disorders secondary

[1] Other than for differential and /or circumstantial purposes.

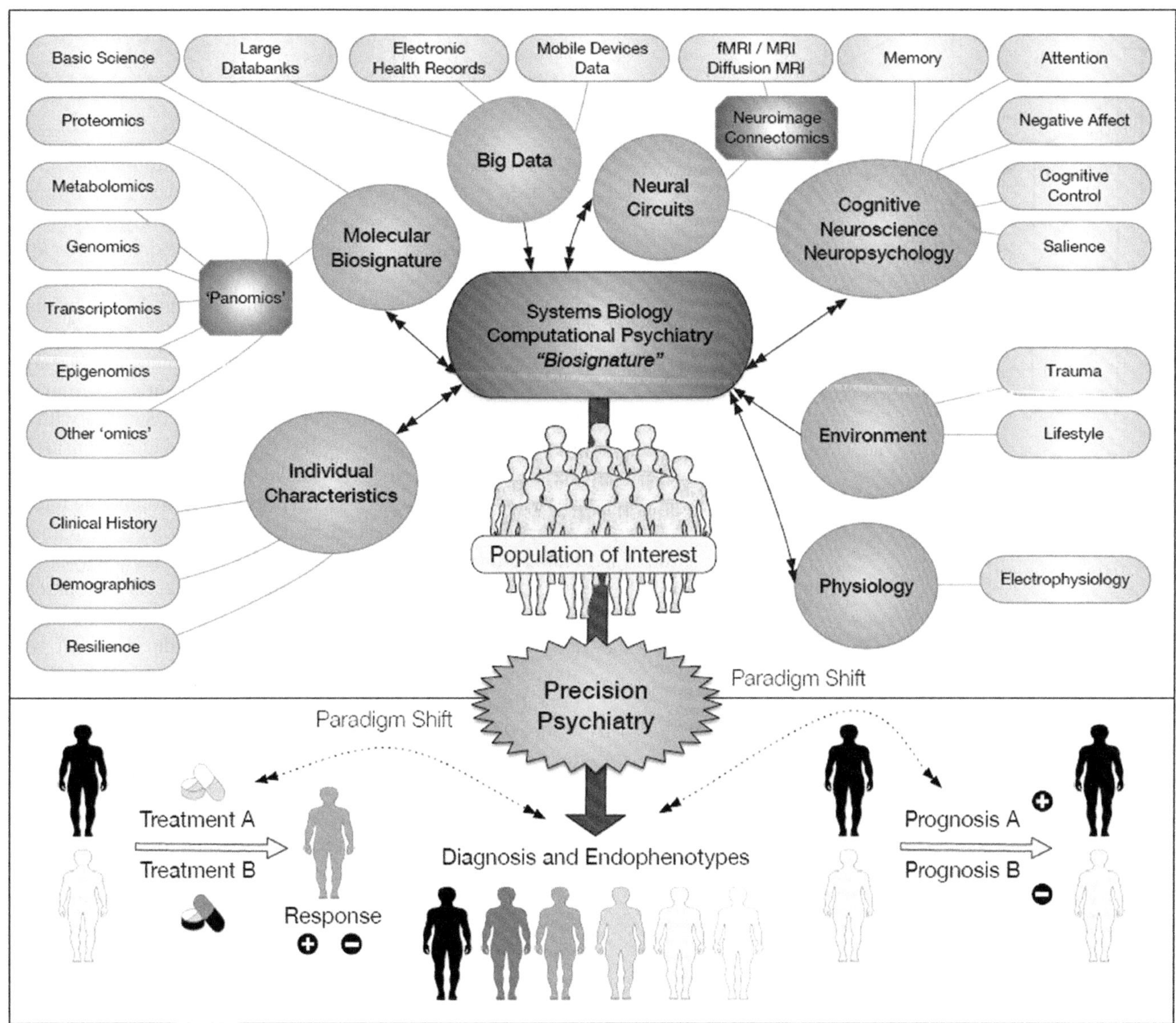

FIGURE 8.6.1 Within precision psychiatry input from diverse domains ("omics"; psychophysiology; others) are incorporated in a biosignature (an integrated set of biomarkers) using systems biology and advanced computational methods. This is presumed to provide better, individualized diagnostics, endophenotypes, classification and prognoses. fMRI, functional MRI. (From Fernandes B.S. et al., *BMC Med.*, 2017, doi: 10.1186/s12916-017-0849-x.)

to organic causes and/or infections (e.g., dementia, delirium, and some stages in syphilis), which are, however, generally delegated toward other medical disciplines (typically neurology).

TRANSDIAGNOSTIC CONSTRUCTS

Interestingly, however, neuroscientific findings have provided convincing arguments for reconsideration and reclassification of some previously well specified mental disorders. For example, the differentiation of DSM-4 anxiety disorders into anxiety disorders; obsessive compulsive disorder(s) (OCD); and post-traumatic stress disorder (PTSD), trauma and stress related disorders in the DSM-5, seems to make sense from a neurobiological perspective. Because OCD primarily reflects dysfunctions in frontal–basal ganglia functioning and anatomy, with other anxiety disorders being associated with disturbances in the amygdala and hippocampal functioning (Trimble & George, 2010).

In line with these developments, ever more scientific evidence points toward the direction of fundamentally more dimensional parameters underlying psychopathology, so-called *Transdiagnostic Constructs*, compared with more conventional

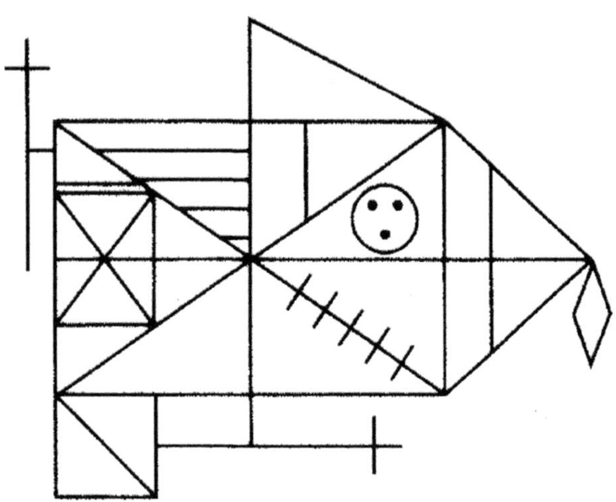

FIGURE 8.6.2 The Rey's complex figure task, an example of neuropsychological assessment applied in, for example, diagnosing autism spectrum disorders.

Transdiagnostic constructs are neurocognitive constructs that cut across different disorders: they express themselves dimensionally and only contribute to the development of psychopathologies when this expression is maladaptive.

Consequently:

1. Delineate a specific psychological process (e.g., working memory, response inhibition, other)
2. Identify the underlying neural system.
3. Bring to bear both animal and human studies
4. Develop paradigms to isolate specific processes
5. Utilize tools that measure the underlying neural and cognitive system

Baker, 2017

and categorical classification methods underlying ICD-11 and DSM-5. Not least because many syndromes present themselves quite heterogeneously (e.g., in symptom progression, severity, and variety), and the generally applied conventional methods do not necessarily acknowledge significant interindividual differences. Typical examples of transdiagnostic constructs are impulsivity; specific executive functions such as inhibitory control & attention; and stress sensitivity. Interacting with other, typically exogenous, factors, deficits in these constructs are associated with quite diverse behavioral and/or psychopathological phenomena. This seems to apply to the molecular and neurotransmitter domain as well, as, for example, dopaminergic deficits are associated with reward feedback and impulse control, and are in this way relevant for dementia, addiction, ADHD, among other disorders.

NEUROIMAGING AND FUNCTIONAL CORTICAL MARKERS

A broad and immensely diverse body of neuroimaging findings has been published over the last three decades. This fortunately provided general outlines related to specific constructs or disorders in an otherwise confusing and inimitable domain that utterly lacks specificity toward diagnostic criteria. However, with a keen eye on discriminative potential, some relevant developments can nevertheless be identified.

Resting state functional MRI (fMRI) data from the inferior frontal gyrus and insula appeared significantly discriminative between ADHD patients and non-affected controls, with data from the frontal and cerebellar regions providing considerable accuracy (~80%) when advanced mathematical analyses were applied (Uddin et al., 2017). In relation to neuroimaging findings, the application of advanced, by default individualized, mathematical and modeling methods seems to be the general way ahead to develop from purely scientific findings based on correlations to applicable and sufficiently discriminative methods relevant for precision psychiatry. Like in autism spectrum disorders where some success has been reported in discriminating patients from non-affected peers, claiming a discriminatory accuracy ranging from 70% up to 90%, typically based on fMRI data, once again modeled using sophisticated mathematical methods ("Machine learning": Guo et al., 2017; Uddin et al., 2017). Hilbert et al. (2017) reported a 90% accuracy to identify individuals suffering from depression or generalized anxiety disorder (compared with healthy subjects) by applying support vector machine algorithms on multifactorial data including self-report questionnaires, cortisol release and MRI data. With a 67% accuracy in classifying the correct disorder within the patient population. Such accuracies are not met in domains where advanced modeling is not applied. As, for example, in relation to conduct and associated disorders (e.g., antisocial personality disorders), where imaging studies have shown deviances in the functional connectivity within and among specific loci in the amygdala, cingulate cortex, prefrontal and limbic structures (e.g., Zhou et al., 2016). However, these findings are typically associated with diverse psychopathic and antisocial traits (like callous/unemotional; Zhou et al., 2016) and are consequently not specific for any of these disorders per se.

Additionally, particularly in mood disorders, approaches based on electroencephalography (EEG) derived parameters have a long history. Event-related potentials (ERP) and error-related negativity (ERN) being in this way established as potent transdiagnostic markers. With an enhanced ERN being typically associated with anxiety, OCD, and neuroticism, and a reduced ERN with a lack of behavioral inhibition, and consequently ADHD, substance abuse, conduct disorders, psychopathy and schizophrenia (Olvet and Hajcak, 2008; Riesel et al., 2017). In combination with other peripheral parameters (such as the startle reflex) such EEG derived parameters seem to effectively discriminate between, for example, phobia, nonphobic anxiety, and depression (Vaidyanathan, et al., 2012).

NEURODEVELOPMENTAL DISORDERS: SCHIZOPHRENIA, PSYCHOSIS AND DEMENTIA

Such findings strongly suggest that ERN related findings are transdiagnostic by nature and may interact with other parameters to provide better accuracy. An interesting finding in this respect is reported by Demro et al. (2017) and Nagai et al. (2017) who report the combination of lowered ERN with higher levels of plasma glutamate in first episode psychosis, a finding that might indicate the prodromal stages of a psychotic disorder.

To capture the progressive neurodegenerative development toward dementia, most prominently Alzheimer's disorder, two potential biomarker complexes have been identified. Being brain amyloid accumulation as imaged by means of positron emission tomography (PET) scans; and neurodegenerative markers such as tau protein increases in the cerebrospinal fluid (CSF), or regional atrophy found using MRI imaging (Morbelli et al., 2017). Additionally, dopaminergic targets are being investigated with interest (Thomas et al., 2017). The latter obviously running into transdiagnostic challenges with syndromes such as schizophrenia and ADHD. Finally, some preliminary findings suggest that models based on resting state EEG connectivity might have a superior classification capability (applying *support vector machine* scripts to accurately discriminate prodromal stages, established Alzheimer's disease and healthy controls) on and above "classical" neuropsychological assessments (Dottori et al., 2017).

IDENTIFIED EPIGENETIC FACTORS

In addition to the above described neuroimaging findings, some studies report an increase in microglial activity in postmortem brain samples in individuals suffering from schizophrenia (Trepanier et al., 2016), selective dysregulation of the endocannabinoid system in psychosis, and CNR1 DNA methylation levels as a potential biomarker for schizophrenia (Addario et al., 2017). Additional targets for schizophrenia might be presented by single nucleotide polymorphisms (SNPs) represented by serum proteins (e.g., chromogranin-A, cystatin-C), though these findings seem to vary with diagnosis and comorbidities (Chan et al., 2017).

Although such findings appear very promising, only very few genetic predispositions can, however, be directly linked to specific mental disorders. As most genetic relations with specific psychopathologies seem to be strongly mediated by environmental factors, typically being stressors of a psychological or physiological nature with a strong developmental character: overtime potentially contributing to the pathogenesis of psychiatric disease (Issler & Chen, 2015, see Figure 8.6.3).

And because epigenetic effects trigger a wide variety of yet indecipherable "downstream" effects, typically related to transdiagnostic constructs such as fear extinction or modified corticosteroid receptor gene expression, their merit as applicable (independent) biomarkers, beyond their undisputed scientific relevance, has yet to be established. Additionally to that

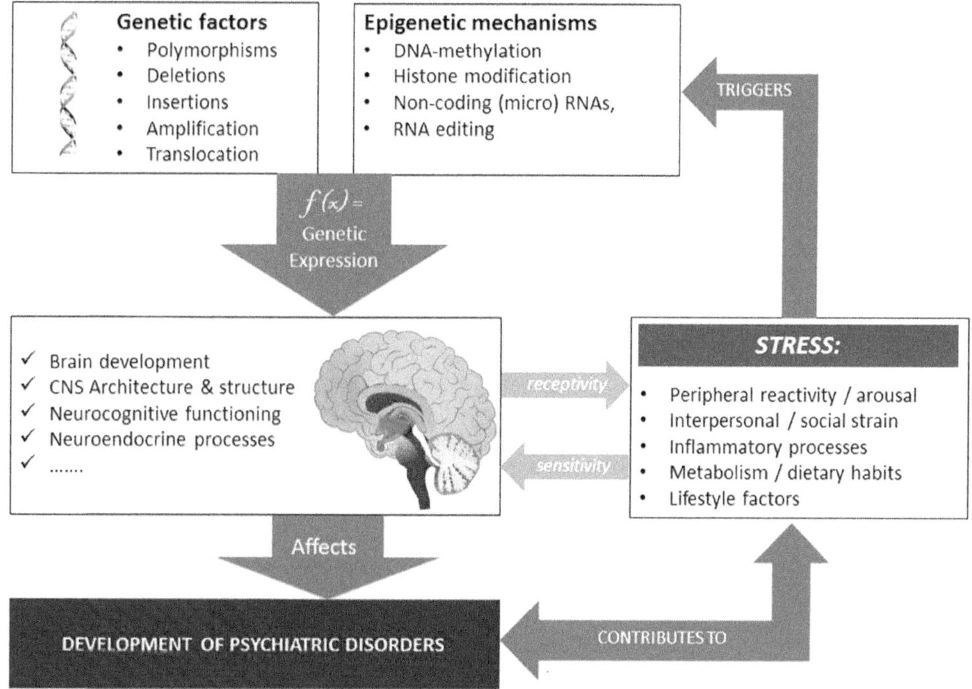

FIGURE 8.6.3 The reciprocal nature of environmental factors ("stress") and genetic predispositions on the etiology of psychiatric disease. With epigenetic processes mediating the impact of environmental stressors on an organism. (Based on Issler, O. and Chen, A. *Nat. Rev. Neurosci.*, 16, 201–212, 2015.)

point, the necessary analytics generally require sophisticated clinical infrastructure, expertise and routine, which are consequently expensive and time consuming, and nevertheless usually error prone (Issler & Chen, 2015). Consequently, presently epigenetic factors can at best be only indirectly linked to specific disorders.

INFLAMMATORY, OXIDATIVE, NEUROENDOCRINE AND NEUROTROPHIC MARKERS OF PSYCHIATRIC DISEASE

Although some research has been done on the potential of inflammatory and oxidative biomarkers in psychiatry, disorder specific markers have not yet been identified. There is evidence suggesting a specific immune-inflammatory biomarker pattern differentiating schizophrenia from bipolar disorders and unaffected individuals (Garcia-Alvarez et al., 2018; Goldsmith et al., 2016). In addition, cytokine levels in patients suffering from major depression seem disturbed, being expressed in increases in interleukin-1 (IL-1), 6 (Il-6), and tumor necrosis factor-α (TNFα) (Dowlati et al., 2010; Goldsmith et al., 2016; Köhler et al., 2017). With increased C-reactive protein (CRP) levels being reported in depression, bipolar disorder, and schizophrenia (Fernandes et al., 2016a, 2016b; Valkanova et al., 2013). In OCD, significantly higher levels of malondialdehyde and lower levels of plasma catalase, superoxide dismutase and glutathione peroxidase are reported (Shrivastava et al., 2017). With the neuropeptide oxytocin being associated with pro/antisocial behavior, and thus being pinpointed as potential biomarker for conduct disorders and psychopathy on one hand, and anxiety disorders on the other. Although additional evidence shows that potential relations seem to be best explained by its influence on traits like callous/unemotional or (the absence of) prosocial behaviors (e.g., Levy et al., 2015). Finally, brain-derived neurotrophic factor (BDNF) has been suggested to be an unspecific state biomarker in bipolar disorder and depression (Fernandes et al., 2009; Molendijk et al., 2014). During acute episodes it is decreased what seems to relate to the severity of the manic and depressive symptoms. However, with reported BDNF alternations in schizophrenia as well, its application as a diagnostic biomarker in any of these disorders seems limited (Fernandes et al., 2014).

Consequently, it appears that the variance of these and other suggested "biomarkers" (such as corticosteroids), seems robust, though rather similar among disorders what renders them less useful as biomarkers for any particular disorder (Pinto et al., 2017). A clear example being dysregulations of many hormones (e.g., glucocorticosteroids, again oxytocin, testosterone, and vasopressin) that have been associated with stress regulation in a social, or even hostile, context, and which reactivity can be mediated by social status, self-esteem, and territorial strain.

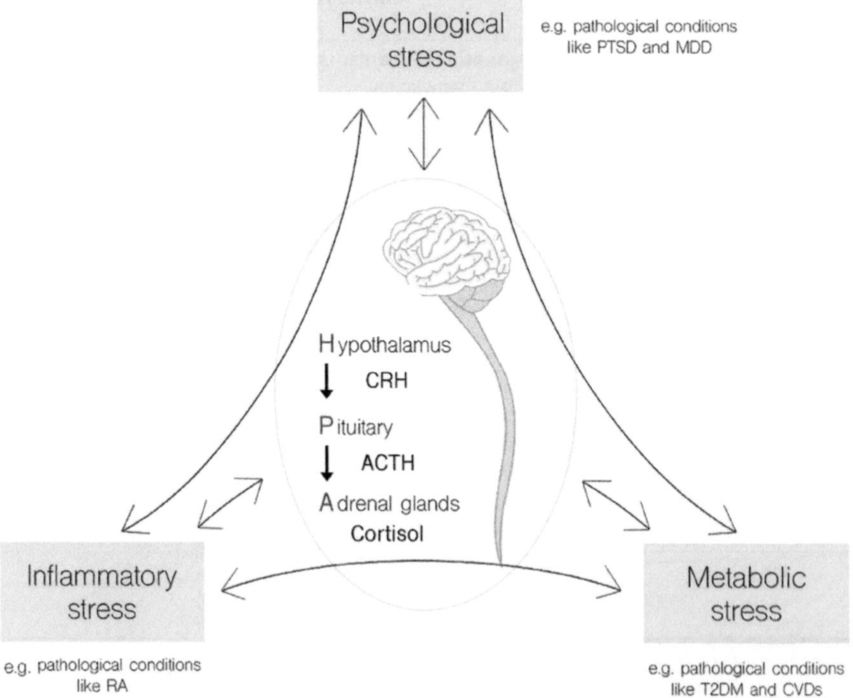

FIGURE 8.6.4 The central role of glucocorticosteroids in stress and psychopathology, inflammatory dysfunctions, and metabolic syndromes. ACTH, adrenocorticotropic hormone; CRH, corticotrophin-releasing hormone; CVD, cardiovascular disease; MDD, major depressive disorder; PTSD, post-traumatic stress disorder; RA, rheumatoid arthritis; T2DM, type 2 diabetes mellitus. (From Bots, S., Stress induced metabolic outcomes: Review on the interplay between psychological stress, metabolic outcomes and altered inflammatory processes. *Master Thesis*, University of Amsterdam, Amsterdam, the Netherlands, 2016.)

All of these factors are associated with a wide variety of disorders, ranging from anxiety to conduct disorders, and from social peculiarities in the autistic or schizotypal domain, to coping deficiencies related to eating disorders. In at least the latter case, hormonal disturbances clearly interact with the metabolic and inflammatory functions of the affected individual (Figure 8.6.4) (Manenschijn et al., 2009). In addition, all of these hormonal disturbances interact with specific receptor complexes, an interaction that appears to be highly sensitive for epigenetic factors (de Kloet et al., 2009, see Figure 8.6.3). Nevertheless, (non-)reactivity challenges are regularly used to assess potential functional disturbances in the hypothalamic-pituitary-adrenal gland (HPA)–axis, which might be associated with psychiatric disorders such as anxiety, PTSD, depression, and conduct disorders. Standardized examples are the generally applied *Cortisol Awakening Response* in cortisol (Stalder et al., 2016); the Dexamethasone Suppression Test (DST: de Kloet et al., 2007); and the Trier Social Stress Test (TSST: Kirschbaum et al., 1993). Cortisol responses to the latter, for example, being associated with anxiety (hypervigilant) and antisocial and/or sensation seeking traits (hypovigilant).

Finally, corticosteroid assessments from hair seem to produce a fairly reliable retrospective marker for experienced stress and/or strain during the previous weeks to months, contrary to established corticoid sources (saliva, blood, and urine) that typically provide momentary or state information (Kirschbaum et al., 2009). Heightened hair cortisol concentrations have been related to depression, bipolar disorder, and specific anxiety disorders. However, recent research suggests that such findings might be best explained by their link with constructs like anhedonia, negative affect and/or arousal (Hillbrandt et al., 2017), whereas meta-analytical findings explicitly report no consistent associations with mood disorders or self-reports of perceived stress, though primarily with more organic "allostatic" markers of stress (Wester and van Rossum, 2015; Stalder et al., 2017).

THE AUTONOMIC NERVOUS SYSTEM (ANS)

Apart from disturbances in the HPA-axis in relation to stress related disorders, subtle deviances in the (non-)reactivity to stress of the autonomic nervous system (ANS) have been pinpointed as potential biomarker for disorders like anxiety, depression, and trauma. A lack of appropriate parasympathetic tone appears to be symptomatic for anxiety related outcomes, even early in development (Beauchaine, 2015).

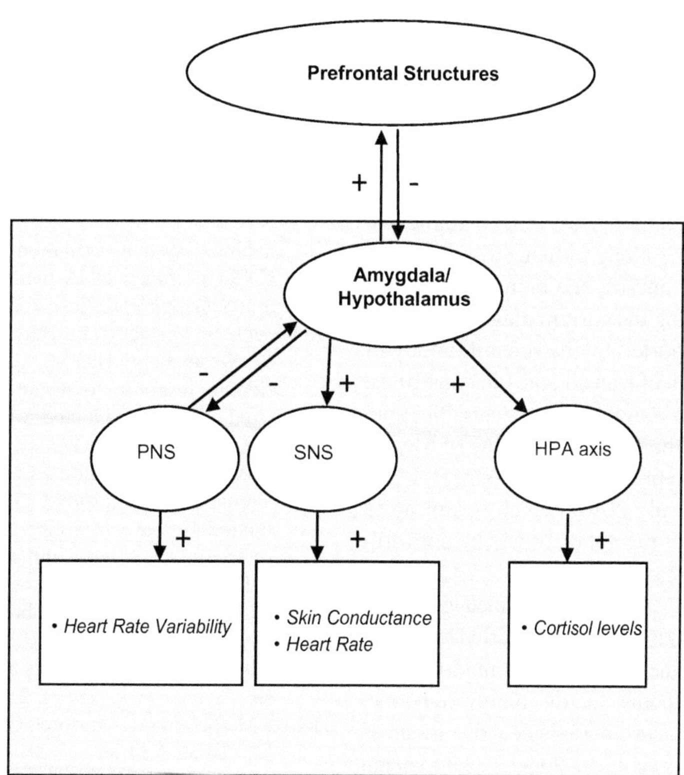

FIGURE 8.6.5 The amygdala and the paraventricular nucleus of the hypothalamus inhibit the parasympathetic nervous system (PNS; causing e.g., lowered heart rate variability), and stimulate the sympathetic nervous system (SNS; increasing skin conductance and heart rate). Meanwhile, nuclei of hypothalamus stimulate the hypothalamic-pituitary-adrenal (HPA)–axis to produce cortisol. Disturbances in amygdala functioning may thus result in a more or less orchestrated overstimulation of the SNS and the HPA-axis combined with a reduction of parasympathetic cardiac control, a typical pattern of responsiveness that might be related to (clinical) anxiety. (From Benarroch, E.E., *Mayo Clin. Proc.*, 68, 988–1001, 1993.)

Whereas heightened parasympathetic tone is reported in youth diagnosed with ADHD, independently from potential comorbidities (like oppositional defiant disorder; or conduct disorder) (Van Lang et al., 2007).

Nonresponsiveness in both the ANS and the HPA-axis on stress (e.g., on social interaction tasks like the TSST) is suggested to underlie the development of conduct disorders and potentially antisocial personality disorder (e.g., Popma et al., 2006). An overall inhibition of peripheral stress-responsiveness (conduct disorders), or an overexcitement (anxiety disorders) at least suggests a disturbed central coordination, with a prominent role for the amygdala and prefrontal structures (see Figure 8.6.5). For example, respiratory sinus arrhythmia (RSA) (as parameter of parasympathetic activity) is indicated as a downstream peripheral marker associated with (prefrontal) cortical dysfunctions and related to diverse psychopathologies (like phobias, attention problems, autism, callousness, and conduct disorders) (Beauchaine, 2015). With preliminary findings suggesting that hypersensitivity for (psychological) stress, potentially progressing into a full-blown anxiety disorder, might be expressed by the *peripheral coherence* of responses between the ANS and the HPA-axis (Kallen et al., 2010).

CONCLUSIONS

Despite a tremendous body of scientific findings on neurophysiological phenomena associated with psychopathology, its suggested potential with regard to applied biomarkers (e.g., in clinical decision making) has not been harvested. In addition, it remains in the balance whether and which articulated "biomarker ambitions" will indeed come into being. The challenge to come to applicable and well-defined biomarkers is definitely enhanced by the fact that the existence of any "simple" one-on-one biomarker for any psychiatric disorder, as defined using the current diagnostic methodologies, is highly unlikely. Supported by the scientific body of evidence pointing into the direction of transdiagnostic constructs, this might call for new avenues for diagnostics and classification, as expressed in initiatives such as the National Institute of Mental Health's Research Domain Criteria project (RDoC, NIMH, USA: Kozak & Cuthbert, 2016: Lang et al., 2016), which is aiming to reorganize the conceptualization of mental disorders by converging neuroscientific and psychological findings.

Secondly, despite the defined caveats and challenges, some potentially very relevant developments have been reported by applying advanced mathematics (Machine Learning) on multifactorial data. A development that does account for the highly dynamic, individualized, and multi-modular, neurophysiological disturbances that may lead to the so typical psychopathological phenomena. The way ahead seems to be to acknowledge these principles and embrace advanced mathematical and modeling methods that aim to provide highly personalized and consequently significantly more valid *biomarker models* to guide and support precision psychiatry.

ACKNOWLEDGMENTS

The authors would like to thank Gregory A. Miller (University of California, Los Angeles) for his inspiration and guidance, Anja Riesel (Humboldt-Universität zu Berlin) for her contribution on transdiagnostic constructs and critically reviewing the manuscript, and Katja Hillbrandt (University of Florida) for her very valuable input on the interpretation of corticosteroid assessments.

REFERENCES

Baker T. (2017). Optimizing multi-modal neuroimaging methods to examine and improve reward functioning in addiction. *Psychophysiology*, 54(1), 102.

Beauchaine T.P. (2015). Respiratory sinus arrhythmia: A transdiagnostic biomarker of emotion dysregulation and psychopathology. *Current Opinion in Psychology*, 3, 43–47.

Benarroch E.E. (1993). The central autonomic network: Functional organization, dysfunction, and perspective. *Mayo Clinical Proceedings*, 68, 988–1001.

Bots S. (2016). Stress induced metabolic outcomes: Review on the interplay between psychological stress, metabolic outcomes and altered inflammatory processes. *Master Thesis*, University of Amsterdam, Amsterdam, the Netherlands.

Chan M.K., Cooper J.D., Heilmann-Heimbach S., Frank J., Witt S.H., Nothen M.M., Steiner J., Rietschel M., & Bahn S. (2017). Associations between SNPs and immune-related circulating proteins in schizophrenia. *Scientific Reports*, 7 (1), 12586.

D'Addario C., Micale V., Di Bartolomeo M., Stark T., Pucci M., Sulcova A., Palazzo M. et al. (2017). A preliminary study of endocannabinoid system regulation in psychosis: Distinct alterations of CNR1 promoter DNA methylation in patients with schizophrenia. *Schizophrenia Research*, 188, 132–140.

De Kloet C.S., Vermetten E., Heijnen C.J., Geuze E., Lentjes E.G.W.M., & Westenberg H.G.M. (2007). Enhanced cortisol suppression in response to dexamethasone administration in traumatized veterans with and without posttraumatic stress disorder. *Psychoneuroendocrinology*, 32 (3), 215–226.

De Kloet E.R., Fitzsimons C.P., Datson N.A., Meijer O.C., & Vreugdenhil E. (2009). Glucocorticoid signaling and stress-related limbic susceptibility pathway: About receptors, transcription machinery and microRNA. *Brain Research*, 1293, 129–141.

Demro C., Rowland L., Wijtenburg S.A., Waltz J., Gold J., Kline E., Thompson E., Reeves G., Hong L.E., & Schiffman J. (2017). Glutamatergic metabolites among adolescents at risk for psychosis. *Psychiatry Research*, 257, 179–185.

Dottori M., Sedeno L., Martorell Caro M., Alifano F., Hesse E., Mikulan E., Garcia A.M. et al. (2017). Towards affordable biomarkers of frontotemporal dementia: A classification study via network's information sharing. *Scientific Reports*, 7 (1), 3822.

Dowlati Y, Herrmann N, Swardfager W, Liu H, Sham L, Reim EK et al. (2010). A meta-analysis of cytokines in major depression. *Biological Psychiatry,* 67(5): 446–457.

Fernandes B.S., Berk M., Turck C.W., Steiner J., & Gonçalves C.A. (2014). Decreased peripheral brain-derived neurotrophic factor levels are a biomarker of disease activity in major psychiatric disorders: A comparative meta-analysis. *Molecular Psychiatry*, 19(7), 750–751. doi: 10.1038/mp.2013.172.

Fernandes B.S., Gama C.S., Kauer-Sant'Anna M., Lobato M.I., Belmonte-de-Abreu P., & Kapczinski F. (2009). Serum brain-derived neurotrophic factor in bipolar and unipolar depression: A potential adjunctive tool for differential diagnosis. *Journal of Psychiatric Research*, 43(15), 1200–1204. doi:10.1016/j.jpsychires.2009.04.010.

Fernandes B.S., Steiner J., Bernstein H.G., Dodd S., Pasco J.A., Dean O.M., Nardin P., Gonçalves C.A., & Berk M. (2016a). C-reactive protein is increased in schizophrenia but is not altered by antipsychotics: Meta analysis and implications. *Molecular Psychiatry*, 21(4):554–564. doi: 10.1038/mp.2015.87.

Fernandes B.S., Steiner J., Molendijk M.L., Dodd S., Nardin P., Gonçalves C.A., Jacka F. et al. (2016b). C-reactive protein concentrations across the mood spectrum in bipolar disorder: A systematic review and meta-analysis. *Lancet Psychiatry*, 3(12), 1147–1156. doi: 10.1016/S2215-0366(16)30370-4.

Fernandes B.S., Williams L.M., Steiner J., Leboyer M., Carvalho A.F., & Berk M. (2017). The new field of "precision psychiatry." *BMC Medicine*. doi: 10.1186/s12916-017-0849-x.

Ford J. (2017). ERP studies of predictive coding deficits in psychosis. *Psychophysiology*, S1, 10.

Garcia-Alvarez L., Caso J.R., Garcia-Portilla M.P., de la Fuente-Tomas L., Gonzalez-Blanco L., Saiz Martinez P., Leza J.C., & Bobes J. (2018). Regulation of inflammatory pathways in schizophrenia: A comparative study with bipolar disorder and healthy controls. *European Psychiatry*, 47, 50–59.

Goldsmith D.R., Rapaport M.H., & Miller B.J. (2016). A meta-analysis of blood cytokine network alterations in psychiatric patients: Comparisons between schizophrenia, bipolar disorder and depression. *Molecular Psychiatry*, 21(12), 1696–1709. doi:10.1038/mp.2016.3.

Guo X., Dominick K.C., Minai A.A., Li H., Erickson C.A., & Lu L.J. (2017). Diagnosing autism spectrum disorder from brain resting-state functional connectivity patterns using a deep neural network with a novel feature selection method. *Frontiers in Neuroscience*, 11, 460. doi: 10.3389/fnins.2017.00460.

Hilbert K., Lueken U., Muehlhan M., & Beesdo-Baum K. (2017). Separating generalized anxiety disorder from major depression using clinical, hormonal, and structural MRI data: A multimodal machine learning study. *Brain and Behavior*, 7(3), e00633. doi:10.1002/brb3.633.

Hillbrandt K., Dominguez V., Keil A., Kirschbaum C., Bradley M., & Lang P. (2017). Hair cortisol concentration in the study of anxiety and depression. *Psychophysiology*, 54(suppl. 1), 165.

Issler O., & Chen A. (2015). Determining the role of microRNAs in psychiatric disorders. *Nature Reviews Neuroscience*, 16, 201–212.

Kallen V.L., Stam J.V., van Pelt J., & Westenberg, P.M. (2010). Associations between ANS and HPA-axis responsiveness to stress in adolescence. *International Journal of Psychophysiology*, 77, 284.

Kirschbaum C., Pirke K.M., & Hellhammer D.H. (1993). The "Trier Social Stress Test" a tool for investigating psychobiological stress responses in a laboratory setting. *Neuropsychobiology*, 28, 76–81.

Kirschbaum C., Tietze A., Skoluda N., & Dettenborn L. (2009). Hair as a retrospective calendar of cortisol production—Increased cortisol incorporation into hair in the third trimester of pregnancy. *Psychoneuroendocrinology*, 34(1), 32–37.

Köhler C.A., Freitas T.H., Maes M., de Andrade N.Q., Liu C.S., Fernandes B.S., Stubbs B. et al. (2017). Peripheral cytokine and chemokine alterations in depression: A meta-analysis of 82 studies. *Acta Psychiatrica* Scandinavica, 135(5), 373–387. doi: 10.1111/acps.12698.

Kozak M.J., & Cuthbert B.N. (2016). The NIMH research domain criteria initiative: Background, issues, and pragmatics. *Psychophysiology*, 53(3), 286–297. doi: 10.1111/psyp.12518.

Lang P.J., McTeague, L.M., & Bradley, M.M. (2016). RDoC, DSM, and the reflex physiology of fear: A biodimensional analysis of the anxiety disorders spectrum. *Psychophysiology*, 53(3), 336–347.

Levy T., Bloch Y., Bar-Maisels M., Gat-Yablonski G., Djalovski A., Borodkin K., & Apter A. (2015). Salivary oxytocin in adolescents with conduct problems and callous-unemotional traits. *European Child and Adolescent Psychiatry*, 24(12), 1543–1551.

Manenschijn L., Van Den Akker E.L.T., Lamberts S.W.J., & Van Rossum E.F.C. (2009). Clinical features associated with glucocorticoid receptor polymorphisms: An overview. *Annals of the New York Academy of Sciences*, 1179, 179–198.

McPartland J.C. (2017). Developing clinically practicable biomarkers for autism spectrum disorder. *Journal of Autism and Developmental Disorders*, 47(9), 2935–2937.

Miller G.A., Rockstroh B.S., Hamilton H.K., & Yee C.M. (2016). Psychophysiology as a core strategy in RDoC. *Psychophysiology*, 53(2016), 410–414.

Molendijk M.L., Spinhoven P., Polak M., Bus B.A., Penninx B.W., & Elzinga B.M. (2014). Serum BDNF concentrations as peripheral manifestations of depression: Evidence from a systematic review and meta-analyses on 179 associations (N=9484). *Molecular Psychiatry*, 19(7), 791–800. doi: 10.1038/mp.2013.105.

Morbelli S., Bauckneht M., & Scheltens P. (2017). Imaging biomarkers in Alzheimer's disease: Added value in the clinical setting. *Quarterly Journal of Nuclear Medicine and Molecular Imaging*, 61(4), 360–371.

Nagai T., Kirihara K., Tada M., Koshiyama D., Koike S., Suga M., Araki T., Hashimoto K., & Kasai K. (2017). Reduced mismatch negativity is associated with increased plasma level of glutamate in first-episode psychosis. *Scientific Reports*, 7 (1), 2258.

National Research Council Committee. (2011). Toward precision medicine. Building a knowledge network for biomedical research and a new taxonomy of disease. *National Research Council Committee on a Framework for Developing a New Taxonomy of Disease*. Washington, DC: National Academies Press.

Olvet D.M., & Hajcak G. (2008). The error-related negativity (ERN) and psychopathology: Toward an endophenotype. *Clinical Psychology Review*, 28(8), 1343–1354.

Pinto J.V., Moulin T.C., & Amaral O.B. (2017). On the transdiagnostic nature of peripheral biomarkers in major psychiatric disorders: A systematic review. *Neuroscience and Biobehavioral Reviews*, 83, 97–108.

Popma A., Jansen L.M., Vermeiren R., Steiner H., Raine A., Van Goozen S.H., van Engeland H., & Doreleijers T.A. (2006). Hypothalamus pituitary adrenal axis and autonomic activity during stress in delinquent male adolescents and controls. *Psychoneuroendocrinology*, 31(8), 948–957.

Riesel A., Klawohn J., & Kathmann N. (2017). The error-related negativity as a transdiagnostic marker for anxiety. *Psychophysiology*, S1, 10.

Shrivastava A., Kar S.K., Sharma E., Mahdi A.A., & Dalal P.K. (2017). A study of oxidative stress biomarkers in obsessive compulsive disorder. *Journal of Obsessive-Compulsive and Related Disorders*, 15, 52–56.

Stalder T., Kirschbaum C., Kudielka B.M., Adam E.K., Pruessner J.C., Wust S., Dockray S. et al. (2016). Assessment of the cortisol awakening response: Expert consensus guidelines. *Psychoneuroendocrinology*, 63, 414–432.

Stalder T., Steudte-Schmiedgen S., Alexander N., Klucken T., Vater A., Wichmann S., Kirschbaum C., & Miller R. (2017). Stress-related and basic determinants of hair cortisol in humans: A meta-analysi. *Psychoneuroendocrinology*, 77, 261–274.

Thomas A.J., Attems J., Colloby S.J., O'Brien J.T., Mckeith I., Walker R., Lee L., Burn D., Lett D.J., & Walker Z. (2017). Autopsy validation of 123 I-FP-CIT dopaminergic neuroimaging for the diagnosis of DLB. *Neurology*, 88(3), 276–283.

Trepanier M.O., Hopperton K.E., Mizrahi R., Mechawar N., & Bazinet R.P. (2016). Postmortem evidence of cerebral inflammation in schizophrenia: A systematic review. *Molecular Psychiatry*, 21(8), 1009–1026.

Trimble M.R., & George M.S. (2010). *Biological Psychiatry*. Chichester, West Sussex: Wiley-Blackwell.

Uddin L.Q., Dajani D.R., Voorhies W., Bednarz H., & Kana R.K. (2017). Progress and roadblocks in the search for brain-based biomarkers of autism and attention-deficit/hyperactivity disorder. *Translational Psychiatry*, 7(8), e1218. doi: 10.1038/tp.2017.

Vaidyanathan U., Nelson L.D., & Patrick CJ. (2012). Clarifying domains of internalizing psychopathology using neurophysiology. *Psychological Medicine*, 42(3), 447–459. 19.

Valkanova V, Ebmeier KP, & Allan CL. (2013). CRP, IL-6 and depression: A systematic review and meta-analysis of longitudinal studies. *Journal of Affective Disorders,* 150(3), 736–744. doi: 10.1016/j.jad.2013.06.004.

Van Lang N.D.J., Tulen J.H.M., Kallen V.L., Rosbergen B., Dieleman G., & Ferdinand, R.F. (2007). Autonomic reactivity in clinically referred children: Attentional deficit Hyperactivity disorder versus anxiety disorder. *European Child and Adolescent Psychiatry*, 16, 71–78.

Wester V.L., & Van Rossum E.F.C. (2015). Clinical applications of cortisol measurements in hair. *European Journal of Endocrinology*, 173(4), M1–M10.

Yee, C.M., Javitt, D.C., & Miller, G.A. (2015). Replacing categorical with dimensional analyses in psychiatry research: The RDoC initiative. *JAMA Psychiatry*, 72, 1159–1160.

Yerys BE, & Pennington BF. (2011). How do we establish a biological marker for a behaviourally defined disorder? Autism as a test case. *Autism Research*, 4, 239–241.

Zhou J., Yao N., Fairchild G., Cao X., Zhang Y., Xiang Y.-T., Zhang L., & Wang X. (2016). Disrupted default mode network connectivity in male adolescents with conduct disorder. *Brain Imaging and Behavior*, 10(4), 995–1003.

Identification of Biomarkers for the Diagnosis, Prediction, and Progression of Alzheimer's Disease 8.7

Bennett P. Greenwood, Vladimir Tolstikov, Paula P. Narain, Jeremy Chaufty, Viatcheslav R. Akmaev, Vivek Vishnudas, Stephane Gesta, Eric J. Nestler, Rangaprasad Sarangarajan, Niven R. Narain, and Michael A. Kiebish

Contents

THE IMPERATIVE NEED FOR PRECISION MEDICINE IN THE TREATMENT OF ALZHEIMER'S DISEASE

Precision medicine is defined as a clinical intervention, using designated therapies based upon a person's or populations' unique genetic or molecular makeup, providing the right treatment, to the right patient, at the right time. This process must be data driven, streamlined through clinical development design, and include characterization of biomarker profiles informing clinical outcomes. Regarding precision medicine, the right time to treat the right patient is established by patient physiology and disease presentation. Toward this, the inherent biology and molecular/physiological presentation of the disease must be defined through either genomic analysis determining relative development risk or characterization of the disease's phenomic signature. Phenomic analysis refers to systematic measurement and subsequent analysis of quantitative traits relying upon comprehensive multi-omic profiling. This includes clinical, behavioral, and dietary data overlaid by proteomic analysis and subsequent posttranslational modifications of proteins within biofluids/cells compared with lipidomic and metabolomic signatures defining a patient's systemic physiological state at a given time. Aligning an individual's or population's disease progression with molecular signatures requires dynamic multiannual sampling of biofluids to track changes within the individual as well as the disease population.

The majority of Alzheimer's disease (AD) cases are idiopathic and difficult to predict during the prodromal phase when symptoms are absent. AD is exemplified by variations in personalized factors including: age of onset, rate of decline, and initial cognitive profile. Such heterogeneity within the disease population further complicates early diagnosis and selection of appropriate treatments, resulting in the need to collect samples from larger population cohorts to detect patients progressing from asymptomatic, to mild cognitive impairment (MCI), to eventual AD diagnosis. Once comprehensive molecular characterization of patient populations defines a molecular signature of AD that changes with disease progression independently from non-diseased controls and other neurological diseases, then clinical development programs can begin to determine if therapeutic intervention works at a particular disease time point.

Following characterization of biomarkers for dynamic progression of MCI toward AD, there will be inherent subpopulations defined by genetic, phenomic, or demographic characteristics that will likely progress more rapidly or have different biological signatures predisposing them to respond to different interventions. These subpopulations will likely need definition using advanced analytical techniques involving various statistical, machine learning, or artificial intelligence approaches incorporating alignment of clinical outcomes with molecular signatures. Currently, the medical field treats AD and MCI similarly; however, if molecular subpopulations are defined, development of therapies targeting specific molecular pathways of unique populations rather than broad disease populations becomes possible. This allows pharmaceutical and diagnostic companies to design clinical trials around subpopulations to increase the likelihood of meeting thresholds of clinical benefit for approval, thus matching the right drug with the right patient.

The critical factor to implementing precision medicine to treat Alzheimer's disease is to have complete participation from patients and acceptance from the broader population that AD is an impending public health concern with broad implications impacting our future. During our assessment, it was clear that alignment of social awareness, government research funding, research publications, and established FDA study designs are not currently structured to impact clinical development of precision medicine in the treatment of Alzheimer's disease.

ALZHEIMER'S DISEASE EPIDEMIOLOGY AND ECONOMIC IMPACT

In 2015, Alzheimer's Disease was the sixth leading cause of death in the United States (Kochanek et al. 2016). Currently, it is an irreversible neurodegenerative disease that is the most common cause of dementia among older people. It slowly destroys memory and cognition until the ability to perform the simplest task is lost. Symptoms typically manifest after 60 years in late-onset Alzheimer's disease (LOAD); however, in rare cases, early-onset Alzheimer's disease (EOAD) symptoms have appeared as early as 30 years of age (Morris 2005; Sperling et al. 2011). In 2016, an estimated 5.4 million Americans were diagnosed with AD, including approximately 200,000 people affected by EOAD; however, according to the Alzheimer's Association these numbers are drastically under representing the total number of people afflicted with AD (Barrett et al. 2006; Zaleta et al. 2012). In 2017, the Alzheimer's Association reported that 3% of people aged 65–74, 17% of people aged 75–84, and about 32% of people 85 and older had AD (Hebert et al. 2013; Alzheimer's Association 2017). Prevalence of AD is likely much higher than these reported numbers for multiple reasons, including diagnostic uncertainty, shortage of disease-modifying treatment, and communication difficulties. Patients and their caregivers are unaware of the diagnosis (Alzheimer's Association 2015a). Based on these findings, the actual number of people affected by AD could be closer to 10–11 million Americans: more than double the population currently thought to be afflicted.

Between 2011 and 2017, AD related spending consistently comprised about 7% of the total national health expenditure (NHE) of the United States (U.S. Centers for Medicare & Medicaid Services). In the same timeframe, yearly expenditures increased by almost 100 billion dollars (Figure 8.7.1).

Because AD generally does not affect people until post-retirement, most of the cost burden falls upon the U.S. government via Medicare. Combining Medicare costs with Medicaid payments, the government pays approximately 66% of the total cost burden associated with AD every year. As such, in 2016 the government spent upward of $160 billion on Alzheimer's care, that is, 13% of the total yearly Medicaid/Medicare budget. The Baby Boomer generation has just reached AD risk ages, meaning that in the foreseeable future there will likely be drastic increases in these costs; the majority are not covered by private insurance.

In 2016, Medicare payments for beneficiaries with dementia averaged over three times more than those without (Alzheimer's Association 2017). The difference is even greater when looking at Medicaid coverage since Medicaid pays for nursing home and other long-term care services for people in low income brackets. In 2016, the average Medicaid payment for Medicare beneficiaries with AD/dementia was 23 times more than payments for those without (Alzheimer's Association 2017). Overall, a person with AD can cost the government over four times more than other beneficiaries in the same age bracket, a daunting amount when anticipating the influx of aging individuals from the baby boomer generation.

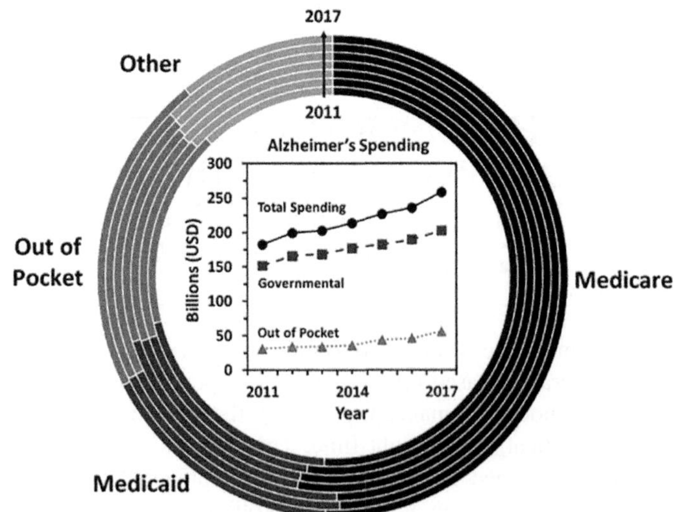

FIGURE 8.7.1 Payer source breakdown of Alzheimer's disease (AD) costs. The U.S. government pays the majority of AD care costs. Roughly 66% of all costs are covered by Medicaid and Medicare and 33% originate out of pocket and from other sources. The inner chart shows total AD spending over time and the allocations from governmental sources and out-of-pocket sources. In 2017, predictions indicated that the total cost burden would exceed $250 billion dollars.

Examination of two other payment sources is important to appreciate the AD/dementia cost burden. Private health insurance payments for beneficiaries with AD/dementia were about double that of healthy patients. The out-of-pocket costs for those with AD/dementia were 5 times greater per person than for those without (Alzheimer's Association 2017). Out-of-pocket costs for AD/dementia patients is a much more significant financial blow than many other diseases because patients with AD can live for many years generating massive total lifetime costs. Adding to this, people with AD/dementia have twice the hospital stays per year as non-dementia patients. Finally, combining these with lost income from family members forced to take time off for patient care, the total impact becomes financially debilitating (Kelley et al. 2015). Further, putting the AD economic burden into perspective is a comparison between yearly costs of Alzheimer's care and yearly costs of cancer care (Figure 8.7.2).

From 2009 to 2014 the yearly cost burden of cancer care is estimated to fluctuate between approximately $70 and $90 billion dollars. Comparatively, the total AD cost burden has consistently increased from approximately $140 billion to over $200 billion dollars (Alzheimer's Association 2015b, 2016). In recent years AD has become more than double the economic burden as cancer and it is only predicted to get worse (Weuve et al. 2014; Alzheimer's Association 2015a; Wimo et al. 2017).

There are two sides to the figurative cost-burden coin. The first, as previously discussed, concerns the overall economic burden in terms of individual and national significance. The second pertains to AD research for academic and industrial labs. The process of drug development for any disease is long and expensive. In academia, this means large competitive grants that are difficult to obtain and typically only fund smaller sized trials. In industry, individual drug development costs exceed $2.5 billion (DiMasi et al. 2016). Studies have shown that on average the probability of a drug successfully passing through all phases and getting FDA approval is only 9.6%. The success rates vary by disease category with oncology lowest at 5.1%, Neurology at 8.4%, and hematology highest at 26.1% (Hay et al. 2014; Biotechnology Innovation Organization 2016). The FDA's approval success rate for drugs used to treat AD is abysmally low; from 2002 to 2012 there were 244 drugs tested in clinical trials. Only one drug was approved by the FDA, leading to a 0.4% approval rate (Cummings et al. 2014). With over 99% of drugs tested to treat AD failing, companies must therefore spend their own budgets developing and testing a drug that, most likely, will fail.

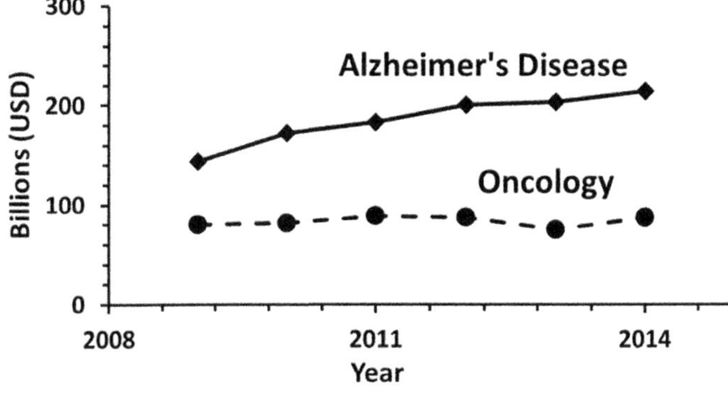

FIGURE 8.7.2 Cost burden comparison with oncology. The Alzheimer's disease (AD) cost burden has steadily increased since 2009. In the past decade, the total AD cost burden has consistently been significantly higher than the total oncological cost burden. These figures do not take into account the total income lost for close family/friends or caretakers of patients. (From Alzheimer's Association, *Alzheimers Dement,* 6, 158–194, 2010; Alzheimer's Association, *Alzheimers Dement,* 7, 208–244, 2011; Alzheimer's Association, *Alzheimers Dement,* 8, 131–168, 2012; Alzheimer's Association, *Alzheimers Dement,* 9, 208–245, 2013; Alzheimer's Association, *Alzheimers Dement,* 10, e47–e92, 2014; Alzheimer's Association, *Alzheimers Dement,* 11, 332–84, 2015a; Alzheimer's Association, *Alzheimers Dement,* 12, 459–509, 2016; Alzheimer's Association, *Alzheimer's Dementia,* 13, 325–373, 2017; U.S. Centers for Medicare & Medicaid Services, OOTA, National Health Statistics Group, *National Health Expenditures by Type of Service and Source of Funds: Calendar Years 1960–2016* [Online], U.S. Centers for Medicare & Medicaid Services, Baltimore, MD. Available: https://www.cms.gov/Research-Statistics-Data-and-Systems/Statistics-Trends-and-Reports/NationalHealthExpendData/NationalHealthAccountsHistorical.html [Accessed January 01, 2018].)

SOCIAL MEDIA/ PUBLIC PERCEPTION OF ALZHEIMER'S DISEASE

In this modern age with information at our fingertips, we have an opportunity to incorporate social media findings pertaining to public interest in any given subject into clinical or scientific research (Brownstein et al. 2009). In fact, in 2009 Google began collaborations with the Centers for Disease Control and Prevention (CDC) to use Internet search engine query data to detect and localize occurrences of influenza epidemics (Ginsberg et al. 2009). Gathering social media data and applying it to clinical research topics such as AD or cancer can help give a societal overview indicating a topic's importance or measuring the general public's interest levels. One such tool, provided free for the public by Google, is Google Trends. It is a public web facility allowing anyone to observe search trends changing over time. Trends are based on terms plugged into Google's search engine showing strong anonymous public interest in the searched term. Searching for a subject provides an interest score calculated by scaling all time points to the date with the highest volume of searches for that term, giving a percentage value interest score. Google Trends data has already been used to study varying topics ranging from cancer screening to the monitoring of non-cigarette tobacco use (Nuti et al. 2014; Cavazos-Rehg et al. 2015; Schootman et al. 2015). Until now, the Google Trends tool has not been applied to studying AD in the United States, though it was touched upon in a study focused on all forms of dementia in Taiwan (Wang et al. 2015). This powerful tool has proven its applicability to the medical field and applying Google Trends to study AD awareness in the United States has produced interesting results.

When plotting interest scores against time, trends of public interest and correlation between events can be seen. For example, in August 2014 a social media craze known as the Ice Bucket Challenge swept the internet, helping raise awareness of ALS. The Google Trends data distinctly shows a massive spike in ALS interest during Q3 of that year, after which interest scores drop back to approximately the same levels as the previous decade (Figure 8.7.3).

Another instance where a public event directly increases public awareness and interest is the annual recurrence of Autism Awareness month (Figure 8.7.4).

When data is plotted by week or month, there is a distinct spike every April indicating the success of the Autism Awareness movement; however, the figures herein show data averaged by quarter causing this trend to be unobservable (Figure 8.7.4).

Social media interest scores indicate that interest in AD declines until about 2010 when it plateaus. There are indications that since 2014, interest has increased and is currently trending upward. Expectations are that interest in AD should increase further over the next decade as more of the Baby Boomer generation enter high-risk ages and more tech-savvy people are affected by the disease, either directly or indirectly, data for EOAD indicates a different trend. Interest was moderately steady until it spiked in 2011 and has since been increasing. A possible explanation is that the younger generation's comfort with technology; namely, they spend much more time on the internet and are far more likely to search for answers via search engines, like Google. As with LOAD, a tech-savvy population reaching prime age will likely lead to spikes and an overall increase in the level of social awareness of EOAD. Overall interest in Parkinson's disease (PD) has shown a steady decrease since 2004 with no indication of this trend changing in recent years (Figure 8.7.3). In comparison with the other neurological disorders, amyotrophic lateral sclerosis (ALS) awareness between 2004 and 2016 is depicted as a flat line indicating no change except for the intense spike attributed to the Ice Bucket Challenge.

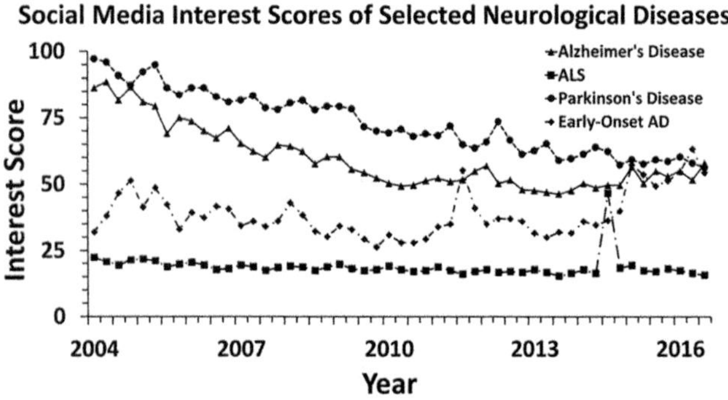

FIGURE 8.7.3 Neurological disease interest scores. Social media search trends show decreased interest in Alzheimer's disease (AD) and Parkinson's disease. An overall increased interest in early-onset AD (EOAD) can be seen. Amyotrophic lateral sclerosis (ALS) shows decreased interest until August 2014 when the Ice Bucket Challenge swept the internet.

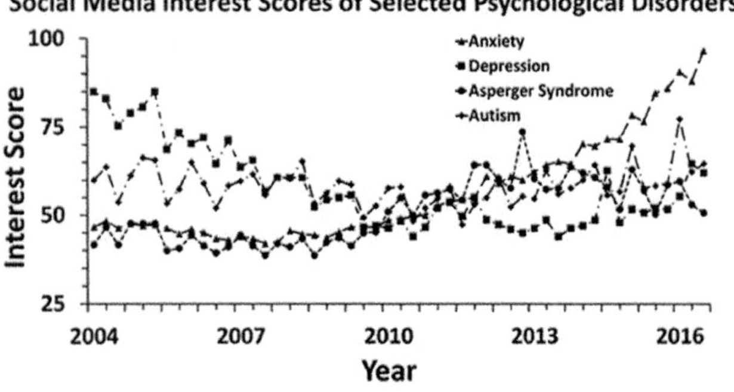

FIGURE 8.7.4 Psychological disorders interest scores. Comparing interest in psychological disorders, anxiety showed the largest increase since 2004. Depression decreased from 2004 to 2016. Neither Asperger syndrome nor autism showed significant interest change, only periodic spikes linked to Autism Awareness Month.

A useful comparison with societal awareness of neurological diseases is public interest in some psychological disorders ranging from anxiety to autism (Figure 8.7.4). Interest in Asperger syndrome and autism both show periodic spiking during Autism Awareness month every year. Otherwise, autism's interest scores are consistent and interest in Asperger syndrome gradually rose in recent years. Even though there are some seemingly random interest spikes, interest in depression has significantly decreased in the last decade. The most interesting score change occurs with anxiety; interest has doubled since 2004. This is the largest social awareness increase observed in any of the sampled diseases/disorders. Several studies suggest a rise in reliance upon technology, particularly social media, induced an increase in anxiety. Alternatively, the enhanced access to information that technology continuously provides could also explain the phenomenon (Shensa et al. 2018; Elhai et al. 2018; Brailovskaia and Margraf 2018). Regardless of cause, anxiety is clearly a growing problem in our society and people are aware.

In terms of social awareness, one of the most-searched disease terms is cancer. Based on 2016 scaled and averaged google trends data, cancer was searched approximately 550 times more often than AD and dementia was searched nearly 50 times more often. Cancer has been one of the most feared diseases worldwide for decades. It can affect any person from any demographic and is so diverse that just about any body part can be affected. Knowing this, it is no surprise that interest in cancer remains consistent. The interest score for cancer dropped in the years after 2004, but it has increased steadily in recent years (Figure 8.7.5).

In comparison, social awareness of dementia steadily increased over the past decade while interest in AD waned, recently plateaued, and is, hopefully, increasing. Incorporating social media data into medical research would do two things. First, it would help direct research efforts toward areas of great social importance and second, it would give a strong indicator of the public's perception of the medical research community. This in turn would help direct the initial research focus of a precision medicine platform.

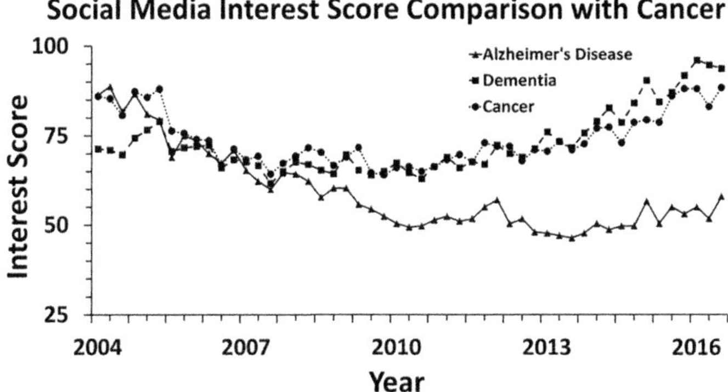

FIGURE 8.7.5 Cancer interest score comparison. Interest in Alzheimer's disease decreased significantly since 2004 but plateaued recently. Dementia and cancer interest initially decreased but eventually increased, exceeding original levels. Dementia and cancer have similar trends but not similar total interest levels.

ALZHEIMER'S RESEARCH AND TREATMENT OPTIONS

Since 2014, over 150,000 works pertaining to diverse cancer research have been published yearly. In 2017, there were over 10,000 publications focusing on AD and over 10,000 mentioning dementia, many likely overlapping (Figure 8.7.6) (NCBI Resource Coordinators 2018). There is no doubt that cancer is costly and has a long way to go until a panacea is developed, but *there are* improved treatment options. Comparatively, AD has no cure and no successful treatments that significantly postpone disease symptoms (Alzheimer's Association 2017). There is a significant lack in the understanding of how AD occurs that translates to failed research toward treatment (Beach 2013).

In the early 1990s, several scientists hypothesized that beta amyloid (Aβ protein deposition in the form of sticky plaques cause AD pathogenesis (Selkoe 1991; Beyreuther and Masters 1991; Hardy and Higgins 1992; Hardy and Selkoe 2002; Selkoe and Hardy 2016). The consequence of Aβ plaque deposition includes twisted tau protein strand formation (neurofibrillary tangles), contributing to neuronal cell death and leading to vascular damage. These biological factors are proposed to underlie severe dementia. This is called the amyloid cascade hypothesis and the majority of research efforts since its publication have focused on Aβ production and deposition in the brain. Supporting this, genetic studies found that mutations occurring in the amyloid precursor protein (APP), presenilin 1, and presenilin 2, lead to Aβ plaque formation and deposition (Goate et al. 1991; Levy-Lahad et al. 1995; Sherrington et al. 1995).

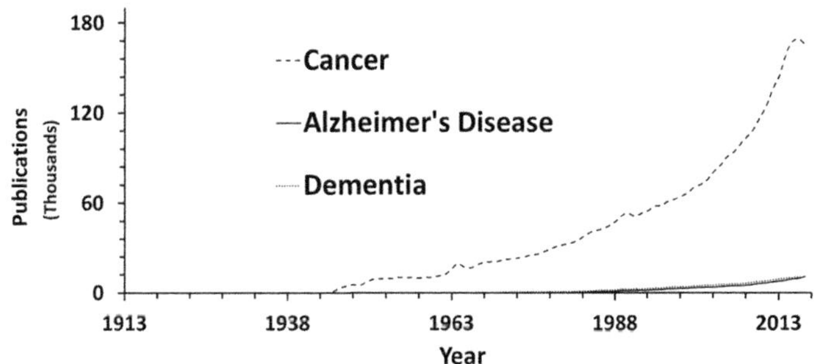

FIGURE 8.7.6 Total PubMed publications. Number of PubMed publications since 1913 mentioning cancer, Alzheimer's disease (AD), or dementia. There is likely significant overlap between the AD and dementia categories.

These genetic mutations are the causative factors in what is known as familial Alzheimer's disease (FAD). FAD is the only form of AD that can be predicted. Due to FAD's similarities with LOAD, many researchers assumed that the genetic mutations and subsequent Aβ deposition were explanations for all AD types (Armstrong 2014). Unfortunately, the disease is not so simple. Over 99% of AD patients do not have genetic links attributing their condition to FAD (Blennow et al. 2006). Therefore, research into non-FAD cases led to the discovery of correlation between individuals carrying the ε4 allele of apolipoprotein E (APOEε4) and LOAD (Corder et al. 1993; Liu et al. 2013; Konishi et al. 2018). About 17% of the worldwide population carries APOEε4, whereas 40% of all LOAD patients are carriers (Farrer et al. 1997). The pathophysiological connection between APOEε4 and LOAD is currently unknown.

Over the past few decades the LOAD progression timeline remained unchanged (Figure 8.7.7) and only recently research efforts investigated the possibility that AD affects the brain 25 years before clinically detectable symptoms appear (Shaw et al. 2007).

There are currently only a small number of biomarkers for AD; all requiring invasive and expensive methods unavailable or not covered by insurance unless deemed medically necessary by symptoms or prior testing (Shaw et al. 2007; Bateman et al. 2011). Additionally to cognitive testing, recent research has used PET scanning to non-invasive screen for AD.

PET scans require PET tracer compounds such as Pittsburgh compound B (PiB), which binds Aβ plaques, allowing observation of cerebral areas with high uptake concentrations (Klunk et al. 2004). PiB research lead to ^{18}F-based PET tracers including Flutemetamol, Florbetapir, and Florbetaben (Morris et al. 2016). This method is accurate enough for diagnosis, but cannot

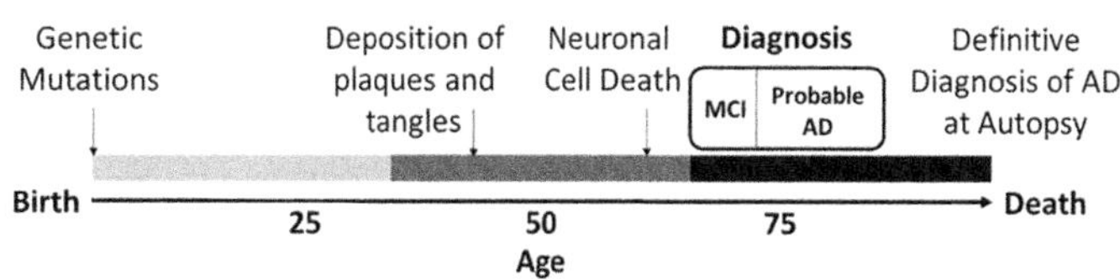

FIGURE 8.7.7 Timeline of late-onset Alzheimer's disease (LOAD). A working disease prevention needs to occur early in life. Once AD is detected it can only be modified, not cured. Based on current treatment options, when symptoms appear it is too late and will be fatal. The only truly definitive diagnosis of AD is postmortem.

detect early stages of AD. Aβ tracer development allows semi-definitive AD confirmation and significantly broadens research capabilities. AD progression research utilizing PET imaging is ongoing through scientific collaborations (Alzheimer's Disease Neuroimaging Initiative 2013). Using PET scans, doctors can be highly confident in their AD diagnosis (Inui et al. 2017). Unfortunately, AD diagnosis confirmation means current disease-modifying therapies will have no effect (Reitz 2012).

Research as far back as 1994 is critical of the amyloid cascade hypothesis (Hardy 2009). Multiple studies found that in head trauma patients, increased APP expression and subsequent Aβ-plaque deposition was a neuronal injury response (Gentleman et al. 1997Roberts 1988; Omalu et al. 2005; Qi et al. 2007). Some hypothesized that poorly controlled βAPP protective response to neuronal injury eventually metamorphosed into neuronal degeneration characteristic of AD (Roberts et al. 1994). Efforts studying early stage AD noted association with mitochondrial dysfunction (Alikhani et al. 2009), others noted a strong insulin resistance correlation and consider AD a neuroendocrine disorder, referring to it as "type 3 diabetes" (Schubert 2005; Steen et al. 2005; Seneff et al. 2011). Even with ample research supporting amyloid involvement, every clinically tested treatment failed to slow cognitive decline (Berk and Sabbagh 2013; Mullard 2017). Full exploration into proposed biomarkers and pathogenesis is beyond the scope of this discussion; however, even though high quality and extensive research was diversely performed, a disparaging lack of understanding into the true origins of AD persists.

In the United States there are currently six pharmacologic treatments for all AD stages approved by the FDA: galantamine, rivastigmine, donepezil, memantine, tacrine, and a combination of donepezil and memantine (Yiannopoulou and Papageorgiou 2013). Temporary symptom improvement by increased neurotransmitter amounts is possible, but no treatment can stop or even slow progression. Donepezil, galantamine, and rivastigmine are cholinesterase inhibitors that stabilize key neurotransmitter levels improving cognitive symptoms (Farlow 2002; Birks 2006). Donepezil is the only approved treatment for all stages. Memantine is an NMDA receptor antagonist that reduces excessive glutamatergic neurotransmission and protects healthy neurons from excitotoxicity and is approved for moderate-to-severe stages (Parsons et al. 1999; Hynd et al. 2004; Olivares et al. 2012). Based on limited understanding of AD pathogenesis and unsuccessful drug development, new methodology is needed to make progress in our understanding and treatment of the disease.

There are some promising new research avenues (Strohle et al. 2015); however, the lack of significant progress in understanding the nature of AD bodes ill for expectations of success. One way to better understand the nature of AD is to accumulate vast clinical data and apply high level statistical comparisons with each individual patient with MCI, dementia, AD, or any other neurological condition. This approach could be vital to identifying biological or chemical indicators, helping differentiate between indistinguishable neurological conditions. Using a precision medicine platform capable of high volume data analysis would produce the desired results.

ALZHEIMER'S DISEASE AND PRECISION MEDICINE

A precision medicine focus is ideal to study AD pathogenesis. Some clinical trials have incorporated aspects of a precision medicine focused platform into their methodology. This is a step in the right direction but further experimental design is needed for significant progress (Reiman et al. 2011; Mills et al. 2013; Sperling et al. 2014). Fully utilizing the power of precision medicine structured clinical trials requires following patients through decades, which would necessitate governmental support. The study enrollment size would be large, ideally with thousands of participants, and ethnically diverse. If feasible, samples of all biofluids—whole blood, plasma, serum, urine, saliva, or cerebrospinal fluid, need be acquired each year. Finally, studies should incorporate multi-omic analysis combining genomic, proteomic, metabolomic, and lipidomic data. Studies thus far have focused on one analysis approach rather than a combination (Mielke and Lyketsos 2006; Finehout et al. 2007; Wilkins and Trushina 2017). In order to fully understand the heterogeneous nature of the diverse population of patients suffering from mild MCI to severe AD, a combination approach should be implemented (Avramouli and Vlamos 2017). Unfortunately, due to length, size, cost and sampling difficulty, a longitudinal study of this magnitude is most likely impossible.

Rather than starting a new long and expensive study, obtaining biofluid samples from existing trials is more realistic. Figure 8.7.8 shows a semi-comprehensive representation of AD clinical trials having collected biofluids run in the United States.

Trials are represented via length, enrollment, and completion date. At the time of data collection, there were over 800 registered clinical trials run in the United States focused on interventional treatment or observational study of AD; however, only 268 trials confirmed biofluid collection. A study using precision medicine techniques would yield best results with high enrollment, thereby providing statistically significant data sets (Kryscio et al. 2013, 2017). A handful of studies fall within parameters ideal for a precision medicine focused clinical trial. These studies are all ongoing or completed in the past decade with enrollment sizes over 200 participants. Of all available trials, these 10 add fit the ideal parameters and could be combined with precision medicine focused analysis.

Due to its slowly progressive nature, AD requires a long observation time to effectively and concisely establish pathogenesis of progression and biomarker identification for each stage (Reitz 2016). Precision medicine focused analysis is ideal for experimental design requiring longitudinal patient observation. Trials thus far relied heavily on LOAD patients and focused on interventional efficacy treating neurological degradation; many fruitlessly conclude early (Ostrowitzki et al. 2017). Current treatment options were designed because symptoms only appear in later stages when intervention is detectable using current cognitive screening methods.

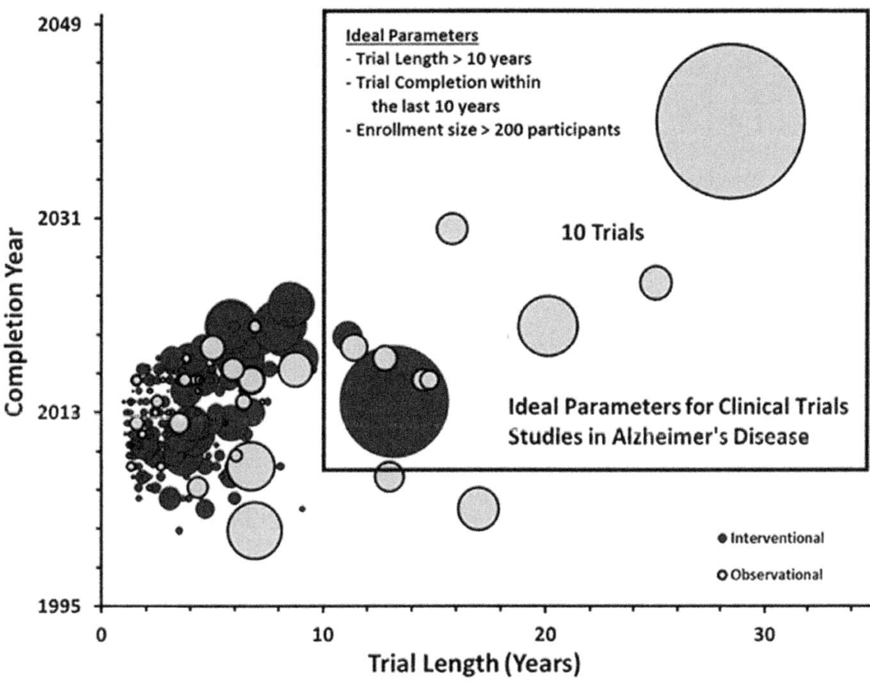

FIGURE 8.7.8 A precision medicine approach to Alzheimer's disease (AD). A semi-comprehensive catalog of clinical trials studying AD in the United States. Bubbles represent individual clinical trials and color represents study type—interventional or observational. Bubble size indicates total enrollment, that is, the largest observational trial (top right) enrolled 20,000 patients, whereas the small trial (bottom left) completed in 2002 with 50 patients. The boxed area indicates clinical trials believed to be most ideal for precision medicine analysis. The parameters for choosing trials are indicated. (All data obtained from clinicaltrials.org.)

There are many hindrances blocking progress toward an AD cure. Combining lengthy observation processes with high developmental costs and difficulties producing specialized small-molecule drugs capable of crossing the blood-brain barrier, leads companies to gravitate toward other research areas with higher chances of success. These reasons contribute to recent AD research platform closures. Companies are understandably justified in their hesitance to spend billions developing drugs likely to fail (Cummings et al. 2014, 2017). Therefore, AD has astronomical yearly costs, nearly stagnant research progress, and no known cure or successful treatment options that effect progression.

The current ecosystem to educate, drive, and implement radical changes in the development of therapeutics for the treatment of Alzheimer's disease is not harmonized to precipitate significant advances that will benefit society. Presently, data exists in a discordant architecture that encompasses patient biology, clinical history, digital social engagement, demographic information, and pharmaceutical and diagnostic development; yet, the utilization of data driven actionable tools have not been fully engaged. Implementation of a synergized analytics and precision medicine driven platform to identify therapeutic signal from population noise will orchestrate the most effective path for harmonization of society, research, and translational medicine that is not effected by human bias. This approach merges biology and technology in a causal manner to direct a revolutionary global impact on health care in Alzheimer's disease and beyond.

REFERENCES

Alikhani, N, Ankarcrona, M & Glaser, E 2009. Mitochondria and Alzheimer's disease: Amyloid-beta peptide uptake and degradation by the presequence protease, hPreP. *J Bioenerg Biomembr,* **41**, 447–51.

Alzheimer's Association 2010. 2010. Alzheimer's disease facts and figures. *Alzheimers Dement,* **6**, 158–94.

Alzheimer's Association 2011. 2011 Alzheimer's disease facts and figures. *Alzheimers Dement,* **7**, 208–44.

Alzheimer's Association 2012. 2012 Alzheimer's disease facts and figures. *Alzheimers Dement,* **8**, 131–68.

Alzheimer's Association 2013. 2013 Alzheimer's disease facts and figures. *Alzheimers Dement,* **9**, 208–45.

Alzheimer's Association 2014. 2014 Alzheimer's disease facts and figures. *Alzheimers Dement,* **10**, e47–92.

Alzheimer's Association 2015a. 2015 Alzheimer's disease facts and figures. *Alzheimers Dement,* **11**, 332–84.

Alzheimer's Association 2015b. *Changing the Trajectory of Alzheimer's Disease: How a Treatment by 2025 Saves Lives and Dollars.* Chicago, IL: Alzheimer's Association.

Alzheimer's Association 2016. 2016 Alzheimer's disease facts and figures. *Alzheimers Dement,* **12**, 459–09.

Alzheimer's Association 2017. 2017 Alzheimer's disease facts and figures. *Alzheimer's & Dementia,* **13**, 325–73.

Alzheimer's Disease Neuroimaging Initiative. 2013. Available: http://www.adni-info.org/Home.html (Accessed February 28, 2018).

Armstrong, RA 2014. A critical analysis of the "amyloid cascade hypothesis". *Folia Neuropathol,* **52**, 211–25.

Avramouli, A & Vlamos, PM 2017. Integrating omic technologies in Alzheimer's Disease. *Adv Exp Med Biol,* **987**, 177–84.

Barrett, AM, Orange, W, Keller, M, Damgaard, P & Swerdlow, RH 2006. Short-term effect of dementia disclosure: How patients and families describe the diagnosis. *J Am Geriatr Soc,* **54**, 1968–70.

Bateman, RJ, Aisen, PS, De Strooper, B, Fox, NC, Lemere, CA, Ringman, JM, Salloway, S, Sperling, RA, Windisch, M & Xiong, C 2011. Autosomal-dominant Alzheimer's disease: A review and proposal for the prevention of Alzheimer's disease. *Alzheimers Res Ther,* **3**, 1.

Beach, TG 2013. Alzheimer's disease and the "Valley Of Death": Not enough guidance from human brain tissue? *J Alzheimers Dis,* 33 Suppl **1**, S219–33.

Berk, C & Sabbagh, MN 2013. Successes and failures for drugs in late-stage development for Alzheimer's disease. *Drugs Aging,* **30**, 783–92.

Beyreuther, K & Masters, CL 1991. Amyloid precursor protein (APP) and beta A4 amyloid in the etiology of Alzheimer's disease: precursor-product relationships in the derangement of neuronal function. *Brain Pathol,* **1**, 241–51.

Biotechnology Innovation Organization, B, Amplion 2016. Clinical development success rates 2006–2015. *BIO Industry Analysis,* **109**, 1–28.

Birks, J 2006. Cholinesterase inhibitors for Alzheimer's disease. *Cochrane Database Syst Rev,* **1**, CD005593.

Blennow, K, De Leon, MJ & Zetterberg, H 2006. Alzheimer's disease. *Lancet,***368**, 387–03.

Brailovskaia, J & Margraf, J 2018. What does media use reveal about personality and mental health? An exploratory investigation among German students. *PLoS One,* **13**, e0191810.

Brownstein, JS, Freifeld, CC & Madoff, LC 2009. Digital disease detection—Harnessing the Web for public health surveillance. *N Engl J Med,* **360**, 2153–5, 2157.

Cavazos-Rehg, PA, Krauss, MJ, Spitznagel, EL, Lowery, A, Grucza, RA, Chaloupka, FJ & Bierut, LJ 2015. Monitoring of non-cigarette tobacco use using Google Trends. *Tob Control,* **24**, 249–55.

Corder, EH, Saunders, AM, Strittmatter, WJ, Schmechel, DE, Gaskell, PC, Small, GW, Roses, AD, Haines, JL & Pericak-Vance, MA 1993. Gene dose of apolipoprotein E type 4 allele and the risk of Alzheimer's disease in late onset families. *Science,* **261**, 921–3.

Cummings, J, Lee, G, Mortsdorf, T, Ritter, A & Zhong, K 2017. Alzheimer's disease drug development pipeline: 2017. *Alzheimers Dement (N Y),* **3**, 367–84.

Cummings, JL, Morstorf, T & Zhong, K 2014. Alzheimer's disease drug-development pipeline: Few candidates, frequent failures. *Alzheimers Res Ther,* **6**, 37.

Dimasi, JA, Grabowski, HG & Hansen, RW 2016. Innovation in the pharmaceutical industry: New estimates of R&D costs. *J Health Econ,* **47**, 20–33.

Elhai, JD, Hall, BJ & Erwin, MC 2018. Emotion regulation's relationships with depression, anxiety and stress due to imagined smartphone and social media loss. *Psychiatry Res,* **261**, 28–34.

Farlow, M 2002. A clinical overview of cholinesterase inhibitors in Alzheimer's disease. *Int Psychogeriatr,* 14 Suppl **1**, 93–126.

Farrer, LA, Cupples, LA, Haines, JL, Hyman, B, Kukull, WA, Mayeux, R, Myers, RH et al 1997. Effects of age, sex, and ethnicity on the association between apolipoprotein E genotype and Alzheimer disease. A meta-analysis. APOE and Alzheimer Disease Meta Analysis Consortium. *JAMA,* **278**, 1349–56.

Finehout, EJ, Franck, Z, Choe, LH, Relkin, N & Lee, KH 2007. Cerebrospinal fluid proteomic biomarkers for Alzheimer's disease. *Ann Neurol,* **61**, 120–9.

Gentleman, SM, Greenberg, BD, Savage, MJ, Noori, M, Newman, SJ, Roberts, GW, Griffin, WS & Graham, DI 1997. A beta 42 is the predominant form of amyloid beta-protein in the brains of short-term survivors of head injury. *Neuroreport,* **8**, 1519–22.

Ginsberg, J, Mohebbi, MH, Patel, RS, Brammer, L, Smolinski, MS & Brilliant, L 2009. Detecting influenza epidemics using search engine query data. *Nature,* **457**, 1012–4.

Goate, A, Chartier-Harlin, MC, Mullan, M, Brown, J, Crawford, F, Fidani, L, Giuffra, L et al. 1991. Segregation of a missense mutation in the amyloid precursor protein gene with familial Alzheimer's disease. *Nature,* **349**, 704–6.

Hardy, J & Selkoe, DJ 2002. The amyloid hypothesis of Alzheimer's disease: Progress and problems on the road to therapeutics. *Science,* **297**, 353–6.

Hardy, J 2009. The amyloid hypothesis for Alzheimer's disease: A critical reappraisal. *J Neurochem,* **110**, 1129–34.

Hardy, JA & Higgins, GA 1992. Alzheimer's disease: The amyloid cascade hypothesis. *Science,* **256**, 184–5.

Hay, M, Thomas, DW, Craighead, JL, Economides, C & Rosenthal, J 2014. Clinical development success rates for investigational drugs. *Nat Biotechnol,* **32**, 40–51.

Hebert, LE, Weuve, J, Scherr, PA & Evans, DA 2013. Alzheimer disease in the United States (2010-2050) estimated using the 2010 census. *Neurology,* **80**, 1778–83.

Hynd, MR, Scott, HL & Dodd, PR 2004. Glutamate-mediated excitotoxicity and neurodegeneration in Alzheimer's disease. *Neurochem Int,* **45**, 583–95.

Inui, Y, Ito, K, Kato, T & Group, S-JS 2017. Longer-term investigation of the value of 18F-FDG-PET and magnetic resonance imaging for predicting the conversion of mild cognitive impairment to Alzheimer's Disease: A multicenter study. *J Alzheimers Dis,* **60**, 877–87.

Kelley, AS, Mcgarry, K, Gorges, R & Skinner, JS 2015. The burden of health care costs for patients with dementia in the last 5 years of life. *Ann Intern Med,* **163**, 729–36.

Klunk, WE, Engler, H, Nordberg, A, Wang, Y, Blomqvist, G, Holt, DP, Bergstrom, M et al 2004. Imaging brain amyloid in Alzheimer's disease with Pittsburgh Compound-B. *Ann Neurol,***55**, 306–19.

Kochanek, KD, Murphy, SL, Xu, J & Tejada-Vera, B 2016. Deaths: Final data for 2014. *Natl Vital Stat Rep,* **65**, 1–122.

Konishi, K, Joober, R, Poirier, J, Macdonald, K, Chakravarty, M, Patel, R, Breitner, J & Bohbot, VD 2018. Healthy versus entorhinal cortical atrophy identification in asymptomatic APOE4 carriers at risk for Alzheimer's Disease. *J Alzheimers Dis,* **61**, 1493–1507.

Kryscio, RJ, Abner, EL, Caban-Holt, A, Lovell, M, Goodman, P, Darke, AK, Yee, M, Crowley, J & Schmitt, FA 2017. Association of antioxidant supplement use and dementia in the prevention of Alzheimer's Disease by vitamin E and selenium trial (PREADViSE). *JAMA Neurol,* **74**, 567–73.

Kryscio, RJ, Abner, EL, Schmitt, FA, Goodman, PJ, Mendiondo, M, Caban-Holt, A, Dennis, BC et al 2013. A randomized controlled Alzheimer's disease prevention trial's evolution into an exposure trial: The PREADViSE Trial. *J Nutr Health Aging,* **17**, 72–5.

Levy-Lahad, E, Wasco, W, Poorkaj, P, Romano, DM, Oshima, J, Pettingell, WH, Yu, CE et al 1995. Candidate gene for the chromosome 1 familial Alzheimer's disease locus. *Science,* **269**, 973–7.

Liu, CC, Liu, CC, Kanekiyo, T, Xu, H & Bu, G 2013. Apolipoprotein E and Alzheimer disease: Risk, mechanisms and therapy. *Nat Rev Neurol,* **9**, 106–18.

Mielke, MM & Lyketsos, CG 2006. Lipids and the pathogenesis of Alzheimer's disease: Is there a link? *Int Rev Psychiatry,* **18**, 173–86.

Mills, SM, Mallmann, J, Santacruz, AM, Fuqua, A, Carril, M, Aisen, PS, Althage, MC et al 2013. Preclinical trials in autosomal dominant AD: Implementation of the DIAN-TU trial. *Rev Neurol (Paris)*, **169**, 737–43.

Morris, E, Chalkidou, A, Hammers, A, Peacock, J, Summers, J & Keevil, S 2016. Diagnostic accuracy of (18)F amyloid PET tracers for the diagnosis of Alzheimer's disease: A systematic review and meta-analysis. *Eur J Nucl Med Mol Imaging*, **43**, 374–85.

Morris, JC 2005. Early-stage and preclinical Alzheimer disease. *Alzheimer Dis Assoc Disord*, **19**, 163–5.

Mullard, A 2017. BACE inhibitor bust in Alzheimer trial. *Nat Rev Drug Discov*, **16**, 155.

Ncbi Resource Coordinators 2018. Database resources of the national center for biotechnology information. *Nucleic Acids Res*, **46**, D8–13.

Nuti, SV, Wayda, B, Ranasinghe, I, Wang, S, Dreyer, RP, Chen, SI & Murugiah, K 2014. The use of google trends in health care research: A systematic review. *PLoS One*, **9**, e109583.

Olivares, D, Deshpande, VK, Shi, Y, Lahiri, DK, Greig, NH, Rogers, JT & Huang, X 2012. N-methyl D-aspartate (NMDA) receptor antagonists and memantine treatment for Alzheimer's disease, vascular dementia and Parkinson's disease. *Curr Alzheimer Res*, **9**, 746–58.

Omalu, BI, Dekosky, ST, Minster, RL, Kamboh, MI, Hamilton, RL & Wecht, CH 2005. Chronic traumatic encephalopathy in a National Football League player. *Neurosurgery*, **57**, 128–34; discussion 128-34.

Ostrowitzki, S, Lasser, RA, Dorflinger, E, Scheltens, P, Barkhof, F, Nikolcheva, T, Ashford, E et al. 2017. A phase III randomized trial of gantenerumab in prodromal Alzheimer's disease. *Alzheimers Res Ther*, **9**, 95.

Parsons, CG, Danysz, W & Quack, G 1999. Memantine is a clinically well tolerated N-methyl-D-aspartate (NMDA) receptor antagonist—A review of preclinical data. *Neuropharmacology*, **38**, 735–67.

Qi, JP, Wu, H, Yang, Y, Wang, DD, Chen, YX, Gu, YH & Liu, T 2007. Cerebral ischemia and Alzheimer's disease: The expression of amyloid-beta and apolipoprotein E in human hippocampus. *J Alzheimers Dis*, **12**, 335–41.

Reiman, EM, Langbaum, JB, Fleisher, AS, Caselli, RJ, Chen, K, Ayutyanont, N, Quiroz, YT, Kosik, KS, Lopera, F & Tariot, PN 2011. Alzheimer's Prevention Initiative: A plan to accelerate the evaluation of presymptomatic treatments. *J Alzheimers Dis*, 26 Suppl **3**, 321–9.

Reitz, C 2012. Alzheimer's disease and the amyloid cascade hypothesis: A critical review. *Int J Alzheimers Dis*, **2012**, 369808.

Reitz, C 2016. Toward precision medicine in Alzheimer's disease. *Ann Transl Med*, **4**, 107.

Roberts, GW 1988. Immunocytochemistry of neurofibrillary tangles in dementia pugilistica and Alzheimer's disease: Evidence for common genesis. *Lancet*, **2**, 1456–8.

Roberts, GW, Gentleman, SM, Lynch, A, Murray, L, Landon, M & Graham, DI 1994. Beta amyloid protein deposition in the brain after severe head injury: Implications for the pathogenesis of Alzheimer's disease. *J Neurol Neurosurg Psychiatry*, **57**, 419–25.

Schootman, M, Toor, A, Cavazos-Rehg, P, Jeffe, DB, Mcqueen, A, Eberth, J & Davidson, NO 2015. The utility of Google Trends data to examine interest in cancer screening. *BMJ Open*, **5**, e006678.

Schubert, D 2005. Glucose metabolism and Alzheimer's disease. *Ageing Res Rev*, **4**, 240–57.

Selkoe, DJ & Hardy, J 2016. The amyloid hypothesis of Alzheimer's disease at 25 years. *EMBO Mol Med*, **8**, 595–08.

Selkoe, DJ 1991. The molecular pathology of Alzheimer's disease. *Neuron*, **6**, 487–98.

Seneff, S, Wainwright, G & Mascitelli, L 2011. Nutrition and Alzheimer's disease: The detrimental role of a high carbohydrate diet. *Eur J Intern Med*, **22**, 134–40.

Shaw, LM, Korecka, M, Clark, CM, Lee, VM & Trojanowski, JQ 2007. Biomarkers of neurodegeneration for diagnosis and monitoring therapeutics. *Nat Rev Drug Discov*, **6**, 295–03.

Shensa, A, Sidani, JE, Dew, MA, Escobar-Viera, CG & Primack, BA 2018. Social media use and depression and anxiety symptoms: A cluster analysis. *Am J Health Behav*, **42**, 116–28.

Sherrington, R, Rogaev, EI, Liang, Y, Rogaeva, EA, Levesque, G, Ikeda, M, Chi, H et al. 1995. Cloning of a gene bearing missense mutations in early-onset familial Alzheimer's disease. *Nature*, **375**, 754–60.

Sperling, RA, Aisen, PS, Beckett, LA, Bennett, DA, Craft, S, Fagan, AM, Iwatsubo, T et al. 2011. Toward defining the preclinical stages of Alzheimer's disease: Recommendations from the National Institute on Aging-Alzheimer's Association workgroups on diagnostic guidelines for Alzheimer's disease. *Alzheimers Dement*, **7**, 280–92.

Sperling, RA, Rentz, DM, Johnson, KA, Karlawish, J, Donohue, M, Salmon, DP & Aisen, P 2014. The A4 study: Stopping AD before symptoms begin? *Sci Transl Med*, **6**, 228fs13.

Steen, E, Terry, BM, Rivera, EJ, Cannon, JL, Neely, TR, Tavares, R, Xu, XJ et al. 2005. Impaired insulin and insulin-like growth factor expression and signaling mechanisms in Alzheimer's disease—Is this type 3 diabetes? *J Alzheimers Dis*, **7**, 63–80.

Strohle, A, Schmidt, DK, Schultz, F, Fricke, N, Staden, T, Hellweg, R, Priller, J, Rapp, MA & Rieckmann, N 2015. Drug and exercise treatment of Alzheimer Disease and mild cognitive impairment: A systematic review and meta-analysis of effects on cognition in randomized controlled trials. *Am J Geriatr Psychiatry*, **23**, 1234–249.

U.S. Centers for Medicare & Medicaid Services, OOTA, National Health Statistics Group. *National Health Expenditures by Type of Service and Source of Funds: Calendar Years 1960-2016* [Online]. Baltimore, MD: U.S. Centers for Medicare & Medicaid Services. Available: https://www.cms.gov/Research-Statistics-Data-and-Systems/Statistics-Trends-and-Reports/NationalHealthExpendData/NationalHealthAccountsHistorical.html [Accessed January 01, 2018 2018].

Wang, HW, Chen, DR, Yu, HW & Chen, YM 2015. Forecasting the incidence of dementia and dementia-related outpatient visits with google trends: Evidence grom Taiwan. *J Med Internet Res*, **17**, e264.

Weuve, J, Hebert, LE, Scherr, PA & Evans, DA 2014. Deaths in the United States among persons with Alzheimer's disease (2010–2050). *Alzheimer's & Dementia*, **10**, e40–46.

Wilkins, JM & Trushina, E 2017. Application of metabolomics in Alzheimer's Disease. *Front Neurol*, **8**, 719.

Wimo, A, Guerchet, M, Ali, GC, Wu, YT, Prina, AM, Winblad, B, Jonsson, L, Liu, Z & Prince, M 2017. The worldwide costs of dementia 2015 and comparisons with 2010. *Alzheimers Dement*, **13**, 1–7.

Yiannopoulou, KG & Papageorgiou, SG 2013. Current and future treatments for Alzheimer's disease. *Ther Adv Neurol Disord*, **6**, 19–33.

Zaleta, AK, Carpenter, BD, Porensky, EK, Xiong, C & Morris, JC 2012. Agreement on diagnosis among patients, companions, and professionals after a dementia evaluation. *Alzheimer Dis Assoc Disord*, **26**, 232–7.

Biomarkers in Lung Cancer

8.8

George Santis, Richard Lee, and Crispin T. Hiley

Contents

INTRODUCTION

Successes in biomarker development in lung cancer have been dominated by companion diagnostics for targeted therapies such as epidermal growth factor receptor (EGFR) inhibitors and more recently anti-PD-L1 immunotherapy in non-small cell lung cancer (NSCLC). Although there is a growing literature of putative prognostic or predictive biomarkers in NSCLC, there are few examples with clinical applications. Major areas of development focus on the following:

1. Early detection and diagnosis, because poor survival is attributed to late presentation with advanced, treatment-refractory disease
2. Improved efficacy and cost-effectiveness of targeted therapies through better subgroup selection
3. Detection of progressive disease and treatment resistance, at single or multiple sites, with the aim of guiding subsequent treatment stratification
4. Improved prognostication through segregation of phenotypes

The latter could enable the patient and clinician to plan for deterioration, to stratify investigations and therapies and potentially to rationalize resources according to most urgent need, although as yet there is little evidence to provide a valid framework through which this is usefully applied.

Challenges derive from the need for clinically applicable biomarkers that discriminate neoplasia from normal with high positive and negative predictive values, in a clinically appropriate timeframe and in a manner that is acceptable to the patient, that is, as defined by the invasiveness, convenience and relative benefit of the diagnostic tests required. In many circumstances, tumor material is also sparse (due to small tumor size or poorly accessible lesions) and signal clarity can be further hindered by high stroma to tumor cell ratio (Hiley et al. 2016). The biomarkers discussed below may be used for more than one application.

BIOMARKERS IN DIAGNOSIS AND EARLY DETECTION

Currently, diagnosis and staging information is derived from imaging using computed tomography (CT)/PET scanning followed by in most cases sampling of primary tumor and/or metastatic sites with minimally invasive techniques such as bronchoscopy, endobronchial ultrasound or CT-guided percutaneous biopsy (Baldwin et al. 2011). Urgent referrals for "suspected cancer" based on abnormal x-rays or symptoms such as hemoptysis can be difficult to discriminate from other pathologies, leading to considerable use of resources and anxiety for the patient.

Although small nodules and tumors may reflect early, potentially curable cancers, a significant proportion also represents nonmalignant disease. Such lesions are technically challenging to biopsy, further limiting material available for analysis in assays. Sampling may be improved by computer-aided navigational tools for small, peripheral lesions difficult to locate by bronchoscopic approaches due to the labyrinthine nature of the bronchial airways (Marino et al. 2016). Similarly improved procedural training may be required as differences in sampling accuracy between trainee and accomplished bronchoscopists become increasingly relevant to provide high quality biopsy samples for molecular testing (Edell and Krier-Morrow 2010, Leong et al. 2013, Silvestri et al. 2015, Kular et al. 2016). Presently, early diagnosis of lung cancer is dominated by CT screening (Ruparel and Janes 2016, Yousaf-Khan et al. 2017). This broad topic is outside the scope of this section, but future screening programs are likely to incorporate the technologies discussed below to enhance their sensitivity and specificity.

Preinvasive vs. Invasive Disease

A particular challenging area is the detection and management of very early stage disease, for example, pulmonary nodules smaller than 10 mm. These are discovered incidentally during investigation for respiratory symptoms; or while imaging other thoracic abnormalities such as aneurysms, cardiac imaging or the peripheries of studies of the head/neck or abdomen; or following a screening CT thorax. The ability to decipher whether such lesions represent invasive or preinvasive disease would represent an important marker of transition in disease phenotype (McCaughan et al. 2011, Koper et al. 2017). Enhancing the accuracy of molecular techniques for such small lesions might be best achieved using localized techniques, for example dedicated bronchoscopic tools to detect endoscopic traits that represent premalignant changes, such as blue-light or autofluorescence bronchoscopy that could additionally be augmented in the future by incorporation of fluorescently tagged marker proteins such as MET (Sun et al. 2011, Burggraaf et al. 2015). The potential to utilize combined histological and genomic analysis to establish if two pulmonary lesions are clonally related and therefore representative of invasive or otherwise independent metachronous tumors and distinguishing whether a lesion is a primary lung cancer or metastatic from another site are also novel possibilities for the future application of these technologies (Girard et al. 2009a, 2009b, Girard et al. 2010, Murphy et al. 2014).

Enhanced Phenotyping of Lung Cancer Subtypes

National guidelines mandate examination of basic immunohistochemical biomarkers such as thyroid transcription factor-1 (TTF1), p63 and cytokeratin 7 (CK7) to differentiate adenocarcinoma from squamous cell carcinoma NSCLC subtypes (Travis et al. 2011, Righi et al. 2014). Such markers are not ubiquitously expressed; for example, TTF1 expression may be positive in an average of 72% of adenocarcinoma specimens in one study (Yatabe et al. 2002). Existing classifications could be enhanced by sequencing data to improve and extend upon existing subtypes (LeBlanc and Marra 2015, Kato et al. 2016). This is based on the fact that distinct genomic aberrations and epigenetic patterns have been observed for the main tumor subtypes. ALK, BRAF, EGFR, ERBB2, KRAS, and STK11 mutations or rearrangements are more common in adenocarcinoma; mutations in DDR2, FGFR3, and NFE2L2 in squamous cell lung cancer; TP53, RB1, MYCN alterations in small cell lung cancer (SCLC); and in mesothelioma, BRCA1-associated protein 1 (BAP1), NF2 (38.1%), and CDKN2A/B loss. Genomic and epigenetic-based classifiers could therefore be used to segregate the major lung cancer subtypes although these approaches have not yet found a role in differentiating clinically relevant phenotypes in routine practice (The Clinical Lung Cancer Genome and Network Genomic 2013, Tomasetti et al. 2016).

Preclinical studies that have explored the role of common mutations such as those in EGFR and KRAS in lung progenitor cells also provide evidence of the importance of these genes in lung cancer evolution that could be relevant to strategies for designing more effective therapies (Salgia and Skarin 1998, Mao 2002). This literature suggests that the cell of origin is an important determinant of the mutational events that lead to the major lung cancer subtypes. Examples include neuroendocrine cells of origin in SCLC and type II or Clara cells in adenocarcinoma (Sutherland et al. 2011, Sutherland et al. 2014). This is of relevance to the scientific and diagnostic principles that differentiate cancer from nonmalignant parenchymal cells and the phenotypic features unique to major lung cancer subtypes. Susceptibility of such originating "cancer stem cells" to therapeutic agents may also be an important determinant of the repertoire of resistant tumor cells primed for later recurrence.

Understanding somatic events in normal tissue is important to avoid confounding the specificity of early detection biomarker, for example, sun-exposed skin may display a range of genomic abnormalities without other evidence of malignancy or genomic damage in other premalignant specimens (Martincorena and Campbell 2015, Martincorena et al. 2015, Campbell et al. 2016, Kadara et al. 2016, Tremblay et al. 2016). Decoding the clinical relevance of early detection biomarkers requires better understanding of the timing and order of genomic events in tumor evolution. Certain aberrations, for example, TP53 mutations, KRAS, p16 promoter hyper-methylation and genome doubling events could occur as early, clonal events and thus serve as earlier markers of a cancer phenotype. On-going studies exploring the role of the evolution of clonal versus subclonal genomic events in determining tumor heterogeneity will help us to understand this theme (Abbosh et al. 2017, Jamal-Hanjani et al. 2017). Data from studies such as TRACERX is providing valuable insights into the role of individual mutations at different stages of tumor evolution—for example, this data suggests that EGFR always arises as an early, clonal event whereas KRAS can be a clonal event in some tumors but subclonal in others. Similarly chromosomal instability and copy number changes appear to be of greater significance than heterogeneity in individual point mutations (Jamal-Hanjani et al. 2017).

The oncogene expediency hypothesis suggests that a single driver mutation is the dominant event yet genetic epistasis leads to gene–gene interactions that result in drivers from mutations that individually would not be considered significant. With this in mind, many genomic events, currently examined individually are likely to be functionally closely related—for example, EGFR and KRAS mutual exclusivity, EGFR and MET crosstalk and promiscuity between HER family dimerization partners. Future approaches to biomarker design and targeted therapy will need to incorporate these principles to be greatest efficacy (Mao 2002, Engelman et al. 2007, The Clinical Lung Cancer Genome and Network Genomic 2013, The Cancer Genome Atlas Research 2014, Rusch 2016, Seifert et al. 2016).

Such findings are likely to challenge the margins drawn between existing histopathological subtypes and the inflexion point delineating normal tissue versus dysplasia and cancer. Although not yet used clinically, genomics based approaches to classifying lung cancer demonstrate good concordance to existing histopathological classifiers and may help to more accurately define those whom can not be subtyped by existing methods (The Clinical Lung Cancer Genome and Network Genomic 2013).

BIOMARKERS TO STRATIFY THERAPY

Biomarkers are also of clear importance in therapeutic trials because they enable medications to be compared more accurately between populations that share mutual genetic aberrations. These enable rationalization of therapy according to the biology of individual tumors and therefore personalized approaches, which should result in increased efficacy and reduced side effects. These "companion" biomarkers are pertinent at the initiation of therapy and for detecting relapse/resistance to therapy but are often easiest to study in the context of selection of later lines of therapy in the metastatic setting. Primarily this evidence hence comes from study of patients with advanced stage disease. As confidence in the use of biomarkers and targeted agents grows there will also be further applications in the context of adjuvant and potentially radical approaches treatment.

Biomarkers in the Radical Setting

These studies predominantly provide observational data that describe the natural history of patients undergoing radical therapy in association with its genetic or molecular correlates in tumor material obtained from biopsy or resected tumor specimens. Biomarkers in this area remain poorly defined. There is mixed data as to the role of EGFR as a prognostic factor in surgically resected patients. Although preclinical indices, suggest that it is biologically plausible that constitutive activation of EGFR by activating mutations would dictate an aggressive clinical phenotype this is not well defined from clinical series (Fang and Wang 2014, Zhang et al. 2014). Correlation is also seen between p53 expression or circulating antibodies against p53 and overall survival (Mattioni et al. 2015), but as yet biomarkers are not available that dictate one preferred radical approach over another. Finally there is the potential to combine radical radiotherapy with targeted agents in patients with actionable driver mutations but there remains a need for further study in each of these areas (Giaccia 2014, McDonald and Popat 2014).

Biomarkers in the Adjuvant Setting

The adjuvant setting provides the opportunity to combine resected tumor laboratory/molecular pathology data with trial data comparing outcomes to drug therapies. Subtype classification based on morphology and immunohistochemistry (i.e., adenocarcinoma vs. squamous cell carcinoma) has been shown to predict disease specific outcomes (Tsao et al. 2015). Although early trials have associated TP53 mutations commutated with KRAS or EGFR with poor survival in adjuvant chemotherapy, analysis from the LACE bio consortium found that TP53 commutation with KRAS or EGFR was not a significant prognostic marker in patients with resected early stage NSCLC. The same group has previously reported similar findings for KRAS and P53 independently (The Clinical Lung Cancer Genome and Network Genomic 2013, Ma et al. 2016, Shepherd et al. 2016).

Another marker, the excision repair cross-complementation group 1 (ERCC1) protein indicates DNA repair capacity and has been studied as a marker of cisplatin sensitivity—on-going challenges with the performance of this marker in clinical applications remain a hurdle (Olaussen et al. 2006, Friboulet et al. 2013). Other putative markers of adjuvant chemotherapy responsiveness include BRCA1, VEGF and beta tubulin (Reiman et al. 2012, Boros et al. 2017, Drilon et al. 2016, Zang et al. 2017).

The application of biomarkers to targeted therapy in the adjuvant setting is limited by the lack of evidence in this setting both in the context of preceding radical chemoradiation and also surgical resection. The on-going ALCHEMIST trial will add further data in this remit by screening resected non-squamous lung cancers for EGFR mutations and ALK translocations to incorporate these patients into randomized trials for Erlotinib or Crizotinib, respectively. An additional arm of this trial will explore the role of PDL1 targeted therapy in the adjuvant setting (NCT02595944) (Kelly et al. 2008, Goss et al. 2013, Govindan et al. 2015).

Biomarkers in Advanced/Metastatic Lung Cancer

Subclassification of patients for targeted molecular agents using "companion tests" allows judicious rationalization and accurate, effective and safe delivery of cutting-edge therapies. Genomic classifications based on current approaches have been able to identify a potentially actionable, oncogenic driver in over half of NSCLC tumors (The Clinical Lung Cancer Genome and Network Genomic 2013). Similarly, next generation sequencing (NGS) in mesothelioma identifies a high proportion of actionable molecular abnormalities (Kato et al. 2016). A range of targeted agents have hence been linked with companion biomarkers (Bansal et al. 2016) yet more pragmatic analyses suggest that these have still not gained widespread use in NSCLC (Shedden et al. 2008).

One of the best-established therapeutic biomarkers of this nature in lung cancer is *EGFR* mutation analysis for the selection of EGFR TKI (Ellison et al. 2012). Activating mutations such as *L858R* or deletions in *Exon 19* determine increased sensitivity to EGFR TKI (Sharma et al. 2007, Shan et al. 2012). Initial development of EGFR TKI identified increased efficacy in young, eastern Asian, female, never smokers with adenocarcinoma in whom such *EGFR* mutations were subsequently found to be enriched. It is likely that EGFR *L858R* and *exon 19*- mutated lung adenocarcinomas represent phenotypically distinct entities that respond differently to TKI therapy. Other mutations, such as in exon 20 have subsequently been found that prevent EGFR TKI activity due to ATP binding pocket conformational changes such as is seen with *T790M*, the major EGFR TKI resistance mutation (Mok 2011, Rosell et al. 2012, Kuan et al. 2015). Although EGFR plays an important role in both NSCLC and glioblastoma multiforme the mutations in the latter are found predominantly in the extracellular domain, compared with the kinase domain mutations seen in NSCLC. This results in a different sensitivity to EGFR inhibitors between these two tumor types, further demonstrating the importance of understanding the target and mechanism by which such agents act (Vivanco et al. 2012). Other common biomarkers under development for use in clinical practice or research trials include ALK, ROS, RET, NTRK and MET (Table 8.8.1).

Predicting Response to Immunotherapy

PDL1 assays

Response to the T-lymphocyte immune checkpoint inhibitors, for example, pembrolizumab, nivolumab, darvalumab or atezolimumab correlate with PDL1 expression on the surface of tumor cells. Staining for PDL1 has therefore been utilized as a biomarker for this approach. The literature remains mixed with a number of studies supporting a predictive value for overall survival and drug efficacy whereas in others, PDL1 was not predictive of benefit (Herbst et al. 2014, Brahmer et al. 2015, Garon et al. 2015, Herbst et al. 2016). This is attributed to difficulties in establishing a consistent performance of biomarkers across trials. Standardization of assay performance, for example, related to inter-operator variability in tissue preparation, antibody selection, staining protocol, thresholds and interpretation are certainly important limitations of the repeatability of this biomarker (McLaughlin et al. 2016, Socinski et al. 2016, Ratcliffe et al. 2017). Percentage and character of T cell infiltration, intratumoral localization of PDL1 positive cells (tumor vs. immune cells) and the role of PD-L2 may also be relevant (Yearley 2015). Furthermore, PDL1 can be upregulated by tumor

TABLE 8.8.1 Genomic biomarkers used in treatment stratification in advanced/metastatic lung cancer

Anaplastic lymphoma kinase (ALK): Fusion genes arise from genomic rearrangement, e.g., ALK-ROS or EML4-ALK resulting in upregulation of the Ras-Raf pathway. Inhibitors: Crizotinib, ceritinib, brigatinib and alectinib (Soda et al. 2007, Mok 2011, Shaw et al. 2014, von Laffert et al. 2014, Nokihara 2016, Soria et al. (2017)

Kirsten rat sarcoma viral oncogene homolog (KRAS): Mutually exclusive with EGFR suggesting inappropriate to target with EGFR TKI. Common mutations seen at G12 and 60/61. Not currently druggable directly although unconfirmed early phase trial evidence that may be accessible to downstream targets, e.g., MEK and potential prognostic significance (Eberhard et al. 2005, Jänne et al. 2013, Abdel-Rahman 2016, Zhang et al. 2016).

MET: Amplification or exon 14 skip mutations potentially in crosstalk with EGFR (Engelman et al. 2007, Tanaka et al. 2012, Vassal et al. 2015, Zhang et al. 2016, Drilon et al. 2017).

Neurotrophic tyrosine kinase receptor 1 (NTKR): Rare rearrangements inhibited by crizotinib or by specific inhibitors entrectinib, AZD7451 (Chong et al. 2017).

ROS1: Fusion gene. ALK inhibitors may also be applicable (Vassal et al. 2015, Shaw et al. 2016, Michels et al. 2017).

RET: Fusion gene: e.g., rearrangements. Inhibitors: Cabozantinib, Vandetanib or Sunitinib (Drilon et al. 2016, Gautschi et al. 2017, Yoh et al. 2017).

Other targets: *BRAF; HER2 (mutation and amplification; VEGF (increased expression); ERCC1; MUC1/16* and *mesothelin* (Kabbinavar et al. 2014, Bansal et al. 2016, Malottki et al. 2016, Zhang et al. 2016, Frezzetti et al. 2017, Kosaka et al. 2017).

intrinsic factors such as MAPK signaling or extrinsic factors such as hypoxia. PDL1 inhibition also performs differently in EGFR mutated versus EGFR wild-type lung cancer. Although EGFR positive tumors typically have a low mutational burden, EGFR can upregulate PDL1—hence such tumors are PDL1 positive but not immunogenic. PD-L1 levels have also been observed to decrease in response to EGFR TKI in cell lines sensitive to TKI and biomarkers that demonstrate interaction between these different pathways may direct a role for co-treatment in such cases. Similarly higher PDL1 expression is observed in ALK positive NSCLC (Azuma et al. 2014, D'Incecco et al. 2015, Garon 2015, Langer et al. 2016).

In other cancers such as melanoma, responses are also seen in PDL1 negative tumors and although biomarkers have been also been developed for PDL1 therapy in such malignancies, it is unclear at present how well these will translate to lung cancer (Weber et al. 2013, Larkin et al. 2015, Schmid 2015).

The combination of anti PD1/PDL1 monoclonal antibodies with ipilimumab also identifies a role for detection of CTLA4 co-stimulation (Hellmann 2016). In addition, other markers quantifying the antitumor immune response are also likely to evolve, for example, serum IL-8, tumor-infiltrating lymphocytes and CD8/CD4 ratio (Kimura 2015, Sanmamed et al. 2016).

Immune infiltrate assays

Tumor infiltrating lymphocytes (TIL) that have migrated into the tumor microenvironment have been considered an important biomarker that could represent the contribution of the host tumor immune response and its susceptibility to treatment. The best means by which to quantify the nature of the antitumor response is not fully elucidated but a number of observations suggest a prognostic role for understanding whether a cytotoxic versus regulatory immune infiltrate is observed: CD8+ effector TIL correlate with survival in NSCLC (Woo et al. 2001, Hiraoka et al. 2006, Tanaka and Sakaguchi 2017) and FoxP3+ T regulatory lymphocytes and neutrophil:lymphocyte ratio have been shown to have prognostic significance in analyses of surgical resection of NSCLC (Petersen et al. 2006, Sarraf et al. 2009, Remark et al. 2015). Immune cell infiltrates thus have a potential role in predicting outcome following lung cancer resection but perhaps one of the most opportune directions of study lies in understanding how the presence of TIL impact the efficacy of immunotherapy—for example higher CD8 + TIL density has been used as a marker of Nivolumab response (Ribas et al. 2016, Haratani et al. 2018). As with prognostic applications, a number of themes remain to be clarified such as which aspect of the tumor is examined—incorporating stromal versus tumor cells; invasive margin versus center and which mechanism can be used for scoring and if this can be digitized/automated (Al-Shibli et al. 2008, Donnem et al. 2015, Donnem et al. 2016, Lizotte et al. 2016).

Immune expression assays and tumor mutational burden

Gene expression analysis to define the tumor-host interaction based on expression patterns can be performed on both tumor and immune cells, for example, through demonstration of altered immune evasion pathways within tumor cells or by representing markers of altered T cell subsets or chemokine profiles (Ulloa-Montoya et al. 2013, Herbst et al. 2014, Yu et al. 2019, Fehrenbacher et al. 2016, Vansteenkiste et al. 2016). By using machine learning approaches it has been demonstrated that tumor genotypes determine immuno-phenotypes in melanoma, which can be used to predict response to CTLA4 and PD1 directed therapy. Similarly other reports of a transcriptomic signal predictive of immunotherapy refractoriness suggest that augmented mutational burden in such patients is specific to particular tumor traits such as mesenchymal transition and angiogenesis (Hugo et al. 2016, Charoentong et al. 2017). Within this, analysis of tumor mutational burden (TMB) is also likely to be an important determinant of susceptibility to PDL1 therapy due to the impact of non-synonymous mutations in creating altered "non-self" peptide on MHC, therefore triggering a T cell response against tumor cells expressing these neoantigens (Rizvi et al. 2015).

In summary, integration of PDL1 status, mutational load and transcriptome analysis are likely of considerable importance to the effective delivery of immunotherapy, but such approaches also demonstrate the significant value of standard IHC and are furthermore, as yet unachievable in a clinically relevant time frame.

Biomarkers to Detect Recurrence or Resistance to Therapy

A number of drug resistance mechanisms are recognized, detection of which is important in determining further therapy (Table 8.8.1). Acquisition of a mutation that changes the efficacy of TKI activity such as *EGFR T790M* is common and accounts for ~50% of acquired resistance to first line EGFR TKI (Yun et al. 2008). Using *T790M* as a biomarker already directs use of later generation EGFR TKI (e.g., Osimertinib) (Mok et al. 2016, Papadimitrakopoulou 2016). Other common mechanisms include "kinase switching" that permits bypass of the inhibited signaling pathway—for example activation of MET can bypass EGFR signaling via HER3 activation in order to sustain downstream PI3K/AKT signaling (Engelman et al. 2007, Campbell, Amin et al. 2010). Finally, phenotypic change, for example, epithelial-mesenchymal transition (EMT) or conversion to SCLC is also recognized (Oser et al. 2015, Morgillo et al. 2016). Understanding these mechanisms is crucial to effective therapy beyond resistance, where sustained treatment can itself risk harm (Soria et al. 2015) (Table 8.8.2).

A common problem with biomarkers for detecting resistance is that current approaches may lack sensitivity. Tumor cell clones possessing the T790M mutation have been detected prior to EGFR TKI exposure, suggesting that such treatment provides a selective environment for such resistance mutations (Su et al. 2012). Deep sequencing of samples from *T790M*-positive NSCLC patients has also shown heterogeneity of resistance mechanism within the same patient suggesting the picture of resistance is more complex than initially appreciated—in one study of patients who had received first-line TKI, almost half had other mechanisms in addition to *T790M*, for example, outgrowth of a competing resistance mechanism (e.g., *MET* or

TABLE 8.8.2 Mechanisms of EGFR TKI resistance and potential biomarkers

MECHANISM	*ESTIMATED FREQUENCY*
EGFR target changes (e.g., T790M)	60%
Small-cell lung cancer conversion	10%
MET amplification	5%–10%
HER2 amplification	8%–13%
Epithelial–mesenchymal transition	1–2%
Others: BRAF, PI3K	2%–3%
Unknown	15%–20%

Source: Camidge, D.R. et al., *Nat. Rev. Clin. Oncol.*, 11, 473–481, 2014.
Note: EGFR, epidermal growth factor receptor; TKI, Tyrosine kinase inhibitor; "MET", MET proto-oncogene; "HER2", Human epidermal growth factor receptor 2; "BRAF", proto-oncogene B-Raf; "PI3K", Phosphoinositide 3-kinase.

HER2 amplification or *BRAF V600E*). Similar investigation of ALK/ROS resistance identified kinase domain mutations in only 33% and 14%–63%, respectively, again highlighting the important role of mechanisms other than kinase domain mutations (Chabon 2016, Gainor 2016, McCoach 2016).

Combined Biomarkers for Multi-target Therapy

In view of this, as with Highly Active Anti-Retroviral Therapy (HAART) for HIV and multiple antibiotic use in severe bacterial infection, it is hypothesized that combined pathway blockade, either through single agents or dual/triple chemotherapy may be more effective than a single agent alone, particularly in the setting of drug resistance, for example combined BRAF/MEK inhibition in *BRAFV600E*-mutant NSCLC, (Planchard et al. 2016). Here biomarkers may allow us to understand the synergy seen between certain therapies used together and the relative contribution of each. This may also allow us to predict which drug combinations are most likely to be effective together. Clearly there will additional complexity to surmount when combining biomarkers compared with those used individually. Although this has been demonstrated in other tumor types such as melanoma, it is not clear if this approach will be valid in lung cancer (Shedden et al. 2008). However, another potential role for combined targeting may lie in conditions with mutual pathogenesis—for example treating idiopathic pulmonary fibrosis and lung cancer simultaneously with Nintedanib may be an important means by which to explore this approach.

CIRCULATING BLOOD BIOMARKERS

Circulating blood or urine biomarkers may provide a sustainable supply of diagnostic clinical and research material as tumor physiology adapts to each round of treatment. These remain in early stages of development but blood, sputum or urine specimens provide a minimally invasive means by which to perform molecular analysis and could plausibly serve as adjuncts for lung cancer diagnosis entirely if a standard biopsy is unsafe, technically challenging or not desired by the patient. Further evolution of these technologies to achieve improved sensitivity and turnaround time remains but these tools will inevitably provide novel approaches in diagnostically challenging patients (Thunnissen 2003, Bordi et al. 2015, Gao et al. 2015, Vargas and Harris 2016).

Detection of circulating tumor DNA (ctDNA) in peripheral blood using technologies such as digital droplet PCR technology allow selective enrichment of common somatic mutations such as *EGFR L858R* or *T790M, KRAS, ROS1, ALK* and *BRAF*, which could be used to stratify therapy (Day et al. 2013). Blood biomarkers such as ctDNA are also good candidates to predict treatment failure. For example, ctDNA *EGFR* mutant alleles are observed to fall following targeted EGFR TKI treatment thus allowing monitoring of therapy. In some circumstances this could be used to preempt drug resistance, which could provide an early signal of relapse. For example, in principal, detection of ctDNA can be used to identify resistance mutations such as *T790M* prior to radiographic progression. As with other markers such as Ca125 in ovarian cancer, it is not clear, however, whether earlier detection and treatment change will have a significant impact on patient outcomes (Oxnard et al. 2014). Whole exome sequencing of plasma specimens has also been shown to identify treatment resistance noninvasively in a number of tumor types including NSCLC (Murtaza et al. 2013). Another technique, Capp-Seq utilizes a library of common mutations derived from population level data for a given type of cancer, to improve the sensitivity of detecting a mutation that denotes recurrence in an individual tumor at the single patient level (Newman et al. 2014). The most desirable approach will depend on the proportion of the circulating free DNA, which originates from tumor but when compared with tissue biopsies, sensitivity and specificity can reach 90% and 100%, respectively (Barrera et al. 2016).

By analyzing intact circulating tumor cells (CTC) it is also feasible to perform *EGFR* mutation detection in single cancer cells, which may allow such samples to be used as a surrogate for intratumoral heterogeneity (Hodgkinson et al. 2014).

Circulating tumor cells (CTC), although more expensive than ctDNA approaches provide a more accurate assessment of gene copy number and also provide details cell surface markers and expression data. These can in principle also be used to establish patient tumor cell culture CTC-derived xenografts (CDX), in mice as a source of tumor material for biomarker and therapy development that can be used to directly test a given patient's tumor against new drugs *ex vivo* (Hodgkinson et al. 2014).

In any case, these approaches may be most practically delivered in situations of high tumor burden, for example in the metastatic setting (Reckamp et al. 2016). In the context of small volume disease (i.e., in the adjuvant setting) the lower disease burden reduces the probability of detecting a variant on random blood draw. Advanced knowledge of the variants in a particular tumor allows a bespoke approach using variants specific to a given patient, sequenced to a very high depths so that there is a chance of finding a particular rare variant as employed in the TRACERx trial. It appears that this is more effective in squamous cell carcinoma where there are greater volumes of ctDNA to track than that derived from adenocarcinoma. The authors of this trial observed patients with lung adenocarcinoma who experienced recurrence without early detection of ctDNA. Using this approach, however, it was possible to track subclonal populations to determine which was responsible for relapse such that they could target therapy accordingly (Abbosh et al. 2017, Jamal-Hanjani et al. 2017).

THE CHALLENGES OF DEVELOPING LUNG CANCER BIOMARKERS

Patient Factors and Limitations of Tissue Sampling and Processing

Current investigations in lung cancer are invasive and not well suited to repeated attempts, particularly given the potential for complications in this highly comorbid group. Comparatively few patients are suitable for surgical resection, further limiting high volume specimens for research. The standard method of processing of lung tissue outside of research orientated academic centers is by "Formalin Fixed Paraffin Embedded" (FFPE) blocks. Technologies for biomarker detection that are amenable to this technique of preservation will be easier to adopt clinically although at present a number of artefacts introduced by the fixation process can hamper downstream assays (Wong et al. 2014). This is important because a new assay must be applicable to large volumes of specimens using high throughput approaches to be able to provide results on a clinically relevant timescale. National efforts are likely to be required to effectively coordinate these approaches on a large scale (Vargas and Harris 2016).

Various assay technologies are in development, many suitable for FFPE or low volume samples that are typical of lung cancer biopsies. The challenge is magnified in cases where even smaller sample sizes derived from minimally invasive approaches such as EBUS are obtained (Lee et al. 2013, Hiley et al. 2016). Techniques with higher sensitivity may be better suited to applications where tumor cell yield is expected to be low or where assay sensitivity is important, for example, smaller or necrotic tumors or those with low tumor cell proportions.

Tissue immunohistochemistry biomarkers give a proteomic viewpoint of cell function and are more readily deployable in clinical laboratory settings. However, many of the studies are early, proof-of-principle studies that lack robust control groups and methodology. These require scrutiny in trials more suited to clinical regulatory approval to ensure that antibodies are comparable between assays and that staining indicates clinically significant biological events (Friboulet et al. 2013). This problem highlights a recurring theme in biomarker development of harmonizing different platforms for analysis of biological pathways, for example, different methods to assess EGFR anomalies and PDL1 assay performance.

Challenges in the Clinical Application of Lung Cancer Genomics

NGS approaches have a higher sensitivity than Sanger sequencing and may also be better suited to smaller quantities of tumor cells. Targeted sequencing panels enable detection of rare variants. Here the constraints are tumor purity (mutant allele frequency in a given sample) and the sequencing coverage required to find it. This can be circumvented by enrichment of the tumor content, and/or by repeated rounds of deep sequencing, albeit with time and cost implications (Takano et al. 2005, Didelot et al. 2012). Coverage is broader with whole genome sequencing (WGS) compared with the capturing of only target coding regions, such as WES, or targeted gene panel sequencing but the latter allows a more focused approach—which either through capture hybridization, or through the use of amplicon based sequencing can further enhance turnaround times by sequencing lower volumes of input DNA. Targeted approaches are also better suited in identifying single nucleotide polymorphisms and insertion/deletions. Although techniques that depend upon DNA amplification by polymerase chain reaction (PCR) versus hybrid capture technology are prone to distortion of sequence quality and potential alteration of sequence proportions, this should be balanced against the ability to detect low frequency variants, time and cost constraints of each approach and the associated labor requirements.

Sequencing of nucleic acid to determine the mutation status has evolved significantly over the last two decades. The established approach has been mutation-specific PCR based tests with or without subsequent Sanger sequencing to confirm mutational status. For example companion diagnostic tests for EGFR TKI are based on PCR assays using primers specific to EGFR exon 19 deletions and L858R mutations. The use of CO-amplification at Lower Denaturation temperature PCR (COLD-PCR) enhances such reactions by preferentially amplifying mutant sequences in the PCR reaction and enhance sensitivity for smaller biopsy specimens including cytological approaches (Li et al. 2008, Santis et al. 2011, Rosell et al. 2012). More recently, clinically used platforms have evolved into multiplexed approaches and targeted NGS panels. These identify the common mutations relevant to NSCLC but also detect a number of additional genomic aberrations in other genes, which are as yet of unknown clinical significance.

Targeted approaches detect a smaller number of clinically relevant mutations, often linked to a targeted therapy for example, panel based approaches allow many DNA fragments relating to genes of interest to be processed in parallel, following either PCR amplification or hybrid capture, allowing analysis of multiple genomic aberrations simultaneously to maximize the information obtained from a small tissue sample (Metzker 2010). It is important that these panels are constructed to include a sufficiently broad coverage, taking into account whether WES or WGS will reveal the required information for the question being addressed. For example WES may not identify aberrations such as MET exon 14 mutations (Schrock et al. 2016). Although sequencing of the entire tumor exome or genome is possible, greater quantities of DNA are required to overcome the high stroma to tumor cell ratio of the typical lung biopsy and necessitate greater sequencing depth resulting in subsequent cost and analysis times that prohibit their clinical utility at present (The Clinical Lung Cancer Genome and Network Genomic 2013, de Bruin et al. 2014, The Cancer Genome Atlas Research 2014, Zhang et al. 2014, McGranahan et al. 2015).

Intratumor heterogeneity and alignment of molecular level data established from primary tumor to sites of metastases adds further complexity. This is of particular relevance to blood biomarkers where the relatively low proportion (<1%) of tumor DNA in approaches such as ctDNA may render this technology difficult to apply to patients with smaller tumors or those with multisite disease (Jr and Bardelli 2014, Gao et al. 2015, Jamal-Hanjani et al. 2017).

SUMMARY AND FUTURE DIRECTIONS

Precision medicine delivered through improved diagnostic, therapeutic and prognostic biomarkers will improve our understanding of how tumors arise, evolve and ultimately lead to clinically detectable disease (Vargas and Harris 2016). In the laboratory setting, surrogates of tumor cell phenotypes such as motility, migration and neo-angiogenesis are compared with true clinical end points of doubling time, tumor burden and evolution of metastasis. At the bedside, biomarker interpretation may require further investigation of anatomically informed approaches such as novel PET tracers or molecular endoscopy to expand upon a growing understanding of the relationship between primary tumor and metastatic lesions, both spatially and longitudinally (Abbosh et al. 2017).

There is a clear need for more robust clinical assessment of existing biomarkers and prompt integration of novel approaches into high quality clinical trials. This will not abrogate the need for novel pharmaceutical agents. Furthermore, bioinformatic approaches are likely to be important for biomarker selection and interpretation in the multiple layers of tumor cell signaling and control. There is increasing focus on big data, artificial intelligence and data mining. Such "biodigital" methods will bring new understanding of lung cancer phenotypes, disease patterns and treatment strategies through programs such as ASCO CancerLinq and Google DeepMind (Armstrong 2016, Miller 2016, Bui et al. 2017).

The use of targeted therapy has captured the imagination of the research community and political leaders internationally with national efforts in the U.S. precision medicine initiative, which plans to study 1 million patients (Vargas and Harris 2016); and similarly ambitious projects of the UK Stratified Medicine Program/National Lung Matrix Trial, U.S. Match trial and French ACSE trial. These programs aim to recruit lung cancer patients nationwide for first-line targeted agents based on molecular profiling (Vassal et al. 2015, Buzyn et al. 2016, Hiley et al. 2016). Approximately half of patients in such trials have exhibited a genetic alteration, which led to use of a targeted agent as first-line therapy in half of these cases (Cabanero and Tsao 2016). Nationwide efforts to deliver biomarker-driven access to targeted therapy intend to coordinate biomarker detection and provide a means of coordinating small patient numbers across a wider patient population (Biankin et al. 2015). This approach is perceived to be the answer to the increasingly complex, time consuming and costly task of bringing novel therapies to clinical practice. However, despite an increasing body of literature on potential prognostic and predictive biomarkers, few have made it into clinical practice and presently genomic changes seldom changes treatments for patients. Ongoing investment in drug development will be an important determinant of improved outcomes (Harris et al. 2018).

Major hurdles include the logistics of large-scale deployment of biomarker driven diagnosis and therapy and infrastructure changes that accelerate the time to access costly, novel treatments. We must ensure that biomarker research keeps pace with developments in targeted therapy and that they are incorporated into drug trials in a manner that the observations made are comparable between studies and different assays—this may require centralization of certain diagnostic services and efforts to align patient subsets across clinical trials. Strategies to account for heterogeneity, within tumors, patients and populations also need to be incorporated into trials and then standard practice. There must be particular emphasis on robustness and translatability to a clinical environment in a means suitable to deliver stratified therapy to the wider population in a cost-effective manner accounting for the differing needs of diverse patient populations.

REFERENCES

Abbosh, C., N. J. Birkbak et al. (2017). "Phylogenetic ctDNA analysis depicts early stage lung cancer evolution." *Nature* **545**(7655): 446.

Abdel-Rahman, O. (2016). "Targeting the MEK signaling pathway in non-small cell lung cancer (NSCLC) patients with RAS aberrations." *Therapeutic Advances in Respiratory Disease* **10**(3): 265–274.

Al-Shibli, K. I., T. Donnem, S. Al-Saad, M. Persson, R. M. Bremnes and L.-T. Busund (2008). "Prognostic effect of epithelial and stromal lymphocyte infiltration in non–small cell lung cancer." *Clinical Cancer Research* **14**(16): 5220–5227.

Armstrong, S (2016). "The computer will assess you now." *British Medical Journal*, **355**: i5680.

Azuma, K., K. Ota et al. (2014). "Association of PD-L1 overexpression with activating EGFR mutations in surgically resected nonsmall-cell lung cancer." *Annals of Oncology* **25**(10): 1935–1940.

Baldwin, D. R., B. White, M. Schmidt-Hansen, A. R. Champion and A. M. Melder (2011). "Diagnosis and treatment of lung cancer: Summary of updated NICE guidance." *British Medical Journal*, **342**: d2110.

Bansal, P., D. Osman, G. N. Gan, G. R. Simon and Y. Boumber (2016). "Recent Advances in targetable therapeutics in metastatic non-squamous NSCLC." *Frontiers in Oncology* **6**: 112.

Barrera, L., E. Montes-Servin, J. R. Borbolla, L. Arnold, J. Poole, V. Alexiadis, V. Singh, B. Gustafson and O. Arrieta (2016).

"Clinical evaluation of the utility of a liquid biopsy (circulating tumoral cells and ctDNA) to determine the mutational profile (EGFR, KRAS, ALK, ROS1 and BRAF) in advanced NSCLC patients." *Annals of Oncology* **27**(suppl_6): 1521PD–1521PD.

Biankin, A. V., S. Piantadosi and S. J. Hollingsworth (2015). "Patient-centric trials for therapeutic development in precision oncology." Nature **526**(7573): 361–370.

Bordi, P., M. Del Re, R. Danesi and M. Tiseo (2015). "Circulating DNA in diagnosis and monitoring EGFR gene mutations in advanced non-small cell lung cancer." *Translational Lung Cancer Research* **4**(5): 584–597.

Boros, A., L. Lacroix et al. (2017). "Prognostic value of tumor mutations in radically treated locally advanced non-small cell lung cancer patients." *Oncotarget* **8**(15): 25189.

Brahmer, J., K. L. Reckamp and P. Baas (2015). "Nivolumab versus docetaxel in advanced squamous-cell non–small-cell lung cancer." *New England Journal of Medicine* **373**: (2):123–135.

Bui, N., S. Henry, D. Wood, H. A. Wakelee and J. W. Neal (2017). "Chart review versus an automated bioinformatic approach to assess real-world crizotinib effectiveness in anaplastic lymphoma kinase–positive non–small-cell lung cancer." *JCO Clinical Cancer Informatics* **1**: 1–6.

Burggraaf, J., I. M. C. Kamerling et al. (2015). "Detection of colorectal polyps in humans using an intravenously administered fluorescent peptide targeted against c-Met." *Nature Medicine* **21**(8): 955–961.

Buzyn, A., J.-Y. Blay, N. Hoog-Labouret, M. Jimenez, F. Nowak, M.-C. L. Deley, D. Perol, C. Cailliot, J. Raynaud and G. Vassal (2016). "Equal access to innovative therapies and precision cancer care." *Nature Reviews Clinical Oncology* **13**(6): 385–393.

Cabanero, M. and M.-S. Tsao (2016). "Taking action on actionable mutations: A French initiative on universality in precision cancer care." *Translational Cancer Research* **5**(1): S35–S39.

Camidge, D. R., W. Pao and L. V. Sequist (2014). "Acquired resistance to TKIs in solid tumours: Learning from lung cancer." *Nature Reviews Clinical Oncology* **11**(8): 473–481.

Campbell, J. D., S. A. Mazzilli, M. E. Reid, S. S. Dhillon, S. Platero, J. Beane and A. E. Spira (2016). "The case for a pre-cancer genome atlas (PCGA)." *Cancer Prevention Research* **9**(2): 119–124.

Campbell, M. R., D. Amin and M. M. Moasser (2010). "HER3 comes of age; New insights into its functions and role in signaling, tumor biology, and cancer therapy." *Clinical Cancer Research: An Official Journal of the American Association for Cancer Research* **16**(5): 1373–1383.

Chabon (2016). "Inter- and intra-patient heterogeneity of resistance mechanisms to the mutant EGFR selective inhibitor rociletinib." *Journal of Clinical Oncology* **34**(suppl): abstr 9000.

Charoentong, P., F. Finotello, M. Angelova, C. Mayer, M. Efremova, D. Rieder, H. Hackl and Z. Trajanoski (2017). "Pan-cancer immunogenomic analyses reveal genotype-immunophenotype relationships and predictors of response to checkpoint blockade." *Cell Reports* **18**(1): 248–262.

Chong, C. R., M. Bahcall et al. (2017). "Identification of existing drugs that effectively target NTRK1 and ROS1 rearrangements in lung cancer." *Clinical Cancer Research* **23**(1): 204–213.

D'Incecco, A., M. Andreozzi et al. (2015). "PD-1 and PD-L1 expression in molecularly selected non-small-cell lung cancer patients." *British Journal of Cancer* **112**(1): 95–102.

Day, E., P. H. Dear and F. McCaughan (2013). "Digital PCR strategies in the development and analysis of molecular biomarkers for personalized medicine." *Methods* **59**(1): 101–107.

de Bruin, E. C., N. McGranahan et al. (2014). "Spatial and temporal diversity in genomic instability processes defines lung cancer evolution." *Science* **346**(6206): 251–256.

Didelot, A., D. Le Corre, A. Luscan, A. Cazes, K. Pallier, J. F. Emile, P. Laurent-Puig and H. Blons (2012). "Competitive allele specific TaqMan PCR for KRAS, BRAF and EGFR mutation detection in clinical formalin fixed paraffin embedded samples." *Experimental and Molecular Pathology* **92**(3): 275–280.

Donnem, T., S. M. Hald et al. (2015). "Stromal CD8+T-cell density—A promising supplement to TNM staging in non–small cell lung cancer." *Clinical Cancer Research* **21**(11): 2635–2643.

Donnem, T., T. K. Kilvaer et al. (2016). "Strategies for clinical implementation of TNM-Immunoscore in resected nonsmall-cell lung cancer." *Annals of Oncology* **27**(2): 225–232.

Drilon, A., F. Cappuzzo, S.-H. I. Ou and D. R. Camidge (2017). "Targeting MET in lung cancer: Will expectations finally be MET?" *Journal of Thoracic Oncology* **12**(1): 15–26.

Drilon, A., I. Bergagnini et al. (2016). "Clinical outcomes with pemetrexed-based systemic therapies in RET-rearranged lung cancers." *Annals of Oncology* **27**(7): 1286–1291.

Drilon, A., N. Rekhtman et al. (2016). "Cabozantinib in patients with advanced RET-rearranged non-small-cell lung cancer: An open-label, single-centre, phase 2, single-arm trial." *The Lancet Oncology* **17**(12): 1653–1660.

Eberhard, D. A., B. E. Johnson et al. (2005). "Mutations in the epidermal growth factor receptor and in KRAS are predictive and prognostic indicators in patients with non–small-cell lung cancer treated with chemotherapy alone and in combination with erlotinib." *Journal of Clinical Oncology* **23**(25): 5900–5909.

Edell, E. and D. Krier-Morrow (2010). "Navigational bronchoscopy: Overview of technology and practical considerations—New current procedural terminology codes effective 2010." *Chest* **137**(2): 450–454.

Ellison, G., G. Zhu, A. Moulis, S. Dearden, G. Speake and R. McCormack (2012). "EGFR mutation testing in lung cancer: A review of available methods and their use for analysis of tumour tissue and cytology samples." *Journal of Clinical Pathology* **66**(2): 79–89.

Engelman, J. A., K. Zejnullahu et al. (2007). "MET amplification leads to gefitinib resistance in lung cancer by activating ERBB3 signaling." *Science* **316**(5827): 1039–1043.

Fang, S. and Z. Wang (2014). "EGFR mutations as a prognostic and predictive marker in non-small-cell lung cancer." *Drug Design, Development and Therapy* **8**: 1595–1611.

Fehrenbacher, L., A. Spira et al. (2016). "Atezolizumab versus docetaxel for patients with previously treated non-small-cell lung cancer (POPLAR): A multicentre, open-label, phase 2 randomised controlled trial." *The Lancet* **387**(10030): 1837–1846.

Frezzetti, D., M. Gallo et al. (2017). "VEGF as a potential target in lung cancer." *Expert Opinion on Therapeutic Targets* **21**(10): 959–966.

Friboulet, L., K. A. Olaussen et al. (2013). ERCC1 isoform expression and DNA repair in non–small-cell lung cancer." *New England Journal of Medicine* **368**(12): 1101–1110.

Gainor, J. F. (2016). "Frequency and spectrum of ROS1 resistance mutations in ROS1-positive lung cancer patients progressing on crizotinib." *Journal of Clinical Oncology* **34**(suppl): abstr 9072.

Gao, F., E. Pfeifer et al. (2015). "Microdroplet digital PCR: Detection and quantitation of biomarkers in archived tissue and serial plasma samples in patients with lung cancer." *Journal of Thoracic Oncology* **10**(1): 212–217.

Garon, E. B. (2015). "Prior TKI therapy in NSCLC EGFR mutant patients associates with lack of response to anti-PD-1 treatment." *Journal of Thoracic Oncology* **10**(suppl 2): MINI03.01.

Garon, E. B., N. A. Rizvi and R. Hui (2015). "Pembrolizumab for the treatment of non-small-cell lung cancer." *The New England Journal of Medicine* 372(21), 2018–2028.

Gautschi, O., J. Milia et al. (2017). "Targeting RET in patients with RET-rearranged lung cancers: Results from the global, multicenter RET registry." *Journal of Clinical Oncology*: JCO.2016.2070.9352.

Giaccia, A. J. (2014). "Molecular radiobiology: The state of the art." *Journal of Clinical Oncology* **32**(26): 2871–2878.

Girard, N., C. Deshpande, C. G. Azzoli, V. W. Rusch, W. D. Travis, M. Ladanyi and W. Pao (2010). "Use of epidermal growth factor receptor/Kirsten rat sarcoma 2 viral oncogene homolog mutation testing to define clonal relationships among multiple lung adenocarcinomas: Comparison with clinical guidelines." *Chest* **137**(1): 46–52.

Girard, N., C. Deshpande, C. Lau, D. Finley, V. Rusch, W. Pao and W. D. Travis (2009a). "Comprehensive histologic assessment helps to differentiate multiple lung primary nonsmall cell carcinomas from metastases." *The American Journal of Surgical Pathology* **33**(12): 1752–1764.

Girard, N., I. Ostrovnaya et al. (2009b). "Genomic and mutational profiling to assess clonal relationships between multiple non–small cell lung cancers." *Clinical Cancer Research* **15**(16): 5184–5190.

Goss, G. D., C. O'Callaghan et al. (2013). "Gefitinib versus placebo in completely resected non–small-cell lung cancer: Results of the NCIC CTG BR19 study." *Journal of Clinical Oncology* **31**(27): 3320–3326.

Govindan, R., S. J. Mandrekar et al. (2015). "ALCHEMIST trials: A golden opportunity to transform outcomes in early-stage non–small cell lung cancer." *Clinical Cancer Research* **21**(24): 5439–5444.

Haratani, K., H. Hayashi et al. (2018). "Tumor immune microenvironment and nivolumab efficacy in EGFR mutation-positive non-small cell lung cancer based on T790M status after disease progression during EGFR-TKI treatment." *Annals of Oncology* **28**(7): 1532–1539.

Harris, L., A. Chen et al. "Abstract B080: Update on the NCI-molecular analysis for therapy choice (NCI-MATCH/EAY131) precision medicine trial." *Molecular Cancer Therapeutics* **17**: B080–B080.

Hellmann M.D., S. N. Gettinger et al. (2016). CheckMate 012: Safety and efficacy of first-line (1 L) nivolumab (nivo; N) and ipilimumab (ipi; I) in advanced (adv) NSCLC (Abstract 3001). *Journal of Clinical Oncology* 34(suppl): abstr 3001.

Herbst, R. S., P. Baas et al. (2016). "Pembrolizumab versus docetaxel for previously treated, PD-L1-positive, advanced non-small-cell lung cancer (KEYNOTE-010): A randomised controlled trial." *The Lancet* **387**(10027): 1540–1550.

Herbst, R.S., J.-C. Soria et al. (2014). "Predictive correlates of response to the anti-PD-L1 antibody MPDL3280A in cancer patients." *Nature* **515**(7528): 563–567.

Hiley, C. T., J. Le Quesne, G. Santis, R. Sharpe, D. G. de Castro, G. Middleton and C. Swanton (2016). "Challenges in molecular testing in non-small-cell lung cancer patients with advanced disease." *The Lancet* **388**(10048): 1002–1011.

Hiraoka, K., M. Miyamoto, Y. Cho, M. Suzuoki, T. Oshikiri, Y. Nakakubo, T. Itoh, T. Ohbuchi, S. Kondo and H. Katoh (2006). "Concurrent infiltration by CD8+ T cells and CD4+ T cells is a favourable prognostic factor in non-small-cell lung carcinoma." *British Journal of Cancer* **94**(2): 275–280.

Hodgkinson, C. L., C. J. Morrow et al. (2014). "Tumorigenicity and genetic profiling of circulating tumor cells in small-cell lung cancer." *Nat Med* **20**(8): 897–903.

Hugo, W., J. M. Zaretsky e t al. (2016). "Genomic and transcriptomic features of response to anti-PD-1 therapy in metastatic melanoma." *Cell* **165**(1): 35–44.

Jamal-Hanjani, M., G. A. Wilson et al. (2017). "Tracking the evolution of non–small-cell lung cancer." *New England Journal of Medicine* **376**(22): 2109–2121.

Jänne, P. A., A. T. Shaw et al. (2013). "Selumetinib plus docetaxel for KRAS-mutant advanced non-small-cell lung cancer: A randomised, multicentre, placebo-controlled, phase 2 study." *The Lancet Oncology* **14**(1): 38–47.

Jr, L. A. D. and A. Bardelli (2014). "Liquid biopsies: Genotyping circulating tumor DNA." *Journal of Clinical Oncology* **32**(6): 579–586.

Kabbinavar, F., L. Fehrenbacher et al. (2014). "Biomarker analyses from a randomized, placebo-controlled, phase IIIb trial comparing bevacizumab with or without erlotinib as maintenance therapy for the treatment of advanced non-small-cell lung cancer (ATLAS)." *Journal of Thoracic Oncology* **9**(9): 1411–1417.

Kadara, H., P. Scheet, I. I. Wistuba and A. E. Spira (2016). "Early events in the molecular pathogenesis of lung cancer." *Cancer Prevention Research* **9**(7): 518–527.

Kato, S., B. N. Tomson, T. P. Buys, S. K. Elkin, J. L. Carter and R. Kurzrock (2016). "Genomic landscape of malignant mesotheliomas." *Molecular Cancer Therapeutics* **15**(10): 2498–2507.

Kelly, K., K. Chansky, L. E. Gaspar, K. S. Albain, J. Jett, Y. C. Ung, D. H. Lau, J. J. Crowley and D. R. Gandara (2008). "Phase III trial of maintenance gefitinib or placebo after concurrent chemoradiotherapy and docetaxel consolidation in inoperable stage III non-small-cell lung cancer: SWOG S0023." *Journal of Clinical Oncology* **26**(15): 2450–2456.

Kimura H, M. Y., Aki Ishikawa, T. lizasa, M. Shingyoji, M. Nakajima, and I. Yoshino (2015). "Final results of phase III trial of adjuvant chemo-immunotherapy in lung cancer." *Journal of Thoracic Oncology* **10**(suppl 2): ORAL04.01.

Koper, A., L. A. H. Zeef, L. Joseph, K. Kerr, J. Gosney, M. A. Lindsay and R. Booton (2017). "Whole transcriptome analysis of pre-invasive and invasive early squamous lung carcinoma in archival laser microdissected samples." *Respiratory Research* **18**: 12.

Kosaka, T., J. Tanizaki et al. (2017). "Response heterogeneity of EGFR and HER2 exon 20 insertions to covalent EGFR and HER2 inhibitors." *Cancer Research.* **77**: 2712–2721.

Kuan, F.-C., L.-T. Kuo, M.-C. Chen, C.-T. Yang, C.-S. Shi, D. Teng and K.-D. Lee (2015). "Overall survival benefits of first-line EGFR tyrosine kinase inhibitors in EGFR-mutated non-small-cell lung cancers: A systematic review and meta-analysis." *British Journal of Cancer* **113**(10): 1519–1528.

Kular, H., L. Mudambi, D. R. Lazarus, L. Cornwell, A. Zhu and R. F. Casal (2016). "Safety and feasibility of prolonged bronchoscopy involving diagnosis of lung cancer, systematic nodal staging, and fiducial marker placement in a high-risk population." *Journal of Thoracic Disease* **8**(6): 1132–1138.

Langer, C. J., S. M. Gadgeel et al. (2016). "Carboplatin and pemetrexed with or without pembrolizumab for advanced, non-squamous non-small-cell lung cancer: A randomised, phase 2 cohort of the open-label KEYNOTE-021 study." *The Lancet Oncology* **17**(11): 1497–1508.

Larkin, J., V. Chiarion-Sileni et al. (2015). "Combined nivolumab and ipilimumab or monotherapy in untreated melanoma." *New England Journal of Medicine* **373**(1): 23–34.

LeBlanc, V. G. and M. A. Marra (2015). "Next-generation sequencing approaches in cancer: Where have they brought us and where will they take us?" *Cancers* **7**(3): 1925–1958.

Lee, R., D. J. Cousins, E. Ortiz-Zapater, R. Breen, E. McLean and G. Santis (2013). "Gene expression profiling of endobronchial ultrasound (EBUS)-derived cytological fine needle aspirates from hilar and mediastinal lymph nodes in non-small cell lung cancer." *Cytopathology* **24**(6): 351–355.

Leong, S., T. Shaipanich, S. Lam and K. Yasufuku (2013). "Diagnostic bronchoscopy--current and future perspectives." *Journal of Thoracic Disease* **5**(Suppl 5): S498–S510.

Li, J., L. Wang, H. Mamon, M. H. Kulke, R. Berbeco and G. M. Makrigiorgos (2008). "Replacing PCR with COLD-PCR enriches variant DNA sequences and redefines the sensitivity of genetic testing." *Nature Medicine* **14**(5): 579–584.

Lizotte, P. H., E. V. Ivanova et al. (2016). "Multiparametric profiling of non–small-cell lung cancers reveals distinct immunophenotypes." *JCI Insight* **1**(14): e89014.

Ma, X., G. Le Teuff et al. (2016). "Prognostic and predictive effect of TP53 mutations in patients with non-small cell lung cancer from adjuvant cisplatin-based therapy randomized trials: A LACE-Bio pooled analysis." *Journal of Thoracic Oncology* **11**(6): 850–861.

Malottki, K., S. Popat, J. J. Deeks, R. D. Riley, A. G. Nicholson and L. Billingham (2016). "Problems of variable biomarker evaluation in stratified medicine research—A case study of ERCC1 in non-small-cell lung cancer." *Lung Cancer* **92**: 1–7.

Mao, L. (2002). "Recent advances in the molecular diagnosis of lung cancer." *Oncogene* **21**(45): 6960–6969.

Marino, K. A., J. L. Sullivan and B. Weksler (2016). "Electromagnetic navigation bronchoscopy for identifying lung nodules for thoracoscopic resection." *The Annals of Thoracic Surgery* **102**(2): 454–457.

Martincorena, I. and P. J. Campbell (2015). "Somatic mutation in cancer and normal cells." *Science* **349**(6255): 1483–1489.

Martincorena, I., A. Roshan et al. (2015). "High burden and pervasive positive selection of somatic mutations in normal human skin." *Science* **348**(6237): 880–886.

Mattioni, M., S. Soddu, A. Prodosmo, P. Visca, S. Conti, G. Alessandrini, F. Facciolo and L. Strigari (2015). "Prognostic role of serum p53 antibodies in lung cancer." *BMC Cancer* **15**(1): 148.

McCaughan, F., C. P. Pipinikas, S. M. Janes, P. J. George, P. H. Rabbitts and P. H. Dear (2011). "Genomic evidence of pre-invasive clonal expansion, dispersal and progression in bronchial dysplasia." *The Journal of Pathology* **224**(2): 153–159.

McCoach (2016). "Resistance mechanisms to targeted therapies in ROS1+ and ALK+ non-small cell lung cancer "*Journal of Clinical Oncology* **34**(suppl): abstr 9065.

McDonald, F. and S. Popat (2014). "Combining targeted agents and hypo- and hyper-fractionated radiotherapy in NSCLC." *Journal of Thoracic Disease* **6**(4): 356–368.

McGranahan, N., F. Favero, E. C. de Bruin, N. J. Birkbak, Z. Szallasi and C. Swanton (2015). "Clonal status of actionable driver events and the timing of mutational processes in cancer evolution." *Science Translational Medicine* **7**(283): 283ra254–283ra254.

McLaughlin, J., G. Han et al. (2016). "Quantitative assessment of the heterogeneity of pd-l1 expression in non–small-cell lung cancer." *JAMA Oncology* **2**(1): 46–54.

Metzker, M. L. (2010). "Sequencing technologies [mdash] the next generation." *Nature Reviews Genetics* **11**(1): 31–46.

Michels, S., M. Gardizi, P. Schmalz, M. Thurat, E. Pereira, M. Sebastian, E. Carcereny et al. (2017). "MA07.05 EUCROSS: A European Phase II trial of crizotinib in advanced adenocarcinoma of the lung harboring ROS1 rearrangements—Preliminary results." *Journal of Thoracic Oncology* **12**(1): S379–S380.

Miller, R. S. (2016). "CancerLinQ update." *Journal of Oncology Practice* **12**(10): 835–837.

Mok, T. S. K. (2011). "Personalized medicine in lung cancer: What we need to know." *Nature Reviews Clinical Oncology* **8**(11): 661–668.

Mok, T. S., Y.-L. Wu et al. (2016). "Osimertinib or platinum–pemetrexed in EGFR T790M–positive lung cancer." *New England Journal of Medicine* **376**(7): 629–640.

Morgillo, F., C. M. Della Corte, M. Fasano and F. Ciardiello (2016). "Mechanisms of resistance to EGFR-targeted drugs: Lung cancer." *ESMO Open* **1**(3): e000060.

Murphy, S. J., M.-C. Aubry et al. (2014). "Identification of independent primary tumors and intrapulmonary metastases using DNA rearrangements in non–small-cell lung cancer." *Journal of Clinical Oncology* **32**(36): 4050–4058.

Murtaza, M., S.-J. Dawson et al. (2013). "Non-invasive analysis of acquired resistance to cancer therapy by sequencing of plasma DNA." *Nature* **497**(7447): 108–112.

Newman, A. M., S. V. Bratman et al. (2014). "An ultrasensitive method for quantitating circulating tumor DNA with broad patient coverage." *Nature Medicine* **20**(5): 548–554.

Nokihara, H. (2016). "Alectinib (ALC) versus crizotinib (CRZ) in ALK-inhibitor naive ALK-positive non-small cell lung cancer (ALK+ NSCLC): Primary results from the J-ALEX study." *Journal of Clinical Oncology* **34**: Abstract 9008.

Olaussen, K. A., A. Dunant et al. (2006). "DNA repair by ERCC1 in non–small-cell lung cancer and cisplatin-based adjuvant chemotherapy." *New England Journal of Medicine* **355**(10): 983–991.

Oser, M. G., M. J. Niederst, L. V. Sequist and J. A. Engelman (2015). "Transformation from non-small-cell lung cancer to small-cell lung cancer: Molecular drivers and cells of origin." *The Lancet Oncology* **16**(4): e165–e172.

Oxnard, G. R., C. P. Paweletz et al. (2014). "Noninvasive detection of response and resistance in EGFR-Mutant lung cancer using quantitative next-generation genotyping of cell-free plasma DNA." *Clinical Cancer Research* **20**(6): 1698–1705.

Papadimitrakopoulou (2016). "Randomised Phase III study of osimertinib versus platinum-pemetrexed for EGFR T790M-positive advanced NSCLC (AURA3)." *Journal of Thoracic Oncology* **11**(suppl): abstr PL03.03.

Petersen, R. P., M. J. Campa, J. Sperlazza, D. Conlon, M.-B. Joshi, D. H. Harpole and E. F. Patz (2006). "Tumor infiltrating Foxp3+ regulatory T-cells are associated with recurrence in pathologic stage I NSCLC patients." *Cancer* **107**(12): 2866–2872.

Planchard, D., B. Besse et al. (2016). "Dabrafenib plus trametinib in patients with previously treated BRAFV600E-mutant metastatic non-small cell lung cancer: An open-label, multicentre phase 2 trial." *The Lancet Oncology* **17**(7): 984–993.

Ratcliffe, M. J., A. Sharpe, A. Midha, C. Barker, M. Scott, P. Scorer, H. Al-Masri, M. Rebelatto and J. Walker (2017). "Agreement between programmed cell death ligand-1 diagnostic assays across multiple protein expression cut-offs in non-small cell lung cancer." *Clinical Cancer Research*. doi:10.1158/1078-0432. CCR-16-2375.

Reckamp, K. L., V. O. Melnikova et al. (2016). "A highly sensitive and quantitative test platform for detection of NSCLC EGFR mutations in urine and plasma." *Journal of Thoracic Oncology* **11**(10): 1690–1700.

Reiman, T., R. Lai et al. (2012). "Cross-validation study of class III beta-tubulin as a predictive marker for benefit from adjuvant chemotherapy in resected non-small-cell lung cancer: Analysis of four randomized trials." *Annals of Oncology* **23**(1): 86–93.

Remark, R., C. Becker et al. (2015). "The non–small cell lung cancer immune contexture. A major determinant of tumor characteristics and patient outcome." *American Journal of Respiratory and Critical Care Medicine* **191**(4): 377–390.

Ribas, A., D. S. Shin et al. (2016). "PD-1 blockade expands intratumoral memory T cells." *Cancer Immunology Research* **4**(3): 194–203.

Righi, L., T. Vavala, I. Rapa, S. Vatrano, J. Giorcelli, G. Rossi, E. Capelletto, S. Novello, G. V. Scagliotti and M. Papotti (2014). "Impact of non-small-cell lung cancer-not otherwise specified immunophenotyping on treatment outcome." *Journal of Thoracic Oncology* **9**(10): 1540–1546.

Rizvi, N. A., M. D. Hellmann et al. (2015). "Mutational landscape determines sensitivity to PD-1 blockade in non–small cell lung cancer." *Science* **348**(6230): 124–128.

Rosell, R., E. Carcereny et al. P.-C. Spanish Lung Cancer Group in collaboration with Groupe Francais de and T. Associazione Italiana Oncologia (2012). "Erlotinib versus standard chemotherapy as first-line treatment for European patients with advanced EGFR mutation-positive non-small-cell lung cancer (EURTAC): A multicentre, open-label, randomised phase 3 trial." *Lancet Oncology* **13**(3): 239–246.

Ruparel, M. and S. M. Janes (2016). "Lung cancer screening: What we can learn from UKLS?" *Thorax* **71**(2): 103–104.

Rusch, V. (2016). "Next generation sequencing (NGS) in resectable non-small cell lung cancer (NSCLC): Therapeutic implications." *Journal of Clinical Oncology* **34**(Suppl): abstr 8541.

Salgia, R. and A. T. Skarin (1998). "Molecular abnormalities in lung cancer." *Journal of Clinical Oncology* **16**(3): 1207–1217.

Sanmamed MF, J. L. Perez-Gracia et al. (2016). "Changes in serum IL8 levels reflect and predict response to anti-PD-1 treatment in melanoma and non-small cell lung cancer patients." *Annals of Oncology* **27**(6): 359–378.

Santis, G., R. Angell, G. Nickless, A. Quinn, A. Herbert, P. Cane, J. Spicer, R. Breen, E. McLean and K. Tobal (2011). "Screening for EGFR and KRAS mutations in endobronchial ultrasound derived transbronchial needle aspirates in non-small cell lung cancer using COLD-PCR." *PLoS One* **6**(9): e25191.

Sarraf, K. M., E. Belcher, E. Raevsky, A. G. Nicholson, P. Goldstraw and E. Lim (2009). "Neutrophil/lymphocyte ratio and its association with survival after complete resection in non–small cell lung cancer." *The Journal of Thoracic and Cardiovascular Surgery* **137**(2): 425–428.

Schmid, P. (2015). "NSCLC with high PD-L1 expression on tumor cells or tumor-infiltrating immune cells represents distinct cancer subtypes." *Annals of Oncology* **26**(suppl 6): abstr 3017.

Schrock, A. B., G. M. Frampton et al. (2016). "Characterization of 298 patients with lung cancer harboring MET exon 14 skipping alterations." *Journal of Thoracic Oncology* **11**(9): 1493–1502.

Seifert, M., B. Friedrich and A. Beyer (2016). "Importance of rare gene copy number alterations for personalized tumor characterization and survival analysis." *Genome Biology* **17**: 204.

Shan, Y., M. P. Eastwood, X. Zhang, E. T. Kim, A. Arkhipov, R. O. Dror, J. Jumper, J. Kuriyan and D. E. Shaw (2012). "Oncogenic mutations counteract intrinsic disorder in the EGFR kinase and promote receptor dimerization." *Cell* **149**(4): 860–870.

Sharma, S. V., D. W. Bell, J. Settleman and D. A. Haber (2007). "Epidermal growth factor receptor mutations in lung cancer." *Cancer* **7**: 169–181.

Shaw, A. T., D.-W. Kim et al. (2014). "Ceritinib in ALK-rearranged non–small-cell lung cancer." *New England Journal of Medicine* **370**(13): 1189–1197.

Shaw, A., G. J. Riley et al. (2016). "Crizotinib in advanced ROS1-rearranged non-small cell lung cancer (NSCLC): Updated results from PROFILE 1001." *Annals of Oncology* **27**(suppl_6): 1206PD–1206PD.

Shedden, K., J. M. G. Taylor et al. (2008). "Gene expression-based survival prediction in lung adenocarcinoma: A multi-site, blinded validation study." *Nature Medicine* **14**(8): 822–827.

Shepherd, F. A., B. Lacas et al. o. b. o. t. L.-B. C. Group (2016). "Pooled analysis of the prognostic and predictive effects of TP53 comutation status combined with KRAS or EGFR mutation in early-stage resected non–small-cell lung cancer in four trials of adjuvant chemotherapy." *Journal of Clinical Oncology* **0**(0): JCO.2016.2071.2893.

Silvestri, G. A., A. Vachani et al. (2015). "A bronchial genomic classifier for the diagnostic evaluation of lung cancer." *New England Journal of Medicine* **373**(3): 243–251.

Socinski, M., B. Creelan et al. (2016). "NSCLC, metastaticCheck-Mate 026: A phase 3 trial of nivolumab vs investigator's choice (IC) of platinum-based doublet chemotherapy (PT-DC) as first-line therapy for stage iv/recurrent programmed death ligand 1 (PD-L1)–positive NSCLC." *Annals of Oncology* **27**(suppl_6): LBA7_PR-LBA7_PR.

Soda, M., Y. L. Choi et al. (2007). "Identification of the transforming EML4-ALK fusion gene in non-small-cell lung cancer." *Nature* **448**(7153): 561–566.

Soria, J.-C., D. S. W. Tan et al. (2017). "First-line ceritinib versus platinum-based chemotherapy in advanced ALK-rearranged non-small-cell lung cancer (ASCEND-4): A randomised, open-label, phase 3 study." *The Lancet* **389**(10072): 917–929.

Soria, J.-C., Y.-L. Wu et al. (2015). "Gefitinib plus chemotherapy versus placebo plus chemotherapy in EGFR-mutation-positive non-small-cell lung cancer after progression on first-line gefitinib (IMPRESS): A phase 3 randomised trial." *The Lancet Oncology* **16**(8). 990–998.

Stahel RA, D. U., Gautschi O et al (2015). "A phase II trial of erlotinib (E) and bevacizumab (B) in patients with advanced non-small-cell lung cancer (NSCLC) with activating epidermal growth factor receptor (EGFR) mutations with and without T790M mutation. The Spanish Lung Cancer Group (SLCG) and the European Thoracic Oncology Platform (ETOP) BELIEF trial." *Annals of Oncology* **26**(Supp 6): abstr 3BA.

Su, K.-Y., H.-Y. Chen et al. (2012). "Pretreatment epidermal growth factor receptor (EGFR) T790M mutation predicts shorter EGFR tyrosine kinase inhibitor response duration in patients with non–small-cell lung cancer." *Journal of Clinical Oncology* **30**(4): 433–440.

Sun, J., D. H. Garfield, B. Lam, J. Yan, A. Gu, J. Shen and B. Han (2011). "The value of autofluorescence bronchoscopy combined with white light bronchoscopy compared with white light alone in the diagnosis of intraepithelial neoplasia and invasive lung cancer: A meta-analysis." *Journal of Thoracic Oncology* **6**(8): 1336–1344.

Sutherland, K. D., J.-Y. Song, M. C. Kwon, N. Proost, J. Zevenhoven and A. Berns (2014). "Multiple cells-of-origin of mutant K-Ras–induced mouse lung adenocarcinoma." *Proceedings of the National Academy of Sciences* **111**(13): 4952–4957.

Sutherland, Kate D., N. Proost, I. Brouns, D. Adriaensen, J.-Y. Song and A. Berns (2011). "Cell of origin of small cell lung cancer: Inactivation of Trp53 and Rb1 in distinct cell types of adult mouse lung." *Cancer Cell* **19**(6): 754–764.

Takano, T., Y. Ohe et al. (2005). "Epidermal growth factor receptor gene mutations and increased copy numbers predict gefitinib sensitivity in patients with recurrent non-small-cell lung cancer." *Journal of Clinical Oncology* **23**(28): 6829–6837.

Tanaka, A. and S. Sakaguchi (2017). "Regulatory T cells in cancer immunotherapy." *Cell Research* **27**(1): 109–118.

Tanaka, A., N. Sueoka-Aragane, T. Nakamura, Y. Takeda, M. Mitsuoka, F. Yamasaki, S. Hayashi, E. Sueoka and S. Kimura (2012). "Co-existence of positive MET FISH status with EGFR mutations signifies poor prognosis in lung adenocarcinoma patients." *Lung Cancer* **75**(1): 89–94.

The Cancer Genome Atlas Research, N. (2014). "Comprehensive molecular profiling of lung adenocarcinoma." *Nature* **511**(7511): 543–550.

The Clinical Lung Cancer Genome, P. and M. Network Genomic (2013). "A genomics-based classification of human lung tumors." *Science Translational Medicine* **5**(209): 209ra153-209ra153.

Thunnissen, F. B. J. M. (2003). "Sputum examination for early detection of lung cancer." *Journal of Clinical Pathology* **56**(11): 805–810.

Tomasetti, M., M. Amati, J. Neuzil and L. Santarelli (2016). "Circulating epigenetic biomarkers in lung malignancies: From early diagnosis to therapy." *Lung Cancer* **107**: 65–72.

Travis, W. D., E. Brambilla, P. Van Schil, G. V. Scagliotti, R. M. Huber, J.-P. Sculier, J. Vansteenkiste and A. G. Nicholson (2011).

"Paradigm shifts in lung cancer as defined in the new IASLC/ATS/ERS lung adenocarcinoma classification." *European Respiratory Journal* **38**(2): 239–243.

Tremblay, A., N. Taghizadeh et al. (2016). "LOw prevalence of high-grade lesions detected with autofluorescence bronchoscopy in the setting of lung cancer screening in the pan-canadian lung cancer screening study." *Chest* **150**(5): 1015–1022.

Tsao, M.-S., S. Marguet et al. (2015). "Subtype classification of lung adenocarcinoma predicts benefit from adjuvant chemotherapy in patients undergoing complete resection." *Journal of Clinical Oncology* **33**(30): 3439–3446.

Ulloa-Montoya, F., J. Louahed et al. (2013). "Predictive gene signature in MAGE-A3 antigen-specific cancer immunotherapy." *Journal of Clinical Oncology* **31**(19): 2388–2395.

Vansteenkiste, J. F., B. C. Cho et al. (2016). "Efficacy of the MAGE-A3 cancer immunotherapeutic as adjuvant therapy in patients with resected MAGE-A3-positive non-small-cell lung cancer (MAGRIT): A randomised, double-blind, placebo-controlled, phase 3 trial." *The Lancet Oncology* **17**(6): 822–835.

Vargas, A. J. and C. C. Harris (2016). "Biomarker development in the precision medicine era: Lung cancer as a case study." *Nature Reviews Cancer* **16**(8): 525–537.

Vassal, G., D. Moro-sibilot et al. (2015). "12LBA Biomarker-driven access to crizotinib in ALK, MET or ROS1 positive (+) malignancies in adults and children: The French national AcSe Program." *European Journal of Cancer* **51**: S715.

Vassal, G., M.-c. Ledeley et al. (2015). "Activity of crizotinib in relapsed MET amplified malignancies: Results of the French AcSe Program." *ASCO Meeting Abstracts* **33**(15_suppl): 2595.

Vivanco, I., H. I. Robins et al. (2012). "Differential sensitivity of glioma- versus lung cancer-specific EGFR mutations to EGFR kinase inhibitors." *Cancer Discovery* **2**(5): 458–471.

von Laffert, M., A. Warth et al. (2014). "Multicenter immunohistochemical ALK-testing of non–small-cell lung cancer shows high concordance after harmonization of techniques and interpretation criteria." *Journal of Thoracic Oncology* **9**(11): 1685–1692.

Weber, J. S., R. R. Kudchadkar et al. (2013). "Safety, efficacy, and biomarkers of nivolumab with vaccine in ipilimumab-refractory or -naive melanoma." *Journal of Clinical Oncology* **31**(34): 4311–4318.

Wong, S. Q., J. Li et al. (2014). "Sequence artefacts in a prospective series of formalin-fixed tumours tested for mutations in hotspot regions by massively parallel sequencing." *BMC Medical Genomics* **7**(1): 23.

Woo, E. Y., C. S. Chu, T. J. Goletz, K. Schlienger, H. Yeh, G. Coukos, S. C. Rubin, L. R. Kaiser and C. H. June (2001). "Regulatory CD4+ CD25+ T cells in tumors from patients with early-stage non-small cell lung cancer and late-stage ovarian cancer." *Cancer Research* **61**(12): 4766.

Yatabe, Y., T. Mitsudomi and T. Takahashi (2002). "TTF-1 expression in pulmonary adenocarcinomas." *The American Journal Surgical Pathology* **26**(6): 767–773.

Yearley (2015). "PD-L2 expression in human tumors: Relevance to anti-PD-1 therapy in cancer." *Annals of Oncology* **26**(suppl 6): abstr 18LBA.

Yoh, K., T. Seto et al. (2017). "Vandetanib in patients with previously treated RET-rearranged advanced non-small-cell lung cancer (LURET): An open-label, multicentre phase 2 trial." *The Lancet Respiratory Medicine* **5**(1): 42–50.

Yousaf-Khan, U., C. van der Aalst et al. (2017). "Final screening round of the NELSON lung cancer screening trial: The effect of a 2.5-year screening interval." *Thorax* **72**(1): 48–56.

Yu, H., Z. Chen et al. (2019). "Correlation of PD-L1 expression with tumor mutation burden and gene signatures for prognosis in early-stage squamous cell lung carcinoma." *Journal of Thoracic Oncology* **14**(1): 25–36.

Yun, C. H., K. E. Mengwasser, A. V. Toms, M. S. Woo, H. Greulich, K. K. Wong, M. Meyerson and M. J. Eck (2008). "The T790M mutation in EGFR kinase causes drug resistance by increasing the affinity for ATP." *Proceedings of the National Academy of Sciences of the USA* **105**(6): 2070–2075.

Zang, J., Y. Hu, X. Xu, J. Ni, D. Yan, S. Liu, J. He, J. Xue, J. Wu and J. Feng (2017). "Elevated serum levels of vascular endothelial growth factor predict a poor prognosis of platinum-based chemotherapy in non-small cell lung cancer." *OncoTargets and Therapy* **10**: 409–415.

Zhang, J., D. Park, D. M. Shin and X. Deng (2016). "Targeting KRAS-mutant non-small cell lung cancer: Challenges and opportunities." *Acta Biochimica et Biophysica Sinica (Shanghai)* **48**(1): 11–16.

Zhang, J., J. Fujimoto et al. (2014). "Intratumor heterogeneity in localized lung adenocarcinomas delineated by multiregion sequencing." *Science* **346**(6206): 256–259.

Zhang, J., S. Khanna, Q. Jiang, C. Alewine, M. Miettinen, I. Pastan and R. Hassan (2016). "Efficacy of anti-mesothelin immunotoxin RG7787 plus nab-paclitaxel against mesothelioma patient derived xenografts and mesothelin as a biomarker of tumor response." *Clinical Cancer Research* **23**(6): 1564–1574..

Zhang, Y., Z. Du and M. Zhang (2016). "Biomarker development in MET-targeted therapy." *Oncotarget* **7**(24): 37370–37389.

Zhang, Z., T. Wang, J. Zhang, X. Cai, C. Pan, Y. Long, J. Chen, C. Zhou and X. Yin (2014). "Prognostic value of epidermal growth factor receptor mutations in resected non-small cell lung cancer: A systematic review with meta-analysis." *PLoS One* **9**(8): e106053.

9

Lessons Learned

What Have We Learned and What Is the Next Challenge?

Translational Science in Modern Pharmaceutical R&D 9.1

The Influence of Organizational Structures on Successes and Failures

Peter M. A. Groenen

Contents

INTRODUCTION

> *"…In a time of drastic change, it is the learners who inherit the future. The learned usually find themselves equipped to live in a world that no longer exists." Eric Hoffer, 1973.*

The successive steps that make up the traditionally linear R&D process for novel first in-class therapeutics have not changed significantly over the last four decades. In principle this process still moves from target selection to chemical or hit finding, hit-to-lead, lead optimization, preclinical development for the best lead and from a first-in-human (FIH), proof of concept (POC) in selected patients finishing with confirmatory clinical trials in phase III (1,2). Some steps can be avoided, for instance developing active molecules extracted from natural sources, using known active substances as a starting point or conditional regulatory approval before finishing phase III trials (3). For biologics (such as antibodies, proteins, gene therapies and cell therapies) the starting point from a target and process up to a FIH is slightly different. The main challenge for the development of a novel therapeutic approach is to successfully predict that the target or pathway addressed plays a pivotal role in the pathology of the disease and that the intended modulation will result in a cure, or symptom relief. For first in class medications by definition, the clinical information is limited or completely absent, and R&D organizations must continuously generate data allowing to take decisions on continuation or stopping of projects. In larger biopharmaceutical companies the processes for resource management and balancing of risks are complicated because of the organization size and global dispersion, whereas smaller (start-up) companies may only just focus on one challenging project. Because of the high degree of uncertainty, managing portfolios to balance benefit and risk has become a highly specialized discipline (4) with estimation of value and probability of success being extremely challenging for early discovery projects (5). Two key factors for both the success of a project and the timely decision making, are the experience level of the project team and their interaction with the management. The following underappreciated factors among others contribute significantly to the innovation gap: competition with other projects for resources, and managing scientific, functional and cultural diversity. This section therefore deals with the generic problems and challenges teams encounter in the biopharmaceutical industry while attempting to implement novel paths to success. Translational Science, for the purpose of this section, will be defined as the particular scientific mindset that deals with the critical elements of disease and treatment in different species with the aim of improving the likelihood of clinical meaningful effects. To obtain early insights in the molecular aspects of the pathology, translation from conceptual *in vitro* to cellular and *in vivo* models, translatable biomarkers for critical aspects of drug/patient interaction (6) have become a major tool to reduce expensive late stage clinical attrition. Implementing biomarker discovery and development with tight budgets and timelines, is a daunting task with many challenges, not in the least caused by the slow reaction time of adaptions in larger organizations.

THE INNOVATION GAP AND THE INDUSTRIAL RESPONSE

The complex nature of biological research and the increased specialization, with new modern technologies such as proteomics, metabolomics, genomics and computational sciences (bio-, chemo- and medical informatics), resulted in a drug discovery and development process that has become very multidisciplinary with a high degree of interaction and many decision points. The process has become prone to errors, tunnel vision and groupthink (7) due to this increased specialization by different members of research teams as well as the complex nature of decision making through portfolio and resource management. Although there was great optimism at the start of the twenty-first century (8) 20 years later none of these newly introduced technologies has significantly improved the output as measured by the number of new medical entities (NMEs). Each discipline, from target discovery in biology to toxicology in preclinical development, is working on a small but essential part. Each step of the process is characterized by a decision point and hand-over as is depicted in Figure 9.1.1. This industrialized R&D is considered to be one of the factors of the current decline in productivity we have been witnessing in the last 25 years (9). Pisano argues that Life sciences, in contrast to other science related industries like microelectronics, appear to be inherently more complex where scaling of data generation does clearly not improve successful market entry of novel drugs.

Since the release of the Critical Path Report in 2004 by the U.S. Food and Drug Administration (FDA) (10,11), which described the innovation gap and offered suggestions to increase the success rates of experimental medicines, two pharmaceutical companies, have independently published insights into the 20–30 years of successes and failures in their drug programs. The AstraZeneca in depth review of successes and failures can be regarded as prototypical for many if not all larger biopharmaceutical companies (12). The paper extensively analyzed its portfolio over the last decades and did not hesitate to touch upon the culture aspects. One of the most prominent errors in R&D decision making was the simple thought that by calculating the attrition rate, simply increasing "shots on goal" would solve the problem. This pressure on R&D teams resulted in lowering the quality criteria for the projects and avoiding critical assessment of possible issues in clinical applications. The 5R framework AstraZeneca synthesized as a requirement for all projects as they progress deals with scientific and medical backing of projects, right target, right tissue, right safety, right patients and finally, the right commercial potential. Nevertheless, the critical part here is obviously the management backing so that teams feel encouraged to fail for the right reasons and avoid groupthink pressured by the need for a positive outcome of a project. The cases and the concept of the 5Rs make this paper therefore an interesting read in the context of this section. Similar conclusions, but not so extensively described were brought forward by a similar analysis from Pfizer coining the "three Pillars of survival": exposure at the target, binding to the pharmacological target and expression of pharmacological activity as a result of the previous two pillars (13).

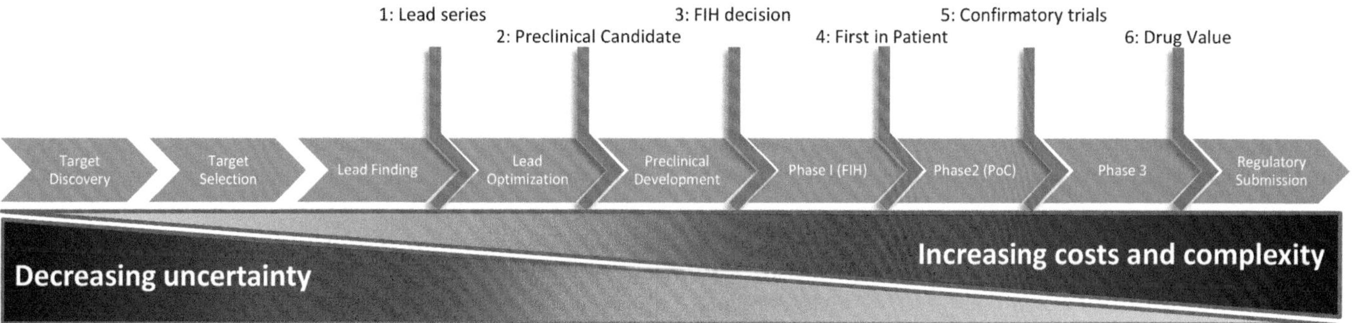

FIGURE 9.1.1 Simplified schematic representation of the current generic pharmaceutical R&D process. Each step is characterized by decision points that in most companies are used for a rigorous review of all data to decide on go/no-go for the next stage or return to further optimize and/or collect more evidence to support the continuation of the project.

The current landscape of organizational structures that companies have developed over many decades to improve success rates has received little attention in the scientific literature searching for explanations of the stagnation of new medical products to the market. Coping with the increasing complexity of the process accompanied with increasing costs and increased attrition companies constantly attempted to adopt new approaches and structurally adapt to the rapid changes. Since most published analyses (14–24) have dealt with the scientific aspects that explain the attrition to some extent, the focus in this chapter will be on the role of personal interactions, where organizational structures, behaviors and decision making are further explored to understand their undeniable influence on success rates (or failures) for pharmaceutical R&D programs.

Over the last 10 years new disciplines have emerged and most companies' project management and life cycle management departments have grown, serving as the glue between the different disciplines within an organization and control the complex process while guarding budget and timelines. Since the publication of the FDA critical path report many companies have also created translational medicine departments, sometimes accompanied by biomarker departments. Translation medicine departments harbor clinicians, biologists, pharmacologists and computational scientists. The key goal of a translational medicine is to break through the linear process and force the different disciplines to circle back whenever needed with a clear mind on bench to bedside and back to translate clinical experiences into the discovery process (6). This is potentially disrupting and leads to frustration. This transition between species and assays has often resulted in the formation of dedicated departments, focusing on biomarker discovery and development. Translational science is a defined discipline rather than just a conceptual way of working in most companies because of the expertise on species differences, assay conversion and translational challenges to get biomarker assays into the clinic and data back into research.

A very important step toward what we now consider a commodity was the foundation of pharmacology as a scientific discipline in 1847 by Rudolf Buchheim, who founded the world's first pharmacology institute. The second important event was the first chemically synthesized drug, amyl nitrite for the treatment of angina, by Guthry in 1859. In Germany and Switzerland, pharma companies started to emerge build on a history of dye chemistry. While in Europe, companies such as Hoffmann-La Roche, Ciba, Merck, Hoechst, Bayer, Schering and Boehringer started to emerge, at the same time in the United States, well-known companies today, like Eli Lilly, Parke Davis and Squibb started to enter into the business of pharmaceuticals (2,27).

The industry for long was very chemistry driven, whereas pharmacology started to pick up later. During the mid-twenties of the last century, both scientific disciplines were of equal importance in the discovery of new medicines. This combination, together with what we today consider as limited regulation, led to period of around 50 years with an increasingly more successful industry. Drugs were mostly discovered either by serendipity after testing drugs on animals or directly on humans or followed a rational approach based on effects seen from plant extracts. In the mid-seventies companies started to move away from a polypharmacology, with unknown targets to more target based drug discovery approaches, preceding the molecular biology revolution. Since the birth of target-based drug discovery, the industry witnessed a decline in the discovery of new drugs where the annual number of NMEs is now already stabilizing over the last 25 years while R&D costs jump as a result of an increase in projects yet with a higher attrition. Certainly, the application of increasingly more advanced technologies has added significantly to the cost increase as well. A very important confounding factor for the decreased productivity has been the dramatically increased regulatory control as a result of a few public health scandals, with still the most prominent one to date, the Vioxx case where Merck & Co. did not report on significant drug-related mortality (28).

A BRIEF HISTORY OF MODERN PHARMACEUTICAL R&D

To understand the organizational evolution of the pharmaceutical R&D in the context of increasing organizational size, regulatory complexity, compliance requirements and the introduction of a vast amount of new, highly sophisticated technologies together with far reaching automation require some historical awareness. For a far more detailed overview the reader is referred to an excellent review and book on the subject (2,25).

The foundations for the modern pharmaceutical industry were laid by three important elements: (1) the chemical industry (26), (2) biomedical science and (3) the emergence of organic chemistry, sometime in the third quarter of the nineteenth century. The combination of these three elements significantly changed the ad-hoc nature of developing medicines into an industrialized coordinated industry.

THE CHANGING STRUCTURE OF PHARMACEUTICAL R&D

This slow paced evolution as is outlined above, was accompanied with a gradual change of organizational structures in the industry (29). Until the late 1950s, the development of new medicines by companies that commercialized the inventions and the fundamental science by academic institutions at the basis of new discoveries, had been strictly separated worlds. Only a few companies existed like Bell labs, Xerox, GE and AT&T, were in engaged in fundamental research (9). Since then the boundaries started to blur and pharmaceutical companies started to be involved in fundamental research. As the industry turned more toward scientific innovations themselves, with new technologies, centralized structures were introduced. In the decades that followed, up until recently the structures adopted were quite similar between different companies seeking a similar level of innovation.

Today, smaller companies mostly adhere to centralized R&D structures whereas bigger, global spanning R&D based pharmaceutical companies have decentralized organizations (29). Size presumably affects the innovation power of companies as industrialization of life sciences is likely not scalable (30). A significant contribution to successes or failures of drug development projects, however, was not detected in a study by Ringel et al. (31) but a more in depth analysis by Backfisch, accounting more for the failures of smaller businesses, actually suggests that the bigger pharmaceutical companies have better success ratios (32). The top 10 biopharmaceutical companies (based on annual revenue) are all global, have a global R&D presence or in the least multiple sites and over 50,000 employees. To have this efficiently working together requires tremendous efforts and is a balancing act between decentralized very fluid organizational structures and strongly centralized, bureaucratic structures (29). Specifically, the latter has been suggested to be one of the culprits of the innovation gap. GSK was one of the first to recognize this and formed so-called "centers of excellence" for which the long term effects are difficult to foresee but the short term effects apparently already measurable (33). However, because it unclear if company size correlates strongly with success in number of NMEs (31) and the biotech myth of being significantly more efficient in bringing new drugs to the market unmasked (2,9), the question arises how much stronger company culture may correlate to successes and failures as compared with company size. Even if size matters, is it not more logical to consider the importance of how the transactions between the different functional areas are organized and how a company culture deals with uncertainty?

The need to rapidly respond to the changing requirements for an R&D organization, as new technologies emerged with an unprecedented speed, forced companies to become much more flexible at the organizational level (34–36). In several published analyses addressing the so-called innovation gap, the exact causes that were synthesized from the different facts remained speculative. In 2004, the FDA launched the critical path initiative and published a report "Innovation or Stagnation: Challenge and Opportunity on the Critical Path to New Medical Products." The report lists a number of attention areas to improve on the R&D process, and solutions to overcome the innovation gap. Interestingly the term "translational medicine" was coined for a research activity to close the gap between discovery and development activities. Directly related to this "a new" discipline was revamping biomarkers as an element of the new pharmaceutical R&D toolkit.

This report, which has been cited by many of the following papers attempting to further define the (technological) needs and improvements, initiated a small revolution in the R&D process and structural organization. Both translational medicine and biomarker discovery and development became widely accepted in the industry as two new and essential disciplines in the classical linear R&D process. However, just as the introduction of molecular biology in the nineties of the last century and genomics and bioinformatics around 2000, this was quite disruptive in the sense of established processes and procedures in pharmaceutical R&D. Suddenly, not just in the established team structure needed changing but more importantly, the decision-making process. The "traditional" information packages used thus far with the accompanied rules for progressing in the R&D pipeline suddenly were not equipped anymore to deal with the vast amount of data, for often it was unclear how they correlated to the "old" parameters.

TEAM STRUCTURE AND THE DECISION-MAKING PROCESS

There is a vast amount of literature that deals with the management of pharmaceutical R&D and even more about R&D in general. Every company has adopted very specific structures and decision gates, but there is a remarkable general pattern that probably emerged around the end of the 1970s with the introduction of recombinant DNA technology (37). As mentioned before, the increased technocracy in pharmaceutical R&D has led to a high degree of specialism in the different steps leading up to final registration of a drug. Decision makers are faced with a huge number of parameters that determine the fate of a drug candidate. From basic physiochemical properties that affect binding to the target, to interactions with the complex physiology of the human body with understandably the limitation in preclinical research of abilities to test this. The reliance on a multitude of animal models, a range of different species and uncertainties on the translatability to humans makes discovering and developing new drugs sometimes look more like professional gambling rather than science, where calculating the odds makes the difference. Ringel et al. performed an analysis to explore which factors most strongly correlate with the pharmaceutical company's successes as measured by the number of NMEs. Most obvious factors like company size or R&D spending did not correlate. Remarkably, only two factors correlate statistically significant with a positive outcome: the level of science in a company as measured by the output in peer reviewed literature and the presence of experienced decision makers with a successful track record (31). That latter is a difficult to quantify aspect or to define in clear words what skills these experienced decision makers have.

The high degree of specialization has led to the formation of functional areas with a certain critical mass (depending on the size of the company). The concentration of experts in defined organizational structures is necessary to assure proper management of technologies and investments both in hardware as well as human resources. The challenge that follows is to have a seamless integration of all the individual functional areas geared toward only one goal: the delivery of a successful drug as measured by a successful clinical outcome followed by a regulatory acceptance. The challenge that most modern organizations are faced with is the integration between functional lines and the matrix. Only those organizations that succeed in mastering this problem of obvious opposing forces and interests in such a matrix organization, are going to be successful. The division of the research and development organizations has been recognized for more than a decade as one of contributions to clinical failures because of inadequate handovers (38).

THE TRANSLATIONAL MEDICINE PARADIGM AND BIOMARKERS

Where there is an increased technological specialization, risks of silos and tunnel vision start to occur, which, as discussed, can be partially overcome in a matrix organization with strong project management. One of the "skills" lost over the last decades is probably that of the traditional pharmacologist, who in the early days of drug discovery played an essential role in "translating" finding into clinical practice. The FDA Critical Path report spurred a number of initiatives (38) that industry embraced such as informatics, modeling and simulation, imaging and biomarkers. There are many good reviews dealing with this topic and one certainly worth reading is one by Martin Wehling (39) proposing a weighted scoring system for translatability.

Several studies have examined (big) clinical trial failures in the last two decades and, while many could probably not have been predicted, a significant number (sometimes with the benefit of the hindsight) could have been prevented had the organization been better prepared to challenge preclinical findings and designed better clinical trials (40). Many projects that finally failed in the most expensive part of the R&D process could have been stopped earlier or diverted to other indications if there would have been more scrutiny on translatability. It's rather ironic that with the increased technocracy in R&D since the early 1990s, the traditional skills of the pharmacologist in the R&D process probably became underappreciated. Prior to the 1990s pharmacologists would view their projects more holistically and drugs were more promiscuous in the pathways modulated (41,42). The clear shortcomings of rodent models in the context of human diseases such as central nervous system (CNS)-, cancer- and immune-related disorders are frequently discussed (24,43,44) and already in 2002 and later 2004, in two landmark comparative genomics analyses between the brown rat (*Rattus norvegicus*), the laboratory mouse (*Mus musculus*) and man, a clear evolutionary driven deviation of the immune and reproductive system at the molecular level was shown (45,46). With the focus on single targets brought about by the advent of recombinant DNA technologies, came the risk that despite the orthology at the single gene level, systems may show a different level of evolutionary conservation.

So, the question arises why an old discipline needed to be resuscitated? Where did we get lost in translation? The answer is not simple but the solution clear. The management for these complex, multidisciplinary projects needs to improve on the de-risking as they move forward and become data heavy. The multitude of parameters tested has increased over decades and an improved overall assessment of translatability needs an absolute anchoring in the decision making. A translational medicine department obviously does not replace the need for educating all scientists in a project team to think about their approaches in a translational context. Translational scientists, however, besides having a strong background in medical sciences, should be comfortable with the modern, data heavy Omics approaches, have strong data analytics skills and moreover should always think about experiments that better assess species similarities or differences to more adequately predict the human physiological response.

THE CHALLENGES OF DECISION MAKING

The challenges faced by an R&D organization are plenty, and those falling within the remit of a Translational Science department are not trivial. A number of clear bottlenecks in the current methods for managing translational science activities can be identified:

Negative decisions. Project discontinuation is hard and progression seeking bias often prevails (47). We also have to consider that we have unknown-unknowns and we realize a vast number of known-unknowns while only having limited knowledge. Stopping a project is considered as negative and the sunk cost fallacy weighs in significantly (48). Decision makers tend to look at past investments and use that to justify future investments. Equally important is the factor of the perception of failure. Due to cognitive biases of the projects value, biomarker data often is perceived as disruptive because the data do not always support the hypotheses of the experimental drugs, possibly even suggesting discontinuation. But also the company culture itself can easily promote a behavior to shy away from negative data as discussed in the AstraZeneca self-analysis paper (12).

Project versus company goals. Per definition, projects and project teams are geared toward completing a specific task. In drug discovery this can be the identification of lead compounds for a given target molecule to modulate disease parameters. In clinical development guiding a compound though safety/tolerability studies to a first clinical proof of concept. No matter what the specific goal is, making binary decisions on complex data sets with a multidisciplinary team is subject to great risks and projects generally do not have to (or want to) take competing projects into account. With the ultimate goal to reach the specific project milestone, within a strict budget and narrow timelines, the risk that the goal justifies all means and teams turn a blind eye to negative data is high. In an R&D organization where there are multiple projects running in parallel tapping into the same resources, portfolio management and overall project management provide the tools to ensure optimal resource use and that company goals are constantly monitored. The long-term profitability of a company depends on critical go/no-go decisions with projects competing for resources. Because time and resources are an issue, anything that lengthens a project critical path is considered to be negative. Time consuming additional studies to better understand mechanism of action and molecule-target interactions have been considered as a delaying and unnecessary cost element. The impact of typical translational research should be considered as being long term and companywide. Project teams may not always benefit from translational medicine and may consider it as having a negative impact if a project is terminated or must regress several steps.

Translational complexity. Despite the huge advances in our ability to measure changes in genes, proteins, metabolites, cells, tissues and so on, our understanding of complex human physiology and disease still remains limited. This is clearly demonstrated by the lack of true innovation in pharmaceutical R&D but also in the slow progress in understanding disease etiology or molecular aspects of common diseases such cardiovascular, inflammatory and CNS disorders. Specifically these disease areas show a significantly lower probability of success (POS) than other areas (21) reflected in the decision of many biopharmaceutical companies to pull-out of the development of psychotropics and anti-infectives. In the rather linear R&D process, certainly before the costly and time dependent clinical development parts starts, our means to accurately predict a clinical outcome are rather limited. The integration of forward and backward translation, or from bench to bedside and backward (49), poses an enormous burden on an R&D organization, where resources are clearly a limiting factor.

High degree of data complexity and specialized functions. Because pharmaceutical R&D is characterized by vast amounts of data produced by state-of-the-art technologies, mostly operated and well understood by experts with a high degree of specialization, the decision making process in teams becomes tedious and depends on high quality project management lead by experienced team leaders (50). Biomarker studies add another complexity as the translational part not only requires a good technical understanding of the platforms used, but also thorough biological and medical knowledge. Yet, some of these molecular views on disease and drug effects deviate from the traditional classifications of disease (51,52) and clinical end points (53). These latter alternative views are often suppressed in favor of timely progress and/or avoiding complicated discussions with regulators. Lastly, the complexity of the data often leads to a behavior where decision makers cannot comprehend the impact and either rely on key opinion leaders and ignore internal expert opinions or ignore the data.

Linear versus nonlinear thinking. High tech industries, where engineering is involved, have paved the way. The automotive industry has come a long way, though the concept of the car they produce is still the same. The process coming from the first Ford T-model construction lines to the modern common platforms with parallel engineering and multiple interactions with part suppliers has dramatically changed the industry, which after the sixties and seventies was in great need for innovation. Even in recent times the companies (like German and Japanese manufacturers) that were more efficient could survive crises much better as those who were in need for drastic measures (like the American car companies). Therefore, also in the pharmaceutical R&D the rather linear process to deliver a new treatment must have more feedback loops and parallel activities to improve efficiency and increase the POS of clinical projects. As an example, the fixed path from preclinical to clinical by animal model testing (partially required by the health authorities) has been suggested as a main reason for attrition and costs while alternative models could be explored and may offer a more successful approach with lower attrition (54).

Uncertainty. As mentioned before, complexity in pharmaceutical R&D projects is quite high and the level of uncertainty significant. We only have a limited amount of data (representing all aspects of disease and treatment), while time and resources are always a limiting factor to explore a wider space. As mentioned earlier, the knowledge as compared with the known-unknowns and undefinable unknown-unknowns are posing a real challenge for making binary decisions on project continuation. Of course, in practice it is not really binary because we constantly seek confirmation and attempt to refine and redo experiments. This uncertainty of all aspects of disease pathophysiology and treatment effects brings in quite some anxiety for project teams as well as decision makers. Portfolio management based on risk assessment of projects is a science on its own (5) and recently a few interesting simulation experiments have been performed to assess the level of error there is in decision making based on limited information (55,56). Interestingly, the number of people required for a good decision (or to increase the chances to make one) was 100 for early projects increasing with project complexity (56).

BIOMARKERS IN THE DECISION PROCESS: "FIT FOR PURPOSE"

The most important message from fit-for-purpose to the organization is: a lean approach. Fit-for-purpose is in principle a valid argument for all the approaches but particularly relevant for the costs a full-fledged clinical biomarker program brings with at the extreme end the development of a companion diagnostic. Fit-for-purpose, or context of use as coined in a recent report by the biomarker working group of the FDA, allows teams to carefully take a stepwise approach with early feasibility of a biomarker in a drug discovery program. Tools like a biomarker development plan through a question based approach (6,57) should help identify gaps in an R&D program and subsequent work to assess the limitation is detailed and planned.

So, it is apparent that the decision process for a pharmaceutical project is complex, requiring both a macroscopic as well as

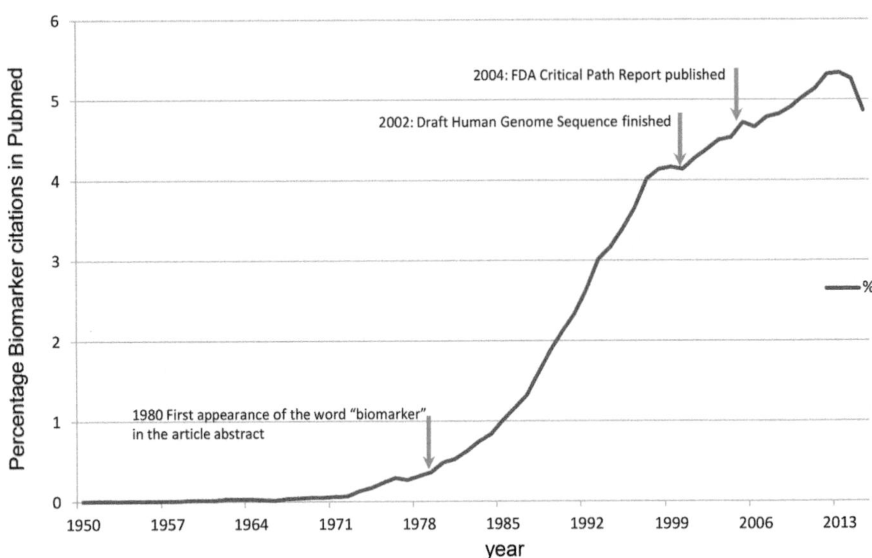

FIGURE 9.1.2 PubMed-indexed articles trend search using Alexandru Dan Corlan. MedLine trend: automated yearly statistics of PubMed results for any query, 2004. The line graph shows the articles published using the word "biomarker" anywhere in the citation as a percentage of the total number of articles published in a given year.

a microscopic view, and requires multiple inputs from different experts. The question that arises for biomarkers and translational medicine is: "Do biomarkers add to the complexity, or lighten the burden in the decision-making process?" The decision maker(s) rely heavily on their experience and lastly intuition, the latter of which is a difficult to quantify subjective element, shaped by experience. Biomarkers represent the "new tool kit" the FDA refers to in its critical path report. The term "biomarker" appears first in PubMed in 1980 (Figure 9.1.2), and the Biomarkers Definitions Working Group has defined a biomarker as a "A characteristic that is objectively measured and evaluated as an indicator of normal biological processes, pathogenic processes, or pharmacologic responses to a therapeutic intervention" (58). Though the name biomarker may be relatively new, the definition clarifies that for centuries we already have been using biomarkers to identify disease and reveal treatment effects, though most often with limited scope and effect. The revisit of the biomarker as an important tool, however, has led to confusion, misplaced expectations and skepticism. Like for the almost forgotten and under-appreciated traditional discipline of pharmacology, biomarkers very quickly were seen as exotic and cost drivers, rather than solutions and enablers of better decision making. Unfortunately, there are no comprehensive reviews dealing with the inevitable cost increase new technologies bring to the whole drug discovery and development process, but it is easy to perceive that they are significant contributors to the increased cost. The challenges of how to use biomarkers for decision making is a multivariate problem to solve, equally difficult as the analysis of biomarkers in the context of the biology we try to understand. But the increases in cost are irrelevant if we consider the benefit of cost savings on halting projects in an early stage that have limited or no potential, or accelerating promising projects by faster assessments. It's clear that the required information on a timescale of just more than a decade is difficult to obtain. Lastly, if biomarkers proof to be essential for the selecting the right patient population (52), we enter the new era of precision medicine where the vision is to be able to treat all patients cost effectively.

In many R&D organizations biomarkers have become a central theme in the decision process as they allow for an objective valuation of the project strength's (12,13,59). Because most biomarkers are also continuous variables, they lend themselves very much for another recent innovation in drug discovery and development projects: Pharmacokinetic/pharmacodynamic (PK/PD) modeling, offering yet another excellent translational tool to allow teams but in the end the decision makers, to assess the likelihood of clinical efficacy within an acceptable safety window (60).

BIOMARKER IMPLEMENTATION IN MODERN PHARMACEUTICAL R&D

Building new capabilities in a constantly changing R&D environment is not trivial and in case of the multiplicity of possibilities for biomarker research and development depending on the available resources and size of the organization. Big organizations may be able to recruit expertise from existing areas (such as liquid chromatography–mass spectrometry [LC-MS] from bioanalytics) or rapidly recruit and deploy new expertise from specialized companies or academia. Obviously, hybrid models are possible as well. Such new dynamic capabilities initially put a burden on an organization and for the reasons laid out in the previous sections, needs newly defined team structures and adaptation and acceptance by decision makers. The latter obviously depends where the initiative was born. If it was a top-down decision, education and restructuring in the respective functional units is normally required.

In an article by Narayanan et al. (35), the processes of how such capabilities may be explored and implemented in R&D organizations is discussed in detail. In the two examples used, "fast cycle capability in the clinical development process" and the "chemical biology in the discovery process" had very different starting points and each experienced a number of clearly recognizable organizational behaviors. In the case of implementing a translational process by means of creating novel biomarker functional units and implant them in the R&D teams, organizations witness a mix of both acceptance problems from project teams and other (more traditional) pharmaceutical R&D units, as well as a struggle by senior management to cope with new attributes to take into account for their decision making as well as often very novel innovative technologies that disrupt the process such as a whole range of Omics technologies. Narayan et al. did not reach a firm conclusion, but from the managerial failure they describe, it becomes clear that with the top down approach stakeholder engagement is the critical element for success. Both the experts in the different functional domains needed to grow a buy-in, as well senior management from the different functional units. This challenge is obviously magnified by the disruptive nature of biomarker R&D and translational medicine together with the absolute requirement of full integration into the existing processes.

IMPLEMENTING TRANSLATIONAL SCIENCE AS A CULTURE OR DISCIPLINE?

This chapter attempts to show the aspects, other than the life science and medical ones, that mostly receive little attention: culture and structure. The impact of mergers and acquisitions and reorganizations that have characterized the pharmaceutical industry for decades did not turn the ship around (61,62). The number of NMEs still lags with the rapidly increasing investments but most importantly with the ever-increasing medical needs because of an aging population. The accumulating literature on management, company structures in highly innovative and technological domains suggest culture, structure, management and decision making have an enormous impact on final successes or failures (63). Is translational science a culture or a discipline? When one thinks of introducing this into the current practice, is education and training sufficient or do we really need to define and implement a new structure? Is it in the end a functional expertise? Clearly looking at the individual elements that contribute to current translational research one is tempted to say no. Strictly spoken it is a mind-set, an umbrella under which many scientific disciplines work. It is not a narrowly defined expertise like toxicology; clinical pharmacology; absorption, distribution, metabolism and excretion (ADME); or medicinal chemistry. Translational medicine or science is an integration of many disciplines, where ideally the adherence to the requirements of clinical translatability is strong. In most translational medicine departments, being academic or commercial in nature, there is a direction and lead of a multiplicity of disciplines toward an improved and sustainable outcome where education and communication are essential. Communication toward the different disciplines and the leadership team, education through internal programs where scientists learn the key elements that drive successful translation.

Many analyses have shown that the key to success is an early assessment of a drug project's viability, to be able allocate or reallocate valuable and mostly competing resources. "Quick win, fail fast" (64) clearly poses a paradigm shift where there is a better cost containment (quick win) and an early halting of likely unsuccessful projects (fail fast) (24). Particularly for "fast fail" the biomarker strategy can help show a concept does not translate very well. In a hospital environment, the challenge is to bring the design thinking closely centered on the patient. For the pharmaceutical industry the answer may lie mainly allowing for the decision making on a continuous process where the interdependencies are carefully evaluated. Once chemical lead-series have been identified, before preclinical development starts, critical steps must be undertaken to test clinical translation. Key questions on target expression in tissue and disease relevance should be answered as well as the definition and the subsequent development or recruitment of tools that will provide the critical readouts for the first clinical studies up to the final proof-of-concept. Literature on disease relation must be critically reviewed and experiments repeated in independent disease cohorts. Several studies have shown the potential delay and costs involved of irreproducible results, but mostly these projects failed early (65,66). Many "traditional disciplines" that have existed for more than half a century are involved but only a common goal and direction that can be provided by a constant testing of translatability.

So, in essence translational science is both a discipline and an organizational model to allow the deepened expertise and knowledge to be orchestrated to find suitable therapies that show commercial viability and allow for precision treatment of disease and or symptoms. It is the absolute and undisputed requirement, but certainly not the only one, to bridge the innovation gap. It prevents decisions being made on pure commercial, emotional or other subjective grounds. But the key role it probably plays either in industry or outside is the integrative power. Because translation relies on deciding on key aspects such as "does the drug get to right target in the right tissue?" and "are the key biomarkers for the perceived mechanism of action really showing the drug has an effect at a tolerable dose?." But not only at the team level should the integration be near perfect but at the senior decision level the decision gates, certainly as projects are getting nearer to FIH, need to be guarded with clear objective guidelines. In a recent paper from Dolgos et al the so called Translational Medicine Guide (TxM) guide at Merck KGaA is discussed as a question based approach to continuously monitor and select the right projects (67). This paper nicely lists the benefits from a tight interplay between project teams, clear guidelines and managerial rigor. Figure 9.1.3 shows a schematic representation of the specific interference of the biomarker guided translational science activities. Although the project generates a continuous stream

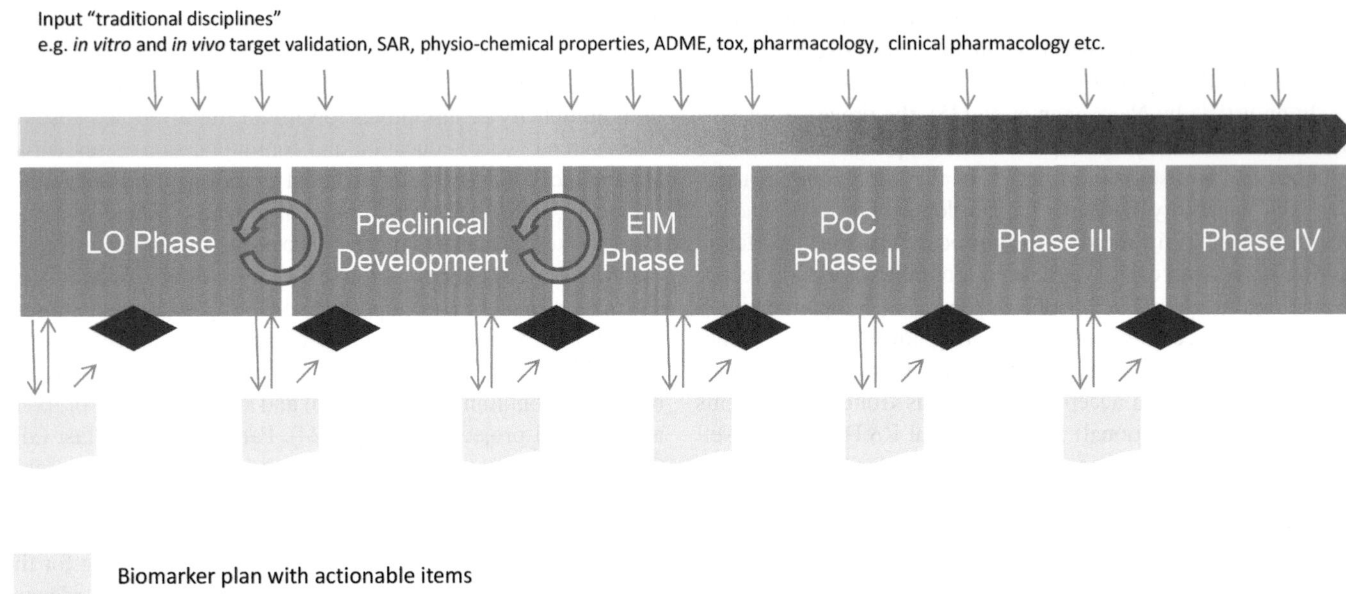

FIGURE 9.1.3 During the lifecycle of a project, many teams will work on obtaining the key parameters that characterize the drugability and drug-like characteristics. The actual number of criteria to take decisions are probably ranging from 10–100, depending on the company and the therapeutic area. Translational science, with the definition of a biomarker plan, containing the key translatable readouts, will enhance or challenge other findings gathered over time. These biomarker readouts are a firm and objective means to allow for informed decision making. During the early phases up to first-in-human (FIH), iterative cycles of compound improvements allow for an increased probability of success (POS). ADME, absorption, distribution, metabolism and excretion; EIM, Entry Into Man; LO, lead optimization; PoC, proof of concept; SAR, structure–activity relationship.

of data to support the decision on compound characteristics, the translational medicine experts challenge team and management at the decision points on key translatability issues and subsequently propose and execute iterative cycles of improvement.

There is neither an absolute success formula nor an absolute definable organization structure that defines the success of a pharmaceutical R&D organization. Based on the limited number of deeper analysis of failures and successes and common logic one can define, however, a best practice that can be readily applied to any R&D organization.

- The first important element for any full-sized biopharmaceutical company is the seamless integration between the discovery and development organization in leadership and project teams.
- For the projects as they progress, translational thinking needs to be embedded as early as possible, whereas decision making should be integral and organization wide.
- The implementation of a translational medicine department guarantees the continuous development of cutting-edge (omics) technologies and drives the mindset for translational decision making.
- Translational medicine as a functional area should be embedded in every project team, from early discovery till early clinical development.
- A full understanding of the Mechanism of Action (MOA) and disease pathophysiology is crucial, therefore, modern data intensive methodologies with "translational informatics" should be the key ingredients of every R&D project.
- Drug project portfolio management needs to be aligned with the biomarker fit-for-purpose biomarker process to guarantee the availability of key decision biomarkers at every stage.
- The full promise of precision medicine can only be realized of we are prepared to invest in precision tools for better and safer treatment options, therefore, we must accept the investments to be high at first, but the return will be there for all stakeholders in health care on the long run.

FUTURE OUTLOOK FOR PHARMACEUTICAL R&D

The unsustainable biopharmaceutical R&D process has been criticized for almost three decades now with many root causes mentioned. However, innovation in the process appears difficult and many interesting and very daring models have been proposed to overcome the innovation gap but most of them seem to lack the incentive or do not really represent thought through viable economic models. Changing the decision-making process and allowing to have new disciplines rule the R&D process may prove to be a small step toward more cost-effective R&D. By ruling out projects that are insufficiently underpinned by evidence to suggest a clinical benefit and concentrating resources on obtaining a maximum of translational evidence to support extensive investments in later stages of development may actually have a significant financial impact. Still with an attrition rate of about 50% (68,69) in phase III clinical trials shows that the POC studies themselves are not equipped to be able to rationalize a true clinical benefit in a representative population.

Ultimately, though it can be disputed that this may not work in larger organizations, small empowered multidisciplinary teams, working within a well-managed portfolio, taking the uncertainty into account but require a rigor on translational aspects, could complement the experienced senior decision makers who on their part take a more coaching role. Faster feedback cycles and allowing more information to flow back from patients, patients experience and patients' expectations could lead to a more sustainable R&D apparatus in the near future.

CLOSING REMARKS

This section carefully reviews the processes from project initiation to product delivery in biopharmaceutical companies and other medical research organizations where the fast and efficient translations from invention to innovation is required to meet the many unmet medical needs and to create a sustainable health care environment where costs are contained and patients become part of the health care chain rather than standing outside as merely consumers. Organizational structures in the industries where advanced technologies and science dominate need to be critically reviewed considering their missions. Standard hierarchal models likely do not work and creativity limited by imposing strict output numbers. Companies should not look for short term benefits as project lifecycles are counted in decades. Finally, this section describes solutions "within," but the last few years many companies, particularly larger ones, have begun to use open innovation models (70,71) to more dynamically face innovation hurdles with wider scientific inputs and locate in so called innovation hubs (72). As was mentioned here, the uncertainty in most multiphase projects is significant as well as the number of experts needed to be involved for better decision making. The future will likely also be affected by using artificial intelligence to aid decision making if the currently predicted digital revolution takes place.

ACKNOWLEDGMENTS

I would like to thank everyone who has inspired me in the last 18 years to write this book section on the so-often underestimated cultural aspects that influence our failures and successes. I would like to in particular thank my coworker Dr. Richard Welford for the critical reading of the manuscript and his thoughtful comments.

REFERENCES

1. Nicolaou, K.C. Advancing the drug discovery and development process. *Angewandte Chemie (International ed in English)*, 2–15 (2014).
2. Malerba, F. & Orsenigo, L. The evolution of the pharmaceutical industry. *Business History* **57**, 664–687 (2015).
3. Leyens, L., Richer, E., Melien, O., Ballensiefen, W. & Brand, A. Available tools to facilitate early patient access to medicines in the EU and the USA: Analysis of conditional approvals and the implications for personalized medicine. *Public Health Genomics* **18**, 249–259 (2015).
4. Jekunen, A. Decision-making in product portfolios of pharmaceutical research and development—Managing streams of innovation in highly regulated markets. *Drug Des Devel Ther* **8**, 2009–2016 (2014).
5. Betz, U.A.K. Portfolio management in early stage drug discovery—A traveler's guide through uncharted territory. *Drug discovery today* **16**, 609–618 (2011).
6. van Gool, A.J., Henry, B. & Sprengers, E.D. From biomarker strategies to biomarker activities and back. *Drug discovery today* **15**, 121–126 (2010).
7. Janis, I.L. *Victims of Groupthink: A Psychological Study of Foreign-Policy Decisions and Fiascoes*. Houghton Mifflin: Boston, MA (1972).
8. Gassmann, O. & Reepmeyer, G. Organizing pharmaceutical innovation: From science-based knowledge creators to drug-oriented knowledge brokers. *Organizing pharmaceutical innovation* **14**, (2005).
9. Pisano, G.P. Can Science Be a Business? Lessons from Biotech. *Harvard Business Review*, **10**, 1–12 (2006).
10. Woodcock, J. & Woosley, R. The FDA critical path initiative and its influence on new drug development. *Annual Review of Medicine* **59**, 1–12 (2008).
11. Administration, U.F.A.D. (2004). *Innovation or Stagnation: Challenge and Opportunity on the Critical Path to New Medical Products*. U.S. Food and Drug Administration: Washington, DC.
12. Cook, D. et al. Lessons learned from the fate of AstraZeneca's drug pipeline: A five-dimensional framework. *Nature Reviews Drug Discovery*, 1–13 (2014).
13. Morgan, P. et al. Can the flow of medicines be improved? Fundamental pharmacokinetic and pharmacological principles toward improving Phase II survival. *Drug Discovery Today* **17**, 419–424 (2012).

14. Scannell, J.W., Blanckley, A., Boldon, H. & Warrington, B. Diagnosing the decline in pharmaceutical R&D efficiency. *Nature reviews Drug discovery* **11**, 191–200 (2012).
15. Scannell, J.W. & Bosley, J. When quality beats quantity: Decision theory, drug discovery, and the reproducibility crisis. *PloS one* **11**, e0147215 (2016).
16. Munos, B. A forensic analysis of drug targets from 2000 through 2012. *Clinical Pharmacology and Therapeutics* **94**, 407–411 (2013).
17. Munos, B. How to handle an industry in disruption: Intervene or laissez-faire? *Science Translational Medicine* **7**, 286ps12 (2015).
18. Munos, B. Lessons from 60 years of pharmaceutical innovation. *Nature Reviews Drug Discovery* **8**, 959–968 (2009).
19. Munos, B.H. & Chin, W.W. How to revive breakthrough innovation in the pharmaceutical industry. *Science Translational Medicine* **3**, 89cm16 (2011).
20. Munos, B.H. & Orloff, J.J. (2016). *Disruptive Innovation and Transformation of the Drug Discovery and Development Enterprise*. US National Academy of Medicine: Boston, MA.
21. Pammolli, F., Magazzini, L. & Riccaboni, M. The productivity crisis in pharmaceutical R&D. *Nature Reviews Drug Discovery* **10**, 428–438 (2011).
22. Sams-Dodd, F. Is poor research the cause of the declining productivity of the pharmaceutical industry? An industry in need of a paradigm shift. *Drug discovery today* **18**, 211–217 (2013).
23. Schmid, E.F. & Smith, D.A. R&D technology investments: Misguided and expensive or a better way to discover medicines? *Drug Discovery Today* **11**, 775–784 (2006).
24. Kola, I. & Landis, J. Can the pharmaceutical industry reduce attrition rates? *Nature Reviews Drug Discovery* **3**, 711–715 (2004).
25. Hill, R.G. & Rang, H.P. *Drug Discovery and Development—E-Book: Technology in Transition*. Elsevier Health Sciences: New York (2012).
26. Murmann, J.P. *Knowledge and Competitive Advantage: The Coevolution of Firms, Technology, and National Institution in the Synthetic Dye Industry, 1850–1914* Cambridge University Press: Cambridge, MA. (2003).
27. Rang, H.P. Chapter 1—The development of the pharmaceutical industry. In: *Drug Discovery and Development (Second Edition)* 2nd ed. Eds. Hill, R.G. and Rang, H.P. 3–18 Churchill Livingstone: London, UK. (2012).
28. Zwillich, T. How Vioxx is changing US drug regulation. *The Lancet* **366**, 1763–1764 (2005).
29. Tirpak TM, M.R., Schwartz L, and Kashdan D. R&D structure in a changing world. *Research Technology Management* **49**, 19–26 (2006).
30. Pisano, G.P. (2010). *The Evolution of ScienceBased Business: Innovating How We Innovate* Harvard Business School: Boston, MA (2010).
31. Ringel, M., Tollman, P., Hersch, G. & Schulze, U. Does size matter in R&D productivity? If not, what does? *Nat Rev Drug Discov* **12**, 901–902 (2013).
32. Martin, B. (2018). *The Development of Firm Size and Innovativeness in the Pharmaceutical industry between 1989 and 2010*. Philipps-Universität Marburg, Faculty of Business Administration and Economics, Department of Economics: Volkswirtschaftliche Abteilung (2018).
33. Garnier, J. Rebuilding the R&D engine in big pharma. *Harvard Business Review*, (2008).
34. Achilladelis, B. & Antonakis, N. The dynamics of technological innovation: The case of the pharmaceutical industry. *Policy Research* **30**, 535–588 (2001).
35. Narayanan, V.K., Colwell, K. & Douglas, F.L. Building organizational and scientific platforms in the pharmaceutical industry: A process perspective on the development of dynamic capabilities. *British Journal of Management* **20**, S25–S40 (2009).
36. Teece, D.J., Pisano, G. & Shuen, A. Dynamic capabilities and strategic management. *Strategic Management Journal* **18**, 509–533 (1997).
37. Cockburn, I.M. The changing structure of The pharmaceutical industry. *Health Affairs* **23**, 10–22 (2004).
38. Butler, D. Crossing the valley of death. *Nature* **453**, 842 (2008).
39. Wehling, M. Assessing the translatability of drug projects: What needs to be scored to predict success? *Nature Reviews Drug Discovery* **8**, 541–546 (2009).
40. Wehling, M. Drug development in the light of translational science: Shine or shade? *Drug Discovery Today* **16**, 1076–1083 (2011).
41. Swinney, D.C. & Anthony, J. How were new medicines discovered? *Nature Reviews Drug Discovery* **10**, 507–519 (2011).
42. Eder, J., Sedrani, R. & Wiesmann, C. The discovery of first-in-class drugs: Origins and evolution. *Nature Reviews Drug Discovery* **13**, 577–587 (2014).
43. Berggren, R., Møller, M., Moss, R., Poda, P. & Smietana, K. Outlook for the next 5 years in drug innovation. *Nature Reviews Drug Discovery* **11**, 435–436 (2012).
44. McGonigle, P. & Ruggeri, B. Animal models of human disease: Challenges in enabling translation. *Biochem Pharmacol* **87**, 162–171 (2014).
45. Gibbs, R.A. et al. Genome sequence of the brown norway rat yields insights into mammalian evolution. *Nature* **428**, 493–521 (2004).
46. Waterston, R.H. et al. Initial sequencing and comparative analysis of the mouse genome. *Nature* **420**, 520–562 (2002).
47. Lendrem, D.W., Lendrem, B.C., Peck, R.W., Senn, S.C., Day, S. & Isaacs, J.D. Progression-seeking bias and rational optimism in research and development. *Nat Rev Drug Discov* **14**, 219–221 (2015).
48. Peck, R.W., Lendrem, D.W., Grant, I., Lendrem, B.C. & Isaacs, J.D. Why is it hard to terminate failing projects in pharmaceutical R&D? *Nature reviews Drug discovery* **14**, 663 (2015).
49. Ledford, H. The full cycle. *Nature* **453**, 843–845 (2008).
50. Bobadilla, N. & Gilbert, P. Managing scientific and technical experts in R&D: Beyond tensions, conflicting logics and orders of worth. *R&D Management* **47**, 223–235 (2017).
51. Smoller, J.W., Andreassen, O.A., Edenberg, H.J., Faraone, S.V., Glatt, S.J. & Kendler, K.S. Psychiatric genetics and the structure of psychopathology. *Mol Psychiatry* 1 (2018).
52. Yan, L. & Zhang, W. Precision medicine becomes reality-tumor type-agnostic therapy. *Cancer Commun (Lond)* **38**, 6 (2018).
53. Parchment, R.E. & Doroshow, J.H. Pharmacodynamic endpoints as clinical trial objectives to answer important questions in oncology drug development. *Semin Oncol* **43**, 514–525 (2016).
54. PWC. Pharma 2020: The vision which path will you take? PWC, (2007).
55. Hedner, T. Change in the Pharmaceutical Industry. Aspects on Innovation, Entrepreneurship, Openness, and Decision Making. Linköping University Electronic Press (2012).

56. Cowlrick I, Hedner T, Wolf R, Olausson M & M, K. Decision-making in the pharmaceutical industry: Analysis of entrepreneurial risk and attitude using uncertain information. *R&D Management* **41**, (2011).
57. Evers, R., Blanchard, R.L., Warner, A.W., Cutler, D., Agrawal, N.G. & Shaw, P.M. A question-based approach to adopting pharmacogenetics to understand risk for clinical variability in pharmacokinetics in early drug development. *Clin Pharmacol Ther* **96**, 291–295 (2014).
58. Definitions, B. & Group, W. Biomarkers and surrogate endpoints: Preferred definitions and conceptual framework. *Clinical pharmacology and therapeutics* **69**, 89–95 (2001).
59. Cohen, A. F., Burggraaf, J., Gerven, J.M.a. V., Moerland, M. & Groeneveld, G.J. The use of biomarkers in human pharmacology (Phase I) studies. *Annual Review of Pharmacology and Toxicology*, 1–20 (2014).
60. Rajman, I. PK/PD modelling and simulations: Utility in drug development. *Drug discovery today* **13**, 341–346 (2008).
61. Goldsmith, A.D. & Varela, F.E. Fragmentation in the biopharmaceutical industry. *Drug Discov Today* **22**, 433–439 (2017).
62. LaMattina, J.L. The impact of mergers on pharmaceutical R&D. *Nature Reviews Drug Discovery* **10**, 559–560 (2011).
63. PWC. (2012). *From vision to decision*. PricewaterhouseCoopers, London, UK (2012).
64. Paul, S.M. et al. How to improve R&D productivity: The pharmaceutical industry's grand challenge. *Nature Reviews Drug Discovery* **9**, 203–214 (2010).
65. Prinz, F., Schlange, T. & Asadullah, K. Believe it or not: How much can we rely on published data on potential drug targets? *Nature Reviews Drug Discovery* **10**, 712 (2011).
66. Begley, C.G. Raise Standards for preclinical cancer research. *Nature* **483**, 8–10 (2012).
67. Dolgos, H. et al. Translational medicine guide transforms drug development processes: The recent merck experience. In: *Drug Discovery Today*, Vol. 21 517–526 Elsevier: Amsterdam, the Netherlands, (2016).
68. Arrowsmith, J. Trial watch: Phase III and submission failures: 2007–2010. *Nature Reviews Drug discovery* **10**, 87 (2011).
69. Harrison, R.K. Phase II and phase III failures: 2013–2015. *Nature Reviews Drug discovery* **15**, 817–818 (2016).
70. Rebhan, M. Towards a systems approach for chronic diseases, based on health state modeling. *F1000Research* **6**, 309 (2017).
71. Yu, H.W. Bridging the translational gap: Collaborative drug development and dispelling the stigma of commercialization. *Drug Discov Today* **21**, 299–305 (2016).
72. Gautam, A. & Pan, X. The changing model of big pharma: Impact of key trends. *Drug Discov Today* **21**, 379–384 (2016).

Biomarker Qualification and Companion Diagnostics 9.2

Failure and Success of Regulatory Processes

Federico Goodsaid

Contents

INTRODUCTION

There are two unique pathways for the regulatory qualification and approval of biomarkers at the FDA: the Biomarker Qualification Pathway (BQP) at the Center for Drug Evaluation and Research (CDER) and the Companion Diagnostic Approval (CDA) at the Center for Devices and Radiological Health (CDRH). Both of these regulatory pathways have witnessed a decade of development and use. Both of these regulatory pathways started with the Pharmacogenomics Guidance [1] and Critical Path for Innovation [2] documents in 2004 and 2005. These documents opened in the FDA both the science and regulatory review of genomic data, as well as the concept of biomarkers as critical components for the development of precision medicine. Although the genesis for both of these pathways has a lot in common, the outcomes for their application and their impact on the development of Precision Medicine have been dramatically different.

Biomarkers can be considered tools to be qualified in close association with individual submissions for drug approvals. The FDA's Drug–Diagnostic Companion Diagnostic Guidance [3] described a specific example of this association, where the approval of a drug and a test are closely linked with each other in both their product concepts and the timelines of their regulatory approvals. Several examples of biomarkers approved through this codevelopment process are shown in the FDA's Table of Pharmacogenomic Biomarkers in Drug Labeling [4]. Biomarkers in this table include both genetic and translational entries. Genetic biomarkers are often integrated in drug development as clinical or nonclinical markers of drug efficacy or safety for the purpose of patient selection in clinical trials, response prediction through stratification or enrichment, or dose optimization. Translational biomarkers have applications similar to those of genetic biomarkers but may also be useful for response monitoring and as early indicators of toxicity or adverse reactions.

Novel approaches are being tested continuously for the successful integration of biomarkers in drug development. Their applications range from early compound selection through post-marketing applications. However, integration of these novel biomarkers into routine nonclinical and clinical practice, and regulatory submissions, have often been slow. Hesitation in the application of these tests is often associated with fear not only about how comprehensive data supporting these applications are but also about the regulatory interpretation of the context of use for these applications [5]. Biomarker tests can be integrated into drug development when we have a consensus about the context in which we are measuring the biomarker and the evidence supporting this measurement. These levels of consensus need to be reflected in the regulatory review of biomarker data.

REGULATORY PATHS IN COMPANION DIAGNOSTIC EVALUATION AND APPROVAL

Regulatory pathways for companion diagnostics have developed in close association with the identification through genetic testing of enrichment biomarkers for selection of specific patient subpopulations. The biomarkers in this case are mostly enrichment biomarkers, and their tests are approved concurrently with therapies developed for the specific patient subpopulations selected by the enrichment biomarkers. The Center for Drugs at the FDA issued in 2014 an enrichment guidance [6] that covers the application of enrichment biomarkers in drug development.

This enrichment guidance is essential to understand the pathway to companion diagnostic approval. A rationale for a biomarker that could be used as a companion diagnostic is possible if the biomarker is a candidate as an effective tool for patient selection to receive targeted therapies. The confirmation of this patient selection marker is obtained in the pivotal study for the therapy. The value of a patient selection marker, therefore, is closely linked to the efficacy of the therapy for which it is used for patient selection.

A major regulatory question that was addressed by the final Companion Diagnostic Guidance was whether an independent clinical utility claim could be made—and should be proven in a pivotal study for a therapy—for a companion diagnostic. Analytical and clinical validity are certainly expected for companion diagnostics, but can their clinical utility claims be defined independently from those of the therapy that they are used to select patients for?

Initial drafts for the Companion Diagnostic Guidance were jointly written by CDER and CDRH around 2005–2006. Proposals at the time [7] were being discussed with relatively modest development and regulatory review burdens for a companion diagnostic by diagnostic companies and regulatory reviewers at CDRH. If only analytical and clinical validity need to be shown for a companion diagnostic, its development costs are independent of those for targeted therapies, and its regulatory reviews only require assessment of how accurate and reproducible the measurement is for a marker that has an independent compelling biological context.

Guidance Documents in the Development of Companion Diagnostics

In vitro diagnostics (IVDs) are regulated by the FDA, Pharmaceuticals and Medical Devices Agency (PMDA) and in the EU. When novel targeted therapies are approved by the FDA, European Medicines Agency (EMA) and PMDA, the companion diagnostic test required to select patients who will receive the therapy is considered for approval concurrently with the therapy. Guidance documents for the submission and review of companion diagnostic products have been developed over the past decade.

The FDA Pharmacogenomics Guidance suggested in 2005 the need for a second document that would cover companion diagnostics. Regulatory documents drafted in this area cover not only the regulatory review process for these companion diagnostic products, but also a continuous expansion of the regulatory space claimed by the FDA CDRH in the United States. These are the key guidance documents in the development of companion diagnostics:

2007: FDA Pharmacogenetic Tests and Genetic Tests for Heritable Markers [8]

Initial, high-level guidance summarizing information needed for regulatory review and approval of pharmacogenetic and genetic tests for heritable markers.

2007: FDA In Vitro Diagnostic Multivariate Index Assays [9]

This was a draft guidance that was not issued in final text. In Vitro Diagnostic Multivariate Index Assays (IVDMIAs) are tests such as those that require microarrays where an algorithm translates the results from multiple variables into a clinically actionable outcome. In oncology, an example of an IVDMIA is the OncotypeDx test. This guidance requested the regulation by the FDA (in addition to Clinical Laboratory Improvement Amendments [CLIA]) of tests that made use of these algorithms. This guidance was the initial attempt for regulation by the FDA of laboratory developed tests (LDTs) (see Timeline to withdrawal of a regulatory guidance).

2011: FDA Commercially Distributed In Vitro Diagnostic Products Labeled for Research Use Only or Investigational Use Only [10]

Frequently Asked Questions (FAQs) in this guidance discuss limits to the use of Research Use Only or Investigational Use Only diagnostic tests in providing clinically actionable data. FAQs B5 and B8 in this guidance specifically addressed the interest of the FDA to regulate these products. This guidance was issued concurrently with white papers aimed at developing guidance for the regulation of LDTs. Although a draft of this guidance [11] was issued in 2014, no final version of this guidance is anticipated in the near future, and the issue of LDT regulation remains an open [12]—and contentious [13]—issue for the application of companion diagnostics.

2011: FDA In Vitro Companion Diagnostic Devices [3]

The Companion Diagnostic Guidance suggested by the Pharmacogenomics Guidance in 2005 was finally issued in 2011. Answers to three key product development questions are covered by this guidance:

- Whose burden is it to show the clinical utility of a companion diagnostics? What determines the clinical utility of a companion diagnostic? *Answer:* The developer of the therapeutic product.

- What determines that a companion diagnostic will require a PMA versus a 510(k)? *Answer:* This is a case-by-case decision for CDRH.
- What will be the goal of a phase III trial for a therapeutic product that requires a companion diagnostic? *Answer:* To prove that both the therapeutic product as well as the companion diagnostic independently show clinical utility.

Over the course of the past decade, an independent clinical utility concept was developed by CDRH for an independent companion diagnostic clinical utility to be shown in the pivotal study for the targeted therapy. This independent clinical utility claim shifts the financial burden for companion diagnostic development primarily to companies testing the therapeutic product because major investments are made by these companies in the clinical development of companion diagnostics. It has also crippled the development of follow-on companion diagnostics because regulatory approval for these requires not only analytical and clinical validity, but also a study with clinical samples (ideally from the original pivotal study) to confirm clinical utility.

The outcome of these strategies has been a limited regulatory impact on day-to-day testing of companion diagnostics. Companion diagnostic products are initially approved by CDRH concurrently with their targeted therapies approved by CDER. However, clinical laboratories around the United States have for the most part ignored the FDA-approved tests and instead run LDT companion diagnostic tests. The lack of an effective internal and/or external reference framework—regulatory or not—to assess the accuracy and reproducibility of LDT tests is a major challenge for the development of accurate and reproducible clinical testing in the United States. In the meantime, FDA—approved companion diagnostic tests will continue to be required for targeted drug approvals, whereas the practice of medicine for targeted drug patient selection will not depend on them.

There are also some unexpected consequences for the current regulatory pathways of companion diagnostics. Although the FDA has not developed yet a guidance for targeted onco next generation sequencing (NGS) panels, the FDA white paper and workshop of February 2016 [14] laid out the expected label claims for targeted NGS onco panels. These include a section on the follow-on companion diagnostics (and corresponding clinical utility claims) and a section on all variants with their analytical and clinical validity claims. The section on the follow-on companion diagnostics is subject to the same problems all other companion diagnostic platforms have. After the launch of the therapeutic product, most testing for the companion diagnostic will not be with FDA-approved tests. It is possible that this may be avoided for sufficiently complex NGS companion diagnostic tests, such as those that require variant signatures across multiple genes. FDA-approved NGS onco panels are likely to succeed in the marketplace for tests whose complexity cannot justify their internal development in individual CLIA labs.

REGULATORY PATHS IN BIOMARKER EVALUATION AND QUALIFICATION

The path from an exploratory biomarker to a biomarker qualified for a specific application context can be long and unpredictable [15]. Application of these biomarkers requires an objective record for their nonclinical or clinical context and supporting qualification evidence. Information from the development of exploratory biomarkers has been shared between the pharmaceutical industry and the FDA through voluntary exploratory data submissions (VXDSs) [16]. Submissions of exploratory biomarker data have allowed reviewers at the FDA to share with scientists in the pharmaceutical industry study designs, sample isolation and storage protocols, technology platforms, analysis algorithms, biological pathway interpretation, and electronic data submission formats. This experience has been valuable in training our reviewers for the analysis and interpretation of biomarker data.

VXDS stressed the need for a regulatory path from exploratory biomarkers to biomarkers qualified for a specific context. Such a path was tested at the FDA through a pilot process for biomarker qualification [17]. This process was focused on the specific needs of the regulatory environment to ensure scientifically accurate and clinically (or preclinically) useful decision making. In the first use of this new joint-agency review process put in place by U.S. and European drug regulators, the FDA and the EMA [18] allow drug companies to submit the results of seven new tests as evidence of nephrotoxicity by new drugs. The qualification of these biomarkers covers voluntary submission of these data for rat studies. On a case-by-case basis, the FDA will also consider possible application of these biomarkers in phase I human trials. The tests measure levels of seven key proteins or biomarkers that scientists from the FDA and EMA believe provide important new safety information about the effect of drugs on the kidney. When reviewing investigational new drug (IND) applications, new drug applications (NDAs), or Biological License Applications (BLAs), both regulatory agencies consider the test results in addition to blood urea nitrogen (BUN) and creatinine.

The development of the new renal toxicity biomarkers was led by the Predictive Safety Testing Consortium (PSTC) [19], whose members include scientists from over a dozen pharmaceutical companies. The PSTC was organized and led by the Critical Path Institute [20]. Researchers from Merck and Novartis identified the nonclinical nephrotoxicity biomarkers, tested them to prove their sensitivity, specificity, and positive and negative predictive value, and then shared their findings with the other consortium members for further study. The consortium then submitted applications for their qualification to FDA and EMA.

This was a unique example of how a group of drug companies can work together to propose and generate qualification data for new safety tests and then present them jointly to the FDA and EMA for qualification. The FDA and EMA laid the groundwork for such joint-agency reviews in 2004 with the development of the VXDS framework. The VXDS review served as the baseline model around which to design the pilot process for biomarker qualification in 2006. A similar biomarker qualification data submission (BQDS) meeting was held in this pilot process to allow an exchange of questions with the sponsor about scientific and clinical information submitted for qualification.

The pilot process for biomarker qualification allowed the PSTC to submit a single application for biomarker qualification to both regulatory agencies, and then to meet jointly with scientists from both agencies to discuss it in detail and to address additional scientific questions posed by the regulators. Each regulatory agency reviewed the application separately and made independent decisions on whether each would allow the new biomarkers to be used. The new biomarkers qualified by FDA and EMA were KIM-1, albumin, total protein, β_2-microglobulin, cystatin C, clusterin, and trefoil factor-3. Testing for these proteins will help scientists assess whether a drug is likely to cause damage to the kidneys, a toxic side effect of some drugs. Both FDA and EMA require drug companies to submit the results of two other tests, BUN and serum creatinine, to show whether such kidney damage has occurred.

The seven new tests have provided important advantages over these two tests. For example, in the rat model, once kidney damage has begun to occur, it takes a week before the two current tests can detect it [21]. The new tests are more sensitive and can reveal cellular damage within hours [22]. BUN and serum creatinine show that damage has occurred somewhere in the kidneys, but the new tests can also pinpoint which parts of the kidney have been affected [23].

Additional studies have been ongoing over the past 7 years to provide clinical data for a qualification that may allow promising drugs to advance into clinical trials that otherwise would have been abandoned, because currently there are no tests available to detect early-onset renal injury. The new tests were developed and will be carried out initially in rats, but they were selected because other studies have shown that similar biomarkers are produced in human kidney cells [24]. If these studies are successful, the PSTC will present a new application seeking acceptance of the human biomarkers over the next 2 years.

The need for an accurate, comprehensive, and efficient process for biomarker qualification is closely linked with our ability to quickly integrate new biomarkers in drug development and regulatory review. The biomarker qualification pilot process at the FDA tested the scientific, clinical, and regulatory components for a biomarker qualification process. Experience gained with the pilot process was useful in the development of a formal regulatory process [25] for biomarker qualification.

Evidentiary Standards

Over the first decade of biomarker qualification at the FDA, three nonclinical and three clinical biomarker qualifications were completed [26]. Of these, all the nonclinical qualification requests were submitted during the pilot phase of this program in 2007–2008. All of the clinical qualifications on record have been for enrichment biomarkers, which the Enrichment Guidance [27] does not require to be qualified. Over 30 qualification submissions have been received at CDER throughout the past decade. Some have been withdrawn, while 25 [28] remain in a prolonged cycle of consultation and advice meetings with the FDA leading to incrementally more complex and expensive qualification studies.

The global futility of biomarker qualification at the FDA is highlighted by the absence over a period of 10 years of a single qualification for a surrogate biomarker. The absence of qualified surrogates has had a major impact on the development of new therapies in the treatment of rare diseases [29], where measurement of conventional end points such as the 6-min walk [30] is particularly difficult for patient populations with severe physiological damage.

The most difficult part of this process has been the definition of incremental contexts of use and the corresponding evidence with which biomarkers may be qualified. We should not misrepresent the industry goal: qualified biomarkers capable of development and approval for new drugs in the clinic. We also cannot misrepresent the goal as far as public health is concerned, which is to obtain better biomarkers for routine clinical use as quickly as the data will allow. Intermediate qualification contexts and data need to be defined so that investment in biomarker qualification studies will be productive both for clinics and for the pharmaceutical industry. Initial studies proposed by consortia are unlikely to match a clear context for qualification for a full surrogate clinical application of biomarkers. What intermediate contexts for qualification can we define, and what study characteristics can we propose for qualification in these intermediate contexts of use? Several authors [31,32] have proposed evidentiary standards for biomarker qualification.

Unlike the incremental process for biomarker qualification embodied in the pilot process for biomarker qualification at the FDA [18], papers on evidentiary standards often propose all-or-nothing qualification contexts, where if the ultimate goal is a clinical qualification, no intermediate qualification contexts are expected to be defined or qualified. This approach is not only time consuming but is also not likely to encourage the investment needed to generate data for biomarker qualification. At each stage, whether the context of use for a biomarker is to be *in vitro*, in a nonclinical animal model, or in the clinic, a company or consortium proposing a qualification will probably seek a quick return on the qualification of a biomarker after data are available to qualify a biomarker in a specific context in drug development. An effective process for biomarker qualification should include incremental application context steps, so that these incremental steps can quickly benefit the drug development process.

Harmonization

The application of biomarkers to drug discovery and development has the potential of improving the efficacy and speed of bringing more effective and safer new drugs to market. This requires that biomarkers that may be applicable to such uses be qualified for a specific application context. To achieve this, both the process of qualification and the evidentiary criteria and standards for qualification will need to be described and defined. The International Committee on Harmonization's (ICH) E16 [33] is a harmonization effort to define the context, structure, and format of the biomarker qualification submission. It is based on the previous experience by the FDA and EMA regarding biomarker qualification. This harmonization effort does not address the evidentiary standards for biomarker qualification.

The structure, format, and content of a submission of biomarker data for qualification depend on the context in which the biomarker is intended to be used. The first step in drafting a submission for qualification of a biomarker is to determine its context of use, preceding specific decisions on applicable structure and format. The context of use for a biomarker is (1) the general area of biomarker application, (2) the specific applications/implementations, and (3) the critical factors that define where a biomarker is to be used and how the information from measurement of this biomarker is to be integrated in drug development and regulatory review. To demonstrate the alignment between proposed context and data, the initial context proposal must be supported by data available at the initial application step or expected to be available throughout the data evaluation process in biomarker qualification. There is a convergent relationship between an initial qualification context and the data supporting it. The initial gap between proposed context and data may need to be filled throughout the qualification process. Initial context proposals, however, should project a significant improvement over currently available biomarkers and/or end points.

The context of a biomarker drives data requirements to demonstrate its qualification for the intended application. The structure of a submission document ensures that the context and data can be submitted in a package consistent for consortia submitting qualifications as well as for reviewers in regulatory agencies evaluating a qualification package. The structure of a qualification submission is independent of the context of this submission, but must also be flexible enough to deal with the specific requirements of each context. On the other hand, the format of data required to qualify a biomarker may vary significantly with the context in which it is to be used. It is therefore only possible to harmonize general regulatory guidelines on data format for biomarker qualification submissions.

SUMMARY

Companion diagnostics and biomarker qualification have followed parallel regulatory policy development paths over the past decade. Companion diagnostics have achieved—at the FDA, at least—a coherent regulatory framework for data review and approval. This regulatory framework has encouraged the development of novel biomarkers and biomarker platforms. Biomarker qualification, however, continues to languish under a regulatory framework that leads to very limited successful qualifications and no impact on the development of better surrogate biomarkers. A major next step in the development of regulatory framework enabling precision medicine will be an effective framework for biomarker qualification. The transfer of the Biomarker Qualification Process to the Office of New Drugs in 2017 is a major step in this direction, as is the development of Type "C" meetings in OND for the discussion of surrogate biomarker proposals for accelerated approvals.

REFERENCES

1. US FDA. Pharmacogenomic Data Submissions. In: HHS, editor: Federal Register; 2005. pp. 14698–14699.
2. US FDA. *Innovation or Stagnation: Challenge and Opportunity on the Critical Path to New Medical Products*. Silver Spring, MD: HHS; 2006 https://www.fda.gov/downloads/ScienceResearch/SpecialTopics/CriticalPathInitiative/CriticalPathOpportunitiesReports/ucm113411.pdf
3. US FDA. In Vitro Companion Diagnostic Devices. In: HHS, editor.: Federal Register; 2014. pp. 45813–45814.
4. US FDA. Table of Pharmacogenomic Biomarkers in Drug Labeling. www.fda.gov/drugs/scienceresearch/researchareas/pharmacogenetics/ucm083378.htm
5. Goodsaid F, Mattes W, editors. *The Path from Biomarker Discovery to Regulatory Qualification*. Amsterdam, the Netherland: Academic Press; 2013.
6. US FDA. Enrichment Strategies for Clinical Trials to Support Approval of Human Drugs and Biological Products. In: HHS, editor.: Federal Register; 2012. pp. 74670–74671.
7. US FDA. Drug-Diagnostic Co-Development Concept Paper. In: HHS, editor.: US Food and Drug Administration; 2005.
8. US FDA. Pharmacogenetic Tests and Genetic Tests for Heritable Markers. In: HHS, editor.: Federal Register; 2007. p. 33765–33766.
9. US FDA. In Vitro Diagnostic Multivariate Index Assays. In: HHS, editor.: Federal Register; 2007. p. 41081–41083.
10. US FDA. Distribution of In Vitro Diagnostic Products Labeled for Research Use Only or Investigational Use Only. In: HHS, editor.: Federal Register; 2013. p. 70306–70307.
11. US FDA. Framework for Regulatory Oversight of Laboratory Developed Tests (LDTs). In: HHS, editor.: Federal Register; 2014. p. 59776–59779.
12. US FDA. Discussion Paper on Laboratory Developed Tests (LDTs). In: HHS, editor.: US Food and Drug Administration; 2017.
13. ACLA. Comments on March 21, 2017 *Discussion Draft of the Diagnostic Accuracy and Innovation Act*. American Clinical Laboratory Association; 2017.
14. US FDA. Public Workshop on Next Generation Sequencing-Based Oncology Panels. In: HHS, editor.: US Food and Drug Administration; 2016.FDA. In Vitro Companion Diagnostic Devices. www.fda.gov/downloads/medicaldevices/deviceregulationandguidance/guidancedocuments/ucm262327.pdf (accessed November. 27, 2016).
15. Wagner JA (2008). Strategic approach to fit-for-purpose biomarkers in drug development. *Annu Rev Pharmacol Toxicol*, 48:631–651.
16. Rifai United States, Gillette MA, Carr SA (2006). Protein biomarker discovery and validation: the long and uncertain path to clinical utility. *Nat Biotechnol*, 24(8):971–983.
17. Goodsaid F, Frueh FW (2007). Implementing the U.S. FDA guidance on pharmacogenomic data submissions. *Environ Mol Mutagen*, 48(5):354–358.
18. Goodsaid F, Frueh F (2006). Process map proposal for the validation of genomic biomarkers. *Pharmacogenomics*, 7(5):773–782.

19. Goodsaid F, Frueh F (2007). Biomarker qualification pilot process at the US Food and Drug Administration. *AAPS J*, 9(1):E105–E108.
20. Goodsaid FM, Frueh FW, Mattes W (2008). Strategic paths for biomarker qualification. *Toxicology*, 245(3):219–223.
21. Anon. (2008). Public consortium efforts in toxicogenomics. *Methods Mol Biol*, 460:221–238.
22. Duarte CG, Preuss HG (1993). Assessment of renal function-glomerular and tubular. *Clin Lab Med*, 13:33–52.
23. Vaidya VS, Ramirez V, Ichimura T, Bobadilla NA, Bonventre JV (2006). Urinary kidney injury molecule: 1. A sensitive quantitative biomarker for early detection of kidney tubular injury. *Am J Physiol Renal Physiol*, 290(2):F517–F529.
24. Zhang J, Brown RP, Shaw M et al. (2008). Immunolocalization of Kim-1, RPA-1, and RPA-2 in kidney of gentamicin-, mercury-, or chromium-treated rats: Relationship to renal distributions of iNOS and nitrotyrosine. *Toxicol Pathol*, 36(3):397–409.
25. Dieterle F, Maurer E, Suzuki E, Grenet O, Cordier A, Vonderscher J (2008). Monitoring kidney safety in drug development: emerging technologies and their implications. *Curr Opin Drug Discov Dev*, 11(1):60–71.
26. US FDA. Guidance for Industry and FDA Staff: Qualification Process for Drug Development Tools. In: HHS, editor.: Federal Register; 2014. p. 831–832.
27. FDA. Enrichment Strategies for Clinical Trials to Support Approval of Human Drugs and Biological Products. www.fda.gov/downloads/drugs/guidancecomplianceregulatoryinformation/guidances/ucm332181.pdf (accessed November 27, 2016).
28. FDA. Biomarker qualification at CDER (presentation) www.fda.gov/downloads/medicaldevices/newsevents/workshopsconferences/ucm490915.pdf (accessed November 27, 2016).
29. FDA. Scientific Dispute Regarding Accelerated Approval of Sarepta Therapeutics' Eteplirsen (NDA 206488)—Commissioner's Decision http://www.accessdata.fda.gov/drugsatfda_docs/nda/2016/206488_summary%20review_Redacted.pdf (accessed November 27, 2016).
30. FDA. Summary Minutes of the Peripheral and Central Nervous System Drugs Advisory Committee Meeting April 25, 2016. http://www.fda.gov/downloads/advisorycommittees/committeesmeetingmaterials/drugs/peripheralandcentralnervoussystemdrugsadvisorycommittee/ucm509870.pdf (accessed November. 27, 2016).
31. Altar CA, Amakye D, Bounos D et al. (2008). A prototypical process for creating evidentiary standards for biomarkers and diagnostics. *Clin Pharmacol Ther*, 83(2): 368–371.
32. Wagner JA, Williams SA, Webster CJ (2007). Biomarkers and surrogate end points for fit-for-purpose development and regulatory evaluation of new drugs. *Clin Pharmacol Ther*, 81(1):104–107.
33. International Committee on Harmonization (ICH) E16. Genomic biomarkers related to drug response: context, structure and format of qualification submissions. http://www.ich.org/cache/html/4773-616-1.html. (accessed October. 19, 2008).

Not All Biomarkers Are Worthwhile

9.3

An Unsuccessful Use and Pitfalls to Avoid

Abdel-Baset Halim

Contents

Incredibly high failure rate in the pharmaceutical industry has been positioning biomarkers and personalized medicine—with its prerequisite drug-diagnostic codevelopment, commonly known as companion diagnostics (CDx)—in the frontline as optimistic rescuers. This hopefulness is potentiated with the recent major advances and competitiveness in molecular diagnostics, which made the laboratory tests widely accessible at affordable prices. If executed right, biomarkers and CDx can potentially help drug industry enhancing the probability of success and, maybe, accelerating time to market; help the diagnostics industry developing tests utilizing precious, clinically annotated, human samples; and, more importantly, benefit patients by supporting accurate diagnosis and selection of the most efficacious and least toxic therapies.

This chapter focuses on some of the critical challenges currently facing biomarkers and precision medicine.

ATTRIBUTES OF A GOOD BIOMARKER

A biomarker and its assay is a complementary process: No good biomarker is without a good assay, and no useful assay is without a good biomarker. Although a good biomarker and its assay can help enhance the probability of success in drug development, the pharmaceutical industry needs to understand the fact that a bad biomarker or its assay can demolish a, potentially, good drug. For a successful course of a biomarker in a clinical trial, the following factors should be fulfilled:

1. The biomarker is properly selected and biologically relevant

2. The right sample type with the right collection frequency at the right times
3. Samples are properly processed, handled, shipped and stored
4. The right assay is properly validated on the right platform using the right reagent with the right quality control
5. Results are properly interpreted

UNSUCCESSFUL USE AND PITFALLS TO AVOID

Although great hopes have been expressed with significant spending on biomarkers, precision medicine and sophisticated technologies including next-generation sequencing, the return on research investments is still questionable. Decline in pharmaceutical productivity is still continuing and it was even worse in 2016 as measured by the number of new molecular entities (NMEs) that received the FDA's and the European Medicines Agency's (EMA) approvals. Only 22 NMEs were approved by the FDA in 2016, compared with 45 NMEs that were approved in 2015, and the EMA reported that it had approved 33 new medicines up until the end of November, compared with 62 at the same period last year (BWT 2016).

Despite all the potential benefits of using biomarkers to advance pharmaceutical research and development, discrepant results can pose a threat to development programs by triggering false go/no-go decisions or enrolling the wrong subjects in targeted therapy. As pharmaceutical companies build their biomarkers and precision medicine strategies, they need to be mindful of some important challenges to be overcome. These include improper planning, intra- (within a pharmaceutical company) and inter-organizational (partners) disconnect or misalignment, biological, pre-analytical, analytical, post-analytical issues (Halim 2014, 2015, Miller and Halim 2016).

Omitting any of the attributes listed above can partially or completely invalidate a biomarker result, which, in turn, can make the efforts be in vain or even enhance probability of failure. Covering all or most of limitations and pitfalls in a book chapter is not possible but here are just some examples.

Improper Selection of a Biomarker

Figure 9.3.1 is a simple schematic presentation of a signaling pathway where a drug can only bind to a wild-type (WT; unaltered) receptor producing an intermediary signal that is required

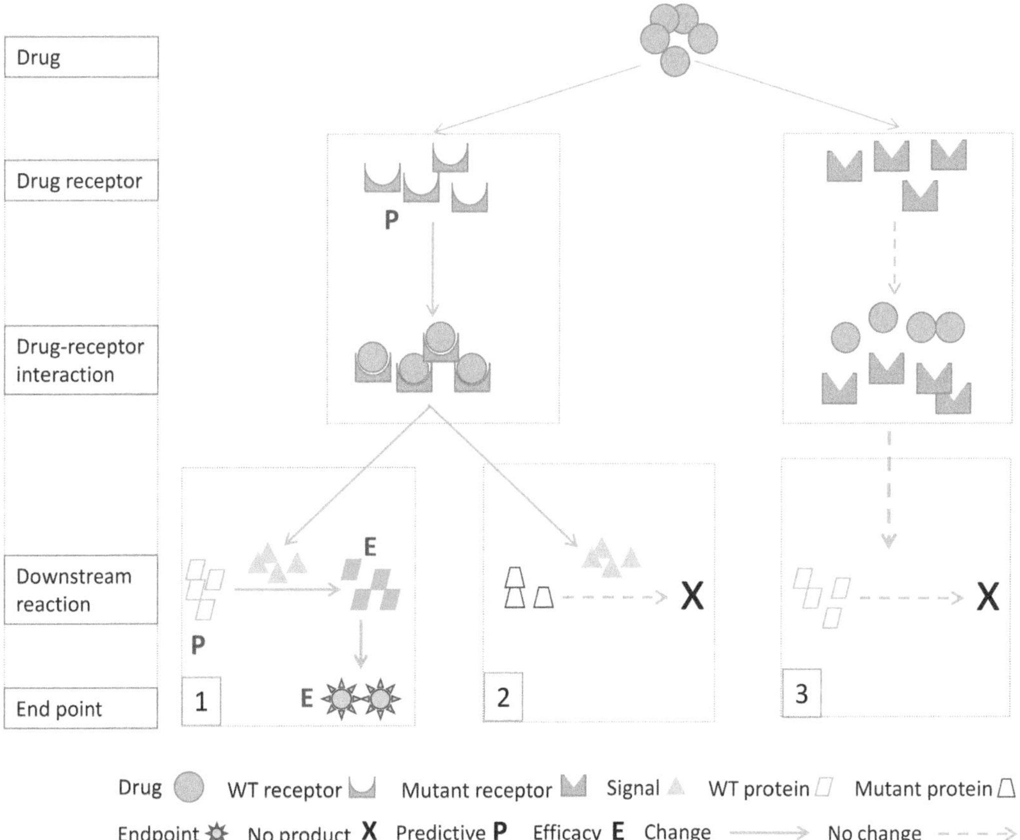

FIGURE 9.3.1 Simple schematic presentation of different mechanistic pathways to demonstrate how a predictive or a pharmacodynamic (PD) biomarker can be hypothesized. WT, wild type.

to change another WT protein into an active form which in turns acts to induce the formation of a final end product. In panel (1), the WT drug's receptor can serve as a predictive biomarker for the drug effect, whereas both the intermediary signal and end product can serve as efficacy biomarkers; to proof the concept or mechanism of drug action or to serve as a pharmacodynamics (PD) biomarker for proof of mechanism (POM), proof of concept (POC) and/or an efficacy biomarker. Altered drug's receptor in panel (3) prevent drug binding and can predict drug inefficacy but panel (2) shows that, whereas drug can hit its target and intermediary signal can be formed, mutant downstream protein can shut down the pathway. For such panel, mutant receptor can indicate drug inefficacy but the WT receptor does not guarantee drug efficacy and the downstream marker has to be genotyped too.

Improper Design and Integration of a Biomarker into a Clinical Trial

For efficacy application, a biomarker has to be well suited within a dosing schedule with, typically, multiple sample collection time points involved. Depending on the biomarker dynamicity and its biological half-life, in addition, to close-to-therapy-initiation, typically within 24 h prior to the first dose, and posttreatment collection, following plasma drug peak, pre-dose sample from each treatment cycle in multi-cycles therapies can be helpful. Also, at or shortly post-event sampling should be considered.

There is a fundamental disconnect usually encountered in biomarker sampling especially in oncology, whereas a trial's objective can be to prove a mechanism or concept or to show efficacy of a compound that needs a reliable biomarker data set, PD are usually considered as exploratory. Listing a biomarker as exploratory on a study protocol is usually automatically misunderstood by a study operational team as an "optional." In my personal experience, by turnover rate I meant percent of samples collected from projected number. Moreover, number of subjects with full sets of samples, baseline and at one or more time points on the treatment schedule, is significantly less. Taking the already small number of subjects in phase I/II oncology trials into consideration, the limited proportion of samples collected make results inconclusive. With the low number of clinical responders usually seen in oncology trials, sometimes only one subject in a therapeutic arm, it is not unlikely to not have samples, at least a complete set, from the responder. Unfortunately, with the highly competitive landscape of clinical research, especially, in oncology, study management is usually put between a rock and a hard place; to either execute the trial on time or satisfy "exploratory" biomarker sample collection and the first option always prevails. This is one of several intra-pharmaceutical organization issues; disconnect between translational medicine and biomarker professionals who put and rationalize the biomarker list to address its objective and clinical operation staff who execute the trial with timelines is the main driver.

For predictive biomarkers, although archived historical samples, for example, formalin-fixed, paraffin-embedded (FFPE) tissue can be used for mutation detection, some mutations can evolve over the course of the disease or refractory to a treatment, for example, EGFR T790M mutation (Kobayashi et al. 2005, Pao et al. 2005). In such cases, fresh sampling at trial entry and at the appearance of new lesion or metastasis can be the only sensible option. However, unfortunately, fresh biopsies are not usually feasible in most of clinical trials and archived historical samples are still used instead from which results can be misleading.

Uncontrolled Pre-analytical Variables

Pre-analytical errors can occur at any time between the test order and the analytical phase, and may affect sample integrity and its suitability for biomarker testing. Sample poor quality has been demonstrated as one of the critical limitations facing biomarkers and companion diagnostics (Halim 2014, 2015). Although the impact of pre-analytical variables on laboratory results has been emphasized for long time (Lippi et al. 2006a, 2006b, Lippi 2009), the topic is still usually overlooked in pharmaceutical trials, especially in the case of molecular testing of tissue samples.

In an article published by Halim in 2015, a biomarker positivity rate by immunohistochemistry (IHC) in 440 samples from a phase III trial, analyzed at a single central lab, declined significantly and proportionally with time elapsed between FFPE tissue section cut and its analysis. The overall average positivity rate, applying a predefined cutoff, was 52.0% but positivity rates were 66.7, 53.3, 41.2, 37.8 and only 30.8% for samples analyzed within 12, 13–24, 25–36, 37–48 and more than 48 weeks respectively. After this data was revealed, separate stability studies were conducted on sectioned tissues at two different laboratories and gradual degradation of the staining signal was seen over time with a stability window of about 12 weeks (data is not published), which confirmed our observation from the clinical trial. Unfortunately, the stability studies were conducted after the fact and data from the phase III trial was disputable. In the same article, samples received from 64 global clinical sites showed different positivity rates ranging from 0.0%–100%. Variability of results from samples contributed by different clinical sites was explained by possible different practices in obtaining and treating the tissues prior to sending to the lab.

In this respect, any of the following factors, if not considered and well controlled, can defeat the whole purpose of biomarkers in a clinical trial:

1. Sample type, for example, plasma versus serum, or FFPE versus fresh-frozen tissues. Sample source is important too, for example, final needle aspirate (FNA) or bronchoalveolar lavage (BAL) should not be automatic alternatives for core needle or surgical biopsy for tissue staining or nucleic acid-based assays without proper validation of an assay for different types of matrices and histological assessment to ensure sample integrity.
2. Sample collection and processing, for example, speed and time of centrifugation to obtain platelet-poor or platelet-rich plasma and type of fixative and length of

time a fresh tissue biopsy is left before it is frozen or fixation time and type of fixative prior to embedding in paraffin.

3. Shipping where samples may be exposed to extreme conditions of heat and/or humidity.
4. Storage conditions and time. Different clinical sites and labs have different practices in storing FFPE specimens where a whole block or cut sections are stored at ambient temperature or in a fridge with some, but not all, paying attention to humidity and light exposure.
5. Inconsistency. Regardless of availability of lab manual for a clinical study, in most of the local practices for tissue sample collection, processing and storage, which are usually varying, prevail. Although it is critical, this point can be beyond a sponsor or CRO's capability to control especially for archived tissue specimens in a global or multi-country trial. However, a sponsor has to consider this variable, among others, in interpreting biomarker results before jumping into conclusions to refer difference in observations to other factors, for example, racial factors.
6. Lack of information. Mandating fresh biopsies is deemed impractical by clinicians or clinical study planners especially from specific tumor sites such as the lungs, and, in the majority of cases, archived FFPE tissues are used. Lack of information on how a sample was obtained, processed, stored or handled exaggerate the issue for not allowing sponsor or lab's to judge viability of the sample for analysis or taking precaution in interpreting results. Typically, people do not pay attention to FFPE sample age but just call it something like, "archived," "historical," "baseline" or "pretreatment" sample. FFPE tissues are usually received from clinical sites in the form of cut sections and, in a large percent of cases, sectioning dates are unknown. Except for real time analysis where biomarker data is used in patient selection or randomization, samples are always archived for analysis in batches. Although archiving samples for batch analysis minimizes lot-to-lot, technician-to-technician and pathologist-to-pathologist variability, archived cut sections are more amenable to degradation of biomarkers compared than whole block. Unfortunately, long-term stability studies in molecular diagnostics are not usually a part of an assay validation.

Assurance of availability of appropriate instruments, for example, centrifuges and freezers and appropriate facility and capabilities to obtain and process fresh frozen and FFPE tissue samples should be on the list during clinical site qualification for a clinical trial. Clear, illustrative lab manuals and procedures, well-trained phlebotomists, and onsite training are essential tools in mitigating the pre-analytical lab errors. The lab requisition and database should be designed to capture all relevant information including date (and time in some cases) of sample collection and FFPE tissue sectioning.

Improper Assay

Biomarker assays range from exploratory type of assays performed on a fit-for-purpose basis to rigorously validated assays when a biomarker is used as a surrogate end point or for patient selection or randomization. Unfortunately, the term "fit-for-purpose" has been used outside its concept where some misinterpret the term as that an assay does not need to be reasonably validated. Fit-for-purpose should mean that an assay is validated and verified for its intended use. For example, in an assay to be used for fresh sample analysis, for example, flow cytometry of whole blood, up to 5–7 days stability check can be enough but for an assay that can be used for batch analysis of isolated peripheral blood mononuclear cells (PBMC), stability period should, typically, cover the longest expected time window from sample processing until analysis, which can be up to a year or more. Under either setup, samples for a stability should be stored under the conditions planned to process, ship and store the clinical samples under. Another example for "fit-for-purpose" assay validation is if a biomarker can be ubiquitous, for example, blood glucose, even if the assay can detect very low levels, for example, 2 mg/dL glucose, no need to spend much time and energy to investigate the assay performance at this low level as it is unlikely to have clinical samples below 30 mg/dL. However, for assays used to "fish out" a biomarker signal, for example, for some cytokines, lower limit of quantification or detection should be verified.

Here are two of many factors that can make an assay as inappropriate:

Improper selection of reagent

Of the tens of thousands of biomarker assays employed for different purposes in drug development, only a fraction apply ready-to-use, regulatory- approved kits and/or systems. The majority are either employing for research-use-only (RUO) kits or "home-brew" assays, commonly known as laboratory developed tests (LDT), for which reagent components are obtained from different sources with no quality or traceability guaranteed. Most of the molecular tests used in oncology, IHC and *in situ* hybridization (ISH) in particular, fall within the "homemade" category. The no-guarantee for traceability or quality is always undermined by assay developers and in most, if not all of time, the manufacturer or distributor's claims about an antibody and ISH probe are taking for granted without verification. The impact of antibodies from different sources on the IHC outcome has been known for decades but the issue is still overlooked. In 1994, Press and colleagues (1994) compared 28 HER2 antibodies on 37 split breast cancer samples and found sensitivities ranging from 6% to 82% and specificities from 92% to 100%. In another study, four commercial antibodies including Dako HercepTest were used to stain split samples from 118 breast cancer samples, where the sensitivity was 59%–77% and specificity was 91%–93% (Thomson et al. 2001). The issue has been a hot topic for PD-L1 recently (Titus 2016).

An IHC assay was developed and validated at a contract research organization (CRO) laboratory for use in a phase I

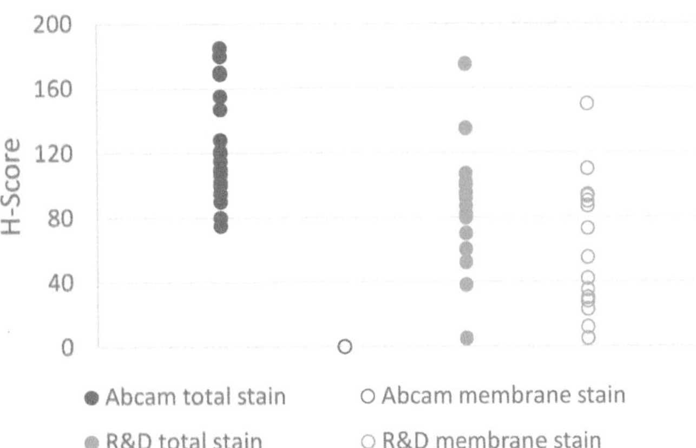

FIGURE 9.3.2 Impact of antibody source on immunohistochemistry results. A biomarker was analyzed in 18 human formalin-fixed, paraffin-embedded colorectal split samples using antibodies from two sources, claimed to target the membrane epitope, generated in rabbit and goat. Whole (cytoplasm + membrane) staining and membrane staining only are presented as H-scores.

clinical trial. The assay utilized a rabbit anti-human DR5 polyclonal antibody, claimed to detect the membrane domain of DR5 but none of the samples tested during validation showed membranous staining. A duplicate assay was developed and validated by the lab using a goat polyclonal antibody from another provider and the two assays were used to analyze split adjacently cut sections from 18 colorectal samples collected in a phase I trial. The goat antibody manufacturer did not list IHC on the antibody package insert but, as shown by Figure 9.3.2, although none of the samples was stained in membrane by the first assay, the second assay detected both cytoplasmic and membranous compartments with membranous staining ranging from 5% to 100% of total stain intensity.

Improper assay performance

As mentioned earlier, an assay should be validated for its intended use. It is not the purpose of this chapter to handle assay validation parameters but draw some lights on how an assay precision and accuracy, and sensitivity and specificity should fit its purpose. Regardless of momentous efforts from lab professionals and *in vitro* diagnostic partners to reduce clinical laboratory errors due to analytical issues via the implementation of a number of quality control and quality assurance (QC&QA) check points, including internal (electronic) QC, liquid QC, calibration, delta checks, method comparison, and proficiency testing, among other measures, the problem is still magnificent.

Precision and accuracy

There are two categories of analytical errors: random and systematic, where random error can affect a sample or a few samples randomly within an analytical run, whereas systematic error (consistent drift) affects all samples analyzed after an error has occurred and until it is fixed.

Each biomarker assay has a "default" imprecision; oscillation of values from the same sample, when measured multiple times, around the average of observations, whereas inaccuracy is the drift of observed (assayed) average from a predefined target (nominal) value. Typically, a clinical lab considers an assay to be well performing if results from a quality control sample are nicely distributed around the average (precision) and within "Average ± 2 Standard Deviation" or within "Average ± a fixed % of Average" (accuracy). In the example depicted by Figure 9.3.3 and Table 9.3.1, both assays in A and B are accurate, with % bias (% difference of observed average from target concentration) of less than 0.5%, and precise but A is significantly more precise than B where results from both oscillate nicely around the "Average" but all results from A are within ±10% of the target, whereas those from B are within ±30% of the target. Assay C is highly precise (all values are within Average ±10%) but the observed average drifts from the target by more than 20%. Assays exemplified by D are neither precise nor accurate where results from the same sample can be randomly scattered beyond ±30% and % bias is close to 20%. Such types of assays can be considered in clinical practice as follows:

1. Assay A is optimal for different purposes of safety and efficacy biomarkers.
2. Assay B can be used as is as long as the biomarker signal can exceed the total allowable error of the assay or if a PD efficacy signal can be looked at as a trend rather than individual sample results. In this respect, an assay acceptable imprecision is not one size fits all but it is determined by the following factors:
   a. Technical complexity—an automated chemistry assay, for example, plasma glucose should be more precise than a multistep assay, for example, reverse transcriptase-polymerase chain reaction (RT-PCR)-based assay preceded by RNA extraction.
   b. The assay intended use—an assay for a surrogate clinical end point should be more precise than an assay for exploratory biomarker.
   c. The expected magnitude of a biomarker signal—Although an assay with imprecision within ±30% cannot be suitable for a biomarker with an expected PD signal of less than 30%, an RT-PCR-based assay with imprecision within, even, ±1-fold (100% change) can still be suitable if the biomarker signal can be in one-half to one order of magnitude or more.
3. Assay C can be used either after correction of the drift, for example, by verifying and adjusting the assay calibration or as is with some precautions, for example, analyzing all samples from a trial in a single lab just to compare posttreatment results with baseline (for PD biomarkers), that is, results will not be used as absolute values but posttreatment results will be calculated as % change from baseline.
4. Assay D should not be used, even for what is called exploratory biomarkers because results will not be conclusive and may lead to a wrong decision.

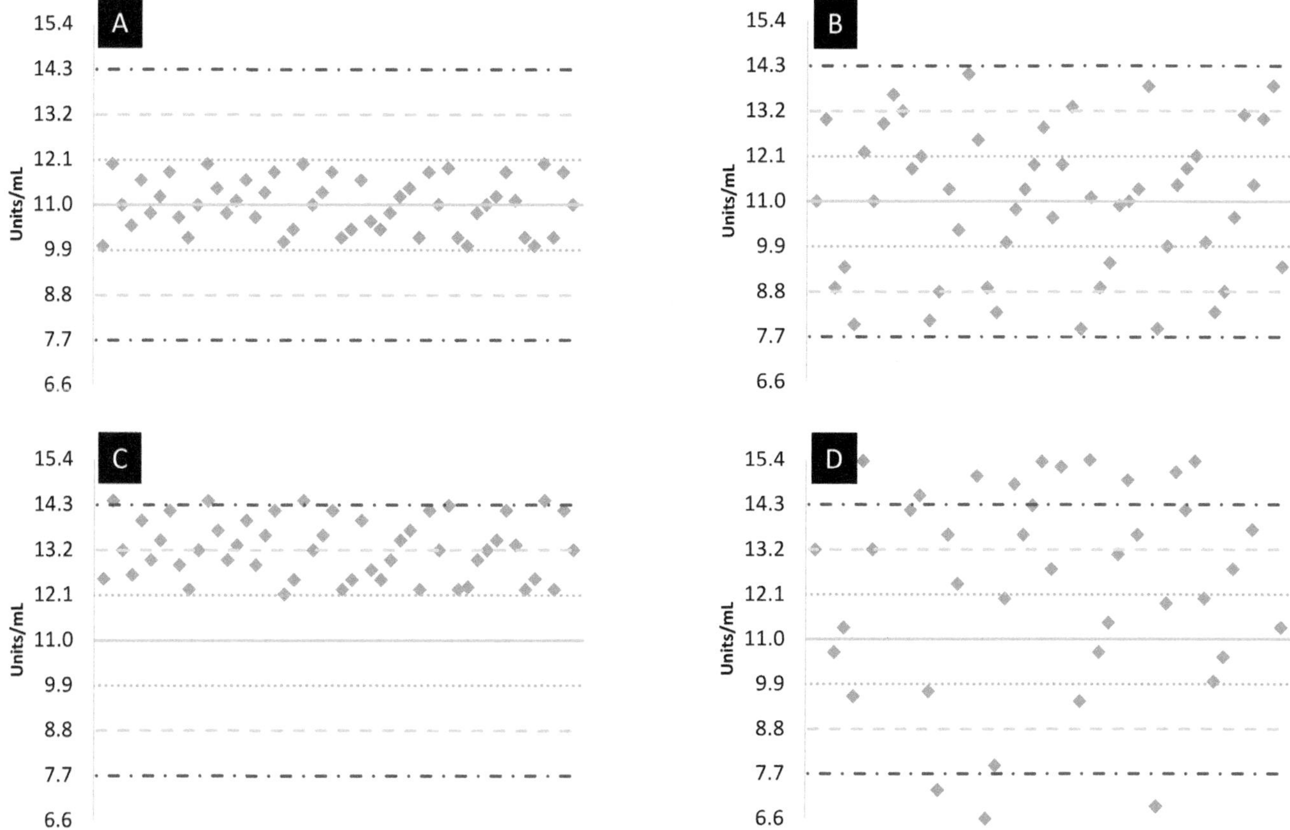

FIGURE 9.3.3 Schematic presentation of precision and accuracy in a laboratory assay. Scatter graphs simulate replicate results from a QC sample analyzed by different assays. A-D represents examples for an assay with acceptable accuracy and precision, imprecise but accurate, precise but inaccurate and neither precise nor accurate assays, respectively.

TABLE 9.3.1 Summary statistics for the different assays depicted in Figure 9.3.3

	ASSAY A	*ASSAY B*	*ASSAY C*	*ASSAY D*
Observed average	11.02	10.96	13.25	13.10
SD	0.64	1.79	0.73	2.69
CV%	5.77	16.31	5.49	20.56
% Bias	0.2	−0.4	20.4	19.1

Note: CV, coefficient of variation; SD, standard deviation.

Sensitivity and specificity

An assay sensitivity is the ability of the test to give a positive result for subjects who have the disease or condition for which they are being tested and specificity is the ability of the test to give a negative result for subjects who do not have the disease or condition. A biomarker prevalence is the proportion of a targeted population who test positive for a specific biomarker. It has been the tradition that an assay sensitivity and specificity are looked at in a general way in isolation from a biomarker prevalence.

Ideally, a diagnostic test should be 100% sensitive and 100% specific, that is, no chance for false negative or false positive results. However, it is unusual or even unlikely to find an assay with these characteristics and each assay always has a certain percentage of false negative and/or false positive and it has been the tradition among laboratory and IVD community to accept assays with sensitivity and/or specificity levels of 90% or even lower in some cases. HercepTest analysis of 60 different gastric cancer specimens, obtained from stomach or gastroesophageal junction performed on 5 nonconsecutive days at three study sites by two observers at each site showed within-site overall day-to-day agreements in the range of 83.1%–98.3% and site-to-site agreement was in the range of 68.3%–90.0%. Details was not given but these agreement figures denote high percentages of falsely positive and false negative results between testing sites, days (analytical runs) and observers (Dako 2014a). This is on top of any built-in reagents-to-reagent false positivity and false negativity (as discussed above). This problem is aggravated by the disconnect between diagnostic and clinical lab industry on one side and clinical and pharmaceutical community on the other side where the latter is usually interpreting results assuming close to 100% specific and 100% sensitive tests. The gap can be obvious when one compares a pharmaceutical report to a related diagnostic claim. Although DAKO kit insert for HercepTest (Dako 2014b), the Herceptin CDx, indicates that roughly 20% of breast cancer is HER2+ and the average false + rate of HercepTest is 18%, Roche announces that Herceptin has been used to treat more than 1.2 million people with HER2-positive breast cancer worldwide from 1998 to 2012 (Roche 2012). With a simple and quick analysis of the two pieces of information, one can easily conclude that about one-half of the 1.2 million women treated with Herceptin because they were HER2+ should not have taken the drug.

The point that is always overlooked is the differential impact of specificity and sensitivity on tests for biomarkers with different prevalence rates. Halim (2015) demonstrated the possible outcomes from assays with different false-positive and false-negative rates in the light of biomarker prevalence in targeted populations. BRAF mutation in melanoma (Colombino et al. 2012, Ascierto et al. 2012), HER2 positivity in breast cancer (Mayo Clinic 2014, Dako 2014b), EGFR mutation in some population with NSCLC (Zhang et al. 2016), and ALK translocation in NSCLC (Bang et al. 2011) were used as examples for target prevalence rates of 50, 20, 10 and 5% respectively. When a 90% specific and 90% sensitive assay was assumed to qualify a patient for treatment, percent of falsely treated from total number of treated patients were 10, 30.8, 50 or 67.9% for the biomarkers with 50, 20, 10 or 5% prevalence respectively. For a biomarker that could be positive in only 5% of patients, if the assay was, even, 95% specific and 95% sensitive, half of the treated subjects would receive a medication wrongly and the only solution is to make the assay as close as possible to 100% sensitive but more importantly, close to 100% specific.

The great impact of an assay low specificity on biomarkers with low prevalence cannot be only dangerous to patients receiving the drug post-approval but can lead to wrong decisions during a drug development where an efficacy signal can be significantly diluted by incorporating a relatively large proportion of false-positive patients in the treatment arm.

Incomparability between different platforms for the same assay

The previous section focused on assay performance within a given methodology in a given laboratory but this section is about the impact of lack of traceability between different labs or between different platforms or methodologies, even, when used by a single lab. This is not only the problem of "sophisticated" biomarker assays, for example, IHC, ISH, genotyping, etc., it also impacts, supposedly, well-standardized chemistry assays that have been used for decades as standard-of-care. Previous reports (Halim 2009, 2011a, 2013) emphasized the impact of this problem on laboratory results and its imminent risk to patients and pharmaceutical development. Results from same sample can be so discrepant if analyzed by different laboratories (even regulated labs) employing different methodologies and/or platforms. INR (international normalized ratio of prothrombin time) values from the same sample were shown to range between 2.9 and 7.6, that is, a results from this sample can be interpreted clinically as within the therapeutic target of Warfarin (the most common anticoagulant medication), slightly anticoagulated, or dramatically anticoagulated. When another sample was analyzed for activated partial thromboplastin time (aPTT), thrombin time (TT), and anti-FXa assay (Heparin test) in different labs employing different platforms and methodologies, the ratio of maximum to minimum reported results was up to 4-, 40-, and 50-fold, respectively at the time 30% or even 20% change can be considered as reliable efficacy or toxicity signal in clinical practice and drug development. When splits from 5 samples were analyzed for alkaline phosphatase, one of the grandfather regulated, safety and efficacy biomarker, at more than 5,000 CLIA-certified, CAP-accredited labs employing different methodologies and instruments, results from same group of laboratories using same methodology and instrument were reasonably acceptable (within-group CV ranged between 3.0% and 12.4%) but the difference between maximum and minimum reported results from the same samples when analyzed on different platforms ranged between 4-fold and 4.6-fold.

Figure 9.3.4 shows results from a CAP-proficiency testing (PT) challenge for lactate dehydrogenase (LDH), another grandfather test commonly used as a safety but also an efficacy biomarker. Splits from a single samples were analyzed by more than 4,000 regulated labs categorized in 21 groups according to the methodologies and instruments; each vertical line on the graph represents minimum, maximum and average results from labs using same methodology and each box represents an instrument used with labs applying same methodology/ies. The difference between maximum and minimum reported results within a group was 10%–23% but the overall drift excluding the last 3 groups on the graph was about 1.5-fold (221.4 vs. 149 U/L) and about 3.8-fold (563.9 vs. 149 U/L) if the last 3 groups were considered. This demonstrate a disconnect between the within-methodology/platform assay performance and the validity of data generated by the assay utilizing different methodologies and/or platforms even within a single lab. LDH has been recognized as a predictive and prognostic marker in different diseases including cancers of the esophagus, stomach, lung, colorectum, pancreas, skin, lung, breast and others where results were used for clinical decision making (Koukourakis et al. 2003, 2006, Yao et al. 2013, Rong et al. 2013, Sun et al. 2014, Zhang et al. 2015, Xiao et al. 2017). 250 U/mL was the most common cutoff cited without any information about the assay(s) and/or platform(s) that were used to generate the data. According to Figure 9.3.4, if the sample was

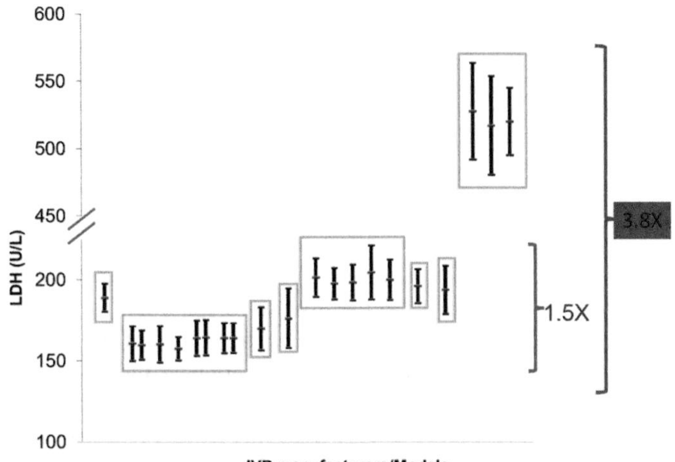

FIGURE 9.3.4 Lactate dehydrogenase (LDH) results (average, minimum and maximum) from a single sample analyzed by more than 4,000 accredited labs categorized in 21 groups depending on methodologies and instruments used. Each vertical line represent a group of labs that used the same methodology and same platform, and each box encompasses the groups of labs that utilized same platform.

analyzed on any but not the last 3 assays, results would be classified as below cutoff but results would be significantly higher than the cutoff if analyzed on any of the last 3 methodologies.

These discrepancies are not necessarily due to different lab practices, as for these automated assays, laboratories are always following manufacturer's instructions but they are caused by the use of different methodologies and platforms. Even if a particular lab uses different platforms, results will be as discrepant as shown above.

These observations demonstrated discrepant results for laboratory tests that are believed to have been standardized for biomarkers that are commonly requested as tools in patient management and used as decision-making in drug development. The discrepancy occurred due to the absence of real standardization or even a reasonable approach to harmonize results from different laboratories or different platforms.

The problem can impact decision making by pharmaceutical developers if they use absolute biomarker values to compare the outcomes of different studies on a drug's efficacy and/or toxicity, or employing biomarkers to bridge between different drug candidates belonging to a particular class of compounds. In addition, global clinical trials may be impacted if different specialty or safety biomarker labs are employed. It is not uncommon for different lab locations within a global organization (or even within one lab location) to use different platforms interchangeably to analyze samples from the same trial.

With these levels of discrepancies in well-trusted assays, which have been used in standard of care for decades, it would not be surprising to see more or, at least, the same level of discrepancy in evolving molecular testing.

Keeping in mind that external quality assessment (EQA) are more controlled as prepared and supplied by an EQA provider and as handled, analyzed and reported by a clinical laboratory, EQA results may not reflect actual inferior practices. In an EQA initiative provided in Europe, 10 pre-characterized FFPE samples made from cell lines with wild type or harboring different EGFR mutations were provided to different laboratories to assess inter-laboratory performance. Seventy-two of the 91 labs returned the results (79.1%) passed the test by having a score ≥18 (considering two points for each successful sample), that is, a lab could fail 1 of the 10 samples and it was stilled classified as successful. Average genotyping error rate was 9.66/sample with up to 35.6% for the sample harboring c.2155G4A, p.(G719S) mutations. False-negative was the main sources of genotyping error where it accounted for 85% (Patton et al. 2014). A similar exercise was attempted earlier for KRAS and involved 59 labs in eight EU countries provided with 10 CRC samples with known KRAS mutations verified by a central lab. Only 70% of the labs identified mutations in all samples with at least 1 sample was wrongly genotyped in >30% of the labs (Bellon et al. 2011).

The inter-laboratory or inter-platform discrepancies in results of evolving biomarkers could be even higher than those mentioned above because data were gathered from "well-controlled" laboratories for supposedly standardized tests used to manage patients' health and as surrogate biomarkers in clinical trials.

A pharmaceutical development program may take as long as 10 years or more, thus switching biomarker vendors is likely, using multiple platforms or changing platforms by a lab is common, and employing different lots of reagents and calibrators is definite. Without paying close attention to these variables, results, even from the same lab, may lead to erroneous go/no-go decisions and make compatibility of results from different studies almost impossible. In addition, unless appropriately understood and interpreted, if such lab tests are used as an efficacy or toxicity biomarker, the drug may be inappropriately labeled.

Until global standardization or harmonization approaches are employed, the pharmaceutical industry needs to monitor biomarker data rigorously and understand these challenges for better interpretation of biomarker results.

Post-analytical Variables

A significant portion of laboratory errors was found in the post-analytical phase. Post-analytical errors include, among others, erroneous verification of analytical data, improper data entry and manual transcription error and incorrect interpretation of test results (Carraro and Plebani 2007, Plebani 2010). Other than automated platforms interfaced with laboratory information system (LIS), most if not all of home-brew assays and technologies rely on manual data recording, transcription and entry, which make them prone to post-analytical errors. A pharmaceutical company needs to take the proper measures to ensure error-free post-analytical phase and reporting including data verification after each step.

REFERENCES

Ascierto PA, Kirkwood, JM, Grob J-J et al. 2012. The role of BRAF V600 mutation in melanoma. *J. Transl. Med.* 10:85.

Bang YJ. 2011. The potential for crizotinib in non-small cell lung cancer: A perspective review. *Ther. Adv. Med. Oncol.* 3: 279–291.

Bellon E, Ligtenberg MJL, Tejpar S et al. 2011. External quality assessment for KRAS testing is needed: Setup of a European program and report of the first joined regional quality assessment rounds. *The Oncologist* 16: 467–478.

BWT. 2016. The biggest news makers and trending stories of 2016–innovation slows: Precision medicine yet to influence output. *BWT* 27(252).

Carraro P and Plebani M. 2007. Errors in a stat laboratory: Types and frequencies 10 years later. *Clin. Chem.* 53: 1338–1342.

Colombino M, Capone M, Lissia A et al. 2012. ABRAF/NRAS mutation frequencies among primary tumors and metastases in patients with melanoma. *J. Clin. Oncol.* 30: 2522–2529.

DAKO. 2014a. HercepTest for gastric cancer, summary of safety and effectiveness (SSED). http://www.accessdata.fda.gov/cdrh_docs/pdf/P980018S010b.pdf (Accessed May 20, 2014).

Dako. 2014b. HercepTest kit insert. http://www.dako.com/28633_04may10_herceptest__brochure-9086.pdf (Accessed May 20, 2014).

Halim AB. 2009. Impact of discrepant results from clinical laboratories on patients and pharmaceutical trials: Evidence from proficiency testing results. *Biomark. Med.* 3(3): 231–238.

Halim AB. 2013. Proficiency testing: A useful tool for monitoring global lab performance and for identifying discordances. *Lab. Med.* 44 (1): e19–e30.

Halim AB. 2014. The biggest challenges currently facing companion diagnostic advancement. *Expert Rev. Mol. Diagn.* 14(1): 27–35.

Halim AB. 2015. Companion diagnostics some imminent but overlooked preanalytical and analytical challenges currently facing biomarkers and companion diagnostics. *Ann. N. Y. Acad. Sci.* 1346(1): 63–70.

Halim AB. 2011a. Discrepant results from clinical laboratories are a potential source of risk to patients under therapy with heparin or anti-thrombin agents: Evidence from proficiency testing data. *Biomark. Med.* 5(2): 211–218.

Kobayashi S, Boggon TJ, Dayaram T et al. 2005. EGFR mutation and resistance of non-small-cell lung cancer to gefitinib. *N. Engl. J. Med.* 352(8): 786–792.

Koukourakis MI, Giatromanolaki A, Sivridis E et al. 2003. Lactate dehydrogenase-5 (LDH-5) overexpression in non-small-cell lung cancer tissues is linked to tumour hypoxia, angiogenic factor production and poor prognosis. *Br. J. Cancer* 89: 877–885.

Koukourakis MI, Giatromanolaki A, Sivridis E, Gatter KC, Harris AL. 2006. Lactate dehydrogenase 5 expression in operable colorectal cancer: Strong association with survival and activated vascular endothelial growth factor pathway—A report of the tumour angiogenesis research group. *J. Clin. Oncol.* 24: 4301–4308.

Lippi, G. 2009. Governance of preanalytical variability: Travelling the right path to the bright side of the moon? *Clin. Chim. Acta.* 404(1): 32–36.

Lippi, G, Guidi, GC, Mattiuzzi, C, and Plebani, M. 2006a. Preanalytical variability: The dark side of the moon in laboratory testing. *Clin. Chem. Lab. Med.* 44(4): 358–365.

Lippi, G, Montagnana, M, and Giavarina, D. 2006b. National survey on the pre-analytical variability in a representative cohort of Italian laboratories. *Clin. Chem. Lab. Med.* 44(12): 1491–1494.

Mayo Clinic. HER2-positive breast cancer: What is it? http://www.mayoclinic.org/breast-cancer/expert-answers/faq-20058066 (Accessed October 14, 2014).

Miller I and Halim A. 2016. Precision Immuno-Oncology Adoption Challenges- Considerations for the Successful Development of Precision Immuno-Oncology Therapies and Their Companion Tests. GEN Exclusives. http://www.genengnews.com/gen-exclusives/precision-immuno-oncology-adoption-challenges/77900814 (Accessed December 26, 2016).

Pao W, Miller VA, Politi KA et al. 2005. Acquired resistance of lung adenocarcinomas to gefitinib or erlotinib is associated with a second mutation in the EGFR kinase domain. *PLoS Med.* 2(3): 225–235.

Patton S, Normanno N, Blackhall F et al. 2014. Assessing standardization of molecular testing for non-small-cell lung cancer: Results of a worldwide external quality assessment (EQA) scheme for EGFR mutation testing. *Br. J. Cancer* 111(2): 413–420.

Plebani M. 2010. The detection and prevention of errors in laboratory medicine. *Ann. Clin. Biochem.* 47: 101–110.

Press MF, Hung G, Godolphin W and Slamon DJ. 1994. Sensitivity of HER-2/neu antibodies in archival tissue samples: Potential source of error in immunohistochemical studies of oncogene expression. *Cancer Res.* 54, 2771–2777.

Roche. 2012. FDA approves Perjeta (pertuzumab) for people with HER2-positive metastatic breast cancer. http://www.roche.com/media/media_releases/med-cor-2012-06-11.htm (Accessed May 20, 2014).

Rong Y, Wu W, Ni X et al. 2013. Lactate dehydrogenase A is overexpressed in pancreatic cancer and promotes the growth of pancreatic cancer cells. *Tumour Biol.* 34: 1523–1530.

Sun X, Sun Z, Zhu Z et al. 2014. Clinicopathological significance and prognostic value of lactate dehydrogenase: A expression in gastric cancer patients. *PLoS One* 9: e91068.

Thomson TA, Hayes MM, Spinelli JJ et al. 2001. HER-2/neu in breast cancer: Interobserver variability and performance of immunohistochemistry with 4 antibodies compared with fluorescent in situ hybridization. *Mod. Pathol.* 14, 1079–1086.

Titus K. 2016. Big hopes, bigger questions with PD-L1. CAP Today 2016, 1.

Xiao Y, Chen W, Xie Z et al. 2017. Prognostic relevance of lactate dehydrogenase in advanced pancreatic ductal adenocarcinoma patients. *BMC Cancer* 17: 25.

Yao F, Zhao T, Zhong C, Zhu J, Zhao H. 2013. LDHA is necessary for the tumorigenicity of esophageal squamous cell carcinoma. *Tumour Biol.* 34: 25–31.

Zhang J, Yao Y-H, Bao-Guo Li et al. 2015. Prognostic value of pretreatment serum lactate dehydrogenase level in patients with solid tumors: A systematic review and meta-analysis. *Sci. Rep.* 5: 9800.

Zhang YL, Yuan JQ, Wang KF et al. 2016. The prevalence of EGFR mutation in patients with non-small cell lung cancer: A systematic review and meta-analysis. *Oncotarget* 29;7(48): 78985–78993.

Research Reproducibility
The Case for Standards

9.4

Leonard P. Freedman and Raymond H. Cypess

Contents

INTRODUCTION

"Standard" is an inflammatory word for many biological scientists, one that conjures up images of bureaucracy, regulation, and paragraphs filled with obscure references and alphanumeric designations. In reality, life science (also known as biological) standards are either highly characterized reagents or written documents that outline community consensus around certain practices (Freedman & Gibson, 2015; ISO-IEC, 2004) Although government agencies can adopt specific standards as regulations, adherence to a majority of standards is voluntary (http://www.ansi.org/about_ansi/faqs/faqs.aspx.)

Standards have extensively benefitted many fields, from the railroad industry to clinical diagnostics, and standards underpin almost all the capabilities of modern life. However, although some life science research standards, such as those dealing with animal welfare, are broadly used, overall the adoption of standards in life science research has been limited. Are more life science standards necessary?

Biomedical research has entered a new realm of complexity. Researchers are discovering targets more quickly, and the increasing understanding of pathway biology has necessitated multidisciplinary approaches to omics. The environment is also highly competitive, with a strong drive to make and report new discoveries. There is a paramount need for each breakthrough to be a quality output that is also timely and economical. This drives much of academic research and sets a landscape of "perverse incentives," as well as commercial research, where the profit potential can be huge.

Over the past several years, concerns about the quality of research have garnered the attention of multiple stakeholders in the scientific community (Ioannidis, 2005; Jasny et al., 2011; Williams, 2015). Specifically, widespread and pervasive irreproducibility of research findings has been documented Prinz et al., (2011); Begley and Ellis, (2012) As a consequence, billions of dollars of research funds have been wasted (Freedman et al., 2015), and researchers must confront a small but growing negative public perception of scientists and the scientific method. Periodically, the scientific community has voiced concerns about irreproducibility, and some stakeholders have started to take action. However, thus far there has been no coordinated global effort to ameliorate these problems.

The scientific community feels the profound effects of research irreproducibility and is ready to find a unifying solution to this problem. Standards can be an intrinsic part of the solution. As this section will describe, standards are a vehicle for reducing irreproducibility by aligning the community around consensus-based methods.

DEFINING STANDARDS

Before a community can internalize issues surrounding standards in life science research, it needs to understand what "standards" are. Standards align people around a consensus-based method for performing important functions. Two commonly described categories of standards are *materials standards* and *written consensus standards* (Figure 9.4.1). Materials standards are highly characterized physical substances, such as chemical or biological reagents (known as reference materials or reference reagents), and are routinely used in several areas of life science research and health care. Written consensus standards are documents that outline community consensus around certain practices. Document (or "paper") standards can cover a broad range of practices, such as designating names, defining common sizes, outlining processes or procedures, and establishing quality systems.

Although government agencies can adopt specific community standards as government regulations, thereby making them mandatory, adherence to standards is most frequently voluntary. Adherence to voluntary standards is commonly checked through accreditation or certification, a process during which an organization is inspected by an independent body to ensure that it is adhering to standards.

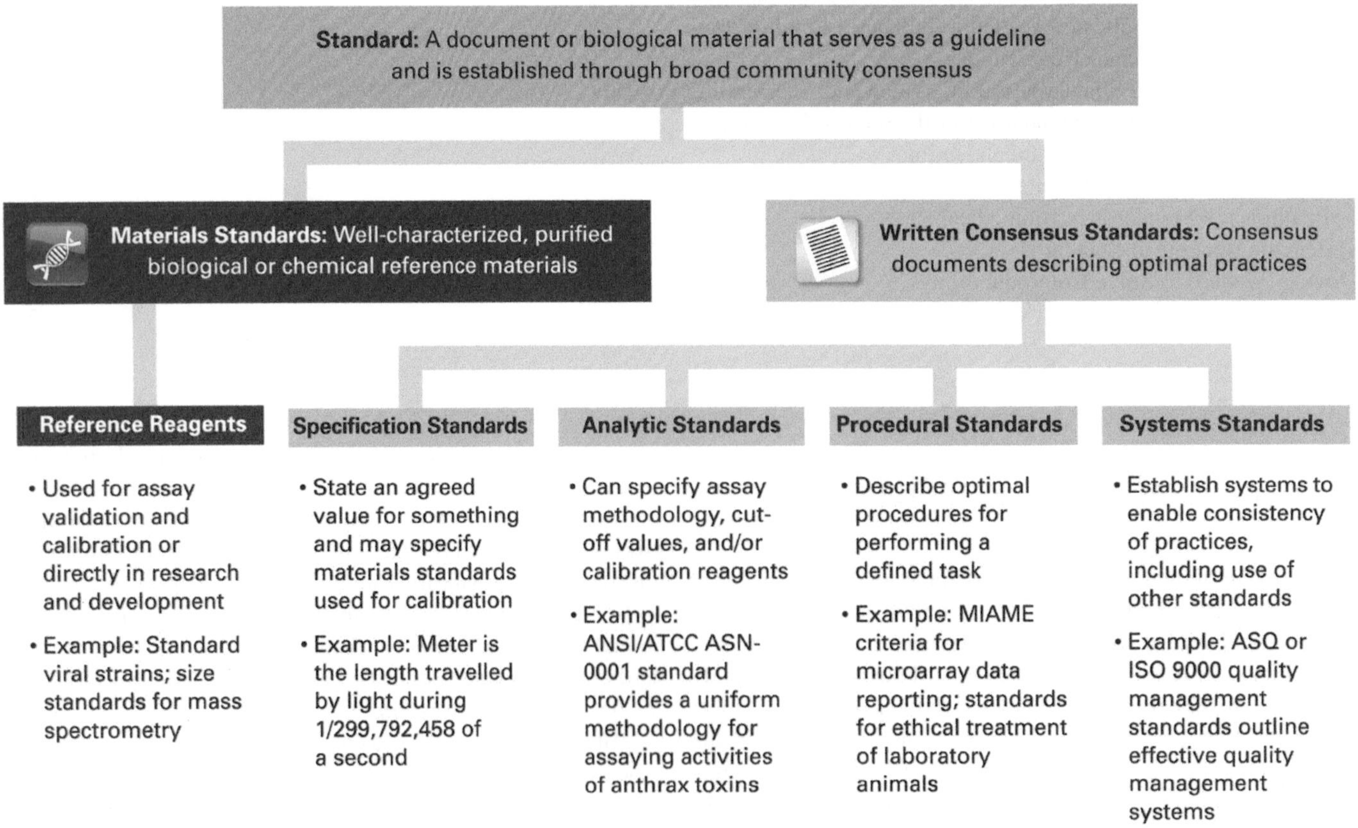

FIGURE 9.4.1 Definition and types of standards.

Standards are broadly used in a variety of fields. Both materials and document standards have been extremely effective in improving reproducibility and quality of laboratory-based fields, such as blood banking and clinical laboratory diagnostics. The fact that standards have been a part of these evolving fields for many years has allowed for documentation of benefits that can be achieved through standards development and implementation over time.

Document Standards

The International Organization for Standardization (ISO) and American National Standards Institute (ANSI) define a standard as "a document, established by consensus and approved by a recognized body, that provides, for common and repeated use, rules, guidelines or characteristics for activities or their results, aimed at the achievement of the optimum degree of order in a given context" (ISO-IEC, 2004).

"Approval by a recognized body" is the main factor that separates voluntary document standards from consensus guidelines or best practices (http://www.ansi.org/about_ansi/faqs/faqs.aspx). Guidelines, recommendations, or best practices can be developed by any group and do not necessarily need to engage all stakeholders that the guideline may affect. In contrast, standards are commonly developed by standards development organizations (SDOs) or consortia. SDOs are collaborative organizations that meet international requirements for openness, balance, consensus, and due process, and provide all affected stakeholders with an opportunity to participate in the development of standards. These requirements are outlined in a system of standards pertaining to SDOs, and SDOs are accredited to follow these standards by national and international standards bodies. ANSI is both the U.S. national standards body and representative to ISO, which is the worldwide federation of national standards bodies. ISO, which was established in 1947 and is based in Geneva, Switzerland, draws members from over 145 countries. There are currently over 200 SDOs in the United States participating in this system, which helps ensure that all major perspectives in a particular field are elicited during the standard development process, and that all the major stakeholders are engaged. Consortia are generally not accredited; instead, they follow the spirit of this system by engaging all key viewpoints in a particular field during standards development.

Document standards have been adopted across multiple industries. For example, the National Electrical Manufacturers Association (NEMA) publishes standards for electrical plugs and sockets in the United States to ensure that all electrical devices intended for home use have similar plugs that are compatible with commonly installed sockets. Standards describing systems that companies need to put in place in order to produce high-quality products and services, such as ISO 9000 series or ASQ Quality Standards, are broadly used in most industries to improve and maintain performance. College of American Pathology (CAP) Standards outline processes and procedures necessary to maintain quality in clinical diagnostic testing and are used by clinical laboratories throughout the United States (CAP, 2016).

Accreditation is one way to encourage compliance with voluntary standards. For example, a company may choose to work only with suppliers that have ISO 9000- or ASQ-accredited quality practices because this accreditation increases the likelihood that the supplier will deliver a high-quality product. As a result of voluntary inducement and community expectations, over one million companies are ISO 9000 series-certified (http://www.iso.org/iso/home/standards/certification/iso-survey.htm). The Centers for Medicare and Medicaid (a government agency) uses accreditation by Joint Commission or CAP (both of which are independent, nonprofit organizations) as a basis for determining whether Medicare will reimburse services from a particular clinical laboratory, resulting in almost universal laboratory certification.

Materials Standards

Materials standards are highly characterized physical substances, such as chemical or biological reagents (known as reference materials or reference reagents), and are routinely used in several areas of life science and health care. Examples of materials standards include reagents for biologics and pharmaceutical development and production, such as well-characterized virus strains to be used for vaccine production (e.g., annual flu vaccines) or in genetic therapy development. Materials standards can also be used to test assays in order to ensure that assays provide accurate measurements and work appropriately. For example, when a sample is tested for a specific infectious organism (e.g., a bacterial pathogen) in a clinical laboratory, infectious organism reference materials (or derivative commercial reagents) can be used to ensure that the assay is sufficiently sensitive and specific to detect the microorganism.

Reference materials can be produced and distributed by several sources, depending on the type of reference material. Sources may include the World Health Organization (WHO); government agencies such as the NIH and the National Institute of Standards and Technology (NIST); nonprofit entities, including biological resource centers like ATCC; and commercial companies. The exact source varies by the specific material.

CURRENT STANDARDS IN LIFE SCIENCE RESEARCH

In life science research, however, standards are not as prolific. The most well-developed standards and accreditation system exist to protect laboratory animal welfare. In the United States, this system is centered on the *Guide for the Care and Use of Laboratory Animals*, 8th edition. National Academies Press, Washington, DC. 2011. REF a collection of standards for appropriate laboratory animal care, initially published in 1963 and revised multiple times thereafter. Adherence to the principles in the guide is required for institutions that receive Public Health Service (NRC, 2011) support (including grants from the National Institutes of Health [NIH] and the Centers for Disease Control and Prevention [CDC]) and is monitored internally by each institution through self-regulating institutional animal care and use committees (http://www.iacuc.org/aboutus.htm).

Materials standards are also used in some research communities, particularly in areas like vaccine and gene therapy development and microbiology. Some research communities have also developed standards that apply to specific fields, such as standards for characterizing and identifying human embryonic stem cells (Loring & Rao, 2006; Shen, 2013a). However, other than in such highly defined areas, few established and broadly-implemented standards exist in life science research.

TACKLING IRREPRODUCIBILITY

Changes in the Life Science Research Landscape Over the past 30 years, life science research advances have benefited the life and health of millions of people. For example, with treatment, HIV patients in the United States and Canada can expect to live an almost a normal life span of over 70 years (Samji et al., 2013). Routinely-recommended genetic testing for diseases such as cystic fibrosis (CF) allows detection and estimation of disease risk prior to pregnancy, enabling couples to make informed decisions, and some CF-affected individuals to live until adulthood (ACOG, 2011). These advances are the result of translating basic research findings into clinical treatments or laboratory tests. In addition, importantly, over 50% of basic research is performed in the academic setting (NSB, 2012).

The current practices used to maintain research quality are deeply entwined with the fundamental tenets of the academic research culture and have successfully produced most biological and treatment advances that we enjoy today. However, changes in the life science landscape, including rising complexity, competition, economic challenges, and translational focus, increase the challenges associated with maintaining life science research quality through traditional systems.

Lack of broadly-accepted standards results in extensive variability in the ways these quality checkpoints are implemented (Figure 9.4.2). For example, academic research leaders and researchers report large differences between laboratories as to which steps of the research process are routinely monitored by the principal investigator (PI). Only a few steps, such as manuscript preparation, are carefully checked for quality.

Peer review of manuscripts prior to publication is a particularly powerful checkpoint that many stakeholders believe fundamentally protects research integrity. However, journals are significantly diverse in their requirements for publication as well as in the criteria that different journals and reviewers use to evaluate work (Bohannon, 2013). Currently, few broadly accepted standards exist to help journals align around requirements for publication. Although some journals may incorporate reporting guidelines and standards into publication decisions, many do not.

Publications release and disseminate results into the broader scientific community where findings and interpretation face additional scrutiny. The feedback from the life science community suggests that researchers believe that when the journal peer-review process fails and erroneous data or interpretations get published, highly important or controversial results will be replicated and either confirmed or disproved by other laboratories within a few years. If findings are reproducible, other laboratories will use them as the basis for additional research, resulting in further publications and the growth of the field. If findings are not reproducible, the situation is more variable (see Allison, et al., 2016). When the original hypothesis is of tremendous significance, the irreproducibility of results is usually published, although this process can take several years. For results of less significance, the inability to replicate findings is frequently not reported.

In some cases, reporting guidelines for certain types of studies or laboratory procedures may be developed by a professional society, and adherence to guidelines may be required by its journal as a condition for publication. For example, *Cytometry A* and *Cytometry B*, the journals of the International Society for Advancement of Cytometry, both require adherence to the MiFlowCyt standard for minimum information required to report the experimental details of flow cytometry experiments ("Cytometry Part A: Author Guidelines," 2016). However, this type of standard is still rare in life science research.

The current research landscape is shaped by multiple changes, including novel technologies, increasing specialization, and imperatives to make a faster impact on patient outcomes. These changes are exerting pressure on traditional paradigms for maintaining research quality (Figure 9.4.2).

Complex and high-throughput technologies and disciplines, such as genomics, proteomics, cellomics, metabolomics, and others, generate increasingly large amounts of biological data (Vucic et al., 2012). For example, the Human Protein Reference Database grew from 2,750 proteins in 2003 (Peri et al., 2003), to over 15,000 in 2016 (www.humanproteinpedia.org). The complexity and breadth of techniques leads to more specialization and a greater reliance on core facilities (centralized, usually institutional, laboratories dedicated to performing one or more complex techniques such as flow cytometry) among researchers as the intricacies of each area present greater challenges and require more focused expertise.

Academic research leaders note that they are relying more on collaboration with experts in complex technologies as a routine part of their work. The trend toward team science is encouraged by the NIH, which funds the majority of academic biomedical research in the United States (Collins, 2016; FASEB, 2015; NSF, 2016). For example, the 2010 NIH *Collaboration and Field Science: A Field Guide* notes that "Innovations and advances that were not possible within one laboratory working in isolation are now emerging from collaborations and research teams that have harnessed techniques, approaches, and perspectives from multiple scientific disciplines and therapeutic areas."

At the same time, multimodal approaches and reliance on team science can undermine the individual investigator's ability

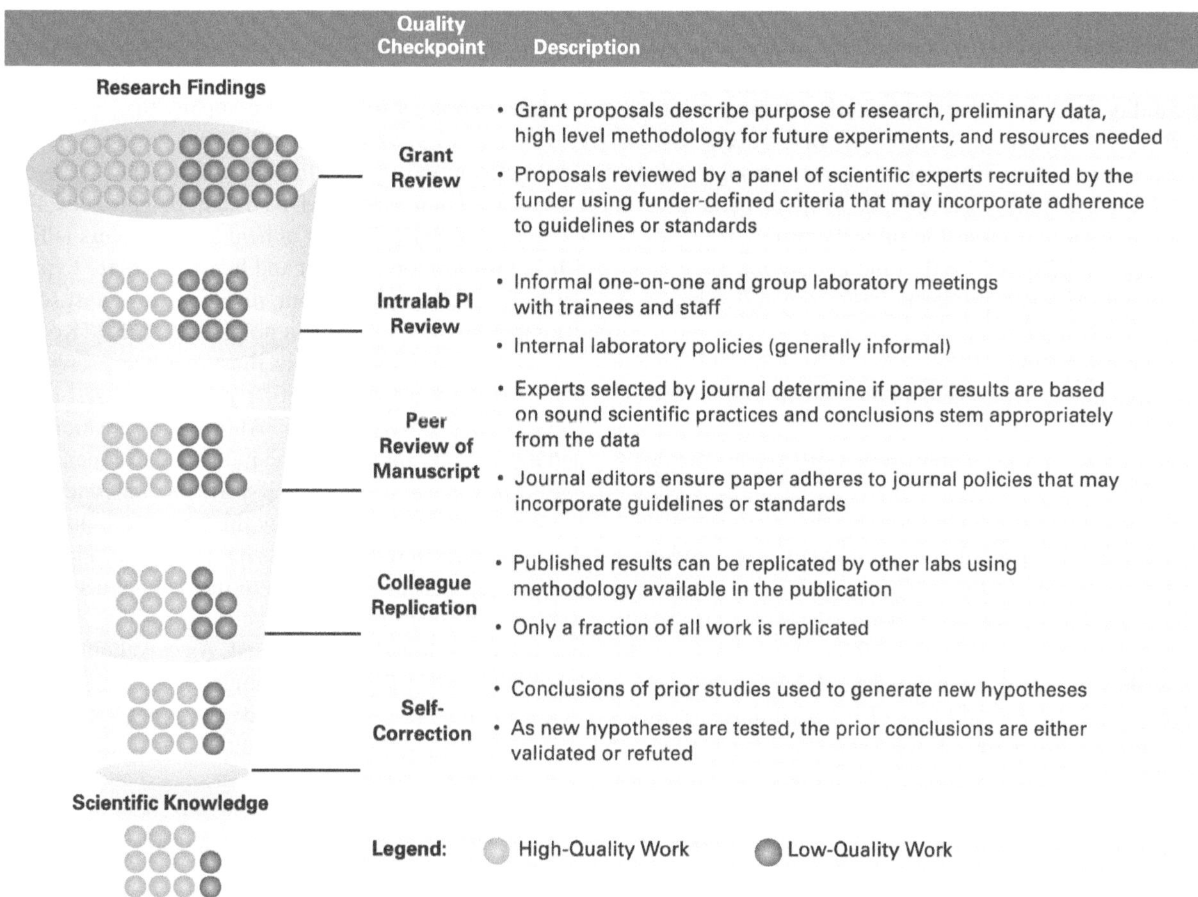

FIGURE 9.4.2 Common academic life science research checkpoints. The funnel represents the overall research process, with circles representing high quality (yellow) or low-quality (blue) work. As the work flows through the checkpoints described on the right, most of the low-quality work is filteredout.

to control the research process and make peer review more challenging. With use of novel, complex technologies to address a broader scope of biological and clinical problems through multimodal and team approaches, these scientists may not have expertise in all the methods being used to generate data for a particular project or paper.

Detailed knowledge of experimental modalities and common pitfalls would allow the PI to maintain research quality by rapidly identifying trainee errors in experimental design or interpretation. However, this checkpoint is weakened when principal investigators must evaluate methods and technologies with which they are less familiar, including interpretation of data obtained from collaborators and core facilities or even routine use of analyte kits. Similarly, journal reviewers selected for their expertise with a particular aspect of a paper might not be familiar with the intricacies of all the technologies used in the research, making it more challenging for reviewers to detect errors and inconsistencies.

The academic research environment has become even more competitive. The percentage of successful NIH grant applications has been in decline for over a decade as NIH funding has flattened (https://report.nih.gov/). Meanwhile, academic research output (e.g., the number of PubMed citations), is steadily increasing. Academic research leaders and researchers report pressure from their institutions and funding agencies to produce more publications. These scientists feel that productivity expectations for advancing professionally and obtaining grants are increasing, particularly because professional advancement decisions frequently take the quantity of published work into consideration.

The dynamic of flattened funding with increasing output leads to fewer available resources for each published paper. Research leaders believe that this resource scarcity leads to the publication of shorter, simpler papers that can be published more rapidly but may not thoroughly explore all aspects of a particular scientific problem, leaving material for subsequent publications. Additionally, the pressure to produce work faster might lead to less thorough experimental design than the PI may otherwise want. These factors can make PI control of research quality less effective.

Life science research is also becoming more translational in nature. Although basic science research has the goal of advancing knowledge about biologic processes, translational research focuses on producing knowledge and products—such as novel drugs or laboratory tests—that can lead to better diagnosis, treatment, prevention, and cures of human diseases. A growing emphasis on producing clinical results can shorten the time between basic research discoveries and use of the discoveries in patients during clinical trials. The increase in importance of translatability is driven in part by NIH funding priorities (Collins, 2010), as well as pressures from institutions, disease foundations, and industry. These pressures are strongly felt by stakeholders. For example, research leaders report that many PIs are incorporating more translatable projects into their laboratories in response to the current NIH funding environment, leading some laboratories with primarily basic research experience to engage in translational research.

Increasing translational focus is also manifested through a growing number of academic-industry collaborations (Schachter, 2012), such as the California Institute for Biomedical Research (Calibr), Johnson & Johnson's Innovation Centers, and Novartis and University of Pennsylvania Center for Advanced Cellular Therapy. Several academic organizations have also established institutes focused on drug discovery such as the Vanderbilt Center for Neuroscience Drug Discovery (http://www.vcndd.com/). These institutes combine academic expertise with structured industrial practices pertaining to processes and reagents to improve translatability and accelerate drug development.

The growing translational focus of research makes it more challenging for sufficient time to pass for "self-correction" to take place before research is adopted by industry or used as a basis for clinical trials or off-label drug use. The shortening of the time available for self-correction as a result of institutional and patient pressures is highlighted by a recent controversy surrounding the reproducibility of results obtained when treating Alzheimer's disease mouse models with a A-approved skin lymphoma drug Targretin® (bexarotene) (Aicardi, 2013).

Bexarotene and AD

Initial results published in *Science* in March 2012 (Cramer et al., 2012) indicated that the drug might have promise in treating Alzheimer's disease (AD) because it increased the clearance of soluble AB amyloid, decreased the area of AB amyloid plaques, and reversed cognitive deficits in putative mouse models of AD. An editorial in NEJM (Lowenthal, et al., 2012) at the time stated, "A single report of this kind of preliminary evidence will require confirmation before Alzheimer's disease investigators even consider launching clinical trials in humans," and cautioned against off-label[2] use of the drug in Alzheimer's disease patients.

Over the next year, multiple laboratories attempted to replicate this high-profile research (Fitz, et al., 2013; Price et al., 2013; Tesseur et al., 2013; Veeraraghavalu et al., 2013). These studies, published in *Science* in May 2013, were more equivocal with the data supporting some but not all of the findings of the original work (Shen, 2013b), dampening the enthusiasm many researchers felt about the promise of this drug.

However, during the year between the publication of the original paper and the subsequent publications, substantial activity to enable the use of bexarotene in AD patients took place. A phase II clinical trial to study the use of bexarotene in AD patients was initiated in January 2013 ("Bexarotene Amyloid Treatment for Alzheimer's Disease (BEAT-AD)," 2016). In April 2013, Case Western Reserve University, where the original research was performed, granted an exclusive license for the treatment strategy to a spinoff company that was cofounded by two of the original study authors and secured $1.4 million in funding from multiple sources, including the Alzheimer's Drug Discovery Foundation (ADDF) and BrightFocus Foundation, to support a phase Ib trial in healthy volunteers (PRNewswire, 2013). Furthermore, patient and family pressure to prescribe the drug off-label was significant, resulting in at least some use of a drug that has side effects but no proven benefit (Koebler, 2013).

The bexarotene example illustrates the growing pressures that make it hard for scientific research to self-correct in an appropriately timely manner. Although replication of key data and discussions about interpretation are a routine part of the scientific process and enable self-correction of scientific findings to take place, the time to allow this process to proceed before a drug is administered to patients has shortened. The example also demonstrates that the changes in the life science research environment are beginning to influence ways that research quality has been traditionally maintained. As stakeholders are affected by these trends, they are beginning to focus attention on quality indicators, like the reproducibility of research findings.

Increasing pressures on traditional systems for maintaining research quality are leading to an increased awareness of irreproducibility of research findings among stakeholders. Reproducibility and reputation are key criteria that stakeholders use to judge the quality of scientific work. Reproducibility can be demonstrated directly when results are replicated externally or expected because of the intrinsic qualities of a paper (Begley, 2013).

Reputation is also viewed as a prominent indicator of quality. This criterion can include the reputation of the PI or laboratory that has produced the work, the institution where the research was performed, or the journal in which the work was published. Over time, the reputations of laboratories, journals, and principal investigators are mainly derived from producing work that is reproducible and stands the test of self-correction.

EXTENT OF IRREPRODUCIBILITY

Irreproducibility is a pervasive, systemic problem experienced by most researchers across life science settings. Discussion about the irreproducibility of many studies, particularly those in genetics, has been ongoing since the early 2000s. This issue was highlighted in a seminal 2005 essay by John Ioannidis, a professor at the Stanford School of Medicine, titled "Why Most Published Research Findings Are False" (Ioannidis, 2005). In this paper, Ioannidis performed mathematical modeling and suggested that "for most study designs and settings, it is more likely for a research claim to be false than true." The article postulated unconscious bias as a key cause.

Several years later, two highly publicized experimental studies showed a significant lack of reproducibility of academic research findings in the industrial setting. In a 2011

Nature Reviews Drug Discovery article, Prinz and colleagues (2011) from Bayer HealthCare reported that in 67 projects over 4 years, published data was reproducible only 21%–32% of the time, depending on the definition of reproducibility (all data vs. main data set) used. A subsequent report coauthored by Begley and Ellis (2012) reported a similar experience at Amgen. Of 53 "landmark" papers that the hematology and oncology department attempted to replicate over 10 years, only 11% were reproducible. Industry stakeholders interviewed for this study also report extensive experiences with lack of reproducibility.

Irreproducibility is not limited to the industrial setting. A recent MD Anderson survey (Mobley et al., 2013) showed that over 65% of senior academic faculty has had the experience of being unable to reproduce a finding from a published paper. Most tried to contact the original authors, but over 60% of the time they received an indifferent or negative response or no response at all.

CAUSES OF IRREPRODUCIBILITY

Irreproducibility is the result of differences in performance of a particular experiment between laboratories. These differences can occur when one laboratory performs the experiment in a more optimal way, or even when both laboratories perform the experiment optimally but with inherent differences in methods or reagents. Multiple systemic causes contribute to variability of both practices to ensure quality and performance of the specific experiment in question, including absence of formal or consistent research quality systems, variable education of staff, and differences in journal review and reporting policies (Figure 9.4.3).

These systemic causes translate into a variety of immediate causes of irreproducibility, including the following:

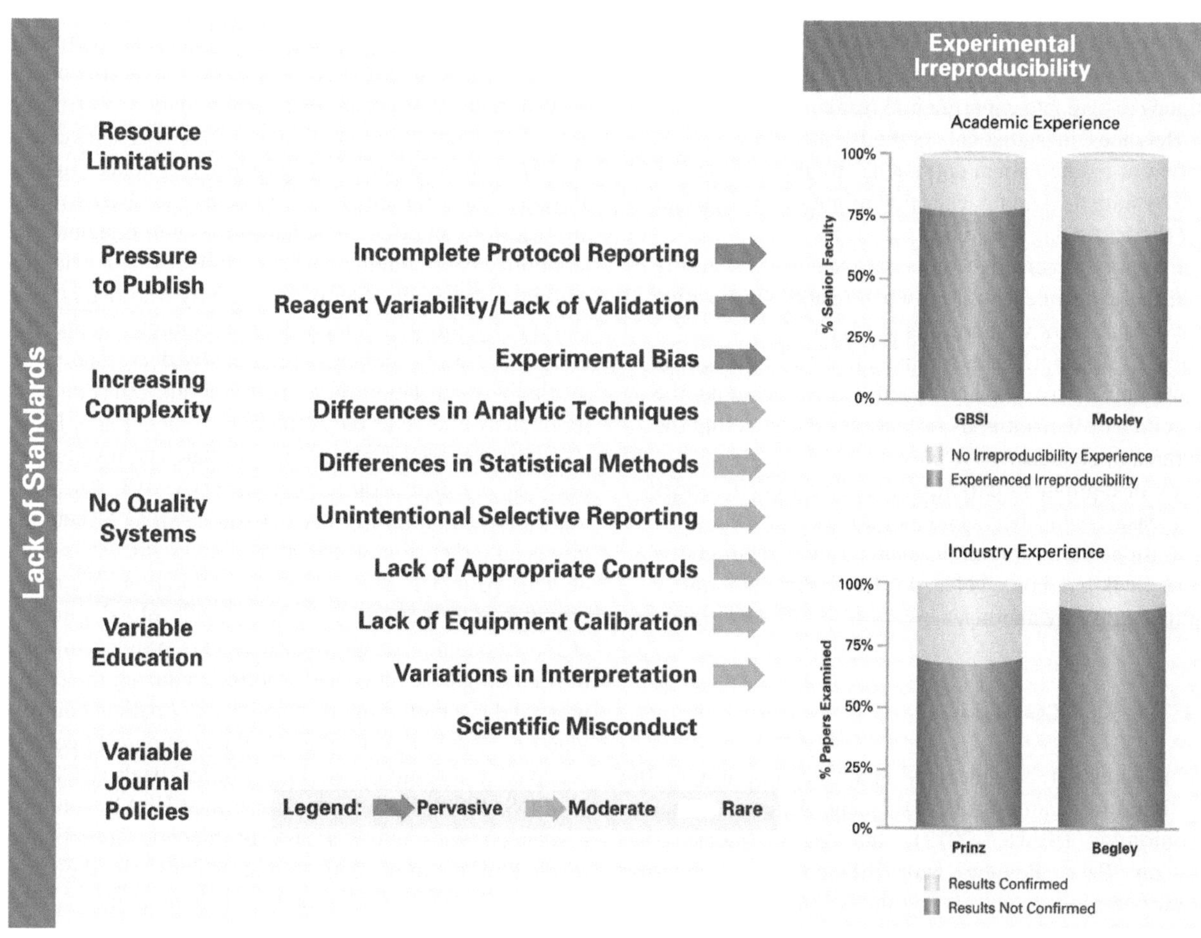

FIGURE 9.4.3 Causes of irreproducibility. Academic experience refers to personal experience of academic senior faculty with irreproducibility. Industry experience refers to percentage of papers reported being able to reproduce. Graphs generated based on responses from study interviews (labeled GBSI, $n = 16$) and Mobley *PLOS One* (2013) ($n = 148$). In industry experience, papers were considered reproducible when literature data was in line with in-house data. (Graphs generated based on Prinz et al. in *Nature Reviews* [2011] and Begley and Ellis in *Nature* [2013]. Causes of irreproducibility are defined and ranked based on interviews for this work, as well as Lozcalzo in *Circulation* [2012] and Begley and Ellis in *Nature* [2013].)

1. Incomplete protocol reporting; for example, when authors leave out certain details that they do not believe to be crucial to the performance of an experiment or that were never recorded as a formal part of the laboratory's procedure for performing the experiment.
2. Reagent variability, such as can occur when two laboratories use what they perceive to be the same reagent, but which are in fact two biologically distinct forms (Vasilevsky et al., 2013). For example, significant differences can exist between antibodies[3] to the same antigen from different manufacturers or even from different lots produced by the same manufacturer.
3. Lack of reagent validation, such as using unauthenticated cell lines (Freedman et al., 2015).
4. Experimental bias (Loscalzo, 2012), such as unconsciously and unintentionally measuring the size of a tumor in an animal from the group an investigator knows to be treated as smaller than in the untreated group. Unblinded experiments where an investigator knows which cell culture dish or group of animals has received treatment are most prone to this type of bias (Vesterinen et al., 2010).
5. Differences in analytic techniques, for example using different assays (tests), imaging, or measurement techniques to evaluate experimental results.
6. Differences in statistical methods, including use of different mathematical approaches to analyze data or use of statistical approaches that might not be optimal for the particular data type.
7. Unintentional selective reporting, such as assuming that the experiment worked when results are positive but did not work when results are negative.
8. Lack of appropriate experimental controls, such as excluding a true negative control to understand the specificity of an antibody or a primer for detecting the antigen or sequence of interest.
9. Lack of equipment calibration leading to differences in measurements between laboratories.
10. Variations in interpretation, such as interpretation of a very weak band on a gel as positive by one laboratory and negative by another.

Scientific Misconduct

Scientific misconduct has recently received attention in the scientific literature (Fang et al., 2012), the American Society of Microbiology (Sliwa, 2012), the 2013 International Congress on Peer Review and Biomedical Publication (http://www.peerreviewcongress.org/index.html), as well as in the popular press (Johnson, 2013; Zimmer, 2012). Particular attention has been brought to the increasing number of peer-reviewed paper retractions and the fact that the majority of retractions result from scientific misconduct (Steen, 2011). Although there is limited overlap, irreproducibility is not equivalent to paper withdrawals or scientific misconduct. Only 0.01% of papers are retracted (Corbyn, 2012), whereas irreproducibility is a pervasive, complex issue that is far more prevalent than paper withdrawals and associated misconduct. Furthermore, most irreproducible papers are never withdrawn (Casadevall et al. 2014). Although scientific misconduct is certainly an issue of grave concern, most interviewed stakeholders view it as an uncommon cause of irreproducibility in the life sciences.

NEGATIVE EFFECTS OF IRREPRODUCIBILITY: THE CASE FOR STANDARDS

Irreproducibility Reflects the Need for Life Science Standards

In its essence, irreproducibility is the result of frequently unrecognized differences in the performance of a particular experiment between laboratories or researchers. Solutions to this pervasive problem can focus on two areas: (1) increasing recognition of intentional differences between experiments conducted by different laboratories (e.g., better data reporting practices), and (2) reducing unintentional differences between laboratories. Standards can effectively reduce differences in practices by aligning the community around consensus-based methods, and are therefore an effective solution for irreproducibility. Multiple laboratory-based fields, such as clinical diagnostics and blood banking, have been able to successfully improve reproducibility and quality through standards. For example, standardization of HbA1c testing in diabetes significantly decreased variability of results between laboratories (Little et al., 2011) and the AABB (previously known as the American Association of Blood Banks) standard for bacterial testing of platelets resulted in a 70% decrease in transfusion-related sepsis (AABB, 2012).

Hemoglobin A1c (HbA1c also known as glycated hemoglobin) is a marker of how effectively blood glucose is controlled in diabetic patients. Although a direct blood glucose measurement provides information about a patient's blood glucose control at that moment, HbA1c measurements provide information on how blood glucose has been controlled during the past month or more and is a more important measure of patient control of blood glucose. HbA1c as a marker of glucose control became broadly accepted as a result of two large clinical trials (DCCTRG, 1993; UKPDS, 1998). Both trials determined HbA1c values that were associated with a lower risk of diabetic complications, and resulted in recommendations that patients' blood glucose should be controlled to a degree that would maintain the patient's HbA1c below these target levels.

At the time these trials came out, laboratories used multiple methods to measure HbA1c that produced results that were not directly comparable to each other (Little et al., 2011). For example, in 1993, only 50% of laboratories were actually reporting results as HbA1c; the remaining were reporting different measurements such as total glycated hemoglobin. Furthermore, laboratories using

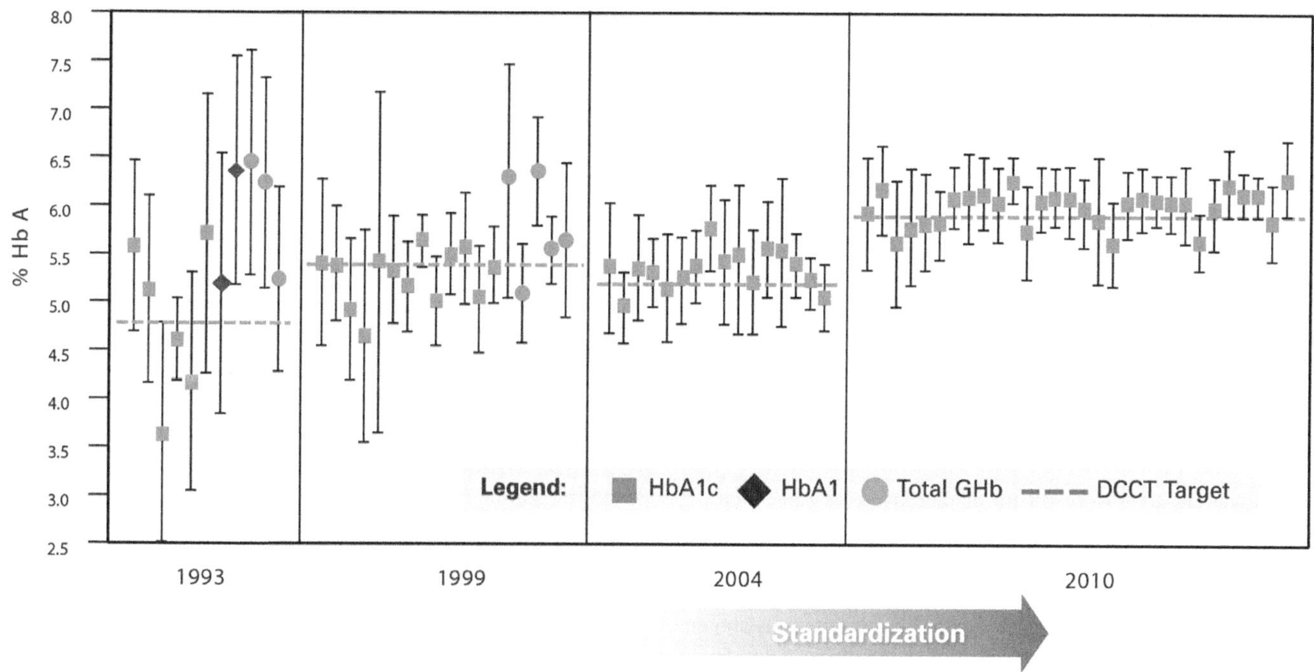

FIGURE 9.4.4 Standardization of HbA1c testing. Each circle, square or rhombus represents the mean value for testing a control material obtained through distinct method of glycated hemoglobin determination and among survey participants. Bars show two standard deviations from the mean. GhB, glycated hemoglobin; DCCT, Diabetes Control and Complications Target. (Reproduced from Little, R.R. et al., *Clin. Chem.*, 57, 205–214, 2011. Data from College of American Pathology surveys 1993–2010.)

the same methods had substantial variability in results. In essence, the results were irreproducible from laboratory to laboratory. This made it difficult or impossible for clinicians to compare the HbA1c levels obtained from their patients to the target levels recommended from the trials or even to compare the results for a single patient over time if they were obtained from different laboratories (Figure 9.4.4). For example, if the same patient's blood sample was sent to two different laboratories, it would not be unusual for one laboratory to produce results that were below the target level established in the clinical trials, indicating that patient treatment was working appropriately, and for a different laboratory to produce results significantly above the target level, indicating that the treatment was not working and should be increased or changed.

Due to the extent of this problem, two professional societies, the American Association of Clinical Chemistry (AACC) and International Federation of Clinical Chemistry (IFCC), created committees to enable the standardization of testing methodologies. These groups created two complementary systems that together serve to standardize manufactured HbA1c assays. The IFCC created reference materials for HbA1c measurement that can be used across different methodologies for calibration and to ensure that findings obtained though all methodologies are comparable. Manufacturers can obtain calibrators and controls with values assigned by the IFCC and use them to determine how their assay methodology compares to the IFCC reference method. The AACC effort developed a standard protocol (Little et al., 2011) and created the National Glycohemoglobin Standardization Program (NGSP), a laboratory-based certification program where manufacturers of HBA1c tests can exchange samples with special NGSP laboratories in order to certify that the results of the test are consistent with the results obtained using the standard protocol. Thus, the IFCC program ensures that each manufacturer can trace its assay to an accurate base through a reagent standard-based system, whereas the NGSP program certifies that the assay accuracy is within an acceptable range of results obtained using a standard protocol.

Together, these standardization systems significantly improved the accuracy, reproducibility, and clinical value of HbA1c testing. As Figure 9.4.4 shows, in 2010 with over 3,000 laboratories participating in the CAP survey for HbA1c testing, the variability between different testing methods, and within each testing method, was significantly lower than in the years before standardization became common (1993 and 1999). Today, clinicians can routinely use and meaningfully compare HbA1c values from different laboratories and be assured that the values they are using to treat their patients are accurately connected to clinical trial results.

IRREPRODUCIBILITY COMPLICATES ACADEMIC-INDUSTRIAL RELATIONSHIPS

The inability to reproduce results also presents a significant challenge for the biomedical research industry. Biopharmaceutical companies might waste resources in attempts to optimize published academic procedures that are not replicable and therefore not translatable, as demonstrated in the Amgen and Bayer experiences described above. Irreproducibility contributes to the unpredictability of development outcomes and failures in target validation and development. Although most companies are highly protective of information regarding costs associated with these failures, one example is provided by Bruce Booth in his blog, LifeSciVC.com: "The company spent $5 million or so trying to validate a platform that didn't exist. When they tried to directly repeat the academic founder's data, it never worked.... Sadly this 'failure to repeat' happens more often than we would like to believe. It has happened to us at Atlas several times in the past decade" (Booth, 2011).

Low reproducibility rates of published data undermine cumulative knowledge production and contribute to both delays and costs of therapeutic drug development. A substantially improved preclinical reproducibility rate would both de-risk preclinical investments and result in an increased hit rate, thereby increasing the productivity of research and improving the speed and efficiency of the therapeutic drug development processes. An analysis by the Global Biological Standards Institute, working with two leading economists from Boston University, has shown that the annual value added to the return on investment from government and private funding would be in the billions in the United States alone (Freedman et al., 2015).

Irreproducibility can significantly complicate academic-industry relationships at a time when both sides are finding these collaborations increasingly desirable. Government funding pressures are increasing the importance and desirability of industry collaborations for both institutions and individual laboratories. Academic research leaders report that in an environment of government funding challenges, industry collaboration can be a highly valuable source of funds and add stability to an otherwise volatile funding environment. Industry is also steadily seeking academic collaborations. Whereas early research and development investment in many companies has flattened, the number of commercial investigational new drug (IND) applications has declined), suggesting decreased R&D efficiency.

To combat these trends, biopharmaceutical companies are increasingly focused on improving the translatability of discoveries and pursuing commercializable opportunities. Academia provides a source of otherwise unobtainable expertise and biological models for biopharmaceutical companies, resulting in a steady stream of collaborative activity. Collaboration with academic laboratories allows biopharmaceutical companies to access leaders in particular scientific disciplines and other experts that might not be available internally. Academic laboratories can also provide valuable resources, such as cell and animal models and active compounds.

Although highly beneficial, the academic-industry relationship is not without impediments; mediating differences between the academic and industrial approaches to quality can be challenging. Generally, Although academic laboratories are focused more on obtaining publishable results, industrial laboratories are more concerned with discovery that occurs within the bounds of replicable processes and can ultimately be transferred into the clinical setting. Whereas maintaining quality in the academic setting is primarily an informal process, many biopharmaceutical companies have formalized some aspects of ensuring quality in the basic research and discovery space. These practices range from basic document control to strict process control systems resembling Good Laboratory Practices that the FDA mandates in later stages of the drug development process (Code of Federal Regulations, Title 21, Part 58). These industry practices can include defined experimental replication procedures or staff training requirements to perform daily research functions. Many academic laboratories and institutions, however, lack such formal quality systems (http://asq.org/glossary/).

Collaborations can rapidly break down when academic results are not reproducible in industrial laboratories because, no matter how promising the academic published results might be, irreproducible results cannot be developed into commercializable products. Industry stakeholders report that differences in quality assurance practices, the quality of reagents, and in the degree of attention paid to the reproducibility of processes can present obstacles to translating academic laboratory results into drug products.

IRREPRODUCIBILITY DAMAGES REPUTATIONS

Irreproducible results can significantly damage the reputations of laboratories, institutions, journals, and funders associated with them. Laboratories whose work cannot be reproduced face reputational concerns that can adversely affect their ability to publish or obtain funding in the future.

Names of institutions where the irreproducible research was produced can be connected indefinitely to irreproducibility. For example, a review article about XMRV (Kakisi et al., 2013) describes the investigators of the original 2006 study as "researchers at the Cleveland Clinic and the University of California," as opposed to referring to the PIs by name. Journals likewise face negative effects on their reputations if a significant amount of published work is shown to be irreproducible. As concerns about irreproducibility gather popular attention, being associated with a publicized incident might worsen public perception of the funding organization or agency and the overall public opinion of life science research.

LIFE SCIENCE RESEARCH STANDARDS: ADDING VALUE AND ACCELERATING PROGRESS

Further Applications of Standards to Life Science Research

Standards can benefit life science research quality and reproducibility by decreasing unrecognized variance between laboratories, and can serve as an effective tool for disseminating and

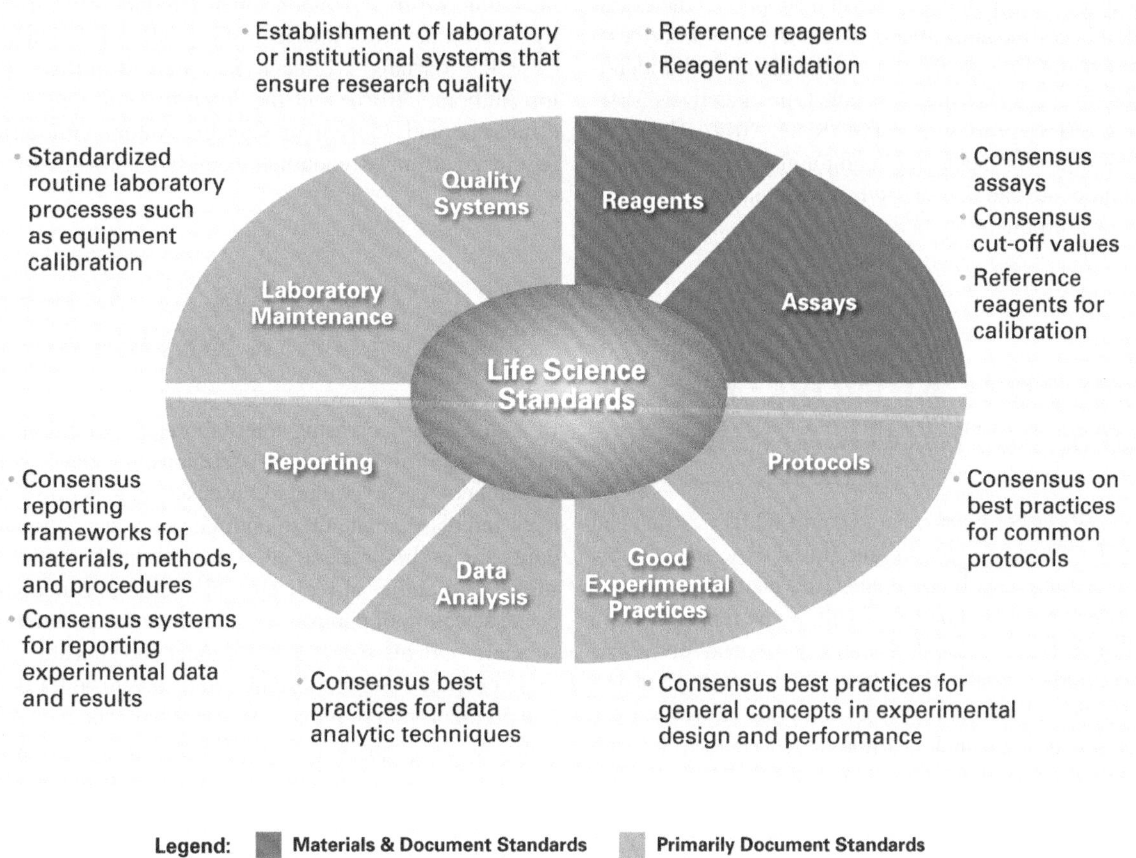

FIGURE 9.4.5 Standards applications in life science research. Standards can have broad applications in life science research. Blue indicates areas where both materials and document standards may be applicable; green indicates where primarily document standards are applicable.

encouraging consensus-based research practices. As illustrated in Figure 9.4.5, it is possible to envision standards pertaining to multiple aspects of life science research.

Standards related to reagents, assays, and laboratory maintenance increase quality and reproducibility by decreasing unintentional (unrecognized) variation between laboratories. Standards related to reagents can include both reference reagents and document standards that describe reagent validation techniques. It is not uncommon for laboratories to use what they believe, based on name or vendor description to be the same antibody or cell line as another laboratory, only to find out later that the particular lot or batch of the reagent had different biological qualities that lead to variability in results. Materials standards for certain research reagents (e.g., antibodies, cell lines, animal models) can be used by laboratories and vendors to decrease these unexpected differences. Document standards pertaining to reagent validation can outline consensus-based ways of ensuring that a particular reagent performs as expected and, likewise, decrease unanticipated variations that lead to irreproducibility.

Assays and measurements present another area where unintended variation can be reduced through biological standards. In many life science research areas, there is little consensus about which assays researchers should use. There might not be reagents available to calibrate the assays in order to ensure that results obtained in one laboratory are comparable to another. Furthermore, there might be variability in cutoffs for considering results positive or negative. Document standards can enable consensus around assays and measurements. For example, a standard can represent the scientific community consensus on which assay will be used to measure activity of a particular bacterial toxin and on the cutoffs for positive and negative results. Additionally, the development of materials standards for assay calibration can particularly benefit fields that use novel and emerging technologies, where such standards are frequently not available.

Standards surrounding laboratory maintenance can define expectations for certain routine "housekeeping" practices. These practices can include the accurate documentation of laboratory protocols and procedures, data recording (e.g., maintenance of laboratory notebooks, deciding how long original data is kept), equipment calibration (e.g., annually ensuring that all pipettes in a laboratory are accurate), checking for reagent expiration, and not using expired reagents. Standards can also outline good practices for common laboratory activities, such as cell or microorganism culture and cell line or microorganism authentication.

A limited number of standards pertaining to reagents, assays, and laboratory maintenance are already available. Some materials standards, primarily microorganisms, are available to researchers through the WHO, NIH, and private bioresource centers.

Document standards are gaining increasing adoption. One example is the ATCC cell line authentication standard that describes a consensus procedure for unambiguous authentication and identification of human cell lines using a specific type of analysis (short tandem repeat [STR] profiling) (ATCC/SDO, 2012). However, many opportunities for achieving community consensus and alignment remain and can serve to reduce unintentional variability and improve reproducibility between laboratories.

STANDARDS AT WORK: CELL LINE AUTHENTICATION

Cell lines are cells either from cancer or normal tissues that can proliferate under laboratory conditions. Initial experiments on a particular tumor type (e.g., breast cancer) are usually performed using these cells. A large number of cells lines representing a variety of tumor types exist. Laboratories depend on visual examination of the cells and vial labels to ensure that they are performing each experiment on the correct cell type. However, even for an expert, it can be difficult or impossible to determine with certainty from which type of tumor a particular cell line originates (Capes-Davis, et al., 2013). Because cells are repeatedly grown, frozen, and stored by laboratories, cross-contamination and errors can occur, resulting in experiments being performed on an incorrect cell type (Freedman et al., 2015).

Examples of using incorrect cell lines in research have been reported for a variety of tumor types. A database maintained by the International Cell Line Authentication Committee (ICLAC) lists over 350 cell lines that have been misidentified or cross-contaminated and publications based on these cell lines (Capes-Davis et al., 2010). Although many of these have been contaminated by the HeLa cell line, the problem is even more pervasive. For example, several common esophageal cell lines that have served as a basis for over 100 publications and several clinical trials have been shown to be from other parts of the body (e.g., lung and colon) (Boonstra et al., 2010). In 2011, a novel gastric MALT lymphoma cell line was described. It was considered groundbreaking because no cell lines were available to model this disease. Recently, however, this cell line was shown to actually be a misidentified, well-known cell line of a different lymphoma type (Capes-Davis, et al., 2013).

Misidentification of cell lines can be prevented by cell line authentication. Authentication is based on determining the genetic signature of a particular cell line and comparing it to established databases to ensure that the cell line used by a laboratory matches the expected signature (Reid, 2011). In 2011 and 2012, an international group of scientists from academia, regulatory agencies, major cell repositories, government agencies, and industry collaborated to develop a standard that describes optimal cell line authentication practices, ANSI/ATCC ASN-0002-2011: Authentication of Human Cell Lines: Standardization of STR Profiling (ATCC/SDO, 2012). More recently, multiple journals including *Nature* ("Announcement: Reducing our irreproducibility," 2013), several American Association for Cancer Research (AACR) journals ("Cancer Research (CanRes): Instructions for Authors," 2016), and the *International Journal of Cancer* ("International Journal of Cancer: Author Guidelines," 2016) require or strongly recommend cell line authentication.

STANDARDS AT WORK: OTHER METRICS

In addition to decreasing inter-laboratory variation in reagents, assays, and daily laboratory operations, standards can also help researchers reach voluntary consensus on general scientific practices and disseminate these optimal practices. For example, standards can outline general good experimental practices, such as a suggested number of replicates, blinded experimentation, appropriate selection of controls, or selection of experimental systems; standards can also serve as a reminder and educational tool about these practices. Standards can also be a vehicle for defining optimal protocols for common experimental procedures. Additionally, standards can apply to data analysis by outlining consensus practices for the use of particular statistical methods, handling of outliers, and even the need for biostatistician involvement.

Standards pertaining to submission of results to peer-reviewed journals or to separate databases can facilitate consistency and completeness of reporting for experimental methods and results. Although some standards already exist in this area, there is tremendous variance between journals and even between reviewers in the same journal about the level of detail reported for many types of experiments (Hirst & Altman, 2012; Tugwell, et al., 2012). Standards could help journals align around the types of data that are necessary to provide about the experimental conditions, analysis, and results (Landis et al., 2012).

Examples of Standards in Biomedical Research

Voluntary standards that have been recently adopted in life science research are focused on establishing data-reporting frameworks. Examples of these include the Minimal Information About a Microarray Experiment (MIAME, developed in 2001) (Brazma et al., 2001), Minimum Information about a MARker gene Sequence (MIMARKS) and Minimum Information about Any (x) Sequence (MIxS, developed in 2011) (Yilmaz et al., 2011), Minimal Information About a Proteomics Experiment (MIAPE, developed in 2007) (Taylor et al., 2007), and Minimum Information about a Flow Cytometry Experiment (MIFlowCyt, developed in 2008) (ISAC, 2008). Multiple other reporting guidelines for a broad range of experimental techniques are also available (DCC, 2016).

These standards, developed by consortia rather than SDOs, apply to high-throughput technologies that generate large data sets of results that are then analyzed using complex bioinformatic techniques. Multiple sources of variability exist in experimental techniques used to generate the data and in assumptions that can be made during the data analysis process (e.g., where to set the cutoff for how much signal intensity is necessary to consider a result positive). These standards encourage more detailed reporting with a particular focus on steps that are common sources of inter-laboratory variability and, as a result, these standards improve reproducibility between laboratories. The MIAME standard has the broadest adoption, and adherence to this reporting framework is currently required by over 50 journals, including high-profile publications such as *Cancer Research*, *Cell*, *Lancet*, and *Nature*.

There are multiple ongoing standards efforts in this area, including Core Information for Metabolomics Reporting (CIMR), a standard to specify minimal guidelines for reporting metabolomics data that is being developed by the Metabolomics Standards Initiative (MSI; http://www.metabolomics-msi.org/); and the Standard Reference System for Information on Bioinformatics Data Structures that is being developed by the Institute of Electrical and Electronics Engineers (IEEE) to help support exchange and comparison of biological data sets in the life sciences (IEEE, 2016). A recently-formed international initiative, the Global Alliance to Enable Responsible Sharing of Genomic and Clinical Data, intends to develop technical and other standards to enable sharing of human genome sequencing data for research purposes in a "secure, controlled and interpretable manner" (Alliance, 2013).

Finally, standards can serve to establish systems that promote and ensure consistent quality and the reproducibility of research results (Freedman et al., 2015). For example, standards may outline quality systems for performing life science research. Standards pertaining to quality systems (such as the International Standards Organization [ISO] 9,000 standards) generally do not explicitly state what constitutes a high-quality outcome (e.g., that every experiment submitted for publication has an appropriate control). Instead, these standards establish institutional frameworks for checking quality consistently based on criteria that may be defined either by other standards or by each institution or laboratory. For instance, such a standard may state that an institution or a laboratory must have a documented internal process for ensuring that all experiments submitted for publication have appropriate controls.

Quality system standards have been successfully and broadly adopted in other industries. For example, over 1 million organizations adhere to quality standards and are ISO-9000 certified (http://asq.org/blog/2012/03/25-years-of-iso-9000/), and over 7,000 laboratories in the United States and internationally adhere to the CAP standards (CAP, 2016). There has also been some progress toward developing these types of standards for life science research. In particular, the American Society for Quality (ASQ) produced a technical report in 2012 titled *ASQ TR1-2012: Best Quality Practices for Biomedical Research in Drug Development* (ASQ, 2012). That document outlines a quality system aimed at nonregulated biomedical research, and could serve as a foundation for standards in this area.

SYSTEMATIC BENEFITS OF STANDARDS

Standards can benefit the quality of the research process by strengthening current checkpoints and counteracting forces that lead to quality challenges (Figure 9.4.6). Document and material standards (e.g., reference reagents) can serve as tools to disseminate best practices to laboratories and institutions, resulting in the overall elevation of minimum experimental quality and facilitate the meaningful comparison of results between different laboratories.

Standards can also ease the evaluation of quality and experimental robustness during the grant review and peer review processes by creating a unifying framework, developed and consented to by experts in the field, and available to a broad range of stakeholders. A standards framework can enable stakeholders to ascertain that best practices are being followed without delving into the details of how they are followed. Over time, adoption of optimal, consensus-based practices and materials standards can lead to increasingly reproducible results and lower reliance on time-consuming self-correction.

By increasing reproducibility, standards can make research results more translatable and facilitate collaboration between industry and academic laboratories. Industry stakeholders report that they would prefer to work with academic laboratories that follow standardized processes and use reliable reagents because this would align these laboratories closer to internal quality processes routinely practiced by industry (described above). Working with laboratories that adopt best practices through consensus standards will increase the reproducibility of their research and reduce technology and process transfer failures. Collaborating with these laboratories could also help to decrease uncertainty of outcomes inherent in academic-industry research partnerships.

Finally, standards can unify and accelerate current and future efforts to improve life science research quality. Stakeholders have started to address irreproducibility in multiple, frequently disjointed ways, including action by journals, funding agencies, and independent organizations. For example, in 2013 *Nature* launched a Reproducibility Initiative ("Announcement: Reducing our irreproducibility," 2013), and the NIH has held several workshops to address the issue of irreproducibility and launched a dedicated website dedicated (NIH, 2016). The 2013 Seventh Peer Review Congress had several presentations and posters dedicated to this and related topics (http://www.peerreviewcongress.org/index.html). Stakeholders interviewed for this Report can also envision multiple other ways of improving reproducibility such as improved education journals, institutions, or organizations and place a significant burden on these stakeholders. Because these efforts are disconnected, each stakeholder must invest anew in identifying best practices and developing their own definitions, guidelines, or systems. By providing a common framework of practices, biological materials, and methods developed by or consented to by key stakeholders in a particular area, standards can facilitate existing efforts to improve life science quality and reproducibility.

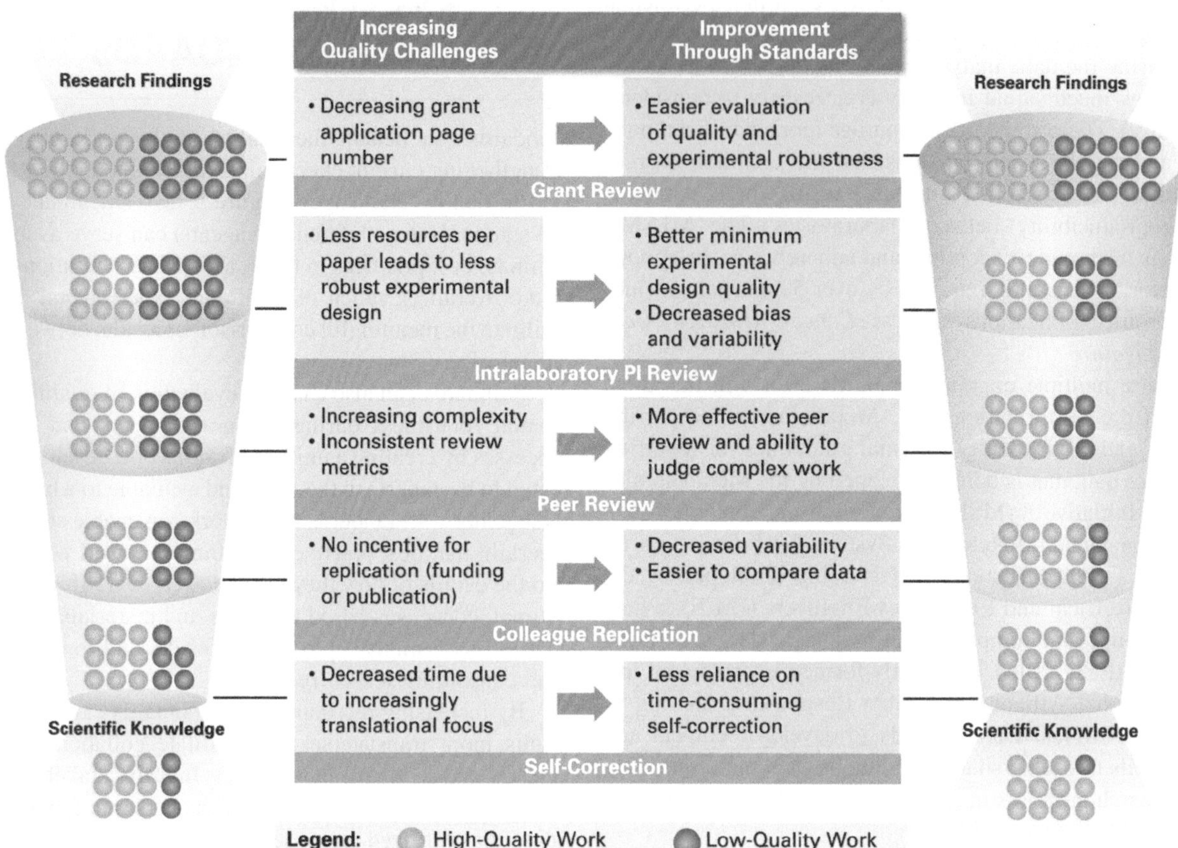

FIGURE 9.4.6 Standards counteract quality challenges. The funnel represents the overall research process, with circles representing high-quality (orange) or low-quality (blue) work. The research flows through the checkpoints described on the right and most of the low-quality work is filtered out. With increasing quality challenges, the filters become more permeable and a lot more low-quality work to pass through. Standards, represented by brown arrows, can counteract the quality challenges and strengthen the current quality checkpoints (orange), resulting in less low-quality research.

Overall, standards are broadly applicable across life science research and can significantly benefit the entire field. Standards can help the life science research community align around consensus-based best practices, reduce uncontrolled variance, and improve reproducibility, resulting in more efficient use of resources, less misinformation, and a more favorable public opinion of research.

MOVING FORWARD TO SOLVE IRREPRODUCIBILITY

Although life science research has made tremendous advances over time, traditional methods for ensuring quality are coming under increasing pressure as the field progresses and the funding environment worsens. Irreproducibility of life science research findings is a pervasive, systemic, and expensive problem that affects those directly involved in research and publication, as well as funders and the general public. Key stakeholders—academic laboratories and institutions, industry and industry associations, investors, government agencies, charitable foundations, journals, and professional societies, standards development organizations, and policy and patient advocacy organizations—associate irreproducibility with poor quality, and are voicing significant concerns about the effects of irreproducibility on their work, on the use of public and private resources, and on public perception of life science research.

Thus far, however, there has been a chasm between the pervasiveness and seriousness of the problem and efforts to find solutions. To date, there have been no coordinated endeavors with a national or international scope to engage all stakeholder groups in order to find effective ways to increase reproducibility across the life sciences research landscape. Emerging efforts, such as the *Nature* Reproducibility Initiative, are important and will surely benefit the field. Nevertheless, the prevalence and magnitude of this problem requires a systemic, unifying effort.

Life science standards can provide the basis for a collaborative, global movement to improve reproducibility.

- Standards are an essential platform that can elevate the quality of the field by encouraging stakeholders to identify and disseminate best practices and optimal materials.
- Standards can serve as a unifying driver for current and future efforts to improve reproducibility by providing stakeholders with a common framework of expert, consensus-based opinions.

Uniting the diverse life science research community in a standardization effort will be challenging. Nonetheless, multiple other fields have successfully incorporated standards to improve quality by associating with core organizations. For example, over a million organizations are ISO 9000-certified and thousands of laboratories adhere to the CAP standards (CAP, 2016). The development and implementation of standards, whether as adherence to written documents or use of reference materials, requires community consensus and alignment around both the necessity for standards and their content.

Ultimately, expanded adoption of life science standards will require:

- Educational initiatives to raise stakeholder awareness of the purpose and benefits of biological standards and understanding of the standards development process;
- Opportunities and forums for stakeholders to identify areas in the life sciences where accelerated standards adoption could provide maximum benefit;
- Engagement of stakeholders with standards development organizations or material reference providers in the development of specific standards;
- Development of effective policies and practices within the life science research community to ensure the proactive development and periodic updating of biological standards.

The life science research community is creative, inventive and innovative. It is a community that has found cures for diseases and extended human life. This community can solve the problem of irreproducibility, and a solid standards framework can unify and accelerate this global effort.

REFERENCES

AABB. (2012). *Public Conference—Secondary Bacterial Screening of Platelet Components*. In AABB (Ed.). Bethesda, MD: AABB.

ACOG. (2011). *Update on Carrier Screening for Cystic Fibrosis* (Vol. 486, pp. 4). Washington, DC: American College of Obstetricians and Gynecologists.

Aicardi, G. (2013). New hope from an old drug: Fighting Alzheimer's disease with the cancer drug bexarotene (targretin)? *Rejuvenation Res, 16*(6), 524–528. doi:10.1089/rej.2013.1497.

Alliance, Global. (2013). Creating a global alliance to enable responsible sharing of genomic and clinical data (pp. 34). Toronto, Ontario, Canada: Global Alliance for Genomics and Health.

Allison, D.B., Brown, A.W., George, B.J., & Kaiser, K.A. (2016). Reproducibility: A tragedy of errors. *Nature, 530*(7588), 27.

ASQ. (2012). ASQ TR1-2012: Best quality practices for biomedical research in drug development (p. 26): American Society for Quality.

ATCC/SDO. (2012). *ASN-0002: Authentication of Human Cell Lines: Standardization of STR Profiling* (A.-S. D. Organization, Trans.). Manassas, Virginia: ATCC-Standards Development Organization (SDO).

Begley, C. G., & Ellis, L. M. (2012). Drug development: Raise standards for preclinical cancer research. *Nature, 483*(7391), 531–533.

Begley, C.G. (2013). Reproducibility: Six red flags for suspect work. *Nature, 497*(7450), 433–434.

Bexarotene Amyloid Treatment for Alzheimer's Disease (BEAT AD) (2016). Retrieved February 29, 2016, from https://http://www.clinicaltrials.gov/ct2/archive/NCT01782742.

Bohannon, John. (2013). Who's afraid of peer review? *Science, 342*(6154), 60–65. doi:10.1126/science.342.6154.60.

Boonstra, J. J., van Marion, R., Beer, D. G., Lin, L., Chaves, P., Ribeiro, C., Dinjens, W. N. (2010). Verification and unmasking of widely used human esophageal adenocarcinoma cell lines. *J Natl Cancer Inst, 102*(4), 271–274.

Booth, B. (2011). Academic Bias & Biotech Failures. *LifeSciVC.com*. Retrieved February 29, 2016, from https://lifescivc.com/2011/03/academic-bias-biotech-failures/.

Brazma, A., Hingamp, P., Quackenbush, J., Sherlock, G., Spellman, P., Stoeckert, C., Vingron, M. (2001). Minimum information about a microarray experiment (MIAME)-toward standards for microarray data. *Nat Genet, 29*(4), 365–371.

Cancer Research (CanRes): Instructions for Authors. (2016). Retrieved February 29, 2016, from http://cancerres.aacrjournals.org/site/misc/ifora.xhtml.

CAP. (2016). Accreditation and Laboratory Improvement. February 29, 2016, from http://www.cap.org/apps/cap.portal?_nfpb=true&_pageLabel=accreditation.

Capes-Davis, A., Alston-Roberts, C., Kerrigan, L., Reid, Y. A., Barrett, T., Burnett, E. C., Drexler, H. G. (2013). Beware imposters: MA-1, a novel MALT lymphoma cell line, is misidentified and corresponds to Pfeiffer, a diffuse large B-cell lymphoma cell line. *Genes Chromosomes Cancer, 52*(10), 986–988.

Capes-Davis, A., Reid, Y. A., Kline, M. C., Storts, D. R., Strauss, E., Dirks, W. G., Kerrigan, L. (2013). Match criteria for human cell line authentication: Where do we draw the line? *Int J Cancer, 132*(11), 2510–2519.

Capes-Davis, A., Theodosopoulos, G., Atkin, I., Drexler, H. G., Kohara, A., MacLeod, R. A., Freshney, R. I. (2010). Check your cultures! A list of cross-contaminated or misidentified cell lines. *Int J Cancer, 127*(1), 1–8. doi:10.1002/ijc.25242.

Casadevall, A., Steen, R. G., & Fang, F. C. (2014). Sources of error in the retracted scientific literature. *FASEB J.* doi:10.1096/fj.14-256735.

Collins, F. (2010). The bridge between lab and clinic. Interview by Meredith Wadman. *Nature, 468*(7326), 877.

Collins, F. (2016). Congressional Justification of the NIH fiscal year (FY) 2017 budget request. Retrieved February 29, 2016, from https://officeofbudget.od.nih.gov/pdfs/FY17/31-Overview.pdf.

Corbyn, Z. (2012). Misconduct is the main cause of life-sciences retractions. *Nature, 490*(7418), 21.

Cramer, P. E., Cirrito, J. R., Wesson, D. W., Lee, C. Y., Karlo, J. C., Zinn, A. E., Landreth, G. E. (2012). ApoE-directed therapeutics rapidly clear beta-amyloid and reverse deficits in AD mouse models. *Science, 335*(6075), 1503–1506. doi:10.1126/science.1217697.

Cytometry Part A: Author Guidelines. (2016). Retrieved February 29, 2016, 2016, from http://onlinelibrary.wiley.com/journal/10.1002/(ISSN)1552-4930/homepage/ForAuthors.html.

DCC. (2016). MIBBI—Minimum Information for Biological and Biomedical Investigations. Retrieved February 29, 2016, from http://www.dcc.ac.uk/resources/metadata-standards/mibbi-minimum-information-biological-and-biomedical-investigations.

DCCTRG. (1993). The effect of intensive treatment of diabetes on the development and progression of long-term complications in insulin-dependent diabetes mellitus. The diabetes control and complications trial research group. *N Engl J Med, 329*(14), 977–986. doi:10.1056/nejm199309303291401.

Fang, F. C., Steen, R. G., & Casadevall, A. (2012). Misconduct accounts for the majority of retracted scientific publications. *Proceedings of the National Academy of Sciences, 109*(42), 17028–17033. doi:10.1073/pnas.1212247109.

FASEB. (2015). *Federal Funding for Biomedical and Related Life Sciences Research: FY 2016* (p. 16). Bethesda, MD: Federation of American Societies for Experimental Biology.

Fitz, N. F., Cronican, A. A., Lefterov, I., & Koldamova, R. (2013). Comment on "ApoE-directed therapeutics rapidly clear beta-amyloid and reverse deficits in AD mouse models." *Science, 340*(6135), 924-c. doi:10.1126/science.1235809.

Freedman, L. P., & Gibson, M. C. (2015). The impact of preclinical irreproducibility on drug development. *Clin Pharmacol Ther, 97*(1), 16–18. doi:10.1002/cpt.9.

Freedman, L. P., Cockburn, I. M., & Simcoe, T. S. (2015). The economics of reproducibility in preclinical research. *PLoS Biol, 13*(6), e1002165.

Freedman, L. P., Gibson, M. C., Ethier, S. P., Soule, H. R., Neve, R. M., & Reid, Y. A. (2015). Reproducibility: Changing the policies and culture of cell line authentication. *Nat Meth, 12*(6), 493–497. doi:10.1038/nmeth.3403.

Hirst, A., & Altman, D. G. (2012). Are peer reviewers encouraged to use reporting guidelines? A survey of 116 health research journals. *PLoS One, 7*(4), e35621. doi:10.1371/journal.pone.0035621.

IEEE. (2016). Standards Development Working Group: 1953 WG—Bioinformatics Standards Working Group. Retrieved February 29, 2016, from http://standards.ieee.org/develop/wg/1953_WG.html.

International Journal of Cancer: Author Guidelines. (2016). Retrieved February 29, 2016, from http://onlinelibrary.wiley.com/journal/10.1002/(ISSN)1097-0215/homepage/ForAuthors.html.

Ioannidis, J. P. (2013). Discussion: Why "An estimate of the science-wise false discovery rate and application to the top medical literature" is false. *Biostatistics, 15*(1), 28–36.

Ioannidis, John P. A. (2005). Why most published research findings are false. *PLoS Med, 2*(8), e124.

ISAC. (2008). Minimum information about a flow cytometry experiment—MIFlowCyt 1.0 (pp. 23): *International Society for Analytical Cytology*.

ISO-IEC. (2004). *ISO/IEC Guide 2:2004: Standardization and Related Activities—General Vocabulary* (8th ed., pp. 26). Geneva, Switzerland: ISO (the International Organization for Standardization) and IEC (the International Electrotechnical Commission).

Jager, L. R., & Leek, J. T. (2014). An estimate of the science-wise false discovery rate and application to the top medical literature. *Biostatistics, 15*(1), 1–12.

Jasny, Barbara R., Chin, Gilbert, Chong, Lisa, & Vignieri, Sacha. (2011). Again, and again, and again. *Science, 334*(6060), 1225. doi:10.1126/science.334.6060.1225.

Johnson, C. (2013). Harvard's Investigation Fills in Picture of Researcher's Misdeeds, *Boston GLobe*. Retrieved from https://http://www.bostonglobe.com/metro/2013/04/14/harvard-investigation-stem-cell-researcher-misconduct-shines-light-secretive-process/uh3aS5FxEQeFAKbREDgiHL/story.html.

Kakisi, O. K., Robinson, M. J., Tettmar, K. I., & Tedder, R. S. (2013). The rise and fall of XMRV. *Transfus Med, 23*(3), 142–151. doi:10.1111/tme.12049.

Koebler, J. (2013). Study finds 'serious problems' with 2012 Alzheimer's breakthrough, *U.S. News and World Report*, p. 2. Retrieved from http://www.usnews.com/news/articles/2013/05/23/study-finds-serious-problems-with-2012-alzheimers-breakthrough.

Landis, S. C., Amara, S. G., Asadullah, K., Austin, C. P., Blumenstein, R., Bradley, E. W., Silberberg, S. D. (2012). A call for transparent reporting to optimize the predictive value of preclinical research. *Nature, 490*(7419), 187–191. doi:10.1038/nature11556.

Little, R. R., Rohlfing, C. L., & Sacks, D. B. (2011). Status of hemoglobin A1c measurement and goals for improvement: From chaos to order for improving diabetes care. *Clin Chem, 57*(2), 205–214.

Loring, J. F., & Rao, M. S. (2006). Establishing standards for the characterization of human embryonic stem cell lines. *Stem Cells, 24*(1), 145–150. doi:10.1634/stemcells.2005-0432.

Loscalzo, J. (2012). Irreproducible experimental results: Causes, (mis)interpretations, and consequences. *Circulation, 125*(10), 1211–1214.

Lowenthal, J., Hull, S. C., & Pearson, S. D. (2012). The ethics of early evidence—preparing for a possible breakthrough in Alzheimer's disease. *N Engl J Med, 367*(6), 488–490.

Mobley, A., Linder, S. K., Braeuer, R., Ellis, L. M., & Zwelling, L. (2013). A survey on data reproducibility in cancer research provides insights into our limited ability to translate findings from the laboratory to the clinic. *PLoS One, 8*(5), e63221.

NIH. (2016). Rigor and Reproducibility. Retrieved February 29, 2016, from https://http://www.nih.gov/research-training/rigor-reproducibility.

NRC. (2011). *Guide for the Care and Use of Laboratory Animals: Eighth Edition*. Washington, DC: The National Academies Press.

NSB. (2012). *Science and Engineering Indicators 2012* (pp. 592). Arlington, VA: National Science Board.

NSF. (2016). Higher Education Research and Development Survey: Fiscal Year 2014 Retrieved February 29, 2016, from http://ncsesdata.nsf.gov/herd/2014/.

Peri, S., Navarro, J. D., Amanchy, R., Kristiansen, T. Z., Jonnalagadda, C. K., Surendranath, V., Pandey, A. (2003). Development of human protein reference database as an initial platform for approaching systems biology in humans. *Genome Res, 13*(10), 2363–2371. doi:10.1101/gr.1680803.

Price, A. R., Xu, G., Siemienski, Z. B., Smithson, L. A., Borchelt, D. R., Golde, T. E., & Felsenstein, K. M. (2013). Comment on "ApoE-directed therapeutics rapidly clear beta-amyloid and reverse deficits in AD mouse models." *Science, 340*(6135), 924-d. doi:10.1126/science.1234089.

Prinz, F., Schlange, T., & Asadullah, K. (2011). Believe it or not: How much can we rely on published data on potential drug targets? *Nat Rev Drug Discov, 10*(9), 712–713.

PRNewswire. (2013). Research foundations collaborate to fund phase 1 study of cancer drug in Alzheimer's disease patients. *PR Newswire, April 24, 2013*(April 24, 2013). http://www.prnewswire.com/news-releases/research-foundations-collaborate-to-fund-phase-1-study-of-cancer-drug-in-alzheimers-disease-patients-204471341.html.

Reid, Y. A. (2011). Characterization and authentication of cancer cell lines: An overview. *Methods Mol Biol, 731*, 35–43.

Samji, H., Cescon, A., Hogg, R. S., Modur, S. P., Althoff, K. N., Buchacz, K., Gange, S. J. (2013). Closing the gap: Increases in life expectancy among treated HIV-positive individuals in the United States and Canada. *PLoS One, 8*(12), e81355. doi:10.1371/journal.pone.0081355.

Schachter, B. (2012). Partnering with the professor. *Nat Biotechnol, 30*(10), 944–952.

Shen, H. (2013a). Stricter standards sought to curb stem-cell confusion. *Nature*, 499, 389. http://www.nature.com/polopoly_fs/1.13434!/menu/main/topColumns/topLeftColumn/pdf/499389a.pdf.

Shen, H. (2013b). Studies cast doubt on cancer drug as Alzheimer's treatment. *Nature* News, May 23, 2013(May).

Sliwa, J. (2012). Has Modern Science Become Dysfunctional? http://www.asm.org/index.php/asm-press-releases/92-news-room/press-releases/1690-release032712b.

Steen, R. G. (2011). Retractions in the scientific literature: Is the incidence of research fraud increasing? *J Med Ethics, 37*(4), 249-253. doi:10.1136/jme.2010.040923.

Taylor, C. F., Paton, N. W., Lilley, K. S., Binz, P. A., Julian, R. K., Jr., Jones, A. R., Hermjakob, H. (2007). The minimum information about a proteomics experiment (MIAPE). *Nat Biotechnol, 25*(8), 887–893. doi:10.1038/nbt1329.

Tesseur, I., Lo, A. C., Roberfroid, A., Dietvorst, S., Van Broeck, B., Borgers, M., De Strooper, B. (2013). Comment on "ApoE-directed therapeutics rapidly clear beta-amyloid and reverse deficits in AD mouse models." *Science, 340*(6135), 924-e. doi:10.1126/science.1233937.

Tugwell, Peter, Knottnerus, Andre, & Idzerda, Leanne. (2012). Why are reporting guidelines not more widely used by journals? *J Clin Epidemiol, 65*(3), 231–233.

UKPDS. (1998). Intensive blood-glucose control with sulphonylureas or insulin compared with conventional treatment and risk of complications in patients with type 2 diabetes (UKPDS 33). UK Prospective Diabetes Study (UKPDS) Group. *Lancet, 352*(9131), 837–853.

Vasilevsky, N. A., Brush, M. H., Paddock, H., Ponting, L., Tripathy, S. J., Larocca, G. M., & Haendel, M. A. (2013). On the reproducibility of science: Unique identification of research resources in the biomedical literature. *PeerJ, 1*, e148. doi:10.7717/peerj.148.

Veeraraghavalu, K., Zhang, C., Miller, S., Hefendehl, J. K., Rajapaksha, T. W., Ulrich, J., Sisodia, S. S. (2013). Comment on "ApoE-directed therapeutics rapidly clear beta-amyloid and reverse deficits in AD mouse models." *Science, 340*(6135), 924-f. doi:10.1126/science.1235505.

Vesterinen, H. M., Sena, E. S., ffrench-Constant, C., Williams, A., Chandran, S., & Macleod, M. R. (2010). Improving the translational hit of experimental treatments in multiple sclerosis. *Mult Scler, 16*(9), 1044–1055. doi:10.1177/1352458510379612.

Vucic, E. A., Thu, K. L., Robison, K., Rybaczyk, L. A., Chari, R., Alvarez, C. E., & Lam, W. L. (2012). Translating cancer "omics" to improved outcomes. *Genome Res, 22*(2), 188–195.

Williams, R (2015). Can't get no reproduction: Leading researchers discuss the problem of irreproducible results. *Circ Res., 117*(8), 667–670. doi:10.1161/circresaha.115.307532.

Yilmaz, P., Kottmann, R., Field, D., Knight, R., Cole, J. R., Amaral-Zettler, L., Glockner, F. O. (2011). Minimum information about a marker gene sequence (MIMARKS) and minimum information about any (x) sequence (MIxS) specifications. *Nat Biotechnol, 29*(5), 415–420. doi:10.1038/nbt.1823.

Zimmer, C. (2012). Misconduct Widespread in Retracted Science Papers, Study Finds, *New York Times*. Retrieved from http://www.nytimes.com/2012/10/02/science/study-finds-fraud-is-widespread-in-retracted-scientific-papers.html?_r=0.

9.5 Can Big Data Analytics Recapitulate Biology? A Survey of Multi-omics Data Integration Approaches

Amrit Singh, Casey P. Shannon, Kim-Anh Lê Cao, and Scott J. Tebbutt

Contents

INTRODUCTION

High-throughput molecular and cellular analytical platforms are inundating researchers with high-dimensional multi-omics data. A major challenge of modern biological research is to integrate within and across these very complex data sets to derive biological insight. Integrating across multiple data sources can improve our understanding of a complex system[1] and may lead to more reliable hypothesis generation. For example, in biology, a given molecular pathway may entail the structural and/or functional interplay between various molecules from messenger RNA transcripts to proteins and metabolites. Biological functions may be modified by genetic (e.g., mutations and polymorphisms), epigenetic (e.g., methylation and microRNA), and environmental regulatory factors. Simultaneous molecular profiling of tens to millions of molecules (later referred to as "variables" or "features") from different biological compartments is necessary to capture this complexity.

Data integration of high-dimensional omics (e.g., transcriptomics, proteomics, metabolomics) data sets may be carried out using *unsupervised analyses* that disregard sample label information (such as disease status). On the other hand, *supervised analyses* can be used to identify patterns, trends, and associations that discriminate between qualitatively distinct groups of specimens. Integrative analyses may also incorporate additional data from curated databases such as protein–protein interactions, canonical pathways, and transcription factor binding data. We refer to approaches that leverage such data as *knowledge-based*, whereas approaches that solely rely on empirical data are referred to as *data-driven*. The biological question, sample size, and types of data available, should inform the type of integrative analysis applied, though many exploratory avenues may be open to the investigators.

We define two broad categories of integrative studies: (1) P-integration and (2) N-integration.[2,3] P-integration integrates multiple data sets that were generated for the same set of P variables. For example, multiple laboratories interested in a specific research area may perform whole genome profiling of different cohorts of individuals. The integrative analysis would constitute a combined analysis of the different data sets for that given omic source. Although this is an important application, we will not discuss P-integration in this chapter. Rather, we will focus specifically on N-integration approaches that are being used to integrate different data sets from multiple omics sources, obtained from the same set of individuals. The need for such methods is timely as multiple omics data sets are becoming increasingly common (e.g., The Cancer Genome Atlas (TCGA); data-types include SNP (Single Nucleotide Polymorphisms), CNV (Copy Number Variants), DNA methylation, gene expression microRNA expression for over 30 cancer types[4]).

We present several approaches for analyzing multi-omics data, including factorization methods, message passing algorithms, methods for multi-block data analysis and generalized canonical correlation analysis (CCA), network based methods, Bayesian methods and classification and regression algorithms (Figure 9.5.1). Many of the techniques covered in this chapter can be considered as belonging to the field of machine learning.[5,6] These integrative techniques are used to "learn" holistic patterns from multi-omics data. We categorize the methods for multi-omics data integration into *unsupervised* and *supervised* analyses, as well as *data-driven* and *knowledge-based* (Figure 9.5.1).

Our survey is not meant to provide a comprehensive review of the literature on multi-omics data integration, but rather to introduce the reader to the different analytical approaches. Table 9.5.1 provides a brief summary including the data-types, number of samples and variables for each method discussed in this chapter.

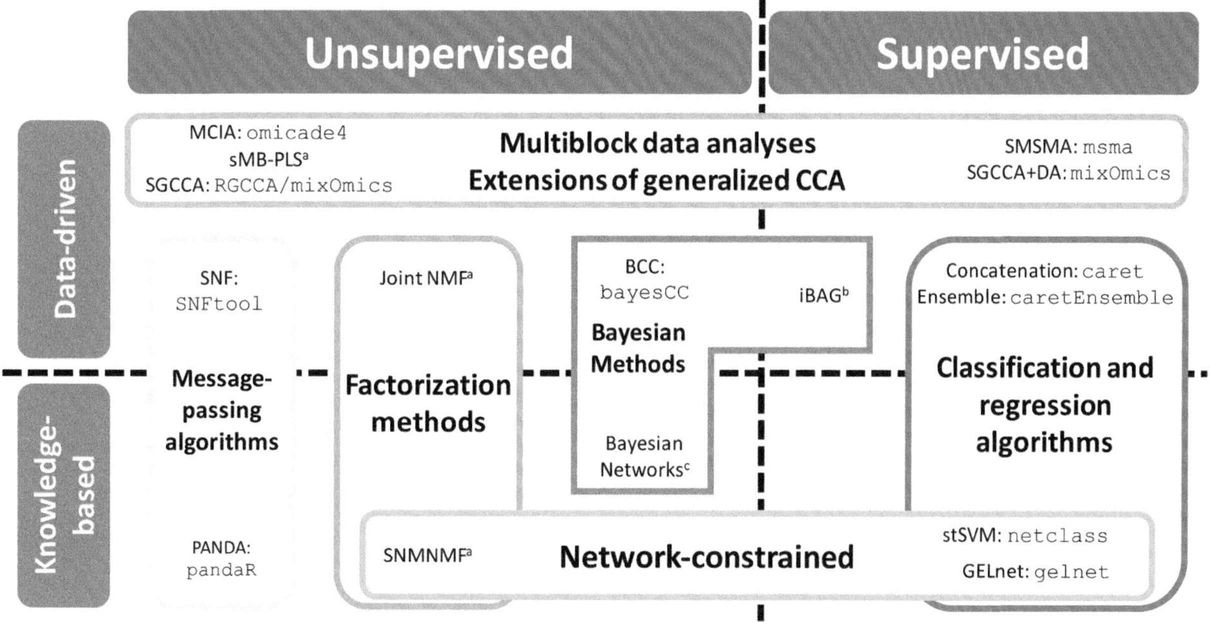

FIGURE 9.5.1 Methods for multi-omics data integration. The integrative methods discussed in this section are based on four major categories, supervised versus unsupervised analyses, and data-driven versus knowledge-based methods. All methods are further grouped into different method-types. For each method-type, specific examples are listed along with the package in the R statistical computing language in the form; Name of method: R-package Methods with a superscript "a" have been implemented in MATLAB® and their source code is linked in their respective publications. iBAG[b] is implemented in R and its associated R-package and R-based Shiny web application can be found on the authors webpage. (From Baladandayuthapani, V., iBAG page, http://odin.mdacc.tmc.edu/~vbaladan/Veera_Home_Page/iBAG_page.html.) The software used to construct Bayesian networks (superscript "c") is called reconstructing integrative molecular Bayesian networks (RIMBANET) implemented in Perl. (From Zhu, J., RIMBANET for Bayesian network reconstruction, http://research.mssm.edu/integrative-networkbiology/RIMBANET/RIMBANET_overview.html.) All other methods have been implemented in R and can be obtained via the comprehensive archive network (CRAN, https://cran.r-project.org/), Bioconductor (https://www.bioconductor.org/), or github (https://github.com/).

TABLE 9.5.1 Summary of studies describing methods for multi-omics data integration covered in this chapter

METHOD	DISEASE	DATA SETS	NO. OF VARIABLES	NO. OF SAMPLES	REFERENCE
Data-driven unsupervised analyses					
Joint non-negative matrix factorization (NMF)	Ovarian cancer	DNA methylation	2,008 probes	385	7
		Gene expression	2,985 genes		
		miRNA expression	270 miRNAs		
Multiple co-inertia analysis (MCIA)	NCI-60: 59 cancer cell lines	Gene expression:	11,051 genes	—	8
		Agilent	8,803 genes		
		HGU95	9,044 genes		
		Hgu133	10,382 genes		
		Hgu133 plus 2.0			
		Protein expression	7,150 proteins	—	
Sparse multi-block partial least squares (sMB-PLS)	Ovarian cancer	Copy number variants	31,324 variants	230	10
		DNA methylation	14,735 probes		
		Gene expression	15,846 genes		
		miRNA expression	799 miRNAs		
Similarity network fusion (SNF)	Glioblastoma multiforme (GBM)	DNA methylation	1,491 probes	215	11
		Gene expression	12,042 genes		
		miRNA expression	534 miRNAs		
	Other cancers: breast, kidney, lung, colon	DNA methylation, gene and miRNA expression	534 miRNAs in GBM to 21,578 methylated genes in lung and colon cancer	92–215	
Bayesian consensus clustering (BCC)	Breast cancer	DNA methylation	574 probes	348	13
		Gene expression	645 genes		
		miRNA expression	423 miRNAs		
		Protein expression	171 proteins		

(Continued)

TABLE 9.5.1 (*Continued*) Summary of studies describing methods for multi-omics data integration covered in this chapter

METHOD	*DISEASE*	*DATA SETS*		*NO. OF VARIABLES*	*NO. OF SAMPLES*	*REFERENCE*
Knowledge-based unsupervised analyses						
Sparse network-regularized joint non-negative factorization (SNMNMF) method	Ovarian cancer	Empirical data	Gene expression	12,456 genes	385	15
			miRNA expression	559 miRNAs		
		Curated data	Protein–protein and DNA–protein interaction data	31,949 gene–gene interactions	—	
			Predicted miRNA–gene interaction	243,331 miRNA–gene interactions	—	
Passing attributes between networks for data assimilation (PANDA)	*Saccharomyces cerevisiae* (yeast)	Empirical data	Gene expression	2555 genes	—	16
		Curated data	Transcription factor (TF) motifs in sequence data	53 TFs	—	
			Protein–protein interactions	135,415 TF–gene interactions	—	
Reconstructing Integrative molecular Bayesian networks (RIMBANET)	*Saccharomyces cerevisiae* (yeast)	Empirical data	DNA variation	—	—	18
			Gene expression	3,662 genes	—	
			Metabolite expression	56 metabolites	120 yeast segregants	
		Curated data	Protein-DNA binding	119 TFs	—	
			Protein–protein interaction	—	—	
			Metabolite–protein interaction	2,252 metabolite–protein interactions	—	
Data-driven supervised analyses						
Integrative Bayesian analysis of genomics (iBAG)	Glioblastoma multiforme	DNA methylation		6,890 probes	201	21
		Gene expression		7,785 genes		
Classification and regression algorithms: Concatenation-based classifiers	Breast cancer	Copy number variation		24,174 variants	77	23
		Gene expression		16,525 genes		
		Protein expression		12,553 proteins		
		Phosphoprotein expression		32,939 phosphoproteins		

(*Continued*)

TABLE 9.5.1 (*Continued*) Summary of studies describing methods for multi-omics data integration covered in this chapter

METHOD	*DISEASE*	*DATA SETS*		*NO. OF VARIABLES*	*NO. OF SAMPLES*	*REFERENCE*
Classification and regression algorithms: Ensemble-based classifiers	Acute kidney rejection	Gene expression		27,306 genes	32	25
		Proteomics		147 protein groups		
Sparse generalized canonical correlation analysis (SGCCA)	Pediatric high-grade gliomas	Copy number variants		1,229 variants	53	32
		Gene expression		15,702 genes		
Supervised multi-block sparse matrix analysis (SMSMA)	Alzheimer's disease	Genotypes		549,709 SNPs	100	35
		Imaging data		2,122,945 features		
Knowledge-based supervised analyses						
Network smoothed T-statistic support vector machine (stSVM)	Breast, ovarian prostate	Empirical data	Gene expression	—	79–228	36
			miRNA expression			
		Curated data	Protein–protein interaction	610,185 gene–gene interactions		
			Pathway data	17,518 gene–gene interactions		
			miRNA–gene interactions	—		
Generalized elastic net (GELnet)	Breast cancer	Empirical data	Gene expression	9,984 genes	54	38
		Curated data	Pathway data	—		
Adaptive group-regularized ridge regression (GRridge)	Cervical cancer	Empirical data	DNA methylation data	40,000 probes	37	39
		Curated data	CpG genomic location	6 probe locations		

The chapter is organized as follows: Section "METHODS FOR UNSUPERVISED MULTI-OMICS DATA INTEGRATION" covers methods for unsupervised analyses, and is further divided into data-driven methods section "Data-Driven Methods for Unsupervised Multi-omics Data Integration" and knowledge-based methods section "Knowledge-Based Methods for Unsupervised Multi-omics Data Integration". Section "METHODS FOR SUPERVISED MULTI-OMICS DATA INTEGRATION" discusses methods for supervised analyses, including data-driven Section "Data-Driven Methods for Supervised Multi-omics Data Integration" and knowledge-based section "Knowledge-Based Methods for Supervised Multi-omics Data Integration" methods. We conclude in section "SUMMARY AND CONCLUDING REMARKS" and address remaining knowledge gaps, limitations and areas for further development.

METHODS FOR UNSUPERVISED MULTI-OMICS DATA INTEGRATION

In this section, we describe methods that integrate multiple high-dimensional omics data sets without using additional phenotypic information on the biological samples.

Data-Driven Methods for Unsupervised Multi-omics Data Integration

Joint nonnegative matrix factorization

Joint non-negative matrix factorization (joint NMF) was proposed to identify subsets of correlated variables across different biological layers within all or a subset of samples.[7] Joint NMF extends NMF by projecting multiple data sets onto the same coordinate axes. Zhang et al.[7] applied Joint NMF to mRNA expression (GE), microRNA expression (ME) and DNA methylation (DM) data from 385 ovarian cancer samples, in order to identify correlated multidimensional (md-) modules. On average, each md-module consisted of 239.6 mRNAs, 13.8 miRNAs and 162.3 methylation markers. Any given md-module was defined as functionally homogeneous if it was enriched in at least one GO biological process category, with a q-value < 0.05. Among the 200 md-modules, 80 GE, 12.5 ME and 62.7% DM were identified as functionally homogeneous. Combining the GE, ME and DM dimensions resulted in 93% of the md-modules being functionally homogenous. Further, a significant proportion of genes were adjacent to DNA methylation markers in the same md-modules as well as genes targeted by miRNAs. Stratifying patients with/without strong association with particular md-modules indicated significant differences with respect to median survival times.

Joint NMF is a useful method to uncover coherent features across multi-omics data while reducing the complexity of the data because it models the global structure of the different data-types. Md-modules consist of regulatory relationships that span different biological layers but an arbitrary threshold is required to determine/define md-module membership. To avoid such ad hoc thresholds, would require the incorporation of sparsity constraints that result in variable selection.

Multiple co-inertia analysis

Similar to joint NMF, multiple co-inertia analysis (MCIA) is a dimension reduction technique that also projects several data sets (also referred to as blocks) onto the same dimensional space in order to improve biological insight. Meng et al.[8] applied MCIA to transcriptomics (from multiple platforms) and proteomics data from 59 cell lines from nine difference tissues as part of the NCI-60 cancer cell line project. Cancer lines from the same tissue of origin clustered closely across all data sets with a stronger similarity between the transcriptomic data sets than between the transcriptomic and proteomic data sets. The first principal component (dimension) separated cell lines based on epithelial and mesenchymal characteristics and the associated loadings (variable importance scores) consisted of larger weights (in absolute value) for epithelial and mesenchymal genes. Integrative analysis of the transcriptomics and proteomics data indicated greater statistical significance for the top identified pathways (e.g., *leukemia extravasation signaling pathway*) compared with the analysis of the transcriptomics data alone.

Because each dimension is associated with a loading vector where the weights for all variables are nonzero, variable selection using penalization constraints can improve biological interpretability. Overall MCIA is a useful method to unravel common trends across multi-omics data sets and may improve biological insights and help generate novel hypotheses.

Sparse multi-block partial squares

Transcriptional output (gene expression) is under strict regulatory control through many factors such as the genome (e.g., copy number variants), and epigenome (miRNA and DNA methylation). Multi-block partial square (MB-PLS) maximizes the covariance between multiple data sets and a response matrix of interest. Sparse Multi-block partial least squares (MB-PLS) (sMB-PLS) extends MB-PLS[9] by allowing for variable selection such that a smaller set of variables from each omic data set can be identified. sMB-PLS was used to identify sets of regulatory factors that affect the expression of a specific collection of mRNA transcripts.[10] Li et al.[10] used sMB-PLS to identify gene expression modules using copy number variants (CNV), DM, ME from 230 ovarian TCGA samples. Md-modules on average consisted of 30 samples, 45 CNVs, 42 DM marks, 5 miRNAs, and 44 genes. Modules obtained using conventional methods such as biclustering algorithms missed variables from one (59% of modules) or two (22% of modules) data sets. Concatenation of the data sets and then applying sPLS resulted in modules that missed at least one data set (47% of modules), where 17% of modules only consisted of variables from one data set. The md-modules consisted of variables with significantly greater functional homogeneity compared with md-modules identified using randomized data. sMB-PLS is a flexible method for the integration of multi-omics data sets as well as performing variable selection from each data set.

Similarity network fusion

Patient phenotyping is useful in identifying subgroups of individuals with similar characteristics in heterogeneous diseases such as cancer and other complex diseases. Similarity network fusion (SNF) was developed to integrate similarity matrices (also called networks) into a "fused" similarity matrix (based on message passing theory), which can then be used to perform sample clustering.[11] Wang et al.[11] applied SNF to mRNA, miRNA and DNA methylation data from five cancer-types from TCGA. Spectral clustering was applied to the patient similarity matrix of each data-type independently and to the patient-similarity network produced by SNF. Patient subgroups identified using SNF were more tightly correlated compared with clusters identified on each data-type separately. The use of local affinity may have led to a reduction in noise (correlations between patient clusters) in SNF compared with spectral clustering applied to each data-type separately. Half of the patient similarities were supported by two data-types, 17.2% by three data-types and one-third by only one data-type. The subtypes identified from the SNF analysis resulted in significantly different survival profiles as compared with the single data set cluster analyses. SNF also outperformed iCluster (a joint latent variable model for integrative clustering) on survival analysis and risk of death prediction using the METABRIC[12] data set.

SNF requires several hyperparameters (similarity index, scaling factors) that must be specified by the user, or tuned (which can be burdensome) and have an impact on the resulting clusters. A limitation of SNF is that the method uses all variables in each data-type, therefore, coherent patterns that exist between subsets of variables from each data-type are missed. SNF would benefit from variable selection and supervised analysis extensions.

Bayesian consensus clustering

Bayesian consensus clustering[13] (BCC) provides a statistical framework for integrative clustering of samples simultaneously using all omics data sets. BCC uses a mixture model to identify a data-specific clustering that is conditional on an overall clustering. Simulation studies were used to demonstrate the reduced error of BCC in identifying correct cluster membership, compared with separate and joint (concatenation of data sets) cluster analysis using a Dirichlet mixture model. The joint method performed well when there was a perfect agreement between data sets, whereas the separate method performed well when there was no agreement between data sets. BCC was also applied to breast cancer data and the identified clusters were different but not independent from those identified using the PAM50 classification (Her2, Basal, LumA and LumB). Most subjects in BCC cluster 1 corresponded to the Basal subtype, whereas BCC cluster 2 consisted of a unique set of LumA samples, which had lower copy number variants (as measured by the fraction of genome altered), and better survival times. This type of LumA cluster has been previously identified by other studies, on independent data sets.[12,14]

BCC is a useful method in identifying a consensus clustering of samples that is represented across the different omic data sets. BCC is able to handle model uncertainty and borrows information across multiple omic data sets. However, because BCC makes prior distributional assumptions, strong deviations from these distributions may result in a consensus clustering with little agreement between omic data sets. Nevertheless, to infer whether or not BCC could be applicable to a given set of multi-omics data, the adherence parameter can be used to determine the level of coherency between each data set-specific clustering and the overall consensus clustering.

Knowledge-Based Methods for Unsupervised Multi-omics Data Integration

The previous section focused primarily on data-driven methods that look for coherent patterns between multi-omics data. This section introduces unsupervised knowledge-based methods.

Sparse network-regularized multiple nonnegative matrix factorization

MicroRNAs play an important regulatory role in mRNA translation by binding to the 3' untranslated regions of their targets. The functional roles of miRNAs may be better understood by incorporating curated interactions with empirical data from expression studies. Sparse network-regularized multiple nonnegative matrix factorization (SNMNMF) is a computational framework for identifying co-modules (sets of miRNAs and mRNAs) by incorporating miRNA and mRNA expression data, as well as PPI and DNA–protein interaction data.[15] Zhang et al.[15] applied SNMNMF to mRNA and miRNA expression form 385 ovarian samples from TCGA. 49 miRNA–gene co-modules were identified, each consisting on average of 3.8 miRNAs and 78 mRNAs per module. The anticorrelation between miRNAs and mRNAs were statistically significant in 69.4% of the modules as compared with randomized miRNA–gene co-modules. 11 of the modules were significantly enriched with at least one miRNA cluster (miRNAs located within 50 kb in the genome). Significantly higher numbers of enriched GO biological processes were identified using the 49 co-modules compared with random modules. Stratifying patients based on co-module activity resulted in three groups that often differed with respect to their survival characteristics.

SNMNMF can functionally annotate groups of miRNA:mRNA pairs, and is a useful hypothesis-generating tool for further mechanistic work using *in vitro* or *in vivo* studies. The method may be extended to other types of data such as drug interaction data and copy number variants. However, the combination of additional sources of data will require the use of additional penalties that may increase the computational cost of tuning additional hyperparameters.

Passing attributes between networks for data assimilation

Passing attributes between networks for data assimilation (PANDA) is a message-passing algorithm that incorporates a cooperativity network (e.g., protein–protein interaction data), coregulatory network (e.g., co-expression network using gene expression) and a regulatory network (e.g., motif data comprising of transcription factor [TF] and gene interactions) in order to determine a consensus regulatory network. Glass et al.[16] applied PANDA to gene expression data in yeast, TF motif (sequence) data and protein–protein interaction data from BIOGRID. The jack knife procedure was used to remove 10% of edges in the cooperativity, regulatory and coregulatory networks in the initial and final networks identified using PANDA, and compared with gold-standard networks in order to compute the performance based on the area under the receiver operating characteristic curve (AUROC). The gold standard for these networks was based on interaction experiments to generate cooperativity (e.g., co-fractionation, colocalization) networks, ChIP-chip data for regulatory networks, and co-targeted in Chip-chip (if both gene pairs had a binding site associated with a particular transcription factor) for coregulatory networks. The resulting PANDA networks were significantly more predictive of all network types as determined using experimental data as compared with the initial networks. Furthermore, PANDA was competitive with other network reconstruction approaches in predicting regulatory networks using regulator knockout, cell-cycle and stress-response data sets.

Significant improvements have recently been made to PANDA for computational efficiency.[17] However, further exploration is needed in order to determine the stability of these networks. Although PANDA was more predictive in estimating networks compared with motif data alone, increases in the AUROCs were modest (between 3% and 8%). Further extensions of PANDA for the identification of differential regulatory networks may provide additional utility in studies with multiple phenotypic groups.

Reconstructing integrative molecular Bayesian networks

Bayesian networks are directed graphs, where the nodes represent random variables (e.g., omic features) and edges represent conditional probabilities between nodes and their parents (also called probabilistic causal networks). Thousands of networks are generated using Monte Carlo simulations (optimized using the Bayesian Information Criterion) and combined into a consensus network (reconstructing integrative molecular Bayesian networks [RIMBANET]). Zhu, Sova and Xu et al.[18] used six data-types to construct probabilistic causal networks. 16 of 59 quantified metabolites had significant log odds (FDR <0.05) and 12 of 16 metQTLs overlapped with four previously identified eQTL hot spot regions.[19] Gene transcripts and metabolites linked to the first two eQTL hot spots recapitulated known biological processes (leucine and pyrimidine biosynthesis pathways). This Bayesian network approach was useful in attributing functional roles to an uncharacterized eQTL hot spot 3, by linking the metabolites isoleucine, threonine and valine as well as the gene *CHA1*, which encodes a serine/threonine deaminase. However, no causal regulator was identified for this hot spot, which may suggest that the genetic variation affected protein levels directly instead of changes in the mRNA levels. Protein-coding variants in genes in this hot spot were identified and tested experimentally by knocking out each candidate gene and comparing its metabolite levels with those in wild-type yeast strains. Knockout of the vacuolar transport regulatory gene *VPS9* resulted in changes to threonine, isoleucine, valine and serine concentrations, suggesting its role as a potential causal regulator of the eQTL hot spot 3.

Bayesian networks have proven useful in accurately predicting complex cell regulatory processes that have been validated experimentally. The use of both high throughput data as well as known interactions from biological databases has proven useful in recapitulating complex cell behavior. However, as these processes are most likely disease-specific, such methods should be applied independently to data from each disease condition.

METHODS FOR SUPERVISED MULTI-OMICS DATA INTEGRATION

In the previous section, the focus was primarily on maximizing the coherence across multi-omics data, regardless of sample groupings. The methods described in this section incorporate phenotypic information in order to identify discriminatory patterns between sample groupings. For example, biological networks across multiple biological compartments that exist in healthy subjects may become dysregulated in cancer subjects.[20] These patterns may result in the identification of biomarker signatures that are predictive of phenotypic traits, either in continuous (e.g., age) or categorical forms (e.g., cancer vs. controls).

Data-Driven Methods for Supervised Multi-omics Data Integration

Integrative Bayesian analysis of genomics

This method uses a two-component hierarchical model to first determine the effects of methylation and other components (e.g., copy number variants) on gene expression and then jointly uses both components to predict a continuous outcome such as survival time. Wang et al.[21] used integrative Bayesian analysis of genomics (iBAG) to integrate gene and methylation data of glioblastoma multiforme (GBM) cancer samples. iBAG was compared with non-integrative (non-INT methods) where a model was constructed using only the gene expression data, and an additive model (ADD), where both gene and methylation data were combined into one joint explanatory matrix in order to predict patient survival times. The C-index (generalization of the AUROC) of the iBAG model was higher than both the non-INT and ADD models, for both training and test data sets, albeit with overlapping 95% confidence intervals. iBAG identified 22 novel genes associated with survival, not previously associated with GBM. The iBAG framework offers the flexibility of incorporating additional sources of information such as pathway information, and additional data-types.

Classification and regression algorithms

Concatenation-based classifiers

A simple approach for multi-omics classification panel is to combine data matrices of different omics into one joint matrix. Classification algorithms such as penalized regression, support vector machines (SVM), and random forest can then be applied to identify a set of molecular features that best predict a given outcome of interest. Although useful, combining data sets of different data-types poses a challenge. For example, different scales of various data-types lead to differences in the relative effect sizes, such that some data-types may over power the signal from other data-types.[22] A recent study using multi-omics data sets (copy number variation, gene expression, and proteomics) from breast cancer tumors demonstrated that a classifier developed using only the proteomics data set outperformed classifiers developed using other omics data sets as well as multi-omics (data-fusion) methods such as concatenation, multiple kernel learning methods, and random forest.[23]

Ensemble classifiers

A predictive model (classifier) is constructed independently for each omic data set and is combined (ensembled) using various classification rules, such as average or majority vote schemes.[6] Ensemble classifiers often have lower error rates compared with single base classifiers. This is possible if the error rate of the single classifiers is less than 50% and if the single classifiers are independent with uncorrelated errors.[24] Gunther et al.[25] developed proteogenomic classifiers that could discriminate acute rejection from non-rejection samples in a kidney transplant study. The ensemble classifiers out-performed the individual classifiers based on the area under the receiver operating characteristic curve using both the average and majority vote aggregation methods. Model stacking is another method to combine model predictions of multiple classifiers using additional classifiers resulting in meta-classifiers.[26]

Multi-block methods and extensions to generalized canonical correlation analysis

The presence of multiple data sets observed on the same set of individuals leads to the natural use of methods for the analysis of multi-block data.[27–31] Sparse generalized canonical correlation analysis (sGCCA) combines the power of multi-block data analysis with well-defined criteria to optimize and the flexibility of Projection to Latent Structures, which incorporates *a priori* relationships between data sets.[30] sGCCA is a dimension reduction approach that maximizes the sum of the pairwise covariances of connected (specified by the user) data sets. Tenenhaus et al.[32] applied sGCCA to genomic data (gene expression, copy number variants and a qualitative variable encoding the location of the tumors; central nuclei or brain stem) from 53 tumor samples of children with pediatric high-grade gliomas. The components generated by sGCCA were used to build a predictive model using Bayesian discriminant analysis. The performance of this approach was comparable to existing approaches such as penalized LDA (linear discriminant analysis),[33] supervised CCA[34] and regularized GCCA.[30]

Supervised multi-block sparse matrix analysis (SMSMA) was recently proposed,[35] and can be viewed as a supervised extension of sMB-PLS in Section 2.1. SMSMA was applied to SNP data, imaging data (magnetic resonance imaging, and positron emission tomography) and a continuous outcome variable measuring dementia in patients with Alzheimer's disease (AD). The components of the SNP and imaging data were used in a multiple logistic regression model to predict AD and healthy controls. Using a 10-fold cross-validation, the AUROC was 0.762 and 0.952 for the SNP and imaging data respectively. Many SNPs with the largest contributions to the principal components were found near previously associated AD risk genes such as *BIN1*, *APOE*, *CCR2*, *LOC651924*, and *IL13*. Therefore, SMSMA is a useful method to obtain a subset of multi-omic features that can discriminate between phenotypic groups. Methods for multi-block data analysis and extensions to generalized CCA offer an extremely flexible methodology for different study designs and are becoming commonly used for the analysis of multi-omics data.

Knowledge-Based Methods for Supervised Multi-omics Data Integration

We describe methods that incorporate experimental data (e.g., gene expression) and data from curated databases such as gene–gene, or miRNA–gene interactions. They may also be augmented using concatenation or ensemble based approaches for the integration of multiple omic data sets.

Network smoothed T-Statistic support vector machines

Biomarker panels (combination of single omic features) can be developed using the top-ranked features based on some univariate ranking procedure such as a *t*-test, or ANOVA. Adjusting this ranking based on biological connectivity between variables may capture the complexity of biological networks that exist in heterogeneous systems. The network smoothed *t*-statistic support vector machines (stSVM) method combines univariate ranking based test statistics with a network graph and selects the top variables from the resulting modified ranking to develop a SVM classifier. stSVM[36] was applied to gene expression data for several cancer types (breast, ovarian, and prostate) in order to predict survival times (dichotomized into 2 classes) and on average outperformed sgSVM (SVM trained using significant genes, false discovery rate (FDR) <5%), aepSVM (average gene expression of KEGG pathways), pathway activity classification (PAC), reweighted recursive feature elimination (RRFE) and netRank (modification of Google's PageRank to rank genes based on differential expression and network connectivity). The features selected using stSVM were consistently selected in the cross-validation folds due to their connectivity in the protein–protein network. Enrichment analysis of the selected features corroborated existing cancer related pathways (e.g., *Pathways in cancer, Prostate Cancer, ERBB signaling*).

Although stSVM had the top consensus ranking across the four analyzed data sets, its cross-validated performance was well in the range of the other methods. Further, feature stability could be due to the strong influence of the networks used, which may overpower the univariate ranking such that the same features were repeatedly selected. However, stSVM provides a biomarker panel that improved interpretability in terms of biological functionality as well as strong classification performance.

Generalized elastic net

Instead of univariate ranking of features prior to classifier development, variable selection can be induced through regularization of the loss function in linear models. Generalized elastic net (GELnet) extends the elastic net penalty[37] to regularize generalized linear models, through the use of penalties for individual features and pairs of features based on domain knowledge. Sokolov et al.[38] demonstrated that GELnet outperformed Elastic Net[37] with respect to the reconstruction error rate, when the same graphical Gaussian model was used to simulate both the gene expression and signaling network. GELnet was applied to mRNA expression data from 54 cancer cell lines as well as their sensitivity profiles to 74 compounds. GELnet achieved a lower root mean square error for 22 of the 74 drugs as compared with Elastic Net, which may suggest the incomplete knowledge of the underlying biological mechanisms that is captured in curated genetic pathway databases. Therefore, GELnet is limited by incomplete pathway information and may not always outperform Elastic Net if the network information is not associated with the expression data.

Adaptive group-regularized ridge regression

Variable selection can significantly improve the predictive performance and interpretability of classification algorithms. This can be further improved by incorporating additional information across groups of omic variables, which together have zero or nonzero coefficients. Adaptive group-regularized ridge regression (GRridge) uses an empirical Bayes method to estimate group-specific penalties based on grouping data (e.g., annotations). This procedure leads to a larger difference between small and large regression coefficients, which may be used to perform variable selection. Furthermore, only one global penalty is needed to be tuned using cross-validation. Van de Wiel et al.[39] applied GRridge to methylation data from 20 normal cervical tissue and 17 CIN3 (cervical intraepithelial neoplasia high-grade precursor lesions) tissue biopsies. Methylation probes were grouped based on their distance from CpG islands: CpG island (CpG), North Shore (NSe), South Shore (SSe), North Shelf (NSf) and South Shelf (SSf) and Distant (D). The estimates of group-specific penalties were large for all groups except for the CpG and SSe group, resulting in a GRridge model containing only CpG and SSe probes. GRridge (AUROC=0.92) outperformed ridge (AUROC=0.86), adaptive ridge (AUROC=0.84) and the group-lasso (AUROC=0.79) in discriminating CIN3 from normal samples. The advantage of tuning a lower number of hyperparameters offers GRridge computational advantages over the group lasso, as well as reduced bias because all variables are kept nonzero.

SUMMARY AND CONCLUDING REMARKS

In this book chapter, we described current approaches for the integration of multi-omics data, focusing on N-integration. We categorized these methods into *unsupervised* and *supervised* analyses, as well as methods that use experimental data only, and those that also incorporate curated data of known molecular relationships. Each of these methods are unique, answering different questions from the data and should be used when appropriate, and in conjunction with one another. Therefore, application of each method will often result in different results and conclusions.

Although the methods surveyed in this chapter are impressive in their ability to model the highly complex regulatory interplay of biological systems, some aspects of this complexity are still underserved. Integration of microbiome,[40] metagenome,[41] environmental exposures,[42] for example, are of particular interest. In addition, methods that can accommodate more complex study designs, such as repeated measures designs for longitudinal studies,[43] and cross-over repeated measure studies[44] would be highly desirable. Finally, because the end user of these algorithms are most likely biomedical researchers, visual aids for the complex outputs of these algorithms would be useful to enable model interpretation and assessment.[45] Last, these algorithms must be user-friendly and be made freely available through open source software.

Each method and general approach has limitations, both at the mathematical and computational level, and all rely on underlying assumptions as to the quality, accuracy and precision of the original data sets, as well as additional issues that plague the phenotyping of individuals in natural populations. Even so, in combination with thoughtful study design, well controlled standard operating protocols for both sample acquisition and processing, and an ability to work in the exploratory space offered by numerous and complementary computational approaches, the era of multi-omics data integration is becoming more established and has tremendous opportunity for increasing biological insights, hypothesis generation, and knowledge. The hypotheses generated through multi-omics data integration must be followed up with experimental studies in order to isolate true biological relationships from purely spurious results.

ACKNOWLEDGMENTS

We thank the biosignatures development team at the PROOF Centre of Excellence for continuous discussions and feedback on the contents of this book chapter.

REFERENCES

1. Gomez-Cabrero, D. et al. Data integration in the era of omics: Current and future challenges. *BMC Syst. Biol.* **8**, I1 (2014).
2. Tseng, G., Ghosh, D. & Zhou, X. J. *Integrating Omics Data.* (Cambridge, UK, University Press, 2015).
3. Rohart, F., Gautier, B., Singh, A. & Le Cao, K.-A. mixOmics: An R package for omics feature selection and multiple data integration. *PLoS Comput Biol.* 13, e1005752 (2017).
4. Weinstein, J. N. et al., The cancer genome atlas pan-cancer analysis project. *Nat Genet.* 45, 1113–1120 (2013).
5. Bishop, C. M. *Pattern Recognition and Machine Learning.* (New York, Springer, 2006).
6. Lin, E. & Lane, H.-Y. Machine learning and systems genomics approaches for multi-omics data. *Biomark. Res.* **5**, 2 (2017).
7. Zhang, S. et al. Discovery of multi-dimensional modules by integrative analysis of cancer genomic data. *Nucleic Acids Res.* **40**, 9379–9391 (2012).
8. Meng, C., Kuster, B., Culhane, A. C. & Gholami, A. M. A multivariate approach to the integration of multi-omics datasets. *BMC Bioinformatics* **15**, 162 (2014).
9. Kowalski, B. R. & Wangen, L. E. A multiblock partial least squares algorithm for investigating complex chemical systems. *J. Chemom.* **3**, 3–20 (1989).
10. Li, W., Zhang, S., Liu, C.-C. & Zhou, X. J. Identifying multi-layer gene regulatory modules from multi-dimensional genomic data. *Bioinformatics* **28**, 2458–2466 (2012).
11. Wang, B. et al. Similarity network fusion for aggregating data types on a genomic scale. *Nat. Methods* **11**, 333–337 (2014).
12. Curtis, C. et al. The genomic and transcriptomic architecture of 2,000 breast tumours reveals novel subgroups. *Nature* (2012). doi:10.1038/nature10983.
13. Lock, E. F. & Dunson, D. B. Bayesian consensus clustering. *Bioinformatics* **29**, 2610–2616 (2013).
14. Jonsson, G. et al. Genomic subtypes of breast cancer identified by array-comparative genomic hybridization display distinct molecular and clinical characteristics. *Breast Cancer Res.* **12**, R42 (2010).
15. Zhang, S., Li, Q., Liu, J. & Zhou, X. J. A novel computational framework for simultaneous integration of multiple types of genomic data to identify microRNA-gene regulatory modules. *Bioinformatics* **27**, i401–i409 (2011).
16. Glass, K., Huttenhower, C., Quackenbush, J. & Yuan, G.-C. Passing messages between biological networks to refine predicted interactions. *PLoS One* **8**, e64832 (2013).
17. van IJzendoorn, D. G. P., Glass, K., Quackenbush, J. & Kuijjer, M. L. PyPanda: A Python package for gene regulatory network reconstruction. *Bioinformatics* **32**, 3363–3365 (2016).
18. Zhu, J. et al. Stitching together multiple data dimensions reveals interacting metabolomic and transcriptomic networks that modulate cell regulation. *PLoS Biol.* **10**, e1001301 (2012).
19. Zhu, J. et al. Integrating large-scale functional genomic data to dissect the complexity of yeast regulatory networks. *Nat. Genet.* **40**, 854–861 (2008).
20. Ha, M. J., Baladandayuthapani, V. & Do, K.-A. DINGO: Differential network analysis in genomics. *Bioinformatics* **31**, 3413–3420 (2015).
21. Wang, W. et al. iBAG: integrative Bayesian analysis of high-dimensional multiplatform genomics data. *Bioinformatics* **29**, 149–159 (2013).
22. Aben, N., Vis, D. J., Michaut, M. & Wessels, L. F. A. TANDEM: A two-stage approach to maximize interpretability of drug response models based on multiple molecular data types. *Bioinformatics* **32**, i413–i420 (2016).
23. Ma, S., Ren, J. & Fenyö, D. Breast cancer prognostics using multi-omics data. *AMIA Summits Transl. Sci. Proc.* **2016**, 52 (2016).
24. Dietterich, T. G. Ensemble methods in machine learning. *In Multiple Classifier Systems* 1–15 (Berlin, Germany, Springer Verlag, 2000). https://link-springer-com.ezproxy.library.ubc.ca/chapter/10.1007/3-540-45014-9_1.
25. Günther, O. et al. A computational pipeline for the development of multi-marker bio-signature panels and ensemble classifiers. **13**, 326 (2012).
26. Whalen, S. & Pandey, G. A comparative analysis of ensemble classifiers: Case studies in genomics. in 807–816 (IEEE, 2013). doi:10.1109/ICDM.2013.21.
27. Lohmöller, J.-B. *Latent Variables Path Modeling with Partial Least Squares.* (Heidelberg, Germany, Physica-Verlag, 1989).
28. Wold, H. Partial least squares. *In Encyclopedia of Statistical Sciences* **6**, 581–591 (New York, Wiley, 1985).
29. Tenenhaus, M., Vinzi, V. E., Chatelin, Y.-M. & Lauro, C. PLS path modeling. *Comput. Stat. Data Anal.* **48**, 159–205 (2005).
30. Tenenhaus, A. & Tenenhaus, M. Regularized generalized canonical correlation analysis. *Psychometrika* **76**, 257–284 (2011).
31. Singh, A., Shannon, C. P., Gautier, B., Rohart, F., Vacher, M., Tebbutt, S. J. & Le Cao, K.-A. DIABLO: An integrative approach for identifying key molecular drivers from multi-omic assays. *Bioinformatics.* In Press (2019).
32. Tenenhaus, A. et al. Variable selection for generalized canonical correlation analysis. *Biostatistics* **15**, 569–583 (2014).
33. Witten, D. M. & Tibshirani, R. Penalized classification using Fisher's linear discriminant. *J. R. Stat. Soc. Ser. B Stat. Methodol.* **73**, 753–772 (2011).
34. Witten, D. M., Tibshirani, R. & Hastie, T. A penalized matrix decomposition, with applications to sparse principal components and canonical correlation analysis. *Biostatistics* **10**, 515–534 (2009).
35. Kawaguchi, A. & Yamashita, F. Supervised multiblock sparse multivariable analysis with application to multimodal brain imaging genetics. *Biostatistics* **00**, 1–15 (2017).
36. Cun, Y. & Fröhlich, H. Network and data integration for biomarker signature discovery via network smoothed t-statistics. *PLoS One* **8**, e73074 (2013).
37. Zou, H. & Hastie, T. Regularization and variable selection via the elastic net. *J. R. Stat. Soc. Ser. B Stat. Methodol.* **67**, 301–320 (2005).

38. Sokolov, A., Carlin, D. E., Paull, E. O., Baertsch, R. & Stuart, J. M. Pathway-based genomics prediction using generalized elastic net. *PLoS Comput Biol.* **12**, e1004790 (2016).
39. van de Wiel, M. A., Lien, T. G., Verlaat, W., van Wieringen, W. N. & Wilting, S. M. Better prediction by use of co-data: Adaptive group-regularized ridge regression. *Stat. Med.* **35**, 368–381 (2016).
40. Lê Cao, K.-A. et al. MixMC: A multivariate statistical framework to gain insight into microbial communities. *PLoS One* **11**, e0160169 (2016).
41. Heintz-Buschart, A. et al. Integrated multi-omics of the human gut microbiome in a case study of familial type 1 diabetes. *Nat. Microbiol.* **2**, 16180 (2016).
42. Robinson, O. et al. The pregnancy exposome: Multiple environmental exposures in the INMA-Sabadell birth cohort. *Environ. Sci. Technol.* **49**, 10632–10641 (2015).
43. Straube, J., Gorse, A.-D., PROOF Centre of Excellence Team, Huang, B. E. & Lê Cao, K.-A. A linear mixed model spline framework for analysing time course 'omics' data. *PLoS One* **10**, e0134540 (2015).
44. Liquet, B., Lê Cao, K.-A., Hocini, H. & Thiébaut, R. A novel approach for biomarker selection and the integration of repeated measures experiments from two assays. *BMC Bioinformatics* **13**, 325 (2012).
45. González, I., Lê Cao, K.-A., Davis, M. J. & Déjean, S. Visualising associations between paired 'omics' data sets. *BioData Min.* **5**, 1–23 (2012).
46. Baladandayuthapani, V. iBAG page. Available at: http://odin.mdacc.tmc.edu/~vbaladan/Veera_Home_Page/iBAG_page.html. (Accessed: August 14 2017).
47. Zhu, J. RIMBANET for Bayesian network reconstruction. Available at: http://research.mssm.edu/integrative-network-biology/RIMBANET/RIMBANET_overview.html. (Accessed: August 14 2017).

Can Health Care Be Made More Affordable by the Use of Biomarkers?

9.6

Reinier L. Sluiter, Wietske Kievit, and Gert Jan van der Wilt

Contents

New knowledge and innovations are rapidly evolving in health care, causing considerable pressure on health care systems' budgets (Sorenson et al. 2008). Decision makers face the challenge of making wise decisions regarding the implementation of these technologies while controlling health care expenditures (Banta 1994). In recent years, we have witnessed a rapid development in biomarkers, which hold the potential of more personalized health care. This raises the question whether the introduction of biomarkers in health care will further contribute to a rise in health care costs, or whether they could be used to make more efficient use of available treatment options. To address this issue, we need to understand how the use of biomarkers may alter the health care trajectories of patients and how this affects patients' outcomes and the associated resource utilization. Whether this renders health care more affordable not only depends on the biomarkers themselves, but also on the context in which they are being used.

COST-EFFECTIVENESS OF BIOMARKERS

A first step in such an inquiry is to explore the extra costs that are associated with the use of a biomarker, the impact on patients' health, and any extra downstream costs or savings (Weinstein and Stason 1977). Use of a biomarker can affect treatment decisions, altering a patient's health prospects and use of health care resources. In theory, a more personalized approach could improve quality of life and save health care resources for instance by preventing adverse drug reactions (ADRs) or avoiding futile treatment. In such cases, the benefits for some patients could outweigh the extra costs of screening all patients. This is expressed in a cost:effectiveness ratio (ICER), representing the extra money that needs to be spent in order to gain one extra unit of effect, for instance a quality-adjusted life year (QALY), when compared with an alternative strategy. To distinguish between what is cost-effective or not, a willingness to pay threshold (WTP) is used, representing the maximum amount of money that is considered acceptable to gain one extra unit of effect. If the cost-effectiveness ratio is below this WTP, the intervention is considered cost-effective. Cost-effectiveness analyses can be performed alongside a clinical trial or conducted using a modeling approach (Drummond 2005).

A lot of research has been conducted to estimate the added value of using various biomarkers compared with current practice. Examples are from cardiac surgery (Koffijberg et al. 2013), coronary artery bypass graft surgery (Henriksson et al. 2010), Alzheimer disease (Lee et al. 2017), non-small cell lung cancer (NSCLC) (Helwick 2012, Romanus et al. 2015), breast and ovarian cancer (Eccleston et al. 2017), cystic fibrosis (van der Ploeg et al. 2015), colorectal screening (Ladabaum et al. 2013) and viral infections (Lubell et al. 2016). These studies looked at the added value of biomarkers by better diagnosing diseases, prioritizing patients for a specific health intervention, or preventing adverse events. Most of these studies concluded that the use of biomarkers was a cost-effectiveness approach compared with current practice, given a specific WTP threshold (Henriksson et al. 2010, Koffijberg et al. 2013, van der Ploeg et al. 2015, Lubell et al. 2016, Eccleston et al. 2017). However, in some cases this conclusion was strongly dependent on a number of model parameters, such as prevalence of the disease. Examples where biomarkers were not found to be cost-effective were related to NSCLC. In these cases, the low biomarker frequency in the target population appeared to be the main reason for this (Helwick 2012, Romanus et al. 2015). However, publication bias might lead to an overly optimistic view of the cost-effectiveness of biomarkers. This concern has been raised for biomarkers in the field of cardiovascular disease, dementia, peripheral depression, and cost-effectiveness studies in general (Bell et al. 2006, Hemingway et al. 2010, Tzoulaki et al. 2013, Wilson et al. 2015, Carvalho et al. 2016). However, these studies also show what factors may be important in determining the cost-effectiveness of biomarkers. In the following, two case studies are presented from the field of pharmacogenetics to explore this issue further.

Key Parameters in Cost-Effectiveness Analysis of Pharmacogenetics

A number of parameters have been reported to be of critical importance on the cost-effectiveness pharmacogenetics (Veenstra et al. 2000, Phillips and Van Bebber 2004, Phillips et al. 2004, Williams et al. 2006). In this paragraph we will highlight these parameters and describe the basic principles how these could affect the cost-effectiveness of pharmacogenetics:

- Prevalence of the genetic mutation

  A higher prevalence of a specific genetic mutation in the general or patient population, the more impact pharmacogenetic testing could have. If more patients have the mutation, the more patients could potentially benefit from the intervention associated with pharmacogenetic testing.
- Prevalence of ADRs

  If an ADR occurs frequently due to treatment with a certain drug, more effect can potentially be gained when optimizing treatment with pharmacogenetic testing. If, however, an ADR is very rare, the potential impact of pharmacogenetic testing is low. Only a few patients would then benefit from it, whereas other patients are screened unnecessarily for a genetic variant.
- Severity of the ADRs

  The more severe an ADR is that can be prevented with an intervention after genotyping, the more impact it can have on the outcomes of genotyping. For instance, by preventing ADRs the quality of life of patients could increase and could also have an influence on the overall health care cost, because costs related to the ADR (i.e., hospital admissions, diagnostic test) and loss of quality of life are prevented.
- Genotype–phenotype association (penetrance)

  If the relationship between genotype and phenotype is weak, testing and adjusting treatment accordingly have little consequences for a patient, because their genetic profile does not results in an expected phenotype. Therefore treatment adjustments were unnecessary and could even result in losses in quality of life or treatment effect due to under treatment.
- Test accuracy of the genetic test

  The test accuracy can be expressed as sensitivity, specificity, positive predictive value and negative predictive value. The positive predictive value is the proportion of patients with a genetic variant that do indeed get for instance an ADR while treated without adjustments. The higher this proportion, the more patients will benefit for the genetic test and the resulting treatment adjustments. The sensitivity is the proportion of patients with an ADR and a genetic variant out of all patients with an ADR. If this proportion is high, it means that only a few patients do get an ADR although they do not have the genetic variant; in other words you will find many of those patients who could benefit the most. If the false negative rate is high, this will result in "undertreatment" because patients still experience side effects. If the test finds patients that do not need treatment adjustments (false positives), then this could lead to "overtreatment" while also resulting in less treatment effect of the pharmacogenetic test.
- Costs of genetic test

  The higher the costs for screening patients, the more costs have to be earned back by preventing ADRs or increasing treatment effect. These same occurs vice versa; how cheaper screening would become the easier these costs are compensated by preventing ADRs or increasing treatment effect.
- The effect and costs of the alternative treatment

  Testing positive on a genetic test will induce an alternative treatment, dose adjustment or another treatment, for those patients with a genetic variant. The higher the costs directly related to this alternative treatment or due to its length, the more money has to be "earned back" by preventing ADRs or gaining more treatment effect. Furthermore, the alternative treatment should really be a better option in terms of effectiveness and toxicity; otherwise, you will gain not enough to outweigh the extra costs of the genetic test.

Illustrative Cases to See the Influence of Key Parameters on the Cost-Effectiveness

The impact of different key parameters described above on the cost-effectiveness of pharmacogenetics will be illustrated by two examples of pharmacogenetic testing; namely (1) *factor V Leiden* in combination with oral contraceptives, and (2) *HLA-B*5701* and antiretroviral therapy of abacavir.

1. *Factor V Leiden* in combination with oral contraceptives

   Factor V Leiden (fVL) mutation leads to a higher chance of venous thromboembolism (VTE), including deep vein thrombosis (DVT) and pulmonary embolism (PE). Another risk factor for VTE is the use of oral contraceptives (OC). Knowledge on the fVL gene activity requires genotyping for this variant allele of all potential OC users. The question arises if the costs of screening outweigh the benefits of preventing VTE, which can be achieved by treatment optimization.

   An earlier cost-effectiveness study by Creinin et al. at testing for fVL before prescribing oral contraceptives concluded that it was not a cost-effectiveness strategy, mainly because of the prevalence of fVL and VTE, and the price of screening (Creinin et al. 1999). Their estimated risk of a VTE in women with fVL using OC was 140 in 100,000. With a prevalence of fVL of almost 5% the ICER was estimated to be $4.8 million per year of life saved when screening all 20-year-old women. When the prevalence would rise to 99%, this ratio would become $404,000 per year of life saved. This shows that even with unrealistic prevalence estimates, testing every 20-year-old woman remains not cost-effective. When the price of screening would be discounted to 34.5% of the current prices, the ICER would still be $1.7 million per life year saved. Screening for fVL can only become cost-effective when genetic testing is almost free of charge.

   This study illustrates that in the general population screening for fVL in order to prevent a first event of VTE is not a very interesting strategy, in terms of cost-effectiveness (Creinin et al. 1999). The reason for this is the low prevalence estimates of fVL and VTE. The prevalence of these genotypes is low being 0%–1% for homozygous fVL variant alleles, and about 3% being heterozygous for an fVL variant allele in the white population. The baseline risk of VTE is in the general population approximately 1:10,000, so for patients with fVL and using OC this is still just 35:10,000. Due to these low prevalences too much patients are screened unnecessary to save a life year by preventing VTE. Even at unrealistic genetic test prices or prevalence rates this strategy would not become cost-effective. However, in high-risk groups, like patients with a history of VTE, screening might become more interesting, due to higher prevalences of fVL and VTE.
2. *HLA-B*5701* and antiretroviral therapy of abacavir

   Abacavir is an antiretroviral therapy often used in the treatment of HIV. One of the adverse reactions of abacavir is a hypersensitivity reaction (HSR) occurring within 6 weeks of treatment, and which in some cases can even be lethal. Different studies showed, however, that these reactions are strongly associated with the HLA-B*5701 allele, which is present in about 5%–8% of the white population. About 50% of the patients that have this allele develop hypersensitive reactions. These ADRs can be prevented by given several other treatment options instead of abacavir.

   A cost-effectiveness analysis by Hughes et al. concluded that pretreatment screening for HLA-B*5701 can be cost-effective depending on the willingness to pay for avoiding a HSR (Hughes et al. 2004). However, the level of cost-effectiveness strongly depended on the treatment costs after genotyping, in other words the alternative treatment patients received instead of abacavir. The cheaper the alternative, the less money have to be spent to avoid a HSR compared with no screening ranging from ICERs of "dominant" (lower costs and more effect gained with testing) for the cheapest alternative up to $36,276 per HSR avoided for the most expensive alternative. Furthermore, test sensitivity and costs associated with HSRs were important parameters. When test sensitivity was higher (80%) compared with the base-case the ICER dropped to $3,941 per HSR avoided compared with no testing, and when costs associated with HSRs was higher ($5,717) the ICER dropped to $5,332. Vice versa, when test sensitivity (40%) and costs associated with HSRs ($572) were lower, ICERs increased respectively up to $12,348 and $11,364 per HSR avoided compared with no testing. As the prevalences of the HLA-B*5701 allele are relatively high (5%–8%), the positive predictive value of the test is high (at least 50% of the patients with a positive test develops a HSR), the costs associated with HSRs could be high $13,249 Hughes et al. [2004]; $32,017 Schackman et al. [2008]), and the consequences of these reactions could even be lethal, screening for this allele preemptive in the treatment with abacavir is already obligatory.

AFFORDABILITY OF BIOMARKERS

Although most biomarkers could be cost-effective, they need not necessarily make health care more affordable. This follows from the fact that interventions can be cost-effective, but still have a considerable budget impact (Gafni and Birch 1993, Sendi and Briggs 2001). In order to become more affordable, it is necessary that when implementing a new innovation like biomarkers, existing health care intervention(s) be disinvested. This disinvestment should then liberate sufficient resources to fund the new innovation, while not resulting in less health benefits (Gafni and Birch 1993). So in case of biomarkers to let health care become more affordable it should first be cost-effective to show the added value compared with standard care. Subsequently, when implementation is considered we must be willing to disinvest current health care programs in order to control health care expenditures (Daniels et al. 2013).

EARLY HEALTH TECHNOLOGY ASSESSMENT (HTA) AND BIOMARKERS

Could we already have predicted in case of fVL that it would not be a cost-effective and affordable approach, although evidence was not completely available? By getting an idea if new strategies concerning personalized medicine are potentially cost effective, we could target the strategies that are most promising instead of researching and developing a biomarker for every disease or drug. Because personalized medicine, including pharmacogenetics, is a relatively new approach in health care, many of the implications are not clear. For example, for pharmacogenetics there is a relatively low amount of scientific evidence concerning the effectiveness of preemptive screening (Alagoz et al. 2016).

Early HTA could offer a solution by saying something about the potential impact of a new innovation or strategy already at the beginning of the development process, so even before considering implementation (where most cost-effectiveness analysis are performed). If the potential impact is small, pharmacogenetics is probably not a very interesting strategy to develop as this will not improve current practice that much, and vice versa. Early HTA also gives an indication about how well the current care is performing, and if there is any room for improvement at all ("effectiveness gap"). Knowing information about the potential impact could support the development of pharmacogenetics (which gene-drug interactions could be interesting to implement?) and also price ceiling (at which price a pharmacogenetic test could be potentially cost-effective?). This could eventually justify the costs of clinical trials and other related research activities in the field of pharmacogenetics. If there is no room for improvement by pharmacogenetics, research activities should preferably be terminated.

In order to show the idea of early HTA, we evaluate the case of HLA-B*5701 and abacavir again with the evidence that was available at the time pharmacogenetic testing was considered.

In the HLA*B5701 case we saw that in about 8% of the patients the genotype was present. Therefore, if all these patients would develop a hypersensitivity reaction, in a cohort of 100 patients you could prevent normally eight cases of hypersensitivity. The median costs associated with mild HSRs were estimated at $2,986, ranging from $572 to $13,558. For fatal reactions costs of $32,017 are mentioned. So preventing 8 cases would result in case of mild hypersensitivity in savings of about $24,013, ranging from $4,574 to $108,628. In case of fatal reactions this could even result in savings of $256,134, when all cases are prevented. So a pharmacogenetic test with 100% sensitivity and specificity could costs around $240 (ranging from $46 to $1,086) when we take the median costs of a mild HSR into account, and even more when all the reactions have a fatal outcome. This seems a very realistic price as the average costs of most pharmacogenetic test lies around the $172.

Subsequently, we know that the sensitivity of the HLA*B5701 screening is about 50%. This means that with screening, of the eight patients potentially developing a HSR by using abacavir in a cohort of 100 patients, half of them could be prevented by a genetic test and change in medication. The specificity of the test is almost 100%, which means that in patients with a HSR, a variant is not present. When we again take into account the median costs of a mild HSR, $11,942 (ranging from $2,287 to $54,232) are the maximum additional costs at which HLA*B5701 screening will be cost-effective compared with current practice. The price of the pharmacogenetic test should therefore not be more then $119 (ranging from $23 to $542) to be cost-effective. However, as we know that the costs of the HSR are much higher with increasing severity or fatal ending, the price of the test can even be much higher. So the average price of $172 per pharmacogenetic test is probably very easily earned back in the case of HLA-B*5701 and abacavir.

These examples shows that with little evidence about the important key parameter, such as prevalences of the allele and ADRs, the specificity and sensitivity of the genetic test, the cost associated with the test and ADRs, this could already tell you something about the potential impact of pharmacogenetic testing. It is important to mention that at the time more evidence is gathered, cost-effectiveness analysis should be performed in order to say something about the "real" impact.

FUTURE DIRECTIONS

The lessons learned from the examples of pharmacogenetics can be translated to various kinds of biomarkers and personalized medicine. Imagine that a new biomarker is discovered that is a predictor for a certain health state. The question then is, is this biomarker (and related interventions) interesting to implement in current practice because it makes health care more affordable? These kinds of evaluations could already be done at a very early stage of the research and development (R&D) process of specific biomarkers, by using methods related to early HTA. These evaluations give a first indication about the potential impact of a new biomarker. For example, the biomarker is probably not becoming a cost-effective approach when evidence surrounding the key parameters is very unfavorable, such as very low prevalences of the biomarker or the disease/ADR it could potentially prevents. This was also the case in the cost-effectiveness studies concerning Alzheimer disease, NSCLC and venous thromboembolism, resulting in not or uncertain cost-effective strategies (Creinin et al. 1999, Helwick 2012, Romanus et al. 2015, Lee et al. 2017). Knowing this information about unfavorable evidence should already mean that these specific biomarkers are not interesting to further investigate, develop and implement in this context. However, for potential interesting biomarkers you can predict the effectiveness gap that is left behind by current treatments. Depending on the effectiveness gap, productivity costs/future research costs can be compensated by gaining more effect and therefore reducing costs compared with current practice. This evaluation shows if it worthwhile spent more time and money in this specific biomarker or that R&D should focus on other areas of biomarkers. For example in areas like BRCA screening in women with ovarian cancer (Eccleston et al. 2017) or screening for the HLA*B5701 gene (Hughes et al. 2004) instead of NSCLC (Helwick 2012, Romanus et al. 2015) or screening for fVL (Creinin et al. 1999). After the decision to develop the intervention, the next step is to gather evidence surrounding at least the most important parameters that determine the cost-effectiveness. This evidence is used to estimate if the new intervention is really going to be cost-effective compared with current practice. In case of missing evidence, threshold analysis could provide information how effective the biomarker should be to become cost-effective. Finally, a decision should be made if the biomarker is interesting enough to implement in current practice, such as in case of the biomarkers for cardiac surgery or cystic fibrosis. Therefore we must decide if we are willing to pay this amount of money to gain health benefits, and which current health care interventions are disinvested to control health expenditures.

It is important to mention that all these evaluations are only useful when we are willing to make choices about terminating R&D processes of certain biomarkers. This, however, raises the question whether it is possible to stop or steer all new innovations. Some innovation processes are started years ago, for example in drug development. Consequently, companies are maybe not able to terminate their R&D processes as a lot of money was already invested. In addition, over time new discoveries concerning the innovation, like new target populations, could make an innovation from unfavorable to favorable in economic terms. All these considerations need to be kept in mind when performing and interpreting results of early HTA evaluations. Therefore these evaluations should preferably start at the beginning of the R&D process and be part of the decision-making process. This will help to allocate resources to more attractive projects, for example when pharmaceutical companies are deciding about which drugs to develop further and which to terminate timely (Grabowski 1997, Miller 2005, Hartz and John 2008). In addition, termination of projects is probably easier when doing these evaluations from the beginning of the R&D process, as investments are low at that stage. This should lead to researching and developing only the most interesting biomarkers, and eventually resulting in a (more) affordable health care.

REFERENCES

Alagoz, O., D. Durham and K. Kasirajan (2016). "Cost-effectiveness of one-time genetic testing to minimize lifetime adverse drug reactions." *Pharmacogenomics J* **16**(2): 129–136.

Banta, H. D. (1994). "Health care technology as a policy issue." *Health Policy* **30**(1–3): 1–21.

Bell, C. M., D. R. Urbach, J. G. Ray, A. Bayoumi, A. B. Rosen, D. Greenberg and P. J. Neumann (2006). "Bias in published cost effectiveness studies: Systematic review." *BMJ* **332**(7543): 699–703.

Carvalho, A. F., C. A. Kohler, A. R. Brunoni, K. W. Miskowiak, N. Herrmann, K. L. Lanctot, T. N. Hyphantis, J. Quevedo, B. S. Fernandes and M. Berk (2016). "Bias in peripheral depression biomarkers." *Psychother Psychosom* **85**(2): 81–90.

Creinin, M. D., R. Lisman and R. C. Strickler (1999). "Screening for factor V Leiden mutation before prescribing combination oral contraceptives." *Fertil Steril* **72**(4): 646–651.

Daniels, T., I. Williams, S. Robinson and K. Spence (2013). "Tackling disinvestment in health care services: The views of resource allocators in the English NHS." *J Health Organ Manag* **27**(6): 762–780.

Drummond, M. (2005). *Methods for the Economic Evaluation of Health Care Programmes.* Oxford Medical Publications. 3rd ed. (New YorkUniversity Press): 379.

Eccleston, A., A. Bentley, M. Dyer, A. Strydom, W. Vereecken, A. George and N. Rahman (2017). "A cost-effectiveness evaluation of germline BRCA1 and BRCA2 testing in UK women with ovarian cancer." *Value Health* **20**(4): 567–576.

Gafni, A. and S. Birch (1993). "Guidelines for the adoption of new technologies: A prescription for uncontrolled growth in expenditures and how to avoid the problem." *CMAJ* **148**(6): 913–917.

Grabowski, H. (1997). "The effect of pharmacoeconomics on company research and development decisions." *Pharmacoeconomics* **11**(5): 389–397.

Hartz, S. and J. John (2008). "Contribution of economic evaluation to decision making in early phases of product development: A methodological and empirical review." *Int J Technol Assess Health Care* **24**(4): 465–472.

Helwick, C. (2012). "Is biomarker testing in NSCLC cost-effective?", 2017, from http://www.ahdbonline.com/issues/2012/august-2012-vol-5-no-5-special-issue-asco-2012-payers-perspective/1066-article-1066.

Hemingway, H., M. Henriksson, R. Chen, J. Damant, N. Fitzpatrick, K. Abrams, A. Hingorani et al. (2010). "The effectiveness and cost-effectiveness of biomarkers for the prioritisation of patients awaiting coronary revascularisation: A systematic review and decision model." *Health Technol Assess* **14**(9): 1–151, iii–iv.

Henriksson, M., S. Palmer, R. Chen, J. Damant, N. K. Fitzpatrick, K. Abrams, A. D. Hingorani et al. (2010). "Assessing the cost effectiveness of using prognostic biomarkers with decision models: Case study in prioritising patients waiting for coronary artery surgery." *BMJ* **340**: b5606.

Hughes, D. A., F. J. Vilar, C. C. Ward, A. Alfirevic, B. K. Park and M. Pirmohamed (2004). "Cost-effectiveness analysis of HLA B*5701 genotyping in preventing abacavir hypersensitivity." *Pharmacogenetics* **14**(6): 335–342.

Koffijberg, H., B. van Zaane and K. G. Moons (2013). "From accuracy to patient outcome and cost-effectiveness evaluations of diagnostic tests and biomarkers: An exemplary modelling study." *BMC Med Res Methodol* **13**: 12.

Ladabaum, U., J. Allen, M. Wandell and S. Ramsey (2013). "Colorectal cancer screening with blood-based biomarkers: Cost-effectiveness of methylated septin 9 DNA versus current strategies." *Cancer Epidemiol Biomarkers Prev* **22**(9): 1567–1576.

Lee, S. A., L. A. Sposato, V. Hachinski and L. E. Cipriano (2017). "Cost-effectiveness of cerebrospinal biomarkers for the diagnosis of Alzheimer's disease." *Alzheimers Res Ther* **9**(1): 18.

Lubell, Y., T. Althaus, S. D. Blacksell, D. H. Paris, M. Mayxay, W. Pan-Ngum, L. J. White, N. P. Day and P. N. Newton (2016). "Modelling the impact and cost-effectiveness of biomarker tests as compared with pathogen-specific diagnostics in the management of undifferentiated fever in remote tropical settings." *PLoS One* **11**(3): e0152420.

Miller, P. (2005). "Role of pharmacoeconomic analysis in R&D decision making: When, where, how?" *Pharmacoeconomics* **23**(1): 1–12.

Phillips, K. A. and S. L. Van Bebber (2004). "A systematic review of cost-effectiveness analyses of pharmacogenomic interventions." *Pharmacogenomics* **5**(8): 1139–1149.

Phillips, K. A., D. L. Veenstra, S. D. Ramsey, S. L. Van Bebber and J. Sakowski (2004). "Genetic testing and pharmacogenomics: Issues for determining the impact to healthcare delivery and costs." *Am J Manag Care* **10**(7 Pt 1): 425–432.

Romanus, D., S. Cardarella, D. Cutler, M. B. Landrum, N. I. Lindeman and G. S. Gazelle (2015). "Cost-effectiveness of multiplexed predictive biomarker screening in non-small-cell lung cancer." *J Thorac Oncol* **10**(4): 586–594.

Schackman, B. R., C. A. Scott, R. P. Walensky, E. Losina, K. A. Freedberg and P. E. Sax (2008). "The cost-effectiveness of HLA-B*5701 genetic screening to guide initial antiretroviral therapy for HIV." *AIDS* **22**(15): 2025–2033.

Sendi, P. P. and A. H. Briggs (2001). "Affordability and cost-effectiveness: Decision-making on the cost-effectiveness plane." *Health Econ* **10**(7): 675–680.

Sorenson, C., M. Drummond and P. Kanavos (2008). Ensuring value for money in health care: the role of health technology assessment in the European Union. Copenhagen, Denmark, S.I. European Observatory on Health Systems and Policies.

Tzoulaki, I., K. C. Siontis, E. Evangelou and J. P. Ioannidis (2013). "Bias in associations of emerging biomarkers with cardiovascular disease." *JAMA Intern Med* **173**(8): 664–671.

van der Ploeg, C. P., M. E. van den Akker-van Marle, A. M. Vernooij-van Langen, L. H. Elvers, J. J. Gille, P. H. Verkerk, J. E. Dankert-Roelse and C. S. Group (2015). "Cost-effectiveness of newborn screening for cystic fibrosis determined with real-life data." *J Cyst Fibros* **14**(2): 194–202.

Veenstra, D. L., M. K. Higashi and K. A. Phillips (2000). "Assessing the cost-effectiveness of pharmacogenomics." *AAPS PharmSci* **2**(3): E29.

Weinstein, M. C. and W. B. Stason (1977). "Foundations of cost-effectiveness analysis for health and medical practices." *N Engl J Med* **296**(13): 716–721.

Williams, S. A., D. E. Slavin, J. A. Wagner and C. J. Webster (2006). "A cost-effectiveness approach to the qualification and acceptance of biomarkers." *Nat Rev Drug Discov* **5**(11): 897–902.

Wilson, C., D. Kerr, A. Noel-Storr and T. J. Quinn (2015). "Associations with publication and assessing publication bias in dementia diagnostic test accuracy studies." *Int J Geriatr Psychiatry* **30**(12): 1250–1256.

Omics Driven Medicine for Democratizing Health Care Delivery and Health Economics

9.7

Michael A. Kiebish, Viatcheslav R. Akmaev, Vivek Vishnudas, Stephane Gesta, Rangaprasad Sarangarajan, and Niven R. Narain

Contents

A NEW ERA OF HEALTH CARE

The health care arena is rapidly evolving and incorporating novel technologies to drive clinical decision support and streamlining uncontrollable costs. Over the next decade, the architecture of health care will require exponential growth and adaption to meet the precision, economic, and outcome based demands of the twenty-first century. These transformative efforts will be heavily driven by both innovative analytics and clinical measurements, which will vastly improve efficacy of current and future interventions for population health. In the past 50 years, we have witnessed an explosive growth in the maturation of clinical technologies to quantitatively measure diverse biomolecules. During this time, there has been an expansion of informative markers for various disease states, including metabolic, oncologic, neurological, cardiovascular, inflammatory, and endocrine disorders (Quinones and Kaddurah-Daouk 2009; Galazis et al. 2012; Martins-de-Souza and Farias 2015; Teixeira et al. 2015; Patel and Ahmed 2015). These clinical assays have evolved from a handful of markers (e.g., high- and low-density lipoprotein cholesterol [HDL-C, LDL-C], creatine, blood urea nitrogen [BUN]) that represent various risk factors to panels of high throughput assays for routine screening. However, the multiplicity of factors that drive human health span far beyond a routine clinical test menu.

In 1953, with the discovery of DNA, we witnessed the emergence of a fundamental breakthrough in the understanding of the building blocks of human biology (Watson and Crick 1953). This revelation sparked a journey toward deconstructing the complexity of physiology as well as the vastness of potential combinations of biomolecules that could exist governing our physiology. The implementation of DNA technology and genomics into translational medicine took several decades to adopt and not until 2012 have the economics and throughput reached a point of broader utility (Simonds et al. 2013; Chrystoja and Diamandis 2014). With this, we have witnessed the launch of several precision medicine initiatives that range in the thousands of participants (Cancer genome project, Human Genome Diversity Project, Personal Genome Project), a hundred thousand participants (Genomics England), as well as a million participants (Precision Medicine Initiative [PMI]-USA) (Jones 2012; Genomes Project et al. 2012; Forbes et al. 2015; Collins and Varmus 2015; England 2018). These projects would have only been possible due to the unprecedented investment in technology and analytics to support these undertakings. Even with the escalating commitment to genomic sequencing as the masthead of precision medicine, genetics does not necessarily provide the full spectrum of potential clinical decision support. Indeed, genomics does provide the blueprint for potential stratification of populations of individuals in a binary manner; however, these can represent only probabilistic risk factors for disease or conditions. It will require the implementation of more adaptive technologies such as proteomics, lipidomics, and metabolomics as well as other quantitative features that will represent the causal quantitative measurements to the clinical phenotype in reengineering population health (Lindon et al. 2007; Khoury et al. 2012; Young 2014; Hawgood et al. 2015).

Proteomics, lipidomics, and metabolomics profiles are highly integrated with an individual's phenotype or current health status. These quantitative measurements are commonly referenced to a person's/patient's phenome due to this integration of clinical phenotypic measurements and broad omic capture of dynamic measurements. Unlike the genome, which sets the fingerprint for the entire body, a person's phenome is distinctly tailored and divergent across cells, tissues and biofluids, allowing for the immense capture of biologically and physiologically significant signatures (Figure 9.7.1).

There is a profound amplification effect and biological prominence of thousands of metabolomic and lipidomic measurements that respond to environmental influence (Kaddurah-Daouk et al. 2008). However, there are more than 10,000 proteins detectable in different tissues and exponentially more posttranslational modifications that are influenced, and to a lesser extent, gene expression and genome changes influenced by environmental impact. Thus, the commanding influence and impact of thousands' of metabolites/lipid species on human health govern the vast majority of clinically relevant signatures and outcomes. Furthermore, the dynamic range of these biomolecules in tissues (e.g., heart, liver, muscle, adipose), biofluids (e.g., blood, serum, plasma, urine, saliva, cerebrospinal fluid [CSF], sweat, tears), cells (e.g., lymphocytes, red and white blood cells, T cells), and particles (e.g., exosomes, microvesicles, secretory vesicles) is immense and can dramatically shift within seconds, minutes, or hours to adapt to physiological perturbants (Clayton et al. 2006; Kim et al. 2014; Wilhelm et al. 2014; Pan et al. 2015; Ivanisevic et al. 2015; Wu et al. 2016).

The physiological imprinting that leads to the susceptibility of disease is embedded in multiple physiologically relevant compartments. In parallel to vast genome projects, the emergence of large-scale and compartmentalized phenome projects will set the tone for inferring causality of molecular signatures with disease stratification (Robinson 2012; Zhang 2015). An individual's molecular system is vastly influenced by their geographical location, pharmacological history, lifestyle, metabolic status, inflammatory state, age, developmental stage, and environmental exposure. All of these factors harmonize a molecular code imprinted in human biology that can be deciphered for determining the stratification of comorbidities as well as inherent population variability. The true artistry and prominence of phenomic stratification is the return on investment on the overall power per analyte in assessing population diversity. Metabolomics, lipidomics, and proteomics measurements predominantly do not take into account race and ethnicity because this information is not the driver of the quantitative measurement, although preliminary reports have identified a few factors that do correlate with these designations (Ishikawa et al. 2013; Saito 2014; Ishikawa et al. 2014; Liu et al. 2015). Thus, instead of deconvoluting the 3 billion base pairs and 34 million variations that exist in the human genome, a more expansive amount of information can be captured based on the dynamic measurements of several thousand analytes in an economical and rapidly precise manner (Genomes Project et al. 2012). Additionally, these thousands of measurements can be more efficiently streamlined to connect to disease pathogenesis and will require less participation from the population because the millions and trillions of variants and

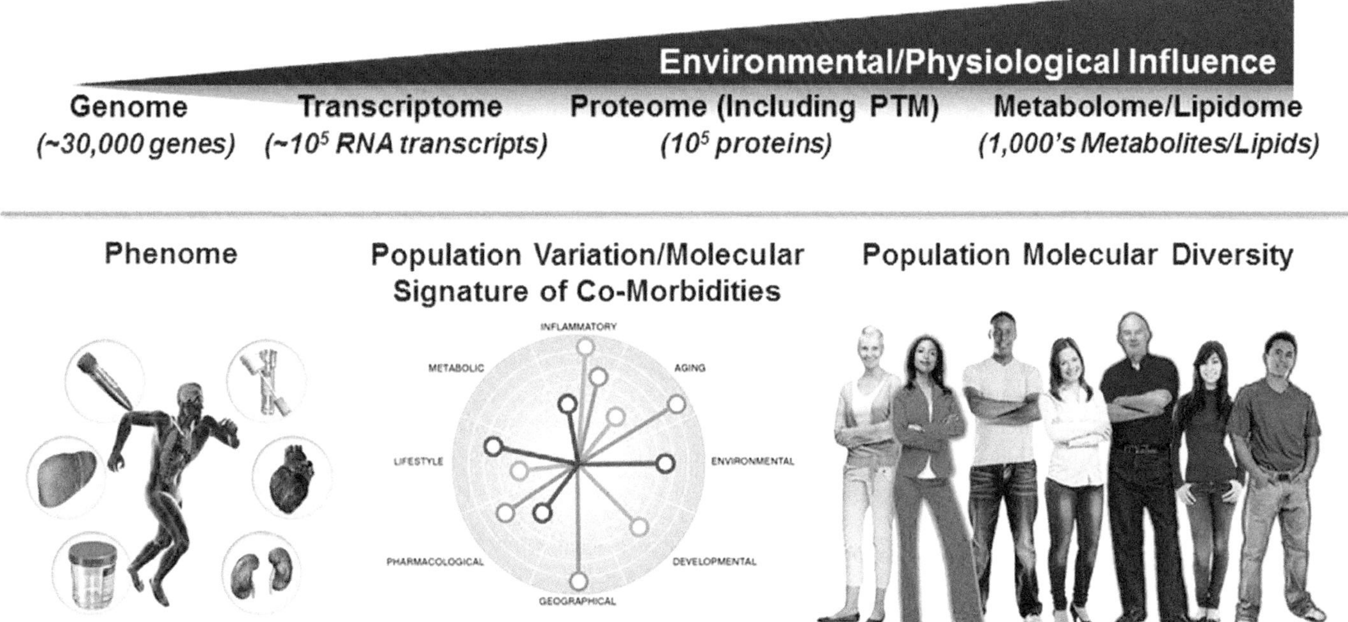

FIGURE 9.7.1 Environmental influence on multi-omic analysis and integration into population health. PTM, posttranslational modification.

sequences, respectively, will complicate the interpretation of the data requiring the need for an exponential increase in participants to infer causality.

EMERGENCE OF A RENAISSANCE IN OMICS MEDICINE

After the emergence of the human genome project, the utilization of genetic mutations or pharmacogenomics has held tremendous promise for the past 15 years. However, the population map of disease stratifying single nucleotide polymorphisms (SNPs) from genome-wide association studies (GWASs) or mutations has not extended for the wider population burdened with disease and multiple comorbidities leading to the Promised Land in health care. Recent analysis of AstraZenenca's pharmaceutical pipeline from 2005 to 2010 demonstrated that association of specific mutations with a particular target of interest during the drug development process was one of the greatest predictors of patient selection for phase IIb clinical trials (Cook et al. 2014). Cook et al illuminates the roadmap of drug development to follow the 5Rs:

- Right target
- Right patient
- Right tissue
- Right safety profile
- Right commercial potential

or on the market, the right patient should be the most important focus. This highlights a critical importance of genetics in some targeted therapy drug development. However, the challenges still exist for the broader population because, due to human heterogeneity and the vast majority of diseases that impact human health, a minuscule number of SNP will be predictive.

A recent commentary by Nicholas Schork in *Nature*, highlights the critical need for stratifying patients to match pharmaceutical intervention not only at the clinical trial stage, but also for agents currently on the market. As referenced, the current landscape is mired in "imprecision medicine" for the top highest grossing pharmaceuticals for schizophrenia, heartburn, arthritis, high cholesterol, depression, asthma, psoriasis, Crohn's disease, multiple sclerosis, and neutropenia. The top 10 drugs for these indications only demonstrated efficacy between 1/3 and 1/24 patients, thus at best 33% efficacy (Schork 2015). In any other industry, this would have been marked as an extreme failure. On the heels of *N*-of-1 trials, precision medicine initiatives, and impending outcome/value based reimbursement models potentially taking suit in 2018, there is a critical need to establish a guided patient selection framework to drive patient care. At the present time, genetic testing might not be a viable option for all indications because a more expansive assessment of the 34 million plus variants will not be powered to handle the 1,500 drugs approved by the FDA in comparison with human diversity and combinatorial efficacy.

The technological revolution that ensued after the sequencing of the human genome spawned several advances that dramatically changed the economics of whole genome sequencing. As society witnessed the costs associated with genome sequencing go from millions to thousands and eventually hundreds of dollars, the recognition of the potential utility of large data capture at a reasonable economic price started to become a reality. The current dogma is that even though the cost has decreased, the time required to perform whole genome sequencing for clinical adoption is not a reality as of yet. Additionally, to exist as a point of care diagnostic, an assay has to be performed within a short timeframe (Chrystoja and Diamandis 2014). At the present time, it has taken several years to accumulate a few thousand genome sequences to reference against and the translational value has not been put into practice for the vast majority of disease indications. All of these factors have a massive effect on the proliferation of genomics as a valuable predictive diagnostic tool. Notably, there will have to be a transition to migrate toward implementation; however, current diagnostics are based on a handful of decades-old biochemical measurements that govern clinical decision-making, thus requiring broader acceptance of large volumes of diagnostic input.

The reality is that adaptive omics platforms have technologically evolved in parallel to the genomics advancements, yet equal attention has not been dedicated to these platforms. The ability to quantify thousands of metabolites, lipids, and proteins within minutes of instrument time has placed metabolomics, lipidomics, and proteomics on the forefront of potential utility for patient care. Each of these omics technologies have witnessed several advancements that have allowed costs to decrease and throughput to increase dramatically. All of these events have had equal or greater impact on the utility of these omics for precision medicine. For proteomics, the streamlining of ultrahigh-resolution, MS^N analysis, automated top-14 depletion workflows, both improvements to two-dimensional (2D) and 1D nano liquid chromatography (nanoLC) and microLC workflows, advancements to data independent acquisition analysis, as well as improved understanding of isobaric multiplexing analysis have all dramatically decreased the cost per sample, increased depth of analysis and the quality of the data (Tu et al. 2010; Wang and Bennett 2013; Zubarev and Makarov 2013; Werner et al. 2014; Liu et al. 2015; Zhang et al. 2015; Kumar et al. 2015; Filip et al. 2015) (Figure 9.7.2). For lipidomics analysis, incorporation of ion mobility, high-resolution MS/MS workflows, increase in detector dynamic range and duty cycles, nano/microLC workflow robustness, MS/MS^{ALL} platforms, as well as direct infusion platforms have all exponentially increased the depth of analysis, decreased the cost per sample, and increased the reproducibility of analysis (Ekroos 2008; Schwudke et al. 2011; Simons et al. 2012; Han et al. 2012; Ekroos 2012; Ståhlman et al. 2012; Paglia et al. 2015; Surma et al. 2015). For metabolomics analysis, improvements with hybrid mass spectrometry workflows, improved accurate mass databases, ion source improvements, streamlined chromatography matched with a database, as well as polarity switching approaches have increased the efficiency, quality, confidence,

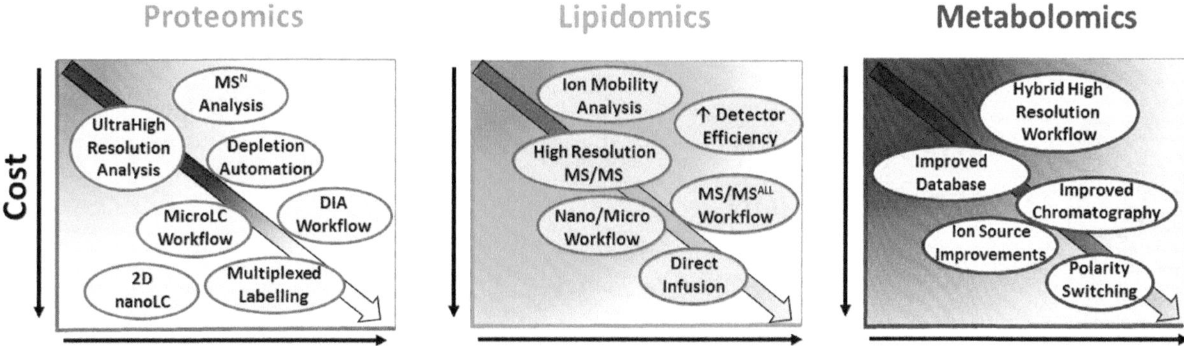

FIGURE 9.7.2 Technical economic advantages in proteomics, lipidomics, and metabolomics analysis for implementation for omic medicine. LC, liquid chromatography; MS, mass spectrometry; DIA, data independent acquisition.

and accuracy of metabolomics analysis (Members et al. 2007; Yuan et al. 2012; Gowda et al. 2014; Liu et al. 2014; Cheema 2015; Fuhrer and Zamboni 2015) (Figure 9.7.2).

Without these advancements, the cost of implementation of these technologies would not justify their potential use in patient care. Further, we are seeing the throughput dramatically increase from 25 samples per day to 200 samples per day on a mass spectrometer, which dramatically changes the economics of cost per sample (Simons et al. 2012; Surma et al. 2015; Fuhrer and Zamboni 2015). Thus, the expansion of clinical patient sample analysis from several informative biomolecules to several thousand in an equal and cost effective time frame is provocative and necessary. Additionally, these omics can provide multidimensional insight and quantitative information in diverse sample types (e.g., urine, saliva, CSF), which reach far beyond genomics capture and engage actionable insight into a patients physiological state. Since each omics analysis is migrating toward hundred dollars per sample and with improvements over the coming years, the economic cost will only decrease as investments are made.

BIOLOGY BASED PHARMACEUTICALS—PUTTING PATIENT BIOLOGY AT THE FOREFRONT

The current dogma in drug development is defining a potential drug target, undergoing high throughput screening, trying to balance toxicity versus efficacy, and attempting to achieve the greatest efficacy in extremely diverse populations. The "red herring" in this dogma is that patient biology and diversity is not at the crux of the equation. The present scenario pushes the limits of Hippocratic morality, socioeconomic doctrine, and sustainable economic modeling. Currently, researchers explore thousands to millions of compounds from chemical libraries for potential "hits." This process can take 3–5 years and cost up to 50 to 150 million dollars. After a compound is discovered, it can take 600–800 million dollars to run through a clinical trial depending on the indication. Ironically, only 1 of every 100 compounds will make it through the process to receive FDA review and monitoring. Thus, the average cost of drug approval can total $2.6 billion and take 12 to 14 years and, even then, the right population of patients is still not defined to demonstrate maximal efficacy (Mullard 2014). The fact is that technology does exist to accelerate trials and after trails stratify patients, as well as decrease the overall cost for drug development.

As mentioned previously, Cook et al highlighted the most prominent positive indicator of clinical trials was stratification of patients (Cook et al. 2014). Granted, in monogenic diseases or diseases with high frequency mutations, this can follow a genomic based stratification. However, in complex diseases with multiple dimensions of comorbidities, this will require a more patient-centric and phenomic strategy (Clayton et al. 2006; Lindon et al. 2007; Khoury et al. 2012; Robinson 2012; Sharma 2013; Zhang 2015; Kamal 2015). A molecular adaptive strategy will have to be incorporated in multiple ways (Cook et al. 2014; Matsui 2014; Simon 2014; Redig and Janne 2015). First, molecular profiling of disease populations will have to take place. Second, multi-omic assessment of phase I clinical trials should examine multiple biofluids and tissues associate with disease progression in Pharmacodynamics (PD) studies during ongoing pharmacokinetic (PK) assessment. Utilizing information gained from defining potential responders to dose escalation studies in phase I, as well as monitoring molecular signatures in phase II and implementing patient stratification in phase III, this overall strategy will rapidly accelerate efficacy assessment as well as dosing profiles for patients (Figure 9.7.3).

This approach also allows for benchmark assessment of the pharmaceuticals in investigation and after investigation potential combination therapies, as well as the current standard of care. Overall, this strategy sets the criteria for future clinical trials in this arena as well as establishes the landscape for pharmacoeconomic advancement and maturity.

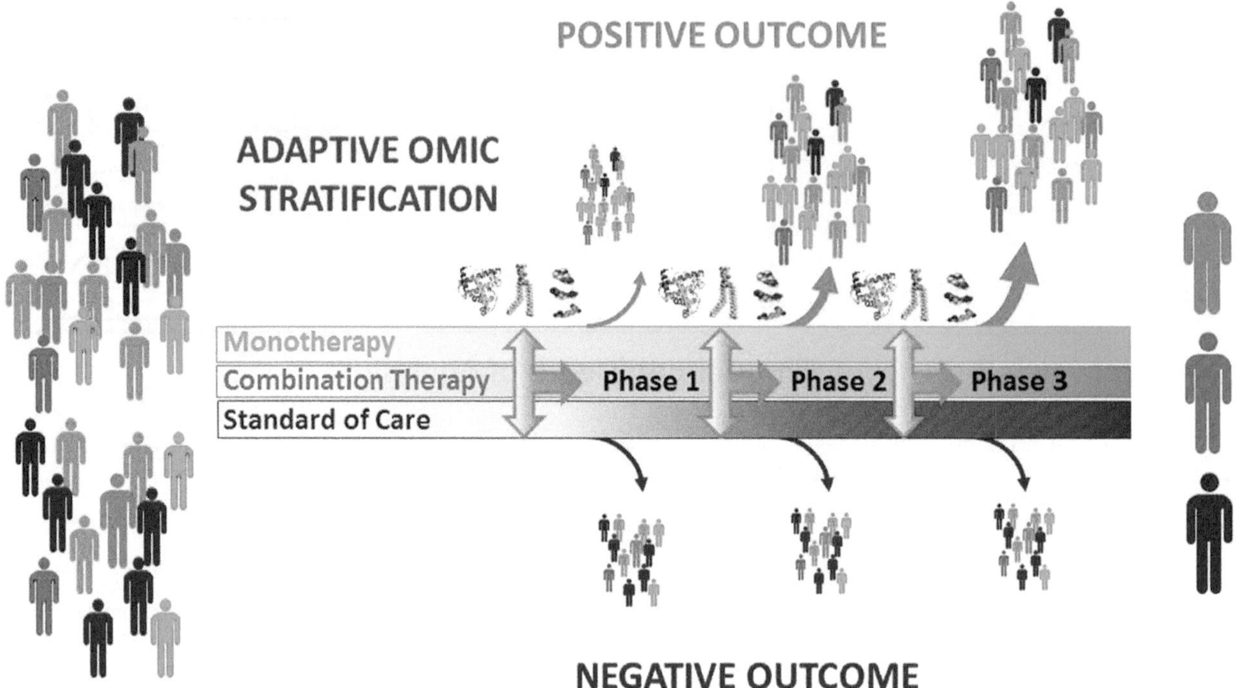

FIGURE 9.7.3 Molecular adaptive omic strategies for clinical trials.

Molecular adaptive clinical development sets the framework for broader population metrics. By employing these phenomic strategies during clinical trial development as well as in critical unmet need areas that exist in the oncology, endocrinology, metabolic, hematologic, inflammatory, as well as neurological space, a more global profile can be obtained for defining molecule signatures of efficacy as well as of avoidance of adverse events. In actuality, economic tallying of the cost for enacting outcome based reimbursement, health disparities, social costs of adverse medical events as well as overall economic impact equals more than 3 trillion dollars annually for the U.S. health care system (Ernst and Grizzle 2001; LaVeist 2011; Goodman et al. 2011; Latkovic 2013). Additionally, the United States spends 572 billion dollars more than any other developed country, stark evidence of opportunities to streamline costs and efficacy of its health care system (Latkovic 2013). Any mechanism that could establish harmonization and balance with patient care, whereas decreasing overall spending would have a massive impact on global economic development and the democratization of health care. With the impending patent cliff existing within the pharmaceutical industry in 2017, the establishment of more refined patient stratification demographics allows for further economic opportunities within the industry while benefiting the patient (Kakkar 2015). Further, it should be emphasized that proteomic, lipidomic, and metabolomic profiles are not dominated by ethnic background as in genomics because a person's genetic composition is a historical framework more than an effective course of action. Thus, these omic technologies are bound by ethical and privacy issues using genomics alone. By stratifying patients based on simplistic biochemical orientation of carbon, nitrogen, hydrogen, phosphorus and oxygen molecules through the heterogeneity of protein, lipids and metabolites beyond the binary orientation of DNA molecules, you essentially deidentify a person's race or sex to allow their current physiologic status to predominate. However, the goal is to deconstruct population variability and assign molecular signatures of response, efficacy, dosing, as well as established safety profiles based on the treatment options that currently exist (Khoury et al. 2012)[16] (Figure 9.7.4). This approach will deidentify a person's genetic associate and instead will link the person's current physiologic status to the most likely intervention with greater success in outcome.

The use of omics technologies should not only be considered for patient stratification, but drug development as well. The acceptance of using proteomic, lipidomic, and metabolomics technologies to identify biological targets for pharmaceutical development has emerged for various disease indications. Thus, through the engagement of diverse patient populations own biology, their clinical adaption to different disease states and immunological response to infections, a wide range of therapeutic agents, antibiotics, and biologics can be developed (Dehairs et al. 2015; Peng et al. 2015; Wu et al. 2016). This approach provides a rapid clinical development roadmap because the major hurdle of toxicity is augmented due to the intervention being biologic in nature.

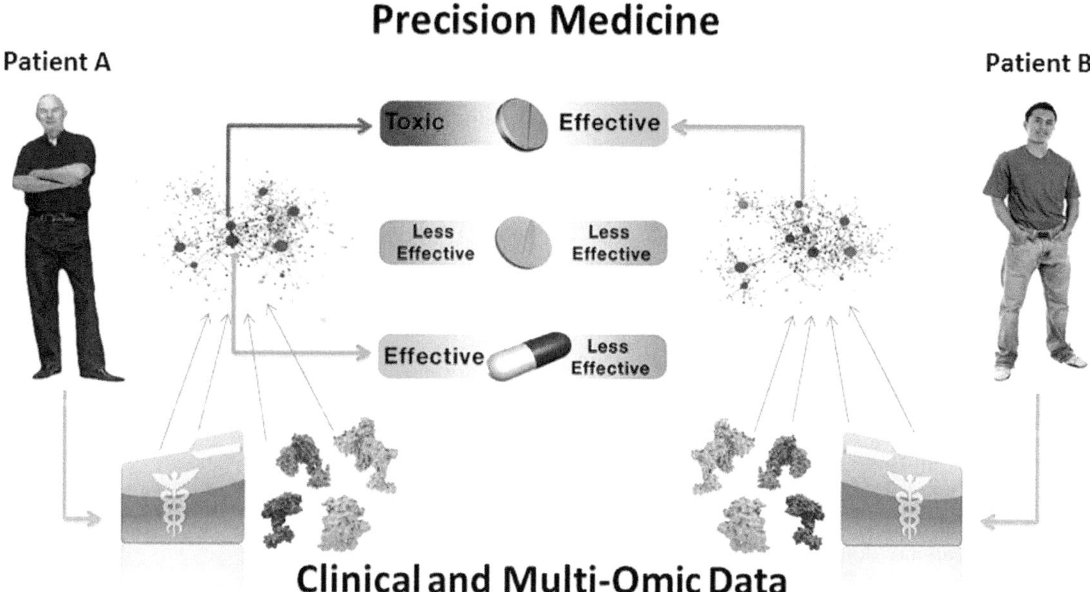

FIGURE 9.7.4 Stratification of patient response for precision medicine and population health.

ESTABLISHING A NEW SOCIAL CONTRACT FOR OMICS MEDICINE AND HEALTH CARE

Upon establishment of the human genome project, the promise of a revolution in new pharmaceutical agents and streamlining of patient care has fallen short of the goal post (Duffy 2016). The current status is that very few drugs have been developed based on our genomic understanding of diseases and even fewer genetic diagnostic tests are utilized to treat and follow disease onset and progression. With proteomics, metabolomics, and lipidomics technologies rapidly evolving into patient care and drug development, the establishment of a new social contract for omics and precision medicine is necessary to allow its broader implementation (Desmond-Hellmann 2012). This will require the broad scale education of the patients, physicians, policy makers, and the public as a whole (Mirnezami et al. 2012). Due to the inherent nature of phenomics being ethnically ungrounded, the opportunity to embrace this approach with underserved populations that utilize the vast majority of health care allows the broader population to feel committed to their own outcome (Laveist 1993; Lillie-Blanton and Laveist 1996; Wells 2004). Because the vast number of clinical trials do not include underserved populations, yet pharmaceuticals will be targeted toward this population, the empowerment of these emerging patient stratification approaches is immense. Additionally, with the goal of these omics technologies to be high throughput and affordable, the adoption will be even greater. The establishment of an omic medicine social contract will create a trust with an individual's own biology as well as their economic well-being. With the public being ever more concerned about the rising cost of health care, they are actively seeking technologies to internally understand their own risk assessment to avoid impending hardship due to health care costs. Thus, the public is now demanding technological advancements and direct-to-consumer options to be proactive in navigating health on their terms.

An expansion of the social contract should recognize that patients must not be fit into parameters or demographics that do not match their current biology or physiology and it is their inherent right to engage their own health for their individual betterment (Hill 1996). There are countless disease indications where race, age, sex, or comorbidities do not dictate the treatment of action and the nonoptimal designation of these parameters have an enormous economic cost and potential lack of efficacy. The treatment of chronic kidney disease, which influences greater than 19 million patients in the United States alone, with a complex disease regimen with multiple comorbidities and greater effect on the African American population is a prime example (Robinson 2006). With total health care spending in excess of 100 billion dollars a year, as well as the complicated array of treatment regiments with mixed efficacy, the need to demonstrate that a patient's biology or omic profile should impart guidance on their treatment regimen is essential (Collins et al. 2010; Jha et al. 2013). Thus, a revolution must occur to employ unbiased quantitative tools to stratify patients based on biology toward the optimal treatment options.

SUMMARY

The embracement of adaptive omics technologies for placing a patient's biology at the forefront for selection of viable and efficacious treatment options requires reengineering of the social health care contract, education of the populous and health care providers, as well as an overall acceptance to empower the patient and allow the physician to truly "deliver" care rather than be put into a position where one "dictates" care. Further, these omics technologies have to be at a stage of maturity, where the proof points for their utility in patient stratification for treatment options exist and the economics of the use of these technologies are at a commercial stage where testing is in the order of a few hundred dollars for all omics per sample. At the consumer level, this economic metric and basepoint will empower patients, entice insurance companies, and encourage physicians to de-risk their clinical guidance with the use of the patients own biology. On the heels of countless precision medicine initiatives throughout the world, every citizen, patient, and physician should question which emerging technology will provide the most accurate guidance for clinical decision-making. The bounty of precision medicine will only be recognized when patients are deidentified by the functional signatures in their biology and not corralled into a "one test fits all" paradigm. To venerate the practice of population health and reconstruction of the current health care climate, we need to further invest in adaptive omics technologies to bring them into mainstream clinical testing while ensuring they are conforming to robust economic stringencies to support their adoption. In the era of connectivity, we should focus on the future of health care being connected with a patient's biology and streamlining sound economic principles to adopt these technologies into the mainstream to adopt Hippocratic policies for the betterment of people. In conclusion, the intersect of technological advancement, complete narrative of patient biology, and the social engineering necessary within the key stakeholders within government, payers, hospitals, pharmaceutical companies, and physicians will drive an era where "patient-centric" medicine is real and sustainable. Twenty-first century medicine has the ability to be dynamic, cost-conscious, and inclusive in delivery of care. As an industry, we have a moral responsibility to develop technologies that will have overarching effects driven by global access and dissemination.

REFERENCES

Cheema, A & Chauthe, S 2015. Quantitative metabolomics. *In:* Hock, F. (ed.) *Drug Discovery and Evaluation: Pharmacological Assays*. Berlin, Germany: Springer.

Chrystoja, CC & Diamandis, EP 2014. Whole genome sequencing as a diagnostic test: Challenges and opportunities. *Clin Chem*, 60, 724–733.

Clayton, TA, Lindon, JC, Cloarec, O, Antti, H, Charuel, C, Hanton, G, Provost, JP et al. 2006. Pharmaco-metabonomic phenotyping and personalized drug treatment. *Nature*, 440, 1073–1077.

Collins, AJ, Foley, RN, Herzog, C, Chavers, BM, Gilbertson, D, Ishani, A, Kasiske, BL et al. 2010. Excerpts from the US renal data system 2009 annual data report. *Am J Kidney Dis*, 55, S1-420, A6–A7.

Collins, FS & Varmus, H 2015. A new initiative on precision medicine. *N Engl J Med*, 372, 793–795.

Cook, D, Brown, D, Alexander, R, March, R, Morgan, P, Satterthwaite, G & Pangalos, MN 2014. Lessons learned from the fate of AstraZeneca's drug pipeline: A five-dimensional framework. *Nat Rev Drug Discov*, 13, 419–431.

Dehairs, J, Derua, R, Rueda-Rincon, N & Swinnen, JV 2015. Lipidomics in drug development. *Drug Discov Today Technol*, 13, 33–38.

Desmond-Hellmann, S 2012. Toward precision medicine: A new social contract? *Sci Transl Med*, 4, 129ed3.

Duffy, DJ 2016. Problems, challenges and promises: Perspectives on precision medicine. *Brief Bioinform*, 17, 494–504.

Ekroos, K 2008. Applications of lipidomics in pharmaceutical R&D. *Eur J Pharm Sci*, 34, S22–S23.

Ekroos, K 2012. Lipidomics perspective: From molecular lipidomics to validated clinical diagnostics. *In*: Ekroos, K (ed.) *Lipidomics: Technologies and Applications*. Weinheim, Germany: Wiley-VCH Verlag GmbH & Co. KGaA.

England, G. 2018. 100,000 Genomes Project. Available: http://www.genomicsengland.co.uk/.

Ernst, FR & Grizzle, AJ 2001. Drug-related morbidity and mortality: Updating the cost-of-illness model. *J Am Pharm Assoc (Wash)*, 41, 192–199.

Filip, S, Vougas, K, Zoidakis, J, Latosinska, A, Mullen, W, Spasovski, G, Mischak, H, Vlahou, A & Jankowski, J 2015. Comparison of depletion strategies for the enrichment of low-abundance proteins in urine. *PLoS One*, 10, e0133773.

Forbes, SA, Beare, D, Gunasekaran, P, Leung, K, Bindal, N, Boutselakis, H, Ding, M et al. 2015. COSMIC: Exploring the world's knowledge of somatic mutations in human cancer. *Nucleic Acids Res*, 43, D805–D811.

Fuhrer, T & Zamboni, N 2015. High-throughput discovery metabolomics. *Curr Opin Biotechnol*, 31, 73–78.

Galazis, N, Iacovou, C, Haoula, Z & Atiomo, W 2012. Metabolomic biomarkers of impaired glucose tolerance and type 2 diabetes mellitus with a potential for risk stratification in women with polycystic ovary syndrome. *Eur J Obstet Gynecol Reprod Biol*, 160, 121–130.

Genomes Project, C, Abecasis, GR, Auton, A, Brooks, LD, Depristo, MA, Durbin, RM, Handsaker, RE, Kang, HM, Marth, GT & Mcvean, GA 2012. An integrated map of genetic variation from 1,092 human genomes. *Nature*, 491, 56–65.

Goodman, JC, Villarreal, P & Jones, B 2011. The social cost of adverse medical events, and what we can do about it. *Health Aff (Millwood)*, 30, 590–595.

Gowda, H, Ivanisevic, J, Johnson, CH, Kurczy, ME, Benton, HP, Rinehart, D, Nguyen, T et al. 2014. Interactive XCMS online: Simplifying advanced metabolomic data processing and subsequent statistical analyses. *Anal Chem*, 86, 6931–6939.

Han, X, Yang, K & Gross, RW 2012. Multi-dimensional mass spectrometry-based shotgun lipidomics and novel strategies for lipidomic analyses. *Mass Spectrom Rev*, 31, 134–178.

Hawgood, S, Hook-Barnard, IG, O'brien, TC & Yamamoto, KR 2015. Precision medicine: Beyond the inflection point. *Sci Transl Med*, 7, 300ps17.

Hill, TP 1996. Health care: A social contract in transition. *Soc Sci Med*, 43, 783–789.

Ishikawa, M, Maekawa, K, Saito, K, Senoo, Y, Urata, M, Murayama, M, Tajima, Y, Kumagai, Y & Saito, Y 2014. Plasma and serum lipidomics of healthy white adults shows characteristic profiles by subjects' gender and age. *PLoS One*, 9, e91806.

Ishikawa, M, Tajima, Y, Murayama, M, Senoo, Y, Maekawa, K & Saito, Y 2013. Plasma and serum from nonfasting men and women differ in their lipidomic profiles. *Biol Pharm Bull*, 36, 682–685.

Ivanisevic, J, Elias, D, Deguchi, H, Averell, PM, Kurczy, M, Johnson, CH, Tautenhahn, R et al. 2015. Arteriovenous blood metabolomics: A readout of intra-tissue metabostasis. *Sci Rep*, 5, 12757.

Jha, V, Garcia-Garcia, G, Iseki, K, Li, Z, Naicker, S, Plattner, B, Saran, R, Wang, AY & Yang, CW 2013. Chronic kidney disease: Global dimension and perspectives. *Lancet*, 382, 260–272.

Jones, B 2012. Genomics: Personal genome project. *Nat Rev Genet*, 13, 599.

Kaddurah-Daouk, R, Kristal, BS & Weinshilboum, RM 2008. Metabolomics: A global biochemical approach to drug response and disease. *Annu Rev Pharmacol Toxicol*, 48, 653–683.

Kakkar, AK 2015. Patent cliff mitigation strategies: Giving new life to blockbusters. *Expert Opin Ther Pat*, 25, 13531359.

Kamal, M 2015. Challenges for the clinical implementation of precision medicine trials. *In:* Le Tourneau, C & Kamal, M (eds.) *Pan-cancer Integrative Molecular Portrait Towards a New Paradigm in Precision Medicine*. Cham, Switzerland: Springer.

Khoury, MJ, Gwinn, ML, Glasgow, RE & Kramer, BS 2012. A population approach to precision medicine. *Am J Prev Med*, 42, 639–645.

Kim, MS, Pinto, SM, Getnet, D, Nirujogi, RS, Manda, SS, Chaerkady, R, Madugundu, AK et al. 2014. A draft map of the human proteome. *Nature*, 509, 575–581.

Kumar, A, Baycin-Hizal, D, Shiloach, J, Bowen, MA & Betenbaugh, MJ 2015. Coupling enrichment methods with proteomics for understanding and treating disease. *Proteomics Clin Appl*, 9, 33–47.

Latkovic, T 2013. The Trillion Dollar Prize: Using outcomes-based payment to address the US healthcare financing crisis. McKinsey & Company. Available: https://healthcare.mckinsey.com/sites/default/files/the-trillion-dollar-prize.pdf.

Laveist, TA 1993. Segregation, poverty, and empowerment: Health consequences for African Americans. *Milbank Q*, 71, 41–64.

Laveist, TA 2011. Perspective: The spectrum of health-care disparities in the USA. *In:* Williams, RA (ed.) *Healthcare Disparities at the Crossroads with Healthcare Reform*. Boston, MA: Springer US.

Lillie-Blanton, M & Laveist, T 1996. Race/ethnicity, the social environment, and health. *Soc Sci Med*, 43, 83–91.

Lindon, JC, Holmes, E & Nicholson, JK 2007. Chapter 19 - Global systems biology through integration of "Omics" results. *In:* Lindon, J (ed.) *The Handbook of Metabonomics and Metabolomics*. Amsterdam, the Netherlands: Elsevier Science B.V.

Liu, X, Ser, Z & Locasale, JW 2014. Development and quantitative evaluation of a high-resolution metabolomics technology. *Anal Chem*, 86, 2175–2184.

Liu, Y, Buil, A, Collins, BC, Gillet, LC, Blum, LC, Cheng, LY, Vitek, O et al. 2015. Quantitative variability of 342 plasma proteins in a human twin population. *Mol Syst Biol*, 11, 786.

Martins-De-Souza, D & Farias, AS 2015. Deciphering the biochemistry and identifying biomarkers to multiple sclerosis. *Proteomics*, 15, 3281–3282.

Matsui, SN, T; Choai, Y 2014. Biomarker-based designs of phase III clinical trials for personalized medicine. *In:* Mcisaac, MCR (ed.) *Developments in Statistical Evaluation of Clinical Trials*. Berlin, Germany: Springer.

Members, MSIB, Sansone, SA, Fan, T, Goodacre, R, Griffin, JL, Hardy, NW, Kaddurah-Daouk, R et al. 2007. The metabolomics standards initiative. *Nat Biotechnol*, 25, 846–848.

Mirnezami, R, Nicholson, J & Darzi, A 2012. Preparing for precision medicine. *N Engl J Med*, 366, 489–491.

Mullard, A 2014. New drugs cost US$2.6 billion to develop. *Nat Rev Drug Discov*, 13, 877.

Paglia, G, Kliman, M, Claude, E, Geromanos, S & Astarita, G 2015. Applications of ion-mobility mass spectrometry for lipid analysis. *Anal Bioanal Chem*, 407, 4995–5007.

Pan, S, Brentnall, TA & Chen, R 2015. Proteomics analysis of bodily fluids in pancreatic cancer. *Proteomics*, 15, 2705–2715.

Patel, S & Ahmed, S 2015. Emerging field of metabolomics: big promise for cancer biomarker identification and drug discovery. *J Pharm Biomed Anal*, 107, 63–74.

Peng, B, Li, H & Peng, XX 2015. Functional metabolomics: From biomarker discovery to metabolome reprogramming. *Protein Cell*, 6, 628–637.

Quinones, MP & Kaddurah-Daouk, R 2009. Metabolomics tools for identifying biomarkers for neuropsychiatric diseases. *Neurobiol Dis*, 35, 165–176.

Redig, AJ & Janne, PA 2015. Basket trials and the evolution of clinical trial design in an era of genomic medicine. *J Clin Oncol*, 33, 975–977.

Robinson, BE 2006. Epidemiology of chronic kidney disease and anemia. *J Am Med Dir Assoc*, 7, S3–S6; quiz S17–S21.

Robinson, PN 2012. Deep phenotyping for precision medicine. *Hum Mutat*, 33, 777–780.

Saito, K, Maekawa, K, Pappan, KL, Urata, M, Ishikawa, M, Kumagai, Y & Saito, Y 2014. Differences in metabolite profiles between blood matrices, ages, and sexes among Caucasian individuals and their inter-individual variations. *Metabolomics*, 10, 402.

Schork, NJ 2015. Personalized medicine: Time for one-person trials. *Nature*, 520, 609–611.

Schwudke, D, Schuhmann, K, Herzog, R, Bornstein, SR & Shevchenko, A 2011. Shotgun lipidomics on high resolution mass spectrometers. *Cold Spring Harb Perspect Biol*, 3, a004614.

Sharma, S & Munshi, A 2013. *Omics Approaches and Applications in Clinical Trials*, New Delhi, India, Springer.

Simon, R 2014. Biomarker based clinical trial design. *Chin Clin Oncol*, 3, 39.

Simonds, NI, Khoury, MJ, Schully, SD, Armstrong, K, Cohn, WF, Fenstermacher, DA, Ginsburg, GS et al. 2013. Comparative effectiveness research in cancer genomics and precision medicine: Current landscape and future prospects. *J Natl Cancer Inst*, 105, 929–936.

Simons, B, Kauhanen, D, Sylvanne, T, Tarasov, K, Duchoslav, E & Ekroos, K 2012. Shotgun lipidomics by sequential precursor ion fragmentation on a hybrid quadrupole time-of-flight mass spectrometer. *Metabolites*, 2, 195–213.

Stâhlman, M, Borén, J & Ekroos, K 2012. High-throughput molecular lipidomics. *In*: Ekroos, K (ed.) *Lipidomics: Technologies and Applications*. Weinheim, Germany: Wiley-VCH Verlag GmbH & Co. KGaA.

Surma, MA, Herzog, R, Vasilj, A, Klose, C, Christinat, N, Morin-Rivron, D, Simons, K, Masoodi, M & Sampaio, JL 2015. An automated shotgun lipidomics platform for high throughput, comprehensive, and quantitative analysis of blood plasma intact lipids. *Eur J Lipid Sci Technol*, 117, 1540–1549.

Teixeira, PC, Ferber, P, Vuilleumier, N & Cutler, P 2015. Biomarkers for cardiovascular risk assessment in autoimmune diseases. *Proteomics Clin Appl*, 9, 48–57.

Tu, C, Rudnick, PA, Martinez, MY, Cheek, KL, Stein, SE, Slebos, RJ & Liebler, DC 2010. Depletion of abundant plasma proteins and limitations of plasma proteomics. *J Proteome Res*, 9, 4982–4991.

Wang, H & Bennett, P 2013. Performance assessment of microflow LC combined with high-resolution MS in bioanalysis. *Bioanalysis*, 5, 1249–1267.

Watson, JD & Crick, FH 1953. Molecular structure of nucleic acids; A structure for deoxyribose nucleic acid. *Nature*, 171, 737–738.

Wells, AL 2004. Reevaluating the social contract in American medicine. *Virtual Mentor*, 6.

Werner, T, Sweetman, G, Savitski, MF, Mathieson, T, Bantschett, M & Savitski, MM 2014. Ion coalescence of neutron encoded TMT 10-plex reporter ions. *Anal Chem*, 86, 3594–3601.

Wilhelm, M, Schlegl, J, Hahne, H, Gholami, AM, Lieberenz, M, Savitski, MM, Ziegler, E et al. 2014. Mass-spectrometry-based draft of the human proteome. *Nature*, 509, 582–587.

Wu, C, Choi, YH & Van Wezel, GP 2016. Metabolic profiling as a tool for prioritizing antimicrobial compounds. *J Ind Microbiol Biotechnol*, 43, 299–312.

Young, HSK 2014. *Health Care*. Burlington, MA: Jones & Bartlett Learning.

Yuan, M, Breitkopf, SB, Yang, X & Asara, JM 2012. A positive/negative ion-switching, targeted mass spectrometry-based metabolomics platform for bodily fluids, cells, and fresh and fixed tissue. *Nat Protoc*, 7, 872–881.

Zhang, X 2015. Precision medicine, personalized medicine, omics and big data: Concepts and relationships. *J Pharmacogenomics Pharmacoproteomics*, 6, e144.

Zhang, Y, Bilbao, A, Bruderer, T, Luban, J, Strambio-De-Castillia, C, Lisacek, F, Hopfgartner, G & Varesio, E 2015. The use of variable Q1 isolation windows improves selectivity in LC-SWATH-MS acquisition. *J Proteome Res*, 14, 4359–4371.

Zubarev, RA & Makarov, A 2013. Orbitrap mass spectrometry. *Anal Chem*, 85, 5288–5296.

Index

Note: Page numbers in italic and bold refer to figures and tables, respectively.